Lecture Notes in Computer Science 3174

Commenced Publication in 1973
Founding and Former Series Editors:
Gerhard Goos, Juris Hartmanis, and Jan van Leeuwen

Editorial Board

Fuliang Yin Jun Wang Chengan Guo (Eds.)

Advances in Neural Networks – ISNN 2004

International Symposium on Neural Networks
Dalian, China, August 19-21, 2004
Proceedings, Part II

Springer

Volume Editors

Fuliang Yin
Chengan Guo
Dalian University of Technology
School of Electronic and Information Engineering
Dalian, Liaoning, China
E-mail: {flyin, cguo}@dlut.edu.cn

Jun Wang
The Chinese University of Hong Kong
Department of Automation and Computer-Aided Engineering
Shatin, New Territories, Hong Kong
E-mail: jwang@acae.cuhk.edu.hk

Library of Congress Control Number: 2004095624

CR Subject Classification (1998): F.1, F.2, D.1, G.2, I.2, C.2

ISSN 0302-9743
ISBN 3-540-22843-8 Springer Berlin Heidelberg New York

Springer is a part of Springer Science+Business Media

springeronline.com

Printed in Germany

Typesetting: Camera-ready by author, data conversion by PTP-Berlin, Protago-TeX-Production GmbH
Printed on acid-free paper SPIN: 11312475 06/3142 5 4 3 2 1 0

Preface

This book constitutes the proceedings of the International Symposium on Neural Networks (ISNN 2004) held in Dalian, Liaoning, China during August 19–21, 2004. ISNN 2004 received over 800 submissions from authors in five continents (Asia, Europe, North America, South America, and Oceania), and 23 countries and regions (mainland China, Hong Kong, Taiwan, South Korea, Japan, Singapore, India, Iran, Israel, Turkey, Hungary, Poland, Germany, France, Belgium, Spain, UK, USA, Canada, Mexico, Venezuela, Chile, and Australia). Based on reviews, the Program Committee selected 329 high-quality papers for presentation at ISNN 2004 and publication in the proceedings. The papers are organized into many topical sections under 11 major categories (theoretical analysis; learning and optimization; support vector machines; blind source separation, independent component analysis, and principal component analysis; clustering and classification; robotics and control; telecommunications; signal, image and time series processing; detection, diagnostics, and computer security; biomedical applications; and other applications) covering the whole spectrum of the recent neural network research and development. In addition to the numerous contributed papers, five distinguished scholars were invited to give plenary speeches at ISNN 2004.

ISNN 2004 was an inaugural event. It brought together a few hundred researchers, educators, scientists, and practitioners to the beautiful coastal city Dalian in northeastern China. It provided an international forum for the participants to present new results, to discuss the state of the art, and to exchange information on emerging areas and future trends of neural network research. It also created a nice opportunity for the participants to meet colleagues and make friends who share similar research interests.

The organizers of ISNN 2004 made great efforts to ensure the success of this event. We would like to thank Dalian University of Technology for the sponsorship, various IEEE organizations (especially the IEEE Circuits and Systems Society) for the technical co-sponsorship, and the members of the ISNN 2004 Advisory Committee for their spiritual support. We would also like to thank the members of the Program Committee and additional referees for reviewing the papers and the Publication Committee for checking and compiling the papers in a very short period of time. We would also like to thank the publisher, Springer-Verlag, for their agreement and cooperation to publish the proceedings as a volume of the Lecture Notes in Computer Science. Finally, we would like to thank all the authors for contributing their papers. Without their high-quality papers, this symposium would not have been possible.

August 2004

Fuliang Yin
Jun Wang
Chengan Guo

Preface

This book constitutes the proceedings of the International Symposium on Neural Networks (ISNN 2004) held in Dalian, Liaoning, China during August 19–21, 2004. ISNN 2004 received over 800 submissions from authors in five continents (Asia, Europe, North America, South America, and Oceania), and 23 countries and regions (mainland China, Hong Kong, Taiwan, South Korea, Japan, Singapore, India, Iran, Israel, Turkey, Hungary, Poland, Germany, France, Belgium, Spain, UK, USA, Canada, Mexico, Venezuela, Chile, and Australia). Based on reviews, the Program Committee selected 329 high-quality papers for presentation at ISNN 2004 and publication in the proceedings. The papers are organized into many topical sections under 11 major categories (theoretical analysis; learning and optimization; support vector machines; blind source separation, independent component analysis, and principal component analysis; clustering and classification; robotics and control; telecommunications; signal, image and time series processing; detection, diagnostics, and computer security; biomedical applications; and other applications) covering the whole spectrum of the recent neural network research and development. In addition to the numerous contributed papers, five distinguished scholars were invited to give plenary speeches at ISNN 2004.

ISNN 2004 was an inaugural event. It brought together a few hundred researchers, educators, scientists, and practitioners to the beautiful coastal city Dalian in northeastern China. It provided an international forum for the participants to present new results, to discuss the state of the art, and to exchange information on emerging areas and future trends of neural network research. It also created a nice opportunity for the participants to meet colleagues and make friends who share similar research interests.

The organizers of ISNN 2004 made great efforts to ensure the success of this event. We would like to thank Dalian University of Technology for the sponsorship, various IEEE organizations (especially the IEEE Circuits and Systems Society) for the technical co-sponsorship, and the members of the ISNN 2004 Advisory Committee for their spiritual support. We would also like to thank the members of the Program Committee and additional referees for reviewing the papers and the Publication Committee for checking and compiling the papers in a very short period of time. We would also like to thank the publisher, Springer-Verlag, for their agreement and cooperation to publish the proceedings as a volume of the Lecture Notes in Computer Science. Finally, we would like to thank all the authors for contributing their papers. Without their high-quality papers, this symposium would not have been possible.

August 2004

Fuliang Yin
Jun Wang
Chengan Guo

ISNN 2004 Organization

ISNN 2004 was organized and sponsored by Dalian University of Technology in cooperation with the Chinese University of Hong Kong. It was technically cosponsored by the IEEE Circuits and Systems Society, IEEE Computational Intelligence Society (Beijing, Hong Kong, and Singapore Chapters), and IEEE Control Systems Society and Robotics and Automation Society (Hong Kong Joint Chapter).

Committees

General Co-chairs

Fuliang Yin, Dalian, China
Jun Wang, Hong Kong

Advisory Committee Co-chairs

Gengdong Cheng, Dalian, China
Yixin Zhong, Beijing, China
Jacek M. Zurada, Louisville, USA

Advisory Committee Members

Shun-ichi Amari, Tokyo, Japan
Zheng Bao, Xi'an, China
Guoliang Chen, Hefei, China
Ruwei Dai, Beijing, China
Anthony Kuh, Honolulu, USA
Chunbo Feng, Nanjing, China
Toshio Fukuda, Nagoya, Japan
Zhenya He, Nanjing, China
Kararo Hirasawa, Fukuoka, Japan
Frank L. Lewis, Fort Worth, USA
Yanda Li, Beijing, China
Erkki Oja, Helsinki, Finland
Zong Sha, Beijing, China
Tzyh-Jong Tarn, St. Louis, USA
Shoujue Wang, Beijing, China
Zhongtuo Wang, Dalian, China
Youshou Wu, Beijing, China
Bo Zhang, Beijing, China

Program Committee Co-chairs

Chengan Guo, Dalian, China
Andrzej Cichocki, Tokyo, Japan
Mingsheng Zhao, Beijing, China

Program Committee Members

Sabri Arik (Istanbul, Turkey), *Amit Bhaya* (Rio de Janeiro, Brazil), *Jinde Cao* (Nanjing, China), *Yijia Cao* (Hangzhou, China), *Laiwan Chan* (Hong Kong), *Ke Chen* (Manchester, UK), *Luonan Chen* (Osaka, Japan), *Tianping Chen* (Shanghai, China), *Yiu Ming Cheung* (Hong Kong), *Chuanyin Dang* (Hong Kong), *Wlodzislaw Duch* (Torun, Poland), *Mauro Forti* (Siena, Italy), *Jun Gao* (Hefei, China), *Shuzhi Sam Ge* (Singapore), *Xinping Guan* (Qinhuangdao, China), *Dewen Hu* (Changsha, China), *DeShuang Huang* (Hefei, China), *Donald L. Hung* (San Jose, USA), *Danchi Jiang* (Canberra, Australia), *Licheng Jiao* (Xi'an, China), *H.K. Kwan* (Windsor, Canada), *Xiaoli Li* (Birmingham, UK), *Yuanqing Li* (Tokyo, Japan), *Xue-Bin Liang* (Baton Rouge, USA), *Lizhi Liao*

(Hong Kong), *Xiaofeng Liao* (Chongqing, China), *Chin-Teng Lin* (Hsingchu, Taiwan), *Derong Liu* (Chicago, USA), *Baoliang Lu* (Shanghai, China), *Hongtao Lu* (Shanghai, China), *Fa-Long Luo* (San Jose, USA), *Qing Ma* (Kyoto, Japan), *Zongyuan Mao* (Guangzhou, China), *Xuemei Ren* (Beijing, China), *Rudy Setiono* (Singapore), *Peter Sincak* (Kosice, Slovakia), *Jianbo Su* (Shanghai, China), *Fuchun Sun* (Beijing, China), *Johan Suykens* (Leuven, Belgium), *Ying Tan* (Hefei, China), *Dan Wang* (Singapore), *Lipo Wang* (Singapore), *Wei Wu* (Dalian, China), *Yousheng Xia* (Hong Kong), *Zhongben Xu* (Xi'an, China), *Simon X. Yang* (Guelph, Canada), *Hujun Yin* (Manchester, UK), *Jianwei Zhang* (Hamburg, Germany), *Liming Zhang* (Shanghai, China), *Liqing Zhang* (Shanghai, China), *Yi Zhang* (Chengdu, China), *Weixin Zheng* (Sydney, Australia)

Organizing Committee Chair	*Min Han* (Dalian, China)
Publication Co-chairs	*Hujun Yin* (Manchester, UK) *Tianshuang Qiu* (Dalian, China)
Publicity Co-chairs	*Tianshuang Qiu* (Dalian, China) *Derong Liu* (Chicago, USA) *Meng Joo Er* (Singapore)

Table of Contents, Part II

Part VI Robotics and Control

Application of RBFNN for Humanoid Robot Real Time Optimal Trajectory Generation in Running ... 1
Xusheng Lei, Jianbo Su

Full-DOF Calibration-Free Robotic Hand-Eye Coordination Based on Fuzzy Neural Network ... 7
Jianbo Su, Qielu Pan, Zhiwei Luo

Neuro-Fuzzy Hybrid Position/Force Control for a Space Robot with Flexible Dual-Arms ... 13
Fuchun Sun, Hao Zhang, Hao Wu

Fuzzy Neural Networks Observer for Robotic Manipulators Based on H_∞ Approach ... 19
Hong-bin Wang, Chun-di Jiang, Hong-rui Wang

Mobile Robot Path-Tracking Using an Adaptive Critic Learning PD Controller ... 25
Xin Xu, Xuening Wang, Dewen Hu

Reinforcement Learning and ART2 Neural Network Based Collision Avoidance System of Mobile Robot ... 35
Jian Fan, GengFeng Wu, Fei Ma, Jian Liu

FEL-Based Adaptive Dynamic Inverse Control for Flexible Spacecraft Attitude Maneuver ... 42
Yaqiu Liu, Guangfu Ma, Qinglei Hu

Multivariable Generalized Minimum Variance Control Based on Artificial Neural Networks and Gaussian Process Models ... 52
Daniel Sbarbaro, Roderick Murray-Smith, Arturo Valdes

A Neural Network Based Method for Solving Discrete-Time Nonlinear Output Regulation Problem in Sampled-Data Systems ... 59
Dan Wang, Jie Huang

The Design of Fuzzy Controller by Means of CI Technologies-Based Estimation Technique ... 65
Sung-Kwun Oh, Seok-Beom Roh, Dong-Yoon Lee, Sung-Whan Jang

A Neural Network Adaptive Controller
for Explicit Congestion Control with Time Delay ... 71
Bo Yang, Xinping Guan

Robust Adaptive Control Using Neural Networks and Projection ... 77
Xiaoou Li, Wen Yu

Design of PID Controllers Using Genetic Algorithms Approach
for Low Damping, Slow Response Plants ... 83
PenChen Chou, TsenJar Hwang

Neural Network Based Fault Tolerant Control of a Class
of Nonlinear Systems with Input Time Delay ... 91
Ming Liu, Peng Liu, Donghua Zhou

Run-to-Run Iterative Optimization Control of Batch Processes ... 97
Zhihua Xiong, Jie Zhang, Xiong Wang, Yongmao Xu

Time-Delay Recurrent Neural Networks for Dynamic Systems Control ... 104
Xu Xu, Yinghua Lu, Yanchun Liang

Feedforward-Feedback Combined Control System
Based on Neural Network ... 110
Weidong Zhang, Fanming Zeng, Guojun Cheng, Shengguang Gong

Online Learning CMAC Neural Network Control Scheme
for Nonlinear Systems ... 117
Yuman Yuan, Wenjin Gu, Jinyong Yu

Pole Placement Control for Nonlinear Systems via Neural Networks ... 123
Fei Liu

RBF NN-Based Backstepping Control
for Strict Feedback Block Nonlinear System and Its Application ... 129
Yunan Hu, Yuqiang Jin, Pingyuan Cui

Model Reference Control Based on SVM ... 138
Junfeng He, Zengke Zhang

PID Controller Based on the Artificial Neural Network ... 144
Jianhua Yang, Wei Lu, Wenqi Liu

Fuzzy Predictive Control Based on PEMFC Stack ... 150
Xi Li, Xiao-wei Fu, Guang-yi Cao, Xin-jian Zhu

Adaptive Control for Induction Servo Motor
Based on Wavelet Neural Networks ... 156
Qinghui Wu, Yi Liu, Dianjun Zhang, Yonghui Zhang

The Application of Single Neuron Adaptive PID Controller in Control System of Triaxial and Torsional Shear Apparatus ... 163
Muguo Li, Zhendong Liu, Jing Wang, Qun Zhang, Hairong Jiang, Hai Du

Ram Velocity Control in Plastic Injection Molding Machines with Neural Network Learning Control ... 169
Gaoxiang Ouyang, Xiaoli Li, Xinping Guan, Zhiqiang Zhang, Xiuling Zhang, Ruxu Du

Multiple Models Neural Network Decoupling Controller for a Nonlinear System ... 175
Xin Wang, Shaoyuan Li, Zhongjie Wang, Heng Yue

Feedback-Assisted Iterative Learning Control for Batch Polymerization Reactor ... 181
Shuchen Li, Xinhe Xu, Ping Li

Recent Developments on Applications of Neural Networks to Power Systems Operation and Control: An Overview ... 188
Chuangxin Guo, Quanyuan Jiang, Xiu Cao, Yijia Cao

A Novel Fermentation Control Method Based on Neural Networks ... 194
Xuhua Yang, Zonghai Sun, Youxian Sun

Modeling Dynamic System by Recurrent Neural Network with State Variables ... 200
Min Han, Zhiwei Shi, Wei Wang

Robust Friction Compensation for Servo System Based on LuGre Model with Uncertain Static Parameters ... 206
Lixin Wei, Xia Wang, Hongrui Wang

System Identification Using Adjustable RBF Neural Network with Stable Learning Algorithms ... 212
Wen Yu, Xiaoou Li

A System Identification Method Based on Multi-layer Perception and Model Extraction ... 218
Chang Hu, Li Cao

Complex Model Identification Based on RBF Neural Network ... 224
Yibin Song, Peijin Wang, Kaili Li

Part VII Telecommunications

A Noisy Chaotic Neural Network Approach to Topological Optimization of a Communication Network with Reliability Constraints 230
Lipo Wang, Haixiang Shi

Space-Time Multiuser Detection Combined with Adaptive Wavelet Networks over Multipath Channels 236
Ling Wang, Licheng Jiao, Haihong Tao, Fang Liu

Optimizing Sensor Node Distribution with Genetic Algorithm in Wireless Sensor Network 242
Jianli Zhao, Yingyou Wen, Ruiqiang Shang, Guangxing Wang

Fast De-hopping and Frequency Hopping Pattern (FHP) Estimation for DS/FHSS Using Neural Networks 248
Tarek Elhabian, Bo Zhang, Dingrong Shao

Autoregressive and Neural Network Model Based Predictions for Downlink Beamforming 254
Halil Yigit, Adnan Kavak, Metin Ertunc

Forecast and Control of Anode Shape in Electrochemical Machining Using Neural Network 262
Guibing Pang, Wenji Xu, Xiaobing Zhai, Jinjin Zhou

A Hybrid Neural Network and Genetic Algorithm Approach for Multicast QoS Routing 269
Daru Pan, Minghui Du, Yukun Wang, Yanbo Yuan

Performance Analysis of Recurrent Neural Networks Based Blind Adaptive Multiuser Detection in Asynchronous DS-CDMA Systems 275
Ling Wang, Haihong Tao, Licheng Jiao, Fang Liu

Neural Direct Sequence Spread Spectrum Acquisition 281
Tarek Elhabian, Bo Zhang, Dingrong Shao

Multi-stage Neural Networks for Channel Assignment in Cellular Radio Networks 287
Hyuk-Soon Lee, Dae-Won Lee,, Jaewook Lee

Experimental Spread Spectrum Communication System Based on CNN 293
Jianye Zhao, Shide Guo, Daoheng Yu

Neural Congestion Control Algorithm in ATM Networks with Multiple Node 299
Ruijun Zhu, Fuliang Yin, Tianshuang Qiu

Neural Compensation of Linear Distortion in Digital Communications 305
Hong Zhou

On the Performance of Space-Time Block Coding
Based on ICA Neural Networks 311
Ju Liu, Hongji Xu, Yong Wan

ICA-Based Beam Space-Time Block Coding
with Transmit Antenna Array Selection 317
Hongji Xu, Ju Liu

Nonlinear Dynamic Method to Suppress Reverberation
Based on RBF Neural Networks 324
Bing Deng, Ran Tao

Part VIII Signal, Image, and Time Series Processing

A New Scheme for Detection and Classification
of Subpixel Spectral Signatures in Multispectral Data 331
Hao Zhou, Bin Wang, Liming Zhang

A Rough-Set-Based Fuzzy-Neural-Network System
for Taste Signal Identification 337
Yan-Xin Huang, Chun-Guang Zhou, Shu-Xue Zou, Yan Wang, Yan-Chun Liang

A Novel Signal Detection Subsystem of Radar Based on HA-CNN 344
Zhangliang Xiong, Xiangquan Shi

Real-Time Detection of Signal in the Noise Based
on the RBF Neural Network and Its Application 350
Minfen Shen, Yuzheng Zhang, Zhancheng Li, Jinyao Yang, Patch Beadle

Classification of EEG Signals Under Different Brain Functional States
Using RBF Neural Network 356
Zhancheng Li, Minfen Shen, Patch Beadle

Application of a Wavelet Adaptive Filter Based on Neural Network
to Minimize Distortion of the Pulsatile Spectrum 362
Xiaoxia Li, Gang Li, Ling Lin, Yuliang Liu, Yan Wang, Yunfeng Zhang

Spectral Analysis and Recognition Using Multi-scale Features
and Neural Networks 369
YuGang Jiang, Ping Guo

A Novel Fuzzy Filter for Impulse Noise Removal 375
Chang-Shing Lee, Shu-Mei Guo, Chin-Yuan Hsu

Neural Network Aided Adaptive Kalman Filter
for Multi-sensors Integrated Navigation 381
Lin Chai, Jianping Yuan, Qun Fang, Zhiyu Kang, Liangwei Huang

Solely Excitatory Oscillator Network for Color Image Segmentation 387
Chung Lam Li, Shu Tak Lee

Image De-noising Using Cross-Validation Method
with RBF Network Representation 393
Ping Guo, Hongzhai Li

Ultrasonic C-scan Image Restoration Using Radial Basis Function Network 399
Zongjie Cao, Huaidong Chen, Jin Xue, Yuwen Wang

Automatic Image Segmentation
Based on a Simplified Pulse Coupled Neural Network 405
Yingwei Bi, Tianshuang Qiu, Xiaobing Li, Ying Guo

Face Pose Estimation Based on Eigenspace Analysis
and Fuzzy Clustering 411
Cheng Du, Guangda Su

Chaotic Time Series Prediction
Based on Local-Region Multi-steps Forecasting Model 418
Minglun Cai, Feng Cai, Aiguo Shi, Bo Zhou, Yongsheng Zhang

Nonlinear Prediction Model Identification and Robust Prediction
of Chaotic Time Series 424
Yuexian Hou, Weidi Dai, Pilian He

Wavelet Neural Networks for Nonlinear Time Series Analysis 430
Bo Zhou, Aiguo Shi, Feng Cai, Yongsheng Zhang

Part IX Biomedical Applications

Neural Networks Determining Injury by Salt Water
for Distribution Lines 436
Lixin Ma, Hiromi Miyajima, Noritaka Shigei, Shuma Kawabata

EEG Source Localization Using Independent Residual Analysis 442
Gang Tan, Liqing Zhang

Classifying G-protein Coupled Receptors with Support Vector Machine 448
Ying Huang, Yanda Li

A Novel Individual Blood Glucose Control Model Based on Mixture of Experts Neural Networks ... 453
Wei Wang, Zheng-Zhong Bian, Lan-Feng Yan, Jing Su

Tracking the Amplitude Variation of Evoked Potential by ICA and WT ... 459
Haiyan Ding, Datian Ye

A Novel Method for Gene Selection and Cancer Classification ... 465
Huajun Yan, Zhang Yi

Nonnegative Matrix Factorization for EEG Signal Classification ... 470
Weixiang Liu, Nanning Zheng, Xi Li

A Novel Clustering Analysis Based on PCA and SOMs for Gene Expression Patterns ... 476
Hong-Qiang Wang, De-Shuang Huang, Xing-Ming Zhao, Xin Huang

Feedback Selective Visual Attention Model Based on Feature Integration Theory ... 482
Lianwei Zhao, Siwei Luo

Realtime Monitoring of Vascular Conditions Using a Probabilistic Neural Network ... 488
Akira Sakane, Toshio Tsuji, Yoshiyuki Tanaka, Kenji Shiba, Noboru Saeki, Masashi Kawamoto

Capturing Long-Term Dependencies for Protein Secondary Structure Prediction ... 494
Jinmiao Chen, Narendra S. Chaudhari

A Method for Fast Estimation of Evoked Potentials Based on Independent Component Analysis ... 501
Ting Li, Tianshuang Qiu, Xuxiu Zhang, Anqing Zhang, Wenhong Liu

Estimation of Pulmonary Elastance Based on RBF Expression ... 507
Shunshoku Kanae, Zi-Jiang Yang, Kiyoshi Wada

Binary Input Encoding Strategy Based Neural Network for Globulin Protein Inter-residue Contacts Map Prediction ... 513
GuangZheng Zhang, DeShuang Huang, Xin Huang

Genetic Regulatory Systems Modeled by Recurrent Neural Network ... 519
Xianhua Dai

A New Computational Model of Biological Vision for Stereopsis ... 525
Baoquan Song, Zongtan Zhou, Dewen Hu, Zhengzhi Wang

A Reconstruction Approach to CT with Cauchy RBFs Network ... 531
Jianzhong Zhang, Haiyang Li

Part X Detection, Diagnostics, and Computer Security

Fault Diagnosis on Satellite Attitude Control with Dynamic Neural Network 537
HanYong Hao, ZengQi Sun, Yu Zhang

A New Strategy for Fault Detection of Nonlinear Systems Based on Neural Networks 543
Linglai Li, Donghua Zhou

Non-stationary Fault Diagnosis Based on Local-Wave Neural Network 549
Zhen Wang, Ji Li, Zijia Ding, Yanhui Song

Hybrid Neural Network Based Gray-Box Approach to Fault Detection of Hybrid Systems 555
Wenhui Wang, Dexi An, Donghua Zhou

Application of Enhanced Independent Component Analysis to Leak Detection in Transport Pipelines 561
Zhengwei Zhang, Hao Ye, Rong Hu

Internet-Based Remote Monitoring and Fault Diagnosis System 567
Xing Wu, Jin Chen, Ruqiang Li, Fucai Li

Rough Sets and Partially-Linearized Neural Network for Structural Fault Diagnosis of Rotating Machinery 574
Peng Chen, Xinying Liang, Takayoshi Yamamoto

Transient Stability Assessment Using Radial Basis Function Networks 581
Yutian Liu, Xiaodong Chu, Yuanyuan Sun, Li Li

An Optimized Shunt Hybrid Power Quality Conditioner Based on an Adaptive Neural Network for Power Quality Improvement in Power Distribution Network 587
Ming Zhang, Hui Sun, Jiyan Zou, Hang Su

Cyclic Statistics Based Neural Network for Early Fault Diagnosis of Rolling Element Bearings 595
Fuchang Zhou, Jin Chen, Jun He, Guo Bi, Guicai Zhang, Fucai Li

Application of BP Neural Network for the Abnormity Monitoring in Slab Continuous Casting 601
Xudong Wang, Man Yao, Xingfu Chen

A TVAR Parametric Model Applying for Detecting Anti-electric-Corona Discharge 607
Zhe Chen, Hongyu Wang, Tianshuang Qiu

The Study on Crack Diagnosis Based on Fuzzy Neural Networks Using Different Indexes 613
Jingfen Zhang, Guang Meng, Deyou Zhao

Blind Fault Diagnosis Algorithm for Integrated Circuit Based on the CPN Neural Networks 619
Daqi Zhu, Yongqing Yang, Wuzhao Li

A Chaotic-Neural-Network-Based Encryption Algorithm for JPEG2000 Encoded Images 627
Shiguo Lian, Guanrong Chen, Albert Cheung, Zhiquan Wang

A Combined Hash and Encryption Scheme by Chaotic Neural Network 633
Di Xiao, Xiaofeng Liao

A Novel Symmetric Cryptography Based on Chaotic Signal Generator and a Clipped Neural Network 639
Tsing Zhou, Xiaofeng Liao, Yong Chen

A Neural Network Based Blind Watermarking Scheme for Digital Images 645
Guoping Tang, Xiaofeng Liao

Color Image Watermarking Based on Neural Networks 651
Wei Lu, Hongtao Lu, Ruiming Shen

A Novel Intrusion Detection Method Based on Principle Component Analysis in Computer Security 657
Wei Wang, Xiaohong Guan, Xiangliang Zhang

A Novel Wavelet Image Watermarking Scheme Combined with Chaos Sequence and Neural Network 663
Jian Zhao, Mingquan Zhou, Hongmei Xie, Jinye Peng, Xin Zhou

Quantum Neural Network for Image Watermarking 669
Shiyan Hu

An NN-Based Malicious Executables Detection Algorithm Based on Immune Principles 675
Zhenhe Guo, Zhengkai Liu, Ying Tan

The Algorithm for Detecting Hiding Information Based on SVM 681
JiFeng Huang, JiaJun Lin, XiaoFu He, Meng Dai

An E-mail Filtering Approach Using Neural Network 688
Yukun Cao, Xiaofeng Liao, Yunfeng Li

Part XI Other Applications

Topography-Enhanced BMU Search in Self-Organizing Maps 695
James S. Kirk, Jacek M. Zurada

Nonlinear Optimal Estimation for Neural Networks Data Fusion 701
Ye Ma, Xiao Tong Wang, Bo Li, JianGuo Fu

Improvement of Data Visualization Based on SOM 707
Chao Shao, Houkuan Huang

Research on Neural Network Method in GPS Data Transformation 713
Min Han, Xue Tian, Shiguo Xu

Dynamic File Allocation in Storage Area Networks
with Neural Network Prediction 719
Weitao Sun, Jiwu Shu, Weimin Zheng

Competitive Algorithms for Online Leasing Problem
in Probabilistic Environments 725
Yinfeng Xu, Weijun Xu

Wavelet Network for Nonlinear Regression Using Probabilistic Framework 731
Shu-Fai Wong, Kwan-Yee Kenneth Wong

A Neural Network Approach for Indirect Shape from Shading 737
Ping Hao, Dongming Guo, Renke Kang

A Hybrid Radial Basis Function Neural Network
for Dimensional Error Prediction in End Milling 743
Xiaoli Li, Xinping Guan, Yan Li

Bayesian Neural Networks for Life Modeling and Prediction
of Dynamically Tuned Gyroscopes 749
Chunling Fan, Feng Gao, Zhihua Jin

Furnace Temperature Modeling for Continuous Annealing Process
Based on Generalized Growing and Pruning RBF Neural Network 755
Qing Chen, Shaoyuan Li, Yugeng Xi, Guangbin Huang

A Learning-Based Contact Model in Belt-Grinding Processes 761
Xiang Zhang, Bernd Kuhlenkötter, Klaus Kneupner

Application of General Regression Neural Network
to Vibration Trend Prediction of Rotating Machinery 767
Zhipeng Feng, Fulei Chu, Xigeng Song

A Novel Nonlinear Projection to Latent Structures Algorithm 773
Shi Jian Zhao, Yong Mao Xu, Jie Zhang

Inversing Reinforced Concrete Beams Flexural Load Rating Using ANN and GA Hybrid Algorithm 779
Zeying Yang, Chengkui Huang, Jianbo Qu

Determining of the Delay Time for a Heating Ventilating and Air-Conditioning Plant Using Two Weighted Neural Network Approach 786
Mengdi Hu, Wenming Cao, Shoujue Wang

Intelligent Forecast Procedures for Slope Stability with Evolutionary Artificial Neural Network 792
Shouju Li, Yingxi Liu

Structural Reliability Analysis via Global Response Surface Method of BP Neural Network 799
Jinsong Gui, Hequan Sun, Haigui Kang

Modeling Temperature Drift of FOG by Improved BP Algorithm and by Gauss-Newton Algorithm 805
Xiyuan Chen

A Novel Chaotic Neural Network for Automatic Material Ratio System 813
Lidan Wang, Shukai Duan

Finite Element Analysis of Structures Based on Linear Saturated System Model 820
Hai-Bin Li, Hong-Zhong Huang, Ming-Yang Zhao

Neural Network Based Fatigue Cracks Evolution 826
Chunsheng Liu, Weidong Wu, Daoheng Sun

Density Prediction of Selective Laser Sintering Parts Based on Artificial Neural Network 832
Xianfeng Shen, Jin Yao, Yang Wang, Jialin Yang

Structure Optimization of Pneumatic Tire Using an Artificial Neural Network ... 841
XuChun Ren, ZhenHan Yao

A Multiple RBF NN Modeling Approach to BOF Endpoint Estimation in Steelmaking Process 848
Xin Wang, Shaoyuan Li, Zhongjie Wang, Jun Tao, Jinxin Liu

Application of Neural Network on Wave Impact Force Prediction 854
Hongyu Zhang, Yongxue Wang, Bing Ren

RBF Neural Networks-Based Software Sensor for Aluminum Powder Granularity Distribution Measurement 860
Yonghui Zhang, Cheng Shao, Qinghui Wu

A Technological Parameter Optimization Approach in Crude Oil Distillation Process Based on Neural Network ... 866
Hao Tang, Quanyi Fan, Bowen Xu, Jianming Wen

A Neural Network Modeling Method for Batch Process ... 874
Yi Liu, XianHui Yang, Jie Zhang

Modelling the Supercritical Fluid Extraction of Lycopene from Tomato Paste Waste Using Neuro-Fuzzy Approaches ... 880
Simon X. Yang, Weiren Shi, Jin Zeng

Detection of Weak Targets with Wavelet and Neural Network ... 886
Changwen Qu, You He, Feng Su, Yong Huang

An Artificial Olfactory System Based on Gas Sensor Array and Back-Propagation Neural Network ... 892
Huiling Tai, Guangzhong Xie, Yadong Jiang

Consumer Oriented Design of Product Forms ... 898
Yang-Cheng Lin, Hsin-Hsi Lai, Chung-Hsing Yeh

An Intelligent Simulation Method Based on Artificial Neural Network for Container Yard Operation ... 904
Chun Jin, Xinlu Liu, Peng Gao

A Freeway Traffic Incident Detection Algorithm Based on Neural Networks ... 912
Xuhua Yang, Zonghai Sun, Youxian Sun

Radial Basis Function Network for Chaos Series Prediction ... 920
Wei Chi, Bo Zhou, Aiguo Shi, Feng Cai, Yongsheng Zhang

Feature Extraction and Identification of Underground Nuclear Explosion and Natural Earthquake Based on FM^{m}let Transform and BP Neural Network ... 925
Xihai Li, Ke Zhao, Daizhi Liu, Bin Zhang

A Boosting-Based Framework for Self-Similar and Non-linear Internet Traffic Prediction ... 931
Hanghang Tong, Chongrong Li, Jingrui He

Traffic Flow Forecasting Based on Parallel Neural Network ... 937
Guozhen Tan, Wenjiang Yuan

A High Precision Prediction Method by Using Combination of ELMAN and SOM Neural Networks ... 943
Jie Wang, Dongwei Yan

Short-Term Traffic Flow Forecasting Using Expanded Bayesian Network for Incomplete Data ... 950
Changshui Zhang, Shiliang Sun, Guoqiang Yu

Highway Traffic Flow Model Using FCM-RBF Neural Network ... 956
Jian-Mei Xiao, Xi-Huai Wang

Earthquake Prediction by RBF Neural Network Ensemble ... 962
Yue Liu, Yuan Wang, Yuan Li, Bofeng Zhang, Gengfeng Wu

Rainfall-Runoff Correlation with Particle Swarm Optimization Algorithm ... 970
Kwokwing Chau

A Study of Portfolio Investment Decision Method Based on Neural Network ... 976
Yongqing Yang, Jinde Cao, Daqi Zhu

Portfolio Optimization for Multi-stage Capital Investment with Neural Networks ... 982
Yanglan Zhang, Yu Hua

Efficient Option Pricing via a Globally Regularized Neural Network ... 988
Hyung-Jun Choi, Hyo-Seok Lee, Gyu-Sik Han, Jaewook Lee

Effectiveness of Neural Networks for Prediction of Corporate Financial Distress in China ... 994
Ji-Gang Xie, Jing Wang, Zheng-Ding Qiu

A Neural Network Model on Solving Multiobjective Conditional Value-at-Risk ... 1000
Min Jiang, Zhiqing Meng, Qiying Hu

DSP Structure Optimizations – A Multirate Signal Flow Graph Approach ... 1007
Ronggang Qi, Zifeng Li, Qing Ma

Author Index ... 1013

Table of Contents, Part I

Part I Theoretical Analysis

Approximation Bounds by Neural Networks in L^p_ω 1
Jianjun Wang, ZongBen Xu, Weijun Xu

Geometric Interpretation of Nonlinear Approximation Capability for Feedforward Neural Networks 7
Bao-Gang Hu, Hong-Jie Xing, Yu-Jiu Yang

Mutual Information and Topology 1: Asymmetric Neural Network 14
David Dominguez, Kostadin Koroutchev, Eduardo Serrano, Francisco B. Rodríguez

Mutual Information and Topology 2: Symmetric Network 20
Kostadin Koroutchev, David Dominguez, Eduardo Serrano, Francisco B. Rodríguez

Simplified PCNN and Its Periodic Solutions 26
Xiaodong Gu, Liming Zhang, Daoheng Yu

On Robust Periodicity of Delayed Dynamical Systems with Time-Varying Parameters 32
Changyin Sun, Xunming Li, Chun-Bo Feng

Delay-Dependent Criteria for Global Stability of Delayed Neural Network System 38
Wenlian Lu, Tianping Chen

Stability Analysis of Uncertain Neural Networks with Delay 44
Zhongsheng Wang, Hanlin He, Xiaoxin Liao

A New Method for Robust Stability Analysis of a Class of Recurrent Neural Networks with Time Delays 49
Huaguang Zhang, Gang Wang

Criteria for Stability in Neural Network Models with Iterative Maps 55
Carlos Aguirre, Doris Campos, Pedro Pascual, Eduardo Serrano

Exponential Stability Analysis for Neural Network with Parameter Fluctuations 61
Haoyang Tang, Chuandong Li, Xiaofeng Liao

Local Stability and Bifurcation in a Model of Delayed Neural Network 67
Yiping Lin, Zengrong Liu

On the Asymptotic Stability of Non-autonomous Delayed Neural Networks 72
Qiang Zhang, Haijun Wang, Dongsheng Zhou, Xiaopeng Wei

Global Exponential Stability of Cohen-Grossberg Neural Networks
with Multiple Time-Varying Delays 78
Kun Yuan, Jinde Cao

Stability of Stochastic Cohen-Grossberg Neural Networks 84
Lin Wang

A Novel Approach to Exponential Stability Analysis of
Cohen-Grossberg Neural Networks 90
Anhua Wan, Weihua Mao, Chun Zhao

Analysis for Global Robust Stability of Cohen-Grossberg Neural
Networks with Multiple Delays 96
Ce Ji, Huaguang Zhang, Huanxin Guan

On Robust Stability of BAM Neural Networks with Constant Delays 102
Chuandong Li, Xiaofeng Liao, Yong Chen

Absolutely Exponential Stability of BAM Neural Networks
with Distributed Delays 108
Wenjun Xiong, Qiuhao Jiang

Stability Analysis of Discrete-Time Cellular Neural Networks 114
Zhigang Zeng, De-Shuang Huang, Zengfu Wang

Novel Exponential Stability Criteria for Fuzzy Cellular Neural
Networks with Time-Varying Delay 120
Yong Chen, Xiaofeng Liao

Stability of Discrete Hopfield Networks with Delay in Serial Mode 126
Runnian Ma, Youmin Xi, Hangshan Gao

Further Results for an Estimation of Upperbound of Delays
for Delayed Neural Networks 132
Xueming Li, Xiaofeng Liao, Tao Xiang

Synchronization in Two Uncoupled Chaotic Neurons 138
Ying Wu, Jianxue Xu, Daihai He, Wuyin Jin, Mi He

Robust Synchronization of Coupled Delayed Recurrent Neural Networks 144
Jin Zhou, Tianping Chen, Xiang Lan

Part II Learning and Optimization

Self-Optimizing Neural Networks ... 150
Adrian Horzyk, Ryszard Tadeusiewicz

Genetically Optimized Self-Organizing Neural Networks Based on PNs and FPNs ... 156
Ho-Sung Park, Sung-Kwun Oh, Witold Pedrycz, Hyun-Ki Kim

A New Approach to Self-Organizing Hybrid Fuzzy Polynomial Neural Networks: Synthesis of Computational Intelligence Technologies ... 162
Ho-Sung Park, Sung-Kwun Oh, Witold Pedrycz, Daehee Park, Yongkab Kim

On Soft Learning Vector Quantization Based on Reformulation ... 168
Jian Yu, Pengwei Hao

A New Approach to Self-Organizing Polynomial Neural Networks by Means of Genetic Algorithms ... 174
Sung-Kwun Oh, Byoung-Jun Park, Witold Pedrycz, Yong-Soo Kim

Fuzzy-Kernel Learning Vector Quantization ... 180
Daoqiang Zhang, Songcan Chen, Zhi-Hua Zhou

Genetic Generation of High-Degree-of-Freedom Feed-Forward Neural Networks ... 186
Yen-Wei Chen, Sulistiyo, Zensho Nakao

Self-Organizing Feature Map Based Data Mining ... 193
Shangming Yang, Yi Zhang

Diffusion and Growing Self-Organizing Map: A Nitric Oxide Based Neural Model ... 199
Shuang Chen, Zongtan Zhou, Dewen Hu

A New Adaptive Self-Organizing Map ... 205
Shifeng Weng, Fai Wong, Changshui Zhang

Evolving Flexible Neural Networks Using Ant Programming and PSO Algorithm ... 211
Yuehui Chen, Bo Yang, Jiwen Dong

Surrogating Neurons in an Associative Chaotic Neural Network ... 217
Masaharu Adachi

Ensembles of RBFs Trained by Gradient Descent ... 223
Carlos Hernández-Espinosa, Mercedes Fernández-Redondo, Joaquín Torres-Sospedra

Gradient Descent Training of Radial Basis Functions 229
Mercedes Fernández-Redondo, Carlos Hernández-Espinosa, Mamen Ortiz-Gómez, Joaquín Torres-Sospedra

Recent Developments on Convergence of Online Gradient Methods for Neural Network Training 235
Wei Wu, Zhengxue Li, Guori Feng, Naimin Zhang, Dong Nan, Zhiqiong Shao, Jie Yang, Liqing Zhang, Yuesheng Xu

A Regularized Line Search Tunneling for Efficient Neural Network Learning 239
Dae-Won Lee, Hyung-Jun Choi, Jaewook Lee

Transductive Learning Machine Based on the Affinity-Rule for Semi-supervised Problems and Its Algorithm 244
Weijiang Long, Wenxiu Zhang

A Learning Algorithm with Gaussian Regularizer for Kernel Neuron 252
Jianhua Xu, Xuegong Zhang

An Effective Learning Algorithm of Synergetic Neural Network 258
Xiuli Ma, Licheng Jiao

Sparse Bayesian Learning Based on an Efficient Subset Selection 264
Liefeng Bo, Ling Wang, Licheng Jiao

A Novel Fuzzy Neural Network with Fast Training and Accurate Generalization 270
Lipo Wang, Bing Liu, Chunru Wan

Tuning Neuro-fuzzy Function Approximator by Tabu Search 276
Guangyuan Liu, Yonghui Fang, Xufei Zheng, Yuhui Qiu

Finite Convergence of MRI Neural Network for Linearly Separable Training Patterns 282
Lijun Liu, Wei Wu

A Rapid Two-Step Learning Algorithm for Spline Activation Function Neural Networks with the Application on Biped Gait Recognition 286
Lingyun Hu, Zengqi Sun

An Online Feature Learning Algorithm Using HCI-Based Reinforcement Learning 293
Fang Liu, Jianbo Su

Optimizing the Weights of Neural Networks Based on Antibody Clonal Simulated Annealing Algorithm 299
Xiaoyi Jin, Haifeng Du, Wuhong He, Licheng Jiao

Backpropagation Analysis of the Limited Precision on High-Order Function Neural Networks 305
Minghu Jiang, Georges Gielen

LMS Adaptive Notch Filter Design Based on Immune Algorithm 311
Xiaoping Chen, Jianfeng Gao

Training Radial Basis Function Networks with Particle Swarms 317
Yu Liu, Qin Zheng, Zhewen Shi, Junying Chen

Optimizing Weights by Genetic Algorithm for Neural Network Ensemble 323
Zhang-Quan Shen, Fan-Sheng Kong

Survival Density Particle Swarm Optimization for Neural Network Training 332
Hongbo Liu, Bo Li, Xiukun Wang, Ye Ji, Yiyuan Tang

Modified Error Function with Added Terms for the Backpropagation Algorithm 338
Weixing Bi, Xugang Wang, Ziliang Zong, Zheng Tang

Robust Constrained-LMS Algorithm 344
Xin Song, Jinkuan Wang, Han Wang

A Novel Three-Phase Algorithm for RBF Neural Network Center Selection 350
Dae-Won Lee, Jaewook Lee

Editing Training Data for *k*NN Classifiers with Neural Network Ensemble 356
Yuan Jiang, Zhi-Hua Zhou

Learning Long-Term Dependencies in Segmented Memory Recurrent Neural Networks 362
Jinmiao Chen, Narendra S. Chaudhari

An Optimal Neural-Network Model for Learning Posterior Probability Functions from Observations 370
Chengan Guo, Anthony Kuh

The Layered Feed-Forward Neural Networks and Its Rule Extraction 377
Ray Tsaih, Chih-Chung Lin

Analysing Contributions of Components and Factors to Pork Odour Using Structural Learning with Forgetting Method 383
Leilei Pan, Simon X. Yang, Fengchun Tian, Lambert Otten, Roger Hacker

On Multivariate Calibration Problems 389
Guo-Zheng Li, Jie Yang, Jun Lu, Wen-Cong Lu, Nian-Yi Chen

Control of Associative Chaotic Neural Networks Using a Reinforcement Learning 395
Norihisa Sato, Masaharu Adachi, Makoto Kotani

A Method to Improve the Transiently Chaotic Neural Network 401
Xinshun Xu, Jiahai Wang, Zheng Tang, Xiaoming Chen, Yong Li, Guangpu Xia

A New Neural Network for Nonlinear Constrained Optimization Problems 406
Zhiqing Meng, Chuangyin Dang, Gengui Zhou, Yihua Zhu, Min Jiang

Delay PCNN and Its Application for Optimization 413
Xiaodong Gu, Liming Zhang, Daoheng Yu

A New Parallel Improvement Algorithm for Maximum Cut Problem 419
Guangpu Xia, Zheng Tang, Jiahai Wang, Ronglong Wang, Yong Li, Guang'an Xia

A New Neural Network Algorithm for Clique Vertex-Partition Problem 425
Jiahai Wang, Xinshun Xu, Zheng Tang, Weixing Bi, Xiaoming Chen, Yong Li

An Algorithm Based on Hopfield Network Learning for Minimum Vertex Cover Problem 430
Xiaoming Chen, Zheng Tang, Xinshun Xu, Songsong Li, Guangpu Xia, Jiahai Wang

A Subgraph Isomorphism Algorithm Based on Hopfield Neural Network 436
Ensheng Yu, Xicheng Wang

A Positively Self-Feedbacked Hopfield Neural Network for N-Queens Problem 442
Yong Li, Zheng Tang, Ronglong Wang, Guangpu Xia, Jiahai Wang

NN-Based GA for Engineering Optimization 448
Ling Wang, Fang Tang

Applying GENET to the JSSCSOP 454
Xin Feng, Lixin Tang, Hofung Leung

Part III Support Vector Machines

Improvements to Bennett's Nearest Point Algorithm for Support Vector Machines 462
Jianmin Li, Jianwei Zhang, Bo Zhang, Fuzong Lin

Distance-Based Selection of Potential Support Vectors by Kernel Matrix 468
Baoqing Li

A Learning Method for Robust Support Vector Machines ... 474
Jun Guo, Norikazu Takahashi, Tetsuo Nishi

A Cascade Method for Reducing Training Time and the Number of Support Vectors ... 480
Yi-Min Wen, Bao-Liang Lu

Minimal Enclosing Sphere Estimation and Its Application to SVMs Model Selection ... 487
Huaqing Li, Shaoyu Wang, Feihu Qi

Constructing Support Vector Classifiers with Unlabeled Data ... 494
Tao Wu, Han-Qing Zhao

Nested Buffer SMO Algorithm for Training Support Vector Classifiers ... 500
Xiang Wu, Wenkai Lu

Support Vector Classifier with a Fuzzy-Value Class Label ... 506
Chan-Yun Yang

RBF Kernel Based Support Vector Machine with Universal Approximation and Its Application ... 512
Junping Wang, Quanshi Chen, Yong Chen

A Practical Parameters Selection Method for SVM ... 518
Yongsheng Zhu, Chunhung Li, Youyun Zhang

Ho–Kashyap with Early Stopping Versus Soft Margin SVM for Linear Classifiers – An Application ... 524
Fabien Lauer, Mohamed Bentoumi, Gérard Bloch, Gilles Millerioux, Patrice Aknin

Radar HRR Profiles Recognition Based on SVM with Power-Transformed-Correlation Kernel ... 531
Hongwei Liu, Zheng Bao

Hydrocarbon Reservoir Prediction Using Support Vector Machines ... 537
Kaifeng Yao, Wenkai Lu, Shanwen Zhang, Huanqin Xiao, Yanda Li

Toxic Vapor Classification and Concentration Estimation for Space Shuttle and International Space Station ... 543
Tao Qian, Roger Xu, Chiman Kwan, Bruce Linnell, Rebecca Young

Optimal Watermark Detection Based on Support Vector Machines ... 552
Yonggang Fu, Ruimin Shen, Hongtao Lu

Online LS-SVM Learning for Classification Problems Based on Incremental Chunk ... 558
Zhifeng Hao, Shu Yu, Xiaowei Yang, Feng Zhao, Rong Hu, Yanchun Liang

A Novel Approach to Clustering Analysis Based on Support Vector Machine 565
Zhonghua Li, ShaoBai Chen, Rirong Zheng, Jianping Wu, Zongyuan Mao

Application of Support Vector Machine in Queuing System 571
Gensheng Hu, Feiqi Deng

Modelling of Chaotic Systems with Novel Weighted Recurrent Least Squares Support Vector Machines 578
Jiancheng Sun, Taiyi Zhang, Haiyuan Liu

Nonlinear System Identification Based on an Improved Support Vector Regression Estimator 586
Li Zhang, Yugeng Xi

Anomaly Detection Using Support Vector Machines 592
Shengfeng Tian, Jian Yu, Chuanhuan Yin

Power Plant Boiler Air Preheater Hot Spots Detection System Based on Least Square Support Vector Machines 598
Liu Han, Liu Ding, Jin Yu, Qi Li, Yanming Liang

Support Vector Machine Multiuser Detector for TD-SCDMA Communication System in Multipath Channels 605
Yonggang Wang, Licheng Jiao, Dongfang Zhao

Eyes Location by Hierarchical SVM Classifiers 611
Yunfeng Li, Zongying Ou

Classification of Stellar Spectral Data Using SVM 616
Fei Xing, Ping Guo

Iris Recognition Using Support Vector Machines 622
Yong Wang, Jiuqiang Han

Heuristic Genetic Algorithm-Based Support Vector Classifier for Recognition of Remote Sensing Images 629
Chunhong Zheng, Guiwen Zheng, Licheng Jiao

Landmine Feature Extraction and Classification of GPR Data Based on SVM Method 636
Jing Zhang, Qun Liu, Baikunth Nath

Occupant Classification for Smart Airbag Using Stereovision and Support Vector Machines 642
Hyun-Gu Lee, Yong-Guk Kim, Min-Soo Jang, Sang-Jun Kim, Soek-Joo Lee, Gwi-Tae Park

Support Vector Machine Committee for Classification 648
Bing-Yu Sun, De-Shuang Huang, Lin Guo, Zhong-Qiu Zhao

Automatic Modulation Classification by Support Vector Machines 654
Zhijin Zhao, Yunshui Zhou, Fei Mei, Jiandong Li

Part IV Blind Source Separation, Independent Component Analysis and Principal Component Analysis

Blind Source Separation Using for Time-Delay Direction Finding 660
Gaoming Huang, Luxi Yang, Zhenya He

Frequency-Domain Separation Algorithms for Instantaneous Mixtures 666
Tiemin Mei, Fuliang Yin

Cumulant-Based Blind Separation of Convolutive Mixtures 672
Tiemin Mei, Fuliang Yin, Jiangtao Xi, Joe F. Chicharo

A New Blind Source Separation Method Based on Fractional Lower Order Statistics and Neural Network 678
Daifeng Zha, Tianshuang Qiu, Hong Tang, Yongmei Sun, Sen Li, Lixin Shen

A Clustering Approach for Blind Source Separation with More Sources than Mixtures 684
Zhenwei Shi, Huanwen Tang, Yiyuan Tang

A Blind Source Separation Algorithm with Linear Prediction Filters 690
Zhijin Zhao, Fei Mei, Jiandong Li

A Blind Source Separation Based Micro Gas Sensor Array Modeling Method 696
Guangfen Wei, Zhenan Tang, Philip C.H. Chan, Jun Yu

A Novel Denoising Algorithm Based on Feedforward Multilayer Blind Separation 702
Xiefeng Cheng, Ju Liu, Jianping Qiao, Yewei Tao

The Existence of Spurious Equilibrium in FastICA 708
Gang Wang, Dewen Hu

Hardware Implementation of Pulsed Neural Networks Based on Delta-Sigma Modulator for ICA 714
Hirohisa Hotta, Yoshimitsu Murahashi, Shinji Doki, Shigeru Okuma

Study on Object Recognition Based on Independent Component Analysis 720
Xuming Huang, Chengming Liu, Liming Zhang

Application of ICA Method for Detecting Functional MRI Activation Data 726
Minfen Shen, Weiling Xu, Jinyao Yang, Patch Beadle

Single-Trial Estimation of Multi-channel VEP Signals Using Independent Component Analysis 732
Lisha Sun, Minfen Shen, Weiling Xu, Francis Chan

Data Hiding in Independent Components of Video 738
Jiande Sun, Ju Liu, Huibo Hu

Multisensor Data Fusion Based on Independent Component Analysis for Fault Diagnosis of Rotor 744
Xiaojiang Ma, Zhihua Hao

Substructural Damage Detection Using Neural Networks and ICA 750
Fuzheng Qu, Dali Zou, Xin Wang

Speech Segregation Using Constrained ICA 755
Qiu-Hua Lin, Yong-Rui Zheng, Fuliang Yin, Hua-Lou Liang

Adaptive RLS Implementation of Non-negative PCA Algorithm for Blind Source Separation 761
Xiao-Long Zhu, Xian-Da Zhang, Ying Jia

Progressive Principal Component Analysis 768
Jun Liu, Songcan Chen, Zhi-Hua Zhou

Extracting Target Information in Multispectral Images Using a Modified KPCA Approach 774
Zhan-Li Sun, De-Shuang Huang

Use PCA Neural Network to Extract the PN Sequence in Lower SNR DS/SS Signals 780
Tianqi Zhang, Xiaokang Lin, Zhengzhong Zhou

Hierarchical PCA-NN for Retrieving the Optical Properties of Two-Layer Tissue Model 786
Yaqin Chen, Ling Lin, Gang Li, Jianming Gao, Qilian Yu

Principal Component Analysis Neural Network Based Probabilistic Tracking of Unpaved Road 792
Qing Li, Nannig Zheng, Lin Ma, Hong Cheng

Chemical Separation Process Monitoring Based on Nonlinear Principal Component Analysis 798
Fei Liu, Zhonggai Zhao

An Adjusted Gaussian Skin-Color Model Based on Principal Component Analysis 804
Zhi-Gang Fan, Bao-Liang Lu

Convergence Analysis for Oja+ MCA Learning Algorithm 810
Jiancheng Lv, Mao Ye, Zhang Yi

On the Discrete Time Dynamics of the MCA Neural Networks 815
Mao Ye, Zhang Yi

The Cook Projection Index Estimation Using the Wavelet Kernel Function 822
Wei Lin, Tian Zheng, Fan He, Xian-bin Wen

Part V Clustering and Classification

Automatic Cluster Number Determination via BYY Harmony Learning 828
Xuelei Hu, Lei Xu

Unsupervised Learning for Hierarchical Clustering Using Statistical Information 834
Masaru Okamoto, Nan Bu, Toshio Tsuji

Document Clustering Algorithm Based on Tree-Structured Growing Self-Organizing Feature Map 840
Xiaoshen Zheng, Wenling Liu, Pilian He, Weidi Dai

Improved SOM Clustering for Software Component Catalogue 846
Zhuo Wang, Daxin Liu, Xiaoning Feng

Classification by Multilayer Feedforward Ensembles 852
Mercedes Fernández-Redondo, Carlos Hernández-Espinosa, Joaquín Torres-Sospedra

Robust Face Recognition from a Single Training Image per Person with Kernel-Based SOM-Face 858
Xiaoyang Tan, Songcan Chen, Zhi-Hua Zhou, Fuyan Zhang

Unsupervised Feature Selection for Multi-class Object Detection Using Convolutional Neural Networks 864
Masakazu Matsugu, Pierre Cardon

Recurrent Network as a Nonlinear Line Attractor for Skin Color Association 870
Ming-Jung Seow, Vijayan K. Asari

A Bayesian Classifier by Using the Adaptive Construct Algorithm of the RBF Networks 876
Minghu Jiang, Dafan Liu, Beixing Deng, Georges Gielen

An Improved Analytical Center Machine for Classification 882
FanZi Zeng, DongSheng Li, ZhengDing Qiu

Analysis of Fault Tolerance of a Combining Classifier 888
Hai Zhao, Bao-Liang Lu

Training Multilayer Perceptron with Multiple Classifier Systems 894
Hui Zhu, Jiafeng Liu, Xianglong Tang, Jianhuan Huang

Learning the Supervised NLDR Mapping for Classification 900
Zhonglin Lin, Shifeng Weng, Changshui Zhang, Naijiang Lu, Zhimin Xia

Comparison with Two Classification Algorithms of Remote Sensing Image Based on Neural Network 906
Yumin Chen, Youchuan Wan, Jianya Gong, Jin Chen

Some Experiments with Ensembles of Neural Networks for Classification of Hyperspectral Images 912
Carlos Hernández-Espinosa, Mercedes Fernández-Redondo, Joaquín Torres-Sospedra

SAR Image Recognition Using Synergetic Neural Networks Based on Immune Clonal Programming 918
Shuiping Gou, Licheng Jiao

Speaker Identification Using Reduced RBF Networks Array 924
Han Lian, Zheng Wang, Jianjun Wang, Liming Zhang

Underwater Acoustic Targets Classification Using Welch Spectrum Estimation and Neural Networks 930
Chunyu Kang, Xinhua Zhang, Anqing Zhang, Hongwen Lin

Automatic Digital Modulation Recognition Using Wavelet Transform and Neural Networks 936
Zhilu Wu, Guanghui Ren, Xuexia Wang, Yaqin Zhao

Local Face Recognition Based on the Combination of ICA and NFL 941
Yisong Ye, Yan Wu, Mingliang Sun, Mingxi Jin

Facial Expression Recognition Using Kernel Discriminant Plane 947
Wenming Zheng, Xiaoyan Zhou, Cairong Zou, Li Zhao

Application of CMAC-Based Networks on Medical Image Classification 953
Weidong Xu, Shunren Xia, Hua Xie

Seismic Pattern Recognition of Nuclear Explosion Based on Generalization Learning Algorithm of BP Network and Genetic Algorithm 959
Daizhi Liu, Renming Wang, Xihai Li, Zhigang Liu

A Remote Sensing Image Classification Method Using Color and Texture Feature 965
Wen Cao, Tian-Qiang Peng, Bi-Cheng Li

Remote Sensing Image Classification Based on Evidence Theory and Neural Networks 971
Gang Chen, Bi-Cheng Li, Zhi-Gang Guo

Neural Networks in Detection and Identification of Littoral Oil Pollution by Remote Sensing 977
Bin Lin, Jubai An, Carl Brown, Hande Zhang

Identifying Pronunciation-Translated Names from Chinese Texts Based on Support Vector Machines 983
Lishuang Li, Chunrong Chen, Degen Huang, Yuansheng Yang

FPGA Implementation of Feature Extraction and Neural Network Classifier for Handwritten Digit Recognition 988
Dongsheng Shen, Lianwen Jin, Xiaobin Ma

Replicator Neural Networks for Outlier Modeling in Segmental Speech Recognition 996
László Tóth, Gábor Gosztolya

A Novel Approach to Stellar Recognition by Combining EKF and RBF Net 1002
Ling Bai, Ping Guo

Pattern Recognition Based on Stability of Discrete Time Cellular Neural Networks 1008
Zhigang Zeng, De-Shuang Huang, Zengfu Wang

The Recognition of the Vehicle License Plate Based on Modular Networks 1015
Guangrong Ji, Hongjie Yi, Bo Qin, Hua Xu

Neural Network Associator for Images and Their Chinese Characters 1021
Zhao Zhang

Exploring Various Features to Optimize Hot Topic Retrieval on WEB 1025
Lan You, Xuanjing Huang, Lide Wu, Hao Yu, Jun Wang, Fumihito Nishino

Author Index 1033

Application of RBFNN for Humanoid Robot Real Time Optimal Trajectory Generation in Running

Xusheng Lei and Jianbo Su

Department of Automation & Research Center of Intelligent Robotics
Shanghai Jiaotong University, Shanghai, 200030, China
{xushenglei, jbsu}@sjtu.edu.cn

Abstract. In this paper, a method for trajectory generation in running is proposed with Radial Basis Function Neural Network, which can generate a series of joint trajectories to adjust humanoid robot step length and step time based on the sensor information. Compared with GA, RBFNN use less time to generate new trajectory to deal with sudden obstacles after thorough training. The performance of the proposed method is validated by simulation of a 28 DOF humanoid robot model with ADAMS.

1 Introduction

There are significant progresses in realization of stable dynamic walking in several full-body humanoid robots [1]. To make humanoid robots run is currently one of the most exciting topics in robotics. Since running has a special phase that whole body of humanoid robot is in the air, which makes high requirement for system energy supply. Furthermore, humanoid robot has to overcome high impact force at the contact phase because of its high speed. To solve these problems, Hodgins developed locomotion control algorithms that allow a humanoid robot model to run at a variety of speeds in an arbitrary direction on the plane [2]. With an aerial posture controller, Hyon and Emura achieved desired posture control for a biped robot [3]. By controlling hip height and feet stride, Kwon and Park realized a fast movement of a biped robot and used impedance control method to absorb impact force at landing moments [4]. Nagasaki and Kajita proposed the method to control total linear and angular momentum of humanoid robot to generate a reliable running pattern [5].

All of the above methods were tested in simulations because running needs so much energy and current humanoid robot systems can not support such motion. As an assistant tool for human beings, humanoid robot should have ability to involve into common living environment. It is hard to use external power supply since power cable will prevent humanoid robot from complex motions. Therefore, it is critical to study humanoid robot running trajectory generation based on minimized energy.

Since a humanoid robot is a complex nonlinear system with too many variables and there are a lot of constraints from mechanical and dynamic point of view, realization of a stable running and further searching for an optimal solution are shown to be computationally expensive. Genetic algorithm has already shown its great searching efficiency in a big range. Hence we use GA as a tool to get optimal running

F. Yin, J. Wang, and C. Guo (Eds.): ISNN 2004, LNCS 3174, pp. 1–6, 2004.

trajectory. Using consumed energy as criteria, the trajectory generated by GA consumes minimal energy in running process. However, GA takes a long time to get optimal solutions, which is not fit for real time motion control. RBFNN has shown strong ability in approximation problems, which provides high accuracy and is computationally simple [6]. It can quickly generate results after thorough training. So we use RBFNN to divide computational expenses by offline training from GA results to decrease online computational costs. Consequently, the humanoid robot can change step length and step time based on sensor information to adapt to environment.

The rest of the paper is organized as follows. In Section 2, Dynamic model of humanoid robot and RBFNN are described. The proposed method is discussed in Section 3. Simulations are given in Section 4, followed by Conclusions in Section 5.

2 Dynamic Model of Humanoid Robot and RBFNN

A humanoid robot is a mechanism that interacts with the environment through its body. Based on this conception, the dynamics of the humanoid robot is normally described as follows:

$$\tau = H(q)\ddot{q} + C(q,\dot{q}) + G(q) + J^T f\,, \tag{1}$$

where τ, q, $C(q,\dot{q})$, $G(q)$ are the vectors of generalized torque, joint coordinates, centrifugal torques, gravity torques, respectively, $H(q)$ is the inertial matrix. Jacobian matrix J^T transfers external force f to humanoid robot system.

The RBFNN is a three-layer forward neural network, consisting of one input layer, one hidden layer and one output layer. The input layer connects the network to the environment. The hidden layer applies a set of nonlinear functions to transfer the input space to hidden space, which processes the input data according to its distance from the associated centers. The output layer is the linear summation of the output of the hidden layer. The output of the RBFNN is computed according to the following equation

$$y = f(x) = \sum_{k=1}^{N} w_k \phi\,(\| x - c_k \|)\,, \tag{2}$$

where $x \in R^m$ is the input vector; $y \in R^n$ is the output vector; w_k is the weights of hidden layers; $\phi(\cdot)$ is the nonlinear function in hidden layer; $\| \cdot \|$ denotes the Euclidean norm; c_k is the center of the k hidden nodes.

3 Trajectory Generation

In this section, we first analyze the characteristics of running, separate the whole process into contacting phase and flying phase and describe them in detail. Then we introduce the energy consumption criteria and use GA to get the optimal trajectory. In the last, we propose method to train RBFNN based on GA results.

A complete running cycle can be divided into contacting phase and flying phase. In the contacting phase, there is only one foot having a contact with the ground, which plays an important role in system stability. The robot has to absorb the impact force at landing moment to avoid slipping, push the ground to get high rebound force and add energy for the continuous flying phase. In the flying phase there is no force except for gravity force on the system. So the trajectory of the hip link in the motion direction is modeled into a ballistic trajectory. The leg swinging forward for touchdown is defined as the main leg and another one as assistant leg. When the main leg touches the ground in one running step, it changes roles with assistant leg in succeeding running period.

Since joint torque is proportional to current, there exists the relationship between the consumed energy and motor torque. So the torque can be used as the criteria for energy consumption. The energy to control the motion of the robot is proportional to the integration of the square of the joint torque, so the cost function J can be defined as:

$$J = \frac{1}{2}\left(\int_0^T \tau' \tau dt + \int_0^T C dt \right), \tag{3}$$

where T is the step time, τ is the torque vector and C is the switch constant given as follows:

$$C = \begin{cases} A_1 & \textit{if the hell and toe stub into the ground} \\ A_2 & \textit{if ZMP exceeds the polygon region in contacting phase} \\ A_3 & \textit{if COG exceeds the knee region in flying phase} \\ 0 & \textit{otherwise} \end{cases}$$

where A_1, A_2 and A_3 are user-defined parameters.

For GA, the most important factor is fitness function, which decides the performance of the whole process. At the same time, consumed energy is the key for trajectory generation, so we choose J in (3) as the fitness function that involves ZMP and COG as constraints to guarantee system's stability. Using (1), we can get corresponding torque to realize planed trajectory, then we can get the index of consumed energy based on (3). The individuals are given different fitness values based on the cost function J. When termination is not met, the individuals with high fitness can be selected as parents to generate offspring. The trajectories breaking constraints will be discarded gradually in GA selection process because they have little chance to generate offspring. GA moves from generation to generation to select suitable solution until the termination constraints are met. Finally, we can get trajectories to realize stable running consuming minimal energy.

The most important factors affecting the performance of RBFNN are the selection of the centers and weights of radical basis function. In order to get a good training result of RBFNN, we use orthogonal least squares learning algorithm [7] to select centers and weights for the RBFNN. The RBFNN can be seen as a linear matrix model:

$$y = p \cdot \theta + E = w \cdot A \cdot \theta + E = w \cdot g + E , \tag{4}$$

where w is an $N \times M$ matrix with orthogonal columns, $\theta = [\theta_1, ..., \theta_M]^T$ is the weight values of RBFNN, $E = [\varepsilon(1), ..., \varepsilon(N)]^T$ represents the errors signal. A is a $M \times M$ triangular matrix :

$$A = \begin{bmatrix} 1 & a_{12} & a_{13} & \cdots & a_{1M} \\ 0 & 1 & a_{23} & \cdots & a_{2M} \\ \cdots & 0 & 1 & \cdots & \cdots \\ \cdots & \cdots & \cdots & \cdots & a_{(M-1)M} \\ 0 & 0 & \cdots & 0 & 1 \end{bmatrix}, \tag{5}$$

and

$$g = A \cdot \theta = w^{-1} y . \tag{6}$$

Based on the Gram-Schmidt method, the values of w and A can be got after computation, then we can calculate the weight values of the neural network.

In order to enable the humanoid robot to adapt to environment changes, its step length and time should be able to adjust online. Thus these variables are chosen to be the input variables of RBFNN and the hip, knee and ankle angles as the output of the RBFNN. The shoulder and elbow angles are synchronized with the angles of leg, which play as compensator to reduce the errors caused by the swing of legs. The selection of transform function is not crucial to the performance of RBFNN, so we choose Gaussian function as transformation function. The orthogonal least squares learning algorithm is used to train RBF model. After thorough training, The RBFNN can generate the new trajectory for step length and step time quickly to adapt to the changes of the environment. As the most weight of humanoid robot is in the upper body and the ankle is the direct mechanism to control foot, the hip and ankle angles are the main factors to affect system stability. So the system use feedback control algorithm to adjust hip and ankle angles to compensate the errors gradually to keep stability based on the sensor information.

Table 1. Parameters of the Humanoid Robot

Link	Weight (kg)	Length (cm)
head	2	10
body	6	30
thigh	1	15
shank	1	15
ankle	0.9	10
foot	0.1	16
arm	0.5	15

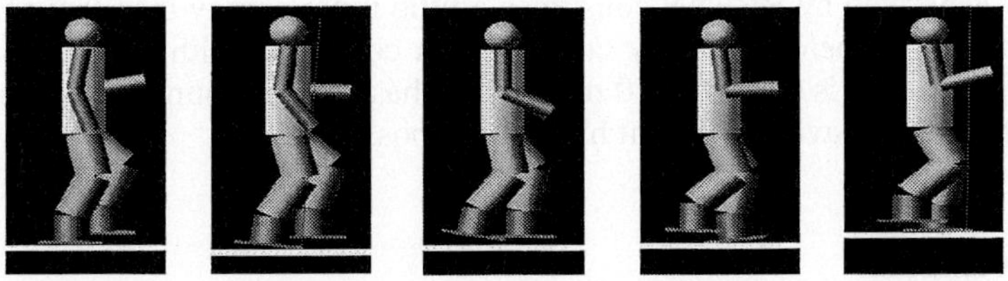

(a)t=6.0s (b)t=6.2s (c)t=6.25s (d)t=6.3s (e)t=6.5s

Fig. 1. Animation of run pattern using ADAMS.

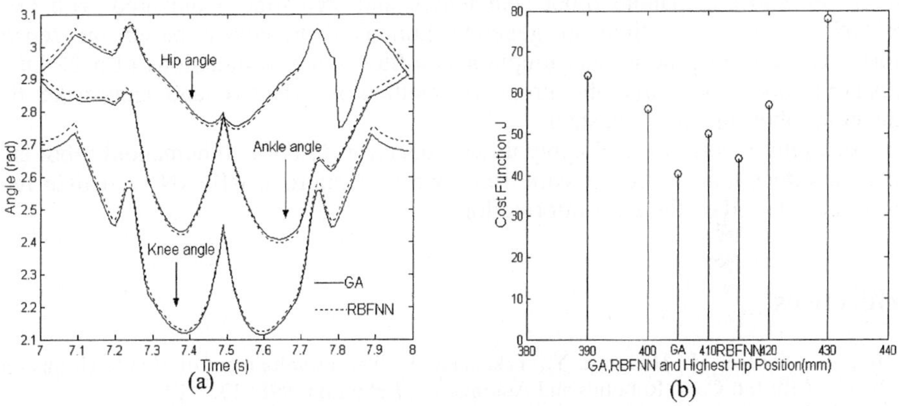

Fig. 2. (a) GA and RBFNN angle trajectory in running, (b) Energy consumption for different methods

4 Simulations

A humanoid robot simulation platform is constructed by ADAMS, whose model has a total of 28 degrees of freedom, i.e. six degrees of freedom for each leg (three for hip joint, one for the knee joint and two for the ankle joint), and six degrees of freedom for each arm (three for shoulder joint, one for the elbow joint and two for the wrist joint). There are two degrees of freedom in the neck joint to allow the head to turn to different directions and two in trunk joint to realize complex turn motion. The model parameters used in simulations are shown in Table 1.

A running pattern is shown in Fig. 1. The flying phase is from (b) to (d), the fly time is $0.1s$ and the fly height is $0.0125\ m$. In order to improve the performance of RBFNN, we get a lot of data using GA method. The optimal trajectory is generated for different step lengths and step times. Step length varies from 0.1 to 0.4 m and step time varies from 0.1 to 1 s. After thorough studying, RBFNN can generate optimal trajectory quickly. The time RBFNN needs is no more than one percent of the time GA needs and the trajectory satisfies the constraints mentioned before. The joint trajectories of GA and RBFNN for the step length 0.22 m and step time 0.5s are shown in Fig. 2(a). It is easy to see that the difference of joint angles between GA and RBFNN is very small. The values of J in running period for GA, RBFNN and the method having different highest hip positions are presented in Fig. 2(b). The trajectory generated by RBFNN consumes a little more energy than that of GA, but it reduces nearly 20 percent energy consumption compared with the trajectory whose highest hip position is equal to 410 mm, which has the minimum energy consumption for the trajectories having different highest hip positions.

5 Conclusions

In this paper, a real time trajectory generation method based on RBFNN is proposed. After thorough training of GA results, RBFNN can generate optimal trajectory quickly to adjust humanoid robot step length and step time. Compared with GA, RBFNN can use less time to generate humanoid trajectory based on sensor information, and the energy consumption is a little more. Simulations of a 28-DOF humanoid robot show that the proposed method is effective and can make the humanoid robot run and walk stably.

As a future work, we will apply these trajectories to our real humanoid robot and test their effectiveness. At the same time, we will improve RBFNN to realize real time control based on the error information.

References

1. Hirai, K., Hirose, M., Haikawa, Y., Takenaka, T.: The Development of Honda Humanoid Robot. IEEE Int. Conf. Robotics and Automation, Belgium (1998) 1321-1326
2. Jessica, K., Hodgins, K.: Three-Dimensional Human Running. IEEE Int. Conf. On Robotics and Automation, Minesota (1996) 3271-3276
3. Hyon, S.H., Rmura, T.: Aerial Posture Control for 3D Biped Running Using Compensator Around Yaw Axis. IEEE Int. Conf. On Robotics and Automation, Taiwan (2003) 57-62
4. Kwon, O., Park, J.H.: Gait Transitions for Walking and Running of Biped Robots. IEEE Int. Conf. On Robotics and Automation, Taiwan (2003) 1350-1355
5. Kajita, S., Nagasaki, T., Yokoi, K., Kaneko, K., Tanie, K.: Running Pattern Generation and Its Evaluation Using A Realistic Humanoid Model. IEEE Int. Conf. On Robotics and Automation, Taiwan (2003) 1336-1342
6. Capi, G., Nasu, Y., Barolli, L., Mitobe K.: Real Time Gait Generation for Autonomous Humanoid Robots: A Case Study for Walking. Vol. 42. Robotics and Autonomous Systems (2003) 107–116
7. Chen, S., Cowan, C.F.N., Grant, P.M.: Orthogonal Least Squares Learning Algorithm for Radial Basis Function Networks. Vol.2. IEEE Transactions on Neural Networks (1991) 302-309

Full-DOF Calibration-Free Robotic Hand-Eye Coordination Based on Fuzzy Neural Network

Jianbo Su[1,2] Qielu Pan[1] and Zhiwei Luo[2]

[1] Department of Automation, Shanghai Jiaotong University
Shanghai, 200030, China
{jbsu,qlpan}@sjtu.edu.cn
[2] Bio-Mimetic Control Research Center, RIKEN
Anagahora, Shimoshidami, Moriyama-ku, Nagoya, 463-0003, Japan
{su,luo}@bmc.riken.jp

Abstract. This paper studies coordination control for an uncalibrated single eye-in-hand robotic system to track an object in 3-D space with simultaneous translations and rotations. A set of object features is properly defined to derive an invertible nonlinear visual mapping model from visual feedback to robot control. A novel fuzzy neural network is proposed to realize the nonlinear mapping model effectively, which is essential to implement six-degree-of-freedom control of robot hand. Simulation results show the performance of the proposed method.

1 Introduction

The robotic hand-eye coordination problem has been studied for a long time. Since the hand-eye relation model is strongly nonlinear, and is closely related to system configurations, it is very difficult to get a model of enough accuracy for control [1]. This motivates researches on uncalibrated hand-eye coordination [2][3], which is to seek robot control directly from visual feedback without knowledge of the hand-eye relation model.

A widely accepted strategy to deal with the nonlinearity is to approximate it with a series of linearities. The image Jacobian matrix, which linearly describes temporary relations of a differential movement in image coordinate system and robotic coordinate system, is involved to address the uncalibrated hand-eye coordination problem and many methods have been developed thereafter. Hand-eye coordination performance strongly relies on performance of online estimation of the image Jacobian matrix, such as accuracy, estimation speed, and data accumulation procedures [4].

There are efforts to utilize the Artificial Neural Network (ANN) to deal with the image Jacobian matrix through offline training [6][7]. But the capacity of the ANN to approach a nonlinear function is not well exploited. In our previous research [5], we proposed a novel nonlinear visual mapping model to extend the image Jacobian matrix model for uncalibrated hand-eye coordination, and then use ANN to realize this nonlinear model. Since the nonlinear model describes the hand-eye relations more completely, the system performance can be

F. Yin, J. Wang, and C. Guo (Eds.): ISNN 2004, LNCS 3174, pp. 7–12, 2004.

enhanced to a great extent. Control of an uncalibrated hand-eye system to track a translationally moving object in 3-D space was analyzed.

An object moving in 3-D space generally has 3-D translations and 3-D rotations. This paper studies the coordination control for an uncalibrated robotic hand-eye system to track a moving object with full six-degree-of-freedom(DOF) motions. The difficulties of the full six-DOF tracking problem lie in that the translation and rotation movements of the object are coupled in image plane. Consequently it is not easy to infer the translation and rotation controls of the robot from visual feedback. Relations of the visual feedback and the robot control are analyzed, which leads to an invertible nonlinear visual mapping model of six dimensions by taking into account the object physical model. A novel fuzzy neural network(FNN) is then proposed to realize the nonlinear mapping model efficiently. Simulations show the validity of the proposed method.

2 System Model of Full-DOF Visual Tracking

2.1 Problem Description

For the full-DOF tracking problem, the robot hand should execute translation and rotation movements at the same time. A typical robot tracking system is shown in Fig.1. The camera is fixed in the robot hand. Relations between the robot (hand) and the camera are unknown. An object is moving in the working space freely (including translations and rotations). The tracking task is finished when the robot hand catches the object or is kept a constant relative pose with respect to the object.

2.2 Image Feature Space

To avoid tracking singularity problem [4], we should employ the same number of object features as the control DOFs of the robot hand, in that an invertible visual mapping model is obtained to uniquely transform errors from image feature space to robot control space, or vice versa. Thus the physical features of the object in image plane must be utilized in addition to its position coordinates in image plane, which implies the object model should be known a prior.

Suppose the object is of a triangular shape, the object feature vector in image plane is defined as: $f = [x, y, \theta, l_1, r_{21}, r_{31}]$, where (x, y) are the coordinates of a corner point of the object in image plane, θ is the angle of one edge rotating counterclockwise from the x axis. The lengths of the three edges are respectively l_1, l_2, l_3, and $r_{21} = l_2/l_1$,$r_{31} = l_3/l_1$. The feature definitions are shown in Fig. 2. Note that the feature definition is not arbitrary. A different feature set may lead to a different structure of the visual mapping model, which is most likely to have different convergence performance in training when the model is realized by ANN. Empirically, features that have different attributes from others are selected to make them as uncorrelated to each other as possible. For example, if the features of the three corner points of the triangle are selected, the resulted visual mapping model converges very slowly when trained by ANN.

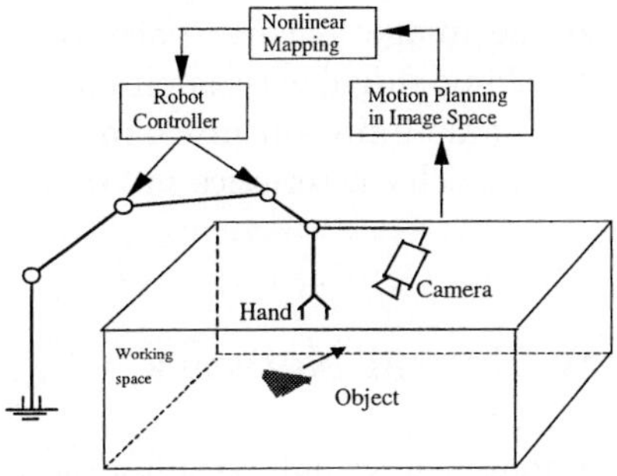

Fig. 1. Overview of the full-DOF robotic hand-eye coordination system

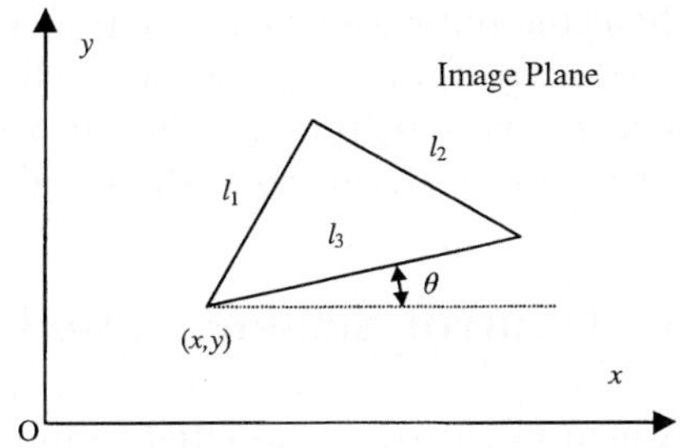

Fig. 2. Feature definitions for a triangular object

2.3 Visual Mapping Model

When the object feature vector is defined, its position and dynamics can be extracted by image processing. Suppose at the k-th visual sampling instant, feature position, velocity and acceleration vectors in image plane are f, $\dot{f}$ and $\ddot{f}$, respectively, we have:

$$\dot{f}(k) = \dot{f}_o(k) + \dot{f}_c(k), \tag{1}$$

where $\dot{f}_o$ is the feature motion in the image plane caused by the object motion, and $\dot{f}_c$ is the feature motion caused by the camera motion.

The feature acceleration $\ddot{f}(k)$ at instant k is decided by the relative movement between the object and the camera. It is surely related to feature's position f, velocity $\dot{f}$ and the hand's acceleration $\ddot{p}(k)$. Thus we obtain the mapping model:

$$\ddot{f}(k) = g'\left(f(k), \dot{f}(k), \ddot{p}(k)\right), \tag{2}$$

where $g'(\cdot)$ is a proper function. Since we have the same numbers of object features and control DOFs of the robot hand, Eq. (2) has the same dimensions of input and output. All coefficient matrices generally have full ranks. Thus the input-output relationship can be exchanged to obtain an inverse mapping model from the image feature space to the robotic movement space, i.e.,

$$\ddot{p}(k) = g\left(f(k), \dot{f}(k), \ddot{f}(k)\right), \tag{3}$$

where $g(\cdot)$ is a properly defined function. This model, called the visual mapping model, nonlinearly relates the hand movement to the motion of the object in image feature space. Notice that this visual mapping model has the same dimensions of the input and output.

The characteristics that the visual mapping model (3) has the same dimensions of input and output spaces is very important for avoiding tracking singularity [4], which means that some DOFs for hand movements might become under-controlled during the tracking procedure. This happens when image features employed are not sufficient to reflect robot motion in some directions [7].

Here the scheme that robot control space and the image feature space are of the same dimensions can guarantee that the tracking singularity is eliminated practically, though not theoretically. In addition, same input-output dimension policy is very helpful for obtaining good convergence characteristics in training if the model is realized with neural networks.

3 Control Scheme Based on Fuzzy Neural Network

A neural network is adopted to implement the nonlinear visual mapping model shown in (3). Since the dimensionality of (3) is high and its input-output relations are complex enough, we propose to use fuzzy neural network(FNN) to realize the model of (3).

Fuzzy neural network is a combination of a fuzzy controller and a neural network. It enables integrating any priori knowledge of the hand-eye coordination system by forming fuzzy rules and membership functions to help enhancing system performance. In addition, the approximating accuracy by a fuzzy neural network could be improved by refining the fuzzy rules involved without increasing much computational expense. This is critical for the control of an uncalibrated visual tracking system of high dimensions.

Fig.3 shows the structure of the fuzzy neural network we adopt in this paper. The network has four layers, i.e., the input layer, the fuzzification layer, the fuzzy inference layer and the defuzzification output layer. Each input $x_i, (i = 1, \cdots, n)$, is fuzzified to l fuzzy sets, which define the accuracy of the system output. Thus the fuzzification layer has $n \times l$ nodes altogether. In the inference layer, there are also l nodes compatible with the number of the membership functions for each input. Following transfer functions are embedded respectively in each layer:

$$u_{ij} = exp\left[-\left[\frac{x_i - m_{ij}}{\sigma_{ij}}\right]^2\right], (i = 1, \cdots, n, \quad j = 1, \cdots, l) \tag{4}$$

$$\Phi_j = u_{1j} \cdot u_{2j} \cdots u_{nj} = \prod_{i=1}^{n} u_{ij}, (j = 1, \cdots, l) \tag{5}$$

$$y = W_1\Phi_1 + W_2\Phi_2 + \cdots + W_l\Phi_l = \sum_{j=1}^{l} W_j\Phi_j . \tag{6}$$

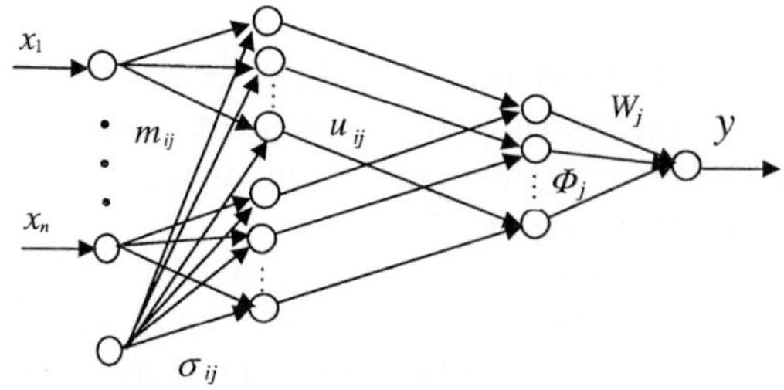

Fig. 3. Structure of the four-layered fuzzy neural network

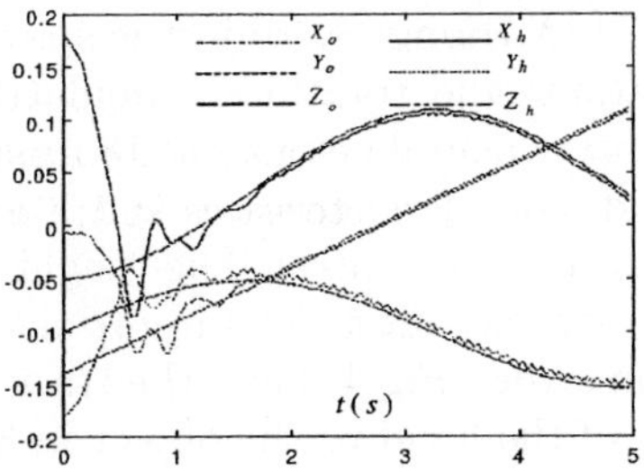

Fig. 4. Tracking procedure of the hand to the object

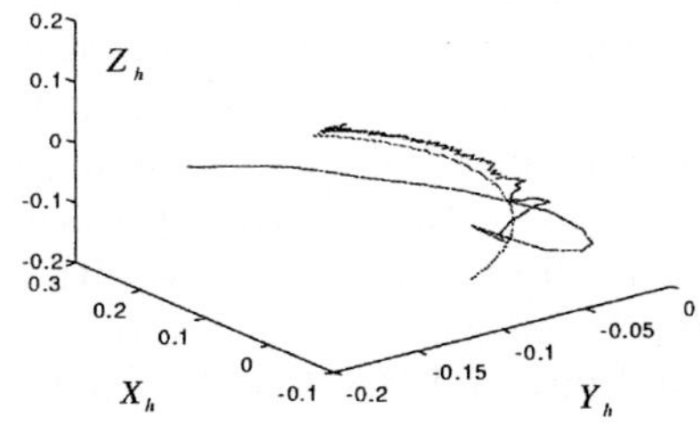

Fig. 5. Tracking trajectory in 3-D space

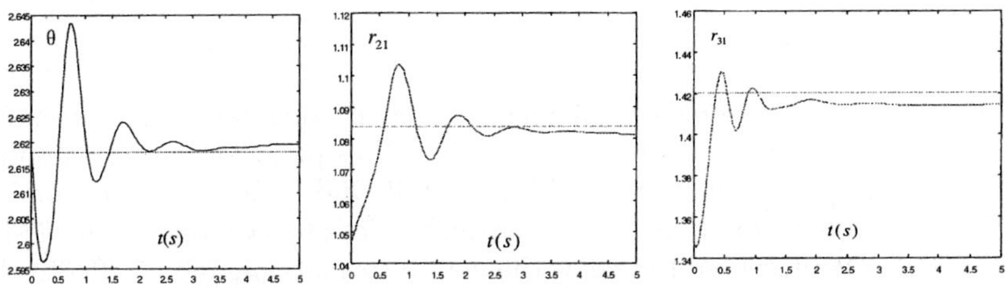

(a) Tracking of θ (b) Tracking of r_{21} (c) Tracking of r_{31}

Fig. 6. Tracking of the pose features of the object

Learning algorithm for training the neural network is based on the back propagation (BP) algorithm. Denote the system output error as

$$E_p = (y - Y)^2/2, \tag{7}$$

where y is the system's actual output and Y is the system's expected output. Then the δ-learning algorithm can be used here to update all weights:

$$m_{ij}(n+1) = m_{ij}(n) - \eta\left(\partial E_p/\partial m_{ij}\right), \tag{8}$$

$$\sigma_{ij}(n+1) = \sigma_{ij}(n) - \eta\left(\partial E_p/\partial \sigma_{ij}\right), \tag{9}$$

$$w_i(n+1) = w_i(n) - \eta\left(\partial E_p/\partial w_i\right), \tag{10}$$

where η is the learning rate. Substituting (4)~(7) into (8)~(10), we have:

$$m_{ij}(n+1) = m_{ij}(n) - 2\eta(y-Y)w_j \cdot \prod_{k=1}^{n} u_{ki} \cdot \frac{x_i - m_{ij}}{\sigma_{ij}^2}, \tag{11}$$

$$\sigma_{ij}(n+1) = \sigma_{ij}(n) - 2\eta(y-Y)w_j \cdot \prod_{k=1}^{n} u_{ki} \cdot \frac{x_i - m_{ij}}{\sigma_{ij}^3}, \tag{12}$$

$$w_i(n+1) = w_i(n) - \eta(y-Y)\Phi_i. \tag{13}$$

4 Simulations

The system configuration shown in Fig.1 is adopted in simulations. A single camera is mounted at the end link of the robot manipulator, with its pose relative to the robot hand $[15cm, 10cm, 3cm, 30°, 20°, 0°]$. A triangular object is spirally moving in robot's 3-D workspace with translations and rotations. Simulations are done by using NN toolbox in Matlab 5.1. A fuzzy neural network of 18 inputs, 54 nodes in membership generation layer, 18 nodes in inference layer, and 1 node in defuzzification output layer is constructed and used in control. The weights of the FNN are trained offline with 5460 training samples after 5000 iterations. A PI controller is then used to converge the system errors. Fig.4 shows the tracking procedure and Fig. 5 is the tracking trajectory of the hand to the object in 3-D space. Fig. 6 is the trajectories of the image features of the object with respective to time. It is easy to see that both the position and pose of the robot hand can successfully track those of the object with satisfactory performance.

5 Conclusions

Full-DOF tracking problem is a challenging topic in visual servoing, especially by an uncalibrated hand-eye coordination system. This paper proposed a nonlinear visual mapping model for the uncalibrated full-DOF visual tracking problem, which is invertible through involving information of object model. A novel fuzzy neural network is presented to realize the nonlinear mapping model effectively by reaching a tradeoff of system accuracy and computational expenses. The prior knowledge of the system can be integrated into system modelling to help the convergence of the training of the neural network. Simulation results verify the satisfactory performance of the proposed method.

Future work lies in studying the relations of the neural network structure with the nonlinear visual mapping model. Convergence of the whole system controller is also under investigation.

References

1. Feddema, J.T., Lee, C.S.G.: Adaptive Image Feature Prediction and Control for Visual Tracking with a Hand-Eye Coordinated Camera. IEEE Trans. S.M.C., **20**, (1990) 1172–1183.
2. Yoshimi, B.H., Allen, P.K.: Alignment Using an Uncalibrated Camera System. IEEE Trans. Robot. Automat., **11**, (1995) 516–521
3. Hespanha, J., Dodds, Z., Hager, G.D., Morse, A.S.: What Can Be Done with an Uncalibrated Stereo Systems?. Proc. 1998 IEEE Inter. Conf. on Robot. Automat., 1366–1372.
4. Sutanto, H., Sharma, R., Varma, V.: Image Based Autodocking without Calibration. Proc. 1997 IEEE Inter. Conf. on Robot. Automat., 974–979
5. Su, J.B., Xi, Y.G., Hanebeck, U.D., Schmidt, G.: Nonlinear Visual Mapping Model for 3-D Visual Tracking with Uncalibrated Eye-in-Hand Robotic System. IEEE Trans. System, Man and Cybernetics, Part B, **34**, (2004) 652–659
6. Hashimoto, H., Kubota, T., Sato, M., Hurashima, F.: Visual Control of Robotic Manipulator Based on Neural Networks. IEEE Trans. on Industrial Electronics, **39**, (1992) 490–496
7. Sharma, R., Srinivasa, N.:A Framework for Active Vision-Based Robot Control Using Neural Networks. Robotica, **16**, (1998) 309–327

Neuro-Fuzzy Hybrid Position/Force Control for a Space Robot with Flexible Dual-Arms*

Fuchun Sun, Hao Zhang, and Hao Wu

Department of Computer Science and Technology, State Key Lab of Intelligent Technology and Systems, Tsinghua University, Beijing 100084, P.R. China
{sfc, zhanghao, wuhao}@s1000e.cs.tsinghua.edu.cn

Abstract. A neuro-fuzzy (NF) hybrid position/force control with vibration suppression is developed in this paper for a space robot with flexible dual-arms handling a rigid object. An impedance force control algorithm is derived using force decomposition, and then singular perturbation method is used to construct the composite control, where an adaptive NF inference system is employed to approximate the inverse dynamics of the space robot. Finally, an example is employed to illustrate the validity of the proposed control scheme.

1 Introduction

Recently, hybrid position/force control has been paid more and more attention for the flexible robot contacting a constrained environment. Chiou [1] discussed the force control stability of the flexible manipulators. Matsuno [2] considered the application of hybrid position/force control strategy to the flexible manipulators. Siciliano and Villani [3] proposed a parallel control method for a two-link flexible manipulator based on singular perturbation decomposition. Sudipto [4] pointed out that the traditional singular perturbation model is not suitable for treating relatively large flexibility, and presented a new singular perturbation model. The Uchiyama Lab research group [5] considered the problems of multiple cooperating flexible manipulators handling an object. They designed control law for force, position and flexible vibration respectively, and composite the control input in frequency domain. However, there is no research on hybrid position/force control for a space robot with flexible dual-arms handling a rigid object.

This paper is concerned with the NF hybrid position/force control for a space robot with flexible dual-arms handling a rigid object. A new singular perturbation model is used to realize the vibration suppression control and NF position/force control with an ANFIS model [6], and the proposed control algorithm is verified by an example.

* This work was jointly supported by the National Excellent Doctoral Dissertation Foundation (Grant No: 200041), the National Science Foundation for Key Technical Research of China (Grant No: 90205008), the National Key Project for Basic Research of China (Grant No: G2002cb312205), the National 863 Project of China (Grant No: 2003AA742016) and Basic Research Foundation of Tsinghua University (Grant No: JC2003028).

F. Yin, J. Wang, and C. Guo (Eds.): ISNN 2004, LNCS 3174, pp. 13–18, 2004.

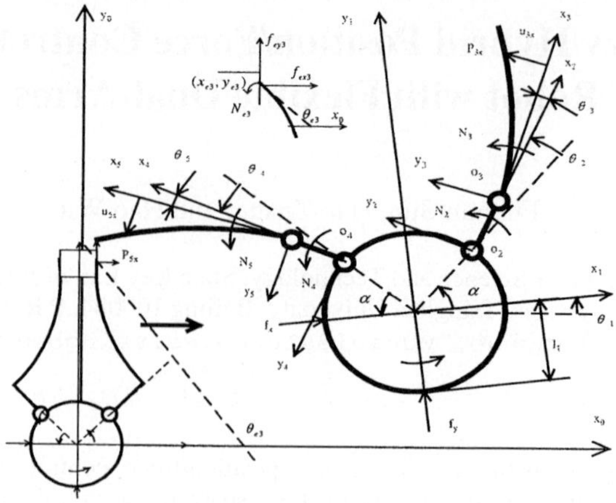

Fig. 1. Coordinates of the Space Robot System

2 Space Robot Modeling

The cooperating system of a space robot with flexible dual-arms handling a rigid object is shown in Fig.1, the front links 3, 5 are flexible. Define the robot's generalized coordinate as $\boldsymbol{p}=(x_{r1}, y_{r1}, \theta_1, \theta_2, \theta_3, \theta_4, \theta_5, \boldsymbol{Q}_3^{\mathrm{T}}, \boldsymbol{Q}_5^{\mathrm{T}})^{\mathrm{T}}$ and task space coordinates as $\boldsymbol{x}_a=(x_{r1}, y_{r1}, \theta_1, x_{e3}, y_{e3}, x_{e5}, y_{e5})^{\mathrm{T}}$, where $(x_{1r}, y_{1r}, \theta_1)^{\mathrm{T}}$ is the position and attitude vector of the vehicle. The contact force is defined as $\tilde{\boldsymbol{f}}=(f_{ex3}, f_{ey3}, f_{ex5}, f_{ey5})^{\mathrm{T}}$, and the tip position vector is $\boldsymbol{p}_e=(x_{e3}, y_{e3}, x_{e5}, y_{e5})^{\mathrm{T}}$.

The dynamics equation of a space robot can be written as follow:

$$\boldsymbol{M}(\boldsymbol{p})\ddot{\boldsymbol{p}}+\boldsymbol{F}(\boldsymbol{p},\dot{\boldsymbol{p}})+\begin{bmatrix} 0 \\ \boldsymbol{D}_p\dot{\boldsymbol{p}}+\boldsymbol{K}_p\boldsymbol{p} \end{bmatrix}=\begin{bmatrix} \boldsymbol{\tau}_{Qa} \\ 0 \end{bmatrix}-\tilde{\boldsymbol{J}}_1^T\tilde{\boldsymbol{f}} \tag{1}$$

where $\boldsymbol{p}_a=(x_1, y_1, \theta_1, \theta_2, \theta_3, \theta_4, \theta_5)^{\mathrm{T}}$ and $\boldsymbol{p}_p=(\boldsymbol{Q}_3^{\mathrm{T}}, \boldsymbol{Q}_5^{\mathrm{T}})^{\mathrm{T}}$ rigid and flexible components of $\boldsymbol{p}$, $\boldsymbol{M}$ inertia matrix, $\boldsymbol{F}$ vector of Coriolis and centrifugal forces, $\boldsymbol{K}_p$ a stiffness matrix determined by stiffness of flexible links EI_3 and EI_5, $\boldsymbol{D}_p$ damping matrix, $\boldsymbol{\tau}_{Qa}$ the active forces and torques, and $\tilde{\boldsymbol{J}}_1$ the robot Jacobian matrix. Define the object's coordinates as $\boldsymbol{x}_o=(x_{o1}, y_{o1}, \theta_{o1})^{\mathrm{T}}$. The object dynamics can be written as $\boldsymbol{M}_o\ddot{\boldsymbol{x}}_o+\boldsymbol{F}_o=\boldsymbol{J}_o^T\tilde{\boldsymbol{f}}$ [7], where $\boldsymbol{M}_o$ and $\boldsymbol{J}_o$ are the object inertial and Jacobian matrices, respectively. The motion of robot and the object should satisfy $\boldsymbol{J}_o\dot{\boldsymbol{x}}_o=\tilde{\boldsymbol{J}}_1\dot{\boldsymbol{p}}$ [7]. Therefore, equation (1) can be expressed as

$$\hat{M}(p)\ddot{p}+\hat{F}(p,\dot{p})+\begin{bmatrix}0\\ D_p\dot{p}+K_pp\end{bmatrix}=\begin{bmatrix}\tau_{Qa}\\ 0\end{bmatrix}-\tilde{J}_1^T\tilde{f}_I \tag{2}$$

where $\hat{M}=M+\tilde{J}_1^{\mathrm{T}}J_o^{+\mathrm{T}}M_oJ_0^+\tilde{J}_1,\hat{F}=F+\tilde{J}_1^{\mathrm{T}}J_0^{+\mathrm{T}}M_oJ_o^+(\dot{\tilde{J}}_1-\dot{J}_oJ_o^+\tilde{J}_1)\dot{p}+\tilde{J}_1^{\mathrm{T}}J_o^{+\mathrm{T}}F_o$, $\tilde{f}_I=\left(I_4-J_o^{+\mathrm{T}}J_o^{\mathrm{T}}\right)\tilde{f}$ denotes "internal force" ("+" denotes Moore-Penrose inverse). By partitioning the matrix and vectors according to the rigid and flexible components, equation (2) can be written as:

$$\begin{bmatrix}\hat{M}_{aa}(p_a,p_p) & \hat{M}_{ap}(p_a,p_p)\\ \hat{M}_{pa}(p_a,p_p) & \hat{M}_{pp}(p_a,p_p)\end{bmatrix}\begin{bmatrix}\ddot{p}_a\\ \ddot{p}_p\end{bmatrix}+\begin{bmatrix}\hat{F}_a(p_a,p_p,\dot{p}_a,\dot{p}_p)\\ \hat{F}_p(p_a,p_p,\dot{p}_a,\dot{p}_p)\end{bmatrix}+$$

$$\begin{bmatrix}0\\ D_p\dot{p}_p+K_pp_p\end{bmatrix}=\begin{bmatrix}\tau_{Qa}\\ 0\end{bmatrix}-\begin{bmatrix}\tilde{J}_{1a}^{\mathrm{T}}(p)\\ \tilde{J}_{1p}^{\mathrm{T}}(p)\end{bmatrix}\tilde{f}_I \tag{3}$$

3 Singular Perturbation Model and Control

The perturbation parameter is defined as [3] $\varepsilon=1/\sqrt{k_m}$, where k_m is the smallest stiffness in K_p. When the robot is static, $p_p=-K_p^{-1}\tilde{J}_{1p}^T\tilde{f}_I$, so we define the new flexible variable as $z=(p_p-p_{p0})/\varepsilon^2, p_{p0}=-K_p^{-1}\tilde{J}_{1p}^T\tilde{f}_I$. In equation (3), assuming that $u=\ddot{p}_a$ is new control input and ignoring the variation of p_{p0}, and $K_{ps}=\varepsilon^2K_p, D_{ps}=\varepsilon D_p$, the slow system can be described as $\ddot{p}_{as}=u_s$ and

$$z_s=-K_{ps}^{-1}\left(\hat{M}_{pa}(p_{as},p_{p0})u_s+\hat{F}_p(p_{as},p_{p0},\dot{p}_a,0)\right) \tag{4}$$

By setting the fast time scale $t_f=t/\varepsilon$, the fast system of $z_f=z-z_s$ is

$$\hat{M}_{pp}(p_{as},p_{p0})\frac{d^2z_f}{dt_f^2}+D_{ps}\frac{dz_f}{dt_f}+K_{ps}z_f=-\hat{M}_{pa}(p_{as},p_{p0})u_f \tag{5}$$

In the slow system, we adopt impedance strategy to translate the internal force to the position error of end effectors as $K_a\ddot{p}_{ec}+K_v\dot{p}_{ec}=\tilde{f}_{Id}-\tilde{f}_I$ [8]. Hence the slow control is $u_s=\ddot{p}_{ar}+K_D(\dot{p}_{ar}-\dot{p}_{as})+K_P(p_{ar}-p_{as})$, where $K_P,K_D>0$, p_{ar} is the reference trajectory of p_{as}, which is solved by inverse kinematics. In fast system, robust H_∞ theory is used to design the linear feedback control law as $u_f=K_{Pf}z_f+K_{Df}\dot{z}_f$. The composite control law is

$$u=u_s(p_{as},\dot{p}_{as})+u_f(z_f,\dot{z}_f). \tag{6}$$

4 Inverse Dynamics Modeling

To approximate an unknown mapping $\boldsymbol{y} = f(\boldsymbol{x})$, $f : R^m \to R^n$, the following ANFIS fuzzy inference system is constituted by l rules of Takagi-Sugeno type

$$R = \{R^i\} = \{\text{IF } \boldsymbol{p} \text{ is } A^i, \text{THEN } \boldsymbol{y} = \boldsymbol{V}_i^{\mathrm{T}}\boldsymbol{x}\}, i = 1, \cdots l \quad (7)$$

where $\boldsymbol{p}(\boldsymbol{x}) \in R^k$ the partition vector, A^i the fuzzy set, $\boldsymbol{V}_i$ the linear coefficient matrix. The ANFIS model can be expressed as

$$\boldsymbol{y} = \hat{f}(\boldsymbol{x}) = \left(\sum_{i=1}^{l}\theta_i(\boldsymbol{p})\boldsymbol{V}_i^{\mathrm{T}}\boldsymbol{x}\right) / \left(\sum_{i=1}^{l}\theta_i(\boldsymbol{p})\right) \quad (8)$$

where $\theta_i(\bullet)$ the membership function of A^i. The data are partitioned to clusters by Hyperplane C-means algorithm [9]. According to the clustering result, the membership functions are constructed by Fuzzy Min-Max network (FMM) [10] as

$$\theta_i(\boldsymbol{p}) = \exp\left(\sum_{i=1}^{k} -\left(Per_{ij}(\boldsymbol{p}) - Per_{ij}\right)^2 / \left(2s_{ij}^2\right)\right) \quad (9)$$

where $Per_{ij} = w_{ij} - v_{ij}$, $Per_{ij}(\boldsymbol{p}) = \max\left(w_{ij} - v_{ij}, w_{ij} - \boldsymbol{p}_j, \boldsymbol{p}_j - v_{ij}\right)$, w_{ij}, v_{ij}, s_{ij} obtained by FMM. The parameters in $\{\theta_i(\bullet)\}$ are trained by steepest descent method and ones in $\{\boldsymbol{V}_i\}$ are trained by least square method.

From equation (3), the actual driving forces and torques can be calculated as

$$\boldsymbol{\tau}_{Qa} = \left(\hat{\boldsymbol{M}}_{aa} - \hat{\boldsymbol{M}}_{ap}\hat{\boldsymbol{M}}_{pp}^{-1}\hat{\boldsymbol{M}}_{pa}\right)\boldsymbol{u} + \left(\hat{\boldsymbol{F}}_a - \hat{\boldsymbol{M}}_{ap}\hat{\boldsymbol{M}}_{pp}^{-1}(\hat{\boldsymbol{F}}_p + \boldsymbol{K}_{pp}\boldsymbol{p}_p)\right) + \left(\tilde{\boldsymbol{J}}_{1a}^{\mathrm{T}} - \hat{\boldsymbol{M}}_{ap}\hat{\boldsymbol{M}}_{pp}^{-1}\tilde{\boldsymbol{J}}_{1p}^{\mathrm{T}}\right)\tilde{\boldsymbol{f}}_{\mathrm{I}} \quad (10)$$

Suppose $\tilde{\boldsymbol{f}}_I \approx \tilde{\boldsymbol{f}}_{Id}$, where $\tilde{\boldsymbol{f}}_{Id}$ is the prescribed internal force. So, an ANFIS network can be built to approximate the inverse dynamics model (10) as

$$\boldsymbol{\tau}_{Qa} = \sum_{i=1}^{l}\mu_{Gi}(\boldsymbol{p}, \dot{\boldsymbol{p}})(\boldsymbol{B}_{1i}\boldsymbol{p} + \boldsymbol{B}_{2i}\dot{\boldsymbol{p}} + \boldsymbol{B}_{3i}\boldsymbol{u} + \boldsymbol{b}_{4i}) \quad (11)$$

The model can be expressed as $\boldsymbol{\tau}_{Qa} = \boldsymbol{W}^{\mathrm{T}}\boldsymbol{X}(\boldsymbol{x})$, where $\boldsymbol{x} = (\boldsymbol{p}, \dot{\boldsymbol{p}}, \boldsymbol{u})^{\mathrm{T}}$, $\boldsymbol{W}$ composed by the elements of $\boldsymbol{B}_{1i}, \boldsymbol{B}_{2i}, \boldsymbol{B}_{3i}, \boldsymbol{b}_{4i}$. We adopt projection method to update $\boldsymbol{W}$ as

$$\boldsymbol{W}_i(t) = \boldsymbol{W}_i(t-\lambda) + \frac{\boldsymbol{X}(\boldsymbol{x}(t))}{\boldsymbol{X}^{\mathrm{T}}(\boldsymbol{x}(t))\boldsymbol{X}(\boldsymbol{x}(t))}\left(\left(\boldsymbol{\tau}_{Qa}(t)\right)_i - \boldsymbol{W}_i^{\mathrm{T}}(t-\lambda)\boldsymbol{X}(\boldsymbol{x}(t))\right) \quad (12)$$

where $\boldsymbol{W}_i$ is the ith column of $\boldsymbol{W}$, λ is the sampling interval.

Remark. A stability proof for the space robot with control law and projection-learning algorithm will be direct. Here the theorem and proof are omitted for paper length.

5 Simulation

5.1 Task Design

The parameters of the space robot with flexible dual arms are shown in Table 1, and EI_3=EI_5=300Nm, $\alpha = \pi / 4$.

Table 1. Space robot parameters

	Link 1	Link 2	Link 3	Link 4	Link 5
Weight(kg)	40	2	2	2	2
Length(m)	2	2	2	2	2

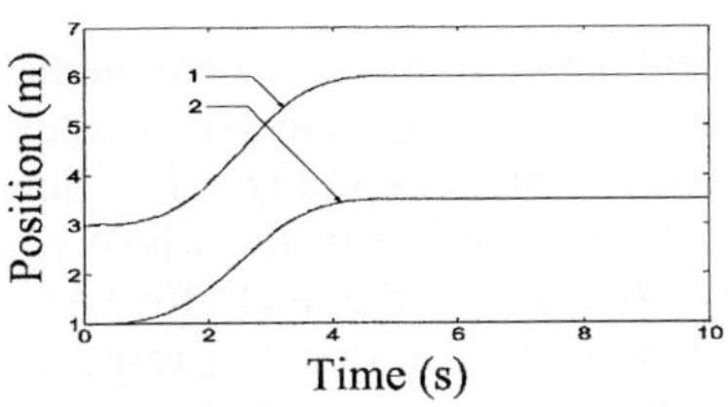

Fig. 2. Tip position of link 3 1 - y_{e3} 2 - x_{e3}

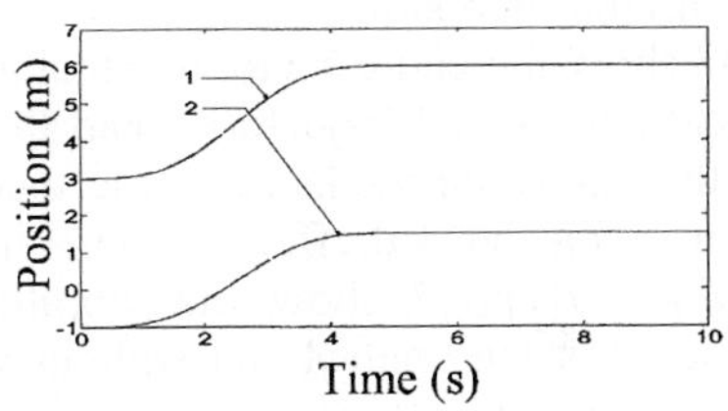

Fig. 3. Tip position of link 5 1 - y_{e5} 2 - x_{e5}

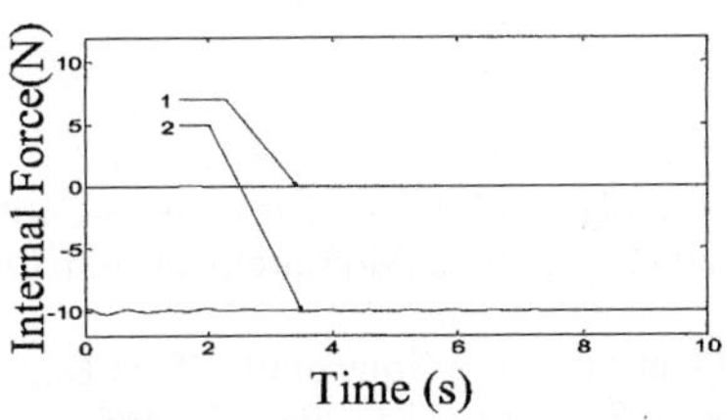

Fig. 4. Internal force at end of link 3 1 - f_{Iey3} 2 - f_{Iex3}

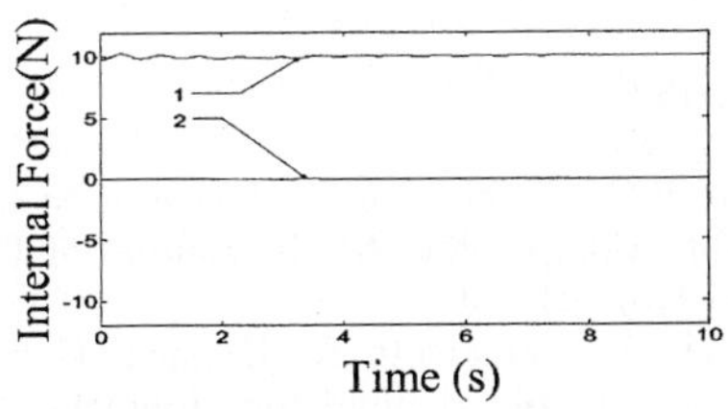

Fig. 5. Internal force at end of link 5 1 - f_{Iex3} 2 - f_{Iey3}

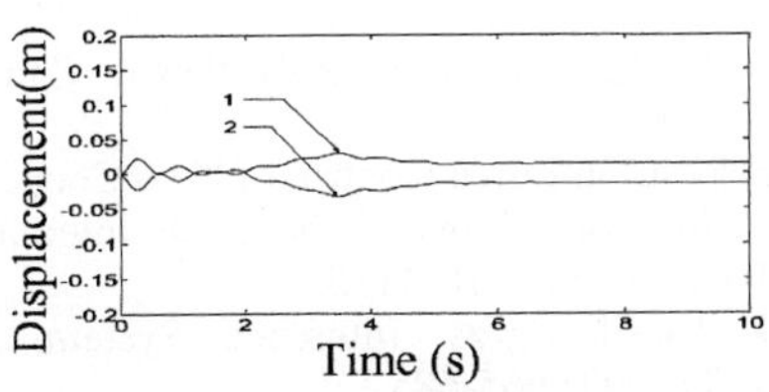

Fig. 6. Tip vibrations 1- u_{3l} 2 – u_{5l}

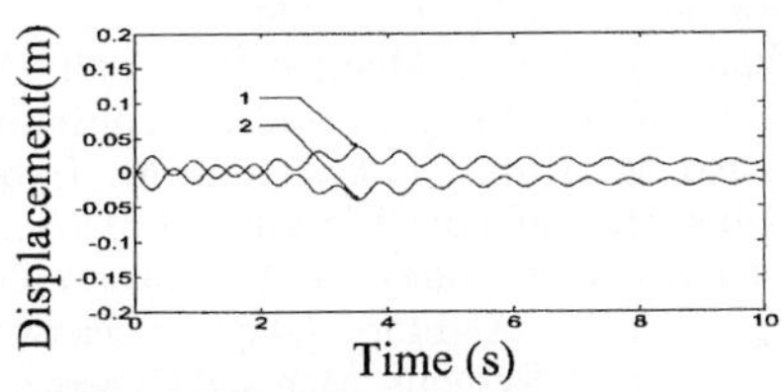

Fig. 7. Tip vibrations 1- u_{3l} 2 – u_{5l}

The target object: $\boldsymbol{M}_o = diag(2,2,4), \boldsymbol{F}_o = \mathbf{0}$.

Trajectory: Set $\boldsymbol{x}_{a0} = (0,0,0,1,3,-1,3)^{\mathrm{T}}$ and $\boldsymbol{x}_{ag} = (2.5,3,0,3.5,6,1.5,6)^{\mathrm{T}}$. Let

$$z(t)=\begin{cases}0.036t^3,\ 0\le t<5/3\\ 1/6+0.3(t-5/3)+0.18(t-5/3)^2-0.072(t-5/3)^3,\ 5/3\le t<10/3\\ 1+0.036(t-5)^3,\ 10/3\le t<5\\ 1,\ t\ge 5\end{cases}$$

The ideal trajectory in task space is $\boldsymbol{x}_{ad}(t)=\boldsymbol{x}_{a0}+z(t)(\boldsymbol{x}_{ag}-\boldsymbol{x}_{a0})$.

Internal force: $\tilde{\boldsymbol{f}}_{Id}=(-10,0,10,0)^{\mathrm{T}}$.

5.2 Simulation Result

The simulation is performed using the parameters presented in Section 5.1. Figs. 2-6 show the simulation result.

Figs.2-3 show the end-effectors' actual and ideal position vector, where the dashed lines indicate the actual trajectories and the full lines the ideal trajectories. It can be observed that the robot tracks the prescribed trajectory accurately. Figs. 4-5 show the internal forces on the end-effectors. The internal force is stable in the whole control process. Figs. 6 and 7 show the vibration amplitudes of the end-effecters with vibration suppression control and without vibration suppression control respectively, where u_{il} denotes the tip vibration amplitudes of link i. A good control performance is illustrated by using the NF hybrid position/force control proposed in this paper.

References

1. Chiou, B.C., Shahinpoor M.: Dynamic Stability Analysis of a Two-Link Force-Controlled Flexible Manipulator, ASME Journal of Dynamics Systems, Measurement and Control, 112 (1990) 661-666
2. Matsuno, F., Yamamoto. K.: Dynamic Hybrid Position/Force Control of a Two Degree-of-Freedom Flexible Manipulator," Journal of Robotic Systems, 11 (1994) 355-366
3. Siciliano, B., Villani, L., Two-time Scale Force and Position Control of Flexible Manipulators," in Proceedings of the IEEE International Conference on Robotics and Automation (2001) 2729-2734
4. Sudipto, S., Robot Manipulation with Flexible Link Fingers, PhD thesis, California Institute of Technology, Pasadena, California (1996)
5. Yamano, M., Kim, J.S., Uchiyama M., Hybrid Position/Force Control of Two Cooperative Flexible Manipulators Working in 3D Space, in Proceedings of the IEEE International Conference on Robotics and Automation, Leuven (1998) 1110-1115
6. Jang, J.–S.R.: ANFIS: Adaptive Network based Fuzzy Inference System, IEEE Transaction on Systems, Man and Cybernetics, 3 (1993) 665-685
7. Bonitz, R., Hsia, T.: Internal Force based Impedance Control for Cooperating Manipulators, in Proceedings of the IEEE International Conference on Robotics and Automation, vol. 3 (1993) 944-949
8. Bruno, S., Villani, L.: Robot Force Control, Kluwer Academic, Boston (1999)
9. Panella, M., Rizzi, A., Mascioli, F.M.F., Martinelli, G.: ANFIS Synthesis by Hyperplane Clustering, in Proceedings of IFSA World Congress and 20th NAFIPS International Conference, vol 1 (2001) 340 -345
10. Chen, X., Jin, D., Li, Z.: Recursive Training for Multi-Resolution Fuzzy Min-Max Neural Network Classfier, in Proceedings of International Conference on Solid-State and Integrated-Circuit Technology, vol.1 (2001) 131-134

Fuzzy Neural Networks Observer for Robotic Manipulators Based on H_∞ Approach

Hong-bin Wang, Chun-di Jiang, and Hong-rui Wang

#524, Yanshan University, Qinhuangdao 066004, China
cat6488@163.com

Abstract. This paper presents an observer for robotic systems using FNN method to estimate the joint velocities of a robot, and then H_∞ approach is embedded to attenuate the effect of external distributes and parametric uncertainties of the robotic systems. Then a simulation example of 2-DOF robotic systems is given at last, from the simulation results, we can see the well performance of the designed observer and the estimation errors of the joint velocities are negligible.

1 Introduction

Robotic manipulator systems are highly nonlinear, couples and time varying, which are subject to uncertainties unavoidably. The uncertainties consist of both structured uncertainties and unstructured uncertainties, which cannot be modeled precisely. For this reason, FNN technique has become a research hotspot in recent years [1,2]. The advantages of FNN technique have two main asides; first, it can provide a solution to robotic control, which deals with high nonlinear, high coupling, and un-modeled uncertainties because fuzzy control is a model-free approach; second one is that the learning ability and optimization ability of neural network.

The problem of nonlinear position control of robotic systems is solved by the passivity backstepping method [3]. The global stability can be attained if the uncertainty can be modeled, but this is not the case. In this paper FNN and H_∞ performance are embedded to solve this problem. The working principles of this FNN control are to realize the process of fuzzy reasoning using the structure of an NN and express the parameters of fuzzy reasoning through the weights of an NN [4].

2 Problem Presentation

The dynamics of a robotic system can be expressed as the following

$$M(q)\ddot{q}+C(q,\dot{q})\dot{q}+G(q)=\tau+\tau_d \quad (1)$$

F. Yin, J. Wang, and C. Guo (Eds.): ISNN 2004, LNCS 3174, pp. 19–24, 2004.

Where $M(q)$ is the inertia matrix, $C(q,\dot{q})$ is the centripetal and Coriolis torques, $G(q)$ is the gravitational torques, q denotes the joint displacements and τ is the vector of applied joint torques, τ_d is the external disturbance.

Considering the 2-DOF (degree of freedom) robotic systems, the dynamic equation (1) can be rewritten using the state space method as

$$\varepsilon_1 = q = [q_1 \quad q_2]^T, \varepsilon_2 = \dot{q} = [\dot{q}_1 \quad \dot{q}_2]^T, \varepsilon = [\varepsilon_1 \quad \varepsilon_2]^T \tag{2}$$

$$\begin{cases} \dot{\varepsilon}_1 = \varepsilon_2 \\ \dot{\varepsilon}_2 = -M^{-1}(\varepsilon_1)(C\varepsilon_2 + G) + M^{-1}(\varepsilon_1)(\tau + \tau_d) \end{cases} \tag{3}$$

It is always the case that the $M(q)$, $C(q,\dot{q})$ and $G(q)$ are bounded matrices containing parametric uncertainty. We assume that the dynamic parameter can be expressed by the nominal parameter and the parameter uncertainty

$$M(q) = M_0(q) + \Delta M(q), C(q,\dot{q}) = C_0(q,\dot{q}) + \Delta C(q,\dot{q}) \tag{4}$$
$$G(q) = G_0(q) + \Delta G(q)$$

These parameters satisfy the following assumptions:

Assumption 1. The bounded of uncertainty parameters is known as follows

$$\|\Delta M(q)\| \le \delta_M, \quad \|\Delta C(q,\dot{q})\| \le \delta_C, \quad \|\Delta G(q)\| \le \delta_G \tag{5}$$

Where δ_M, δ_C and δ_G are positive constants, and $\delta_M \le \lambda_{\min}(M_0)$.

The FNN observer is used to approximate the nonlinear function as

$$M_0^{-1}(\varepsilon_1)[C_0(\varepsilon_1,\varepsilon_2)\varepsilon_2 + G_0(\varepsilon_1)] = W_0^{*T}\theta_0(\varepsilon_1,\varepsilon_2) + \varepsilon_r(\varepsilon_1,\varepsilon_2) \tag{6}$$

Where $\theta_0(\varepsilon_1,\varepsilon_2)$ is the FNN generalize vector, and ε_r denotes the reconstruction error of FNN. We assume that there exists $\Omega_{W_0} = \{W_0 / \|W_0\| \le M_{W_0}\}$. W_0^* Is the optimal weight matrix, which can be expressed as

$$W_0^* = \arg \min_{W_0 \in \Omega_{W_0}} \{\sup |M_0^{-1}(\varepsilon_1)[C_0(\varepsilon_1,\varepsilon_2)\varepsilon_2 + G_0(\varepsilon_1)] - W_0^T\theta_0(\varepsilon_1,\varepsilon_2)|\} \tag{7}$$

Since the speed vector $\dot{q}$ is unknown, the estimation of robotic dynamic parameters can only be expressed in term of ε_1 and $\hat{\varepsilon}_2$, where $\hat{\varepsilon}_2$ denotes the estimation of ε_2. The approximate FNN has the form

$$W_0^T\theta_0(\varepsilon_1,\hat{\varepsilon}_2) = W_0^{*T}\theta_0(\varepsilon_1,\varepsilon_2) - \tilde{W}_0^T\theta_0(\varepsilon_1,\hat{\varepsilon}_2) - W_0^{*T}\tilde{\theta}(\varepsilon_1,\varepsilon_2,\hat{\varepsilon}_2) \tag{8}$$

$$\tilde{\theta}_0(\varepsilon_1,\varepsilon_2,\hat{\varepsilon}_2) = \theta_0(\varepsilon_1,\varepsilon_2) - \theta_0(\varepsilon_1,\hat{\varepsilon}_2) \tag{9}$$

Assumption 2. The FNN reconstruction error and estimation error are bounded by $\|\varepsilon_r(\varepsilon_1,\varepsilon_2)\| \le \delta_r$ and $\|\tilde{\theta}_0(\varepsilon_1,\varepsilon_2,\hat{\varepsilon}_2)\| \le \delta_\theta$, where δ_r and δ_θ are positive constants.

The proposed robust FNN observer has the form as the following

$$\dot{z}_1 = \hat{\varepsilon}_2 + k_3\tilde{\varepsilon}_1 \qquad (10)$$
$$\dot{z}_2 = -W_0^T\theta_0(\varepsilon_1,\hat{\varepsilon}_2) + M_0^{-1}(\varepsilon_1)\tau + k_4\tilde{\varepsilon}_1 - v_0(\varepsilon_1,\varepsilon_2,\hat{\varepsilon}_2)$$

Where $\tilde{\varepsilon}_i = \varepsilon_i - \hat{\varepsilon}_i, i = 1,2$, $k_i = k_i^T > 0, i = 3,4$. And v_0 is a robust compensate term for the proposed observer. Then the estimated dynamic can be expressed by $\hat{\varepsilon}_1$ and $\hat{\varepsilon}_2$

$$\hat{\varepsilon}_1 = z_1, \hat{\varepsilon}_2 = z_2 + k_5\tilde{\varepsilon}_1 \qquad (11)$$

Therefore the estimated dynamic can be rewritten as

$$\dot{\hat{\varepsilon}}_1 = \hat{\varepsilon}_2 + k_3\tilde{\varepsilon}_1 \qquad (12)$$
$$\dot{\hat{\varepsilon}}_2 = -W_0^T\theta_0(\varepsilon_1,\hat{\varepsilon}_2) + M_0^{-1}(\varepsilon_1)\tau + k_4\tilde{\varepsilon}_1 - v_0(\varepsilon_1,\varepsilon_2,\hat{\varepsilon}_2)$$

Then the observer error dynamic is obtained

$$\dot{\tilde{\varepsilon}}_1 = \tilde{\varepsilon}_2 - k_3\tilde{\varepsilon}_1 \qquad (13)$$
$$\dot{\tilde{\varepsilon}}_2 = -M^{-1}(\varepsilon_1)\Delta M(\varepsilon_1)M_0^{-1}(\varepsilon_1)\tau + M^{-1}(\varepsilon_1)\tau_d - h(\varepsilon_1,\varepsilon_2) - \varepsilon_r(\varepsilon_1,\varepsilon_2) -$$
$$\tilde{W}_0^T\theta_0(\varepsilon_1,\hat{\varepsilon}_2) - W_0^{*T}\tilde{\theta}_0(\varepsilon_1,\varepsilon_2,\hat{\varepsilon}_2) + (k_5k_3 - k_4)\tilde{\varepsilon}_1 - k_5\tilde{\varepsilon}_2 + v_0$$

$$h(\varepsilon_1,\varepsilon_2) \qquad (14)$$
$$= M^{-1}(\varepsilon_1)[C(\varepsilon_1,\varepsilon_2)\varepsilon_2 + G(\varepsilon_1)] - M_0^{-1}(\varepsilon_1)[C_0(\varepsilon_1,\varepsilon_2) + G_0(\varepsilon_1)] =$$
$$-M^{-1}(\varepsilon_1)\Delta M(\varepsilon_1)\left[W_0^T\theta_0(\varepsilon_1,\hat{\varepsilon}_2) + \varepsilon_r(\varepsilon_1,\varepsilon_2) + \tilde{W}_0^T\theta_0(\varepsilon_1,\hat{\varepsilon}_2) + W_0^{*T}\tilde{\theta}_0(\varepsilon_1,\varepsilon_2,\hat{\varepsilon}_2)\right]$$
$$+M^{-1}(\varepsilon_1)\Delta C(\varepsilon_1,\varepsilon_2)\tilde{\varepsilon}_2 + M^{-1}(\varepsilon_1)[\Delta C(\varepsilon_1,\varepsilon_2)\hat{\varepsilon}_2 + \Delta G(\varepsilon_1)]$$

Then the state equation of estimation error dynamic is attained

$$\dot{\tilde{\varepsilon}} = (A_0 + \Delta A_0)\tilde{\varepsilon} + B_{01}\{v_0 - M^{-1}(\varepsilon_1)\Delta M(\varepsilon_1)M_0^{-1}\tau \qquad (15)$$
$$-\varepsilon_r(\varepsilon_1,\varepsilon_2) - \tilde{W}_0^T\theta_0(\varepsilon_1,\hat{\varepsilon}_2) - W_0^{*T}\tilde{\theta}_0(\varepsilon_1,\varepsilon_2,\hat{\varepsilon}_2)$$
$$+M^{-1}(\varepsilon_1)\Delta M(\varepsilon_1)\left[W_0^T\theta_0(\varepsilon_1,\hat{\varepsilon}_2) + \varepsilon_r(\varepsilon_1,\varepsilon_2) + \tilde{W}_0^T\theta_0(\varepsilon_1,\hat{\varepsilon}_2) + W_0^{*T}\tilde{\theta}_0(\varepsilon_1,\varepsilon_2,\hat{\varepsilon}_2)\right]$$
$$-M^{-1}(\varepsilon_1)[\Delta C(\varepsilon_1,\varepsilon_2)\hat{\varepsilon}_2 + \Delta G(\varepsilon_1)]\} + B_{02}\tau_d$$

$$A_0 = \begin{bmatrix} -k_3 & I_{2\times2} \\ k_5k_3 - k_2 & -k_5 \end{bmatrix}, \Delta A_0 = \begin{bmatrix} 0_{2\times2} & 0_{2\times2} \\ 0_{2\times2} & -M^{-1}(\varepsilon_1)\Delta C(\varepsilon_1,\varepsilon_2) \end{bmatrix} \qquad (16)$$

$$B_{01} = \begin{bmatrix} 0_{2\times2} \\ I_{2\times2} \end{bmatrix}, B_{02} = \begin{bmatrix} 0_{2\times2} \\ M^{-1}(\varepsilon_1) \end{bmatrix} \qquad (17)$$

Assumption 3. There exist constant matrices E_0 and F_0, satisfy $\Delta A_0 = E_0\Sigma_0F_0$, where $\Sigma_0 \in R^{2\times2}$ and $\Sigma_0^T\Sigma_0 \le I_{2\times2}$, then considering the nominal parameters, yields

$$\|M^{-1}(\varepsilon_1)\| \le \frac{1}{\lambda_{\min}(M_0(\varepsilon_1)) - \delta_M} \qquad (18)$$

$$\|a_1\| = \|-I_2 + M^{-1}(\varepsilon_1)\Delta M(\varepsilon_1)\| \le \frac{\lambda_{\min}(M_0(\varepsilon_1))}{\lambda_{\min}(M_0(\varepsilon_1)) - \delta_M} \tag{19}$$

$$\|a_2\| = \|-M^{-1}(\varepsilon_1)\Delta C(\varepsilon_1, \varepsilon_2)\| \le \frac{\delta_C}{\lambda_{\min}(M_0(\varepsilon_1)) - \delta_M}$$

$$\|a_3\| = \|M^{-1}(\varepsilon_1)\Delta M(\varepsilon_1)\| \le \frac{\delta_M}{\lambda_{\min}(M_0(\varepsilon_1)) - \delta_M}$$

$$\|a_4\| = \|-M^{-1}(\varepsilon_1)\Delta G(\varepsilon_1)\| \le \frac{\delta_G}{\lambda_{\min}(M_0(\varepsilon_1)) - \delta_M}$$

Theorem 1. Considering the whole dynamic system (14), with $\tau_{\hat{d}} \in L_2[0 \quad \infty)$ and the assumptions 1-3 are satisfied. Suppose for any given $Q_0 = Q_0^T > 0$, there exists a positive matrix $P_0 = P_0^T > 0$, and the Riccati equation holds.

$$P_0A_0 + A_0^T P_0 + \frac{1}{\varepsilon_0} P_0\overline{B}_{02}\overline{B}_{02}^T P_0 + \lambda_0^2 P_0 E_0 E_0^T P_0 - P_0 B_{01} R_0^{-1} B_{01}^T P_0 + Q_0 + \frac{1}{\lambda_0^2} F_0 F_0^T = 0 \tag{20}$$

Where $P_0B_{01} = C_{01}^T$, and $\overline{B}_{02}$ is the upper bound of B_{02} and satisfy $B_{02}B_{02}^T \le \overline{B}_{02}\overline{B}_{02}^T$, the robust compensate term is defined as the follows

$$v_0 = v_{oh} + v_{os} + v_{ow} \tag{21}$$

$$\operatorname{sgn}(\varepsilon_i) = \begin{cases} 1 & if\ \varepsilon_i > 0 \\ 0 & otherwise \end{cases} \quad i = 1,2 \tag{22}$$

$$v_{oh} = -R_0^{-1} B_{01}^T P_0 \tilde{\varepsilon} = -R_0^{-1} C_{01} \tilde{\varepsilon} = -R_0^{-1} \tilde{\varepsilon}_1 \tag{23}$$

$$v_{os} = -\{\frac{\delta_C\|\hat{\varepsilon}_2\| + \delta_G + \delta_M \|M_0^{-1}(\varepsilon_1)\tau + W_0^T\theta_0(\varepsilon_1, \hat{\varepsilon}_2)\|}{\lambda_{\min}(M_0(\varepsilon_1)) - \delta_M} + \frac{\lambda_{\min}(M_0(\varepsilon_1))[\delta_r + M_{W_0}\delta_\theta]}{\lambda_{\min}(M_0(\varepsilon_1)) - \delta_M}\}\operatorname{sgn}(\tilde{\varepsilon}_1)$$

$$v_{ow} = -\frac{2\delta_M M_{W_0} \|\theta_0(\varepsilon_1, \hat{\varepsilon}_2)\|}{\lambda_{\min}(M(\varepsilon_1)) - \delta_M} \operatorname{sgn}(\tilde{\varepsilon}_1)$$

The direct adaptive training law is shown in (23)

$$\dot{W}_0 = \begin{cases} -S_0\tilde{\theta}_0(\varepsilon_1, \hat{\varepsilon}_2)\varepsilon_1^T, & if\ \|W_0\| < M_{W_0} or \|W_0\| = M_{W_0} and \tilde{\varepsilon}_1^T W_0^T \theta_0(\varepsilon_1, \hat{\varepsilon}_2) \ge 0 \\ \Pr(\bullet), & if \|W_0\| = M_{W_0} or \|W_0\| = M_{W_0} and \tilde{\varepsilon}_1^T W_0^T \theta_0(\varepsilon_1, \hat{\varepsilon}_2) < 0 \end{cases} \tag{24}$$

$$\Pr(\bullet) = -S_0\theta_0(\varepsilon_1, \hat{\varepsilon}_2)\varepsilon_1^T + \frac{S_0\varepsilon_1^T W_0^T \theta_0(\varepsilon_1, \hat{\varepsilon}_2)}{\|W_0\|^2} W_0$$

Where S_0 is a constant positive matrix, then the system (23) satisfies the following H_∞ tracking performance

$$\int_0^T \left\|\tilde{\varepsilon}(t)\right\|_{Q_0}^2 dt + \int_0^T \left\|v_0(t)\right\|_{R_0}^2 dt \le \beta_0 + \varepsilon_0 \int_0^T \left\|\tau_d\right\|^2 dt \tag{25}$$

Where $\beta_0 \in R$ and $\|W_0\| \le M_{W_0}$, state estimation error $\tilde{\varepsilon}_1$, $\tilde{\varepsilon}_2$ are bounded.

Proof. Choosing a Lyapunov function as

$$V(\tilde{\varepsilon},t) = \frac{1}{2}\tilde{\varepsilon}^T P_0 \tilde{\varepsilon} + \frac{1}{2} tr\{\tilde{W}_0^T S_0^{-1} \tilde{W}_0\} \tag{26}$$

Considering equation (19) and (21) and the norm inequality (18), we can attain the differential of the above equation

$$\begin{aligned}
\dot{V}(\tilde{\varepsilon},t) \le & -\frac{1}{2\varepsilon_0}\tilde{\varepsilon}^T P_0 \left(\overline{B}_{02}\overline{B}_{02}^T - B_{02}B_{02}^T\right)P_0\tilde{\varepsilon} - \frac{1}{2\lambda_0^2}\tilde{\varepsilon}^T F_0^T \left(\Sigma_0^T\Sigma_0 - I_{2\times 2}\right)F_0\tilde{\varepsilon} - \\
& \frac{1}{2}\left(\frac{1}{\sqrt{\varepsilon_0}}B_{02}^T P_0\tilde{\varepsilon} - \sqrt{\varepsilon_0}\,\tau_d\right)^T \left(\frac{1}{\sqrt{\varepsilon_0}}B_{02}^T P_0\tilde{\varepsilon} - \sqrt{\varepsilon_0}\,\tau_d\right) - \\
& \frac{1}{2}\left(\lambda_0 E_0^T P_0^T - \frac{1}{\lambda_0}\Sigma_0 F_0\tilde{\varepsilon}\right)^T \left(\lambda_0 E_0^T P_0^T - \frac{1}{\lambda_0}\Sigma_0 F_0\tilde{\varepsilon}\right) + \\
& \frac{1}{2}\varepsilon_0 \tau_d^T \tau_d - \frac{1}{2}\tilde{\varepsilon}^T Q_0 \tilde{\varepsilon} - \frac{1}{2} v_{oh}^T R_0 v_{oh} \\
\le & \frac{1}{2}\varepsilon_0 \tau_d^T \tau_d - \frac{1}{2}\tilde{\varepsilon}^T Q_0 \tilde{\varepsilon} - \frac{1}{2} v_{oh}^T R_0 v_{oh}
\end{aligned} \tag{27}$$

Integrating the above inequality from $t=0$ to $t=T$ and considering $V(\tilde{\varepsilon}(t),t) \ge 0$, the following result can be gained from the above inequality

$$\int_0^T \tilde{\varepsilon}^T Q_0 \tilde{\varepsilon}\, dt + \int_0^T v_{oh}^T R_0 v_{oh}\, dt \le V(\tilde{\varepsilon}(0),0) + \varepsilon_0 \int_0^T \tau_d^T \tau_d\, dt \tag{28}$$

$$\dot{V}(\tilde{\varepsilon}(t),t) < 0$$

Define $\beta_0 = V(\tilde{\varepsilon}(0),0)$, then the system satisfies the H_∞ attenuate performance and the system is stable.

3 Simulation Experiment

To show the feasibility and performance of FNN observer, a simulation study of a two links robotic system has been carried out. Here choose the exogenous disturbances $\tau_d = \{\exp(-0.1t)\}_{2\times 1}$, the 2-DOF robotic system choose from the reference [2], and the other parameters choose as follows

$k_3 = 20I_{2\times 2}$, $k_4 = 125I_{2\times 2}$, $k_5 = 125I_{2\times 2}$ So that A_0 is a Hurwitz matrix.

$Q_0 = \begin{bmatrix} I_{2\times 2} & 0_{2\times 2} \\ 0_{2\times 2} & 2I_{2\times 2} \end{bmatrix}$, $R_0 = 10I_{2\times 2}$, $\beta_0 = 0.1$, $\lambda_0 = 1$, $E_0 = [0,0,2,2]^T$, $F_0 = [0,0,1,1]^T$.

Thus the simulation figure 1 are gained as following:

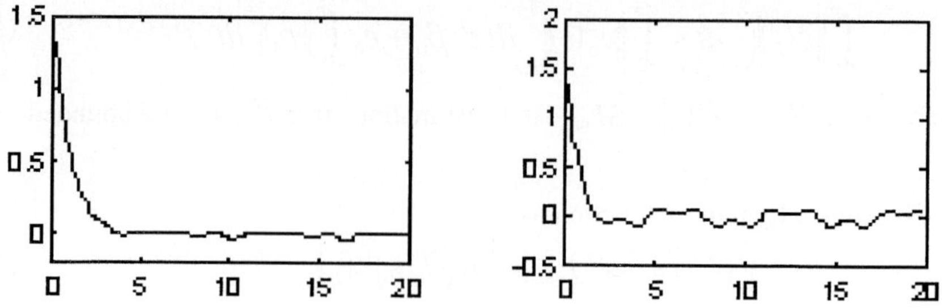

(a) (b)

Fig. 1. The observer error of joint velocities: (a) is the observer error of joint 1, (b) is the observer error of joint 2. From the figures, we know the estimation error is negligible and the observer is effectively

4 Conclusion

In this paper, FNN observer combined with H_∞ tracking performance guarantee the stability and the attenuate of external disturbance. The simulation example testifies the feasibility and the practicability of the designed observer.

References

1. Zhi, L., Chunwen, L.: Fuzzy Neural Networks Quadratic Stabilization Output Feedback Control for Biped Robots via H_∞ Approach. IEEE Transactions on System, Man and Cybernetics-PART B: Cybernetics, Vol. 33. IEEE (2003) 67–84
2. Han, L., Ding, L., Yanxi, Y., Xin, J.: Robot Manipulator Controller Based on Fuzzy Neural and CMAC Network. Proceedings of the Second International Conference on Machine Learning and Cybernetics, Xi An, China (2003) 525–529
3. Nganga-Kouya, D., Saad, M., Lamarche, L.: Backstepping Passivity Adaptive Position Control for Robotic Manipulators. Proceeding of the American Control Conference. Anchorage AK (2002) 4607–4611
4. Minpeng, L., Pengyung, W.: Neural-Fuzzy Control System for Robotic Manipulators. IEEE Control Systems Magazine (2002) 53–63

Mobile Robot Path-Tracking Using an Adaptive Critic Learning PD Controller*

Xin Xu[1,2], Xuening Wang[2], Dewen Hu[2]

[1] School of Computer, National University of Defense Technology, 410073, Changsha, P.R. China
xuxin_mail@263.net

[2] Department of Automatic Control, National University of Defense Technology, 410073, Changsha, P.R. China

Abstract. This paper proposes a novel self-learning PD (Proportional-Derivative) control method for mobile robot path-tracking problems. In the self-learning PD control method, a reinforcement-learning (RL) module is used to automatically fine-tune the PD coefficients with only evaluative feedback. The optimization of the PD coefficients is modeled as a Markov decision problem (MDP) with continuous state space. Using an improved AHC (Adaptive Heuristic Critic) learning control method based on recursive least-squares algorithms, the near-optimal control policies of the MDP are approximated efficiently. Besides its simplicity, the self-learning PD controller can be adaptive to uncertainties in the environment as well as the mobile robot dynamics. Simulation and experimental results on a real mobile robot illustrate the effectiveness of the proposed method.

1 Introduction

The path-tracking problem of wheeled mobile robots (WMRs) has been an important research area due to its theoretical nature as well as its practical importance. In theory, the design of path tracking controller for mobile robots is a difficult problem since there are not only non-holonomic constraints on robot kinematics but also various uncertainties in robot dynamics. In practice, wheeled mobile robots are more and more widely applied in industry, transportation and space exploration, which will lead to dynamic environment for mobile robot controllers. Thus, to design adaptive and robust path tracking controllers for mobile robots has become a challenging problem in the literature.

Until now, previous path-tracking control methods for mobile robots can be mainly classified into three categories, i.e., PID (Proportional-Integral-Derivative) control [1][2], nonlinear feedback control [3] and intelligent control methods [4-5]. Despite of its simplicity, the performance of PID control is heavily determined by the tuning of PID coefficients, which is a very difficult task for time-varying and nonlinear plants. Nonlinear feedback control methods are based on nonlinear dynamic models of

* Supported by the National Natural Science Foundation of China Under Grant 60303012, 0234030, & 60225015

F. Yin, J. Wang, and C. Guo (Eds.): ISNN 2004, LNCS 3174, pp. 25–34, 2004.

mobile robots so that explicit and exact modeling of robot dynamics is required. Intelligent control methods for mobile robots have attracted lots of research interests in recent years since intelligent controllers do not need explicit and exact model of robot dynamics and machine learning algorithms can be used to realize adaptivity to uncertainties. However, most existing intelligent controllers for mobile robots either require human knowledge for constructing control rules or collecting labeled samples for supervised learning [4]. In this paper, we propose a novel intelligent control method that combines the advantages of PID control and a class of machine learning method called reinforcement learning (RL) control [6] so that the design and optimization of path tracking controllers can be implemented only by evaluative feedback during the interaction between robot and the environment.

As a class of adaptive optimal control methods, reinforcement learning provides a flexible approach to the design of intelligent agents in situations for which both planning and supervised learning methods are impractical or difficult to be employed. Unlike supervised learning methods, RL methods need only evaluative feedback from the environment to optimize system performance so that it can be applied to problems where significant domain knowledge is either unavailable or costly to obtain. Due to the above merits of RL, in recent years, lots of research work has been done in the theory and applications of RL. For a comprehensive survey, please refer to [6].

Although many applications of RL, such as elevator control, call admission control and routing in communication networks, etc., have been reported in the literature, there are very few results on applying RL to path-tracking control of mobile robots. In this paper, we propose an adaptive critic learning PD control scheme for the path-tracking problem of mobile robots, where a RL module is combined with a gain-variable PD controller to fine-tune the PD gains automatically only based on evaluative feedback from the environment. In the RL module, an improved adaptive heuristic critic (AHC) algorithm based on the recursive least squares temporal difference method [7] is employed to optimize the controller performance using value function approximation in continuous state and action space. Compared with previous supervised learning methods for learning PD control [8], the proposed self-learning PD controller realizes the auto-tuning of PD gains without any exemplary signals and it is adaptive to uncertainties and disturbances in robot dynamics. Moreover, the performance of the closed-loop path-tracking control system can be improved via online learning.

The rest of the paper is organized as follows. In Section 2, the mobile robot path-tracking problem is described. In Section 3, the architecture and the learning algorithm of the self-learning PD controller are presented. Simulation and real-time experimental results are provided to illustrate the effectiveness of the proposed method in Section 4 and Section 5, respectively. Section 6 summarizes the main contributions of the paper and discusses future work.

2 The Mobile Robot Path-Tracking Problem

As was extensively studied in [9], the kinematics and dynamics of wheeled mobile robots can be divided into several categories due to different kinds of mechanical configurations. Since the adaptive critic learning PD control method proposed in this paper does not rely on explicit mathematical model and it has online learning ability to be adaptive to uncertainties, in the following, we will only focus on a class of mobile robots with two differentially driven wheels and one or more castor wheels. Nevertheless, the method presented in this paper can be easily applied to other kinds of WMRs with different kinematic and dynamic properties.

As discussed in [9], the kinematics of mobile robots can be described as follows:

$$\begin{cases} \dot{x} = v\cos\theta \\ \dot{y} = v\sin\theta \\ \dot{\theta} = \varpi \end{cases} \tag{1}$$

where (x, y) are the coordinates of robot mass center, θ is the heading angle, v is the velocity and ω is the angular velocity.

Let v_c and ω_c denote the desired forwarding velocity and angular velocity of the robot mass center, respectively. Since the position control and velocity control of mobile robots are usually studied separately in order to simplify controller design, in the following, we will only focus on the lateral position control problem with constant forwarding velocities so that the desired angular velocity serves as the control input of the system. For velocity control problem, the proposed learning PD control method can also be applied.

For lateral position control, the tracking errors are defined as follows.

$$e_1 = y_d - y \tag{2}$$

$$e_2 = \theta_d - \theta \tag{3}$$

where e_1 is the lateral error and e_2 is the orientation error.

The objective of controller design is to specify a control law ω_c such that the lateral tracking error e_1 and the orientation error e_2 are minimized.

To solve the above path-tracking problem, a conventional PD controller computes the control input ω_c according to the following equation.

$$\omega_c = k_1 e_1 + k_2 e_2 \tag{4}$$

where k_1 and k_2 are the gain coefficients. k_1 is the proportional gain and k_2 can be viewed as the derivative gain since the orientation error determines the variation direction of the lateral error.

3 The Adaptive-Critic Learning PD Controller

3.1 Architecture of the Adaptive-Critic Learning PD Controller

The path-tracking control system based on the proposed adaptive critic PD controller is depicted in Fig.1.

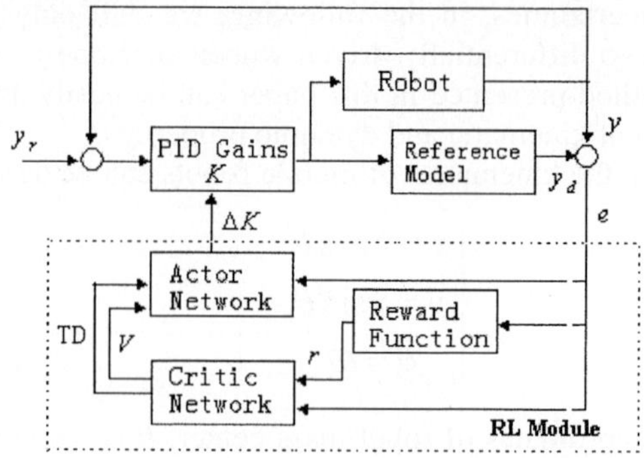

Fig. 1. Architecture of the self-learning PD controller

In Fig.1, y_r is the desired output vector, y is the output vector of the robot system and y_d is the output of a reference model. The controller is composed of two main parts, i.e., the PD gain module and the RL module.

The RL module in Fig.1 is an actor-critic learning control structure, which was early studied in [11] and later in [7] and [12], etc. There are three parts in the actor-critic structure, i.e., an actor network, a critic network and a reward function. The actor network is used to fine-tune the PD gains through its output vector ΔK. The error signal *e* serves as the inputs of the actor network as well as the critic network and the reward function. The critic network has another input which is the reward r determined by the reward function. The outputs of the critic network include an evaluation function of the current state and a temporal difference (TD) signal. The control output of the self-learning PD controller is given by

$$u = (K + \Delta K)^T (y_r - y) \tag{5}$$

where K and ΔK are the fixed gain vector and the variable gain vector, respectively.

The reward function and the objective function for optimization are designed as follows.

$$r_t = c|e_1| \tag{6}$$

$$J = \sum_{t=0}^{T} \gamma^t r_t \tag{7}$$

where e_1 is the lateral error defined in Section 2, c is a constant or piece-wise constant negative number and γ is a positive discount factor which is close to 1. Thus, when the objective function J is maximized, the tracking error e_1 will be minimized.

In the self-learning PD controller, the fixed gains may be selected based on *a priori* information of the system so that existing parameter tuning methods for PID controllers can be used and the learning process will be accelerated. In addition, better choice of the fixed PD part will greatly improve the stability and robustness of the system since parameter tuning only takes place in the vicinity of the fixed gains.

3.2 MDPs and the Adaptive Critic Learning Method

To realize the above optimization objective without model information, the whole system is modeled as an MDP and an improved adaptive critic learning method called Fast-AHC [7] is used to learn a near-optimal control policy for the MDP.

A Markov decision process (MDP) is an elegant mathematical model for sequential decision-making. To solve the optimal control problem of MDPs with model information, lots of methods have been studied in operations research [10]. However, in reinforcement learning, the model information of MDPs is assumed to be unknown, which is more realistic in many complex applications such as mobile robot control.

To solve the MDP, the adaptive critic method uses the critic network to evaluate the policy or predict the value functions and the actor network is employed to estimate the policy gradient using a stochastic action exploration method. For the formal definition of value function and policy evaluation, please refer to [10].

In adaptive critic learning, the policy evaluation in the critic is usually accomplished by temporal-difference (TD) learning methods such as TD(λ) algorithms [13]. The stochastic action exploration in the actor network can be described as follows.

Let $\Delta K'$ denote the output of the actor network. The actual action ΔK to be performed is determined by the following Gaussian probability distribution

$$\Delta K \sim N(\Delta K', \sigma(e)) \tag{8}$$

where $\Delta K'$ is the mean value and $\sigma(e)$ is the variance which is computed by

$$\sigma = \frac{\sigma_1}{(1+\exp(\sigma_2 V(e))} \tag{9}$$

where σ_1 and σ_2 are two positive constants and V is the value function estimation from the critic network.

3.3 The Fast-AHC Algorithm for the Learning PD Controller

For the path-tracking control problem, the underlying MDP has continuous state space so that neural networks need to be used for generalization and value function approximation. In the proposed adaptive-critic learning controller, a neural network with linear basis functions and two multi-layer neural networks with general sigmoid functions are selected as the critic network and the actor networks, respectively. The Fast-AHC algorithm proposed in [7] is applied to estimate a near optimal policy for the path-tracking problem by tuning PD parameters online. As discussed in [7], the Fast-AHC algorithm is an adaptive heuristic critic method using recursive least-squares temporal-difference (RLS-TD) learning techniques. By making use of the RLS-TD(λ) algorithm [7] in the critic, the efficiency of learning prediction for value functions is improved so that better control performance can be achieved in the closed-loop learning control system.

Let x denote the state of the MDP and $\varphi(x)$ denote the feature vector determined by the critic network. The estimated value function using linear basis functions has the following form:

$$V(x) = \varphi^T(x)W \tag{10}$$

where W is the weight vector.

The RLS-TD(λ) algorithm used in the critic network has the following update rules:

$$K_{t+1} = P_t\vec{z}_t / (1 + (\varphi^T(x_t) - \gamma\varphi^T(x_{t+1}))P_t\vec{z}_t) \tag{11}$$

$$W_{t+1} = W_t + K_{t+1}(r_t - (\varphi^T(x_t) - \gamma\varphi^T(x_{t+1}))W_t) \tag{12}$$

$$P_{t+1} = P_t - P_t\vec{z}_t[1 + (\varphi^T(x_t) - \gamma\varphi^T(x_{t+1}))P_t\vec{z}_t)]^{-1}(\varphi^T(x_t) - \gamma\varphi^T(x_{t+1}))P_t \tag{13}$$

where $P_0=\delta I$, δ is a positive constant, I is the identity matrix and

$$\vec{z}_{t+1} = \gamma\lambda\vec{z}_t + \varphi(x_t) \tag{14}$$

When the stochastic action exploration method in (8) is employed, the policy gradient in the actor network is estimated by

$$\frac{\partial J_\pi}{\partial u} = \frac{\partial J_\pi}{\partial \bar{y}_t}\frac{\partial \bar{y}_t}{\partial u} \approx \hat{r}_t \frac{y_t - \bar{y}_t}{\sigma_t}\frac{\partial \bar{y}_t}{\partial u} \tag{15}$$

where

$$\hat{r}_t = r_t + \gamma V(x_{t+1}) - V(x_t) \tag{16}$$

The learning algorithm for the adaptive-critic learning PD controller is summarized as follows.

(**Algorithm 1**)

Given: A stop criterion, maximal learning step T for one episode, learning parameters including γ, λ, δ, β, σ_1, σ_2, and fixed PD gains K.

(1) Initialize the weights of the critic and actor networks.

(2) While the stop criterion is not satisfied,

(a) Initialize the states of the system, set time step t=0.

(b) For the current state s_t, compute the outputs of the actor network, determine the variable PD gains ΔK according to (8).

(c) Compute the control output of the PD controller based on the fixed PD gains and variable PD gains.

(d) Observe the next state of the system and the reward of the state transition.

(e) Update the weights of the critic network according to (11)-(13).

(f) Update the actor weights according to the following equation:

$$\Delta u = -\beta_t \frac{\partial J_\pi}{\partial u} \quad (17)$$

where β_t is a positive learning factor.

(g) If $t<T$, $t=t+1$, return to (b),
Else return to (2).

4 Simulation

To illustrate the effectiveness of the proposed adaptive critic learning PD controller, computer simulation for the path tracking of a wheeled mobile robot was conducted. In the simulation, to take the robot dynamics into consideration, the relationships between the desired velocities and the real velocities are described by the following first-order differential equations.

$$\tau_a \dot{\varpi} + \varpi(t) = \varpi_c(t) \quad (18)$$

$$\tau_v \dot{v} + v(t) = v_c(t) \quad (19)$$

where τ_a, τ_v are two positive constants.

A time step t=0.05s is selected for the simulation and the time step for learning control is 0.2s. Since only lateral control of the vehicle is considered, the robot forwarding velocity is set to a constant v=0.4m/s.

The reference model is chosen as

$$\dot{y}_d = -by_d \quad (20)$$

where b=0.2. The reward function has the following form:

$$r_t = \begin{cases} -2|y - y_d|, & |y - y_d| > 0.1 \\ -0.5, & 0.05 \le |y - y_d| \le 0.1 \\ 0, & |y - y_d| < 0.05 \end{cases} \quad (21)$$

The other parameters of the learning PD controller are selected as: λ=0.6,P_0=0.1I, β=0.2. A CMAC network is used in the critic, which has 4 tilings and 7 partitions for each input. The number of physical memory units in the CMAC is set as 40 so that a linear basis function with 40 features is employed in the critic network. Two multi-layer neural networks are used as the actor networks which have 6 sigmoid neurons in the middle layer and one output neuron in the output layer. There are 4 inputs for the critic and the actor networks, which are the lateral error e_1, the direction error e_2 and their derivatives. The variable PD gains determined by the actor networks have the following form:

$$\Delta k_i = c_i \times (\varpi_i - 0.5), \quad i=1,2 \quad (22)$$

where ω_1 and ω_2 are the stochastic exploration actions given by (8), c_1=0.4, c_2=0.8.

In the simulation, the path-tracking performance of conventional PD controllers with fixed PD gains is also studied. The reference path is a straight line with initial lateral error e_1=2.0m and initial direction error e_2=0. The simulation results are shown in Fig.2 and Fig.3. In Fig.2, the lateral tracking errors of a conventional PD controller and the adaptive-critic learning PD controller are compared. The conventional PD controller has fixed gains k_1=0.2, k_2=0.4, which are manually optimized. The learning PD controller has the same initial gains as the conventional PD controller and its performance is measured after a learning process of 10 trials. For each trial, the state of the control system is re-initialized. In Fig.2, it is shown that the learning PD controller obtains better tracking performance with shorter rising time and smaller overshooting. In Fig.3, the variation of the lateral gain of the learning PD controller is depicted, which clearly shows that the learning controller has the ability to adjust its PD gains automatically so that better performance can be obtained.

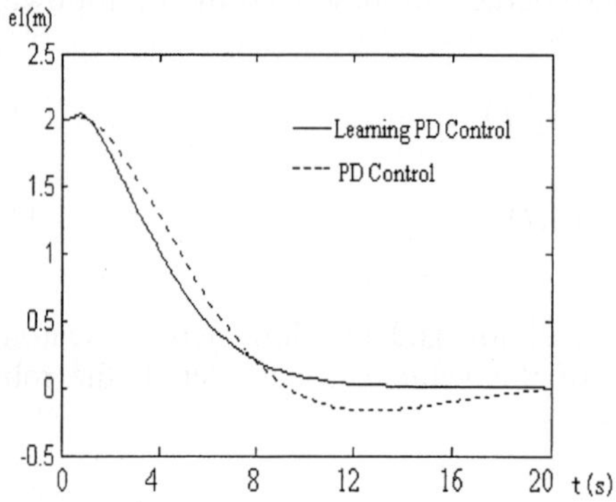

Fig. 2. Lateral tracking errors after 10 trials

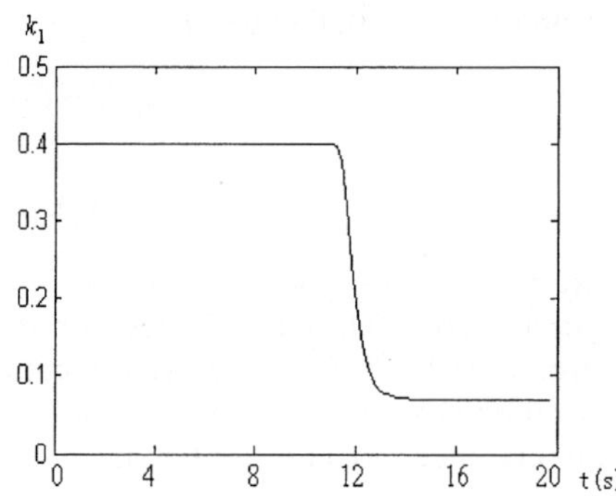

Fig.3. Gain variation of learning controller

5 Experiments on a Real Robot

In this section, we will present experimental results on a real wheeled mobile robot to study the performance of the proposed adaptive-critic learning PD controller. The mobile robot is a six-wheeled robot based on the LABMATE platform manufactured by TRC Inc. Fig.4 shows the closed-loop path-tracking control system of the robot. There are two control loops in the system. One is the inner loop of the HP motion controller for the motor speed and the other is the outer loop of the path-tracking PID controller.

The path-tracking task of the robot is as same as that in the simulation. The initial position error e_1=0.25m. The velocity of the robot in the experiment is set as a constant v=0.35m/s. The learning algorithm and parameters are almost as same as those employed in the simulation except that the sampling time for the controller is 400ms. The experimental results are shown in Fig.5 and Fig.6, where the solid line in Fig.5 shows the tracking error of the learning PD controller after 2 trials and the solid line in Fig.6 shows the learning tracking error after 4 trials. The dotted lines in Fig.5

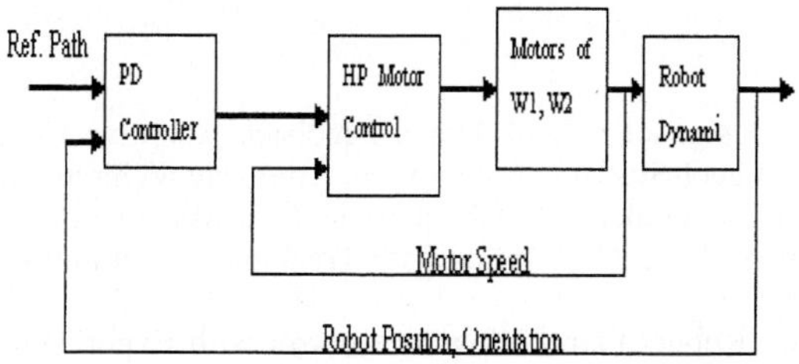

Fig. 4. The closed-loop control system of the robot

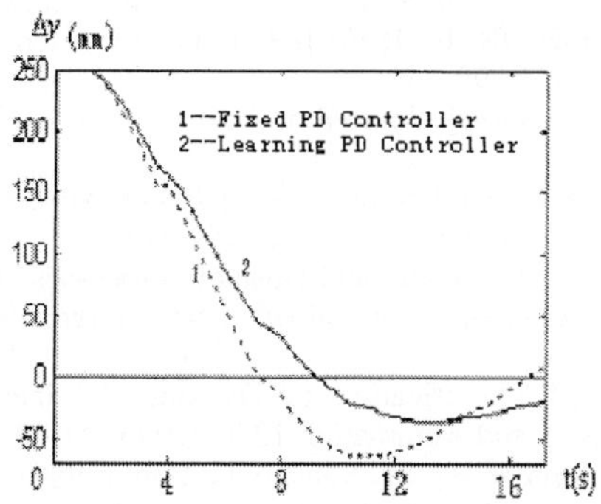

Fig. 5. Tracking errors after 2 trials

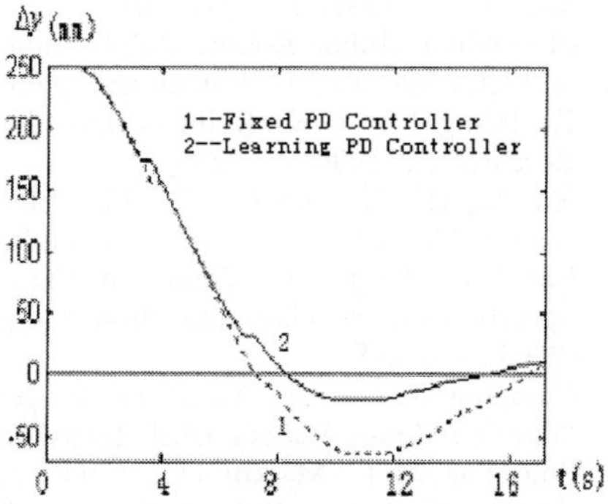

Fig. 6. Tracking errors after 4 trials

and Fig.6 are the tracking errors of a conventional PD controller. The results clearly illustrate the adaptive optimization ability of the reinforcement learning PD control method.

6 Conclusion and Future Work

In this paper, an adaptive-critic learning PD control method is proposed for mobile robot path tracking problems so that the uncertainties in robot control can be compensated automatically by reinforcement learning. By modeling the performance optimization of PD controller as an MDP, the adaptive critic learning structure using the Fast-AHC algorithm based on RLS-TD learning prediction is employed to estimate the near optimal policy of the MDP efficiently. Simulation and experimental results illustrate the effectiveness of the proposed method. Since stability is one of main concerns of controller design, further work needs to be done to develop the stability mechanism as well as theoretical analysis of the self-learning PD controller.

References

1. Choi, S. B.: The Design of a Look-Down Feedback Adaptive Controller for the Lateral Control of Front-Wheel-Steering Autonomous Highway Vehicles. In: Proceedings of the American Control Conference, v3, Albuquerque, New Mexico (1997) 1603-1607
2. Shin, D.H., Singh, S., Lee, J.J.: Explicit Path Tracking by Autonomous Vehicles. Robotica 10 (1992) 539-554
3. Aguilar, M., et al.: Robust Path Following Control with Exponential Stability for Mobile Robots. In: IEEE International Conference on Robotics and Automation, v4, Leuven, Belgium (1998) 3279-3284
4. Fierro, R., Lewis, F.L.: Control of a Nonholonomic Mobile Robot Using Neural Networks. IEEE Trans. on Neural Networks 9 (1998) 589-600
5. Sanchez, O., Ollero, A., Heredia, G.: Adaptive Fuzzy Control for Automatic Path Tracking of Outdoor Mobile Robots: Application to Romeo 3R. In: IEEE International Conference on Fuzzy Systems, v1, Barcelona, Spain (1997) 593-599
6. Kaelbling, L.P., et al.: Reinforcement Learning: a Survey. Journal of Artificial Intelligence Research 12 (1996) 237-285
7. Xu, X., He, H., and Hu, D.: Efficient Reinforcement Learning using Recursive Least-squares Methods. Journal of Artificial Intelligence Research 16 (2002) 259-292
8. Lee, C.H., Teng, C.C.: Tuning of PID Controllers for Stable and Unstable Processes with Specifications on Gain and Phase Margins. International Journal of Fuzzy Systems 3(1) (2001) 346-355
9. Campion, G., et al.: Structural Properties and Classification of Dynamic Models of Wheeled Mobile Robots. IEEE Trans. on Robotics and Automation 12(1) (1996) 47-62
10. Puterman, M. L.: Markov Decision Processes. John Wiley and Sons, New York (1994)
11. Barto, A.G., Sutton, R.S., Anderson, C.W.: Neuron-like Adaptive Elements that can Solve Difficult Learning Control Problems. IEEE Transactions on System, Man and Cybernetics 13 (1983) 834-846
12. Berenji, H.R., Khedkar, P.: Learning and Tuning Fuzzy Logic Controllers Through Reinforcement. IEEE Trans.on Neural Networks 3(5) (1992) 724-740
13. Sutton, R.: Learning to Predict by the Method of Temporal Differences. Machine Learning 3(1) (1988) 9-44

Reinforcement Learning and ART2 Neural Network Based Collision Avoidance System of Mobile Robot

Jian Fan, GengFeng Wu, Fei Ma, and Jian Liu

School of Computer Engineering and Science, Shanghai University,
Shanghai, China
{jfan, gfwu}@mail.shu.edu.cn

Abstract. In view of the collision avoidance problem of multi-moving-obstacles in path planning of mobile robot, we present a solution based on reinforcement learning and ART2 (Adaptive Resonance Theory 2) neural network as well as the method of rule-based collision avoidance. The simulation experiment shows that the solution is of good flexibility and can solve the problem on random moving obstacles.

1 Introduction

Recently, one of the hottest topics in robotics is mobile robot system. A key point in the research of mobile robot system is collision avoidance. Mobile robot can get the obstacles position by sensors, also can plan path using graphic method, free space method, grid method, artificial potential field method and so on. These methods are effective in static environments, but are difficult to solve the problem on dynamic environments.

Chun-Ta Chen [1] presented a collision avoidance system (CAS) based on the research of cockroach avoiding to be caught. He solved the problems by recognizing moving-obstacle and selecting collision avoidance behavior (for short as CAB), but the system failed to solve the problem of avoiding multi-obstacles.

In this paper, we propose a method of rule-based robot collision avoidance. We use reinforcement learning (RL) to acquire collision avoidance rules (for short as CAR) automatically and adopt ART2 neural network to integrate RL and neural network to decrease the memory required for rule storing.

2 Rules-Based Collision Avoidance Method of Mobile Robot

2.1 The Acquisition of Obstacle Moving States

According to the mechanism of collision avoidance proposed by [1], we suppose that the moving velocity and direction of mobile robot (R) are known. According to the information of obstacle (O) position relative to R, CAS can compute the moving state of O, such as moving direction, velocity, present position and so on (shown as Fig. 1).

F. Yin, J. Wang, and C. Guo (Eds.): ISNN 2004, LNCS 3174, pp. 35–41, 2004.

The velocity V^R and direction θ^R of R relative to Cartesian coordinate are known. The angel (θ_{old}, θ_{new}) and the displacement (d_{old}, d_{new}) of O relative to R in time t and (t+Δt) can be measured by sensors, the velocity V^o and direction θ^o of O relative to Cartesian coordinate can be computed by regular mathematic formulation.

2.2 The Definition of Collision Avoidance Rules (CAR)

R acquires the O's position information through sensors and adopts corresponding CAB (collision avoidance behavior) based on the position state. But the behaviors of R (including acceleration, deceleration, deflection and so on) is limited. We disperse CABs of R into a set of behavior space. A CAB is composed of deflection angle and motion acceleration. The problem can be simplified as the selection of a CAB from behavior table according to O and R's state. A CAR (collision avoidance rule) is composed of O state, R state and a behavior of robot collision avoidance, which can be formally defined as:

R_{ij}: IF state parameter of R, state parameter of O, THEN deflection angle of R, acceleration of R.

When R collides with O, R selects a proper CAB from behavior space table based on rules stored in system. R will select a new rule according to new state. In this way, R can arrive at destination without collision by repeating this process.

2.3 Evaluation and Selection of Reinforcement Learning (RL) Based CAB

RL is another machine learning method different from supervised learning and unsupervised learning. It can optimize the decision of behavior via estimating feedback signals got from interactive process with environment, so it is more valuable in solving complex optimization control problem. We use it to automatically evaluate and select CAB. The flow chat of RL is shown as Fig. 2.

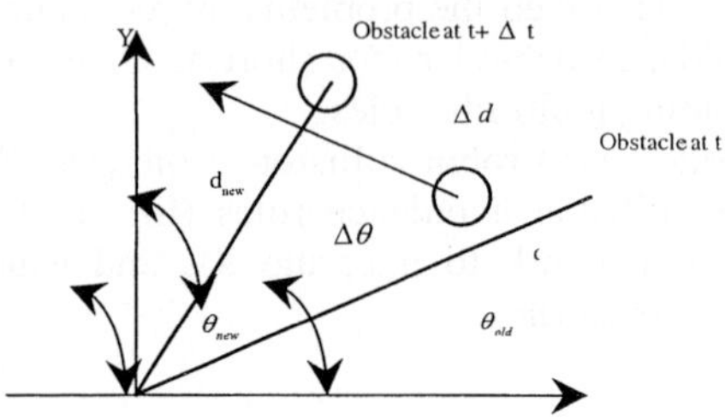

Fig. 1. System State of CAS

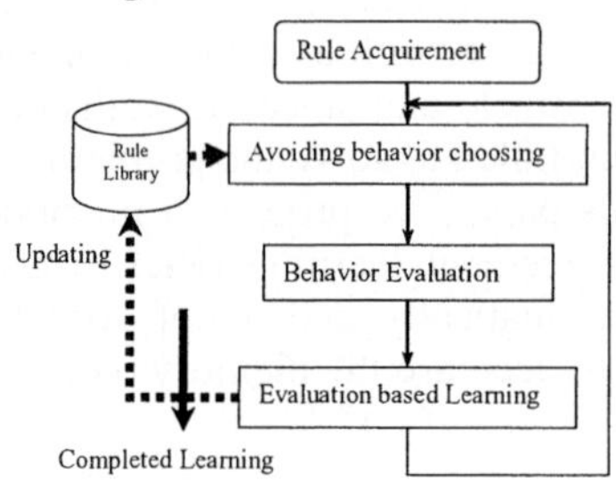

Fig. 2. Flow chat of RL

We define an evaluation score table composed of an evaluation score S_{ij} for each possible CAB in the behavior space. Where i represents the ith motion deflection angle, j represents the jth acceleration. Mobile robot selects a minimum PB_{ij} of CAB in present state according to formula (1).

$$PB_{ij} = \delta\Delta\, dir_i + \varphi\Delta\, sp_j + \eta / S_{ij} \tag{1}$$

Where $\triangle dir_i$ is the change value of present collision avoidance acceleration, $\triangle sp_j$ is the change value of deflection angle; S_{ij} is the evaluation score of CAB, δ, φ, η are their corresponding weight vector, respectively. PB_{ij} is the integrated evaluation value of CAB in present state.

In the state of initial learning, each CAB has the same initial evaluation score. Every R selects a CAB with minimum PB_{ij} in both initial learning and subsequent learning, and uses evaluation function E to evaluate the behavior based on feedback information of environment. E is defined as formula (2).

$$E=\begin{cases}\lambda & |\Delta dir|*|\Delta sp|=0,\text{No Collision}\\ \alpha/|\Delta dir|+\beta/|\Delta sp| & |\Delta dir|*|\Delta sp|\neq 0,\text{No Collision}\\ -\gamma & \text{Collision}\end{cases} \tag{2}$$

Where α and β are the weight parameters of each standard. $|\triangle dir|$ and $|\triangle sp|$ are the change values of present acceleration and present deflection angle, γ and λ are positive constant.

When a CAB is selected and executed, S_{ij} is recomputed as formula (3).

$$S_{new}=S_{old}+E \tag{3}$$

If collision happens, E is negative which decreases the score of present behavior; otherwise E is positive which increases the score. So the meaning of E is that if there is no collision happens, the behavior with minimum change of motion state will get maximal score.

3 ART2 Neural Network Based Collision Avoidance System

With the increase of obstacles and parameters, the number of control rules increases exponentially. Therefore, large memory is needed to store rules if we use traditional way to implement above RL solution. In order to decrease memory space, we propose an ART2 neural network based CAS with RL (for short as ART2CAS) to realize machine learning and store CARs.

3.1 The Structure of ART2CAS

ART2 is designed for classifying arbitrary sequence model and simulating input model. It can classify input vector in any precision. The structure of ART2CAS is shown as Fig. 3.

Neural network controlling: It is composed of an ART2 neural network shown in Fig. 4, whose main function is to store control rules and do the inference based on input state. The inputs of F1 layer represent linguistic variables of precondition, namely the state of robot and obstacles, and each class of F2 layer represents linguistic variable of conclusion for each rule, namely a group of CABs.

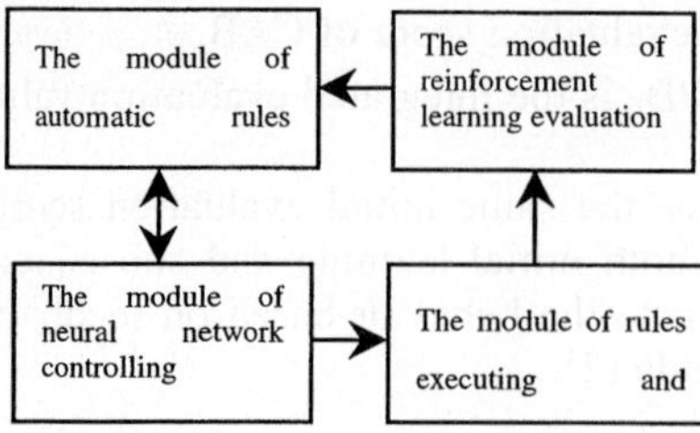

Fig. 3. System structure of ART2CAS

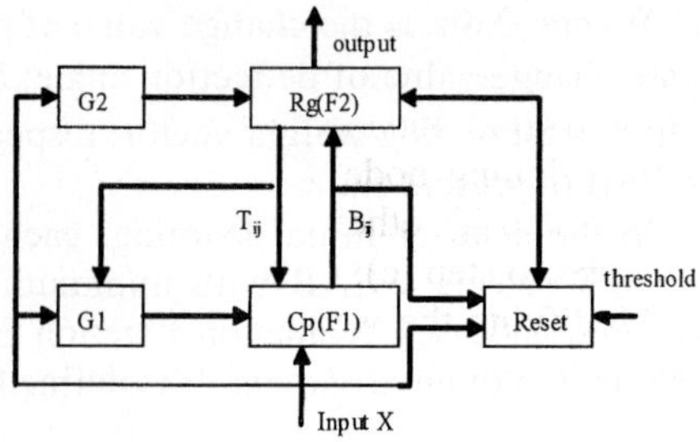

Fig. 4. ART2 neural network structure

Rule executing and state collecting: Executing the rule and taking charge to collect system state information for the use of other modules.
Reinforcement learning evaluation: The main function of this module is to provide the evidence of evaluation for RL algorithm. It analyzes the conclusion of present rule and gives the value of evaluation function E to other modules based on formula (2). The RL algorithm needs to continuously evaluate the system capability to confirm the intensity of award or punishment; thereby the optimal control output is acquired.
Automatic rules selecting: This module provides a RL algorithm based neural network learning process without input and output samples knowledge. The input of module is the system state and the value of evaluation function E. If E is larger than some given threshold, the system selects the present rule. Otherwise, the system does the reasoning again and selects a group of CABs by neural network, moreover executes the CABs by the executing module. Then the module of RL evaluation evaluates system to get E. The process is repeated.

3.2 The Learning Algorithm of ART2 Neural Network

The typical learning algorithm of ART2 neural network is unsupervised. After given a uniform threshold, neural network shown in Fig 4 can automatically classify the inputs. In order to automatic acquire CARs, we modify the learning algorithm. For each group of input state, ART2 neural network relearns the state no longer based on uniform threshold, but based on a changeable evaluation function E for each input. This is the process of the rule generation of collision avoidance.

The modified learning algorithm of ART2 neural network is as follows:

(a) Initializing the linked weight matrix and gain control G1.
(b)Given a input state, namely given the neural network input X=(x1, x2,…,xn), x∈ {R}n.
(c) The input layer Cp exports vector C=X and weights the weight vector of competitive layer Rg from bottom to top, gets the total input uj=CBj,j∈ {1,2, …,m} of jth node in Rg layer.
(d) Nodes of Rg layer compete based on the competitive learning rules. The system selects the node with the maximal input value $u_{j^*} = \max_{j\in\{1,2,\ldots,m\}}\{u_j\}$, i.e. the jth node of Rg layer win the competition.

(e) Information evaluation. The system selects the CAB of present winning node as system output, the module of RL evaluation computes the E based on the information acquired from outside. If the result is larger than given threshold, the system takes j* as the winning node, (i.e. a rule defined by present state and CAB), the algorithm turns to step (f), otherwise system sends reset signal, sets j* node to be inhibition state and turns to step (c).
(f) Modifying the weight vector of network. The system modifies the weight vector sent by the winning node of Rg layer from top to bottom and from bottom to top. When the input which is similar to X appears again, it will win easily.
(g) Getting rid of the inhibition state of the node setting in step (e) and turn to step (b) to wait for the next input.

In the simulation program of our research, evaluation function E has been simplified to be two kinds of evaluation, one for successful collision avoidance and the other was for failure one. If no collision, the neural network modified weight vector in order that present rule is more possible to be selected when similar input state happens next time, otherwise restrain present optimal matching rule and select a new optimal rule which can avoid collision successfully.

4 Simulation Experiment and Analysis

In our simulation environment, mobile robot R and moving obstacles O1, O2 are moving on a 20×20 unit plane. The R's velocity is 0.5 units, and its moving direction is $\pi/2$ in polar coordinates and keeps constantly. The position of O1 and O2 are distributed at random, and their velocity and direction value are also produced at random.

In simulation one, we place two obstacles, the maximum deflection angle is $\pm 30^0$, and the maximum acceleration is ±0.02 unit. In the behavior space table, Robot can only change its angle in 5^0, and acceleration is 0.005 units. Each obstacle's state can be described as four parameters (velocity, direction, x coordinate, y coordinate). The F1 layer of ART2 neural network has 12 nodes corresponding to 12 state variables of robot and obstacles. The F2 layer has 117 nodes corresponding to robot's 117 CABs. Fig. 5(a) shows that collision is inevitable when ART2CAS is not active. After using our ART2CAS, robot chooses the optimal behavior after 6 times RL shown as Fig. 5(b). At that time robot's deflection angle is 25^0, and its acceleration is zero. The robot avoids the obstacles O1 and O2 successfully with the CAR selected by RL and stores CAR in neural network (shown as Fig. 5(c)).

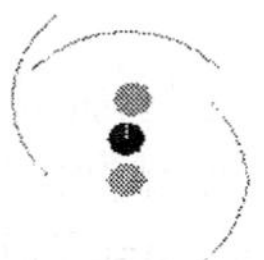

Fig. 5.(a) Situation of Collision

Fig. 5.(b) Learning of ART2CAS

In simulation two (shown as Fig. 6), there are two obstacles, the nodes of each layer in neural network is the same as simulation one. The motion states of obstacles (initial position, velocity and direction) are at random in order to produce enough

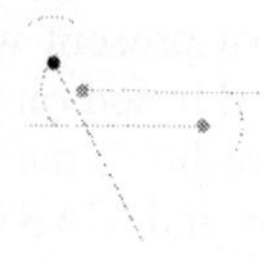

Fig. 5.(c) Successful learning of ART2CAS

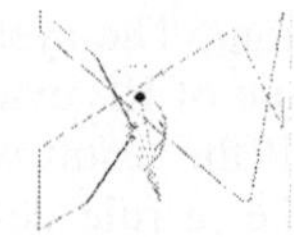

Fig. 6. Learning of Multi-obstacles collision avoidance

collision states. After using our ART2CAS to learn and train repeatedly, the robot stores enough CARs so as to avoid two obstacles with random velocity and random position successful.

In case of more obstacles or mobile robots, we just need to respectively modify the node number of F1 layer based on the obstacles and mobile robots as well as the node number of F2 layer based on the redefined CAB space table of robot in order to set up new neural network structure. Our ART2CAS can handle the collision avoidance problem with the corresponding number of obstacles completely.

5 Conclusion

The paper has given a collision avoidance system based on RL algorithm and ART2 neural network (ART2CAS) against the collision avoidance problem of robot on the condition of multi-motion-obstacle. The simulation experiment indicated that appropriative memory space of the system was far decreased and the system could solve the problem of collision avoidance of any motion obstacles. The system can deal with the collision avoidance problem on the complex condition of having static and dynamic obstacles at the same time, since the static obstacle can be viewed as a dynamic obstacle with zero velocity.

Acknowledgement. The research was supported by Science and Technology Committee of Shanghai Municipality (Project no: 015115042) and the fourth key subject construction of Education Committee of Shanghai Municipality (Project no: B682), we are greatly grateful to the foundation.

References

1. Chun-Ta Chen , Quinn, R.D., Ritzmann, R.E.: A Crash Avoidance System Based Upon the Cockroach Escape Response Circuit. Proceeding of the IEEE International Conference on Robotics and Automation (1997) 2007-2012
2. Arai, Y., Fujii, T.: Collision avoidance in multi-robot systems based on multi-layered Reinforcement. Robotics and Autonomous Systems 29. Elsevier Science B.V. (1999) 21-32
3. Ping Tang: Dynamic Obstacle Avoidance Based on Fuzzy Inference and Transposition Principle for Soccer Robots. IEEE International Fuzzy System Conference (2001)1062-1064
4. Fujimori, A., Tani, S.: A Navigation of Mobile robot with Collision Avoidance for Moving Obstacles. IEEE ICIT '02, Bangkok, Thailand (2002) 1-6

5. Suwimonteerabuth, D. , Chongstitvatana, P.: Online Robot Learning By Reward and Punishment for a mobile robot. Proceedings of the IEEE/RSJ International Conference on Intelligent Robots and Systems. (2002) 921-926
6. Smart, W.D., Pack Kaelbling, L.: Effective Reinforcement learning for Mobile Robots. Proceedings of the IEEE International Conference of Robotics and Automation, Washington D.C. (2002) 3404-3410
7. Xu, H. , Yang, S.X.: Real-time Collision-free Motion Planning of Nonholonomic Robots using a Neural Dynamics based Approach. Proceedings of the IEEE International Conference of Robotics and Automation, Washington D.C. (2002) 3087-3092

FEL-Based Adaptive Dynamic Inverse Control for Flexible Spacecraft Attitude Maneuver

Yaqiu Liu, Guangfu Ma, and Qinglei Hu

Harbin Institute of Technology, School of Astronautics
150001 Harbin, Heilongjiang Province, China
{yqliu, magf, huqinglei}@hit.edu.cn

Abstract. Based on Feedback-Error-Learning (FEL), an adaptive dynamic inverse control approach for single-axis rotational maneuver of spacecraft with flexible appendages by use of on-off thrusters is discussed. This approach uses a conventional feedback controller (CFC) concurrently with a Nonlinear Auto-Regressive Exogenous Input (NARX) neural network, and the NARX neural network can act dynamic adaptive inverse feed-forward controller, which is adapted on-line using the output of the CFC as its error signal, to improve the performance of a conventional non-adaptive feedback controller. The neural network (NN) does not need training in advance, and can utilize input and output on-line information to learn the systematic parameter change and unmodeled dynamics, so that the self-adaptation of control parameter is adjusted. However, the CFC should at least be able to stabilize the system under control when used without the neural network. The numerical simulations have shown that the control strategy is effective and feasible.

1 Introduction

Modern spacecrafts often employ large flexible structures such as solar arrays, while requirement for attitude control performance becomes more stringent. For attitude operations that require small control actions, reaction or momentum wheels are used. However, during orbital correction maneuver, the required torque is normally too high for reaction wheels. Therefore, thrusters are normally used for attitude control during these maneuvers. Thrusters are on-off devices and are normally capable of providing only fixed torque. Achieving high attitude control performance using thrusters is a challenging task for flexible spacecraft, where thruster firings could excite modes resulting attitude control instability or limit cycles. Research toward this end has been focused mainly to seek efficient methods to convert continuous input commands to on-off signals suitable for controlling on-of thruster. Pulse modulators are commonly employed due to their advantages of reduced propellant consumption and near linear duty cycle in Refs. [1-2]. On-off thruster firing, no matter the method of modulation, will introduce vibration to the flexible structures to some degree, which will reduce the precision pointing of the onboard payloads. The investigation of methodologies using artificial neural networks for control of light weight materials

F. Yin, J. Wang, and C. Guo (Eds.): ISNN 2004, LNCS 3174, pp. 42–51, 2004.

with distributed flexibility in advanced space applications has been the subject of intensive research in the recent years in Ref. [3-5]. Based on their inherent learning ability and the massively parallel architecture, neural networks are considered promising controller candidates, particularly for nonlinear and uncertain systems. These characteristics have motivated the present research towards the development of neural network control methodologies that make use of real time controller tuning and of output measurements to cooperate the CFC to increase performance. Song et. al. [6] have proposed a hybrid controller for gas tubine applications by using Multilayer Perceptron (MLP) neural network with classical LQG/LTR controller. Apolloni et. al. [7] have proposed a hybrid control strategy in satellite attitude control using a PID and MLP neural network. But in their approach a neural network was being trained in an off-line manner. Kawato et. al [5] proposed a feed-forward controller learning scheme, which uses a feedback signal as the error for training a NN model. They call this learning method feedback-error-learning. As for the structure of the neural network, the dynamic neural network NARX neural network in Ref. [8] is general enough to approximate any nonlinear dynamical system in Ref. [9], and has self-feedback to be adapted using a method such as Real-Time Recurrent Learning [10] or Back-Propagation Through Time [11]. A feed-forward neural network is one whose input contains no self-feedback of its previous outputs. Its weights may be adapted using the BP algorithm.

In this paper, based on NARX neural network acting as a feed-forward controller, we propose an adaptive inverse control approach using FEL that is based on the output of a previously adjusted conventional feedback controller (CFC) with fixed parameters. The neural network aims to improve the performance of a conventional non-adaptive feedback controller. Nevertheless, The CFC should at least be able to stabilize the system under control when used without the neural network.

2 Mathematical Modeling

The slewing motion of a rigid hub with appendages attached to the hub is graphically presented in Fig.1. [12]. The rotational motion only without any translation of the center of mass of the whole structure is considered in this paper. The center rotation is denoted as θ, and the deflection of the appendage is represented by $w(x,t)$. The governing differential equations of motion can be derived from the extended Hamilton's principle which states that

$$\int(\delta L + \delta W)\mathrm{d}t = 0 \tag{1}$$

Where $L=T-U$ represents system Lagrangian as a difference between kinetic and potential energy of the system, and $\delta W=\delta\theta u$ is the virtual work by non-conservative control torque applied at the center body. Therefore, the governing equations of motion for the spacecraft model are given by

$$I_c\ddot{\theta} + \int_b^l \rho x\left(x\ddot{\theta} + \frac{\partial^2 w}{\partial t^2}\right)dx + m_t l\left(l\ddot{\theta} + \left.\frac{\partial^2 w}{\partial t^2}\right|_l\right) = u \quad ; \quad \rho\left(x\ddot{\theta} + \frac{\partial^2 w}{\partial t^2}\right) + EI\frac{\partial^4 w}{\partial x^4} = 0 \tag{2}$$

where I_c is the moment of inertia of the center body, ρ is linear mass density of the appendage, EI is the elastic rigidity of the flexible structure, m_t is the tip mass, b is the radius of sphere l is the distance from the center origin to the tip of the flexible structure, and u is the control torque applied by the actuator located at the center hub.

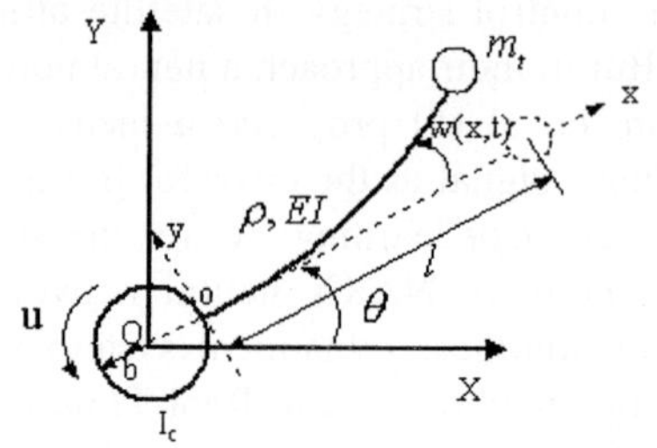

Fig. 1. Spacecraft model with single-axis rotation

The flexible dynamics are also governed by the boundary conditions at the root and tip of the structure such as

$$w(x,t) = \frac{\partial w(x,t)}{\partial t} = 0 \quad \text{at } x = b \tag{3a}$$

$$EI\frac{\partial^2 w(x,t)}{\partial x^2} = 0 \quad EI\frac{\partial^3 w(x,t)}{\partial x^3} = m_t\left(l\ddot{\theta} + \frac{\partial^2 w(x,t)}{\partial t^2}\right) \quad \text{at } x = l \tag{3b}$$

The target maneuver is single axis rotational rest-to-rest maneuver about body axis with simultaneous vibration control. The solar array is under bending deflection only and tends to vibrate due to the coupling effect with the rigid body rotation. The dynamics are therefore coupled motion between the center body rotation and flexible deflection of the solar array model.

3 Control Strategy

In this section, a dynamic adaptive inverse feed-forward controller design method using FEL algorithm, based on NARX neural network, is presented for the single-axis rotational maneuver control of spacecraft with flexible appendages. This control law design needs two steps. The first step is to design a classical controller, PD controller, which should be able to stabilize the system under control when used without the neural network. Then a dynamic adaptive inverse feed-forward controller is to design to improve the performance of a conventional non-adaptive PD controller.

3.1 PD Controller Design Based on Lyapunov Technique

The Lyapunov control design process starts with a choice of a candidate Lyapunov function. The choice here is taken as the following form, which is a modification of the form used in Refs. [13,14].

$$V = \frac{1}{2} I_c \ddot{\theta} + \frac{a_1}{2} \{ \int_b^l \rho(x\dot{\theta} + \frac{\partial w(x,t)}{\partial t})^2 dx + \int_b^l EI \left(\frac{\partial^2 w(x,t)}{\partial t^2} \right)^2 dx \} + \frac{a_1}{2} m_t \left(l\dot{\theta} + \frac{\partial w(x,t)}{\partial t}\Big|_l \right)^2 + \frac{a_2}{2} (\theta - \theta_f)^2 \tag{4}$$

where θ_f is a final target attitude, and it is obvious by inspection that imposing a_1,a_2>0, guarantees that $V \geq 0$ and that the global minimum of $V = 0$ occurs only at the zero state. The Lyapunov function is a combination of the structure energy and error energy with respect to an equilibrium point

$$\left(w(x,t), \frac{\partial w(x,t)}{\partial t}, \theta, \dot{\theta} \right) = \left(0, 0, \theta_f, 0 \right) \tag{5}$$

Time derivative of the Lyaounov function with the governing equations and boundary conditions yields

$$\dot{V} = \left[u + a_2(\theta - \theta_f) + a_3(bM_1 - M_2) \right] \dot{\theta} \tag{6}$$

where $a_3 = a_1 - 2 > -2$ is parameters to guarantee positiveness of the Lyapunov function, and

$$bM_1 - M_2 = \int_b^l \rho x \left(x\ddot{\theta} + \frac{\partial^2 w}{\partial t^2} \right) dx + m_t l \left(l\ddot{\theta} + \frac{\partial^2 w}{\partial t^2}\Big|_l \right) \tag{7}$$

A simple choice of the controller design can be written the following form

$$u = -a_4 \dot{\theta} - a_2(\theta - \theta_f) \tag{8}$$

where $a_3 = 0$,and $a_4 > 0$ is another design parameter for which $\dot{V} = -a_4 \dot{\theta}^2 \leq 0$. The control law in equation (8) is a typical PD controller using center body angular information The controller is therefore robust, and easy to implement. Stability of the closed-loop system is guaranteed irrespective of mathematical modeling errors. However, even though stability is guaranteed by the control law with two feedback gains, the performance requirement of the closed-loop system may not be satisfied automatically, which will be discussed later on.

3.2 Adaptive Dynamic Inverse Feed-Forward Controller

The neural network controller adopts three layers NARX neural network, which is general enough to approximate any nonlinear dynamical system in Ref. [13]. The control strategy structure is shown in Fig.2. Generally speaking, a feed-forward neural network is one whose input contains no self-feedback of its previous outputs. Its weights may be adapted using the popular back-propagation algorithm. For the NARX neural network, it has self-feedback and must be adapted using a method such as RTRL, or BPTT. Although compelling arguments may be made supporting either algorithm, we have chosen to use RTRL in this work, since it easily generalizes as is required later and is able to adapt neural network used in an implementation of adaptive inverse control in real time. In term of the results in Ref. [3] how to adapt a feed-forward neural network and how to compute Jacobians of a neural network, it is a simple matter to extend the back-propagation algorithm to adapt NARX filters. The presentation is as follows. An NARX filter computes a function of the following form:

$$y_k = f(x_k, x_{k-1}, \cdots, x_{k-n}, y_{k-1}, y_{k-2}, \cdots, y_{k-m}, W) \tag{9}$$

To adapt using the familiar "sum of squared error" cost function, we need calculate

$$\frac{dJ_k}{dW} = \frac{d\,\frac{1}{2}\|e_k\|^2}{dW} = -e_k^T \frac{dy_k}{dW} = -e_k^T\left(\frac{\partial y_k}{\partial W} + \sum_{i=0}^{n}\frac{\partial y_k}{\partial x_{k-i}}\frac{dx_{k-i}}{dW} + \sum_{i=1}^{m}\frac{\partial y_k}{\partial y_{k-i}}\frac{dy_{k-i}}{dW}\right). \tag{10}$$

where the first term $\partial y_k/\partial W$ is the direct effect of a change in the weights on y_k, and is one of the Jacobians calculated by the back-propagation algorithm. The second term is zero, since dx_{k-1}/dW is zero for all k. The final term may be broken up into two parts. The first, $\partial y_k/\partial y_{k-1}$, is a component of the matrix $\partial y_k/\partial X$, as delayed versions of y_k are part of the network's input vector X. The back-propagation algorithm may be used to compute this. The second part, dy_{k-1}/dW, is simply a previously calculated and stored value of dy_k/dW. When the system is "turned on," dy_{k-i}/dW are set to zero for i = 0,-1,-2,…, and the rest of the terms are calculated recursively from that point on.

Note that the back-propagation procedure naturally calculates the Jacobians in such a way that the weight update is done with simple matrix multiplication. Let

$$(d_w y)_k = \left[\left(\frac{dy_{k-1}}{dW}\right)^T \cdots \left(\frac{dy_{k-m}}{dW}\right)^T\right]^T, \quad (d_x y)_k = \left[\left(\frac{\partial y_k}{\partial y_{k-1}}\right) \cdots \left(\frac{\partial y_k}{\partial y_{k-m}}\right)\right]. \tag{11}$$

The latter is simply the columns of corresponding to the feedback inputs to the network, and is directly calculated by the back-propagation algorithm. Then, the weight update is efficiently calculated as

$$\Delta W_k = \left(\eta e_k^T\left[\frac{\partial y_k}{\partial W} + (d_x y)_k (d_w y)_k\right]\right)^T. \tag{12}$$

According to the general back-propagation algorithm, the adaptation signal used to adjust neural network controller is error signal of input to plant. Search suitable adaptation signals used to train network being important problem. The advantage of the BPTM (back-propagation through model) approach in adaptive inverse control theory is that nonlinear dynamic inversion may be computationally intensive and precise dynamic models that may not be available in Ref. [3]. However, it is convenient for FEL to adopt directly output of a feedback controller with fixed parameters to adapt a NARX neural network, without the need of the plant's model. Combined these characteristic, adaptive dynamics inverse control scheme based on the FEL algorithm using NARX network is proposed in this paper.

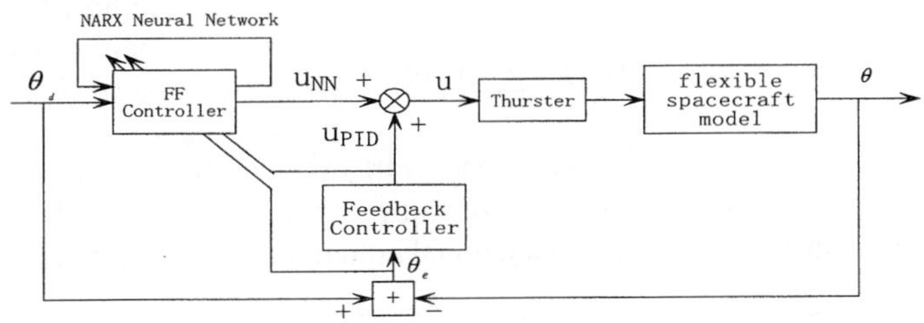

Fig. 2. Adaptive inverse control approach *Feedback-Error-Learning*

In Fig.2, let $\Phi(.)$ be the function implemented by the NARX network controller

$$u_k = \Phi(u_{k-1}, u_{k-2} \cdots u_{k-m}, r_k, r_{k-1} \cdots r_{k-q}, W_C) \tag{13}$$

The learning rule used is based on Hebbian learning scheme in the following form:

$$\Delta W_c = \eta (\frac{d\Phi}{dW_c})^T u_{\text{FB}} \tag{14}$$

where η is the adaptive learning rate, u_{FB} is the learning error. The NARX neural network can self-adapt to minimize the feedback error u_{FB}. Using (14) and the chain rule for total derivatives, the learning algorithm adopting RTRL can be made the following form

$$\frac{du_k}{dW_C} = \frac{\partial u_k}{\partial W_C} + \sum_{j=1}^{m} (\frac{\partial u_k}{\partial u_{k-j}})(\frac{du_{k-j}}{dW_C}) \tag{15}$$

where the first term is the Jacobian $\partial u_k / \partial u_{k-j}$, which may be computed via the back-propagation algorithm. The next term, du_{k-j}/dW_c is the current or a previously computed and saved version of du_k/dW_c, computed via (15). Here du_k/dW_c as expanded in (15) has the same in form as (10) and can be computed in the same way.

4 Simulation Results

In order to test the proposed control schemes, numerical simulations have been performed. The numerical model of the spacecraft is from the Ref. [12]. The low-frequency modes are generally dominant in a flexible system, in this paper, and the first two modes frequency are 3.161rad/s and 16.954rad/s, respectively.

The model is placed into state space in preparation for digital simulation using the MATLAB/Simulink software package. The original hybrid differential equation of the motion can be discretized into a finite dimensional mathematical model. The mathematical model is developed for simulation study and the model-based control law design. The flexible displacement is approximated as

$$w(x,t) = \sum_{i=1}^{2} \phi_i(x)\eta_i(t) \tag{16}$$

where $\phi_i(x)$ i=1,2 is the *ith* shape functions to be obtained by solving a characteristic equation for a cantilevered beam problem, and η_i is the *ith* generalized coordinates for the flexible deflection. The finite dimensional equations of motion in matrix form can be written as

$$\begin{bmatrix} I_c & M_{\theta\eta} \\ M_{\eta\theta} & M_{\eta\eta} \end{bmatrix} \begin{bmatrix} \ddot{\theta} \\ \ddot{\eta} \end{bmatrix} + \begin{bmatrix} 0 & 0 \\ 0 & K_{\eta\eta} \end{bmatrix} \begin{bmatrix} \theta \\ \eta \end{bmatrix} = \begin{bmatrix} 1 \\ 0 \end{bmatrix} u \tag{17}$$

element mass and stiffness matrices ($M_{\theta\eta}$, $M_{\eta\eta}$, $K_{\eta\eta}$) are computed by using Lagrange's equation, and $\eta=[\eta_1\ \ \eta_2]^T$ is the flexible coordinate vector. The state-space representation of the system equation is

$$\dot{x} = Ax + Bu, \qquad y = Cx + Du. \tag{18}$$

where the state vector is defined as $x = [\theta \quad \eta \quad \dot{\theta} \quad \dot{\eta}]$. The output y is the vector of the states; hence, C is an 6×6 identity matrix, and the feedback values of angular position and rate are measured exactly.

In the simulation, the neural network controller adopts three layers NARX network with with 3 regressive exogenous input and 50 delay input version. The hidden and output network layers used hyperbolic tangent and linear functions respectively. The adaptive learning rate in (14) is of the following form in the simulation:

$$\eta_k = \eta_0 / (1 + K_\eta \,|\tanh(\sum_{i=0}^{n} e_{k-i})|) \tag{19}$$

where K_η=10, n=3 and η_0=0.001.

These parameters were adjusted empirically. And the choice of the PD parameter is based on Lyapunov technique considering the stability of the system, the time of convergence and the final performance. The single-axis rotational maneuver command is 60 degree. Fig.3(a) shows the angle slew of the flexible spacecraft when the PD controller is used alone, and the amplitude of the angle error at the end of maneuver is about 1.6 degree, and the angle velocity is about 0.2deg/s shown in Fig.4(a), which is a poor results and can not satisfy the requirement of the precision pointing of the onboard payloads. This results from the coupled motion between the center body rotation and flexible deflection of the solar array model, and the vibration of the modes will reduce the precision pointing. It should be noted that there still exists some room for improvement with different design parameter sets of PD. But there is not much improvement in the maneuver angle and angle velocity response using on-off thruster.

In order to improve the precision pointing, we use the PD controller concurrently with a NARX neural network acting dynamic adaptive inverse feed-forward controller, which is adapted on-line using the output of the PD controller as its error signal. The angle slew of the flexible spacecraft with PD and NN controller is shown in Fig.3(b), and the error of the angle at the end of maneuver is about 0.06 degree, and the angle velocity is about 0.02deg/s shown in Fig.4(b).

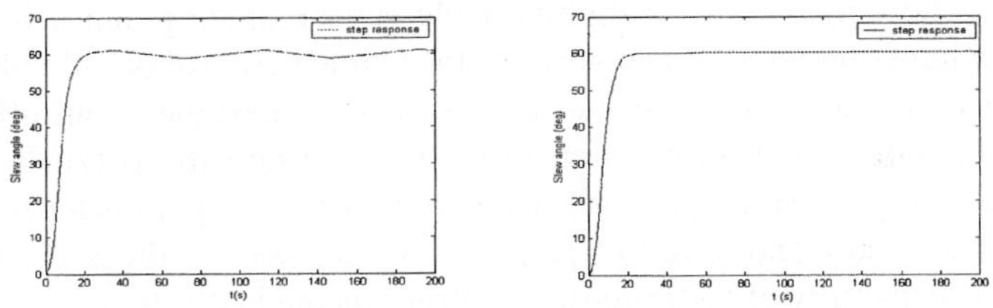

Fig. 3. (a) 60° Slew with PD control (left) and (b) with PD+NN control (right)

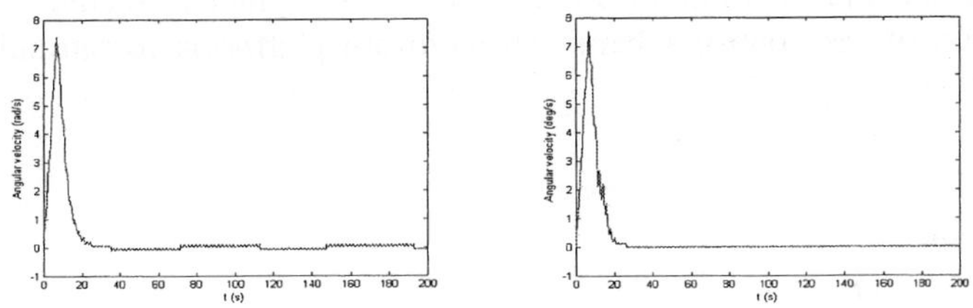

Fig. 4. (a) Angular velocity response with PD control (left) and (b) with PD+NN control (right)

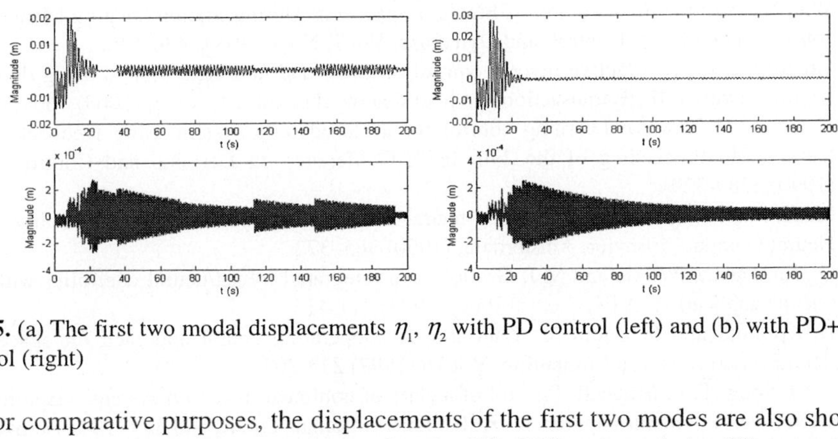

Fig. 5. (a) The first two modal displacements η_1, η_2 with PD control (left) and (b) with PD+NN control (right)

For comparative purposes, the displacements of the first two modes are also shown in Fig.5(b) and Fig.5(a), respectively, for the PD+NN case and the PD controller alone. The PD+NN control strategy can effectively suppress the vibration of the flexible modes, which mean that the NN was able to filter out the higher frequency mode vibration of the flexible solar array.

5 Conclusions

A single-axis rotational maneuver control of spacecraft with flexible appendages using a conventional feedback controller (CFC) concurrently with a NARX neural network is discussed in this paper. Based on Feedback-Error-Learning scheme, the NARX neural network acts dynamic adaptive inverse feed-forward controller, which is adapted on-line using the output of the CFC as its error signal, to improve the performance of a conventional non-adaptive feedback controller. It is shown how the network can be employed as a multivariable self-organizing and self-learning controller in conjunction with a PD controller for attitude control of a flexible spacecraft. In contrast to the previous classical approaches, the designed controller can utilize input and output on-line information to learn the systematic parameter change and unmodeled dynamics, so that the self-adaptation of control parameter is adjusted. The problem of accuracy and speed of response has all been addressed and through an example, the capability of the proposed control scheme is illustrated.

Future research will investigate the performance of this neural control approach when used to suppress vibration in real time, and is planted to study the digital implementation of the control scheme on hardware platforms for attitude control experimentation.

References

1. Anthony, T., Wei, B., and Carroll, S.: Pulse Modulated Control Synthesis for a Spacecraft. Journal of Guidance, Control, and Dynamics, Vol.13, No.6, (1990) 1014-1015.
2. Wie, B.: Attitude Stabilization of Flexible Spacecraft During Station-keeping Manuevers. Journal of Guidance, Control, and Dynamics, Vol.7, No.4, (1984) 430-436.
3. Gregory L. Plett: Adaptive inverse control of linear and nonlinear system using dynamic neural networks. IEEE transactions on neural networks, vol. 14, No. 2, (2003) 360-376
4. H Gomi, M Kawato: Learning control for a closed loop system using feedback-error-learning. In Proceeding of the 29th IEEE Conference on Decision and Control, vol.6 (1990) 3289-3294
5. M. Kawato: Feedback-error-learning neural network for supervised motor learning. Adv. Neural Comput., Elsevier, Amsterdam, (1990) 365–372
6. Q. Song and J. Wilkie and M.J. Grimble: An integrated robust/neural controller with gas turbine application. IEEE Control Proc, (1994) 411-415
7. B. Apolloni and F. Battini C. Lucisano: A cooperating neural approach for spacecraft attitude control. Neuro Computing, Vol.16 (1997).279-307
8. Y. D. Song, T. L. Mitchell: Control of a class of nonlinear uncertain systems via compensated inverse dynamics approach. IEEE Trans. Automat. Contr., Vol.39 (1994) 1866-1871

9. H. T. Siegelman, B. B. Horne, and C. L. Giles: Computational capabilities of recurrent NARX neural networks. IEEE Trans. Syst., Man, Cybern. B, Vol.27 (1997) 208-215.
10. R. J. Williams and D. Zipser: experimental analysis of the real-time recurrent learing algorithm. Connection Sci., Vol.1, No.1 (1989) 87-111.
11. D.H. Neguyen and B. Widrow: Neural networks for self-learning control systems. IEEE Trans. Neural networks, Vol.10 (1990) 18-23.
12. Jin Jun, Shan: Study on CSVS Method for the Flexible Spacecraft. Ph.D thesis, Harbin Institute of Technology (2002)
13. Junkins, J. L. and Bang, H. C.: Maneuver and Vibration Control of Hybird Coordinate Systems using Lyapunov Stability Theory. Journal of Guidance, Control, and Dynamics, Vol.16, No.4, (1993) 668-676
14. Byers, R. M., Vadali, S.R: Near-Minimum-time Closed-loop Slewing of Flexible Spacecraft. Journal of Guidance, Control, and Dynamics, Vol.12, No.6 (1989) 858-865

Multivariable Generalized Minimum Variance Control Based on Artificial Neural Networks and Gaussian Process Models

Daniel Sbarbaro[1]*, Roderick Murray-Smith[2], and Arturo Valdes[1]

[1] Department of Electrical Engineering, Universidad de Concepcion, Chile
{dsbarbar, avald}@die.udec.cl
[2] Department of Computer Science, University of Glasgow, Glasgow, U.K. & Hamilton Institute, NUI Maynooth, Ireland.
rod@dcs.gla.ac.uk

Abstract. The control of an unknown multivariable nonlinear process represents a challenging problem. Model based approaches, like Generalized Minimum Variance, provide a flexible framework for addressing the main issues arising in the control of complex nonlinear systems. However, the final performance will depend heavily on the models representing the system. This work presents a comparative analysis of two modelling approaches for nonlinear systems, namely Artificial Neural Network (ANN) and Gaussian processes. Their advantages and disadvantages as building blocks of a GMV controller are illustrated by simulation.

1 Introduction

The control of a nonlinear multivariable system is a difficult task due to the number of variables involved and the complex coupling among inputs and outputs. This problem is even worse if the model of the system is unknown.

The use of a flexible modelling technique, such as artificial neural networks (ANN), for controlling Multivariable Nonlinear systems has been addressed from a theoretical point of view and illustrated by simulation in [4]. A design methodology based on the idea of a model reference and ANN is also developed. Generalized Minimum Variance control approach is a versatile and very popular formalism for designing linear self-tuning controllers [1], Gawthrop has also shown that model reference and self-tuning controllers are closely related. The GMV controllers can be enhanced to deal with nonlinear systems if nonlinear models are considered. ANN models, in this context, were first proposed by [2]. In [3] an extension to a multivariable system is presented, but in this case a linear model extended by a ANN was considered. A typical assumption used in these approaches is the "Certainty Equivalence principle" under which the controller is designed in terms of the identified model as if the model were the real process.

There has been an increase of interest in the use of Gaussian Process priors as alternative to ANN [9,5,8]. Mackay describes in his seminal paper the similarity

* This work was supported by Fondecyt project 1040486

F. Yin, J. Wang, and C. Guo (Eds.): ISNN 2004, LNCS 3174, pp. 52–58, 2004.

and differences of both approaches from a statistical perspective [5]. In [6] a MV adaptive controller for single-input single-output system based on GP is presented. One of the main advantages of this modelling approach is that provides a direct estimation of the output variance, which is normally ignored by other methods. This is a relevant information for control design, since it allows the design of self-tuning controllers without resorting to the Certainty Equivalence principle.

Without losing generality, this work considers the MIMO systems as multiloop systems, consisting of several SISO systems where coupling effects are considered as disturbances.

This paper is organized as follows: section 2 describes the GMV approach for control design. Section 3 presents briefly models used to approximate the system and the resulting expression for GMV controllers. Section 4 illustrates by performing simulations, the main characteristics of both modelling approaches. Some final remarks are given in section 5.

2 Multivariable Generalized Minimum Variance Controller

Let us consider the Multivariable nonlinear system described by the following affine in the control structure:

$$\mathbf{Y}(k+1) = F(\mathbf{X}(k)) + G(\mathbf{X}(k))\mathbf{U}(k) + \Xi(k) \tag{1}$$

where Ξ is a vector of Gaussian zero mean random noise, $\mathbf{Y}(k) = [y_1(k) \ldots y_n(k)]^T$, $\mathbf{U}(k) = [u_1(k) \ldots u_m(k)]^T$, n the number of outputs and m the number of inputs. The vector $\mathbf{X}$ represents a vector with all the past information as described by:

$$\mathbf{X}^T(k) = [\mathbf{Y}(k)^T \mathbf{Y}(k-1)^T \ldots \mathbf{Y}(k-r)^T \mathbf{U}(k-1)^T \mathbf{U}(k-2)^T \ldots \mathbf{U}(k-s)^T] \tag{2}$$

The integers r and s are the number of delayed inputs and outputs respectively necessary for describing the system. As pointed out in [2] the affinity property represents more an advantage for designing the controller, rather than a modelling constraint. In the generalized minimum variance control approach, the performance index is:

$$J = E\{\mathbf{E}(k+1)^T \mathbf{Q} \mathbf{E}(k+1)\} + \mathbf{U}(k)^T \mathbf{R}(q^{-1}) \mathbf{U}(k) \tag{3}$$

where $\mathbf{E}(k+1) = \mathbf{Y}_d(k+1) - \mathbf{Y}(k+1)$, the matrix $\mathbf{Q}$ is definite positive matrix and $\mathbf{R}$ is polynomial matrix in the backward shift operator q^{-1} . The expected value of the quadratic error term can be expanded in terms of the variance and the mean values as follows:

$$\begin{aligned} E\{\mathbf{E}(k+1)^T \mathbf{Q} \mathbf{E}(k+1)\} &= \\ (\mathbf{Y}_d(k+1) - \mu_Y(k+1))^T \mathbf{Q} (\mathbf{Y}_d(k+1) - \mu_Y(k+1)) & \\ + \sum_{j=1}^{n} q_{jj} var\{y_j(k)\} + \sum_{i=1}^{n} \sum_{j=1}^{n} q_{ij} covar\{y_i(k) y_j(k)\} & \end{aligned} \tag{4}$$

where μ_Y represents a vector having the mean value of each output; i.e. $\mu_Y = [\mu_{y_1} \dots \mu_{y_n}]^T$.

Without losing generality we will consider $\mathbf{Q} = diag[q_1 \dots q_n]$ and $\mathbf{R}(q^{-1}) = diag[R_1(q^{-1}) \dots R_n(q^{-1})]$. The necessary condition for $u_i(k)$ being a minimum of the cost function 3 is:

$$\mathbf{E}(k+1)^T \mathbf{Q} \frac{\partial \mathbf{E}(k+1)}{\partial u_i(k)} + R_i(q^{-1}) u_i(k) + \sum_{j=1}^{n} q_{jj} \frac{\partial var\{y_j(k)\}}{\partial u_i(k)} = 0. \quad (5)$$

3 Modelling Approaches

3.1 ANN Models

Every output of the nonlinear system is represented by a set of ANNs structured as follows:

$$y_j(k+1) = NN_{j,0}(k) + \sum_{l=1}^{m} NN_{j,l}(k) u_l(k) \quad (6)$$

where

$$NN_{j,l}(k) = \mathbf{W}_{j,l} \sigma(\mathbf{V}_{j,l} \mathbf{X}(k)). \quad (7)$$

The uncertainty associated with each output can be estimated by several standard confidence bound estimation algorithms [7], [10] which normally gave satisfactory results. However, the size of the estimated confidence interval could be biased [11]. A more important issue that hamper the application of these estimates in a control algorithm is the fact that they depends on a Jacobian matrix in a very complex form, making impossible their use in the control algorithm. In addition, they are usually expressed as uncertainty of parameters (even though the parameters often have no physical interpretation), and do not take into account uncertainty about model structure, or distance of current prediction point from training data used to estimate parameters. Thus, we will consider the outputs of the ANN as if they were the real output; i.e. we will apply the "certainty equivalence principle". In this way, equation (5):

$$\sum_{j=1}^{n} q_{jj} \left[y_j^d(k+1) - NN_{j,0}(k) + \sum_{l=1}^{m} NN_{j,l}(k) u_l(k) \right] NN_{j,i} + R_i(q^{-1}) u_i(k) = 0 \quad (8)$$

From this set of equations we can obtain the control as follows:

$$\mathbf{U}(k) = \mathbf{M}(k)^{-1} \begin{bmatrix} \sum_{j=1}^{n} q_{jj} \left[y_j^d(k+1) - NN_{j,0}(k) \right] NN_{j,i}(k) - \bar{R}_1(q^{-1}) u_1(k) \\ \vdots \\ \sum_{j=1}^{n} q_{jj} \left[y_j^d(k+1) - NN_{j,0}(k) \right] NN_{j,m}(k) - \bar{R}_m(q^{-1}) u_m(k) \end{bmatrix} \quad (9)$$

$$\mathbf{M}(k) = \begin{bmatrix} r_{10} + \sum_{j=1}^{m} q_{jj} NN_{j,1}(k)^2 & \dots & \sum_{j=1}^{m} q_{jj} NN_{j,m}(k) NN_{j,1}(k) \\ & \vdots & \\ \sum_{j=1}^{m} q_{jj} NN_{j,1}(k) NN_{j,m}(k) & \dots & r_{m0} + \sum_{j=1}^{m} q_{jj} NN_{j,m}(k)^2 \end{bmatrix} \tag{10}$$

where $\bar{R}_i(q^{-1}) = r_{i0} + r_{i1}q^{-1}...$ is a polynomial of order p.

3.2 GP Models

Let us consider the jth output of the system be described as:

$$y_j(k+1) = F_j(\phi(k)) + \epsilon_i(k) \tag{11}$$

where $\phi(k)^T = [\mathbf{X}^T(k) \;\; \mathbf{U}^T(k)]^T$. Instead of parameterising the system (1) as a parametric model, we can place a prior directly on the space of functions where F_j is assumed to belong. A Gaussian process represents the simplest form of prior over functions, we assume that any p points have a p-dimensional multivariate Normal distribution. Let the input and output sequence be stacked in the following matrix Φ_N and vector $\mathbf{y}_N$, respectively. We will assume zero mean, so for the case with partitioned data $\mathbf{y}_N$ and $y(k+1)$ we will have the multivariate Normal distribution,

$$\begin{bmatrix} \mathbf{y}_N \\ y(k+1) \end{bmatrix} \sim \mathcal{N}(0, \mathbf{C}_{N+1}), \quad \mathbf{C}_{N+1} = \begin{bmatrix} \mathbf{C}_N & \mathbf{K} \\ \mathbf{K}^T & \kappa \end{bmatrix}. \tag{12}$$

where $\mathbf{C}_{N+1}$ is the full covariance matrix. Like the Gaussian distribution, the Gaussian Process is fully specified by a mean and its covariance function, so we denote the distribution $GP(\mu, C)$. The covariance function $C(\phi(k), \phi(l))$ expresses the expected covariance between $y(k+1)$ and $y(l+1)$. We can therefore, infer $y(l+1)$'s from the N data pairs Φ_N and $\mathbf{y}_N$'s rather than building explicit parametric models.

As in the multinormal case, we can divide the joint probability into a marginal Gaussian process and a conditional Gaussian process. The marginal term gives us the likelihood of the training data,

$$P(\mathbf{y}_N|\Phi_N) = (2\pi)^{-\frac{N}{2}} |\mathbf{C}_N|^{-\frac{1}{2}} e^{\left(-\frac{1}{2}\mathbf{y}_N^T \mathbf{C}_N^{-1} \mathbf{y}_N\right)}. \tag{13}$$

The conditional part of the model, which best relates to a traditional regression model is therefore, the Gaussian process, which gives us the output posterior density function conditional on the training data $\Phi_N, \mathbf{y}_N$ and the test points $\phi(k)$,

$$P(y(k+1)|\Phi_N, \mathbf{y}_N, \phi(k)) = \frac{P(y(k+1), \mathbf{y}_N)}{P(\mathbf{y}_N)} = \frac{1}{(2\pi\hat{\sigma}_y^2)^{\frac{1}{2}}} e^{-\frac{(y(k+1)-\hat{\mu}_y)^2}{2\hat{\sigma}_y^2}},$$

where, as in the straightforward multinormal case described earlier,

$$\hat{\mu}_{y_i} = \mathbf{K}^T \mathbf{C}_N^{-1} \mathbf{y}_N \qquad var\{y_i\} = \hat{\sigma}_y^2 = \kappa - \mathbf{K}^T \mathbf{C}_N^{-1} \mathbf{K}, \tag{14}$$

so we can use $\hat{\mu}(\phi(k))$ as the expected model output, with a variance of $\hat{\sigma}(\phi(k))^2$. These expressions can be simplified [6] to obtain:

$$\hat{\mu}_{y_j} = \omega_{j,0} + \sum_{l=1}^{m} \omega_{j,l} u_l(k) = \omega_{j,0} + \Omega_j^T \mathbf{U}(k) \tag{15}$$

$$var\{y_j\} = \sigma_{j,0} + \sum_{l=1}^{m} \gamma_{j,l} u(l) + \sum_{i=1}^{m}\sum_{l=1}^{m} \sigma_{j,il} u_l(k) u_i(k) \tag{16}$$

$$= \sigma_{j,0} + \mathbf{U}(k)^T \Gamma_j + \mathbf{U}(k)^T \Sigma_j \mathbf{U}(k). \tag{17}$$

where $\sigma_{j,0}, \omega_{j,0}, \Gamma_j$, and Σ_j are all functions of time depending on the data gathered up to time k and the set of parameters associated to the covariance functions. Thus, the control signal is obtained as:

$$\mathbf{U}(k) = \mathbf{M}(k)^{-1} \begin{bmatrix} \sum_{j=1}^{n} q_{jj} \left[y_j^d(k+1) - \omega_{j,0}(k) \right] \omega_{j,1}(k) + \gamma_{j,1} - \bar{R}_1(q^{-1}) u_1(k) \\ \vdots \\ \sum_{j=1}^{n} q_{jj} \left[y_j^d(k+1) - \omega_{j,0}(k) \right] \omega_{j,m}(k) + \gamma_{j,m} - \bar{R}_m(q^{-1}) u_m(k) \end{bmatrix} \tag{18}$$

$$\mathbf{M}(k) = \begin{bmatrix} r_{10} + \sum_{j=1}^{m} q_{jj} \omega_{j,1}(k)^2 + \sigma_{j,11} & \dots & \sum_{j=1}^{m} q_{jj} \omega_{j,m}(k) \omega_{j,1}(k) + \sigma_{j,1m} \\ \vdots & & \\ \sum_{j=1}^{m} q_{jj} \omega_{j,1}(k) \omega_{j,m}(k) + \sigma_{j,m1} & \dots & r_{m0} + \sum_{j=1}^{m} q_{jj} \omega_{j,m}(k)^2 + \sigma_{j,mm} \end{bmatrix} \tag{19}$$

Notice that equations (18) and (19) have additional terms compared to (9) and (10). As pointed out in [6] these terms provide extra robustness in the face of uncertainty in the output estimation.

4 Simulations

In order to illustrate the main characteristics of both models, the following multivariable nonlinear system was considered:

$$y_1(k) = \frac{0.7 y_1(k-1) y_1(k-2)}{1 + y_1(k-1)^2 + y_2(k-2)^2} + \frac{10^{-4} u_2(k-1)}{1 + 3y_1(k-2)^2 + y_2(k-1)^2} + u_1(k-1) + .25u_1(k-2) + 0.5u_2(k-2) \tag{20}$$

$$y_2(k) = \frac{0.5 y_2(k-1) sin(y_2(k-2))}{1 + y_2(k-1)^2 + y_1(k-2)^2} + 0.5u_2(k-2) + .3u_1(k-2) + +u_2(k-1)(0.1u_2(k-2) - 1.5) \tag{21}$$

The nonlinear functions were approximated by four ANNs with 8 units in the hidden layer and the following structure :

$$y_i(k) = NN_{i,0}(k-1) + NN_{i,1}(k-1) u_i(k-1) \; i = 1, 2 \tag{22}$$

where

$$NN_{j,l}(k) = \mathbf{W}_{j,l} \tanh(\mathbf{V}_{j,l}\mathbf{X}(k)) \quad l = 0,1 \quad j = 1,2 \tag{23}$$

and $\mathbf{X} = [y_1(k)\ y_1(k-1)\ y_2(k)\ y_2(k-1)\ u_2(k-1)\ u_1(k-1)]^T$. The initial parameters were all set to small positive random numbers. The algorithm for updating the parameters was a standard backpropagation algorithm. The weighting polynomial associated with the control signal was selected as: $R(q^{-1}) = \lambda(1 - q^{-1})$; i.e. weights the control deviations, the parameter was $\lambda = .05$. This model structure is suitable for on-line learning, as shown in the simulations. The polynomial $R(q^{-1})$ introduces a certain degree of regularization on the control signal, avoiding extremely large excursions of the system output. Nevertheless, we can still see in figure 1(a) some large values. The convergence is quite slow and after 800 sampling steps the controller does not quite compensate for the coupling between the systems.

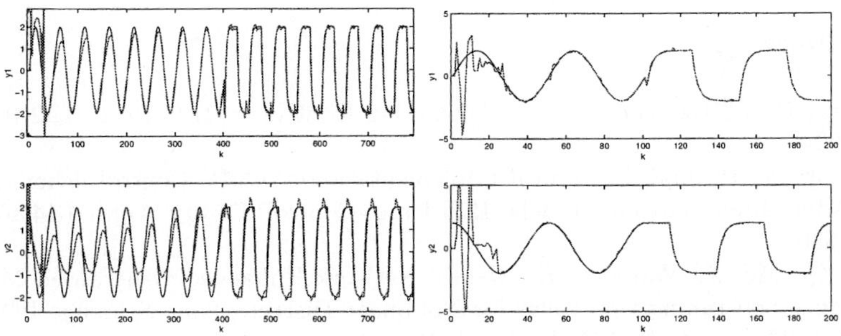

(a) ANN based controller (b) GP-based controller

Fig. 1. Simulation results for both models

The GP model considered two covariance function with the following structure for each output:

$$\begin{aligned}\mathbf{C}_1(\phi_1(i), \phi_1(j)) &= C(\mathbf{X}(i), \mathbf{X}(j); \Theta_{1,0}) + u_1(i)C(\mathbf{X}(i), \mathbf{X}(j); \Theta_{1,1})u_1(j) \\ \mathbf{C}_2(\phi_2(i), \phi_2(j)) &= C(\mathbf{X}(i), \mathbf{X}(j); \Theta_{2,0}) + u_2(i)C(\mathbf{X}(i), \mathbf{X}(j); \Theta_{2,1})u_2(j)\end{aligned} \tag{24}$$

where $\phi_1(k) = [y_1(k)\ y_1(k-1)\ y_2(k)\ y_2(k-1)\ u_2(k-1)\ u_1(k-1)\ u_1(k)]^T$, $\phi_2(k) = [y_1(k)\ y_1(k-1)\ y_2(k)\ y_2(k-1)\ u_2(k-1)\ u_1(k-1)\ u_2(k)]^T$, and

$$C(\mathbf{X}(i), \mathbf{X}(j); \Theta_{lz}) = v_{lz} e^{-\frac{1}{2}\sum_{k=1}^{p} \alpha_{k,lz}(x_k(i) - x_k(j))^2} + a_{lz}. \tag{25}$$

The parameters of the covariance function were adjusted by maximizing a likelihood function two times during the simulation, at time $k = 15$ and $k = 25$. In addition, the first ten samples were used to do an off-line identification and after that the control loop was closed. Figure 1(b) shows these initial steps. It is worth pointing out that, in this case, with very few observations, the controller is able to handle the system quite well. It can also be noticed that there are no abrupt changes when the controller enters in operation; this is due to the extra terms that provide a regularization of the control signal. As seen in figure 1(b) the long run behaviour of the controller is also quite acceptable.

5 Conclusions

The comparative study carried out in this paper shows that GP models are suitable models to be used as building blocks of a GMV controller. The use of ANN, in this context, is limited since it is not possible to have a simple expression of the uncertainty bounds associated with the output. The use of the weighting polynomial plays an important role in smoothing the control action in both approaches.

References

1. Clarke, D.W., Gawthrop, P.G.: Self-tuning Controller. IEE Proc. **122**(9) (1975) 929-934
2. Bittanti, S., Piroddi, L.: Neural Implementation of GMV Control Schemes Based on Affine Input/output Models. IEE Proc.-Control Theory Appl. **144**(6) (1997) 521–530
3. Zhu, Q., Ma, Z., Warwick, K.: Neural Network Enhanced Generalised Minimum Variance Self-tuning Controller for Nonlinear Discrete-Time Systems. IEE Proc.-Control Theory Appl. **146**(4) (1999) 319–326
4. Narendra, K., Mukhopadhyay, S.: Adaptive Control of Nonlinear Multivariable Systems Using Neural Networks. Neural Networks **7** (1994) 737–752
5. MacKay, D.: Gaussian Processes - A Replacement for Supervised Neural Networks?. Lecture Notes for a Tutorial at NIPS (1997)
6. Murray-Smith, R., Sbarbaro, D.: Nonlinear Adaptive Control Using Non-Parametric Gaussian Process Prior Models. Proceedings of the 15th IFAC World Congress (2003)
7. Chryssolouris, G., Lee, M., Ramsey, A.: Confidence Interval Prediction for Neural Network Models. IEEE Transactions Neural Networks **7**(1)(1996) 229-232
8. Williams, C. K. I.: Prediction with Gaussian Processes: From Linear Regression to Linear Prediction and Beyond. In: Learning and Inference in Graphical Models, Ed. M. I. Jordan, Kluwer. (1988) 599–621
9. Williams, C. K. I., Rasmussen, C. E.: Gaussian Processes for Regression. Neural Information Processing Systems, Vol. 8. MIT Press (1996) 514-520
10. De Veaux, R. D., Schumi, J., Schweinsberg, J., Ungar, L. H.: Prediction Intervals for Neural Networks via Nonlinear Regression. Technometrics **40**(4)(1998) 273-282
11. Yang, L., Kavli, T., Carlin, M., Clausen, S., de Groot P. F. M.: An Evaluation of Confidence Bound Estimation Methods for Neural Networks. In (H.-J. Zimmermann et. al. ed.) Advances in Computational Intelligence and Learning: Methods and Applications. Kluwer Academic Publishers (2002) 71-84

A Neural Network Based Method for Solving Discrete-Time Nonlinear Output Regulation Problem in Sampled-Data Systems*

Dan Wang[1,2] and Jie Huang[1]

[1] Department of Automation and Computer-Aided Engineering, The Chinese University of Hong Kong, jhuang@acae.cuhk.edu.hk,
[2] Current affiliation: Temasek Laboratories, National University of Singapore, tslwangd@nus.edu.sg

Abstract. Many of nonlinear control systems are sampled-data system, i.e. the continuous-time nonlinear plants are controlled by digital controllers. So it is important to investigate that if the solution of the discrete-time output regulation problem is effective to sampled-data nonlinear control systems. Recently a feedforward neural network based approach to solving the discrete regulator equations has been presented. This approach leads to an effective way to practically solve the discrete nonlinear output regulation problem. In this paper the approach is used to sampled-data nonlinear control system. The simulation of the sampled-data system shows that the control law designed by the proposed approach performs much better than the linear control law does.

1 Introduction

We consider the control problem of a benchmark system, inverted-pendulum on a cart that is a well known unstable nonlinear system. The motion of the system can be described by [1]

$$(M+m)\ddot{x} + ml(\ddot{\theta}cos\theta - \dot{\theta}^2 sin\theta) + b\dot{x} = u$$
$$m(\ddot{x}cos\theta + l\ddot{\theta} - gsin\theta) = 0$$

where M is the mass of the cart, m the mass of the block on the pendulum, l the length of the pendulum, g the acceleration due to gravity, b the coefficient of viscous friction for motion of the cart, θ the angle the pendulum makes with vertical, x the position of the cart, and u the applied force. With the choice of the state variables $x_1 = x, \quad x_2 = \dot{x}, \quad x_3 = \theta, \quad x_4 = \dot{\theta}$, and using Euler's method with T as sampling period, the inverted pendulum on a cart system can be discretized into a discrete time nonlinear system as follows.

$$x_1(k+1) = x_1(k) + Tx_2(k)$$

* The work described in this paper was partially supported by a grant from the Research Grants Council of the Hong Kong Special Administration Region (Project No. CUHK4209/00E).

F. Yin, J. Wang, and C. Guo (Eds.): ISNN 2004, LNCS 3174, pp. 59–64, 2004.

$$
\begin{aligned}
x_2(k+1) &= x_2(k) + \frac{T}{M + m(sinx_3(k))^2}(u + mlx_4^2(k)sinx_3(k) \\
&\quad - bx_2(k) - mgcosx_3(k)sinx_3(k)) \\
x_3(k+1) &= x_3(k) + Tx_4(k) \\
x_4(k+1) &= x_4(k) + \frac{T}{l(M + m(sinx_3(k))^2)}((M+m)gsinx_3(k) \\
&\quad - ucosx_3(k) + bx_2(k)cosx_3(k) - mlx_4^2(k)sinx_3(k)cosx_3(k)) \\
y(k) &= x_1(k)
\end{aligned} \tag{1}
$$

where the parameters are given by $b = 12.98Kg/sec$, $M = 1.378Kg$, $l = 0.325m$, $g = 9.8m/sec^2$, $m = 0.051Kg$.
The control objective is to design a state feedback control law for this system such that the position of the cart can asymptotically track a reference input y_d. Since this system is nonminimum-phase, it is impossible to solve this problem by using conventional inversion based control methods such as input-output linearization[1]. We have to solve the problem using output regulation method. Output regulation problem has been extensively studied since 1970's (interested readers may refer to references in [2]). It is known that the solvability of the discrete time nonlinear output regulation problem relies on a set of algebraic functional equations called discrete regulator equations. Due to the nonlinear nature, the discrete regulator equations cannot be solved exactly. Many efforts have been made on developing numerical methods to solve the regulator equations approximately. Recently, we proposed a neural network approach to solve the discrete output regulation problem [2]. Since most of the nonlinear control systems are sampled-data control systems, it is important to consider if the method can achieve good performance for sampled-data systems. In this paper, we will design a neural network based discrete-time control law using the approach given in [2] and conduct a series simulations to investigate if this approach is effective to sampled-data systems.

2 The General Formulation of the Discrete-Time Output Regulation Problem

Consider a class of discrete-time nonlinear plant described below,

$$
\begin{aligned}
x(k+1) &= f(x(k), u(k), v(k)), \quad x(0) = x_0,\ k \geq 0 \\
e(k) &= h(x(k), u(k), v(k))
\end{aligned} \tag{2}
$$

where x is the state, u the input, e the error output representing the tracking error, and v the reference input and/or a class of disturbances generated by the following exogenous system.

$$
v(k+1) = a(v(k)), \quad v(0) = v_0 \tag{3}
$$

In the sequel, we assume all the functions $f(\cdot,\cdot,\cdot)$, $h(\cdot,\cdot,\cdot)$ and $a(\cdot)$ are sufficiently smooth for $x \in X \subset R^n$, $u \in U \subset R^m$, and $v \in V \subset R^q$ and satisfy $f(0,0,0) = 0$,

$h(0,0,0) = 0$, and $a(0) = 0$. Here X, U, and V are open neighborhoods of the respective Euclidean spaces.

The output regulation problem aims to design a feedback control law such that the equilibrium of the closed-loop system is locally asymptotically stable, and the output $e(k)$ of the plant vanishes asymptotically, i.e., $\lim_{k\to\infty} e(k) = 0$ for all sufficiently small initial conditions. Under some standard assumptions, the problem is solvable if and only if the following algebraic functional equations are solvable.

$$\mathbf{x}(a(v)) = f(\mathbf{x}(v), \mathbf{u}(v), v), \tag{4}$$

$$0 = h(\mathbf{x}(v), \mathbf{u}(v), v) \tag{5}$$

Equation (4) and (5) are known as the discrete regulator equations. If the solutions of equations (4) and (5) are found out, either state feedback or output feedback control law can be readily synthesized. However, due to the nonlinear nature of equation (4) and (5), it is almost impossible to obtain the exact solution for the discrete regulator equations. In this paper, we will use the approximation method given in [2] to solve the discrete nonlinear functional equation (4) and (5).

The approach will be based on the well known universal approximation theorem. Let the pth component of $\mathbf{x}(v)$ and $\mathbf{u}(v)$ be denoted as $\mathbf{x}_p(v)$ and $\mathbf{u}_p(v)$, respectively. The neural network representations of $\mathbf{x}_p(v)$ and $\mathbf{u}_p(v)$ can be given as follows:

$$\hat{\mathbf{x}}_p(W^{xp}, v) = \sum_{j=1}^{N_{xp}} w_j^{xp} \phi\left(\sum_{i=1}^{r} w_{ji}^{xp} v_i + w_{j0}^{xp}\right), \tag{6}$$

$$\hat{\mathbf{u}}_p(W^{up}, v) = \sum_{j=1}^{N_{up}} w_j^{up} \phi\left(\sum_{i=1}^{r} w_{ji}^{up} v_i + w_{j0}^{up}\right) \tag{7}$$

Thus the solutions of the regulator equations (4) and (5) can be approximately represented as follows:

$$\hat{\mathbf{x}}(W_x, v) = \begin{bmatrix} \hat{\mathbf{x}}_1(W^{x1}, v) \\ \vdots \\ \hat{\mathbf{x}}_n(W^{xn}, v) \end{bmatrix}, \quad \hat{\mathbf{u}}(W_u, v) = \begin{bmatrix} \hat{\mathbf{u}}_1(W^{u1}, v) \\ \vdots \\ \hat{\mathbf{u}}_m(W^{um}, v) \end{bmatrix} \tag{8}$$

where W_x, W_u are the weight vectors consisting of all W^{xp} and W^{up} respectively. Our goal is to find integers N_{xp}, N_{up}, and weight vector W such that, for a given compact set $\Gamma \subset V$ and a given $\gamma > 0$,

$$\sup_{v\in\Gamma} ||\hat{\mathbf{x}}(a(v)) - f(\hat{\mathbf{x}}(W_x, v), \hat{\mathbf{u}}(W_u, v), v)|| < \gamma \tag{9}$$

$$\sup_{v\in\Gamma} ||h(\hat{\mathbf{x}}(W_x, v), \hat{\mathbf{u}}(W_u, v), v)|| < \gamma \tag{10}$$

Next, let

$$J(W, v) = \frac{1}{2}||\hat{\mathbf{x}}(a(v)) - f(\hat{\mathbf{x}}(W_x, v), \hat{\mathbf{u}}(W_u, v), v)||^2 + \frac{1}{2}||h(\hat{\mathbf{x}}(W_x, v), \hat{\mathbf{u}}(W_u, v), v)||^2 \tag{11}$$

and

$$J(W) = \sum_{v \in \Gamma} J(W, v) \tag{12}$$

Finally we can find the desirable weight W by solving the parameter optimization problem, $\min_W J(W)$. Existence of such a weight is guaranteed by the Universal Approximation Theorem.

3 Solution to the Inverted Pendulum on a Cart System

Given $y_d(k) = Asin\omega kT$ as the reference input for the control problem of the inverted pendulum on a cart system given in Section 1, we introduce the following exosystem

$$v(k+1) = Sv(k), v(0) = [0, A]^T \tag{13}$$

where

$$S = \begin{bmatrix} cos\omega T & sin\omega T \\ -sin\omega T & cos\omega T \end{bmatrix}, \quad v(k) = \begin{bmatrix} v_1(k) \\ v_2(k) \end{bmatrix}$$

Clearly, $v_1(k) = Asin\omega kT$. Thus, we can define the error equation as follows

$$e(k) = y(k) - v_1(k) = x_1(k) - v_1(k)$$

The discrete regulator equations associated with this tracking problem is given as follows:

$$\begin{aligned}
\mathbf{x}_1(Sv) &= \mathbf{x}_1(v) + T\mathbf{x}_2(v) \\
\mathbf{x}_2(Sv) &= \mathbf{x}_2(v) + \frac{T}{M + m(sin\mathbf{x}_3(v))^2}(\mathbf{u} + ml\mathbf{x}_4^2(v)sin\mathbf{x}_3(v) \\
&\quad - b\mathbf{x}_2(v) - mgcos\mathbf{x}_3(v)sin\mathbf{x}_3(v)) \\
\mathbf{x}_3(Sv) &= \mathbf{x}_3(v) + T\mathbf{x}_4(v) \\
\mathbf{x}_4(Sv) &= \mathbf{x}_4(v) + \frac{T}{l(M + m(sin\mathbf{x}_3(v))^2)}((M+m)gsin\mathbf{x}_3(v) \\
&\quad - \mathbf{u}cos\mathbf{x}_3(v) + b\mathbf{x}_2(v)cos\mathbf{x}_3(v) - ml\mathbf{x}_4^2(v)sin\mathbf{x}_3(v)cos\mathbf{x}_3(v)) \\
0 &= \mathbf{x}_1(v) - v_1
\end{aligned} \tag{14}$$

With equation (14) ready, we can define $J(W)$ as in (12). But for this example, we can further simplify our problem. In fact, by inspection, we can partially solve (14) as follows

$$\mathbf{x}_1(v) = v_1 \tag{15}$$

$$\mathbf{x}_2(v) = [(cos\omega T - 1)v_1 + sin\omega T * v_2]/T \tag{16}$$

$$\mathbf{x}_4(v) = \frac{1}{T}(\mathbf{x}_3(Sv) - \mathbf{x}_3(v)) \tag{17}$$

$$\begin{aligned}
\mathbf{u} &= \frac{M + m(sin\mathbf{x}_3(v))^2}{T^2}[cos\omega T - 1, sin\omega T](Sv - v) - ml\mathbf{x}_4^2(v)sin\mathbf{x}_3(v) \\
&\quad + \frac{b}{T}[cos\omega T - 1, sin\omega T]v + mgcos\mathbf{x}_3(v)sin\mathbf{x}_3(v))
\end{aligned} \tag{18}$$

with $\mathbf{x}_3(v)$ satisfying following equation.

$$\mathbf{x}_3(S^2v) = 2\mathbf{x}_3(Sv) - x_3(v) + \frac{T^2}{l(M+m(sin\mathbf{x}_3(v))^2)}((M+m)gsin\mathbf{x}_3(v) \\ - mgcos\mathbf{x}_3(v)^2 sin\mathbf{x}_3(v)) + \frac{cos\mathbf{x}_3(v)}{l}[cos\omega T - 1, sin\omega T](v - Sv) \quad (19)$$

Therefore the only unknown function is $\mathbf{x}_3(v)$. Assume $\hat{\mathbf{x}}_3(W,v)$ is the neural network representation of $\mathbf{x}_3(v)$, and define

$$F(W,v) = -\hat{\mathbf{x}}_3(W,S^2v) + 2\hat{\mathbf{x}}_3(W,Sv) - \hat{\mathbf{x}}_3(W,v) + \frac{T^2}{l(M+m(sin\hat{\mathbf{x}}_3(W,v))^2)} \\ (\ (M+m)gsin\hat{\mathbf{x}}_3(W,v) - mg(cos\hat{\mathbf{x}}_3(W,v))^2 sin\hat{\mathbf{x}}_3(W,v)) \\ + \frac{cos\hat{\mathbf{x}}_3(W,v)}{l}[cos\omega T - 1, sin\omega T](v - Sv) \quad (20)$$

and let

$$J(W) = \sum_{v\in\Gamma} \frac{1}{2}(F(W,v))^2 \quad (21)$$

Then we can minimize (21) using the gradient method. After some trial and error, we have obtained a feedforward neural network $\hat{\mathbf{x}}_3(W,v)$ with $r = 2$ and $N = 15$, respectively. The resulting weight vector W satisfies $J(W) < 10^{-4}$. In our design, we have selected the circumference ${v_1}^2 + {v_2}^2 = A^2$ as the domain Γ of v_1,v_2. In terms of $\hat{\mathbf{x}}_3(W,v)$, we can get $\hat{\mathbf{x}}_4(W,v)$ and $\hat{\mathbf{u}}(W,v)$ by (19) and (18). Let

$$\hat{\mathbf{x}}(W,v) = [\mathbf{x}_1(v), \mathbf{x}_2(v), \hat{\mathbf{x}}_3(W,v), \hat{\mathbf{x}}_4(W,v)]^T \quad (22)$$

Then the control law can be approximated as follows

$$u(k) = \hat{\mathbf{u}}(W,v(k)) + L[x(k) - \hat{\mathbf{x}}(W,v(k))] \quad (23)$$

where L is such that all the eigenvalues of $\frac{\partial f(0,0,0)}{\partial x} + \frac{\partial f(0,0,0)}{\partial u}L$ are inside the unit circle. Now we can test the corresponding control law with the sinusoidal input $Asin\omega kT$. We select all the eigenvalues of $\frac{\partial f(0,0,0)}{\partial x} + \frac{\partial f(0,0,0)}{\partial u}L$ at $Z = 0.8187$ or $Z = 0.9048$, where $Z = 0.8187$ and $Z1 = 0.9048$ correspond to the point $s = -4$ on the S-plane for $T = 0.05$ and $T = 0.025$ respectively. Thus we get

$$L = [7.8997\ \ 21.6948\ \ 51.9021\ \ 9.3280];\ \ for\ \ T = 0.05 \quad (24)$$

$$L = [9.6092\ \ 23.0741\ \ 56.0933\ \ 10.1023];\ \ for\ \ T = 0.025 \quad (25)$$

In our simulations, the discrete-time control law is used to control the sampled-data inverted pendulum on a cart system. To assess the performance of the control law, we make a comparison with the performance of the control law obtained by the conventional discrete-time linear controller design (linear controller). Both the control laws designed by nonlinear approximation approach and linearization method can track the reference input but the nonlinear control law can achieve much smaller tracking error than the linear control law does. As ω increases, the tracking error of linear controller increases drastically. But the

Table 1. Maximal steady state tracking errors

ω	T	A	NN (abs)	NN (%)	linear (abs)	linear (%)
π	.1	0.6	.4076	67.94	.7186	119.77
$\pi/2$	.05	2.5	.2663	10.65	.4943	19.77
2.5	.025	1.0	.0872	8.72	.2273	22.73
3	.025	1.0	.0999	9.99	.4958	49.58
3	.01	0.6	.0152	2.54	.0919	15.31
3	.01	0.8	.0183	2.29	.2074	25.92
3	.01	1.0	.0329	3.29	.4055	40.55

control law designed by nonlinear approximation approach can keep the tracking errors at a relatively small level. On the other hand, as expected, reducing the sample period T makes control performance better.

Table 1 lists the maximal steady state tracking errors of the closed-loop system for the reference inputs at several different amplitudes with different ω and T.

4 Conclusions

A simulation study is conducted on the output regulation problem of the benchmark system, inverted-pendulum on a cart system, in sampled-data case. The discrete-time control law is designed using the neural network based approach proposed in previous papers. Then the controller is used to control the sampled-data inverted-pendulum on a cart system at different sampling period T to study its effectiveness to sampled-data nonlinear systems. The simulations demonstrate that the control law designed by the proposed approach can achieve a much better performance than the linear control law does. This investigation not only gives a practical solution to one of the most important control problems, but also shows the great potential of the neural networks techniques in engineering applications.

References

1. R. Gurumoorthy and S.R. Sanders: Controlling Nonminimum Phase Nonlinear Systems- the Inverted Pendulum on a Cart Example. Proceedings of American Control Conference, San Francisco, California, USA (1993) 680-685
2. Dan Wang and Jie Huang: A Neural Network Based Approximation Method for Discrete-Time Nonlinear Servomechanism Problem. IEEE Transactions on Neural Networks, Vol.12, No.3 (2001) 591-597

The Design of Fuzzy Controller by Means of CI Technologies-Based Estimation Technique

Sung-Kwun Oh[1], Seok-Beom Roh[1], Dong-Yoon Lee[2], and Sung-Whan Jang[1]

[1] School of Electrical, Electronic and Information Engineering, Wonkwang University, South Korea
{ohsk, nado, swhjang}@wonkwang.ac.kr

[2] Department of Information and Communication Engineering, Joongbu University, South Korea
dylee@joongbu.ac.kr

Abstract. In this study, we introduce a new neurogenetic approach to the design of fuzzy controller. The development process exploits the key technologies of Computational Intelligence (CI), namely genetic algorithms and neurofuzzy networks. The crux of the design methodology is based on the selection and determination of optimal values of the scaling factors of the fuzzy controllers, which are essential to the entire optimization process. First, the tuning of the scaling factors of the fuzzy controller is carried out, and then the development of a nonlinear mapping for the scaling factors is realized by using GA based Neuro-Fuzzy networks (NFN). The developed approach is applied to a nonlinear system such as an inverted pendulum where we show the results of comprehensive numerical studies and carry out a detailed comparative analysis.

1 Introduction

Fuzzy control is one of the useful control techniques for uncertain and ill-defined nonlinear systems. Control actions of a fuzzy controller are described by some linguistic rules. This property makes the control algorithm easy to understand. Heuristic fuzzy controllers incorporate the experience of knowledge into rules, which are fine-tuned based on trial and error. The intent of this study is to develop, optimize and experiment with the fuzzy controllers (fuzzy PD controller or fuzzy PID controller) when developing a general design scheme of Computational Intelligence. One of the difficulties in the construction of the fuzzy controller is to derive a set of optimal control parameters of the controller such as linguistic control rules, scaling factors, and membership functions of the fuzzy controller. Genetic algorithms (GAs) can be used to find the optimal control parameters. However, evolutionary computing is computationally intensive and this may be a point of concern when dealing with amount of time available to such search. For instance, when controlling a nonlinear plant such as an inverted pendulum of which initial states vary in each case, the search time required by GAs could be prohibitively high when dealing with dynamic systems. As a consequence, the parameters of the fuzzy controller cannot be easily adapted to the changing initial state of this system such as an angular position of the pendulum. To alleviate this shortcoming, we introduce a nonlinear mapping from the

F. Yin, J. Wang, and C. Guo (Eds.): ISNN 2004, LNCS 3174, pp. 65–70, 2004.

initial states of the system and the corresponding optimal values of the parameters. With anticipation of the nonlinearity residing within such transformation, in its realization we consider GA-based Neurofuzzy networks. Bearing this mind, the development process consists of two main phases. First, using genetic optimization we_determine optimal parameters of the fuzzy controller for various initial states (conditions) of the dynamic system. Second, we build up a nonlinear model that captures a relationship between the initial states of the system and the corresponding genetically optimized control parameters. The paper includes the experimental study dealing the inverted pendulum with the initial states changed.

2 Design Methodology of Fuzzy Controller

The block diagram of fuzzy PID controller is shown in Fig. 1. Note that the input variables to the fuzzy controller are transformed by the scaling factors (GE, GD, GH, and GC) whose role is to allow the fuzzy controller to properly "perceive" the external world to be controlled.

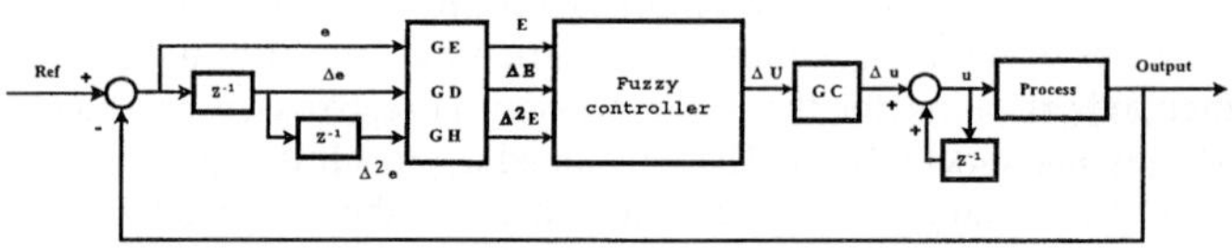

Fig. 1. An overall architecture of the fuzzy PID controller

The above fuzzy PID controller consists of rules of the following form, cf. [3]
R_j : if E is A_{1j} and ΔE is A_{2j} and Δ^2E is A_{3j} then ΔU_j is D_j
The capital letters standing in the rule (R_j) denote fuzzy variables whereas D is a numeric value of the control action.

3 Auto-tuning of the Fuzzy Controller by GAs

Genetic algorithms (GAs) are the search algorithms inspired by nature in the sense that we exploit a fundamental concept of a survival of the fittest as being encountered in selection mechanisms among species[4]. In GAs, the search variables are encoded in bit strings called chromosomes. They deal with a population of chromosomes with each representing a possible solution for a given problem. A chromosome has a fitness value that indicates how good a solution represented by it is. In control applications, the chromosome represents the controller's adjustable parameters and fitness value is a quantitative measure of the performance of the controller. In general, the population size, a number of bits used for binary coding, crossover rate, and mutation rate are essential parameters whose values are specified in advance. The genetic search is guided by a reproduction, mutation, and crossover. The standard ITAE expressed for the reference and the output of the system under control is treated as a fitness function [2].

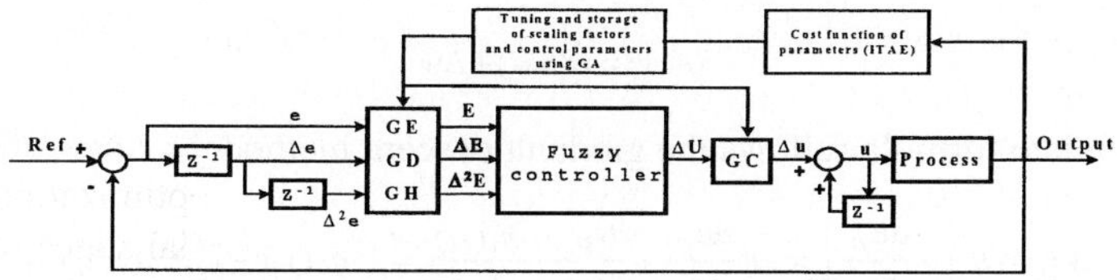

Fig. 2. The scheme of auto-tuning of the fuzzy PID controller involving estimation of the scaling factors

The overall design procedure of the fuzzy PID controller realized by means of GAs is illustrated in Fig. 2. It consists of the following steps

[Step 1.] Select the general structure of the fuzzy controller

[Step 2.] Define the number of fuzzy sets for each variable and set up initial control rules.

[Step 3.] Form a collection of initial individuals of GAs. We set the initial individuals of GAs for the scaling factor of fuzzy controller.

[step 4] Here, all the control parameters such as the scaling factors GE, GD, GH and GC are tuned at the same time.

4 The Estimation Algorithm by Means of GA-Based Neurofuzzy Networks (NFN)

As visualized in Fig. 3, NFN can be designed by using space partitioning in terms of individual input variables. Its topology is concerned with a granulation carried out in terms of fuzzy sets being defined for each input variable.

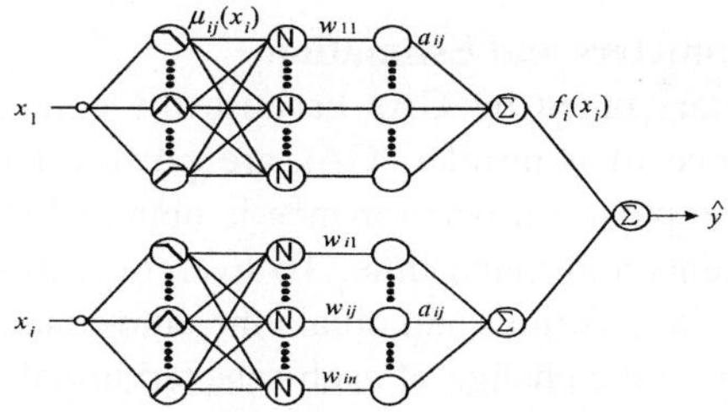

Fig. 3. NFN architecture

The output of the NFN $\hat{y}$ reads as

$$\hat{y} = f_1(x_1) + f_2(x_2) + \cdots + f_m(x_m) = \sum_{i=1}^{m} f_i(x_i) \tag{1}$$

with m being the number of the input variables(viz. the number of the outputs f_i's of the "Σ" neurons in the network). The learning of the NFN is realized by adjusting connections of the neurons and as such it follows a standard Back-Propagation (BP) algorithm.

As far as learning is concerned, the connections change as follows

$$w(new) = w(old) + \Delta w \tag{2}$$

where the update formula follows the gradient descent method

$$\Delta w_{ij} = \eta \cdot \left(-\frac{\partial E_p}{\partial w_{ij}} \right) = -\eta \frac{\partial E_p}{\partial \hat{y}_p} \cdot \frac{\partial \hat{y}_p}{\partial f_i(x_i)} \cdot \frac{\partial f_i(x_i)}{\partial w_{ij}} = 2 \cdot \eta \cdot (y_p - \hat{y}_p) \cdot \mu_{ij}(x_i) \tag{3}$$

with η being a positive learning rate.
Quite commonly to accelerate convergence, a momentum term is being added to the learning expression. Combining (3) and a momentum term, the complete update formula combining the already discussed components is[5,6]

$$\Delta w_{ij} = 2 \cdot \eta \cdot (y_p - \hat{y}_p) \cdot \mu_{ij}(x_i) + \alpha(w_{ij}(t) - w_{ij}(t-1)) \tag{4}$$

(Here the momentum coefficient, α, is constrained to the unit interval).
In this algorithm, to optimize the learning rate, momentum term and fuzzy membership function of the above NFN we use the genetic algorithm.

5 Experimental Studies

In this section, we demonstrate the effectiveness of the fuzzy PID controller by applying it to the inverted pendulum system. Our control goal here is to balance the pole without regard to the cart's position and velocity[7], and we compare the fuzzy PID controller and the fuzzy PD controller with the conventional PID controller under same conditions to validate the fuzzy PID controller and the fuzzy PD controller.

Tuning of Control Parameters and Estimation
We get control parameters tuned by GAs because the control parameters have an effect on the performance of controller. GAs are powerful nonlinear optimization techniques. However, the powerful performance is obtained at the cost of expensive computational requirements and much time. To overcome this weak point, first, we select several initial angular positions and obtain the auto-tuned control parameters by means of GAs according to the change of each selected initial angular positions, then build a table. Secondly, we use GA-based NFN to estimate the control parameters in the case that the initial angular positions of the inverted pendulum are selected arbitrarily within the given range. Here we show that the control parameters under the arbitrarily selected initial angular position are not tuned by the GAs but estimated by the estimation algorithm of GA-based NFN. Table 1 shows the estimated scaling factors of the fuzzy PID controller and describes performance index (ITAE, Overshoot(%), Rising time(sec)) of the fuzzy PID controller with the estimated scaling factors in case that the initial angles of inverted pendulum are 0.1849(rad), 0.4854(rad), 0.38878(rad) and 0.7133(rad) respectively.
In case of the fuzzy PD controller, the estimated scaling factors and performance index are shown Table 2.

In case of the PID controller, the estimated scaling factors by means of neuro-fuzzy model are shown in Table 3 .

Table 1. The estimated parameters by means of the GA-based NFN and performance index(ITAE, Overshoot(%), Rising time(sec)) of the fuzzy PID controller

Case	Initial Angle(rad)	GE	GD	GH	GC	ITAE	Over Shoot(%)	Rising Time(sec)
1	0.184900	4.107681	65.87493	203.5847	1.936434	0.069752	0.000000	0.091611
2	0.485400	1.840316	60.66174	181.5137	1.892224	0.553420	2.293559	0.144283
3	0.388780	2.230318	62.79905	183.0733	1.867487	0.329704	2.612525	0.126850
4	0.713300	1.303365	49.17086	159.8879	1.742495	1.496100	2.290066	0.186604

Table 2. The estimated parameters by means of the GA-based NFN and performance index(ITAE, Overshoot(%), Rising time(sec)) of the fuzzy PD controller

Case	Initial Angle (rad)	GE	GD	GC	ITAE	Over Shoot(%)	Rising Time(sec)
1	0.184900	8.840099	0.399971	3.912999	0.060869	0.000000	0.08198996
2	0.485400	4.566366	0.278876	3.872337	0.518902	0.035730	0.14340339
3	0.388780	5.315691	0.308264	3.776772	0.312420	0.000000	0.12621812
4	0.713300	3.064440	0.211538	3.837129	1.466210	2.028139	0.18146067

Table 3. The estimated parameters, ITAE, Overshoot(%) and Rising time of the PID controller

Case	Initial Angle(rad)	K	Ti	Td	ITAE	Over Shoot(%)	Rising Time(sec)
1	0.184900	129.29765	253.00385	0.098317	0.182383	3.127633	0.15356993
2	0.485400	129.29129	248.09918	0.101459	0.759869	2.092925	0.17871029
3	0.388780	129.52347	251.77655	0.095975	0.512037	3.624664	0.16010507
4	0.713300	128.20747	249.20142	0.093086	1.822047	3.968492	0.19895169

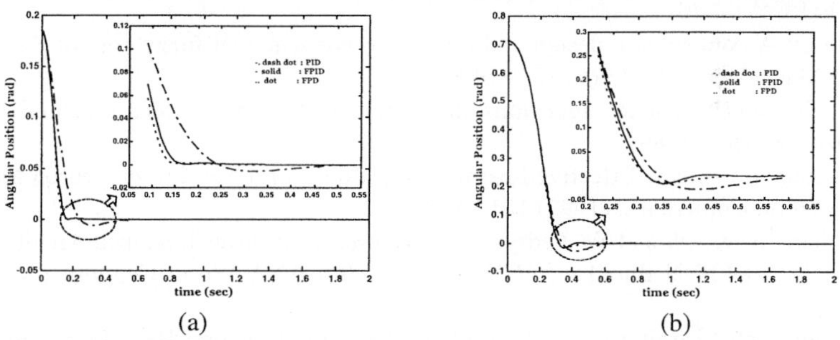

Fig. 4. Pole angle for initial angle (a)θ= 0.1849(rad) (Case 1) and (b)θ= 0.7133(rad) (Case 4).

Fig. 4 demonstrates pole angle for initial angle (a)θ= 0.1849(rad) (Case 1) and (b) θ = 0.7133(rad) (Case 4)

From the above Figures 4, we know that the fuzzy PD and fuzzy PID are superior to the PID controller from the viewpoint of performance index. The output performance of the fuzzy controllers such as the fuzzy PD and the fuzzy PID controller including nonlinear characteristics are superior to that of the PID controller especially when using the nonlinear dynamic equation of the inverted pendulum.

6 Conclusions

In this paper, we have proposed a two-phase optimization scheme of the fuzzy PID and PD controllers. The parameters under optimization concern scaling factors of the input and output variables of the controller that are known to exhibit an immense impact on its quality. The first phase of the design of the controller uses genetic computing that aims at the global optimization of its scaling factors where they are optimized with regard to a finite collection of initial conditions of the system under control. In the second phase, we construct a nonlinear mapping between the initial conditions of the system and the corresponding values of the scaling factors. We investigated a number of alternatives there by looking at GA-based Neuro-fuzzy networks model. While the study showed the development of the controller in the experimental framework of control of a specific dynamic system (inverted pendulum), this methodology is general and can be directly utilized to any other system. Similarly, one can envision a number of modifications that are worth investigating. For instance, a design of systems exhibiting a significant level of variability could benefit from the approach pursued in this study.

Acknowledgements. This paper was supported by Wonkwang University in 2004.

References

1. S.-K. Oh, W. Pedrycz: The Design of Hybrid Fuzzy Controllers Based on Genetic Algorithms and Estimation Techniques. Kybernetes 31(2002) 909-917
2. S.K. Oh, T. Ahn, H. Hwang, J. Park, K. Woo: Design of a Hybrid Fuzzy Controller with the Optimal Auto-tuning Method. Journal of Control, Automation and Systems Engineering. 1 (1995) 63-70
3. H. X. Li: A comparative design and tuning for conventional fuzzy control. IEEE Trans. Syst., Man, Cybern. B. 27 (1997) 884-889
4. D.E. Goldberg: Genetic algorithms in Search, Optimization, and Machine Learning. Addison-Wesley (1989)
5. T. Yamakawa: A New Effective Learning Algorithm for a Neo Fuzzy Neuron Model. 5th IFSA World Conference, (1993) 1017-1020
6. B.-J. Park, S.-K. Oh and W. Pedrycz: The Hybrid Multi-layer Inference Architecture and Algorithm of FPNN Based on FNN and PNN. 9th IFSA World Congress,(2001) 1361-1366.
7. J.R. Jang: Self-Learning Fuzzy Controllers Based on Temporal Back Propagation. IEEE Trans. on Neural Networks. 3(1992) 714-723
8. H.-S. Park, S.-K. Oh: Multi-FNN identification Based on HCM Clustering and Evolutionary Fuzzy Granulation. International Journal of Control, Automation, and Systems. 1(2) (2003) 194-202
9. H.-S. Park, S.-K. Oh: Rule-based Fuzzy-Neural Networks Using the Identification Algorithm of GA hybrid Scheme. International Journal of Control, Automation and Systems, 1(1) (2003) 101-110

A Neural Network Adaptive Controller for Explicit Congestion Control with Time Delay*

Bo Yang and Xinping Guan

Institute of Electrical Engineering, Yanshan University,
Qinhuangdao 066004, P. R. China
{yangbo, xpguan}@ysu.edu.cn

Abstract. This paper examines explicit rate congestion control for data networks. A neural network (NN) adaptive controller is developed to control traffic where sources regulate their transmission rates in response to the feedback information from network switches. Particularly, the queue length dynamics at a given switch is modeled as an unknown nonlinear discrete time system with cell propagation delay and bounded disturbances. To overcome the effects of delay an iterative transformation is introduced for the future queue length prediction. Then based on the causal form of the dynamics in buffer an adaptive NN controller is designed to regulate the queue length to track a desired value. The convergence of our scheme is derived mathematically. Finally, the performance of the proposed congestion control scheme is also evaluated in the presence of propagation delays and time-vary available bandwidth for robustness considerations.

1 Introduction

Asynchronous Transfer Mode (ATM) as the paradigm for broadband integrated service digital networks (B-ISDN) provides different services to multimedia sources with different traffic characteristics by statistically multiplexing cells of fixed length packets with specified quality of service (QoS). However the flexibility to accommodate these sources may cause serious network congestion and result the severe buffer over flow. In this case the cells, which are stored in the buffer and are intended to transmit through it may by discarded. Then this causes the degradation of QoS such as cell loss, excessive queuing delay. Thus an appropriate congestion control is required. In order to avoid congestion the queue length should be controlled around a desired level, which is related to cell queuing delay and resource utilization.

The above queue management strategy can be carried out as a feedback scheme. In the context of ATM available bit rate (ABR) service, the sources send a control cell (resource management cell) once every *N* data cells, which can be used by switches to

* Supported in part by the NSFC (No. 60174010) and the Key Research Project of Ministry of Education (No. 204014). This work was also supported in part by Yanshan University (No. YDJJ2003010).

F. Yin, J. Wang, and C. Guo (Eds.): ISNN 2004, LNCS 3174, pp. 71–76, 2004.

convey feedback. The control cells travel to the destination and are returned to the sources in the same path. Then the sources adjust their sending rate according to the feedback signal in the form of binary or an explicit rate (ER) value. The rate-based control with explicit feedback is chosen as the standard flow control scheme in ATM networks Forum.

The challenging aspects of congestion control are the no stationary, uncertain, and even nonlinear characteristics of network flow with long transmission delays. In most cases, only parts of the problem are taken into account. The direct method to overcome the effects of time delay in congestion control is predictive principle (see for example [5]). However the similar literatures are based on the known linear queue length dynamic differential equation. In facts the linear model is not accurate system model in congestion control. To surmount the nonlinear characteristics of queue length dynamics in [1] the authors divided the global nonlinear dynamics into several local linear operating regions using fuzzy IF-THEN rules, and then they derived the optimal ABR traffic fuzzy adaptive predictive control. In [3] Jagannathan et al. also considered the buffer dynamics at the network switch as a nonlinear process. Then a novel adaptive congestion control scheme was developed using a one layer neural networks (NN). But they assumed the round trip time (RTT) delay to be zero for simplicity. Ignoring RTT is usually not practical in high-speed networks, where the bandwidth-delay product can be very large. This characteristic will crash some idealized algorithms in some severe operations.

In this paper, despite the RTT delay, nonlinear dynamics in buffer and uncontrollable traffic we aim to achieve high utilization of resources while maintaining QoS in terms of user requirements (cell-loss, cell-delay and cell-delay variation). We propose to achieve this by controlling queue levels to stationary values by making use of output feedback with NN due to its universal approximation capability. We particularly focus on controlling the ABR sources to response the congestion and idle resource states. Since the desired the feedback control contains future information due to time delay, we transform this unreachable information into causal forms. Then we use current and previous queue length as well as previous ER values to tune the weights of high order NN (HONN) to generate an adaptive explicit congestion feedback control. The closed-loop convergence of ER and queue length as well as the weights is proven using Lyapunov analysis.

The paper is organized as follows. In section 2 we describe the congestion control problem in a single buffer. In section 3 we design the adaptive NN flow controller and discuss the closed-loop convergence mathematically. In section 4 we demonstrate the functionality of the proposed solution through simulations. The concluding remarks come from section 5.

2 ATM Switch Nonlinear Dynamic Model

Now that physical systems are inherently nonlinear, we model the dynamic characteristics in the single buffer as an unknown nonlinear discrete model.

$$q_{k+1} = f(q_k, \cdots, q_{k-n+1}, u_{k-\tau}, \cdots, u_{k-\tau-m+1}, c_k) + Tu_{k-\tau+1}, \tag{1}$$

where $n \geq m$, $n \geq \tau \geq 1$, $q_k \in \Re$ is the queue length at time instant k, T is the sampling time, and $u_k, c_k \in \Re$ is the controllable traffic accumulation rate and service rate respectively. τ denotes the average round trip propagation delay between sources and destinations. $f(\cdot)$ is an unknown smooth nonlinear function of queue length, traffic input rate and available capacity for ABR traffic.

Remark 1. The nonlinear queue model (1) is similar with the one in [3], except for our concern with delay. Besides, (1) can be viewed as a general form of M/M/1 queue model in [6].

In order to get the explicit form of desired flow control, which can drive queue length to desired value q_d. Moving (1) (τ-1) steps ahead results in

$$u_k^* = (q_d - f(q_{k+\tau-1}, \cdots, q_{k-n+\tau}, u_{k-1}, \cdots, u_{k-m+1}, c_{k+\tau-1}))/T, \tag{2}$$

By means of (2) the desired control includes unknown function $f(\cdot)$, unknown output $q_{k+\tau-1}, \cdots, q_{k+1}$ and future available bandwidth for ABR sources. In the following, the HONN will be applied for dealing with this problem.

3 HONN Traffic Rate Controller Design

In order to predict the future queue length and ER value in (2), it is necessary to select a large number of past values of input and output. The number of past values is carefully selected based on (2)

$$q(k) = [q_k \quad q_{k-1} \quad \cdots \quad q_{k-n+1}]^T, u_{k-1}(k) = [u_{k-1} \quad \cdots \quad u_{k-\tau-m+1}]^T$$

$$c(k) = [c_{k+\tau-1} \quad c_{k+\tau-2} \quad \cdots \quad c_k]^T$$

$$\eta(k) = [q^T(k) \quad u_{k-1}^T(k)]^T, \bar{\eta}(k) = [\eta^T(k) \quad q_d]^T$$

Following (1) we know q_{k+1} is a function of q(k), $u_{k-\tau+1}, \cdots, u_{k-\tau-m+1}$ and c_k. Similarly, q_{k+2} can be represented as a function of q(k) , $u_{k-\tau+2}, \cdots, u_{k-\tau-m+1}$, c_k and c_{k+1}. Continue the iteration recursively, it's easy to prove $q_{k+\tau-1}$ is a function of η(k) and $c(k)$. Finally, $q_{k+\tau}$ can be expressed as

$$q_{k+\tau} = G_\tau(\eta(k), c(k)) + Tu_k, \tag{3}$$

where $G_\tau(\cdot)$ is a smooth function. Using mean value theorem (3) can be further written as Taylor's formula

$$q_{k+\tau} = G_\tau(\eta(k), 0) + \phi(k) + Tu_k, \tag{4}$$

where $\phi(k)=\left(\partial G_{\tau}/\partial c\,|c(k)=\varphi_1(k)\right)^T c(k),\ \varphi_1(k)\in L(0,c(k))$ and $L(0,c(k))$ is a line defined by

$$L(0,c(k))=\{\varphi(k)|\varphi(k)=\ell c(k),\ 0<\ell<1\}, \tag{5}$$

Since $G_\tau(\cdot)$ is smooth and $c(k)$ is physically limited, we can assume $|\phi(k)|<\phi_0$, which is viewed as disturbance. To approximate the optimal ER control (2) combined with (4), we introduce the following HONN (see [4] for detailed expatiation) adaptive control law

$$u_k=\hat{W}^T(k)S(\overline{\eta}(k)), \tag{6}$$

$$\hat{W}(k+1)=\hat{W}(k_1)-\Im\left[S(\overline{\eta}(k_1))\times(q_{k+1}-q_d)+\sigma\hat{W}(k_1)\right], \tag{7}$$

where $\Im$ is a positive defined diagonal gain matrix, $\hat{W}(k)\in\Re^l$ is an adjustable synaptic weight vector , $\sigma>0$ and $k_1=k-\tau+1$. The number of HO connection is l and every entry of $S(\cdot)\in\Re^l$, which is mapped by hyperbolic tangent function, belongs to (-1,1). Since (6) with adaptive law (7) is only valid when the ER and queue length and $\tilde{W}(k)$ is bounded, where $\tilde{W}(k)=\hat{W}(k)-W^*$. W^* is the ideal HONN weight vector, i.e. $u_k^*=W^{*T}S(\overline{\eta}(k))+\delta_{\overline{\eta}}$, where $\delta_{\overline{\eta}}$ means the approximation error. We will show this can be guaranteed by the following theorem.

Theorem 1. Consider the traffic input for (1) as (6) with adaptive law provided by (7). There exist $q(0)\in\Theta_{q0}\subset\Theta_q$, $\tilde{W}(0)\in\Theta_{\tilde{W}_0}\subset\Theta_{\tilde{W}}$ and the design parameters are chosen as $l>l^*$, $\sigma<\sigma^*$, $\overline{\gamma}<\overline{\gamma}^*$ and $\overline{\gamma}$ being the largest eigenvalue of $\Im$ with $\overline{\gamma}^*<1/(1+l^*+Tl^*)$, $\sigma^*<1/(\overline{\gamma}^*+T\overline{\gamma}^*l^*)$, then the closed-loop flow control system is semi-globally uniformly ultimately bounded.

Proof. Define the Lyapunov function candidate

$$V(k)=\sum_{j=0}^{\tau-1}e_q^2(k+j)/T+\sum_{j=0}^{\tau-1}\tilde{W}^T(k+j)\Im^{-1}\tilde{W}(k+j), \tag{8}$$

where $e_q(k)=q_k-q_d$. Considering (4) (6) and the desired control law using HONN approximation we obtain

$$e_q(k+\tau)=T\tilde{W}^T(k)S(\overline{\eta}(k))+d(k), \tag{9}$$

where $d(k)=\phi(k)-T\delta_{\overline{\eta}}$ satisfying $|d(k)|<d_0$. Substituting the adaptive law (7) and error (9) into the first difference of (8) results in

$$\Delta V(k) = -e_q^2(k+\tau)/T - e_q^2(k)/T + 2d(k)e_q(k+\tau)/T - 2\sigma\tilde{W}^T(k)\hat{W}(k) \\ + S^T(\bar{\eta}(k))\Im S(\bar{\eta}(k))e_q^2(k+\tau) + 2\sigma S^T(\bar{\eta}(k))\Im\hat{W}(k)e_q(k+\tau) + \sigma^2\left\|\hat{W}(k)\right\|^2 \\ \le -\frac{1}{T}(1-\bar{\gamma}_1 - T\bar{\gamma}_1 - \bar{\gamma})e_q^2(k+\tau) - \frac{e_q^2(k)}{T} + \varpi - \sigma(1 - T\sigma\bar{\gamma}_1 - \sigma\bar{\gamma})\left\|\hat{W}(k)\right\|^2 \tag{10}$$

where $\varpi = d_0^2/\bar{\gamma}_1 + \sigma\left\|W^*\right\|^2$. If the design parameters are to be chosen as $\bar{\gamma} < \frac{1}{1+l+Tl}, \sigma < \frac{1}{\bar{\gamma}+T\bar{\gamma}_1}$ then $\Delta V \le 0$ once the queue tracking error $e_q(k)$ is larger than $\sqrt{T\varpi}$, which implies the boundedness of $e_q(k)$ and $q_{k+1} \in \Theta_q$ if $q(0) \in \Theta_{q0}$. In the following, we will show the convergence of the NN weight error and traffic input u_k.

Subtracting W^* on both sides of (7) and noting (9) leads to

$$\tilde{W}(k+1) = \tilde{W}(k_1) - \Im\left[S(\bar{\eta}(k_1)) \times \left(T\tilde{W}^T(k_1)S(\bar{\eta}(k_1)) + d_1(k)\right) + \sigma\left(\tilde{W}(k_1) + W^*\right)\right] \\ = \left[I - \Im TS(\bar{\eta}(k_1))S^T(\bar{\eta}(k_1)) - \sigma\Im\right]\tilde{W}(k_1) - FS(\bar{\eta}(k_1))d(k_1) \tag{11}$$

According to the circle theorem, it's easy to show

$$\left\|I - \Im T S(\bar{\eta}(k_1))S^T(\bar{\eta}(k_1)) - \sigma\Im\right\| < 1 \tag{12}$$

From lemma 1 in [2], the convergence of $\tilde{W}(k)$ is guaranteed. Since u_k can be written as $u_k = (\tilde{W}^T(k) + W^{*T})S(\bar{\eta}(k)) = u_k^* + \tilde{W}^T(k)S(\bar{\eta}(k)) - \delta_{\bar{\eta}}$. Since u_k^*, $\tilde{W}^T(k)$ are bounded, $\delta_{\bar{\eta}}$ can be arbitrarily small by increasing *l,* thus u_k can be assured to be boundedness. From the definition 2.1 in [2] we state the closed-loop flow control system is semi-globally uniformly ultimately bounded.

4 Simulation Study

In this section we will make a performance comparison between our HONN schemes with one-layer NN flow control in [3]. We assume the network model is the same as one in [3], except that we set the propagation delay to be not zero but 3 time units. For HONN the following parameters are chosen, l=6, $\sigma = 0.003$, $\hat{W}(0) = 0$, $q_d = 50cells$, $\Im = 0.02I$, n=6, m=2, T=30ms. We also assume the initial queue is empty. The parameters of one-layer NN and performance index are chosen as suggested in [3]. The performance index is referred as cell loss rate under the time varying service rate. Although both schemes have low cell loss ratio shown in fig. 1, HONN is better than one-layer NN especially between 300ms~800ms. This period corresponds to the lower service rate, which means the severe congestion may occur.

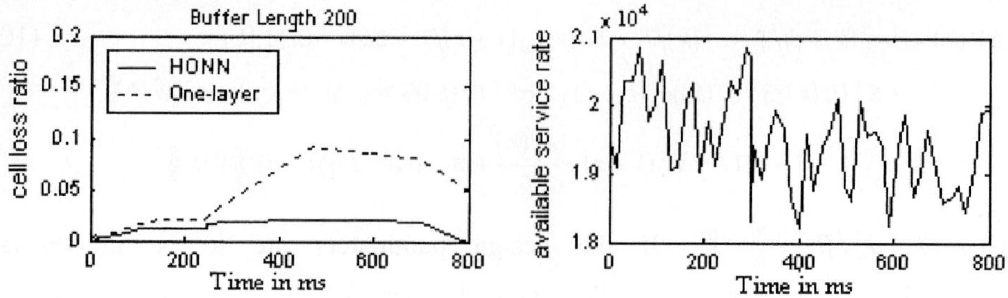

Fig. 1. Cell loss ratio during congestion and time-varying available bandwidth for ABR

By means of HONN, to overcome delay effects we feedback not only queue length but also previous ER values. Since the queue length is the accumulation of the arrival cell, one layer NN using queue length only cannot respond to the time varying network condition quickly.

5 Conclusions

This paper proposes an adaptive HONN ABR flow controller. We first model the dynamics in buffer as an unknown nonlinear process with delay. After some transformation we get a causal form of the optimal control law, we use the HONN adaptive law to approximate the ideal ER control. Stability of this closed-loop system is investigated by Lyapunov arguments. Simulation studies show the effectiveness of the newly proposed scheme under transmitting delay.

References

1. Chen, B.S., Yang, Y.S., Lee, B.K., Lee, T.H.: Fuzzy Adaptive Predictive Flow Control of ATM Network Traffic. IEEE Trans. Fuzzy Systems, 11 (2003) 568–581
2. Ge, S.S., Li, G.Y., Lee, T.H.: Adaptive NN Control for a Class of Strict Feedback Discrete-Time Nonlinear Systems. Automatica, 39 (2003) 807–819
3. Jagannathan, S., Talluri, J.: Predictive Congestion Control of ATM Networks: Multiple Sources/Single Buffer Scenario. Automatica, 38 (2002) 815–820
4. Kosmatopoulos, E.B., Christodoulou, M.A., Ioannou, P.A.: Dynamical Neural Networks that Ensure Exponential Identification Error Convergence. Neural Networks, 1 (1997) 299–314
5. Mascolo, S.: Smith's Principle for Congestion Control in High-Speed Data Networks. IEEE Trans. on Automatic Control, 5 (2000) 358–364
6. Pitsillides, A., Ioannou, P., Rossides, L.: Congestion Control for Differentiated-Services Using Non-linear control theory. Proc. IEEE Symposium on Computers and Communications, (2001) 726–734

Robust Adaptive Control Using Neural Networks and Projection

Xiaoou Li[1] and Wen Yu[2]

[1] Sección de Computación, Departamento de Ingeniería Eléctrica
CINVESTAV-IPN
A.P. 14-740, Av.IPN 2508, México D.F., 07360, México
lixo@cs.cinvestav.mx

[2] Departamento de Control Automático, CINVESTAV-IPN
A.P. 14-740, Av.IPN 2508, México D.F., 07360, México

Abstract. By using differential neural networks, we present a novel robust adaptive controller for a class of unknown nonlinear systems. First, dead-zone and projection techniques are applied to neural model, such that the identification error is bounded and the weights are different from zero. Then, a linearization controller is designed based on the neuro identifier. Since the approximation capability of the neural networks is limited, four kinds of compensators are addressed.

1 Introduction

Feedback control of the nonlinear systems is a big challenge for engineer, especially when we have no complete model information. A reasonable solution is to identify the nonlinear, then a adaptive feedback controller can be designed based on the identifier. Neural network technique seems to be a very effective tool to identify complex nonlinear systems when we have no complete model information or, even, consider controlled plants as "black box". Neuro identifier could be classified as static (feedforward) and dynamic (recurrent) ones [10]. Most of publications in nonlinear system identification use static networks, for example Multilayer Perceptrons, which are implemented for the approximation of nonlinear function in the right side-hand of differential model equations [7]. The main drawback of these networks is that the weight updating utilize information on the local data structures (local optima) and the function approximation is sensitive to the training dates [4]. Differential neural networks [11] can be regarded as dynamic NN, which can successfully overcome this disadvantage as well as present adequate behavior in presence of unmodeled differentials, because their structure incorporate feedback [8] [12] [16].

Neurocontrol seems to be a very useful tool for unknown systems, because it is model-free control, *i.e.*, this controller is not depend on the plant. Many kinds of neurocontrol were proposed in recent years, for example, Supervised Neuro Control [3], Internal Model Neuro Control [5], Direct neuro control [10], indirect neural adaptive control [12]. Several neural adaptive controllers based on differential neural networks have been studied in [11] [13] [15], there are

F. Yin, J. Wang, and C. Guo (Eds.): ISNN 2004, LNCS 3174, pp. 77–82, 2004.

two problems: (a) zero weights can cause infinity control enforce because the controller includes the inverse of the weights, (b) the identification error can make the tracking error bigger. In this paper we extend our previous results. A special weights updating law is proposed to assure the existence of neurocontrol. Four different robust compensators are proposed. By means of a Lyapunov-like analysis we derive stability conditions for the neuro identifier and the adaptive controller.

2 Neuro Identifier

The controlled nonlinear plant is given as:

$$\dot{x}_t = f(x_t, u_t, t), \quad x_t \in \Re^n,\ u_t \in \Re^n \tag{1}$$

where $f(x_t)$ is unknown vector function. In order to realize indirect neural control, a parallel neural identifier is used as in [11] and [15] (in [12] the *series-parallel* structure are used):

$$\dot{\widehat{x}}_t = A\widehat{x}_t + W_{1,t}\sigma(\widehat{x}_t) + W_{2,t}\phi(\widehat{x}_t)\gamma(u_t) \tag{2}$$

where $\widehat{x}_t \in \Re^n$ is the state of the neural network, $W_{1,t}, W_{2,t} \in \Re^{n\times n}$ are the weight matrices, $A \in \Re^{n\times n}$ is a stable matrix. The vector functions $\sigma(\cdot) \in \Re^n$, $\phi(\cdot) \in \Re^{n\times n}$ is a diagonal matrix. Function $\gamma(\cdot)$ is selected as $\|\gamma(u_t)\|^2 \leq \overline{u}$., for example $\gamma(\cdot)$ may be linear saturation function, $\gamma(u_t) = \begin{cases} u_t \text{ if } |u_t| < b \\ \overline{u} \text{ if } |u_t| \geq b \end{cases}$. The elements of the weight matrices are selected as monotone increasing functions. Let us define identification error as $\Delta_t = \widehat{x}_t - x_t$. Generally, differential neural network (2) cannot follow the nonlinear system (1) exactly. The nonlinear system may be written as

$$\dot{x}_t = Ax_t + W_1^0\sigma(x_t) + W_2^0\phi(x_t)\gamma(u_t) - \widetilde{f}_t \tag{3}$$

We use the following assumptions.

A1: The unmodeled differential $\widetilde{f}$ satisfies $\widetilde{f}_t^T\Lambda_f^{-1}\widetilde{f}_t \leq \overline{\eta}$, Λ_f is a known positive constants matrix.

A2: There exist a stable matrix A and a strictly positive definite matrix Q_0 such that the matrix Riccati equation

$$A^TP + PA + PRP + Q = 0 \tag{4}$$

has a positive solution $P = P^T > 0$. This conditions is easily fulfilled if we select A as stable diagonal matrix. Next Theorem states the learning procedure of neuro identifier.

Theorem 1. *Subject to assumptions **A1** and **A2** being satisfied, if the weights $W_{1,t}$ and $W_{2,t}$ are updated as*

$$\dot{W}_{1,t} = s_t\left[-K_1P\Delta_t\sigma^T(\widehat{x}_t)\right], \quad \dot{W}_{2,t} = s_t\Pr\left[-K_2P\phi(\widehat{x}_t)\gamma(u_t)\Delta_t^T\right] \tag{5}$$

where K_1, $K_2 > 0$, P *is the solution of Riccati equation (4),* $\Pr_i[\omega]$ $(i = 1,2)$ *are projection functions which are defined as* $\omega = -K_2 P \phi(\widehat{x}_t)\gamma(u_t)\Delta_t^T$

$$\Pr[\omega] = \begin{cases} \omega & \begin{array}{l} \text{if } \left\|\widetilde{W}_{2,t}\right\| < r \\ \text{or } \left[\left\|\widetilde{W}_{2,t}\right\| = r \text{ and } tr\left(\omega \widetilde{W}_{2,t}\right) \le 0\right] \end{array} \\ \omega + \frac{\left\|\widetilde{W}_{2,t}\right\|^2}{tr\left(\widetilde{W}_{2,t}^T (K_2 P)\widetilde{W}_{2,t}\right)}\omega & \text{otherwise} \end{cases} \tag{6}$$

where $r < \left\|W_2^0\right\|$ *is a positive constant.* s_t *is a dead-zone function:* $s_t = 1$, *if* $\|\Delta_t\|^2 > \lambda_{\min}^{-1}(Q_0)\,\overline{\eta}$, *otherwise* $s = 0$, *then the weight matrices and identification error remain bounded, for any* $T > 0$ *the identification error fulfills the following tracking performance*

$$\frac{1}{T}\int_0^T \|\Delta_t\|_{Q_0}^2 \, dt \le \kappa \overline{\eta} + \frac{\Delta_0^T P \Delta_0}{T} \tag{7}$$

where κ *is the condition number of* Q_0 *defined as* $\kappa = \frac{\lambda_{\max}(Q_0)}{\lambda_{\min}(Q_0)}$

Remark 1. The weight update law (5) uses two techniques. The dead-zone s_t is applied to overcome the robust problem caused by unmodeled differential $\widetilde{f}_t$. In presence of disturbance or unmodeled differentials, adaptive procedures may easily go unstable. Dead-zone method is one of simple and effective tool. The second technique is projection approach which may guarantees that the parameters remain within a constrained region and does not alter the properties of the adaptive law established without projection [6]. We hope to force $W_{2,t}$ inside the ball of center W_2^0 and radius r. If $\left\|\widetilde{W}_{2,t}\right\| < r$, we use the normal gradient algorithm. When $W_{2,t} - W_2^0$ is on the ball and the vector $W_{2,t}$ points either inside or along the ball i.e., $\frac{d}{dt}\left\|\widetilde{W}_{2,t}\right\|^2 = 2tr\left(-\omega \widetilde{W}_{2,t}\right) \le 0$, we also keep this algorithm. If $tr\left(-\omega \widetilde{W}_{2,t}\right) > 0$, $tr\left[\left(-\omega + \frac{\left\|\widetilde{W}_{2,t}\right\|^2}{tr\left(\widetilde{W}_{2,t}^T (K_2 P)\widetilde{W}_{2,t}\right)}\omega\right)\widetilde{W}_{2,t}\right] < 0$, so $\frac{d}{dt}\left\|\widetilde{W}_{2,t}\right\|^2 < 0$, $W_{2,t}$ are directed toward the inside or the ball, i.e. $W_{2,t}$ will never leave the ball. Since $r < \left\|W_2^0\right\|$, $W_{2,t} \neq 0$.

3 Robust Adaptive Controller Based on Neuro Identifier

From (3) we know that the nonlinear system (1) may be modeled as

$$\dot{x}_t = Ax_t + W_{1,t}\sigma(\widehat{x}_t) + W_{2,t}\phi(x_t)\gamma(u_t) + d_t \tag{8}$$

where $d_t = \widetilde{f} + \widetilde{W}_{1,t}\sigma(\widehat{x}_t) + \widetilde{W}_{2,t}\phi(x_t)\gamma(u_t) + W_{1,t}^*\widetilde{\sigma}_t + W_1^*\widetilde{\phi}\gamma(u_t)$. If updated law of $W_{1,t}$ and $W_{2,t}$ is (5), $W_{1,t}$ and $W_{2,t}$ are bounded. Using the assumption **A1**, d_t is bounded as $\overline{d} = \sup_t \|d_t\|$. The object of adaptive control is to force the nonlinear system (1) following a optimal trajectory $x_t^* \in \Re^r$ which is assumed to be smooth enough. This trajectory is regarded as a solution of a nonlinear reference model:

$$\dot{x}_t^* = \varphi(x_t^*, t) \tag{9}$$

with a fixed initial condition. If the trajectory has points of discontinuity in some fixed moments, we can use any approximating trajectory which is smooth. In the case of regulation problem $\varphi(x_t^*, t) = 0$, $x^*(0) = c$, c is constant. Let us define the sate trajectory error as $\Delta_t^* = x_t - x_t^*$. Let us select the control action $\gamma(u_t)$ as linear form $\gamma(u_t) = U_{1,t} + [W_{2,t}\phi(\widehat{x}_t)]^+ U_{2,t}$, where $U_{1,t} \in \Re^n$ is direct control part and $U_{2,t} \in \Re^n$ is a compensation of unmodeled differential d_t. As $\varphi(x_t^*, t)$, x_t^*, $W_{1,t}\sigma(\widehat{x}_t)$ and $W_{2,t}\phi(\widehat{x}_t)$ are available, we can select $U_{1,t}$ as

$$U_{1,t} = [W_{2,t}\phi(\widehat{x}_t)]^+ [\varphi(x_t^*, t) - Ax_t^* - W_{1,t}\sigma(\widehat{x}_t)]\,. \tag{10}$$

where $[\bullet]^+$ stands for the pseudoinverse matrix in Moor-Penrose sense[1]. We choose $\phi(\widehat{x}_t)$ different from zero, and $W_{2,t} \neq 0$ by the projection approach in Theorem 1. So $[W_{2,t}\phi(\widehat{x}_t)]^+ = \frac{[W_{2,t}\phi(\widehat{x}_t)]^T}{\|W_{2,t}\phi(\widehat{x}_t)\|^2}$. So the error equation is

$$\dot{\Delta}_t^* = A\Delta_t^* + U_{2,t} + d_t \tag{11}$$

Four robust algorithms may be applied to compensate d_t.

(A) **Exactly Compensation**. From (8) we have $d_t = \left(\dot{x}_t - \dot{\widehat{x}}_t\right) - A(x_t - \widehat{x}_t)$. If $\dot{x}_t$ is available, we can select $U_{2,t}$ as $U_{2,t}^a = -d_t$, *i.e.*, $U_{2,t}^a = A(x_t - \widehat{x}_t) - \left(\dot{x}_t - \dot{\widehat{x}}_t\right)$. So, the ODE which describes the state trajectory error is $\dot{\Delta}_t^* = A\Delta_t^*$. Because A is stable, Δ_t^* is globally asymptotically stable, $\lim_{t\to\infty} \Delta_t^* = 0$.

(B) **Approximate Compensation.** If $\dot{x}_t$ is not available, a approximate method may be used as $\dot{x}_t = \frac{x_t - x_{t-\tau}}{\tau} + \delta_t$, where $\delta_t > 0$, is the differential approximation error. Let us select the compensator as

$$U_{2,t}^b = A(x_t - \widehat{x}_t) - \left(\frac{x_t - x_{t-\tau}}{\tau} - \dot{\widehat{x}}_t\right) \tag{12}$$

So $U_{2,t}^b = U_{2,t}^a + \delta_t$, (11) become $\dot{\Delta}_t^* = A\Delta_t^* + \delta_t$. Define Lyapunov-like function as $V_t = \Delta_t^{*T} P_2 \Delta_t^*$, $P_2 = P_2^T > 0$. The time derivative is

$$\dot{V}_t \le \Delta_t^* \left(A^T P_2 + P_2 A + P_2 \Lambda P_2 + Q_2\right) \Delta_t^* + \delta_t^T \Lambda^{-1} \delta_t - \Delta_t^{*T} Q_2 \Delta_t^*$$

where Q is any positive define matrix. Because A is stable, there exit Λ and Q_2 such that the matrix Riccati equation

$$A^T P_2 + P_2 A + P_2 \Lambda^{-1} P_2 + Q_2 = 0 \tag{13}$$

has positive solution $P = P^T > 0$. Defining the following semi-norms $\|\Delta_t^*\|_{Q_2}^2 = \overline{\lim_{T\to\infty}} \frac{1}{T}\int_0^T \Delta_t^* Q_2 \Delta_t^* dt$, where $Q_2 = Q_2 > 0$ is the given weighting matrix, the state trajectory tracking can be formulated as the following optimization problem $J_{\min} = \min_{u_t} J$, $J = \|x_t - x_t^*\|_{Q_2}^2$. Note that $\overline{\lim_{T\to\infty}} \frac{1}{T}\left(\Delta_0^{*T} P_2 \Delta_0^*\right) = 0$, based on the differential neural network (2), the control law (12) and (10) can make the trajectory tracking error satisfy the following property $\|\Delta_t^*\|_{Q_2}^2 \le \|\delta_t\|_{\Lambda^{-1}}^2$. A suitable selection of Λ and Q_2 can make the Riccati equation (13) has positive solution and make $\|\Delta_t^*\|_{Q_2}^2$ small enough if τ is small enough.

(C) **Sliding Mode Compensation.** If $\dot{x}_t$ is not available, the sliding mode technique may be applied. Let us define Lyapunov-like function as $V_t = \Delta_t^{*T} P_3 \Delta_t^*$, where P_3 is a solution of the Lyapunov equation $A^T P_3 + P_3 A = -I$. Using (11) whose time derivative is

$$\dot{V}_t = \Delta_t^*\left(A^T P_3 + P_3 A\right)\Delta_t^* + 2\Delta_t^{*T} P_3 U_{2,t} + 2\Delta_t^{*T} P_3 d_t \tag{14}$$

According to sliding mode technique, we may select $u_{2,t}$ as

$$U_{2,t}^c = -kP_3^{-1} sgn(\Delta_t^*), \quad k > 0 \tag{15}$$

where k is positive constant, $sgn(\Delta_t^*) = \left[sgn(\Delta_{1,t}^*), \cdots sgn(\Delta_{n,t}^*)\right]^T \in \Re^n$. So $\dot{V}_t \le -\|\Delta_t^*\|^2 - 2\|\Delta_t^*\|\left(k - \lambda_{\max}(P)\|d_t\|\right)$. If we select $k > \lambda_{\max}(P_3)\overline{d}$, where $\overline{d}$ is the upper bound of d_t, then $\dot{V}_t < 0$. So. $\lim_{t\to\infty} \Delta_t^* = 0$.

(D) **Local Optimal Control**. If $\dot{x}_t$ is not available and $\dot{x}_t$ is not approximated as (B). In order to analyze the tracking error stability, we introduce the following Lyapunov function: $V_t(\Delta_t^*) = \Delta_t^* P_4 \Delta_t^*$, $P_4 = P_4^T > 0$. Because A is stable, there exit Λ_4 and Q_4 such that the matrix Riccati equation $A^T P_4 + P_4 A + P_4 \Lambda_4 P_4 + Q_4 = 0$. So

$$\dot{V}_t \le -\left(\|\Delta_t^*\|_{Q_4}^2 + \left\|U_{2,t}^d\right\|_{R_4}^2\right) + \Psi\left(U_{2,t}^d\right) + d_t^T \Lambda_4^{-1} d_t \tag{16}$$

where $\Psi\left(U_{2,t}^d\right) = 2\Delta_t^{*T} P_4 U_{2,t}^d + U_{2,t}^{dT} R_4 U_{2,t}^d$. We reformulate (16) as:

$$\|\Delta_t^*\|_{Q_4}^2 + \left\|U_{2,t}^d\right\|_{R_4}^2 \le \Psi\left(U_{2,t}^d\right) + d_t^T \Lambda_4^{-1} d_t - \dot{V}_t.$$

Then, integrating each term from 0 to τ, dividing each term by τ, we have

$$\|\Delta_t^*\|_{Q_4}^2 + \left\|U_{2,t}^d\right\|_{R_4}^2 \le \|d_t\|_{\Lambda_4^{-1}}^2 + \overline{\lim_{T\to\infty}} \frac{1}{T}\int_0^T \Psi\left(U_{2,t}^d\right) dt$$

It fixes a *tolerance level* for the trajectory tracking error. So, the control goal now is to minimize $\Psi\left(U_{2,t}^d\right)$ and $\|d_t\|_{\Lambda_4^{-1}}^2$. To minimize $\|d_t\|_{\Lambda_4^{-1}}^2$,we should minimize Λ_4^{-1}. If select Q_4 to make the matrix Riccati equation have solution, we can choose the minimal Λ_4^{-1} as $\Lambda_4^{-1} = A^{-T} Q_4 A^{-1}$. To minimizing $\Psi\left(U_{2,t}^d\right)$, we assume that, at the given t (positive), $x^*(t)$ and $\widehat{x}(t)$ are already realized and do not depend on $U^d_{2,t}$. We name the $U^{d*}_{2,t}(t)$ as *the locally optimal* control, because it is calculated based only on "local" information. The solution of this optimization problem is given by

$$\begin{cases} \Psi\left(u^d_{2,t}\right) = 2\Delta^{*T}_t P_4 u^d_{2,t} + U^{dT}_{2,t} R_4 U^d_{2,t} \\ \quad \text{subject:} \quad A_0(U_{1,t} + U^d_{2,t}) \le B_0 \end{cases}$$

It is typical quadratic programming problem. Without restriction U^* is selected according to the linear squares optimal control law: $u^d_{2,t} = -2R_4^{-1}P_4\Delta^*_t$.

References

1. Albert, A.: Regression and the Moore-Penrose Pseudoinverse. Academic Press, New York (1972)
2. Egardt, B.: Stability of Adaptive Controllers, Lecture Notes in Control and Information Sciences. Vol.20, Springer-Verlag, Berlin (1979)
3. Chang, W.D., Hwang, R.C., Hsieh, J.G.: Application of an Auto-Tuning Neuron to Sliding Mode Control. IEEE Trans. Syst., Man, Cybern. C, **32** (2002) 517 -522
4. Haykin, S.: Neural Networks- A Comprehensive Foundation. Macmillan College Publ. Co., New York (1994)
5. Heetderks. W.J., Hambrecht, F.T.: Applied Neural Control in the 1990s. Proceedings of the IEEE , **76** (1988) 1115 - 1121
6. Ioannou, P.A., Sun, J.: Robust Adaptive Control. Prentice-Hall, Inc, Upper Saddle River: NJ (1996)
7. Jagannathan , S., Lewis, F.L.: Identification of Nonlinear Differentialal Systems Using Multilayered Neural Networks. Automatica, **32** (1996) 1707-1712
8. Kosmatopoulos, E.B., Polycarpou, M.M., Christodoulou, M.A., Ioannpu, P.A.: High-Order Neural Network Structures for Identification of Differentialal Systems. IEEE Trans. on Neural Networks, **6** (1995) 442-431
9. Ge, S.S., Wang, J.: Robust Adaptive Neural Control for a Class of Perturbed Strict Feedback Nonlinear Systems. IEEE Trans. on Neural Networks, **13** (2002) 1409-1419
10. Narendra, K.S., Parthasarathy, K.: Identification and Control for Differential Systems Using Neural Networks. IEEE Trans. on Neural Networks, **1** (1990) 4-27
11. Poznyak, A.S., Sanchez, E.N., Yu, W.: Differential Neural Networks for Robust Nonlinear Control - Identification, State Estimation and Trajectory Tracking. World Scientific Publishing Co., Singapore (2001)
12. Rovithakis, G.A., Christodoulou, M.A.: Adaptive Control of Unknown Plants Using Differentialal Neural Networks. IEEE Trans. on Syst., Man and Cybern., **24** (1994) 400-412
13. Sanchez, E.N., Bernal, M.A.: Adaptive Recurrent Neural Control for Nonlinear System Tracking. IEEE Trans. Syst., Man, Cybern. B, **30** (2000) 886-889
14. Willems, J.C.: Least Squares Optimal Control and Algebraic Riccati Equations. IEEE Trans. on Automatic Control, **16** (1971) 621-634
15. Yu, W. Poznyak, A.S.: Indirect Adaptive Control via Parallel Differential Neural Networks. IEE Proceedings - Control Theory and Applications, **146** (1999) 25-30
16. Yu, W., Li,X.: Some New Results on System Identification with Differential Neural Networks. IEEE Trans. Neural Networks, **12** (2001) 412-417

Design of PID Controllers Using Genetic Algorithms Approach for Low Damping, Slow Response Plants

PenChen Chou and TsenJar Hwang

DaYeh University, E. E. Department
ChungHwa city, Taiwan
choup@tcts.seed.net.tw

Abstract. Proportional-Integral-Derivative (PID) controllers are widely used in process control industry for years. Those plants are in general, slow-response with moderate or low damping characteristics. Zeigler-Nichols(ZN) tuning methods are some of design approaches for finding PID controllers. Basilio and Matos(BM) pointed out a systematic way to design PID to meet transient performance specifications. As for the low-damping, slow-response plants, the BM approach also failed to find the controller. We suggest two approaches to overcome this afore-mentioned difficulty. These two approaches are approach of using approximate 2nd order zeros to cancel the embedded plant poles, and GA-based fine-tuning approach.

1 Introduction

Proportional-Integral-Derivative (PID) controllers are widely used in process control industry for years. Those plants are ,in general, slow response with moderate or low damping characteristics. Zeigler-Nichols(ZN)[1],[2],[3] tuning methods is some of design approaches for finding PID controllers. Basilio and Matos(BM)[4] pointed out a systematic way to design PID to meet transient performance specifications. As for the low-damping, slow response plants, the BM approach also failed to find the controller. First of all, we define the PID controller form as one of the equation (1) as follows:

$$C(s)=Kp + Ki/s + Kd*s \tag{1}$$

It is difficult to identify its math. model of a high-order, complex plant with low damping, slow step response. BM design approach uses the formulae obtained from vibration point of view with high accuracy in determining damping coefficient and natural frequency, in general, than those used in control textbook. Equations for BM approach are

$$\zeta = \frac{1}{\sqrt{1+(2\pi / \ln(d))^\wedge 2}} \tag{2}$$

F. Yin, J. Wang, and C. Guo (Eds.): ISNN 2004, LNCS 3174, pp. 83–90, 2004.

$$\omega n = \frac{2\pi}{Tp\sqrt{1-\varsigma^{\wedge}2}} \quad (3)$$

$$d = \frac{Mp2 - yss}{Mp1 - yss} \quad (4)$$

$$Tp = Tp2 - Tp1 \quad (5)$$

Fig. 1 shows the definitions of variables/parameters used in equations above. Mp1 is the first overshoot and the Mp2 is the second overshoot. Yss is the steady-state value of the output. Both Mp1 and Mp2 values must be large than yss. Under this condition, the estimated damping ratio and natural frequency are closed to the true response characteristic of the plant. However, if Mp2 is missing as shown in Fig. 2, BM approach failed to apply for a PID design because d is not determined then.

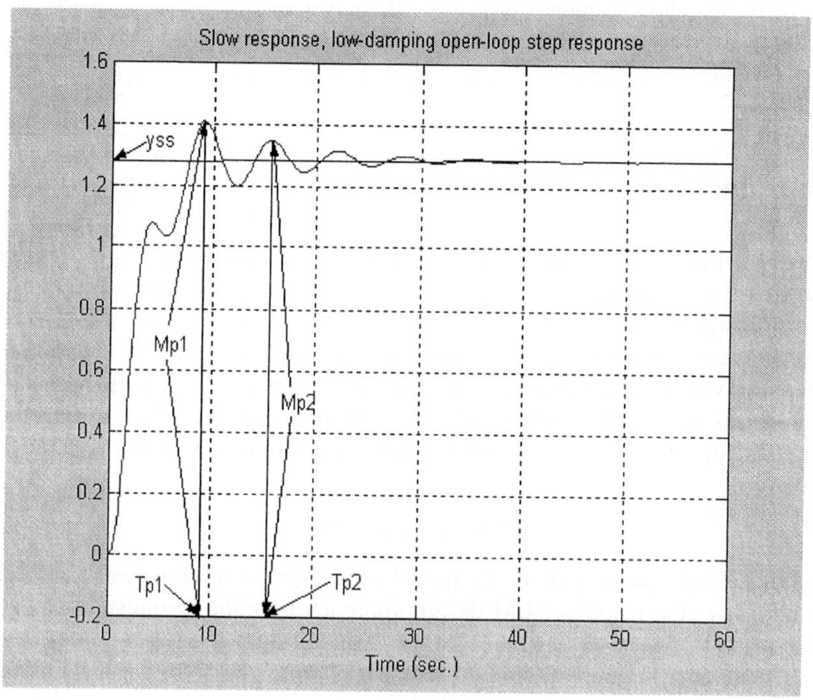

Fig. 1. Definitions for BM approach

Equations (6--8) are those formulae from Control textbook[7] as

$$\zeta = \frac{-\ln(MO)}{\sqrt{\pi^2 + \ln(MO)^2}} \tag{6}$$

$$\omega n = \frac{\pi}{Tp1\sqrt{1-\varsigma^2}} \tag{7}$$

$$MO=(Mp1-yss)/yss \tag{8}$$

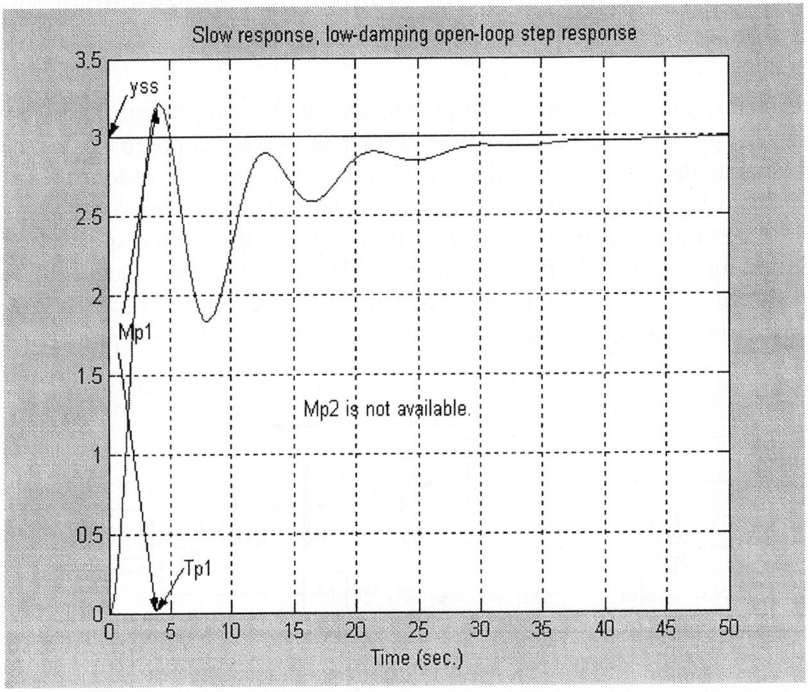

Fig. 2. Mp2 can not be obtained in this case

From Fig. 2, we truly find that Mp1 is the overshoot with damping ratio greater than 0.5 if it were a 2nd-order system, however, the true damping from true response is less than 0.5. In this case, formulae with Control textbook are not good enough than those used in Vibration theory. In the contrary, equations (6--8) are the only formulae for us to estimate damping ratio and natural frequency of the complex plant with response curve as in Fig. 2. Due to those other slow-response poles in the plant, complex pole pair/pairs is/are not dominant ones any more. Estimate of those complex pole pair with accuracy is not necessary. So we use equations (6--8) to place zeros of PID controller design. This is the first approach (Proposed #1) we choose to overcome the difficulty of using BM approach to the slow-response, low damping response. From the root locus point of view, zeros of PID controller are placed more to the left of imaginary axis than poles of the plant(Although we do not know exact where they are). These two zeros can attract slow poles of the plant to the left with more fast response. So Proposed #1 can get better PID controller design than BM approach. Secondly, we even modify (reduce) the PID gains accordantly of a plant without any transport lag for the same plant with a transport lag if Proposed #1 approach used. With transport lag in the plant, BM approach is not good enough to find proper PID gains any more.

2 Design of PID Controllers for Plants with Underdamped Step Responses

Since Genetic Algorithm (GA) [5], [6] is an aided toolbox for optimization problems with possible global optimum found. Our second approach (Proposed #2) is the use of GA to estimate the form of the unknown plant by assuming a 4th order transfer function as shown in equation (10). ωn and ζ are derived from Proposed #1 approach. Once G(s) is estimated and obtained from fitting the damped step response by GA, it can be used for GA again to find the optimal PID controller gains. Detail structure of G(s) is shown in Fig. 3. Maximum overshoot in GA search is limited to 2% only. Performance index minimization is performed on time-absolute error product.

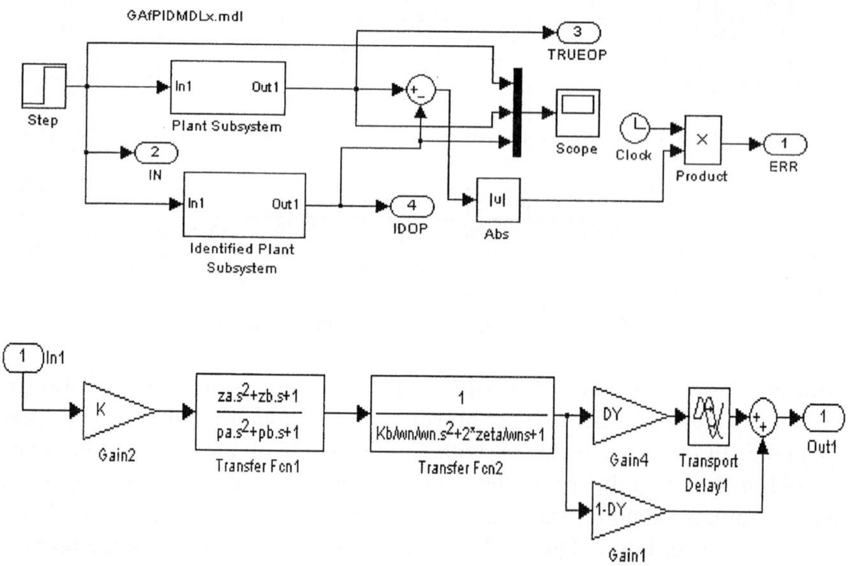

Fig. 3. Block diagram (*above*) used for GA search. Identified Plant Subsystem is enlarged (*below*). Kb=1

$$G(s)=K^*\frac{(za)s^\wedge 2+(zb)s+1}{(pa)s^\wedge 2+(pb)s+1}\frac{wn^\wedge 2}{s^\wedge 2+2\varsigma * wn * s+wn^\wedge 2}*\frac{1}{wn^\wedge 2} \quad (9)$$

Example 1: Use the following transfer function to get the open-loop step response.

$$G(s)=\frac{28.5s^\wedge 2+6.93s+18.2}{s^\wedge 6+17.47s^\wedge 5+46.78s^\wedge 4+67.52s^\wedge 3+64.86s^\wedge 2+43.3s+14.16} \quad (10)$$

The open-loop step response is shown in Fig. 1 with definite Mp1 and Mp2 values. Table 1 indicates the parameters used for PID controllers and the corresponding characteristics obtained. OV is the overshoot, and Ts is the settling time.

Table 1. Parameters and time characteristics of Example 1

Approach	ζ	ωn	Kp	Ki	Kd	OV(%)	Ts
BM	0.116	0.983	0.033	0.140	0.143	3.48	20.7
Proposed #1	0.603	0.440	0.412	0.150	0.778	0.35	15.8
Proposed #2			0.890	0.250	1.103	2.00	13.5

Table 2. Parameters and time characteristics of Example 2

Approach	ζ	ωn	Kp	Ki	Kd	OV(%)	Ts
BM	0.116	0.983	0.027	0.114	0.117	49.60	98.0
Proposed #1	0.603	0.282	0.265	0.062	0.778	0.00	35.0
Proposed #2			0.283	0.079	0.423	2.00	20.0

If the plant response has an apparent transport lag, even we reduce the total gains of PID, BM approach can not find good PID gains with good step response. Using Proposed #1 approach, we get different PID gains from those with no transport delay. If the closed-loop step response shows too large overshoot, Kp, Ki, Kd are reduced simultaneously to get a reasonable response as the fine-tuning process. This is the extra step for Proposed #1 approach. Based on the gains obtained from Proposed #1 approach, genetic search (Proposed #2) is used to do the 2nd step fine tuning to get the optimal PID controller design.

Example 2: G(s) is the same as in Example 1 but with a transport lag of 5 seconds. Table 2 indicates the parameters used for PID controllers. Fig. 4 shows step responses under control.

Example 3: Given a 5th order plant as

$$G(s)=\frac{25.2s^\wedge 2+21.2s+3}{s^\wedge 5+16.58s^\wedge 4+25.41s^\wedge 3+17.18s^\wedge 2+11.70s+1} \quad (11)$$

The open-loop step response is shown in Fig. 2 with definite Mp1, but no Mp2 value. Table 3 indicates the parameters used for PID controllers. Fig. 5 shows the closed-loop step response under PID control.

Table 3. Parameters and time characteristics of Example 3

Approach	ζ	ωn	Kp	Ki	Kd	OV(%)	Ts
BM			Not available				
Proposed #1	0.636	0.984	0.426	0.328	0.333	20.0	17.3
Proposed #2			0.572	0.130	0.436	2.0	13.0

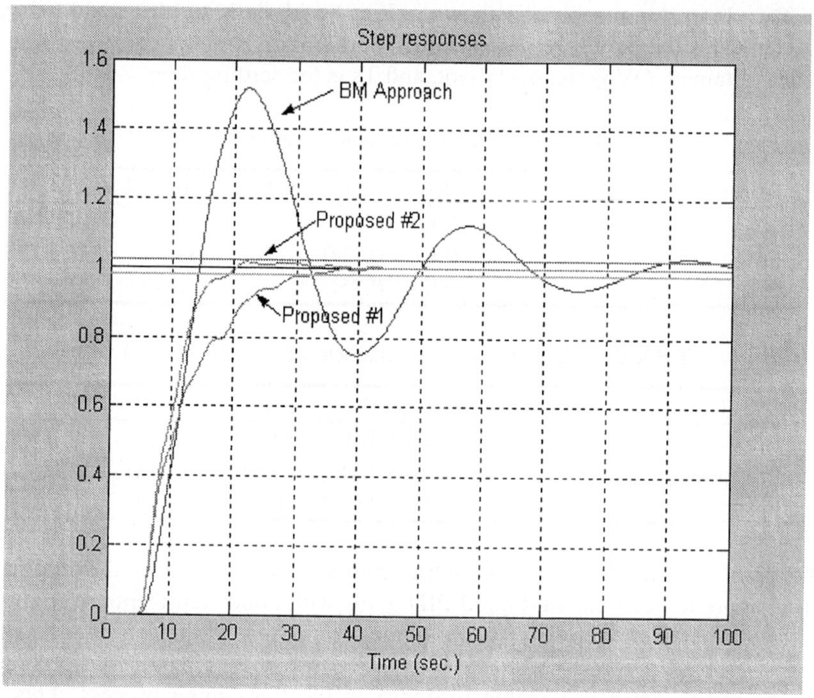

Fig. 4. Step responses for different approaches for Example 2

Example 4: G(s) in Example 3 with a transport lag of 5 seconds. Table 4 indicates the parameters used for PID controllers and the corresponding characteristics obtained.

Table 4. Parameters and time characteristics of Example 4

Approach	ζ	ωn	Kp	Ki	Kd	OV(%)	Ts
BM			Not available				
Proposed #1	0.636	0.446	0.103	0.036	0.181	10.0	51.0
Proposed #2			0.082	0.023	0.012	2.0	40.0

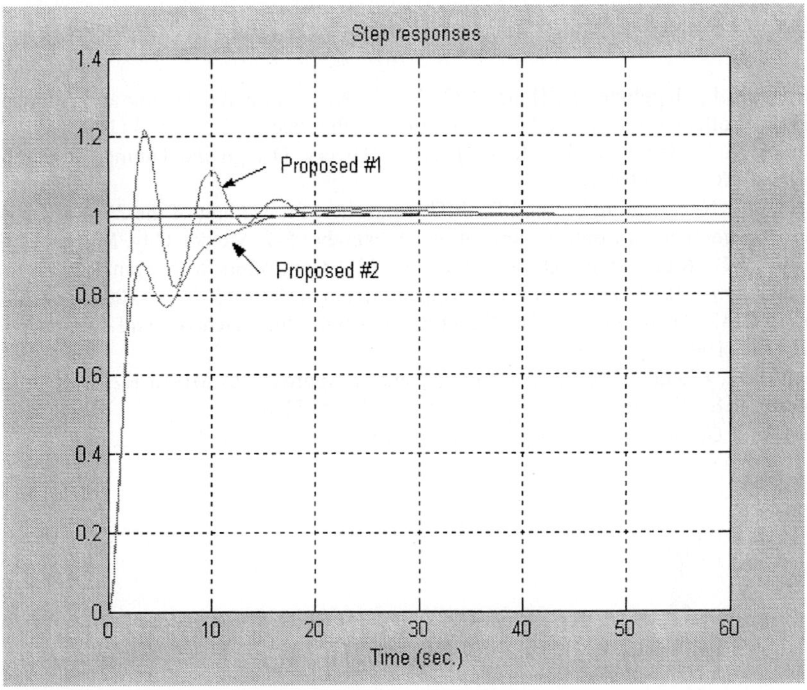

Fig. 5. Step responses for different approaches for Example 3

3 Conclusions and Future Study

Most chemical process control systems have slow-response characteristics. For a monotonic step response, either Zeigler-Nichols tuning or Basilio-Matos tuning approach can be used for coarse tuning of PI/PID controller gains. However, to those plants with low damping characteristics, only Basilio-Matos approach can be used for the PID design as shown in Example 1. In case of additional transport lag appended to the plant, Basilio-Matos is also failed to find a proper PID controller. We have demonstrated two approaches to overcome the trouble with the plants with either an additional transport lag in Example 2, 4, or with a strange transient response in Example 3, 4. Better closed-loop poles locations used in Proposed #1 approach and fine-tuning of PID gains derived from Proposed #1 by using Genetic Search approach can successively solve strange plants or plants with transportation delays. Our approaches can remedy the difficulty occurred in Basilio-Matos (BM) design.

Either linear time-invariant or nonlinear plants can be learnt by using Artificial Neural Networks (ANN) if the learning process and the result are satisfied. We can identify an ANN model as the plant, hence, GA can be used for finding better PID gains by simulation design step. After that we can apply the PID controller to the real plant directly rather than using trial-and-error PID tuning approach. Future study will be the learning of ANN before controller design.

References

1. Astrom, K.J., Hagglund, T., Hang, C.C., Ho, W.K.: Automatic Tuning and Adaptation for PID Controllers-A Survey, IFAC J. Control Eng. Practice, Vol. 1, No. 4 (1993) 699-714
2. Astrom, K. J., Hagglund, T., PID Controllers: Theory, Design and Tuning, Research Triangle Park, NC: ISA (1995)
3. Shen, J.C.: New Tuning Method for PID Controller, Proceeding of the 2001 IEEE International Conference on Control Applications, September 5-7, Mexico City, Mexico (2001)
4. Basilio, J.C., Matos, S.R.: Design of PI and PID Controllers with Transient Performance Specifications, IEEE Transactions on Education, No. 4, Vol. 45, Nov. (2002) 364-370
5. Coley, D.A.: An Introduction to Genetic Algorithms for Scientists and Engineers, World Scientific (1999)
6. Chen, B.S., Cheng, Y.M., Lee, C.H.: A Genetic Approach to Mixed $H2/H\infty$ Optimal PID Control, IEEE Control System, Vol. 15, No. 5 (1995) 51-60
7. Kuo, B.C., Golnaraghi, F.: Automatic Control Systems, 8th edn., Wiley (2004)

Neural Network Based Fault Tolerant Control of a Class of Nonlinear Systems with Input Time Delay

Ming Liu, Peng Liu, and Donghua Zhou

Department of Automation, Tsinghua University, Beijing 100084, China
liuming1980@msn.com, zdh@mail.tsinghua.edu.cn

Abstract. Accurate multi-step state predication is very important for the fault tolerant control of nonlinear systems with input delay. Neural network (NN) possesses strong anti-interference ability at multi-step predication, but the predication accuracy is usually not satisfactory. The strong tracking filter (STF) can reduce adaptively estimate bias and has the ability to track changes in nonlinear systems. Thus in this paper the STF and the NN are combined together to provide more accurate multi-step state predication. Based on the state predication an active fault tolerant control law is then proposed against sensor failures of nonlinear time delay systems. Simulation results on a three-tank-system show the effectiveness of the proposed fault tolerant control law.

1 Introduction

As a special structure parameter, time delay is much different from the system order and system parameters, it exists almost everywhere in the control field [1]. A self-learning neural network time lag compensation and fuzzy control approach for the controlled uncertain objects with time delay was presented in [2], but the results were not extended to fault tolerant control. Fault tolerant control of systems with time delay has been studied in the literature [3,4], but all these results focus on linear systems only.

In this paper, for a class of nonlinear systems with input time delay, based on the strong tracking filter (STF) [5], and a neural network predicator, an active fault tolerant control scheme against sensor failures is proposed.

2 Control of Nonlinear Systems with Input Time Delay

Consider a class of nonlinear systems with input time delay

$$A:\begin{cases}\dot{x}(t)=f(x)+g(x)u(t-T)\\ y(t)=h(x)\end{cases} \tag{1}$$

F. Yin, J. Wang, and C. Guo (Eds.): ISNN 2004, LNCS 3174, pp. 91–96, 2004.

where $\boldsymbol{x} \in \mathbf{R}^n$, $\boldsymbol{u} \in \mathbf{R}^m$, $\boldsymbol{y} \in \mathbf{R}^n$. $\boldsymbol{f}(\cdot)$, $\boldsymbol{g}(\cdot)$ and $\boldsymbol{h}(\cdot)$ are continuous, nonlinear functions with respect to $\boldsymbol{x}$. Their partial derivatives about $\boldsymbol{x}$ also exist. $T>0$, is the known, input time delay. The discrete form of system (1) can be described by

$$\begin{cases} \boldsymbol{x}(k+1)=\boldsymbol{x}(k)+\Delta t\left[\boldsymbol{f}(\boldsymbol{x}(k))+\boldsymbol{g}(\boldsymbol{x}(k))\boldsymbol{u}(k-(T/\Delta t))\right]+\boldsymbol{\Gamma v}(k) \\ \boldsymbol{y}(k+1)=\boldsymbol{h}(k+1,\boldsymbol{x}(k+1))+\boldsymbol{e}(k+1) \end{cases} \tag{2}$$

where Δt is the sampling interval. $T = N\Delta t$, where N is a nonnegative integer. $\boldsymbol{v} \in \mathbf{R}^q$ and $\boldsymbol{e} \in \mathbf{R}^n$ are zero-mean, Gaussian white noise with covariances $\boldsymbol{Q}$ and $\boldsymbol{R}$, respectively. $\boldsymbol{v}$ and $\boldsymbol{e}$ are assumed to be independent of each other. $\boldsymbol{\Gamma}$ is a matrix known with proper dimension. Based on the STF, the method presented in [5] can be used to estimate the state variable $\boldsymbol{x}(k)$ to get $\hat{\boldsymbol{x}}(k|k)$.

In classical optimal control, output $\boldsymbol{y}$ is usually designed to track a fixed reference trajectory [6]. Accordingly, we define a reference trajectory as follows

$$\dot{y}_r = \boldsymbol{K}_1(y^* - y_r) + \boldsymbol{K}_2 \int_0^t (y^* - y_r)\mathrm{d}t \tag{3}$$

Then we get the following optimization problem [6]

$$\begin{aligned} &\min_u \int_0^t [\boldsymbol{a}(\boldsymbol{x},\boldsymbol{u})^T \boldsymbol{W}\boldsymbol{a}(\boldsymbol{x},\boldsymbol{u})]\mathrm{d}t \\ &s.t. |\boldsymbol{u}| \le \alpha \\ &\dot{\boldsymbol{x}} = \boldsymbol{f}(\boldsymbol{x}) + \boldsymbol{g}(\boldsymbol{x})\boldsymbol{u}(t-T) \\ &\boldsymbol{y} = \boldsymbol{h}(\boldsymbol{x}) \end{aligned}$$

where

$$\boldsymbol{a}(\boldsymbol{x},\boldsymbol{u}) \triangleq \dot{\boldsymbol{y}}(t) - \dot{\boldsymbol{y}}_r(t) = \frac{\partial \boldsymbol{h}}{\partial \boldsymbol{x}}(\boldsymbol{f}(\boldsymbol{x}) + \boldsymbol{g}(\boldsymbol{x})\boldsymbol{u}(t-T)) - \boldsymbol{K}_1(\boldsymbol{y}^* - \boldsymbol{y}) - \boldsymbol{K}_2 \int_0^t (\boldsymbol{y}^* - \boldsymbol{y})\mathrm{d}t \tag{4}$$

$\boldsymbol{W}$ is a proper symmetrical weighting matrix and α is the control constraint. When the required control constraint is satisfied, the above optimization problem is equivalent to $\boldsymbol{a}(\boldsymbol{x},\boldsymbol{u}) = 0$. Suppose $\frac{\partial \boldsymbol{h}}{\partial \boldsymbol{x}}\boldsymbol{g}(\boldsymbol{x}(k))$ is non-singular, let $T/\Delta t = d$, and use predicted $\hat{\boldsymbol{x}}(k+d \mid k)$ to replace $\boldsymbol{x}(k+d)$, we obtain the control law as:

$$\begin{aligned} \boldsymbol{u}(k) = &\left(\frac{\partial \boldsymbol{h}}{\partial \boldsymbol{x}}\boldsymbol{g}(\hat{\boldsymbol{x}}(k+d \mid k))\right)^{-1} \cdot \Bigg\{ \boldsymbol{K}_1\left(\boldsymbol{y}^*(k+d) - \boldsymbol{y}(k+d)\right) + \boldsymbol{K}_2 \Delta t \sum_{j=1}^{k+d} \left[\boldsymbol{y}^*(j)\right. \\ &\left. - \boldsymbol{y}(j)\right] - \frac{\partial \boldsymbol{h}}{\partial \boldsymbol{x}} \boldsymbol{f}(\hat{\boldsymbol{x}}(k+d \mid k)) \Bigg\} \end{aligned} \tag{5}$$

To realize this control law, we need to predict the state and output at $k+1, \cdots, k+d$.

3 Neural Network Based State Predication of Nonlinear Systems with Input Time Delay

Since neural network (NN) can possess strong anti-interference ability [7] in multi-step predication at time $k+j$ $(1 \le j \le d)$, we propose

$$\hat{\boldsymbol{x}}_p\left(k+d\middle|k\right)=\boldsymbol{x}_m\left(k+d\mid k\right)+\hat{\boldsymbol{x}}\left(k\middle|k\right)-\boldsymbol{x}_m\left(k\mid k-d\right) \tag{6}$$

where $\boldsymbol{x}_m\left(k+d\mid k\right)$ is the predicted $\boldsymbol{x}\left(k+d\right)$ at time k by NN, $\hat{\boldsymbol{x}}_p\left(k+d\middle|k\right)$ is the corrected $\boldsymbol{x}_m\left(k+d\mid k\right)$. $\hat{\boldsymbol{x}}\left(k\middle|k\right)$ is the state estimation at time k by the STF.

$$\boldsymbol{x}_m\left(k+d\mid k\right)=NNP\left[\hat{\boldsymbol{x}}\left(k\middle|k\right),\boldsymbol{u}\left(k-1\right),\boldsymbol{u}\left(k-2\right),\ldots,\boldsymbol{u}\left(k-M\right)\right] \tag{7}$$

where NNP means neural network predicator.

Substitute $\hat{\boldsymbol{y}}_p\left(k+d\right)=\boldsymbol{h}\left(\hat{\boldsymbol{x}}_p\left(k+d\middle|k\right)\right)$ into (5), the control law can be realized finally by

$$\begin{aligned}\boldsymbol{u}\left(k\right)=&\left(\frac{\partial \boldsymbol{h}}{\partial \boldsymbol{x}}\boldsymbol{g}\left(\hat{\boldsymbol{x}}_p\left(k+d\middle|k\right)\right)\right)^{-1}\cdot\Big\{\boldsymbol{K}_1\left(\boldsymbol{y}^*\left(k+d\right)-\hat{\boldsymbol{y}}_p\left(k+d\right)\right)+\boldsymbol{K}_2\Delta t\\ &\sum_{j=1}^{k}\left[\boldsymbol{y}^*(j)-\boldsymbol{y}(j)\right]+\boldsymbol{K}_2\Delta t\sum_{j=k+1}^{k+d}\left[\boldsymbol{y}^*(j)-\hat{\boldsymbol{y}}_p(j)\right]-\frac{\partial \boldsymbol{h}}{\partial \boldsymbol{x}}\boldsymbol{f}\left(\hat{\boldsymbol{x}}_p\left(k+d\middle|k\right)\right)\Big\}\end{aligned} \tag{8}$$

Feed-forward NN is used for predication in (7), which can be trained with the improved BP algorithm. Samples are given as $\{(\boldsymbol{X}_i,\boldsymbol{Y}_i)\}$ $(i=1,2,\cdots,N)$. To accelerate the convergence of the BP algorithm, momentum factor is introduced as

$$\boldsymbol{W}(k+1)=\boldsymbol{W}(k)+\eta\boldsymbol{G}(k)+\alpha\cdot\Delta\boldsymbol{W}(k) \tag{9}$$

$\boldsymbol{W}\in R^l$ is the weights vector of the NN. η is a constant representing the step size, which can be regulated. Momentum factor α is between 0 and 1, $0<\alpha<1$. $\boldsymbol{G}(k)$ is the negative gradient of error function $E(\boldsymbol{W})$.

$$E\left(\boldsymbol{W}\right)=\frac{1}{2}\sum_{i=1}^{N}\left\|\boldsymbol{Y}_i-\hat{\boldsymbol{Y}}_i\right\|^2,\Delta\boldsymbol{W}(k)=\boldsymbol{W}(k)-\boldsymbol{W}(k-1) \tag{10}$$

4 Fault Tolerant Control of Nonlinear Systems with Input Time Delay

With state augmentation techniques, the system (2) with sensor failures can be modeled by

$$\begin{cases} \boldsymbol{x}_e(k+1) = \boldsymbol{f}_e(k, \boldsymbol{u}(k), \boldsymbol{x}(k), \boldsymbol{y}_0(k)) + \boldsymbol{\Gamma}_e \boldsymbol{v}(k), \\ \boldsymbol{y}(k+1) = \boldsymbol{h}(k+1, \boldsymbol{x}(k+1)) + \boldsymbol{y}_0(k+1) + \boldsymbol{e}(k+1) \end{cases} \tag{11}$$

$$\boldsymbol{f}_e(k, \boldsymbol{u}(k), \boldsymbol{x}(k), \boldsymbol{y}_0(k)) = \begin{bmatrix} \boldsymbol{x}(k) + \Delta t\left[\boldsymbol{f}(\boldsymbol{x}(k)) + \boldsymbol{g}(\boldsymbol{x}(k))\boldsymbol{u}(k-d)\right] \\ \boldsymbol{y}_0(k) \end{bmatrix}$$
$$\boldsymbol{\Gamma}_e = \begin{bmatrix} \boldsymbol{\Gamma} \\ \boldsymbol{0}_{m\times q} \end{bmatrix} \tag{12}$$

$\boldsymbol{y}_0 \in \mathbf{R}^m$ is the deviation of the sensor, is unknown and time-varying, and it can be estimated by the STF. And $\boldsymbol{x}_e(k+1) = \left[\boldsymbol{x}(k+1)^T \quad \boldsymbol{y}_0(k+1)^T\right]^T$.

Based on (11), we can obtain the estimated sensor deviation $\hat{y}_0(k)$ online with the STF [5]. Comparing $\hat{y}_0(k)$ with a threshold vector ε, we can detect possible sensor failures. When a sensor failure is detected at $k = \lambda$, the output of the sensor is not valid anymore. Substitute $\boldsymbol{y}(k)$ in (8) by $\hat{\boldsymbol{y}}(k) = \boldsymbol{h}(\hat{\boldsymbol{x}}(k|k))$, we can finally get the following fault tolerant control against sensor failures

$$\boldsymbol{u}(k) = \left(\frac{\partial \boldsymbol{h}}{\partial \boldsymbol{x}} \boldsymbol{g}(\hat{\boldsymbol{x}}_p(k+d|k))\right)^{-1} \cdot \Big\{\boldsymbol{K}_1(\boldsymbol{y}^*(k+d) - \hat{\boldsymbol{y}}_p(k+d)) + \boldsymbol{K}_2 \Delta t \sum_{j=1}^{\lambda-1}[\boldsymbol{y}^*(j) - \boldsymbol{y}(j)] + \boldsymbol{K}_2 \Delta t \sum_{j=\lambda}^{k}[\boldsymbol{y}^*(j) - \hat{\boldsymbol{y}}(j)] + \boldsymbol{K}_2 \Delta t \sum_{j=k+1}^{k+d}[\boldsymbol{y}^*(j) - \hat{\boldsymbol{y}}_p(j)] - \frac{\partial \boldsymbol{h}}{\partial \boldsymbol{x}} \boldsymbol{f}(\hat{\boldsymbol{x}}_p(k+d|k))\Big\} \tag{13}$$

After all the elements of the sensor deviation vector are less than their thresholds, the control law will be switched back to (8).

5 Simulation Study

A modified mathematical model, coming from the setup DTS200, is adopted for computer simulations. For details, see [8].

The system can be modeled by the following discrete model

$$\begin{cases} \boldsymbol{x}(k+1) = \boldsymbol{x}(k) + \Delta t\left[\boldsymbol{A}(\boldsymbol{x}(k)) + \boldsymbol{B} \cdot \boldsymbol{u}(k - (\tau/\Delta t))\right] \\ \boldsymbol{y}(k+1) = \left[x_1(k+1) \quad x_2(k+1)\right]^{\mathrm{T}} \end{cases} \tag{14}$$

In the control algorithm, we choose $\boldsymbol{K}_1 = 0.05\boldsymbol{I}_2$, $\boldsymbol{K}_2 = 0.0015\boldsymbol{I}_2$. In the STF algorithm [5], we choose $\boldsymbol{Q} = 2\times 0.01\boldsymbol{I}_3$, $\boldsymbol{R} = 0.01\boldsymbol{I}_2$, $\beta = 5$, $\boldsymbol{\alpha} = diag[1,1,1]$. The initial liquid levels $h_1 = 25\text{cm}$, $h_2 = h_3 = 10\text{cm}$; time-delay $\tau = 10\text{s}$; sampling interval is $\Delta t = 1\text{s}$, and the simulation time is 800s. At the time of 550s, change the set-point of T2 to 14cm; at the time of 600s, change the set-point of T1 to 18cm.

The neural network used for predication is described as (7), where M=3. The input layer has 9 neurons, the hidden layer has 7 neurons and the output layer has 3 neurons. We adopt improved BP algorithm as NN's training algorithm. This BP NN has 3 layers. Choose the Sigmoid function $f\left(x\right)=1/\left(1+e^{-x}\right)$ to describe the neurons.

When $1 \le k < 10$, $\boldsymbol{u}(k)=0$; $10 < k \le 100$, $\boldsymbol{u}(k)=[100*rand(1),100*rand(1)]^{T}$, where $rand(1)$ represent a random number between 0 and 1. With open-loop system (14), we can get 100 samples for the NN's training.

Two initial weights vectors are produced randomly after we get the samples for training. Adjust the weights with (9), in which $\eta=0.7$, $\alpha=0.6$.

Train the NN with the algorithm introduced in Section 3 until $E(\boldsymbol{W})<0.1$. Then we get two weighting matrices, they are not shown due to the limit of space.

When the system runs normally, the simulation results is shown in Fig.1, where curves 1 and 3 represent the real liquid levels of T1 and T2 respectively and curves 2 and 4 represent the predicted results of the liquid levels. This simulation result shows that we can predict the system's state quite well with the NN algorithm.

When the system is under sensor failure, $\boldsymbol{y}\left(k+1\right)=\boldsymbol{F}\left[\boldsymbol{x}_1\left(k+1\right)\ \ \boldsymbol{x}_2\left(k+1\right)\right]^{\mathrm{T}}$, where $\boldsymbol{F}=\begin{bmatrix}0 & 0\\ 0 & 1\end{bmatrix}$ is a matrix representing the sensor failure. As an example, we suppose the sensor failure happens between the time of 250s and 350s. The sensor failure can be detected by comparing $\left|\hat{y}_0\left(k\right)\right|$ with the threshold ε_1=1, ε_2=0.5.

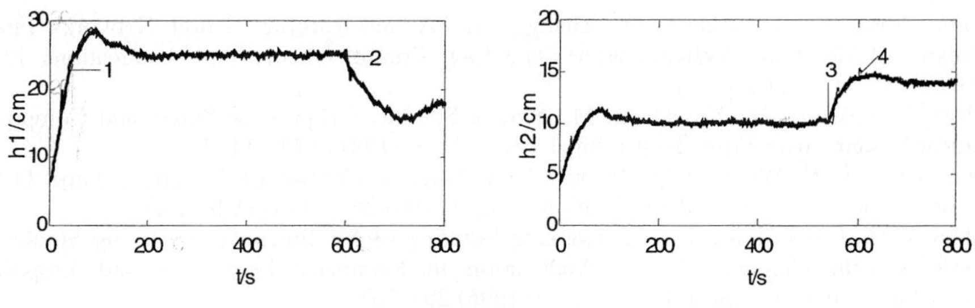

Fig. 1. NN prediction results in normal case

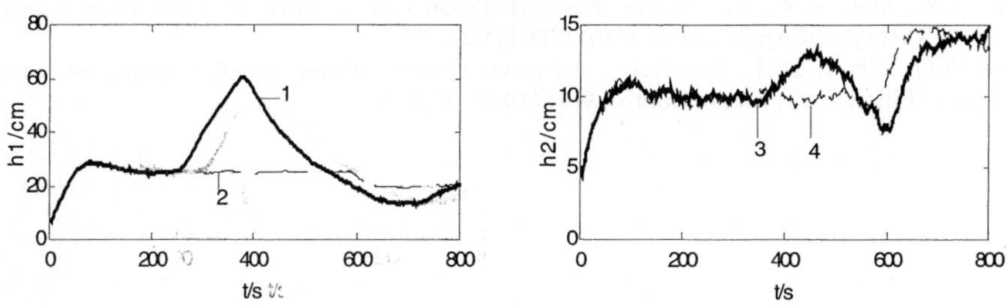

Fig. 2. Simulation results under sensor failure F

In Fig.2, curves 1 and 3 show the results when the proposed fault tolerant control law is not adopted under sensor failure $\boldsymbol{F}$. It is shown that great fluctuation of the liquid level of T1 occurs. Curves 2 and 4 show the results when the proposed fault tolerant control law is adopted under sensor failure $\boldsymbol{F}$. It is clearly that failure $\boldsymbol{F}$ does not have much effect on the liquid levels of the tanks.

Simulation results are also satisfactory under sensor failure on $\boldsymbol{x}_2$, the figures are omitted due to space limitation.

6 Conclusions

In this paper, an active fault tolerant control law against sensor failures for a class of nonlinear time delay systems is proposed on the basis of the STF and a NN predictor. Simulation results illustrate the effectiveness of the proposed approach.

Acknowledgement. This work was supported by the NSFC (Grant No. 60025307, 60234010), and the national 973 program (Grant No. 2002CB312200) of China.

References

1. Zhou, D.H., Frank, P.M.: A Real-time Estimation Approach to Time-varying Time Delay and Parameters of NARX Processes. Computers and Chemical Engineering 23(11-12) (2000) 1763–1772
2. Chen, S.B., Wu, L., Zhang, Q., Zhang, F.E.: A Self-learning Neural Networks Fuzzy Control of Uncertain Systems with Time Lag. Control Theory and Applications 13(3) (1996) 347-355 (in Chinese)
3. Mori, T., Fukuma, N., Kuwahara, M.: Simple Stability Criteria for Single and Composite Linear Systems with Time-delays. Int. J. contr. 34(6) (1981) 1175-1184
4. Sun, J.S., Li, J., Wang, Z.Q.: Robust Fault-Tolerant Control of Uncertain Time Delay System. Control Theory and Applications 15(2) (1998) 267-271 (in Chinese)
5. Zhou, D.H., Frank, P.M.: Strong Tracking Filtering of Nonlinear Time-varying Stochastic Systems with Coloured Noise: Application to Parameter Estimation and Empirical Robustness Analysis. Int. J. Contr. 65(2) (1996) 295-307
6. Lee, P.L., Sullivan, G.R.: Generic Model Control (GMC). Comput. Chem. Engng. 12(6) (1998) 573-580
7. Xu, X.Y., Mao, Z.Y.: The Neural Network Predicative Control of Time-delay Systems. Control Theory and Applications 18(6) (2001) 932-935
8. Xie, X.Q., Zhou, D.H., Jin, Y.H.: Adaptive Generic Model Control Based on Strong Tracking Filter. J. of Process Control 9(4) (1999) 337-350

Run-to-Run Iterative Optimization Control of Batch Processes Based on Recurrent Neural Networks

Zhihua Xiong[1], Jie Zhang[2], Xiong Wang[1], and Yongmao Xu[1]

[1] Department of Automation, Tsinghua University, Beiijng, 100084, P.R. China,
{zhxiong, wangxiong, xym-dau}@tsinghua.edu.cn

[2] School of Chemical Engineering and Advanced Materials, University of Newcastle, Newcastle upon Tyne, NE1 7RU, UK
jie.zhang@ncl.ac.uk

Abstract. Recurrent neural network (RNN) is used to model product quality of batch processes from process operational data. Due to model-plant mismatches and unmeasured disturbances, the calculated control policy based on the RNN model may not be optimal when applied to the actual process. Model prediction errors from previous runs are used to improve RNN model predictions for the current run. It is proved that the modified model errors are reduced from run to run. Consequently control trajectory gradually approaches the optimal control policy. The proposed scheme is illustrated on a simulated batch reactor.

1 Introduction

It is very important for good optimal control performance of batch processes to obtain an accurate model capable of providing accurate long range predictions [1]. To overcome the difficulties in developing mechanistic models, empirical models based on process operational data can be utilized. Neural networks have been shown to be able to approximate any continuous nonlinear functions and have been proposed as a promising tool for identifying empirical models [2]. If properly trained and validated, these neural network models can be used to predict steady-state and dynamic process behavior reasonably well [3]. As a non-linear regression tool, neural networks have been increasingly used in modeling and control of chemical processes, especially complex non-linear processes where process understanding is limited. In this paper, recurrent neural networks (RNN) [4] are used to represent the non-linear relationship between product qualities and control trajectory of batch processes.

In practice, the effort of batch process optimization is often hampered by model-plant mismatches and unknown disturbances. "Optimal on the model" may not necessarily mean "optimal on the process". Since batch processes are of repetitive nature, it would be possible to use information of previous runs to improve the operation of the current run. This is referred to as "run-to-run" or "batch-to-batch" optimization [5,6]. Recently, iterative learning control (ILC) has been introduced to directly update input trajectory [7,8]. But the ILC type approach can not be used directly to product quality control of batch processes, because it is usually more difficult to set the reference trajectories of product qualities during a whole batch run. However, the main interest in batch process is on the final product qualities and/or quantities, and the desired values of product qualities at the end of a batch are usually

F. Yin, J. Wang, and C. Guo (Eds.): ISNN 2004, LNCS 3174, pp. 97–103, 2004.

known. This makes it still possible to improve the product qualities from run to run. In this study, the idea of ILC is combined with iterative optimization to improve product quality from run to run. RNN model predictions are iteratively modified by using errors of RNN model during previous runs. Updated control policy is calculated for each run using the modified model predictions. By such a means, model errors based on modified predictions are gradually reduced with respect to the number of batch runs and control policy gradually approaches the optimal trajectory.

2 Batch Process Modeling Using Recurrent Neural Networks

A batch process is considered here where its run length (t_f) is fixed and divided into N equal intervals, and input and product quality sequences are defined as $\mathbf{U}_k=[u_k(0),\ldots, u_k(N\text{-}1)]^T$, $\mathbf{Y}_k=[y_k(1),\ldots, y_k(N)]^T$, where k is the batch run index, $y\in R^n$ are product quality variables, $u\in R^m$ are the input (manipulted) variables for the product quality. In this study, RNN models are used to model the non-linear relationship between u_k and y_k as shown in Fig 1 for modeling a second order nonlinear system.

$$\hat{y}_k(t+1) = \text{RNN}\,[\,\hat{y}_k(t), \hat{y}_k(t\text{-}1), u_k(t), u_k(t\text{-}1)\,] \tag{1}$$

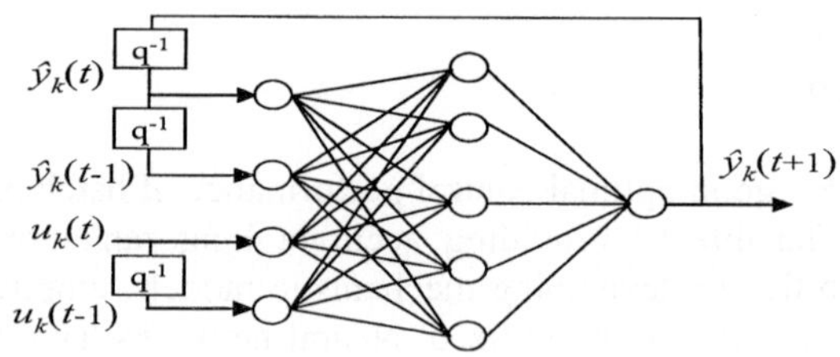

Fig. 1. A globally recurrent neural network. The neuron activation function in the hidden layer is sigmoidal function whilst in the output layer it is linear. The lagged network output $\bullet_k(t+1)$ is fed back to network input nodes as indicated by the back-shift operator q^{-1}. In this way, dynamics are introduced into the network.

Given the initial conditions and the input sequence $\mathbf{U}_k$, a RNN model can predict recursively output at the end of a batch, which can be formulated as $\hat{y}_k(t_f) = f_{RNN}(\mathbf{U}_k)$, where f_{RNN} is a nonlinear function related to the RNN model. Thus the predictions from a RNN model are long range or multi-step-ahead predictions. The RNN network can be trained in the "back-propagation through time" fasion [9] using the Levenberg-Marquart optimization algorithm to minimize its long-range prediction errors. Therefore a RNN model can usually offer much better long-range predictions than a feed-forward neural network [4]. However, due to model-plant mismatches, prediction offsets of RNN models can occur. Since batch processes are intended to be run repeatedly, it is reasonable to correct the RNN model predictions by adding filtered model errors of previous run [10]. As the dynamics of product quality variables are usually non-linear and measurement noises always exist, information of prediction errors in all previous runs should be utilized in deciding the input change for the next run. In this study, the average model errors $\bar{\hat{e}}_k(t_f)$ of all previous runs are used to modify the RNN model predictions for the current run, which is defined as

$$\bar{\hat{e}}_k(t_f)=\frac{1}{k}\sum_{i=1}^{k}\hat{e}_i(t_f)=\frac{1}{k}\sum_{i=1}^{k}(y_i(t_f)-\hat{y}_i(t_f)) \tag{2}$$

where $\hat{e}_i(t_f)$ is RNN model error, $y_i(t_f)$ and $\hat{y}_i(t_f)$ are the measured and predicted product qualities at the end time t_f of the ith batch, respectively. By filtering this average model error, the modified prediction of a RNN model is defined as

$$\tilde{y}_{k+1}(t_f)=\hat{y}_{k+1}(t_f)+\alpha\bar{\hat{e}}_k(t_f) \tag{3}$$

where α is an adjustable filter parameter.

3 Run-to-Run Iterative Optimization Control

The aim of run-to-run iterative optimization control is to find an update mechanism for the input sequence $\mathbf{U}_{k+1}$ of a new run based on the information from previous runs so that the measured $y_{k+1}(t_f)$ at the batch end converges asymptotically towards the desired $y_d(t_f)$. When the input degrees of freedom are deficient, the offsets are minimized as dictated by a quadratic objective for run-to-run optimization [7,8]. The run-to-run iterative optimization for product quality control can be formulated as

$$\min_{\mathbf{U}_{k+1}} J=\| y_d(t_f)-\tilde{y}_{k+1}(t_f)\|^2_Q+\|\mathbf{U}_{k+1}-\mathbf{U}_k\|^2_R \tag{4}$$

where $Q=\lambda_q\cdot\mathbf{I}_n$ and $R=\lambda_r\cdot\mathbf{I}_N$ are weighting matrices. The quadratic objective in Eq(4) penalizes the input change instead of the input, and it has an integral action for the input with respect to the run index k and achieves the minimum achievable error in the limit [8].

The modified prediction error at the end time t_f between measured product quality and modified prediction is calculated by $\varepsilon_{k+1}(t_f)=y_{k+1}(t_f)-\tilde{y}_{k+1}(t_f)$. Considering Eq(3), it can be rewritten as $\varepsilon_{k+1}(t_f)=\hat{e}_{k+1}(t_f)-\alpha\bar{\hat{e}}_k(t_f)$. Eq(2) can be reformulated as

$$\bar{\hat{e}}_k(t_f)=\frac{k-1}{k}\bar{\hat{e}}_{k-1}(t_f)+\frac{1}{k}\hat{e}_k(t_f) \tag{5}$$

Then $\varepsilon_{k+1}(t_f)$ can be rewritten further as

$$\varepsilon_{k+1}(t_f)=\frac{k-1}{k}\varepsilon_k(t_f)+\eta_{k+1}(t_f) \tag{6}$$

$$\eta_{k+1}(t_f)=\hat{e}_{k+1}(t_f)-\frac{k-1+\alpha}{k}\hat{e}_k(t_f) \tag{7}$$

As k approaches infinity, it indicates in Eq(5) that

$$\lim_{k\to\infty}\bar{\hat{\mathbf{e}}}_k(t)=\lim_{k\to\infty}\frac{k-1}{k}\bar{\hat{\mathbf{e}}}_{k-1}(t)+\lim_{k\to\infty}\frac{1}{k}\hat{\mathbf{e}}_k(t)=\bar{\hat{\mathbf{e}}}_\infty(t) \tag{8}$$

The RNN model predictions at the end of the (k+1)th batch run can be approximated by the first order Taylor series approximation as

$$\hat{y}_{k+1}(t_f) = f_{RNN}(\mathbf{U}_{k+1}) = f_{RNN}(\mathbf{U}_k) + \left[\left. \frac{\partial f_{RNN}}{\partial \mathbf{U}} \right|_{\mathbf{U}_k} \right]^T \Delta \mathbf{U}_{k+1} \tag{9}$$

where $\Delta\mathbf{U}_{k+1}=\mathbf{U}_{k+1}-\mathbf{U}_k$. It is reasonable to assume that the iterative optimization produces no worse solutions since the trivial solution $\Delta\mathbf{U}_{k+1}=0$ can ensue this. Under this assumption, it holds $0 \le \|y_d(t_f) - \tilde{y}_{k+1}(t_f)\|^2_Q \le \|y_d(t_f) - \tilde{y}_k(t_f)\|^2_Q$. As k approaches infinity, it follows $\lim_{k\to\infty} \tilde{y}_{k+1}(t_f) = \lim_{k\to\infty} \tilde{y}_k(t_f)$. From Eq(3) and Eq(9), it holds that

$$\begin{aligned} \lim_{k\to\infty} \tilde{y}_{k+1}(t_f) &= \lim_{k\to\infty} [\hat{y}_{k+1}(t_f) + \alpha \bar{\hat{e}}_k(t_f)] = \lim_{k\to\infty} [f_{RNN}(\mathbf{U}_k) + \left[\left. \frac{\partial f_{RNN}}{\partial \mathbf{U}} \right|_{\mathbf{U}_k} \right]^T \Delta\mathbf{U}_{k+1} + \alpha \bar{\hat{e}}_k(t_f)] \\ &= \lim_{k\to\infty} \tilde{y}_k(t_f) + \lim_{k\to\infty} \left[\left. \frac{\partial f_{RNN}}{\partial \mathbf{U}} \right|_{\mathbf{U}_k} \right]^T \Delta\mathbf{U}_{k+1} \end{aligned} \tag{10}$$

It follows from the above equation that

$$\lim_{k\to\infty} \Delta\mathbf{U}_{k+1} = 0, \quad \text{i.e.} \quad \lim_{k\to\infty} \mathbf{U}_{k+1} = \lim_{k\to\infty} \mathbf{U}_k = \mathbf{U}_\infty \tag{11}$$

Therefore the iterative optimal control scheme converges with respect to the batch run number k. It also indicates from Eq(7) that

$$\lim_{k\to\infty} \eta_{k+1}(t_f) = \lim_{k\to\infty} \hat{e}_{k+1}(t_f) - \lim_{k\to\infty} \frac{k-1+\alpha}{k} \hat{e}_k(t_f) = \hat{e}_\infty(t_f) - \hat{e}_\infty(t_f) = 0 \tag{12}$$

Therefore from Eq(6), it holds that

$$\lim_{k\to\infty} \varepsilon_{k+1}(t_f) = \lim_{k\to\infty} \frac{k-1}{k} \varepsilon_k(t_f) + \lim_{k\to\infty} \eta_{k+1}(t_f) = \varepsilon_\infty(t_f) \tag{13}$$

The procedure of run-to-run model-based iterative optimization is outlined as follows: At the current run k, input trajectory $\mathbf{U}_k$ is implemented into the batch process and outputs $y_k(t_f)$ are obtained after the completion of batch. RNN model predictions for the next batch are modified by using prediction errors of all previous runs. Based on the modified predictions, the optimization problem is solved again and a new control policy $\mathbf{U}_{k+1}$ for the next run is calculated. This procedure is repeated from run to run.

4 Simulation on a Batch Reactor

The typical batch reactor is taken from [11]. Its reaction scheme is $A \xrightarrow{k_1} B \xrightarrow{k_2} C$, and the equations describing the reactor are:

$$\begin{aligned} \frac{dx_1}{dt} &= -4000\exp(-2500/T)x_1^2 \\ \frac{dx_2}{dt} &= 4000\exp(-2500/T)x_1^2 - 6.2\times10^5\exp(-5000/T)x_2 \end{aligned} \tag{14}$$

where x_1 and x_2 represent the dimensionless concentrations of A and B, T is the temperature of the reactor and is scaled to dimensionless value as $T_d = (T-T_{min})/(T_{max}-T_{min})$, t_f is fixed to be 1.0 h, and the values of $T_{\min}$ and $T_{\max}$ are 298K and 398K, respectively. The initial conditions are $x_1(0)=1$ and $x_2(0)=0$, and the constraint on the control $u=T_d$ is $0 \le T_d \le 1$. The performance index is to maximise the concentration of B at the end of batch $x_2(t_f)$. In this study, the above mechanistic model, Eq(14), is assumed to be not available for process optimization control and a recurrent neural network model is utilized to model the non-linear relationship between $y=x_2$ and u.

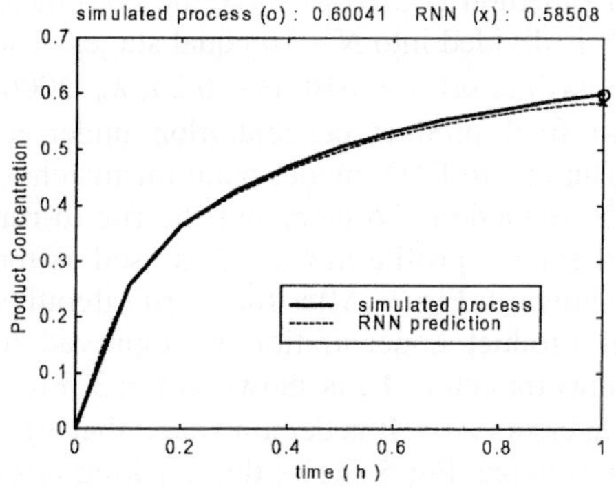

Fig. 2. Long-range predictions of RNN model

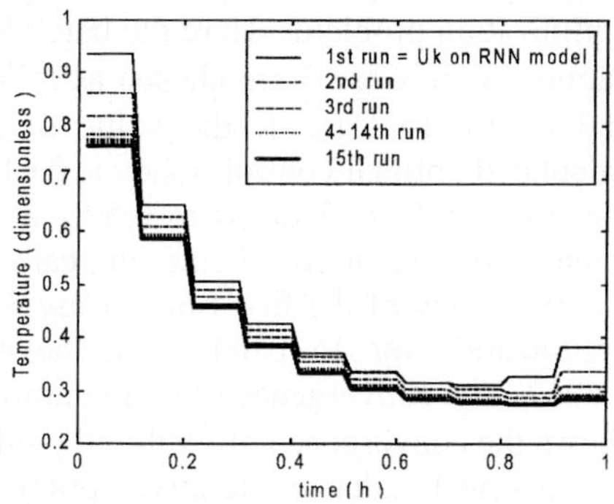

Fig. 3. Convergence of input trajectory $\mathbf{U}_k$

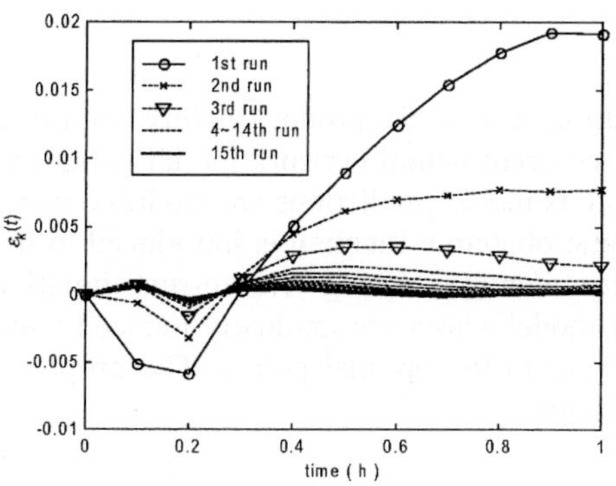

Fig. 4. Convergence of modified prediction error $\varepsilon_k(t)$

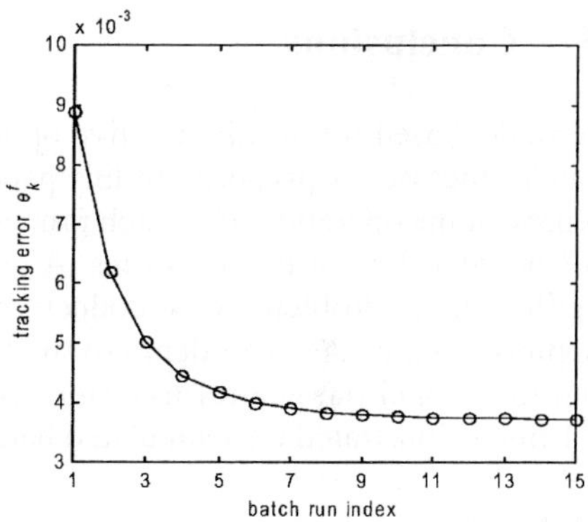

Fig. 5. Convergence of tracking error e_k^f

Fifteen batches of process operation under different feeding policies were simulated from the mechanistic model. Normally distributed random noises with zero means were added to all the "measurements" to simulate the effects of measurement noises. Among these data, 10 batches were used as training data, 4 batches were used as validation data, and the remaining 1 batch was used as unseen testing data to evaluate the developed network. Different networks were trained on the training data and determined by cross-validation. The network with the least sum of squared errors on the validation data was chosen as the best network. After RNN training and validation, the final selected model is of the form as Eq (1). The appropriate number of hidden neurons was found through cross-validation as 8. Long-range predictions of y from this RNN model on the unseen testing batch are shown in Fig 2. It can be seen that the prediction errors increase as the prediction horizon increases. Neverthless, the

RNN model prediction at the end of the batch can be considered as being quite accurate.

To investigate the performance of the proposed control strategy, three cases are studied: Case 1 – mechanistic model based optimization; Case 2 – only RNN model based optimization; and Case 3 – RNN model based run-to-run iterative optimization. The sequential quadratic programming (SQP) method is used to solve the non-linear optimization problems. Here the batch length is divided into $N = 10$ equal stages. The parameters in Case 3 are chosen as follows: $y_d(t_f)$ is set to 0.610, $\alpha = 0.25$, λ_q=2000, and λ_r=10. In Case 1, the value of actual final product concentration under its calculated optimal control policy is 0.6105. But due to RNN model-plant mismatches, the value in Case 2 drops to 0.596, a 2.38% reduction. To carry out the run-to-run iterative optimization scheme, the calculated control profile in Case 2 is used as the control policy of the first run in Case 3, as shown in Fig 3. After run-to-run iterative optimization for 15 batch runs, the actual product concentration is improved to 0.6062. The convergence of temperature (input) trajectory $\mathbf{U}_k$ is shown in Fig 3. Fig 4 shows the convergence of modified prediction errors $\varepsilon_k(t)$. It is demonstrated that $\varepsilon_k(t_f)$ is converged under the iterative optimization scheme. Fig 5 shows that tracking error $e_k^f = y_d(t_f) - y_k(t_f)$ is gradually reduced from run to run. It can be seen that after 9 batches e_k^f is lower than 0.0038.

5 Conclusions

A model-based run-to-run iterative optimization scheme for product quality control in batch processes is proposed in this paper. Recurrent neural network model is built to represent the operation of a batch process and its model predictions are modified using prediction errors of previous runs. A quadratic objective function is introduced to the optimization problem of product quality control. Using run-to-run iterative optimization, it has been demonstrated that model errors are gradually reduced from run to run and the control trajectory converges to the optimal policy. The proposed scheme is illustrated on a simulated batch reactor.

References

1. Bonvin, D.: Optimal Operation of Batch Reactors – a Personal View. J. Proc. Cont. **8** (1998) 355-368
2. Cybenko, G.: Approximation by Superpositions of a Sigmoidal Function. Math. Control Signal Systems, **2** (1989) 303-314
3. Sjoberg, J., Zhang, Q., Ljung, L., et al.: Nonlinear Black-box Modeling in System Identification: a Unified Overview. Automatica **31** (1995) 1691-1724
4. Tian, Y., Zhang, J., Morris, A.J.: Optimal Control of a Batch Emulsion Copolymerisation Reactor Based on Recurrent Neural Network. Chem. Engng. Processing **41** (2002) 531-538
5. Srinivasan B., Primus, C.J., Bonvin, D., Ricker, N.L.: Run-To-Run Optimization via Control of Generalized Constraints. *Control Engineering Practice* **9** (2001) 911-919
6. Dong, D., McAvoy, T.J., Zafiriou, E.: Run-to-Run Optimization Using Neural Network Models. *Ind. Eng. Chem. Res.* **35** (1996) 2269-2276
7. Amann, N., Owens, D.H., Rogers, E.: Iterative Learning Control for Discrete-time System with Exponential Rate of Convergence. IEE Proc.D-Contr. Theo. Appl. **143** (1996)217-224

8. Lee, J.H., Lee, K.S., Kim, W.C.: Model-Based Iterative Learning Control with a Quadratic Criterion for Time-Varying Linear Systems. *Automatica* **36** (2000) 641-657
9. Werbos, P.J.: Backpropagation through Time: What It Does and How To Do It. Proc. IEEE. **78** (1990) 1550-1560
10. Doyle III, F.J., Harrison, C.A., Crowley, T.J.: Hybrid Model-based Approach to Batch-to-Batch Control of Particle Size Distribution in Emulsion Polymerization. Computers Chem. Engng. **27** (2003) 1153-1163
11. Ray, W.H.: Advanced Process Control. McGraw-Hill: New York (1981)

Time-Delay Recurrent Neural Networks for Dynamic Systems Control

Xu Xu[1,2], Yinghua Lu[3,4], and Yanchun Liang[3*]

[1] College of Mathematics, Jilin University, Changchun 130012, China
[2] Institute of Vibration Engineering Research, Nanjing University of Aeronautics and Astronautics, Nanjing 210016, China
[3] College of Computer Science and Technology, Jilin University, Changchun 130012, China
*Corresponding author: ycliang@email.jlu.edu.cn
[4] College of Computer Science, Northeast Normal University, Changchun 130024, China

Abstract. A time-delay recurrent neural network (TDRNN) model is presented. TDRNN has a simple structure but far more "depth" and "resolution ratio" in memory. A TDRNN controller for dynamic systems is proposed. A dynamic recurrent back-propagation algorithm is developed and the optimal adaptive learning rates are also proposed to guarantee the global convergence. Numeral experiments for controlling speeds of ultrasonic motors show that the TDRNN has good effectiveness in identification and control for dynamic systems.

1 Introduction

Artificial neural network (ANN) has been applied widely in many fields [1,2]. "Depth" and "resolution ratio" in memory are main characteristics to scale dynamic performance of neural network [3]. "Depth" denotes how far information can be memorized; "resolution ratio" denotes how much information in input series of neural network can be held.The popular neural networks have much defect on dynamic performance [4]. This paper proposed a novel time-delay recurrent network model which has far more "depth" and "resolution ratio" in memory for dynamic systems control. The proposed control scheme is examined by the numerical experiments for controlling speeds of ultrasonic motors.

2 Time-Delay Recurrent Neural Network (TDRNN)

Denote the outputs of input nodes as $\boldsymbol{I} = \{I_i\}_{i=1}^{N}$, the outputs of the context neurons as $\mathbf{z} = \{z_i\}_{i=1}^{N}$, the outputs of the hidden neurons as $\boldsymbol{x} = \{x_i\}_{i=1}^{H}$, and the output of the neural network as $\mathbf{u} = \{u_i\}_{i=1}^{R}$. The output of the context neuron in time k is

$$z_i(k) = b_i(k) + \lambda z_i(k-1), \quad z_i(0) = 0, \qquad (i = 1, 2, ..., N) \tag{1}$$

F. Yin, J. Wang, and C. Guo (Eds.): ISNN 2004, LNCS 3174, pp. 104–109, 2004.

where $b_i(k) = I_i(k) + \alpha I_i(k-1) + ... + \alpha I_i(k-M)$, $\alpha + \lambda = 1$ and M is the step number of time delay. Taking expansion for $z_i(k-1)$, $z_i(k-2)$, ... etc. by using Eq. (1), we have

$$z_i(k) = b_i(k) + \lambda b_i(k-1) + ... + \lambda^{k-1} b_i(1) = A(k) + B(k) \tag{2}$$

where $A(k) = I_i(k) + I_i(k-1) + ... + I_i(k-M)$, $B(k) = \sum_{l=n}^{k-2} \lambda^{l-M+1} b_i(k-l-1)$.

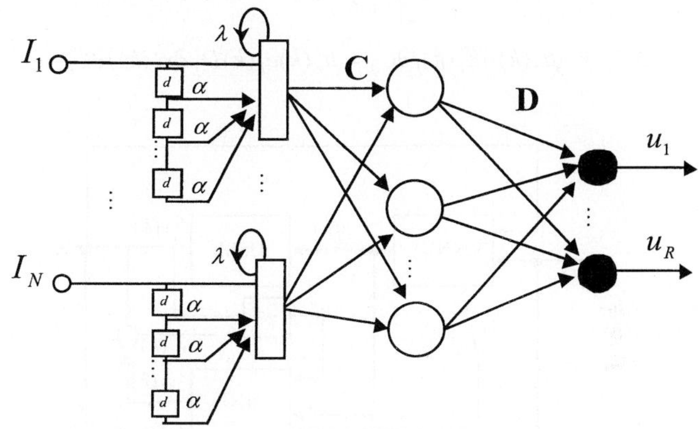

Fig. 1. Time delay recurrent neural network structure proposed in this paper

Although introducing some context and delay neurons in TDRNN, the weights have not increased. Furthermore, from the term $A(k)$ and $B(k)$ in Eq. (2) it can be seen that the context neurons includes all previous inputs, it has far more memory "depth". In additon, context neurons memory accurately the inputs from *k-M* to *k* time, this is quite different from the memory performance of popular recurrent neural network. If *M* is moderate large, the TDRNN has high more memory "resolution ratio". Good dynamic memory performances of TDRNN render it fit to apply to dynamic systems.

3 Time-Delay Recurrent Neural Networks Based Control System

In this paper, an approach of identification and control of dynamic system using TDRNN is presented. For simplicity, the dynamic system is assumed to be a single input-output system. Both NNI and NNC use the same TDRNN architecture shown in Fig.1. The block diagram of dynamic system control proposed in this paper is shown in Fig. 2.

3.1 Dynamic Training Algorithm of NNI

The outputs of the neurons in the hidden and output layers of the NNI are

$$x_j(k) = f(\sum_{i=1}^{n} C_{ij}(k) z_i(k)) \ (j = 1, 2, ..., H), \quad y_I(k) = \sum_{j=1}^{H} D_j(k) x_j(k) \tag{3}$$

The error function is defined as $E_I(k) = e_I^2(k)/2$, let $\mu_D(k)$ be the learning rate for the weight vector $\mathbf{D}$ and $\mu_c(k)$ be the learning rate for the weights $\mathbf{C}$. The modification of the weight of NNI are

$$\Delta D_j = -\mu_d(k) \partial E_I(k)/\partial D_j = \mu_d(k) e_I(k) \partial y_I(k)/\partial D_j \tag{4}$$

$$\Delta C_{ij} = -\mu_c(k) \partial E_I(k)/\partial C_{ij} = \mu_c(k) e_I(k) D_j \partial x_j(k)/\partial C_{ij} \tag{5}$$

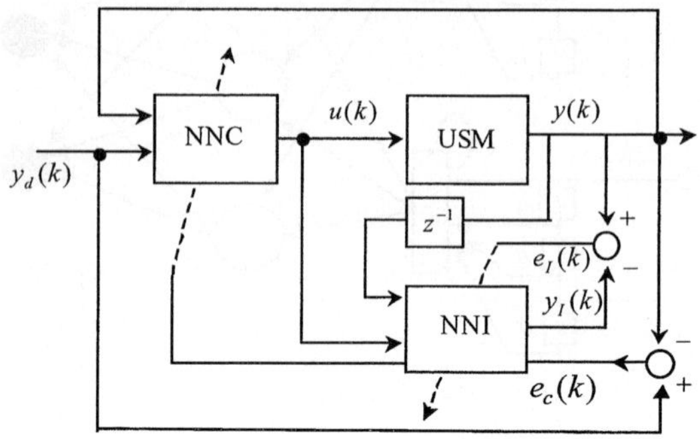

Fig. 2. Block diagram of the control scheme

Supposing that the sampling interval is small enough, then we have

$$\partial x_j(k)/\partial C_{ij} = f_j'(k) z_i(k) \approx f_j'(k) b_i(k) + \lambda \partial x_j(k-1)/\partial C_{ij} \tag{6}$$

According to the B-P algorithm, the update rules of the weights of the NNI are

$$D_j(k) = D_j(k-1) + \Delta D_j(k),\ C_{ij}(k) = C_{ij}(k-1) + \Delta C_{ij}(k),\ (\ i = 1,..., N;\ j = 1,..., H) \tag{7}$$

3.2 Dynamic Training Algorithm of NNC

Denote the outputs of the input nodes in the NNC as $\boldsymbol{I}^c(k) = \{I_i^c(k)\}_{i=1}^{N^c}$, the outputs of the context neurons as $z^c(k) = \{z_i^c(k)\}_{i=1}^{N^c}$, the input of the ith context neuron as $\mathbf{b}_i^c(k) = \{b_i^c(k)\}_{i=1}^{N^c}$, and the outputs of the hidden neurons as $\boldsymbol{x}^c = \{x_i^c(k)\}_{i=1}^{H^C}$. The inputs to the NNC are the reference input $y_d(k)$ and the previous system output $y(k-1)$. According the the B-P algorithm, we have

$$B_j(k) = B_j(k-1) + \Delta B_j(k),\ A_{ij}(k) = A_{ij}(k-1) + \Delta A_{ij}(k),\ (\ i = 1,...N^c;\ j = 1,..., H^c) \tag{8}$$

where $\Delta B_j(k) = \mu_B(k)e_c(k)Y_u x_j^c(k)$, $\Delta A_{ij}(k) = -\mu_A(k)e_c(k)Y_u B_j(k)\partial x_j^c(k)/\partial A_{ij}$, $\mu_B(k)$ is the learning rate for the weight vector B associated with the hidden and output neurons, $\mu_A(k)$ is the learning rate for the weights A associated with the context and hidden neurons. $Y_u = \sum_{j=1}^{H} D_j(k)f_j'(k)C_{1j}(k)$, and the value of $\partial x_j^c(k)/\partial A_{ij}$ satisfies that $\partial x_j^c(k)/\partial A_{ij} = f_j'(k)b_i^c(k) + \lambda \partial x_j^c(k-1)/\partial A_{ij}$.

3.3 Convergence Analysis

Theorem 1: Suppose that the modification of the weights of the NNI is determined by Eq.(7). If learning rate $\mu_D(k)$ is adopted as $\mu_D = \lambda_D E_I(1+E_I)/\|\partial E_I/\partial \mathbf{D}\|^2$ and $\mu_c(k)$ is adopted as $\mu_c = \lambda_c E_I(1+E_I)/\|\partial E_I/\partial \mathbf{C}\|^2$, then global convergence of the update rules (7) is guaranteed. Where $\lambda_D > 0$ $\lambda_C > 0$ are the minimal learning rates of weights **D** and **C,** respectively. $\|\cdot\|$ represents the Euclidean norm.

Proof: Because the error of identification is determined by the weights **D** and **C**, the error during the learning process can be represented as $E_I = E_I(\mathbf{C},\mathbf{D})$, then we have $d\mathbf{C}/dt = -\mu_c \partial E_I/\partial \mathbf{C}$ and $d\mathbf{D}/dt = -\mu_D \partial E_I/\partial \mathbf{D}$, so

$$\begin{aligned}\frac{dE_I}{dt} &= \left[\frac{\partial E_I}{\partial \mathbf{C}}\right]^T \frac{d\mathbf{C}}{dt} + \left[\frac{\partial E_I}{\partial \mathbf{D}}\right]^T \frac{d\mathbf{D}}{dt} = -\mu_C\left\|\frac{\partial E_I}{\partial \mathbf{C}}\right\|^2 - \mu_D\left\|\frac{\partial E_I}{\partial \mathbf{D}}\right\|^2 \\ &= -(\lambda_c + \lambda_D)E_I(1+E_I) = -\lambda_I E_I(1+E_I)\end{aligned} \tag{9}$$

Eq. (9) can be written as $dE_I/(E_I(1+E_I)) = -\lambda_I dt$. Let E_0 is the identified error at initial time, then integrating it we obtain that $E_I/(1+E_I) = E_0\exp(-\lambda_I t)$. It can be seen that when $t \to \infty$, E_I is such that $E_I \to 0$. According to the Lyapunov stability theory, we have shown that the identified error converges to the zero point as $t \to \infty$. This completes the proof of Theorem 1.

Theorem 2: Suppose that the modification of the weights of the NNC is determined by Eq. (8). If $\mu_B = \lambda_B E_c(1+E_c)/\|\partial E_c/\partial \mathbf{B}\|^2$ and $\mu_A = \lambda_A E_c(1+E_c)/\|\partial E_c/\partial \mathbf{A}\|^2$, global convergence of the update rules (8) is guaranteed.

Proof: Define control error of system is $E_c(k) = e_c^2(k)/2$, in a similar way to the proof of Theorem 1, we can complete the proof of this theorem.

4 Numerical Simulation Results and Discussions

Numerical simulations are performed based on the proposed scheme in this paper for the speed control of a longitudinal oscillation USM [5].

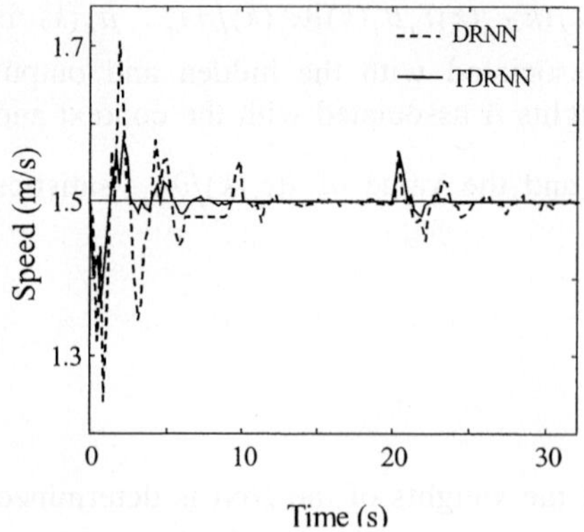

Fig. 3. Control curve comparison

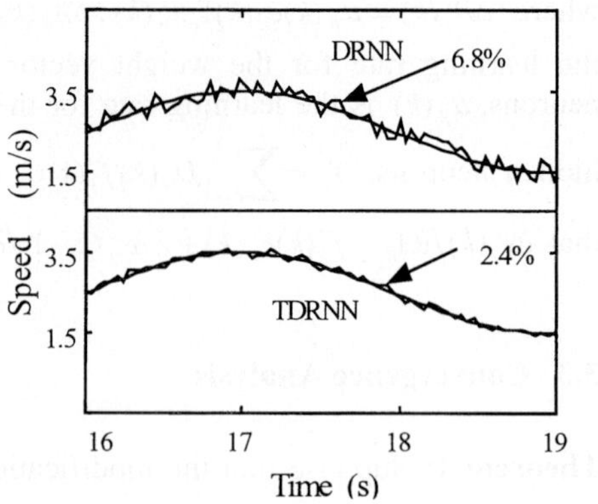

Fig. 4. Fluctuation comparison of control curves

Figure 3 shows the speed control curves using the scheme base on DRNN [6] and the proposed scheme in this paper. From the figure it can be seen that the time of convergence using the proposed method is much less than that using the method base on DRNN. In addition when a parameter of the USM varies suddenly, the large fluctuation can be eliminated after a short time control using this scheme.

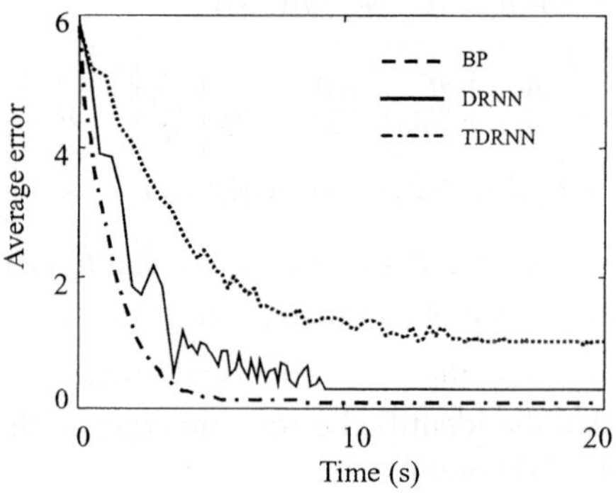

Fig. 5. Average errors using different methods

Figures 4 show the case of the fluctuation using the method base on DRNN [6] and the proposed scheme in this paper, where the fluctuation is defined as $\zeta = (V_{max} - V_{min}) / V_{aver} \times 100\%$, and V_{max}, V_{min} and V_{aver} represent the maximum, minimum and average values of the speeds. It can be seen that the control precision can be increased around 3 times when the proposed method is employed.

Figure 5 shows the comparison of the average errors using different neural networks in the control system. From the comparison it can be seen that the time and precision of convergence using the proposed method is much superior to that obtained using the methods based on DRNN and BP network.

5 Conclusions

This paper constructs a time-delay recurrent neural network with better performance in memory than popular delay neural networks and recurrent networks. A TDRNN controller is utilized for controlling dynamic systems. A dynamic recurrent back propagation algorithm and the optimal adaptive learning rates are presented to guarantee the global convergence. Numeral experiments for controlling speeds of ultrasonic motors show that the TDRNN has good effectiveness in the identification and control for dynamic systems.

Acknowledgments. The authors are grateful to the support of the science-technology development project of Jilin Province of China under Grant No. 20030520, the key science-technology project of the National Education Ministry of China under Grant No. 02090, and the doctoral funds of the National Education Ministry of China under Grant No. 20030183060. The first author would also like to thank the support of the young scholarship of Mathematics College of Jilin University of China.

References

1. Liang, Y.C., Lin, W.Z., Lee, H.P., Lim, S.P., Lee, K.H., Feng, D.P.: A Neural-network-based Method of Model Reduction For Dynamic Simulation of MEMS. Journal of Micromechanics and Microengineering, 11(2001) 226-233
2. Xu, X., Liang, Y.C., Lee, H.P. , Lin W.Z., Lim, S.P., Lee, K.H., Shi, X.H. : Identification and Speed Control of Ultrasonic Motors Based On Neural Networks. Journal of Micromechanics and Microengineering, 13(2003) 104-114
3. Yan, P. F., Zhang, C. S.: Artificial Neural Network and Simulated Evolutionary Computation. Thinghua University Press, Beijing China (2000)
4. Haykin, S.: Neural Networks: A Comprehensive Foundation. NJ: Prentice Hall International (1999)
5. Xu, X., Liang, Y.C., Lee, H.P., Lin, W.Z., Lim, S.P., Lee, K.H.: Mechanical Modeling of A Longitudinal Oscillation Ultrasonic Motor and Temperature Effect Analysis. Smart Materials and Structures, 12(2003) 514-523.
6. Ku, C.C., Lee, K.Y.: Diagonal Recurrent Neural Networks For Dynamic Systems Control. IEEE Transactions On Neural Networks, 6(1995)144-156

Feedforward-Feedback Combined Control System Based on Neural Network

Weidong Zhang, Fanming Zeng, Guojun Cheng, and Shengguang Gong

Naval University of Engineering, 430033 Wuhan, China
zweid7@263.net

Abstract. A neural network is used as the feedforward controller in a feedforward-feedback combined system. The network is trained by the feedback output that is minimized during training and most control action for disturbance rejection is finally performed by the rapid feedforward action of the network. The neural feedforward controller is independent of the model of plant and self-adaptive to time-variable system. The dynamic architecture of the neural controller is chosen, and the methods for delay time treatment and training network on line are investigated. An application to oxygen replenishment of an underwater plant is used to prove the effectiveness of the scheme and the simulation shows that the dynamic performance of the oxygen control is greatly improved by this neural combined control system.

1 Introduction

In the industrial plant, the feedforward-feedback combined control system is often applied to achieve good control effect. When the disturbances are measurable, the feedforward control is rapid, intellect, and sensitive, which can greatly improve the dynamic control performance of the plant. Meanwhile, the uncertain disturbances and model mismatch can be compensated by the feedback control.

However, lack of precision of the model makes it difficult to design the model-based feedforward controller for plants with time delays, strong interactions and nonlinearities. Normally, it is necessary to adjust the dynamic parameters of the feedforward controller on industrial site. However, a perfect and simple method to correct the parameters has still not been found up to now.

Artificial neural network can be used to provide promising control solution to those very complicated systems [1,2,3]. The dynamic relationship between the disturbances and feedforward actions can be learned by neural network, and the network can be used directly as the feedforward controller once the training is accomplished. In addition, the change in time-variable system can also be adapted by neural network. And fault tolerance is also provided since damage to a few links of the network would not significantly impair the overall performance of the network.

Generally, Using neural network to identify a nonlinear model of the plant requires a large number of input-output training pattern, the training procedure may be very

F. Yin, J. Wang, and C. Guo (Eds.): ISNN 2004, LNCS 3174, pp. 110–116, 2004.

complicated and time-consuming for a neural feedforward controller. In this paper, a more convenient method is used. The neural network is trained by minimizing the feedback control output. As we know it, the feedback output would be zero if the plant has been compensated perfectly by the feedforward action. Conversely, if the neural network is trained and generates an output to force the feedback output to approach zero, the trained network will achieve the characteristics of the feedforward controller.

The neural architecture suitable for dynamic system identification and training methods on-line are discussed. The delay time estimation for the neural feedforward controller is researched. Finally, this neural feedforward algorithm is applied to closed cycle diesel (CCD) oxygen replenishment and the effectiveness of the scheme is proved.

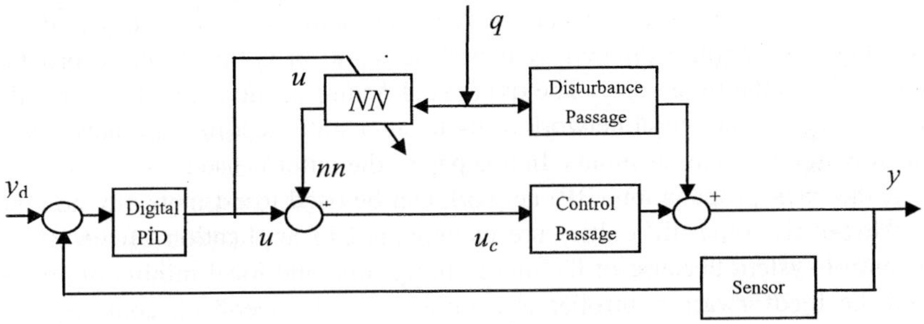

Fig. 1. Neural feedforward control learning system

2 Neural Feedforward Control Learning System

Based on a feedback control system, the neural network is incorporated as the feed-forward controller to form a combined control system that is shown in Figure.1.

2.1 Neural Feedforward Controller

The disturbance q imposed on the plant is measured and used as the input signal of the network *NN* (see Figure.1). Then, the neural network output *nn* is added to the PID output u to form the manipulated variable u_c to regulate the process. The purpose of training the network is to find the connection weights with which the network produces the control action to minimize the PID output, that is, to make the u close to zero. To achieve the purpose, some schemes have to be found to train the network. In fact, the feedback output itself or the deviation y-y_d could be used as the error signal to train the network. Because the aim of the training is to reduce the training error, if the training could be accomplished successfully, the feedback output used as the training error would be decreased and approximate to zero. Thus, in the proposed neural combined control system, the PID output *u* is used to train the network *NN*, and finally the output *nn* of the network will perform the major control action.

2.2 Architecture of Neural Network and Learning Algorithm

Generally, the dynamic relationship between disturbance patterns and feedforward control actions learned by neural network can be characterized by nonlinearities and time delays. Hence the dynamic characteristics of the feedforward compensation model can be expressed by a finite impulse response (FIR) system such as:

$$nn(k) = f(q(k-1-d), \cdots, q(k-m-d)) \tag{1}$$

where $nn(k)$ is the output of the system, $q(k)$ is the input of the system, f is the unknown nonlinear function estimated by neural network, m is the structure orders of the system and d is the time delay. The dynamical nature of the network can be identified either by recurrent neural network or by using a static feedforward network that is presented with past states as inputs. In this paper, the latter is used.

Usually, the back-propagation (BP) network can be used to estimate any nonlinear function. However, some difficulties are encountered in application of the BP network to control system because of its long training time and local minimum problem etc. When the feedforward controller is coupled with the feedback controller, it is possible to use an approximate model for feedforward controller because the model mismatch could be compensated by the feedback control. Based on that, the adaptive linear element (Adaline), a linear two-layered neural network with one neuron can serve as an alternative of the BP network that is more suitable for real-time training and control.

If the feedforward compensation model consists of large delay, the application of feedforward control scheme is more difficult because the delay time discrimination is troublesome. So a method is used to deal with the time delay in this neural-network-based feedforward control scheme. The transport delay time is covered by the input sample sequence of the network. For instance, if the delay time is 2s, and the sampling time interval is 0.1s, then the input of the network can be: $q(k\text{-}14)$, $q(k\text{-}15)$, $\cdots$, $q(k\text{-}23)$, $q(k\text{-}24)$. The simulation shows that a good compensation effect can be achieved by this scheme supported by the adaptability of neural network although inaccurate input signals are supplied.

The network is trained on line with feedback controller output and disturbance patterns, and the connection weights of the network are updated (accordingly the feedforward output of the network is also updated) at each sample time. That implies training and control are simultaneously carried out during the training period.

As Adaline network is used and trained on-line, the weights at $(k+1)$ *th* learning step could be updated by using the Widrow-Hoff rule:

$$w_j(k+1) = w_j(k) + \eta \cdot u(k) \cdot q_j \tag{2}$$

Where $u(k)$ is the feedback controller output, η is the learning rate coefficient, q is the input of the network.

The following steps are taken as training is performed:

1. Initializing the connection weights W with sufficiently small random numbers.
2. Taking a sample $q(k)$ from the disturbance.
3. Calculating the digital PID controller output $u(k)$ according to the deviation between the plant output and desired value.
4. Renewing the connecting weights W and the output $nn(k)$ of the network with $u(k)$ and the past values of the disturbance (d steps time delay) that have been stored.
5. Adding $nn(k)$ to $u(k)$to form the manipulated variable, and applying it to the process.
6. Return to 2

3 Application on Closed Cycle Diesel Oxygen Replenishment

The closed cycle diesel (CCD) is developed for extended endurance submarines. To finish the closed cycle, a certain number of CO_2 is removed from the exhaust gas and the oxygen consumed during combustion is replenished. Since the oxygen consumption is related to fuel provision and fuel rack is measurable, the feedforward compensation can be used to improve the CCD oxygen control. However, the CCD system is very complex because of the nonlinear effects associated with the combustion and mass transfer processes, and the transport delay associated with the respective pipes connecting the parts of the system. For this reason, the conventional model-based feedforward control algorithm is difficult to be used

The neural feedforward control is used to overcome the difficulties discussed above and the simulation is carried out to verify the effectiveness of the method

A lumped parameter model [4,5] is established by the gas thermodynamics, mass conservation, fluid flow characteristics and mass transfer formulas. The engine thermodynamic subsystem, scrubber mass transfer subsystem and mixing chamber subsystem with the additional associated control models are involved which reflect the innate characters of nonlinearities, time delays and complex interaction of the CCD system. The MATLAB SIMULINK tool is used to accomplish the simulation and the s-function is used to create a block for neural feedforward controller.

A (12,1) two-layered Adaline network is used and the size of the network is selected according to a rough estimate of the complexity of the problem and simulation results. The initial connection weights of the network and the learning rate coefficient are chosen sufficiently small to ensure the stability of the system. In the simulation, the sampling time interval is 0.1s, and the learning rate coefficient is 0.005. During the training period, the step load change is imposed upon the closed system, and repeated until the desired performance is achieved.

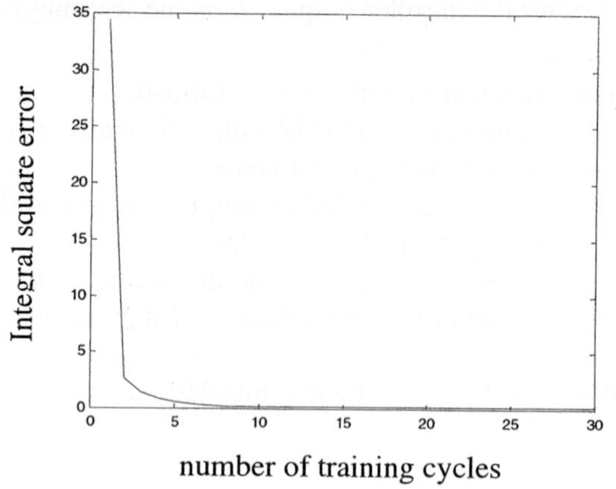

Fig. 2. Change of integral square error

The training period including one step load change is considered one training cycle. The integral square error (ISE) of training for one training cycle is recorded at each training cycle. Figure 2 shows changes in the integral square error vs. the number of training cycles. As the outputs of PID are used as error signals for training the network, the decrement of the integral square error indicates that PID outputs are minimized during training process, which also proves that the PID outputs can provide the correct gradient direction for training. The integral square error of training decreases greatly during the period of only a small number of training cycles. After 10~20 times' training, the integral square error are nearly approach fixed which implies that the training can be completed within limited training cycles and makes it convenient to apply this neural feedforward strategy in practice

The changes in the output of Adaline network and that of the PID are shown in Figure.3. As training proceeds, the output of PID becomes small and small. Meanwhile the network keeps configuring its connecting weights and its output becomes large gradually. The feedback control action, which initially performs the major control, is gradually shifted to the feedforward action of the network.

The changes in oxygen volumetric concentration in response to the step load are shown in Figure.4, in which the control effect of the combined neural control system is compared to that of the PID controller. It can be seen that when a fast great load change is imposed upon the CCD system, the dynamic response of the oxygen volumetric concentration is unsatisfactory when only the feedback controller is used (dash line), because of the slow response of the oxygen sensor and nonlinear factors in valves. However, as the feedforward action of the network is formed after training, the dynamic performance of the oxygen control is greatly improved (solid line).

The results of simulation prove that the neural scheme is workable.

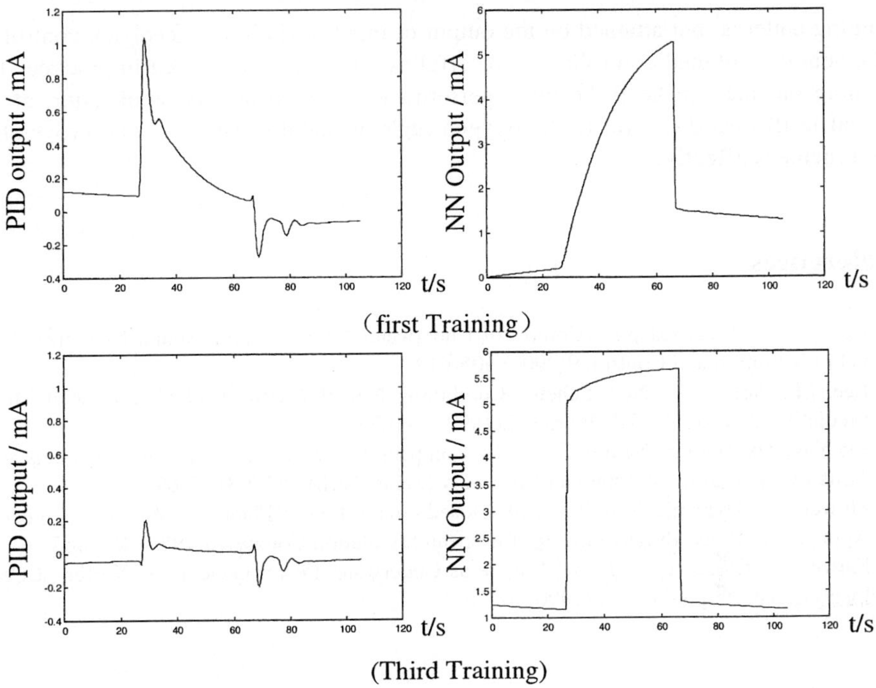

Fig. 3. PID, NN Output as Training Proceeds

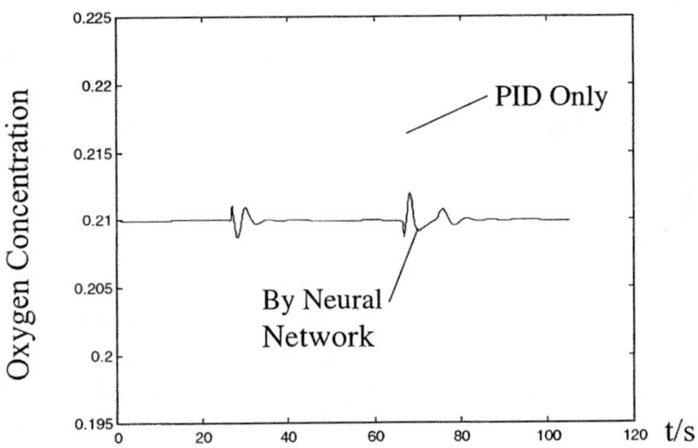

Fig. 4. Response of Oxygen Concentration

4 Conclusion

The feedforward-feedback control is popular in industry. This paper tries to solve its application problems using neural network. The dynamic relationship between disturbance and feedforward control action is not trained by large number of input-output training patterns, but attained by the output or input deviation of feedback controller. The neural combined controller is powerful as tool but easy as used in practice. The training on line can keep the good performance of control. The application of the neural feedforward control to the oxygen replenishment of CCD system proves that the scheme is effective.

References

1. Daniel, J.I.: Direct Adaptive Control of Underwater Vehicles using Neural Networks. Journal of Vibration and Control. 9(2003) 605-619
2. Lee, M., Park, S.A.: New Scheme Combining Neural Feedforward Control with Model Predictive Control. AICHE Journal. 2(1992) 193-200
3. Lu Yan, Du Jihong: Neural Networks Compensate Control for Unknown Systems with Time Delay. Journal of Tsinghua University (Sci & Tech). 9(1998) 67-69
4. Flower, A.: Dynamic Simulation of Closed-circuit Power Plant for Marine Engineering Application. Proceedings of the 3rd European Simulation Congress. (1989) 457-463
5. Flower, A.: Closed-cycle Diesel Engine as Underwater Power Generators. Northeast Coast Institution & Shipbuilders. 2(1990) 67-76

Online Learning CMAC Neural Network Control Scheme for Nonlinear Systems

Yuman Yuan, Wenjin Gu, and Jinyong Yu

Department of Automatic Control Naval Aeronautical Engineering Institute
Yantai, Shandong Province, P.R. China
arman2002@sohu.com

Abstract. The cerebella model articulation controller (CMAC) neural network control scheme is a powerful tool for practical real-time nonlinear control applications. The conventional leaning controller based on CMAC can effectively reduce tracking error, but the CMAC control system can suddenly diverge after a long period of stable tracking, due to the influence of accumulative errors when tracking continuous variable signals such as sinusoidal wave. A new self-learning controller based on CMAC is proposed. It uses the dynamic errors of the system as input to the CMAC. This feature helps the controller to avoid the influence of the accumulative errors and the stability of the system is ensured. The simulation results show that the proposed controller is not only effective but also of good robustness. Moreover, it has a high learning rate, which is important to online learning.

1 Introduction

The cerebella model articulation controller (CMAC) was proposed by Albus [1]. CMAC has many excellent properties, which make it a real alternative to backpropagated multilayer networks (MLPs). The speed of learning of CMAC may be much higher then that of a corresponding MLP using backpropagation training algorithm. The classical binary CMAC can be implemented without using multipliers, so it is especially suited to digital hardware implementation, ideal network architecture for embedded applications. This neural network is capable of learning nonlinear functions extremely quickly due to the nature of its weight updating, so it is a powerful and practical tool for real-time control. Miller et al [2,3,4] proposed a feasible control scheme to combine the CMAC and a traditional controller for robot manipulator control. However, the Miller scheme can be unstable[4],the reason is that the structure of the controller is imperfect. A new control scheme for CMAC control is present in this paper, which improve the stability of the CMAC control system significantly.

2 CMAC Neural Network

The structure of CMAC neural network [4] is shown in Fig.1. The input vectors in the input space s are a number of sensors in real world. Input space consists of all

F. Yin, J. Wang, and C. Guo (Eds.): ISNN 2004, LNCS 3174, pp. 117–122, 2004.

possible input vectors. CMAC maps the input vector into C points in the conceptual memory A. As shown in Fig.1, two "close" inputs will have overlaps in A, the closer the more overlapped, and two "far" inputs will have no overlap. Since practical input space is extremely large, in order to reduce the memory requirement, A is mapped onto a much smaller physical memory A' through hash coding. So, any input presented to CMAC will generate C physical memory locations. The output Y will be the summation of the content of the C locations in A'.

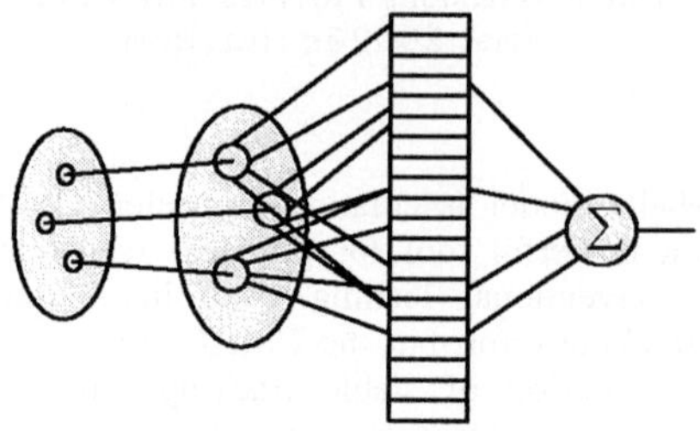

Fig. 1. Block diagram of CMAC

From the above, we can see that the associative mapping within the CMAC network assures that nearby points in the input space generalize while distant points do not generalize. Moreover, since from A' to Y is a linear relationship but from s to A' is a nonlinear relationship, the nonlinear nature of the CMAC network perform a fixed nonlinear mapping from the input vector to a many-dimensional output vector.
Network training is typically based on observed training data pairs *s* and Y_d, where Y_d is the desired network output corresponding to the input s, using the least mean square (LMS) training rule. The weight can be calculated by:

$$w(k+1) = w(k) + \frac{\eta(Y_d - F(s))}{C} \quad (1)$$

where η is the learning step length, *F(s)* is the output of the CMAC network.
Thus, if we define an acceptable error ε, no changes needed for the weights when $|Y_d - F(s)| \le \varepsilon$. The training can be done after a set of training samples being tested or after each training sample being tested.

3 Control Problem and Schemes

The old CMAC control system proposed by Miller[3] is shown in Fig. 2, which is a class of supervisor control structure. The CMAC network is used as a feedforward controller, which learns the inverse dynamics of the plant. The system to be controlled is

$$Y(k+1) = G(Y(k), U(k)) \quad (2)$$

The control *U(k)* is generated by

$$U(k) = U_c(k) + U_p(k) \tag{3}$$

where $U_p(k) = P(r(k) - Y(k))$ is the output of the traditional constant gain controller, P is the proportional gain and r is the desired system output. $U_c(k)$ is the output of the CMAC module for the input vector $(r(k+1), Y(k))$. Then the CMAC is updated by the following gradient-type learning rule:

$$W_i(k+1) = W_i(k) + \frac{\eta}{C}(U(k) - U_c(k)) \tag{4}$$

where C is the size of generalization, *Wi* is the content of the *i*th memory location, and η is the learning rate. In essence, in this scheme, the constant gain controller helps the CMAC to learning the inverse of the plant. However, these two controllers are independent, and the conflicts between them are the source of the instability[4].

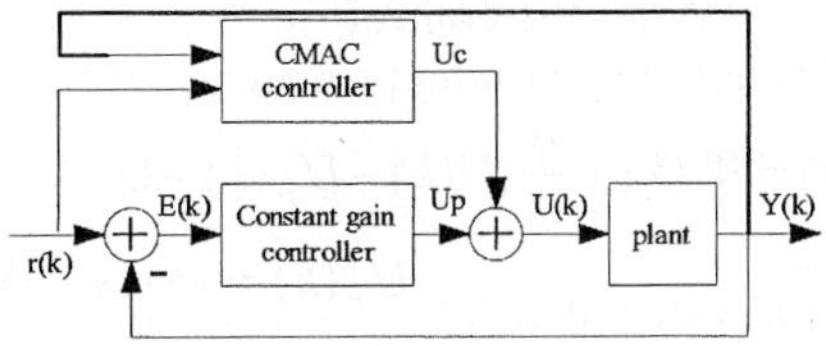

Fig. 2. Original scheme of CMAC controller

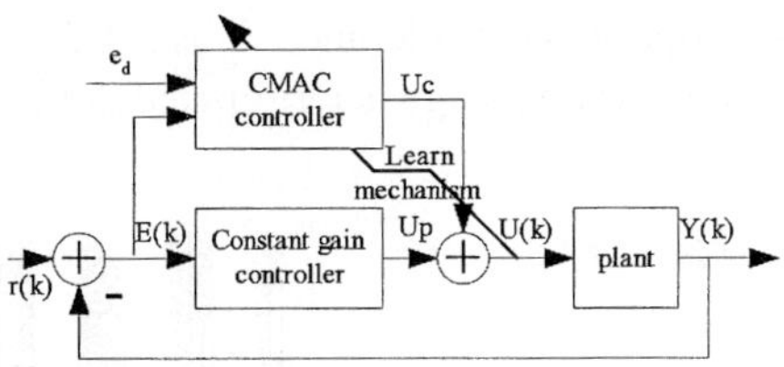

Fig. 3. New scheme of CMAC controller

The structure of the new scheme, as shown in Fig. 3, where e_d is the desired error, the structure is similar to which is proposed by Miller[3], the difference between them is that the input of CMAC in new scheme is the tacking error of the system rather than the desired output and the system output. Instead of having the system inverse solely learned by the CMAC, the constant gain controller and the CMAC are integrated, and will be used to approximate the system inverse together. In other words, the CMAC will be used to learn the difference between the system inverse and the constant gain control.

There are tow schemes for CMAC network controller input vector as follow:

a) using vector ($e(k+1)$, $e(k)$) as CMAC input

In this case, all value of CMAC weight is set to zero at first. At control stage, the desired error of next step $e_d(k+1)$, which is zero usually, and the system error $e(k)$

are quantized and input to the CMAC to get the output of the CMAC network $U_c(k)$, plus the constant gain controller output $U_p(k)$,i.e. $U(k)=U_p(k)+U_c(k)$, is applied to the plant. Then, at learning stage, the actual error of system $e(k+1)$ and $e(k)$ are input to the CMAC network to calculate the output $U_c^{'}(k)$, and to adjust its weight according to:

$$W_i(k+1)=W_i(k)+\frac{\eta}{C}(U(k)-U_c^{'}(k)) \tag{5}$$

b) using vector $(Y(k+1),Y(k))$ as CMAC input

In this case, the constant gain control has to be calculated based on the same input vector as is applied to the CMAC. Since the input vector to the CMAC is $(r(k+1),Y(k))$, the proportional control U_P is modified to be $U_p(k)=P(r(k+1)-Y(K))$. After the control $U(k)=U_p(k)+U_c(k)$ is applied to the plant, the CMAC goes through the learning process according to

$$W_i(k+1)=W_i(k)+\frac{\eta}{C}(U(k)-U_p(k)-U_c(k)) \tag{6}$$

Note that, in (6), $U_p(k)$ is calculated by $U_p(k)=P(r(k+1)-Y(K))$, and $U_c(k)$ is generated from the input vector $(Y(k+1),Y(k))$, where $Y(k+1)$ is the actual system output at time step $k+1$.

In both two cases, an appropriate cost tolerance ε is set for cost function. When the cost is less than ε, then the learning stage is terminated , otherwise, it is started.

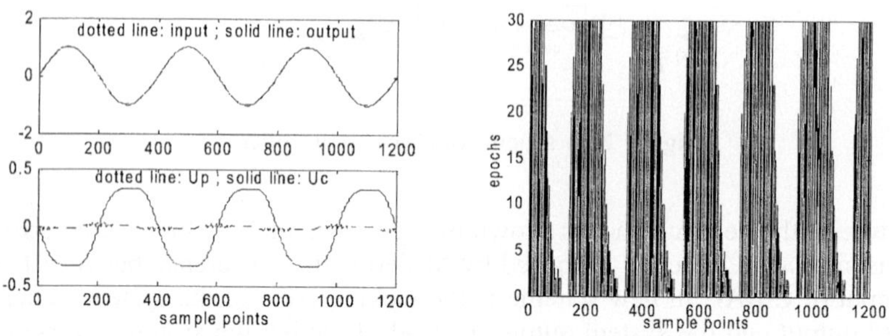

Fig. 4. Result using old scheme with $\eta=0.02$

4 Simulation Results

As in [5], the nonlinear system used in the simulation is $Y(k+1)=0.5Y(k)+\sin(Y(k))+U(k)$. The reference input signal is the command $\sin(2\pi k/400)$. The

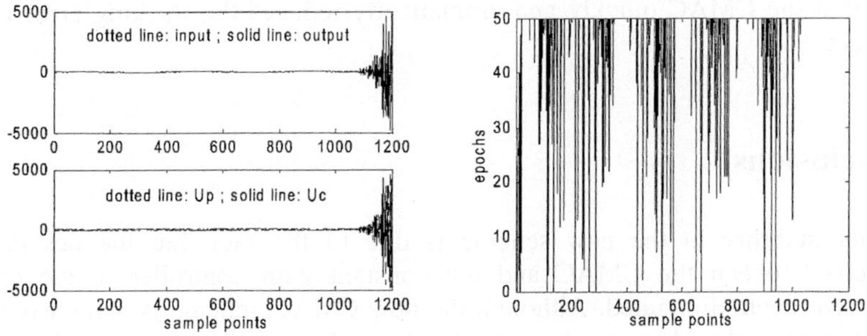

Fig. 5. Result using old scheme with $\eta = 0.05$

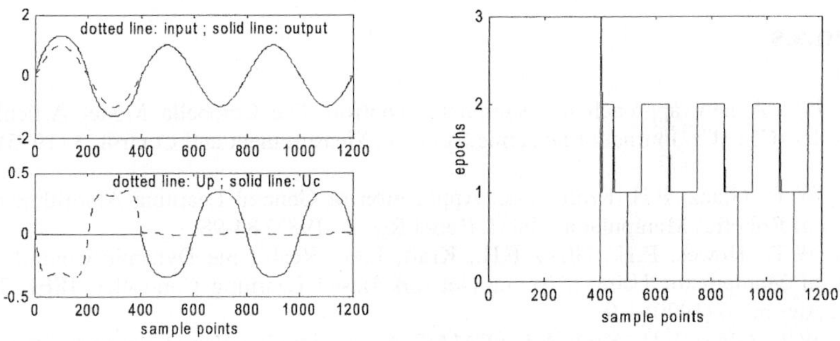

Fig. 6. Result of scheme (a)

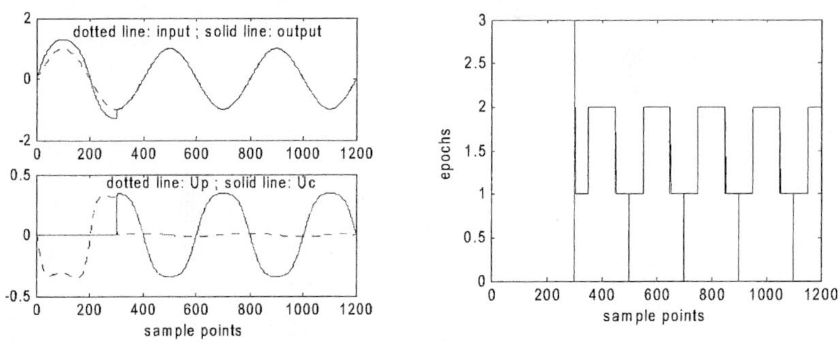

Fig. 7. Result of scheme (b)

parameters of CMAC network are: active elements, C = 15; Learning rate η =0.9; memory size = 2000; tolerance of the cost function for LMS learning scheme is 1e-6. Using the old scheme, when $\eta > 0.05$, the system becomes unstable. The results using the old scheme are shown in Fig 4 and Fig 5, with maximum epoch is 30. The results using the two new schemes are shown in Figs.6 and 7, respectively. In Figs 6 and 7, the CMAC network is added into the system at 400th and 300th sample point, respectively. Before that, only the constant gain controller is in charge. One can observe that the CMAC quickly and dramatically reduces the tracking error once it takes effect.

5 Conclusions

The better stability in the new scheme is due to the fact that the destabilizing interactions between the CMAC and the constant gain controller in the original scheme have been eliminated. Although the new CMAC scheme is demonstrated by simple examples, the ideas can be extended to sophisticated and practical nonlinear control problems.

References

1. Albus J.S.: A new approach to manipulator control: The Cerebella Model Articulation Controller (CMAC). Journal of Dynamic Systems, Measurement and Control. 3 (1975) 220-227
2. Miller, W.T., Glanz, F.H. Kraft, L.G.: Application of General Learning Algorithm to the Control of Robotics Manipulators. Int. J. Robot Res. 6 (1987) 84-98
3. Miller, W.T., Hewes, R.H., Glanz F.H., Kraft, L.G.: Real-Time Dynamic Control of an Industrial Manipulator Using a Neural-Network-Based Learning Controller. IEEE Trans. Robot. Autom. 6 (1990) 1-9
4. Miller, W.T., Glanz, F.H., Kraft, L.G.: CMAC: An Associative Neural Network Alternative to Backpropagation. Proceedings of the IEEE. 10 (1990) 1561-1567
5. Chen F.C. and Chang C.H.: Practical Stability Issues in CMAC Neural Network Control System. IEEE Trans. Control Syst. Technol. 1 (1996) 86-91

Pole Placement Control for Nonlinear Systems via Neural Networks

Fei Liu

Institute of Automation, Southern Yangtze University,
Wuxi, Jiangsu, 214036, P. R. China
Dr_fliu@hotmail.com

Abstract. This paper extends pole placement control of conventional linear systems to a class of nonlinear dynamical systems via neural networks. An application of typical inverted pendulum illustrates the design method. Multi-layer neural networks are selected to approach nonlinear components arbitrarily, and then are represented by linear difference inclusion (LDI) format. With pole placement regions formed in linear matrix inequalities (LMIs), quadratic stability theory is used as a basic analysis and synthesis methodology. Pole placement controllers via state feedback are derived by numerical solutions of a set of coupled LMIs. Applying common back propagation algorithm (BP) for networks training and interior point computation for LMI solving, some simulation results show the validity of pole placement control.

1 Introduction

Pole placement is the most popular technique in control system design. It specifies not only the stability but also the transient responses of closed-loop systems. For linear time-invariant systems, pole locations bound the maximum overshoot, the delay time, the rise time, the settling time, and etc. For linear time-varying uncertain systems, instead of exact pole location, it is often sufficient to place poles in a prescribed region of complex plane [1]. Since linear matrix inequality (LMI) technique is widely used as a power tool for controller design, pole regions are significantly defined by the LMI formulation [2], which cover half-plane, disks, sectors, and vertical strips as special cases.

However, how about nonlinear systems? In general, pole placement is only the notion of linear systems. Motivated by robust pole placement of linear time-varying systems [3], a feasible method dealing with pole placement design of nonlinear systems is based on Takagi-Sugeno fuzzy models [4], [5]. In this paper, another method will be presented, which applies neural networks to model nonlinear systems. It is well known that neural network has universal approximation capability and self-learning ability [6], [7]. By now, neural networks were already used in closed-loop control systems by different ways [8], [9]. Recent years, one of interesting neural network-based design approach for nonlinear systems was by means of linear difference inclusion (LDI) representation [10], [11].

In this paper, a typical inverted pendulum is used to illustrate pole placement method via neural networks. Firstly, nonlinearity of inverted pendulum is

F. Yin, J. Wang, and C. Guo (Eds.): ISNN 2004, LNCS 3174, pp. 123–128, 2004.

approximated by multi-layer neural networks in LDI representations. Then, based Lyapunov quadratic stability theory, sufficient conditions for existence of state feedback controller with specified pole placement region constraint are given in terms of a set of LMIs. The controller can be directly obtained by searching the common solutions among the set of LMIs, and the simulation results show its feasibility and validity.

2 Pole Placement and Nonlinear Systems

In control engineering practices, satisfactory transient response or performance requirements can often be guaranteed by forcing the closed-loop poles into a suitable region. In general, good damping, fast decay and settling time may be imposed by limiting the poles in a disk, a conic sector, a vertical strip or their intersection [12]. For linear time-invariant dynamical system $\dot{x} = Ax$, while given a poles region D of the complex left-half plane, pole placement design means that all eigenvalues of the matrix A lie in the region D. For example, condition of all closed-loop poles on the left-half plane within vertical strip $D = \{z \in C : \mathrm{Re}(z) < -\alpha, \alpha > 0\}$, will force the closed-loop transient response not slower than $e^{-\alpha t}$, and the larger the value of α is, the faster the transient response will be. A more general region D [2] can be defined as $D = \{z \in C : f_D(z) < 0\}$, where $f_D(z) = L + zM + \bar{z}M^T$ is characteristic function and L, M are real matrices with $L^T = L$. For given L and M, the LMI condition for closed-loop systems to satisfy region D are

$$L \otimes P + M \otimes PA + M^T \otimes A^T P < 0 . \tag{1}$$

where $\otimes$ denotes the Kronecker product of matrices and P is positive symmetric matrix .

While a class of nonlinear dynamical systems is modeled by LDI, it appears in uncertain linear system form. This gives a way to present so-called pole placement concerning nonlinear systems. Although the analytical relations between transient behaviors and above so-called pole placement remain open by now, pole placement control, from the point of view of engineering, provides some useful help to adjust system behaviors. Considering nonlinear dynamics $\dot{x} = f(x)$, the nonlinear functions $f(x)$ can be approximated arbitrarily by multi-layered neural networks in compact representation described as following [11]: $f(x) \approx \Psi_L[W_L \cdots \Psi_2[W_2 \Psi_1[W_1 x] \cdots]$, where the activation function vector of i-th layer has the form $\Psi_i[\cdot] : R^{n_i} \mapsto R^{n_i}$ defined as $\Psi_i[\xi] = [\psi_1(\xi_1), \cdots, \psi_{n_i}(\xi_{n_i})]^T$, $i = 1, \cdots, L$. All the connecting weight matrices $W_i \in R^{n_i \times n_{i-1}}$ from the $i-1$th layer to the i-th layer will be determined via learning algorithm, such as back propagation (BP). In most cases, activation functions ψ_j, $j = 1, \cdots, n_i$, of neurons in all L layers adopt sigmoid type or line type. Denoting the minimum and maximum values of ψ_j as δ_{j0} and δ_{j1} respectively, it follows

$\psi_j = \sum_{k=0}^{1} h_{jk}\delta_{jk}$, which is so-called min-max form of activation functions, where h_{jk} is real number associated with ψ_j satisfying $h_{jk} > 0$ and $\sum_{k=0}^{1} h_{jk} = 1$. Define a set of index vector of the i-th layer as $\Gamma_{n_i} = \{ \nu \in R^{n_i} \mid \nu_j \in \{0,1\}, j = 1,\cdots,n_i \}$, which indicates that the i-th layer with n_i neurons has 2^{n_i} combinations of binary indicator with $k = 0,1$. Obviously, all the elements of index vectors for all L-layers neural network have $2^{n_L} \cdots 2^{n_2} \times 2^{n_1}$ combinations in the set $\Omega = \Gamma_{n_1} \cdots \oplus \Gamma_{n_2} \oplus \Gamma_{n_L}$. Substitute in the min-max formed Ψ_i, nonlinear functions $f(x)$ can be rewritten in linear difference inclusion (LDI) form as

$$f(x) \approx \sum_{\nu \in \Omega} \mu(\nu) A_\nu(\nu, W_i^*, \Psi_i)x = \sum_{\nu} \mu_\nu A_\nu x , \tag{2}$$

where A_ν is a set of the matrices of appropriate dimensions, which can be computed by optimal weighs W_i^*, activation functions Ψ_i and all index vectors ν. Coefficient μ_ν is a product of real number h_{jk} of all neurons in whole neural network, obviously, $\mu_\nu > 0$, $\sum_{\nu} \mu_\nu = 1$, for $\nu \in \Omega$.

For dynamical systems with nonlinear function $f(x)$ as in (2), considering state feedback controller $u = Kx$, we introduce following definition based on (1): Given region D with selected gain matrices L, M, the poles of dynamical systems with nonlinear $f(x)$ will be restricted in D, if there exists a positive symmetric matrix P satisfying

$$L \otimes P + M \otimes P(\sum_{\nu} \mu_\nu A_\nu + BK) + M^T \otimes (\sum_{\nu} \mu_\nu A_\nu^T + K^T B^T)P < 0 , \tag{3}$$

where B is the gain matrix of control input. Pre- and post-multiply above (3) by $I \otimes P^{-1}$, let $P^{-1} = X$, $KP^{-1} = Y$, and keep in mind the coefficient μ_ν defined in (2), above inequality (3) is guaranteed by following condition in term of a set of LMIs

$$L \otimes X + M \otimes (A_\nu X + BY) + M^T \otimes (A_\nu X + BY)^T < 0 . \tag{4}$$

Different from condition (1), which deals with linear systems, above LMIs (4) can deal with nonlinear systems. In other words, for nonlinear dynamical systems formulated in LDI, if there exists positive symmetric matrix X and matrix Y satisfying LMIs (4), we can always find a feedback controller $u = YX^{-1}x$ such that the so-called poles of resulting closed-loop systems are forced into a region D with gain matrices L, M.

3 Application Study

Consider a simple model of inverted pendulum given by [13]

$$\begin{bmatrix} \dot{x}_1 \\ \dot{x}_2 \end{bmatrix} = \begin{bmatrix} x_2 \\ \frac{g}{l}\sin(x_1) - \frac{k}{m}x_2 + \frac{1}{ml^2}u \end{bmatrix}, \tag{5}$$

where $g = 9.81$, $l = 1.0$, $m = 1.0$, $k = 2.0$, $x_1 = \theta$ is the angle (in radians), $x_2 = \dot{\theta}$ is the angular velocity (in radians per second), and u is the control input. The model is easily rewritten as $\dot{x} = Ax + Bu + \bar{f}(x)$, where $A = \begin{bmatrix} 0 & 1 \\ 0 & -2 \end{bmatrix}$, $\bar{f}(x) = 9.81\begin{bmatrix} 0 \\ f(x_1) \end{bmatrix}$, $B = [0 \quad 1]^T$, $f(x_1) = \sin(x_1)$.

Introduce a 3-layered neural network to approximate nonlinear function $f(x_1)$, and its middle hidden layer has three neurons z_1, z_2 and z_3. From input layer x_1 to output layer $f(x_1)$, the weight matrices are given by $W_{zi} = [w_{11} \quad w_{12} \quad w_{13}]^T$, $W_{oz} = [w_{21} \quad w_{22} \quad w_{23}]$. Here, the hidden layer and output layer define the same sigmoid type activation function, i.e. $\psi_j(\xi) = \frac{2}{1+e^{-2\xi}} - 1$, $j = 1, \cdots, n_i$, $\forall i, j$. The maximum and minimum values of ψ_j are $\delta_{j1} = 1$, $\delta_{j0} = 0$, with hidden layers ($j = 1,2,3$) and output layer ($j = 4$). Applying basic back propagation learning algorithm, for input/output data $x_1 \to \sin(x_1)$, all the weights among the layers are obtained as $w_{11} = 0.2282$, $w_{12} = 2.0345$, $w_{13} = 0.8581$, $w_{21} = -8.4883$, $w_{22} = -1.6492$, $w_{33} = 6.9209$ and $f(x_1) = \Psi_2[W_{oz}\Psi_1[W_{zi}x_1]]$. Using min-max form of activation functions, the three neurons of hidden layer can be formulated as

$$z_j = (h_{j1}\delta_{j1} + h_{j0}\delta_{j0})w_{1j}x_1 = \sum_{i_j=0}^{1} h_{ji_j}\delta_{ji_j}w_{1j}x_1,\ j = 1,2,3. \tag{6}$$

Similarly, the output layer of neural network follows

$$\begin{aligned} f(x_1) &= \sum_{i_4=0}^{1} h_{4i_4}\delta_{4i_4}W_{oz}[z_1\ z_2\ z_3]^T \\ &= \sum_{v}\mu_v A_v x_1 = \sum_{i_4=0}^{1}\sum_{i_3=0}^{1}\sum_{i_2=0}^{1}\sum_{i_1=0}^{1} h_{4i_4}h_{3i_3}h_{2i_2}h_{1i_1}(\delta_{4i_4}[\delta_{1i_1}\quad \delta_{2i_2}\quad \delta_{3i_3}]W_{zi} \bullet W_{oz}^T)x_1. \end{aligned} \tag{7}$$

Obviously, $\sum_v \mu_v = \sum_{i_4=0}^{1}\sum_{i_3=0}^{1}\sum_{i_2=0}^{1}\sum_{i_1=0}^{1} h_{4i_4}h_{3i_3}h_{2i_2}h_{1i_1}$, $A_v = \delta_{4i_4}[\delta_{1i_1}\quad \delta_{2i_2}\quad \delta_{3i_3}]W_{zi} \bullet W_{oz}^T$, where $\bullet$ denotes array multiplication and the subscript v is given by

$$v = \{[i_1\ i_2\ i_3\ i_4] \in R^4 \mid i_j \in \{0, 1\}, j = 1, 2, 3, 4\}.$$

Thus, by computing, we get $A_{0001}=0$, $A_{0011}=5.9388$, $A_{0101}=-3.3553$, $A_{0111}=2.5835$, $A_{1001}=-1.9370$, $A_{1011}=4.0018$, $A_{1101}=-5.2923$, $A_{1111}=0.6465$.

As simulation examples, consider a disk of radius d and center $(-q,0)$ with characteristic function $f_{dk}(z)=\begin{bmatrix} -d & q+z \\ q+\bar{z} & -d \end{bmatrix}$, assume $d=q=9$, by solving LMIs (4) with $L=\begin{bmatrix} -9 & 9 \\ 9 & -9 \end{bmatrix}$ and $M=\begin{bmatrix} 0 & 1 \\ 0 & 0 \end{bmatrix}$, a pole placement controller is obtained as $K=[-61.8317 \quad -13.4424]$. The transient response of the inverted pendulum displays in Fig.1, compared with cases $d=q=20$ and $d=q=50$. Consider vertical strip $\mathrm{Re}(z)<-\alpha$ with characteristic function $f_{vs}(z)=z+\bar{z}+2\alpha<0$. Assume $\alpha=0.2$, by solving LMIs (4) with $L=0.4$ and $M=1$, the transient response displays in Fig.2, compared with cases $\alpha=0$ and $\alpha=1$. From Fig.2, obviously, the large the value of α is, the faster the transient response is.

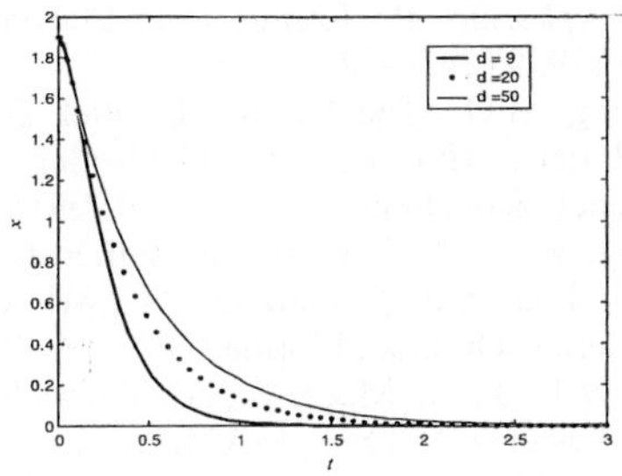

Fig. 1. Disk constraints and transient response

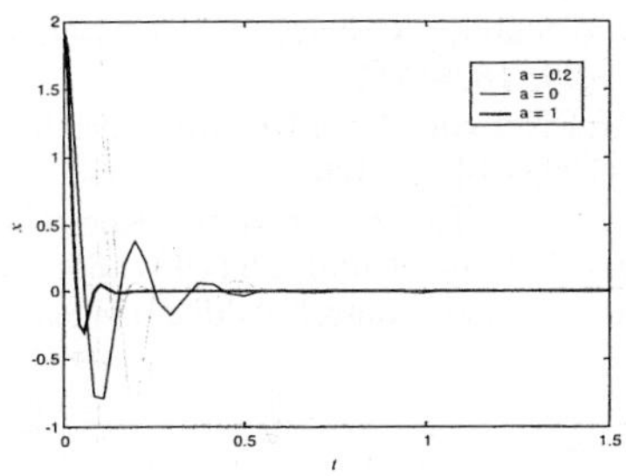

Fig. 2. Vertical strip constraints and transient response

Furthermore, by intersecting above special regions, the poles may be restricted in more general region. For example, for region $D=\{z\in C: f_{dk}(z)<0, f_{vs}(z)<0\}$, the pole placement controller may be obtained by coupling LMIs (4) with different M, L on common matrices X, Y.

4 Conclusion

By means of linear difference inclusion (LDI), the design methodology of pole placement with LMI region is applied to nonlinear dynamical systems approximated by neural networks. Compared with similar T-S fuzzy-based approach, neural networks-based approach needs less information about nonlinearity. In other ways, neural networks in LDI representation requires common positive symmetric matrix and controller gain. This brings biggish conservation.

References

1. Garcia, G., Bernussou, J.: Pole Assignment for Uncertain Systems in a Specified Disk by State Feedback. IEEE Trans. Automat. Contr. 40 (1995) 184-190
2. Chilali, M., Gahinet, P.: H_∞ Design with Pole Placement Constraints: An LMI Approach. IEEE Trans. Automat. Contr. 41 (1996) 358-367
3. Chilali, M., Gahinet, P., Apkarian, P.: Robust Pole Placement in LMI Regions. IEEE Trans. Automat. Contr. 44 (1999) 2257-2270
4. Li, J., Niemann, D., Wang, H.O., Tanaka, K.: Parallel Distributed Compensation for Takagi-Sugeno Fuzzy Models: Multiobjective Controller Design. Proceedings of the American Control Conference, San Diego, California (1999) 1832-1836
5. Joh, J., Langari, R., Chung, W. J.: A New Design Method for Continuous Takagi-Sugeno Fuzzy Controller with Pole Placement Constraints: A LMI Approach. IEEE Int. Conf. on Systems, Man, and Cybernetics, Orlando, Florida (1997) 2969-2974
6. Suykens, J. A. K., Vandewalle, J., De Moor, B.: Artificial Neural Networks for Modeling and Control of Non-linear Systems. Norwell, MA, Kluwer (1996)
7. Hornik, K.: Approximation Capabilities of Multilayer Feedforward Network. Neural Networks (1991) 251-257
8. Psaltis, D., Sideris, A., Yamamura, A.: A Multilayered Neural Network Controller. IEEE Contr. Syst. Mag. 8 (1988) 17-21
9. Narenda, K. S., Parthasarathy, K.: Identification and Control of Dynamical System Using Neural Networks. IEEE Trans. Neural Networks 1 (1990) 4-27
10. Tanaka, K.: An Approach to Stability Criteria of Neural-Network Control Systems. IEEE Trans. Neural Networks 7 (1996) 629-642
11. Limanond, S., Si, J.: Neural-Network-Based Control Design: An LMI Approach. IEEE Trans. Neural Networks 9 (1998) 1422-1429
12. Gutman, S., Jury, E. J.: A General Theory for Matrix Root Clustering in Subregions of the Complex Plane. IEEE Trans. Automat. Contr. 26 (1981) 853-863
13. Passino, K. M., Yurkovich, S.: Fuzzy Control. Addison Wesley Longman, Inc (1995)

RBF NN-Based Backstepping Control for Strict Feedback Block Nonlinear System and Its Application

Yunan Hu[1,2], Yuqiang Jin[2], and Pingyuan Cui[1]

[1] Department of Astronautics and Mechanics, Harbin Institute of Technology, Harbin 150001, P.R.China
cui@astro.hit.edu.cn

[2] Department of Automatic Control, Naval Aeronautical Engineering Academy, Yan tai 264001, P.R.China
hya507@sina.com

Abstract. Based on neural networks, a robust control design method is proposed for strict-feedback block nonlinear systems with mismatched uncertainties. Firstly, Radial-Basis-Function (RBF) neural networks are used to identify the nonlinear parametric uncertainties of the system, and the adaptive tuning rules for updating all the parameters of the RBF neural networks are derived using the Lyapunov stability theorem to improve the approximation ability of RBF neural networks on-line. Considering the known information, neural network and robust control are used to deal with the design problem when control coefficient matrices are unknown and avoid the possible singularities of the controller. For every subsystem, a nonlinear tracking differentiator is introduced to solve the "computer explosion" problem in backstepping design. It is proved that all the signals of the closed-loop system are uniform ultimate bounded.

1 Introduction

Nonlinear adaptive control has been developed rapidly during last ten years [1~3],and mostly results limited to the systems in which parametric uncertainties are linear , can be linearized and the uncertainties satisfy the matched conditions. As a breakthrough of the difficulties, adaptive backstepping design method can deal with the so-called mismatched uncertainties [4]. For strict-feedback nonlinear systems, backstepping is a systematic design method. However, for a class of nonlinear systems in block standard forms, it is more difficult, fewer results are obtained up to now. Block control principle is developed on the basis of "block control standard form". Based on the block control principle, some design methods are proposed in recent years [5-7].

RBF NN is a typical local approximator and finds wide applications in control system design. Generally, only the weight values are updated in most applications, which is difficult to get ideal approximation when one doesn't know much about prior knowledge on the system. In order to solve the problem, reference [8] made some modifications on the NN updated laws. The author derived the tuning rules for all the parameters of RBF NN (weights, width and centers of Gaussian functions), but its implementation of the tuning rules is on the basis of gradient optimization, which cannot guarantee the stability of the overall system [3]. References [10] and [11] pro-

F. Yin, J. Wang, and C. Guo (Eds.): ISNN 2004, LNCS 3174, pp. 129–137, 2004.

posed design methods of neural network adaptive control for a class of nonlinear systems with general uncertainties. But there exist drawbacks in their design method: a) only can deal with scalar systems at each step of backstepping, cannot be extended to multi-variable sub system, and controller design of multi-variable system with unknown control coefficient matrix is a difficult problem; b) highly rely on the approximation ability of NN and don't take use of known information efficiently.

In this paper, we discuss the design of Backstepping for a class of nonlinear systems in block control form and with mismatched uncertainties, solve the problems mentioned above.

2 Fully Tuned RBF NN

Defination 1. Fully tuned RBF NN: all the parameters of RBF NN (including weights, widths and centers) are updated.

Assumption 1. Function vector $\Delta \boldsymbol{f}:\Omega \mapsto R^r$, Ω belongs to a sub-compact set of R^n. $\forall\ \boldsymbol{\varepsilon}=[\varepsilon_1 \quad \varepsilon_2 \quad \cdots \quad \varepsilon_r]^T$, $\varepsilon_i>0$, there always exist an optimal Gaussian base function vector $\boldsymbol{\varphi}^*:R^n \mapsto R^l$ and weight matrix $\boldsymbol{W}^* \in R^{l\times r}$ such that

$$\Delta \boldsymbol{f}=\boldsymbol{W}^{*T}\boldsymbol{\varphi}^*+\boldsymbol{\varepsilon},\forall \boldsymbol{x}\in\Omega, \tag{1}$$

where $\boldsymbol{\varphi}^*=\left[\exp\left(-\left\|\varsigma-\boldsymbol{\mu}_1^*\right\|^2/\sigma_1^{*2}\right) \quad \cdots \quad \exp\left(-\left\|\varsigma-\boldsymbol{\mu}_l^*\right\|^2/\sigma_l^{*2}\right)\right]^T$, $\boldsymbol{\mu}_i^*, i=1,2,\ldots,l$ is the optimal center, l is the number of hidden layer nodes, $\sigma_i^*, i=1,2,\ldots,l$ is the optimal affect size, $\varsigma\in R^n$ is the input of RBF NN, and $\boldsymbol{\varepsilon}$ is the construction error of NN.

The errors between the optimal values and estimated values will have influence on the system. The influence is stated in Theorem 1.

Theorem 1. Define $\tilde{\boldsymbol{W}}=\hat{\boldsymbol{W}}-\boldsymbol{W}^*, \tilde{\boldsymbol{\varphi}}=\hat{\boldsymbol{\varphi}}-\boldsymbol{\varphi}^*$, $\tilde{\boldsymbol{\mu}}_i=\hat{\boldsymbol{\mu}}_i-\boldsymbol{\mu}_i^*$, $\tilde{\sigma}_i=\hat{\sigma}_i-\sigma_i^*$, $i=1,2,\ldots,l$. The output approximate error of fully tuned RBF NN can be expressed as:

$$\hat{\boldsymbol{W}}^T\hat{\boldsymbol{\varphi}}-\boldsymbol{W}^{*T}\boldsymbol{\varphi}^*=\tilde{\boldsymbol{W}}^T\left(\hat{\boldsymbol{\varphi}}-\hat{\boldsymbol{\varphi}}_\mu'.*\hat{\boldsymbol{\mu}}-\hat{\boldsymbol{\varphi}}_\sigma'.*\hat{\boldsymbol{\sigma}}\right)+\hat{\boldsymbol{W}}^T\left(\hat{\boldsymbol{\varphi}}_\mu'.*\tilde{\boldsymbol{\mu}}+\hat{\boldsymbol{\varphi}}_\sigma'.*\tilde{\boldsymbol{\sigma}}\right)+\boldsymbol{d}_u, \tag{2}$$

where the up bound of $\boldsymbol{d}_u$ is:

$$\|\boldsymbol{d}_u\|\le\left\|\boldsymbol{\mu}^*\right\|\left\|\hat{\boldsymbol{W}}^T\hat{\boldsymbol{\varphi}}_\mu'\right\|+\left\|\boldsymbol{\sigma}^*\right\|\left\|\hat{\boldsymbol{W}}^T\hat{\boldsymbol{\varphi}}_\sigma'\right\|+\left\|\boldsymbol{W}^{*T}\right\|\left\|\hat{\boldsymbol{\varphi}}_\mu'.*\hat{\boldsymbol{\mu}}\right\|+\left\|\boldsymbol{W}^{*T}\right\|\left\|\hat{\boldsymbol{\varphi}}_\sigma'.*\hat{\boldsymbol{\sigma}}\right\|+\sqrt{l}\left\|\boldsymbol{W}^{*T}\right\|. \tag{3}$$

The meaning of the symbols used in (3) can be founded in reference [12].

3 NN-Based Adaptive Control

3.1 System Description

Consider the following nonlinear system

$$\dot{X}_1 = f_1(X_1) + g_1(X_1)X_2, \dot{X}_2 = f_2(\overline{X}_2) + g_2(\overline{X}_2)X_3, \tag{4}$$

$$\cdots\cdots, \dot{X}_n = f_n(\overline{X}_n) + g_n(\overline{X}_n)u, \tag{5}$$

where $\overline{X}_i = \left[X_1^T \quad \cdots \quad X_i^T\right]^T$, $f_i(\overline{X}_i^T), g_i(\overline{X}_i^T)$ are uncertain functions in the system, $X_i = \left[x_{i1} \cdots x_{in_i}\right]^T$, $i = 1, \cdots, n$, n_i is the order of the i th sub-block, and $u \in R^p$. In fact, we can divide the uncertain function into two parts, namely, the nominal part and the uncertain part: $f_i(\overline{X}_i) = f_{i0}(\overline{X}_i) + \Delta f_i(\overline{X}_i)$, $g_i(\overline{X}_i) = g_{i0}(\overline{X}_i) + \Delta g_i(\overline{X}_i)$, where $g_i(\overline{X}_i)$ is invertible, $f_{i0}(\overline{X}_i), g_{i0}(\overline{X}_i)$ are the nominal values, and $\Delta f_i(\overline{X}_i), \Delta g_i(\overline{X}_i)$ are uncertain terms.

3.2 Controller Design and Stability Analysis

The goal of the controller design is to cancel the influence of uncertainties on system performance and track the desired signal X_{id}.

Introduce a new error state $z_i \in R^{n_i}$, and define

$$z_i = X_i - X_{id}, i = 1, \cdots, n, \tag{6}$$

where X_{id} is the desired state trajectory. From (4) - (6), the error state dynamics can be expressed as

$$\dot{z}_1 = f_1(X_1) + g_1(X_1)X_2 - \dot{X}_{1d}. \tag{7}$$

Step 1: Considering and rearranging the system (7), we have

$$\dot{z}_1 = f_{10}(X_1) + g_{10}(X_1)X_2 - \dot{X}_{1d} - {}_1, \tag{8}$$

where ${}_1 = -(\Delta f_1(X_1) + \Delta g_1(X_1)X_2)$ is introduced by general uncertainties. We apply a fully tuned RBF NN to compensate the affect. Assume that

$$_1 = W_1^{*T} \varphi_1^*(Z_1) + \varepsilon_1, \tag{9}$$

where $Z_1 = \left[X_1^T \quad X_2^T\right]^T$ is the input of NN, subscript i in W_i, φ_i, μ_{ij}, σ_{ij} denotes the NN parameters of the i th sub-system, and ε_1 is the construction error of NN. Taking X_2 as a virtual control of (7), there exists an ideal virtual control

$$X_{2d}^* = -g_{10}^{-1}(X_1)\left[f_{10}(X_1) - \dot{X}_{1d} + k_1 z_1 - W_1^{*T}\varphi_1^*(Z_1) - \varepsilon_1\right], \tag{10}$$

such that $\dot{z}_1 = -k_1 z_1 + g_{10}(X_1)(X_2 - X_2^*)$, where $k_1 > 0$ is a design parameter. Because X_2^* is not available, we choose the desired virtual control as

$$X_{2d} = -g_{10}^{-1}(X_1)\left[f_{10}(X_1) - \dot{X}_1^d + k_1 z_1 - \hat{W}_1^T\hat{\varphi}_1(Z_1) - v_1\right], \tag{11}$$

where $\hat{W}_1, \hat{\varphi}_1$ are estimates of W_1^* and φ_1^* respectively, and v_1 is a robust term, defined by (22). Choose Lyapunov function as

$$V_1 = \frac{1}{2}z_1^T z_1 + \frac{1}{2}tr\left[\tilde{W}_1^T \Gamma_{W_1}^{-1}\tilde{W}_1\right] + \frac{1}{2}\sum_{i=1}^{l}\left(\tilde{\mu}_{1i}^T\Gamma_{1\mu}^{-1}\tilde{\mu}_{1i}\right) + \frac{1}{2}\sum_{i=1}^{l}\Gamma_{1\sigma}^{-1}\tilde{\sigma}_{1i}^2, \tag{12}$$

where $\tilde{W}_1 = \hat{W}_1 - W_1^*$, $\tilde{\mu}_{1i} = \hat{\mu}_{1i} - \mu_{1i}^*$, $\tilde{\sigma}_{1i} = \hat{\sigma}_{1i} - \sigma_{1i}^*$, $\hat{\mu}_{1i}$, $\hat{\sigma}_{1i}$ are the estimates of μ_{1i}^* and σ_{1i}^* respectively, $i = 1,\cdots,l$, $\Gamma_{W_1} = \Gamma_{W_1}^T > 0$, $\Gamma_{1\mu} = \Gamma_{1\mu}^T > 0, \Gamma_{1\sigma} > 0$ are design parameters. According to Theorem 1, the derivative of V_1 is given by

$$\begin{aligned}\dot{V}_1 = &-k_1\|z_1\|^2 + z_1^T g_{10}(X_1)(X_2 - X_{2d}) + z_1^T d_{u1} - z_1^T\varepsilon_1 + tr\left[\tilde{W}_1^T\Gamma_{W_1}^{-1}\dot{\tilde{W}}_1\right] \\ &+ z_1^T\left[\tilde{W}_1^T\left(\hat{\varphi}_1 - \hat{\varphi}_{1\mu}^{'}.*\hat{\mu}_1 - \hat{\varphi}_{1\sigma}^{'}.*\hat{\sigma}_1\right) + \hat{W}_1^T\left(\hat{\varphi}_{1\mu}^{'}.*\tilde{\mu}_1 + \hat{\varphi}_{1\sigma}^{'}.*\tilde{\sigma}_1\right)\right] \\ &+ \sum_{i=1}^{l}\left(\tilde{\mu}_{1i}^T\Gamma_{1\mu}^{-1}\dot{\tilde{\mu}}_{1i}\right) + z_1^T v_1 + \sum_{i=1}^{l}\Gamma_{1\sigma}^{-1}\tilde{\sigma}_{1i}\dot{\tilde{\sigma}}_{1i}\end{aligned} \tag{13}$$

Let the parameter adaptive laws of NN be

$$\dot{\hat{W}}_1 = -\Gamma_{W_1}\left(\hat{\varphi}_1 - \hat{\varphi}_{1\mu}^{'}.*\hat{\mu}_1 - \hat{\varphi}_{1\sigma}^{'}.*\hat{\sigma}_1\right)z_1^T, \quad \dot{\hat{\mu}}_{1i} = -\Gamma_{1\mu}\hat{\varphi}_{1\mu_i}^{'}\left(\hat{W}_1 z_1\right)_i, \tag{14}$$

$$\dot{\hat{\sigma}}_{1i} = -\Gamma_{1\sigma}\hat{\varphi}_{1\sigma i}^{'}\left(\hat{W}_1 z_1\right)_i, \quad i = 1,\cdots,l, \tag{15}$$

where $\delta_{W_1}, \delta_{1\mu}, \delta_{1\sigma} > 0$ are parameter to be designed, $\hat{\varphi}_{1\mu_i}^{'} = \dfrac{\partial\varphi_{1i}}{\partial\mu_{1i}}$, $\hat{\varphi}_{1\sigma i}^{'}$ and $\left(\hat{W}_1 z_1\right)_i$ denote the i th element of $\hat{\varphi}_{1\mu}^{'}, \hat{\varphi}_{1\sigma}^{'}$ and $\hat{W}_1 z_1$. Choose the robust term as

$$v_1 = -z_1\left(\left\|\hat{W}_1^T\hat{\varphi}_{1\mu}^{'}\right\|^2 + \left\|\hat{W}_1^T\hat{\varphi}_{1\sigma}^{'}\right\|^2 + \left\|\hat{\varphi}_{1\mu}^{'}.*\hat{\mu}_1\right\|^2 + \left\|\hat{\varphi}_{1\sigma}^{'}.*\hat{\sigma}_1\right\|^2\right)/\eta_1, \tag{16}$$

with $\eta_1 > 0$. Substituting (14)-(16) into (13) and assuming that $\|\varepsilon_1\| \le \varepsilon_{1H}$, we can get that

$$\dot{V}_1 \le -k_1 \|z_1\|^2 + z_1^T g_{10}(X_1) z_2 + c_1 , \tag{17}$$

where $c_1 = \frac{[(2+l)\eta_1]}{4}\|W_1^{*T}\|^2 + \frac{\eta_1}{4}\|\mu_1^*\|^2 + \frac{\eta_1}{4}\|\sigma_1^*\|^2 + \frac{\eta_1}{4}\varepsilon_{1H}^2$.

Step 2: Choosing Lyapunov Function as

$$V_i = V_{i-1} + \frac{1}{2} z_i^T z_i + \frac{1}{2} tr\left[\tilde{W}_i^T \Gamma_{W_i}^{-1} \tilde{W}_i\right] + \frac{1}{2}\sum_{j=1}^{l}\left(\tilde{\mu}_{ij}^T \Gamma_{i\mu}^{-1} \tilde{\mu}_{ij}\right) + \frac{1}{2}\sum_{j=1}^{l} \Gamma_{i\sigma}^{-1} \tilde{\sigma}_{ij}^2 \tag{18}$$

Similar to the Step 1, we have

$$\dot{V}_i \le -\sum_{j=1}^{i} k_j \|z_j\|^2 + z_i^T g_{i0}(\overline{X}_i) z_{i+1} + \|z_i\| \varepsilon_{iH} + z_i^T v_i + \|z_i^T\| \|d_{ui}\| + \sum_{j=1}^{i-1} (c_j) \tag{19}$$

$$\le -\sum_{j=1} k_j \|z_j\| + z_i\, g_{i0}(X_i) z_{i+1} + \sum (c_j),$$

where $c_i = \frac{[(2+l)\eta_i]}{4}\|W_i^{*T}\|^2 + \frac{\eta_i}{4}\|\mu_i^*\|^2 + \frac{\eta_i}{4}\|\sigma_i^*\|^2 + \frac{j\eta_i}{4}\varepsilon_{iH}^2$.

Step n: Choosing Lyapunov function as

$$V_n = V_{n-1} + \frac{1}{2} z_n^T z_n + \frac{1}{2} tr\left[\tilde{W}_n^T \Gamma_{W_n}^{-1} \tilde{W}_n\right] + \frac{1}{2} tr\left[\tilde{W}_g^T \Gamma_g^{-1} \tilde{W}_g\right] + tr\left[\tilde{W}_{gn}^T \Gamma_{gn}^{-1} \dot{\tilde{W}}_{gn}\right]. \tag{20}$$

Taking its time derivative

$$\dot{V}_n = \dot{V}_{n-1} + z_n^T\left(f_{n0}(\overline{X}_n) + g_n(\overline{X}_n) u - \dot{\hat{X}}_{nd} - \quad_n \right) + tr\left[\tilde{W}_n^T \Gamma_{W_n}^{-1} \dot{\tilde{W}}_n\right] \tag{21}$$

where $\quad_n = -\left[\Delta f_{n0}(\overline{X}_n) + \dot{\hat{X}}_{nd} - \dot{X}_{nd}\right]$. Denoting $\hat{g}_n(\overline{X}_n) := g_{n0}(\overline{X}_n) + \Delta\hat{g}_n(\overline{X}_n)$ and assuming that it is invertible. Denoting $[b_{ij}] = \hat{g}_n^{-1}(\overline{X}_n), i, j = 1, \cdots, n_n$. For the simplicity, uncertainties in the sub-system are approximated by RBF NN that only weight matrix is updated

$$\Delta\hat{f}_n(\overline{X}_n) + \dot{\hat{X}}_{nd} - \dot{X}_{nd} = W_n^* \varphi_n + \varepsilon_{Wn}, \Delta g_n(\overline{X}_n) = W_g^* \varphi_g + \varepsilon_g,$$

$W_g = \left[\left(W_{gij}^T\right)_{1\times l}\right]_{n_n \times (n_n \times l)}$, $\varphi_{gn} = diag(\varphi_n, \cdots, \varphi_n)$. Choosing control law as

$$u = -\left[g_{n0}(\overline{X}_n) + \hat{W}_g \varphi_g(Z_n)\right]^{-1}\left[f_{n0}(\overline{X}_n) - \dot{\hat{X}}_{nd} + k_n z_n + g_{(n-1)0}^T(\overline{X}_{(n-1)}) z_{(n-1)} - \hat{W}_n \varphi_n(Z_n) - v_n\right], \tag{22}$$

where $v_n = \left[v_{n1}, \cdots, v_{nn_n}\right]^T$, $v_{ni} = D_{\varepsilon fi}\operatorname{sgn}(z_{ni}) + U_{\max}\operatorname{sgn}(z_{ni})\sum_{j=1}^{n_n} D_{\varepsilon gij}$, $D_{\varepsilon fi}$,

$D_{\varepsilon gij}, i, j = 1, \cdots, n_n$ are the up bounds of the NN approximated errors, and $U_{\max} > 0$. From (22), we have

$$u_i = \sum_{j=1}^{n_n} b_{ij} \left[f_{n0}(\overline{X}_n) - \dot{\hat{X}}_{nd} + k_n z_n + g_{(n-1)0}^T (\overline{X}_{(n-1)}) z_{(n-1)} - \hat{W}_n \varphi_n (Z_n) - v_n \right]_j . \quad (23)$$

Let

$$a_i(\overline{X}_n) = \sum_{j=1}^{n_n} |b_{ij}| \left[|f_{n0}(\overline{X}_n)|_j + |\dot{\hat{X}}_{nd}|_j + k_n |z_n|_j + |g_{(n-1)0}^T (\overline{X}_{(n-1)}) z_{(n-1)}|_j \right]$$

$$+ \sum_{j=1}^{n_n} \left[|b_{ij}| \|\hat{W}_n^T \varphi_n (Z_n)\|_j + D_{\varepsilon fi} \right], \quad h_i(\overline{X}_n) = \sum_{j=1}^{n_n} |b_{ij}| \sum_{l=1}^{n_n} D_{\varepsilon gjl}$$

, where $|\cdot|_j$ denotes the absolute of the j th element. Because $D_{\varepsilon gjl}$ can be arbitrary small, following the idea of reference [14], we can assume that $0 \le h_i(\overline{X}_n) < 1$. With (23), we can get

$$|u_i| \le a_i(\overline{X}_n) + U_{\max}(\overline{X}_n) h_i(\overline{X}_n) \le U_{\max}(\overline{X}_n),$$

$$U_{\max}(\overline{X}_n) \ge \max_{i=1,\cdots,n_n} \left[\frac{a_i(\overline{X}_n)}{1 - h_i(\overline{X}_n)} \right]. \quad (24)$$

Equation (21)can be rewritten as

$$\dot{V}_n \le \dot{V}_{n-1} + z_n^T \left(f_{n0}(\overline{X}_n) + [g_{n0}(\overline{X}_n) + \Delta \hat{g}_n(\overline{X}_n)] u - \dot{\hat{X}}_{nd} - \quad _n + [\Delta g_n(\overline{X}_n) - \Delta \hat{g}_n(\overline{X}_n)] u \right),$$

$$+ tr\left[\tilde{W}_n^T \Gamma_{W_n}^{-1} \dot{\tilde{W}}_n \right] + tr\left[\tilde{W}_g^T \Gamma_g^{-1} \dot{\tilde{W}}_g \right] \quad (25)$$

Because $\hat{W}_n$, $\hat{W}_{gn}$ cannot be guaranteed in the bounded closed set in control process, we apply the project operator in [15]. Assuming that the feasibility sets of the parameter space are

$$\Omega_{Wn0} \triangleq \left\{ \hat{W}_n \,\middle|\, \|\hat{W}_n^T \hat{W}_n\|^2 \le \beta_{Wn} \right\}, \Omega_{Wn} \triangleq \left\{ \hat{W}_n \,\middle|\, \|\hat{W}_n^T \hat{W}_n\|^2 \le \beta_{Wn} + \lambda_{Wn} \right\},$$

where, $\beta_{Wn} > 0$, $\lambda_{Wn} > 0$, $\Omega_{gn0} \triangleq \left\{ \hat{W}_g \,\middle|\, \|\hat{W}_g^T \hat{W}_g\|^2 \le \beta_{Wg} \right\}$,

$\Omega_{gn} \triangleq \left\{ \hat{W}_g \,\middle|\, \|\hat{W}_g^T \hat{W}_g\|^2 \le \beta_{Wg} + \lambda_{Wg} \right\}$, with $\beta_{Wg} > 0$, $\lambda_{Wg} > 0$.

The parameter adaptive laws are chosen as:

$$\dot{\hat{W}}_n = \Gamma_{W n} \mathrm{proj}(\hat{W}_n, \Phi_W), \; \dot{\hat{W}}_g = \Gamma_g \mathrm{proj}(\hat{W}_g, \Phi_g) \quad (26)$$

Defining the project operator as

$$\operatorname{proj}\left(\hat{\boldsymbol{M}},\boldsymbol{\Phi}_M\right)=\begin{cases}\boldsymbol{\Phi}_M-\dfrac{\left(\left\|\hat{\boldsymbol{M}}\right\|^2-\beta_M\right)\boldsymbol{\Phi}_M^T\hat{\boldsymbol{M}}}{\delta_M\left\|\hat{\boldsymbol{M}}\right\|^2}\hat{\boldsymbol{M}}, if\ \left\|\hat{\boldsymbol{M}}\right\|^2>\beta_M \text{ and } \boldsymbol{\Phi}_M^T\hat{\boldsymbol{M}}>0,\\ \boldsymbol{\Phi}_M, elsewise\end{cases}$$

where $\boldsymbol{M}=\boldsymbol{W}_n,\boldsymbol{W}_g$, $\boldsymbol{\Phi}_{Wn}=\boldsymbol{\Gamma}_{W_n}z_n\phi_{Wn}^T$ and $\boldsymbol{\Phi}_{Wg}=-\boldsymbol{\Gamma}_g z_n\boldsymbol{u}^T\varphi_g^T$. Using the results of [15], we have

$$\dot{V}\le-\sum_{j=1}^{n}k_n\left\|z_n\right\|^2+\sum_{j=1}^{n-1}\left(c_j\right), \tag{27}$$

where $k=\min_{j=1,\cdots,n}\left\{2k_j\right\}$. From (27), we can conclude that all the signals in closed-loop system are UUB. So we can assume that $c_0=\sup_{\hat{W}_1,\cdots,\hat{W}_n,\hat{W}_g}\left(\frac{1}{2}\sum_{j=1}^{n}tr\left[\tilde{\boldsymbol{W}}_j^T\boldsymbol{\Gamma}_{W_j}^{-1}\tilde{\boldsymbol{W}}_j\right]+\frac{1}{2}tr\left[\tilde{\boldsymbol{W}}_g^T\boldsymbol{\Gamma}_g^{-1}\tilde{\boldsymbol{W}}_g\right]\right)<\infty$. Let $c_n=kc_0$ and $c=\sum_{j=1}^{n}c_j$. (27) can be rewritten as $V(t)\le V(0)e^{-kt}+\frac{1}{k}c$, $\forall t\ge 0$. Consider (20), we can conclude that

$$\sum_{j=1}^{n}\left\|\tilde{\boldsymbol{W}}_j\right\|^2\le\frac{2V(t)}{\lambda_{\min}\left(\boldsymbol{\Gamma}_{W_j}^{-1}\right)},\ \sum_{j=1}^{n}\left\|\tilde{\boldsymbol{\mu}}_j\right\|^2\le\frac{2V(t)}{\lambda_{\min}\left(\boldsymbol{\Gamma}_{j\mu}^{-1}\right)},\ \sum_{j=1}^{n}\left\|\tilde{\boldsymbol{\sigma}}_j\right\|^2\le\frac{2V(t)}{\lambda_{\min}\left(\Gamma_{j\sigma}^{-1}\right)}, \tag{28}$$

$$V(t)\ge\frac{1}{2}\sum_{j=1}^{n}\left\|z_j\right\|^2 \tag{29}$$

From (27) and (29), we have $\dot{V}(t)\le-\frac{k}{2}\sum_{j=1}^{n}\left\|z_j\right\|^2+c$. Integrating it yields

$$\int_0^t\left\|z_j(\tau)\right\|^2d\tau\le 2\left[V(0)+tc\right]/k,\ j=1,\cdots,n. \tag{30}$$

From the discussion above, we have the following result,

Theorem 2. Considering the system (4)-(5), if Assumptions 1 holds, using the proposed approach, the following results hold,

(1)The state tracking error of the system z_j and the parameter estimated error of NN are bounded and converge to the neighborhoods of the origins exponentially,

$$\Omega_j=\left\{z_j,\tilde{\boldsymbol{W}}_j,\tilde{\boldsymbol{\mu}}_j,\tilde{\boldsymbol{\sigma}}_j\left|\sum_{j=1}^{n}\left\|z_j\right\|^2\le 2\left[V(0)+\frac{1}{k}c\right],\sum_{j=1}^{n}\left\|\tilde{\boldsymbol{W}}_j\right\|^2\le\frac{2\left[V(0)+\frac{1}{k}c\right]}{\lambda_{\min}\left(\boldsymbol{\Gamma}_{W_j}^{-1}\right)},\right.\right.$$

$$\left.\sum_{j=1}^{n}\left\|\tilde{\boldsymbol{\mu}}_j\right\|^2 \le \frac{2\left[V(0)+\frac{1}{k}c\right]}{\lambda_{\min}\left(\boldsymbol{\Gamma}_{j\mu}^{-1}\right)}, \sum_{j=1}^{n}\left\|\tilde{\sigma}_j\right\|^2 \le \frac{2\left[V(0)+\frac{1}{k}c\right]}{\lambda_{\min}\left(\Gamma_{j\sigma}^{-1}\right)}\right\}, \; j=1,\cdots,n$$

(2)The following inequalities hold,

$$\lim_{t\to\infty}\frac{1}{t}\int_0^t\left\|z_j(\tau)\right\|^2 d\tau \le 2c/k\,, \; \sum_{j=1}^{n}\left\|z_j\right\|^2 \le 2\left[V(0)+\frac{1}{k}c\right].$$

Remark 1. From the result (1) of the Theorem 2, we know that adjusting the values of $k_i, \delta_{W_i}, \delta_{i\mu}, \delta_{i\sigma}, \boldsymbol{\Gamma}_{W_i}, \boldsymbol{\Gamma}_{i\mu}, \Gamma_{i\sigma}, \eta_i$, we can control the convergence rate and the size of the convergence region.

4 Simulation Example

The missile model with general set of uncertainties considered hereafter is a three-axes , highly nonlinear model of a BTT missile, The nonlinear dynamic equations are given as follows:

$$\dot{x}_1 = f_1(x_1)+b_1(x_1)x_2+h_1(x_1)u\,,\; \dot{x}_2 = f_2(x_1,x_2)+b_2u\,, \tag{31}$$

where $x_1=[\alpha \;\; \beta \;\; \phi]^T$, $x_2=[p \;\; q \;\; r]^T$, $u=[\delta_x \;\; \delta_y \;\; \delta_z]^T$. The meaning of the symbols used in this paper can be founded in [12].

In order to validate the effectiveness and validity of the proposed method, nonlinear 6-DOF simulation results for a BTT missile model are presented, The solid lines represent the command signals, The dotted lines and the dash-dotted lines represent the simulation results of the adaptive controller under uncertain air parameters with 50 percent (up and down) uncertainty, It is shown that the control system has good stability, performance and robustness even if the large model uncertainties exist.

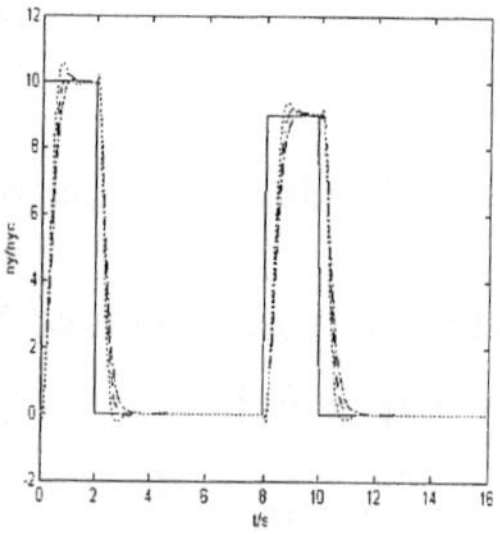

Fig. 1. n_y tracking curves

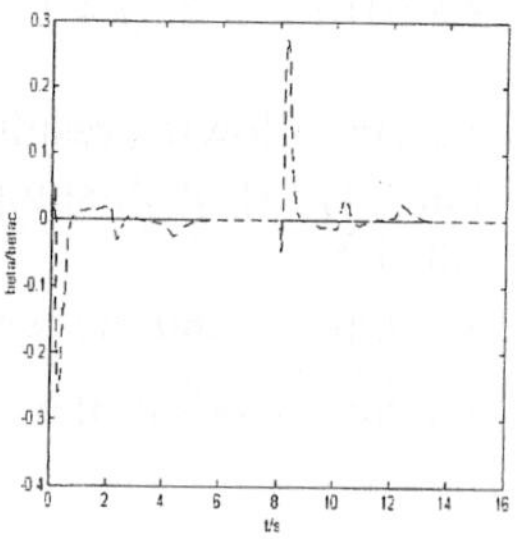

Fig. 2. β tracking curves

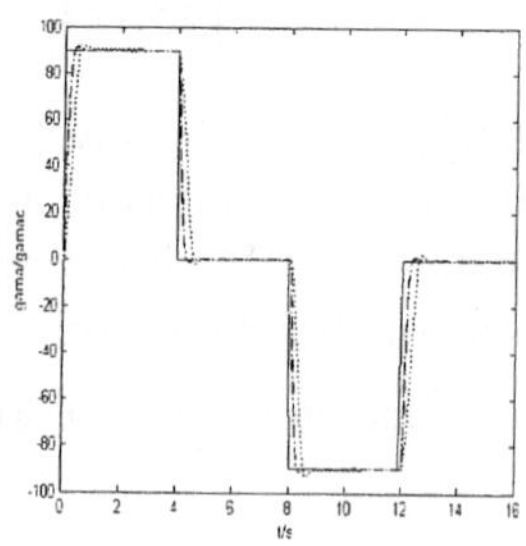

Fig. 3. ϕ tracking curves

5 Conclusions

The main contribution of this paper are: a) relaxed the requirement of SISO in references [10] and [11]; b)took the prior knowledge of the system into consideration and successfully solved the design problem for a class of multi-input multi-output nonlinear systems when the control coefficient matrix is unknown; c)avoided the possible singularities of the controller; d)Applied Lyapunov stability theory, the parameter updating laws of the fully tuned RBF NN was derived, which guaranteed the stability of the overall systems; e) by introducing nonlinear tracking differentiators and NNs, the "computation explosion" is reduced.

References

1. Sastry, S. S., Isidori, A.: Adaptive Control of Linearizable System. IEEE Translations on Automatic Control, 11 (1989) 1123-1131
2. Seto, D., Annaswamy, A. M., Baillieul, J.: Adaptive Control of Nonlinear Systems with a Triangular Structure. IEEE Translations on Automatic Control, 7 (1994) 1411-1428
3. Polycarpou, M. M.: Stable Adaptive Neural Control Scheme for Nonlinear Systems. IEEE Translations on Automatic Control, 3 (1996) 447-451
4. Krstic, M., Kanellakopoulos, I., Kokotovic, P.: Nonlinear and Adptive Control Design. Wiley–Interscience Publication (1995)
5. Vadim, I. Utkin., De-Shiou, C., Hao-Chi, C.: Block Control Principle for Mechanical Systems. Journal of Dynamic Systems, Measurement, and Control, 1 (2000) 1-10
6. Loukianov, A., Toledo, B.C., Dodds, S.J.: Nonlinear Sliding Surface Design in the Presence of Uncertainty. Proceedings of the 14th IFAC, Beijing, P.R.China (1999) 55-60
7. Jagannathan, S., Lewis, F.L.: Robust Backstepping Control of a Class of Nonlinear Systems Using Fuzzy Logic. Information Sciences, 2 (2000) 223-240
8. Yan, L., Sundararajan, N., Saratchandran, P.: Neuro-controller Design for Nonlinear Fighter Aircraft Maneuver Using Fully Tuned RBF Networks. Automatica, 8 (2001) 1293-1301
9. Park, J., Sandberg, I. W.: Universal Approximation Using Radial Basis Function Networks. Neural Computation, 2 (1991) 246-257
10. Zhang, T., Ge, S.S., Hang, C.C.: Adaptive Neural Network Control for Strict-Feedback Nonlinear Systems Using Backstepping Design. Automatica, 12 (2000) 1835-1846
11. Ge, S. S., Wang, C.: Adaptive NN Control of Uncertain Nonlinear Pure-Feedback Systems. Automatica, 4 (2002) 671-682
12. Jin, Y.Q.: Nonlinear Adaptive Control System Design for Missile. Yantai, P.R.China (2003)
13. Han, J., Wang, W.: Nonlinear Tracking Differentiator. System Science and Mathematics, 2 (1994) 177-183
14. Ordonez, R., Spooner, J.T.: Stable Multi-input Multi-output Adaptive Fuzzy Control. Proceedings of the 35th CDC, Japan (1996) 610-615
15. Khalil, H.K.: Adaptive Output Feedback Control of Nonlinear Systems Represented by Input-Output Models. IEEE Transactions on Automatic Control, 2 (1996) 177-188

Model Reference Control Based on SVM

Junfeng He and Zengke Zhang

Department of Automation, Tsinghua University, Beijing, PR China
heroson98@mails.tsinghua.edu.cn

Abstract. A model reference control algorithm based on Support Vector Machine (SVM) for discrete nonlinear systems is proposed in this article. It uses SVM as regression tools to learn the feed forward controller from input and output data. Further more, a compensator is used as the feed back controller to make the whole system more robust. Advantages of SVM and the robust compensator help the method perform excellently as shown in the experiments.

1 Introduction

Model Reference Control (MRC) is a very important and practical method in nonlinear control field. The popular way of MRC nowadays might be learning the controller from input and output data with neural networks. However, despite many advances, there are numerous weak points for this classic neural networks approach, for example, the existence of many local minima solutions, the problem to select hidden units and so on. Major breakthroughs at this point might be Support Vector Machine (SVM) [6], successfully both theoretically and in applications. There are a lot of examples [1,3,6], which reported excellent performances of SVM in regression and nonlinear modeling area. In this paper, we propose a MRC method based on SVM, which would show that SVM is also a wonderful choice in MRC field.

2 Model Reference Control

For MRC as shown in figure 1, plant P is the system to be controlled, and M is the reference model (always designed as a simple linear system) to generate reference output signal d, which satisfies our requirements under the input of r. Controller C is to be learned to generate proper signal u for the plant, so that the output y of the plant could track reference output d. More details about MRC could be found in references [4,5]. A standard assumption in MRC is that: Plant P and the inverse system P^{-1} are both stable. Another assumption is to assume the relative degree of the reference model M is greater than or equal to that of the plant P.
If we know the exact model P, we could compute the controller as:

$$C = M \circ P^{-1} \quad . \tag{1}$$

F. Yin, J. Wang, and C. Guo (Eds.): ISNN 2004, LNCS 3174, pp. 138–143, 2004.

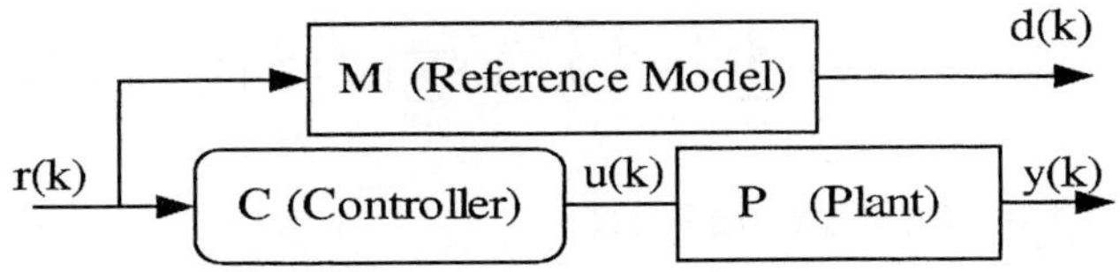

Fig. 1. Model Reference Control Scheme

It guarantees the satisfaction of the MRC objective. However, in many applications we might not get the exact model P, or could not compute the controller C from model P because of intensive computation loads, so a more practical way is to learn the controller from the data. Often, researchers use BP neural networks to do that. The weights of neural networks are adapted to minimize the risk function:

$$R(w_{i,j}) = \sum_{t=1}^{N} (y(k) - d(k))^2 \quad . \tag{2}$$

3 SVM for Regression

In this section, we would introduce only the basic idea of SVM for regression. For complete details, please refer to reference [6].
Consider regression with the following function set:

$$f(x) = (w \cdot \phi(x)) + b \ , \ x \in R^n, \phi : R^n \to R^{n_h}, w \in R^{n_h}, b \in R \quad . \tag{3}$$

with given training data $\{x_i, y_i\}$ (i=1...*l*) , ($x_i \in R^n$ is input data; $y_i \in R$ is output data). The nonlinear mapping $\phi : R^n \to R^{n_h}$ maps the input data to a so-called high-dimensional feature space. In fact, with some kinds of mapping ϕ, the function set described by equation (3) could approximate all the continuous function up to any desired accuracy [6]. SVM tries to solve the optimization problem according to the SRM (structure risk minimization) rule:

$$\min \qquad R_{emp}(w,b) = \frac{1}{l}\sum_{i=1}^{l} | y_i - (w \cdot \phi(x_i)) - b |_\varepsilon \quad , \tag{4}$$

$$st \ : (w \cdot w) \le C \quad .$$

Function $|\cdot|_\varepsilon$ is called $\varepsilon-insensitive$ function, defined as:

$$| y_i - (w \cdot \phi(x_i)) - b |_\varepsilon = 0 \ , \qquad \text{if} \ | y_i - (w \cdot \phi(x_i)) - b | \le \varepsilon \ ; \tag{5}$$

$$= | y_i - (w \cdot \phi(x_i)) - b | - \varepsilon \ , \quad \text{otherwise.}$$

After rewrite the optimization problem and use the Lagrangian method [1,6], we could get the final dual problem:

$$\max W(\partial,\partial^*) = -\varepsilon\sum_{i=1}^{l}(\partial_i^* + \partial_i) + \sum_{i=1}^{l} y_i(\partial_i^* - \partial_i) - 0.5\sum_{i,j=1}^{l}(\partial_i^* - \partial_i)(\partial_j^* - \partial_j)K(x_i, x_j) \ , \tag{6}$$

$$\text{st}: \quad \sum_{i=1}^{l}\partial_i^* = \sum_{i=1}^{l}\partial_i, \quad 0 \le \partial_i^* \le C, 0 \le \partial_i \le C \ .$$

For any given input data x, the output is:

$$f(x) = w \cdot \phi(x) + b = \sum_{i=1}^{N}(\partial_i^* - \partial_i)K(x_i, x) + b \ . \tag{7}$$

The kernel function K corresponds to:

$$K(x_i, x_j) = \phi(x_i)^T \phi(x_j) \ . \tag{8}$$

4 MRC Based on SVM

4.1 Use SVM in MRC

The MRC method introduced in section 2 could not directly use SVM to learn the controller, since SVM would use the "target" ("labeled") output of the controller, which is unknown in that scheme. However, for linear SISO system, Widrow [7] once proposed an alternative method, which could be developed to use SVM to learn the controller: the plant is placed in front of the controller when learning, but placed after the controller when controlling. For linear control, exchanging their places would theoretically not change the control result. However, if P is nonlinear, theoretically we could exchange the place of C and that of P to get the same control result if only the reference model is a unit mapping or pure time delay system. So another scheme of model reference control both suitable for linear and nonlinear system, is provided in figure 2.

In figure 2, the parallel reference model M_0 is a unit mapping or delay system so that the controller C learns to approximate the inverse system $P^{-1}$of the plant P. Moreover, a forward reference model M is used to generate the reference output satisfying our requirements. To learn the controller in figure 2, we could use the learning scheme shown in figure 3. Since the NARMA model of the controller could be described as:

$$z(k+n) = f(z(k+n-1), \ldots, z(k), u(k+m), \ldots, u(k)) \ . \tag{9}$$

Our task is to find the function f with the data: z(k) = q(k) and u (k)=s (k), k=1…N(Here, q is the output of the parallel reference model M_0 and s is the output of plant under the same input v, as shown in figure3. The input v is usually selected to cover a wide range in frequency domain). The order of the controller (m and n) is often appointed by prior knowledge, posterior error or other methods. Let :

$$y(k) = q(k+n) \ ; \tag{10}$$

$$x(k) = [q(k+n-1), \ldots, q(k), s(k+m), \ldots, s(k)]^T \quad (k = 1 \ldots N-n) \ . \tag{11}$$

The problem of learning the controller equals to find the regression function y = f (x) with data set:{ x (k), y (k)},k=1…N-n. It could be solved with SVM directly in the same form as introduced in section 3.

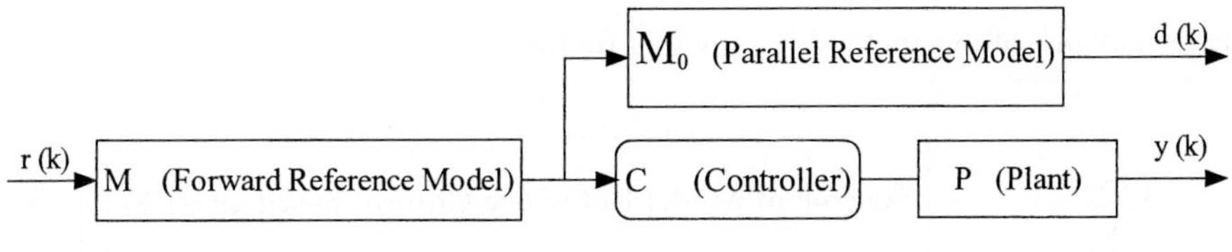

Fig. 2. New Model Reference Control Scheme

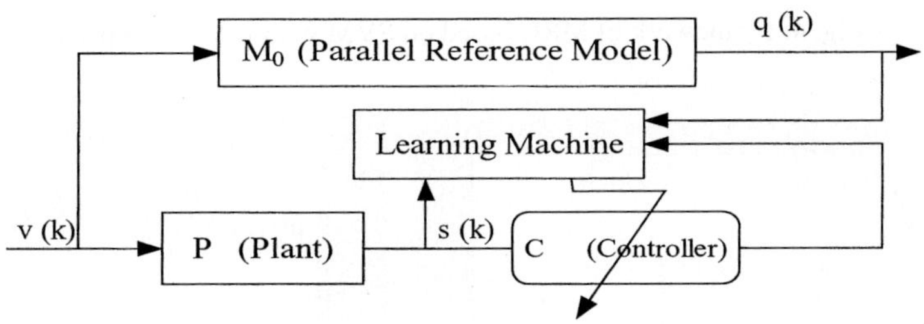

Fig. 3. Scheme to Learn the Controller in MRC

4.2 Robust Compensator

Since MRC based on SVM introduced above is open-loop control, to make it more robust, we use a robust compensator [2] as the feedback controller. As shown in figure 4, suppose G is the real transfer function from r to y in MRC control scheme of figure 2, and suppose G_0 is the transfer function of M. In our method, G is expected to approximate G_0, but there might be differences between them because of learning error. Moreover, noises might exist in the system. Concerning all the noises, errors and disturbances, y could be viewed as the output of G_0 with the artificial equivalent noise e:

$$y = G_0 r + e \,. \tag{12}$$

All the impacts of errors, disturbances and noises are included in the artificial equivalent noise e. The main idea of the robust compensator is to compensate or counteract e as follows:

$$r = u - v \,. \tag{13}$$

$$v = G_0^{-1} e = G_0^{-1} \,(y - y_0) = G_0^{-1} \,(y - G_0 r) = G_0^{-1} y - r \,. \tag{14}$$

However G_0^{-1} is non-causal and could not be realized. A way out to this problem is to use a linear low-pass filter F to make it causal .The relative order of F is designed to be higher than that of G_0:

$$v = F\,G_0^{-1}\,e = FG_0^{-1}\,y - Fr\,. \tag{15}$$

The framework of our method is shown in figure 4.

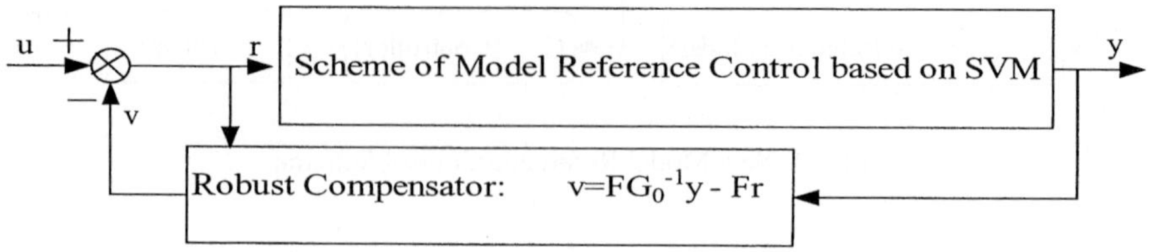

Fig. 4. Framework of MRC based on SVM with robust compensator

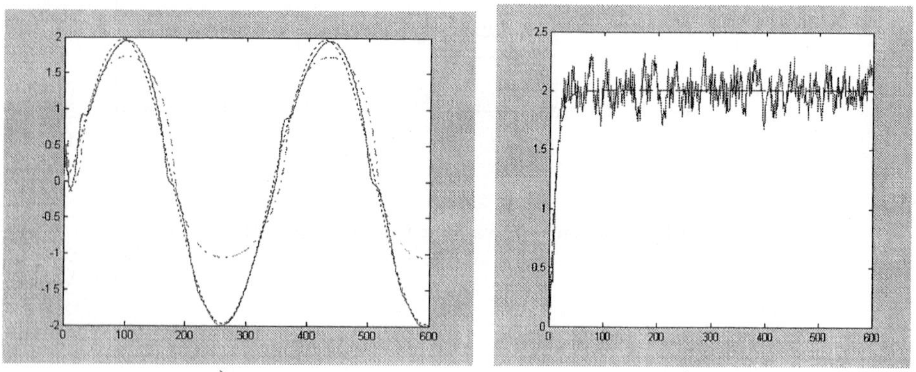

a) b)

Fig. 5. The simulations of control8ß results. a) Comparing the control result of MRC based on SVM with the control result of MRC based on neural networks under the same input. The dotted line is the reference output; the dash-dotted line is the output of MRC based on neural network; and the solid line is the output of MRC based on SVM, which approximate the reference output more accurately. b) The result of our control method with uniform noises of $SNR \approx 25$ in the input of the plant. The dotted line is the reference output and the solid line is the real output of plant.

5 Simulations

The plant to be controlled is a nonlinear system as follows:

$$y(k+1) = 2.5y(k)y(k-1)/(1+ y^2(k)+ y^2(k-1)) + 0.3\cos(y(k) + y(k-1))+1.2u(k). \tag{16}$$

When learning, 50 data is used as training set and another 50 data as the test set. In our method, the kernel type of SVM is RBF kernel with σ =5, and C= 20. The order of controller is determined by learning error on the test set.
As shown in graph a) of figure 5, we compare our control method with the classic MRC method based on neural networks. Obviously, MRC based on SVM works more outstandingly than MRC based on neural networks.
As shown in graph b) of figure 5, even when strong noises exist, our new method still works quite well.

6 Conclusions

A new MRC method based on SVM for linear and nonlinear system is investigated and some exciting results are achieved. Though we only discuss the method for SISO system in this article, it is easy to extend this algorithm to make it suitable for MIMO system. However, some future work must be done to make the method more sound in theory and practical in applications. For example, the methods for non-minimum phase system still need more investigation.

References

1. J.A.K, Suykens, et al.: Optimal Control by Least Squares Support Vector Machine. Neural Networks, Vol. 14. Pergamon, New York (2001) 23-35
2. Yi-Sheng Zhong. : Robust output tracking control for SISO plants with multiple operating points and with parametric and unstructured uncertainties. INT.J.CONTROL. Vol. 75. Taylor and Francis, London (2002) 219-241
3. De Kruif, B.J. De vries, T.J.A. :On using a support vector machine in Learning Feed-Forward Control. IEEE/ASME International Conference on Advanced Intelligent Mechatronics, Vol. 1. Como Italy (2001) 272-277
4. G.W. Irwin, Hunt K J et al. (eds.): Neural network Applications In Control. The institution of Electrical Engineers. (1992)
5. Hans butler. (ed.): Model Reference Adaptive Control- from theory to practice. Prentice hall (1992)
6. Vapnik, V.: Statistical learning theory. Wiley New York (1998)
7. Bernad Widrow. : Adaptive Inverse Control. Prentice-Hall. (1996)

PID Controller Based on the Artificial Neural Network

Jianhua Yang, Wei Lu, and Wenqi Liu

School of Electr. & Inf. Eng., Dalian Univ. Of Technol., Dalian 116024, China
{jianhuay,wqliu}@dlut.edu.cn
luweimail@126.com

Abstract. The paper provides a new style of PID controller that is based on neural network according to the traditional one's mathematical formula and neural network's ability of nonlinear approximation. It also discusses the corresponding learning algorithm and realizing method. This new controller is proven highly practical and effective in the simulation test. This new controller has more advantage than the traditional one, such as more convenient in parameter regulating, better robust, more independence and adaptability on the plant, etc.

1 Introduction

The traditional PID control is widely applied to industrial produce process, for its control mode is direct, simply and robust. But, there are some disadvantages of PID control. Firstly, it is difficult to regulate the three parameters of PID controller: KP, KI and KD in some control systems. Secondly, the traditional PID control used in nonlinear, time varying or large inertia systems will be not very effective.

Artificial neural networks (ANN) have been used in various control field of with the characteristics of self-adapting and learning, nonlinear mapping, strong robustness and fault-tolerance. In this paper, an ANN PID controller which is based both ANN and the traditional PID control is presented. The simulation proves the controller is applicable, and it is easily realized and more convenient to regulate parameters.

2 Traditional PID Control with Artificial Neural Network

It is well known, there are two traditional PID controller modes, one is locational mode, and the other is incremental mode. The ANN realization of locational mode PID is shown below. It can be referenced to get the one of incremental mode.

$$u(k) = k\left[e(k) + \frac{T}{T_i}\sum_{j=0}^{k} e_j(k) + \frac{T_d}{T}(e(k) - e(k-1))\right] = k_p e(k) + k_i T\sum_{j=0}^{k} e_j(k) + k_d\frac{\Delta e(k)}{T} \tag{1}$$

Where { $k_p = k$, $k_i = \frac{k}{T_i}$, $k_d = kT_d$ }, T is the sampling period, $u(k)$ is the output of the controller, $e(k)$ is the deviation.

F. Yin, J. Wang, and C. Guo (Eds.): ISNN 2004, LNCS 3174, pp. 144–149, 2004.

For equation (1), $u(k)$ is the linear combination of $e(k)$, $T\sum_{j=0}^{k} e_j(k)$ and $\frac{\Delta e(k)}{T}$, i.e.

$$u(k) = f(e(k), T\sum_{j=0}^{k} e_j(k), \frac{\Delta e(k)}{T}) \tag{2}$$

Feed-forward ANN is used to construct an ANN PID controller. Generally, a three-layered feed-forward ANN with appropriate network structure and weights can approach to any random continuous function. The ANN is designed with three layers in consideration of the control system real time requirement.

Obviously, there are three nodes in input layer, the deviation $e(k)$, the cumulation of deviation $\sum_{j=0}^{k} e_j(k)$ and the variety of deviation $\Delta e(k)$. Only one node in output layer, that is, the output of the controller $u(k)$. In order to simplify the structure of the ANN, hidden layer nodes, which can correctly reflect the relationship between the input and the output of the ANN, are designed as few as possible, and 8 nodes is assumed in this paper.In practice, hidden layer nodes which can be acquired by test method or experience. The neuron activation function of input layer is assumed linear function $f_i(x) = x$; that of hidden layer is assumed the Sigmoid function $f_h(x) = \frac{1}{1+e^{-x}}$; and that of output layer is assumed linear function $f_o(x) = x$. So a 3-8-1 network is constructed, which can take the place of traditional PID controller.

3 Rule of Neutral Networks PID Weights Adjusting

The capability error function J of the network constructed above can be written as follows:

$$J = \frac{1}{2}[y_p(k+1) - y(k+1)]^2 \tag{3}$$

Steepest descent method is used to regulate weights to make J least.

The networks inputs of input layer are $\{e(k), \sum_{j=0}^{k} e_j(k), \Delta e(k)\}$. Corresponding weights are $\{1, T, 1/T\}$.

Input of hidden layer's *j*th neuron can be written as follows:

$$I_{hj} = \sum_{i=1}^{3} \omega_{ij} X_i \tag{4}$$

Where ω_{ij} is the weight which connect *j*th(j=1,2…8) neuron in hidden layer with *i*th(i=1,2,3) neuron in input layer; X_i (i=1,2,3) are the outputs of input layer, that is: $X_1 = e(k)$, $X_2 = T\sum_{j=0}^{k} e_j(k)$, $X_3 = \frac{\Delta e(k)}{T}$.

Output of the *j*th neuron in hidden layer is

$$O_{hj} = f_h(I_{hj}) \tag{5}$$

Input of the neuron in output layer is

$$I_{o1} = \sum_{j=1}^{8} \omega_{j1} O_{hj} \tag{6}$$

ω_{jl} (j=1,2…8)is the weight which connects the only neuron in output layer with the *j*th neuron in hidden layer.

Output of the neuron in output layer is

$$O_{ol} = f_o(I_{ol}) = I_{ol} \tag{7}$$

Output of the neutral network PID controller is

$$u(k) = O_{ol} = f(I_{ol}) = I_{ol} \tag{8}$$

3.1 Adjusting of Output Layer's Weights

Weight learning rule is

$$\Delta\omega_{jl} = -\eta\frac{\partial J}{\partial\omega_{jl}} \tag{9}$$

Where η is learning speed, it commonly is (0, 1).

$$\frac{\partial J}{\partial\omega_{jl}} = \frac{\partial J}{\partial y(k+1)}\frac{\partial y(k+1)}{\partial O_{ol}}\frac{\partial O_{ol}}{\partial\omega_{jl}} = -(y_p(k+1) - y(k+1))O_{hj}\frac{\partial y(k+1)}{\partial u(k)} \tag{10}$$

Where $\frac{\partial y(k+1)}{\partial u(k)}$ is unknown which denotes system's input-output relationship. For most systems, its signal is definite. $\frac{\partial y(k+1)}{\partial u(k)}$ is replaced with $\mathrm{sgn}(\frac{\partial y(k+1)}{\partial u(k)})$, and learning speed η is used to equalize the calculate error. Adjusting rule of output layer's weights ω_{jl} is:

$$\omega_{jl}(k+1) = \omega_{jl}(k) + \eta(y_p(k+1) - y(k+1))O_{kj}\cdot\mathrm{sgn}(\frac{\partial y(k+1)}{\partial u(k)}) \tag{11}$$

3.2 Adjusting of Hidden Layer's Weights

Weight learning rule is

$$\Delta\omega_{ij} = -\eta\frac{\partial J}{\partial\omega_{ij}} = -\eta\frac{\partial J}{\partial I_{hj}}\frac{\partial I_{hj}}{\partial\omega_{ij}} = -\eta\frac{\partial J}{\partial O_{hj}}\frac{\partial O_{hj}}{\partial I_{hj}}X_i = -\eta\frac{\partial J}{\partial O_{hj}}O_{hj}(1-O_{hj})X_i \tag{12}$$

$$\begin{aligned}\frac{\partial J}{\partial O_{hj}} &= \frac{\partial J}{\partial I_{ol}}\frac{\partial I_{ol}}{\partial O_{hj}} = \frac{\partial J}{\partial I_{ol}}\frac{\partial}{\partial O_{hj}}\sum_{j=1}^{8}\omega_{jl}O_{hj} = \frac{\partial J}{\partial I_{ol}}\omega_{jl} = \frac{\partial J}{\partial y(k+1)}\frac{\partial y(k+1)}{\partial O_{ol}}\omega_{jl} \\ &= -(y_p(k+1) - y(k+1))\omega_{jl}\frac{\partial y(k+1)}{\partial u(k)}\end{aligned} \tag{13}$$

Where $\frac{\partial y(k+1)}{\partial u(k)}$ is replaced by $\mathrm{sgn}(\frac{\partial y(k+1)}{\partial u(k)})$, learning speed η is used to equalize the calculate error. Corresponding adjusting rule of hidden layer's weights ω_{ij} is:

$$\omega_{ij}(k+1) = \omega_{ij}(k) + \eta(y_p(k+1) - y(k+1))\omega_{jl}O_{hj}(1-O_{hj})X_i\cdot\mathrm{sgn}(\frac{\partial y(k+1)}{\partial u(k)}) \tag{14}$$

4 PID Control System Based on Neural Network

As shown in the following figure, r is system referenced input, y is system output, u is controlling output of neural network PID. Deviation e, cumulation of deviation $\sum e$ and variety of deviation Δe are applied as the inputs of the network in neural network PID controller which adapt the weights according to the following steps:

1. Decide network structure at first. Because the nodes of network input layers and output layers are known, only the nodes of hidden layers remained undecided.
2. Initialize the weights of hidden layers ω_{ij} and the ones of output layers ω_{jl} with less random number, select the speed of learning η;
3. Sample the system, get $e(k)$, calculate $\sum_{j=0}^{k} e_j(k)$ and $\Delta e(k)$, which are the network inputs;
4. Calculate the outputs of hidden layer and output layer, get the controlling amount u;
5. Calculate the system output and get $y(k+1)$;
6. According to the weights adapting rule of output layer and hidden layer, regulate each connection weight of output layer and hidden layer.
7. Go to (3).

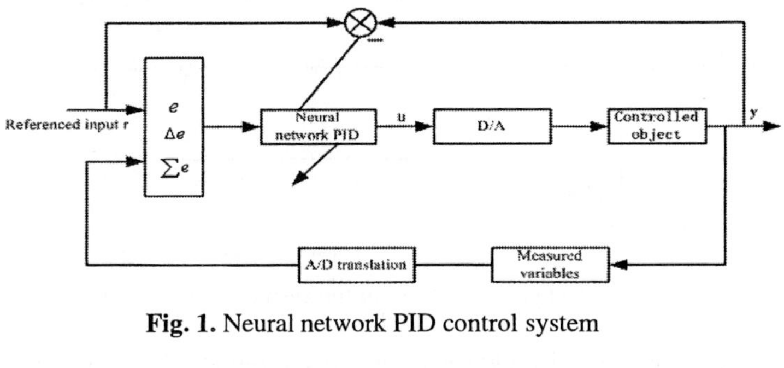

Fig. 1. Neural network PID control system

5 Simulation Results and Conclusion

During the simulation, a typical second-order system, which model is $G(s)=\frac{1}{s(s+1)}$, is taken as the controlled object. The NN PID controller with the 3-8-1 network structure and the traditional controller are applied respectively. When the input signal is unit step signal and the sampling period is T=0.05s, the corresponding response curves are shown in Fig.2.

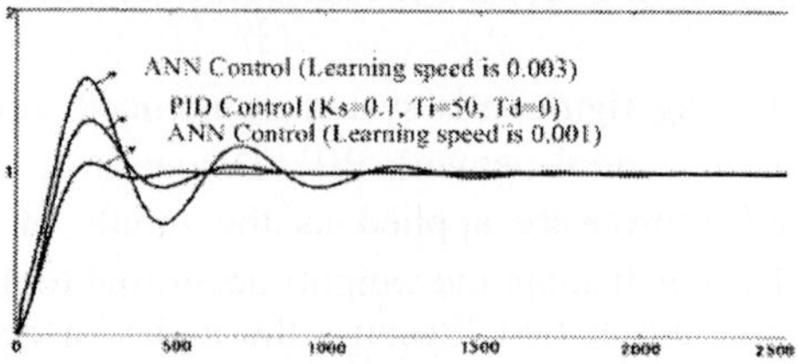

Fig. 2. The curves of simulation system

It can be drawn that the NN PID controller is obviously applicable. Its number of the adaptable parameter is only one, i.e. the learning speed η of network. So the parameter regulating of this controller is much more convenient than traditional PID controller. From the figure, it is obvious that the learning speed η have a strong impact on the system. The response curves with different characters can be achieved by adapting the learning speed η, because the regulation range of the weights depends on it. The regulation range of the weights is smaller, when η is small. While, it is bigger, when η is bigger. In the figure following, the influence is well shown.

The model of the controlled object is $G(s)=\dfrac{1}{0.0001s^2+0.0002s+1}$ The input signal is unit step signal, the sampling period is T=0.05s, the learning speed is η=0.01 and η=0.03 respectively. And the response curves for the unit step signal are shown in Fig.3.

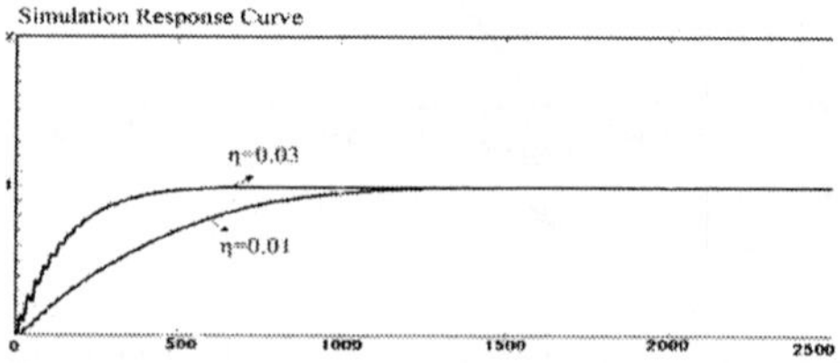

Fig. 3. The response curves of the system with different η

Obviously, in the case of bigger learning speed, the regulation range of weights is greater, and systematic adapting time would be relatively shorter. While the appropriate learning speed is selected, a satisfying control quality would be obtained.

The time-varying nonlinear model of the controlled object is

$$y(k)=\frac{a_0(k)y(k-1)+u(k-1)}{1+y^2(k-1)}$$

Where $a_0(k)$ is $a_0(k)=1+0.1\sin(\frac{k\pi}{25})$ .With the controlled object used above, the systemic response to the square wave signal is also investigated. The amplitude of square wave signal is 1, and the period is 10s. The response curve of the neural network PID controller with 3-8-1 structure is shown in Fig.4. The learning speed η=0.03, sampling period T=0.05s. The system has better input signal trace ability and excellent robustness.

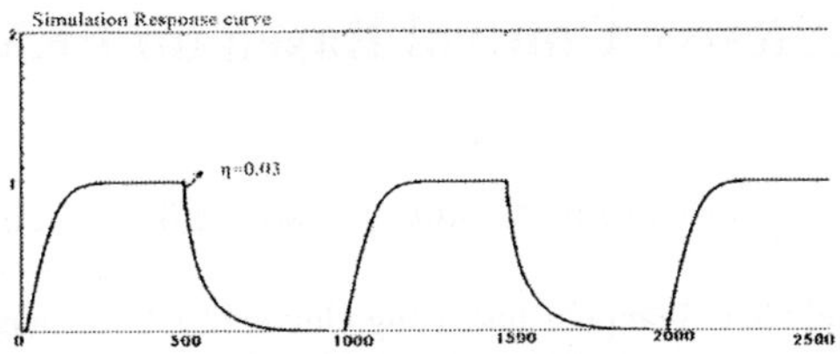

Fig. 4. Response curve of second-order system for square wave signal

According to the above curves, the NN PID controller is more independent and adaptability on the model of the controlled object. This also exactly reflects the outstanding learning and retaining ability and self-adapting ability of neural network. In addition, the initial weights of network also have great influence on the performance of the system, because they not only impact on whether the network output could reach minimum, but also affect learning time of the network heavily.

6 Conclusion

The advantages of the ANN PID controller are summarized as following:

1. The mathematics model for the controlled object is not necessary;
2. It is easy to regulate the parameters;
3. The structure of the controller and the algorithm are simple, it is convenient to apply them to online real-time control;
4. It is more robust. The method can be applied to industrial process control, and take full advantage of both neural network and traditional PID.
5. The initial weights of network have great influence on the performance of the system, which is selected by test or experience in the actual control.

References

1. Yu Yongquan, Huang Ying, Zeng Bi: A PID Neural Network Controller. Proceeding of the International Joint Conference on Neural NetWorks, vol.3. IEEE Copmuter Society Press, California(2003)1933-1938
2. Iwasa,T., Morizumi,N., Ormatu,S.: Temperature Control in a Batch Process by Neural Networks.Proceeding of IEEE World Congress on Computational Intelligence, vol.2. IEEE Press,New York(1992)992-995
3. Li qi, Li shihua: Analysis and Improvement of a Kind of neural Networks Intelligent PID Control Alogrithm. Control and Decision,vol.13. Control and Decision Editorial Department,ShenYang China(1998)311-316
4. Moradi,M.H.: New Techniques for PID Controller Design. Proceeding of 2003 IEEE Conference on Control Applications,vol.2. IEEE Press,New York(2003)903-908

Fuzzy Predictive Control Based on PEMFC Stack

Xi Li[1], Xiao-wei Fu[2], Guang-yi Cao[1], and Xin-jian Zhu[1]

[1]Institute of Fuel Cell, Shanghai Jiao Tong University, Shanghai 200030, China
{lixi,gycao,xjzhu}@sjtu.edu.cn
http://www.automation.sjtu.edu.cn
[2]Institute of Computer and Science, Wuhan University of Science and Technology, Wuhan 430081, China
fxw_wh0409@mail.wust.edu.cn

Abstract. A nonlinear predictive control algorithm based on fuzzy model is presented for a family of complex system with severe nonlinearity such as Proton exchange membrane fuel cell (PEMFC). In order to implement nonlinear predictive control of the plant, the fuzzy model is identified by learning offline and rectified online. The model parameters are initialized by fuzzy clustering, and learned using back-propagation algorithm offline. If necessary, it can be rectified online to improve the predictive precision in the process of real-time control. Based on the obtained model, discrete optimization of the control action is carried out according to the principle of Branch and Bound (B&B) method. The test results demonstrate the effectiveness and advantage of this approach.

1 Introduction

The nominal operating temperature of PEMFC is about 80℃,which impacts the important index such as the traits of the voltage and current density. Consequently, it's crucial that the temperature can be controlled in an appropriate range. However, it is known that the dynamics of PEMFC system is a nonlinear system with complex chemical and electro-chemical reactions. And it is very difficult to model PEMFC system using the traditional method. In last several decades, fruits of the PEMFC stack model have been obtained. However, most of them are based on mass, energy and momentum conservation laws. And their expressions are too complicated to be used to design a control system, especially in the design of the on-line control [5].

A fuzzy predictive control algorithm, based upon discrete optimization and fuzzy model identified, is a relatively simple and effective identification approach and is proposed as a nonlinear model predictive control (NMPC) strategy in this paper [1], which can fit and predict nonlinear system characteristics of complicated industrial process with severe nonlinearity such as PEMFC, so it can be applied to complex nonlinear systems effectively. A fuzzy Takagi-Sugeno(T-S)[4]predictive model is viewed as a feed-forward network, whose weight coefficients are directly correlated

F. Yin, J. Wang, and C. Guo (Eds.): ISNN 2004, LNCS 3174, pp. 150–155, 2004.

with the parameters of the fuzzy model, and then have clear physical meanings. The fuzzy rule parameters are learned offline through back-propagation (BP) algorithm. To lessen the possibility of the parameters trapping into local optimum, the initial value of the premise parameters is set by fuzzy clustering, which is more reasonable than the initializing method used by Wang [7]. When the obtained offline model is used in real-time control, consequent parameters of the rules, but not all the parameters, can be rectified online to insure the real-time predictive precision.

Based on the predictive model and the principle of the B&B method, a structured search technology is used to search optimal control action in the discrete space for the modeling and control of the stack of the PEMFC avoiding the internal complicity [1,2].

The paper is organized as follows: Section 2 introduces the method of fuzzy modeling PEMFC. Section 3 presents the discrete optimization and application in the control of the PEMFC stack. The final conclusion is given in the section4.

2 Structures and Algorithm of Identification of the PEMFC Stack

2.1 Descriptions and Analysis of PEMFC Stack

The operating process of fuel cell is the reverse reaction of electrolyzed water. According to the PEMFC dynamic characteristic, the temperature of the stack can be impacted by the adjustment of the speed of the hydrogen and cooling water. To facilitate model and control design, we use the form as a model of the stack:

$$\overrightarrow{T(k+1)} = \phi[\overrightarrow{T(k)}, \overrightarrow{V_w(k)}, \overrightarrow{V_a(k)}, \overrightarrow{V_c(t)}] \tag{1}$$

where $V_w(k)$, $V_a(k)$ and $V_c(t)$ denote the speed of the cooling water , hydrogen, and the air respectively, and $T(k)$ denotes the operating temperature of the stack, ϕ denotes the nonlinear relation between $T(k)$, $V_w(k)$, $V_a(k)$ and $V_c(t)$ with K being discrete variant[6].

2.2 Structure of the Multi-input and Multi-output Fuzzy Model

Consider a multi-input and multi-output (MIMO) system $y = F(\boldsymbol{u})$ with input vector $\boldsymbol{u} \in U \subseteq R^{nj}$ and output vector $\boldsymbol{y} \in Y \subseteq R^{no}$. It can be regarded as a system consisting of n_o coupled (multi-input and single-output) MISO subsystems, each of which can be fit by a fuzzy T-S model independently. We define

$$\{u_j(k)\}_0^{nuj} = [u_j(k), .., u_j(k - nuj + 1)](j = 1, ..., n_j), \tag{2}$$

$$\{y_l(k)\}_0^{nyl} = [y_l(k),..., y_l(k-nyl+1)](l=1,...,n_o) \quad (3)$$

with nuj and nyl being the order of u_j and y_l respectively, and then each multi-input and single-output (MISO) subsystem can be denoted as:

$$y_l(k+1) = f(x(k)), l=1,...,n_o. \quad (4)$$

$$x(k) = [\{u_1(k)\}_0^{nu1},...,\{u_{n_j}(k)\}_0^{nun_j}, \{y_1(k)\}_0^{ny1},...,\{y_{n_o}(k)\}_0^{nyn_o}] \quad (5)$$

$x(k) = [x_1(k),..., x_n(k)]$ is the regression data vector consisting of input/output data at the kth instant and before. The T-S model employed to fit the MISO subsystem in this paper is a collection of fuzzy rules, which is in the form of "if…then…" The ith rule of the lth output is given by $R_{l,i}$: If $x(k)$ is $A_{l,i}$ Then

$$\hat{y}_{l,i}(k+1) = p_{i,0}^l + p_{i,1}^l x_1(k) + ... + p_{i,n}^l x_n(k), i=1,...,c \quad (6)$$

In this model, what is to be partitioned is the whole space spanned by regression data vectors, but not the single variable space presented by Takagiet al [4].

From Eq (7), the membership function of fuzzy set A_j is defined as the Gaussian product function. The weighted average defuzzifier is in the form of Eq (8) and Eq (9).

$$\mu_{Ai}(x_m) = \prod_{p=1}^{n} (\exp(-\frac{1}{2}(\frac{x_{mp} - \upsilon_{ip}}{\sigma_{ip}})^2)) \quad (7)$$

$$\hat{y}_l(k+1) = \sum_{i=1}^{c} w_i([1 \ x_m] * p_i^{\mathrm{T}}) \Big/ \sum_{i=1}^{c} w_i \quad (8)$$

$$w_i = \prod_{p=1}^{n} (\exp(-\frac{1}{2}(\frac{x_{mp} - \upsilon_{ip}}{\sigma_{ip}})^2)) \quad (9)$$

where, x_m is the mth data vector, $p = 1,...n$, n is the dimension of the data vector, $\upsilon_i = [\upsilon_{i1} \ \upsilon_{ip} \ ... \ \upsilon_{in}]$ and $\sigma_i = [\sigma_{i1} \ \sigma_{ip} \ ... \ \sigma_{in}]$ are the membership function parameters to be identified. Here, $\theta = [\ p_{1,0}^l \cdots p_{c,0}^l \ \ p_{1,1}^l \cdots p_{c,1}^l \cdots p_{1,n}^l \cdots p_{c,n}^l\ ]^T$ is the consequent parameter vector to be identified, c is the number of fuzzy rules and w_i is match degree of the $x_m = [x_{m1} ... x_{mn}]$ with respect to the ith rule.

2.3 BP Learning of Model Parameters

The above fuzzy logic system can be represented as a feed-forward network, whose weight coefficients have clear physical meanings and can be leaned by back-propagation of output predictive error [7]. A collection of training data $X=\{x_1,\ldots,x_m,\ldots,x_t\}$ (t is number of training data) produced by the controlled plant, find a parameter set of fuzzy system that minimizes the output predictive error:

$$E=\sum_{m=1}^{t}(y_{l,m}-\hat{y}_{l,m})^2/2 \tag{10}$$

where, $y_{l,m}$ is the actual output of system with respect to the x_m, $\hat{y}_{l,m}$ is the predictive output of fuzzy system. From the above, we can determine gradient of E with respect to design parameters of the fuzzy system and obtain the learning algorithms of υ_i, σ_i and p_i.

$$\upsilon_{i,p}(k+1)=\upsilon_{i,p}(k)-\alpha\frac{\partial E}{\partial \upsilon_{i,p}}(k),\sigma_{i,p}(k+1)=\sigma_{i,p}(k)-\alpha\frac{\partial E}{\partial \sigma_{i,p}}(k) \tag{11}$$

$$p_{i,0}^l(k+1)=p_{i,0}^l(k)-\alpha\frac{\partial E}{\partial p_{i,0}^l}(k),p_{i,p}^l(k+1)=p_{i,p}^l(k)-\alpha\frac{\partial E}{\partial p_{i,p}^l}(k) \tag{12}$$

Here, α is the learning factor, and $p=1,\ldots n$.

The BP tends to fall into local optimum, so the initialization of the design parameter is very important. Here, the fuzzy c means (FCM) [3] clustering is used to initialize the premise parameters υ_i and σ_i. Using FCM, we can obtain the cluster centers $c_i=[c_{i1}\ldots.c_{in}](i=1,\ldots,c)$ and fuzzy partition matrix $U_{c\times t}$ (U_{im} is the match degree of x_m to the tth cluster.), which are used to determine the initial value:

$$\upsilon_i=c_i,\sigma_i=\beta*\sum_{m=1}^{t}(U_{im}*\|x_m-c_i\|)\Big/\sum_{m=1}^{t}U_{im} \qquad i=1,\ldots c \tag{13}$$

where, β is a coefficient and should be determined in advance. Apparently, this initializing method is more reasonable than the one used by Wang [7], in which the first c data vectors in training data set are used as initial parameters.

2.4 Consequent Parameters Rectification Online

For a complex system that operates over a wide range of operating conditions, if the predictive precision of the model can't satisfy the need of real-time control, the consequent parameters p_i^l should be rectified online. From Eq. (8), the Recursive Least Squares (RLS) is used to rectify the parameter vector θ.

The modeling algorithm is employed to identify the PEMFC stack by T-S fuzzy model. Fig.1 shows that the fuzzy model can imitate the dynamic temperature response of actual stack, and the maximal error is not beyond 0.8℃.

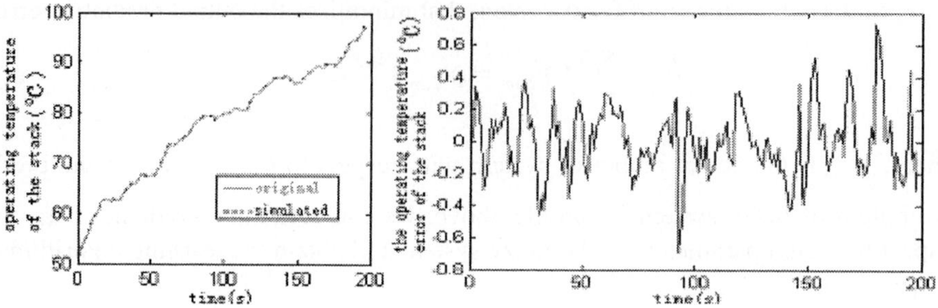

Fig. 1. Schematic of variant operating temperature

According to the requirement of the technics, the test results have shown that the modeling accuracy is high and the fuzzy model can be established fast avoiding complicated physical conversation law and differential equations groups to describe the stack. The T-S fuzzy model can be used to predict the temperature responses on-line that make it possible to design online controller of PEMFC stack.

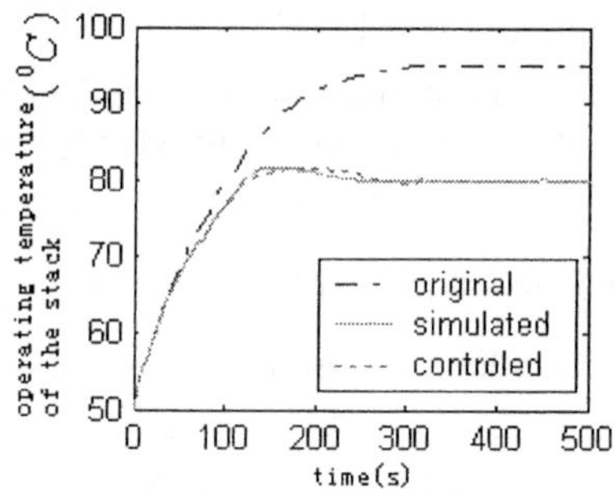

Fig. 2. Schematic of the obtained tracking curves

3 Control Based on Branch and Bound of PEMFC Stack

Based on the fuzzy model and discrete search space of the B&B principle, an optimal control sequence is found to minimize the objective function [2]. The PEMFC stack is controlled by the algorithm above. And the obtained tracking curves are shown in Fig.2. From the simulation and effect, the results of the test enter the stable state much more quickly than the factual control, which is perhaps due to the disturbance, lag and other uncertain factors in the factual process. In a word, the control algorithm can adjust the operating temperature to the value set, make the fluctuation of the temperature least, and obtain the satisfied controlled effectiveness.

4 Conclusions

In this paper, a nonlinear predictive control algorithm based on fuzzy model is presented for the modeling and control of PEMFC stack. Fast convergence of the model identified and effectiveness of the optimization method make the system output track the reference trajectory steadily and quickly The test results show the validity and advantage of this approach avoiding the complexity of the internal system, which lays for the on-line control of PEMFC system well. Moreover, it can be extended to the state of the multivariate and constrained system conveniently.

Acknowledgements. This work is supported by National 863 Scientific Project Development Funds (No. 2002AA517020), PR China.

References

1. Cheng, K.H., Wang, Q.: A Branch and Bound Algorithm for the Traveling Purchaser Problem. European Journal of Operational Research, Vol. 97, No. 3, March 16 (1997) 571-579
2. Liu, B., Shen, Q., Su, H.Y., Chu, J.: A Nonlinear Predictive Control Algorithm Based on Fuzzy Online Modeling and Discrete Optimization. In: IEEE International Conference on Systems, Man & Cybernetics. October 5–8, Washington, D.C., USA (2003)
3. Wolkenhauer, O.: Fuzzy Clustering [Online]. Available: http://www.sbi.uni-rostock.De/data_engineering/clustering.pdf
4. Tomohiro, T., Michio, S.: Fuzzy Identification of System and Its Applications to Modeling and Control [J]. IEEE Transaction on Systems Man Cybernet (1985) 15(1) 16-32
5. Costamagna, P., Srinivasan, S.: Quantum Jumps in the PEMFC Science and Technology from the 1960s to the Year 2000[J], J. Power Sources 102 (2001) 253-269
6. Berning, T., Lu, D.M., Djiali, N.: Three-Dimensional Computation Analysis of Transport Phenomena in a PEM Fuel Cell [J], J. Power Sources 106 (2002) 284-294
7. Wang, L.X.: Adaptive Fuzzy Systems and Control: Design and Stability. Prentice-Hall, Englewood Cliffs, NJ (1994)

Adaptive Control for Induction Servo Motor Based on Wavelet Neural Networks

Qinghui Wu[1], Yi Liu[2], Dianjun Zhang[2], and Yonghui Zhang[1]

[1] Institute of Advanced Control Technology,
Dalian University of Technology, Dalian, China, 116024
qinghuiwu@tom.com
[2] School of Civil and Hydraulic Engineering,
Dalian University of Technology, Dalian, China, 116024
taylor_power@163.com

Abstract. This paper presents an adaptive control system using wavelet neural networks (WNN) for an induction servo drive motor with complex of non-linear, multivariable, strong coupling, slow time-varying properties, etc., and many uncertainties such as mechanical parametric variation and external disturbance. The motivation of developing a new method is to overcome the limitation of conventional control methods which depends on the accurate model and cannot guarantee satisfactory control performance. The proposed scheme with on-line learning has the good tracking and dynamic performance, the ability of adaptive learning from the process and good robustness to uncertainties. Simulation results demonstrate the effectiveness of the proposed method.

Keywords: Servo induction motor; Wavelet neural network; Adaptive learning control.

1 Introduction

The induction motor is a multivariable nonlinear coupling system. In the last three decades, different vector control methods have been proposed for better dynamic performance during transient process, such as field-oriented control (FOC), field acceleration method (FAM), universal field orientation (FUO), direct self control (DSC) and so forth [1]. However, because of computation complex, these methods are very difficult to be realized in real application. Direct torque control (DTC) is a high performance control method after vector control and used widely [2]. It directly decouples rotor flux and torque fully by detecting stator voltage and stator current. Differential geometry is introduced into electric drive field. But its decoupling control theory based on differential geometry needs complex mathematics knowledge, such as differential geometry, Lee algebra and etc. It cannot be extended to use. It uses dynamic feedback linearization method to decouple the system in the global range. Under parametric variations and uncertain disturbance, in order to guarantee system robustness, sliding mode control is also introduced in electric drive. Once the system

F. Yin, J. Wang, and C. Guo (Eds.): ISNN 2004, LNCS 3174, pp. 156–162, 2004.

states enter into sliding mode moving, the system will be insensitive to any disturbance [3]. The merits of the new technique will be valued in the control fields. However its inherent chattering problems are not solved very well, it is difficult to be used in real applications.

In the past few years, the neural network control has been also studied in electric drive. Much research has been done on wavelet neural networks combining the capability of neural networks in learning from processes and the ability of wavelet multi-resolution [4, 5].

A new control scheme based on wavelet neural networks for induction servo motor is proposed in this paper. The proposed scheme with on-line learning has the good tracking performance, the ability of adaptive learning from the process and good robustness to uncertainties. Simulation results demonstrate the effectiveness of the method.

2 Model of Induction Servo Motor Drive

The Laplace transform of the mechanical equation for an induction servo drive described in Ref. [5] as follows

$$\theta(s)=\frac{1}{s}\left[\frac{1}{s+B/J}\left(\frac{K_t}{J}U(s)-\frac{1}{J}T_l+\xi\right)\right] \tag{1}$$

where J is the moment of inertia, B is the damping coefficient, θ is the rotor position, T_l represents the external load disturbance, ξ represents the effects of parametric variation, external disturbance, U is torque current as the control input and K_t is the torque constant, respectively. The nominal model of the plant can be depicted in Fig.1.

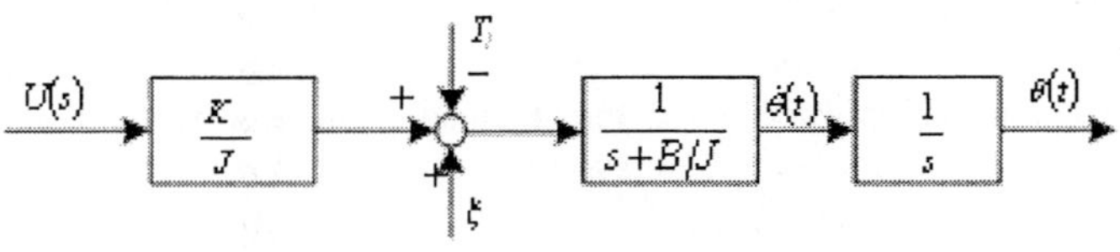

Fig. 1. Block diagram of the plant

3 Design of an Adaptive Controller Based on WNN

3.1 Wavelet Neurocontroller

The wavelet neurocontroller (WNC) is used to drive the unknown dynamic system so that the error between the actual output of the plant and the desired output is minimized gradually. A three-layer WNN shown in Fig.2 is comprised of an input layer, a hidden layer and an output layer. Action functions in hidden layer neurons and output layer adopt wavelet functions and linear functions, respectively. The block diagram of the WNN based control system is shown in Fig.3. In the hidden layer, each node performs a wavelet from the translations and dilations of the mother wavelet, which is the second derivative of a Gaussian function, *i.e.*

$$\psi(x) = (1 - x^2)e^{-\frac{x^2}{2}} \tag{2}$$

So the active function of the neuron in the hidden layer can be gained by the translations and dilations of the mother wavelet in the following

$$\psi_j(x) = \psi\left(\frac{x - b_j}{a_j}\right) = \left[1 - \left(\frac{x - b_j}{a_j}\right)^2\right] e^{-\left(\frac{x-b_j}{a_j}\right)^2 / 2}, \quad j = 1, \cdots 5 \tag{3}$$

The number of neurons in the hidden layer is five. To realize the control object, define the tracking error as $e(k) = r(k) - y(k)$, where $r(k)$ and $y(k)$ represent a desired trajectory and the plant output, respectively. The inputs to the WNC are the tracking error and the increment of the error, *i.e.* $x = [e(k), e(k)(1 - z^{-1})]$.

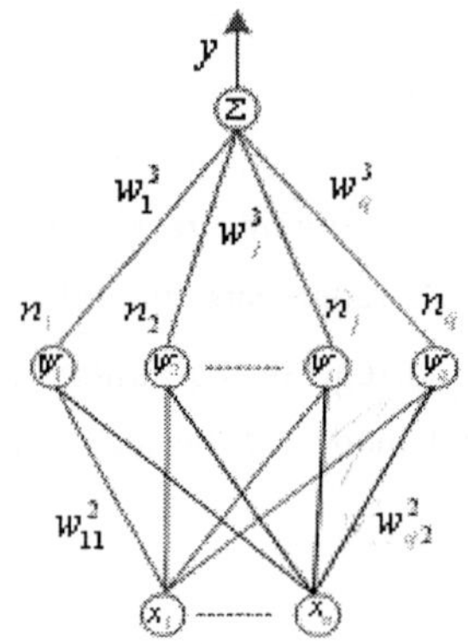

Fig. 2. Wavelet neural network architecture

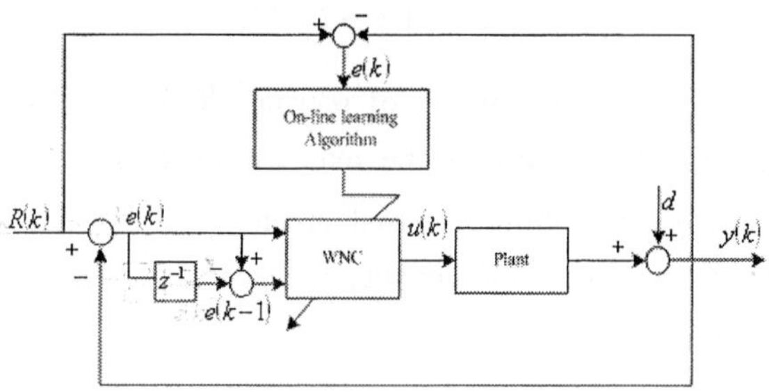

Fig. 3. Block diagram of WNN based control system

3.2 Online Learning Mechanism

The parameters of WNN needing to learn consist of dilations, translations and the weights between hidden layer and output layer. To describe the on-line learning mechanism, the energy function is defined as

$$e = r - y, \qquad J = \frac{1}{2}e^2 \tag{4}$$

Each parameter is adjusted from back layer to fore one by the negative gradient of the energy function in (4). To describe on-line learning mechanism, define the neurons in every layer as follows

The first layer (*i.e.* the input layer):

$$net_i^1 = x_i^1, \quad y_i^1 = f_i^1(net_i^1) = net_i^1, \quad i=1,2 \tag{5}$$

The second layer (*i.e.* the hidden layer):

$$net_k^2 = \sum_{i=1}^{2} w_{ki}^2 x_i^2, \quad y_k^2 = f_k^2(net_k^2) = \psi_k(net_k^2), \quad k = 1,\ldots,5 \tag{6}$$

The third layer (*i.e.* the output layer):

$$net_o^3 = \sum_{j=1}^{5} w_j^3 x_j^3, \quad y_o^3 = f_o^3(net_o^3) = net_o^3, \quad o = 1 \tag{7}$$

In the output layer, the weight is updated by the amount

$$\Delta w_j^3(t) = -\eta_w^3 \frac{\partial J}{\partial w_j^3} = -\left(\eta_w^3 \frac{\partial J}{\partial y_o^3}\right)\left(\frac{\partial y_o^3}{\partial net_o^3}\frac{\partial net_o^3}{\partial w_j^3}\right) \tag{8}$$

and the error term to be propagated is given by

$$\delta_o^3 = -\frac{\partial J}{\partial y_o^3} = -\frac{\partial J}{\partial e}\frac{\partial e}{\partial y_o^3} = -\frac{\partial J}{\partial e}\frac{\partial e}{\partial y}\frac{\partial y}{\partial y_o^3} \tag{9}$$

In the same way, the weight in the hidden layer is updated by the amount

$$\Delta w_{ki}^2(t) = -\eta_w^2 \frac{\partial J}{\partial w_{ki}^2} = -\eta_w^2 \frac{\partial J}{\partial net_k^2}\frac{\partial net_k^2}{\partial w_{ki}^2} = -\eta_w^2\left(\frac{\partial J}{\partial y_k^2}\frac{\partial y_k^2}{\partial net_k^2}\right)\frac{\partial net_k^2}{\partial w_{ki}^2} \tag{10}$$

where

$$\frac{\partial J}{\partial y_k^2} = \frac{\partial J}{\partial net_o^3}\frac{\partial net_o^3}{\partial y_k^2} = \frac{\partial J}{\partial net_o^3}\frac{\partial}{\partial y_k^2}\sum_j y_j^2 w_j^3 = \frac{\partial J}{\partial net_o^3}\cdot w_k^3 \tag{11}$$

where

$$\frac{\partial J}{\partial net_o^3} = \frac{\partial J}{\partial y_o^3}\frac{\partial y_o^3}{\partial net_o^3} = -\delta_o^3 \tag{12}$$

In addition, the parameters of wavelets are also adjusted equally according to the negative gradient method. First the update law of the dilations is

$$\Delta a_j(t) = -\eta_a \frac{\partial J}{\partial a_j} = -\eta_a \frac{\partial J}{\partial y_j^2}\left(\frac{\partial y_j^2}{\partial net_j^2}\frac{\partial net_j^2}{\partial a_j}\right) \tag{13}$$

and the update law of the translations is

$$\Delta b_j(t) = -\eta_b \frac{\partial J}{\partial b_j} = -\eta_b \frac{\partial J}{\partial y_j^2}\left(\frac{\partial y_j^2}{\partial net_j^2}\frac{\partial net_j^2}{\partial b_j}\right) \tag{14}$$

where

$$\frac{\partial J}{\partial y_j^2} = \frac{\partial J}{\partial net_o^3}\frac{\partial net_o^3}{\partial y_j^2} = -\delta_o^3 w_j^3 \tag{15}$$

where η_a and η_b are the learning rates of the dilation and the translation, respectively.

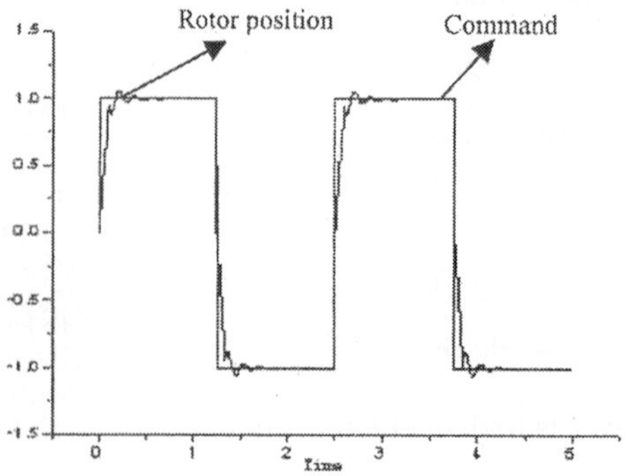

Fig. 4. All the learning rates are set 0.1 without any disturbance.

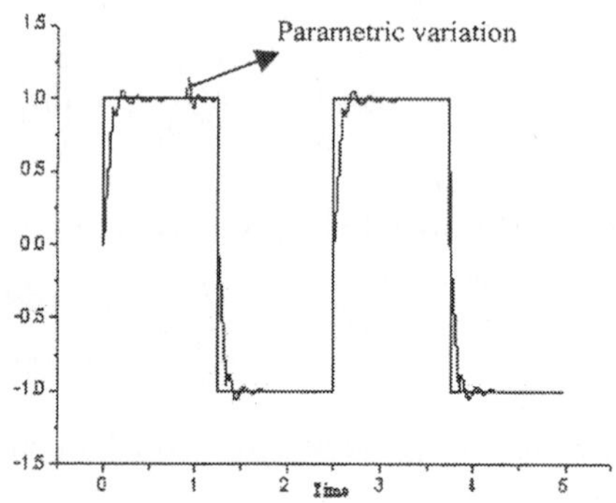

Fig. 5. A variation is inflicted at 0.8 second

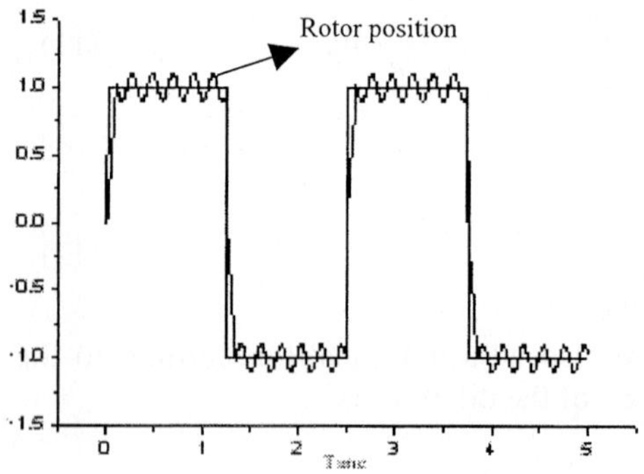

Fig. 6. All the learning rates are set at 0.9

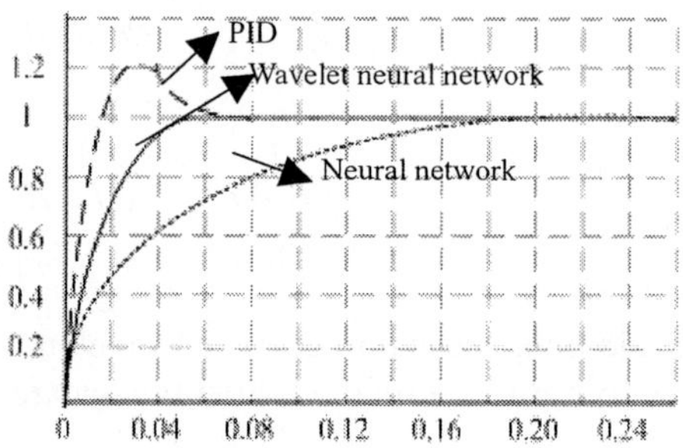

Fig. 7. Dynamic response curve of the control system with different control methods

Since the plant is normally unknown, the sensitivity term $\partial y / \partial y_o^3$ is unknown. Though this unknown value can be estimated by using the identifier, the amount of the computation is required [6]. To solve this problem, an approximation law is adopted as follows:

$$\delta_o^3 \approx e \tag{16}$$

4 Simulated Results

The parameters of the plant model follow as

$$K_t = 0.5 N\, m/A \qquad J = 5\times10^{3} N\cdot m\cdot s^2 \qquad B = 5\times10^{-3}\, N\cdot m\cdot s/rad \tag{17}$$

Case 1: The on-line adaptive ability of WNN based control system: In this simulation, all the learning rates are 0.1 without any disturbance. The output of the system tracking the square wave as the reference input is shown in Fig.4.
Case 2: The robust to disturbances: In this example, all learning rates are the same as in Case 1, except a disturbance is inflicted on the input at 0.8 second, shown in Fig.5.
Case 3: The relations between the performances of the WNN based control system and the learning rates. In this study, all learning rates are set 0.9. The effect is shown in Fig.6.
Case 4: Dynamic response characteristics of the system with WNNC, PID and NNC. The compared effectiveness is shown in Fig.7.

5 Conclusions

This paper presented a wavelet neural network based control system, *i.e.* wavelet neurocontroller. Due to the orthogonality of wavelets, this design is more efficient than conventional controllers based on neural networks. During learning process, the error between the reference output and the real output is chosen as the unknown sensitivity, which lessens the amount of computation.

Simulation shows the effectiveness of the proposed control system by tracking square wave trajectory and the better dynamics by comparing with other control methods. The virtues of this design are that the system still keeps steady under disturbances and that convergence of the tracking error can be guaranteed without prior knowledge of the controlled plant. Otherwise, the update law is more adaptive to on-line applications and the proposed method is easy to be extended for any nonlinear control problem.

References

1. José Andrés , Richard M:Vector control methods for induction machines: an overview IEEE Trans. Education, 44 (2) (2001): 170-176.
2. Ortega, R., Barabanov, N.,et.:Direct torque control of induction motors: stability analysis and performance improvement, IEEE Trans. Automatic Control 46 (8) (2001) 1209-1222.
3. Wai, R.-J.:Adaptive sliding-mode control for induction servomotor drive. IEEE Proc. Electric Power Applications, 147 (6) (2000) 553-562.
4. Ciftcioglu, O.:From neural to wavelet network, NAFIPS(1999) 894-898.
5. Wai, R.-J., Lin, C.-M., Hsu, C.-F.:Hybrid control for induction servomotor drive. IEEE Proc. Control Theory and Application, 149 (6) (2002) 555-562.
6. Chao-chee Ku, Lee, K.Y.:Diagonal recurrent neural networks for dynamic systems control, IEEE Trans. Neural Networks, 6(1) (1995).144-156.

The Application of Single Neuron Adaptive PID Controller in Control System of Triaxial and Torsional Shear Apparatus

Muguo Li, Zhendong Liu, Jing Wang, Qun Zhang, Hairong Jiang, and Hai Du

The State Key Laboratory of Coastal and Offshore Engineering,
Dalian University of Technology, 116024, Dalian, P.R.China
lmguo@dlut.edu.cn

Abstract. In this paper, an adaptive single neuron-based PID controller for the Triaxial and Torsional Shear Apparatus is proposed. Considering variations of the control precision, the single neuron adaptive PID controller is used to construct the control system to achieve the adaptive control of triaxial and torsion testing. The single neuron adaptive PID controller using hybrid Supervised-Hebb rule is proposed to tune control parameters. The testing results of actual application show that the single neuron adaptive PID controller makes better improvement in the control precision and robustness of control system.

1 Introduction

Triaxial experimental technique is significant in the soil mechanics, especially in the study of the strength of the soil and constitutive relationship. It is an important means in the determination of the strength, stress-strain feature, and other mechanical characteristics of soil. Most of the existing dynamic triaxial apparatus are single torsion or single axle direction apparatus, which do not reflect the influence of initial principal stress coefficient and initial principal stress direction. They can not realize many complicated stressing condition, such as the successive rotation of the principal stress direction. A Soil Static and Dynamic Universal Triaxial and Torsional Shear Apparatus was imported from SEIKEN INC, Japan in 2001 by Dalian University of Technology. It overcomes some of the disadvantages of the past testing apparatus. For the reason that the PID parameters are fixed, the apparatus does not work well for different frequency control signal. It is an urgently concerned problem to improve control level of the triaxial apparatus. Various types of modified PID controllers have been developed, such as self-tuning PID controllers [1], and fuzzy PID controllers [2]. Self-tuning provides a systematic and flexible approach for deal with uncertain, nonlinear, and time-varying parameters.

The triaxial apparatus is a complex, multivariable and uncertain with time-varying plant. How to get appropriate control parameters is very difficult. In this paper, we use neural network theory to improve the precision and robustness of the system and make the system gain exact control effect under different frequency control signals. The

F. Yin, J. Wang, and C. Guo (Eds.): ISNN 2004, LNCS 3174, pp. 163–168, 2004.

core of principle for single neuron adaptive PID control is based on PID control, adaptive and self-learning of single neuron. Based on adaptive neural network, we designed a control system, which can real-time tune the PID parameters.

2 Brief Introduction of the Control System

The control system of the improved triaxial testing apparatus has two parts. One controls the axial loading, the other controls the torsion loading. Each part consists of two loop closed control loop: (1) Hydraulic power system with the control of servo-valve. This system controls the servo-valve based on the signals from the control system, and then controls the hydraulic power system to execute axial loading and torque onto the samples. (2) Analog PID closed control loop. The loop is a main control loop of existing system. As it has perfect hardware protection circuit, emergency chain protection design, signal anti-interference design and outstanding power amplification circuit design, this circuit is kept. (3) Adaptive PID control loop of single neuron. It is the main control loop of the improved system. It performs the adaptive control of parameter and makes the system gain better control precision and robustness. As an identifying model, the neural network is a physical implement for actual system. The algorithm of single neuron is simple and can be used for on-line real-time control.

3 The Design of Adaptive PID Control System of Single Neuron

Traditional PID control algorithms have the advantages of intuitive structure and algorithm. However, they are not very efficient for nonlinear and time-varying systems, since the PID coefficients are often fixed. Adaptive PID control system of single neuron is the main part of the testing apparatus. Preserving analog PID closed-loop control and adding an adaptive closed-loop control of single neuron to make it become a pair of closed-loop control system. The self-tuning of system parameters are realized using the neural network.

3.1 The Principle of the Algorithm of Adaptive PID Control System of Single Neuron

In recent years, the current interest has been focused on design of self-tuning controller by using neural network. Neural network is an information processing paradigm that is inspired by the way of biological nervous systems, such as the brain, process information. The key element of this paradigm is the novel structure of the information processing system. Several dozen kinds of network models, such as BP net, Hopfield net, etc. have already been proposed. The models have been applied widely in pattern-recognition and artificial intelligence. In the automatic control field, adaptive

PID control with neural network has got very great development during the past few years [3-8]. These algorithms synthesize the advantages of the neural network and PID control, keep some traits of traditional PID control structure and overcome some shortcomings of traditional PID. Single neuron is the basic unit of neural network, and has the similar capability of adaptation like neural network. From the structure and function points of view, an artificial neuron can be regarded as a nonlinear multi-input and multi-output processing unit. In addition, the structure of a single neuron is simple, and it is therefore easy to realize by software in real-time cases. Because the network is made up of single neuron, the computation load is low and the structure is simple and the convergence speed can be accelerated greatly, which assures the system to realize online self-tuning.

3.2 The Proposed Algorithm

We propose an adaptive single neuron-based PID controller, which has the inputs: X_1 is system-error $e(k)$, X_2 is first order difference of system-error $\Delta e(k)$, and X_3 is second order difference of system error $\Delta e^2(k)$:

$$X_1 = r(k) - y(k) \tag{1}$$

$$X_2 = e(k) - e(k-1) \tag{2}$$

$$X_3 = e(k) - 2e(k-1) + e(k-2) \tag{3}$$

where $r(k)$ is the reference enactment signal of the system, $y(k)$ is the real-time signal that the sensor gathers. Neuron controller output $U(k)$ is:

$$U(k) = U(k-1) + K(k)\sum_{i=1}^{3} W_i(k)X_i(k) \tag{4}$$

where W_i is the input weight and $K(k)$ is proportional coefficient of neuron. In order to ensure convergence and robustness, standardization learning algorithm can be adopted:

$$U(k) = U(k-1) + k\sum_{i=1}^{3} W_i^{'}(k) * X_i(k) \tag{5}$$

$$W_i^{'}(k) = \frac{W_i(k)}{\sum_{i=1}^{3} |W(k)_i|} \tag{6}$$

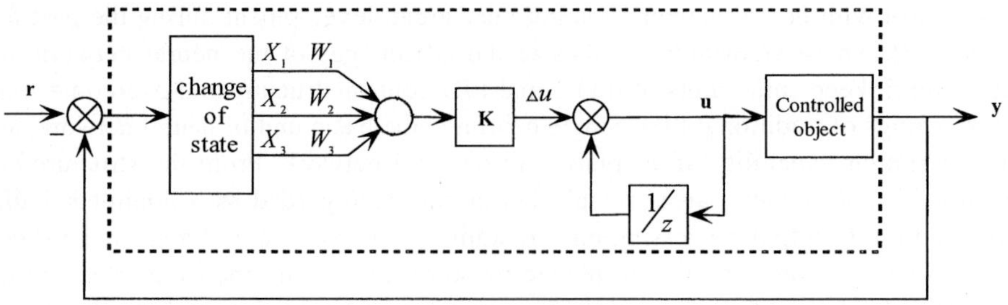

Fig. 1. Structure of single neuron adaptive PID controller

Fig. 1 is the structure of single neuron adaptive PID controller. In the single neuron adaptive PID control algorithm, the key is the learning rule of input weight. This trait is applied to self-tuning of the parameters of adaptive PID control system. Considering W_i should be concerned with the correlative functions of input, output and output error of neuron, hybrid Supervised-Hebb rule is adopted:

$$W_i(k) = W_i(k-1) + \eta_i * X_i(k) * e(k) * U(k) \tag{7}$$

where η_i is the learning rate of the system.

In the single neuron adaptive PID control algorithm, the value of η_i decides directly the convergence speed of control system. The bigger learning rate is, the faster convergence speed is, but the system is apt to over-adjustment. However, if the learning rate is too small, the system response will be slow. At the same time, if the learning rate is a fixed value during the whole learning process, the system control performance will be limited. So the following algorithm is introduced into the system to optimize the enactment of the learning rate and to make it change with the system state. The adjustment criterion is to check whether learning rate η_i will reduce the error for the revision value of weighting value $W_i(k)$. If the value is reduced, it means that learning rate is small, an additional value needs to be added.

In single neuron adaptive PID control algorithm, optimizing the proportion parameter *K* can improve control performance. The optimizing function $K(k)$ is the key of the whole algorithm. If $\mathrm{sgn}[e(k)] = \mathrm{sgn}[e(k-1)]$, then

$$K(k) = K(k-1) + c\frac{K(k-1)}{T_v(k-1)} \tag{8}$$

$$T_v(k) = T_v(k-1) + L * \mathrm{sgn}[|\Delta e(k)| - T_v(k-1)|\Delta^2 e(k)|] \tag{9}$$

where $0.05 \le L \le 0.1$, $0.025 \le c \le 0.05$. If $\mathrm{sgn}[e(k)] \ne \mathrm{sgn}[e(k-1)]$

$$K(k) = 0.75K(k-1) \tag{10}$$

4 Operation Result and Analysis of Performance

The single neuron adaptive PID control algorithm is applied to the Triaxial and Torsion Shear Apparatus. We keep hardware analog PID closed control loop, software closed control loop based on single neuron adaptive control algorithm is outside analog PID closed control loop. The single neuron adaptive control algorithm is realized in MS VC++. Fig.2 is the comparison of responding curves under the function of step signal before upgrading and after upgrading. Fig.3 is the comparison of responding curves of the testing system under the function of dynamic enactment signal before upgrading and after upgrading. Initial parameters of the single neuron adaptive PID control system are:

$$x_I = 0.000001, x_P = 0.00004, x_D = 0 \tag{11}$$

$$w_1(0) = 1, w_2(0) = 3, w_3(0) = 0, K(0) = 2 \tag{12}$$

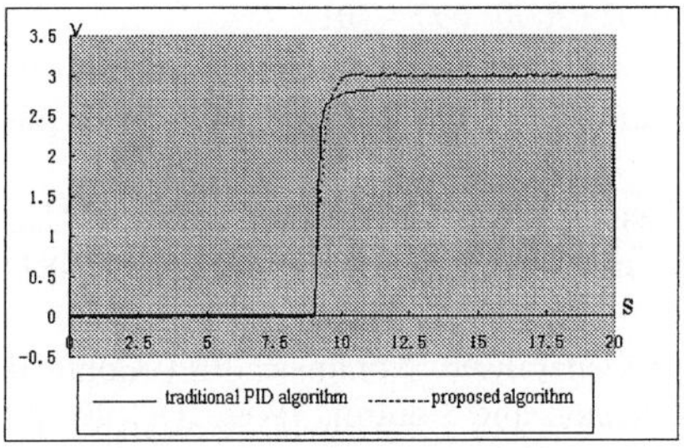

Fig. 2. The step response of system

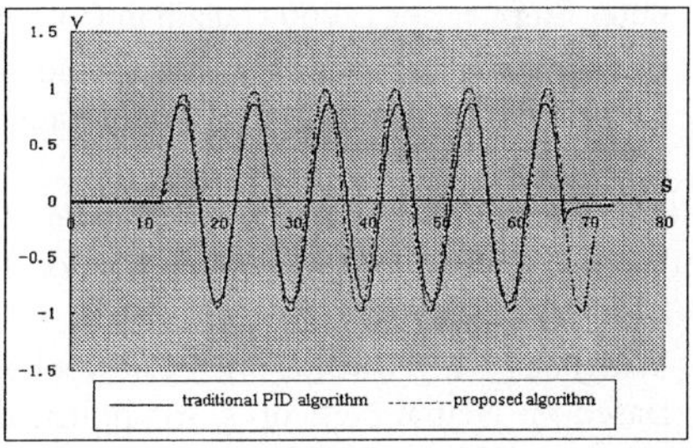

Fig. 3. The dynamic response of system

From the diagrams above, we can see that responding speed is accelerated for the same input of step signal after adding single neuron adaptive PID control closed loop. There is 2.5 seconds at least from step change to steady state before upgrading. But there is only 1.0 to 1.5 seconds after upgrading. So the systematic residual error before upgrading is removed totally after upgrading. In the comparison of responding curves under the same dynamic enactment signal, improved ability of dynamic tracking can be shown after adding single neuron adaptive PID controller.

5 Conclusions

In this paper, single neuron adaptive algorithm has been proposed to improve the precision and robustness of control system. The algorithm takes the advantage of the neuron ability of self-organizing and self-learning to optimize parameters of controller. The control system based on neural network has the ability to tune the parameters of the controller. The improved Triaxial and Torsion Shear Apparatus has already been in operation and better operation results have been achieved.

References

1. Tan, K., Huang, S.,Ferdous, R.: Robust Self-tuning PID Controller for Nonlinear Systems. In: Industrial Electronics Society, 2001, IECON '01, The 27th Annual Conference of the IEEE, **1** (2001) 758–763
2. Lo, K.L., Sadegh, M.O.: Systematic Method for the Design of a Full-scale Fuzzy PID Controller for SVC to Control Power System Stability. In: Generation, Transmission and Distribution, IEE Proceedings, **150** (2003) 297–304
3. Huailin, S., Xiucai, G., Hua, S.: PID Neural Networks in Multivariable Systems. In: Intelligent Control, 2002, Proceedings of the 2002 IEEE International Symposiumon (2002) 440–444
4. Hassanzadeh, I., Khanmohammadi, S., Sharifian, M.B.B, Jiang, J.: A SISO Discrete Control System Containing Neural Estimator and Neural Controller. In: Electrical and Computer Engineering, 2000 Canadian Conference on, **2** (2000) 697 - 701
5. Chen, C. L., Chang, F. Y.: Design and Analysis of Neural/Fuzzy Variable Structural PID Control Systems. In: Control Theory and Applications, IEE Proceedings, **143** (1996) 200–208
6. Matsukuma, T., Fujiwara, A., Namba, M., Ishida, Y.: Non-linear PID Controller Using Neural Networks. In: Neural Networks, 1997, International Conference on, **2** (1997) 811–814
7. Yonghong, T., Xuanju, D., Cauwenberghe,A: Generalised Nonlinear PID Controller Based on Neural Networks. In: Information, Decision and Control, 1999, IDC 99, Proceedings (1999) 519–524
8. Ohnishi,Y., Yamamoto,T., Yamada,T., Nanno,L., Tanaka,M.: A Design of Nonlinear PID Control Systems with a Neural-net Based System Estimator. In: SICE 2002, Proceedings of the 41st SICE Annual Conference, **4** (2002) 2272–2273

Ram Velocity Control in Plastic Injection Molding Machines with Neural Network Learning Control

Gaoxiang Ouyang[1], Xiaoli Li[1], Xinping Guan[1],
Zhiqiang Zhang[1], Xiuling Zhang[1], and Ruxu Du[2]

[1] Institute of Electrical Engineering, Yanshan University, Hebei,
066004 Qinhuangdao, China
{xlli, xpguan}@ysu.edu.cn

[2] Dept. Automation & Computer Aided Engg., Chinese University of Hong Kong, China
rdu@acae.cuhk.edu.hk

Abstract. In plastic injection molding, the ram velocity plays an important role in production quality. This paper introduces a new method, which is a combination of the current cycle feedback control and neural network (NN) learning, to control the ram velocity in injection process. It consists of two parts: a PD controller (current cycle feedback control) is used to stabilize the system, and the feedforward NN learning is used to compensate for nonlinear/unknown dynamics and disturbances, thereby enhancing the performance achievable with feedback control alone. The simulation results indicate that the proposed NN learning control scheme outperforms the conventional PD controller and can greatly reduce tracking errors as the iteration number increase.

1 Introduction

Recently, a number of new control techniques were applied for plastic injection molding processes. It is agreed that the control of the injection is the most important. In particular, since the injection ram velocity controls the rate of the molten material entering the mold cavity, controlling the ram velocity is very important [1-2]. In fact, many part quality problems, such as weld line, burn, jetting, short-shot and flash, can be eliminated by properly controlling the injection velocity [3-4]. Learning control methodologies make use of the repetitive nature of tasks to obtain improved performance without the necessity of a parametric model of the system. Hence, this technique is suitable for plastic injection molding process, which essentially carry out repetitive tasks and the accurate model of injection molding machines (IMM) is normally unavailable. The ability to learn a non-linear mapping and the non-requirement of a priori information about the system model make the NN an attractive alternative to the conventional learning control scheme [5]. The most popular control scheme is one that utilizes the learning ability of NN to identify the system inverse and thus generate control input to the plant [6-7]. As for the training method, some theoretical results are examined in [7-9] on the weight updating law and the convergence performance of NN controller.

F. Yin, J. Wang, and C. Guo (Eds.): ISNN 2004, LNCS 3174, pp. 169–174, 2004.

In this paper, we propose a new NN learning control scheme based on a feedback-feedforward configuration. The feedforward component applies the multi-layer NN to estimate the inverse dynamics of the IMM. Thus, when it is employed in the neural learning control scheme, the repeatable structured and unstructured uncertainties of IMM can be compensated. The rest of the paper is organized as follows: Section 2 is a study on the IMM. Section 3 describes the proposed NN learning control scheme. Section 4 shows the simulation results. Finally, Section 5 contains the conclusions.

2 A Study on the IMM

Detailed discussions on injection molding process can be found in many manufacturing technology monographs, such as [1]. Briefly, an injection molding cycle contains three stages: injection, holding and plasticating, as well as ejection of the part. The injection stage starts when the molds are closed and the plastic melt is being injected into the mold. The screw, not rotating at this point, is pushed forward so that the plastic melt in front of the screw is forced into the mold. In the holding and plasticizing stage, two actions occur at the same time. In this study, we will focus on the injection phase, in particular, the ram position control in the injection phase. However, controlling the ram position is rather difficult. Experiments indicate that even under a same working condition, the ram velocity may be different. In other words, the ram position must be precisely controlled to ensure the part quality.

In order to control the ram trajectory, first it is necessary to develop the dynamic model. In this study, the model developed in [2,3] is adopted with a slight modification, as detailed in Appendix. Fig. 1 shows the block diagram for the ram position control, where P_1 is the hydraulic pressure, P_2 is the actuator pressure, and Q_1 is the material flow from control valve to the line, Q_2 is the material flow from the line to the actuator, and X_{ram} is the ram position. From the figure, it is seen that the inputs include the pressure of the pump (P_{sup}), the nominal material flow (Q_{nom}) and the initial ram position (X_0).

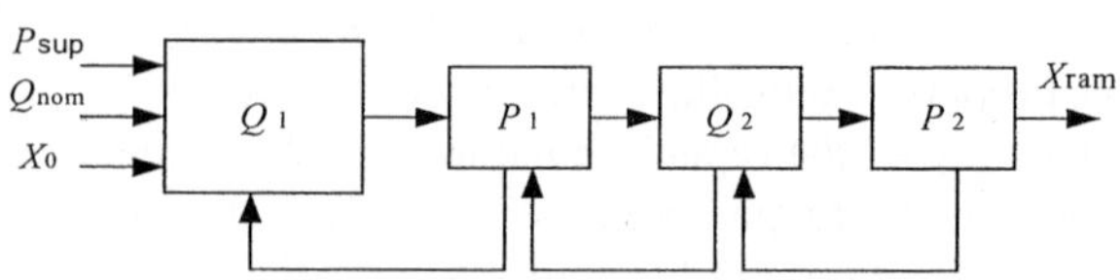

Fig. 1. The block diagram for the ram position control

3 NN Learning Control Scheme

This new control method is illustrated in Fig. 2(left). From the figure, it is seen that the two controllers work in parallel. The feedback PD controller is used to stabilize the system, which takes the system output to be near the desired trajectory; then the feedforward NN controller is used to predict the desired control input and achieve precision tracking. Assuming $y_d(t)$ is the desired ram trajectory, then the control scheme can be described as

$$\begin{aligned} u_k(t) &= u^{ff}{}_k(t) + u^{fb}{}_k(t) \\ &= u_{k-1}{}^{nn} + K_p(y_d(t) - y_k(t)) + K_d(\dot{y}_d(t) - \dot{y}_k(t)) \end{aligned} . \tag{1}$$

where the subscript k of each variable indicates its value at the kth iteration. $u_k{}^{fb} = K_p(y_d(t) - y_k(t)) + K_d(\dot{y}_d(t) - \dot{y}_k(t))$ is the feedback term. The iterative neural learning controller is operated in two process periods: training period and control period. During the training period, universal approximation property of the NN is exploited to estimate the IMM inverse dynamics off-line based on input-output pairs gathered in the $(k-1)$th operation cycle [5]. During the control period, the weights of the well-trained NN controller are fixed and provide feedforward control signal, $u_k{}^{ff}(t)$, to compensate for the repeatable uncertainty and nonlinear effects. After each operation of the IMM, the NN will be retrained with the operation data obtained from previous iteration.

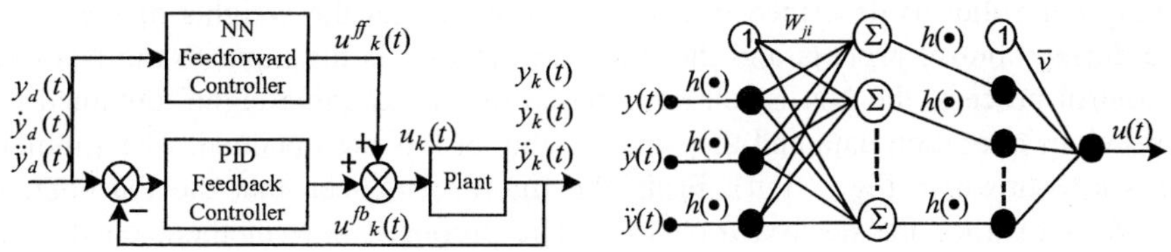

Fig. 2. The proposed NN learning control scheme (left) and the multi-layer NN (right)

Fig. 2(right) shows the structure of the multi-layer NN. During the training period, the input to the NN is the actual operation vector $x = [y\ \dot{y}\ \ddot{y}]^T$. Each input state passes through a nonlinear threshold activation function as

$$h(x) = 1/(1 + e^{-4\lambda \cdot x}) . \tag{2}$$

where $\lambda > 0$ is a constant parameter. The overall control system has a learning-then-control feature, which may look similar to the other learning control schemes [5]. However, the difference lies in the NN training method. To ensure the convergence of the NN, we apply a three-layer NN and a modified weight-tuning algorithm using dead-zone technique to reject noise referring to [5,7]. Note that since the teaching data of NN are recorded from the actual operation, such input-output pairs definitely represent the actual inverse dynamics of IMM, which include repeatable structured and unstructured uncertainties. The success of the iterative neural learning control scheme depends largely on whether the NN model can converge the actual system inverse dynamic model. This condition can be guaranteed by the well-known universal approximation property of NN [8]. Thus repeatable uncertainties, which can be learned during the training process, can be compensated.

4 Simulation Results

A simulation study is conducted using the proposed new method to control the ram position of a plastic IMM. Many plastic IMM have build-in ram position sensors, but not ram velocity sensors. Therefore, the control focuses on the ram position. Herein, as shown in Fig. 3, the desired ram trajectory is given below:

$$x(t) = \begin{cases} 60 - 20\left(t - \sin(2\pi t)/2\pi\right) & 0 < t \le 1 \\ 40 - 12.5(t-1)^2 & 1 < t \le 2 \\ 2.5 + 25(t-1)^{-1} & 2 < t \le 6 \end{cases} . \tag{3}$$

Then, we select $K_p = -100, K_d = -5$ as PD controller gains in simulation. The control period for each iteration is 6s with sampling time selected as 5ms. First, we employ the PD controller alone to bring the ram position into the neighbourhood of the desired trajectory. Using the operation data obtained from the results of the PD controller we employed a three-layer NN with three input neurons, seven hidden neurons and one output neuron to estimate the inverse dynamics of the IMM off-line. The referred weight-tuning algorithm with dead-zone is employed in the training process [5,7]. After finishing the training of the NN, we implement the iterative learning controller as described in Fig. 2(left). Note that the weights of the NN are fixed during control process and the desired trajectory is the input to the NN. After the control process, the NN is retrained based on the teaching signal obtained from the previous operation data and the new weights of NN are obtained. The simulation results are shown in Fig. 3(left). From the figure, it is seen that the PD controller alone fails to trace the desired trajectory. The proposed new method, on the other hand, is able to minimize the tracking error step by step. In particular, at the fifth iteration, the tracking error is very closely to zero.

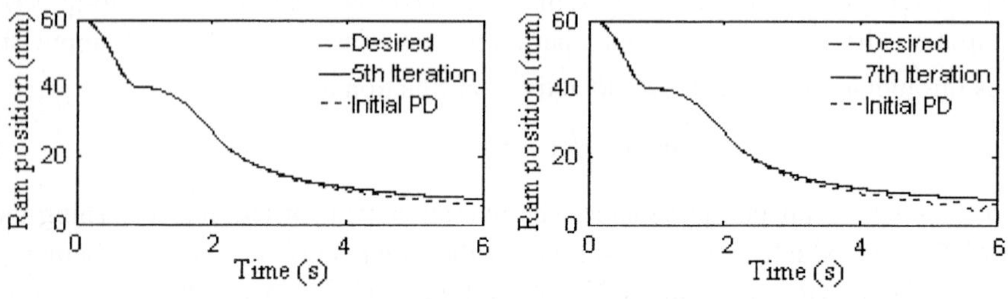

Fig. 3. Ram position tracking performance

To illustrate the robustness of the new control method, we assume that the effective bulk modulus of the plastic material, $\beta_1(t) = 1.5\times10^5 \sin(12t)$, is changing with respect to time. This is rather common for smaller plastic IMM, since the heating would take longer time. Fig. 3(right) shows the simulation results. From the figure, it is seen that the use PD controller alone cannot eliminate the tracking error. The new method, on the other hand, can greatly reduce tracking errors as the iteration number increase. In particular, at the seventh iteration, the tracking error is very closely to zero.

5 Conclusions

This paper presents the new method for controlling the ram position of IMM. The new method is based on a combination of the conventional current cycle PD control and feedforward NN learning control. The scheme applies a NN to construct feedforward control signal by making use of the previous operation input-output data. The computer simulation indicates:

1. The new method can greatly reduce the tracking errors, and significantly outperform the conventional PD controller, which achieves satisfactory tracking performance after a few iterations;
2. The new method has a learning-then-control feature thus it avoids the disadvantage of heavy on-line computation.

References

1. Dominick, V.R., Donald, V.R., Marlene, G.R.: Injection Molding Handbook. 3rd edn. Kluwer Academic Publishers, New Jersey (2000)
2. Havlicsek, H., Alleyne, A.: Nonlinear Control of an Electrohydraulic Injection Molding Machine via Iterative Adaptive Learning. IEEE-ASME T. Mech. 4 (1999) 312-323
3. Zheng, D.N., Alleyne, A.: Modeling and Control of an Electro-hydraulic Injection Molding Machine with Smoothed Fill-to-pack Transition. J. Manuf. Sci. E.–T. ASME. 125 (2003) 154-163
4. Tan, K.K., Tang, J.C.: Learning-enhanced PI Control of Ram Velocity in Injection Molding Machines. Eng. Appl. Artif. Intel. 15 (2002) 65-72
5. Xiao, J.Z., Song, Q., Wang, D.W.: A Learning Control Scheme Based on Neural Network for Repeatable Robot Trajectory Tracking. 14th IEEE Symposium on Intelligent Control/Intelligent Systems and Semiotics, ISIC/ISAS'99, USA (1999)
6. Psaltis, D., Sideris, A., Yamamura, A.A.: A Multilayered Neural Network Controller. IEEE Contr. Syst. Mag. 8 (1988) 17-20
7. Song, Q., Xiao, J.Z., Soh, Y.C.: Robust Back Propagation Training Algorithm for Multilayered Neural Tracking Controller. IEEE T. Neural Network. 10 (1999) 1133-1141
8. Hornik, K., Stinchcombe, M., White, H.: Multilayer Feedforward Networks are Universal Approximators. Neural Networks, 2 (1989) 359-366
9. Polycarpou, M.M., Ioannou, P.A.: Learning and Convergence Analysis of Neural-type Structured Networks. IEEE T. Neural Network. 3 (1992) 39-50

Appendix: The Dynamic Model for the Ram Position Control

In this paper, the simulation is conducted based on the dynamic model developed in [2,3]. The model is for ram position control in plastic IMM. As shown in Fig. 1, the injection molding process model can be decomposed into four functional blocks.

1. The flow from the pump to the valve, Q_1:

$$Q_1 = Q_{nom}\sqrt{\frac{P_{sup} - P_1}{C}}(t-\tau)\,. \tag{A.1}$$

where, Q_1 is the material flow through flow control valve into the line; $C = 1/7.997$ is a constant related to the plastic fluid density and discharge coefficient; P_1 is the hydraulic pressure; P_{sup} is the pump supply pressure and 2500 psi at maximum; Q_{nom} is the nominal flow, it is a function of flow control voltage.

2. The pressure in the line, P_1:

$$P_1 = \int \frac{\beta_1}{V_1}(Q_1 - Q_2) * saturation\,. \tag{A.2}$$

where, β_1 is the bulk modulus (= $5.75\times 10^7\ P_a$ in this study) and V_1 is the volume (= 0.002 m^3); Q_2 is the material flow from line into the actuator.

3. The flow to the cylinder, Q_2:

$$Q_2 = Cv\sqrt{P_1 - P_2}\,. \tag{A.3}$$

where, Cv is the orifice coefficient (= 6.0×10^{-6}).

4. The pressure in the actuator, P_2:

$$P_2 = \int \frac{\beta_1}{V_2}(Q_2 - C_L P_2 - A_a \frac{dX_{ram}}{dt})\,. \tag{A.4}$$

where, P_2 is the actuator pressure and X_{ram} is the position of the ram; A_a is the area of the ram cylinder (= 0.02 m^2); C_L is the leakage coefficient (= 3.0×10^{-11}); V_2 the actuator volume (=0.003m^3).

In addition, the pressure in the nozzle is:

$$P_n = \int \frac{\beta_2}{V_2 - A_n X_{ram}}(A_n v_r - Q_p)\,. \tag{A.5}$$

where P_n is the pressure of barrel nozzle and A_n is the area of nozzle; Q_p is the polymer melt flow rate (=10 L/min); β_2 is the nozzle bulk modulus (=$5.0\times 10^7\ P_a$).

And the ram velocity is:

$$\dot{v}_r = \frac{1}{M}\left[P_2 A_2 - P_n A_n - 2\pi\eta R_n^{1-n}(X_0 + X_{ram})\left(\frac{(1/n-1)\,v_r}{K_r^{1-(1/n)} - 1}\right)^n\right]. \tag{A.6}$$

where, v_r is the ram velocity; M is the actuator mass (= 80 kg); R_n is the nozzle radius (=0.0175 m); n is the power law index for polymer melt (= 0.8); η is the viscosity of the material (= 460); K_r is the ratio between screw radius and nozzle radius (= 0.9); X_0 is the initial length of screw (= 0.1 m).

Multiple Models Neural Network Decoupling Controller for a Nonlinear System*

Xin Wang[1], Shaoyuan Li[1], Zhongjie Wang[2], and Heng Yue[3]

[1] Institute of Automation, Shanghai Jiao Tong University, 200030
wangxin26@sjtu.edu.cn
[2] Depart. of Control Science & Engineering, Tongji University, 200092
[3] Research Center of Automation, Northeastern University, 110004

Abstract. For a discrete-time nonlinear MIMO system, a multiple models neural network decoupling controller is designed in this paper. At each equilibrium point, the system is expanded into a linear and nonlinear term. These two terms are identified using two neural networkss, which compose one system model. Then, all models, which are got at all equilibrium points, compose the multiple models set. At each instant, the best model is chosen as the system model according to the switching index. To design the controller accordingly, the nonlinear term and the interactions of the best model is viewed as measurable disturbance and eliminated by the use of the feedforward strategy. The simulation example shows that the better system response can be got even when the system is changed around these equilibrium points.

1 Introduction

In recent years, for linear multivariable systems, researches on adaptive decoupling controller have made much success. As for nonlinear Multi-Input Multi-Output (MIMO) systems, few works have been observed. Ansari *et al.* simplified a nonlinear system into a linear system by using Taylor's expansion at the equilibrium point and controlled it using a linear adaptive decoupling controller accordingly [1]. However, for a system with strong nonlinearity, it can not get good performance. In [2], an exact linear system can be produced utilizing a feedback linearization approach. But accurate information, such as the parameters of the system, must be known precisely. Furthermore, a variable structure controller with sliding mode was proposed [3] and industrial experiment in binary distillation columns was presented [4], which requires the system was an affine system. Although the design methods above can realize nonlinear decoupling control, there were too many assumptions required on the system so that they can not be used in the industrial process directly. To solve this problem, Neural Network (NN) decoupling controller was proposed. In [5], two NNs were needed to identify the linear and nonlinear term expanded using Taylor's formula at the origin. Unfortunately, when the equilibrium point was far from the origin, the system lost its stability.

---

* This work is supported by 863 High Technology Project (2002AA412130, 2003AA412310).

F. Yin, J. Wang, and C. Guo (Eds.): ISNN 2004, LNCS 3174, pp. 175–180, 2004.

In this paper, a multiple models NN decoupling controller (MMNNDC) is designed. At each equilibrium point, by using Taylor's formula, the system is expanded into linear and nonlinear terms. These two terms are identified using two NNs respectively, which compose one system model. All models, which are got at all equilibrium points, compose the multiple models set. At each instant, one best model is chosen out as the system model. To control the system, the nonlinear term and the interactions of the above model is viewed as measurable disturbance and eliminated by the choice of the polynomial matrices. The simulations illustrate the effectiveness of the method.

2 Description of the System

The system is a discrete-time nonlinear MIMO system of the form

$$y(t+1) = f[y(t),\cdots,u(t),\cdots], \tag{1}$$

where $u(t)$, $y(t)$ are $n\times 1$ input, output vectors respectively and $f[\cdot]$ is a vector-based nonlinear function which is continuously differentiable and Lipshitz.

Suppose that $(u_1, y_1),\cdots(u_l, y_l),\cdots(u_m, y_m)$ are m equilibrium points. At each equilibrium point (u_l, y_l), using Taylor's formula, it obtains

$$y(t+1) = y_l + \sum_{n_1=1}^{n_a} f'_{n_1}\Big|_{\substack{u=u_l\\ y=y_l}} \cdot [y(t-n_a+n_1)-y_l] + \sum_{n_2=0}^{n_b} f'_{n_2}\Big|_{\substack{u=u_l\\ y=y_l}} \cdot [u(t-n_b+n_2)-u_l] \tag{2}$$
$$+\, o[x(t)],$$

where $f'_{n_1} = \dfrac{\partial f}{\partial y(t-n_a+n_1)}$, $f'_{n_2} = \dfrac{\partial f}{\partial u(t-n_b+n_2)}$, $x(t) = [y(t)-y_l,\cdots; u(t)-u_l,\cdots]$,

$o[x(t)]$ satisfies $\lim\limits_{\|x(t)\|\to 0} \dfrac{\|o[x(t)]\|}{\|x(t)\|} = 0$, where $\|\cdot\|$ is the Euclidean norm operator.

Define

$$\bar{y}(t) = y(t) - y_l, \tag{3}$$

$$\bar{u}(t) = u(t) - u_l, \tag{4}$$

$$v(t) = o[x(t)], \tag{5}$$

$$A^l_{n_1} = (-1)\cdot f'_{n_1}\Big|_{\substack{u=u_l\\ y=y_l}}, n_1 = 1,\cdots,n_a, \tag{6}$$

$$B^l_{n_2} = f'_{n_2}\Big|_{\substack{u=u_l\\ y=y_l}}, n_2 = 0,\cdots,n_b, \tag{7}$$

$$A^l(z^{-1}) = I + A^l_1 z^{-1} + \cdots + A^l_{n_a} z^{-n_a}, \tag{8}$$

$$B^l(z^{-1}) = B^l_0 + B^l_1 z^{-1} + \cdots + B^l_{n_b} z^{-n_b}. \tag{9}$$

Then system (2) can be rewritten as

$$A^l(z^{-1})\bar{y}(t+1) = B^l(z^{-1})\bar{u}(t) + v(t) . \tag{10}$$

3 Design of MMNNDC

For unknown system, two NNs are employed to identify the linear and nonlinear term of the system (10). These two NNs compose one system model. Similarly, at all equilibrium points, all NNs employed compose the multiple models set. At each instant, the best model is chosen out according to the switching index and the corresponding controller is designed, in which the nonlinear term and the interactions is viewed as measurable disturbance and eliminated using the feedforward strategy.

3.1 Foundation of Multiple Models Set

At each equilibrium point (u_l, y_l), the system (10) is excited using white noise. One BP network NN_l is trained off-line to approximate the system's input-output mapping. So $\hat{A}^l(z^{-1})$ and $\hat{B}^l(z^{-1})$, the estimation of the $A^l(z^{-1})$ and $B^l(z^{-1})$, are obtained. Then another BP network, NN_2, is employed to approximate $v(t)$ online, *i.e.*

$$\hat{v}(t) = NN[W, x(t)] , \tag{11}$$

where $NN[\cdot]$ means the structure of the neural network and W is the weighting value. So the model at the equilibrium point (u_l, y_l) is obtained. Similarly, the models at all equilibrium points can be set up and compose the multiple models set.

3.2 The Switching Index

To the models in the multiple models set, at each instant, only one model is chosen as the system model according to the switching index, which has the form

$$J_l = \left\| e^l(t) \right\|^2 = \left\| y(t) - y^l(t) \right\|^2 , \tag{12}$$

where $e^l(t)$ is the output error between the real system and the model l. $y^l(t)$ is the output of the model l. Let $j = \arg\min(J_l), l = 1, \cdots, m$ correspond to the model whose output error is minimum, then it is chosen to be the best model.

3.3 Multiple Models Neural Network Decoupling Controller Design

For the best model, to realize the decoupling control, the interaction between the input $\bar{u}_j(t)$ and the output $\bar{y}_i(t)$, $(j \neq i)$ is viewed as measurable disturbance. Then (10) can be rewritten as

$$A^l(z^{-1})\bar{y}(t+1) = \bar{B}^l(z^{-1})\bar{u}(t) + \bar{\bar{B}}^l(z^{-1})\bar{u}(t) + v(t) , \tag{13}$$

where $\overline{\boldsymbol{B}}'(z^{-1}) = \mathrm{diag}[\boldsymbol{B}'_{ii}(z^{-1})]$ is a diagonal polynomial matrix with a known nonsingular matrix $\overline{\boldsymbol{B}}'_0$, $\overline{\overline{\boldsymbol{B}}}'(z^{-1}) = \boldsymbol{B}'(z^{-1}) - \overline{\boldsymbol{B}}'(z^{-1})$. For a simple case, $\boldsymbol{A}'(z^{-1})$ is assumed to be a diagonal matrix.

Because the nonlinear system is rewritten as a linear equation in (13), the linear adaptive decoupling controller can be designed to control the system, in which the nonlinear term is viewed as measurable disturbance and eliminated with the interaction by the choice of the polynomial matrices. Like the conventional optimal controller design, for the model j, the cost function is of the form

$$J_c = \left\| \boldsymbol{P}(z^{-1})\overline{\boldsymbol{y}}(t+k) - \boldsymbol{R}(z^{-1})\boldsymbol{w}(t) + \boldsymbol{Q}(z^{-1})\overline{\boldsymbol{u}}(t) + \boldsymbol{S}_1(z^{-1})\overline{\boldsymbol{u}}(t) + \boldsymbol{S}_2(z^{-1})\boldsymbol{v}(t) \right\|^2, \tag{14}$$

where $\boldsymbol{w}(t)$ is the known reference signal, $\boldsymbol{P}(z^{-1}), \boldsymbol{Q}(z^{-1}), \boldsymbol{R}(z^{-1}), \boldsymbol{S}_1(z^{-1}), \boldsymbol{S}_2(z^{-1})$ are weighting polynomial matrices respectively. Introduce the identity as

$$\boldsymbol{P}(z^{-1}) = \boldsymbol{F}(z^{-1})\boldsymbol{A}(z^{-1}) + z^{-1}\boldsymbol{G}(z^{-1}). \tag{15}$$

Multiplying (13) by $\boldsymbol{F}(z^{-1})$ from left and using (15), the optimal control law can be derived as follows

$$\begin{gathered}\boldsymbol{G}(z^{-1})\overline{\boldsymbol{y}}(t) + [\boldsymbol{F}(z^{-1})\overline{\boldsymbol{B}}(z^{-1}) + \boldsymbol{Q}(z^{-1})]\overline{\boldsymbol{u}}(t) + [\boldsymbol{F}(z^{-1})\overline{\overline{\boldsymbol{B}}}(z^{-1}) + \boldsymbol{S}_1(z^{-1})]\overline{\boldsymbol{u}}(t) \\ + [\boldsymbol{F}(z^{-1}) + \boldsymbol{S}_2(z^{-1})]\boldsymbol{v}(t) = \boldsymbol{R}\boldsymbol{w}(t),\end{gathered} \tag{16}$$

combing (16) with (13), the closed loop system equation is obtained as follows

$$\begin{gathered}\left[\boldsymbol{P}(z^{-1}) + \boldsymbol{Q}(z^{-1})\overline{\boldsymbol{B}}^{-1}(z^{-1})\boldsymbol{A}(z^{-1})\right]\overline{\boldsymbol{y}}(t+1) = \left[\boldsymbol{Q}(z^{-1})\overline{\boldsymbol{B}}(z^{-1})^{-1}\overline{\overline{\boldsymbol{B}}}(z^{-1}) - \boldsymbol{S}_1(z^{-1})\right]\overline{\boldsymbol{u}}(t) \\ \left[\boldsymbol{Q}(z^{-1})\overline{\boldsymbol{B}}(z^{-1})^{-1} - \boldsymbol{S}_2(z^{-1})\right]\boldsymbol{v}(t) + \boldsymbol{R}(z^{-1})\boldsymbol{w}(t).\end{gathered} \tag{17}$$

To eliminate the nonlinear form and the interactions of the system exactly, let

$$\boldsymbol{Q}(z^{-1}) = \boldsymbol{R}_1\overline{\boldsymbol{B}}(1), \tag{18}$$

$$\boldsymbol{S}_1(z^{-1}) = \boldsymbol{R}_1\overline{\overline{\boldsymbol{B}}}(1), \tag{19}$$

$$\boldsymbol{S}_2(z^{-1}) = \boldsymbol{R}_1, \tag{20}$$

$$\boldsymbol{R}(z^{-1}) = \boldsymbol{P}(1) + \boldsymbol{R}_1\boldsymbol{A}(1), \tag{21}$$

where $\boldsymbol{R}_1$ is a constant matrix and decided by the designer to guarantee the stability of the closed loop system. So $\boldsymbol{P}(z^{-1}), \boldsymbol{R}_1$ are selected off-line to satisfy

$$\left|\boldsymbol{B}(z^{-1})\boldsymbol{B}^{-1}(1)\boldsymbol{R}_1^{-1}\boldsymbol{P}(z^{-1}) + \boldsymbol{A}(z^{-1})\right| \neq 0 \quad |z| > 1. \tag{22}$$

Although the second $\overline{\boldsymbol{u}}(t)$ is the interaction of the system and viewed as measurable disturbance, to obtain the control input, it must be included. So the control law is rewritten from (16) as

$$\begin{gathered}[\boldsymbol{F}(z^{-1})\boldsymbol{B}(z^{-1}) + \boldsymbol{Q}(z^{-1}) + \boldsymbol{S}_1(z^{-1})]\overline{\boldsymbol{u}}(t) + \\ \boldsymbol{G}(z^{-1})\overline{\boldsymbol{y}}(t) + [\boldsymbol{F}(z^{-1}) + \boldsymbol{S}_2(z^{-1})]\boldsymbol{v}(t) = \boldsymbol{R}\boldsymbol{w}(t).\end{gathered} \tag{23}$$

Remark 1. Equation (23) is a nonlinear equation because $\overline{\boldsymbol{u}}(t)$ is include into the nonlinear term $\boldsymbol{v}(t)$. Considering $\overline{\boldsymbol{u}}(t)$ will converge to a constant vector in steady state, then substitute $\overline{\boldsymbol{u}}(t)$ in the nonlinear term $\boldsymbol{v}(t)$ with $\overline{\boldsymbol{u}}(t-1)$ and solve (23).

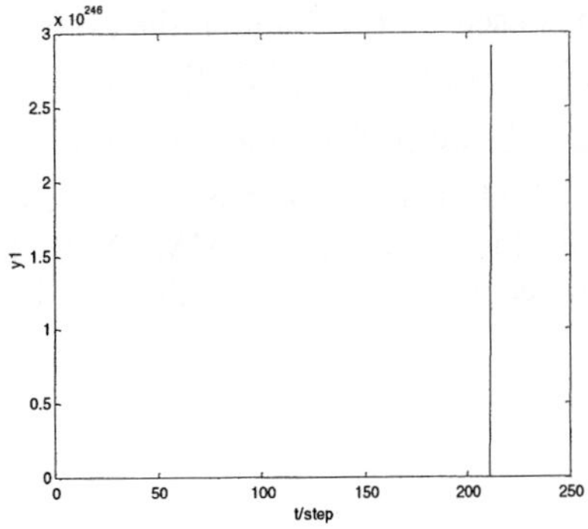

Fig. 1. The output $y_1(t)$ of NNDC

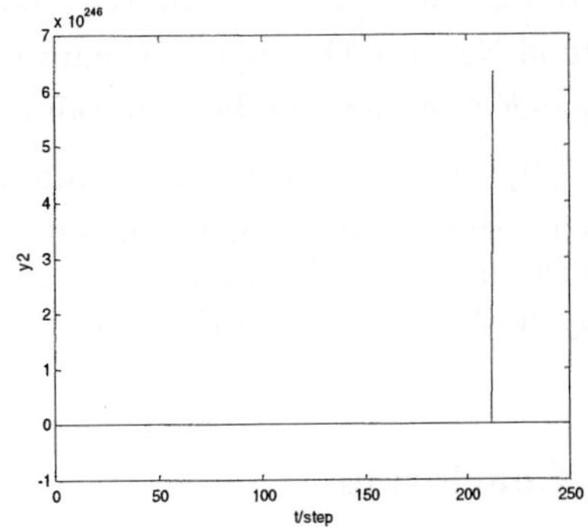

Fig. 2. The output $y_2(t)$ of NNDC

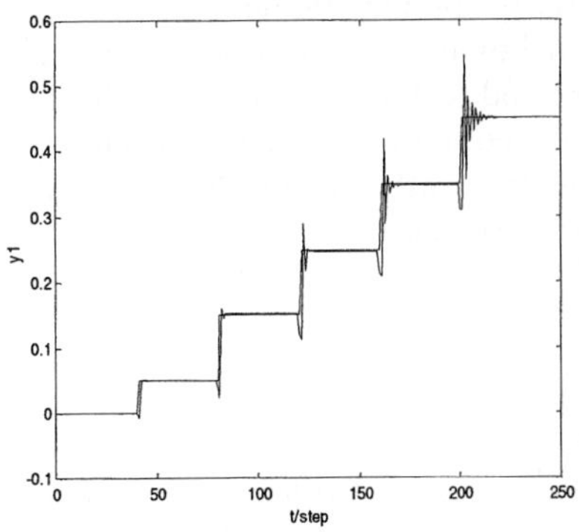

Fig. 3. The output $y_1(t)$ of MMNNDC

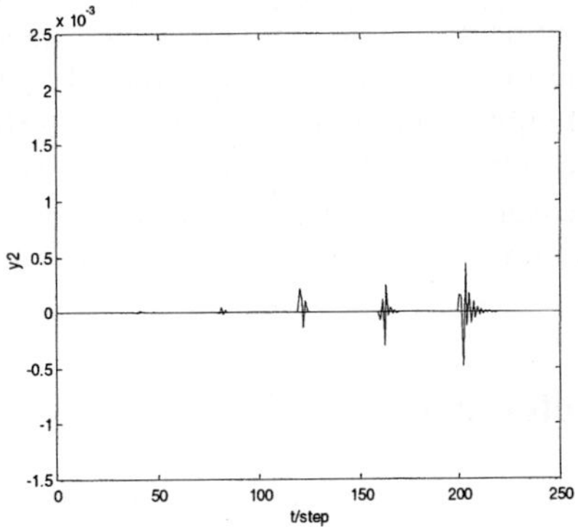

Fig. 4. The output $y_2(t)$ of MMNNDC

4 Simulation Studies

A discrete-time nonlinear multivariable system is described as follows

$$\begin{cases} y_1(t+1) = \dfrac{-0.2y_1(t)}{1+y_1^2(t)} + \sin[u_1(t)] - 0.5\sin[u_1(t-1)] + 1.5u_2(t) + 0.2u_2(t-1) \\ y_2(t+1) = 0.6y_2(t) + 0.2u_1(t) + 1.3u_1(t-1) + u_2(t) + u_2^2(t) + \dfrac{1.5u_2(t-1)}{1+u_2^2(t-1)} \end{cases}, \quad (24)$$

which is the same as the simulation example in [5]. The known reference signal $\boldsymbol{w}$ is set to be a time-varying signal. When $t=0$, w_1 equals to 0 and when t is 40, 80, 120, 160, 200, it changed into 0.05, 0.15, 0.25, 0.35, 0.45 respectively, while w_2 equals to 0 all the time.

In Fig.1 and 2, the system (24) is expanded only at the original point (0,0) and a Neural Network Decoupling Controller (NNDC) is used. In Fig.3 and 4, the system is expanded at six equilibrium points, *i.e.* $[0,0]^T$, $[0.1,0]^T$, $[0.2,0]^T$, $[0.3,0]^T$, $[0.4,0]^T$ and. Note that the equilibrium points are far away from the set points. The results show that although the same NNDC method is adopted, the system using NNDC loses its stability (see Fig.1 and 2), while the system using MMNNDC not only gets the good performance but also has good decoupling result (see Fig.3 and 4).

5 Conclusion

A MMNNDC is designed to control the discrete-time nonlinear multivariable system. At each equilibrium point, one NN is trained offline to identify the linear term of the nonlinear system and the other neural network is trained online to identify the nonlinear one. The multiple models set is composed of all models, which are got from all equilibrium points. According to the switching index, the best model is chosen as the system model. The nonlinear term and the interaction of the system are viewed as measurable disturbance and eliminated using feedforward strategy. The simulation example shows that the effectiveness of the controller proposed.

References

1. Ansari R.M., Tade M.O.: Nonlinear Model-based Process Control: Applications in Petroleum Refining. Springer, London (2000)
2. Wang W.J., Wang C.C.: Composite Adaptive Position Controller for Induction Motor Using Feedback Linearization, IEE Proceedings D Control Theory and Applications, 45 (1998) 25–32
3. Wai R.J., Liu W.K.: Nonlinear Decoupled Control for Linear Induction Motor Servo-Drive Using The Sliding-Mode Technique. IEE Proceedings D Control Theory and Applications, 148 (2001) 217–231
4. Balchen J.G., Sandrib B.: Elementary Nonlinear Decoupling Control of Composition in Binary Distillation Columns. Journal of Process Control, 5 (1995) 241–247
5. Yue H., Chai T.Y.: Adaptive Decoupling Control of Multivariable Nonlinear Non-Minimum Phase Systems Using Neural Networks. Proceedings of the American Control Conference. (1998) 513–514

Feedback-Assisted Iterative Learning Control for Batch Polymerization Reactor

Shuchen Li[1,2], Xinhe Xu[1], and Ping Li[2]

[1] School of Information Science & Engineering, Northeastern University, 110004 Shenyang, China
lishuchen@lnpu.edu.cn
[2] School of Information Engineering ,Liaoning University of Petroleum & Chemical Technology, 113001 Fushun, China

Abstract. An algorithm of the feedback-assisted iterative learning control (FBAILC) was proposed for a batch repeatable operation process. Control law of FBAILC was based on the inverse of process model, added the filter polynomial in iterative learning and analyzed the convergence of FBAILC algorithm. On-line estimator method of the process parameters was introduced in application, which achieved the parameter self-tuning of controller. The effectiveness of the proposed method was demonstrated by simulation results.

1 Introduction

A batch polymerization reactor is a typical device of petroleum chemical process. Dual-model control is used in the traditional industry, which heats up system by the maximum heat flow at initial state, and switches to the PID control when it reaches preset point temperature. Dual-model control can be considered to be an industrial application of bang-bang time-optimal control, however, it exists in a common drawback, i.e. it lacks adaptability to the process changes. Antonio [1] introduced an adaptive control technique to heat-up control, indicated identifying the process model on-line, and calculating the optimal switching time on-line. This approach hinders its industrial application because the repeatable process leaves behind the lead-time for the identification in each repeatable process is insufficient. Applying the predictive control to the polymerization reactor had been employed in numerous researches and applications [2], but the predictive control depends on the predictive model which is significant different at each batch, such as the difference of catalyst activation and fouling products on the reactor wall will influence on the performance of the predictive control.

Iterative learning control (ILC) is a novel control algorithm, which handles especially the partial repeatable operation process, has been successful used in the industrial robot, machine tool of numerical control and batch chemical device [3~5]. It makes use of the previous process data of control system to attain an ideal control input trajectory through ILC algorithm according to the process output and desire reference trajectory. ILC was open-loop control at original, then combined with feedback control and better control performance was obtained [6~7].

F. Yin, J. Wang, and C. Guo (Eds.): ISNN 2004, LNCS 3174, pp. 181–187, 2004.

In this paper, Feedback-assisted iterative learning control is applied to temperature control of a batch polymerization reactor. The following improvements on control algorithms: Firstly, the process model is identified on-line through the maximum initial heat-up process, and realizes self-tuning on-line for the parameters of controller. Secondly, the reference trajectory is built on-line to make sure the asymptotic convergence of iterative learning. Thirdly, the filter is introduced in FBAILC algorithm to improve robustness of the control algorithm.

2 Process Description

Fig.1 shows the procedure of the heat-up. The temperature in the reactor will be close to the circumstance's temperature (the temperature is difference with the season) after the injected material is finished. Ratio of heat-up indicates system properties of this repeatable process, such as material quality and catalyst activation, however, these properties are indefinite and relevant to partial repeat of the previous process. The system switches to the feedback control (the controller is usually PID) when temperature reaches some range. PID controller parameters by initial value are tuned formidably because the properties of each batch process are different.

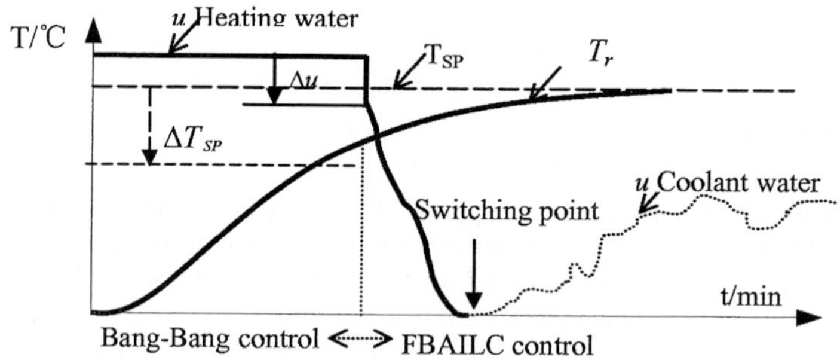

Fig. 1. Diagram of heat-up process control

3 Feedback-Assisted Iterative Learning Control

3.1 FBAILC Law

Consider a minimum-phase SISO discrete-time system defined:

$$y(t) = p(q^{-1})u(t) + d(t) \ , t \in (0,1,\dots,n) \tag{1}$$

where $u \bullet y$ and d denote the input, output of the process and unknown disturbance, respectively. The first-order FBALC [8] algorithm can be written as follows:

$$u_k(t) = u_{k-1}(t) + H_1(q^{-1})[r(t) - y_{k-1}(t)] + H_0(q^{-1})[r(t) - y_k(t)] \tag{2}$$

where the subscript k , r denote the batch numbers and the reference trajectory of the process output, and H_0 and H_1 are transfer impulse functions for the feedback controller and the learning, respectively. In (2) the first two terms on the right side constitute the learning part, which generates an current bias control signal from the previous batch data, and the last term represents the feedback control signal on this batch process. For an unknown repetitive disturbance, to make sure process (1) convergence conditions based on (2) control law is satisfied:

$$\left|1 - P(q^{-1})H_1(q^{-1})\right| < \left|1 + P(q^{-1})H_0(q^{-1})\right| \tag{3}$$

more precisely, we can have:

$$\lim_{k\to\infty}\left\|r - y_k\right\|_2 = 0 \tag{4}$$

where $q = e^{jwh}$, $wh \in [0,\pi]$, h is the sampling interval.

There are many possibilities for choosing H_1. Among them H_1=P^{-1} gives the fastest convergence, obtaining zero tracking error at all frequencies, however, the exact process model is practically unavailable, so the inverse of an approximate model of process $\overline{P}$ substitutes for H_1. (3) can be rewritten as:

$$\left|1 - \frac{P(q^{-1})}{\overline{P}(q^{-1})}\right| < \left|1 + P(q^{-1})H_0(q^{-1})\right| \tag{5}$$

We can see that the process can make sure convergence of algorithm if the relative model uncertainty is smaller in (5). When the high frequency noise is present in the output y_{k-1}(t), u_k(t) may have large spikes due to the learning control. At the same time, depending on the process, (5) may pose strict restrictions on the allowable model uncertainty. To assure stabilization of algorithm is to introduce a low-pass filter $F(q^{-1})$, control law (2) can be rewritten as:

$$u_k(t) = F(q^{-1})\{u_{k-1}(t) + \overline{P}^{-1}[r(t) - y_{k-1}(t)]\} + H_0(q^{-1})[r(t) - y_k(t)] \tag{6}$$

Note that $F(q^{-1})$ is be introduced, not only enlarges stabilization region, but also reduces the convergence speed of iterative learning.

3.2 Convergence Analysis

The process model is:

$$y_k(t) = P_k(q^{-1})u_k(t) \tag{7}$$

Rewritten (6) by the subscript k of reference trajectory, we have:

$$u_k(t) = F(q^{-1})\{u_{k-1}(t) + \overline{P}^{-1}[r_{k-1}(t) - y_{k-1}(t)]\} + H_0(q^{-1})[r_k(t) - y_k(t)] \tag{8}$$

Substituting (7) into (8), omitting $(\cdot)$ and attaining equation:

$$e_k = \frac{F}{1+P_k H_0}\left[\frac{P_k}{P_{k-1}} - \frac{P_k}{\overline{P}}\right] e_{k-1} + \frac{1}{1+P_k H_0}\left[r_k - \frac{FP_k}{P_{k-1}} r_{k-1}\right] \tag{9}$$

where $e_k=r_k-y_k$, $e_{k-1}=r_{k-1}-y_{k-1}$. For $k \to \infty$ as $e_k(t) \to 0$, the following two conditions should be satisfied:

$$\left|\frac{F}{1+P_k H_0}\left[\frac{P_k}{P_{k-1}} - \frac{P_k}{\overline{P}}\right]\right| < 1 \tag{10}$$

$$r_k = \frac{FP_k}{P_{k-1}} r_{k-1} \tag{11}$$

3.3 On-Line Parameters Estimation for Process Model

Condition (10) may be satisfied if the process does not change much, but convergence condition (11) cannot be satisfied unless the process model and reference trajectory are suited. We can obtain Fig.2*a* curve for heat-up process, the reactor temperature is related to the jacket temperature by a first-order delay model with unit steady state gain. Pure delay time of process is $\tau=t_1-t_0$, we have:

$$y(s) = \frac{K_P}{T_P S+1} e^{-\tau s} u(s) \quad \text{or} \quad y(s) = \frac{1}{T_P S+1} u(s) \tag{12}$$

where T_P denotes time constant of the process, $y=T_r-T_{ro}$, $u=T_j-T_{ro}$, T_r and T_j denote the reactor and jacket temperature, T_{ro} is lower than the reactor target temperature T_r^{SP} at the FBALC is started. If the process gain and pure delay time are invariable, $K_p=1$, dynamic properties of the process are realized by a parameter T_P, Rearranging (12), we can have:

$$\hat{T}_P = \frac{u(t)-y(t)}{dy(t)/dt} = \frac{T_j(t)-T_r(t)}{dT_r(t)/dt} \tag{13}$$

Choosing reference trajectory as follows:

$$T_r^{ref}(t) = (T_r^{sp} - T_{r0}) + \frac{\alpha t}{\left\{\left[\frac{\alpha}{T_{r0}} t\right]^N + 1\right\}^{1/N}} \tag{14}$$

where N is a constant, α is the slope of the tangent line that is denoted by the following equation:

$$\alpha = \frac{T_{r0} - T_{r1}}{t_2 - t_1} \tag{15}$$

Because of $|F(j\omega)| = 1$, condition(11) can be rewritten:

$$r_k(s)=\frac{T_{k-1}^P s+1}{T_k^P s+1} r_{k-1}(s) \tag{16}$$

Applying the initial and final value theorems to (16):

$$\left[\frac{dr_k}{dt}\right]_{t=0}=\left[\frac{T_{k-1}^P}{T_k^P}\right]\left[\frac{dr_{k-1}}{dt}\right]_t=0\bullet\left[\frac{dr_k}{dt}\right]_{t\to\infty}=\left[\frac{dr_{k-1}}{dt}\right]_{t\to\infty} \tag{17}$$

where the reference trajectory is $[dr_k/dt]_{t\to\infty}=0$. The above equation is illustrated that the reference trajectory is satisfied with convergent condition (11) at the initial and finial time.

3.4 Parameters Self-Tuning for Feedback Controller

The feedback controller is used PID control law, u_0, K_c, T_i, T_d need be tuned, that are the initial value of output, proportional gain, integral time and differential time, respectively. The tuning process is:

$$\Delta u=\Delta u_0+k_{u0}(\alpha-\alpha_0),\ u_0=u_{\max}-\Delta u \tag{18}$$

where $\Delta u_0, \alpha_0$ are the initial value change of controller output in the normal circumstance and the slope of the heat-up tangent line, respectively. k_{uo} is a given tuning coefficient. In Fig.1, Δu is tuned automatically according to α. The formulation of K_c, T_i, T_d self-tuning are:

$$Kc=\frac{T_P}{\tau K_P}\left(\frac{4}{3}+\frac{\tau}{4T_P}\right),\quad T_i=\tau\frac{32+6\tau/T_P}{13+8\tau/T_P},\quad T_d=\tau\frac{4}{11+2\tau/T_P} \tag{19}$$

The above equations on the right side are from estimate values of section 3.3.

4 Simulation Analysis

The model of simulation process is used by follows: $1.2e^{-5s}/(20s+1)$, the iterative learning filter is: $F(q^{-1})=(1+0.5q^{-1}+0.5q^{-2})/2$, $u_{\max}$=85, $\Delta u_0=10$, T_r^{SP}=75, T_{r0}=65, the sharpness coefficient of reference trajectory N=1, the slope of heat-up mass flow α_0=3.19, the tuning coefficient of PID initial value k_{uo}=4.5. The curves of ideal control are shown in Fig.2*b*.

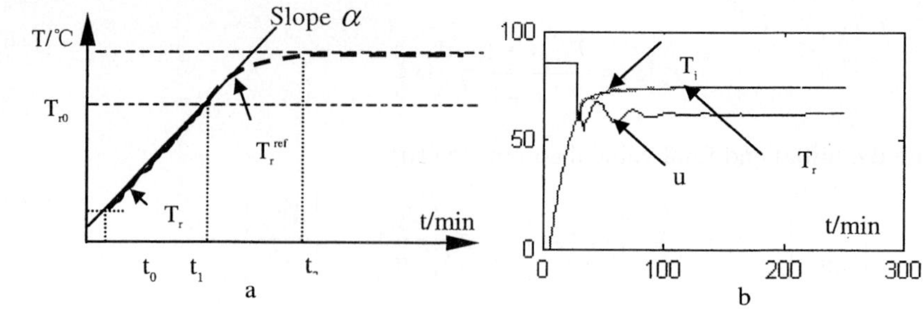

Fig. 2. Diagram of process parameters estimation and control

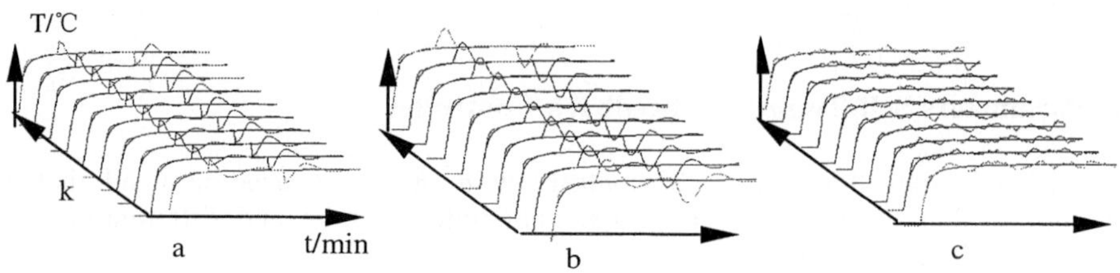

Fig. 3. Control curve of FBAILC for measuring noise

4.1 Simulation Analysis for Disturbance

Control curves of FBAILC for measuring noise are given in Fig.3. Fig.3 curve of k=1 is PID control, curves of k=2~10 are FBAILC. Fig.3*a*, the constant disturbances (+5℃ and -5℃, respectively) are added to the measurement at 70min and 160min.With the increase of repetitive number, the tuning time is gradually fast to the constant disturbances of the same magnitude. Fig.3*b*, the increasing disturbance by 1℃/min is added to the measurement at the beginning of 70min, but the descending

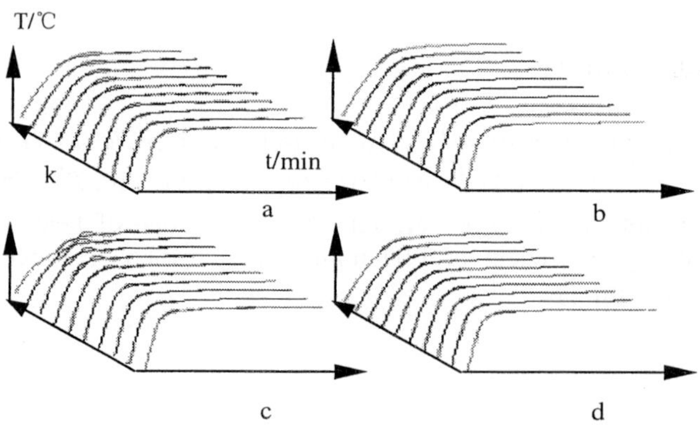

Fig. 4. Control comparison of time-varying parameters

disturbance is added to the measurement at 160min in the reverse way. Thought it is exaggerated to added disturbance, the magnitude of control variable reduces with the increase of trial numbers. Fig.3*c*, ±1℃ magnitude of random disturbances are added in measurement, however, FBAILC algorithm is the same as PID control, cannot improve on random disturbance.

4.2 Simulation Analysis for Moved Parameters of Process

Fig.4 is shown the simulation curves of time constant and gain moving for process. Fig.4*a*, the time constant of process is increasing at magnitude 1+2+...+k of PID(k_c=2.5, t_i=6.0, t_d=1.0) control curve, Fig.4*b* is shown about corresponding FBAILC curve. Fig.4*c*, the gain of process is increasing at magnitude 1+2+...+k of PID (k_c=2.5, t_i=6.0, t_d=1.0) control curve, Fig.4*d* is shown about corresponding FBAILC curve. Form simulation curve, we can attain that FBAILC algorithm is preferable adaptive ability and robustness.

5 Conclusion

FBAILC improves the control performances on feedback control for the repetitive disturbance, makes sure convergence of algorithm by the reference trajectory build on-line. Furthermore, estimation of the process parameters on-line and parameters of controller self-tuning on-line are introduced to improve stabilization and robustness of algorithm. Simulation analyzes the performance of FBAILC, and the effectiveness of FBAILC was demonstrated by simulation results.

References

1. Antonio, V.: Time-Optimal Plug & Control for Integrating and FOPDT Processes. Journal of Process Control. 3 (2003) 195-202
2. Andrew, W.D., Jay, H. L.: Building Inferential Prediction Models of Batch Processes Using Subspace Identification. Journal of Process Control .5 (2003) 397–406
3. Y.-S.X., J.-Y.S.: Performance Evaluation and Optimization of Human Control Strategy. Robotics and Autonomous Systems . 1 (2002) 19–36
4. F.-R.G., Y.-G.Y. , C. S.: Robust Iterative Learning Control with Applications to Injection Molding Process. Chemical Engineering Science.24 (2001) 7025–7034
5. Mezghani, M., Rouxb, G., Cabassud, M.: Robust Iterative Learning Control of an Exothermic Semi-Batch Chemical Reactor•Mathematics and Computers in Simulation. 6 (2001) 367–385
6. Peter, B., Gold S.: On the Equivalence of Causal LTI Iterative Learning Control and Feedback Control. Automatica.4 (2002) 703–708
7. J.-X.X., T.-H.L., Ying ,T.: Enhancing Trajectory Tracking for a Class of Process Control Problems Using Iterative Learning. Engineering Applications of Artificial Intelligence, 1(2002) 53–64
8. Lee, K.S., Bang, S.H.: Iterative Learning Control of Heat-up Phase for a Batch Polymerization Reactor. Journal of Process Control. 4 (1996) 255-262

Recent Developments on Applications of Neural Networks to Power Systems Operation and Control: An Overview

Chuangxin Guo[1], Quanyuan Jiang[1], Xiu Cao[2], and Yijia Cao[1]

[1] College of Electrical Engineering
Zhejiang University, Hangzhou 310027, Zhejiang, China
yijiacao@zju.edu.cn

[2] Department of Computer Science and Engineering,
Fudan University, Shanghai 200433 , China
xiucao@fudan.edu.cn

Abstract. Artificial neural networks (ANNs) have found many potential applications in power systems operation and control recently. This paper presents a categorization of the main significant applications of neural networks, which includes power system controller design, power system security assessment and load forecasting. It is desired that they are helpful to the construction of more efficient, robust ANNs to solve a broader range of problems in power systems.

1 Introduction

Power system blackouts are rare events. However, when they occur, the effects on commerce, industry, and everyday life of the general population can be quite severe. In order to prevent a blackout, it is important to develop some strategies to meet the load demand and satisfy the stability and reliability criteria. Due to the increasing complexity of modern power systems, it is always attractive to employ intelligent system technology. The term "intelligent system" is often used to represent any combination of artificial neural networks (ANN's), expert systems, fuzzy-logic systems, and other emerging technologies, such as genetic algorithms. This paper concentrates on applications of ANNs to power systems and presents an overview on main significant applications of neural networks in recent years, which include power system controller design, power system security assessment and load forecasting.

2 An Introduction to Artificial Neural Networks

An ANN can be seen as a union of simple processing units, based on neurons that are linked to each other through connections similar to synapses. Many different types of ANN's are described in the technical literature. The main types of networks that have

F. Yin, J. Wang, and C. Guo (Eds.): ISNN 2004, LNCS 3174, pp. 188–193, 2004.

been used in power system applications are classified as: 1) Multilayer perceptrons; 2) Hopfield networks; 3) Kohonen networks; 4) Other networks. In this section, due to its very widespread application (over 80%), we concentrate on the overview of applications of the multi-layer perceptron (MLP) network. A multilayered perceptron (MLP) was used and trained with a supervised learning algorithm called back-propagation. A MLP consists of several layers of processing units that compute a nonlinear function of the internal product of the weighted input patterns. These types of networks can deal with nonlinear relations between the variables; however, the existence of more than one layer makes the weight adjustment process for problem solution difficult. In the original version, the back-propagation learning algorithm adjusts the weight of the connections one pattern at a time. The input patterns are represented by a vector $X_p = (x_{p1}, x_{p2}, \cdots, x_{pN})$ and are submitted to the ANN through the input layer that simply redistributes them to the following hidden layer. Each neuron of the following layer receives the weighted signals (signal multiplied by a weight) and generates an output signal to the following layer. This process is repeated until the output layer is reached, where the neurons will generate the output of the ANN for the given input vector. With the output of the ANN obtained, the weight adjustment of the connections will begin in the direction from output layer to input layer. The weight adjustments are realized in order to minimize the error function for a certain pattern. Equation (1) illustrates the error function

$$E_p = \frac{1}{2}\sum_j (d_{pj} - y_{pj})^2, \qquad (1)$$

where d_p is the desired output for input pattern p and y_p is the actual output pattern. Equation (2) mathematically defines how the connection weights of the network are modified

$$\omega_{ij}(t+1) = \omega_{ij}(t) + \eta\delta_j y_i\ , \qquad (2)$$

where ω_{ji} is the weight of the connection between neurons i and j, and η is the learning rate. $\delta_j = (d_{pj} - y_{pj})$ if neuron j is an output layer unit and $\delta_j = \sum_k \delta_k \omega_{jk}$ if neuron j is a hidden layer unit. The choice of an appropriate learning parameter η will considerably influence the convergence rate of the algorithm [1]. For more details about neural networks, one can refer to [1].

3 An Overview on Recent Applications

ANN applications can be broken into two main categories: 1) curve fitting and regression and 2) pattern recognition and classification. While in many ways these two categories are the same, regression analyses often require some type of *a priori* knowledge of the general trends involved and the variables that significantly influence the trends. Examples of ANN applications to a variety of power system problems that are pertinent to industrial and commercial power systems can be found in

[2]–[18]. Note that these references are intended to be representative only; a very large number of successful applications are omitted for the sake of brevity.

3.1 Power System Controller Design

A synchronous generator in a power system is a nonlinear fast-acting multiple-input–multiple-output (MIMO) device. Conventional linear controllers (CONVCs) for the synchronous generator consist of the automatic voltage regulator (AVR) to maintain constant terminal voltage and the turbine governor to maintain constant speed and power at some set point. These controllers are designed to control, in some optimal fashion, the generator around one particular operating point; and at any other point the generator's damping performance is degraded. Artificial neural networks (ANNs) offer an alternative for the CONVC as nonlinear adaptive controllers. Researchers in the field of electrical power engineering have until now used two different types of neural networks, namely, a multilayer perceptron network (MLPN), or a radial basis function network (RBFN), both in single and multimachine power system studies [3]–[7]. Proponents of each type of neural network have claimed advantages for their choice of ANN, without comparing the performance of the other type for the same study. The applications of ANNs in the power industry are expanding, Park *et al* [3] compared the performances of a multilayer neuralcontroller (MLPNC) and a radial basis function neuralcontroller (RBFNC) for back-propagation via time based indirect adaptive control of the synchronous generator. The results show that the RBFNC improves the system damping and dynamic transient stability more efficiently than the MLPNC. They also further proposed a novel optimal neurocontroller that replaces the conventional controller (CONVC)[4], which consists of the automatic voltage regulator and turbine governor, to control a synchronous generator in a power system using a multilayer perceptron neural network (MLPN) and a radial basis function neural network (RBFN). Changaroon *et al* [5] presents the development of a neural network based power system stabilizer (PSS) designed to enhance the damping characteristics of a practical power system network representing a part of Electricity Generating Authority of Thailand (EGAT) system. The proposed PSS consists of a neuro-identifier and a neuro-controller which have been developed based on Functional Link Network (FLN) model. In addition, Lim [6] also proposed a neural network-based control scheme in conjunction with a coherency recognition technique to enhance the damping characteristics of multi-machine power systems.

3.2 Power System Security Assessment

Power system security assessment deals with the system's ability to continue to provide service in the event of an unforeseen contingency. Neural networks have shown great promise as a means of predicting the security of large electric power systems [8]–[12]. Neural networks offer several advantages over traditional techniques including the ability to learn from examples. The first step in applying neural networks to power system security assessment is the creation of an appropriate training data set.

A common approach is to simulate the system in response to various disturbances and then collect a set of pre-disturbance system features along with the corresponding system security index. Possible security indices include the Critical Clearing Time (CCT) and the system Energy Margin (EM). Paucar and Fernandes [8] also proposed an ANN methodology to compute the critical clearing time (CCT) of power system transient stability. Simulation results show that the maximum testing error in the calculation of the critical clearing time is below 5%. One of the most important aspects of achieving good neural network performance has proven to be the proper selection of training features. Proposed methods for selecting an appropriate subset of features are numerous. For example, Jensen et al investigated the use of Fisher's linear discriminant function, coupled with feature selection techniques as a means for selecting neural network training features for power system security assessment. A case study was performed on the IEEE 50-generator system to illustrate the effectiveness of the proposed techniques [9]. An alternate to feature selection is feature extraction. Feature extraction can offer sensitivity information to help the identification of input features best suited for control action. Moulin *et al* [10] presented a new learning-based nonlinear classifier, the Support Vector Machines (SVMs) NNs, showing its suitability for power system transient stability analysis (TSA). It can be seen as a different approach fo cope with the problem of high dimensionality due to its fast training capability, which can be combined with existing feature extraction techniques. SVMs' theoretical motivation is conceptually explained and they are applied to the IEEE 50 generator system TSA problem. Aspects of model adequacy, training time and classification accuracy are discussed and compared to stability classifications obtained by Multi-Layer Perceptrons (MLPs). A novel two-layer fuzzy hyper-rectangular composite neural network (FHRCNN) was developed [11]. The process of presenting input data to each hidden node in a FHRCNN is equivalent to firing a fuzzy rule. An efficient learning algorithm was developed to adjust the weights of an FHRCNN. From simulation tests on the IEEE 39-bus system, it reveals that the proposed novel FHRCNN can yield a much better performance than that of conventional multilayer perceptrons (MLP's) in terms of computational burden and classification rate. Also a hybrid pattern classifier that combines neural networks and decision trees has been proposed to assess a multi-machine power system based on phasor measurements with classification rates of over 99% for the training set and over 94% for the test set [12].

3.3 Load Forecasting

Short-load forecasting has always been the essential part of an efficient power system planning and operation. Now artificial neural network becomes to be the most popular approach for its self-learning, generalization and non-linear mapping abilities. Several approaches have been studied using neural networks for load forecasting [13]–[19]. Senjyu *et al* proposed a one-hour-ahead load forecasting method using the correction of similar day data [13]. In the proposed prediction method, the forecasted load power is obtained by adding a correction to the selected similar day data. The correction is yielded from the neural network. Since the neural network yields the correction which is a simple data, it is not necessary for the neural network to learn all similar day's data. Therefore, the neural network can forecast load power by simple learning. If the forecast day is changed, the neural network is retrained and it can obtain the relationship between load and temperature around the forecast day. Therefore, it is possible to deal with seasonal change by using the proposed neural network. Neural network are suitable for load forecasting because of their approximation ability for nonlinear mapping and generalization. Most of these neural network based methods, reported so far uses weather information, load power of the day before forecast day and day type for input variables [14]. However there are many factors that influenced the precision of load forecasting directly or indirectly, and it is hard to determine the uncertainty between load and various factors. Therefore, the forecasting process has become even more complex, and more accurate forecasts are needed. Now data mining, a newly emerging technology, can solve the uncertainty that arises from inexact, noisy, or incomplete information. It provided an effective way for us to solve the difficulties. Rough set theory [15] as the most typical algorithm of data mining has been applied in expert system, decision support systems, and machine learning. In machine learning, rough set is used to extract decision rules from operation data; in neural network, rough set is used in knowledge discovery, data pre-processing and modeling knowledge-based neural network [16-17]. In order to improve the performance of load forecasting, a novel model integrated with a fuzzy neural network has been presented for short-term load forecasting [18]. In the model fuzzy sets are employed to handle linguistic input information and ambiguity in output decision, while rough set extracts the relevant domain knowledge to the load, and then the network structure and initial weights are auto-adjusted by the knowledge encoded in the neural network. Data mining, integrated with neural networks, could be a promising approach for accurate load forecasting.

Acknowledgment. The authors would like appreciate the financial support from the NSF of Zhejiang Province, China (No. 502053) and Doctoral Program Fund in Higher Education Institutions, China.

References

1. Nauck, D., Klawonn, F. Kruse R.: Foundations of Neuro-Fuzzy Systems, John Wiley & Sons (1997)
2. Halpin, S. M., Burch, R. F. IV.: Applicability of Neural Networks to Industrial and Commercial Power Systems: A Tutorial Overview. IEEE Trans. on Industr. App. 33 (1997) 1355-1363
3. Park, J., Harley, R.G., Venayagamoorthy, G.K.: Comparison of MLP and RBF Neural Networks using Deviation Signals for Indirect Adaptive Control of a Synchronous Generator. Proceedings of the 2002 International Joint Conference on Neural Networks, 1 (2002) 919 - 924

4. Park, J., Harley, R.G., Venayagamoorthy, G.K.: Adaptive-critic-based Optimal Neurocontrol for Synchronous Generators in a Power System using MLP/RBF Neural Networks. IEEE Trans. on Industr. App. 39 (2003) 1529 - 1540
5. Changaroon, B., Srivastava, C. S., Thukaram, D.: A Neural Network Based Power System Stabilizer Suitable for On-Line Training—A Practical Case Study for EGAT System. IEEE Trans. on Energ. Conv. 15 (2000) 103-110
6. Lim, C.M.: Neural Network Control for Damping of Multi-mode Oscillations in a Power System. Proceedings of IEEE Region 10 International Conference on Electrical and Electronic Technology. 2 (2001) 658 – 663
7. Venayagamoorthy, G.K., Harley, R.G., Wunsch, D.C.: Adaptive Neural Network Identifiers for Effective Control of Turbo-generators in a Multimachine Power System. Proc. IEEE Power Engineering Society Winter Meeting. 3 (2001) 1294 – 1299
8. Paucar, V.L., Fernandes, F.C.: A Methodology based on Neural Networks for the Determination of the Critical Clearing Time of Power Systems Transient Stability. Proceedings of International Conference on Power System Technology. 4 (2002) 2669 – 2673
9. Jensen, C.A., El-Sharkawi, M.A., Marks, R.J., II.: Power System Security Assessment using Neural Networks: Feature Selection using Fisher Discrimination. IEEE Transactions on Power Syst. 16 (2001) 757 – 763
10. Moulin, L.S., da Silva, A.P.A., El-Sharkawi, M.A., Marks, R.J.: Support Vector and Multilayer Perceptron Neural Networks Applied to Power Systems Transient Stability Analysis with Input Dimensionality Reduction. Proc. of IEEE Power Engineering Society Summer Meeting. 3 (2002) 1308 – 1313
11. Su, M., Liu, C., Tsay, S.: Neural-Network-Based Fuzzy Model and its Application to Transient Stability Prediction in Power Systems. IEEE Trans. on Syst. Man and Cyber., Part C. 29 (1999) 149—157
12. Ah King, R. T .F., Rughooputh, H. C. S.: A Hybrid Neural Network-Decision Tree-based Method for Transient Stability Assessment. Proc. 6th IEEE AFRICON. 2 (2002) 947 – 952
13. Senjyu, T., Takara, H., Uezato, K., Funabashi, T.: One-hour-ahead Load Forecasting using Neural Network. IEEE Transactions on Power Syst. 17 (2002) 113 – 118
14. Osowski, S., Siwek, K.: Regularisation of Neural Networks for Improved Load Forecasting in the Power System. IEE Proceedings- Generation Transmission and Distribution 149 (2002) 340 – 344
15. Slowinski, R. (ed.): Intelligent Decision Support, Handbook of Applications and Advances of the Rough Sets Theory. Dordrecht, Kluwer, The Netherlands (1992)
16. Guo, C. X, Hu, J. S., Ye, B., Cao, Y. J.: Swarm Intelligence for Mixed-variable Design Optimization, Journal of Zhejiang University SCIENCE. 5(2004) 851—860
17. Lambert-Torres, G.: Application of Rough Sets in Power System Control Center Data Mining. Proc 2002 IEEE Power Engineering Society Winter Meeting. (2002) 627 –631
18. Feng L., Qiu J. and Cao, Y. J.: Performance and Complexity of the Rough Fuzzy-Neural Network for Short-Term Load Forecasting. Proc 2004 IEEE Power Systems Conference and Expo PSCE , New York (2004)

A Novel Fermentation Control Method Based on Neural Networks

Xuhua Yang, Zonghai Sun, and Youxian Sun

National Laboratory of Industrial Control Technology, Institute of Modern Control Engineering, Zhejiang University, Hangzhou, Zhejiang, P.R. China. 310027
xhyang@iipc.zju.edu.cn

Abstract. This paper proposes a novel fermentation control method. Two stages are involved. First, propose the fermentation time model and the optimal fermentation temperature model based on RBF Neural networks. Second, on the base of the two models, propose the novel fermentation control method by which different fermentation batch can adopt different optimal fermentation temperature trajectory which fits itself. Using this method, each fermentation batch can be fermented at optimal fermentation temperature trajectory and will improve average product proportion. The practical application showed that this method can improve average product proportion 3% effectively.

1 Introduction

Control to fed-batch microbiological fermentation process is a very difficult task since many uncertainties and nonlinearities are encountered. Neural networks have super capability in the aspect of identification and approximation to the nonlinear system. Because of this reason, neural networks' application[1,2] in the field of fermentation process is getting more and more.

This paper proposes a novel fermentation control method on the base of RBF neural networks. This method can simply the modeling greatly and avoids the theoretical derivation and the identification of the complex fermentation model. The application of this method have gotten success and showed that this method has great application and promotion value. This research to microbiological fermentation is carried out with avermectin's fermentation process in Shenghua Biok Biology Co., Ltd.

2 The Fermentation Time Length and the Optimal Fermentation Temperature Trajectory

In the microbiological fermentation process, there exists such phenomenon: in different fermentation batch, though all fermentation technical parameters were controlled primarily, the fermentation time length and the product proportion are

F. Yin, J. Wang, and C. Guo (Eds.): ISNN 2004, LNCS 3174, pp. 194–199, 2004.

different from each other.(The technical parameters are temperature, *pH* value, pressure, ventilation quantity, electromotor rotate speed and so on.)

In fermentation factory, all fermentation technical parameters are constant in different fermentation batch. We knew:

(1)The fermentation time length of the avermectin is about $10 \sim 15$ days.

(2)Looking over avermectin's statistic among manufacture, there are one average product proportion and one maximal product proportion.

maximal product proportion - average product proportion $\in 15\% \sim 20\%$.

Why does this thing happen?

After analyzing the whole fermentation process carefully, we can see that the main difference of different fermentation batch is that there exist a little difference of fermentation preliminary parameters in different fermentation batch. The abamectin's fermentation preliminary parameters of a fermentation batch contain seeds' preliminary parameters and the culture medium's preliminary ingredient proportion in this fermentation batch.

So, we can conclude that the difference of fermentation preliminary parameters leads to the different fermentation time length and the different product proportion in different fermentation batch.

On the base of the conclusion, we can get two deductions as following:

(1)The fermentation time length of a fermentation batch is only determined by preliminary parameters in this fermentation batch.

(2)There exists optimal fermentation technical parameters to a fermentation batch. A fermentation batch's optimal fermentation technical parameters are optimal to itself and not optimal to other fermentation batches. A fermentation batch's optimal fermentation technical parameters are only determined by preliminary parameters in this fermentation batch.

This paper only researches the temperature technical parameters.

The abamectin's fermentation time length and the optimal fermentation temperature trajectory varying with time of a fermentation batch are only determined by preliminary parameters in this fermentation batch.

3 The Fermentation Time Model and the Optimal Fermentation Temperature Model

In fermentation process, suppose the temperature is T , the time variable is t ,the fermentation time length is $TimeLength$, seeds' preliminary parameters are $z_1, z_2, \cdots, z_m$, culture medium preliminary ingredient proportion are $k_1, k_2, \cdots, k_n$, then the fermentation preliminary parameters are $z_1, z_2, \cdots, z_m, k_1, k_2, \cdots, k_n$.

According above discussion, there exist mapping relationship as following :

The fermentation time length is

$$TimeLength = Function\_TL(z_1, z_2, \cdots, z_m, k_1, k_2, \cdots, k_n) \quad (1)$$

The optimal fermentation temperature trajectory varying with time is

$$T(t) = Function\_T(z_1, z_2, \cdots, z_m, k_1, k_2, \cdots, k_n, t) \quad (2)$$

$$t \in [0, TimeLength]$$

Formula(1) is the fermentation time model and Formula(2) is the optimal fermentation temperature model.

4 The Principle of the Novel Fermentation Control Method

The novel fermentation control method is that different fermentation batch should adopt different optimal fermentation temperature which fits itself. Using this method, each fermentation batch can be fermented at optimal fermentation temperature and will improve average product proportion.

Two stages are involved. First, build the fermentation time model and the optimal fermentation temperature model based on RBF neural networks. Second, on the base of the two models, get the optimal fermentation temperature trajectory varying with time before the fermentation's beginning of a fermentation batch. So, different fermentation batch can adopt different optimal fermentation temperature which fits itself and the fermentation temperature will always be optimal in the fermentation process of a fermentation batch and will improve average product proportion.

The specific operation procedure of the novel fermentation control method is as following:

(1)Get the fermentation time model based on RBF neural networks and historical fermentation data.

(2)Get the optimal fermentation temperature model base on RBF neural networks and historical fermentation data.

(3)Before the fermentation's beginning of a fermentation batch, using the preliminary fermentation parameters and the fermentation time model, we can get this fermentation batch's fermentation time length. Using the preliminary fermentation parameters, the fermentation time length and the optimal fermentation temperature model, we can get the optimal fermentation temperature trajectory varying with time. The fermentation time length will be used to be the practical fermentation time length. The temperature trajectory will be used to be the optimal fermentation temperature setting value trajectory in the temperature control. So, during the practical fermentation process, the practical fermentation temperature trajectory varying with time will be the optimal fermentation temperature setting value trajectory and the fermentation temperature will always be optimal and will improve average product proportion.

5 Application Method

In a fermentation batch, suppose the fermentation time length is *TimeLength* and the fermentation preliminary parameters are $z_1, z_2, \cdots, z_m, k_1, k_2, \cdots, k_n$.

(1)Get the fermentation time model $Function\_TL$

We use a RBF neural networks to get the fermentation time model.

Let the preliminary fermentation parameters $z_1, z_2, \cdots, z_m, k_1, k_2, \cdots, k_n$ as input and the fermentation time length as output to construct a RBF neural networks which has $(m+n)$ inputs and single output.(This paper adopts the RBF neural networks' learning algorithm in reference 3.)

We adopt supervised training method to train this RBF neural networks to get the fermentation time model. Training data are from historical batch reports. The training input sample set adopts preliminary fermentation parameters data of each fermentation. The training output sample set adopts each fermentation's time length. After training, we can get the fermentation time model $Function\_TL$.

(2)Get the optimal fermentation temperature model $Function\_T$

Looking over all fermentation batches process or experimentation process records up to now, the fermentation process which has maximal product proportion is considered as the optimal fermentation process. Namely, the function relationship between the preliminary fermentation parameters and the fermentation temperature trajectory varying with time in this fermentation batch is considered as optimal fermentation model. If finding a fermentation process which has more product proportion, we should adopt data in this fermentation batch.

We use a RBF neural networks to get the optimal fermentation temperature model. Let the preliminary fermentation parameters and time variable t ($t \in [0, TimeLength]$) as inputs and the optimal fermentation temperature trajectory varying with time as output, construct a RBF neural networks which has $(m+n+1)$ inputs and single output.

We adopt supervised training method to train the RBF neural networks to get the optimal fermentation temperature model. Training data are from the optimal fermentation process. The training input sample set adopts preliminary fermentation parameters and time variable t ($t \in [0, TimeLength]$). The training output sample set adopts the temperature curve data in the optimal fermentation process. After training, we can get the optimal fermentation temperature model $Function\_T$.

(3)Get the fermentation time length

$$TimeLength = Function\_TL(z_1, z_2, \cdots, z_m, k_1, k_2, \cdots, k_n)$$

(4)Get the optimal fermentation temperature trajectory varying with time

$$T(t) = Function\_T(z_1, z_2, \cdots, z_m, k_1, k_2, \cdots, k_n, t) \quad (t \in [0, TimeLength])$$

So, before the fermentation's beginning of a fermentation batch, we can obtain the fermentation time length and the optimal fermentation temperature trajectory varying with time. The fermentation time length will be used to be the practical fermentation time length. The temperature trajectory will be used to be the optimal fermentation temperature setting value trajectory in the temperature control. So, during the practical fermentation process, the practical fermentation temperature varying with time will be the optimal fermentation temperature setting value trajectory and the fermentation temperature will always be optimal and will improve average product proportion.

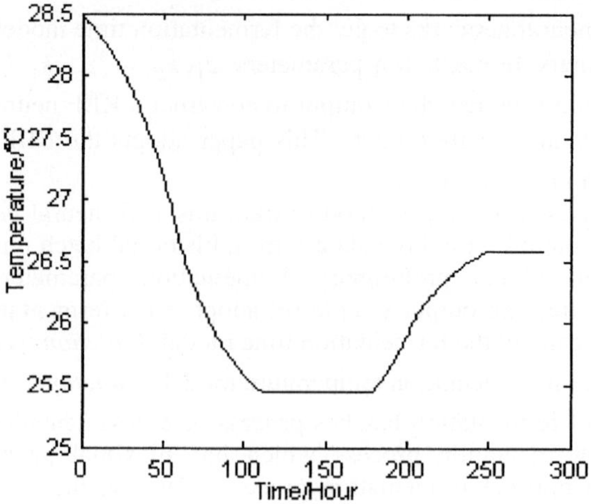

Fig. 1. Fermentation temperature curve varying with time in the optimal fermentation process

6 Practical Application

The avermectin's preliminary fermentation parameters which is in the optimal fermentation process are as following:

the preliminary seeds' quality parameters and their real values:

z_1 seed cell shape index[1] 0.8

z_2 seed culture medium pH 6.9

z_3 seed culture medium miscellaneous bacteria contamination and phagocytosis index[2] 0.01

culture medium preliminary ingredient proportion(fermentation tank's effective capacity is 38 tons)

k_1 yeast 1.169%

k_2 amylum 7.792%

k_3 bean cake powder 1.169%

k_4 corn plasm 0.91%

k_5 earthnut cake powder 1.2%

The fermentation temperature curve varying with time in the optimal fermentation process is showed as figure 1.

---

1 Indicate the growth status of seeds. Its range is from 0 to 1 and indicate the growth status of seeds from bad to good.

2 Indicate the extent that seed culture medium is contaminated by miscellaneous bacteria and phagocytosis. Its range is from 0 to 1 and indicate the extent from bad to good.

The abamectin fermentation time in optimal fermentation process is 285 hours. Namely, fermentation time $TimeLength = 285$.

Above all parameters are gotten by measure instrument.

We controlled the fermentation process according to the application method in 5 practically and the practical application showed that this method can improve average abamectin product proportion 3% .

7 Conclusion

This paper proposes a novel fermentation control method by which different fermentation batch can adopt different optimal fermentation temperature which fits itself. Using this method, each fermentation batch can be fermented at optimal fermentation temperature and will improve average product proportion. The practical application showed that this method can improve average product proportion 3% effectively and has great application and promotion value.

References

1. Leal Ascencio, R.R.; Reynaga, F.; Herrera, E.; Gschaedler, A.: Artificial neural networks as a biomass virtual sensor for a batch process, Intelligent Control Proceedings of the 2001 IEEE International Symposium on. (2001) 246 –251
2. Zhihua Xiong; Jie Zhang: Modeling and optimal control of fed-batch processes using control affine feedforward neural networks, American Control Conference. Vol. 6 (2002) 5025 -5030
3. Chen, S.; Cowan, C.F.N.; Grant, P.M.: Orthogonal least squares learning algorithm for radial basis function networks, IEEE Transactions on Neural Networks. Vol.2 , No. 2 (1991) 302 - 309

Modeling Dynamic System by Recurrent Neural Network with State Variables

Min Han, Zhiwei Shi, and Wei Wang

School of Electronic and Information Engineering, Dalian University of Technology, Dalian 116023, China
minhan@dlut.edu.cn

Abstract. A study is performed to investigate the state evolution of a kind of recurrent neural network. The state variable in the neural system summarize the information of external excitation and initial state, and determine its future response. The recurrent neural network is trained by the data from a dynamic system so that it can behave like the dynamic system. The dynamic systems include both input-output black-box system and autonomous chaotic system. It is found that the state variables in neural system differ from the state variable in the black-box system identified, this case often appears when the network is trained with input-output data of the system. The recurrent neural system learning from chaotic system exhibits an expected chaotic character, its state variable is the same as the system identified at the first period of evolution and its state evolution is sensitive to its initial state.

1 Introduction

In recent years, RNN (Recurrent Neural Network) has been introduced to model and control dynamic system, there are typically two types RNN model for dynamic system modeling, the first one is input-output based RNN model, which is well known as NARX (Nonlinear AutoRegresisve eXogenous) neural network or parallel model [1] or output error model, the second one for dynamic system modeling is state-based RNN, this kind of the RNN use state feedback instead of output feedback. To make a multi-step prediction using NARX neural network, the output is feedback [2]. The well known state-based RNN is Elman network [3][4]. In Elman network the hidden-layer output is feedback and delayed as the input of the hidden-layer. Up to now, Elman type network is a popular network model in the field of control and modeling.

Each state neuron is computed with a single hidden neuron in Elman network, and this constraint leads generally to a non-minimal state-space representation [5]. J.M. Zamarreno proposes a 5-layer state space neural network [6], the model only contains one time delay in the state layer, which is similar to the hidden layer delay in the Elman structure. The neural state space network with 3-layer weights is extensively studied by J.A. Suykens [7]. The state equation and output equation are respectively implemented by two MLPs(Multi-Layer Perceptron) in [6][7], and J.M. Zamarreno's network is a cascaded MLPs state space neural network and the network in the [7] is a parallel MLPs state space neural network. When the two mappings are implemented by a single MLP, the network becomes the model discussed by I. Rivals in [5].

F. Yin, J. Wang, and C. Guo (Eds.): ISNN 2004, LNCS 3174, pp. 200–205, 2004.

The state of a dynamic system is defined as a set of quantities that summarizes all the information about the past behavior of the system that is needed to uniquely describe its future behavior, except for the pure external effects arising from the applied input (excitation). In this paper, an empirical study is preformed on the state variable of the neural system, and the emphasis is focused on the relationship between the neural system and the system identified, and the influence of the initial condition on the evolution of system. The state space neural network is implemented by a single MLP, and the systems under investigation include both input-output black-box system and autonomous chaotic system.

Section 2 gives the structure of neural network and the dynamic system for further discussion. Section 3 is our discussion on the evolution of the state variable, and we draw our conclusion in Section 4.

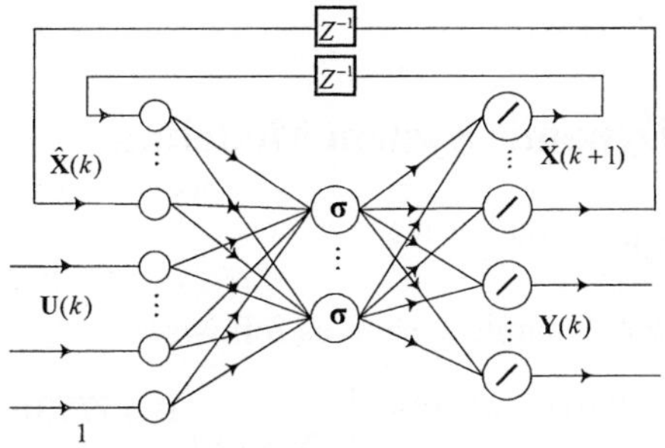

Fig. 1. Structure of the Recurrent Neural Network

2 The Recurrent Network and Dynamic System

The model to be discussed can be described by equation (1), its input $\mathbf{U} \in \mathbb{R}^m$ and its output $\mathbf{Y} \in \mathbb{R}^q$, $\hat{\mathbf{X}} \in \mathbb{R}^n$ acts as state variable (see Fig.1).

$$\begin{bmatrix} \hat{\mathbf{X}}(k+1) \\ \mathbf{Y}(k) \end{bmatrix} = \mathbf{W}_2 \cdot \boldsymbol{\sigma}(\mathbf{W}_1 \cdot \begin{bmatrix} \hat{\mathbf{X}}(\mathrm{k}) \\ \mathbf{U}(\mathrm{k}) \\ 1 \end{bmatrix}) \tag{1}$$

where $\mathbf{W}_2 \in \mathbb{R}^{(n+q)\times N}$, $\mathbf{W}_1 \in \mathbb{R}^{N\times(n+m+1)}$, and $\boldsymbol{\sigma}(\mathbf{X}) = diag\{\sigma(x_1), \sigma(x_2), \ldots, \ \sigma(x_N)\}$. when $\sigma(x)$ is a sigmoid function, It is called nonlinear state space neural network. Let $\mathbf{W}_1 \in \mathbb{R}^{N\times(n+1)}$ and $\mathbf{W}_2 \in \mathbb{R}^{n\times N}$, an autonomous neural system is produced and given by:

$$\hat{\mathbf{X}}(k+1) = \mathbf{W}_2 \cdot \boldsymbol{\sigma}(\mathbf{W}_1 \cdot \begin{bmatrix} \hat{\mathbf{X}}(k) \\ 1 \end{bmatrix}) \tag{2}$$

The model is called linear state space network when transfer function $\sigma(\cdot)$ is linear function, and the linear neural system is given by:

$$\begin{bmatrix} \hat{\mathbf{X}}(k+1) \\ \mathbf{Y}(k) \end{bmatrix} = \mathbf{W}_2 \cdot \mathbf{W}_1 \cdot \begin{bmatrix} \hat{\mathbf{X}}(k) \\ \mathbf{U}(k) \end{bmatrix} \tag{3}$$

For modeling black-box dynamic system, the equation (1) or (3) is used, and the system input output data are presented for network learning. An autonomous neural system (2) can be used to model a chaotic system such as Rossler attractor. In the first example, a predefined linear system is learned by the equation (3), and the system input output data are used to train the neural model, and then the obtained neural system is compared with the predefined system. In the second example, Rossler attractor is learned by a neural network (2), the training data are the three state variables sampled at fixed time intervals in the Rossler equation from a specified initial condition. We focus our attention on the evolution of the state variable of the neural system.

3 Simulation of Dynamic System Modeling

3.1 Simulation Example 1

For a given linear discrete dynamic system as follows:

$$\begin{aligned} \mathbf{X}(k+1) &= \begin{bmatrix} 2.8201 & -1.9981 \\ 4.2227 & -3.0201 \end{bmatrix} \cdot \mathbf{X}(k) + \begin{bmatrix} 0.057736 \\ 0.081220 \end{bmatrix} \cdot \mathbf{U}(k) \\ \mathbf{Y}(k) &= [-0.49571 \quad 0.62325] \cdot \mathbf{X}(k) + 0.040001 \cdot \mathbf{U}(k) \end{aligned} \tag{4}$$

A second order state-space neural network model with $\mathbf{W}_1(2\times3)$, $\mathbf{W}_2(3\times2)$, is built to model the dynamic system. By using numerical optimization algorithm to minimize the error between the network output and the ideal output, $\mathbf{W}_1$, W_2 are found as follows:

$$\mathbf{W}_1 = \begin{bmatrix} 0.63934 & -0.69824 & 0.20667 \\ 0.52390 & 0.72704 & 0.13112 \end{bmatrix}, \quad \mathbf{W}_2 = \begin{bmatrix} 0.67903 & -0.19202 \\ 0.47575 & -0.27582 \\ 0.074888 & 0.18704 \end{bmatrix}$$

The obtained state-space neural network is:

$$\begin{aligned} \hat{\mathbf{X}}(k+1) &= \begin{bmatrix} 0.33353 & -0.61373 \\ 0.15967 & -0.53272 \end{bmatrix} \cdot \hat{\mathbf{X}}(k) + \begin{bmatrix} 0.11516 \\ 0.062158 \end{bmatrix} \cdot \mathbf{U}(k) \\ \mathbf{Y}(k) &= \begin{bmatrix} 0.14587 & 0.083697 \end{bmatrix} \cdot \hat{\mathbf{X}}(k) + 0.040001 \cdot \mathbf{U}(k) \end{aligned} \tag{5}$$

From the obtained state-space representation, it can be found that the state-variables are different from the state-variables of the predefined state-space representation. The weight connections are trained to make the network approximate the target, while the state variables of the neural network are unconstrained, so the state variables of the network go freely in the state-space to model the input-output relationship until it finds a proper organization to fit the input-output data. The pulse-transfer function and eigenvalues of the neural network are calculated to have a comparison with the system identified (as shown in Table 1).

From the comparison of the eigenvalues and pulse-transfer function between the predefined system and the neural network model, it can be found that the eigenvalues and pulse-transfer function of the original dynamic system are almost exactly found back. In [6], J.M. Zamarreno also obtains the eigenvalues of the system identified, but the pulse-transfer function is not considered, which describes the relationship between the system input and system output.

Table 1. The comparison of Eigenvalues and Transfer-Function

Items	Eigenvalues	Transfer-Function
System identified	[0.19934, -0.39934]	$\frac{0.04000Z^2+0.03000Z+2.105e-5}{Z^2+0.20000Z-0.07961}$
Neural network	[0.19974, -0.39893]	$\frac{0.04001Z^2+0.02997Z+5.402e-7}{Z^2+0.19920Z-0.07969}$

3.2 Simulation Example 2

Equation (2) is now used to learn from Rossler equation (6). The variables *x*, *y*, *z* in the equation are the state variables of the autonomous dynamic system.

$$\begin{aligned} x &= -(y+z) \\ y &= x + a\cdot y \\ z &= b + z\cdot(x-c) \end{aligned} \tag{6}$$

In the Rossler equation (6), let a=0.398, b=2 and c=4. A trajectory has been generated from the initial condition [-2.2, 1.1, 3.3] by using Runge-Kutta method with fixed step-length (0.1). The first 250 data points are used for trajectory learning. The number of the hidden layer neurons is 10, and the number of the state layer neurons is 3 according to the Rossler equations.

The evolutions of the z-component of the Rossler attractor (solid line) and the RNN (doted line) are shown in Fig.2, from the time point 0 to time point 250, the neural network has well learned the evolution of the state variable, and from the time point 251 to about 800, the neural network gives satisfying prediction result.

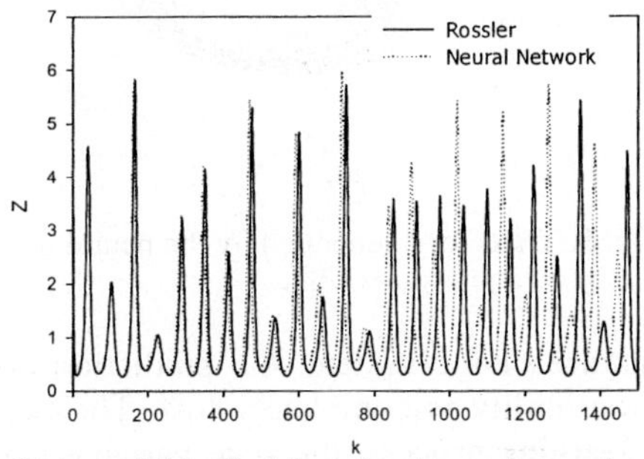

Fig. 2. State variable Z of the attractor and neural network.

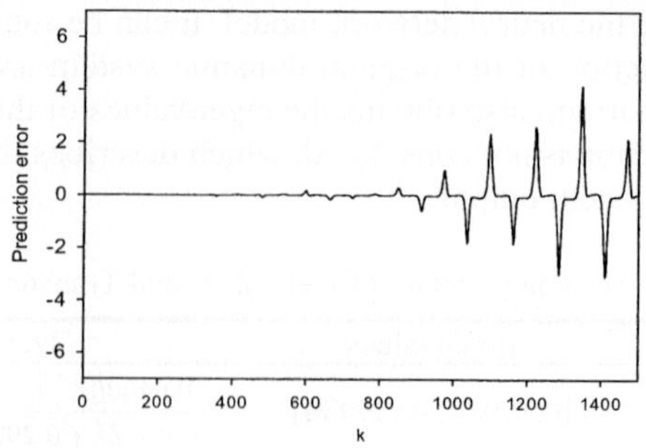

Fig. 3. Prediction error of the neural network about state variable Z.

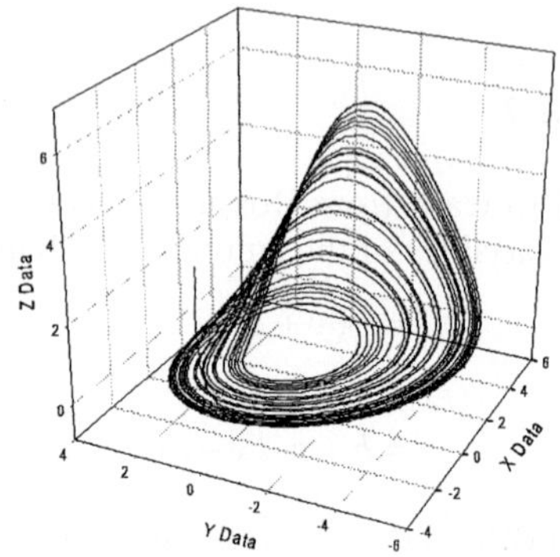

Fig. 4. Trajectory of the Rossler attractor.

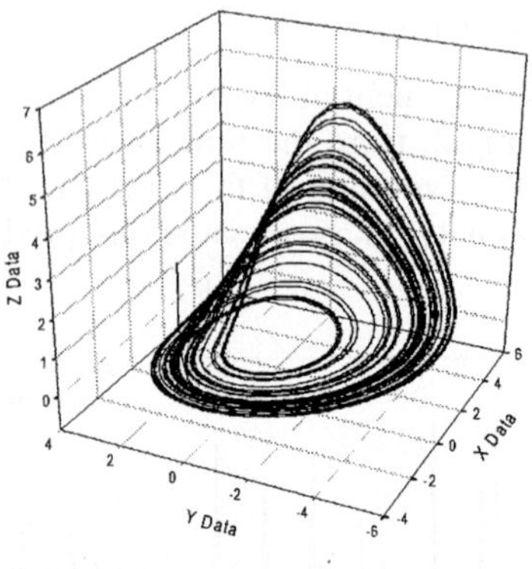

Fig. 5. Trajectory generated by the neural network

Due to the chaotic nature of the system, the prediction error become larger and larger until the prediction ability is completely lost. The prediction error is shown from Fig.3. During the learning process, the state variable is constrained by the network training because the object function is defined on the state variable. Fig.4 and Fig.5 are respectively the trajectory of the Rossler attractor and the trajectory generated by the neural network from the initial condition [-2.2, 1.1, 3.3], both trajectory contain 2000 time points.

4 Conclusion

The state variable of a kind of RNN is studied in this paper. For a black-box system, the RNN is trained with system input-output data, and its state variable has no direct constraint by the object function. It is found that the evolution of the state variables of the two systems differs from each other, but the eigenvalue and the pulse-transfer function of the two systems are exactly the same in our simulation. The recurrent neural system learning from chaotic system exhibits an expected chaotic character, its state variable is the same as the system identified at the first period of evolution and its state evolution is sensitive to its initial state.

Acknowledgements. This research is supported by the project (60374064) of the National Nature Science Foundation of China; it is also supported by the project (50139020) of the National Natural Science Foundation of China. All of these supports are appreciated.

References

1. Narendra, K.S., Parthasarathy, K.: Identification and Control of Dynamical Systems Using Neural Networks. IEEE Transactions on Neural networks **1** (1990) 4-27.
2. Suykens, J.A.K., Vandewalle, J.: Recurrent Least Squares Support Vector Machines. IEEE Transactions on Circuits and Systems I: Fundamental Theory and Applications **47** (2000) 1109-1114.
3. Elman, J. L.: Finding Structure in Time. Cognitive Science **14** (1990) 179-211.
4. Pham, D.T., Liu, X., Training of Elman Networks and Dynamic System Modeling. International Journal of Systems Science **27** (1996) 221-226.
5. Rivals, I., Personnaz, L.: Black-box Modeling with State-space Neural Networks. In: Zbikowski, R., Hunt, K. J. (eds.): Neural Adaptive Control Technology, World Scientific, Singapore (1996) 237-264.
6. Zamarreno, J.M., Vega, P.: State Space Neural Network. Properties and Application. Neural Networks **11** (1998) 1099-1112.
7. Suykens, J.A. De Moor, B.L.R., Vandewalle, J.: Nonlinear System Identification Using Neural State Space Models, Applicable to Robust Control Design. International Journal of Control **62** (1995) 129-152.

Robust Friction Compensation for Servo System Based on LuGre Model with Uncertain Static Parameters

LixinWei[1], Xia Wang[2], and Hongrui Wang[2]

[1] Yanshan University, Qinhuangdao 066004, China
gglcc2002@yahoo.com.cn
[2] Hebei University, Baoding 071002, China
gglcc2002@yahoo.com.cn

Abstract. A new adaptive robust friction compensation for servo system based on luGre model is proposed. Considered the uncertainty of steady state parameters in the friction model, a RBF neural network is adopted to learn the nonlinear friction-velocity relationship in steady state. The bristle displacement is observed using the output of the network. Nonlinear adaptive robust control laws are designed based on backstepping theory to compensate the unknown system parameters. System robustness and asymptotic results is proved and shown in simulation results.

1 Introduction

Friction in servomechanisms is a nonlinear phenomenon that reduces positioning and tracking accuracy. Friction is usually composed of static, coulomb, viscous friction and Stribeck effect. There are also some dynamics characters such as varying breakaway force and frictional lag. Counteracting the effects of friction requires a suitable friction model. Much work has been done in paper [1][2] to characterize nonlinear friction behavior and static models expressed as algebraic equation and dynamic models expressed as differential equation are proposed. Static exponential model depict most character between friction and steady velocity. To achieve higher performance, frictional dynamics need to be considered. Canudas et al. proposed a new analytic friction model that captures most of the frictional behaviors observed experimentally. The friction interface is modeled as elastic bristles. Deflection of the bristles caused by a relative motion of two surfaces gives rise to the friction force. The steady state of this luGre model is a nonlinear function of velocity, which is equal to exponential model.

Friction characteristics are known to change easily with environment. In applications, friction parameters usually need to be identified, which leads to a vast use of adaptive control schemes. Traditional adaptive control schemes can't cope with nonlinear parameter in the models. Some linearization methods are proposed for exponential model in document [3]. In luGre model, the nonlinearity of friction is represented by the nonlinear function in steady state. The static friction parameters, how

F. Yin, J. Wang, and C. Guo (Eds.): ISNN 2004, LNCS 3174, pp. 206–211, 2004.

ever, is always assumed to be known via experimental in so many compensations as in [4] and [5]. Observer is constructed under this ideal presupposition, which makes the compensation sensitive to the changes of environment. Therefore, how to handle the uncertainty of static parameters in luGre model becomes important.

2 LuGre Friction Model and the Observer Based on Network

The form of LuGre model is

$$\dot{z} = v - \frac{|v|}{g(v)} z \tag{1}$$

$$\sigma_0 g(v) = F_c + (F_s - F_c)\exp(-(v/v_s)^2) \tag{2}$$

$$F = \sigma_0 z + \sigma_1 \dot{z} + \sigma_2 v \tag{3}$$

where v is the relative velocity between tow surfaces, z is the average deflection of bristles, $\sigma_0 > 0$ is the stiffness, $\sigma_1 > 0$ is a damping coefficient, and $\sigma_2 > 0$ is the viscous coefficient. The function $\sigma_0 g(v)$ determines the steady nonlinear relation between velocity and friction. Static parameters $F_c > 0$, $F_s > 0$, $v_s > 0$, represent coulomb friction, static friction and Stribeck velocity. Normally, they are determined by measuring the steady-state friction when the velocity is constant and are taken to be changeless for long. Actually, these parameters always change with environment and should be identified momentarily. We can rewrite equation (3) as follows

$$F = \beta_0 z - \beta_1 f(v) z + \beta_2 v \tag{4}$$

where $f(v) = \frac{|v|}{g(v)} \geq 0$, $\beta_0 = \sigma_0 > 0$, $\beta_1 = \sigma_1 > 0$, $\beta_2 = \sigma_1 + \sigma_2 > 0$. $f(v)$ is a nonlinear function learned by the RBF neural network. We train the network with the measured data beforehand and adjust its weights while working.

2.1 Static Nonlinear Friction Estimation Based on RBF Neural Network

Figure 1 shows the structure of the RBF network in which $\varphi = [G_1(v), \cdots G_5(v)]^T$, $G_j(v) = \exp[-(v - c_j)^2 / 2\delta_j^2]$, $j = 1 \cdots 5$.

$$\hat{f} = \theta^T \varphi \tag{5}$$

Define the optimum estimation and minimum estimation error

$$\hat{f}(v \mid \theta^*) = \min_{\theta} \left| \hat{f}(v \mid \theta) - f \right| \tag{6}$$

$$E = f(v) - \hat{f}(v \mid \theta^*) \tag{7}$$

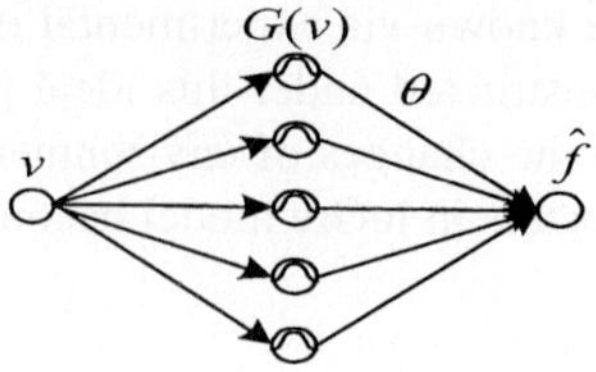

Fig. 1. Structure of RBF neural network

2.2 Observer of Bristle Deflection

It is well known that the friction state z is not measurable. In order to handle different nonlinearities of z, two nonlinear observers are constructed with the output of network. The dynamics of its estimations $\hat{z}_1$ and $\hat{z}_2$ are given as

$$\dot{\hat{z}}_1 = v - \bar{f}\hat{z}_1 - k_1\varepsilon_2 \tag{8}$$

$$\dot{\hat{z}}_2 = v - \bar{f}\hat{z}_2 + k_2\varepsilon_2 \tag{9}$$

where $\bar{f} = \hat{f} + E$. The dynamics of observation errors are

$$\dot{\tilde{z}}_1 = -f\tilde{z}_1 + \tilde{f}\hat{z}_1 + k_1\varepsilon_2 \tag{10}$$

$$\dot{\tilde{z}}_2 = -f\tilde{z}_2 + \tilde{f}\hat{z}_2 - k_2\varepsilon_2 \tag{11}$$

where $\tilde{z}_i = z - \hat{z}_i$, $i = 1, 2$, $\tilde{f} = \bar{f} - f$, $k_1, k_2 > 0$.

3 Controller Design Based on Backstepping Theory

Servo system is usually described as

$$J\ddot{x} = u - F - T_l \tag{12}$$

where J , u , F , T_l are the system inertia, control torque, friction force and load torque.

3.1 Some Hypothesis

(1) $\beta_0 \le g_0$, $\beta_1 \le g_1$, $f \le l$.

(2) The observation error is bounded and satisfies $|\tilde{z}_1| \le r_1$, $|\tilde{z}_2| \le r_2$.

(3) The estimation error of neural network is bounded and satisfies $|\tilde{f}| \le s$.

3.2 Controller Design

The problem derive from systematic unknown parameters can be conquered by adaptations. Because we considered the uncertainties of the static parameters in the friction model, a robust term is employed to ensure the stability of the whole system. The adaptive robust control law is deduced step by step through backstepping design.

We rewrite the system in state space form

$$\begin{aligned} \dot{x}_1 &= x_2 \\ \dot{x}_2 &= \frac{1}{J}(u - F - T_l) \end{aligned} \tag{13}$$

Define the position error

$$\varepsilon_1 = x - x_d = x_1 - x_d \tag{14}$$

Define a positive function $V_1 = \varepsilon_1^{\,2}/2$. So the stabilizing function is

$$\alpha_1 = \dot{x}_d - h_1\varepsilon_1 \tag{15}$$

where h_1 is a positive constant. Introduce the error

$$\varepsilon_2 = x_2 - \alpha_1 \tag{16}$$

Define another positive function $V_2 = V_1 + \varepsilon_2^{\,2}/2$. Take the derivative of V_2 and use (13)(15)(16)

$$\begin{aligned} \dot{V}_2 &= \dot{V}_1 + \varepsilon_2\dot{\varepsilon}_2 = -h_1\varepsilon_1^{\,2} + \varepsilon_2(\varepsilon_1 - \dot{\alpha}_1 + \dot{x}_2) \\ &= -h_1\varepsilon_1^{\,2} + \varepsilon_2[\varepsilon_1 - \dot{\alpha}_1 + \frac{1}{J}(u - F - T_l)] \end{aligned} \tag{17}$$

Design the control law as $u = \hat{F} + \hat{T}_l - \hat{J}(\varepsilon_1 - \dot{\alpha}_1 + h_2\varepsilon_2) + \Delta/\varepsilon_2$. where $\hat{J}$, $\hat{T}_l$, $\hat{\beta}_0$, $\hat{\beta}_1$, $\hat{\beta}_2$ are estimated values whose adaptive law is to be deduced later and $\hat{F} = \hat{\beta}_0\hat{z}_1 - \hat{\beta}_1\bar{f}\hat{z}_2 + \hat{\beta}_2 v$ is to counteract the friction F, $\Delta = \Delta_1 + \Delta_2$ is a robust term.

It follows

$$\begin{aligned} \hat{F} - F &= \hat{\beta}_0\hat{z}_1 - \hat{\beta}_1\bar{f}\hat{z}_2 + \hat{\beta}_2 v - \beta_0 z + \beta_1 f z - \beta_2 v \\ &= \tilde{\beta}_0\hat{z}_1 - \beta_0\tilde{z}_1 + \tilde{\beta}_2 v - \tilde{\beta}_1\bar{f}\hat{z}_2 + \beta_1 f\tilde{z}_2 + \beta_1(f - \bar{f})\hat{z}_2 \\ &= \tilde{\beta}_0\hat{z}_1 + \tilde{\beta}_2 v - \tilde{\beta}_1\bar{f}\hat{z}_2 - \beta_0\tilde{z}_1 + \beta_1 f\tilde{z}_2 + \beta_1(\theta^* - \theta)^T\varphi\hat{z}_2 + \beta_1\tilde{E}\hat{z}_2 \end{aligned} \tag{18}$$

where $\tilde{\beta}_0 = \hat{\beta}_0 - \beta_0$, $\tilde{\beta}_1 = \hat{\beta}_1 - \beta_1$, $\tilde{\beta}_2 = \hat{\beta}_2 - \beta_2$, $\tilde{E} = E - \hat{E}$.

Let V_3 be a positive define function of the form

$$\begin{aligned} V_3 = V_2 + \frac{1}{J}\{&\frac{\tilde{J}^2}{2a_1} + \frac{\tilde{T}_l^{\,2}}{2a_2} + \frac{\tilde{\beta}_0^{\,2}}{2a_3} + \frac{\tilde{\beta}_2^{\,2}}{2a_4} + \frac{\tilde{\beta}_1^{\,2}}{2a_5} + \frac{\beta_1}{2a_6}\tilde{E}^2 + \\ &\frac{\beta_1}{2a_7}(\theta^* - \theta)^T(\theta^* - \theta) + \frac{\beta_0}{2k_1}\tilde{z}_1^{\,2} + \frac{\beta_1 f}{2k_2}\tilde{z}_2^{\,2}\} \end{aligned} \tag{19}$$

Take its derivation and use equation (17) and (18), we get

$$\begin{aligned}\dot V_3 = {} & -h_1\varepsilon_1^2 - h_2\varepsilon_2^2 + \frac{1}{J}\{\tilde J[\varepsilon_2(\varepsilon_1-\dot\alpha_1+h_2\varepsilon_2)-\frac{\dot{\hat J}}{a_1}]+\tilde T_l(\varepsilon_2+\frac{\dot{\hat T}_l}{a_2}) \\ & + \beta_0(\varepsilon_2\hat z_1+\frac{\dot{\hat\beta}_0}{a_3})+\tilde\beta_2(\varepsilon_2 v+\frac{\dot{\hat\beta}_2}{a_4})+\tilde\beta_1(\varepsilon_2\bar f\hat z_2+\frac{\dot{\hat\beta}_1}{a_5})+\beta_0\tilde z_1(-\varepsilon_2+\frac{\dot{\tilde z}_1}{k_1}) \\ & + \beta_1 f\tilde z_2(\varepsilon_2+\frac{\dot{\tilde z}_2}{k_2})+\beta_1(\theta^*-\theta)^T(\varepsilon_2\varphi\hat z_2-\frac{\dot\theta}{a_7})+\beta_1\tilde E(\varepsilon_2\hat z_2-\frac{\dot{\hat E}}{a_6})+\Delta\}\end{aligned} \tag{20}$$

Take adaptive laws and the correction term in observer as follows

$$\begin{aligned}&\dot{\hat J}=a_1\varepsilon_2(\varepsilon_1-\dot\alpha_1+h_2\varepsilon_2),\ \dot{\hat T}_l=-a_2\varepsilon_2,\ \dot{\hat\beta}_0=-a_3\varepsilon_2\hat z_1 \\ &\dot{\hat\beta}_1=-a_5\varepsilon_2\bar f\hat z_2,\ \dot{\hat\beta}_2=-a_4\varepsilon_2 v,\ \dot\theta=a_7\varepsilon_2\varphi\hat z_2,\ \dot{\hat E}=a_6\varepsilon_2\hat z_2\end{aligned} \tag{21}$$

and substitute equation (10) and (11) into equation (20), we obtains

$$\dot V_3 \le -h_1\varepsilon_1^2-h_2\varepsilon_2^2+\frac{1}{J}[-\frac{\beta_0 f\tilde z_1^2}{k_1}-\frac{\beta_1 f^2\tilde z_2^2}{k_2}]+\frac{1}{J}[\frac{\beta_0|\tilde z_1||\tilde f|\hat z_1}{k_1}+\Delta_1+\frac{\beta_1 f|\tilde z_2||\tilde f|\hat z_2}{k_2}+\Delta_2] \tag{22}$$

Since β_0, β_1, $|\tilde z_i|$ and $\tilde f$ are all bounded, define

$$\Delta_1=-\frac{\hat z_1}{k_1}g_0 r_1 s,\ \Delta_2=-\frac{\hat z_2}{k_2}g_1 l r_2 s \tag{23}$$

we get

$$\dot V_3 \le -h_1\varepsilon_1^2-h_2\varepsilon_2^2-\frac{1}{J}[\frac{\beta_0 f\tilde z_1^2}{k_1}+\frac{\beta_1 f^2\tilde z_2^2}{k_2}]\le 0 \tag{24}$$

Therefore, the system defined by equation (1)-(3) and (12) is globally stable. It follows from Barbalart's lemma that $e(t)$ and $\tilde z$ asymptotically converge to zero.

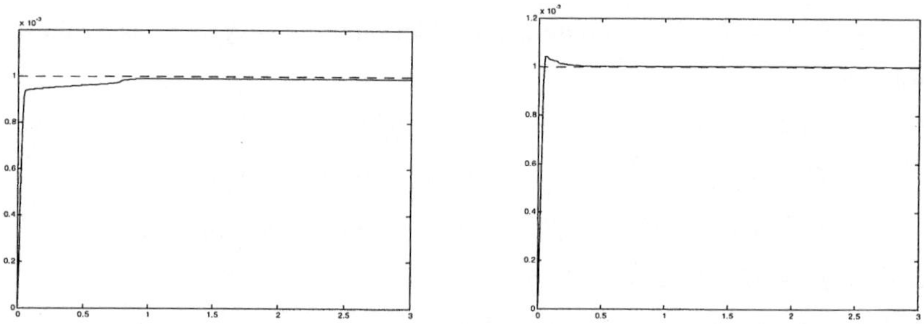

(a) (b)

Fig. 2. (a) Result of normal compensation Based on LuGre model. (b) Result with the steady friction identified by neural network

4 Simulation Results

Simulation is conducted to show the superiority of our compensation compared to the usual compensation based on LuGre Model. First, we adjust control parameters to achieve satisfying results for both methods when the steady friction is $F = \mathrm{sgn}(1+0.5\exp(-(v/0.001)^2)) + 0.4v$. Then, the steady friction is turns to $F = \mathrm{sgn}(1.2+0.6\exp(-(v/0.0015)^2)) + 0.4v$. Results of two different methods are presented in figure 2. The dotted line is the reference signal. We can see that our compensation appears robust to the change in the model.

5 Conclusion

A RBF neural network is proposed to learn the steady nonlinear function of LuGre friction model. A nonlinear adaptive robust controller is designed to handle the unknown parameters of servo system based on backstepping theory. Asymptotic positioning results are proved in the article and robust performance is shown in simulation results.

References

1. Armstrong-Helouvry, B., Dupont, P., Canudas De Wit, C.: A Survey of Models, Analysis Tools and Compensation Methods for the Control of Machines with Friction. Automatica, Vol. 30. Elsevier Science Ltd (1994) 1083–1138
2. Canudas de Wit, C., Olsson, H., Astrom, K. J., Lischinsky, P.: A New Model for Control of Systems with Friction. IEEE Transactions on Automatic Control, Vol. 40. IEEE (1995) 419–425
3. Taware, A., Tao, G., Pradhan, N., Teolis, C.: Friction Compensation for a Sandwich Dynamic System. Brief Papers in Automatica, Vol. 39. Elsevier Science Ltd (2003) 481–488
4. Ro, P.-I., Shim, W., Jeong, S.: Robust Friction Compensation for Submicrometer Positioning and Tracking for a Ball-Screw-Driven Slide System. Precision Engineering, Vol. 24. Elsevier Science Inc (2000) 160–173
5. Tan, Y.L., Chang, J., Tan, H.L.: Adaptive Backstepping Control and Friction Compensation for AC Servo With Inertia and Load Uncertainties. IEEE Transactions on Industrial Electronics, Vol. 50. IEEE (2003) 944–952

System Identification Using Adjustable RBF Neural Network with Stable Learning Algorithms

Wen Yu[1] and Xiaoou Li[2]

[1] Departamento de Control Automático, CINVESTAV-IPN
A.P. 14-740, Av.IPN 2508, México D.F., 07360, México
yuw@ctrl.cinvestav.mx

[2] Sección de Computación, Departamento de Ingeniería Eléctrica, CINVESTAV-IPN
A.P. 14-740, Av.IPN 2508, México D.F., 07360, México

Abstract. In general, RBF neural network cannot match nonlinear systems exactly. Unmodeled dynamic leads parameters drift and even instability problem. According to system identification theory, robust modification terms must be included in order to guarantee Lyapunov stability. This paper suggests new learning laws for normal and adjustable RBF neural networks based on Input-to-State Stability (ISS) approach. The new learning schemes employ a time-varying learning rate that is determined from input-output data and model structure. The calculation of the learning rate does not need any prior information such as estimation of the modeling error bounds.

1 Introduction

Resent results show that RBF neural network seems to be very effective to identify a broad category of complex nonlinear systems when we do not have complete model information [10]. It is well known that normal identification algorithms are stable for ideal plants [4]. In the presence of disturbance or unmodeled dynamics, these adaptive procedures can go to instability easily. The lack of robustness in parameters identification was demonstrated in [2] and became a hot issue in 1980s. Several robust modification techniques were proposed in [4]. The weight adjusting algorithms of neural networks is a type of parameters identification, the normal gradient algorithm is stable when neural network model can match the nonlinear plant exactly [11]. Generally, we have to make some modifications to the normal gradient algorithm or backpropagation such that the learning process is stable. For example, in [6] some hard restrictions were added in the learning law, in [13] the dynamic backpropagation was modified with NLq stability constraints. Another generalized method is to use robust modification techniques of robust adaptive control [4]. [8] applied $\sigma-$modification, [5] used modified $\delta-$rule, and [15] used dead-zone in the weight tuning algorithms. The motivation of this paper is to prove that the normal gradient law and backpropagation-like algorithm without robust modifications are L_∞ stable for identification error.

F. Yin, J. Wang, and C. Guo (Eds.): ISNN 2004, LNCS 3174, pp. 212–217, 2004.

Input-to-state stability (ISS) is an elegant approach to analyze stability besides Lyapunov method. It can lead to general conclusions on the stability by using input and state characteristics. We will use input-to-state stability approach to obtain some new learning laws that do not need robust modifications. A simple simulation gives the effectiveness of the suggested algorithm. To the best of our knowledge, ISS approach for RBF neural network was not still applied in the literature. The adjustable RBF neural network is referred to the activation function of RBF neural network can be updated by a learning algorithm.

In this ISS approach is applied to system identification via RBF neural network. Two cases are considered: (1) For normal RBF neural network, the activation functions are assumed to be known, and learning is carried on the weights, (2) Learning algorithm concerns both the activation functions and the weights. The new stable algorithms with time-varying learning rates are applied.

2 Identification Using Normal RBF Neural Network

Consider following discrete-time nonlinear system in NARMA form

$$y(k) = f\left[y\left(k-1\right), y\left(k-2\right), \cdots u\left(k-1\right), u\left(k-2\right), \cdots\right] = f\left[X\left(k\right)\right] \tag{1}$$

where $X\left(k\right) = \left[y\left(k-1\right), y\left(k-2\right), \cdots u\left(k-d\right), u\left(k-d-1\right), \cdots\right]^T \in \Re^n$, $f\left(\cdot\right)$ is an unknown nonlinear difference equation representing the plant dynamics, $u\left(k\right)$ and $y\left(k\right)$ are measurable scalar input and output, d is time delay. A normal RBF neural network can be expressed as

$$\widehat{y}\left(k\right) = W\left(k\right) \Phi\left[X\left(k\right)\right] \tag{2}$$

where the weights $W\left(k\right) = \left[w_1 \cdots w_n\right]$, the activation function $\Phi\left[X\left(k\right)\right] = \left[\phi_1 \cdots \phi_n\right]^T$. The Gaussian function is $\phi_i = \exp\left[-\left(\frac{x_i - c_i}{\sigma_i}\right)^2\right]$, c_i and σ_i are the center and width parameters of the activation function of ϕ_i. When we have some prior information of the identified plant, we can construct the activation function $\phi_1 \cdots \phi_n$. In this section we assume $\phi_1 \cdots \phi_n$ are given by prior knowledge. The object of RBF neural modeling is to find the weights $W\left(k\right)$, such that the output $\widehat{y}\left(k\right)$ of RBF neural networks (2) can follow the output $y\left(k\right)$ of nonlinear plant (1). Let us define identification error as $e\left(k\right) = \widehat{y}\left(k\right) - y\left(k\right)$. We will use the modeling error $e\left(k\right)$ to train the RBF neural networks (2) online such that $\widehat{y}\left(k\right)$ can approximate $y(k)$. According to function approximation theories of RBF neural networks [3], the identified nonlinear process (1) can be represented as

$$y\left(k\right) = W^* \Phi\left[X\left(k\right)\right] - \mu_1\left(k\right) \tag{3}$$

where W^* is unknown weights which can minimize the unmodeled dynamic $\mu_1\left(k\right)$. The identification error can be represented by (2) and (3)

$$e\left(k\right) = \widetilde{W}\left(k\right) \Phi\left[X\left(k\right)\right] + \mu_1\left(k\right) \tag{4}$$

where $\widetilde{W}(k) = W(k) - W^*$. In this paper we are only interested in open-loop identification, we assume that the plant (1) is bounded-input and bounded-output (BIBO) stable, *i.e.*, $y(k)$ and $u(k)$ in (1) are bounded. By the bound of the activation function Φ, $\mu_1(k)$ in (3) is bounded. The following theorem gives a stable gradient descent algorithm for RBF neural modeling.

Theorem 1. *If we use the RBF neural networks (2) to identify nonlinear plant (1), the following gradient descent algorithm with a time-varying learning rate can make identification error* $e(k)$ *bounded*

$$W(k+1) = W(k) - \eta_k e(k) \Phi^T [X(k)] \tag{5}$$

where the scalar $\eta_k = \frac{\eta}{1 + \|\Phi[X(k)]\|^2}$, $0 < \eta \leq 1$. *The normalized identification error* $e_N(k) = \frac{e(k)}{1+\max_k\left(\|\Phi[X(k)]\|^2\right)}$ *satisfies the following average performance*

$$\limsup_{T\to\infty} \frac{1}{T} \sum_{k=1}^{T} \|e_N(k)\|^2 \leq \overline{\mu}_1 \tag{6}$$

where $\overline{\mu}_1 = \max_k \left[\|\mu_1(k)\|^2\right]$.

Proof. We selected a positive defined scalar L_k as

$$L_k = \left\|\widetilde{W}(k)\right\|^2 \tag{7}$$

By the updating law (5), we have

$$\widetilde{W}(k+1) = \widetilde{W}(k) - \eta_k e(k) \Phi^T [X(k)]$$

Using the inequalities

$$\|a - b\| \geq \|a\| - \|b\|, 2\|ab\| \leq a^2 + b^2$$

for any a and b. By using (4) and $0 \leq \eta_k \leq \eta \leq 1$, we have

$$\begin{array}{l} \Delta L_k = L_{k+1} - L_k = \left\|\widetilde{W}(k) - \eta_k e(k) \Phi^T(X)\right\|^2 - \left\|\widetilde{W}(k)\right\|^2 \\ = \eta_k^2 \|e(k)\|^2 \|\Phi[X(k)]\|^2 - 2\eta_k \|e(k)[e(k) - \mu_1(k)]\| \\ \leq \eta_k^2 \|e(k)\|^2 \|\Phi[X(k)]\|^2 - 2\eta_k \|e(k)\|^2 + 2\eta_k \|e(k)\mu_1(k)\| \\ \leq -\eta_k \|e(k)\|^2 \left(1 - \eta_k \left\|\Phi^T(X)\right\|^2\right) + \eta_k \|\mu_1(k)\|^2 \end{array} \tag{8}$$

Since $\eta_k = \frac{\eta}{1 + \|\Phi[X(k)]\|^2}$,

$$\begin{array}{l} \eta_k \left(1 - \eta_k \|\Phi[X(k)]\|^2\right) = \eta_k \left(1 - \frac{\eta}{1 + \|\Phi[X(k)]\|^2} \|\Phi[X(k)]\|^2\right) \\ \geq \frac{\eta_k}{1+\max_k\left(\|\Phi[X(k)]\|^2\right)} \geq \frac{\eta}{\left[1+\max_k\left(\|\Phi[X(k)]\|^2\right)\right]^2} \end{array}$$

So

$$\Delta L_k \leq -\pi \left\| e\left(k\right)\right\|^2 + \eta \left\| \mu_1\left(k\right)\right\|^2 \tag{9}$$

where π is defined as $\pi = \frac{\eta}{\left[1+\max\limits_k\left(\left\|\Phi[X(k)]\right\|^2\right)\right]^2}$, Because $n \min\left(\widetilde{w}_i^2\right) \leq L_k \leq n \max\left(\widetilde{w}_i^2\right)$, where $n \min\left(\widetilde{w}_i^2\right)$ and $n \max\left(\widetilde{w}_i^2\right)$ are $\mathcal{K}_\infty$-functions, and $\pi \left\| e\left(k\right)\right\|^2$ is an $\mathcal{K}_\infty$-function, $\eta \left\| \mu\left(k\right)\right\|^2$ is a $\mathcal{K}$-function. So L_k admits a ISS-Lyapunov function. By [7], the dynamic of the identification error is input-to-state stable. From (4) and (7) we know L_k is the function of $e\left(k\right)$ and $\mu_1\left(k\right)$. The "INPUT" corresponds to the second term of (9), *i.e.*, the modeling error $\mu_1\left(k\right)$. The "STATE" corresponds to the first term of (8), *i.e.*, the identification error $e\left(k\right)$. Because the "INPUT" $\mu_1\left(k\right)$ is bounded and the dynamic is ISS, the "STATE" $e\left(k\right)$ is bounded. (8) can be rewritten as

$$\Delta L_k \leq -\eta \frac{\left\| e\left(k\right)\right\|^2}{\left[1+\max\limits_k\left(\left\|\Phi\left[X\left(k\right)\right]\right\|^2\right)\right]^2} + \eta_1 \overline{\mu} \tag{10}$$

Summarizing (10) from 1 up to T, and by using $L_T > 0$ and L_1 is a constant, we obtain

$$\eta \sum_{k=1}^{T} \left\| e_N\left(k\right)\right\|^2 \leq L_1 - L_T + T\eta_1\overline{\mu} \leq L_1 + T\eta_1\overline{\mu}$$

(6) is established.

3 Identification Using Adjustable RBF Neural Network

When we regard the plant as a black-box, neither the weight nor the activation function are known. Now the object of the RBF neural modeling is to find the weights, as well as the activation functions $\phi_1 \cdots \phi_n$, such that the RBF neural networks (2) can follow the nonlinear plant (1). Similar as (3), (1) can be represented as

$$y\left(k\right) = W^* \Phi^* \left[X\left(k\right)\right] - \mu_2\left(k\right) \tag{11}$$

where $\Phi\left[X\left(k\right)\right] = \left[\phi_1^* \cdots \phi_n^*\right]^T$, $\phi_i^* = \exp\left[-\left(\frac{x_i - c_i^*}{\sigma_i^*}\right)^2\right]$. In the case of three independent variables, a smooth function f has Taylor formula as

$$f\left(x_1, x_2, x_3\right) = \sum_{k=0}^{l-1} \frac{1}{k!}\left[\left(x_1 - x_1^0\right)\frac{\partial}{\partial x_1} + \left(x_2 - x_2^0\right)\frac{\partial}{\partial x_2} + \left(x_3 - x_3^0\right)\frac{\partial}{\partial x_3}\right]_0^k f + R_l$$

where R_l is the remainder of the Taylor formula. If we let x_1, x_2, x_3 correspond w_i, c_i and σ_i, x_1^0, x_2^0, x_3^0 correspond w_i^*, c_i^* and σ_i^*,

$$\begin{aligned} \widehat{y}\left(k\right) &= y\left(k\right) + \mu_2 + \sum_{i=1}^{n} \left(w_i - w_i^*\right)\phi_i \\ &+ \sum_{i=1}^{n} \frac{\partial\left(w_i\phi_i\right)}{\partial c_i}\left(c_i - c_i^*\right) + \sum_{i=1}^{n} \frac{\partial\left(w_i\phi_i\right)}{\partial \sigma_i}\left(\sigma_i - \sigma_i^*\right) + R \end{aligned} \tag{12}$$

where R is second order approximation error of the Taylor series. Using the chain rule, we get

$$\frac{\partial(w_i\phi_i)}{\partial c_i} = \frac{\partial(w_i\phi_i)}{\partial \phi_i}\frac{\partial \phi_i}{\partial c_i} = w_i 2\phi_i \frac{x_i - c_i}{\sigma_i^2}$$
$$\frac{\partial(w_i\phi_i)}{\partial \sigma_i} = \frac{\partial(w_i\phi_i)}{\partial \phi_i}\frac{\partial \phi_i}{\partial \sigma_i} = w_i 2\phi_i \frac{(x_i - c_i)^2}{\sigma_i^3}$$

The identification error is

$$\begin{gathered} e_k = \widetilde{W}\Phi + 2WD_1\Phi_1\widetilde{C} + 2WD_2\Phi_1\widetilde{\Omega} + \zeta_k \\ \widetilde{W} = W(k) - W^*, \Phi = [\phi_1 \cdots \phi_n]^T, \zeta_k = R + \mu_2, \widetilde{C} = [(c_1 - c_1^*), \cdots (c_n - c_n^*)]^T, \\ \Phi_1 = diag[\phi_1 \cdots \phi_n], D_1 = diag\left[\frac{x_1 - c_1}{\sigma_1^2}, \cdots \frac{x_n - c_n}{\sigma_n^2}\right], \\ D_2 = diag\left[\frac{(x_1 - c_1)^2}{\sigma_1^3}, \cdots \frac{(x_n - c_n)^2}{\sigma_n^3}\right], \widetilde{\Omega} = [(\sigma_1 - \sigma_1^*), \cdots (\sigma_n - \sigma_n^*)]^T \end{gathered} \tag{13}$$

Theorem 2. *If we use the adjustable RBF neural network (2) to identify nonlinear plant (1), the following backpropagation algorithm makes identification error $e(k)$ bounded*

$$\begin{aligned} W(k+1) &= W(k) - e_k\eta_k\Phi \\ C(k+1) &= C(k) - 2e_k\eta_k W(k) D_1\Phi_1 \\ \Omega(k+1) &= \Omega(k) - 2e_k\eta_k W(k) D_2\Phi_1 \end{aligned} \tag{14}$$

where $C = [c_1, \cdots c_n]^T$, $\Omega = [\sigma_1, \cdots \sigma_n]^T$, $\eta_k = \frac{\eta}{1+\Psi_k}$, $\Psi_k = \|\Phi\|^2 + 4\|WD_1\Phi_1\|^2 + 4\|WD_2\Phi_1\|^2$, $0 < \eta \leq 1$. *The average of the identification error satisfies*

$$J = \limsup_{T\to\infty} \frac{1}{T}\sum_{k=1}^{T} e_k^2 \leq \frac{\eta}{\pi}\overline{\zeta} \tag{15}$$

where $\pi = \frac{\eta}{(1+\Psi_k)^2} > 0$, $\overline{\zeta} = \max_k\left[\zeta_k^2\right]$

Proof. We selected a positive defined scalar L_k as

$$L_k = \left\|\widetilde{W}(k)\right\|^2 + \left\|\widetilde{C}(k)\right\|^2 + \left\|\widetilde{\Omega}(k)\right\|^2 \tag{16}$$

So we have

$$\begin{aligned} \Delta L_k &= \left\|\widetilde{W}(k) - \eta_k e_k\Phi\right\|^2 - \left\|\widetilde{W}(k)\right\|^2 + \left\|\widetilde{C}(k) - 2\eta_k e_k W D_1\Phi_1\right\|^2 \\ &- \left\|\widetilde{C}(k)\right\|^2 + \left\|\widetilde{\Omega}(k) - 2\eta_k e_k W D_2\Phi_1\right\|^2 - \left\|\widetilde{\Omega}(k)\right\|^2 \\ &= \eta_k^2\|\Phi\|^2 e_k^2 - 2\eta_k e_k\widetilde{W}(k)\Phi + 4\eta_k^2\|WD_1\Phi_1\|^2 e_k^2 \\ &-4\eta_k e_k WD_1\Phi_1\widetilde{C}(k) + 4\eta_k^2\|WD_2\Phi_1\|^2 e_k^2 - 4\eta_k e_k WD_2\Phi_1\widetilde{\Omega}(k) \\ &= \eta_k^2 e_k^2\left(\|\Phi\|^2 + 4\|WD_1\Phi_1\|^2 + 4\|WD_2\Phi_1\|^2\right) \\ &-2\eta_k e_k\left(\widetilde{W}\Phi + 2WD_1\Phi\widetilde{C} + 2WD_2\Phi\widetilde{\Omega}\right) \end{aligned} \tag{17}$$

The remaining parts are similar to the proof of Theorem 1.

4 Conclusion

This paper applies input-to-state stability approach to adjustable RBF neural networks and proposes robust learning algorithms which can guarantee the stability of training process. The proposed algorithms are effective. The main contributions are: (1) By using ISS approach, we conclude that the commonly-used robustifying techniques in discrete-time neural modeling, such as projection and dead-zone, are not necessary. (2) New algorithms with time-varying learning rates are proposed, which are robust to any bounded uncertainty.

References

1. Cybenko, G.: Approximation by Superposition of Sigmoidal Activation Function. Math.Control Sig. Syst., **2** (1989) 303-314
2. Egardt, B.: Stability of Adaptive Controllers. Lecture Notes in Control and Information Sciences, Vol.20, Springer-Verlag, Berlin (1979)
3. Haykin,S.: Neural Networks- A Comprehensive Foundation. Macmillan College Publ. Co., New York (1994)
4. Ioannou, P.A., Sun, J.: Robust Adaptive Control. Prentice-Hall, Inc, Upper Saddle River: NJ (1996)
5. Jagannathan, S., Lewis, F.L.: Identification of Nonlinear Dynamical Systems Using Multilayered Neural Networks. Automatica, **32** (1996) 1707-1712
6. Jin, L., Gupta, M.M.: Stable Dynamic Backpropagation Learning in Recurrent Neural Networks. IEEE Trans. Neural Networks, **10** (1999) 1321-1334
7. Jiang, Z.P., Wang, Y.: Input-to-State Stability for Discrete-Time Nonlinear Systems. Automatica, **37** (2001) 857-869
8. Kosmatopoulos, E.B., Polycarpou, M.M., Christodoulou, M.A., Ioannou, P.A.: High-Order Neural Network Structures for Identification of Dynamical Systems. IEEE Trans. on Neural Networks, **6** (1995) 442-431
9. Mandic, D.P., Hanna, A.I., Razaz, M.A.: Normalized Gradient Descent Algorithm for Nonlinear Adaptive Filters Using a Gradient Adaptive Step Size. IEEE Signal Processing Letters, **8** (2001) 295-297
10. Peng,H., Ozaki,T., Ozaki, V.H., Toyoda, Y.: A Parameter Optimization Method for Radial Basis Function Type Models. IEEE Trans. Neural Networks, **14** (2003) 432-438
11. Polycarpou, M.M., Ioannou, P.A.: Learning and Convergence Analysis of Neural-Type Structured Networks. IEEE Trans. Neural Networks, **3** (1992) 39-50
12. Song, Q., Xiao, J., Soh, Y.C.: Robust Backpropagation Training Algorithm for Multilayered Neural Tracking Controller. IEEE Trans. Neural Networks, **10** (1999) 1133-1141
13. Suykens, J.A.K., Vandewalle, J., De Moor, B.: NLq Theory: Checking and Imposing Stability of Recurrent Neural Networks for Nonlinear Modelling. IEEE Transactions on Signal Processing, **45** (1997) 2682-2691
14. Yu, W., Li, X.: Some New Results on System Identification with Dynamic Neural Networks. IEEE Trans. Neural Networks, **12** (2001) 412-417
15. Yu, W., Poznyak , A.S., Li, X.: Multilayer Dynamic Neural Networks for Nonlinear System On-line Identification. International Journal of Control, **74** (2001) 1858-1864

A System Identification Method Based on Multi-layer Perception and Model Extraction

Chang Hu and Li Cao

Department of Automation, Tsinghua University, 100084 Beijing P.R. China
huchang@mails.tsinghua.edu.cn
caoli@mail.tsinghua.edu.cn

Abstract. Artificial Neural Networks (ANNs) have provided an interesting and labor-saving approach for system identification. However, ANNs fall short when an explicit model is needed. In this paper, a method of getting the explicit model by extracting it from a trained ANN is proposed. To identify a system, a Multi-Layer Perceptron (MLP) is constructed, trained and a polynomial model is extracted from the trained network. This method is tested in the experiments and shows its capability for system identification, compared with the Least Squared method.

1 Introduction

The artificial neural network (ANN) has provided us with an interesting and useful approach for system identification. When dealing with system identification problems one has to choose among a huge number of algorithms and thereby implement them. With the help of ANNs the designer can be relieved from the implementation of algorithms, therefore concentrate on the structure of the model itself[1]. The ANN is irrelevant with the training algorithm adopted, that provides adaptability to some extent.

When ANNs are to be adopted for system identification, the result is usually a black-box model, with the parameters unknown. Therefore, ANN falls short when an explicit model is necessary. To compensate with this problem a number of works concerning model extraction from ANNs is proposed[2].

Among various types of ANNs, Multi-layer Perception (MLP) has proved itself as an efficient tool for non-linear regression[3]. What is more, it is well fit for model extraction due to its simplicity on structure.

This paper is to propose a new system identification method using ANNs and model extraction. Section 2 discusses the auto-regressive moving average (ARMA) model for system identification. Section 3 and 4 discuss extracting ARMA from a trained MLP.

In Section 5, MLP is compared with Least Square (LS) though two experiments. In the first experiment, the validity of MLP-identified ARMA is shown by comparison with LS-identified ARMA. In the second experiment, MLP-identified ARMA has shown its advantage to be reasonably interpreted.

F. Yin, J. Wang, and C. Guo (Eds.): ISNN 2004, LNCS 3174, pp. 218–223, 2004.

2 The ARMA Model for System Identification

In system identification, the auto-regressive moving average (ARMA) model is often adopted. The ARMA model can be defined as:

$$y(t)+a_1 y(t-1)+\cdots\cdots+a_{d_y} y(t-d_y) \\ =b_0 u(t)+b_1 u(t-1)+\cdots\cdots+b_{d_u} u(t-d_u)+e(t) \tag{2-1}$$

Where $y(t)$ and $u(t)$ are the output and input of the system, respectively.

3 MLP model

A MLP is a feed-forward network.(as shown in Fig.3.1). When the network has only one output it can be viewed as a Multi-Input-Single-Output (MISO) system:

$$\hat{y}=f(\vec{u}) \tag{3-1}$$

where $\vec{u} =[u_1 ,u_2 ,\cdots\cdots,u_m]^T$ is the input vector.

The output of the network is denoted as $\hat{y}$ to show the difference between network output and the output y out of the actual system .

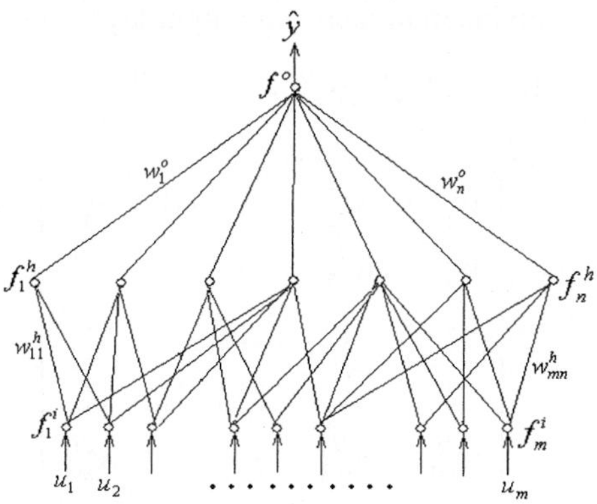

Fig. 3.1. Network structure

When it comes to system identification, the input and output of a neural network is usually adjusted to meet the requirements of the model. In case of an ARMA model, the input and output of the network is configured as follows:

$$\vec{u}=[y(t-1),....,y(t-d_y),u(t),....,u(t-d_u)]^T \tag{3-2}$$

4 Extracting ARMA Model from MLP

As[4] has pointed out, a MLP with sigmoid transfer function can be viewed as ARMA when every neuron is working near zero.

Similarly, consider a MLP in Fig.3.1, which has transfer functions f^I in input layer, f^H in hidden layer and f^O in output layer.

For input layer,

$$y_1^I = f^I(u_1), \quad \cdots\cdots\cdots \quad , y_m^I = f^I(u_m) \tag{4-1}$$

for hidden layer:

$$y_1^H = f^H(\sum_{i=1}^{m} w_{i1}^H y_i^I), \quad \cdots \quad , y_n^H = f^H(\sum_{i=1}^{m} w_{in}^H y_i^I) \tag{4-2}$$

for output layer:

$$\hat{y} = f^O(\sum_{j=1}^{n} w_j^O y_j^H) \tag{4-3}$$

when f^I, f^H and f^O are sigmoid functions, with input near zero, they could be approximated by linear functions $f(x) = x$. Substituting into Eq.4-1 to Eq.4-3, yields and putting the weights in hidden and output layer into matrices,

$$W^O = [w_1^O, w_2^O, \cdots\cdots, w_n^O] \tag{4-4}$$

$$W^H = \begin{bmatrix} w_{11}^H & w_{12}^H & \cdots & \cdots & w_{1n}^H \\ w_{21}^H & w_{22}^H & \cdots & \cdots & w_{2n}^H \\ \vdots & \vdots & \ddots & & \vdots \\ \vdots & \vdots & & \ddots & \vdots \\ w_{m1}^H & w_{m2}^H & \cdots & \cdots & w_{mn}^H \end{bmatrix} \tag{4-5}$$

Eq.4-3 is then

$$\hat{y} = W^O (W^H)^T \vec{u} \tag{4-6}$$

As a result, if the input and output of the network are set properly, a polynomial model could be extracted from the trained network, by multiplying the weight matrices of the hidden and output layers.

5 Experiments

In order to evaluate the technique to extract model from MLP as an ARMA model, experiments have been conducted. They are a simulated one and a real-world one, respectively.

In the following experiments, the proposed method is compared with the system identification method Least Square (LS) [5]. LS is designed to minimize the MSE criterion, Thus we use MSE to evaluate the performance of MLP, too.

$$mse(\vec{z},\hat{\vec{z}}) = \sum_i (z_i - \hat{z}_i)^2 \tag{5-1}$$

5.1 Experiment 1: A Simulation

The proposed method and LS are used to identify data generated by the ARMA in

$$z(k) = -0.8z(k-1) + 0.05z(k-2) + 0.05u(k) - 0.4u(k-1) + 0.4u(k-2) \tag{5-2}$$

Where *u(k)* is Gaussian white noise for the convenience of system identification.

In order to minimize the effect brought by different initial states in MLP, a number of initial network states have been tried out, and the ARMA models from those MLPs are averaged to produce the final model.

ARMA models identified by both MLP and LS are compared in Table.5.1.1 and Table 5.1.2. The tables have also shown a comparison of estimated output of both ARMA models. Both noisy data (with noise amplitude 10% of amplitude of $u(k)$) and noise-free data are used. In the noisy case, the noisy output $\tilde{z} = z + e$ is used for identification and the true output z is used to calculate MSE.

Table 5.1.1. ARMA model and MSE comparison. (Noise-free data)

	ARMA.	MSE.
Original	-0.8000 0.0500 0.0500 -0.4000 0.4000	
LS	-0.8000 0.0500 0.0500 -0.4000 0.4000	0
MLP	-0.8124 0.0511 0.0499 -0.4041 0.4055	4.67e-5

Table 5.1.2. ARMA model and MSE comparison. (noisy data)

	ARMA.	MSE.
Original	-0.8000 0.0500 0.0500 -0.4000 0.4000	
LS	-0.6020 0.0376 0.0707 -0.3988 0.3397	0.0051
MLP	-0.6267 0.0217 0.0542 -0.4237 0.3859	0.0049

In the noise-free case, although ARMA model extracted from MLP is not as good as that from LS in terms of MSE, it is still a reasonable one because it's similar to that from LS. In the noisy case, the performance of MLP-estimated ARMA is comparable to that of LS-estimated ARMA.

5.2 Experiment 2: A Real-World Problem

To evaluate the performance of the proposed method, experiment upon a real-world data set is conducted. The data set is time series representing the load of air-conditioning system in a building.

The model to be identified is

$$z(k) = a_1 z(k-1) + a_2 z(k-2) + a_3 z(k-3) + bu(k) \tag{5-3}$$

Where *u(k)* is temperature at time *k* and *z(k)* is current load. The ARMA models identified by both methods are normalized for interpretation. They are listed in Table 5.2.1 and their absolute values are plot as stems in Fig. 5.2.2.

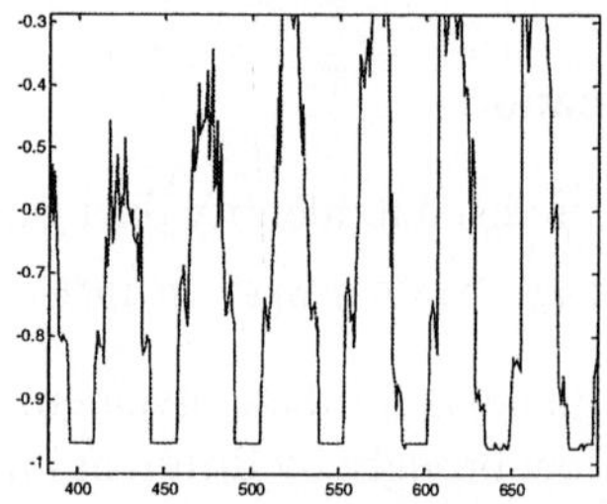

Fig. 5.2.1 Real-world data (x-axis: time in 0.5hour; y-axis: load in MW)

Table 5.2.1. Normalized ARMA models

No.	ARMA.
LS	0.8565 1.0000 -0.3382 0.0894
MLP	0.9886 0.3562 -0.1530 1.0000

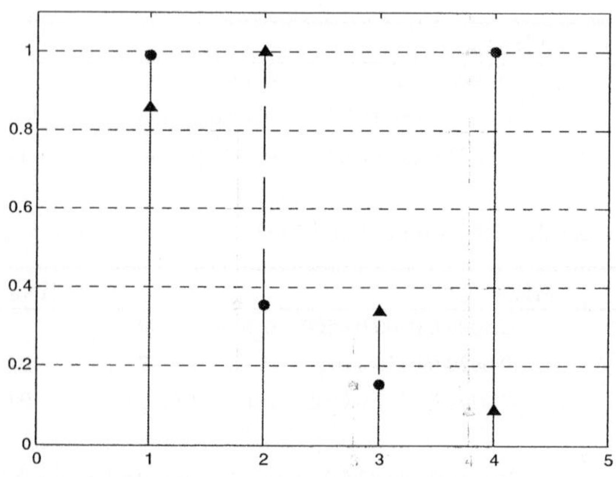

Fig. 5.2.2 Absolute coefficients of Normalized ARMA models (solid: MLP-ARMA; dashed: LS-ARMA)

In MLP-identified ARMA, *z(k-1)* and *u(k)* are given almost the same weight. However, LS- identified ARMA emphasizes *z(k-2)*. According to ground-true knowledge, the load of an air-conditioning system is determined by its load in the last half hour and current temperature instead of load one hour before. Therefore, the MLP-identified ARMA appears to be more reasonable.

6 Conclusions

In this paper, a system identification method by extracting ARMA model from trained MLP is proposed and evaluated. The performance of the proposed method has been evaluated through comparison with the Least Square method.

Experiments with simulated data validate MLP-identified ARMA by showing its similarity to LS-identified ARMA. What is more, MLP-identified ARMA has shown its advantage over LS-identified ARMA, when both models are to be interpreted in real-world case.

References

1. McLoone, S.: Neural Network Identification: a Survey of Gradient Based Methods; Optimisation in Control: Methods and Applications (Ref. No. 1998/521), IEE Colloquium on , 10 Nov. (1998) 4/1 -4/4
2. Setiono, R., Wee Kheng Leow, Zurada, J.M.: Extraction of Rules from Artificial Neural Networks for Nonlinear Regression; Neural Networks, IEEE Transactions on, Volume: 13 Issue: 3, May (2002) 564 -577
3. Stubberud, A., Wabgaonkar, H., Stubberud, S.: A Neural-Network-Based System Identification Technique, Decision and Control, 1991, Proceedings of the 30th IEEE Conference on, 11-13, Dec., vol.1. (1991) 869 -870
4. Muller, C., Mangeas, M.: Neural Networks and Times Series Forecasting: a Theoretical Approach, Systems, Man and Cybernetics, 1993. 'Systems Engineering in the Service of Humans', Conference Proceedings, International Conference on , 17-20 Oct. vol.2. (1993) 590 -594
5. Xiao, D.Y., Fang C. Z.: Process Identification; Tsinghua University Press, Aug (1988) 160-162

Complex Model Identification Based on RBF Neural Network

Yibin Song, Peijin Wang, and Kaili Li

School of Computer, Yantai University, Shandong, China 264005
sybw@ytu.edu.cn

Abstract. Based on the principle of Radial Basis Function (RBF) Neural Network, a learning method is presented for the identification of a complex system model. The RBF algorithm is employed on the learning and identifying process of the nonlinear model. The simulation results show that the presented method has good effect on speeding up the learning and approaching process of the nonlinear complex model, and has an excellent performance on learning convergence.

Keywords: RBF Neural Networks learning algorithm model identification.

1 Introduction

RBF Neural Network (RBF-NN) is usually to take three-layer structure. The mapping from the input layer to the hidden layer is nonlinear and the linear mapping is adopted between the hidden layer and the output layer. In the RBF Network, the radial basis function is adopted as the mapping function. The main function of RBF is taking as the "Base" for hidden layers so as to construct the hidden space. The input vector can map to hidden space directly but not to use connecting weights. The mapping relation will be confirmed as soon as the center point confirmed. But the mapping between the hidden layer and the output layer is linear and the output of Neural Network (NN) is the linearly weighted sum of hidden layers. The connecting weights are adjustable parameters for Network. So, the connecting weights of network can be evaluated by sets of linear equation or by LMS method directly. The learning speed of NN will be more quickly. Meanwhile, the fussy calculation and the local minima can be avoided.

Usually, the multiplayer NN needs a significant amount of calculation of weights for getting the global approach. As a result, some problems, including low learning speed and local minima problem, might emerge. In contract, the RBF-NN is a typical local approaching network. Only a few weights need to be adjusted for the output of NN. Therefore, the RBF-NN could complete the learning process more quickly than that of the BP-NN. It is obviously the RBF-NN is more effective than normal BP-NN in approaching ability, sorting ability and identification speed. It is a powerful tool for realizing the modeling and identifying process of the complex plants or systems.

Based on the principle of RBF-NN, this paper presents a method for the modeling and identification of the structure-unknown complex plants or systems. An actual identification of nonlinear model is applied to validate the effects of algorithm. The simulation results show that the presented method has good effect on speeding up the

F. Yin, J. Wang, and C. Guo (Eds.): ISNN 2004, LNCS 3174, pp. 224–229, 2004.

training and modeling process, especially suitable for the real-time request of complex system control. The algorithm also shows an excellent performance on learning convergence.

2 Learning Principle of RBF-NN

2.1 RBF-NN Structure Analysis

In the structural characteristics of network, RBF-NN adopts direct mapping between input layer and hidden layer. But the mapping between hidden layer and output layer adopts the linearly weighted sum of hidden layers as the mapping mode (shown as in Fig. 1). This structure of NN can reduce the complexity of computational problems so as to speed up the learning process. It is suitable especially for accomplishing the function approaching, model identification and sorting process quickly.

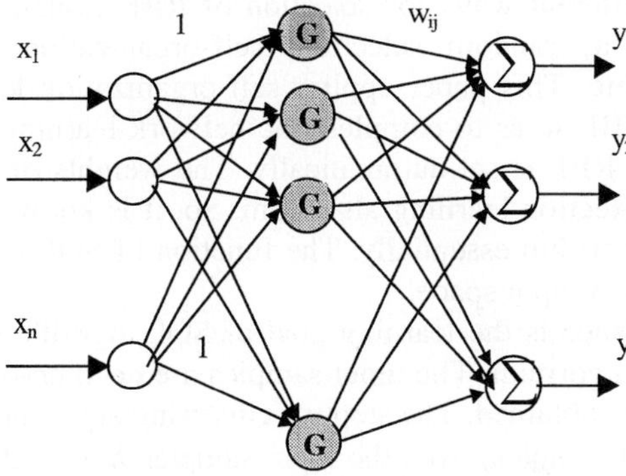

Fig. 1. Structure of RBF-NN

In the RBF-NN, the task of the hidden layer is different from that of the output layer. So does the learning strategy. The hidden layer adopts the nonlinear optimizing strategy. So, the learning speed of the function parameters will be slower accordingly. Oppositely, the output layer adopts the linear optimizing strategy for the adjusting of weights. The learning speed is quicker than that of the hidden layer. It is known there are two hierarchy of learning processes for RBF-NN.

2.2 Learning Algorithm of RBF-NN

In RBF-NN, the input vectors of lower dimensions are mapped to the hidden space of higher dimensions first. The hidden cells select the basis function to realize the vector conversion, and then, sorted or identified by output layer. The form of base function ***F*** can be described as:

$$F(X)=\sum_{i=1}^{N} w_i\varphi(\| X-X_i \|) \tag{1}$$

where: $\varphi(\| \mathbf{X}-\mathbf{X}_i \|)$ is the Radial Basis Function (RBF), usually uses Gaussian function, $\| \bullet \|$ denotes the Euclid Norm.

To take known data $X_i \in R^{n_i}$ as the center of RBF and φ is radial-symmetrical to the center point. If Gauss function is adopted as the Radial Basis Function (RBF), it can be described as:

$$G(\| X-t_i \|^2)=\exp(-\frac{M}{d_m^2}\| X-t_i \|^2) \tag{2}$$

where, **M** is the number of center, $\mathbf{t}_i$ is the center of RBF and $\mathbf{d}_m$ is the maximal distance between selected centers. The mean-square-error of RBF takes $\sigma=d_m/\sqrt{2M}$.

According to the different ways for selection of RBF center, RBF-NN is usually applied the methods as random selection, self-organization learning (SOL) or supervisory learning, etc. This paper applies self-organization learning algorithm to choose the center of RBF so as to complete the network learning. In SOL algorithm, the central position of RBF is set automatically. The weights of output layer can be calculated by error-correction-learning algorithm. So, it is known the SOL algorithm is a mixed learning algorithm essentially. The function of SOL is to adjust the center of RBF to the key area of input space.

Clustering least distance is the learning goal of SOL algorithm. This paper adopts the k-mean clustering algorithm. The input samples are decomposed to M classes and M clustering centers are obtained. The steps of clustering algorithm are:

1 Choose randomly M samples, from the input samples X_j (j=1,2,⋯,N), as the initial clustering centers t_i (i=1,2,⋯,M),

2 Distribute input samples X_j (j=1,2,⋯,N) to every t_i and to form clustering sets θ_i (i=1,2,⋯,M). The following condition are met:

$$d_i=\min_i\| X_j-t_i \| \tag{3}$$

(j=1,2,⋯,N i=1,2,⋯,M)

where d_i is the minimal Euclidean distance.

3 Calculate the sample's mean (i.e. clustering center $\mathbf{t}_i$) in θ_i:

$$t_i=\frac{1}{M_i}\sum_{X_j\in\theta_j} X_j \tag{4}$$

where $\mathbf{M}_i$ is the number of input sample in θ_i

4 Repeat above calculation till to get the change of distribution of clustering center less than designed value ε .

5 After the decision of RBF center, the mean square error σ can be calculated by equation $\sigma=\mathbf{d}_m/\sqrt{2M}$ and then to get the output of hidden layer by equation (2).

The error-correction-learning algorithm can calculate the linear weights between the hidden layer and the output layer. Suppose that the output of the *kth* neuron of output layer is $\hat{\mathbf{y}}_{\mathbf{k}}$ and that of the *jth* neuron of hidden layer is g_j. The relation is:

$$\hat{y}_k = \sum_j w_{kj} \cdot g_j \tag{5}$$

If the actual output is y_k, the error is: $\varepsilon_k = y_k - \hat{y}_k$.

The purpose of BP learning is to amend connecting weights so as to minimize the error index $E = f(\sum \varepsilon^2_k)$ and to meet the desired performance **J**. When input mode is $\mathbf{X_p}$, the correcting value of w_{kj} should be:

$$\Delta_p w_{kj} = -\alpha \frac{\partial E_p}{\partial w_{kj}} \tag{6}$$

where, α is adjusting factor for learning rate. The updating equation is:

$$w_{kj}(k+1) = w_{kj}(k) + \Delta_p w_{kj} \tag{7}$$

In order to improve the convergence of algorithm and its learning effects, a dynamic correcting factor can be set in the correcting equation, that is:

$$w(k+1) = w(k) + \alpha[(1-\eta)T(k) + \eta T(k-1)] \tag{8}$$

where, $0<\alpha<1$, $0\leqslant\eta<1$ is a dynamic factor and $T(k)=-\partial \mathbf{E}/\partial \mathbf{w}(\mathbf{k})$ is the direction of negative grads in the **k***th* time learning.

However, simulation results show that the effect of this method is not perfect for improving the learning process. The key question is how to select the learning rate. Sometimes it is a very difficult decision for the algorithm. The difficulty is how to pay attention to both learning speed and the convergence of algorithm. This paper presents an improved algorithm. It adds a rate-adaptive factor $\delta(k)$ to the learning equation, that is:

$$\begin{cases} w(k+1) = w(k) + \alpha(k)T(k) \\ \alpha(k) = \delta(k)\alpha(k-1) \\ \delta(k) = \varepsilon \cdot 2^{\lambda} \end{cases} \tag{9}$$

where λ= Sign [T(k)T(k-1)] is the direction of grads, $\delta(k)$ is the adaptive factor associated with learning error ε.[5]

Simulation results show that the improved algorithm can speed up the convergent process ($\Delta_p w_{kj}$ or $\Delta_p w_{ji} \rightarrow \mathbf{0}$) of weights learning effectively (shown as in table 1).

3 Simulation Results

From the characteristics of RBF-NN, this paper applies RBF-NN to identify the nonlinear model via learning about the sampled data. The performance of designed

NN is analyzed. Suppose that a nonlinear system is described as the following equation:

Input: $$u(k) = 0.25\left[Sin^2(\frac{2k\pi}{25}) + Sin^3(\frac{2k\pi}{21}) \right] \quad (10)$$

Nonlinear plant: $$y(k+1) = \frac{y(k)}{1+y^2(k)} + [u(k+1)]u(k)[u(k-1)] \quad (11)$$

For the unknown - structure system, an RBF-NN needs to model the unknown object and identify the behaviors of the plant by learning the sampling data. As an example, 60 pairs of input data ***u(k)*** and output data ***y(k)*** are sampled as the learning information. The network parameters are selected as: spread coefficient Sc=1.5, learning error $\varepsilon \leq 0.001$。

The simulation results show that the RBF-NN accomplished the desired identification accurately only after training for 37 epochs. The learning and identifying processes are shown in Fig. 2 and Fig. 3. In Figure 2, '+' stands for sampling data, '—' shows the results after learning by the RBF-NN.

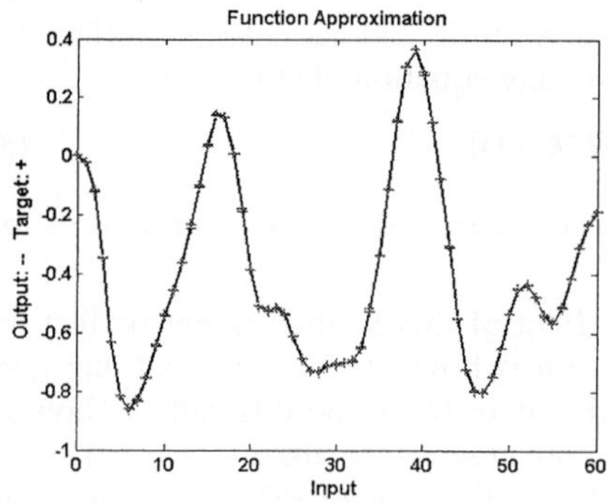

Fig. 2. Learning result of RBF-NN

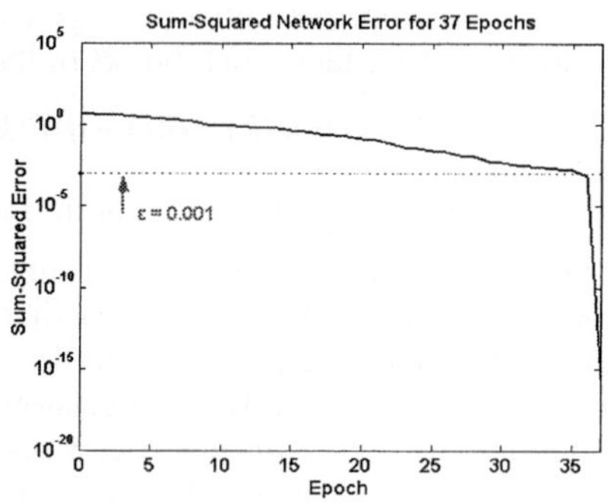

Fig. 3. Convergent process of Learning error of RBF-NN

Table 1. Comparisons of the training performance

Algorithm	Pairs of samples	Training epochs	Sum-squared error (ε ≦)
BP	30	15000	0.3
Improved BP	30	450	0.1
RBF	60	37	0.001

The training results of the normal BP-NN for the same plant are shown in Fig. 4 and Fig. 5. There are only 30 pair of samples for the learning of the BP algorithm and the learning error is only $\varepsilon \leq 0.3$. But the sum-squared learning error cannot be attained until 15000 epochs of training by the normal BP algorithm. The improved BP

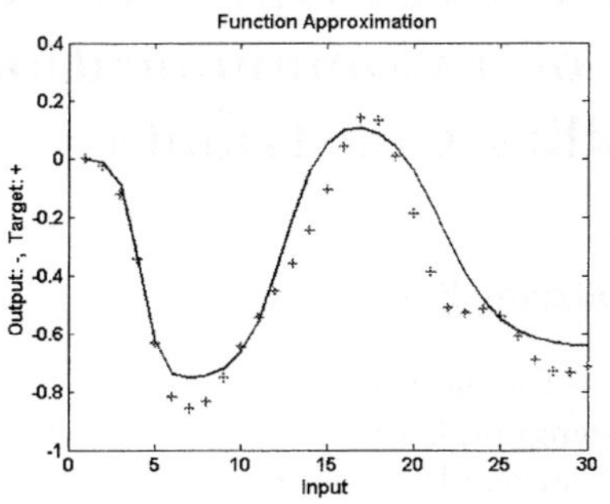

Fig. 4. Learning result of normal BP-NN

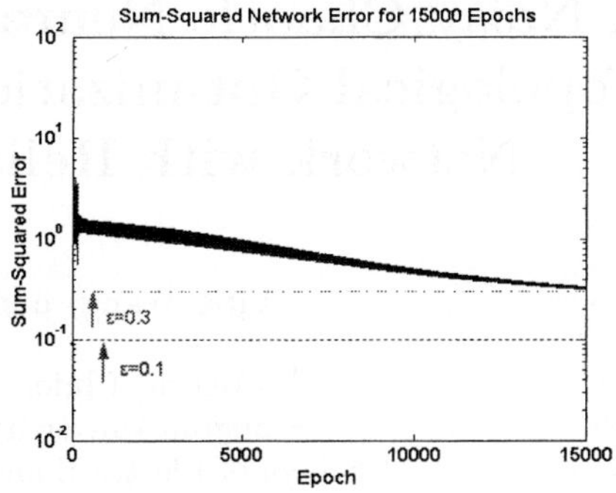

Fig. 5. Convergent process of Learning error of normal BP-NN

algorithm still needs 450 epochs training to attain error goal of $\varepsilon \leqslant 0.1$. It is obviously that the RBF-NN accomplishes the training more quickly. Detail results are shown in Table 1.

Simulation results show clearly that the performance of RBF-NN is superior to the BP-NN not only in training speed but also in identifying accuracy. It proved that the RBF-NN is more suitable for the modeling and identifying of complex plants. It is an effective method for the prediction and control of complex systems.

4 Conclusions

The excellent characteristics of RBF algorithm are its quick convergence and accurate learning results. This paper discussed different training algorithms respectively, i.e. RBF-NN, BP-NN and improved BP-NN. After applying to the identification of a typical nonlinear system and simulations, some conclusions can be obtained.

Although the improved BP-NN can speed up the learning process, the RBF-NN has much quicker training process and much excellent learning performance. It is more suitable for modeling and identifying of complex plants. Therefore, RBF-NN is an effective way for the realization of the real-time control of complex systems.

References

1. Delgado A., Kambhampati C., Warwick K.: Dynamic Recurrent Neural Network for System Iidentification and Control. IEE Proc. Control Theory Application, 142(6), (1995), 307-314,.
2. Norio Baba: A New Approach for Finding the Global Minimum of Error Function of Neural Networks. Neural Networks, 3(1990). 535-549.
3. Kosmatopulos E B: High-Order Neural Network Structures for Identification of Dynamical Systems. *IEEE Trans. On Neural Networks.* 6(2), (1995) 422-431.
4. M. Bianchini, P. Frasconi and M. Gori: Learning Without Local Minima in Radial Basis Function Networks, *IEEE Trans. on Neural Networks*, 6(3), (1995) 749-756.
5. Song Yibin: Analysis and Application of Neuron Learning Rules with adaptive Accelerating Factors. Proc. of IEEE ICNN&B'98, (1998) 274-277.

A Noisy Chaotic Neural Network Approach to Topological Optimization of a Communication Network with Reliability Constraints

Lipo Wang[1] and Haixiang Shi[2]

[1] College of Information Engineering
Xiangtan University, Xiangtan, China
[2] School of Electrical and Electronic Engineering
Nanyang Technological University
Block S1, Nanyang Avenue, Singapore 639798
{elpwang, pg02782641}@ntu.edu.sg

Abstract. Network topological optimization in communication network is to find the topological layout of network links with the minimal cost under the constraint that all-terminal reliability of network is not less than a given level of system reliability. The all-terminal reliability is defined as the probability that every pair of nodes in the network can communicate with each other. The topological optimization problem is an NP-hard combinatorial problem. In this paper, a noisy chaotic neural network model is adopted to solve the all-terminal network design problem when considering cost and reliability. Two sets of problems are tested and the results show better performance compared to previous methods, especially when the network size is large.

1 Introduction

An important stage of network design is to find the best way to layout all the components to optimize costs while meeting a performance criterion, such as transmission delay, throughput, or reliability [1]. This paper focuses on network design of large backbone communication networks with an all-terminal reliability (ATR) constraint, i.e., an acceptable probability that every pair of nodes can communicate with each other [1]. The network topology problem can be formulated as a combinatorial optimization problem which is NP-hard [2]. Previous work on this problem can be categorized to enumerative-based and heuristic methods. Jan *et al* [1] developed an algorithm using decomposition based on branch-and-bound to find the exact, optimal solution. They divided the problem into several subproblems by the number of links of subnetworks. The maximum network size of fully connected graphs for which a solution has been found is 12 nodes (60 links) [4]. Due to the NP-hard nature of the problem, heuristic approaches are often adopted to solve this problem. Berna *et al* [3] used genetic algorithms to solve the all-terminal network design problem. They formulate the network design as an integer vector which is the chromosome. Each gene of the

F. Yin, J. Wang, and C. Guo (Eds.): ISNN 2004, LNCS 3174, pp. 230–235, 2004.

chromosome represents a possible link of the network. Because the network reliability measure is also an NP-hard problem, instead of the exact calculation of reliability, the network reliability is estimated through three steps. First step is the connectivity check for a spanning tree. The next step is the evaluation of a 2-connectivity measure and the last step is to compute Jan's upper bound [1]. The precise estimation of the reliability of the network can be obtained through Monte Carlo simulations. Hosam *et al* [4] adopted a Hopfield neural network called optimized neural network (OPTI-nets) to solve this problem. The links in the backbone network are represented by neurons, where each neuron represents a link. A neuron is set if its output is a nonzero value and the corresponding link is hence selected [4]. They tested their OPTI-nets on a large network size with 50 nodes and 1225 edges.

In this paper, we use a noisy chaotic neural network (NCNN) [7]-[9] to solve this NP-hard problem. In section II, we briefly introduce the NCNN model [7]-[9]. Section III presents the NCNN solution to the topological optimization problem. Section IV includes the results. We draw conclusions in Section V.

2 Noisy Chaotic Neural Network

Chen and Aihara [5][6] proposed chaotic simulated annealing (CSA) by starting with a sufficiently large negative self-coupling in the neurons and then gradually decreasing the self-coupling to stabilize the network. They called this model the transiently chaotic neural network (TCNN). By adding decaying stochastic noise into the TCNN, Wang and Tian [7] proposed a new approach to simulated annealing using a noisy chaotic neural network (NCNN). This novel model has been applied successfully in solving several challenging optimization problems including the traveling salesman problem (TSP) and the channel assignment problem (CAP) [7] [8] [9]. The NCNN model is described as follows [7]:

$$x_{jk}(t) = \frac{1}{1+e^{-y_{jk}(t)/\varepsilon}} \quad , \tag{1}$$

$$y_{jk}(t+1) = ky_{jk}(t) + \alpha \left[\sum_{i=1, i\neq j}^{N} \sum_{l=1, l\neq k}^{N} w_{jkil} x_{jk}(t) + I_{ij} \right] - z(t)\left[x_{jk}(t) - I_0\right] + n(t) \, , \tag{2}$$

$$z(t+1) = (1-\beta_1)z(t) \quad , \tag{3}$$

$$A[n(t+1)] = (1-\beta_2)A[n(t)] \quad , \tag{4}$$

where

x_{jk} : output of neuron jk ;
y_{jk} : input of neuron jk ;

w_{jkil}: connection weight from neuron jk to neuron il, with $w_{jkil} = w_{iljk}$ and $w_{jkjk} = 0$;

$$\sum_{i=1,i\neq j}^{N} \sum_{l=1,l\neq k}^{N} w_{jkil} x_{jk} + I_{ij} = -\partial E / \partial x_{jk} : \quad \text{input to neuron } jk \tag{5}$$

E : energy function;
t : time steps;
N : number of network nodes;
I_{jk} : input bias of neuron jk ;
k : damping factor of nerve membrane $(0 \leq k \leq 1)$;
α : positive scaling parameter for inputs ;
β_1 : damping factor for neuronal self-coupling $(0 \leq \beta_1 \leq 1)$;
β_2 : damping factor for stochastic noise $(0 \leq \beta_2 \leq 1)$;
$z(t)$: self-feedback connection weight or refractory strength $(z(t) \geq 0)$, $z(0)$ is a constant;
I_0 : positive parameter;
ε : steepness parameter of the output function $(\varepsilon > 0)$;
$n(t)$: random noise injected into the neurons, in $[-A, A]$ with a uniform distribution;
$A[n]$: amplitude of noise n.

3 Applying NCNN to Topological Optimization Problem

3.1 Problem Formulation

A communication network can be modeled by a graph $G = (N, L, P, C)$. Here N and L are network sites and communication links, respectively. P is the set of reliability for all links and C is the set of costs for all links. The optimization problem is [3]:

$$\text{Minimize:} \quad Z = \sum_{i=1}^{N-1} \sum_{j=i+1}^{N} c_{ij} v_{ij} \tag{6}$$

$$\text{Subject to:} \quad R \geq R_0$$

where the $C = \{c_{ij}\}$ is the cost of link (i, j) and $V = \{v_{ij}\}$ is the neuron binary output which is defined as:

$$v_{ij} = \begin{cases} 1, & \text{if link } (i,j) \text{ is selected for the optimized network design ;} \\ 0, & \text{otherwise.} \end{cases}$$

R is the all-terminal reliability of the network and R_0 is the network reliability requirement.

We formulate the energy function of the NCNN as follows:

$$E = -W_1 R + W_2 \sum_{i=1}^{N} \sum_{j=i+1}^{N} v_{ij} c_{ij} + W_3 \eta^{\varsigma} \times \left| 1 - \frac{R}{R_0} \right| + W_4 \sum_{i=1}^{N} \sum_{j=1,j\neq i}^{N} v_{ij}(1 - v_{ij}) \,, \tag{7}$$

where $\varsigma = u(1 - \frac{R}{R_0})$ and $u(x)$ is a unit step function, i.e., $u(x) = 0$ for $x < 0$ and $u(x) = 1$ for $x \geq 0$.

W_1, W_2, W_3, and W_4 are weight factors. The W_1 term encourages the network to increase the network reliability. The W_2 term is the costs of network links to be minimized. The W_3 term is the constraint term which discourages the network from adding more links to increase the network reliability unnecessarily over R_0 and η is the penalty factor. The W_4 term is used to help the convergence of neurons. When the energy goes toa minimum, the W_4 term forces the output of neurons to a value of 0 or 1.

The reliability term in energy function R is replaced with a upper bound instead of exact calculation due to the NP-hard complexity for an exact calculation [4]. The upper bound on the ATR is [4]:

$$R(G) \leq \prod_{i=1, i \neq s}^{N} \left[1 - \prod_{j \in A(i)} (1 - p_{ij}) \right] , \tag{8}$$

where $A(x)$ is defined as the set of vertexes connected to vertex x, p_{ij} is the reliability of link (i, j). Node s is selected as the node with maximum node degree [4].

From dynamic equations (2), (5) and (7), the motion equation of the NCNN is:

$$\begin{aligned} y_{jk}(t+1) = k y_{jk}(t) + \alpha \, \{ W_1 \frac{\partial R}{\partial v_{ij}} - W_2 c_{ij} - \frac{W_3}{R_0} (-1)^{1-\varsigma} \\ \eta^{\varsigma} \frac{\partial R}{\partial v_{ij}} - W_4 (1 - 2 v_{ij}) \} - z(t) \left[x_{jk}(t) - I_0 \right] + n(t) \, , \end{aligned} \tag{9}$$

where from equation (8) [4],

$$\begin{aligned} \frac{\partial R}{\partial v_{ij}} = \prod_{k=1, k \neq s, i, j}^{N} \left[1 - \prod_{l=1, l \neq k}^{N} \psi_{kl} \right] p_{ij} \\ \left[\prod_{l=1, l \neq i}^{N} \psi_{jl} + \prod_{l=1, l \neq j}^{N} \psi_{il} - 2 \prod_{l=1, l \neq j}^{N} \psi_{lj} \prod_{l=1, l \neq i, j}^{N} \psi_{li} \right] \end{aligned} \tag{10}$$

and

$$\psi_{ab} \equiv 1 - p_{ab} \times v_{ab} \, . \tag{11}$$

4 Results

The parameters for the energy function is determined empirically as below: $W_1 = 1.5, W_2 = 0.0001, w_3 = 1.2, w_4 = 1, \eta = 10$.

The benchmark problems are adopted from [3] in which the results for $n < 10$ are verified by using the exact branch-and-bound method [1]. The problems contains 20 different cases. Case 1-17 are problems concerning fully connected networks and case 18-20 are for non-fully connected networks. The comparison of results is listed in Table 1, where p is the link probability for each link and R_0 is the objective network reliability. The simulation of each case was run 10 times and only the best results are listed as in the references to which we shall compare our results [1][3][4]. We compare the results with three other methods as showed in Table 1. From the results, it can be seen that branch-and-bound cannot be applied to problems with larger node sizes but it can obtain optimal solution. Genetic algorithms are effective but failed to find good solutions when the node size becomes large as showed in case 15-17. And it also can be seen that our NCNN can find better solutions than OPTI-net does on these cases.

Table 1. Comparison of results of test cases from [3]

Case	# of Nodes	# of edges	p	R_0	BnB[1][3]	GA[3]	OPTI-net[4]	NCNN
1	5	10	0.8	0.9	255	255	255	255
2	5	10	0.9	0.95	201	201	201	201
3	7	21	0.9	0.9	720	720	720	720
4	7	21	0.9	0.95	845	845	845	845
5	7	21	0.95	0.95	630	630	630	630
6	8	28	0.9	0.9	208	208	208	208
7	8	28	0.9	0.95	247	247	247	247
8	8	28	0.95	0.95	179	179	179	179
9	9	36	0.9	0.9	239	239	239	239
10	9	36	0.9	0.95	286	286	308	286
11	9	36	0.95	0.95	209	209	209	209
12	10	45	0.9	0.9	154	156	154	154
13	10	45	0.9	0.95	197	205	197	197
14	10	45	0.95	0.95	136	136	136	136
15	15	105	0.9	0.95		317	304	304
16	20	190	0.95	0.95		926	270	202
17	25	300	0.95	0.9		1606	402	377
18	14	21	0.9	0.9	1063	1063	1063	1063
19	16	24	0.9	0.95	1022	1022	1077	1022
20	20	30	0.95	0.9	596	596	596	596

5 Conclusion

In this paper we apply the noisy chaotic neural network to solve the topological optimization problem in backbone networks with all-terminal reliability constraint. The results on 20 benchmark problems show that our NCNN outperforms other methods, especially in large problems.

References

1. Jan, R.H., Hwang, F.J., Chen, S.T.: Topological Optimization of a Communication Network Subject to a Reliability Constraint. IEEE Trans. Reliability, Vol. 42. (1993) 63–70
2. Garey, M.R., Johnson, D.S.: Computers and Intractability: A Guide to the Theory of NP-Completeness. W. H. Freeman & Co. (1979)
3. Dengiz, B., Altiparmak, F., Smith, A.E.: Efficient Optimization of All-terminal Reliability Networks, Using an Evolutionary Approach. IEEE Trans. Reliability, Vol. 46. (1997) 18–26
4. AboElFotoh, Hosam M.F., Al-Sumait, Loulwa S.: A Neural Approach to Topological Optimization of Communication Networks, with Reliability Constraints. IEEE Trans. Reliability, Vol. 50. (2001) 397–408
5. Chen, L., Aihara, K.: Transient Chaotic Neural Networks and Chaotic Simulated Annealing, M. Yamguti(ed.). Towards the Harnessing of Chaos. Amsterdam, Elsevier Science Publishers B.V. (1994) 347–352
6. Chen, L., Aihara, K.: Chaotic Simulated Annealing by a Neural Network Model with Transient Chaos. Neural Networks, Vol. 8. (1995) 915–930
7. Wang, L., Tian, F.: Noisy Chaotic Neural Networks for Solving Combinatorial Optimization Problems. Proc. International Joint Conference on Neural Networks. IJCNN (2000)
8. Li, S., Wang, L.: Channel Assignment for Mobile Communications Using Stochastic Chaotic Simulated Annealing. The 6th International Workshop-Conference on Artificial and Natural Neural Networks. IWANN (2001)
9. Wang, L., Li, S., Tian, F., Fu, X.: A Noisy Chaotic Neural Network for Solving Combinatorial Optimization Problems: Stochastic Chaotic Simulated Annealing. IEEE Trans. System, Man, Cybern, Part B - Cybernetics, August, (2004)

Space-Time Multiuser Detection Combined with Adaptive Wavelet Networks over Multipath Channels

Ling Wang[1, 2], Licheng Jiao[1, 2], Haihong Tao[2], and Fang Liu[3]

[1] Institute of Intelligent Information Processing, Xidian University, 710071 Xi'an, China

[2] Key Lab for Radar Signal Processing, Xidian University, 710071 Xi'an, China

wanglingboy@yahoo.com.cn, {lchjiao, hhtao}@xidian.edu.cn

[3] School of Computer Science and Technology, Xidian University, 710071 Xi'an, China

f63liu@163.com

Abstract. The capacity and performance of code division multiple access (CDMA) systems are limited by multiple access interference (MAI) and "near-far" problem. Space-time multiuer detection combined with adaptive wavelet networks over frequency selective fading multipath channels is proposed in this paper. The structure of the multiuser detector is simple and its computational complexity mostly lies on that of wavelets networks. With numerical simulations and performance analysis, it is shown that the proposed detector can converge at the steady state rapidly and offer significant performance improvement over the space-time matched filtering detector, the conventional RAKE receiver and matched filter detector only at the time domain. Therefore, it can suppress the MAI and solve the "near-far" problem effectively.

1 Introduction

The future generation of wireless networks will support multiple classes of traffic with different quality of service requirements such as data rate and bit-error rate(BER). DS-CDMA has been emerging as a popular multiple access technology for wireless communication systems. But There exist two major obstacles in the system: MAI and near-far problem. Recently, multiuser detection and space-time processing techniques have been a growing interest for improving the receiver performance and network capacity[1, 2]. By combining the two techniques, the resulting space-time multiuser detectors can further improve the system capacity over traditional time-domain detectors and enhance the quality of service provided to all the users[3, 4].

Multiuser detection can be regarded as the classification and approximation problems. Aazhang and Kechiotis *et al.* proposed multiuser detectors based on BP neural networks and Hopfield neural networks, respectively[5, 6]. But these detectors still exist some inherent problems, *e.g.* slower convergence speed, more difficult choice of bases function, and more probability of being trapped in local minima. Wavelets show promise for both signal representation and classification[7, 8], both of which can be viewed as feature extraction problems in which the goal is to find a set

F. Yin, J. Wang, and C. Guo (Eds.): ISNN 2004, LNCS 3174, pp. 236–241, 2004.

of daughter wavelets that either best represent the signal or best separate various signal classes in the resulting feature space. H. H. Szu proposed a kind of feedforward adaptive wavelets networks(AWN)[7], parameters of which are iteratively computed to minimize an energy function for representation or classification.

In this paper, we consider the space-time multiuser detection combined with adaptive wavelet networks(STMUDCAWN) over frequency selective fading channels. Due to the excellent classification and approximation ability of the AWN with highly parallel structure and adaptability to system parameters, we choose the space-time matched filter to reduce the computational complexity of the proposed detector. It will be shown that the STMUDCAWN can suppress the MAI and solve the "near-far" problem effectively. On the other hand, the computational complexity of the AWN component of the detector is linear with the number of its hidden layer.

2 Signal Model

It is assumed that there are K simultaneous users with common carrier frequency in a cellular and the digital modulation technique is BPSK. If antenna arrays have P elements, the receiver can be modeled as a single input and multiple output system. So the received signal vector over frequency selective fading multipath channels is

$$\mathbf{r}(t) = \sum_{i=0}^{M-1}\sum_{k=1}^{K}\sqrt{w_k}\, b_k(i)\sum_{l=1}^{L}\mathbf{a}_{kl}\beta_{kl}s_k(t - iT - \tau_k - lT_c) + \sigma\mathbf{n}(t) , \quad (1)$$

where M is the number of data symbols per user. $\sqrt{w_k}$, $s_k(t)$, and τ_k denote the amplitude, normalized signature waveform, and relative delay of the kth user, respectively. L is the number of resolvable paths for each user. $b_k(i) \in \{\pm 1\}$ is the ith transmitted symbol by user k. T is symbol interval. $\mathbf{a}_{kl} = [a_{kl1}, \cdots, a_{klP}]^T$ and β_{kl} is, respectively, the array response vector and zero-mean complex Gaussian fading gain. $\mathbf{n}(t) \sim \mathcal{N}\left(\mathbf{0}, \sigma^2\mathbf{I}\right)$, where $\mathbf{I}$ denotes the identity matrix.

3 Space-Time Multiuser Detection Combined with Adaptive Wavelet Networks (STMUDCAWN)

The optimal multiuser detector is a NP-complete problem[1]. With more users, it will confront the "exponential expansion" and can hardly applied in the practical CDMA system. Many researchers on neural networks show that neural networks can solve the problem excellently and is easy to implement by ASIC. Adaptive wavelets networks are highly interconnected networks of relatively simple processing units operating in parallel[7]. The massive parallelism of neural nets and their robustness to the characteristics of the problems under consideration make them desirable for solving various complex tasks. The AWN with the simple structure contains the nonlinearity of wavelets function in the artificial neurons rather than that of the standard sigmoidal function. We consider the space-time multiuser detector combined with the AWN for approximating the solution to the optimal detector over frequency selective fading channels. To reduce the computational complexity of the STMUDCAWN, we choose the space-time matched filter with low computational load as the pretreatment. The block diagram of the STMUDCAWN is illustrated in Fig. 1.

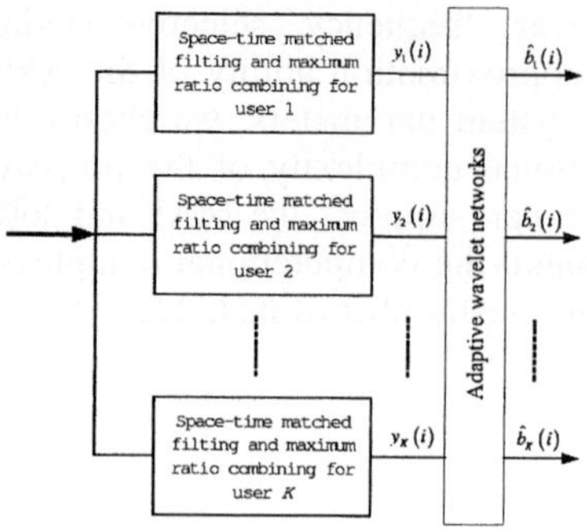

Fig. 1. Block diagram of the STMUDCAWN

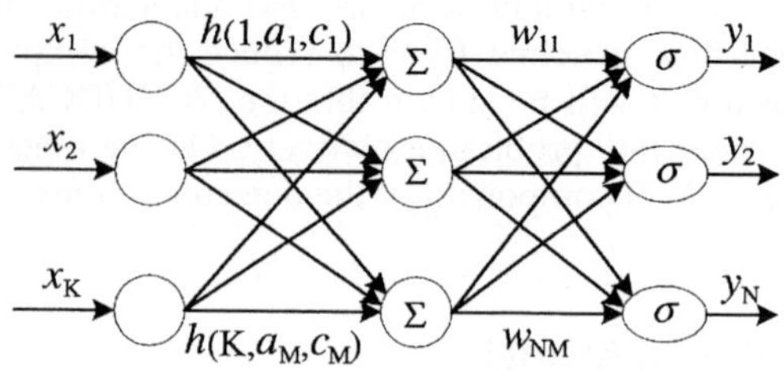

Fig. 2. Adaptive wavelets networks classifier

3.1 Space-Time Matched Filtering and Maximum Ratio Combining

The received signal is first matched with the estimated direction vector corresponding to L propagation paths at the space domain. We can obtain

$$z_{kl}(t) = \mathbf{a}_{kl}^H \mathbf{r}(t) \ . \tag{2}$$

And then the signal from L branches is correlated with respective delay version of the signature waveform of the kth user at the time domain. It operates as the following formula.

$$q_{kl}(i) = \int_{-\infty}^{\infty} z_{kl}(t) s_k(t - iT - \tau_k - lT_c) dt \ . \tag{3}$$

By maximal ratio combining technique, we can obtain the output of the space-time matched filter and maximum ratio combiner.

$$y_k(i) = \sum_{l=1}^{L} \beta_{kl}^* q_{kl}(i) \ . \tag{4}$$

3.2 Adaptive Wavelets Networks(AWN) and Learning Algorithm

Representation and classification can be viewed as the feature extraction problem in which the goal is to find a set of daughter wavelets that either best represent the signal or best separate various signal classes in the resulting feature space. Specially, the extraction of features is the inner products of a set of wavelets with the input signal. And then these features can be input to classifier. The key problem is which wavelets should be selected and how to select. We extend the AWN classifier proposed by

H. H. Szu [7] to the multiple classification network illustrated in Fig. 2, the wavelets feature component of which uses wavelets weights rather than wavelets nonlinearity of the representation network.

The ith output of the network is

$$y_i(t)=\sigma\left[\sum_{j=1}^{M} w_{ij}\sum_{k=1}^{K} x_k(t)h\left(\left(k-c_j\right)/a_j\right)\right] , \quad j=1,\ldots,N , \tag{5}$$

where w_{ij}, a_i, and c_i are the weight coefficients, dilations, and translations for each daughter wavelet which can be optimized by minimizing an energy function. K, M, and N are the node number of input, hidden, and output layer, respectively. $x_k(t)$ and $y_i(t)$ is the kth input and the ith output element for networks, respectively. $h(t)$ is the mother wavelet function. And $\sigma(x)=1/[1+\exp(-x)]$ is a sigmoidal function. The wavelets network classifier parameters w_{ij}, a_i, and c_i can be optimized by minimizing the total error energy function as follows.

$$E=\frac{1}{2}\sum_{q=1}^{P}\sum_{i=1}^{N}\left(d_i^q-y_i^q\right)^2 , \tag{6}$$

where P is the number of the training samples, d_i^q is the desired value of y_i^q .We choose $h(t)=\cos(1.75t)\exp(-t^2/2)$ as the mother wavelet for classification. It can be proved that this function confirms the frame condition[7]. Let $f(t)=\sin(1.75t)\exp(-t^2/2)$ and $t'=(k-c_j)/a_j$, then the gradients of E become

$$\partial E/\partial w_{ij}=-\sum_{q=1}^{P}\sum_{i=1}^{N}(d_i^q-y_i^q)y_i(1-y_i)w_{ij}\left(\sum_{k=1}^{K}x_k h(t')\right) , \tag{7}$$

$$\partial E/\partial c_j=-\sum_{q=1}^{P}\sum_{i=1}^{N}(d_i^q-y_i^q)y_i(1-y_i)w_{ij}\sum_{k=1}^{K}x_k\cdot\left\{\left[1.75f(t')+h(t')\right]t'/a_j\right\} , \tag{8}$$

$$\partial E/\partial a_j=t'g(c_j) . \tag{9}$$

Based on the formulas above, it is straightforward to deduce the conjugate gradient method for iteratively updating the parameters of the AWN.

4 Simulation Results and Performance Analysis

Several simulation experiments are presented to illustrate the effectiveness of the STMUDCAWN in comparison with the space-time multiuser detector combined with the conventional BP neural networks employing the sigmoidal function (STMUDCBPNN), the space-time matched filter(STMF), the conventional RAKE receiver(RAKE) which combines the multipath signal of the interesting user received by single antenna to improve signal interference plus noise ratio (SINR) over multipath fading channel, and the conventional detector (CD) which directly decides on the output of the matched filter at time domain in single antenna over single path channels.

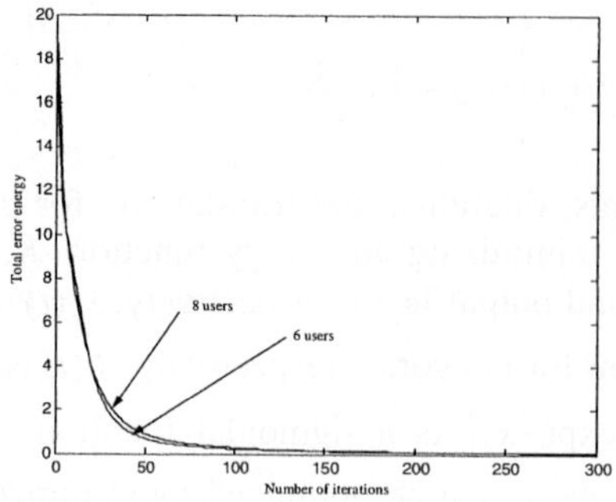

Fig. 3. Total error energy versus iterations number

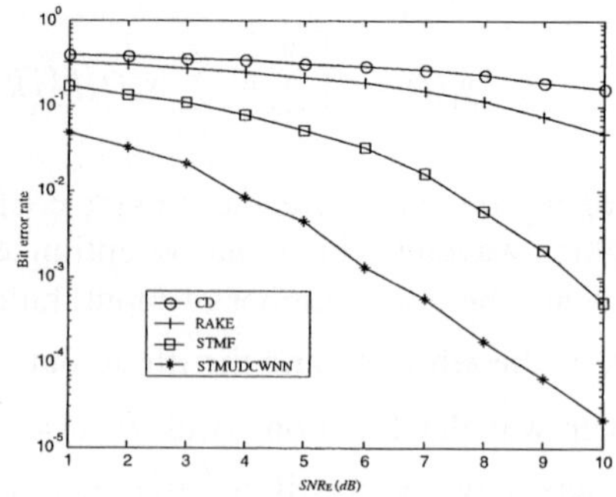

Fig. 4. BER versus SNR with 6 users

In the following simulation, the signature waveform of each user is Gold sequence with the length of 31. The number of paths of each user, of antenna elements, and of nodes in hidden layer, respectively, is 3, 3,10. The fading gain and the DOA of each path of each user are randomly generated and keep unchanged in all experiments. Define Signal-to-Noise Ratio(SNR) of user k as SNR_k. The interesting user is user 1 while all other users are the interfering users, that's SNR_E=SNR_1, SNR_I=$SNR_{2\sim K}$. Without loss of generalization, we only demodulate the data bits of the desired user thus letting the node number of output layer of the AWN N=1.

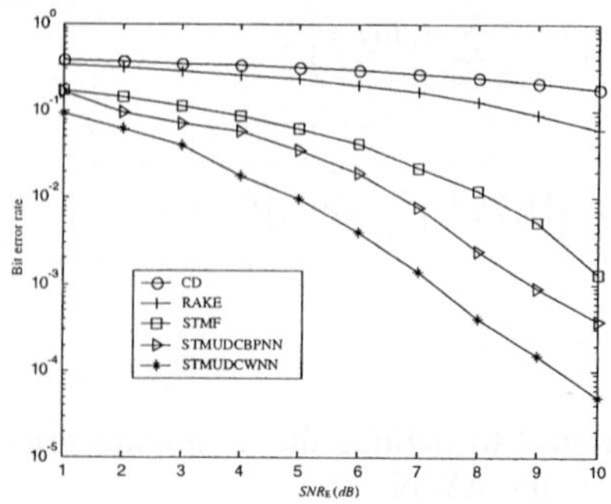

Fig. 5. BER versus SNR with 8 users

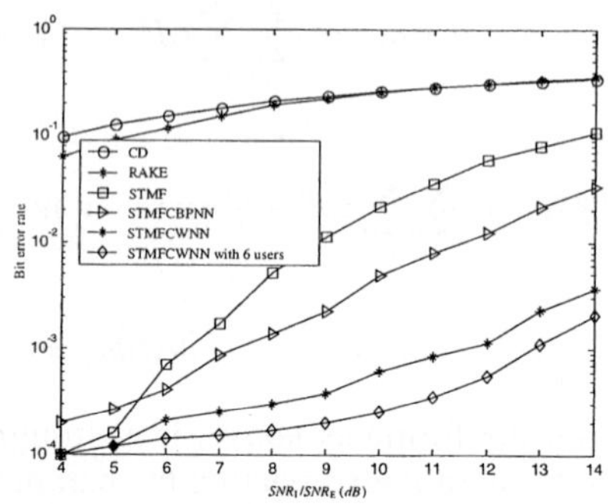

Fig. 6. BER versus "Far-to-Near ratio"

The convergence performance of the STMUDCAWN in the system with 8 users is shown in Fig. 3. The SNR of the interesting user and all interfering users are 6dB and 17dB, respectively. The number of the training samples is Q=100. From the figure, the STMUDCAWN converge at the steady state and the precision of the network is very high after 100 iterations.

The performance of bit error rate (BER) versus SNR in the systems with 6 and 8 users, shown in Fig. 4 and 5, respectively. SNR_I=17dB keeps unchanged while SNR_E changes from 1 to 10dB. Under the environment, the MAI is very strong.

The performance of suppressing multiple access interference(MAI) and near-far resistance in the system with 6 and 8 users is stimulated in Fig. 6, where SNR_E=8dB keeps unchanged while the "Far-to-Near ratio"(FNR), defined as SNR_I / SNR_E, changes from 4 to 14dB.

From Fig. 4 to 6, we can see that the BER performance of suppressing MAI and background noise of the STMUDCAWN is far better than that of the STMUDCBPNN, CD, RAKE and STMF. On the other side, when the number of users is different, we also can see that the BER performance of the proposed method will become worse slightly as the number of users increases.

5 Conclusions

The space-time multiuser detection combined with adaptive wavelet networks over frequency selective fading multipath channels have been presented in this paper. The parameters of wavelet and the weight efficient of the output layer of the network can be adaptively computed by minimizing the energy function. The structure of the multiuser detector is simple and the computational complexity of the detector mostly lies on that of wavelets networks. Once the parameters of the AWN are obtained by training, its complexity is linear with the number of the nodes of the hidden layer. Simulation results show that the proposed space-time multiuser detector can converge at the steady state rapidly and can suppress the MAI and solve the "near-far" problem effectively.

References

1. Verdu, S.: Multiuser Detection. Cambridge Univ. Press, New York (1998)
2. Liberti, J.C., Jr., Rappaport, T.S.: Smart Antennas for Wireless Communications: IS-95 and Third Generation CDMA Applications. Prentice Hall PTR, New Jersey (1999)
3. Wang, X., Poor, H.V.: Space-Time Multiuser Detection in Multipath CDMA Channels. IEEE Trans. on Signal Processing. 9 (1999) 2356–2374
4. Sfar, S., Murch, R. D., Letaief, K. B.: Layered Space-Time Multiuser Detection over Wireless Uplink Systems. IEEE Trans. on Wireless Communication. 4 (2003) 653–668
5. Aazhang, B., Paris, B.P., Orsak, G.C.: Neural Networks for Multiuser Detection in CDMA Communications. IEEE Trans. on Communication. 7 (1992) 1212–1222
6. Kechriotis, G., Manolakos, E.S.: Hopfield Neural Network Implementation of the Optima CDMA Multi-user Detector. IEEE Trans. on Neural Networks. 1 (1996) 131–141
7. Szu, H.H., Telfer, B., Kadambe, S.: Neural Network Adaptive Wavelets for Signal Representation and Classification. Optical Engineering. 9 (1992) 1907–1916
8. Reyneri, L.M.: Unification of Neural and Wavelet Networks and Fuzzy Systems. IEEE Trans. on Neural Networks. 4 (1999) 801–814

Optimizing Sensor Node Distribution with Genetic Algorithm in Wireless Sensor Network

Jianli Zhao, Yingyou Wen, Ruiqiang Shang, and Guangxing Wang

School of Information Science & Engineering, Northeastern University, Shenyang, China
tsingdao@hotmail.com

Abstract. Wireless sensor networks have recently emerged as a premier research topic. They have great longterm economic potential, ability to transform our lives, and pose many new system-building challenges. One of the fundamental problems in sensor networks is the calculation of the coverage. In this paper, an optimal distribution based on Genetic Algorithm is proposed in the initial planning of sensor network. Moreover, a new optimizing algorithm of sensor node distribution is designed by utilizing node topology in sensor network, which provides a sound effective means for the topology management of sensor network. Simulation shows that this efficient algorithm optimally solved the best-coverage problem raised in [1].

1 Introduction

Wireless sensor network [2][3] is a network that is composed of mobile portable equipments, which has the ability of computing, storage and wireless communication. The nodes in wireless sensor network are self-organize, which can collect information or send data by multi-hop routing in special case. Because of the wireless sensor network's speediness, agility, robust and adaptive survival, it will be used to military and civil field widely. In wireless sensor network, node distribution and topology are important to improve the network's survival. In a large-scale random scattering network, for getting a better node distribution, more redundancy sensor nodes are needed which can make sure a best coverage. But in some sensor network nodes can be set specially, which can not only reduce the redundancy nodes, routing request and maintenance overhead, power consuming but also expend the network's sensing range. So how to get the optimized node distribution is an important problem in the wireless sensor network. And by the mobile sensor node's adjusting to the network topology, the network topology discovery and node position's obtaining can effectively avoid the shadow and blind point in sensing area. Then the optimization of node distribution is an effective approach to improve the sensor network performance. Although there are a lot research about sensor network's topology and orientation, few are about distribution optimization algorithm by the current topology. Aimed at this problem, we give the optimal sensor node distribution and distribution optimization algorithm in sensor network using Genetic Algorithm advantage to resolve the problem.

F. Yin, J. Wang, and C. Guo (Eds.): ISNN 2004, LNCS 3174, pp. 242–247, 2004.

2 Optimal Sensor Node Distribution in Sensor Network

As the limited sensing rage of nodes in wireless sensor network, obtaining a better network coverage area must depend on reasonable node distribution. In a given object area, how to get the better node distribution belongs to the optimal problem. Genetic Algorithm has been attached importance to this research because of its advantage [4][5].

Convenient for problem's description, we define position area sensing parameter. Firstly object area is to be discrete by some precision, a point represents a little local area. According to geographical environment parameter and corresponding radio propagation mode, every node is assigned a sensing parameter; different sensing parameter means that the sensor node at this point has different available sensing range. Optimal node distribution problem translates into that how to gain the best coverage in the object area.

2.1 Coding Mapping of Optimal Sensor Node Distribution in Genetic Algorithm

Aimed at the problem of sensor node distribution in wireless sensor network, we adopt a coding scheme based on node coordinate. Sensor node position in object area is replaced by its coordinate. Using GPS, sensor network can be organized by the translating from node coordinate to longitude and latitude. These gene expressions match the practice project and can reflect applied environment peculiarity of sensor node intuitively. If every chromosome of individual is composed of N genes in Genetic Algorithm solution space, every gene represents the position of a sensor node. Two genes of gene group express the X-axis direction and Y-axis direction of the node in a coordinate respectively.

Using coding mapping, we set up the initial population in Genetic Algorithm. Given object area:

$$\{\ S(x,y):\quad 0 \le x \le Length \quad 0 \le y \le Width\ \} \tag{1}$$

Here, Length and Width are the length and width of the object area respectively. The population is composed of T individuals. There are N sensor nodes, and then the initial population is as follows:

$$\begin{cases} Pop_1 = \{g_{11}, g_{12}, g_{13}, \cdots\cdots g_{1j}, \cdots g_{1N}\} \\ Pop_2 = \{g_{21}, g_{22}, g_{23}, \cdots\cdots g_{2j}, \cdots g_{2N}\} \\ Pop_3 = \{g_{31}, g_{32}, g_{33}, \cdots\cdots g_{3j}, \cdots g_{3N}\} \\ \quad\vdots \\ Pop_i = \{g_{i1}, g_{i2}, g_{i3}, \cdots\cdots g_{ij}, \cdots g_{iN}\} \\ \quad\vdots \\ Pop_T = \{g_{T1}, g_{T2}, g_{T3}, \cdots\cdots g_{Tj}, \cdots g_{TN}\} \end{cases} \tag{2}$$

Here, $g_{ij} = (x_{ij}, y_{ij})\ (1 \le i \le T, 1 \le j \le N)$ is the node coordinate of the No.j gene in No.i individual.

2.2 Crossover, Mutation, and Selection in Genetic Algorithm

Crossover can bring new population in Genetic Algorithm. To make the gene in parents surviving in the offspring as possible and avoid the obtained optimal solution losing, we adopt one approach of circle crossover. If there are K individuals to propagate in parents: (Pop_1, Pop_2 , Pop_3,··· Pop_K); for the first time selected (Pop_1, Pop_2) to be parents, the second time selected (Pop_2, Pop_3) to be parents,··· the ith time selected (Pop_i, Pop_{i+1}), then (K-1) offspring come forth. In this way, the gene of every parent will survive in offspring as possible. The detailed process of crossover is as follows: Select K individuals to propagate using crossover probability P_c in parents. Given parents are (Pop_i, Pop_{i+1}), equipotent gene groups are g_{ik}, $g_{(i+1)k}$ respectively, then the node coordinate which the equipotent gene group Chd_{ik} of offspring individual represents is as follows:

$$\begin{cases} x_{Chd_{ik}} = (x_{Pop_{ik}} + x_{Pop_{(i+1)k}})/2 + \psi_x & \psi_x \sim N(0, x_{Pop_{(i+1)k}} - x_{Pop_{ik}}) \\ y_{Chd_{ik}} = (y_{Pop_{ik}} + y_{Pop_{(i+1)k}})/2 + \psi_y & \psi_y \sim N(0, y_{Pop_{(i+1)k}} - y_{Pop_{ik}}) \end{cases} \quad (3)$$

Here, ($x_{Pop_{ik}}$, $y_{Pop_{ik}}$) and ($x_{Pop_{(i+1)k}}$, $y_{Pop_{(i+1)k}}$) is the coordinate of parents node; ψ_x, ψ_y obey Gauss Distribution that average value is 0.

Mutation can bring new content for population. To avoid a big random alteration and keep the obtained solution of optimal node distribution, we just only use mutation to the offspring individuals which parents propagate. Select gene group to mutate by probability P_m from gene groups which offspring individuals make up of. If the sensor node coordinate of which the primary gene group represents is (v1, v2), it is mutated to $(v_1 + \psi_1, v_2 + \psi_2)$, here ψ_1, ψ_2 obey Gauss Distribution.

Selection generates next generation population based on all parents and offspring, i.e. expands sampling space. This sampling space allows improving Genetic Algorithm performance using bigger crossover probability and mutation probability and won't bring too much random alteration.

To avoid being limited to local optimal solution because the algorithm converges too fast, we expand the offspring sampling space to the sampling space that is composed of all parents and offspring every several generations. By this transforming, one side the better genes of parents will be kept in the offspring; on the other hand, define the searching space newly to avoid the local optimal solution. In the initial running stages of the algorithm, it needs a bigger searching space, but in late it needs to reduce searching space for a fast convergence. We adopt the subsection function to ascertain the step length of sampling space.

$$Length(n) = \begin{cases} ([n/phase]+1) * step_1 & n \le n_{phase} \\ step_2 & n > n_{phase} \end{cases} \quad (4)$$

Here, n is the generation number that the algorithm runs, n_{phase} is the generation number of phase running, *phase* is the generation number of sub-phase running; *step1*, *step2* are the step length at different phase.

3 Optimal Sensor Node Distribution in Sensor Network

In the course of forming of the sensor network, especially in some emergency or the situation with limited condition, such as military field and remote area, usually adopt random scattering way to finish the sensor nodes distribution of the network. Though having a better network sensing range by a lot of redundancy nodes, to the application of sensor network, it can avoid shadow and blind point effectively in sensing area by the discovery of network topology, management and the localization technologies of nodes, utilizing mobile sensor node adjustment ability to the topological structure of the network. This is an effective approach to improve the whole performance of sensor network.

3.1 Topology Discovery and Node Localization in Sensor Network

The network topological structure management of sensor network has always been a research topic of the sensor network. Its data-centric application requires that the nodes that obtains effective information must understand the sensor distribution centered at sink node, which makes that the gained information parameter is corresponding to the practical sense environment. Traditional algorithm of sensor network topology discovery gets the information of neighbors by radio broadcast between nodes, and completes the topological structure by the diffusion of topology discovery grouping of sink node [6][7].

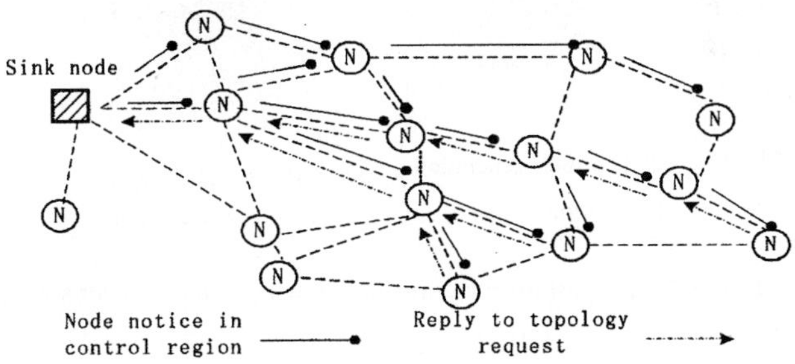

Fig. 1. The topology discovery centered at sink node in wireless sensor network

The process of topology discovery is described in Fig.1, firstly sink node broadcasts topology discovery group, and sensor nodes that received the group memorize the address of sink and notice the source address, setting the timeouts at the same time, which makes the node confirming its logical address in the topological tree. And then transmits the broadcast notice group. If a node receives notice group from different nodes, it adjusts the record of the upstream node to gain the minimum spanning topological structure according to the number of hops in the notice. Leaf node notices its node Id to upstream node. Every node transmits the information that is collected form downstream nodes to its upstream nodes respectively. Thus sink node gains the minimum spanning topological tree.

Localization technology is popular especially in military fields. In sensor network node self-localization adopts global positioning system technology. To some application that needn't a high precision, it can use local position algorithm to realize localization [8][9].

3.2 Optimal Sensor Node Distribution Based on Topology Discovery and Node Localization

It can effectively adjust network structure and realize reconfiguration topology by the aforesaid topological discovery and localization technology of the node because of the mobile characteristic of some nodes in sensor network, which can make the network having an optimal coverage. This is an important part in wireless sensor network management. So, we adjust the aforementioned algorithm of sensor node distribution: in the initial stage of the algorithm the stable node position or low mobility node position is wrote down to gain the optimal node distribution solution under some stable nodes and low mobility nodes through this algorithm. In general, the more stable nodes exist in the network, the smaller the optimal space is and the less the difficulty of optimal adjustment is. In extreme case, optimizing the network that has only one stable node is very difficult. So, we simulate this case by distribution optimal algorithm.

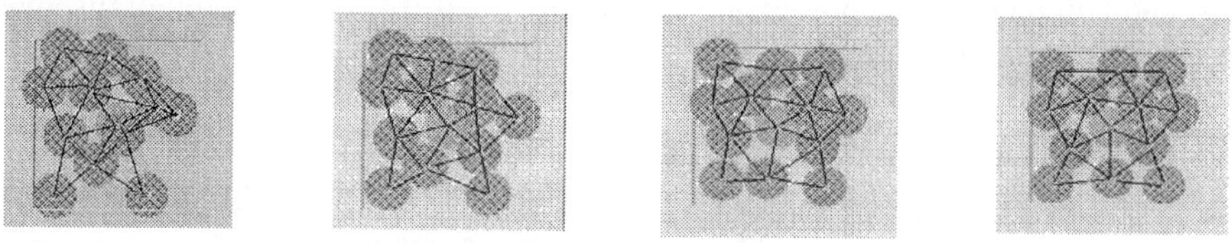

(1) Initial distribution (2) The 100th generation distribution (3) The 200th generation distribution (4) The 400th generation distribution

Fig. 2. The simulation of optimal distribution of 13 sensor nodes

There are 13-sensor nodes in a 1.5×1.5Km square object area. The initial distribution is as Fig 2(1). The sensor node in the object area centre can't move, but other nodes can. The precision of object area dispersing is 10 m, for simplifying, we suppose that the node have the same sensing parameter, namely in any discrete area, the effective sensing radius of each sensor node is 200 meters. Using the same genetic operation parameter, the algorithm runs for 400 generations. In Fig 2, the sensing range of sensor node in the centre of square area is showed by undertone shadow. Simulation shows, the algorithm can effectively adjust mobile node position to get an optimal network topological structure and best network coverage according to stable node distribution. So, it can provide reliable and exact information to the management and adjustment of practical wireless sensor network.

4 Conclusion

In sensor networks, a reasonable distribution of sensor nodes and dynamic topology adjustment will do much good to information aggregation and network survival. To solve this problem, an optimal distribution based on genetic algorithm is proposed in the initial planning of sensor network. Moreover, a new optimizing algorithm of sensor node distribution is designed by utilizing topology management and node localization in sensor network, which provides a sound effective means for the topology management of sensor network. Simulation shows that this algorithm can achieve optimal node distribution in the object area. The research in the future will concentrates on the optimization of the route algorithm of the sensor network mainly.

References

1. Seapahn Meguerdichian, Farinaz Koushanfar, Miodrag Potkonjak, Mani Srivastava: Coverage Problems in Wireless Ad-hoc Sensor Networks. IEEE INFOCOM 2001, Vol 3. IEEE Press, Washington DC（2001）1380–387
2. Hill J., R., Szewczyk, A. Woo, Hollar, S. Culler, D. and Pister, K. System Architecture Directions for Networked Sensors. Proceedings of the 9th ACM International Conference on Architectural Support for Programming Languages and operating Systems. ACM Press, New York N.Y. (2000) 93-104
3. Pattie, G., Kaiser, W.: Wireless Sensor Networks. Communications of the ACM, Vol.43. ACM Press, New York N.Y. (2000) 51–58
4. Lawrence, Handbook of Genetic Algorithms.Van Nostrand Reinhold, New York N.Y. (1996)
5. Calegari ,P, Cuidec F: Genetic Approach to Radio Network Optimization for Mobile Systems. Proceedings of the 47th IEEE VTC. IEEE Press, Washington DC (1997) 755-759
6. Deb, B., Bhatangar, S., Nath, B. : A Topology Discovery Algorithm for Sensor Networks with Applications to Network Management. Proceedings of the IEEE CAS Workshop on wireless Communications and Networking. IEEE Press, Washington DC (2002)
7. Chih-fan Hsin, Mingyan Liu: A Distributed Monitoring Mechanism for Wireless Sensor networks. Proceedings of the ACM workshop on Wireless security. ACM Press, New York N.Y. (2002) 57 - 66
8. Joe Albowicz, Alvin Chen, and Lixia Zhang: Recursive Position Estimation in Sensor Networks. Proceedings of the International Conference on Network protocols (ICNP '01). IEEE Computer Society, Washington, DC, Washington DC (2001) 35–41
9. Nirupama Bulusu, Deborah Estrin, Lewis Girod, John Heidemann: Scalable Coordination for Wireless Sensor Networks: Self-Configuration Localization Systems. Proceedings of the 6th IEEE International Symposium on Communication Theory and Application. IEEE Press, Washington DC (2001)

Fast De-hopping and Frequency Hopping Pattern (FHP) Estimation for DS/FHSS Using Neural Networks

Tarek Elhabian, Bo Zhang, and Dingrong Shao

School of Electronics & Information, Beijing University of Aeronautics & Astronautics, China.
{Tareksalah2002, shaodingrong}@hotmail.com

Abstract. A Fast de-hopping and FHP estimation model for Direct Sequence/Frequency Hopping Spread Spectrum (DS/FHSS) system is proposed. The Neural Networks (NNs) were used to mimic the Parallel Matched Filtering (PMF). The signal samples and its Fast Fourier Transform (FFT) were used for Back propagation Neural Network (BNN) training. The FH patterns designated as concatenated prime codes [8] were used for the Radial Basis Function (RBF) training. Computer simulations show that the proposed method can effectively identify the frequency and estimate its pattern. Small hardware resources compared with PMF hardware.

Keywords. Acquisition, Direct Sequence, Frequency Hopping Pattern, Neural Network, Matched Filter, Fourier transform.

1 Introduction

Restricted for a long time to the military domain, SS techniques are now used in more non-military applications. They are also proposed for many digital communication systems. The most widely techniques used being DS [1] and FH [5]. DS techniques use a pseudo-noise sequence PN, which is multiplied by the signal to spread it directly. FH techniques the available channel bandwidth is divided into a large number of contiguous frequency slots, and among them one is chosen for the transmission in any signal interval. The choice is done by a PN code. The decision of switching between acquisitions and tracking mode in DS or FH [7] receiver system is very difficult problem. FH has been extensively studied as SS technique for interference avoidance as opposed to interference attenuation achieved by other SS techniques [5] using NN [3]. The DS/FH signal can be expressed as follows:

$$s(t)=\sum_{n} d_n(t)\cdot e^{j(2\pi f_n+\varphi_n)}\sum_{m=1}^{N_c} c_{nm}\cdot p(t-mT_c-T_h)\quad , T_h=N_cT_c . \tag{1}$$

Where: f_n is the nth hopping frequency, φ_n is the phase, T_h is the hopping time; $C_{nm}(t)$ is the DS code, T_c is the code chip duration of PN code, $d_n(t)$ is the information data.

The received DS/FH signal as follows:

F. Yin, J. Wang, and C. Guo (Eds.): ISNN 2004, LNCS 3174, pp. 248–253, 2004.

$$s_r(t) = s_1(t) + \sum_{i=2}^{k} s_i(t) + n(t) + j(t), \qquad s_1(t) = m(t)\cos(2\pi f_n t + \varphi_n) \tag{2}$$

Where, *i* is the user number of the system, *m(t)* is the DS signal (*m(t)= d(t)·c(t), c(t)* is PN code and *d(t)* is the information signal); φ_n is the phase; *n(t)* is the noise; *j(t)* is the jammer interferences; $s_1(t)$ is the desired signal. So the $s_r(t)$ must be multiplied by the local frequency to get the de-hopped signal information. The paper is organized as follows: The FH conventional acquisition techniques and model description are presented in Sections 2 and 3. In Sections 4 and 5, the signal frequency recognition and FH pattern estimation subsystems are presented. The experiments of simulations and results are presented in Section 6. The conclusions are mentioned in Section 7.

2 Conventional Techniques for FH Acquisition

The most techniques used for FH acquisition are: (1) Serial search, which the receiver generates the PN code that represent one of the frequency values and wait until one of the coming frequency hop match with it. (2) Parallel search, which the receiver generates all frequencies (n hopping frequency) using n DDS and tests them with the coming frequency at the same time using separate matched filters. Each of the matched filters is adjusted on certain code. Often, the serial search is used although it takes very long acquisition time. The better way to see the frequency hops at the receiver is to use a time-frequency representation by using the FFT [3].

3 Model Description

Fig. 1 contains two subsystems for de-hopping and pattern estimation. The first one is Signal Frequency Recognition (SFR) using the BNN, the second one for the Frequency Hopping Pattern Estimation (FHPE) using the RBF.

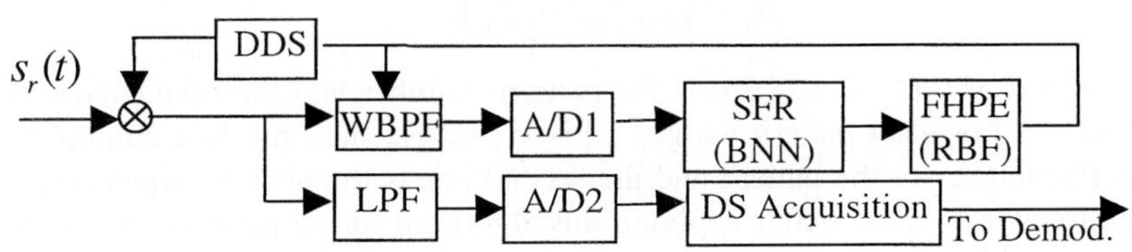

Fig. 1. DS/FH receiver

At starting time the received signal is down converted to IF using the Wide Band Pass Filter (WBPF) until the FHPE estimate the frequency pattern and then control the DDS to generate the corresponding frequency of the pattern.

4 Signal Frequency Recognition

For each signal, there are two vectors (input vector V^i and desired vector FFT^i). The vector V^i has a set of 30 sampled values and the vector FFT^i has the corresponding FFT points as described in Equations 3 and 4.

$$V^i = (v_1^i, v_2^i, \cdots, v_m^i). \tag{3}$$

$$FFT^i = ((r_1^i + jc_1^i), (r_2^i + jc_2^i), \cdots, (r_m^i + jc_m^i)). \tag{4}$$

Where $i = (1,2,\ldots,25)$ is the signal number and $m = (1,2,\ldots,30)$ is the sample number or the FFT points. Each sampled value at vector V^i was represented by 8 bits (A/D1 output) as in Equation 5. Also the FFT^i vectors were rearranged such that the real values and then corresponding imaginary values in one vector as in Equation 6.

$$IS_m^i = (bv_1^i, bv_2^i, \cdots, bv_m^i),\ bv_m^i = (b_0, b_1, \cdots, b_7)_m^i. \tag{5}$$

$$IFFT^i = (r_1^i, r_2^i, \cdots, r_m^i, c_1^i, c_2^i, \cdots, c_m^i). \tag{6}$$

The matrixes ISS^i and FFT^i as in Equations 7 and 6 are used as input/desired vectors for BNN training to recognize at any phase shift the corresponding FFT point's vector. The BNN was trained using the normal training method, Momentum learning method and VLR or adaptive method to improve the convergence time.

$$ISS^i = \left[IS_1^{iT}\ IS_2^{iT}\ \cdots\ IS_m^{iT} \right],\ IS_m^{iT} \text{ is the transpose of } IS_m^i. \tag{7}$$

5 Frequency Hopping Pattern Estimation

The codes from concatenated prime code $GF(5^2)$ [8] were used as FH patterns for the simulation. Each pattern has different frequency values as in Equation (8).

$$p_n = \left[f_1, f_2, \cdots, f_m \right]. \tag{8}$$

Where n and $m = (1, 2, \ldots, 25)$ is the patterns number and the frequencies number respectively. For each pattern we got a [5×25] matrix that, the first column has the first 5 frequencies of the pattern and the second one is the next 5 frequencies shifted by one frequency value. After applying this algorithm for all patterns we got the matrix ipp [5×625] as in Equations 9 and 10.

$$pp_n = \left[p_1^T\ p_2^T\ \cdots\ p_n^T \right],\ p_n^T \text{ is the transpose of } p_n. \tag{9}$$

$$ipp = \left[pp_1\ \ pp_2\ \cdots\ pp_n \right],\ n=1,2,\ldots,25. \tag{10}$$

So the number of columns are 625, each 25 columns related to one frequency pattern as described in Equation (10) are used to train the RBF network. If the RBF network got one of these columns values, the FHPE will estimate that, it relates to the pattern number n and generate the next frequency.

6 Experiments and Results

Using 30 shifted sampled values for 25 signals to test the BNN trained by different algorithm to recognize the signal. Other experiments were done to test the performance of RBF, by using *ipp* matrix vectors. The other RBF experiments were done with different parameter for FHP estimation purpose. The objectives of experiments were measure the performance of the BNN network for frequency identification when change the number of hidden layer neurons and changing the training method and measure the performance the RBF for FHP estimation with signal interferences, study the effect of using Euclidean Distance (ED) and Hamming Distance (HD) on the RBF performance with changing the frequency space that the frequencies values of the patterns were selected.

6.1 Experiments and Results of SFR

Table 1 shows the recognition rate against the number of hidden neurons at different training methods. All BNN training methods made a high recognition rate at small number of hidden neurons. The VLR with momentum algorithm is superior over the normal, VLR without momentum and momentum algorithm.

Table 1. Recognition rate at different BNN training algorithms

No of neuron / Rec rate%	5	7	10	30	50	100	150	200	400	500	600
Normal	80.3%	86.0%	87.0%	100%	100%	100%	100%	100%	100%	100%	100%
VLR	90.0%	93.0%	94.0%	100%	100%	100%	100%	100%	100%	100%	100%
Momentum	84.0%	87.0%	89.0%	100%	100%	100%	100%	100%	100%	100%	100%
VLR+Mom	91.0%	94.0%	98.0%	100%	100%	100%	100%	100%	100%	100%	100%

Table 2. The time consumed in the BNN training methods using the tanh sigmoid function

No of neuro / Time in min.	5	7	10	30	50	100	150	200	400	500	600
Normal	3.08	3.24	3.39	5.31	6.75	11.13	12.72	9.68	4.46	4.11	3.87
VLR	3.15	3.39	3.48	5.14	6.64	10.8	9.76	3.35	2.53	2.89	3.22
Momentum	3.06	3.29	3.58	5.05	6.68	11.18	16.10	11.2	6.23	7.34	6.88
VLR+Mom	3.04	3.18	3.36	4.95	6.48	10.84	1.97	1.14	0.64	0.69	0.81

Table 2 shows the VLR without and with momentum training methods were speed up the convergence time especially the second one was the most speed. This speed up convergence is noticeable at large numbers of hidden neurons.

Fig. 2 illustrates the relation between the Mean Square Error (MSE) and time for each BNN training method. From left to right and up to down the 4th figure show how fast decreasing of the MSE when applying the VLR with momentum training algorithm compared to the others training methods because of at this algorithm the momentum

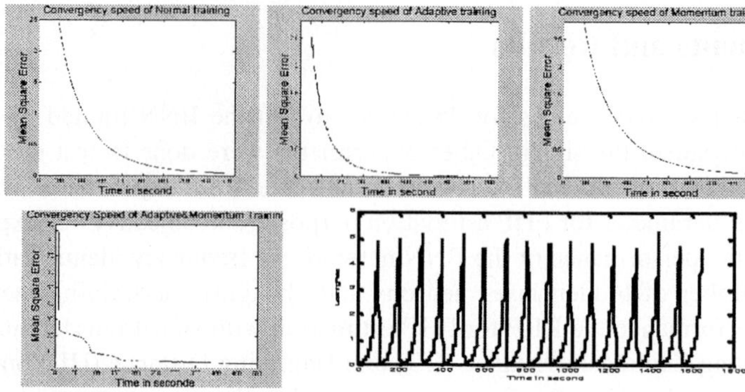

coefficient and the learning rate are adapted continuously according to the MSE value at previous iteration as shown in 5th figure.

Fig. 2. MSE against time at different training methods and the learning coefficient against time

6.2 Experiments and Results of FHPE

625 five-element vectors were used to test the RBF to estimate the FH pattern. Fig. 3 shows that, when using large space of frequency to select the frequency values of the FH patterns gave a lower probability of error especially if using the HD rather than the ED and how much the probability of error was decreased when the HD rule were used.

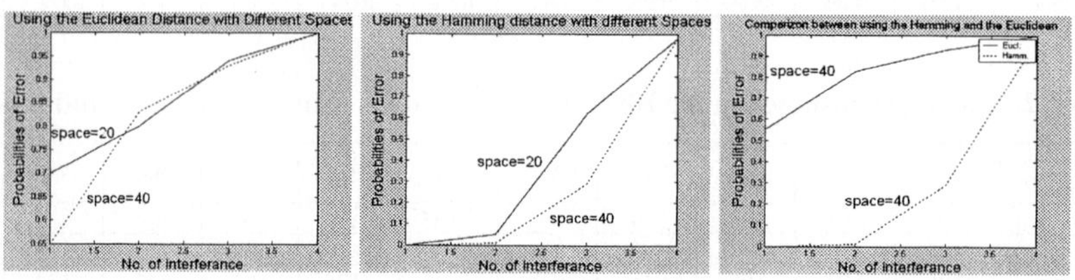

Fig. 3. Probability of error against the interference when using ED and HD

At 600hop/sec, the experiments have been shown that, without interference, the FHPE can correctly estimate the FH pattern after $2T_h$, and with interference is $5T_h$.

7 Conclusions

An efficient and fast FH acquisition and FHP estimation techniques have been presented. First the BNN was introduced as a technique for de-hopping SS. The BNN was trained on the samples values of 25 different signals using different learning algorithms. Secondly, the RBF was trained by using 25 FH patterns to estimate the pattern after five hopping. After sufficient training the experiments and results it has been shown that the performance of the system simulates the conventional matched filter receiver. It has been shown through simulation that the NNs for synchronization is comparable to the parallel search synchronization. The proposed model can acquire acquisition and estimates the FHP pattern within 5 hopping times ($5T_h$), which are much shorter than that of serial search technique. Using simple hardware, its realization is simpler than parallel search technique.

References

1. Bouder, C., Burel, G.: Spread Spectrum Codes Identification by Neural Networks. In: Systems and Control: Theory and Applications, ISBN 960-8052-11-4, WSES (2000) 257–262
2. Day, P.S.: Handbook of Neural Computation. IOP Publishing Ltd and Oxford University Press (1997)
3. Dolent, M., Lorenz, D.R.: Recurrent Neural Network Topologies for Spectral State Estimation and Differentiation. ANNIE 2000 Conference, St. Louse MO (2000)
4. Chau, A.Y., Wang, J.K.: Spectral-Estimation-Based Acquisition for Frequency-Hopping Spread-Spectrum Communications in Nonfading or Rayleigh-Fading Channel. IEEE Transactions on Communications, VOL. 45, IEEE (1997)
5. Berder, O., Bouder, C., Burel, G.: Identification of Frequency Hopping Communications. Problems in Modern Applied Mathematics, ISBN 960-8052-15-7, WSES (2000) 259-264
6. Hagan, T.M., Demuth, B.H., Beale, M.: Neural Network Design, PWS Publishing Co., (1996)
7. Fan, C.C., Tsai, Z.H.: A Differentially Coherent Delay-Locked Loop for Spread-Spectrum Tracking Receiver. IEEE Communication Letters, VOL. 3, IEEE (1999)
8. Hong, C.F., Yang G.C.: Concatenated Prime Codes. IEEE Communications Letters, Vol. 3, IEEE (1999) 260-262

Tarek Elhabian was born in 1966. He received the B.S. degree in computer science form Military Technical College, Egypt in July 1988 and M.S. degree, in EE & Sys. Eng., from Ain Shams University, Egypt, 2000. He is now a Ph.D. student at Beijing University of Aeronautics & Astronautics (BUAA), China.

Bo Zhang was born in 1972. He received the M.S. degree of EE in BUAA. he is studying in BUAA for Ph.D. now.

Dingrong Shao graduated from the Dept. of E. E. of Beijing Aeronautical Institute, Beijing, China in 1962. From 1980 to 1983, he worked on carrier free radar at the Catholic University of America, Washington, DC, USA. 1992 he became a Professor of BUAA.

Autoregressive and Neural Network Model Based Predictions for Downlink Beamforming

Halil Yigit[1], Adnan Kavak[2], and Metin Ertunc[3]

[1] Kocaeli University, Dept. of Electronics and Computer Ed., 41300, Izmit, Kocaeli, Turkey
halilyigit@kou.edu.tr

[2] Kocaeli University, Dept. of Computer Engineering, 41040, Izmit, Kocaeli, Turkey
akavak@kou.edu.tr

[3] Kocaeli University, Dept. of Mechatronics Engineering, 41040, Izmit, Kocaeli, Turkey
hmertunc@kou.edu.tr

Abstract. In Time-Division-Duplex (TDD) wireless communications, downlink beamforming performance of a smart antenna system can be degraded due to variation of spatial signature vectors in vehicular scenarios. To mitigate this, downlink beams must be adjusted according to changing propagation dynamics. This can be achieved by modeling spatial signature vectors in the uplink period and then predicting them for new mobile position in the downlink period. This paper examines time delay feedforward neural network (TDFN), adaptive linear neuron (ADALINE) network and autoregressive (AR) filter to predict spatial signature vectors. We show that predictions of spatial signatures using these models provide certain level of performance improvement compared to conventional beamforming method under varying mobile speed and filter (delay) order conditions. We observe that TDFN outperforms ADALINE and AR modeling for downlink SNR improvement and relative error improvement with high mobile speed and higher filter order/delay conditions in fixed Doppler case in multipaths.

1 Introduction

Smart Antenna Systems (SAS) are proven to provide significant capacity increase and performance enhancement at the base station of wireless communications [1-3]. Spatial signature vector or channel vector describes the propagation characteristics of the signals present at an antenna array of a smart antenna system (SAS). For the downlink beamforming in Time Division Duplex (TDD) system, the SAS conventionally uses the last known spatial signature estimated during the uplink interval as the weight vector. This conventional approach [4] performs well as long as channel characteristics remain the same between consecutive time intervals. When the mobile terminal is stationary or moving a small distance of two wavelengths, spatial signature variations are not significant and direction of arrivals (DOAs) are almost unchanged [5]. However, if the mobile user moves at a relatively high speed, spatial signature vectors can change rapidly due to fast fading effects induced by Doppler shift at each multipath [6] and each element of spatial signatures can be modeled as the sum of sinusoids [7,8]. Under such circumstances, employing the spatial signature

F. Yin, J. Wang, and C. Guo (Eds.): ISNN 2004, LNCS 3174, pp. 254–261, 2004.

of the previous uplink time slot as the downlink weight vector for the new mobile position may result in performance degradation. This can be avoided by accurately predicting the spatial signatures in the downlink interval and using predicted spatial signatures as transmission weight vectors to control downlink beams.

The aim of present work is to predict downlink weight vectors via time delay feedforward neural network (TDFN) modeling, adaptive linear neuron (ADALINE) network modeling [9-12], and autoregressive (AR) modeling of uplink spatial signature vectors. We compare the performance of these models under varying mobile speed (V) and prediction filter (delay) order (P). The performance metrics used are downlink SNR of the received power (SNR improvement, Δ_{SNR}) and norm of error vector between actual and predicted spatial signatures (relative error improvement, $\Delta\varepsilon$), which indicate the accuracy of predictions and mobile user tracking. The simulation results show that the TDFN performs better than ADALINE network and conventional AR filter with both Δ_{SNR} and $\Delta\varepsilon$ under high mobile speed and prediction filter order/delay in fixed Doppler case.

The remainder of this paper is organized as follows: Section 2 describes a spatial signature model that is used in the predictions. In Section 3, prediction methods based on AR modeling and neural network modelings are explained. Section 4 presents simulation conditions and results, and concluding remarks are given in Section 5.

2 Spatial Signature Vector Model

The transmitted band-limited baseband signal $s(t)$ from a mobile unit is received by an M element antenna array at the base station as

$$\mathbf{x}(\mathrm{t}) = s(t)\mathbf{a}(t) + \mathbf{I}(t) + \mathbf{n}(t) \tag{1}$$

where $\mathbf{a}(t)$ is the spatial signature vector or the channel vector corresponding to the mobile user, $\mathbf{I}(t)$ is the multiple-access interference, $\mathbf{n}(t)$ is the complex-valued noise vector. Transmissions from the mobile user to the antenna array typically occur with multipaths, each associated with a direction of arrival (DOA) that forms a steering vector. The spatial signature vector determines the physical propagation characteristics in the wireless environment and is expressed as a weighted sum of steering vectors [5]. The spatial signature vector $\boldsymbol{a}(\mathrm{t})$ corresponding to the mobile user is given by,

$$\boldsymbol{a}(\mathrm{t}) = \sum_{i=1}^{L} r_i \alpha_i e^{j\varphi_i(t)} \boldsymbol{v}(\theta_i) \tag{2}$$

where L is the number of multipaths; $\mathbf{r}_i$ is the large-scale propagation path loss; α_i is the complex attenuation caused by reflections from local scatterers; $\varphi_i(\mathrm{t})$ is the phase shift induced by varying path lengths due to mobile movement (Doppler effect); $\boldsymbol{v}(\theta_i)$ is the Mx1 steering vector for the ith multipath component which arrives at an angle of θ_i. For downlink transmission, the SAS conventionally employs spatial signature based beamforming method in which the spatial signature vector $\boldsymbol{a}(\mathrm{t}\text{-}1)$ obtained during the previous time slot is used as a weight vector $\mathbf{w}(\mathrm{t})$ at current time slot,

$$\mathbf{w}(t) = \mathbf{a}(t-1) \tag{3}$$

In fast fading mobile environments, using the weight vector in (3) for downlink transmission might lead to inaccurate beams, which is directed towards the previous mobile position, and therefore, cause degradation in the received power at the mobile terminal. This degradation can be eliminated by updating (predicting) the downlink weight vector and therefore steering the downlink beam for the new mobile position. To accomplish this, the smart antenna system must observe spatial signature samples over several uplink time slots, then construct a model for these spatial signatures in the uplink period, and then predict the spatial signature vector to be used as a weight vector for the downlink beamforming at current time interval.

3 Prediction Models

3.1 Autoregressive (AR) Modeling

Small movements of the mobile user cause changes in the relative phase of multipath components due to Doppler shift, elements of the spatial signature vectors are expressed as the sum of sinusoids. Hence, each element of the spatial signature vector can be treated as time varying autoregressive (AR) process [13,14] given by,

$$a_k(t) = -\sum_{j=1}^{P} b_{k,j}^{*}\, a_{k,j}(t-j) \qquad k = 1, 2, ..., M \tag{4}$$

where P is the model or prediction filter order and $b_{k,j}$ are the filter coefficients. The prediction filter coefficients are calculated by solving the Yule-Walker equations [14].

3.2 Neural Network Modelings

In this study, we use time delay feedforward neural network (TDFN) and adaptive linear neuron (ADALINE) network. TDFN and ADALINE network [11-13] use tapped delay line to perform temporal processing. Both neural network structures are used to predict the next value of the spatial signature vector as the downlink weight vector.

The difference between two structures is that ADALINE network predicts the vectors adaptively. It doesn't include any training process. The weights and biases are updated sequentially based on the new input and target vectors for each time step. TDFN uses a training set obtained from the data set. A model is constructed by training network on training set until minimum error is reached after a maximum of training epochs and then the vectors are predicted using this model. The structure of TDFN and ADALINE network structures are as illustrated in Fig.1 and Fig. 2, respectively.

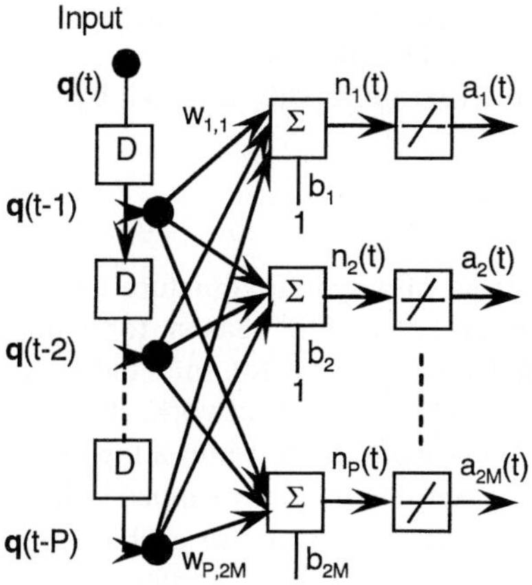

Fig. 1. ADALINE network structure

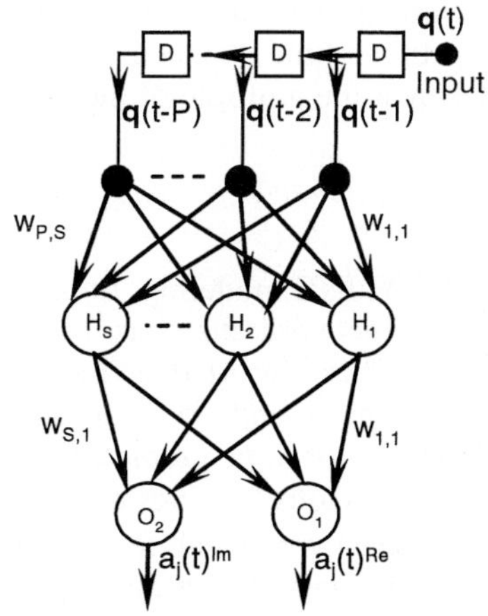

Fig. 2. TDFN structure without bias

Since the spatial signature $\boldsymbol{a}(t)$ is Mx1 complex valued vector, it is decomposed into real $\boldsymbol{a}^{\mathrm{Re}}(t)$ and imaginary $\boldsymbol{a}^{\mathrm{Im}}(t)$ parts. In ADALINE modeling, they are recombined to input to the network as 2Mx1 vector at each sampling time t given by,

$$\mathbf{q}(t) = [q_1(t)\ q_2(t)\ \cdots\ q_{2M-1}(t)\ q_{2M}(t)]^T = [a_1^{\mathrm{Re}}(t)\ a_1^{\mathrm{Im}}(t)\ \cdots\ a_M^{\mathrm{Re}}(t)\ a_M^{\mathrm{Im}}(t)]^T \tag{5}$$

In TDFN modeling, the weights and biases are adjusted using Levenberg-Marquardt learning algorithm to train the network in the training process. Input to TDFN as 2x1 dimensional vectors can be shown for each elements of spatial signature vector as follows,

$$\mathbf{q}(t) = \begin{bmatrix} \mathrm{Re}\{a_j(t)\} \\ \mathrm{Im}\{a_j(t)\} \end{bmatrix} \qquad \mathrm{j} = 1,\cdots,\mathrm{M} \tag{6}$$

4 Simulations and Results

4.1 Simulation Setup

We perform computer simulations to evaluate the performance of prediction based downlink beamforming of a SAS in TDD mode using ADALINE network modeling, TDFN modeling, and AR modeling. Each spatial signature vector is 7x1 complex valued vectors because the base station has 7-element uniform linear array antenna. We generate Q=30 spatial signature samples to be compatible with UTRA-TDD standard [15] using the model in Equation (2) for a given mobile speed (V), the number (L) and mean DOA (θ) of multipaths, and angle spread (Δθ). We take duplexing time (time between consecutive samples) as 10/15 ms and the carrier frequency is as 1.8 GHz frequency. Within each 20 ms TDD frame, we assume that L and θ are constant, and other parameters such as Doppler shift $f_{d,i}$ parameters of each multipath are time dependent variables. We consider Doppler shifts in the multipaths are taken as the multiple of the Doppler shift in the first multipath, as given by

$$f_{d,i} = i\frac{v}{\lambda} \quad i = 1, \cdots, L \tag{7}$$

In ADALINE modeling, to predict the next value of a spatial signature vector, we employ a network that has 14 neurons at output layer. The reason for using 14 neurons is because of decomposing each spatial signature vector into the real and imaginary parts and constituting a 14x1 input vector to the network. The spatial signature vector to be predicted at time t, **q**(t), enters from the left into a tapped delay line. The previous P values of **q**(t) are available as outputs from the tapped delay line. We have used very small learning rate (η=0.04). In AR modeling, for each element of the spatial signature, an autoregressive model with order P is constructed from the first Q/2 spatial signature samples during the uplink interval. In the downlink interval, next Q/2 spatial signature vectors are predicted using these AR model coefficients. In TDFN modeling, the number of hidden and output neurons is 2P and 2, respectively. The reason for using two output neurons is that we obtain real and imaginary part of each element of predicted spatial signature vector at the time interval. We first identify the available parameters as learning rate (η=0.05), step size (0.005), and the number of epochs (5000). We construct a training set and a comprehensive testing set from the first decomposed Q/2 spatial signature samples during the uplink interval. After training process, next Q/2 spatial signature vectors are tested (predicted) with updated weights and biases.

4.2 Performance Metrics

We use two different measures to test the accuracy of the above prediction models for the spatial signatures. The first measure, which indicates the proximity of the predicted and conventional spatial signatures to the actual ones in length, is defined as the relative error improvement (•ε) given by,

$$\Delta\varepsilon = 20\log_{10}\frac{\|\varepsilon_{conv.}\|}{\|\varepsilon_{pred.}\|} = 20\log_{10}\frac{\|\boldsymbol{a}_{actual} - \boldsymbol{a}_{conv.}\|}{\|\boldsymbol{a}_{actual} - \boldsymbol{a}_{pred.}\|} \tag{8}$$

where $\boldsymbol{a}_{actual}$ and $\boldsymbol{a}_{pred}$ are the actual and predicted spatial signatures, respectively, for the current time slot (mobile position) in the downlink interval, and $\boldsymbol{a}_{conv}$ is the spatial signature of the previous time slot. The second measure is the signal to noise ratio (SNR) improvement (Δ_{SNR}). It can also be viewed as the measure of how accurately the downlink beam is pointed in the direction of the new mobile position. By substituting predicted spatial signature in place of the conventional spatial signature in (3), we obtain some improvement in the downlink received power (SNR) given by,

$$\Delta_{SNR}(dB) = 20\log_{10}\left(\frac{\left|\mathbf{a}^{H}_{pred.}\mathbf{a}_{actual}\right|}{\left\|\mathbf{a}^{H}_{pred.}\right\|} \frac{\left\|\mathbf{a}^{H}_{conv.}\right\|}{\left|\mathbf{a}^{H}_{conv.}\mathbf{a}_{actual}\right|} \right) \tag{9}$$

To obtain statistically conclusive results, we performed 500 simulation runs for a given angle spread ($\Delta\theta$), mobile speed (V), prediction filters order (P), and multipath number (L). In each simulation run, the only changing parameter in propagation environment is DOAs of multipaths, hence mobile start point randomly change within scattering environment. The prediction performance is averaged over 10 ms downlink period. Thus, we have obtained 500 values for Δ_{SNR} and $\Delta\varepsilon$ values.

4.3 Results

The performance of TDFN, ADALINE network, and AR filter prediction models are examined for various mobile speeds (V) and filter/delay order (P). Fig. 3 represents the cumulative distributions of simulation results which are obtained for V=100 km/h, P=5, $\Delta\theta$=60°, L=6. Obviously, $\Delta\varepsilon$ performances of TDFN based prediction are above those of ADALINE and AR based prediction.

Note that for each simulation run, we have averaged Δ_{SNR} and $\Delta\varepsilon$ values over downlink period. We observe that at the 50 percentiles (in the mean value), TDFN has Δ_{SNR} and $\Delta\varepsilon$ values approximately 34 dB and 3.6 dB, respectively and these values are greater than the AR (8.5 dB-3 dB) and ADALINE (18 dB-3.4 dB) prediction does. When the prediction filter order or delay in the neural models is varied from 4 to 6 and mobile speed is fixed as V=100 km/h, similar statistics are obtained for Δ_{SNR} and $\Delta\varepsilon$ values over 500 simulation run. In other words, TDFN performance is always above the AR and ADALINE performance for V=100 km/h. The results for mean values are tabulated in Table 1. Performance of three methods increases with increasing order/delay values compared to each others. But the rate of increase in $\Delta\varepsilon$ for TDFN modeling is larger than that for ADALINE and AR modeling.

Table 1. Mean values of Δ_{SNR} and $\Delta\varepsilon$ for various mobile speeds for fixed Doppler case and random Doppler case

Model	V(km/h)	Δ_{SNR} (dB)	$\Delta\varepsilon$ (dB)
TDFN network	30	-2.620	8.066
	100	3.560	34.268
ADALINE network	30	-0.096	2.873
	100	3.373	17.886
AR filter	30	0.110	6.938
	100	2.9177	8.593

The mean values obtained from cumulative distributions of three methods under fixed filter order/delay (P=5), and low and high mobile speed conditions are given in Table 2. We observe that for low speed (V=30 km/h) and high speed (V=100 km/h) conditions in fixed Doppler case, $\Delta\varepsilon$ performances of these methods increase with increasing the mobile speed. But TDFN based prediction are better than those of ADALINE network and AR based prediction; that is, TDFN network achieves better approximation of predicted spatial signatures to the actual ones in length than other models do. Δ_{SNR} performance of TDFN is affected significantly by mobile speed. As seen from Table 2, in low mobile speed, its performance gets worse, while in high speed Δ_{SNR} performance becomes better.

Table 2. Mean values of Δ_{SNR} and $\Delta\varepsilon$ for various filter order/delay for fixed Doppler case and random Doppler case

Model	P	Δ_{SNR} (dB)	$\Delta\varepsilon$ (dB)
TDFN network	4	3.644	33.258
	5	3.500	34.268
	6	3.547	34.792
ADALINE network	4	3.340	15.949
	5	3.373	17.886
	6	3.363	15.460
AR filter	4	2.905	8.398
	5	2.918	8.593
	6	2.985	8.946

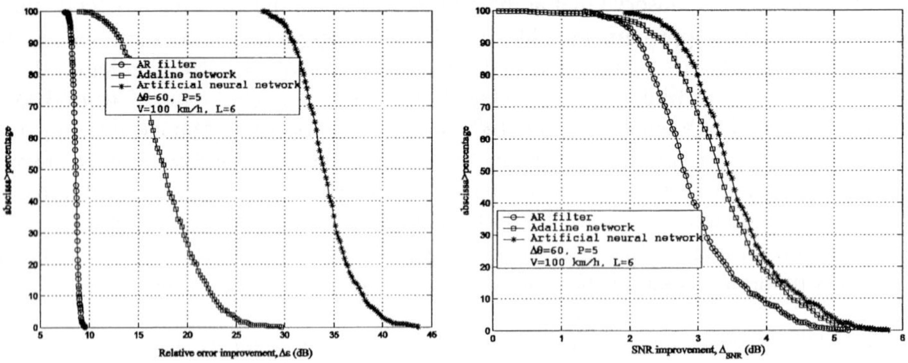

(a) (b)

Fig. 3. (a) and (b) Cumulative distribution of $\Delta\varepsilon$ and Δ_{SNR} in fixed Doppler case under V=100 km/h, P=5, L=6, and $\Delta\theta$ =60°

5 Conclusion

For smart antenna systems that operate in Time-Division-Duplex (TDD) mode, TDFN network, ADALINE network, and AR modeling based predictions of uplink spatial signatures for controlling downlink beams in fast fading wireless scenarios are studied. Performances of three models are investigated for varying mobile speed (V) and filter order/delay (P) conditions. We found that for low mobile speed (V=30 km/h), using TDFN modeling for spatial signature predictions has no advantage in terms Δ_{SNR} (or downlink beamforming accuracy). However, TDFN, ADALINE, and AR modeling can achieve certain level of relative error improvement ($\Delta\varepsilon$) for all mobile speed and varying filter order. But TDFN performance is always above for all conditions. Thus, it is feasible to employ TDFN, ADALINE, or AR modeling based prediction of spatial signatures depending on the propagation conditions and link budget requirements.

References

1. Godara, L.C.: Applications Of Antenna Arrays To Mobile. Communications, Part I: Performance Improvement, Feasibility, And System Considerations. IEEE Proceedings. Vol. 85(7) (1997) 1031-1060
2. Kavak, A.: Adaptive Antenna Arrays for Downlink Capacity Increase in Third Generation Wireless CDMA. In Proc. IEEE Radio and Wireless Conf. (2001) 77-80
3. Winters, J.H.: Smart Antennas for WirelessSystems. IEEE Personal Comm. Mag., Vol. 5(1) (1998) 23-27
4. Krim, H., Viberg, M.: Two Decades ofArray signal Processing Research. IEEE Signal Proc. Mag., Vol. 13(4) (1996) 67-94
5. Kavak, A., Yang, W., Xu, G., Vogel, W.J.: Characteristics of Vector Propagation Channels in dynamic mobile scenarios. IEEE Trans. on Ant. Prop., Vol. 39(12) (2001) 1695-1702
6. Yang, W., Kavak, A., Kim, S., Xu, G., Hansen, L.: Evaluation of Spatially Sselective Receiving/transmission Techniques for a Smart Antenna System Operating at 1.8 GHz in non-stationary scenarios. In Proc. IEEE VTC'99, Vol. 1. (1999) 862-866
7. Hwang, J.K., Winters, J.H.: Sinusoidal Modeling and Prediction of Fast Fading Processes. In Proc. IEEE GLOBECOM, Vol. 1. Sydney, Australia (1998) 892-897
8. Duel-Hallen, A., Hu, S., Hallen, H.: Long-range prediction of fading signals. IEEE Signal Proc. Mag., Vol.17(3) (2000) 62-75
9. Haykin, S.: Neural networks: A Aomprehensive Foundation. 2nd edition. Prentice Hall, New Jersey (1999)
10. Yigit, H., Kavak, A., Ertunc, H.M.: Using Autoregressive and Adaline Neural Network Modeling to Improve downlink performance of smart antennas. ICM'04 (2004) accepted.
11. Freeman, J.A., Skapura, D.M.: Neural Networks Algorithms, Applications, and Programming Techniques. Addison-Wesley (1991)
12. Du, K.-L., Lai, A.K.Y., Cheng, K.K.M., Swamy, M.N.S.: Neural Methods for Antenna Array signal processing: a review. Elsevier Sig. Proc. Vol. 82 (2002) 547-561
13. Dong, L., Xu, G., Ling, H.: Prediction of Fast Fading Mobile radio Channels in wideband communication systems. In Proc. IEEE GLOBECOM (2001) 3287-3291
14. Haykin, S.: Adaptive Filter Theory. Prentice Hall, NJ(1996)
15. Haardt, M., et al.: The TD-CDMA based UTRA TDD mode. IEEE J. Sel. Areas in Commun., Vol.18(8) (2000) 1375-1385.

Forecast and Control of Anode Shape in Electrochemical Machining Using Neural Network

Guibing Pang, Wenji Xu, Xiaobing Zhai, and Jinjin Zhou

School of Mechanical Engineering, Dalian University of Technology, Dalian 116023, China
{pangguibingsx, zxbsx}@163.com, {wenjixu, zhoujj}@dlut.edu.cn

Abstract. It is difficult for numerical method to forecast and control the anode shape in Electrochemical Machining (ECM) with an uneven interelectrode gap, so this paper introduces Artificial Neural Network (NN) to solve this problem. The experiments with different cathode shapes and minimal interelectrode gaps are carried out and the corresponding anode shapes are obtained. Those cathode and anode shapes are discretized and taken as the input samples of a B-P network. Quasi-Newton algorithm is used to train this network. To verify the validity of the trained network, results obtained by NN are compared with that obtained by the experiments, and the results show that the former is close to the later, which indicates it is feasible to apply NN to solve this problem.

1 Introduction

ECM has been applied in the machining of many metal parts for its advantages. In ECM [1], the workpiece is shaped by the different metal removal rate in different area of the anode surface, which is caused by the uneven distribution of current on the anode surface. In the general ECM process, the cathode moves towards the anode, so the balanced state of interelectrode gap can be obtained after a period of machining and the final shape of the anode is determined by superposing the interelectrode gap over the cathode shape. In this case, some numerical or empirical methods can be used to forecast and control the anode shape [2-4]. However, the anode needs to be shaped with a static cathode sometimes [5], in other words, there doesn't exist the equilibrium interelectrode gap in the machining process and the anode shape entirely relies on the difference of the metal removal rate on its surface. Therefore, it is necessary to study two problems under the condition of uneven interelectrode gap. One is the forecast-problem in which the anode shape corresponding to a certain cathode shape needs to be forecasted (in this paper, it is also named as direct-problem), and the other is the control-problem, in which the cathode shape should be decided and this cathode is used to obtain an expectant anode shape (in this paper, it is also named as reverse-problem). It is difficult for the existing theory of ECM to solve the above problems, especially the control-problem. So, it is required to introduce new methods. NN possesses a better adaptability to the complicated circumstance and can approach an arbitrary non-linear function with any precision, so this paper tries to use NN to forecast and control the anode shape with an uneven interelectrode gap.

F. Yin, J. Wang, and C. Guo (Eds.): ISNN 2004, LNCS 3174, pp. 262–268, 2004.

2 Forecast and Control of Anode Shape

Fig.1 illustrates the machining process with an uneven interelectrode gap in two dimensions. In Fig.1, $z=h(x)$ denotes the cathode shape, $z=g(x, t)$ denotes the anode shape, t_0 is the initial time of machining, t_i is an arbitary time of machining, t_n is the end time of machining and the curves pointed by t_0, t_i and t_n denote the anode shape at that time, respectively.

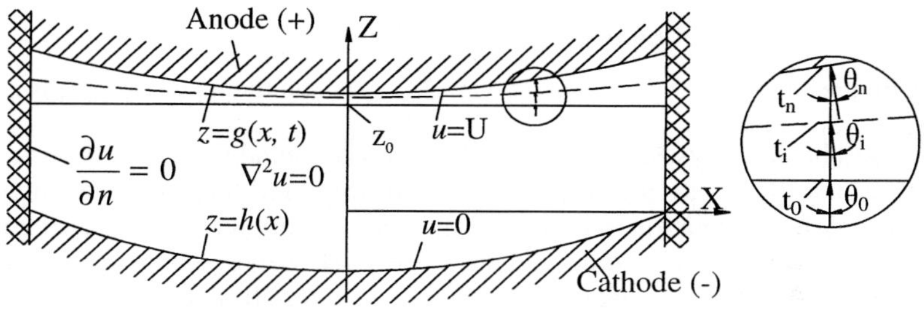

Fig. 1. Machining process with an uneven interelectrode gap.

Supposing that the function of electric field distribution in Fig.1 is $u=u(z, x)$, the function of current density distribution in the normal direction of the anode curve at time t is:

$$i=\lambda\cdot\frac{\partial u}{\partial n}=\lambda\cdot(\frac{\partial u}{\partial z}+\frac{\partial u}{\partial x})=\lambda\cdot(u'_z(z,x)+u'_x(z,x)+u'_z(z,x)\cdot g'_x(x,t)) \tag{1}$$

where λ is electrolytic conductivity.

According to Faraday law, the removal depth of the anode surface is:

$$V=\eta\omega i\Delta t \tag{2}$$

where η is current efficiency, ω is volumetric electrochemical equivalent of the workpiece material, i is current density and Δt is machining time.

By taking Eq.1 to Eq.2, the following equation can be obtained:

$$V=\eta\omega\Delta t\lambda(u'_z(z,x)+u'_x(z,x)+u'_z(z,x)\cdot g'_x(x,t)) \tag{3}$$

When $\Delta t\to 0$, Eq.3 denotes the removal depth distribution in the normal direction of anode curve at time t. If Eq.3 is taken as the removal depth distribution in Z direction, the anode curve at time t_i can be described as:

$$z\big|_{t=t_i}=z(x,t_{i-1})+\eta\omega\Delta t\lambda(u'_z(z,x)+u'_x(z,x)+u'_z(z,x)\cdot g'_x(x,t_{i-1})) \tag{4}$$

Where $\Delta t=t_i-t_{i-1}$ and i=1,2,3…n. Owing to the substitution of the removal depth distribution in the normal direction for that in Z direction, an error will occur inevitably and its value depends on Δt. The less Δt, the less the error.

At the end time of machining t_n, the anode curve can be described as:

$$z\Big|_{t=t_n} = z_0 + \eta\omega\Delta t\lambda\sum_{i=1}^{n}(u'_z(z,x) + u'_x(z,x) + u'_z(z,x)\cdot g'_x(x,t_{i-1})) \tag{5}$$

where $g(x, t_0)=z_0$ is the initial condition and the function of u is:

$$\nabla^2 u = \frac{\partial^2 u}{\partial x^2} + \frac{\partial^2 u}{\partial z^2} = 0 \tag{6}$$

Boundary conditions of Eq.6 are shown as follows:

U=0, on the cathode; u =U, on the anode; $\frac{\partial u}{\partial n} = 0$, on the insulating walls.

From the above analyses, it is known that the final anode shape is determined by the electric field distribution in the machining process and the total machining time. The electric field distribution is related to the cathode and anode shape, cathode and anode position, and interelectrode voltage closely. However, anode shape is always changing in the machining process, which leads to a real-time change of the electric field distribution. Namely, the anode shape and the electric field distribution interact with each other in the machining process. When Eq.5 is solved, the electric field distribution at time t_i needs to be decided according to the anode shape at that time, and then the anode shape at time t_{i+1} is decided according to the electric field distribution at time t_i. This process is cycled continually and the final anode shape is obtained. The solving precision relies on the interval from t_i to t_{i-1}, and the less the interval, the higher the precision. Therefore, the calculation will be complicated if a higher precision is required.

For the control-problem, it is more difficult to be solved. When the cathode shape is uncertain, the boundary condition of the electric field distribution on the cathode surface couldn't be decided and the function of electric field distribution couldn't be established. So it is very difficult or nearly impossible for numerical method to solve the cathode shape.

3 Application of B-P Network

Based on the above analyses, the author introduces NN to solve this problem. NN can avoid the complicated modeling and numerical calculation. It is especially in favor of the reverse-problem, because the solving process of the reverse-problem is essentially a converse mapping process of the direct-problem.

3.1 Research Method

This research is made up of three sections, ie: 1) basic experiments in which lots of training data are obtained; 2) network training including the training of direct-problem and that of reverse-problem; 3) random verification in which the random input data are taken into the network and the output data are obtained, and then the reliability of the output data is validated by the experiments.

3.2 Data Processing

According to Eq.5 and Eq.6, the final anode shape is determined by the conjunct actions of the cathode shape, the interelectrode gap, the interelectrode voltage and the machining time. The interelectrode voltage and the machining time are generally decided by the practical requirement, and in this paper, the interelectrode voltage is 20V and the machining time is 4s. For satisfying the requirements of the input and output in NN, it is necessary to discretize the cathode and anode shape. Fig.2 and Table.1 illustrate the method.

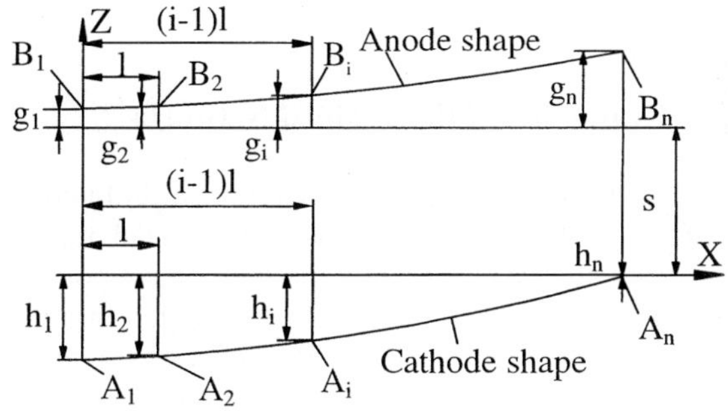

Fig. 2. Method to describe the cathode and anode shape.

Table 1. Data processing method

	Forecast	Control
Input	$H=[h_1, \ldots h_i, \ldots h_n]$, s	$G=[g_1, \ldots g_i, \ldots g_n]$, s
Output	$G=[g_1, \ldots g_i, \ldots g_n]$	$H=[h_1, \ldots h_i, \ldots h_n]$

In Fig.2, A_1, $\ldots A_i$, $\ldots A_n$ are the control points of the cathode shape, so the cathode shape can be described by $H=[h_1, \ldots h_i, \ldots h_n]$ and l. The rule to decide H is that A_1 and A_n are the boundary points, h_i is a random number from 0 to h_{i+1} and the max of h_i is no more than 0.3 mm which is decided by the practical requirement. By above rule, the randomicity of the samples can be ensured. The value of n is decided according to the width of the cathode and it is an integer more than 1. Likewise, B_1, $\ldots B_i$, $\ldots B_n$ are the control points of the anode shape corresponding to that of the cathode shape and the anode shape can be described by $G=[g_1, \ldots g_i, \ldots g_n]$ and l. s denotes the minimum of the interelectrode gap and it is a random number from 0 to 0.3 mm, which is decided by the practical requirement.

In the experiments, the width of the cathode is 6 mm, so n is 7, namely l equals 1 mm. By the above method, 80 groups of combinations of different cathode shapes (H) and the minimum of interelectrode gaps (s) are constructed randomly. The corresponding anode shapes are obtained by the experiments and the control points of the anode shapes (G) are measured. After normalized, the data of H, G and s are taken as the training samples of the network.

3.3 Network Training

This paper adopts the network structure with four layers (2×20×41×1), which include an input layer, an output layer and two hidden layers. Quasi-Newton algorithm is used to train the network owing to its good convergence. There is the program of Quasi-Newton algorithm in the MATLAB software and the training is accomplished by taking the samples to this program. In this way, the trained networks of forecast-problem and control-problem are obtained.

3.4 Verification and Evaluation of the Trained Networks

To verify the validity of the trained network of forecast-problem, twenty groups of new combinations of the different H and s are constructed randomly. With those combinations, the experiments are carried out and the control points of the real anode shapes (G′) are obtained. The same combinations are inputed into the trained network and the control points of the simulated anode shapes (G) are worked out. By comparing G with G′, it is found that eighteen relative errors are less than 5% and the mean of the relative errors is 2.3%.

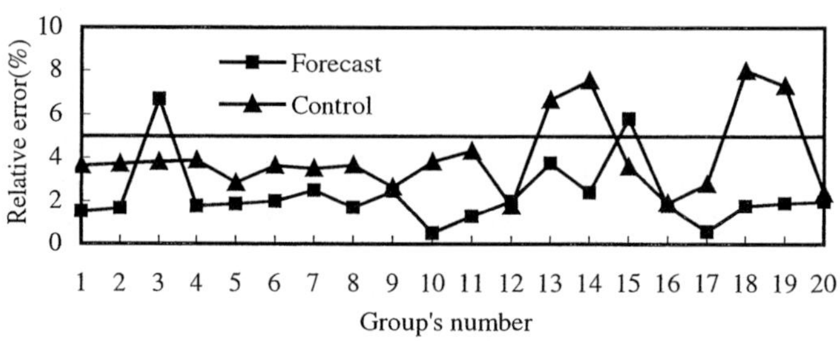

Fig. 3. Relative errors in the forecast and control problems.

The cathode shapes (H) corresponding to twenty groups of the combinations of the expectant anode shapes (G) and s are worked out by the trained network of the control-problem, and then those cathodes are manufactured. The experiments with those cathodes and s are carried out and the control points of the real anode shapes (G•) are obtained. By comparing G with G•, it is found that sixteen relative errors are less than 5% and the mean of the relative errors is 4.07%. The relative errors in the forecast-problem and control-problem are shown in Fig.3. It needs to be pointed out that the relative error in Fig.3 is the maximum of the errors in all control points for every anode shape. The above results indicate that the relative errors in the forecast and control problems could satisfy the practical requirements in ECM to a certain extent.

In Table.2, Table.3 and Fig.4, two instances of the forecast-problem and the control-problem are illustrated.

Table 2. An instance of the forecast-problem

i	1	2	3	4	5	6	7
Simulated anode shape (g_i /mm)	0.039	0.040	0.042	0.046	0.051	0.063	0.088
Real anode shape (g_i' /mm)	0.040	0.041	0.043	0.045	0.053	0.065	0.085
Relative error (%)	2.5	2.44	2.33	2.22	3.77*	3.1	3.53

3.77* is the maximal error in this instance and is shown in Fig.3 (group 13).

Table 3. An instance of the control-problem

i	1	2	3	4	5	6	7
Expectant anode shape (g_i/mm)	0.034	0.038	0.040	0.042	0.050	0.062	0.067
Real anode shape (g_i'/mm)	0.035	0.036	0.041	0.045	0.053	0.063	0.070
Relative error (%)	2.86	5.56	2.44	6.67*	5.66	1.59	4.29

6.67* is the maximal error in this instance and is shown in Fig.3 (group 13).

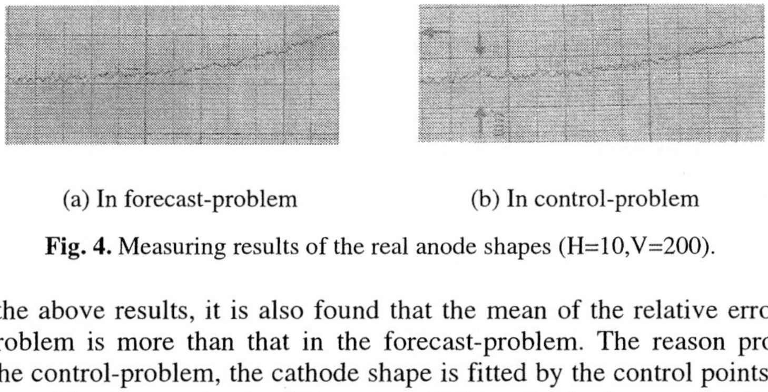

(a) In forecast-problem (b) In control-problem

Fig. 4. Measuring results of the real anode shapes (H=10,V=200).

From the above results, it is also found that the mean of the relative errors in the control-problem is more than that in the forecast-problem. The reason probably is that, for the control-problem, the cathode shape is fitted by the control points, when it is used to machine the anode, the fitting error of the cathode shape will influence the anode shape. But, how to improve the control precision is still a problem worth to be further studied.

4 Conclusions

- It is difficult for the numerical method to forecast and control the anode shape in ECM with an uneven interelectrode gap. The application of NN could avoid the complicated mathematic modeling and solving, so this problem could be simplified largely.

- In this research, the anode and cathode shapes are discretized by the control points which are taken as the input and output data of the network. This method provides a reference for the similar problems.
- The comparison of the results obtained by NN and that obtained by experiments shows that the former is close to the later, which indicates it is feasible to apply NN to forecast and control the anode shape in ECM with an uneven interelectrode gap and this paper's work lays a foundation for the further study.

Acknowledgements. This study gets support from Dalian Science and Technology Project Foundation under the contract 2003A1GX061. We would like to appreciate all the teachers and students who have discussed with us about this topic and given us good advice.

References

1. Rajurkar, K.P., Zhu, D., Mcgeough, J.A., Kozak, J., Silva, A.De.: New Development in Electro-Chemical Machining. Ann. CIRP. 2 (1999) 567-579
2. Jain, V.K., Rajurkar, K.P.: Integrated Approach for Tool Design in ECM. Prec. Eng. 2 (1991) 111-124
3. Bhattacharyya, S., Ghosh, A., Mallik, A.K.: Cathode Shape Prediction in Electrochemical Machining Using a Simulated Cut-and-Try Procedure. J. Mater. Proc. Technol. 1 (1997) 146-152
4. Kozak, J.: Mathematical Models for Computer Simulation of Electrochemical Machining Processes. J. Mater. Proc. Technol. 1 (1998) 170-175
5. Zhang, M., Zhou, J.J., Wang, L.M., Yi, J.J.: Numerical Simulation of Electric Field for Internal Gear Modification with ECM. J. Dalian Univ. Technol. 3 (2000) 323-325 (in Chinese)

A Hybrid Neural Network and Genetic Algorithm Approach for Multicast QoS Routing

Daru Pan[1], Minghui Du[1], Yukun Wang[2], and Yanbo Yuan[2]

[1] School of Electronic & Information Engineering, South China Univ. of Tech., Guangzhou 510640, China
Pdr168@163.net, ecmhdu@scut.edu.cn
[2] JinPeng Group, Guangzhou 510665, China
{Wangyk, Yuanyb}@gzjpg.com

Abstract. Computing the Multicast QoS routing is an NP-complete problem. Generally, it was solved by heuristic algorithms, which include tabu search, simulated annealing, genetic algorithms (GA), neural networks (NN), etc. In this paper, a hybrid neural network and genetic algorithm approach is described to compute the multicast QoS routing tree. The integration of neural network and genetic algorithm can overcome the premature and increase the convergence speed. The simulation results show that the proposed approach outperforms the traditional GA and NN algorithm in terms of both solution accuracy and convergence speed.

1 Introduction

As the rapid development of the internet, the interest and demands of distributed multimedia applications are increasing. These applications involve the transmission of multimedia information and therefore it is essential to satisfy QoS constraints. The challenge to multicast routing is how to build multicast tree subjected to the QoS constraints, for building a constrained multicast tree has been proved to be an NP-completed problem [1]. Generally, these problems are solved by heuristic method. Several deterministic heuristic methods to solve this problem have been proposed [2], but they are computationally expensive and cannot be practiced for large sized networks. Methods based on computational intelligence such as neural networks (NN) and genetic algorithms (GA) may be a good way to solve these problems.

There are already some NN-based algorithms and GA-based algorithms to solve the QoS-based routing problem. Chotipat et al. [3] have proposed a modified version of Hopfield neural network model to solve delay constrained multicast routing. It can find near optimal multicast route very fast, when implemented by hardware, while it can't overcome the drawback of easily getting stuck in local minim . In order to improve it, Jzau-Sheng Lin et al.[4] have used an annealed Hopfield neural network with a new cooling schedule, and Su-Bing Zhang et al. [5] have proposed a dynamic routing algorithm based on chaotic neural network, but all these methods have a drawback: they only care about one QoS

F. Yin, J. Wang, and C. Guo (Eds.): ISNN 2004, LNCS 3174, pp. 269–274, 2004.

parameter, the energy function is hard to construct, when more than two QoS parameters are considered. Xiang et al. [6] have presented a GA-based algorithm for QoS routing in multicast. When the size of the destination group is large, the representation becomes very complicated, and it is hard to practice. Wang et al. [7] have proposed a tree data structure for genotype representation in their GA-based algorithm, and A.T. haghighat et al. [8] have extended the algorithm of Wang et al., but all of these GA-based approaches have a weakness: It is easy to get into local convergence and sometimes its convergence rate is too low.

In the remainder of the paper, we propose a novel hybrid neural network and genetic algorithm approach for the QoS multicast routing problem. Some novel schemes are also proposed for coding, crossover, mutation and back-propagation.

2 Problem Formulations

We model the network system as a directed connected graph $G(V, E)$, where V is a finite set of nodes and E is a finite set of links. Each link $e(u, v) \in E$ is associated with several nonnegative additive and subtractive QoS values: cost, delay, available bandwidth, jitter, etc. Cost is represented a function, C(l), all the addictive QoS parameters are denoted by the QoS function, $Q_i(e)$, and the subtractive ones are denoted by $QM_i(e)$.

For a multicast connection, the problem of QoS routing is to find a multicast Steiner tree $T(s, D)$, where s is the source of a multicast request and D is the set of destinations D $= \{d_1, \ldots, d_n\}$.

Given a multicast Steiner tree, let $PT_e = \sum_{i=1}^{n} P(s, d_i)$denote all the path of the tree. The total cost of the tree $T(s, D)$ is defined as $\sum_{e \in PT_e} C(e)$, the total addictive value of the tree $T(s, D)$ is represented as the sum of all the links in the tree: $\sum_{e \in PT_e} Q_i(e)$, and the bottleneck value of the tree is defined as the minimum available value at any link along the path: $\min(QM_i(e), e \in PT(s, d))$. Let Δ_i be corresponding addictive QoS constraints and δ_i be corresponding subtractive QoS constraints that need to be satisfied. The multiple-constraint multicast routing problem is defined as follows:

$$\begin{cases} \sum\limits_{e \in PT_e} Q_i(e) \leq \Delta_i \\ \min(\sum\limits_{e \in PT_e} C(e)) \\ \min(QM_i(e), e \in PT(s, d)) \geq \delta_i \end{cases} . \quad (1)$$

3 The Proposed Algorithm

The genetic algorithm is a highly parallel randomly searching algorithm that imitates life evolution in the nature, its global searching performance is very good, while the BP algorithm does quite well in local searching, we combine these two algorithms to yield a better performance. Generally, to solve the combinatorial optimization problems, there are two methods to combine the GA with the NN: 1) The GA acts as the major method, and the NN is used to help the GA to gain the good initial population or the better offspring. 2) The NN is the major method, the GA is used as a common training method for the NN, or is used to help to evaluate and optimize the neural network's parameters and structure.

Most of the studies concentrate on the latter and use the Hopfield neural network. In this paper, we deal with the former, and our algorithm is described as follows:

1: Code and initialize a population of chromosomes.
2: Evaluate each chromosome in the population by the fitness.
3: Generate offspring by selecting parents, applying crossover and mutation.
4: Evaluate the new chromosomes and insert them into the population.
5: Apply the back-propagation operator on the offspring, and replace the original one with the better solution.
6: If stopping criteria are met, return the best solution. Otherwise, go to 2.

3.1 Coding and Initial Population

In the hybrid algorithm, we have used tow types of coding scheme: a floating-point number coding scheme and a nodes-sequence coding scheme. In the nodes-sequence coding scheme, each chromosome is coded as several nodes sequences between every pair of source and destination node of a multicast tree. At first, we use the nodes-sequence coding scheme for the mutation and crossover operations, after each iteration of the evolution, we use a floating-point number to represent this chromosome, and then we apply the modified back-propagation neural networks into the population with these floating-point numbers.

3.2 Fitness Function

GA adopts fitness value to direct the search. It means, for an individual with larger fitness value (the multicast tree's delay is bounded with a low cost and other QoS metrics constraints are satisfied), it will appear in the next generation with larger possibility. In our algorithm, the constraints may be handled through penalty function. Then the constrained optimization problem may be transformed to unconstrained optimization. Therefore, the fitness function is designed as follows:

$$F(T(s,M)) = \frac{\alpha(\beta_i(\prod_{d\in M}\phi(Q_i(P(s,d)-\Delta_{id}))+\chi_j(\prod_{d\in M}\phi(QM_j(P(s,d)-\delta_{jd})))}{\sum_{e\in T(s,M)} C(e)}$$

$$\phi(Z) = \begin{cases} 1 & Z\leq 0 \\ r & Z>0 \end{cases} . \quad (2)$$

Where α, β_i, χ_j are the positive real coefficients, which reflect how important the cost, the addictive QoS metrics and the subtractive QoS metrics are respectively, $\phi(Z)$the penalty function and r the degree of penalty.

3.3 Crossover and Mutation Strategy

For each path, $P(s, d_i)$, based on the total cost, we use the same penalty method in the path as in the whole multicast tree:

$$F(P(s,d_i)) = \frac{\alpha(\beta\phi(Q_j(P(s,d_i)) - \Delta_{jd_i}) + \chi(\phi(QM_k(P(s,d_i)) - \delta_{kd_i}))}{\sum\limits_{e\in P(s,d_i)} C(e)}. \quad (3)$$

We compute the fitness of two paths in an individual and select the better path, we get the set of the paths: $TP_O = \{P_O(s,d_1), P_O(s,d_2), ..., P_O(s,d_m)\}$, then we use one of the following well-known crossover schemes for it:

One point crossover operator

One point crossover operator, with fixed probability $P_c(\approx 0.6 - 0.9)$

Two point crossover operator

The mutation operation is applied with the fixed probability (0.005-0.1), Firstly, select tow nodes in the $P_O(s, d_i)$randomly, say n and m, and then search a path from n to m with the well-known randomized depth-first search algorithm. Secondly, insert this path in to the $P_O(s, d_i)$, if there are any same nodes in the new $P_O(s, d_i)$, delete the link between the same nodes.

3.4 Back Propagation

After the iteration of the evolution, we reconstruct every tree and code it in floating numbers. Then we apply the modified back-propagation to it. Back-propagation needs a teacher that knows the correct output for any input ("supervised learning"). We don't have the exact result here, so we modify the BP a little bit, we use the difference between the best individual and every individual in this generation to direct the error back-propagation. We also apply momentum and variable learning rate to the BP neural network, which help to accelerate convergence and to keep the evolution trajectory moving in a consistent direction. We obtain the following equation for it:

$$\begin{cases} T_j(k+1) = T_j(k) + \eta\Delta(T_j) + \alpha(k)E_j(k) \\ \alpha(k) = 2^{\lambda}\alpha(k-1) \\ \lambda = sgn[E_j(k) - E_j(k-1)] \\ \Delta(T_j) = [T_{bigest}(k) - T_j(k)]\ |N(0,1)| \\ E_j(k+1) = \eta acc_j(k)\Delta(F_j) + \alpha(k)E_j(k) \\ acc_j(k) = \begin{cases} 1\,if\ the\ update\ improved\ \cos t \\ 0 \end{cases} \end{cases} \quad . \quad (4)$$

where $T_j(k)$ is the variable of an jth individual at the kth generation. η is the learning rate, $\alpha$the momentum rate, $N(0, 1)$ denotes the normal distribution, $|\cdot|$ denotes an absolute value, $\Delta(T_j)$ is the amount of change of an individual, $E_i(k)$ is the evolution tendency of previous evolution.

4 Simulation Results

In this section, performance of our hybrid algorithm and other GA, NN algorithms is evaluated via simulation under various random graphs. The random graphs used in the experiments are generated by the Salama[9] graph generator. The source and the destinations are randomly generated . We have implemented the proposed hybrid algorithm on a P4 (1.5G), 256 MB RAM, PC in c program.

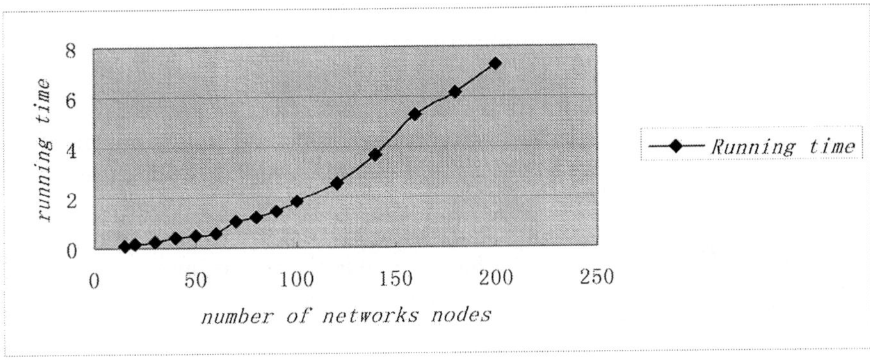

Fig. 1. Running time of the proposed Algorithm

We carried out the proposed hybrid algorithm for different kinds of networks with different nodes as shown in Fig. 1. It shows that by increasing the network scale, the algorithm still yields good performance: the running time of the proposed hybrid algorithm grows slowly with the size of the network. It can satisfy the real time request of the running real-time networks.

By simulations, we compare the results of our hybrid algorithm with the traditional GA and NN method. For the sake of comparison, we use the same parameters and set the same delay bounds and bandwidth constraints to the multicast tree. The experiments mainly concern the convergence ability and the convergence speed.

Table 1. Comparison of the proposed algorithm, GA and NN. CIA:Convergence iteration of average;P(400), P(1000): Possibility of convergence within 400 and 1000 iterations P(loc): Possibility of getting stuck into the local convergence

Algorithm	CIA	P(400)	P(1000)	P(loc)
Proposed Algorithm	353	0.78	1	0.001
GA	1602	0	0.2	0.08
NN	2116	0	0.08	0.11

Performance behavior of proposed algorithm and GA methods for a network with 30 nodes is shown in tabl.1. All the algorithms can converge to a good routing, while the proposed method has found a feasible route by few generations(or iterations), and also, the propose algorithm's possibility of getting stuck into the local convergence is much lower than the others. Therefore, the search efficiency of the hybrid method is better than GA and NN method.

5 Conclusions

In this paper, we have proposed a novel QoS multicasting routing approach by combining back-propagation neural networks and genetic algorithm. The performance evaluation via simulations shows that the proposed hybrid algorithm has better behavior than GA and NN. From the simulation results, we conclude:

The back-propagation speeds up the convergence and jumps out of the premature convergence effectively; the proposed algorithm can find a feasible QoS route faster than GA and NN.

The nodes sequences coding scheme simplifies the mutation and crossover operation. The proposed algorithm has easy genetic operations;

The proposed algorithm is easy to be extended to applications with more than two QoS constraints.

References

1. Wang, Z., Crowcroft, J.: Quality of service for supporting multimedia applications. IEEE JSAC, 14 (1996) 1228-1234
2. Wu, J.J., Hwang, R.H., Liu, H.I.: Multicast routing with multiple QoS constraints in ATM networks. Information Sciences, 124 (2000) 29–57
3. Chotipat, P., Goutam, C., Norio, S.: A Neural Network Approach to Multicast Routing in Real-Time Communication Networks. Network Protocols, IEEE CNF, 11 (1995) 332-339
4. Lin, J.S., Liu, M., Huang, N.F.: The shortest path computation in MOSPF protocol using an annealed Hopfield neural network with a new cooling schedule, Information Sciences, 129 (2000) 17-30
5. Zhang, S., Liu, Z.: A new dynamic routing algorithm based on chaotic neural networks. Journal of china instituation of communication 12(2001) 1-7
6. Xiang, F., Junzhou, L., Jieyi, W.: G. Guanqun. QoS routing based on genetic algorithm. Computer Communication., 22 (1999) 1394-1399
7. Wang, Z., Shi, B., Zhao, E.: Bandwidth-delay-constrained least-cost multicast routing based on heuristic genetic algorithm. Computer Communication., 24 (2001) 685-692.
8. Haghighatab, A.T., Faezb, K., Dehghan, M.A.: Mowlaeib, Y.: GA-Based heuristic algorithms for QoS based multicast routing. Knowledge-Based Systems, 16 (2003) 305-312
9. Salama, H.F., Reeves, D.S., Viniotis, Y. : Evaluation of multicast routing algorithms for real-time communication on high-speed networks. IEEE JSAC, 15 (1997) 332-345

Performance Analysis of Recurrent Neural Networks Based Blind Adaptive Multiuser Detection in Asynchronous DS-CDMA Systems

Ling Wang[1, 2], Haihong Tao[2], Licheng Jiao[1, 2], and Fang Liu[3]

[1] Institute of Intelligent Information Processing, Xidian University,
710071 Xi'an, China
[2] Key Lab for Radar Signal Processing, Xidian University,
710071 Xi'an, China
wanglingboy@yahoo.com.cn, {hhtao, lchjiao}@xidian.edu.cn
[3] School of Computer Science and Technology, Xidian University,
710071 Xi'an, China
f63liu@163.com

Abstract. With dynamics property and highly parallel mechanism, recurrent neural networks (RNN) can effectively implement blind adaptive multiuser detection at the circuit time constant level. In this paper, the RNN based blind adaptive multiuser detection is extended to ubiquitous asynchronous DS-CDMA systems, and the performance of the output signal to interference plus noise ratio, asymptotic multiuser efficiency, computational complexity, operating time, and mismatch of the detector are quantitatively analyzed. With performance analysis and numerical simulations, it is shown that RNN based blind adaptive multiuser detection can converge at the steady quickly and offer significant performance improvement over some existing popular detectors in eliminating multiple access interference and "near-far" resistance.

1 Introduction

Direct-sequences code-division multiple access (DS-CDMA) schemes have been considered as the attractive multiple access technique in third-generation (3G) and future beyond 3G or 4G systems [1, 2]. In the DS-CDMA framework, multiple-access interference (MAI) existing at the received signal creates "near-far" effects and constitutes the main limitation. Multiuser detection (MUD) techniques can efficiently suppress MAI and substantially increase the capacity of systems.

Recently, considerable attention has been focused on blind adaptive multiuser detection (BAMUD) [3-6] only requiring the knowledge of the signature waveform and timing of the desired user. BAMUD based on least mean squares (LMS), recursive least squares (RLS), and Kalman algorithms were proposed in [3, 4, 5]. Despite their simple structures it is still difficult to implement them in real time. A circuit implementation for BAMUD based on the recurrent neural networks (RNN) proposed by Tank *et al* [7, 8] was developed in [6], which only discussed the synchronous system and was not be involved in the quantitative performance metrics.

F. Yin, J. Wang, and C. Guo (Eds.): ISNN 2004, LNCS 3174, pp. 275–280, 2004.

In this paper, we extend the recurrent neural networks based blind adaptive multiuser detector to asynchronous systems and its performance of the output signal to interference plus noise ratio, asymptotic multiuser efficiency, computational complexity, operating time, and mismatch effect is quantitatively analyzed .

2 Discrete Asynchronous Signal Model

Consider an antipodal K-user asynchronous DS-CDMA system signaling through an additive white Gaussian noise channel. Without loss of generality, let user 1 be the desired user, and the delay of user 1 to K satisfy $0=\tau_1 \le \tau_2 \le \cdots \le \tau_k < T$, where $\tau_k = p_k T_c$, $p_k \in \{0,1,...,G-1\}$. T and T_c, respectively, is the symbol and chip interval, satisfying $T=GT_c$, where G is the spreading gain. Assumed that $b_k(i)$ and A_k, respectively, is the data bit at the ith interval and its amplitude transmitted by user k. By passing through a chip-matched filter, followed by a chip-rate sampler, the discrete-time output of the receiver is

$$\mathbf{r}(i) = \mathbf{SAb}(i) + \mathbf{n}(i) \ , \tag{1}$$

where $\mathbf{s}_k$ is the signature waveform vector with size of $G\times 1$ corresponding to user k, $\mathbf{b}_I(i) = [d_2(i), b_2(i), ..., d_K(i), b_K(i)]^T$, $\mathbf{b}(i) = [b_1(i), \mathbf{b}_I^T(i)]^T$, $d_k(i) = b_k(i-1)$, $\mathbf{A} = diag(A_1, \mathbf{A}_I)$, $\mathbf{A}_I = diag(A_2, A_2, ..., A_K, A_K)$, $\mathbf{S}_I = [\mathbf{s}_{21}, \mathbf{s}_{20}, ..., \mathbf{s}_{K1}, \mathbf{s}_{K0}]$, $\mathbf{S} = [\mathbf{s}_1, \mathbf{S}_I]$, $\mathbf{s}_{k0} = \mathbf{T}_0 \mathbf{s}_k$, $\mathbf{s}_{k1} = \mathbf{T}_1 \mathbf{s}_k$, $\mathbf{T}_0 = \begin{bmatrix} \mathbf{0} & \mathbf{0} \\ \mathbf{I}_{(G-p_k)\times(G-p_k)} & \mathbf{0} \end{bmatrix}$, $\mathbf{T}_1 = \begin{bmatrix} \mathbf{0} & \mathbf{I}_{p_k \times p_k} \\ \mathbf{0} & \mathbf{0} \end{bmatrix}$, $\mathbf{n}(t) \sim \mathcal{N}\left(\mathbf{0}, \sigma^2 \mathbf{I}\right)$, and $\mathbf{I}$ is the identity matrix.

3 RNN Based Blind Adaptive Multiuser Detection

The structure of the RNN for solving the linear programming problem [7] with M=3 inputs and N=3 outputs is illustrated in Fig. 1. When $f(z)=K_1 z$ and $g(u)=K_2 u$, where K_1 and K_2 are amplifying gain, the obtained RNN is always stable. It can be used to solve the Least Square (LS) problem with infinite precision in real time without the "programming complexity" and local minimum problem inherent in Hopfield networks [7, 8]. The blind multiuser detector based on the minimum output energy (MOE) is $\mathbf{w}_{1MOE} = \beta \mathbf{R}^{-1} \mathbf{s}_1$, where $\beta = \mathbf{s}_1^T \mathbf{R}^{-1} \mathbf{s}_1$, $\mathbf{R} = E[\mathbf{r}\mathbf{r}^T] = \mathbf{SA}^2\mathbf{S}^T + \sigma^2 \mathbf{I}$ [3, 4]. Letting the connected strength matrix $\mathbf{D}$=$\mathbf{R}$ and bias current $\mathbf{p}$=$\mathbf{s}_1$ of the RNN, we can obtain

$$d\mathbf{u}/dt = -\left(\mathbf{I}/RC + K_1 K_2 \mathbf{R}^T \mathbf{R}/C\right)\mathbf{u} + K_1 \mathbf{R}^T \mathbf{s}_1 / C \ , \tag{2}$$

where R and C is the resistance and capacitance. At the stable point, its output is

$$\mathbf{v} = \left[(1/RK_1K_2)\mathbf{I} + \mathbf{R}^T\mathbf{R}\right]^{-1} \mathbf{R}^T \mathbf{s}_1 \approx \mathbf{R}^{-1}\mathbf{s}_1 \ . \tag{3}$$

The RNN based blind adaptive multiuser detector (RNNBBAMUD) can be get by

$$\mathbf{w}_{1RNN} = \mathbf{v}/\mathbf{s}_1^T \mathbf{v} \approx \beta \mathbf{R}^{-1} \mathbf{s}_1 = \beta [\mathbf{SA}^2\mathbf{S}^T + \sigma^2 \mathbf{I}]^{-1} \mathbf{s}_1 \ . \tag{4}$$

If R, K_1, and K_2 are appropriately chosen, the obtained detector can approach to that based on the MOE criteria at any precision. Letting λ be the forgetting factor, the correlation matrix can be iteratively updated to track dynamic channels as follows.

$$\mathbf{R}(n) = \lambda\mathbf{R}(n-1) + \mathbf{r}(n)\mathbf{r}^T(n) \ . \tag{5}$$

4 Performance Analysis

4.1 Output Signal to Interference plus Noise Ratio (SINR), Bit Error Rate (BER), and Asymptotic Multiuser Efficiency (AME)

After the RNNBBAMUD converges, the detector will keep unchanged and approach to the MOE based detector. The output signal of the detector of the desired user

$$y_1(i) = \mathbf{w}_1^T \mathbf{s}_1 A_1 b_1(i) + \mathbf{w}_1^T [\mathbf{S}_I \mathbf{A}_I \mathbf{b}_I(i) + \mathbf{n}(i)] \ . \tag{6}$$

According to the formula above, we can obtain the output SINR of user 1.

$$SINR_{1theo} \approx A_1^2 / \left\{ \beta^2 [\left\| \mathbf{s}_1^T [\mathbf{S}\mathbf{A}^2\mathbf{S}^T + \sigma^2 \mathbf{I}]^{-T} \mathbf{S}_I \mathbf{A}_I \right\|^2 + \left\| [\mathbf{S}\mathbf{A}^2\mathbf{S}^T + \sigma^2\mathbf{I}]^{-1} \mathbf{s}_1 \right\|^2 \right\} \ . \tag{7}$$

It is obvious to write out the expression of BER at steady state by (6)

$$P_1(\sigma) = 2^{2(1-k)} \sum_{\mathbf{b}_I \in \{-1,1\}^{2(K-1)}} Q\left[\left(A_1 - \mathbf{w}_1^T \mathbf{S}_I \mathbf{A}_I \mathbf{b}_I \right) / \sigma \left\| \mathbf{w}_1 \right\| \right] \ , \tag{8}$$

where $Q(x) = (1/\sqrt{2\pi}) \int_x^{\infty} e^{-(x^2/2)} dx$. Under the high signal to noise ratio condition ($\sigma \to 0$), the error probability is mainly decided by the maximal term of sum in (8). Therefore, the AME of the desired user in asynchronous systems is

$$\eta_{1asyn} = \max{}^2 \left\{ 0, 1 - \sum_{k=2}^{K} (A_k / A_1) \left[\left| \mathbf{w}_1^T \mathbf{s}_{k0} \right| + \left| \mathbf{w}_1^T \mathbf{s}_{k1} \right| \right] \right\} / \left\| \mathbf{w}_1 \right\|^2 \ . \tag{9}$$

4.2 Computational Complexity and Operating Time

By (2), we can know that the convergence time of the RNN is dependent on the minimal eigenvalue $\lambda_{\min}$ of the matrix $\mathbf{I}/RC + K_1 K_2 \mathbf{R}^T \mathbf{R}/C$. Although the accurate value is related to the background noise level, we can choose the lower bound as $1/RC$. The corresponding circuit time constant (CTC) is equal to RC in second. Therefore, the approximate convergence time can be obtained by $5RC$. In actual systems, since $\mathbf{R}^T\mathbf{R}$ and $\mathbf{R}$ have the same signal subspace, $\lambda_{\min} = 1/RC + \sigma^4 K_1 K_2 / C$ for the synchronous system with $K < G$ and it is possible that $\lambda_{\min} >> 1/RC$ for the asynchronous system with $K \geq G/2$. Therefore, the RNN can converge at the steady solution with faster speed in asynchronous systems than that in synchronous systems.

Apart from the convergence time needed by the RNN, the iterative updating for the correlation matrix by (5) will also consume processing time. We compare the computational complexity and operating time of several detectors in Table 1.

Table 1. Computational complexity (CC) and operating time (OT) needed by the LMS, RLS, KALMAN, and RNN based blind adaptive multiuser detectors by using a digital signal processor with the instruction period of 25 nanoseconds and programming efficiency of 50%.

Detector	LMS	RLS	KALMAN	RNN
CC	$6G$	$4G^2+7G$	$4G^2-3G$	$2G^2+2G$
OT (in sec.)	$300G$	$200G^2+350G$	$200G^2-150G$	$<100G^2+100G+5RC$

4.3 Mismatch Effect

Due to multipath propagation, timing error, the deviation between **D** and **R**, and the estimation error of **R**, mismatch always exists in actual systems. Suppose that $\hat{\mathbf{s}}_1=\mathbf{s}_1+\boldsymbol{\delta}_1$, $\mathbf{D}_g=\hat{\mathbf{R}}^T+\boldsymbol{\Delta}_g$, $\mathbf{D}_f=\hat{\mathbf{R}}+\boldsymbol{\Delta}_f$, and $\hat{\mathbf{R}}=\mathbf{R}+\boldsymbol{\Delta}_\mathbf{R}$, where $\boldsymbol{\delta}_1$, $\boldsymbol{\Delta}_g$, $\boldsymbol{\Delta}_f$, and $\boldsymbol{\Delta}_\mathbf{R}$ are corresponding mismatch level. Furthermore, let $\boldsymbol{\Delta}'_g=\boldsymbol{\Delta}^T_\mathbf{R}+\boldsymbol{\Delta}_g$ and $\boldsymbol{\Delta}'_f=\boldsymbol{\Delta}_\mathbf{R}+\boldsymbol{\Delta}_f$, it follows that $\mathbf{D}_g=\mathbf{R}^T+\boldsymbol{\Delta}'_g$, $\mathbf{D}_f=\mathbf{R}+\boldsymbol{\Delta}'_f$, and $\boldsymbol{\Delta}=\mathbf{R}^T\boldsymbol{\Delta}'_f+\boldsymbol{\Delta}'_g\mathbf{R}$. According to the dynamic equation in (2), we can get the error of the RNNBBAMUD.

$$\left\|\mathbf{w}_{opt}-\mathbf{v}\right\|\le\left\|\mathbf{R}^{-1}\right\|^2\{\alpha+\zeta+\chi\}\ , \tag{10}$$

where $\mathbf{w}_{opt}=\mathbf{R}^{-1}\mathbf{s}_1$, $\alpha=\left\|\mathbf{w}_{opt}\right\|/RK_1K_2$ can appropriately be adjusted by R, K_1, and K_2. $\zeta=\left\|\boldsymbol{\Delta}\right\|\left\|\mathbf{w}_{opt}\right\|+\left[\left\|\mathbf{R}^{-1}\right\|^2/RK_1K_2+1\right]\left\|\boldsymbol{\Delta}'_g\right\|$ is induced by the estimation error of correlation matrix and connection strength error. $\chi=\left[\left\|\mathbf{R}^{-1}\right\|/RK_1K_2+1/\left\|\mathbf{R}^{-1}\right\|\right]\left\|\boldsymbol{\delta}_1\right\|$ is brought about by the signature waveform mismatch.

5 Numerical Simulation

The performance of the RNNBBAMUD in asynchronous systems with 6 users is verified by Monte Carlo numerical simulation compared with that of the blind adaptive multiuser detectors based on, respectively, the LMS [3], RLS [4], and KALMAN [5] algorithms, the conventional matched filter (MF), MMSE detector, and the performance bound of the MF in single user system, namely by single user bound (SUB). The forgetting factor is equal to 0.997. The signature waveform of each user is Gold sequence with G=7. $R=10k\Omega$ and $C=10pF$ in the RNN. Define Signal-to-Noise Ratio(SNR) of user k as SNR_k. The desired user is user 1 while all other users are the interfering users, that's $SNR_E=SNR_1$, $SNR_I=SNR_{2\text{-}K}$.

The average output SINR performance versus iteration number in the dynamical environment changing number of users is illustrated in Fig. 2. While $n<300$, user 1 to 4 are active users. At n=300, user 5 and 6 are added to the system. At n=700, user 3 and 4 leave the system. Furthermore, SNR_E=15dB, and SNR_I=25dB thus existing strong MAI. From the figure, the RNNBBAMUD takes on faster convergence rate and is closer to the theoretic value than other mentioned detectors.

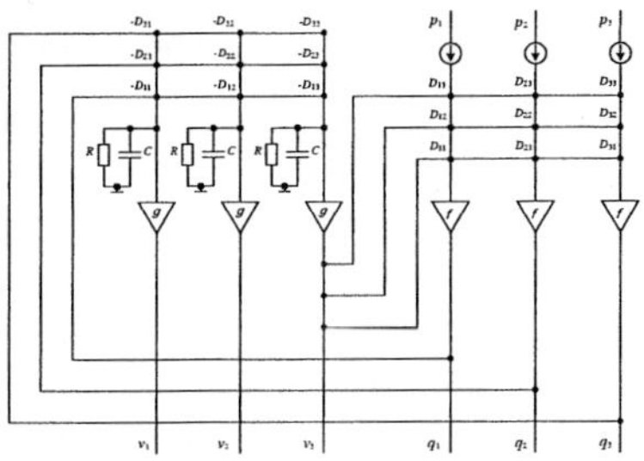

Fig. 1. RNN with M=3 and N=3

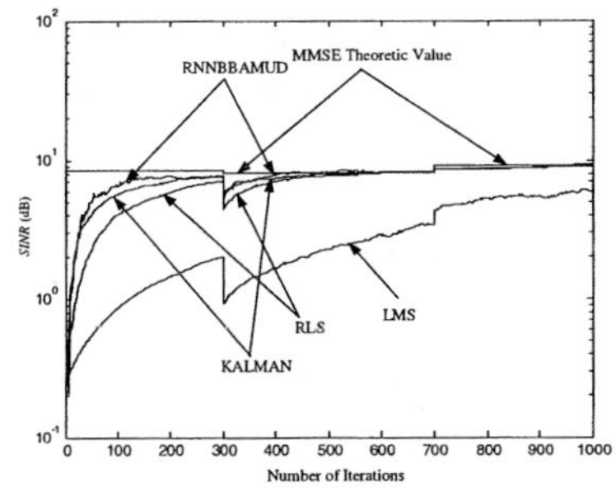

Fig. 2. Average SINR versus number of iterations

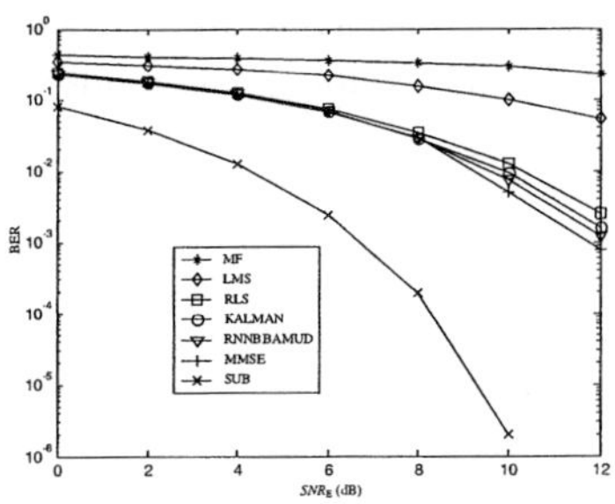

Fig. 3. BER versus SNR

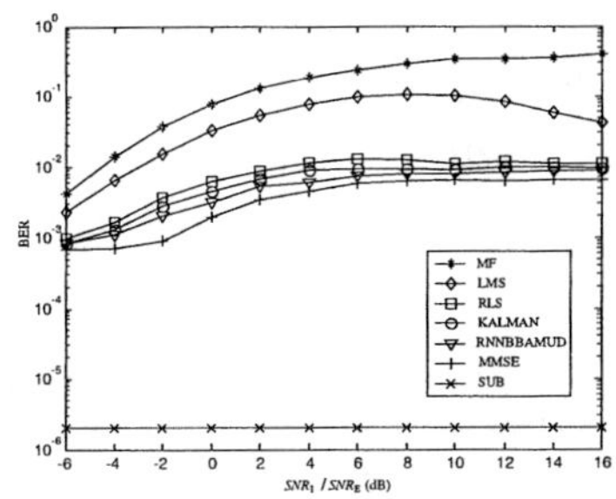

Fig. 4. BER versus RNR

The BER performance versus SNR at the steady state, assumed n=1000, is demonstrated in Fig. 3, where SNR_I=18dB keeps unchanged while SNR_E changes from 0 to 12dB. The performance of suppressing MAI and near-far resistance at the steady state is stimulated in Fig. 4, where SNR_E=8dB keeps unchanged while the "Far-to-Near ratio"(FNR), defined as SNR_I / SNR_E, changes from -6 to 16dB. From Fig. 3 and 4, it is shown that the RNNBBAMUD demonstrates the superior ability of suppressing MAI and background noise.

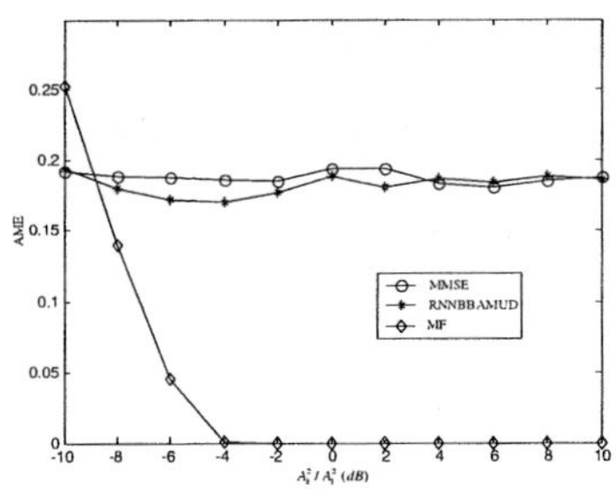

Fig. 5. AME versus FNR

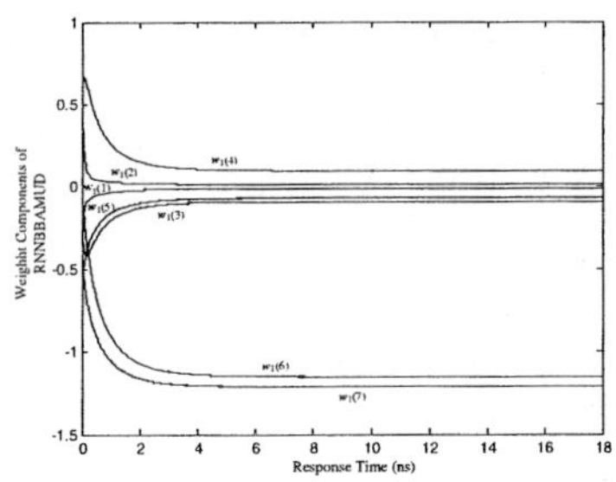

Fig. 6. Transient response of RNNBBAMUD

The AME performance versus FNR at the steady state is presented in Fig. 5. From the figure, we can see that the AME of the RNNBBAMUD approaches to that of the MMSE based detector in asynchronous channel.

The dynamics property of the RNN is simulated in Fig. 6, where SNR_E=10dB, SNR_I=18dB, n=1000, and approximate theoretic CTC is 1.08*ns*. The figure shows that the RNN can reach the steady state at CTC level.

6 Conclusions

In this paper, we extend the RNN based blind adaptive multiuser detection in asynchronous DS-CDMA systems, and quantitatively analyze the performance of the output SINR, AME, computational complexity, operating time, and mismatch of the detector. With performance analysis and numerical simulation, it is shown that the RNN based blind adaptive multiuser detector is superior to some famous existing popular detectors in eliminating MAI and "near-far" resistance. Furthermore, the detector can converge at the steady state quickly and can be implemented in real time.

References

1. Hanzo, L., Yang, L.L., Kuan, E.L., Yen, K.: Single- and Multi-Carrier DS-CDMA: Multi-User Detection, Space-Time Spreading, Synchronization, Networking and Standards. IEEE Press-Wiley, New York (2003)
2. Yang, L.L., Hanzo, L.: Multicarrier DS-CDMA: A Multiple Access Scheme for Ubiquitous Broadband Wireless Communications. IEEE Commun. Magazine. 10 (2003) 116–124
3. Honig, M., Poor, H.V.: Blind Multiuser Detection. IEEE Trans. on Information Theory. 41 (1995) 944–960
4. Poor, H.V., Wang, X.D.: Code-Aided Interference Suppression for DS/CDMA Communications, Part II: Parallel Blind Adaptive Implementations. IEEE Trans. on Commun. 9 (1997) 1112–1122
5. Zhang, X.D., Wei, W.: Blind Adaptive Multiuser Detection Based on Kalman Filtering. IEEE Trans. on Signal Processing. 1 (2002) 87–95
6. Wang, N., Zheng, B.Y., Zhu, W.P.: A Blind Multiuser Detector Based on Neural Network. Int. Conf. on Commun. Technology. IEEE Press, New York (1998) 22–24
7. Tank, D.W., Hopfield, J.J.: Simple "Neural" Optimization Networks: an A/D Converter Signal Decision Circuit, and Linear Programming Circuit. IEEE Trans. on Circuit and Systems. 5 (1986) 533–541
8. Luo, F.L., Unbehauen, R.: Applied Neural Networks for Signal Processing. Cambridge University Press, Cambridge (1997)

Neural Direct Sequence Spread Spectrum Acquisition

Tarek Elhabian, Bo Zhang, and Dingrong Shao

School of Electronics & Information, Beijing University of Aeronautics & Astronautics, China.
{Tareksalah2002, shaodingrong}@hotmail.com

Abstract. A neural network model is described which simulate the Parallel Matching Filtering (PMF) for Direct Sequence Spread Spectrum (DSSS) signal acquisition. This system is based on training the Counter Propagation Network (CPN) in all half chip phase shifts of the Pseudo Noise (PN) code. The trained network can be used at the receiver for the signal acquisition. The CPN performance in Additive Wight Gaussian Noise (AWGN) channel is evaluated. Computer simulations carried on maximal length sequences of length N=256, show that the proposed system can effectively decide the half chip phase shift of the received code even at much lower Signal to Noise ration (S/N) until S/N = -27.74dB. This model has a simple architecture, so can be realized in a simple hardware. This makes the neural network based acquisition technique faster and more robust than the other conventional acquisition techniques.

Keywords. Neural Network, Matching Filtering, Direct Sequence Spread Spectrum Acquisition, Counter Propagation Network

1 Introduction

The code synchronization is required for implementation of spread spectrum systems. In the DSSS system [1], the message signal is multiplied by a high-rate digital signal (called spreading code) which widening the signal bandwidth. This signal then is modulated on a carrier. The rate of the spreading code is called chip rate and it is usually much higher than the data rate. Hence, the energy of the signal is spread over a much higher bandwidth and the spectral density is lowered. At the receiver, the spreading code is removed by multiplication of the received signal with a local replica of the code. Hence, the spectral density of the information signal is raised to its original level and the interfering signals are spread and removed using a filter. The overall synchronization process can be divided into two stages initial synchronization (acquisition), and tracking [2], [3]. The acquisition is more difficult than tracking [1]. Typical acquisition methods include serial, parallel and hybrid systems. Since a common serial acquisition time is rather long [4] and detection probability also low, so the parallel method was proposed for DSSS acquisition [5]. The purpose of the DSSS receiver is to de-spread the received signal and to extract the information content from the de-spread signal. The received signal can be expressed as:

$$r(t) = sig(t) + n(t) \tag{1}$$

F. Yin, J. Wang, and C. Guo (Eds.): ISNN 2004, LNCS 3174, pp. 281–286, 2004.

$$sig(t) = \sqrt{2S}\, d(t)C(t + \xi T_c)\cos(w_0 t + w_d t + \theta). \tag{2}$$

Where: S is the signal power, $d(t)$ is the data, w_0, w_d and θ are the carrier frequency, the Doppler frequency and carrier phase respectively. The expression $C(t+\xi T_c)$ is a ± 1 valued, with respect to an arbitrary code delayed ξT_c time. The noise component $n(t)$ represents the AWGN. Recent years the applications of neural networks in pattern recognition, digital signal processing, as well as in digital communications, have been studied increasingly [6], [7]. In this paper, a neural direct sequence acquisition system is presented using the CPN.

2 PN Code Phase Shifts Descriptions

The spreading PN code is assumed to be the maximal length sequences of period N = 256 as:

$$PN = [c_0, c_1, c_2, \ldots, c_{N-1}] \tag{3}$$

Each PN code bit was duplicated to make a vector of 512 bits as in equation 4. This duplication done to make the proposed system recognize each half chip phase shift of the PN code. The vector represents the zero phase shift sequence as:

$$C = [c_0, c_0, c_1, c_1 \ldots, c_{N-1}, c_{2N-1}] \tag{4}$$

The next phase shift sequences can be formed by taking the $2N$ consecutive values starting from the second value of the previous sequence as:

$$C^1 = [c_0, c_1, c_1 \ldots, c_{2N-1}, c_{2N-1}, c_0] \tag{5}$$

All the 512 PN code phase shifts C^{2N} were utilized for training and testing the CPN. Also they added with AWGN in different power levels for testing the CPN. The aim of the detector is to detect the presence of the signal $sig(t)$ in the received signal $r(t)$.

3 Neural DSSS Acquisition Model

The neural network models divided mainly into two categories: (1) current networks or feed forward; (2) recurrent networks or feedback networks. The Hopfield net is considered as a good example for the recurrent networks. The response of such networks is dynamic since this output feedback to modify the input. However, the stability could be problem for this network [8]. The CPN is a combination of two well-known networks, which are the Kohonen and Grooberg networks. Fig. 1 shows the structure of CPN. The neurons in layer 1 act as distributors to Kohonen's layer and perform no computation.

The Kohonen's layer functions in a winner take-all manner [8]; that is, for a given input vector, only one Kohonen's neuron outputs a logical one and all others output a

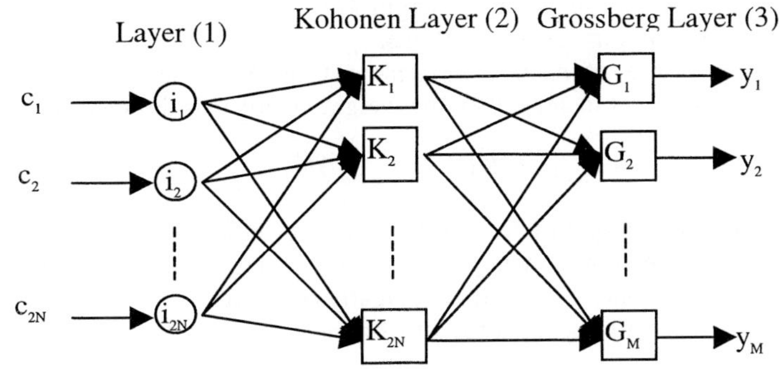

Fig. 1. CPN structure

zero. Each neuron in layer 1 connects to every neuron in the layer 2 through a separate weight w_{ij}, where i=j=1,2, …, 2N. The Kohonen's neuron output K_i is simply the summation of the weight inputs and can be defined in matrix form as:

$$\begin{bmatrix} K_1 \\ K_2 \\ \vdots \\ K_{2N} \end{bmatrix} = \begin{bmatrix} w_{1,1} & w_{1,2} & \cdots & w_{1,2N} \\ w_{2,1} & w_{2,2} & \cdots & w_{2,2N} \\ \vdots & \vdots & \vdots & \vdots \\ w_{2N,1} & w_{2N,2} & \cdots & w_{2N,2N} \end{bmatrix} \begin{bmatrix} c_1 \\ c_2 \\ \vdots \\ c_{2N} \end{bmatrix}. \tag{5}$$

$$K = WC. \tag{6}$$

Where W is the weight matrix after training. The Kohonen's neuron K_i with the largest output value should be the "winner" and its value is set to one, and all others are set to zeros. In other words, the activation function of the Kohonen's neurons is represented by a step function. The learning approach of this layer is a self-organizing algorithm that operates in the unsupervised mode. The training method can be described as:

1. Initialize the weight matrix W of order $2N\times 2N$ to small random values.
2. For each PN shift, the input pattern is represented by vector of $2N$ elements.
3. Normalizing each input vector C_i before applying to the Kohonen's layer.
4. Adjust the weight vector such that

$$W_i = W_{i-1} + \alpha(C_i - W_{i-1}). \tag{7}$$

   Where: α is a training rate coefficient that usually ≤ 0.7. It is apparent from equation (7) that the training process actual selects the Kohonen's neuron K_i whose weight vector is most similar to the input vector.

5. The above steps are repeated to each PN code phase shift to calculate W.

The only action of each neuron in the Grossberg layer is to output the value of the weight that connects it to the single nonzero Kohonen's neuron. The job of this layer is to mapping the winner neuron number at Kohonen layer to the PN code phase shift number. The training algorithm of this layer can be described as:

1. Initialize the weight matrix V of order $2N\times M$ to zero values.

2. The desired output vector Y_i of length M=9 which the PN code phase shift numbers 512 can represented in 9 bits.
3. Let the response of each Kohonen's neuron K_i is set to one and other components are set to zero; then the weight vector V_i is defined as:

$$v_{ij\ new} = v_{ij\ old} + \beta(y_i - v_{ij\ old})K_i \tag{8}$$

Where: K_i is the output of Kohonen's neuron i (only one Kohonen's neuron is nonzero), y_j is the component j of the vector of desired outputs and β initially is set to approximately 0.1 and is gradually reduced as training progresses.

4. The above steps are repeated for all Kohonen's neurons to calculate V.

4 Experiments and Results

The proposed model was tested using 512 vectors, all the PN code half chip phase shifts. The experiments were done when applying all the PN half chip code shifts and found that, the recognition rate was 100% until S/N = -27.74dB. The Kohonen training network has useful and interesting advantage that, the ability to extract the statistical properties of the input data (PN code shifts) [8].

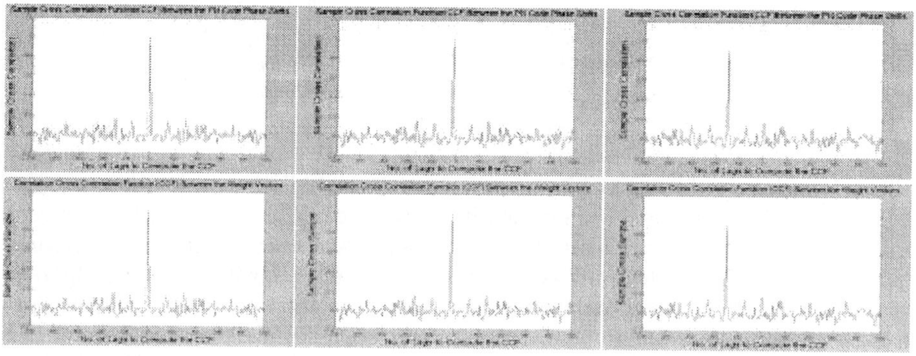

Fig. 2. From left to right and upper to down are the Cross Correlations of the PN code shifts and weight vectors respectivly

This advantage matching with the proposed model simulations which the cross correlations between the weight vectors after training are the same cross correlations between the PN code shift vectors as shown in Fig. 2. From left to right and upper to down, the fist three figures are the cross correlation of the zero half chip phase shift with itself, 5th and 50th half chip phase shifts respectively. The second three figure related to the cross correlation of the first weight vector of the W matrix with itself, 5th and 50th weight vector respectively.

It is well known that circular cross correlation of two discrete sequences in the time domain equivalent to multiplication of one sequence by the conjugate of the other sequence in the frequency domain. Fig. 3 from left to right shows that, the output result of the Kohonen layer when applying the PN half chip code shifts number 10, 50 and 200 without noise. Which the output of the Kohonen layer is the cross correlation between the PN half chip code shift and the trained weight matrix vectors.

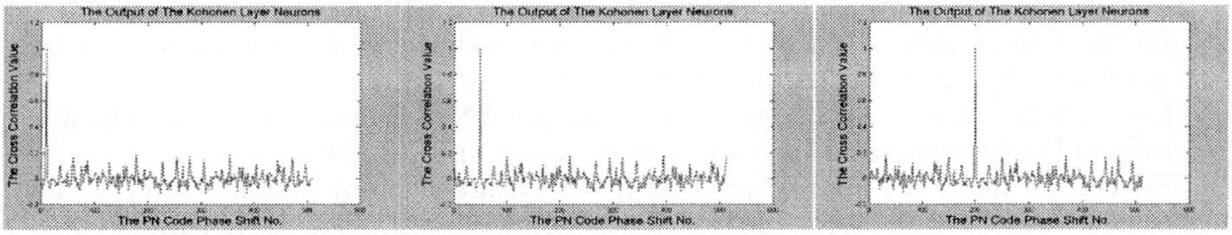

Fig. 3. From left to right the output of Kohonen layer when applying the PN code shifts No. 10, 50 and 200 respectivly

Fig. 4 from left to right shows that, the output of the Kohonen layer when applying the PN half chip shift number 50 with different AWGN, which showed the corresponding cross correlations between each of them with the trained weight matrix vectors respectively.

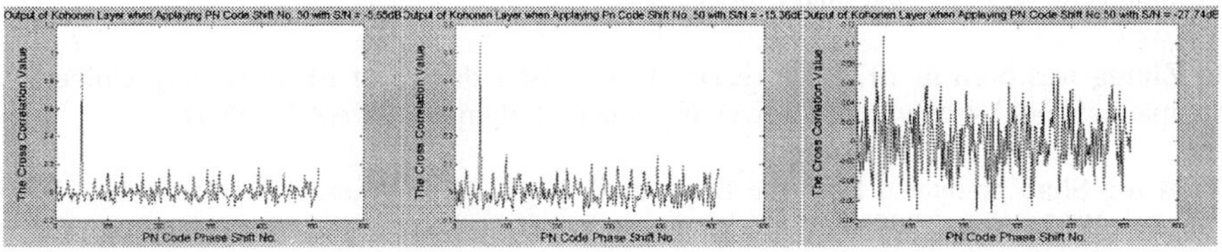

Fig. 4. From left to rigyyht the output of Kohonen layer when applying the PN code shifts number 50 at S/N= -5.65dB, -15.36dB and -27.74dB

5 Conclusions

A neural network detector has been described which can be used for non-coherent PN code acquisition in a DSSS system. First the CPN was introduced as an approach for DSSS acquisition. The CPN is trained on the PN half chip code shifts under consideration using the self-organizing algorithm. The neural network uses 256×2 shifts as its inputs of the received sequences for training to get the weights and test. It has been shown through simulation that the performance of the CPN is suitable for the DSSS acquisition even at lower signal to noise ration, until S/N = -27.74dB.

References

1. Simon, K.M., Omura, K.J., Scholtz, A.R., Levitt, K.B.: Spread Spectrum Communications Handbook. McGraw Hill Companies Inc., (2002)
2. Thompson, W.M., Dianat, S.: Non-coherent PN Code Acquisition in Direct Sequence Spread Spectrum Systems Using a Neural Nework. IEEE (2000)

3. Fan, C.C., Tsai, Z.H.: A Differentially Coherent Delay-Locked Loop for Spread-Spectrum Tracking Receivers. IEEE Journal on selected areas in Commun., VOL. 18, IEEE (2000)
4. Katz, D.M., Glisic, S.: Modeling of Code Acquisition Process in CDMA Networks-Asynchronous Systems. IEEE Commun. Letters, VOL. 3, IEEE (1999)
5. Jordan, T.M., Luecke, R.J.: A Rapid-Acquisition Architecture for Advanced Avionics and Spread-Spectrum Applications. IEEE (1990)
6. Leung, S.H., Weng, J.F.: Neural Networks for PN Synchronization and Optimal Symbol Decision in DS/SS Indoor Communications. IEEE (1994)
7. Bouder, C., Burel, G.: Spread Spectrum Codes Identification by Neural Networks. In: Systems and Control: Theory and Applications, ISBN 960-8052-11-4, WSES (2000) 257-262
8. Wasserman, D.P.: Neural Computing Theory and Practice.Van Nostrand Reinhold, ANZA (1989)

Tarek Elhabian was born in Cairo, Egypt, on March 6, 1966. He received the B.S. degree in computer science form Military Technical College, Cairo, Egypt in July 1988 and M.S. degree, in electrical computer engineering & system engineering, from Ain Shams University, Cairo, Egypt in July 2000. He is now a Ph.D. student at Beijing University of Aeronautics & Astronautics, China. His research interest is in neural network applications, spread spectrum communications, signal and image processing.

Bo Zhang was born in 1972. He received the master degree of EE in Beijing University of Aeronautics and Astronautics (BUAA). Now he is studying in BUAA for Ph.D.

Dingrong Shao graduated from the Dept. of E. E. of Beijing Aeronautical Institute, Beijing, China in 1962. From 1980 to 1983, he was engaged in the research on carrier free radar as a visiting scholar at the Catholic University of America, Washington, DC, USA. Since 1981, he works with the dept. of E. E. of Beijing University of Aeronautics and Astronautics (BUAA), 1992 he became a Professor. His main interests are in the fields of communication and signal processing. He published more then 60 papers in the fields of communication and signal processing. He received 5 scientific and technological awards at the ministerial level of China.

Multi-stage Neural Networks for Channel Assignment in Cellular Radio Networks

Hyuk-Soon Lee, Dae-Won Lee, and Jaewook Lee

Department of Industrial Engineering,
Pohang University of Science and Technology,
Pohang, Kyungbuk 790-784, Korea.
{fnledo,woosuhan,jaewookl}@postech.ac.kr

Abstract. In this paper, we consider the channel assignment problem in cellular mobile communication systems, which assigns a channel to every requested call with the minimum span of channels subject to interference constraints. In this paper, a multi-stage heuristic algorithm using neural networks for the channel assignment problem is proposed. The proposed algorithm is devised to find first a good initial feasible assignment, then a locally optimal assignment, and finally a overall best quality assignment. The proposed algorithm has been applied to the Philadelphia benchmark instances. Experimental results show that the proposed method is competitive with the other existing algorithms.

1 Introduction

In the last few years, demands for the cellular mobile service are growing rapidly. Since the available electromagnetic frequency spectrum is limited, the efficient use of frequency spectrum is regarded as of major importance. In a cellular mobile network, the service area of the system is divided into a large number of cells. When a user requests a call for this system, a channel is assigned there to provide the communication service. However, due to the nature of radio transmissions, calls generated in a cell may cause interference with calls generated in the other cells.

The objective of the channel assignment problem (CAP) is to assign a channel to every customer's call while satisfying the constraints imposed to avoid the radio interference among the channels assigned to the same cell or in relatively adjacent cells. The mathematical model for the CAP considered in this paper is as follows:

$$\begin{aligned}&\text{minimize} \max_{i,k} f_{ik}\\ &\text{subject to } \ |f_{ik} - f_{jl}| \geq c_{ij} \quad \forall i,k,j,l (i \neq j, k \neq l)\\ &f_{ik} \ \text{ is nonnegative integer}\end{aligned} \tag{1}$$

where N is the number of cells in the system, m_i, $1 \leq i \leq N$ is the number of channels required in cell i, c_{ij}, $1 \leq i,j \leq N$ is the channel separation constraint between a call in cell i and a call in cell j, and f_{ik}, $1 \leq i \leq N$, $1 \leq k \leq m_i$ is the frequency assigned to the kth call in the ith cell.

F. Yin, J. Wang, and C. Guo (Eds.): ISNN 2004, LNCS 3174, pp. 287–292, 2004.

In this paper, we propose a novel multi-stage algorithm to find a conflict-free frequency assignment with the minimum number of total frequencies. In the first stage, a good initial assignment is found by using a so-called frequency insertion strategy (FIS). In the second stage, this initial assignment is adapted to a locally optimal assignment using an adaptive local search. In the final stage, a parallel neural-network algorithm is employed to find a better optimal assignment than the assignment obtained from the second stage. This paper is organized as follows. In Section 2, a novel multi-stage algorithm is proposed. In Section 3, some experimental results and comparison for the benchmark problems are discussed. Section 4 contains a conclusion.

2 The Proposed Method

2.1 Initial Assignment Stage

In the first stage, a good initial feasible assignment is found using a frequency insertion strategy (FIS) [4]. The FIS initially permits constraint-violating assignment pretending that we do not have enough frequencies and then inserts necessary frequencies to resolve the violation by sliding the relevant calls, increasing the number of frequencies required. The basic procedure of FIS is as follows:

Step 1. Assign the frequency to a call that results in least increment of maximum frequency.
Step 2. If the call to which the frequency assigned causes constraint violation, this violation may be avoided by using frequency rearrangement called frequency iteration.

In step 2, to avoid constraint violations, we calculate conflicts from frequency 1 to maximum frequency and assign the calls to the minimum-conflict channel as widening bandwidth following conflict size. A simple example of FIS is given for illustration in Fig. 1.

2.2 Local Search Stage

The basic idea behind using local search for CAP is as follows; Given a current ordered list of calls x_p, we construct a new neighbor x'_p by selecting two calls and swapping their positions in the call ordering list. The neighborhood $\mathcal{N}(x)$ is searched in this way for a new configuration x_{p+1} for which $f(x_{p+1}) < f(x_p)$. During the local search stage, frequency exhaustive strategy (FES), which assigns each call to the least possible frequency [2], is used instead of FIS. When it converges to the local minimum assignment or fails to find such a solution before limit is reached, the local-search algorithm terminates.

One distinguished feature of the proposed local search distinguished from existing ones such as CAP3 [3] is that it obtains new call ordering from the result of FIS and uses it as an initial call list of a local search in the Stage

Calls : {4, 4, 4, 1, 2, 3}

Iteration 1

Freq.	1	2	3	4	5	6	7	8	9	10	11
Calls	4					4					4

Remaining Calls : {1, 2, 3}

Freq.	1	2	3	4	5	6	7	8	9	10	11
Calls	4		1			4					4
Conflicts	2	1	0	0	1	2	1	0	0	1	2

Remaining Calls : {2,3}

Iteration 2

Freq.	1	2	3	4	5	6	7	8	9	10	11
Calls	4		1			4		2			4
Conflicts	2	2	2	1	1	2	1	0	0	1	2

Remaining Calls : {3}

Iteration 3

Freq.	1	2	3	4	5	6	7	8	9	10	11
Calls	4		1	3		4		2			4
Conflicts	2	2	2	1	1	2	2	2	1	1	2

Freq.	1	2	3	4	5	6	7	8	9	10	11
Calls	4		1	3		4		2			4

Freq.	1	2	3	4	5	6	7	8	9	10	11	12
Calls	4		1		3		4		2			4

Fig. 1. A simple example of frequency assignment using FIS

II. Existing local search algorithms for CAP mostly use a node-degree ordering heuristics [2] 2] to generate an initial list of calls, whereas the proposed method obtains a new call list according to the order of calls assigned to ascending frequency order obtained by FIS of Stage I. In the case of example in Fig. 1, an initial call list for local search is not [4 4 4 1 2 3],but [4 1 3 4 2 4].

2.3 Neural Network Assignment Stage

In the final stage, neural-network assignment is employed to find a better solution using the result of assignment in the Stage II. After releasing calls in some cells, the calls are assigned to channels once again by using neural-network. To find the assignment that contains no violate constraint within the M channels, we use a hystereses McCulloch-Pitts neuron model where the output of the ith neuron V_i as follows [6]:

$$V_i(t+\Delta t) = \begin{cases} 1 & \text{if } U_i > UTP \text{ (upper trip point)} \\ 0 & \text{if } U_i < LTP \text{ (lower trip point)} \\ V_i(t) & \text{if } LTP \le U_i \le UTP \end{cases} \tag{2}$$

where U_i is the input of ith neuron. To solve N-cell-M-channel CAP, a total $N \times M$ binary neuron is used. The motion equation of the ijth neuron in the N-cell-M-channel problem is given by :

$$\frac{dU_{ij}}{dt} = -A\left(\sum_{q=1}^{m} V_{iq} - d_i\right) - B\left(\sum_{\substack{q=j-(c_{ii}-1)\\ q\neq j, 1\le q\le M}}^{j+(c_{ii}-1)} V_{iq} + \sum_{\substack{p=1\\ p\neq i\\ c_{ip}>0}}^{N} \sum_{\substack{q=j-(c_{ip}-1)\\ 1\le q\le M}}^{j+(c_{ip}-1)} V_{pq}\right) \tag{3}$$

where A and B are coefficients, V_{ij} is output of neuron#ij that represents the assignment of channel #j to cell #i, and d_i is the demand of cell #i. The first

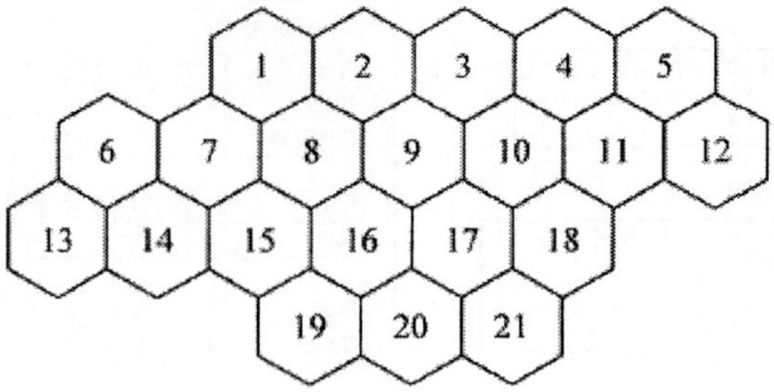

Fig. 2. The 21-cell cellular networks of the Philadelphia problem

term has nonzero output, if a total of d_i frequencies are not required for cell $\#i$. The second term has nonzero output if the assignment of frequency violates the constraints. All the input values U_{ij} are updated while all the outputs V_{ij} are fixed. Then all the outputs V_{ij} are updated while all the input values U_{ij} are fixed. The energy function E corresponding to Eq.(3) is defined as follows:

$$E = \frac{A}{2}(\sum_{q=1}^{m} V_{iq} - d_i) + \frac{B}{2}\sum_{i=1}^{n}\sum_{j=1}^{m}(\sum_{\substack{q=j-(c_{ii}-1)\\ q\neq j, 1\leq q\leq M}}^{j+(c_{ii}-1)} V_{iq} + \sum_{\substack{p=1\\ p\neq i\\ c_{ip}>0}}^{N} \sum_{\substack{q=j-(c_{ip}-1)\\ 1\leq q\leq M}}^{j+(c_{ip}-1)} V_{pq})V_{ij} \tag{4}$$

The energy function E is zero when no constraints are violated. And the goal of neural network model described by Eq.(3) is to find zero point of the energy function E.

2.4 Algorithm

Algorithm 1 *(Multi-stage algorithm using neural-network algorithm for CAP)*

% Initial assignment stage using FIS
1) *Set lower bound of maximum frequency, max iterate* $i_{\max}$
2) *Assign frequency using FIS algorithm. And* X *is a new call ordering list obtained from the result of FIS, and* $f_{\max}$ *is max frequency.*

% Local search stage
3) *Select a call* a_{ik} *in maximum frequency cell, and a call* a_{jl} *randomly* $(j \neq i)$.
4) *Make new call ordering list* X' *by swapping* a_{ik} *and* a_{jl} *from* X.
5) *Assign frequency to call ordering list* X' *using FES.* $f'_{\max}$ *is max frequency.*
6) *If* $f'_{\max} < f_{\max}$, *then* $X \leftarrow X'$, $f_{\max} \leftarrow f'_{\max}$.
Go to 3) until iteration number is equal to the max iterate $i_{\max}$, *or* $f'_{\max}$ = *lower bound*

Table 1. Demands of Philadelphia instances

#	Reuse distance	Demand vector
1	$(2\sqrt{3}, \sqrt{3}, 1, 1, 1, 0)$	(8,25,8,8,8,15,18,52,77,28,13,15,31,15,36,57,28,8,10,13,8)
2	$(\sqrt{7}, \sqrt{3}, 1, 1, 1, 0)$	(8,25,8,8,8,15,18,52,77,28,13,15,31,15,36,57,28,8,10,13,8)
3	$(2\sqrt{3}, \sqrt{3}, 1, 1, 1, 0)$	(5,5,5,8,12,25,30,25,30,40,40,45,20,30,25,15,15,30,20,20,25)
4	$(\sqrt{7}, \sqrt{3}, 1, 1, 1, 0)$	(5,5,5,8,12,25,30,25,30,40,40,45,20,30,25,15,15,30,20,20,25)
5	$(2\sqrt{3}, \sqrt{3}, 1, 1, 1, 0)$	(20,20,20,20,20,20,20,20,20,20,20,20,20,20,20,20,20,20,20,20,20)
6	$(\sqrt{7}, \sqrt{3}, 1, 1, 1, 0)$	(20,20,20,20,20,20,20,20,20,20,20,20,20,20,20,20,20,20,20,20,20)
7	$(2\sqrt{3}, \sqrt{3}, 1, 1, 1, 0)$	(16,50,16,16,16,30,36,104,154,56,26,30,62,30,72,114,56,16,20,26,16)
8	$(2\sqrt{3}, 2, 1, 1, 1, 0)$	(8,25,8,8,8,15,18,52,77,28,13,15,31,15,36,57,28,8,10,13,8)
9	$(2\sqrt{3}, \sqrt{3}, 1, 1, 1, 0)$	(32,100,32,32,32,60,72,208,308,112,52,60,124,60,144,228,112, 32,40,52,32)

Table 2. Minimum bandwidth of algorithms tested on benchmark problem

	LB	CRF	CRR	CCF	CCR	DRF	DRR	DCF	DCR	RSD	CAP3	Proposed
1	427	543	464	460	476	543	521	475	504	485	463	<u>**427**</u>
2	427	543	468	451	501	543	466	447	495	486	463	<u>**427**</u>
3	258	360	345	296	296	346	296	304	297	299	302	<u>284</u>
4	258	347	285	285	273	346	270	280	269	296	271	<u>261</u>
5	240	302	299	275	261	295	298	280	251	296	264	<u>249</u>
6	179∼189	289	223	206	197	295	220	220	<u>197</u>	226	232	209
7	856∼857	1088	928	922	939	1088	1046	952	1003	1010	945	<u>**856**</u>
8	525∼528	646	547	582	588	646	<u>544</u>	586	577	626	584	<u>544</u>
9	1714∼1725	2178	1856	1846	1889	2178	2096	1905	2016	1956	1893	<u>**1714**</u>

Note: The boldface represent the values to reach lower bounds and underlines represent the best solutions among the 12 algorithms.

% Neural-network assignment stage

7) Release calls in some cells which include max frequency cell.

8) Assign neural-network assignment within the $f_{\max}$. If the assignment is success, then go to 9). If the assignment fails, then stop.

9) Decrease $f_{\max}$ by 1 and go to 7)

3 Simulation Results

In this section, we conduct several experiments for the purpose of illustrating the results of our paper. The proposed algorithm described in the previous section is tested on Philadelphia benchmark problem [1] as shown in Fig. 2. Table 1 represents the demand vectors and reuse distances of all instances. d_j in the reuse distance $(d_0, d_1, ..., d_k)$ means that at least d_j unit distance is needed to avoid interference supposing that the difference between all pairs of assigned frequencies is more than j. And the ith element of demand vector shows the demand of cell i.

Table 2 compares the quality of the solutions obtained from Sivarajan's 8 algorithms [2], RSD [5], CAP3 [3], and our proposed method. As shown in Table 2, our method finds optimum solutions (i.e. lower bounds) for four out of nine cases and yields the best solutions among the 11 algorithms in all the cases except only one cases. We released 7 cells in order according to the closeness to the maximum frequency.

4 Conclusion

In this paper, a novel multi-stage algorithm for CAP has been proposed. Firstly, the proposed method is performed by means of FIS and local search for a good initial point. And then after releasing the calls in cells which include maximum frequency cell, channels are assigned using neural network algorithm. The performance is verified through Philadelphia benchmark problems. An application of the method to more large-scale benchmark problems remains to be investigated.

Acknowledgement. This work was supported by by Com^{2}Mac-KOSEF.

References

1. Anderson, L.G: A Simulation Study of Some Dynamic Channel Assignment Algorithms in a High Capacity Mobile Telecommunications System. IEEE Transactions on Communications, 21 (1973) 1294-1301
2. Sivarajan, K.S., McEliece, R.J., Ketchum, J.W.: Channel Assignment in Cellular Radio. in Proc. 39th IEEE Vehicular Tech. Conf. May1-3 (1989) 846-850
3. Wang, W., Rushfold, C.K.: An Adaptive Local-Search Algorithm for the Channel-Assignment Problem. IEEE Trans. Veh. Tech., Vol. 45, No.3 (1996)
4. Shin, W.-Y.: Frequency Insertion Strategy for Channel Assignment Problem. M.S. Dissertation, Dept.I.E. POSTECH (2003)
5. Battiti, R., Bertossi, A., Cavallaro, D.: A Randomized Saturation Degree Heuristic for Channel Assignment in Cellular Radio Network. IEEE Trans. Veh. Tech., Vol. 50, No. 2 (2001)
6. Funabiki, N., Okutani, N.: A Three-Stage Heuristic Combined Neural-Network Algorithm for Channel Assignment in Cellular Mobile Systems. IEEE Trans. Veh. Tech., Vol. 49, No. 2 (2000)

Experimental Spread Spectrum Communication System Based on CNN

Jianye Zhao, Shide Guo, and Daoheng Yu

Department of Electronics, Peking University, Beijing 100871, China
phdzjy@263.net

Abstract. A new spread spectrum communication system based on CNN is proposed in this paper. Chaos is generated with three cell CNN, then it's transferred to a digital sequence. The chaotic sequence is better than gold sequence when they are utilized in direct sequence spread spectrum system. Compared with the traditional gold sequence system, there is 2dB improvement in CNN chaotic sequence system when the channel is additive white Gaussian noise channel, and there is more improvement in CNN chaotic sequence system when the channel is multi-path channel. The structure of hardware CNN spread spectrum system is also shown at last.

1 Introduction

Cellular Neural Network (CNN) is a powerful nonlinear processor array which is easily realized with circuits. CNN[1,2] is locally interconnected, each cell is connected only to its neighbor, so VLSI CNN chips is easily implemented . Most of CNN's application is in the field of image processing[2], the reason is the difficulty of real-time image processing with traditional computer. There are a great deal of data to be processed, and only parallel processing algorithm can be adopted for real-time processing. But making a traditional parallel processing array is very difficult and expensive. The stability of CNN is studied, and such complex phenomena as chaos can be also performed by CNN[3]. On the other hand, study of chaotic behavior is attractive in the field of wireless communications for its security [4,5]. Digital spread spectrum communication is widely applied, and the spread sequence utilized in the system is very important. The chaotic sequence has better random characteristic, and it is employed in spread spectrum communication system[5], but it's difficult to get qualified real-time chaotic sequence. In this paper, a new spread spectrum communication system based on CNN is proposed. Real-time chaotic sequence is generated with CNN, then the sequence is modulated with DSP modulator. The experimental results confirm the good quality of CNN chaotic sequence.

This paper is organized as follows: Section 2 describes chaos in CNN system. We discuss the stability of CNN's with opposite-sign template in this section, and the chaotic attractor is shown. In section 3, the random characteristic of CNN sequence is analyzed. An experimental spread spectrum is shown in section 4, the experimental result is also given in this section. Section 5 shows some concluding remarks.

F. Yin, J. Wang, and C. Guo (Eds.): ISNN 2004, LNCS 3174, pp. 293–298, 2004.

2 Chaos in CNN

Usually representative small networks with only three cells are taken as objects for discussing chaotic behavior in the CNN. When the characteristic of chaotic behavior in CNN system, the *Lyapunov analysis* approach is employed, and cellular neural network is regarded as a dynamic system . The CNN system can be described as :

$$C\frac{dV_{Xij}(t)}{dt}=-\frac{1}{R_X}V_{Xij}(t)+\sum_{(k,l)\in N_r(i,j)}A_{ijkl}V_{Ykl}(t)+\sum_{(k,l)\in N_r(i,j)}B_{ijkl}u_{kl}(t)+\frac{1}{\sigma^2}I_{ij} \quad (1)$$

$V_{Xij}(t)$ is the state of cell C(i,j),

$$V_{Yij}(t)=\frac{1}{2}(|V_{Xij}(t)+1|-|V_{Xij}(t)-1|) \quad (2)$$

When three cells CNN is studied, it is described as:

$$\begin{aligned}\frac{dx_1}{dt}&=-x_1+p_1f(x_1)-sf(x_2)-tf(x_3)\\ \frac{dx_2}{dt}&=-x_1-sf(x_1)-p_2f(x_2)-rf(x_3)\\ \frac{dx_3}{dt}&=-x_1-tf(x_1)+rf(x_2)-p_3f(x_3)\end{aligned} \quad (3)$$

When p_1=1.25, p_2=1.1, p_3=1.0,t=3.2,r=4.4,we get the strange attractor in Fig.1.

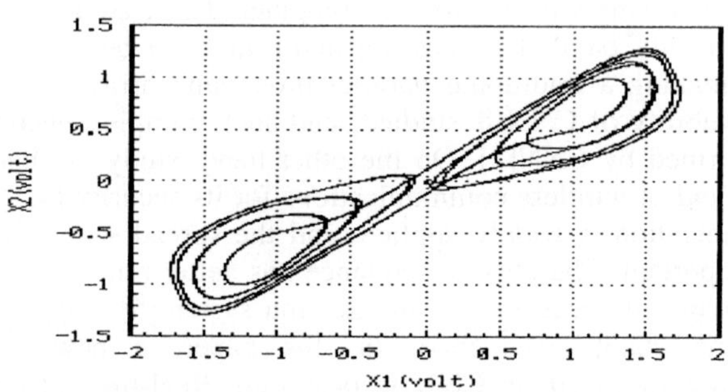

Fig. 1. The chaotic attractor of CNN

CNN work in linear region, partial saturation region or saturation region, and there is very rich dynamical behavior in small sized networks, such as three cell CNN. These dynamical behaviors can also be utilized. The synchronization of chaos is attractive to wireless communication, because secure communication can be realized with a synchronizing chaos system. If we put the CNN chaos into an A/D converter, we will get the digital CNN chaotic sequence.

3 Random Characteristic of CNN Chaotic Sequence

Chaos generated with CNN have complex dynamic behavior, so its random characteristic is suitable for spread spectrum. We can get binary CNN chaotic with a simple A/D converter. The value of chaos is x: if x>0, the corresponding value of binary sequence is *x*, then *x*=1;

if x⩽0, the corresponding value of binary sequence is *x*, then *x*=-1;

The random characteristic can be described as following context:

At first, the mean value of binary sequence is zero. Because each mean value of the cell is zero. After sampled with A/D, and the sampling frequency(Dt=1 μ s) is very fast, the sequence do not change its mean value, the mean value is described as:

$$\bar{x} = \lim_{N \to \infty} \frac{1}{N} \sum_{i=0}^{N-1} x_i = 0 \tag{4}$$

The next equation defines the auto-correlation function of the CNN sequence:

$$ac(m) = \lim_{N \to \infty} \frac{1}{N} \sum_{i=0}^{N-1} (x_i - \bar{x})(x_{i+m} - \bar{x}) = \lim_{N \to \infty} \frac{1}{N} \sum_{i=0}^{N-1} x_i x_{i+m} \tag{5}$$

the cross-correlation function of the CNN chaotic sequence is:

$$cc_{12}(m) = \lim_{N \to \infty} \frac{1}{N} \sum_{i=0}^{N-1} (x_{1i} - \bar{x})(x_{2(i+m)} - \bar{x}) = \lim_{N \to \infty} \frac{1}{N} \sum_{i=0}^{N-1} x_{1i} x_{2(i+m)} \tag{6}$$

We simulates the CNN chaotic sequence with TMS320C54 DSK tools, the following figures show the simulation results: the auto-correlation function is almost the δ function ,the cross-correlation function is almost equal to 0.

4 The CNN Chaotic Spread Spectrum Communication System

Because the statistical characteristic of CNN chaotic sequence is similar with that of Gold sequence: the auto-correlation function is almost the δ function, the cross-correlation function is almost equal to 0, the CNN chaotic sequence can be employed in direct sequence spread spectrum (DSSS). The CNN chaotic sequence is sensitive to initial value, so lots of different sequences can be generated. It's attractive to DSSS system. The transmitter of DSSS system can be described as:

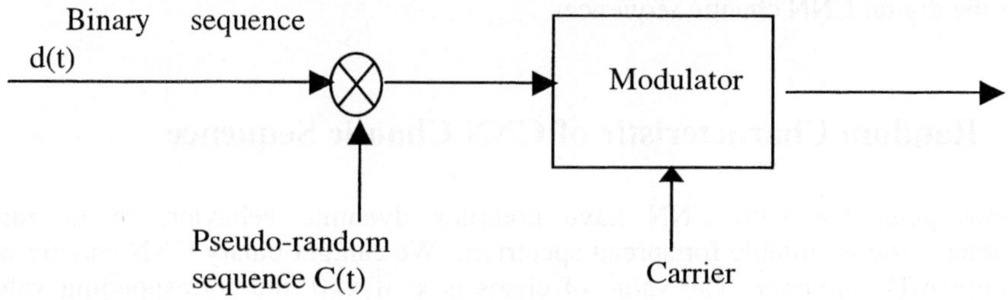

Fig. 2. DS-BPSK Modulator

In Fig.2, C(t) is spread spectrum signal. Usually C(t) is named as direct sequence. Both gold sequence and CNN chaotic sequence are good direct sequences, but the quality of CNN sequence is better.

We assume the signal pass the additive white Gaussian noise channel, so we can describe receiver as Fig 3. Fading channel will studied in the following works:

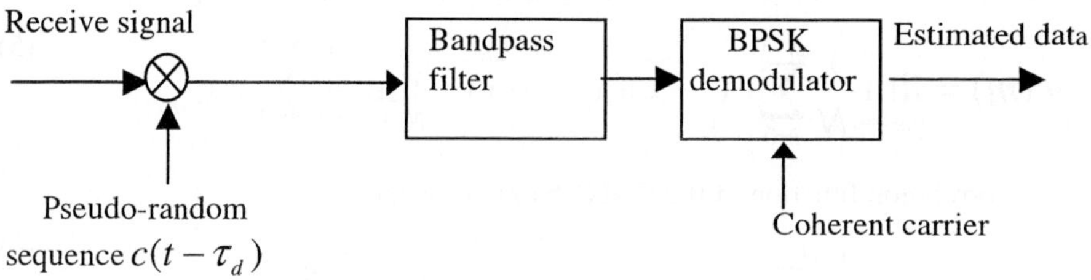

Fig. 3. DS-BPSK demodulator

Different users employ different sequences, so CDMA system could be set up. We set up an experimental CNN-DSSS system based on TMS320C5416 DSP. Two kinds of sequence are utilized in our experiment:1024 bit Gold sequence, and CNN sequence which is divided into 1024 bit segment. The Bit error rate(BER) curve is shown in Fig.4.

In Fig 4,* represents that DS is Gold sequence;o represents that DS is CNN chaotic sequence. Compared the traditional gold sequence system, there is 2dB improvement in CNN chaotic sequence system. The CNN chaotic sequence is more qualified than Gold sequence. The channel of Fig.4 is additive Gaussian white noise channel, the results confirm that the BER of CNN-DSSS system is better than that of traditional Gold sequence system. We also simulate a multi-path channel, the channel model for this simulation is the familiar delay line model as follows:

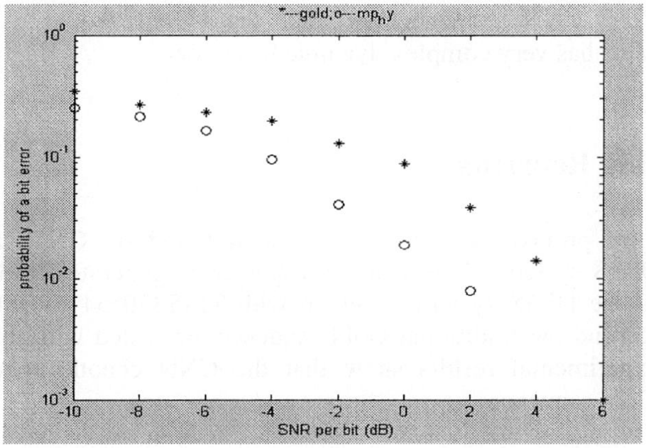

Fig. 4. BER of DSSS system in additive white Gaussian noise channel

$$r(n) = \frac{1}{\sqrt{2}}\left[t(n) + t(n-1)\right] + \eta(n) \tag{7}$$

In equation 7,r(n) is the receive PN code, t(n) is the transmitted PN code, and $\eta(n)$ is additive white Gaussian noise. Fig.5 shows the simulation results:

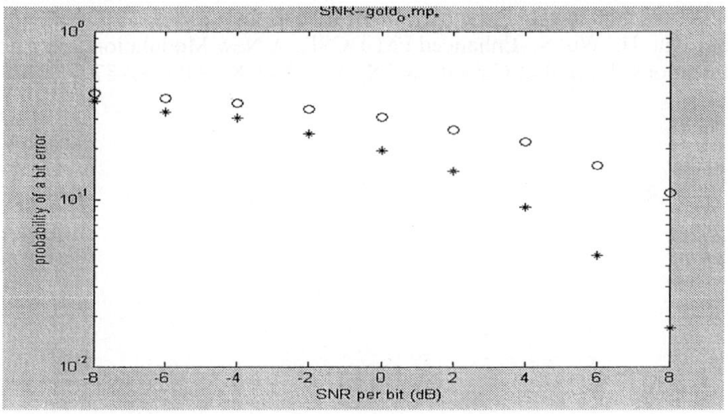

Fig. 5. BER of DSSS system in multi-path channel

In Fig 5, o represents that DS is Gold sequence, * represents that DS is CNN chaotic sequence. The marks in these two figures are quite different because the simulation results are gotten in different channel. Compared with the traditional gold sequence system, there is almost 4dB improvement in CNN chaotic sequence system when the channel described with equation (6) is utilized. The CNN chaotic sequence is excellent in the two-path channel. The dynamic behavior of CNN is more complex than that of traditional system, so the CNN sequence has excellent random characteristic. The CNN sequence is better than Gold sequence because CNN sequence generator has very complex dynamic behaviors.

5 Concluding Remarks

In this paper, we propose a new DSSS system based on CNN, and set up an experimental DSSS system, CNN chaotic sequence is generated with TMS320C54 DSK tools, and the DSSS system is set up with TMS320c5416 DSP. Both CNN chaotic sequence and the traditional Gold sequence are tested with the experimental system. The experimental results show that the CNN chaotic sequence is more qualified than Gold sequence

References

1. Chua, L.O., Yang, L.: Cellular Neural Networks: Theory. IEEE Trans. Circuits Syst.. Vol. 35 (1988) 1257-1272
2. Chua, L.O., Yang, L.: Cellular Neural Networks: Applications. IEEE Trans. Circuits Syst. Vol. 35 (1988) 1273-1290
3. Zou, F., Nossek, J.: Bifurcation and Chaos in Cellular Neural Networks. IEEE Trans. Circuits Syst. Vol. 40 (1993) 166-173
4. Kapitaniak, T., Chua, L.O., Zhong, G.Q.: Experiment Hyperchaos in Coupled Chua's Circuits. IEEE Trans. on CAS, Part I, Vol. 41 (1994) 499-503
5. Wang, M., Qiu, H., Wu, S.: Enhanced FM-DCSK: A New Modulation Scheme for Chaotic Communications. Journal of Circuits and System, Vol. 8 (2003) 82-87

Neural Congestion Control Algorithm in ATM Networks with Multiple Node

Ruijun Zhu, Fuliang Yin, and Tianshuang Qiu

College of Informatics and Electrical Engineering,
Dalian Univ. of Technology, Dalian,116023,China.
{zhurj,flyin,qiutsh}@dlut.edu.cn

Abstract. In ATM networks, congestion control is a distributed algorithm to share network resources among competing users.It is important in situation where the availability of resources and the set of competing users vary over time unpredictably, round trip delay is uncertain and constraints on queue, rate and bandwidth are saturated, which results in wasted bandwidth and performance degradation. A neural congestion control algorithm is proposed by real-time scheduling between the self-tuning neural controller and the modified EFCI algorithm, which makes the closed-loop systems more stable and robust with respect to uncertainties and more fairness in resources allocation. Simulation results demonstrated the effectiveness of the proposed controller.

1 Introduction

Congestion control has special significance in the performance of ATM networks. ATM Forum traffic management specification version 4.1 [1] adopted rate-based flow control as the standard methods to control congestion in which no detailed specifications for the end system and switch was established. So many engineers put forward to a great deal of practical methods[2] [3], and also performed the experimental analysis by simulation [4][5][9], However, no systematic theories is proposed for the networks in the rapid change environment and limited resource so that there are great oscillation in cell rate, utilization and buffer queue.

In the practical networks, the round trip delay is time-varying, the accessed users are stochastic and the high-priority traffic, such as VBR is heavily burst and unpredicted, and the cell rate, buffer size and link bandwidth are limited by the physical devices, therefore it is very difficult to obtain the global performance even for single-node networks. And these factors are interconnected and mixed in multiple-node networks, which make it much harder to optimize the global performance. Kolarov[6] proposed a dual PD controller, which improved the performance and obtained the global stability and steady fairness, but it didn't increase the computational complexity because both EFCI and high-gain controller as well as initial rate recovery controller must be used at the same time. Wang[7] et.al. presented a predictive controller, and they takes the CBR and VBR into account, Benmohaned[8] extended the Kolarov's single-node works [6] to multiple-node. But these works may only guarantee the local

F. Yin, J. Wang, and C. Guo (Eds.): ISNN 2004, LNCS 3174, pp. 299–304, 2004.

stability because they didn't consider the nonlinearity. Then we[9][10][11] proposed a series of predictive congestion controllers by using dual PD controller or modified EFCI algorithm, and guarantee the global stability and steady fairness. in which the discrete-event simulation is also used to demonstrate the effectiveness of the proposed controller, the extension to multiple-node can see [12] for the detail. In this paper, A neural congestion control algorithm is proposed based on self-tuning technique for multiple-node networks to compensate the unmodeled nonlinear dynamics, which guarantee the local stability of the closed-loop systems, then modified EFCI is also presented to cancel the saturated nonlinearity. The real-time scheduling controllers are given to make the closed-loop systems more stable, robustness and steady fairness. The simulation results show that the proposed controller is robust to uncertainties and steadily fair in the resource allocation.

2 Neural Congestion Control Algorithm

2.1 Queueing Model of Multiple-Node Networks

In multiple-node networks, the nonlinear queueing model of each switch can be described as follows[8] [12]

$$x_i(n+1) = SAT_{X^0}\{x_i(n) + T_s(\sum_{k=1}^{D_i} \bar{l}_i^k q_k(n+1-k) - C) + m_i(n)\} \quad (1)$$

where X^0 denotes the buffer size, and $x_i(n)$ is the incoming packets at time slot $[n, n-1]$ in switch i. $m_i(n) = f_{ii}(n) + d_i(n) + \nu_i(n)$, in which $d_i(n)$ denotes quantized errors and high-priority traffic, $\nu_i(n)$ stands for measured disturbance, and $f_{ii}(n)$ is the nonlinear function of the networks states $Y_j(n)$ of link j $(j = 1, 2, ..., i-1)$, and

$$Y_j(n) = (x_j(n) - x_0, ..., x_j(n-D_j) - x_0, q_j(n), ..., q_j(n-K_j)) \quad (2)$$

and the saturation function is defined by

$$Sat_a(z) = \begin{cases} 0 & z \leq 0 \\ a & z \geq a \\ z & other \end{cases} \quad (3)$$

2.2 Neural Networks[15]

Neural networks is widely used in many research area, such as signal processing, control engineering etc.. And the approximate and generalized property is very important as a modelling tool. The networks which is most popularly used is backward propagation (BP) neural networks. The neuron usually takes the sigmoid function as

$$f(x) = \frac{1}{1 + e^{-x/Q}} \quad (4)$$

when there are enough neurons, the three-layer networks can approximate any piecewise continuous function by using backward propagation algorithm. Next, we will use this kind of networks to modelling the nonlinear dynamics $m_i(n)$ as in (1) and design the controller.

2.3 Neural Congestion Controller

For simplicity, we only consider the one-node queueing model (1) in the following form

$$x(n+1) = x(n) + (\sum_{k=1}^{D_i} l_k q(n+1-k) - C) + m(n)\} \tag{5}$$

Where we omit the index i and saturation function, and $l_k = T_s \bar{l}^k$.

The equation (5) can be rewritten in input-output form as

$$\Delta A(z^{-1})x(n) = B(z^{-1})\Delta q(n) + \Delta m(n) \tag{6}$$

where $\Delta = A(z^{-1}) = 1 - z^{-1}, B(z^{-1}) = \sum_{k=0}^{D_i-1} l_k z^{-k}$ and $\Delta m(n) = m(n) - m(n-1)$.

Define the following performance index

$$J = |E(z^{-1})x(n+1) - Rx_0(n) - Q(z^{-1})q(n) - H\Delta m(n)|^2 \tag{7}$$

and use the Diophantine equation

$$\begin{aligned} T(z^{-1}) &= B(z^{-1})E(z^{-1}) - \Delta A(z^{-1})Q(z^{-1}) \\ E(z^{-1}) &= z^{-1}G(z^{-1}) + \Delta A(z^{-1})C(z^{-1}) \end{aligned} \tag{8}$$

when $A(z^{-1}), B(z^{-1})$, Δm are known, the controller is given by

$$q(n) = \frac{Rx_0(n) - F\Delta m(n) - G(z^{-1})x(n) - C(z^{-1})\Delta m(n)}{-G(z^{-1}) + C(z^{-1})B(z^{-1})} \tag{9}$$

and the closed-loop system is

$$T(z^{-1})x(n+1) = B(z^{-1})[Rx_0(n) - F\Delta m(n) - Q(z^{-1})\Delta m(n)] \tag{10}$$

To cancel the static offset and the effect of nonlinear part in the steady state, R, H may be chosen in the most simple form $R = T(1)/B(1), H = Q(1)/B(1)$.

when $A(z^{-1}), B(z^{-1})$, Δm are unknown, we can use the identification algorithm given in [13] to online estimate the controller parameters $G(z^{-1}), C(z^{-1})$ $Q(z^{-1})$ and nonlinear function $\Delta m(n)$. Using the estimated parameter and nonlinear function, the neural congestion control algorithm may be rearranged by

$$q(n) = \frac{Rx_0(n) - F\widehat{\Delta m}(n) - \widehat{G}(z^{-1})x(n) - \widehat{C}(z^{-1})\widehat{\Delta m}(n)}{-\widehat{G}(z^{-1}) + \widehat{C}(z^{-1})\widehat{B}(z^{-1})} \tag{11}$$

The overall structure of the closed-loop systems is shown in Fig. 1. Similar to [12],[13], it easily known that the closed-loop system (6) and (11) is bounded input and bounded output stable (BIBO).

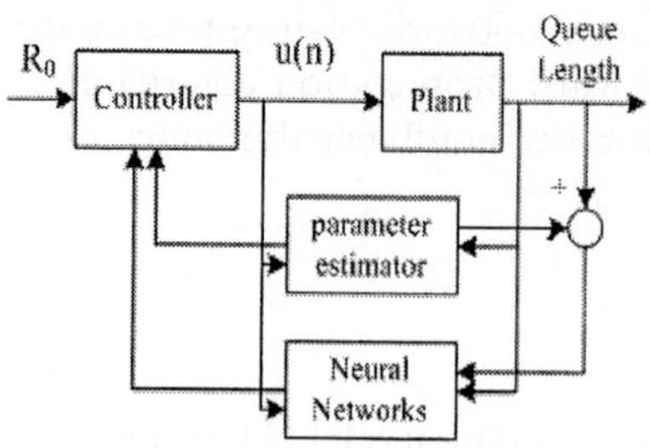

Fig. 1. Neural Control system structure

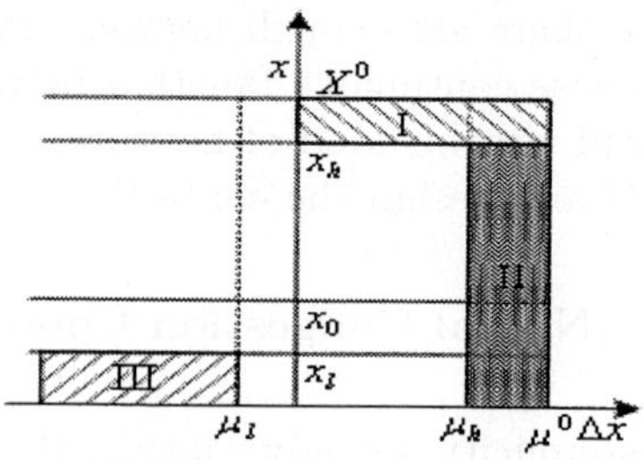

Fig. 2. Phasic map of x and Δx

2.4 Real-Time Scheduling Controller

We can conclude that the controller (11) may only guarantee the controlled system locally stable, and once the users rapidly change and the system will run away the equilibrium, the buffer queue will enter the nonlinear zone, and cause the system unstable. In this section, we modify EFCI algorithm by setting the EFCI bit not only using queue but also the rate of queue change. That is, when the system goes into the area saturated by lower bound of the buffer and the average rate of queue may go down (area I, II in Fig. 2), the increase algorithm in the modified EFCI works, i.e., If $x \le x_l$ and $\Delta x = (x(k) - x(k-1))/T \le \mu_l$, then $q(n) = (1 + RIF) * q(n-1)$

When the system comes near the area saturated by the upper bound of the buffer and the average rate of queue may go down (area III in Fig. 2) , the decrease algorithm in the modified EFCI starts, i.e., If $x > x_h$and $\Delta x > \mu_h$, then $q(n) = q(n-1) - RDF * PCR$. The real-time scheduling controller may be written by

$$q(n) = \begin{cases} (1 + RIF) * q(n-1) & as \;\; x \le x_l \;\; and \;\; \Delta x \le \mu_l \\ q(n-1) - RDF * PCR & as \;\; x > x_h \;\; and \;\; \Delta x > \mu_h \\ (11) & others \end{cases} \tag{12}$$

where RDF and RIF are the constants. So the closed-loop system will be asymptotically approach to the nonzero equilibrium point.

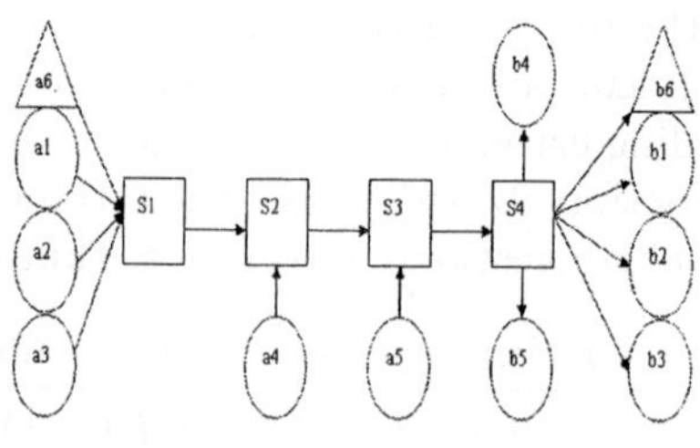

Fig. 3. Multiple-node Network topology

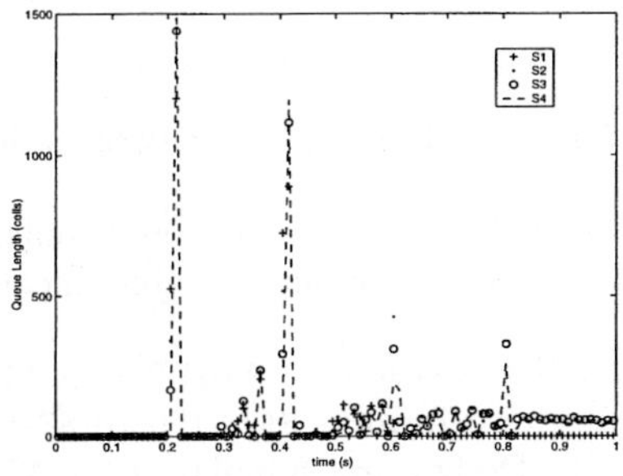

Fig. 4. Buffer Queue length

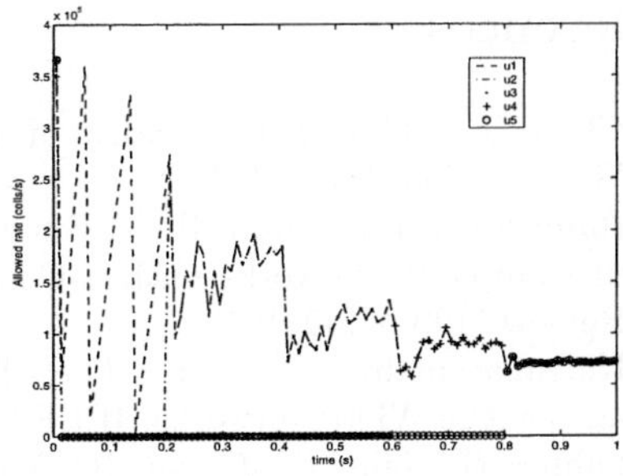

Fig. 5. ACR of each ABR source

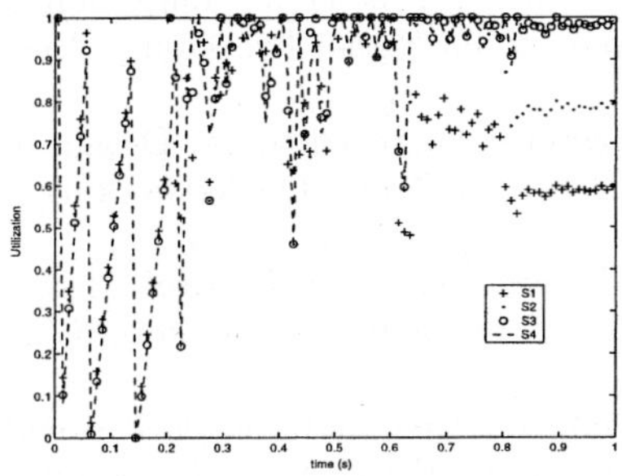

Fig. 6. Link Utilization

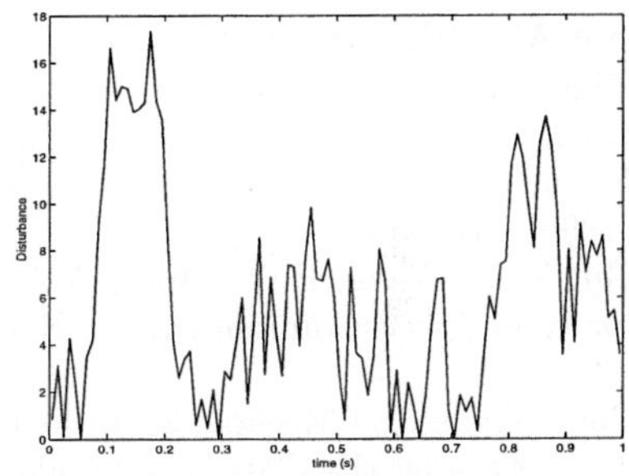

Fig. 7. Measured noise

3 Simulation

In this section, we test our proposed controller by simulation. The networks topology used in the simulation is shown in Fig. 3, which consists of four ABR sources, one VBR source and four switches, and the capacity of each output port is set 10000cells, link capacity is set to 155.52Mb/s. The parameters of every ABR source are: PCR=318cells/ms(135Mb/s), ICR=318cells/ms(135Mb/s), MCR=1 cells/ms(Mb/s), RIF=1/16, RDF=1/16. The sample time is taken $T_s = 1ms$, the RTT of each hop is set to $d_s = 1.$ and $\lambda_1 = 0.99, \lambda_2 = 1.5$, $x^0 = 50$ cells, $x_h = 1050$cells, $x_l = 4$cells, $\mu_h = 9500$cells/s,$\mu_l = 1000$cells/s. The queue length of the buffer is shown in Fig. 4, The sending rate for four ABR sources are given in Fig. 5, the link utilization is plotted in Fig. 6, the traffic rate of VBR service is pictured in Fig. 7. From the simulation results, we can see that the proposed controller has good performance as well as fairness and rapidly respond to time-varying conditions.

Acknowledgement. This work is supported by grant of 2002AA412010 of Key project of State 863 Plan of China and grant of 01053 of Key Scientific Research of Ministry of Education of China.

References

1. Chao, H.J., Guo, X.L.: Quality of Service Control in High-Speed Networks. John Wiley & Sons,Inc. (2002)
2. Ramkrishnan, K., Jain, R.: A Binary Feedback Scheme for Congestion Avoidance in Computer Networks with a Connectionless Layer. Computer Communication Review (1995) 139–155
3. Kalyanaraman, S., Jain, R.: The ERICA Switch Algorithm for ABR Traffic Management in ATM Networks. IEEE/ACM Trans. on Networks **8(1)** (2000) 87–98
4. Fahmy, S., Jain, R., Goyal, R., Vandalore, B.: Design and Simulation of ATM-ABR End System Congestion Control. Transactions of the Society for Computer Simulation **78(3)** (2002) 150–162
5. Zhu, R.J., Teng, H.T., Yin, F.L.: Simulation Study of Predictive Congestion Control Algorithm. Seventh International Conference on Signal Processing. Beijing, China (2004)
6. Kolarov A., Ramanurthy, G.: A Control-Theoretic Approach to the Design of an Explicit Rate Controller for ABR Service. IEEE/ACMTrans. On Networking **7(5)** (1999) 741–753
7. Gu, Y.R., Wang, H.O., Hong, Y.G.: A Predictive Congestion Control Algorithm for High Speed Communication Networks. American Control Conference **5** (2001) 3779–3780
8. Benmohaned L., Meekov, S.M.: Feedback Control of Congestion in Packet Switching Networks: the Case of a Single Congested Node. IEEE/ACM Trans. on Networks **1** (1993) 693–707
9. Zhu, R.J., Ma, J.R.: Discrete-Event-Based Simulation Methods for ATM Flow Control. Control and Decision (in chinese) **18** (2003) 284–286
10. Zhu, R.J, Suo, D.H.: ATM Predictive Congestion Controller. Control and Decision (in chinese) **19** (2004) 61–65
11. Zhu, R.J., Ma, J.R.: Stable Congestion Control Algorithm with Maxmin Fairness. Journal of Dalian University of Technology **44** (2004) 309–312
12. Zhu, R.J., Wang, J.W.: Predictive Congestion Control Methods of ATM Networks with Multiple Nodes. Journal of Dalian University of Technology **43** (2003) 81–83
13. Zhu, Q.M., Ma, Z., Warwick, K.: Neural Network Enhanced Generalised Minimum Variance Self-Tuning Controller for Nonlinear Discrete-Time Systems.IEE Porc-D **146** (1999) 319–326
14. Guo, J., Chen, Q.W.: Adaptive Predictive Control of a Class of Nonlinear Systems. Control Theory and Application (in Chinese) **19(1)** (2002) 68–72
15. Sun, Z.Q.: Theory and Technology of Intelligent Control. Tsinghua Press Beijing, China (1992)

Neural Compensation of Linear Distortion in Digital Communications

Hong Zhou

Dept. of Elect. Commun. & Inform. Eng., Osaka Institute of Technology
5-16-1, Asahi-ku, Ohmiya, Osaka,
535-8585, Japan
zhou@elc.oit.ac.jp

Abstract. In this paper, we proposed a novel and simple linear distortion compensation scheme utilizing Kohonen's neural network for digital communications. The scheme compensates the distortions that exist in modulator and demodulator, at the decision stage of the receiver at the same time of data transmission and decision process. The scheme can compensate not only static but also changing distortions. Computer simulations using QPSK signal have confirmed the effectiveness of the proposed scheme.

1 Introduction

In advanced digital communication systems, quadrature modulator and demodulator are necessary system devices [1]. However, in real circuits of quadrature modulator and demodulator, the linear distortions degrade greatly the transmission quality. Thus far, Affine transformation method has been reported to compensate such distortions [2]. In this paper, we propose a novel linear distortion compensation method based on Kohonen's Self-Organization Map (SOM) neural network. Using the proposed method, the linear distortion can be compensated by a very simple system in high precision. The method does not need any pilot symbols and carries out the compensation processing at the same time of information data transmission. The method can also adaptively compensate the distortion changing.

2 The Linear Distortion

Figure 1 (a) and (b) respectively show the block diagram of a quadrature modulator and a coherent quadrature demodulator. In its ideal type, the upper (in-phase channel, I-ch) part and lower (quadrature channel, Q-ch) part of the modulator (demodulator) are symmetry and no any linear distortion exists. In real circuits, however, because of imperfect selection of electronics components, incomplete adjustment of the circuits and so on, the symmetry no longer stands and the linear distortion is caused. There are basically three kinds of linear distortion: 1) DC level offset δ_i in I-ch part and δ_q in

F. Yin, J. Wang, and C. Guo (Eds.): ISNN 2004, LNCS 3174, pp. 305–310, 2004.

Q-ch part, 2) amplitude level imbalance between I-ch and Q-ch (the gain of the multiplier for I-ch $Ai \neq$ that of the multiplier for Q-ch Aq), 3) quadrature reference phase offset θ due to imperfect orthogonally of the $\pi/2$ shifter (carrier $\cos\omega_c t$ for I-ch while $\sin(\omega_c t+\theta)$ for Q-ch).

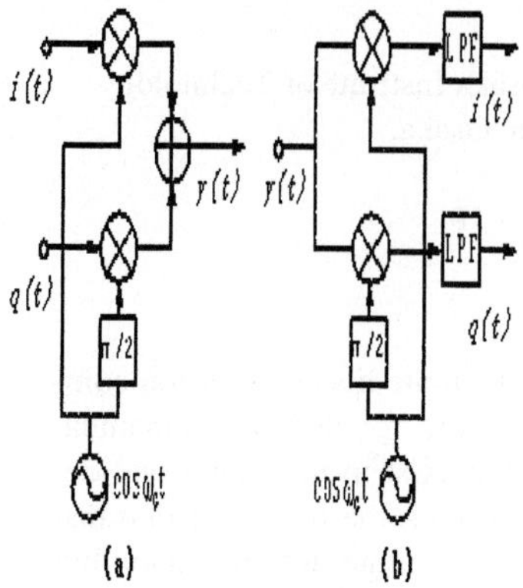

Fig. 1. Quadrature modulator and Demodulator

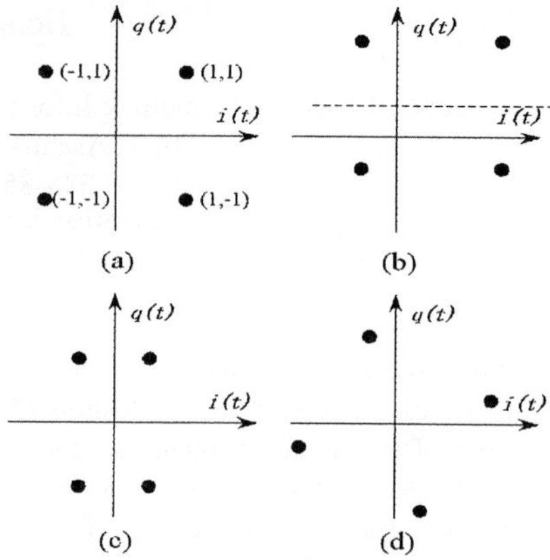

Fig. 2. Various signal constellation of QPSK

The effect of the linear distortions is that the final demodulator outputs *i(t)* and *q(t)* at the receiver deviate from their ideal ones. Hereafter in this paper, we take QPSK modulation scheme as an example. The results and conclusions for QPSK can be easily expended to other modulation schemes. Assuming that the linear distortions only exist in modulator and no any noise exists, Fig. 2 shows the effect of the linear distortions to signal constellations. Fig.2 (a) is the constellation for no distortion case. Each signal point is decided by the value of *i(t)* and *q(t),* and corresponds to a digit pair, for example, the point in the first quadrant corresponding to (1,1). Fig.2 (b)(c)(d) are respectively the constellation for DC level offset only case, amplitude level imbalance only case and imperfect orthogonality of the $\pi/2$ shifter only case. When three kinds of distortion exist at the same time, the constellation is more complicated. In the case of Additive White Gaussian Noise (AWGN) presence, the received signal point will scatter around the points without noise in Fig. 2. Hereafter, we call signal point without noise as center point.

In the receiver, the demodulator is followed by a decision system. The decision system recovers digital signal based on the received noisy signal point *i(t)* and *q(t)* according to the maximum likelihood decision rule which decides the digit pair corresponding to the closest center point. If the linear distortions exist, the decision system may make a wrong decision according to the ideal center points, causing performance degradation. For example, in the case of DC level offset as shown in Fig. 3, although the received signal point *X* should be belong to the distorted center point *A* and (-1, 1) should be recovered, the decision system will make a wrong decision (1, 1) as *X* is most close to the ideal center point *a*.

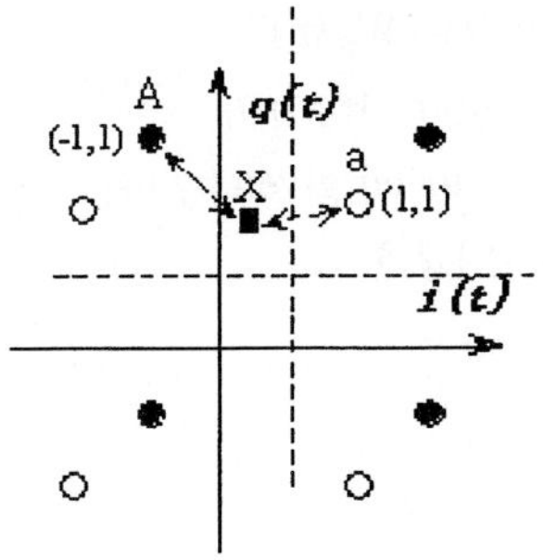

Fig. 3. Effect of DC level offset

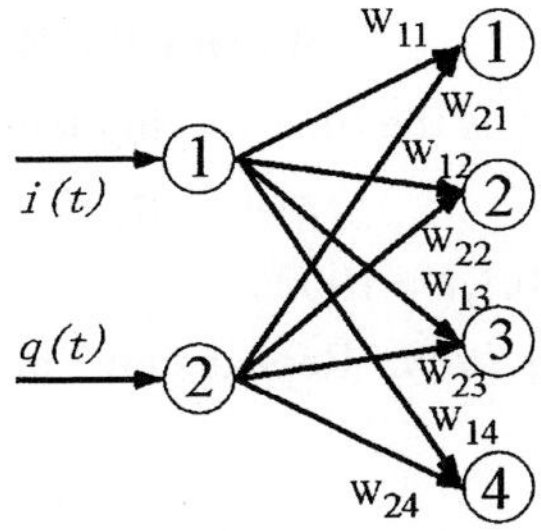

Fig. 4. Kohonen's SOM network for distortion compensation and decision

3 Distortion Compensation Utilizing Neural Network

The task of linear distortion compensation, i.e., finding out the positions of the distorted center points based on continuously received noisy signals and, at the same time, deciding the distorted center point which the recently received noisy signal is belong to, is a typical problem of self organizing pattern recognition. This problem can be resolved by Kohonen's SOM neural network [3]. The proposed scheme introduces Kohonen's network into the decision system after the demodulator and concentrates compensation of all linear distortions appearing in the modulator and demodulator at one place.

Fig. 4 is the proposed decision system based on Kohonen's network. The network has two input nodes j (j=1,2), respectively connecting to $i(t)$ and $q(t)$ outputs of the demodulator, and four output nodes k (k=1,2,3,4) respectively corresponding to four center signal points. Each input node is fully connected to all output nodes with synaptic weights W_{jk} that in fact means coordinates of the corresponding center signal points. For example, W_{11} and W_{21} are respectively the in-phase and quadrature coordinates of center point 1. The proposed network does compensation and decision processing at the same time of information data transmission in the following steps.

Step 1 Initialization: Initialize the values of the synaptic weights W_{jk} by using the coordinates of ideal center points of QPSK, i.e., $W_{jk}=\{(1,1),(1-1),(-1,-1),(-1,1)\}$.
Step 2 Competition and decision: Determine the best-matching output node l to the recently received signal $(i(t), q(t))$ by

$$l\{(i(t),q(t))\} = \arg\min_{k}\{\| (i(t),q(t)) - (W_{1k}, W_{2k}) \|\} \tag{1}$$

where $\| (i(t),q(t)) - (W_{1k}, W_{2k}) \|$ means the Euclidean distance between $(i(t), q(t))$ and (W_{1k}, W_{2k}). Eq. (2) can be equivalently calculated by selecting the output node with the largest inner product $(i(t),q(t))^T (W_{1k}, W_{2k})$. Then a digit pair is recovered corresponding to the best-matching output node l.
Step 3 Adaptation: The synaptic weights related to the best-matching output node l (W_{1l}, W_{2l}) is updated by

$$\begin{aligned} W_{1l}(n+1) &= W_{1l}(n) + \eta(n)(i(t) - W_{1l}(n)) \\ W_{2l}(n+1) &= W_{2l}(n) + \eta(n)(q(t) - W_{2l}(n)) \end{aligned} \quad (2)$$

where $\eta(n)$ is the time varying learning rate parameter given by [4]

$$\eta(n) = \eta_0 \exp(-n/\tau) \quad n=0,1,2,3,.... \quad (3)$$

with $\eta_0 = 0.1, \tau = 1000$.

Step 4 Repetition: Repeat step 2 and 3.

4 Computer Simulations

Computer simulation has been carried out using the communication model shown in Fig. 5 to evaluate the performance of the proposed compensation scheme. In simulation, it is assumed that there is only AWGN and only modulator has the linear distortions.

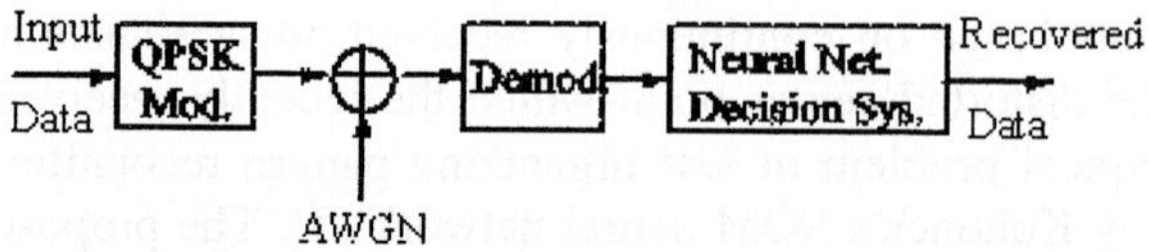

Fig. 5. Communication system for computer simulations

Fig. 6 shows the average Bit Error Rate (BER) performance of the QPSK transmission system with and without compensation for DC level offset only case ($\delta_i = \delta_q = 0.5$) and ($\delta_i = \delta_q = -0.2$). For comparison, the results of no distortion case are also given in the figure. Without compensation, the BER performance shows a big fall from that of no distortion After utilizing the proposed compensation scheme, the effect of the distortion is totally removed and the BER performance is identical to that of no distortion case.

Fig. 7 depicts BER improvement with the proposed scheme for various relative DC level offset values at average signal to noise power ration (SNR) = 14dB. As relative DC offset changes from 0 to ± 0.5, although average BER degrades from 2.4×10^{-4} to 1.8×10^{-2} without compensation, it remains almost no change at about 2×10^{-4} with the proposed compensation scheme.

Fig. 8 shows BER performance for various quadrature reference phase offset values at average SNR = 14dB. As phase offset varies from 0° to ±30°, with compensation the average BER remains almost no change, while average BER degrades from about 2×10^{-4} to about 6×10^{-2} without compensation.. Fig. 9 shows BER performance when three kinds of distortion exist simultaneously ($\delta_i = \delta_q = 0.3, A_i / A_q = 1.2, \theta = 10°$). The proposed scheme is still effective even under such severe conditions

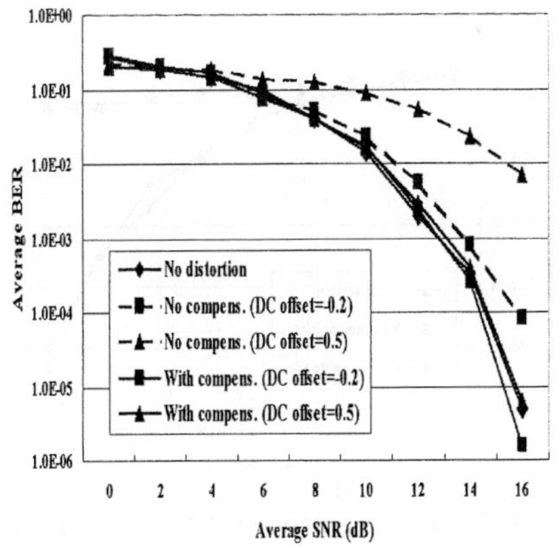

Fig. 6. BER performance in DC level offset only case

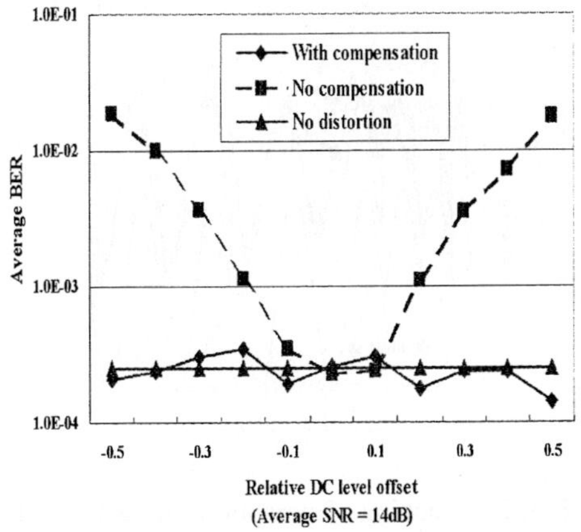

Fig. 7. BER performance improvement in DC level offset only case

In some cases, the linear distortions may not be fixed values and may be time varying due to, for example, changing of the environment temperature. In order to evaluate the performance of the proposed scheme in such cases, simulation is also done assuming that the relative DC level offset varies in a sine curve with amplitude of 1 and period of 1000 transmission bits as shown in Fig. 10. Fig. 10 shows the variation of the I-ch and Q-ch coordinates W_{11}, W_{21} for center point 1 at average SNR=6dB. They track the variation of DC level offset very well and vary around 1 from 0 to 2.

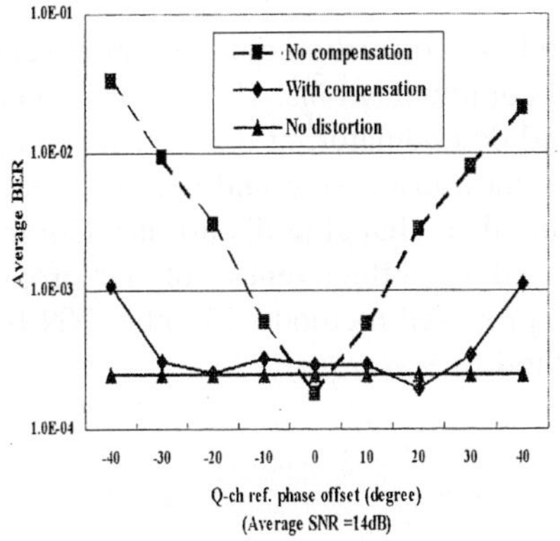

Fig. 8. BER performance improvement in quadrature reference phase offset only case

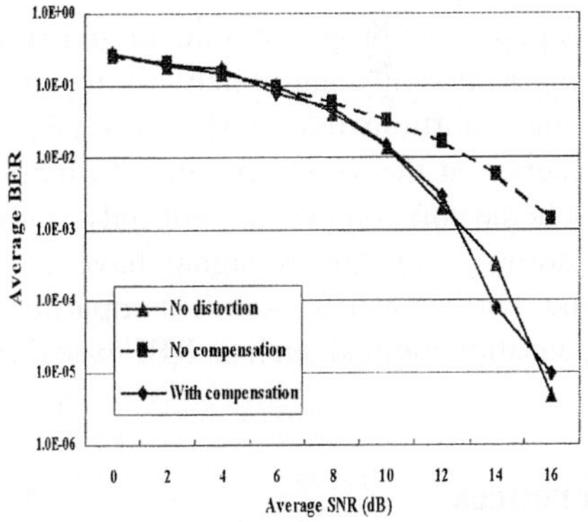

Fig. 9. BER performance in three kinds of linear distortion case

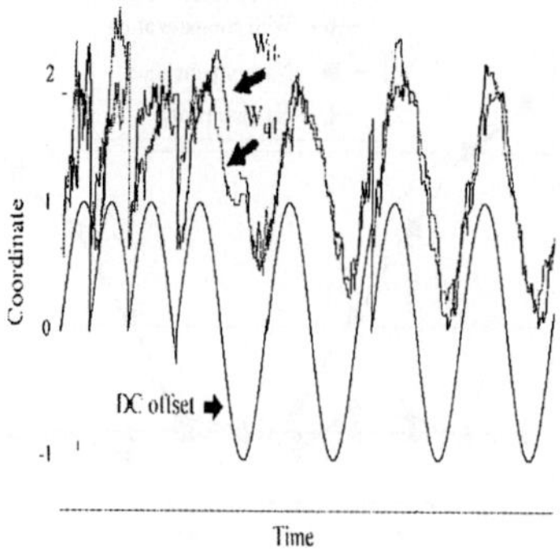

Fig. 10. Tracking of synaptic weights to the changing of DC level offset

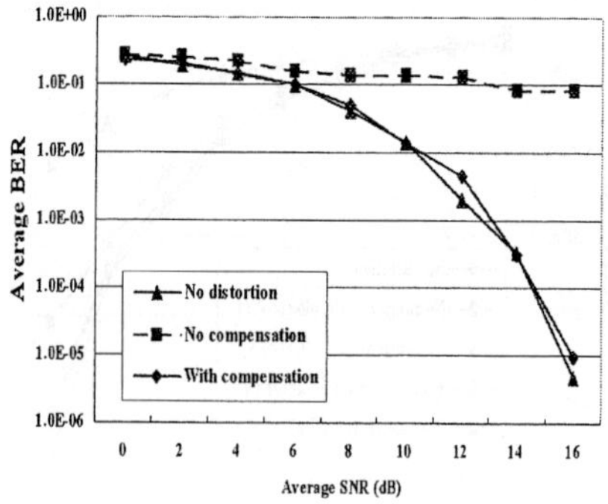

Fig. 11. BER performance in DC level offset changing case

Fig.11 gives the BER performance for this case. Although the system without compensation is thoroughly out of order, the system with the proposed scheme works perfectly, having identical BER performance to that of the system without distortion. The proposed scheme can even adaptively compensate the changing linear distortion. This important characteristic means that it is possible using the proposed neural network type decision system to compensate power attenuation of the received signal due to transmission distance, fading due to multipath propagation in mobile radio.

5 Conclusions

In this paper, we proposed a novel and simple linear distortion compensation scheme utilizing Kohonen's neural network for digital communications. The scheme compensates the distortions that exist in modulator and demodulator, at the decision stage in the receiver at the same time of information data transmission and decision process. The scheme can compensate not only static but also changing distortions. Computer simulations using QPSK signal have confirmed the effectiveness of the proposed scheme. Further works include comparing the proposed method with other NN based compensation method such as RBF based method.

References

1. Proakis, J. G.: Digital Communications. Third ed., Prentice Hall(1995)
2. Suzuki, H., Yoshino, H.: Affine Transformation for Linear Distortion Compensation - An Application to Linear Signalling with Equalization in Mobile Radio. Trans. IEICE, Vol.J75-B-II, No.1(1992) 1-9
3. Kohonen, T.: The Self-organizing Map. Proc. IEEE, Vol.78 (1990)1464-1480
4. Haykin, S.: Neural Networks. Second ed., Prentice Hall(1999)

On the Performance of Space-Time Block Coding Based on ICA Neural Networks*

Ju Liu[1,2], Hongji Xu[1], and Yong Wan[3]

[1]School of Information Science and Engineering, Shandong University
250100 Jinan, China
[2] State Key Lab. of Mobile Communications, Southeast University
210096 Nanjing, China
{Juliu, Hongjixu}@sdu.edu.cn
[3]Department of Physics, Qingdao University, 266071 Qingdao, China
{zhangjianping}@tv.hisense.com

Abstract. For conventional space-time block coding (STBC), the decoding usually requires accurate channel state estimation. However, the accuracy of the channel estimation strongly determines the system performance. Independent component analysis (ICA) techniques can be applied to perform blind detection so as to detect the transmitted symbols without any channel information. In this paper, we establish the special ICA model for the STBC system and study the performance of STBC schemes based on ICA neural networks; what is more, several different ICA algorithms of blind separation are used for performance evaluation. By using the ICA based schemes, the good robustness against channel estimation errors and time variation of the fading channels can be acquired. The computer simulation analyzes the bit error rate (BER) performance of these methods and indicates the optimal separation algorithm suitable for STBC scheme.

1 Introduction

In recent years, blind source separation (BSS) techniques have attracted special attention in the fields of image processing, wireless communication etc. Independent component analysis (ICA) is a signal processing and data analysis method within the family of BSS. Even without any information of the transmission channel, ICA can recover the transmitted symbols only from the observations.

ICA techniques have been applied into many fields of communications. Different methods for blind beamforming without a prior information about sensor location and response patterns have been proposed [1][2]. In addition, many ICA algorithms have been used to the blind interference suppression of the CDMA systems [3][4].

Space-time block coding (STBC) proposed by Alamouti is a powerful transmit diversity scheme, which can achieve the full diversity without bandwidth expansion [5]. However, the decoding of STBC scheme usually requires accurate channel state in-

* This work was supported by National Natural Science Foundation of China (No. 60372031) and Open Foundation of State Key Laboratory of Mobile Communications (No. A0401).

F. Yin, J. Wang, and C. Guo (Eds.): ISNN 2004, LNCS 3174, pp. 311–316, 2004.

formation (CSI) at the receiver. The performance of STBC strongly depends on the channel estimation, but if the system is in a rapidly changing mobile environment or there is an estimation error, the system performance will be degraded seriously. However, due to the characteristic of ICA, it could be used in blind detecting scheme of conventional STBC to detect the transmitted symbols without channel estimation.

In this paper, the methods combining STBC and ICA-based blind detection algorithms are studied. Three ICA algorithms including joint approximate diagonalization of eigen-matrices (JADE), equivariant adaptive separation via independence (EASI) and FastICA are used to perform the blind symbol detection, and the performances of these different algorithms are also analyzed.

2 STBC System Model

Consider a STBC system with n_T transmit antennas and n_R receive antennas. For simplicity, the modulation scheme is assumed to be BPSK. A frame of transmitted symbols is divided into many blocks. Let $\mathbf{s}(k)=[s_1(k) \quad s_2(k) \quad \cdots \quad s_N(k)]^T$ be the kth block of N symbols to be transmitted, and it is encoded by the STBC encoder and mapped onto an $n_T \times N$ code matrix $\mathbf{C}(k)=[\mathbf{c}_1(k) \quad \mathbf{c}_2(k) \quad \cdots \quad \mathbf{c}_N(k)]$ by [6]:

$$\mathbf{C}(k)=\sum_{n=1}^{N}\mathbf{A}_n s_n(k) \tag{1}$$

where $\mathbf{A}_n$ is the orthogonal coding matrix corresponding to the nth symbol $s_n(k)$, and $\mathbf{c}_n(k)$ is the $n_T \times 1$ coded vector at the nth time instant of the kth block. We can model the transmission during the nth time instant as follows:

$$\mathbf{c}_n(k)=\mathbf{P}_n\mathbf{s}(k) \quad n=1,\ldots,N \tag{2}$$

and therefore

$$\mathbf{C}(k)=[\mathbf{P}_1\mathbf{s}(k) \quad \mathbf{P}_2\mathbf{s}(k) \quad \cdots \quad \mathbf{P}_N\mathbf{s}(k)] \tag{3}$$

where $\mathbf{P}_n$ is a permutation matrix corresponding to $\mathbf{c}_n(k)$.

Assume that $\mathbf{H}$ is a $n_R \times n_T$ complex channel response matrix, which remains constant during one detection period, and then the received signal can be written as:

$$\mathbf{Y}(k)=\mathbf{H}\mathbf{C}(k)+\mathbf{V}(k) \tag{4}$$

where $\mathbf{Y}(k)$ is an $n_R \times N$ received signal matrix and $\mathbf{V}(k)$ is an $n_R \times N$ noise matrix whose elements are Gaussian random variables with mean zero and variance σ^2.

Denote $\mathbf{Y}(k)=[\mathbf{y}_1(k) \quad \mathbf{y}_2(k) \quad \cdots \quad \mathbf{y}_N(k)]$ and $\mathbf{V}(k)=[\mathbf{v}_1(k) \quad \mathbf{v}_2(k) \quad \cdots \quad \mathbf{v}_N(k)]$, and we can obtain a new expression:

$$\mathbf{y}_n(k)=\mathbf{H}\mathbf{c}_n(k)+\mathbf{v}_n(k)=\mathbf{H}\mathbf{P}_n\mathbf{s}(k)+\mathbf{v}_n(k) \tag{5}$$

where both $\mathbf{y}_n(k)$ and $\mathbf{v}_n(k)$ are $n_R \times 1$ complex vectors, and they represent the received signal and noise during the nth time instant in the kth block, respectively.

3 The Blind Detection Schemes Based on ICA

3.1 ICA-Based System Model

ICA can be used to enhance the flexibility of the communication system. Here we consider applying the ICA algorithms into the receiver to perform the blind detection.

In order to separate the source signals effectively, we transform (5) into the following expression

$$\begin{bmatrix} \mathbf{y}_1^R(k) \\ \vdots \\ \mathbf{y}_N^R(k) \\ \mathbf{y}_1^I(k) \\ \vdots \\ \mathbf{y}_N^I(k) \end{bmatrix} = \begin{bmatrix} \mathbf{H}^R\mathbf{P}_1 \\ \vdots \\ \mathbf{H}^R\mathbf{P}_N \\ \mathbf{H}^I\mathbf{P}_1 \\ \vdots \\ \mathbf{H}^I\mathbf{P}_N \end{bmatrix} \mathbf{s}(k) + \begin{bmatrix} \mathbf{v}_1^R(k) \\ \vdots \\ \mathbf{v}_N^R(k) \\ \mathbf{v}_1^I(k) \\ \vdots \\ \mathbf{v}_N^I(k) \end{bmatrix} \tag{6}$$

where $(\mathrm{A})^R$ and $(\mathrm{A})^I$ represent the real and the imaginary component of A, respectively. The above expression can also be denoted as:

$$\tilde{\mathbf{y}}(k) = \tilde{\mathbf{H}}\mathbf{s}(k) + \tilde{\mathbf{v}}(k) \tag{7}$$

where both $\tilde{\mathbf{y}}(k)$ and $\tilde{\mathbf{v}}(k)$ are $2n_R N \times 1$ vectors and $\tilde{\mathbf{H}}$ is a $2n_R N \times n_T$ matrix.

Note that Equation (7) can be regarded as a linear mixing model of ICA, and in this model, the received signals $\tilde{\mathbf{y}}(k)$ are $2n_R N \times 1$ matrices. This is an overdetermined blind source separation problem [7]. Firstly, the received signals can be projected onto an N-dimensional block representing the signal-plus-noise subspace and a $(2n_R N - N)$-dimensional block representing the noise subspace. This process is called pre-whitening, which may be written by

$$\mathbf{x}(k) = \mathbf{V}_S \tilde{\mathbf{y}}(k) \tag{8}$$

where $\mathbf{V}_S$ is the pre-whitening matrix with the decreased dimension N.

Secondly, we should find a separating matrix $\overline{\mathbf{W}}$ to apply to the vector $\mathbf{x}(k)$, and then the source signal can be recovered:

$$\mathbf{z}(k) = \overline{\mathbf{W}}\mathbf{x}(k) = \hat{\mathbf{s}}(k) \tag{9}$$

3.2 Three ICA Algorithms for Recovering Source Signals

In this section, we choose three algorithms to recover original signals: Cardoso and Souloumiac's JADE algorithm [1], Cardoso and Laheld's EASI algorithm [2] and Hyvärinen's FastICA algorithm [8].

For the JADE algorithm, it finds a unitary matrix that approximately diagonalizes the most significant eigenmatrices of the fourth-order cumulants of the whitened signals. The criterion of JADE is defined as [1]

$$C_{JADE} = \sum_{i,k,l=1}^{N} \left| cum[\mathbf{z}_i, \mathbf{z}_i^*, \mathbf{z}_k, \mathbf{z}_l^*] \right|^2 \qquad (10)$$

The main reason for considering this criterion is its link to underlying eigenstructures which allows for an efficient optimization of it by using joint diagonalization.

The EASI algorithm was originally developed to separate signals when no noise was present. It is a serial updating adaptive blind separating algorithm based on kurtosis contrast. The training formulation of the separating matrix is as follow

$$\overline{\mathbf{W}}(k+1) = \overline{\mathbf{W}}(k) - \lambda(k)\mathbf{F}(\mathbf{z}(k))\overline{\mathbf{W}}(k) \qquad (11)$$

where $\lambda(k)$ is a step size for training and $\mathbf{F}(\mathbf{z}) = \mathbf{z}\mathbf{z}^T - \mathbf{I} + \mathbf{g}(\mathbf{z})\mathbf{z}^T - \mathbf{z}\mathbf{g}^T(\mathbf{z})$ is a function of the nonlinear function $\mathbf{g}(\mathbf{z})$ [2].

FastICA is a computationally optimized ICA algorithm and has a relatively fast convergence performance. The basic updating step can be expressed as

$$\overline{\mathbf{w}}_i^+ = E\{\mathbf{x}g(\overline{\mathbf{w}}_i^T\mathbf{x})\} - E\{g'(\overline{\mathbf{w}}_i^T\mathbf{x})\}\overline{\mathbf{w}}_i, i = 1, \ldots, N \qquad (12)$$

where $\overline{\mathbf{w}}_i^T$ is one row of the separating matrix $\overline{\mathbf{W}}$, and $\overline{\mathbf{w}}_i^+$, the new value of $\overline{\mathbf{w}}_i$, is normalized to unit norm after every iteration. Function $g(\cdot)$ is the derivative of a sufficiently regular non-quadratic function [8].

4 Simulations

In this section, we will investigate the performances of Alamouti scheme with different ICA blind detecting algorithms. A communication system with two transmit antennas and two receive antennas is considered. Assume that the modulation scheme is BPSK. The quasi-static flat Rayleigh fading channel is assumed.

Fig.1 shows the received signals and the corresponding separated signals on two receive antennas for SNR=5dB. In this simulation, we chose JADE algorithm to show the effect of BSS and the results are basically the same as the other ICA algorithms. It is clear that the received data streams can be separated effectively.

Fig.2-Fig.4 show the BER performance comparison for the ICA methods with the different numbers of samples in source signals. In Fig.2, we can find that when the source has a few samples the ICA-based detecting methods have a performance gap compared with the standard Alamouti scheme, and the EASI algorithm has the worst BER performance and slowest convergence rate. The performances of the JADE and FastICA algorithm are the same basically. After analyzing Fig.3 and Fig.4, we know that the BER performance and the convergence rate of the ICA methods will improve when the length of the source signal increases. While the number of the transmitted symbols gets to about 800, the performance of the EASI will improved obviously; what is more, all the three ICA algorithms have the approximate performances with the scheme using ideal channel estimation.

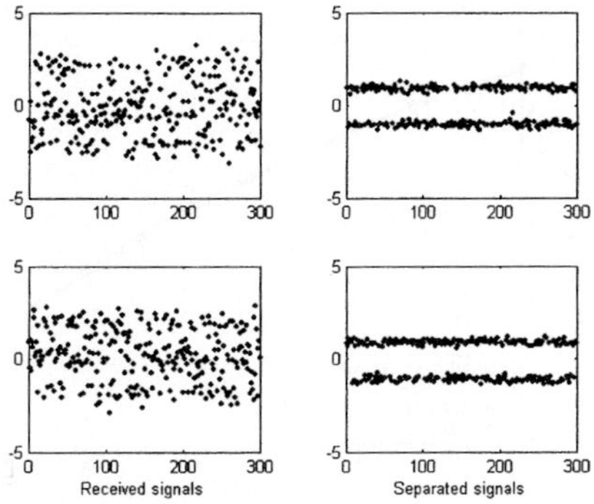

Fig. 1. Comparison between received and separated signals

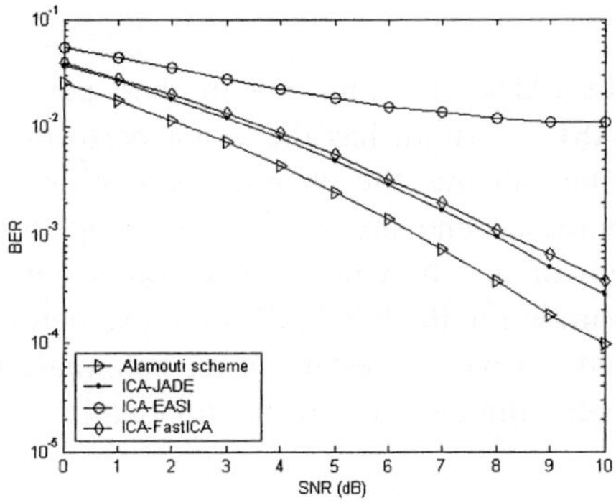

Fig. 2. BER performance comparison for the transmitted signal with 100 bits

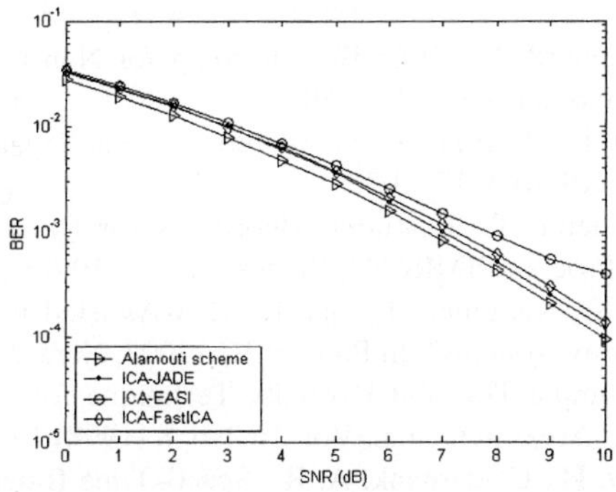

Fig. 3. BER performance comparison for the transmitted signal with 300 bits

5 Conclusions

In this paper, the performances of STBC system based on ICA network are investigated, and three different ICA methods of blind detection are used for performance comparison. The simulation validates the effectiveness of the ICA-based schemes,

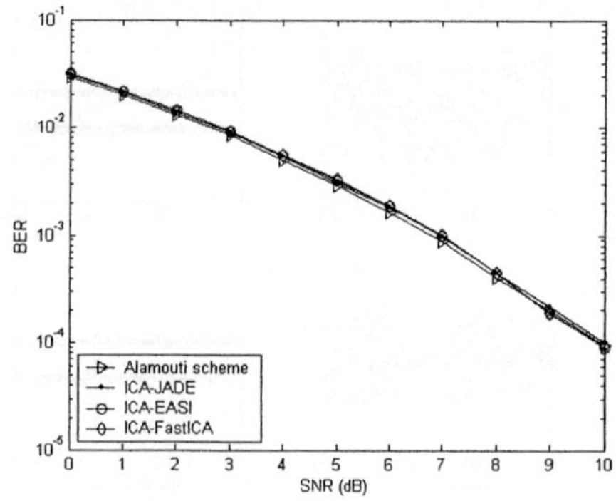

Fig. 4. BER performance comparison for the transmitted signal with 800 bits

and it also shows that the JADE algorithm has the best performance compared with the others, while the EASI algorithm has the worst performance. By exploiting the ICA-based blind receiving scheme, the channel estimation can be avoided, so the spectral efficiency is increased and the system can adapt to more rapidly varying environments. A problem of this ICA-based detection scheme is that it has failure problem of the blind separation in the low SNR region sometimes. If this happens, the retransmission is required, so we can establish an automatic retransmission request (ARQ) mechanism to processing these events automatically.

References

1. Cardoso, J. F., Souloumiac, A.: Blind Beamforming for Non Gaussian Signals, IEE Proceedings-F, Vol. 140. No. 6 (1993) 362–370
2. Cardoso, J. F., Laheld, B. H.: Equivariant Adaptive Source Separation, IEEE Trans. Signal Proc., Vol. 44. No. 12 (1996) 3017–3030
3. Joutsensalo, J., Ristaniemi, T.: Learning Algorithms for Blind Multi-User Detection in CDMA Downlink, In Proc. of PIMRC'98, Boston, U.S.A (1998) 1040–1044
4. Ristaniemi, T., Raju, K., Karhunen, J., Oja, E.: ICA-Assisted Inter-Cell-Interference Cancellation in CDMA Array Systems", In Proc. of ICA2003, Nara, Japan (2003)
5. Alamouti, S. M.: A Simple Transmit Diversity Technique for Wireless Communications, IEEE Journal on Selec. Areas in Comm, Vol. 16. No. 8 (1998) 1451–1458
6. Tarokh, V., Jafarkhani, H., Calderbank, A. R.: Space–Time Block Codes from Orthogonal Designs, IEEE Trans. Inform. Theory, Vol. 45 (1999) 1456–1467
7. Liu, J., Iserte, A. P., Lagunas, M. A.: Blind Separation of OSTBC Signals Using ICA Neural Networks, in Proc. of ISSPIT, Darmstadt, Germany (2003)
8. Hyvärinen, A.: Fast and Robust Fixed-Point Algorithms for Independent Component Analysis. IEEE Trans. on Neural Networks, Vol. 10. No. 3 (1990) 626–634

ICA-Based Beam Space-Time Block Coding with Transmit Antenna Array Selection*

Hongji Xu[1] and Ju Liu[1, 2]

[1]School of Information Science and Engineering, Shandong University
250100 Jinan, China
[2] State Key Lab. of Mobile Communications, Southeast University
210096 Nanjing, China
{Hongjixu, Juliu}@sdu.edu.cn

Abstract. Space-time block coding (STBC) can provide a fairly good diversity advantage over the Rayleigh fading channel, and the beam space-time block coding (BSTBC) scheme combining STBC with the transmit beamforming (TBF) can achieve both diversity gains and beamforming gains. In this paper, we establish a BSTBC model firstly, and then present a method which considers transmit antenna array selection (TAAS) used for BSTBC at the transmitter side and applies independent component analysis (ICA) techniques to assist the channel estimation and achieve blind detection at the receiver side. The utility of TAAS can obtain higher diversity gains as well as keep the original number of the radio-frequency chains. The ICA-based blind scheme can enhance the flexibility of the communication system and adapt to more rapidly varying channels. Simulation results for Rayleigh channel demonstrate the validity and significant performance improvement of the proposed scheme.

1 Introduction

Transmit antenna array (TAA) can be used at the base station (BS) to improve the performance of wireless communication systems. The beam space-time block coding (BSTBC) scheme utilizes smart beamforming arrays to replace the conventional isotropic transmit antennas, which can combine the benefits of beamforming with those given by space-time block coding (STBC).

For conventional STBC scheme, the transmit antenna selection (TAS) can be considered to achieve high diversity order. The idea of TAS can also be utilized into the BSTBC system, therefore we can exploit the transmit antenna array selection (TAAS) scheme to choose the optimal array subset for the downlink transmission.

The decoding of STBC usually requires accurate channel estimation [1], but it will lead to a loss of spectral efficiency due to the use of pilot sequences. Independent

* This work was supported by National Natural Science Foundation of China (No. 60372031) and Open Foundation of State Key Laboratory of Mobile Communications (No. A0401).

F. Yin, J. Wang, and C. Guo (Eds.): ISNN 2004, LNCS 3174, pp. 317–323, 2004.

component analysis (ICA) technique can recover the transmitted symbols without any information of the spatial channel, so it could be exploited in the blind receiver.

In this paper, transmit beamforming (TBF) and STBC are coupled to construct the BSTBC model, and then BSTBC is combined with TAAS at the transmitter; at the same time ICA techniques are applied to assist the channel estimation and symbol detection so that blind receiving scheme is achieved at the receiver.

2 Downlink Transmission System Model

2.1 System Model

As shown in Fig.1, a wireless communication system combining BSTBC with TAAS at the transmitter is considered. In BSTBC scheme, several beamforming sub-arrays (BFSAs) are used for transmission, and each BFSA is an M-element uniform linear array (ULA). Suppose that BS is equipped with n_T BFSAs and the mobile station (MS) is equipped with n_R receive antennas. If N_T BFSAs are chosen for linking the transmit radio-frequency (RF) chains and all the other BFSAs are kept silent, we refer to it as an (n_T, N_T; n_R) system. Let $P_i, i=1,\cdots,N_T$ be the same transmit power for each BFSA.

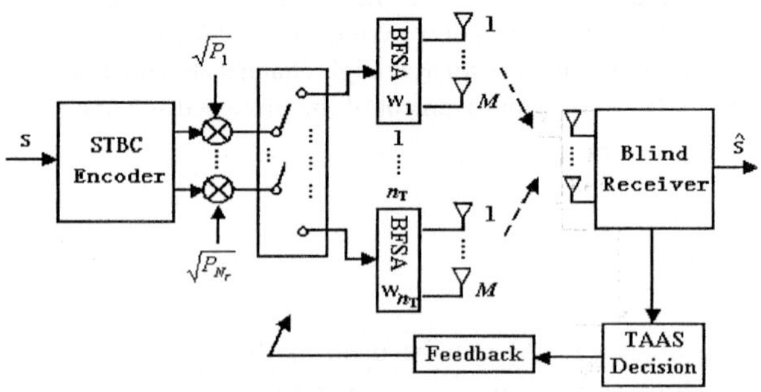

Fig. 1. The structure of the proposed communication system

For the STBC scheme using BPSK modulation, assume that $\mathbf{s}(k)$ is the kth block transmitted signal vector of N symbols. $\mathbf{c}_n(k)$ is the $N_T \times 1$ coded vector at the nth time instant of the kth block, and it can be expressed as:

$$\mathbf{c}_n(k) = \mathbf{P}_n \mathbf{s}(k) \quad n = 1,...,N \tag{1}$$

where $\mathbf{P}_n$ is a permutation matrix corresponding to $\mathbf{c}_n(k)$.

For the (n_T, N_T; n_R) wireless system in this paper, we may define an $n_R \times n_T$ channel matrix $\mathbf{H}$. After the process of TAAS, the optimal N_T TAAs will be used for signal transmission. Suppose that each branch of the encoded signals is sent to a transmit BFSA and weighted by an $M \times 1$ weight vector $\mathbf{w}_i$, $i=1,...,N_T$ ($\|\mathbf{w}_i\|=1$), and then we can define a weight matrix $\mathbf{W}$ containing the diagonal components $\mathbf{w}_i^H$.

For the convenience of the following expression, we define a new $N_T \times n_R$ complex channel response matrix $\mathbf{H}'$, its component $\mathbf{h}_{ij}$ is an $M \times 1$ complex channel vector between the *i*th transmit BFSA and the *j*th receive antenna, and it can be written by

$$\mathbf{h}_{ij}(t) = \sum_{l=1}^{L} \alpha_{l,ij}(t) \cdot \mathbf{a}_{ij}(\theta_l) \tag{2}$$

where, for a certain i and j, $\alpha_{l,ij}(t)$ is the complex channel gain of the *l*th path, and $\mathbf{a}_{ij}(\theta_l)$ is the $M \times 1$ downlink array steer vector at the angle of departure (AOD) θ_l, and L is the number of the independent fading paths.

The weight matrix $\mathbf{W}$ and the channel response matrix $\mathbf{H}'$ can be combined into an $n_R \times N_T$ "effective channel " $\overline{\mathbf{H}} = (\mathbf{W}\mathbf{H}')^T$. So the received signal can be written as

$$\mathbf{y}_n(k) = \overline{\mathbf{H}}\mathbf{c}_n(k) + \mathbf{v}_n(k) = \overline{\mathbf{H}}\mathbf{P}_n\mathbf{s}(k) + \mathbf{v}_n(k) \tag{3}$$

where both $\mathbf{y}_n(k)$ and $\mathbf{v}_n(k)$ are $n_R \times 1$ complex valued vectors, and they represent the received signal and noise during the *n*th time instant in the *k*th block, respectively.

2.2 Optimum TBF Scheme

We choose the optimum TBF algorithm that maximizes the receive signal-to-noise ratio (SNR) at MS. The optimum weights of each transmit BFSA can be described as:

$$\mathbf{w}_{i\_opt} = \max_{\mathbf{w}_i : \|\mathbf{w}_i\|=1} \mathbf{w}_i^H \mathbf{R}_{d,i} \mathbf{w}_i \tag{4}$$

where $\mathbf{R}_{d,i}$ is the $M \times M$ downlink spatial covariance matrix (DSCM) corresponding to the *i*th BFSA, and $\mathbf{w}_{i\_opt}$ is the eigenvector associated with the maximum eigenvalue of $\mathbf{R}_{d,i}$.

For a usual closed-loop frequency-division duplex system, the optimum weights may be obtained by feedback. But in this paper, in order to calculate the weights we exploit spatial covariance matrix transformation (SCMT) to estimate the DSCM $\mathbf{R}_{d,i}$ in virtue of the uplink spatial covariance matrix (USCM) $\hat{\mathbf{R}}_{u,i}$ calculated at the BS,

$$\hat{\mathbf{R}}_{d,i} = \mathbf{T} \cdot \hat{\mathbf{R}}_{u,i} \cdot \mathbf{T}^H \tag{5}$$

where $\mathbf{T}$ is a linear transformation matrix which relates to the spatial manifold of uplink and downlink [2]. Hence the weight feedback can be avoided.

2.3 TAAS Scheme

For the TAAS scheme, only N_T out of n_T TAAs are chosen for transmission. Here Let N_T=2, and then the aim of TAAS is to transmit Alamouti codes at the best TAA pair.

Let $\beta_{ji} = \sum_{l=1}^{L} \left|\alpha_{l,ij}\right|^2$ be the squared sum of absolute value of the L fading path gain in $\mathbf{h}_{ij}$, so we can obtain an $n_R \times n_T$ channel squared gain matrix $\mathbf{B}$ whose elements are β_{ji}. Assume that the uth and vth TAA are chosen for transmission, and then the receive SNR γ at the output data stream is

$$\gamma = \gamma_0 \sum_{j=1}^{n_R} (\beta_{j,u} + \beta_{j,v}) \tag{6}$$

where γ_0 is the transmit SNR.

We should choose two columns of the gain matrix $\mathbf{B}$ with the largest and the second largest sums. Clearly, it is based on the rule of optimal array selection:

$$\text{TAAS}(u,v) = \underset{1 \le u,v \le n_T, u \ne v}{\arg\max} \left\{ \sum_{j=1}^{n_R} (\beta_{j,u} + \beta_{j,v}) \right\}. \tag{7}$$

We can apply the conclusion of TAS into the analysis of TAAS. Therefore, for the (n_T, 2; n_R) TAAS scheme, the gain from the TAAS technique in expected receive SNR over the standard Alamouti scheme can be expressed as [3]

$$G_{TAAS}(dB) = 10\log_{10}\left(\frac{E(\overline{\beta}_{n_T} + \overline{\beta}_{n_{T-1}})}{2n_R}\right) \tag{8}$$

where $\overline{\beta}_{n_T}$ and $\overline{\beta}_{n_{T-1}}$ are the largest and the second largest column sums of matrix $\mathbf{B}$.

3 ICA-Based Blind Receiving Scheme

In this paper, ICA is regarded as an assistant approach for channel estimation and symbol detection in the traditional receiver. We use ICA network as front-end processing module to achieve blind symbol detection. The symbols detected by the ICA network are used as the reference signal in order to estimate the channel responses. Fig. 2 shows the diagram of the ICA-based receiver.

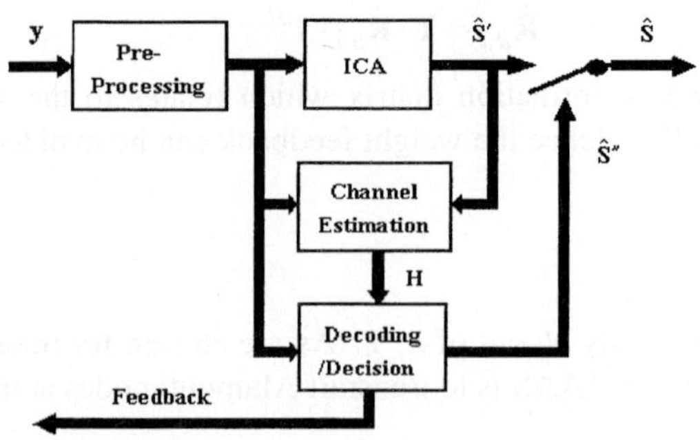

Fig. 2. The diagram of the ICA-based channel estimation and detection scheme

For the system mentioned in Section 2, define $\mathbf{Y}(k)=[\mathbf{y}_1(k) \quad \mathbf{y}_2(k) \quad \cdots \quad \mathbf{y}_N(k)]$, $\mathbf{V}(k)=[\mathbf{v}_1(k) \quad \mathbf{v}_2(k) \quad \cdots \quad \mathbf{v}_N(k)]$ and $\overline{\overline{\mathbf{H}}}(k)=[\overline{\mathbf{H}}\mathbf{P}_1 \quad \overline{\mathbf{H}}\mathbf{P}_2 \quad \cdots \quad \overline{\mathbf{H}}\mathbf{P}_N]$. To separate the source signals effectively, we transform (3) into the following expression

$$\tilde{\mathbf{y}}(k)=\tilde{\mathbf{H}}\mathbf{s}(k)+\tilde{\mathbf{v}}(k) \tag{9}$$

where

$$\tilde{\mathbf{y}}(k)=\begin{bmatrix}\mathrm{Re}(\mathbf{Y}^T(k))\\ \mathrm{Im}(\mathbf{Y}^T(k))\end{bmatrix},\ \tilde{\mathbf{v}}(k)=\begin{bmatrix}\mathrm{Re}(\mathbf{V}^T(k))\\ \mathrm{Im}(\mathbf{V}^T(k))\end{bmatrix},\ \tilde{\mathbf{H}}=\begin{bmatrix}\mathrm{Re}(\overline{\overline{\mathbf{H}}}^T(k))\\ \mathrm{Im}(\overline{\overline{\mathbf{H}}}^T(k))\end{bmatrix}, \tag{10}$$

$\tilde{\mathbf{y}}(k)$ and $\tilde{\mathbf{v}}(k)$ are $2n_RN\times 1$ vectors, and $\tilde{\mathbf{H}}$ is a $2n_RN\times N_T$ matrix. $\mathrm{Re}(\cdot)$ and $\mathrm{Im}(\cdot)$ represent the function of extracting real and imaginary components, respectively.

Note that (9) can be regarded as a linear mixing model of ICA. As the received signals $\tilde{\mathbf{y}}(k)$ are $2n_RN\times 1$ matrices, this is an overdetermined blind source separation problem [4]. Firstly, the received signals can be projected onto an N-dimensional signal-plus-noise subspace by using a pre-whitening matrix $\mathbf{V}_S$:

$$\mathbf{x}(k)=\mathbf{V}_S\tilde{\mathbf{y}}(k). \tag{11}$$

Secondly, we should find a separating matrix $\overline{\mathbf{W}}$ to apply to the vector $\mathbf{x}(k)$, and then the source signal can be recovered:

$$\mathbf{z}(k)=\overline{\mathbf{W}}\mathbf{x}(k)=\hat{\mathbf{s}}(k). \tag{12}$$

In this step, we choose FastICA algorithm, a computationally optimized ICA algorithm, to obtain the separating matrix $\overline{\mathbf{W}}$. The basic updating step is

$$\overline{\mathbf{w}}_i^+=E\{\mathbf{x}g(\overline{\mathbf{w}}_i^T\mathbf{x})\}-E\{g'(\overline{\mathbf{w}}_i^T\mathbf{x})\}\overline{\mathbf{w}}_i,\ i=1,\ldots,N \tag{13}$$

where $\overline{\mathbf{w}}_i^T$ is one row of the separating matrix $\overline{\mathbf{W}}$, and $\overline{\mathbf{w}}_i^+$, the new value of $\overline{\mathbf{w}}_i$, is normalized to unit norm after every iteration. Function $g(\cdot)$ is the derivative of a sufficiently regular non-quadratic function [5].

The estimate of source signal $\hat{\mathbf{s}}(k)$ will be used as the reference signal to estimate spatial channels so as to direct the process of TAAS by feedback. The estimated CSI can be used to decoding after the array selection, but we also can exploit directly the ICA network to detect the transmitted signals in a blind mode. Therefore, we can choose the best output within the two detection schemes according to actual channel states, and the structure is shown in Fig. 2.

4 Simulation Results

Consider a system with n_T transmit BFSAs at BS and two receive antennas at MS. Each BFSA is a 4-element ULA. The original data stream is 300 blocks with two BPSK symbols per block. The quasi-static flat Rayleigh fading channel is assumed.

Fig.3 (a) shows BER performance comparison between the FastICA-based scheme and the conventional STBC scheme. In the simulation, we do not consider TAAS scheme and exploit only two transmit BFSAs. Clearly, the BSTBC scheme has a better performance than the original STBC. The FastICA-based BSTBC scheme has approximate performance compared with the BSTBC scheme having the perfect CSI, which can also demonstrate the validity of channel estimation using ICA network.

Fig.3 (b) shows the performance of the TAAS considering BSTBC and FastICA-based blind detection scheme. It is clear that the BER performance is better and better by increasing the number of BFSA.

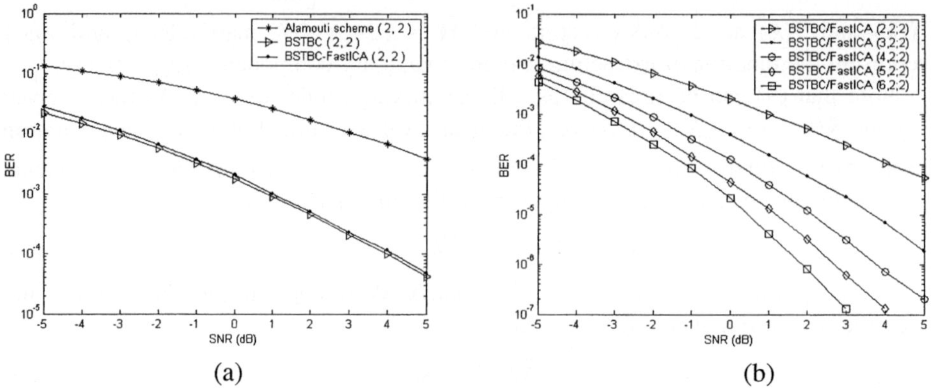

Fig. 3. Performance of the proposed method combining BSTBC, TAAS and FastICA-based blind detection scheme. (a) Combined BSTBC and FastICA-based detection scheme without TAAS; (b) Combined BSTBC and FastICA-based detection scheme with TAAS

5 Conclusions

In this paper, we investigate a method which combines BSTBC and TAAS at the transmitter and applies ICA technique to assist the channel estimation as well as achieves blind detection at the receiver. The full diversity gains, beamforming gains and the multi-access interference cancellation performances are achieved by using the BSTBC. The utility of TAAS obtains full diversity order as if all the transmit BFSAs are exploited, and it improves effectively the performance of the transmission system without increasing the complexity of the system obviously. The ICA-based channel estimation and blind detection scheme can estimate wireless channel without using pilot sequences and it can provide a high degree of robustness against channel estimation errors and time variation of the fading channels.

References

1. Alamouti, S. M.: A Simple Transmit Diversity Technique for Wireless Communications, IEEE Journal on Selec. Areas in Comm, Vol. 16. No. 8 (1998) 1451–1458
2. Aste, T., Forster, P., Fety, L., Mayrargue, S.: Downlink Beamforming Avoiding DOA Estimation for Cellular Mobile Communications, in Proc. IEEE International Conference on Acoustics, Speech, and Signal Processing (ICASSP), Vol. 6 (1998) 3313–331
3. Gore, D., Paulraj, A.: Space-Time Block Coding with Optimal Antenna Selection, in Proc. IEEE International Conference on Acoustics, Speech, and Signal Processing (ICASSP), Vol. 4. Salt Lake City, UT (2001) 2441–2444
4. Liu, J., Iserte, A. P., Lagunas, M. A.: Blind Separation of OSTBC Signals Using ICA Neural Networks, in Proc. IEEE International Symposium on Signal Processing and Information Technology (ISSPIT), Darmstadt, Germany (2003)
5. Hyvärinen, A.: Fast and Robust Fixed-Point Algorithms for Independent Component Analysis. IEEE Trans. on Neural Networks, Vol. 10. No. 3 (1990) 626–634

Nonlinear Dynamic Method to Suppress Reverberation Based on RBF Neural Networks

Bing Deng[1,2] and Ran Tao[1]

[1] Department of Electronic engineering, School of Information Science and Technology, Beijing Institute of Technology, 100081 Beijing, China
{dengbing, rantao}@bit.edu.cn

[2] Department of Electronic engineering, Naval Aeronautical Engineering Institute, 264001, Yantai, Shandong, China

Abstract. On condition that the noise, such as the reverberation, could be modeled as a dynamical model with lower dimensions, it would be picked out from its mixture with a useful signal by using the nonlinear dynamic method proposed in this paper. In other words, the useful signal could be separated from the noise with this method, which constructs the nonlinear inverse to linear filter, based on the spectrum difference between the noise and the useful signal, in virtue of successive approximation with Radial Basis Function Neural Networks. Two examples, with the sine pulse as the useful signal, are displayed. The artificial chaotic signal plays the role of the noise in one example, and the actual reverberation in another. These examples confirm the feasibility of this method.

1 Introduction

In underwater acoustics and active sonar, the reverberation becomes the dominant noise. Usually the reverberation is modeled as stochastic model, and its non-correlation is used to improve SRR (Signal Reverberation Ratio). Since the reverberation could be modeled as a dynamical model, the traditional method won't make full use of the structural detail of the reverberation. In fact, Frison et al have determined the chaos property of the ocean ambient noise and the continuous wave signal influenced by the sea channel in 1996 [1], [2]. And Broomhead et al have explained that fluid systems are the origin of many nonlinear disorganized phenomena [3], such as the sea clutter in radar, some noise in sonar, which was backed up by Zhiming Cai [4]. So a heuristic nonlinear dynamic method is proposed to suppress reverberation in this paper.

In Section II, we describe the principle and algorithm of this method. Section III displays the simulation results, which are analyzed, as well. And the conclusion is presented in Section IV.

F. Yin, J. Wang, and C. Guo (Eds.): ISNN 2004, LNCS 3174, pp. 324–330, 2004.

2 Principles and Algorithm

2.1 Principles

Let the receiving modal be as follows:

$$z_n = x_n + y_n \tag{1}$$

where x_n represents the observation of a nonlinear system (may be but need not be a chaos system), i.e., $x_n = h(\vec{s}_n)$, $h(\bullet)$ denotes the observation function of the nonlinear system, $\vec{s}_n \in M$, M is the state vector set of the nonlinear system; And y_n is the useful signal, whose bandwidth is much narrower than x_n. Then y_n can be filtered through a narrow-band FIR filter. Because feedback may produce false dynamic feature, which is unavoidable in IIR filters, so FIR filter is preferable. Let the transfer function of the narrow-band FIR filter be $F(\cdot)$. Naturally the outcome of z_n through the filter, z'_n, is another observation of this nonlinear system.

$$\begin{aligned} z'_n = F(x_n + y_n) = F(x_n) = x'_n \\ = h(\vec{s}'_n),\ \vec{s}'_n \in \mathrm{M} \end{aligned} \tag{2}$$

Since x_n and x'_n are both the observation of this nonlinear system, we can model this system by using them respectively, i.e., dynamic reconstruction. The normal way of dynamic reconstruction is delay embedding [5], which convert the scalar time series $x_n, n = 1, \cdots, \mathrm{N}$ to the vector series $\vec{x}_{n,\mathrm{D},\tau}$ according to (3).

$$\begin{aligned} \vec{x}_{n,\mathrm{D},\tau} &= \Phi_{\mathrm{D},\tau}(\vec{s}_n) = [x_n, x_{n+\tau}, \cdots, x_{n+(\mathrm{D}-1)\tau}]^{\mathrm{T}} \\ &= [h(\vec{s}_n), h(\vec{s}_{n+\tau}), \cdots, h(\vec{s}_{n+(\mathrm{D}-1)\tau})]^{\mathrm{T}},\quad n = 1, \cdots, \mathrm{N} - (\mathrm{D}-1)\tau \end{aligned} \tag{3}$$

where D, τ is the embedding dimension and the embedding delay respectively. The vector series $\vec{x}_n$ can reconstruct the state space of the nonlinear system and keep all dynamic information. Dynamic reconstruction of x'_n is the same as x_n.

$$\vec{x}'_{n,\mathrm{D},\tau} = \Phi_{\mathrm{D},\tau}(\vec{s}'_n) = [x'_n, x'_{n+\tau}, \cdots, x'_{n+(\mathrm{D}-1)\tau}]^{\mathrm{T}} \tag{4}$$

Aiming to simplify the description of formula, we might as well let $\tau = 1$. If the coefficients of the FIR filter are $\{c_j, j = 0, \cdots, \mathrm{q}-1\}$, then

$$x'_n = \sum_{j=0}^{q-1} c_j x_{n+j} \tag{5}$$

Obviously the relation between $\vec{x}'_{n,\mathrm{D},1}$ and $\vec{x}_{n,\mathrm{D+q-1},1}$ is as follows:

$$\vec{x}'_{n,\mathrm{D},1} = \phi_{\mathrm{q}} \vec{x}_{n,\mathrm{D+q-1},1} \tag{6}$$

where ϕ_q is a $D\times(D+q-1)$ matrix,

$$\phi_q = \begin{bmatrix} c_0 & c_1 & c_2 & \cdots & c_{q-1} & 0 & 0 & \cdots & 0 \\ 0 & c_0 & c_1 & \cdots & c_{q-2} & c_{q-1} & 0 & \cdots & 0 \\ 0 & 0 & c_0 & \cdots & c_{q-3} & c_{q-2} & c_{q-1} & \cdots & 0 \\ \vdots & \vdots & \vdots & \ddots & \cdots & \cdots & \cdots & \ddots & \vdots \\ 0 & 0 & 0 & \cdots & c_0 & c_1 & c_2 & \cdots & c_{q-1} \end{bmatrix} \tag{7}$$

Eq.(6) can be rewritten as:

$$\vec{x}'_{n,D,1} = \phi_q \circ \Phi_{D+q-1,1}(\vec{s}_n) \tag{8}$$

Since ϕ_q is an injection Based on the set $\Phi_{D+q-1,1}(M)$, i.e., $\forall \vec{x}'_{n,D,1}$, one and only $\vec{x}_{n,D+q-1,1}$ can be found certainly to satisfy (9).

$$\vec{x}_{n,D+q-1,1} \in \Phi_{D+q-1,1}(M), \quad \vec{x}'_{n,D,1} = \phi_q \vec{x}_{n,D+q-1,1} \tag{9}$$

Thus inverse transform must exist between $\vec{x}'_{n,D,1}$ and $\vec{x}_{n,D+q-1,1}$. Let ϕ_q^{-1} represents the inverse of ϕ_q. We can utilize RBF neural networks to approximate ϕ_q^{-1} by means of learning input/output data $\vec{x}'_{n,D,1}$ and $\vec{x}_{n,D+q-1,1}$. Then $\vec{x}_{n,D+q-1,1}$ can be recovered through (10), which means the estimation of x_n, $\hat{x}_n$, is obtained.

$$\hat{\vec{x}}_{n,D+q-1,1} = \hat{\phi}_q^{-1} \vec{x}'_{n,D,1} \tag{10}$$

where $\hat{\phi}_q^{-1}$ denotes the approximation result.

Finally y_n can be recovered by z_n subtracting $\hat{x}_n$, i.e., $\hat{y}_n = z_n - \hat{x}_n$.

2.2 Algorithm

This method is used in 'blind' signal separation in this paper (only the spectrum difference between x_n and y_n is known), which means the learning data to train RBF NN come from the observation data, polluted by the useful signal. Therefore it is important that the part of the observation data is picked out as the learning data to ensure accurate approximation to ϕ_q^{-1}. As a result, the following two simulations both use the sine pulse as the useful signal, which make some part of the observation data not polluted by the sine pulse in the time domain. Some actual observation models accord with these simulations, for example, the active sonar for detecting mines lying on the seabed, the echo of the target is much shorter than the reverberation in the time domain.

The algorithm is as follows:

Step 1. To construct the FIR filter based on the spectrum of the useful signal y_n, and filter the useful signal from z_n to get z'_n, i.e., x'_n.

Step 2. Dynamic reconstruction. First, we determine the embedding dimension D and the embedding delay τ in virtue of x'_n [6], [7], which are needed in dynamic reconstruction. Then we model x'_n and z_n with the dynamic model.

Step 3. Constructing RBF Neural Networks to approximate ϕ_q^{-1} by means of learning P pairs of input/output data $\vec{x}'_{n,D,\tau}$ and $\vec{z}_{n,D+q-1,\tau}$, $n = 1, \cdots, P$. In this paper, two layer RBF NN is used, in which a group of coarse approximation is made at first and combined linearly to obtain the optimal mode at last [8]. The two-layer RBF NN, compared with the RBF NN using strict interpolation, reduces the computation cost and improves prediction accuracy.

Step 4. To estimate x_n by Inputting $\vec{x}'_{n,D,\tau}$ to the trained two layer RBF NN and obtain $\hat{y}_n$ through z_n subtracting $\hat{x}_n$.

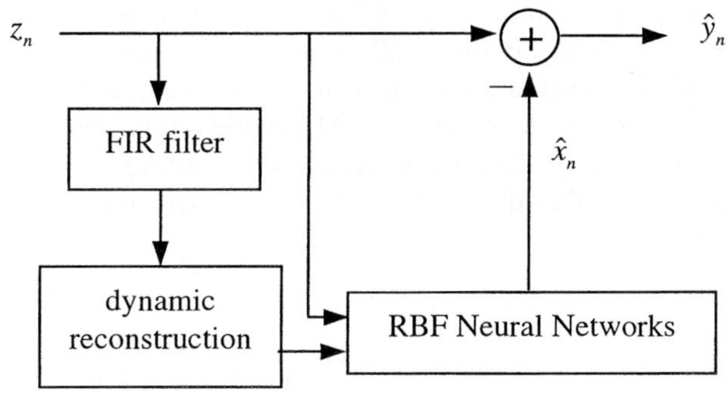

Fig. 1. Block diagram of this proposed Nonlinear Dynamic Method

3 Simulations

Two examples, with the sine pulse as the useful signal, are followed. In the first one the artificial chaotic signal is used as the noise, and the actual reverberation in the second one, respectively.

In the first example, the artificial chaotic signal is Lorenz series. The frequency and width of the sine pulse is 175KHz, 50 μs respectively. The sample frequency is 1MHz, the sample duration is 1ms, and SNR is -8.8093dB. The sine pulse exists from 200th to 250th μs. Totally 300 pairs of data are used, which are divided into 60 groups, 5 radial basis functions in each group, to train the RBF NN. $D=4$, $\tau=1$. After the sine pulse is separated from Lorenz series by this proposed method, SNR is increased to 11.7264dB.

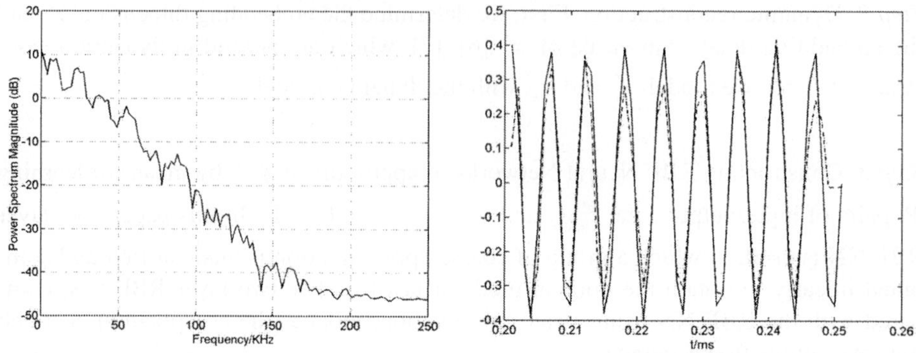

(a) (b)

Fig. 2. Example with Lorenz series as the noise, (a) PSD of the observation data; (b) the original pulse (*solid line*) compared with the recovered pulse (*dash-dot line*)

In the second example, the actual reverberation is used, which is from the experiment made in Songhua Lake, China. When the transmitting signal is a mono-frequency pulse with 3ms duration, 20° transmitting angle and 100KHz mid-frequency, we call the reverberation mono-frequency reverberation in this paper. Respectively the reverberation is called LFM reverberation when the transmitting signal is a LFM pulse with 7.2ms duration, 20° transmitting angle and 90-180KHz band. In the example, two kinds of reverberation are both used. And the sampling

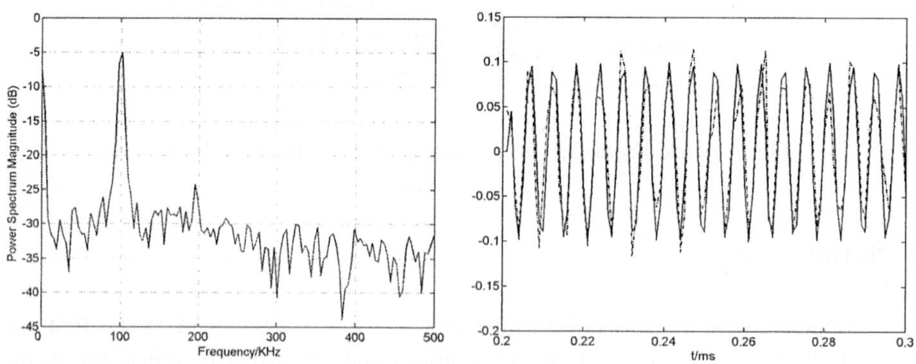

(a) (b)

Fig. 3. Example with mono-frequency reverberation as the noise, (a) PSD of the observation data; (b) the original pulse (*solid line*) compared with the recovered pulse (*dash-dot line*)

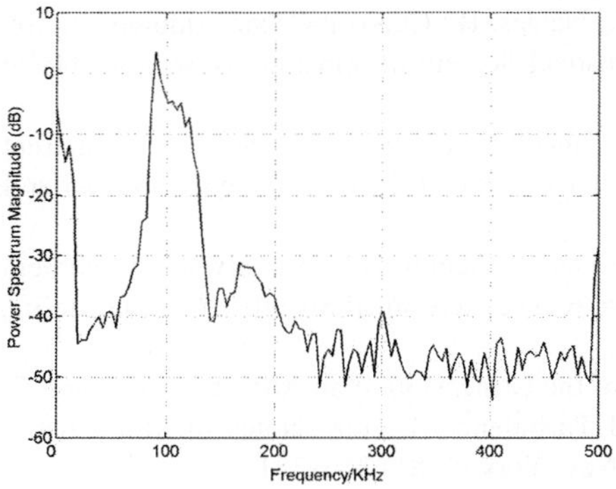

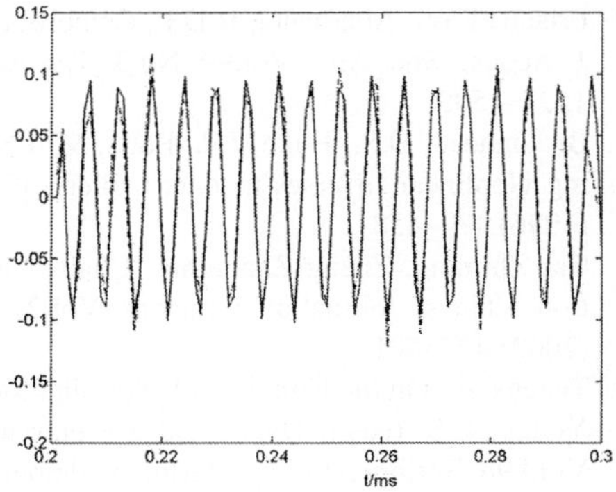

Fig. 4. Example with LFM reverberation as the noise, (a) PSD of the observation data; (b) the original pulse (*solid line*) compared with the recovered pulse (*dash-dot line*)

frequency is still 1MHz, but only 1000 samples are used. The useful signal is still the sine pulse, which is similar to the one used in the first example, but the amplitude descend and the duration expand to $100\,\mu s$. In this example 600 pairs of data are used, which are divided into 100 groups, 6 radial basis functions in each group, to train the RBF NN. When mono-frequency reverberation is used, $D = 10$, $\tau = 2$, SRR is increased to 8.6159dB from -7.3593dB; When LFM reverberation is used, $D = 7$, $\tau = 1$, SRR is increased to 13.1992dB from -6.0387dB. These results show that this proposed method can separate the useful signal from the reverberation effectively (even if SRR is very low), and obtain better results with LFM reverberation as ambient noise than mono-frequency reverberation.

4 Conclusion

In this paper, a nonlinear dynamics noise-reduction method based on RBF NN is proposed. And its feasibility is testified by two examples, in which simulation chaos data and actual reverberation data are used as the noise respectively. Though the theoretical base of this method is chaos theory, its application isn't confined to process chaos signal. Only if the noise could be modeled as a dynamical model with lower dimensions and its bandwidth is wider than the useful signal, this method would have effect to separate them.

References

1. Frison, T.W., Abarbanel, H.D.I., Cembrola, J., Katz, R.: Nonlinear analysis of environmental distortions of continuous wave signals in the ocean. J. Acoust. Soc. Am, Vol.99, No.1. The Acoustical Society of America, New York (1996) 139–146

2. Frison, T.W., Abarbanel, H.D.I., Cembrola, J., Neales, B.: Chaos inOcean Ambient 'noise'. J. Acoust. Soc. Am., Vol.99, No.3. The Acoustical Society of America, New York (1996) 1527–1539
3. Broomhead, D.S., Huke, J.P., Potts, M.A.S.: Cancelling Deterministic Noise by Constructing Nonlinear Inverses to linear filters. Physica D, Vol.89. Elsevier Science, Netherlands (1996) 439–458
4. Cai Zhiming, Zheng Zhaoning.: Chaos Characteristic Analysis of Underwater Reverberation. Chinese Journal of Acoustics, Vol.27, No.6. The Acoustical Society of China, Beijing (2002) 497–501
5. Takens, F.: On the Numerical Determination of the Dimension of an Attractor. In: Rand, D., Young, L.S. (eds.): Dynamical Systems and Turbulence. Lecture Notes in Mathematics, Vol.898. Springer-Verlag, Berlin Heidelberg New York (1981) 366–381
6. Kim, H. S., Eykholt, R., Salas, J. D.: Nonlinear Dynamics, Delay Times, and Embedding Windows. Physica D, Vol.127. Elsevier Science, Netherlands (1999) 48–60
7. Liangyue Cao.: Practical Method for Determining the Minimum Embed–ding Dimension of a scalar time series. Physica D, Vol.110. Elsevier Science, Netherlands (1997) 43–50
8. Xiangdong He, Alan Lapedes: Nonlinear Modeling and Prediction by Successive Approxi-Mation Using Radial Basis Functions. Physica D, Vol.70. Elsevier Science, Netherlands (1993) 289–301

A New Scheme for Detection and Classification of Subpixel Spectral Signatures in Multispectral Data*

Hao Zhou[1], Bin Wang[1,2], and Liming Zhang[1]

[1] Department of Electronics Engineering, Fudan University, Shanghai 200433, China
{032021038,wangbin,lmzhang}@fudan.edu.cn
[2] The Key Laboratory of Wave Scattering and Remote Sensing Information, Fudan University, Shanghai 200433, China

Abstract. Mixed pixels exist in almost all multispectral and hyperspectral remote sensing images. Their existence impedes the quantification analysis of remote sensing images. This paper proposes a new scheme for detection and classification of subpixel spectral signatures in multispectral remote sensing images. By minimizing the energy function with two special constraints, the mixed pixels can be decomposed more precisely. Further, we point out that our scheme can also be used for the decomposition of mixed pixels in hyperspectral remote sensing imagery. Finally, the performances of the proposed scheme are demonstrated experimentally and the comparisons of the performances with conventional methods such as back-propagation (BP) neural network are made.

1 Introduction

Usually, ground objects in remote sensing images are detected by units of the pixels. Due to the limit of spatial resolution, in most cases, one pixel may cover hundreds of square meters with various ground objects and becomes a mixed pixel. The mixed pixel problem not only influences the precision of object recognition and classification, but also becomes an obstacle to quantification analysis of remote sensing images. This problem can be overcome by obtaining the percentages of interesting object precisely [1][2].

A kind of widely used methods for the decomposition of mixed pixels are neural network methods such as back-propagation (BP) and radial basis function (RBF) neural networks [3][4]. They are useful for detection and classification of subpixel spectral signatures in multispectral remote sensing images. However, their decomposition precisions are usually low and their performances decrease quickly in the presence of noise. This kind of methods based on neural networks cannot meet the constraint conditions such as the sum of percentages of decomposition results should be 1 and the percentage of each decomposition result should fall into the range [0, 1]. In this paper, we propose a scheme which can produce more precise results than the

* This research was supported by the grants from the National Natural Science Foundation of China (No. 30370392 and No. 60171036) and the Major State Basic Research Development Program of China (No. 2001CB309401).

F. Yin, J. Wang, and C. Guo (Eds.): ISNN 2004, LNCS 3174, pp. 331–336, 2004.

BP or RBF networks by minimizing a new objective function with special constraints. Especially, our scheme is more robust and can still obtain better results even through there exists stronger noise.

The remainder of this paper is organized as follows. Section 2 is used to describe the proposed scheme. Some experimental results are shown in section 3. Conclusion is given in section 4.

2 The Proposed Scheme

2.1 Linear Spectral Mixture Analysis

Over the past years, linear spectral mixture analysis (LSMA) has been widely used for mixed pixel decomposition. It assumes that the spectral signature of an image pixel is linearly mixed by the spectral signatures of objects present in the image. Define $\boldsymbol{X}$ as a multispectral vector of a single pixel in multispectral remote sensing images, and $\boldsymbol{A}$ as a reflectance characteristic matrix composed of reflectance of each object in each spectral band, and $\boldsymbol{S}$ as a vector composed of the percentage of each object. So we can obtain the equation

$$\boldsymbol{X} = \boldsymbol{A} \times \boldsymbol{S} . \tag{1}$$

If the multispectral remote sensing images have n bands and m sorts of interesting objects, then $\boldsymbol{X}$ is a $n \times 1$ vector, $\boldsymbol{A}$ is a $n \times m$ matrix and $\boldsymbol{S}$ is a $m \times 1$ vector. In this model, the selection of matrix $\boldsymbol{A}$ is important to the precision of decomposition results.

2.2 Constraint Conditions

The decomposition results $\boldsymbol{S}$ based on LSMA should satisfy the following two constraints.

(a) The sum of percentages s_i of interesting objects in every single pixel should be 1, *i.e.*

$$\sum_{i=1}^{m} s_i = 1 \cdot \tag{2}$$

(b) The percentages s_i of interesting objects should fall into the range from 0 to 1, *i.e.*

$$0 \le s_i \le 1, \ (i = 1,2,\ldots,m) . \tag{3}$$

In addition, the decomposition results $\boldsymbol{S}$ should satisfy the condition of LSMA model, *i.e.*

$$\boldsymbol{X} = \boldsymbol{A} \times \boldsymbol{S} . \tag{4}$$

2.3 The Proposed Scheme Based on Constraint Conditions

If the reflectance characteristic matrix A is selected precisely and there is no noise in the images, the decomposition results should satisfy the Eq. (4). But because of measure noise, weather condition and atmosphere dispersion condition, *etc.*, it is impossible to get an ideal situation mentioned above. So precise decomposition results should be obtained by minimizing the following energy function

$$E = (X - AS)^T (X - AS). \tag{5}$$

In order to make the results satisfy the Eq. (2) and (3), we introduce them to Eq. (5) and form the objective function

$$E=(X-AS)^T(X-AS)+M\left(\sum_{i=1}^{m} s_i -1\right)^2 +\sum_{i=1}^{m}[c_1 s_i^{2l} +c_2(1-s_i)^{2l}]. \tag{6}$$

The Eq. (6) uses the second part to constrain the sum of percentages s_i to 1. M is a Lagrange variable and can be obtained by iterative operation. For the consideration of simplicity, we set M as a large number, and the algorithm can still work well without the precision loss of decomposition results.

The Eq. (6) uses the third part to constrain the percentages s_i in the range from 0 to 1. l is a large positive integer. When $0\le s_i \le 1$, s_i^{2l} and $(1- s_i)^{2l}$ are close to zero, but when $s_i<0$ or $s_i>1$, at least one of s_i^{2l} and $(1- s_i)^2$ is much greater than 1, so through iterative operation, s_i can be constrained back to [0,1]. c_1 and c_2 are adjustment factors. By changing the values of M, c_1 and c_2, we can adjust the proportion of the two constraints. Here, we make the two constraints equally important.

In order to minimize the objective function E, calculate its differential coefficient

$$\frac{\partial E}{\partial S} = -2A^T(X-AS)+2M\left(\sum_{i=1}^{m} s_i -1\right)\boldsymbol{I} + 2lc_1\left(s_1^{2l-1}, s_2^{2l-1},\dots, s_m^{2l-1}\right)^T$$
$$- 2lc_2\left((1-s_1)^{2l-1},(1-s_2)^{2l-1},\dots,(1-s_m)^{2l-1}\right)^T. \tag{7}$$

$\boldsymbol{I}$ is a unit column vector. From the Eq. (7), we can get the following iterative algorithm

$$S(k+1)=S(k)-\eta\Bigg\{-A^T(X-AS(k))+M\left(\sum_{i=1}^{m} s_i(k)-1\right)I$$
$$+c_1'(s_1(k)^{2l-1}, s_2(k)^{2l-1},\dots, s_m(k)^{2l-1})^T$$
$$-c_2'\left((1-s_1(k))^{2l-1},(1-s_2(k))^{2l-1},\dots,(1-s_m(k))^{2l-1}\right)^T\Bigg\}. \tag{8}$$

η is the iterative step size. c_1' and c_2' are constants calculated from c_1 and c_2, respectively.

3 Experiment Results

In this section, we decompose the mixed pixels with two algorithms, BP neural network and the proposed algorithm, and then compare their results. The experiments are performed on artificial simulation images and Landsat multispectral images.

3.1 Experiments for Simulation Images

In order to avoid the influence of imprecise selection of reflectance characteristic matrix ***A***, first, the experiment is carried out for simulated images. We assume that artificial images include two kinds of ground objects. The percentages of them in the image are shown in Fig. 1.(a). In the Fig. 1.(a), pure black denotes that the percentage of a certain sort of object in this pixel is 0, while pure white denotes 1. Define 4×2 matrix ***A*** as below

$$\boldsymbol{A} = \begin{bmatrix} 20 & 150 \\ 50 & 100 \\ 150 & 50 \\ 160 & 70 \end{bmatrix}.$$

We can obtain 4 simulated images shown in Fig. 1 (b),(c),(d) and (e) as 4 bands of multispecrtral images. The performance of BP network depends on the quality of training set to a great extent. In our experiment, we randomly produce a training set containing one thousand two-dimensional vectors with 1 as the sum of every vector's two elements. Training error is set as 10^{-6}. The decomposition results of BP network are shown in Fig. 2.

(a) (b) (c) (d) (e)

Fig. 1. Simulation images. (a) The percentages of standard ground objects. (The image size is 100×100 .); (b) (c) (d) (e) 4 mixed simulation images.

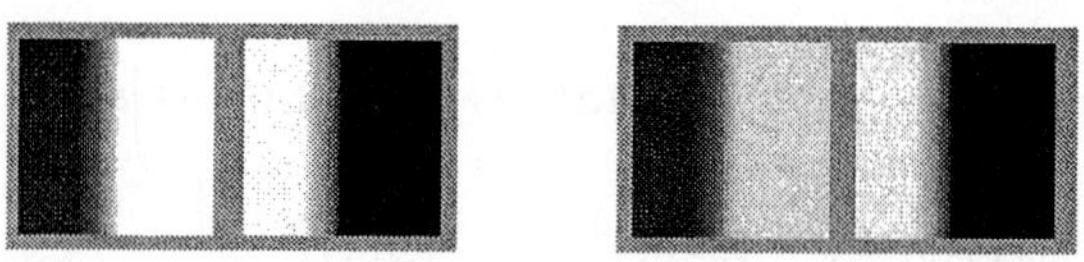

(a) (b)

Fig. 2. The decomposition results of BP network. (a) The results in the absence of noise, (b) The results in the presence of 20db noise.

In the proposed algorithm, the program iterates 500 times for every pixel or jumps to the next pixel when $\Delta E \leq 0.01$. Here, we set $M = 1000$, $c_1' = c_2' = 1$, and l=25. The decomposition results of our method are shown in Fig. 3.

Fig. 3. The decomposition results of our method. (a) The results in the absence of noise, (b) The results in the presence of 20db noise.

From the above results, we can see that the results of our method are more precise than that of BP neural network. For example, in a pixel, the percentages of two sorts of objects are 2% and 98%, the decomposition results of our method are 2.781% and 97.401% when BP are 22.201% and 104.956%. Obviously, the precision is improved. More examples are given in Table 1.

Table 1. The result comparison of the BP network and our method

Percentages of two objects in source images		Results of the BP network		Results of our method	
		No noise	Add noise	No noise	Add noise
1	2%	22.201%	32.063%	2.781%	4.048%
	98%	104.956%	81.232%	97.401%	99.016%
2	12%	28.347%	33.151%	12.787%	12.272%
	88%	94.574%	72.523	88.010%	90.520%
3	20%	37.877%	39.386%	19.215%	20.753%
	80%	86.058%	68.188%	80.676%	81.047%
4	32%	50.686%	44.843%	31.254%	30.266%
	68%	78.216%	56.957%	68.216%	71.394%
5	45%	55.468%	50.177%	43.856%	42.059%
	55%	72.503%	72.158%	56.021%	56.764%

The Table 1. presents 5 mixed pixels selected from artificial images and their decomposition results by the two algorithms: the BP network and the proposed algorithm. From the Fig. 2. the Fig. 3. and the Table 1. we can conclude that

(a) In no noise situation, all two algorithms can decompose the mixed pixel, but the results of our method are more precise.

(b) When noise is added to the source images, the results of BP network become worse, but our method demonstrates the stronger ability of noise suppression.

3.2 Experiments for Landsat Multispectral Images

In this part, we used 1st – 5th and 7th bands of Landsat multispectral images covering an area of Shanghai as experimental data (image size is 256×256). This area mainly includes 3 sorts of typical ground objects: water, plant and soil (include artificial architectures). The reflectance characteristic matrix $\boldsymbol{A}$ is composed of the mean of 20 samples selected in each sort of object in every band. In the experiment, we apply separately the BP network and our method to Landsat multispectral images. From the experimental results, we can find that there are many noisy pixels in the decomposition results of BP network and most of its results do not satisfy the constraints mentioned above. In addition, we find out this problem also exists in the RBF networks. The proposed method solved this problem. By adding two constraints to the energy function, this proposed algorithm can control the percentages of interesting objects in most pixels within the range [0, 1], which makes the results more precise.

Finally, we have to point out that our method can also be used for the decomposition of mixed pixels of hyperspectral remote sensing images although only the experiment results for multispectral remote sensing images have been shown in this paper.

4 Conclusions

In this paper, we presented a scheme which can be used to decompose the mixed pixel in multispectral and hyperspectral remote sensing images. By introducing constraints into the objective function, the proposed method can satisfy the request of pixel decomposition: the sum of percentage of each interesting object in every single pixel should be 1, and the percentage of each interesting object should fall into the range from 0 to 1. By these constraints, our method obtained more precise results, and stronger ability of noise suppression. Its performance is better than the conventional BP and RBF neural networks.

References

1. Cheng-I Chang, Xiao-Li Zhao: Least Squares Subspace Projection Approach to Mixed Pixel Classification for Hyperspectral Images. IEEE Transactions on Geoscience and Remote Sensing, vol. 36, no. 3(1998), 898-912
2. Muramatsu K., Furumi S., Fujiwara N., Hayashi A., Daigo M., andOchiai F.: Pattern Decomposition Method in the Albedo Space for Landsat TM and MSS Data Analysis. International Journal of Remote Sensing, vol.21, no.1(2000), 99-119
3. Tanqiang Peng, Bicheng Li, and Huan Su: A Remote Sensing Iimage Classification Method Based on Evidence Theory and Neural Networks. Proceedings of the 2003 International Conference on Neural Networks and Signal Processing, Vol. 1, (2003), 240-244
4. Heermann P. D., and Khazenie N.: Classification of Multispectral Remote Sensing Data Using a Back-propagation Neural Network. IEEE Transactions on Geoscience and Remote Sensing, vol. 30, no. 1(1992), 81-88

A Rough-Set-Based Fuzzy-Neural-Network System for Taste Signal Identification*

Yan-Xin Huang, Chun-Guang Zhou, Shu-Xue Zou, Yan Wang, and Yan-Chun Liang

College of Computer Science and Technology, Jilin University, Changchun, 130012, P.R. China
jluhyx89@mail.jl.cn

Abstract. A voting-mechanism-based fuzzy neural network model for identifying 11 kinds of mineral waters by its taste signals is proposed. In the model, A classification rule extracting algorithm based on discretization methods in rough sets is developed to extract fewer but robust classification rules, which are ease to be translated to fuzzy if-then rules to construct a fuzzy neural network system. Finally, the particle swarm optimization is adopted to refine network parameters. Experimental results show that the system is feasible and effective.

1 Introduction

The first step in designing a fuzzy inference system is the partition of fuzzy input space, i.e., the number of fuzzy subsets in each coordinate, the type and the parameters of the membership function corresponding to each fuzzy subset are determined. A good partitioning method can create a small rule base with robust rules. General methods for partitioning the fuzzy input space include the C-means clustering algorithm ([1]) and the subtractive clustering algorithm ([1],[2]) and so on. Despite of the popularity of the C-means, one of its major drawbacks is very sensitive to the values of the initial cluster centers ([3]). Similarly the subtractive clustering algorithm often suffers from determining its parameters ([1]). A classification rule extracting algorithm in rough sets has been presented in [4] to identify tea taste signals. Further a voting-mechanism-based fuzzy neural network model is proposed in this paper. Based on the discretization method by Nguyen, H.S ([5]), a new classification rule extracting algorithm is proposed to implement partitioning fuzzy input space. Finally the particle swarm optimization (PSO) is adopted to refine network parameters. Experimental results show that the system is feasible and effective.

* This paper is supported by the National Natural Science Foundation of China under Grant No. 60175024 and the Key Laboratory for Symbolic Computation and Knowledge Engineering of Ministry of Education of China.

F. Yin, J. Wang, and C. Guo (Eds.): ISNN 2004, LNCS 3174, pp. 337–343, 2004.

2 Discretization of Continuous Signals

A set of training samples in classification systems can be represented as a decision table $A = (U, C \cup \{d\})$ in rough sets, where $U = \{x_1, x_2, ..., x_n\}$, is the universe; $C = \{C_1, C_2, ..., C_s\}$ is a set of condition attributes; and $\{d\}$ is a set of decision attributes. The classification rules in rough sets are in the form as follows:
$R: \, if \; C_1(x) = c_1^{(j)} \wedge C_2(x) = c_2^{(j)} \wedge ... \wedge C_s(x) = c_s^{(j)} \; then \; d(x) = v^{(j)}$
where $c_i^{(j)} \in V_{c_i}$, V_{c_i} represents the value set of condition attribute $C_i, i \in \{1, 2, ..., s\}$; $v^{(j)} \in V_d, j \in \{1, 2, ..., r\}$, V_d represents the value set of decision attribute d. Such a decision table in which the condition attributes have real values and the decision attributes have discrete values is mainly concerned in this paper, and it is assumed that no any two training samples with the same condition attribute valves have the different decision attribute values.
The discretization decision table of $A = (U, C \cup \{d\})$ is represented as $A_{|D} = (U, C_D \cup \{d\})$, where $C_D = \{C_1^{D_{C_1}}, C_2^{D_{C_2}}, ..., C_s^{D_{C_s}}\}$ represents discretization condition attribute set, and getting a optimal discretization decision table is a NP-hard problem ([5]). Nguyen, H.S ([5]) proposed a heuristic searching for approximately optimal discretizations with the computational complexity of $O(sn|D| + log^n)$. The algorithm is adopted in this paper.

3 Extraction of Classification Rules

Some classification rules extracted directly from $A_{|D}$ are not robust enough. Below an algorithm is proposed to extract fewer but robust classification rules.

Definition 1. *(The strength of the classification rules) The strength of a classification rule R is defined as:*
Strength(R)$=|\{x | x \in U$*,and* x *matches the antecedent of the rule* $R\}|$.

Definition 2. *(The neighbor rules): Any two classification rules R1 and R2 extracted from* $A_{|D}$ *are called neighbor with regard to* j *if* $\exists j \in \{1, 2, ..., s\}$*, the values of* $C_j^{D_{C_j}}(x)$ *in R1 and R2 are two consecutive discrete values, and* $\forall i \neq j, i \in \{1, 2, ..., s\}$*, the values of* $C_i^{D_{C_i}}(x)$ *in R1 and R2 are the same.*

Definition 3. *(The incorporable rules): Let a classification rule R in the conjunctive form be* $if \; C_1^{D_{C_1}}(x) = (c_1^1 \vee c_1^2 \vee ... \vee c_1^{k_1}) \wedge C_2^{D_{C_2}}(x) = (c_2^1 \vee c_2^2 \vee ... \vee c_2^{k_2}) \wedge ... \wedge C_s^{D_{C_s}}(x) = (c_s^1 \vee c_s^2 \vee ... \vee c_s^{k_s}) \; then \; d(x) = v^{(j)}$. *Without losing generality, assume* $c_i^1 < c_i^2 < ... c_i^{k_i}, \forall i \in \{1, 2, ..., s\}$. *The rule R is called incorporable if* $c_i^1, c_i^2, ..., c_i^{k_i}$ *are all the consecutive discrete values,*$\forall i \in \{1, 2, ..., s\}$.

Definition 4. *(The neighbor incorporable rules): Any two incorporable rules* $R1: \; if \; C_1^{D_{C_1}}(x) = (c_{11}^1 \vee c_{11}^2 \vee ... \vee c_{11}^{k_{11}}) \wedge C_2^{D_{C_2}}(x) = (c_{12}^1 \vee c_{12}^2 \vee ... \vee c_{12}^{k_{12}}) \wedge ... \wedge C_s^{D_{C_s}}(x) = (c_{1s}^1 \vee c_{1s}^2 \vee ... \vee c_{1s}^{k_{1s}}) \; then \; d(x) = v^{(j_1)}$ *and* $R2: \; if \; C_1^{D_{C_1}}(x) = (c_{21}^1 \vee c_{21}^2 \vee ... \vee c_{21}^{k_{21}}) \wedge C_2^{D_{C_2}}(x) = (c_{22}^1 \vee c_{22}^2 \vee ... \vee c_{22}^{k_{22}}) \wedge ... \wedge C_s^{D_{C_s}}(x) =$

$(c_{2s}^1 \vee c_{2s}^2 \vee ... \vee c_{2s}^{k_{2s}})$ *then* $d(x) = v^{(j_2)}$ *are called neighbor with regard to* j *if the following conditions are satisfied for R2 (or R1):*
(1)$\exists j \in \{1, 2, ..., s\}$*, the value of* $C_j^{D_{C_j}}(x)$ *in R2 (or R1) is a single discrete value* c_{2j}^1 *(or* c_{1j}^1*), and the following relation between* c_{2j}^1 *in R2 (or* c_{1j}^1 *in R1) and* $c_{1j}^1 \vee c_{1j}^2 \vee ... \vee c_{1j}^{k_{1j}}$ *in R1 (or* $c_{2j}^1 \vee c_{2j}^2 \vee ... \vee c_{2j}^{k_{2j}}$ *in R2) are held:* $c_{2j}^1 = c_{1j}^{k_{1j}} + 1$ *(or* $c_{1j}^1 = c_{2j}^{k_{2j}} + 1$*), or* $c_{2j}^1 = c_{1j}^{k_{1j}} - 1$ *(or* $c_{1j}^1 = c_{2j}^{k_{2j}} - 1$*);*
(2)$\forall i \neq j, i \in \{1, 2, ..., s\}$*, the values of* $C_i^{D_{C_i}}(x)$ *in R1 and R2 are the same.*

Theorem 1. *(1)Any two neighbor rules with the same rule consequent can be represented as a incorporable rule; (2)Any two neighbor incorporable rules with the same rule consequent can be represented as a incorporable rule.*
Proof: (1)Let two neighbor rules with regard to j *be* $R1$: *if* $C_1^{D_{C_1}}(x) = c_1^1 \wedge C_2^{D_{C_2}}(x) = c_2^1 \wedge ... \wedge C_j^{D_{C_j}}(x) = c_j^1 \wedge ... \wedge C_s^{D_{C_s}}(x) = c_s^1$ *then* $d(x) = v^{(j_{i_1})}$ *and* $R2$: *if* $C_1^{D_{C_1}}(x) = c_1^1 \wedge C_2^{D_{C_2}}(x) = c_2^1 \wedge ... \wedge C_j^{D_{C_j}}(x) = c_j^2 \wedge ... \wedge C_s^{D_{C_s}}(x) = c_s^1$ *then* $d(x) = v^{(j_{i_2})}$*. Since* $v^{(j_{i_1})} = v^{(j_{i_2})}$*, they can be represented as a classification rule R in the conjunctive form: if* $C_1^{D_{C_1}}(x) = c_1^1 \wedge C_2^{D_{C_2}}(x) = c_2^1 \wedge ... \wedge C_j^{D_{C_j}}(x) = (c_j^1 \vee c_j^2) \wedge ... \wedge C_s^{D_{C_s}}(x) = c_s^1$ *then* $d(x) = v^{(j_i)}$*, where* $v^{(j_i)} = v^{(j_{i_1})} = v^{(j_{i_2})}$*. Since* c_j^1 *and* c_j^2 *are two consecutive discrete values, from Definition 3 we have that R is a incorporable rule. (2)can analogously be proved.*

A s-dimensional array RUL is used to keep all possible classification rules in the discrete space $C_1^{D_{C_1}} \times C_2^{D_{C_2}} \times ... \times C_s^{D_{C_s}}$, e.g., a classification rule $R:$ *if* $C_1^{D_{C_1}}(x) = c_1 \wedge C_2^{D_{C_2}}(x) = c_2 \wedge ... \wedge C_s^{D_{C_s}}(x) = c_s$ *then* $d(x) = v^{(i)}$ can be kept as $RUL(c_1, c_2, ..., c_s) = v^{(i)}$ in RUL. RUL is created as follows:
(1)Initialize RUL by *, where * represents an arbitrary decision attribute value;
(2)Produce rule base $RUL(R^{(j)}), j = 1, 2, ..., r$, by extracting the classification rules with rule consequent $d(x) = v^{(j)}$ from $A_{|D}$. Then Update RUL by each rule base $RUL(R^{(j)})$.
Note that, in step (2) any classification rule R extracted from $A_{|D}$ is required satisfying the condition: Strength(R)$\geq \lambda n$, where n is the number of the training samples; and $0 \leq \lambda \leq 1$, is a threshold. λ is used to obtain fewer but robust classification rules. Below Algorithm I is used to obtain an incorporable rule base $RUL\_Incor(R^{(j)})$ from RUL and $RUL(R^{(j)}), j = 1, 2, ..., r$.
Algorithm I:
For $j = 1, 2, ..., r$, $RUL\_Incor(R^{(j)})$ is obtained as follows:
(1)if $RUL(R^{(j)}) = \emptyset$, output $RUL\_Incor(R^{(j)})$ and stop, otherwise a classification rule R is chosen randomly from $RUL(R^{(j)})$ and set $RUL(R^{(j)}) = RUL(R^{(j)}) - \{R\}$;
(2)Set $L = \{1, 2, ..., s\}$;
(3)if $L = \emptyset$, then set $RUL\_Incor(R^{(j)}) \cup \{R\}$, and go to (1), otherwise go to (4);
(4)An element i in L is chosen randomly, and the neighbor classification rules with i to R in RUL are searched. Those classification rules are required satisfying the conditions: $d(x) = v^{(j)}$ *or* $d(x) = *$. If no such rules can be sought, go to

(6), else go to (5);
(5)The neighbor classification rules with i to R in $RUL(R^{(j)})$ are removed from $RUL(R^{(j)})$. Combine R and its neighbor rule in RUL to a incorporable rule. Set the new incorporable rule be R as well. go to (2);
(6)Set $L = L - \{i\}$, go to (3).
Note that, in step (5) if R is the incorporable, then the neighbor rule to R may consists of many classification rules in $RUL(R^{(j)})$, and they all should be removed.

4 Translation from the Classification Rules to the Fuzzy if-then Rules

Fuzzy if-then rules are in the form as follows:
$R: \; if \; C_1(x) \; is \; A_1 \; and \; C_2(x) \; is \; A_2 \; and ...and \; C_s(x) \; is \; A_s \; then \; d(x) \; is \; v^{(j)}$
where $x \in U$; A_i is a fuzzy subset (linguistic term) with membership function $a_i : R \to [0,1], i \in \{1,2,...,s\}; v^{(j)} \in V_d$. From section 2 and section 3, the condition attribute values of the classification rules in $RUL\_Incor(R^{(j)})$ are all in the form of $(c_i^1 \vee c_i^2 \vee ... \vee c_i^{k_i})$, where $(c_i^1, c_i^2, ..., c_i^{k_i})$ are consecutive discrete values, $j = 1,2,...,r; i \in \{1,2,...,s\}$, and $(c_i^1 \vee c_i^2 \vee ... \vee c_i^{k_i})$ can be mapped to $[l_{C_i}, r_{C_i}) \subset R$. Then $[l_{C_i}, r_{C_i})$ can be translated to a fuzzy subset with a Gaussian membership function as follows:

$$a_i(y) = e^{-\frac{(y-c)^2}{\sigma^2}} \tag{1}$$

where $c = \frac{1}{2}(r_{Ci}+l_{Ci}), \sigma = \frac{1}{3}(r_{Ci}-l_{Ci})$. So, any classification rule in RUL_Incor $(R^{(j)}), j = 1,2,...,r$, can be translated to a fuzzy rule by the above method.

5 Voting-Mechanism-Based Fuzzy Neural Network

5.1 Fuzzy Neural Network Architecture

Based on the fact that an object with unknown class tag is generally close to those samples whose class tags are the same with the one while far from the samples that have different ones, a voting-mechanism-based fuzzy neural network system (VMFNN) is proposed.
Assume the number of fuzzy if-then rules translated from $RUL\_Incor(R^{(j)})$ is $a_j, j = 1,2,...,r$, and those fuzzy rules are represented as:
$R_{i,j}: \; if \; C_1(x) \; is \; A_{1,i} \; and \; C_2(x) \; is \; A_{2,i} \; and...and \; C_s(x) \; is \; A_{s,i} \; then \; d(x) \; is \; v^{(j)}$
where $i = 1,2,...,a_j, j = 1,2,...,r$. The matching degree of the fuzzy rule antecedent for an object x is computed by the multiplication T-norm ([6]) as follows:

$$A_T(R_{i,j}) = \prod_{k=1}^{s} A_{k,i}(C_k(x)) \tag{2}$$

Then a subsystem S_j can be constructed using all $R_{i,j}, i = 1, 2, ..., a_j$. Its output is defined as:

$$O_j = 1 - exp^{(-\sum_{i=1}^{a_j} A_T(R_{i,j}))} \tag{3}$$

Finally, the output of the fuzzy neural network system is defined as:

$$Y = v^{(j)}, \quad subject \quad to \quad O_j = \max\{O_1, O_2, ..., O_r\} \tag{4}$$

According to the above, the VMFNN model with 5 layers is derived as follows:
[A] Input layer. $O_j = I_j = C_j(x), j = 1, 2, ..., s$.
[B] Fuzzification layer. The node parameters include the centers and the radiuses of membership functions in the fuzzy rule antecedent. The output of the nodes in the layer are the membership degree values calculated by Eq. (1).
[C] Fuzzy rule layer. By Eq. (2), every circle node in the rectangles multiplies the incoming signals and sends the product out. The outputs of the layer make up the input parts of the corresponding subsystems.
[D] Subsystem output layer. The outputs of the layer are calculated by Eq.(3).
[E] Voting output layer. The output of the layer is calculated by Eq.(4).
Note that Node functions in the same layers are of the same form as described above. And the nodes in [B] layer have parameters, while the ones in other layers have none.

5.2 Optimization for the Fuzzy Neural Network

Assume the system contains r subsystems, subsystem i contains a_i fuzzy rules, $i = 1, 2, ..., r$, and each of which has $2s$ parameters (i.e., the centers and the radiuses in the Gaussian function), where s is the dimensionality of input feature vectors. Then the number of the system parameters is totally $M = 2s \sum_{i=1}^{r} a_i$.
The PSO is used to refine the system parameters. The PSO algorithm flow can refer to [7]. Let Q be the size of the particle swarm (generally set $Q = 20$). The initial velocity of the particles are initialized by random numbers uniformly distributed on $[-0.3, +0.3]$, and the initial positions of the particles are initialized by the system parameters with 15% noise. Taking use of information included by the particle i, a fuzzy system as described in section 4 can be constructed. The misclassification rates of the system constructed by particle i are defined as:

$$E_i = \frac{err_i}{n} \tag{5}$$

where $i = 1, 2, ..., Q$, n is the number of the training samples, and err_i is the misclassification number of the system constructed by particle i. Eq.(5) can be used as the fitness of the particle i. Set acceleration coefficients $W = 0.7298,\ C_1 = 1.42,\ and\ C_2 = 1.57$, which satisfy the convergence condition of the particles: $W > (C_1 + C_2)/2 - 1$ ([7]). Since $C_2 > C_1$, the particles will faster converge to the global optimal position of the swarm than the local optimal position of each particle. To avoid the premature convergence of the particles, an inactive particle, whose position unchanged in consecutive S epochs (set $S = 10$ in the paper), will be reinitialized.

6 Experimental Results

6.1 Taste Signals

With the developments in studying taste sensors and recognition systems, many favorable outcomes were obtained with respect to the basic taste signals and some mixed taste signals. Moreover successes in extracting taste signals from rice, sauce, beverages and alcoholic drinks are reported recently ([8]). Taste signals of 11 kinds of mineral waters by Saitama University in Japan are used in our experiment. Experimental data consist of 1100 points with 100 points for each taste signal. Original taste signals are 5-dimensional data that are outputs of $Na^+, K^+, pH, Cl^-, Ca^{++}$ sensor electrodes. To eliminate the correlation among the taste signals and decrease the dimensions, the Principal Component Analysis (PCA) is used for original signals and resultant 2-dimensional data are obtained.

6.2 Discretization of the Taste Signals and Extraction of the Classification Rules

The discretization algorithm of continuous signals by Nguyen,H.S ([5]) is used for the taste signals so that 30 cuts on c_1-coordinate and 39 cuts on c_2-coordinate are obtained. Obviously $n = 1100$. Set parameter $\lambda = 0$ and $\lambda = 0.003$ respectively, Algorithm I is used to extract the incorporable rules from the discrete taste signals. Comparison of the number of the resultant incorporable rules with different parameter λ is given in Tab.1.

Table 1. Comparison of the number of the incorporable rules with different parameter λ

No.	Taste Signals	Num.of rules ($\lambda = 0$)	Num.of rules ($\lambda = 0.003$)	No.	Taste Signals	Num.of rules ($\lambda = 0$)	Num.of rules ($\lambda = 0.003$)
1	Vital	7	2	7	Crystal	3	1
2	Valvet	4	2	8	Minmi-Alps	3	1
3	Yourou	5	3	9	Pierval	9	2
4	Rokkou	13	3	10	Fujisan	2	1
5	Volvic	4	2	11	Shimanto	4	1
6	Kireira-mizu	3	2				

6.3 Identification of Taste Signals

For $\lambda = 0$ and $\lambda = 0.003$, the fuzzy neural network systems I and II (below which are abridged as Sys I and Sys II) can be constructed respectively. The original taste signals and the ones polluted by the noise are used for our identification experiment. Note that, assume original signal is A , then the signal polluted by

the noise is $A' = A + A \times \eta \times rand$, where $0 \le \eta \le 1$, is the noise level (set $\eta = 0.2$ in the experiment), and $rand$ is a random number uniformly distributed on $[-1, +1]$.
The PSO is used to train Sys I and Sys II respectively, and after 100 training epochs, Sys I got the misclassification rate of 94.3% for the original taste signals and 78.4% for the polluted taste signals, while Sys II got the misclassification rate of 97.4% for the original taste signals and 84.6% for the polluted taste signals. Moreover, because Sys II is constructed by fewer fuzzy if-then rules than Sys I, it needs only 51.363 sec. per 100 training epochs, while Sys I needs 116.127 sec. accordingly. Therefore Sys II is better than Sys I in terms of learning capability, error tolerance and running speed in our experiment.

7 Conclusions and Discussions

Choosing a proper parameter λ in Algorithm I can obtain robust fuzzy if-then rules, then those rules can be used to construct a fuzzy neural network system with many perfect properties, such as robustness, learning capability, simplicity and running speed. Generally set $\lambda \in [0, 0.05]$. Applying the methods to high-dimensional complex data and attempting other optimization methods are future works.

References

1. J S R Jang, C T Sun, E Mizutani: Neuro-Fuzzy and Soft computing. Xi An, China: Xi An Jiaotong University Press, 1996.
2. S. Chiu: Fuzzy Model Identification Based on Cluster Estimation. Journal of Intelligent and Fuzzy Systems, Vol. 2, No. 3, (1994) 267-278.
3. L. Bottou and Y. Bengio: Convergence properties of the k-means algorithms. In G. Tesauro and D. Touretzky, (eds.): Advances in Neural Information Processing Systems 7, The MIT Press, Boston, (1995)585-592.
4. Yan-Xin Huang, Chun-Guang Zhou, Guo-Hui Yang et al. A study of identification of tea taste signals based on rough set theory (in Chinese). Journal of Jilin University (Information science Edition), Vol. 20, No. 3, (2002)73-77.
5. Nguyen H.S.: Discretization of Real Value Attributes: A Boolean Reasoning Approach [PH.D thesis]. Warsaw: Computer Science Department, University of Warsaw, 1997.
6. Ludmila I Kuncheva. How Good are Fuzzy if-then Classifiers? IEEE Transaction on Systems, Man, and Cybernetics, Part B: Cybernetics, Vol. 30, No. 4, (2000)501-509.
7. Frans van den Bergh. An Analysis of Particle Swarm Optimizers [PH.D thesis]. Pretoria: Natural and Agricultural Science Department, University of Pretoria, 2001.
8. Chun-Guang Zhou, Yan-Chun Liang, Yong Tian et al. A study of identification of taste signals based on fuzzy neural networks (in Chinese). Journal of Computer Research & Development, Vol. 36, No. 4, (1999)401-409.

A Novel Signal Detection Subsystem of Radar Based on HA-CNN

Zhangliang Xiong and Xiangquan Shi

School of Electronic Engineering and Photo-electronic Technology,
Nanjing University of Science & Technology, Postal code 210094,
200 Xiaolingwei, Nanjing, Jiangsu, P.R. China
xdz3701@yahoo.com.cn
http://www.njust.edu.cn

Abstract. According to the recently neurophysiology research results, a novel signal detection subsystem of radar based on HA-CNN is proposed in this paper. With a kind of improved chaotic neuron that is based on discrete chaotic map and Aihara model, Hierarchical-Associative Chaotic Neural Network (HA-CNN) exhibits promising chaotic characteristics. The function of HA-CNN in the signal detection subsystem of radar is to reduce the influence of the environmental strong noisy chaotic clutter and distill the useful signal. The systematic scheme of signal detection with HA-CNN and the detailed chaotic parameter region of HA-CNN applied in signal detection are given and the results of analysis and simulation both show this kind of signal detection subsystem has good detecting ability and fine noise immunity.

1 Introduction

In some cases, useful signal is embedded in strong noisy clutter background. Thus the radar's ability to detect targets embedded in strong noisy clutter background depends on the Signal-to-Clutter Ratio (SCR) rather than the Signal-to-Noise Ratio (SNR). A great deal of studies has shown that the classical stochastic analysis methods ignored the inherent physical characteristics of strong noisy clutter and brought loss in signal processing. It is more precise to use nonlinear model such as chaos and fractal to analyze it in many cases.

The dynamics of all kinds of traditional neural network in essence can be analyzed and explained by using the fixed-point theory [1] in the traditional discrete dynamical system and in the mean time the explanation and classification of chaos can also be expressed in terms of its relationship with fixed points. Based on the fixed-point theory, it is naturally to expand either hierarchical or associative traditional neural network in the symmetrical system into chaos theory in the asymmetrical system.

In this paper, we construct a novel signal detection subsystem of radar with HA-CNN that is based on fixed-point theory. The function of HA-CNN in the signal detection subsystem is to reduce the influence of the environmental strong noisy chaotic clutter and distill the useful signal.

F. Yin, J. Wang, and C. Guo (Eds.): ISNN 2004, LNCS 3174, pp. 344–349, 2004.

2 System Model

H. Leung presents one kind of signal detection method based on nonlinear prediction [2]. Unfortunately, the H. Leung's method ignores the disturbance of noise to the predictive ability of the chaos model and does not utilize pre-information of the useful signal sufficiently. This paper develops H. Leung 's theory and proposes a system model as shown in Fig. 1.

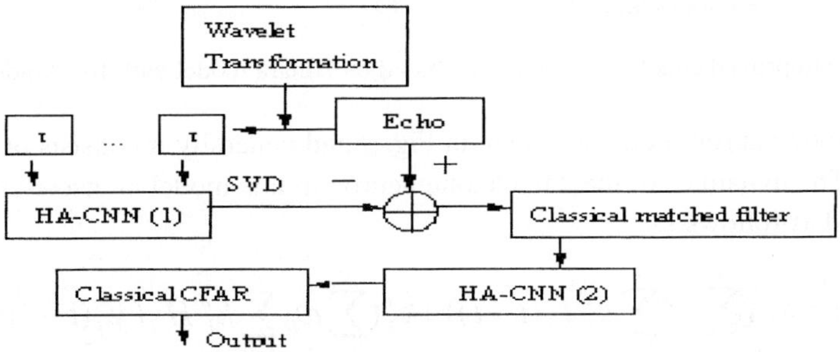

Fig. 1. The block diagram of signal detection subsystem of radar that is based on HA-CNN

The flow of signal detection with HA-CNN in the background of chaotic clutter is as follows:

- The function of first HA-CNN in the subsystem is to construct the phase space of the strong noisy chaotic clutter. Wavelet transformation is applied here to eliminate the influence of noise on the phase space reconstruction. We adopt Daubechies wavelet in this step. After training, the network can be considered as alternative model of the real noisy chaotic clutter. To avoid few points of the predictive series departure from the truth seriously, the Singular Value Decomposition (SVD) arithmetic is applied to smoothen the output.
- The posterior observational echoes including target signals subtract the predictive values of the trained CNN to obtain the output with small clutter and then the output pass through the classical matched filter to assemble the energy of signal.
- Considering the condition that the output of the classical matched filter is still submerged in system noise, the second HA-CNN is applied to distill the useful signal waveform with Chaotic Resonance (CR) phenomenon and then by some criterion function we can get anticipant target detection results.

3 Theory Analysis

The classical Aihara chaotic neuron model is extension of the Caianiello neuron equation (1961) and the Nagumo-Sato neuron model (1972) that includes a refractory effect. In this paper, we constructed an improved chaotic neuron model based on Aihara model, which combined chaos internal generation approach with external generation approach. It exhibits promising chaotic characteristics with adequate parameters.

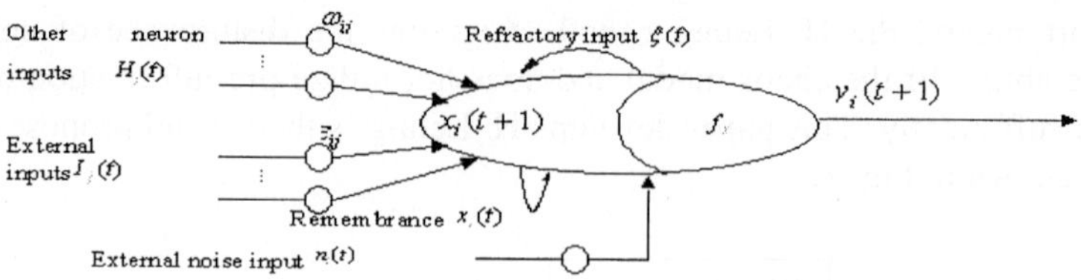

Fig. 2. The improved chaotic neuron model based on Aihara model with five kinds of inputs

The chaotic neuron model is shown in Fig. 2 and generally it consists of five kinds of inputs. The dynamics of the i th chaotic neuron in this model at discrete time $t+1$ is described as follows:

$$x_i(t+1) = k_1^s(\sum_{j=1}^{N}\varepsilon_{ij}\sum_{l=0}^{t}k_e^d I_j(t-l)) + k_2^s(\sum_{j=1}^{N}\omega_{ij}\sum_{l=0}^{t}k_i^d H_j(y_j(t-l)) + B_i) \tag{1}$$
$$+ k_3 x_i(t) - \alpha(t)(\sum_{l=0}^{t}k_r^d \xi(y_i(t-l)) - B_0) + \mu_i n_i(t) - \theta_i(t) .$$

$$y_i(t+1) = f_i(x_i(t+1)) . \tag{2}$$

Where $x_i(t+1)$ and $y_i(t+1)$ are the internal state and output of the i th neuron at the time $t+1$ respectively. $I_j(t-l)$ is the strength of the j th external input at time $t+1$. μ_i is the coefficient of the external noise. $\alpha(t)$ is the positive refractory strength. It can be time-dependent on demand. $\theta_i(t)$ is the dynamical threshold of i th neuron [3]. Refractory function $\xi(\cdot)$ is the determinative factor in (1) to induce nonlinear character into neuron. The $\xi(\cdot)$ is defined as [4]:

$$\varphi_{t+1} = \xi(\varphi_t) = A\left|\sin(a/\varphi_t)\right|, \varphi_t \in [-A,0) \cup (0,A], a \in (0,+\infty) . \tag{3}$$

As shown in Fig. 3, we think it is adequate to apply $0.4 < \alpha < 0.6$, $N=4, a=100$, $\theta = 0.05 + 0.1t$, $1.6e-3 < |n(t)| < 1.8e-3$, $\theta_{\max} < 0.8$ in the chaotic neuron to utilize the effect of discrete chaotic map to obtain favorable chaotic property.

Since the number of environmental variables is large, a statistical study of the variables is ought to carry out as a previous step to the analysis of the chaotic time series. We take the environmental variables in pairs and test their lineal correlation and Spearman rank correlation [5]. The latter are calculated using (4):

$$R_s = (\sum_{i=1}^{n}(x_i - \bar{x})(y_i - \bar{y}))/(\sqrt{\sum_{i=1}^{n}(x_i - \bar{x})^2 \sum_{i=1}^{n}(y_i - \bar{y})^2}), \quad x_i \in X, y_i \in Y . \tag{4}$$

Takens Theorem is the academic base of the phase space reconstruction and the delay-time coordinates (DTC) method is chosen to realize it in this section. The vectors of constructed phase space Ω are defined as

$$\Omega_i = (x_i(t_i), x_i(t_i + \tau), \cdots, x_i(t_i + (d_e - 1)\tau) . \tag{5}$$

We generalize the Cross-Validated Subspace (CVS) method [6] to determine the optimum number of hidden units of HA-CNN. The dimension of the subspace spanned by the echo eigenvectors is used to ascertain suitable number of hidden units and the cross validation method is applied to prevent over-fitting.

Stochastic resonance (SR) is known as a phenomenon in which the presence of noise helps a nonlinear system in distilling a weak signal. It plays positive role in the information processes. Literature [7] shows that the SR-like behavior, i.e. CR, can be observed in deterministic dynamic systems by using the intrinsic chaotic dynamics of a nonlinear map. With $\alpha = 0.5, n = 2e-3, \theta = 0.05 + 0.1t$ and $\theta_{max} < 0.3$ the CR phenomenon can be observed in HA-CNN and we will apply this useful characteristic to distill the output of classical matched filter from system noise.

4 Simulation

To carry out the signal detection simulation, we construct simulated echo signal composed of environmental strong noisy chaotic clutter and target echo according to the historical observed data. HA-CNN in this application owns two hidden layers. As shown in Fig. 4, we think it is appropriate to set the number of units H in hidden layers of HA-CNN as $H = 12$ by the computation with CVS method. The embedding dimension d_e and time delay τ are determined as $d_e = 4, \tau = 4$. According to this result, the number of neurons in the input layer is eight: four is the circumstance parameters and the other four is the historical data. The output layer in the simulation only comprises one neuron. The parameters of first HA-CNN is fixed as: $0.4 < \alpha < 0.6, N = 4, a = 100, \theta = 0.05 + 0.1t, 1.6e-3 < |n(t)| < 1.8e-3, \theta_{max} < 0.8$. The parameters of second HA-CNN is fixed as $\alpha = 0.5, n = 2e-3$, $\theta = 0.05 + 0.1t$ and $\theta_{max} < 0.3$. The 13bit barker code {1,1,1,1,1,0,0,1,1,0,1,0,1} is applied in the simulation as phase modulation code. Fig. 5 shows the simulation of the signal detection courses. As shown in the figures, because SCR_{in} is less than -30dB, the object signal is submerged in the noisy clutter completely, but after processed by HA-CNN system, the useful correlation output is detected effectively.

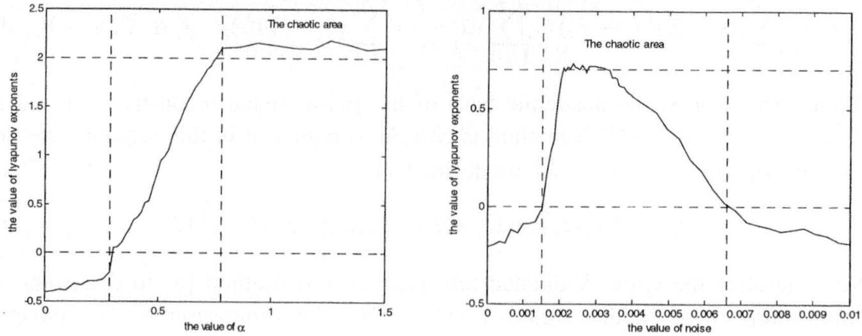

Fig. 3. λ_1 plotted against the refractory strength $\alpha(t)$ ($|n(t)=2e-3|$) and the strength of external noise $n(t)$ ($\alpha(t)=0.1$) respectively

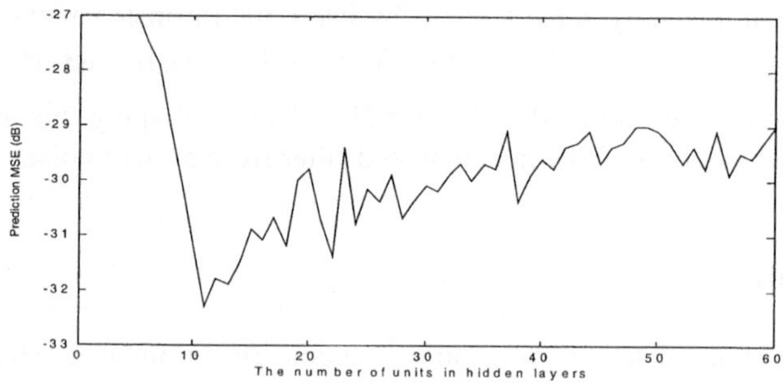

Fig. 4. The number of optimum units in hidden layers plotted against the prediction MSE

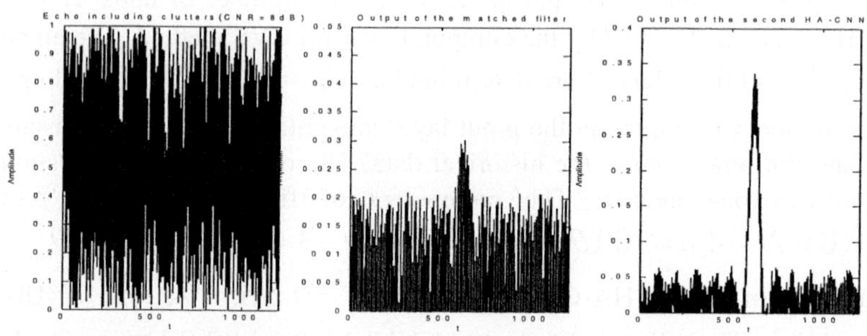

Fig. 5. In normal direction flow, the figure is echo including noisy clutter (CNR=8dB), the output of the classical matched filter and the output of the second HA-CNN sequentially

The signal detection system is immune to the noise mixed in the chaotic clutter unless CNR <5dB and is also not sensitive to the category of noise. Table 1 shows the comparison of different signal detection method, as we can see, signal detection

Table 1. Comparison of different signal detection methods

Category	$P_f = 10^{-6}$	$P_f = 10^{-5}$	$P_f = 10^{-4}$
Classical method	$P_d = 0.37$	$P_d = 0.50$	$P_d = 0.58$
H.Leung method	$P_d = 0.46$	$P_d = 0.60$	$P_d = 0.69$
HA-CNN method	$P_d = 0.59$	$P_d = 0.77$	$P_d = 0.87$

with HA-CNN proposed in this paper can bring obviously increases in the probability of detection compared with old methods.

5 Conclusions

In this paper we developed a novel signal detection subsystem of radar based on HA-CNN. With theoretical analysis and simulation we think it is effective in weak signal detection and in strong chaotic clutter background it can reach a good improvement factor $\delta(=SCR_{out}/SCR_{in})>50dB$. HA-CNN exhibits fine properties and the applications of it are in no way restricted to the proposed. It owns the potential to be widely applied in nonlinear series prediction, pattern recognition and combinatorial optimization etc. The further researches of HA-CNN on theory and practice are absorbing.

References

1. D. Zhou, K. Yasuda, R. Yokoyama.: A Method to Combine Chaos and Neural Network Based on the Fixed Point Theory. Proceedings of IEEE International Symposium on Circuits and Systems. 1 (1997) 645 –648
2. Leung H.: Appling Chaos to Radar Detection in an Ocean Environment: An Experimental Study. Journal of Oceanic Engineering. 1 (1995) 56-64
3. Y. Yao, W. J. Freeman.: Model of Biological Pattern Recognition with Spatially Chaotic Dynamics. Neural Networks. 2 (1990) 153-170
4. He Di, He Chen: Correlation Characteristics of Binary-phase and Quadri-phase ICMIC spreading sequences. Journal of China Institute of Communications. 3 (2001) 13-19
5. C. Pérez-Lleraa, M.C. Fernández-Baizánb, J.L. Feitoc, V. González del Vallea.: Local Short-Term Prediction of Wind Speed: A Neural Network Analysis. Proceedings of the International Environmental Modeling and Software Society, Paris (2002) 381-389
6. Henry Leung, Titus Lo.: Prediction of Noisy Chaotic Time Seires Using an Optimal Radial Basis Function Neural Network. IEEE Trans on Neural Networks. 12 (2001) 1163-1172
7. Haruhiko Nishimura, Naofumi Katada, Kazuyuki Aihara.: Coherent Response in A Chaotic Neural Network. Neural Processing Letters. 1 (2000) 49-58

Real-Time Detection of Signal in the Noise Based on the RBF Neural Network and Its Application

Minfen Shen[1], Yuzheng Zhang[1], Zhancheng Li[1], Jinyao Yang[2], and Patch Beadle[3]

[1] Key Lab. of Guangdong, Shantou University, Guangdong 515063, China
mfshen@stu.edu.cn
[2] Ultrasonic Institute of Shantou, Shantou, Guangdong, China
[3] School of System Engineering, Portsmouth University, Portsmouth, U.K.

Abstract. The problem of real time signal detection in the noise and its applications to the denoising single-trial evoked potentials (EP) was investigated. The main objective is to estimate the amplitude and the latency of the single trail EP response without losing the individual properties of each epoch, which is important for practical clinical applications. Based on the radial basis function neural network (RBFNN), a method in terms of normalised RBFNN was proposed to obtain preferable results against other nonlinear methods such as ANC with RBFNN prefilter and RBFNN. The performance of the proposed methods was also evaluated with MSE and the ability of tracking peaks. The experimental results provide convergent evidence that the NRBFNN can significantly attenuate the noise and successfully identify the variance between trials. Both simulations and real signal analysis show the applicability and the effectiveness of the proposed algorithm.

1 Introduction

In practice, the measurement of the signals we are interested in is often corrupted by noise as a result of many other kinds of background activities. Moreover, majority of signals turn to be typical non-stationary and nonlinear. Conventional signal analysis for these problems has long been dominated by FFT or the traditional ensemble averaging (EA). However, due to the non-stationarity and nonlinearity of the noisy signals, the Fourier analysis fails to investigate a class of signals whose frequency contents change with time and the EA method cannot be used to real-time track the signal in noise. For example, in medical applications, evoked potentials (EPs) represent the weak electrical activity measured from specific regions of brain usually resulting from sensory stimulation. Under some conditions the problem of tracking EPs changes is quite important and practical significant: e.g., for patients under critical care when their physiological conditions vary with time. The measurement of the time-varying EPs is always buried in relatively large background noise which is the on-going electrical activity of other brain cells known as electroencephalographic activity (EEG). To extract the real-time EPs more effectively from the noise, advanced signal processing technique is required. Thus, the problem of dealing with the real-time detection of signal in noise becomes more and more important in many practical measurement environments. Our main task is to design a real-time estimator

F. Yin, J. Wang, and C. Guo (Eds.): ISNN 2004, LNCS 3174, pp. 350–355, 2004.

with which the unwanted contribution of the on-going background noise can be filtered out from the observations as much as possible.

Adaptive signal processing, such as adaptive noise canceller (ANC), has been widely used to improve the estimate result of transient noisy signals [1, 2]. Many adaptive filters need a meaningful reference signal for its good tracking. Taking the nonlinear nature of the recognition signal into consideration, radial basis function neural network (RBFNN) can be adopted as the prefilter to obtain more meaningful reference for ANC [3] since RBFNN is capable of dealing with any nonlinear multidimensional continuous functions [4, 5]. This ANC method with RBFNN prefilter is much better than EA in the case of tracking real-time response and can extract the temporal information of the measurement.

However, when the SNR is very low and the response is fast transient, the methods discussed above may not be valid. The structure of the RBFNN is relatively simple with an output node that has a liner-in-weight property that can be directly employed to track the single trial signal, but the Radial Basis Function (RBF) is able to cover the whole input space if a RBFNN is normalized and applied. NRBFNN, called normalized RBFNN, can be obtained by dividing each radial function in RBF network by the sum of all radial functions. This new technique is helpful to significantly improve the capability of signal detection. In this contribution, the modified NRBFNN procedure is proposed to carry out the real-time detection of signals contaminated with noise. As the application in medicine, we aim at obtaining fast EPs measurement and tracking EPs' variations across trials on time. The propose scheme is also compared with other common used methods such as RBFNN and the nonlinear ANC with RBFNN prefilter in the case of single-trial estimation.

2 Proposed Approach

NRBFNN is used to obtain the fast responses measurement and track the variations across the trials. Fig.1 shows the structure of a general RBF neural network which is a multiplayer feed-forward neural network consisting of input layer, kernel (hidden) layer and output layer. The units in the kernel layer provide an array of nonlinear radial basis function ϕ_i, which are usually selected as Gaussian functions and defined as $y = f(x) = \sum_{i=1}^{M} w_i \phi_i (x_i ; c_i \sigma_i)$.

where x is the input vector, c_i represent the location with kernel unit i, while σ_i models the shape of the activation function. w_i denote the weight from kernel unit to output unit. In the output layer, the value of a unit is obtained through a linear combination of the nonlinear outputs from the kernel layer. w_i are trained in a supervised fashion using an appropriate linear learning method.

Local basis functions are advantageous because of the increased interpretability of the network, the ability to produce locally accurate confidence limits, and also the locality, which can be utilized to improve computational efficiency. Normalization

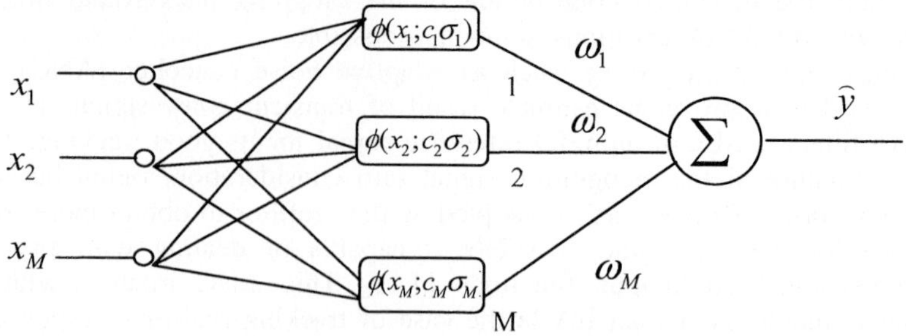

Fig. 1. Structure diagram of RBF neural network

is common practice in local linear modeling. The normalization of the basis functions in such a network is proposed that is motivated by the desire to achieve a partition of unity across the input space. It results in the basis functions covering the whole of the input space to the same degree, i.e. the basis functions sum to unity at every point. By partition of unity it is meant that at any point in the input space the sum of the normalized basis functions equals unity. Partitioning of unity is an important property for basis function networks in many applications, such as noisy data interpolation and regression. It often results in a structure which can be less sensitive to poor center selection and in cases where the network is used within a local model structure. Also since the basis function activations are bounded between 0 and 1, they can be interpreted as probability values. Covered the space between training vectors without excessive overlap, the RBF eventually approach either zero or one at the extremes of the input. The effect of the normalization also improves the interpolation properties and makes the network less sensitive to the choice of the widths σ_i.

To evaluate the performance of real-time detecting signal in the noise, both RBFNN and the nonlinear ANC with RBFNN prefilter are computed and compared with our proposed NRBFNN method. By employing the simulated data and real signals, we carry out the evaluations in the following three aspects: (a) comparing the performance with relative mean square error (MSE), (b) evaluating the ability of tracking signal variation, and (c) testing the signal detection ability at different SNR conditions. All significant results are provided to show the effectiveness and the advantage of the presented NRBFNN method.

3 Results

First of all, we generate 50 epochs of 500 samples with input SNR at –20 dB and –40 dB, respectively, for evaluating the behaviors. All three models, ANC with RBFNN prefilter, RBFNN and NRBFNN, were tested. Fig. 2 shows the corresponding real-time detecting performances at different SNR. Though being able to track single-trial data rapidly, RBFNN can only achieve it under certain circumstance, such as a higher SNR input. In all situations, the normalized RBFNN performs best among all the structures proposed in the fitting ability, as well as in the rate of convergence. Next, the performance of the estimation is also evaluated by calculating the mean square error (MSE) and shown in terms of the MSE versus 3 methods in Fig.3. In terms of MSE comparison, NRBFNN is also found to be the best as compared to the other models. In addition, the ability of tracking signal's variation was investigated. 1000 trials input signals was generated. The first 500 trials remain waveform S1, and other 500 trials adopt a different waveform S2. There existed a jump at the 500th trial. Fig.4 illustrates the error curves comparison in abrupt changing simulation. Finally, we compared the performance of signal detection at different SNR. We consider the local SNR, varying from –40 dB to 0 dB with an increment of –5dB. Fig.5 illustrated the results. It's clear that the MSE of NRBFNN is the smallest at all noise levels. The NRBFNN method effectively eliminates the bad input data in all three models. As an example in medical application, the visual EPs (VEP) on scale of visual spatial attention was studied. Fig. 6 shows the real-time detected VEP responses by using the proposed approach.

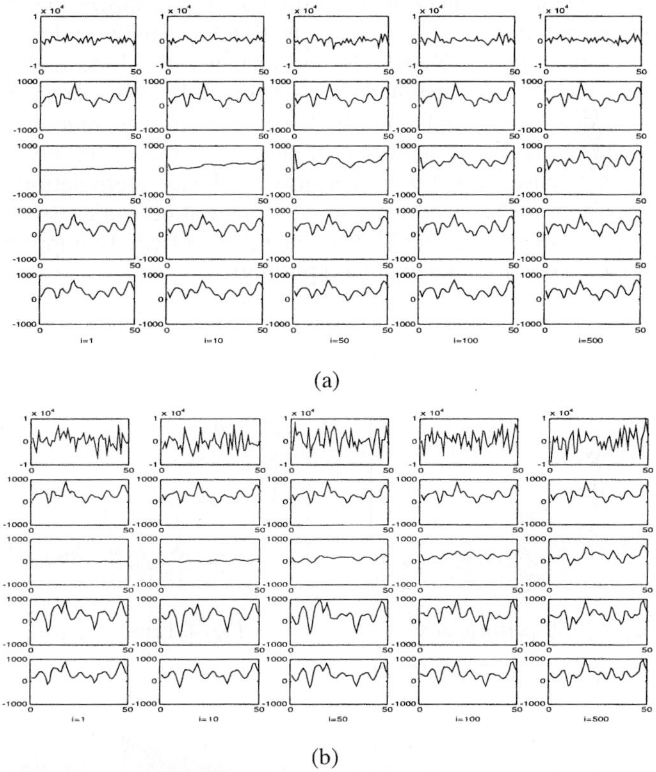

Fig. 2. Performance comparison. (a) SNR=-20dB and (b) SNR=40dB. From top to below: corrupted signal, pure VEPs, output of the nonlinear ANC, results of RBFNN and output of the NRBFNN, in which i indicates the trial number

Based on the analysis and the experimental results of the proposed modified RBFNN method above, several significant conclusions can be drawn: (a) To overcome the drawbacks of conventional method, we proposed NRBFNN to obtain preferable results for real-time signal detection. The structure and the characteristics of NRBFNN were described. By using singular value decomposition algorithm, we investigated and optimized the normalized RBF neural network, which enables to eliminate the redundant hidden nodes of the network and to obtain a reasonable network structure. The results in this contribution showed that the presented normalized RBFNN technique is effectively used for the real-time detection of signal in noise. (b) The performances of real-time detecting signal in the noise were evaluated. We compared our method with both RBFNN and the nonlinear ANC with RBFNN using the simulated data and real signals in three aspects: (a) the performance of the relative mean square error (MSE), (b) the ability of tracking signal variation, and (c) the behaviors of signal detection at different SNR conditions. All significant results are provided to show the effectiveness and the applicability of the presented NRBFNN model.

As an application in medicine, we investigated the problem of obtaining real-time EPs measurement and tracking EPs' variations across the trials. Our particular interest aimed at determining the temporal relationship of variance between trials and measuring response synchronously to each stimulus. It has shown that NRBFNN can significantly improve the behavior of EPs detection, especially under very low SNR conditions. By employing the NRBFNN, the changes of whole real-time VEP for each trial can be clearly demonstrated that significantly help our understanding of many practical problems in medicine.(d) It has been proved that the NRBFNN is more applicable to the real-time detection of signals in noise than other conventional methods. The main reason for that is because the local interpolation properties of NRBFNN is greatly improved , which makes the neural network less sensitive to the choice of other parameters. The NRBFNN scheme significantly improved the ability with respect to the responding speed and output SNR.

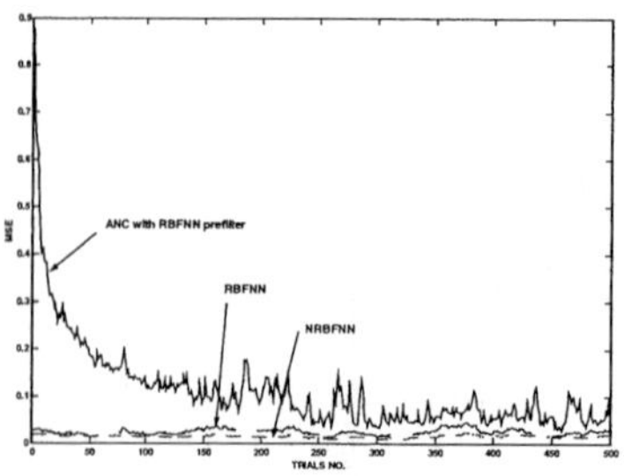

Fig. 3. MSE performance

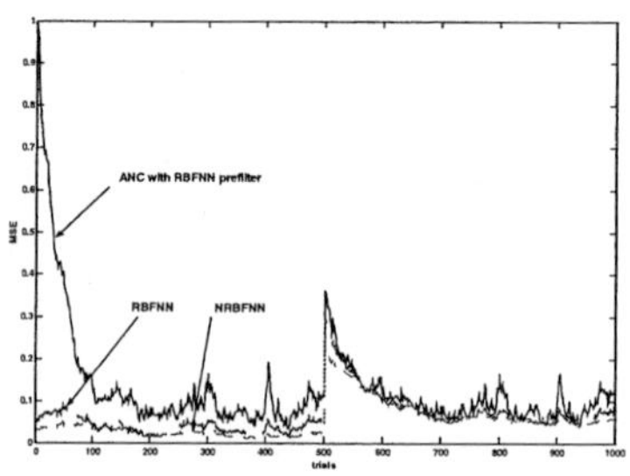

Fig. 4. Errors comparison

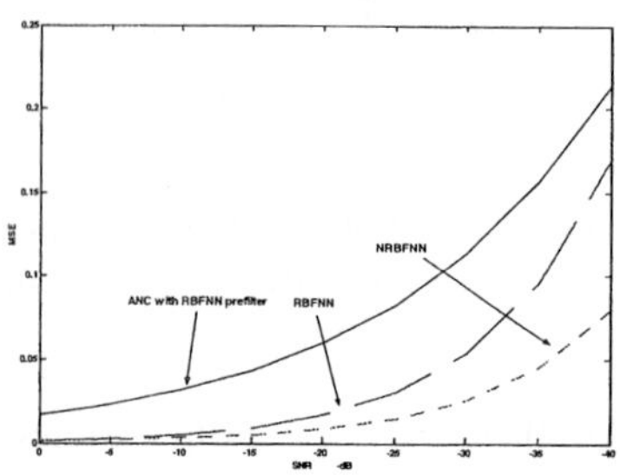

Fig. 5. MSE vs SNR comparison

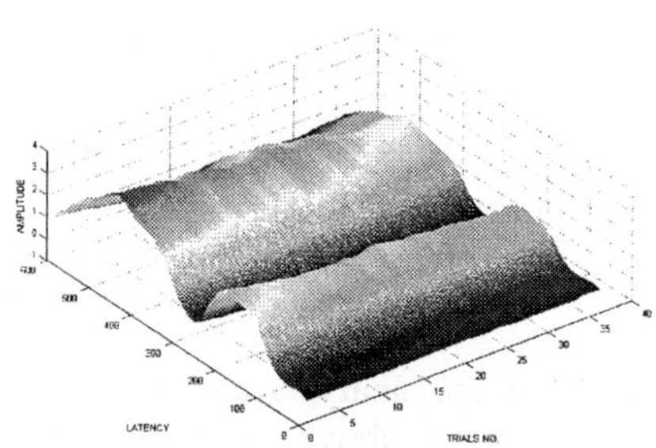

Fig. 6. One ERPs detected via NRBFNN

4 Conclusions

A method of signal detection based on NRBFNN was proposed for dealing with the problem of extracting the time-varying responses in the noise. We also focused on its application to the extraction of the real ERP signal. It has been shown that the NRBFNN is more applicable to the real-time detection of single-trial VEPs than other conventional methods, such as the nonlinear ANC with RBFNN prefilter or the common RBFNN methods.

Acknowledgements. The research was supported by the Natural Science Foundation of China (60271023), the Natural Science Foundation of Guangdong (021264 and 32025).

References

1. Hartman E. J., Keeler J. D. and Kowalski J. M.: Layered Neural Networks with Gaussian Hidden Units as Universal Approximation. Neural Computation, Vol. 2. MIT Press, Cambridge (1989) 210-215
2. Widrow Bernard, Glover John R. Jr., McCool John M., Kaunitz John, Williams Charles S., Hearn Robert H., Zeidler James R., Dong Eugene Jr., Goodlin Robert C.: Adaptive Noise Canceling: Principles and Applications. Proceeding IEEE, Vol. 63.IEEE Press, Piscataway NJ (1975) 1692-1716
3. Z. Y. Zhang: Nonlinear ANC Based on RBF Neural Networks. Journal of ShangHai Jiaotong University Vol.32. Shanghai Jiaotong University Press, Shanghai (1998) 63-65
4. J. C. Platt: A Resource Allocating Network for Function Interpolation. Neural Computation, Vol. 3. MIT Press, Cambridge (1991) 213-225
5. John Moody and Christain J. Darken: Fast Learning in Network of Locally-tuned Processing Units. Neural Computation, Vol. 1. MIT Press, Cambridge (1989) 281-294

Classification of EEG Signals Under Different Brain Functional States Using RBF Neural Network

Zhancheng Li[1], Minfen Shen[1], and Patch Beadle[2]

[1] Key Lab. of Guangdong, Shantou University, Guangdong 515063, China
mfshen57@vip.163.com

[2] School of System Engineering, Portsmouth University, Portsmouth, U.K.

Abstract. Investigation of the states of human brain through the electroencephalograph (EEG) is an important application of EEG signals. This paper describes the application of an artificial neural network technique together with a feature extraction technique, the wavelet packet transformation, in classification of EEG signals. Feature vector is extracted by wavelet packet transform. Artificial neural network is used to recognize the brain statues. After training, the BP and RBF neural network are able to correctly classify the brain states, respectively. This method is potentially powerful for brain states classification.

1 Introduction

EEG signals are the electrical activities in the cortex or on the surface of scalp, caused by the physiological activities of the brain. EEGs reflect the activity of the brain. EEG monitoring has many advantages such as no-invasive and real-time, which make it important in research of brain functioning. Classifying the states of the brain by EEG monitoring is one of the most important clinical applications. For example, assess the depth of anesthesia of the patient during surgery by EEG. In one aspect, it makes sure that the patient gets enough anesthesia, in the other, prevent from over drugged [1]. Dividing different sleep depth by the dynamic change of EEG rhythm energy is another example [2].

EEG rhythms play an important part in the research of brain states. Each rhythm relates to some certain brain function states [3]. As a result, to extract rhythm from EEG precisely is another important research region. Research of brain function states should focus on feature extraction and pattern recognition.

Over these years, various digital signal processing techniques have been widely applied to the analysis of clinical EEG signals such as frequency spectrum and AR model. But these methods assume that the EEG is stationary. However, in clinical practice, the assumption is not satisfied [4].

Short Time Fourier Transform (STFT) solves the problem in a certain extent. STFT assume the stationary of the signal within a temporal window and calculate the Fourier transform in the window. But the length of the window brings another problem. If the window is too narrow, the frequency resolution will be poor and if the window is too wide, the time localization will be less precise. This limitation restricts its application.

F. Yin, J. Wang, and C. Guo (Eds.): ISNN 2004, LNCS 3174, pp. 356–361, 2004.

To overcome these limitations, this paper uses wavelet packet decomposition (WPD) to extract EEG rhythms. WPD is a multi-resolution method which has good resolution in both frequency and time domains. WPD make no assumption of the stationary of the signal. It is a robust tool in non-stationary signal processing. The precise time location ability makes WPD suitable in EEG analysis.

One of the first attempts to apply artificial neural network technique to the problem of EEG signal classification in psychiatric disorders was the pilot work by Tsoi et al. This study set out to extend this initial finding [5].

This paper proposes that use WPD to extract EEG rhythms precisely, and then set up feature vector by that. The feature vector is applied to construct and train the neural network. The neural was then used to classify the functional states of brain. This method is useful in the brain activity monitoring. It will also be useful in the biomedical signal, speech processing and other time-varying signal processing.

2 Method

2.1 Wavelet Packet Transform

Wavelet transform is a powerful tool for non-stationary signal processing. One of its drawbacks is that the frequency resolution is poor in the high frequency region. In certain applications, the wavelet transform may not generate a spectral resolution fine enough to meet the problem requirement. Wavelet packet transform is invited to solve this problem. The use of wavelet packet is a generalization of a wavelet in that each octave frequency band of wavelet spectrum is further subdivided into finer frequency band by using the tow-scale relations repeatedly.

2.2 Construction of Feature Vector

EEG signal contain frequency ranging form 0.5 to 60 Hz, it can be divided into different rhythm with frequency bands (see Table 1).

Table 1. EEG rhythms (Hz)

Delta	Theta	Alpha	Beta	Gamma
0.5-3.5	4-7	8-13	14-25	Above 26

The frequencies of EEG are different based on brain functional states. For example, β wave (20-25Hz) will be domain frequency when object is excited. When released, α wave (8-12Hz) will become domination and α wave will weaken when asleep.

Feature vector should be constructed according to the research subject. This paper concerns about the states of eyes open and eyes closed. These two conditions were chosen for representing physiological brain states differing in their degree of EEG synchronization. It has been well established that EEG was characterized with high amplitude α activity. With WPD technique, EEG is decomposed to extract the α rhythm. We define the node coefficient function as:

$$x(j,n)=\sqrt{\sum_{k=1:N}\omega_{j,n}^{2}(k)/N} \tag{1}$$

in which $\omega_{j,n}$ is the wavelet packet coefficient, N is the length of $\omega_{j,n}$ .WPD can decompose the frequency sub-band into the nodes. α rhythm changes with different eyes states. The nodes which cover α rhythm contain this information. As result, take the node function of the 11-16 nodes of the 6th level WPD as the feature vector. It covers the frequency region of 7.81-13.28 Hz.

$$F=[x(6,11),x(6,12),x(6,13),x(6,14),x(6,15),x(6,16)] \tag{2}$$

The other rhythms have not been taken into the feature vector because they have no fixed tie with the eyes states.

2.3 Neural Network Construction and Training

The structure of the feature vector indicates that there should be 6 nodes in the input layer. The output of the network is used to determine the brain states. Output layer contains 1 node. 0 represent eyes open and 1 indicates eyes closed. The number of the neurons in the hidden layer has great impact in the convergence speed and generalization capability. If there are too few neurons, the network need more training epoch and the precision will decrease. If there are too many, the large scale of neurons will increase the train time. Set the hidden neurons s1=3,6,8,15 and 30, train the BP network. The result can be seen from Fig 1.

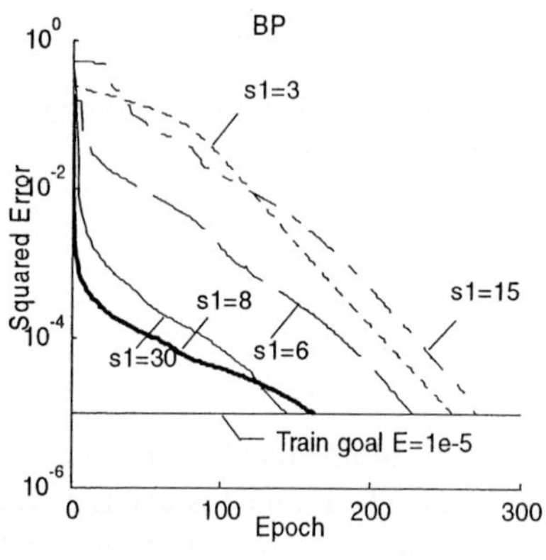

Fig. 1. Training curve of BP ANN

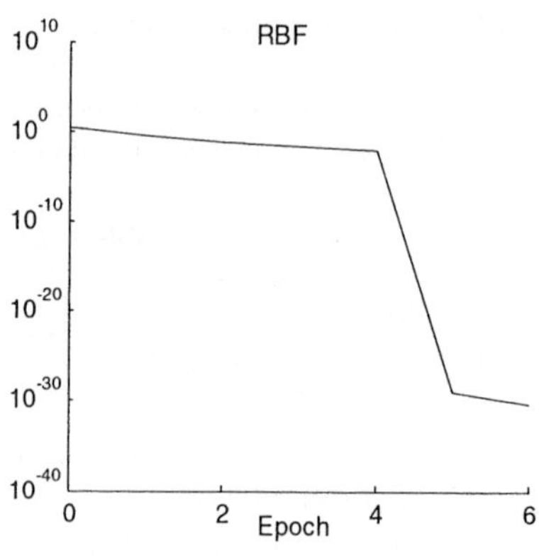

Fig. 2. Training curve RBF network

In this paper, the hidden layer was set as 8 nodes. To set the output value between [0 1] for states determining, the hidden layer uses Log-sigmoid transform function. Six samples are used to train the neural network in which the first three with eyes open and the last three with eyes closed. The output target is [0 0 0 1 1 1]. Adaptive training algorithm is used in the training. The train target is $E=10^{-5}$. The network converges after 163 epochs. RBF neural network's characters of local response make it less sensitive to the input value outside specified region. Hence, input data should be generalized into [-1 1] before training. The number of hidden layer is determined by the training process. Construct and train the RBF neural network by the input sample and target sample. Training progress can be seen from fig 2. It is obvious that RBF converge much faster than the BP network. After 6 epochs, it reaches the MSE of 10^{-30}. Because the output nodes of RBF neural network are linear function, the output of it may be of any value.

Table 2. Samples for network training

Sample	Vector Value						Output	Notes
1	7.6456	4.1828	6.0641	5.5876	8.55	7.1971	0.0019602	Open
2	6.786	19.244	10.217	7.493	3.9024	4.8504	0.0024404	Open
3	15.037	7.037	14.403	11.138	4.3095	11.574	0.0017106	Open
4	20.132	38.1	42.058	29.431	14.873	2.5855	0.99608	Closed
5	12.618	33.171	36.361	35.36	8.1687	7.8465	0.99604	Closed
6	25.796	49.25	66.137	27.016	21.417	10.421	0.99607	Closed

3 Result

It can be seen from Table 3 that both BP and BRB neural network can classify EEG efficiently by taking the threshold value $\lambda=0.5$.

Table 3. Testing samples and output

Sample	Vector Value						BP output	RBF output	Notes
1	6.6659	4.491	3.7405	11.904	9.1182	3.9671	0.021	0.0885	Open
2	10.837	12.657	6.6429	5.2605	8.333	6.5333	0.0015	0.0119	Open
3	15.037	7.037	14.403	11.138	4.3095	11.574	0.0017	0.1691	Open
4	14.171	17.259	45.413	41.219	7.1534	7.8166	0.9962	1.0484	Closed
5	10.903	19.513	36.337	28.404	14.362	5.5268	0.9961	1.0800	Closed
6	12.83	46.11	27.735	29.587	17.245	5.3035	0.9681	1.0294	Closed

To evaluate the performance of the network, we test the networks with 50 eyes-open samples and 50 eyes-closed samples. Table 4 shows that both of them have high accuracy rate, RBF neural network even reach 100% accuracy.

To investigate the distribution of network output of the test samples, we draw the histograms of the output value of the 100 test samples in fig 3.

Table 4. Recognition Rate

Eyes States	Number of samples	BP			RBF		
		Open	Closed	Rate	Open	Closed	Rate
Open	50	50	0	100%	50	0	100%
Closed	50	3	47	94%	0	50	100%
Total	100	53	47	97%	50	50	100%

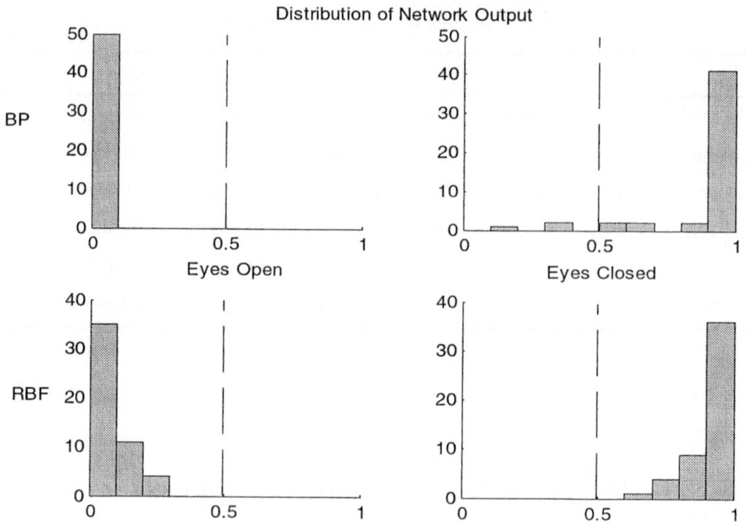

Fig. 3. Histograms of BP and RBF output

4 Conclusion

This paper proposes the method that combines wavelet packet decomposition and artificial neural network in the EEG signal processing. The result demonstrates that wavelet packet transform has good feature extraction ability. It also prove that, after proper training, artificial neural network can research high pattern recognize rate. Experiment shows that EEG rhythm can be accurately extract with wavelet packet decomposition. Both BP and RBF neural network have good generalization capability. Moreover, RBF neural network are better than BP neural network in training rate, approach ability and recognize ability. This method will meaningful in other EEG research. The change of feature vector and threshold will make it suitable for other usages.

There are still some problems that need more advanced investigation. The choice of wavelet subspace will be an important factor. How to choose the optimum wavelet scale or specify scale for specific problem will need more research.

References

1. Sumathy S., Krishnan C. N.: Automatic Machine Classification of Patient Anaesthesia Levels Using EEG Signals. IECON Proceedings, Vol. 3. IEEE, Los Alamitos CA, (1991) 2349-2351
2. Shimada, Takamasa, Shiina, Tsuyoshi, Saito, Yoichi: Detection of Characteristic Waves of Sleep EEG by Neural Network Analysis. IEEE Transactions on Biomedical Engineering, IEEE, Piscataway NJ Vol. 47. (2000) 369-379
3. O.A. Rosso et al: Brain Electrical Activity Analysis Using Wavelet-based Informational Tools. Physica A, Vol. 313. Elsevier, Amsterdam (2002) 587-608
4. O.A. Rosso, S. Blanco, J. Yordanova, V. Kolev, A. Figliola,M. SchWurmann, E. BaHsar, Wavelet Entropy: a New Tool for Analysis of Short Duration Brain Electrical Signals. Journal of Neuroscience Methods, Vol. 105. Elsevier, Amsterdam (2001), 65–75.
5. Hazarika, Neep, Chen, Jean Zhu, Tsoi, Ah Chung, Sergejew, Alex: Classification of EEG Signals using the Wavelet Transform. Signal Processing, Vol. 59. Elsevier, Amsterdam (1997) 61-72
7. Tong, S., Bezerianos, A Paul J., Zhu Y. Thakor, N: Nonextensive Entropy Measure of EEG Following Brain Injury from Cardiac Arrest. Physica A: Statistical Mechanics and its Applications, Vol. 305. (2002) 619-628
8. Yu Ali, Yu Ke, etc: Application of Wavelet Neural Network in EEG Signals Data Compression Representation and Spikes Recognition. Chinese Journal of Biomedical Engineering, Vol. 18. (1999) 142-148
9. Hou Xinguo, Xie Li, Wu Zhengguo, Zhao Yong-ling: Fault Diagnosis Method for Induction Motor Based on Wavelet Transformation and Neural Network, Journal of Data Acquisition & Processing. Vol.19 (2004) 32-36

Application of a Wavelet Adaptive Filter Based on Neural Network to Minimize Distortion of the Pulsatile Spectrum

Xiaoxia Li[1,2], Gang Li[1], Ling Lin[1], Yuliang Liu[1],
Yan Wang[1], and Yunfeng Zhang[2]

[1] College of Precision Instrument and Opto-Electronics Engineering, Tianjin University, Tianjin, 300072, China
[2] School of electrical and automatic engineering, Hebei University of Technology, Tianjin, 300130, China
lxxlj@eyou.com

Abstract. The approach of the Dynamic Spectrum evaluates the pulsatile part of the entire optical signal at different wavelength. In the course of collecting the pulsatile spectrum signal in vivo, it is inevitable to be interfused with yawp signals as high frequency interference, baseline drift and so on. Using the traditional adaptive filter, it is very difficult to collect the reference signal from the in vivo experiment. In this paper, Daubechies wavelet adaptive filter based on Adaptive Linear Neuron networks is used to extract the signal of the pulse wave. Wavelet transform is a powerful tool to disclose transient information in signals. The wavelet used is adaptive because the parameters are variable, and the neural network based adaptive matched filtering has the capability to "learn" and to become time-varying. So this filter estimates the deterministic signal and removes the uncorrelated noises with the deterministic signal. This method can get better result than nonparametric results. This filter is found to be very effective in detection of symptoms from pulsatile part of the entire optical signal.

1 Introduction

Over the last years spectroscopy in the wavelength range between 600 and 1100 nm has been developed to manifold applications in biomedical sensing and diagnostics [1]. The methods for non-invasive concentration measurements of clinically relevant blood and tissue components play an important role. The main difficulty for determining absolute or even exact relative concentrations is the scattering behaviour of the tissue. This leads to significant differences in the ideal Lambert Beer's law.

Several investigations have been published to obtain a quantification of the optical signal. In this context the pulse oximetry is the most established clinical method. Evaluating only the pulsatile part of the entire optical signal, this method is rather

F. Yin, J. Wang, and C. Guo (Eds.): ISNN 2004, LNCS 3174, pp. 362–368, 2004.

independent of individual or time changes in scattering or absorption characteristics of the tissue. But it was only used for double wavelength evaluation.

The approach of the Dynamic Spectrum, proposed by Dr. LI Gang etc, is based on Photo-plethysmography (PPG) with fast Fourier transforms. The Dynamic Spectrum is composed of the magnitude of fundamental wave of the pulse at different wavelength.

In the course of collecting the pulsatile spectrum signal, it is inevitable to be interfused with yawp signals as high frequency interference, baseline drift and so on. It is very difficult to get the clear signal of the pulse wave. The traditional filter such as FIR and IIR, the cut-off frequency is changeless, the filter can't work when the frequency of the noise beyond the cut-off frequency. For the common adaptive filter, it is very difficult to collect the reference signal from the in vivo experiment. In this paper, wavelet adaptive filter based on neural network is used to collect the signal of the pulse wave. Wavelet transform is a powerful tool to disclose transient information in signals [2]. The wavelet used is adaptive because the parameters are not fixed. The neural network based adaptive matched filtering has the capability to "learn" and to become time-varying [3]. In this paper, an adaptive filter based on Daubechies wavelet and Adaptive Linear Neuron networks is introduced. This filter estimates the deterministic signal and removes the noises uncorrelated with the deterministic signal. The method can perform better result than nonparametric results. This filter is found to be very effective in detection of symptoms from pulsatile part of the entire optical signals.

2 Wavelet Transform

Wavelet transforms are inner products between signals and the wavelet family, which are derived from the mother wavelet by dilation and translation. Let $\psi(t)$ be the mother wavelet, the daughter wavelet will be:

$$\psi_{a,b}(t) = \psi((t-b)/a) \tag{1}$$

Where a is the scale parameter and b is the time translation. By varying the parameters a and b, we can obtain different daughter wavelets that constitute a wavelet family. Wavelet transform is to perform the following operation:

$$W(a,b) = \frac{1}{\sqrt{a}} \int x(t)\psi^{*}_{a,b}(t)dt \tag{2}$$

Where '*' stands for complex conjugation

Practically, when computing the wavelet transform, dyadic scales and positions are chosen. The computation can be realized by the discrete wavelet transform. The dyadic wavelet is determined using a scale $a = 2^j$, where $j \in Z$ and Z is the integral set. The WT at scale $a = 2^j$ is obtained by

$$W(2^j,b)=\frac{1}{2^j}\int_{-\infty}^{\infty}x(t)\psi^*_{2^j,b}(t)dt \tag{3}$$

Signal reconstruction in the case of the orthonormal **W** matrix is described by the simple formula

$$f=W^T w \tag{4}$$

For the decomposition and reconstruction of the signal whose length equals an integer power of two ($N=2^p$), the fast pyramid algorithm can be applied. Each row represents a level of decomposition.

3 Wavelet Adaptive Filter Based on Neural Network

3.1 Adaptive Filters and Adaptive Linear Neuron Networks

Adaptive signal processing is a signal processing method developed recently [4]. Adaptive filters are digital filters capable of self adjustment. An adaptive filter is used in applications that require differing filter characteristics in response to variable signal conditions.

In this paper, an adaptive Filter based on Neural Network is used. The Adaptive Linear Neuron network is chosen. The ADALINE has been and is today one of the most widely used neural networks found in practical applications. Adaptive filtering is one of its major application areas. The ADALINE network has one layer of S neurons connected to R inputs through a matrix of weights W. R is the number of elements in input vector. S is the number of neutrons in layer. The networks will use the LMS algorithm or Widrow-Hoff learning algorithm based on an approximate steepest descent procedure. The LMS or Widrow-Hoff learning rule minimizes the mean square error and, thus, moves the decision boundaries as far as it can from the training patterns.

Here we adaptively train the neural linear network to predict the combined pulse wave and noise signal m from an noise signal n. Notice that the noise signal n does not tell the adaptive network anything about the pulse wave signal contained in m. However, the noise signal n. does give the network information it can use to predict the noise's contribution to the signal m. The network will do its best to adaptively output m. In this case, the network can only predict the noise in the signal m. The network error e is equal to m, the combined signal, minus the predicted contaminating noise signal. Thus, e contains only the pulse wave! Our linear adaptive network adaptively learns to cancel the noise. The principle of minimizing noise is shown in Fig.1.

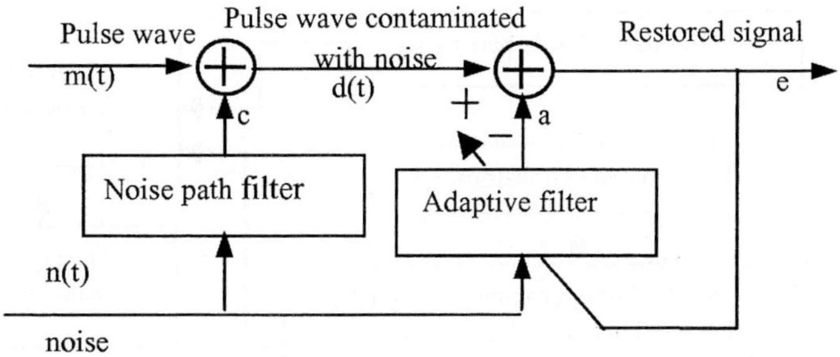

Fig. 1. The principle of minimizing noise

3.2 Procedure to Perform the Adaptive Wavelet Filtering Based on Neural Network

The designed filter, the input signal $d(t)$ is the original spectrum signal, and $d(t) = m(t) + n(t)$, it includes the signal of pulse wave and other noise signals. The reference input signal must be independent of the pulse wave signal and correlated with the original input signal. It is very difficult to get the reference signal in the in vivo experiment. The wavelet transform is concerned. The reconstruction of some levels of decomposed signal can be the reference signal. The wavelet used is adaptive because the parameters are not fixed.

Choice of a wavelet function is very important for any processing application in the wavelet domain. For noise attenuation, a wavelet should produce an optimal separation of signal from noise. The wavelet function must have enough vanishing moments, have good pass band performance, and must have small enough support length to decrease the redundancy of the information. That is constructing the orthogonal wavelet. So, The Daubechies wavelet is chosen in this paper.

The procedure to perform the adaptive wavelet filtering based on neural network is shown in Fig.2.

Decomposition of the original signal with the Daubechies 4 wavelet to 4 levels, select the approximation signal at level 4 contain the information of the baseline drift and the detail signal at level 1,2 contain the information of the high-frequency noise. Get rid of the information of the other level.

Reconstruction the signal X based on this decomposition structure, put signal X as the reference input signal, the original signal as the input signal of the ADALINE network. Training the ADALINE network with the reference input signal.

Then the error signal is approximate the pulse wave signal which is eliminated the high-frequency noise and the baseline drift.

Practically, the original data is preprocessed so that minimum is -1 and maximum is 1.

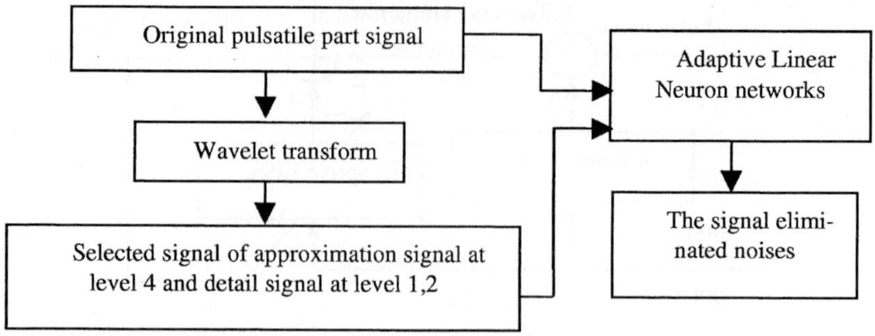

Fig. 2. Diagram of the adaptive wavelet filtering based on neural network

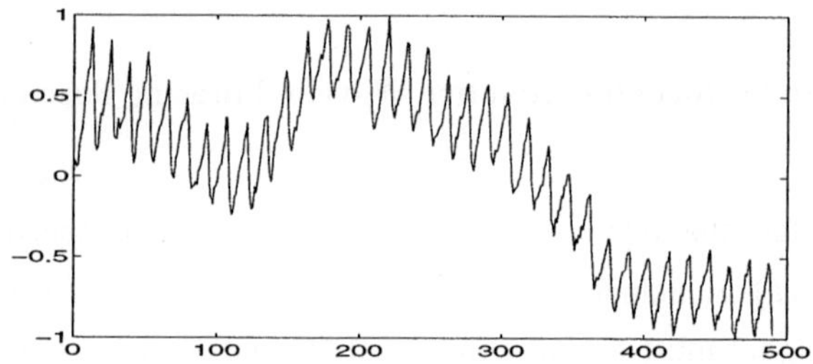

Fig. 3. The original signal of pulse wave

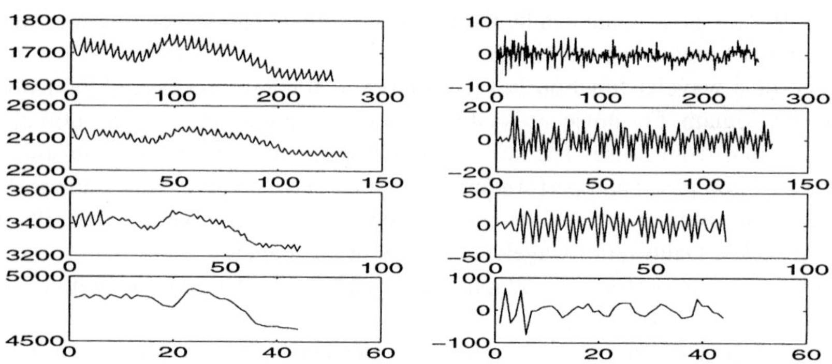

Fig. 4. The approximation signal (*left column*) and the detail signal (*right column*)

4 Experimental Result

For in vivo transmission measurements an experimental set-up was applied. The light from a stabilized tungsten halogen light source was guided to the surface of the top of the finger tip. The transmitted light from the bottom side of the finger was collected by a single optical fiber. This fiber was directly connected to a common portable spectrometer (the USB2000 portable spectrometer of the OCEAN OPTIC Company). The spectrometer scans over a wavelength range from 462.87 to 1136.16nm. For this measurement, spectra were taken with an integration time of 60 ms. The total measurement time was approximately 30 seconds. The original signal is shown in Fig.3.

The approximation signal and the detail signal are shown in Fig.4.

The reference signal is shown in Fig.5. It is reconstructed with the 4th level of approximation signal and 1st, 2nd level of detail signals.

After the procedure of removing noise, the output and the error of the network are shown in Fig.6. The pulse wave signal which is minimized noise is got, shown in Fig.7.

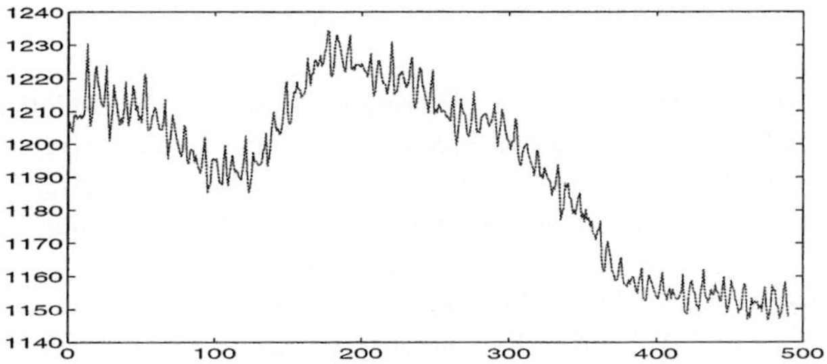

Fig. 5. The reference signal

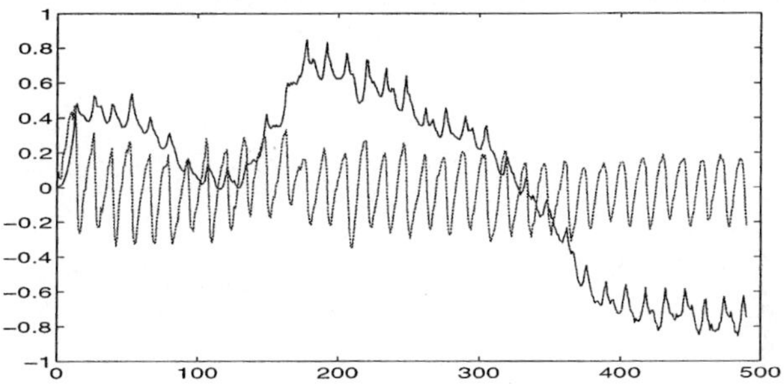

Fig. 6. The output (*the upper one*) and the error (*the middle one*) of the network

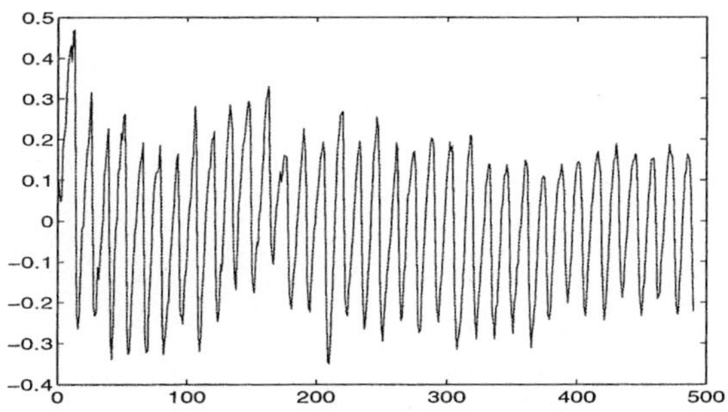

Fig. 7. The pulse wave signal minimized noise

5 Result

The experimental result shows that, the signal quality of pulse wave which extracted from the spectrum is improved greatly with this method. The wavelet adaptive filter based on Neural Network can minimize the noise as high frequency interference, baseline drift and so on.

References

1. Torella. Francesco, Cowley. Richard, Thorniley. Maureen S., McCollum. Charles N.: Monitoring Blood Loss With Near Infrared Spectroscopy. Comparative Biochemistry and Physiology-Part A: Molecular & Integrative Physiology, Vol. 132. (2002) 199-203
2. Park.K.L, Lee.K.J, Yoon.H.R.: Application of a Wavelet Adaptive Filter to Minimize Distortion of the ST-Setment. Biol.Eng.Comput, (1998)581~586
3. Ping. Li, Fang. M.T.C., Lucas. J.: Modeling of Submerged Arc Weld Beads Using Self-Adaptive Offset Neural Networks. Journal of Materials Processing Technology, Vol. 71. (1997) 288-298
4. Lin. J., Zuo, M. J.: Gearbox Fault Diagnosis Using Adaptive Wavelet Filter. Mechanical Systems and Signal Processing , Vol. 17. (2003) 1259-1269

Spectral Analysis and Recognition Using Multi-scale Features and Neural Networks

YuGang Jiang and Ping Guo

Department of Computer Science
Beijing Normal University, Beijing, 100875, P.R.China
jiangyg81@sohu.com; pguo@ieee.org

Abstract. This paper presents a novel spectral analysis and classification technique, which is based on multi-scale feature extraction and neural networks. We propose two feature extraction methods in wavelet domain to implement de-noising process and construct feature spectra. Then a radial basis function network is employed for classifying spectral lines. The input of the neural network is the feature spectra, which is produced by the proposed methods. Real world data experimental results show that our technique is robust and efficient. The classification results are much better than the best results obtained by principle component analysis feature extraction method.

1 Introduction

Current and future large astronomical surveys will yield very large number of fiber spectra. Once the huge number of spectra have been collected, the study of the distribution of spectral types and the analysis of spectral data can help to understand the temporary change of the physical conditions of stars from a statistical point of view, and therefore, to learn about their evolution. This is why spectral recognition is one of the fundamental aspects of the evolutionary study of stars, and a phase that must be carried out in a fast, efficient and accurate way.

Computational artificial neural networks (ANN) are known to have the capability for performing complex mappings between input and output data. They can be applied to different types of problems: classification of objects, modelling the functional relationships and representation of large amount of data. In the classification of astronomical objects the ANN was adopted by Odewahn *et al* in 1992 [1]. The supervised learning networks, such as the back propagation (BP) neural network are used in their research. Radial basis function (RBF) neural network is also applied to automatic classification of spectra [2].

However, stellar spectra are extremely noisy and voluminous. If we use the spectral lines directly as input of a neural network without pre-processing, the classification results are very poor using either RBF network or BP network. In order to obtain a high correct classification rate (CCR), one possible approach is to use multi-scale analysis.

F. Yin, J. Wang, and C. Guo (Eds.): ISNN 2004, LNCS 3174, pp. 369–374, 2004.

The wavelet transform provides a powerful and versatile framework for astronomical signal processing. Each scale of resolution may pick up different types of structure in the signals. It is particularly appropriate for spectral data since the structures we want to recognize have different scales and resolutions. In this paper we propose two feature extraction methods in wavelet domain. One is from scales, and the other is from positions. After getting the feature spectra, we adopt a RBF neural network to complete the classification.

The organization of this paper is as follows: In Section 2, we briefly introduce the wavelet transform and RBF neural networks. Section 3 describes the experiments and gives the results. The conclusions are presented in the last Section.

2 Background

2.1 Briefly Review of Wavelet Transform and Multi-scale Analysis

The wavelet transform is a very powerful tool for signal processing. It is a linear operation that decomposes a signal into components that appear at different scales. A wavelet is a function $\psi(x) \in L^2$ such that $\int_{-\infty}^{+\infty} |\psi(x)|^2 dx < \infty$. Let us denote the dilation and translation of $\psi(x)$ by $\psi_{s,p}(x)$ at scale s and position p as:

$$\psi_{s,p}(x) = \frac{1}{s}\psi(\frac{x-p}{s}). \quad (1)$$

The wavelet transform of a function $f(x)$ at the scale s and position p is defined as:

$$W_f(s,p) = \langle \psi_{s,p}(x), f(x) \rangle. \quad (2)$$

If $\psi_{s,p}(x)$ is orthonormal, $f(x)$ can be reconstructed by:

$$f(x) = \sum_{s,p} W_f(s,p)\psi_{s,p}(x). \quad (3)$$

When the scale s decreases, the support of $\psi_{s,p}(x)$ decreases so the wavelet transform $W_f(s,p)$ is sensitive to finer details. The scale s characterizes the size and regularity of the signal features extracted by the wavelet transform. In this paper, we adopt widely used Daubechies-4 as wavelet basis function.

2.2 RBF Neural Networks

RBF neural network is composed of three layers. The neurons of the input layer only transfer input signal to hidden layer, the neurons of hidden layer are composed of radiant basis functions, and the neurons of output layer are usually simple linear functions [3]. The RBF network is built by considering the basis function as an neuronal activation function and the w_{kj} parameters as weights.

The output of the network looks like:

$$y_k(\mathbf{x}) = \sum_{j=1}^{H} w_{kj}\varphi_j(\mathbf{x}) + w_{k0} \Leftrightarrow \mathbf{y}(\mathbf{x}) = \mathbf{W}\varphi(\mathbf{x}), \quad (4)$$

where $\mathbf{x}$ is a p dimensional input vector, H is the number of RBF hidden neurons and $\mathbf{W}$ holds both weights and bias. In the experiments, we build the radial basis functions as Gaussians:

$$\varphi_j(\|\mathbf{x}-\mu_j\|) = \exp[-\frac{1}{2\gamma_j^2}\|\mathbf{x}-\mu_j\|^2], \tag{5}$$

where μ_j is the center and γ_j is the standard deviation of the Gaussian function.

When training and target samples $\{\mathbf{x}_i, \mathbf{t}_i\}_{i=1}^N$ are given, the weights matrix $\mathbf{W}$ can be obtained as $\mathbf{W} = \mathbf{T}\mathbf{\Phi}^\dagger$, $\mathbf{\Phi}^\dagger$ is the pseudo-inverse of $\mathbf{\Phi}$, where $\mathbf{\Phi}$ is a matrix:

$$\mathbf{\Phi} = \begin{pmatrix} \varphi(\|\mathbf{x}_1-\mu_1\|) & \dots & \varphi(\|\mathbf{x}_N-\mu_1\|) \\ \vdots & \ddots & \vdots \\ \varphi(\|\mathbf{x}_1-\mu_H\|) & \cdots & \varphi(\|\mathbf{x}_N-\mu_H\|) \end{pmatrix}. \tag{6}$$

The network learning procedure consists of two steps. At first, the basis functions are established and their parameters are found. Then, having the basis functions properly established, the weights on output layer can be obtained.

3 Experiments

3.1 The Data Sets and Pre-treatment

Two stellar spectrum libraries selected from Astronomical Data Center (ADC) are used in the experiments. One is contributed by Jacoby [4] and the other is contributed by Pickles [5].

MK classification system of stellar spectra classifies stars into seven main spectral types in the order of decreasing temperature, namely: O, B, A, F, G, K, M [6]. In the library observed by Jacoby, there are 161 spectral samples from the seven classes. These spectra cover the wavelength ranges 351.0 to 742.7 nm with 0.14 nm of spectral resolution. Pickles' library contains a total of 131 spectra covering spectral types ranging from O to M. The wavelength is from 115.0 to 2500.0 nm and the resolution is 0.5 nm.

Before analyzing the spectra, for the construction of the different pattern sets, all spectra are linearly interpolated to the wavelength range of 380.0-742.0 nm with a step of 0.5 nm. Then the spectra of the two catalogues are scaled to flux 1 at wavelength 545.0 nm in order to normalize the flux values.

3.2 Multi-scale Analysis

We perform wavelet decomposition at seven scales of all the spectra by Mallat algorithm [7]. The decomposition of each spectral line produces a wavelet coefficient set W_s at each scale s and one smoothed array at the last scale as shown in Fig. 1.

Here two approaches are proposed to implement de-noising process and construct feature spectra:

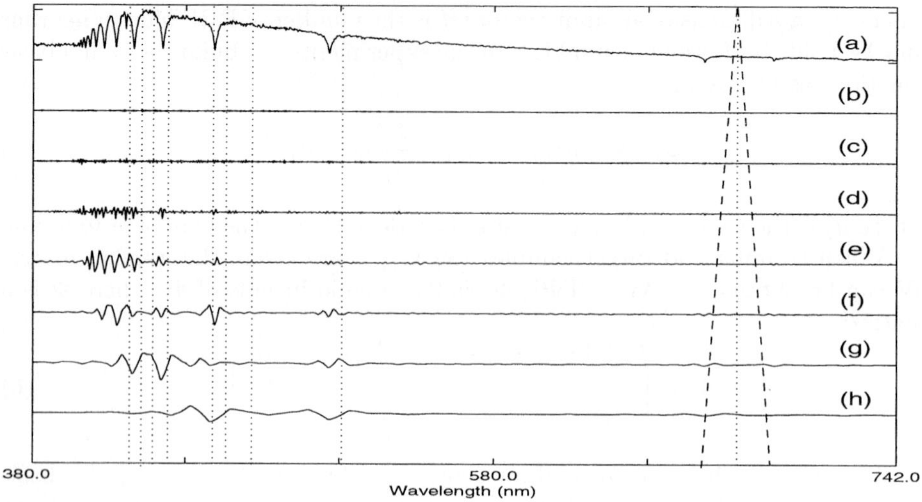

Fig. 1. (*a*) An example of spectrum. (*b*) Wavelet scale 1 (*solid line*). (*c-h*) Same as (*b*), but for wavelet scales from 2 to 7, respectively

Approach I: Extract Features from Scales. When wavelet scale s increases, the frequency represented in the scale decreases.

Due to the very low signal-to-noise ratio (SNR) of the spectra, the traditional threshold de-noising technique [8] can not obtain satisfying result. In the experiment, we find that most of the coefficients in the first four wavelet scales arise from noise. In order to suppress the noise, we omit the wavelet coefficients belonged to higher frequency and reconstruct the feature spectra using wavelet coefficients at scale five, six and seven.

Because we discard wavelet coefficients at four scales, the reconstructed feature spectra with Mallat algorithm are of 54 dimensions, which is about 1/16 of original Dimensions. Fig. 2 shows the feature spectra extracted by this approach.

Approach II: Extract Features from Positions. Knowledge-based system can reproduce the reasoning of experts in the field to classify spectra. Here we propose a knowledge-based feature extraction method in wavelet domain, that is extracting feature coefficients at some fixed wavelengths. Table 1 shows a complete description of the parameters' positions according to astrophysicist's knowledge [9]. They are shown in Fig. 1 by dotted lines.

Through extracting the coefficients at the wavelengths shown in Table 1 and reconstruct them, we can get the feature spectra. When scale s increases, the wavelet function $\psi_{s,p}(x)$ becomes widen. So more coefficients have to be extracted at a wavelength while scale increases, as shown in Fig. 1 (Between the two dashed lines). Then the feature spectra are sampled to 54 dimensions in order to make a comparison with other feature extraction methods. An example of extracted feature spectrum is shown in Fig. 2.

Table 1. Spectral parameters used for feature extraction

Line	Wavelength (nm)
Ca II (K)	393.3
Ca II (H)	396.8
CH band	430.0
H I γ	434.0
H I δ	410.2
He I	402.6
He I	447.1
H I β	486.1
H I α	656.3

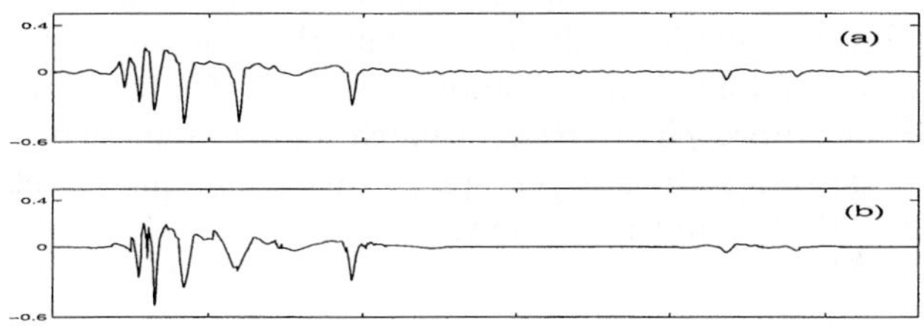

Fig. 2. Feature spectrum. (*a*) First approach. (*b*) Second approach

3.3 Classification

After constructing the feature spectra, we can adopt ANN method to classify them. Usually in ANN classification method, two data sets namely training and testing set are used. In our experiments, the feature spectra data constructed from Jacoby's library are used as training set and data in Pickles' library are used as testing set.

The feature vectors extracted in Section 3.2 are chosen as the input of a RBF neural network. For the network output vector, we use one-of-k encoding method. The classification results are shown in Table 2. RBF network as a final classifier performs differently with different number of hidden neurons. Table 3 shows the CCR with various number of RBF net hidden neurons.

As a comparison, we also adopt a BP net to do the final classification. The inputs of the BP net are obtained by approach I. Principle component analysis (PCA) is a widely used tool for dimension reduction and feature extraction [10]. Here we use first 54 principal components as input of a RBF network. The result of Approach I+BP and PCA+RBF are also shown in Table 2, which illustrate that the accuracy of these methods is not as good as our proposed methods.

Table 2. Comparison of classification accuracy using different methods

	Approach I+RBF	Approach II+RBF	Approach I+BP	PCA+RBF
CCR	94.6%	93.7%	78.3%	75.7%

Table 3. Comparison of RBF net with different number of hidden neurons

Number of Neurons	10	20	30	60
CCR	73.0%	90.5%	94.6%	91.9%

4 Conclusions

In this paper, we present a hybrid technique which combine multi-scale analysis and RBF network to classify the spectra of stars. In order to obtain the input patterns of the neural networks, two feature extraction methods in wavelet domain are proposed. Experimental results show that the feature extraction methods are robust to noise and efficient to computation. The CCR of the proposed Approach I+RBF or Approach II+RBF is much higher than that of BP network or PCA+RBF. Therefore it is a very promising technique for classification of spectra with low SNR from sky surveys.

Acknowledgements. This work was fully supported by a grant from the National Natural Science Foundation of China (Project No. 60275002) and by the "863" Program of China (Project No. 2003AA133060).

References

1. Odewahn, S. C., Stockwell, E. B., Pennington, R. L., Humphreys, R. M., Zumach, W. A.: Automated Star/Galaxy Discrimination with Neural Networks. Astronomical Journal, **103** (1992) 318–331
2. Bai, L., Li, Z. B., Guo, P.: Classification of Stellar Spectral Data Based on Kalman Filter and RBF Neural Networks. Proceedings of IEEE Inter. Conf. on Sys. Man & Cyber., IEEE Press, Piscataway NJ (2003) 274–279
3. Bishop, C. M.: Neural Networks for Pattern Recognition. Oxford University Press, London (1995)
4. Jacoby, G. H., Hunter, D. A., Christian, C. A.: A Library of Stellar Spectra. The Astrophysical Journal Supplement Series, **56** (1984) 257–281
5. Pickles, A. J.: A Stellar Spectral Flux Library: 1150–25000 Å. The Publications of the Astronomical Society of the Pacific, **110** (1998) 863–878
6. Kurtz, M. J.: The MK Process and Stellar Classification. David Dunlap Observatory, Toronto (1984)
7. Mallat, S. G.: A Theory for Multiresolution Signal Decomposition: The Wavelet Representation. IEEE Trans. on Pattern Analysis and Machine Intelligence, **11** (1989) 674–693
8. Donoho, D. L.: De-noising by Soft-Thresholding. IEEE Trans. on Information Theory, **41** (1995) 613–627
9. Zombeck, M. V.: Handbook of Space Astronomy and Astrophysics. Cambridge University Press, London (1990)
10. Jolliffe, I. T.: Principal Component Analysis. Springer-Verlag, New York (1986)

A Novel Fuzzy Filter for Impulse Noise Removal

Chang-Shing Lee[1], Shu-Mei Guo[2], and Chin-Yuan Hsu[2]

[1]Department of Information Management, Chang Jung University, Tainan, Taiwan
leecs@mail.cju.edu.tw
leecs@cad.csie.ncku.edu.tw
[2] Department of Computer Science & Information Engineering,
National Cheng Kung University, Tainan, Taiwan

Abstract. In this paper, we propose a novel *Neural Fuzzy Filter* (*NFF*) to remove impulse noise from highly corrupted images. The proposed filter consists of a *fuzzy number construction process*, a *neural fuzzy filtering process* and an *image knowledge base*. First, the *fuzzy number construction process* will receive sample images or the noise-free image, then construct an *image knowledge base* for the *neural fuzzy filtering process*. Second, the *neural fuzzy filtering process* contains of a *neural fuzzy mechanism*, a *fuzzy mean process*, and a *fuzzy decision process* to perform the task of impulse noise removing. By the experimental results, *NFF* achieves better performance than the state-of-the-art filters based on the criteria of Mean-Square-Error (MSE). On the subjective evaluation of those filtered images, *NFF* also results in a higher quality of global restoration.

1 Introduction

Nowadays, the techniques of image processing have been well developed, but there are still some bottlenecks that are not solved. Many image processing algorithms can't work well in a noisy environment, so the image filter is adopted as a preprocessing module. For example, a median filter [1] is the most used method, but it will not work efficiently when the noise rate is above 0.5. E. Abreu et al. [2] propose an efficient nonlinear algorithm to suppress impulse noise from highly corrupted images while preserving details and features. It is called Signal-Dependent Rank Ordered Mean (SD-ROM) filter. SD-ROM filter can achieve an excellent tradeoff between noise suppression and detail preservation. Adaptive Weighted Fuzzy Mean (AWFM)[3] can improve the WFM filter's incapability in a low noisy environment, and still retains its capability of processing in the heavily noisy environment. F. Russo [4] presents the hybrid neuro-fuzzy filters for images which are highly corrupted by impulse noise. The network structure of the filter is specifically designed to detect different patterns of noisy pixels typically occurring in highly corrupted data. The proposed filters are able to yield a very effective noise cancellation and to perform significantly better than the other approaches. J. H. Wang et al. [5] present a histogram-based fuzzy filter (HFF) to the restoration of noise-corrupted images. G. Pok et al. [6] propose a decision-based, signal-adaptive median filtering algorithm for removal of impulse noise.

F. Yin, J. Wang, and C. Guo (Eds.): ISNN 2004, LNCS 3174, pp. 375–380, 2004.

In this paper, we propose a novel Neural Fuzzy Filter (*NFF*) to removal impulse noise from highly corrupted images. The proposed filter consists of a *fuzzy number construction process*, a *neural fuzzy filtering process*, and an *image knowledge base*. The rest of this paper is organized as follows. In Section 2, we propose the structure of neural fuzzy filter. Section 3 describes the neural fuzzy process for impulse noise removal. The experimental results for the *NFF* are described in Section 4. Finally, we make the conclusion in Section 5.

2 The Structure of Fuzzy Neural Filter

In this paper, we propose a novel Neural Fuzzy Filter (*NFF*) to remove impulse noise from highly corrupted images. Fig. 1 shows the structure of *NFF*. There are two main processes including a *fuzzy number construction process* and a *neural fuzzy filtering process* in Fig. 1. The image knowledge base should be constructed before performing the *neural fuzzy filtering process*. In this paper, we use the trapezoidal function $f_A(x)$ to be membership function of fuzzy set A [3]. The fuzzy set A is denoted by the parameter set $A=[a_A, b_A, c_A, d_A]$. Fig. 2 illustrates an example for *luminance* fuzzy variable with five linguistic terms. The image knowledge base consists of the parameters of the membership functions. In this paper, we define five fuzzy sets for an image including *very dark* (VDK), *dark* (DK), *medium* (MD), *bright* (BR), and *very bright* (VBR) shown in Fig. 2. The fuzzy sets VDK, DK, MD, BR, and VBR are denoted as $VDK=[a_{VDK}, b_{VDK}, c_{VDK}, d_{VDK}]$, $DK=[a_{DK}, b_{DK}, c_{DK}, d_{DK}]$, $MD=[a_{MD}, b_{MD}, c_{MD}, d_{MD}]$, $BR=[a_{BR}, b_{BR}, c_{BR}, d_{BR}]$, and $VBR=[a_{VBR}, b_{VBR}, c_{VBR}, d_{VBR}]$, respectively. The fuzzy sets describing the intensity feature of a noise-free image can be derived from the histogram of the source image. The histogram of a digital image with gray levels in the range [0, L-1] is a discrete function $h(s_k)=g_k$ where s_k is the kth gray level of image S and g_k is the number of pixels with the kth gray level in S. Then the algorithm for *fuzzy number construction process* is as follows:

Algorithm for Fuzzy Number Construction Process:

Step 1: Decide the overlap range of the fuzzy sets, respectively.

Step 1.1: Set c_{VDK} be the first s_k such that $g_k>0$, $a_{DK}\leftarrow c_{VDK}$.

Step 1.2: Set b_{VBR} be the last s_k such that $g_k>0$, $d_{BR}\leftarrow b_{VBR}$.

Step 1.3: Set $range \leftarrow \left\lfloor \frac{(b_{VBR}-c_{VDK})}{2\times N_f-3} \right\rfloor$ where N_f is the number of fuzzy sets.

Step 1.4: Set $a_{VDK}\leftarrow 0$, $b_{VDK}\leftarrow 0$, $c_{VBR}\leftarrow 0$, $d_{VBR}\leftarrow L-1$.

Step 2: Decide the parameter values of the membership function f_{VDK} of fuzzy set VDK as follows: $d_{VDK}\leftarrow c_{VDK}+range$.

Step 3: Decide the parameter values of the membership function f_{DK} of fuzzy set DK by the following sub-step:

Step 3.1: Set $b_{DK}\leftarrow d_{VDK}$.

Step 3.2: Set $c_{DK} \leftarrow range + b_{DK}$, $d_{DK} \leftarrow range + c_{DK}$.

Step 4: Decide the parameter values of the membership function f_{MD} of fuzzy set MD by the following sub-step:

Step 4.1: Set $a_{MD} \leftarrow c_{DK}$, $b_{MD} \leftarrow d_{DK}$.

Step 4.2: Set $c_{MD} \leftarrow range + b_{MD}$, $d_{MD} \leftarrow range + c_{MD}$.

Step 5: Decide the parameter values of the membership function f_{BR} of fuzzy set BR by the following sub-step:

Step 5.1: Set $a_{BR} \leftarrow c_{MD}$, $b_{BR} \leftarrow d_{MD}$.

Step 5.2: Set $c_{BR} \leftarrow b_{BR} + range$.

Step 6: Decide the parameter values of the membership function f_{VBR} of fuzzy set VBR as follows: $a_{VBR} \leftarrow c_{BR}$.

Step 7: Stop.

3 Neural Fuzzy Filtering Process for Impulse Noise Removal

The *neural fuzzy mechanism* consists of three layers including input linguistic neuron layer, input term neuron layer, and rule neuron layer. Now we describe them as follows.

Layer 1(Input linguistic neuron layer): The neurons in first layer just directly transmit input values to next layer. If the input vector is $x = (x_1, x_2, \ldots, x_9)$, then output vector of the input for this layer will be

$$\mu^1 = ((x_{11}, x_{12}, \ldots, x_{15}), (x_{21}, x_{22}, \ldots, x_{25}), \ldots, (x_{91}, x_{92}, \ldots, x_{95})) \tag{1}$$

where x_{ik} is input value of the *k*th linguistic term in the *i*th fuzzy variable.

Layer 2(Input term neuron layer): Each fuzzy variable of the second layer appearing in the premise part is represented with a condition neuron. This layer performs the first inference step to compute matching degrees. If the input vector of this layer is $\mu^1 = ((x_{11}, x_{12}, \ldots, x_{15}), (x_{21}, x_{22}, \ldots, x_{25}), \ldots, (x_{91}, x_{92}, \ldots, x_{95}))$, then the output vector will be

$$\begin{aligned}\mu^2 = ((&f_{A1}(x_{11}), f_{A2}(x_{12}), \ldots, f_{A5}(x_{15})), (f_{A1}(x_{21}), f_{A2}(x_{22}), \ldots,\\ &f_{A5}(x_{25})), \ldots, (f_{A1}(x_{91}), f_{A2}(x_{92}), \ldots, f_{A5}(x_{95})))\end{aligned} \tag{2}$$

Layer 3(Rule neuron layer): The neuron in this layer is called a rule neuron. Eq. (3) shows the matching degree μ_j^3 of Rule neuron *j*, where *j*=1,…5.

$$\mu_j^3 = \begin{cases} \dfrac{\sum_{i=1}^{n} (x_i f_{Aj}(x_{ij}))}{\sum_{i=1}^{n} f_{Aj}(x_{ij})} & , if (\sum_{i=1}^{n} f_{Aj}(x_{ij})) > 0 \\ 0 & , otherwise \end{cases} \tag{3}$$

The *fuzzy mean process* performs the computing of fuzzy mean for input variables. Eq. (4) denotes the computing process with fuzzy set F_mean for *fuzzy mean process*, respectively.

$$y_{mean} = \begin{cases} \dfrac{\sum_{i=1}^{n} \left(f_{F\_mean}(x_i) \times x_i \right)}{\sum_{i=1}^{n} f_{F\_mean}(x_i)} & , if (\sum_{i=1}^{n} f_{F\_mean}(x_i)) > 0 \\ 0 & , otherwise \end{cases} \quad (4)$$

where $F\_mean = [0, \alpha, \beta, L-1]$.

There are five computation functions including $f_{comp}(\cdot)$, $f_{-}(\cdot)$, $f_{\times_1}(\cdot)$, $f_{\times_2}(\cdot)$, and $f_{+}(\cdot)$ and two membership functions including $f_{small}(\cdot)$ and $f_{l\arg e}(\cdot)$ utilized in fuzzy decision process. Now we briefly describe them as follows:

1. $f_{comp}(\mu_1^3(\cdot), \ldots, \mu_5^3(\cdot); y_{mean}) = \mu_j^3$ where the index j makes $|\mu_j^3 - y_{mean}|$ be the minimum value for j=1, …, 5.
2. $f_{-}(f_{comp}(\cdot), x_5) = | f_{comp}(\cdot) - x_5 |$, $f_{\times_1}(f_{comp}(\cdot), f_{l\arg e}(\cdot)) = f_{comp}(\cdot) * f_{l\arg e}(\cdot)$
3. $f_{\times_2}(x_5, f_{small}(\cdot)) = x_5 * f_{small}(\cdot)$, $f_{+}(f_{\times_1}(\cdot), f_{\times_2}(\cdot)) = f_{\times_1}(\cdot) + f_{\times_2}(\cdot)$

We define fuzzy sets $small = [0, 0, s, l]$ and $l\arg e = [s, l, 255, 255]$ for the fuzzy decision process. The final output y of fuzzy decision process is the computing result of $f_{+}(\cdot)$.

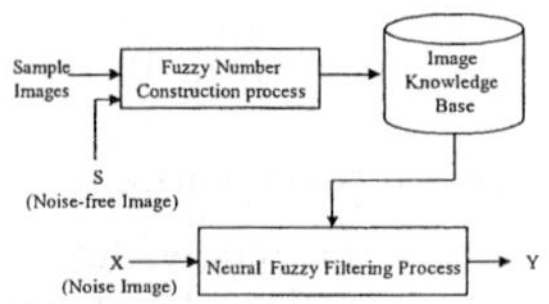

Fig. 1. The structure of Neural Fuzzy Filter.

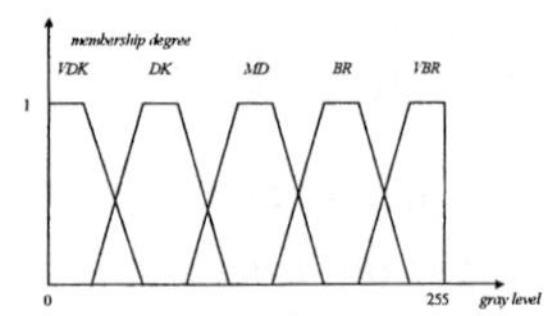

Fig. 2. The *luminance* fuzzy variable with five linguistic terms.

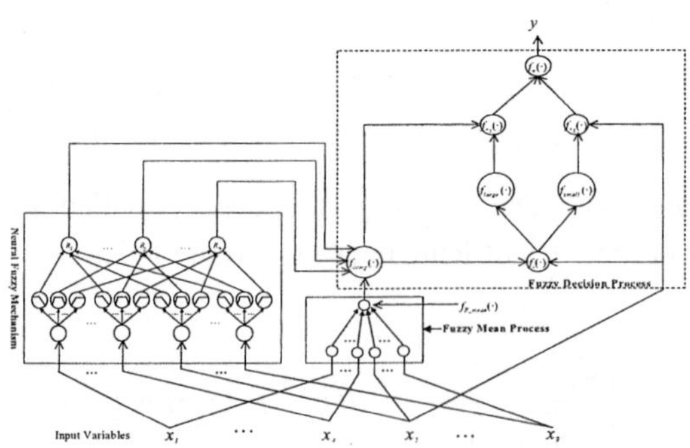

Fig. 3. The architecture of neural fuzzy filtering process.

4 Experimental Results

To test the performance of *NFF*, we compare our approach with the famous filters including Russo, SD-ROM, AWFM, and median filters. The experiments are performed on "Albert" image which is shown in Fig. 4. Fig. 5 shows the MSE curves of "Albert" image with noise probability p, where p=0~0.9. Fig. 6 shows the subjective evaluation with noise probability p=0.8 the experimental results exhibit that *NFF* has the better performance than the other approaches for impulse noise removal.

Fig. 4. "Albert" image.

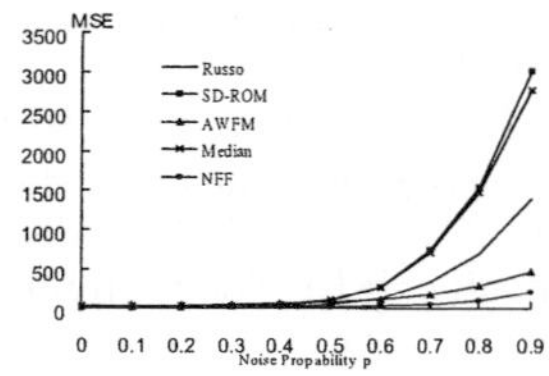

Fig. 5. MSE curves of the proposed method and the compared approaches.

(a) Noise rate 0.8 (b) Russo result (c) SD-ROM result

(d) AWFM result (e) Median result (f) *NFF* result

Fig. 6. Results of gray image Albert with 0.8 impulse noise.

5 Conclusions

In this paper, a novel neural fuzzy filter for impulse noise removal is presented. The proposed filter consists of a fuzzy number construction process, a neural fuzzy filtering process, and an image knowledge base. From the experimental results, we observe that the MSE values of *NFF* are better than the other approaches. Subjective evaluation of *NFF* also shows high-quality restoration of filtered images. In the future, we will extend *NFF* to process color images.

Acknowledgement. This work was partially supported by the National Science Council, R.O.C., under Grant NSC 92-2213-E-309-005.

References

1. Arakawa, K.: Median Filter Based on Fuzzy Rules and Its Application to Image Restoration. Fuzzy Sets and Systems. Vol. 77 (1996) 3-13
2. Abreu, E., Mitra, S.K.: A Signal-Dependent Rank Ordered Mean (SD-ROM) Filter. Speech and Signal Processing. ICASSP-95. Detroit. Michigan (1995)
3. Lee, C.S., Kuo, Y.H.: The Important Properties and Applications of AWFM Filter. International Journal of Intelligent Systems. Vol. 14 (1999) 253-274
4. Russo, F.: Hybrid Neuro-fuzzy Filter for Impulse Noise Removal. Pattern Recognition. Vol. 32 (1999) 307-314
5. Wang, J.H., Liu, W.J., Lin, L.D.: Histogram-Based Fuzzy Filter for Image Restoration. IEEE Trans. on System, Man and Cybernetics, Vol. 32 (2002) 230-238
6. Pok, G., Liu, J.C., Nair, A.S.: Selective Removal of Impulse Noise Based on Homogeneity Level Information. IEEE Trans. on Image Processing, Vol. 12. No. 1 (2003) 85-92

Neural Network Aided Adaptive Kalman Filter for Multi-sensors Integrated Navigation

Lin Chai[1], Jianping Yuan[1], Qun Fang[1], Zhiyu Kang[2], and Liangwei Huang[1]

[1] College of Astronautics, Northwestern Polytechnical University, Xi'an 710072, P. R. China
l.chai@126.com, {jyuan,qfang,lwhuang}@nwpu.edu.cn

[2] No. 805 Institute, Shanghai Academy of Spaceflight Technology, Shanghai 201108, P. R. China
zykang@126.com

Abstract. The normal Kalman filter (KF) is deficient in adaptive capability, at the same time, the estimation accuracy of the neural network (NN) filter is not very well and the performance depends on the artificial experience excessively. It is proposed to incorporate a back-propagation (BP) neural network into the adaptive federal KF configuration for the SINS/GPS/TAN (Terrain Auxiliary Navigation)/SAR (Synthetic Aperture Radar) integrated navigation system. The proposed scheme combines the estimation capability of adaptive KF and the learning capability of BP NN thus resulting in improved adaptive and estimation performance. This paper addresses operation principle, algorithm and key techniques. The simulation results show that the performance of the BP NN aided filter is better than the stand-alone adaptive Kalman filter's.

1 Introduction

More and more different approaches have been introduced in response to the ever-increasing interest of the multi-sensors information fusion field. Among these approaches are the classical and Bayesian inference theories, evidential reasoning, cluster analysis, fuzzy logics, and recently neural networks, wavelet transform [1]. In general these approaches can be partitioned into two schools of thoughts, viz., algorithmic and nonalgorithmic. The representative algorithmic method is Kalman filtering and the representative nonalgorithmic method is neural network. Both methods have their shortcomings. The Kalman filter requires: (1) the system dynamics is completely known and modeled as a Gauss-Markov stochastic process, and (2) the statistics of the system error and the observation error are known to be normally distributed. But many actual systems can hardly meet these requires absolutely. The shortcomings with the neural network approach are: (1) the implementation depends on the artificial experience excessively, (2) the pure neural network system is not easy to be carried out in hardware, and (3) the estimation accuracy is not ideal.

Since both algorithmic and nonalgorithmic approaches are facing some unresolvable implementation problem, another approach must be sought. A hybrid

F. Yin, J. Wang, and C. Guo (Eds.): ISNN 2004, LNCS 3174, pp. 381–386, 2004.

neural network approach is proposed by Chin L. for tracking a non-manoeuvring target [2]. Vaidehi V. extends this approach for tracking multi-targets that are highly manoeuvring [3]. In 1998, a Kalman filter aided neural network scheme for multi-targets tracking is proposed by Wong Y. C. [4]. In 2003, a filtering method of colored noise based on the Kalman filter structure using neural network is proposed for image fusion [5]. In this paper, the BP neural network aided adaptive Kalman filter scheme for SINS/GPS/TAN/SAR integrated navigation system is proposed. The adaptive capability and navigation accuracy is improved by incorporating a BP NN into the adaptive KF in the proposed scheme.

2 The Operation Principle of Integrated Navigation System Based on NN Aided Filtering

2.1 The Concept of the NN Aided Adaptive Kalman Filter

Research works in the past adopted the approach of incorporating a neural network into the classical Kalman filter (see [2,3,4,5]). In order to enhance the adaptive capability, the BP NN aided adaptive KF approach is proposed in this paper.

The basic concept of the proposed method is shown in Fig. 1, in which a BP NN is employed to aid the adaptive KF to reduce the estimation error due to, among other imperfections, highly manoeuvring, model varying effect. The scheme combines the estimation capability of the adaptive KF and the learning capability of the BP NN. The implementation of NN aided adaptive KF has three stages, namely, (1) architecture, (2) training and testing, and (3) recall [3]. The BP NN is trained off line to learn the errors in navigation. The complex modelling requirement in the KF is reduced by the NN because errors in navigation are eliminated during the recall phase.

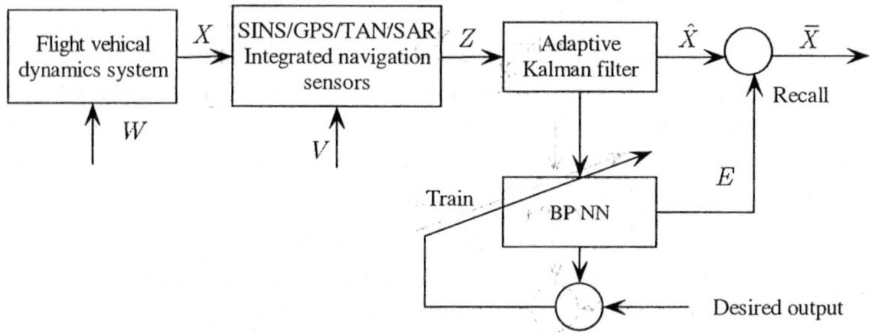

Fig. 1. The principle block diagram of the BP NN aided adaptive Kalman filter

2.2 The Architecture of SINS/GPS/TAN/SAR Integrated Navigation System

In this paper, the proposed SINS/GPS/TAN/SAR fault-tolerant integrated navigation system is based on federated filter structure, as illustrated in Fig. 2. Each local filter is dedicated to a separate sensor subsystem. Before inputting to the master filter, the data of local filters have been detected (and isolated) by BP NN detectors. The global optimal estimation $\hat{X}_k$ from master filter is revised by the data fusion BP NN. Then, the navigation parameter relative real value $\bar{X}_k$ is obtained.

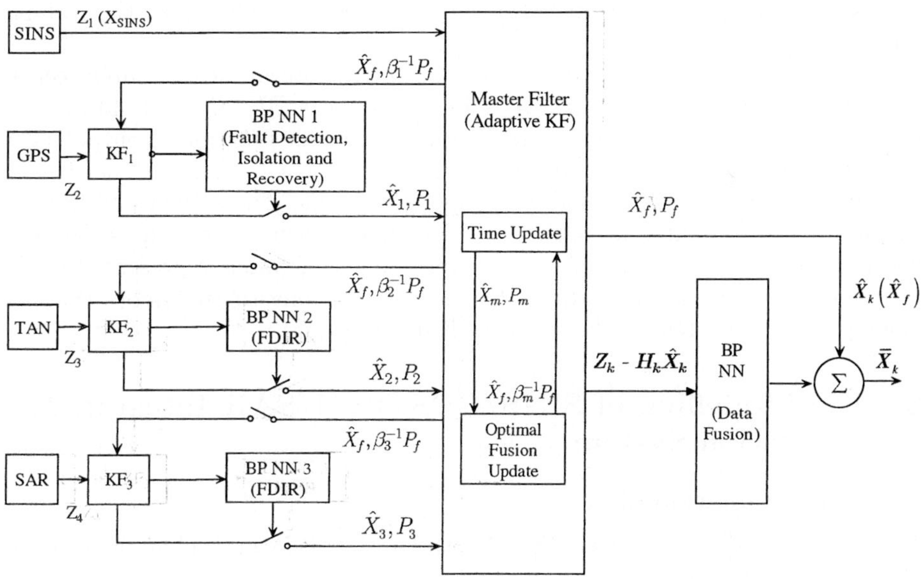

Fig. 2. SINS/GPS/TAN/SAR fault-tolerant integrated navigation system structure

2.3 The Dynamic Information-Sharing Algorithm

The key technique of federated filter is to determine the information-sharing factor β_i. It is always expected that the high navigation precision subsystem has bigger β_i value than that of the low navigation precision subsystem. In design of classical federated filter, β_i is rigid and based on designer's experience. So it is not adapted to highly manoeuvring environment. In this paper, a real-time on-line revised β_i method is proposed, as shown in (1):

$$\beta_i = \frac{trace_i}{trace_1 + trace_2 + \cdots + trace_n + trace_m} \tag{1}$$

Where, $trace_i$ is the trace of covariance matrix P_i; m is master filter; n is the number of local filters.

2.4 The BP NN for Data Fusion

The objective of the proposed scheme is to reduce the error in navigation. So the parameters which have direct influence over the error are chosen as inputs. In this paper, the input signal is the difference between the actual measurement vector and estimated measurement vector of adaptive Kalman filter: $Z_k - H_k\hat{X}_k$. Since the exact state X_k is known (as far as simulation is concerned), the supervised learning algorithm can use the error (difference between known state and estimated state) to train the network, i.e., $Error = X_k - \hat{X}_k$. The measure of inaccuracy in navigation denoted by *Error* in the above equation have to be corrected using the neural network. The neural network output vectors ideally describe the exact difference between the state vector predicted by the Kalman filter and the actual state.

Simulation have been conducted to obtain the neural network architecture that gives the best convergence. It is found that the BP NN with 40 neurons input layer, 1 hidden layer of tangent-sigmoid transfer function, 1 output layer of purelin transfer function, 30 neurons in the hidden layer and 2 neurons in the output layer is suitable for the proposed scheme. The synaptic weights of the network are updated according to the standard Delta rule under supervised learning mode. At the time of recall, when the actual value is presented to the neural network, it has to output the ideal corrections to the errors that are inadequately predicted by the adaptive Kalman filter.

3 The Modelling of SINS/GPS/TAN/SAR Integrated Navigation System

3.1 State Equation

The navigation system works in East-North-Up (ENU) local-level mechanization. The state equation is

$$\Delta\dot{X}_g(t) = F_g(t)X_g(t) + G_g(t)W_g(t) \tag{2}$$

Where, $X_g = [\varphi_e\, \varphi_n\, \varphi_u\, \delta V_e\, \delta V_n\, \delta V_u\, \delta\varphi\, \delta\lambda\, \delta h\, \varepsilon_{bx}\, \varepsilon_{by}\, \varepsilon_{bz}\, \varepsilon_{rx}\, \varepsilon_{ry}\, \varepsilon_{rz}\, \nabla_x\, \nabla_y\, \nabla_z]^{\top}$ is the state vector; $\varphi_e, \varphi_n, \varphi_u$ are the attitude errors and azimuth error of SINS; $\delta V_e, \delta V_n, \delta V_u$ are the velocity errors; $\delta\varphi$, $\delta\lambda$, δh are the position errors; $\varepsilon_{bx}, \varepsilon_{by}, \varepsilon_{bz}$ are the SINS gyro constant drift errors; $\varepsilon_{rx}, \varepsilon_{ry}, \varepsilon_{rz}$ are the SINS gyro first-order Markov process drift errors; $\nabla_x, \nabla_y, \nabla_z$ are the accelerometer errors; $W_g(t)$ is the modelling error (system disturbance) vector; $F_g(t)$ is the state transition matrix; $G_g(t)$ is the error transition matrix.

3.2 Observation Equation

Considering the estimated accuracy and real-time requirement, the position and velocity alternation integration model is adopted in GPS/SINS integrated subsystem. The velocity information integrate by per second, and position information integrate every 20 seconds once. The observation equation of GPS/SINS is

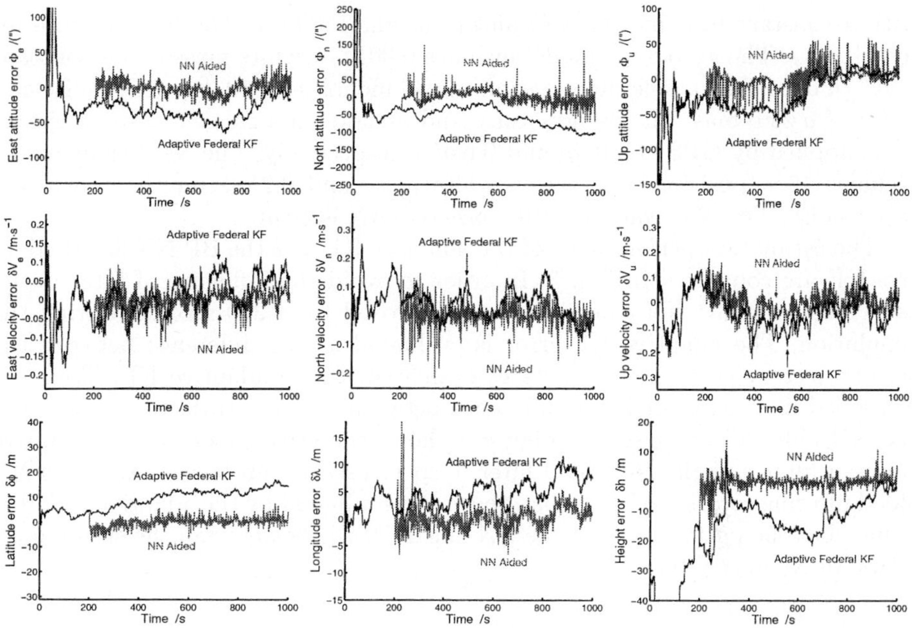

Fig. 3. SINS/GPS/TAN/SAR system estimate errors of two schemes

$$Z_1 = Z_P = \begin{bmatrix} \varphi_I - \varphi_G \\ \lambda_I - \lambda_G \\ h_I - h_G \end{bmatrix} + V_{1P} \quad \text{or} \quad Z_1 = Z_V = \begin{bmatrix} V_{IE} - V_{GE} \\ V_{IN} - V_{GN} \\ V_{IU} - V_{GU} \end{bmatrix} + V_{1V} \tag{3}$$

Where, φ, λ, h are the position outputs; V_E, V_N, V_U are the velocity outputs; V_{1P}, V_{1V} are the observation noise. The observation equation of TAN is

$$Z_2 = h_i - h_t - h_r = H_2 X_g + V_2 \tag{4}$$

Where, h_i is absolute height; h_t is terrain height; h_r is relative height; V_2 is the observation noise. The observation equation of SAR is

$$Z_3 = [\delta\varphi, \delta\lambda, \varphi_u]^\top + V_3 \tag{5}$$

Where, V_3 is the observation noise. Combining (3), (4), and (5), the observation equation of SINS/GPS/TAN/SAR integrated system is shown as follow:

$$Z = [Z_1, Z_2, Z_3]^\top \tag{6}$$

4 Simulation

The typical flight track in simulation is horizontal and fight to the east. The starting position is latitude 34° N and longitude 180° E. Flight time is 1000 s, with a constant velocity $200\,m/s$ and in height $15000\,m$. The filter update interval is $1\,s$. The gyro error model includes $0.001\,^{\circ}/h$ white noise and a constant $0.01\,^{\circ}/h$ drift. The accelerometer error model includes $5 \times 10^{-6}\,g$ white noise and $5 \times 10^{-4}\,g$ zero bias. The pseudo range and pseudo delta-range measurement errors supplied by GPS are $10\,m$ and $0.6\,m/s$ respectively. The SAR error model includes $15\,m$ position observation white noise and 0.2° azimuth observation white noise. The observation white noise of TAN is $30\,m$.

The estimation performance of the adaptive KF and the BP NN aided adaptive KF are compared in Fig. 3. In order to avoid the influence of filter initial oscillation, the BP NN aided filtering is invoked from 200 seconds during the simulation. The adaptive KF error is evaluated as the difference between the true trajectory states and the states evaluated by the adaptive KF. Similarly, NN aided filtering error is the difference between the true trajectory states and the NN aided filtering states. Figure 3 shows the error plots for the adaptive Kalman filter and the NN aided filter, respectively. It can be seen that the NN aided filtering errors are lower than the errors evaluated by the adaptive KF. It is found that the performance of the NN aided filter is better than the stand-alone adaptive Kalman filter.

5 Conclusion

A BP neural network aided adaptive federal Kalman filter scheme for multi-sensors integrated navigation system is proposed. The inaccuracies in the estimation are corrected by the BP NN. The complex modelling requirement in the classical Kalman filter for navigation system is also eliminated by the BP NN. The estimation accuracy improvement due to the proposed scheme is presented. In view of these improvements, it can be concluded that the proposed method is more superior than the adaptive (and conventional) Kalman filtering approach.

References

1. Xu S. H., Wu J. H., Hu Z. Q.: Research on Wavelet Transform Method for the Fault-detection of the Integrated Navigation System. Journal of Astronautics **24** (2003) 111–114
2. Chin L.: Application of Neural Networks in Target Tracking Data Fusion. IEEE Trans. on Aerospace and Electronic Systems **30** (1994) 281–287
3. Vaidehi V., Chitra N., Krishnan C. N.: Neural Network Aided Kalman Filtering for Multitarget Tracking Application. IEEE Proc. of the Radar Conference (1999) 160–165
4. Wong Y. C., Sundareshan M. K.: Data Fusion and Tracking of Complex Target Maneuvers with a Simplex-trained Neural Network-based Architecture. IEEE Proc. of the World Congress on Computational Intelligence **2** (1998) 1024–1029
5. Xiong S. S., Zhou Z. Y.: Neural Filtering of Colored Noise Based on Kalman Filter Structure. IEEE Trans. on Instrumentation and Measurement **52** (2003) 742–747

Solely Excitatory Oscillator Network for Color Image Segmentation

Chung Lam Li and Shu Tak Lee

Department of Computing, The Hong Kong Polytechnic University, Hung Hom, Kowloon, Hong Kong
{cslamli, csstlee}@comp.polyu.edu.hk

Abstract. A solely excitatory oscillator network (SEON) is proposed for color image segmentation. SEON utilizes its parallel nature to reliably segment images in parallel. The segmentation speed does not decrease in a very large network. Using NBS distance, SEON effectively segments color images in term of human perceptual similarity. Our model obtains an average segmentation rate of over 98.5%. It detects vague boundaries very efficiently. Experiments show that it segments faster and more accurately than other contemporary segmentation methods. The improvement in speed is more significant for large images.

1 Introduction

One major problem of traditional color image segmentation is their sequential nature [1]. Being inherently parallel, neural network can be an efficient method for parallel segmentation. However, very little researches have been conducted in this area [2]. One major reason is the lack of an encoding scheme to represent multiple segments.

One way to represent multiple segments in neural network is using oscillatory correlation theory [3]. This theory is supported by many physiological experiments of visual cortex of cats [4]. Wang and Terman proposed an oscillatory network, LEGION (locally excitatory, globally inhibitory oscillator networks), to implement theory [5]. LEGION has been applied to parallel image segmentation.

2 Solely Excitatory Oscillator Network

This paper proposes a solely excitatory oscillator network (SEON) for color segmentation. SEON consists of a 2D matrix of locally connected Peskin oscillators (with its eight nearest neighbours) [6] and a globally connected excitatory separator (ES).

2.1 A Single Oscillator

The activity x_i of a Peskin oscillator i[6] obeys

$$\dot{x}_i = S_i - \gamma x_i, \quad i = 1,\ldots,n \tag{1}$$

F. Yin, J. Wang, and C. Guo (Eds.): ISNN 2004, LNCS 3174, pp. 387–392, 2004.

and assume the values

$$0 \le x_i \le 1 \tag{2}$$

where S_i is the stimulus of the oscillator and γ is constant.

Initially, individual values x_i obeying (2) is prescribed by (1). When the time passes, assuming that S_i>0, x_i will increase and reach the value 1. Then, the oscillator fires and x_i jumps back to zero. At an infinitesimally later time t^+, the values x_j of all coupled oscillators are increased by ε or the other oscillators fire when reaching x_j=1.

$$x_i(t) = 1 \Rightarrow x_j(t^+) = \min(1, x_j(t) + \varepsilon), \quad \forall j \in N(i) \tag{3}$$

where *N(i)* is the nearest eight neigbhors of oscillator *i*. A pulse is transmitted instantaneously with no conduction delays. If S_o=0, x_i decays exponentially toward zero.

2.2 Synchronization and Desynchronization Mechanisms

The following algorithm is used to synchronize oscillators within the same segment:

```
1.The values xi are randomly chosen from [0,1].

2.All oscillators are set to an unfired state.

3.Repeat until no oscillators reach threshold.

          If any oscillator reaches threshold θi =1

           The oscillator fires by setting the oscilla-
           tor to a fired state. Its unfired neighbour-
           ing oscillators are fired (using (3)).

4.Go to step 2.
```

Fig. 1(a) show shows the temporal evolution of 25 oscillators with $\varepsilon \ge 1$. Given that $\varepsilon \ge 1$, the firing of oscillator *i* will always cause the neighbouring oscillators to fire. The process repeats, leading to a chain reaction of firings to all coupled oscillators.

The following algorithm is used to desynchronize oscillators in different segments:

```
1.The ES is set to the ready state.

2.Repeat this step until the end of simulation

          If ES is in a busy state and Tbusy=Tseperation

             ES is set to the ready state.

          If any oscillator reaches the ES threshold θES
          where θES < θi  and ES is in the ready state
```

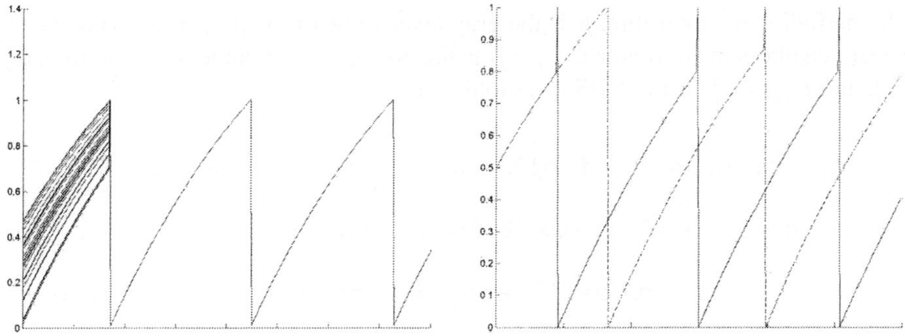

(**a**) Synchronization of 25 oscillators (**b**) Desynchronization of two segments

Fig. 1. Synchronization and desynchronization of oscillators. (a) Synchronization of 25 oscillators (b) Desynchronization of two segments (two groups of oscillators)

```
        ES sends an excitation to the oscillator.
        The state is set to busy and T_busy are set
        to zero. The value of the oscillator is in-
        creased to 1. This causes the oscillator to
        fire.

    If ES is in a busy state

        T_busy = T_busy + 1
```

Fig. 1(b) shows the temporal evolution of two accidentally synchronized segments. When an oscillator in a segment reaches the ES threshold (θ_{ES} = 0.8), ES will send a pulse to that oscillator, leading the oscillator to fire, and instantaneously triggering the chain reaction of firing. This pre-empts the synchronization of oscillators within the same segment, which effectively desynchronizes them.

2.3 Color Image Segmentation

SEON uses National Bureau of Standards (NBS) unit color differenc to measure the color difference in HVC color space [7], which describes color closes to human perception. Table 1 shows the relationship between human perception and NBS distance [7].

Table 1. Human perception and NBS value Table [7]

NBS Value	Human Perception
0 – 1.5	Almost the same
1.5 – 3.0	Slightly different
3.0 – 6.0	Remarkably different
6.0 – 12.0	Very different
> 12.0	Different color

In the following algorithm, p_i is the gray level value of the ith pixel. $T(i)$ is the four nearest neighbors of oscillator i. $d_{NBS}(i,j)$ is the NBS distance value between ith and jth pixel, and θ_{NBS} (=1.5) is the NBS threshold value.

```
1.The stimulus Sj of all oscillators are set to 1.

2.For gray level value k from 0 to 255

        For any oscillator i with gray level value k

          If the oscillator i has not been marked

          For j ∈ T(i)//Each j runs concurrently

            If dNBS(i,j) > θNBS

              Mark the neighbors i as boundary region by
              setting Si to 0.

            Else if no neighbors j with pj ≤ k belongs
            to any region

              Create a new regional minimum by marking
              the oscillator to belong to a new region
              Mr.

            Else if the neighbors j with pj ≤ k belongs
            to only one region

              Mark the oscillator i to belong to that
              region. Create connection between the os-
              cillator i and neighbors j.

            Else if the neighbors j with pj ≤ k belongs
            to more than one region

              If one of the neighbors j is marked as
              boundary region or there exist some neigh-
              bors j with pj < k

                Mark the oscillator i as boundary region
                by setting Si to 0.

              If all neigbhors j satisfy pj = k
```

```
Mark the oscillator i to belong to any
one of the regions and merge the re-
gions. Create connections between the
oscillator i and neighbors j.
```

3 Experiments

120 images are tested to compare the segmentation accuracy between SEON, color RGB vector gradient [8] and gray-level sequential watershed [8]. SEON, RGB gradient and watershed method obtained an average correct segmentation rate of 98.6%, 91.6% and 96.4%, respectively. Thus, SEON achieves the highest rate.

Four methods including SEON, sequential watershed segmentation [8], parallel watershed transformation [9] and sequential RGB vector gradient [8] have been tested to compare the segmentation speed. Table 2 shows the average execution times of the four methods. On average, SEON is the fastest, especially for large images.

Table 2. Average segmentation times. 120 images are tested with 20 images for each size X.

Image Size X	Average segmentation times (in seconds)			
	SEON	Par. Water.	RGB grad.	Seq. Water.
128	0.183	0.19	0.092	0.167
256	0.352	0.547	0.561	0.793
512	0.697	1.618	2.038	2.673
1024	1.389	4.729	9.278	11.563
2048	2.746	14.694	32.542	38.792
4096	5.397	43.619	108.375	119.376

4 Conclusion

In this paper, a solely excitatory oscillator network (SEON) is proposed for color image segmentation. The model effectively segments color image in parallel. Experiments show that the model segments faster and more accurately than other segmentation methods. The improvement of segmentation speed increases with the image size.

References

1. Cheng, H.D., Jiang, X.H., Sun, Y., Wang, J.L.: Color image segmentation: Advances and prospects. Pattern Recognit. 34 (2001) 2259-2281
2. Pal, N., Pal, S.: A review on image segmentation techniques. Pattern Recognit. 26 (1993) 1277-1294

3. von der Malsburg, C.: The correlation theory of brain function. Internal report, 81-2. Max-Planck-Institute for Biophysical Chemistry, Gottingen, FRG (1981)
4. Gray, C.M., Konig, P., Engel, A. K., Singer, W.: Oscillatory responses in cat visual cortex exhibit intercolumnar synchronization with reflects global stimulus properties. Nature. 338 (1989) 334-337
5. Wang, D.L., Terman, D.: Locally excitatory globally inhibitory oscillator networks. IEEE Trans. on Neural Network. 6 (1995) 283-286
6. Peskin, C.S.: Mathematical aspects of heart physiology. Courant Institute of Mathematical Sciences, New York University, New York (1975) 268-278
7. Judd, D.B., Wyszeki, G.: Color in Business, Science, and Industry. Wiley, New York (1963)
8. Gonzalez, R.C., Woods, R.E.: Digital Image Processing. 2nd edn. Prentice Hall, New Jersey (2002)
9. Alina, N.M., Moncef, G.: Parallel image component labeling with watershed transformation. IEEE Trans. on Pattern Anal. Machine Intell. 19 (1997) 441-450

Image De-noising Using Cross-Validation Method with RBF Network Representation

Ping Guo and Hongzhai Li

Department of Computer Science
Beijing Normal University, Beijing, 100875, P.R.China
pguo@ieee.org; lihongzhai@163.com

Abstract. This paper presents a new image de-noising method, which based on the image representation model of radial basis function neural network. In this model, the number and distribution of the centers (which are set to the pixels of the observed image) are fixed, and the model parameters of the image representation are chosen by cross-validation method. Experimental results show that the model can represent the image well, and proposed method can reduce the noises in images without need any noise knowledge *in priori*.

1 Introduction

Image de-noising is an important issue in digital image processing [1]. Most image de-noising methods need the prior knowledge of the noise. Generally speaking, natural noises can be regarded as random signals, which make smooth image fluctuated sharply. The frequency filter methods, as most used image de-noising methods, reduce noises by restrain the high-frequency part of the degraded image. As the result, the image become blurred for some of its details are reduced at the same time [2].

Unlike the frequency filter, the neural network can deal the image in nonlinear and local ways. Egmont-Petersen [3] reviewed more than 200 applications of neural networks in image processing and discussed the present and possible future role of neural networks, especially feed-forward neural networks, Kohonen's feature maps and Hop1eld neural networks. Guo *et al* propose a novel technique for blind image restoration and resolution enhancement based on radial basis function neural network (RBFNN) which designed to represent the observed image [4]. The RBFNN has a universal approximation capability and has been successfully applied to many signal and image processing problems.

In this paper, taking given degraded image as training data set, we propose to build RBFNNs model to represent the underlying image with cross-validation method. The parameters in the network are determined based on training error and validation error.

F. Yin, J. Wang, and C. Guo (Eds.): ISNN 2004, LNCS 3174, pp. 393–398, 2004.

2 RBFNN

RBFNN is a most commonly-used feed-forward network. It usually has one hidden layer, and the basis functions is radial symmetry. The RBFNN can be presented as follow [5]:

$$g(\mathbf{x}) = \sum_{i=1}^{N} w_i \varphi_i(||\mathbf{x} - \mu_i||) + w_0. \tag{1}$$

Where $\mathbf{x}$ is input vector, N is the number of basis functions in RBFNN. μ_i is the centers of each basis function, $||\cdot||$ denotes the Euclidean norm. $\mathbf{w} = \{w_i | i = 1, 2, \cdots, N\}$ is the weight vector between the hidden layer and the output layer. w_0 is the bias neuron, and $g(\mathbf{x})$ stands for output of the network. $\varphi(\mathbf{x})$ stands for the basis function, here the Gaussian function is used as basis function associated with each neuron:

$$\varphi_i(||\mathbf{x} - \mu_i||) = \exp[-\frac{1}{2\sigma^2}||\mathbf{x} - \mu_i||^2]. \tag{2}$$

The parameter σ represents the standard deviation of a Gaussian function. When this Gaussian function is chosen as the RBFNN basis, the degree of generalization becomes rotationally symmetric.

Generally speaking, a RBFNN is trained in two steps: a). Determining the RBFNN center and the standard deviation, and choosing the hidden layer neuron number according to the practical problem to achieve some special effects. And, b). Adopting some optimization techniques such as gradient descent methods to find the RBFNN weight parameters which connect hidden layer and output layer.

In this paper, the sum-of-squares error function is adopted, and the weight vector $\mathbf{w}$ is updated according to the delta learning rule.

Park *et al* proved that RBFNN with one hidden layer is capable of universal function approximation [6,7]. Under certain mild conditions on radial basis functions, RBFNN is capable of approximating arbitrarily well any functions. RBFNN can also approach well to an image at any precision, if we consider an image as a two-dimensional function.

The bias, the number of hidden neurons and the basis functions are keys to efficiency of the approximation processing. The bias is absent for the situation of there are enough hidden neurons. The centers of the basis functions are commonly depend on the distribution of the input. We have analyzed other basis functions also, but it seems that there is no essential distinction compared with Gaussian function.

3 Image Representation

3.1 RBFNN Architecture

Assume the degradation of a digital image is modelled as: $\mathbf{G} = \mathbf{F} + \mathbf{V}$, where $\mathbf{G}$ is the degraded image, $\mathbf{F}$ is the underlying image, and $\mathbf{V}$ is linear space invariant and zero mean additive noise, the size of $\mathbf{G}$, $\mathbf{F}$ and $\mathbf{V}$ are assumed as $L \times L$ scale. On considering the approximation capability of the RBFNN, in this paper, we use the RBFNN to approximate a degraded image $\mathbf{G}$, and try to get the underlying image $\mathbf{F}$.

For image de-noising problem, a elaborating designed RBFNN is taken to represent the degraded image $\mathbf{G}$. Considering the RBFNN in Eq. 1, the basis function center μ_j is fixed and set to the coordinates of the pixels of the images, the output is set to the degraded image pixel gray scale value. Then $||\mathbf{x}_i - \mu_j||$ stands for the Euclidean distance of the coordinates of two pixels $\mathbf{x}_i$ and μ_j.

3.2 The Cross-Validation Method

When a degraded image is given, we can use it to train a RBFNN with the data set $D = \{\mathbf{x}_i, g_i | i = mL + n + 1, \text{with } m, n = 0, 1, \cdots, L-1\}$, (m, n) is the image pixel $\mathbf{x}_i$ integer coordinate. The Eq. 2 can be expressed as $\varphi_{ij} = \exp(-\frac{1}{2\sigma^2}||\mathbf{x}_i - \mu_j||^2)$. Let $\mathbf{\Phi}$ be a matrix with elements $\{\varphi_{ij} | i, j = 1, 2, \cdots, N\}$, according to the properties of the Gaussian basis function, $\mathbf{\Phi}$ can be approximated as a block Toeplitz matrix if σ is small. $\mathbf{\Phi}$ is called as "Basis Function Matrix" (BFM) in this work. The Eq. 1 can be written as a matrix form with training data, $\mathbf{g} = \mathbf{\Phi}\mathbf{w}$, where $\mathbf{g} = [g_1, g_2, \ldots, g_N]^T$, is a $N \times 1$ vector standing for the degraded image grey value in lexicographic orders, $\mathbf{w} = [w_1, w_2, \ldots, w_N]^T$, is a $N \times 1$ weight vector.

With the RBFNN's architecture discussed in Sect. 3.1, the cross-validation method is used to select the parameters of the model for underlying image.

Firstly, the data set D is partitioned into several subsets, four subsets are suggested for the sake of image can be easily divided into even number subsets. In this work,

$$D = \cup(D^1, D^2, D^3, D^4) \tag{3}$$

means D was divided into four subsets: $D^k, k = 1, 2, 3, 4$. Where training data in subset D^1 are those $\{\mathbf{x}_i, g_i | i = (2m)L + 2n + 1\}$. And $D^2 = \{\mathbf{x}_i, g_i | i = (2m)L + (2n+1) + 1, \}$, $D^3 = \{\mathbf{x}_i, g_i | i = (2m+1)L + 2n + 1, \}$, and $D^4 = \{\mathbf{x}_i, g_i | i = (2m+1)L + (2n+1) + 1\}$. In this case $m, n = 0, 1, \cdots, (L-1)/2$.

Secondly, for each k, D^k is picked up as training set in turn, and the remains subsets $\cup_{j \neq k} D^j$ as validation sets. Then a RBFNN output is established as $\mathbf{g}^k = \mathbf{\Phi}^{kk}\mathbf{w}^k$, where $\mathbf{\Phi}^{kk}$ is computed from training set D^k, the σ in $\mathbf{\Phi}^{kk}$ is set to a certain value, $\mathbf{w}^k$ now is the weight vector to this RBFNN.

Here we defined Sub Basis Function Matrix (SBFM), $\mathbf{\Phi}^{kl} = \mathbf{\Phi}(x_k, \mu_l)$, $x_k \in D^k$, $\mu_l \in D^l, k, l = 1, 2, 3, 4$. It is easy to find that there are 16 SBFMs altogether with our data partition method, and it can be proved that each SBFM is approximately a block Toeplitz matrix.

Now for each k, the training error can be computed as:

$$E_t^k = ||\mathbf{g}^k - \mathbf{\Phi}^{kk}\mathbf{w}^k||^2, \qquad k = 1, 2, 3, 4. \tag{4}$$

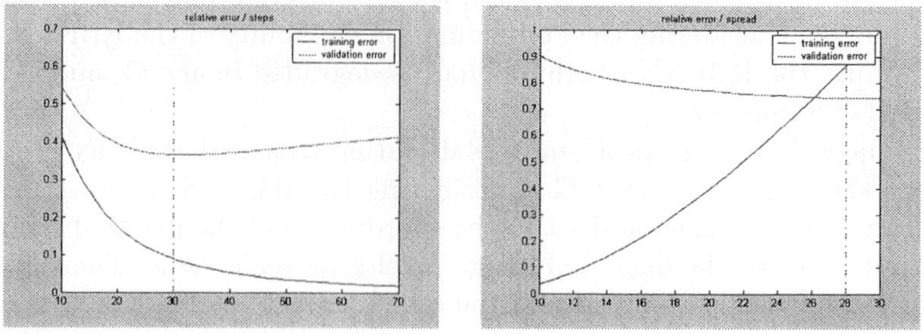

(a) (b)

Fig. 1. (*a*) The relationship of errors and steps. (*b*) The relationship of errors and the σ of SBFM

$\mathbf{w}^k$ is updated in the training procedure. At the same time, the validation error is computed as:

$$E_v^k = \sum_{l \neq k} ||\mathbf{g}^l - \mathbf{\Phi}^{kl}\mathbf{w}^k||^2, \qquad k = 1, 2, 3, 4. \tag{5}$$

At last, the suitable network parameters is selected based on the minimal validation error.

4 Experiments

In experiments, both the training errors and the validation errors are computed. Experimental results are shown in Fig. 1.

Fig. 1(a) show the relation of training error and validation error in one RBFNN training procedure. The training error goes downward near to zero when the training steps is big enough and σ is small. That means the network output gradually approximate the degraded image and fit to noise. While the validation error goes downward first and goes upward after a certain steps.

Considering the situation of the image without noise, then the image is regarded as samples of relative smooth function. If training error is small, the RBFNN can approximate the given image quite well. In this time the validation error is small also. However, when noise is added on the image, the image becomes turbulent. From Eq. 5, the validation error can be regarded as a measure of the difference between the (smooth) network output and the degraded image, that is, a measure of the difference between the underlying image and the degraded image.

Experimental results show that the measure that take one subset as the training set and all remains as validation set can emphases the difference between the underlying image and the degraded image. The position where the validation error turn upward means the RBFNN is approximating to the underlying image mostly, which can be detect in the curves. In this work, this position is found

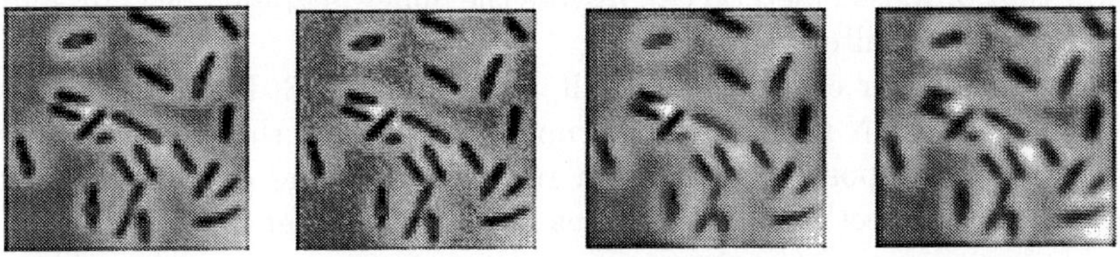

(a) (b) (c) (d)

Fig. 2. Example of image de-noising. (*a*) Original image, resolution 64 × 64, (*b*) image degraded by Gaussian noise, (*c*) restored image by Wiener filter. (*d*) Restored image by RBFNN method

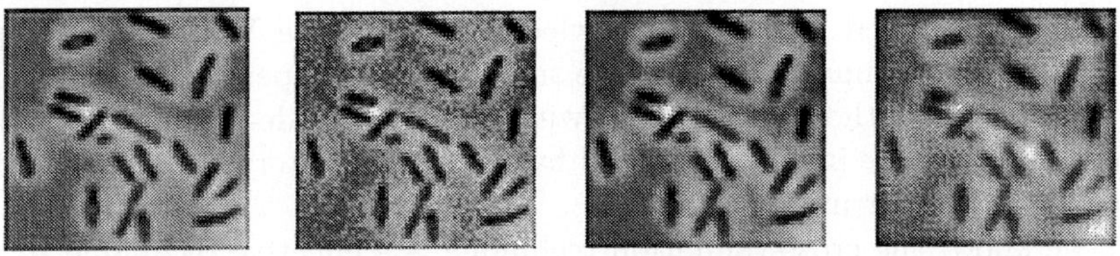

(a) (b) (c) (d)

Fig. 3. Example of image de-noising. (*a*)Original image, (*b*) image degraded by Poisson noise, (*c*) restored image by Wiener filter. (*d*) Restored image by the RBFNN method

adaptively, and the RBFNN's output is simultaneously recorded as restored image. A point should be mentioned here is the selection of the deviation σ of the SBFM. The deviation σ represents the range in which a neuron affecting other pixels from its center. If σ is big, the outputs of the RBFNN become more smooth.

Fig. 1(b) shows the relationship between the errors and the deviation. The errors are computed for given σ from 0.5 to 3 and the RBFNN is trained for 100 times. It is clear that the training error increases stably from near zero, that is, when σ is too small, the RBFNN approximates the degraded image exactly, and it is hardly to get the smooth output as underlying image. When σ is too large, the RBFNN is hardly to approximate the degraded image, the restored image under this situation is also valueless.

In experiment, an exhaustive search method is used to find suitable σ. We assume the range of σ first, then a series of σ are tested, finally we select the σ with minimal validation error.

In order to demonstrate the performance of proposed RBFNN image de-noising method, we show some simulation results now. Two kinds of noises are simulated, namely Gaussian noise and Poisson noise.

One experiment result is shown in Fig. 2. Fig. 2(d) show the restored image by RBFNN method proposed in this paper (30 learning steps). From this experiment we know that proposed method can restore the image degraded by Gaussian noise as same as Wiener filter.

In Fig. 3 another experiment result is shown. Fig. 3(d) shows the restored image by the RBFNN method (30 learning steps). From this experiment we can know also that proposed method can restore the image degraded by Poisson noise, but visual effect is not as good as using the Wiener filter.

5 Discussions and Conclusions

In this paper, we present a new image de-noising method. In this method, the image was represented by a RBFNN model associated with cross-validation method. Unlike other traditional methods such as Wiener filter require to know the noise variance in advance, the main advantage of the proposed method is that it does not need any *prior* knowledge of the noise **V**, and only suppose that the underlying image **F** is relative smooth. From experiments it has proved that if proposed method is combined with other image de-noising methods, for example, the degrade image is restored by the median filter first, then the finally result will be further improved.

As we know, the cross-validation technique is a effective technique to select model of the image representation, however it is computationally expensive. In the further work, we will find some solutions. For example, one possible solution is on considering the properties of block Toeplitz matrix. The block Toeplitz has special structure similar block circular matrix, which can be restored from their first row or first column with simple method, and some algorithms will be developed for the fast computation of SBFM.

Acknowledgement. The research work described in this paper was fully supported by a grant from the National Natural Science Foundation of China (Project No. 60275002).

References

1. Castleman, K. R.: Digital Image Processing. Prentice-Hall, Englewood Cliffs NJ (1996)
2. Sugiyama, M., Ogawa, H.: A Unified Method for Optimizing Linear Image Restoration Filters. Signal Processing, **82** (2002) 1773-1787
3. Egmont-Petersen, M., Ridder, D., Handels, H.: Image Processing with Neural Networks - A Review. Pattern Recognition, **35** (2003) 2279-2301
4. Guo, P., Xing, L.: Blind Image Restoration Based on RBF Neural Networks. Proceedings of SPIE, Vol. 5298. SPIE, Bellingham WA (2004) 259–266
5. Bishop, C. M.: Neural Networks for Pattern Recognition. Oxford University Press, London (1995)
6. Park, J., Sandberg, I.W.: Universal Approximation Using Radial-Basis-Function Networks. Neural Computation, **3** (1991) 246-257
7. Park, J., Sandberg, I.W.: Approximation and Radial Basis-Function Networks. Neural Computation, **5** (1993) 305-316

Ultrasonic C-scan Image Restoration Using Radial Basis Function Network

Zongjie Cao, Huaidong Chen, Jin Xue, and Yuwen Wang

Welding research Institute, Xi'an Jiaotong University, Xi'an 710049, China
{czj, infrinst}@mail.xjtu.edu.cn

Abstract. A method for restoration of ultrasonic C-scan images is presented by using a radial basis function network. The method attempts to reproduce the mapping between the degraded C-scan image and the high quality one by training a RBF network. The inputs for training are the sub-images divided from C-scan image of flat-bottom hole of size 3mm and the output is the corresponding center in high quality image. After the network was trained, the other C-scan images were used to verify the network. The results show that the network produces good restored results, in which the noise is removed and the edges are deblurred. Comparing the restored results by the networks trained by the different sub-images, the sub-images with size 7×7, scanning step of 3 are determined as the optimal inputs for training.

1 Introduction

Ultrasonic C-scanning is the most widely used method for the nondestructive evaluation of materials and structures [1]. The C-scan image is often degraded with blurriness and noise due to the physical characteristics of the transducer and the inhomogeneities in tested materials. It is difficult to obtain the information of the defects from the degraded C-scan image and thus it results in the decrease in the accuracy of nondestructive evaluation. One way to deal with the problem is to use a focused transducer to increase the resolution of images. Unfortunately, the focused transducer cannot be used in any situation and the parameters of transducer have to be considered with some care. In the same time, there are also many image processing methods based on Weiner filter to restore the C-scan image from the degraded one [2]-[4]. However, the Weiner filters are hard to be determined and implemented since most of them are ill-conditioned problems.

In this paper, a radial basis function (RBF) network [5] was trained to approximate the inverse degrade function of C-scan image to restore the high quality C-scan image from the degraded version.

F. Yin, J. Wang, and C. Guo (Eds.): ISNN 2004, LNCS 3174, pp. 399–404, 2004.

2 Model of Degraded C-scan Image

A physical model of C-scan is presented by using the linear shift invariant model. The model gives the following relation:

$$g(x, y) = h(x, y) * f(x, y) + n(x, y) \quad (1)$$

where f is the distribution of defects to be imaged within the interface, g is the degraded C-scan image of f, h is the image degraded function (i.e., the PSF), and n is the additive noise and the other physical effect not included in the convolution model. In frequency domain, this equation can be expressed as follow:

$$G(u,v) = H(u,v) \cdot F(u,v) + N(u,v) \quad (2)$$

Knowing h and n in advance, it is easy to retrieve f from g. Unfortunately, sufficient knowledge about the PSF that depends not only on the characters of transducer but also the tissues in the tested material can hardly be obtained, and the inverse operation is sensitive to noise thus a large error is usually associated with it.

We assume that the degraded function is the integrated results of the transducer parameters with the additive noise. Mathematically, it satisfies the next equation:

$$g(x, y) = D \cdot f(x, y) \quad (3)$$

where D is the integrated degraded function of C-scan image in the spatial domain. Simply, D can be seemed as a nonlinear mapping from f to g. We can build an inverse mapping Q and the restoration of C-scan image is expressed as the relation:

$$f'(x, y) = Q \cdot g(x, y) \quad (4)$$

where $f'(x, y)$ is the optimal approximate image of $f(x, y)$.

It is well known that ANN method is a good tool to deal with this type of problem [6]-[9]. In this paper a radial basis function network was utilized to reproduce the mapping between the degraded C-scan image and the high quality one.

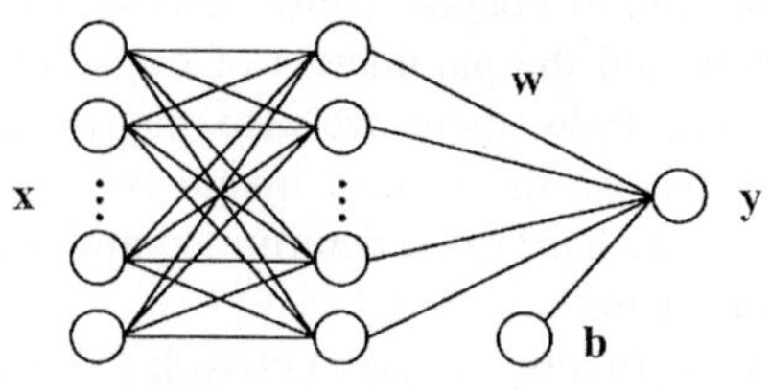

Fig. 1. Structure of radial basis function network

3 Radial Basis Function Network

The radial basis function network is a feedforward structure with a radial basis function in the hidden layer. The character is that its response decreases monotonically with distance from a center point. The center, the distance scale, and the precise shape of the radial basis function are the parameters of the model. Gaussian function in the hidden layer is used in this paper. The structure of the radial basis function network is illustrated in Fig. 1.

There are three layers in RBF network. x is the input, y is the output that is connected with the radial basis neurons in the hidden layer by the weight W. It represents a mapping from $x \in R^n$ to $y \in R$:

$$y = \sum_{i=1}^{m} w_i g_i(\|x - c_i\|) + b \tag{5}$$

where m is the number of neurons in the hidden layer, w_i is the weight associated between the ith neuron in the hidden layer and the output, and g_i is the Gaussian function represented by:

$$g_i(\|x - x_i\|) = \exp\left(\frac{\|x - x_i\|^2}{2\sigma^2}\right) \tag{6}$$

The parameter σ and x_i are the standard deviation and the center of Gaussian function, which are determined with the weight w_i during the training of network. The radial basis function network has a universal approximation capability [10]. Theoretically, it can approximate any nonlinear function with sufficient RBF neurons in the hidden layer.

4 Experiments and Discussion

We have used MATLAB Neural Network ToolBox to implement RBF network. Since the flat-bottom hole is usually employed for reference standards of defect size in ultrasonic nondestructive evaluation, the ultrasonic C-scan images of blind flat-bottom hole were utilized to train and verify the network in the experiment. All the C-scan images have been obtained by JTUIS ultrasonic NDT image system, which is made by Xi'an Jiaotong University. A 5 MHz non-focused transducer of 10 mm diameter was used to scan the samples.

It is assumed that the gray value of a given pixel in the restored image is related to the gray value of a neighborhood of the corresponding pixel in the degraded image. The shape of the neighborhood of the given pixel is assumed to be a rectangular window. During the training phase, a window of $n \times n$ pixels scans and samples the degraded image with a scanning step of d pixels that is not larger than n, so the image is completely covered. All the sub-images with size of $n \times n$ in the degraded image are converted to input vectors of size $n^2 \times 1$ for the learning and the corresponding center pixels of the sub-images in the high quality image are the desired outputs of the network.

Figure 2 shows the input and output ultrasonic C-scan images for the training of RBF network. Fig. 2(a) is an input C-scan image of flat-bottom hole of 3mm diameter, which is blurred at the edge of the hole and stained by the noise, and Fig. 2(b) is the desired output of the network. The size of the images is 128×128, and the distance between the two adjacent pixels is 0.075mm. To study the network ability, three RBF networks were trained by the sub-images with different n and d. The parameters of the networks are in Table 1.

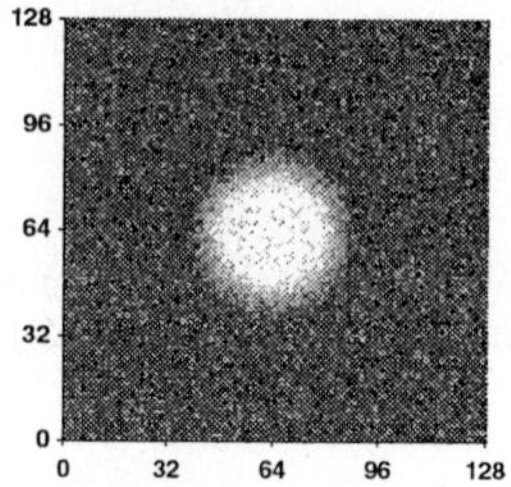

(a)

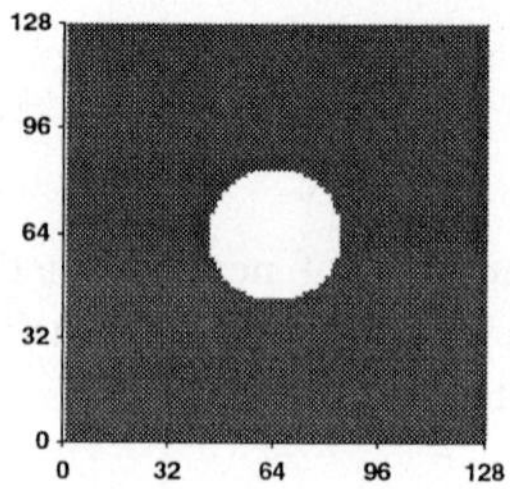

(b)

Fig. 2. C-scan images utilized in training of network: (a) input image; (b) desired output image

Table 1. Parameters of the networks

net	$n \times n$	d
NET1	5×5	5
NET2	7×7	3
NET3	9×9	5

After the training was completed, the mapping between the high quality C-scan image and the degraded one was built. Then the C-scan image of flat-bottom hole of 5mm diameter shown in Fig. 3 was used to test the networks. Fig. 4(a), (b) and (c) are the computed results by the three networks. It is evident that the networks produce an enhancement of the C-scan image. In Fig. 4(a), (b) and (c), most of the noise in Fig. 3 is removed and the edges of the flat-bottom hole are deblurred. The approximate results better represent the actual size of flat-bottom hole.

Comparing the three computed results, the edge of Fig. 4(b) is the most clear and regular. The noise around the edge is not removed in Fig. 4(a) and there is still severe blurriness in Fig. 4(c). To show the effectiveness of the networks more clearly, the edges of the results were detected using Sobel operator. The edge in Fig. 4(e) is more accurate and round than Fig. 4(d) and (f). The diameters of flat-bottom hole calculated from Fig. 4(d), (e) and (f) are 4.75mm, 5.04mm and 4.58mm. It is concluded that NET2 is more efficient than the other two networks thus the sub-images with size 7×7 and scanning step of 3 pixels on the original image can be determined as the optimal parameters for training.

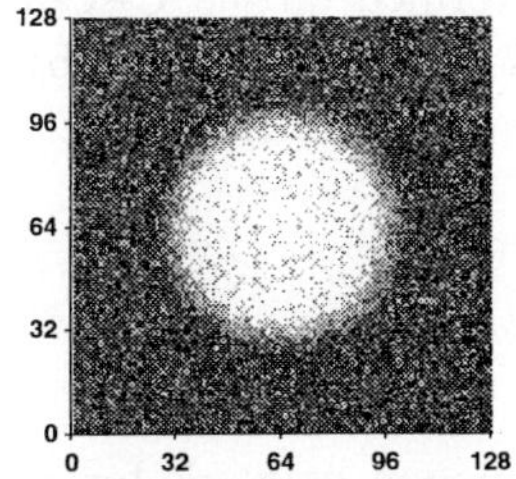

Fig. 3. C-scan image of flat-bottom hole of 5mm diameter

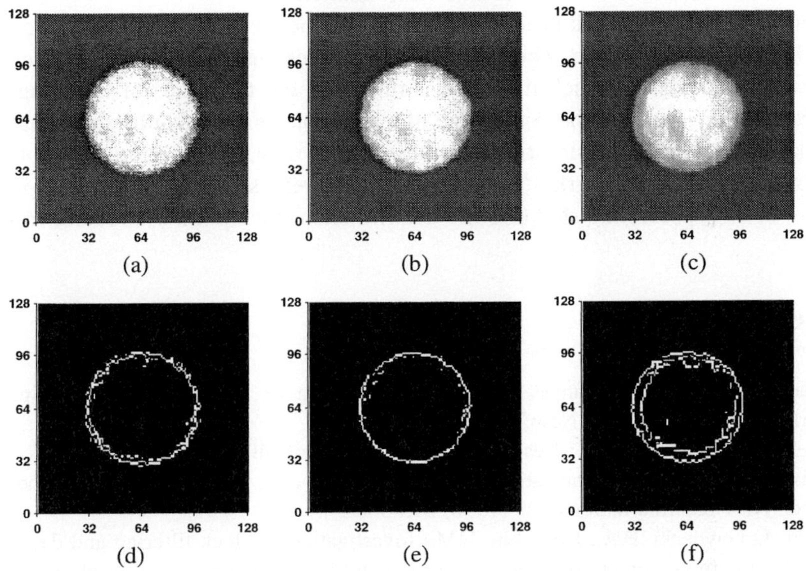

Fig. 4. Tested results of the C-scan image in Fig.3 by the networks: (a) the result calculated by NET1; (b) the result calculated by NET2; (c) the result calculated by NET3; (d) edge detection of (a); (e) edge detection of (b); (f) edge detection of (c)

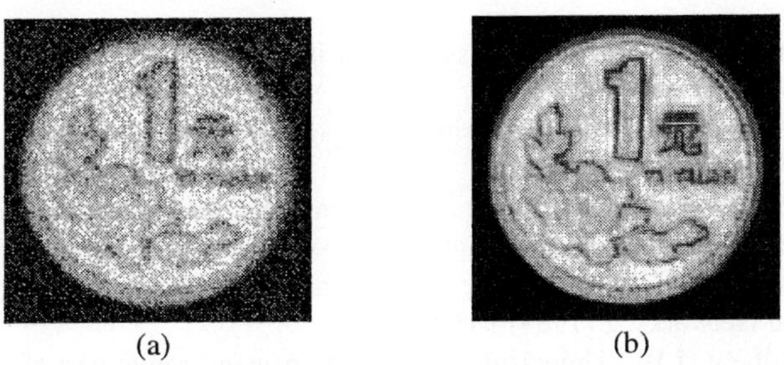

Fig. 5. C-scan images of coin: (a) original image; (b) restored result

Another experiment was performed on the C-scan image of coin shown in Fig. 5(a). Its restored result by NET2 is shown in Fig. 5(b). The noise is removed and the details on the surface of coin are enhanced efficiently.

5 Conclusion

A method using radial basis function network for restoration of ultrasonic C-scan image is presented. The network was trained by different sub-images divided from a C-scan image of flat-bottom hole of size 3mm diameter.

After trained, the network produced the mapping between the degraded C-scan image and the high quality one. The other C-scan images were used to test the network. The restored images, in which the noise is removed and the edges of the images are deblurred, better represent the actual size of tested objects. The results show that RBF network is efficient for restoration of ultrasonic C-scan image. Comparing the restored results by the networks trained by the different sub-images, the sub-images with size of 7×7, scanning step of 3 are determined as the optimal inputs for training.

References

1. Krautkramer, J., Krautkramer, H.: Ultrasonic testing of materials, 3rd edn. Springer-Verlag, Berlin Heidelberg New York (1983)
2. Cheng, S.W., Chao, M.K.: Resolution improvement of ultrasonic C-scan images by deconvolution using the monostatic point-reflector spreading function (MPSF) of the transducer. NDT &E International. 29 (1996) 293-300
3. Zhao, J., Gaydecki, P.A., Burdekin, F.M.: Investigation of block filtering and deconvolution for the improvement of lateral resolution and flaw sizing accuracy in ultrasonic testing. Ultrasonics. 33 (1996) 187-194
4. Karpur, P., Frock, B.G., Bhagat, P.K.: Weiner filtering for image enhancement in ultrasonic nondestructive evaluation. Material Evaluation. 48 (1990) 1374-1379
5. Poggio, T., Girosi, F.: Networks for approximation and learning. Proc. IEEE 78 (1990) 1481-1497
6. Su, M., Basu, M.: A hybrid learning system for image deblurring. Pattern Recognition. 35 (2002) 2881-2894
7. Dunstone, E.S.: Image processing using an image approximation neural networks. Proc. IEEE ICIP-94. 3 (1994) 912-916
8. Pattichis, C.S., Constantinides, A.G.: Medical imaging with neural networks. Proc. IEEE Workshop on Neural Networks for Signal Processing (1994) 431-440
9. Funahashi, K.I.: On the approximate realization of continuous mappings by neural networks. Neural Networks. 2 (1989) 183-192
10. Park, J., Sandberg, I.W.: Universal approximation using radial-basis-function networks. Neural Compution. 3 (1991) 246-257

Automatic Image Segmentation Based on a Simplified Pulse Coupled Neural Network

Yingwei Bi, Tianshuang Qiu, Xiaobing Li, and Ying Guo

School of Electronic and Information Engineering, Dalian University of Technology, Dalian 116024, Liaoning Province, P.R. China
yingweibi@21cn.com, qiutsh@dlut.edu.cn

Abstract. Recent researches indicate that pulse coupled neural network (PCNN) can be effectively utilized in image segmentation. However, the near optimal parameter set should always be predetermined to achieve desired segmentation result for different images, which impedes its application for segmentation of various images. So far as that is concerned, there is no method of adaptive parameter determination for automatic real-time image segmentation. To solve the problem, this paper brings forward a new automatic segmentation method based on a simplified PCNN with the parameters determined by images' spatial and grey characteristics adaptively. The proposed algorithm is applied to different images and the experimental results demonstrate its validity.

1 Introduction

Pulse coupled neural network (PCNN) is formed by the modifications and variations of the neuron model developed by Eckhorn and his co-workers[1] in 1990. Since it can ignore minor intensity variations, bridge small spatial gaps, and produce perfect segmentation results when the intensity ranges of the object and the background overlap[2,3], PCNN is considered greatly suitable for image processing. While used for image segmentation, PCNN is easily implemented. However, for different images, the parameters of the PCNN should always be adjusted to achieve satisfactory results, which impedes its application in automatic segmentation of various images. To solve this problem, Szekely and Lindblad [4] train the parameters with the desired images to achieve optimal ones. Obviously such method is not suitable for automatic real-time segmentation. In this paper, we adopt a simplified PCNN and bring forward a method to determine the parameters by images' own characteristics to achieve automatic image segmentation.

The rest of the paper is organized as follows. Section 2 briefly introduces the preliminary of PCNN neuron model and its application in image processing. In section 3, the method for determination of parameters and the new segmentation method are proposed. Section 4 presents the experimental results and provides some brief discussion on them. Finally, conclusions are given in section 5.

F. Yin, J. Wang, and C. Guo (Eds.): ISNN 2004, LNCS 3174, pp. 405–410, 2004.

2 PCNN Neuron Model

Fig.1 shows a mathematical model of PCNN neuron. It consists of three parts, i.e., the receptive field, the internal activity section and the spike generator. The receptive field is comprised of feeding field and linking field. Let N_{ij} denote the *ij*th neuron. The feeding field of N_{ij} receives the external stimulus S_{ij} and the pulses Y from the neighboring neurons. A signal denoted by F_{ij} is outputted. The linking field receives the pulses from the neighboring neurons and outputs the signal denoted by L_{ij}. In the internal activity section, F_{ij} and L_{ij} are inputted and modulated. The modulation result U_{ij} called internal activity signal is sent to the spike generator, in which U_{ij} is compared with the dynamic threshold θ_{ij} from the threshold signal generator (TSG). If U_{ij} exceeds θ_{ij}, the spike generator outputs one. Then the dynamic threshold is enlarged greatly so that it exceeds U_{ij} and the spike generator outputs zero. Thus a pulse burst is generated.

In the feeding field and linking field, there are six parameters, i.e., three time decay constants ($\alpha_F, \alpha_L, \alpha_\theta$) and three amplification factors (V_F, V_L, V_θ). m_{ijkl} and w_{ijkl} are the weighting factors between the two neighboring neurons N_{ij} and $N_{i+k,j+l}$. β_{ij} is the modulation parameter. $step(\cdot)$ is the unit step function. A more detailed description of PCNN is referred to Johnson and Padgett's paper [5].

In the application of image segmentation, each pixel corresponds to a single PCNN neuron. The neurons are organized in a single layer network to perform the segmentation task.

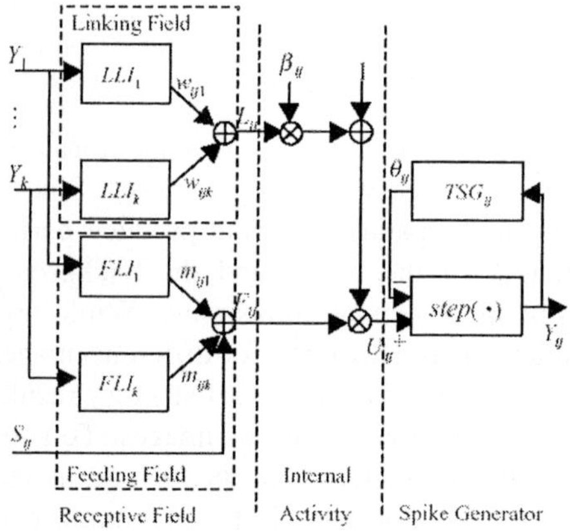

Fig. 1. PCNN neuron model

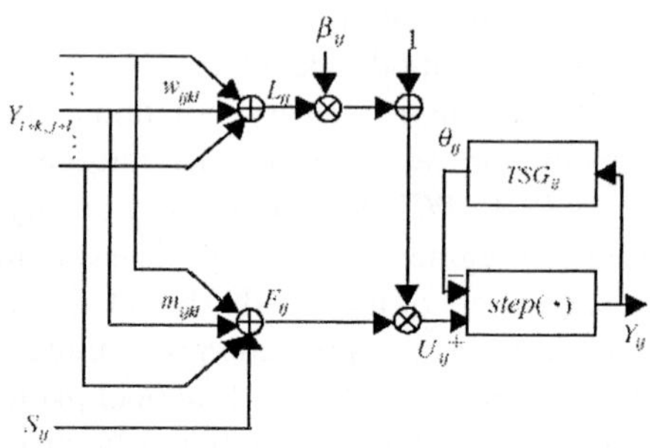

Fig. 2. Simplified PCNN neuron model

3 Algorithm

When PCNN is used for image segmentation, the parameters required to be adjusted are $\alpha_F, \alpha_L, \alpha_\theta, V_F, V_L, V_\theta, m_{ijkl}, w_{ijkl}$ and β_{ij}. Obviously it's time consuming and difficult. To solve the problem, we bring forward an automatic segmentation method based on a simplified PCNN with the parameters determined by the images' own characteristics. The simplified PCNN neuron model is shown in Fig.2.

The set of describing equations for the simplified neuron N_{ij} are listed as

$$F_{ij}(n) = S_{ij} + \sum_{k,l} m_{ijkl} Y_{i+k,j+l}(n-1) \tag{1}$$

$$L_{ij}(n) = \sum_{k,l} w_{ijkl} Y_{i+k,j+l}(n-1) \tag{2}$$

$$U_{ij}(n) = F_{ij}(n)(1+\beta_{ij} L_{ij}(n)) \tag{3}$$

$$\theta_{ij}(n) = \theta_{ij}(n-1) - \Delta + V_\theta Y_{ij}(n-1) \tag{4}$$

$$Y_{ij}(n) = step(U_{ij}(n) - \theta_{ij}(n)) \ . \tag{5}$$

In the equations, Δ is the adjusting step of the dynamic threshold.

In most previous studies, the feeding field always receives only the external stimulus[3,6]. It's effective in most cases except when S_{ij} is zero. If $S_{ij} = 0$, whatever the parameters are, the neuron corresponding to the 'black' pixel will never pulse in advance with the neighboring neurons if they belong to the same region. Hereby, the simplified neuron receives not only the external stimulus but also the outputs of the neighboring neurons. On the other hand, θ_{ij} decays in term of exponential, which is consistent with human visual characteristic but not necessary for computer processing. Therefore, to reduce the complexity, we decrease θ_{ij} linearly. In order to achieve automatic segmentation, the proposed algorithm determines the optimal segmentation result in the binary image series with the method that Ma [7] developed.

In the simplified PCNN model, the parameters to be determined are reduced to m_{ijkl}, w_{ijkl}, β_{ij}, Δ and V_θ. V_θ is always set a large value to increase the dynamic threshold rapidly after the neuron has just pulsed, so we set $V_\theta = 50$.

In an $M \times N$ image, the grey level of pixel (i, j) is denoted by $G(i, j)$. The maximal iteration number $n_{\max}$ is set 50 since the effective segmentation results always come from the 10th-40th iteration by experiments. Let $d_{space}(i+k, j+l)$ and $d_{grey}(i+k, j+l)$ denote the spatial and grey distance between pixel (i, j) and $(i+k, j+l)$. They are computed as follows.

$$d_{space}(i+k, j+l) = \max(|l|, |k|) \tag{6}$$

$$d_{grey}(i+k, j+l) = G(i, j) - G(i+k, j+l) \ . \tag{7}$$

Considering that m_{ijkl} and w_{ijkl} are used to control the influence of $N_{i+k,j+l}$ on N_{ij}, we use the spatial distance and the grey distance to determine them. Since β_{ij} is a parameter to control the degree of enlargement of F_{ij}, we define it as the coefficient of variation (CV), which is a statistical measure of the deviation of variables from its mean. When it's small, the grey level range in the connection neighborhood is small and the corresponding region is relatively homogeneous, so small enlargement can make them pulse simultaneously. Δ is set to make the dynamic threshold cover all possible grey level ranges. So the parameters can be determined as

$$m_{ijkl} = \begin{cases} \dfrac{1/(\left|d_{grey}(i+k, j+l) \times d_{space}(i+k, j+l)\right| + 1)}{\sum_{k,l}(1/(\left|d_{grey}(i+k, j+l) \times d_{space}(i+k, j+l)\right| + 1))} & \text{if } G(i,j) = 0 \\ 0 & \text{otherwise} \end{cases} \quad (8)$$

$$w_{ijkl} = \frac{1/(\left|d_{grey}(i+k, j+l) \times d_{space}(i+k, j+l)\right| + 1)}{\sum_{k,l}(1/(\left|d_{grey}(i+k, j+l) \times d_{space}(i+k, j+l)\right| + 1))} \quad (9)$$

$$\beta_{ij} = \frac{\sqrt{V_{ij}}}{M_{ij}} \quad (10)$$

$$\Delta = \frac{1}{n_{\max}} . \quad (11)$$

M_{ij} and V_{ij} correspond to the mean and variance of the grey levels in the neighborhood of pixel (i, j).

The proposed algorithm can be implemented with the following steps.

Step1 Input the normalized grey levels of the image to the network as the external stimulus signal S_{ij} ($0 \le i \le M-1, 0 \le j \le N-1$).

Step2 Let the segmented binary image of each iteration be stored in $T_{M \times N}$. $T(0) = 0$. Set the maximal entropy $H_{\max} = 0$.

Step3 Set $n_{\max} = 50$, $V_\theta = 50$, $F_{ij}(0) = S_{ij}$, $\theta_{ij}(0) = 1$, $L_{ij}(0) = 0$, $U_{ij}(0) = 0$, $Y_{ij}(0) = 0$, and compute the parameters m, w, β and Δ.

Step4 Set iteration number n as 1.

Step5 Compute $Y_{ij}(n)$. If $Y_{ij}(n) = 1$, $T_{ij}(n) = 1$.

Step6 Compute the entropy, i.e. H, of the image denoted by $T(n)$. If $H > H_{\max}$, $H_{\max} = H$, and store $T(n)$ to T_{save}.

Step7 $n = n + 1$.

Step8 If $n < n_{\max}$, go to *Step5*, otherwise output T_{save} as the final segmentation result.

4 Experimental Results

We implement image segmentation with the proposed algorithm on various images. Fig.3 shows part of the results. It is obvious that the proposed algorithm can effectively segment various images automatically, without predetermining and adjusting the network's parameters. To verify the validity of the algorithm, we also perform medical ultrasound image segmentation without preprocessing, as we know medical ultrasound image segmentation is a difficult problem due to the intrinsic speckle noise and tissue related texture. The result in Fig.4 shows that the proposed algorithm deals well with images of low quality such as ultrasound images. For comparison, segmentation results of PCNN with near optimal parameters adjusted manually are presented in Fig.3 and Fig.4. It's obvious that the proposed method yields a similar result to that of PCNN with near optimal parameters adjusted manually.

(a) (b) (c)

Fig. 3. Segmentation results of 'Lena'. (a) Initial image; (b) Result of PCNN with parameters as: $V_L = 0.2$, $V_\theta = 50$, $V_F = 0$, $\alpha_L = 1$, $\alpha_\theta = 0.05$, $\alpha_F = +\infty$, $\beta = 0.1$; (c) Result of the proposed method

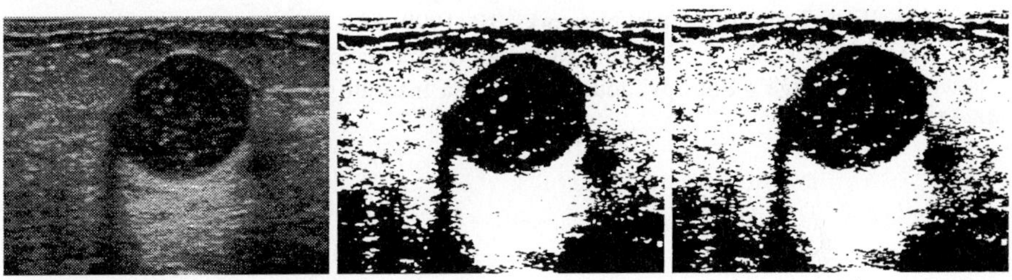

(a) (b) (c)

Fig. 4. Segmentation result of medical ultrasound image. (a) Initial 'Breast Cyst' image; (b) Result of PCNN with parameters as: $V_L = 0.4$, $V_\theta = 50$, $V_F = 0$, $\alpha_L = 1$, $\alpha_\theta = 0.1$, $\alpha_F = +\infty$, $\beta = 0.06$; (c) Result of the proposed method

5 Conclusions

Pulse coupled neural network is based on the study of biological neural behavior. It's ability of ignoring minor intensity variations, bridging small spatial gaps, and producing perfect segmentation results when the intensity ranges of the object and the background overlap adapts it to image segmentation. However the necessity to determine network's parameters impedes its application in automatic segmentation of various images. To solve the problem, this paper proposed a method to determine the parameters by images' own characteristics automatically. The experimental results demonstrate the validity of the proposed algorithm.

In the proposed algorithm, the parameters are set for general images. For specific image segmentation, other characteristics such as texture can be considered to determine the parameters to achieve more accurate segmentation. Further research on this may be regarded.

References

1. Eckhorn, R., ReitBoeck, H.J., Arndt, M., Dicke, P.: Feature Linking via Synchronization Among Distributed Assemblies: Simulation of Results from Cat Visual Cortex. Neural Comput., 2 (1990) 293-307
2. Ranganath, H.S., Kuntimad, G., Johnson, J.L.: Pulse Coupled Neural Networks for Image Processing. Southeastcon '95. 'Visualize the Future'., Proc., IEEE, 26-29 (1995) 37-43
3. Ranganath, H.S., Kuntimad, G.: Image Segmentation Using Pulse Coupled Neural Networks. Proc. IEEE Int. Conference on Neural Networks, Orlando, FL. 2 (1994) 1285-1290
4. Szekely, G., Lindblad, T.: Parameter Adaptation in a Simplified Pulse Coupled Neural Network. Proc. Wkshp. Virtual Intell./DYNN, VI-DYNN'98, Stockholm, Sweden, Vol.3728, SPIE (1999)
5. Johnson, J.L., Padgett, M.L.: PCNN Models and Applications. IEEE Trans. on Neural Networks, 10 (1999) 480-498
6. Kuntimad, G., Ranganath, H.S.: Perfect Image Segmentation Using Pulse Coupled Neural Networks. IEEE Trans. on Neural Networks, 10 (1999) 591-598
7. Ma, Y.D., Dai, R.L., Li, L.: Automated Image Segmentation Using Pulse Coupled Neural Networks and Image's Entropy. Journal of China Institute of Communications, 23 (2002) 46-51

Face Pose Estimation Based on Eigenspace Analysis and Fuzzy Clustering

Cheng Du and Guangda Su

Electronic Engineering Department, State Key Laboratory of Intelligent Technology and Systems, Tsinghua University, Beijing, P.R. China, 100084
ducheng00@mails.tsinghua.edu.cn

Abstract. In this paper, a new method is proposed to estimate the accurate pose of face images. A face image collection device is designed to collect face image with accurate pose. Principle Component Analysis (PCA) is used to set up the eigenspace of face pose, then a fuzzy C-means clustering method is applied to divide the train samples into several classes, and to decide the centers of the classes. When a face image is presented to our system, we first project the image into the eigenspace, then the subordinate degree of the input face image to each class is calculated, finally the pose of the input face image is calculated combining the subordinate degrees. Experiments show that the proposed method can effectively estimate the accurate pose of face image.

1 Introduction

Face pose estimation is very important in the research areas of human computer interaction and face recognition. Many human computer interaction applications like virtual reality and video teleconferencing require the information of where a person is looking at and what he or she is paying attention to. Also, most face recognition and expression recognition algorithms have higher recognition rate when the frontal view of the face is used as an input. Therefore, as a preprocessing step, it is very important to estimate the posture of the human face from the input face image, and reproject it to generate the frontal view face image.

Conventionally there are many methods for estimating the pose of face images which can be mainly categorized into two classes: the first class is based on the detection of facial features such as eyes, nose, mouth, etc. [1-3]. The second class is based on the intensity distribution of face image [4-8]. The former methods generally involve 3D position estimation of the facial features, thus accurate camera parameters must be known for face pose estimation. Additionally facial feature detection by the image understanding techniques is still a hard problem under arbitrary conditions, such as illumination and background scenes; the latter methods are basically model based method in which the image models obtained previously are used for face pose estimation. Those methods do not need feature detection, but it takes much labor to train the models before estimation of the face pose.

F. Yin, J. Wang, and C. Guo (Eds.): ISNN 2004, LNCS 3174, pp. 411–417, 2004.

Gao, Y. et al [1] estimate face pose by locations of the two irises and the symmetric axis of face. After that they synthesized the frontal view face image of the input face, the recognition rate increased greatly after the synthesis. But they didn't report how accurate their pose estimation method is. Their pose estimation method is very sensitive to the location error of the facial features and facial expression. Srinivasan S. and Boyer K.L [7] proposed a view based eigenspace method to estimate pose of face image. They constructed different eigenspaces for face images of different poses. The input image is projected to all the eigenspaces. They argued that the norms of the projections represent the similarity between the image and the corresponding eigenspace. Based on this, they proposed a refined estimation method and an extreme estimation method. The RMS is 12.9° for the refined method and 26.0° for the extreme method. Blanz V. and Vetter T. [5] proposed a pose estimation method based on fitting a 3D morphable model, the average RMS of their method is smaller than 5°, but their method needs 3D scans of head to set up shape model of human face, and their algorithm is time consuming. In this paper, we combine the eigenspace analysis and fuzzy clustering to estimate pose of face image. We first construct face eigenspace using face images of different pose, then we project face images of different pose to the eigenspace, then we divide the projections into several classes using fuzzy C-means clustering algorithm. When a face image is presented to our system, we first project the face image to the eigenspace, then we calculate the subordinate degree of the projection to all the classes, finally the rotation degree of the input face image is calculated according to the subordinate degrees.

In this paper, we only consider face images rotating from left to right with no tilt, we believe that our method can be extended to the more generalized case of tilt and roll. The paper is organized as follow: section 2 is the introduction to our face image collection device, by this device we can collect face image with accurate pose, section 3 is construction of face eigenspace, section 4 is our face pose estimation algorithm, section 5 are experiments and conclusions.

2 Construction of Face Eigenspace

The eigenspace analysis is also called principal component analysis (PCA) which is first introduced to the face recognition area by Turk M. and Pentland A. [9]. The main purpose of PCA is to reduce the dimension of a face image.

We first compute the covariance matrix of the train images. The covariance matrix is computed using face images of all the poses in the train database. The eigenvectors $e_1, e_2, \cdots e_m$ of the covariance matrix are then computed and sorted according to the eigenvalues. The eigenvectors corresponding to the top M eigenvalues are stored. These eigenvectors are typically called eigenfaces. The space spanned by eigenfaces is called eigenspace. For each image x_i in the database which dimension is n, an eigen representation is then obtained by projecting the database image onto each of the eigenface vectors $w_{ij} = e_j^T x_i, j = 1, 2, \cdots M$. So each face image is now represented by an M dimensional vector of eigen coefficients. Here $M \ll n$, this amounts to a significant dimensionality reduction.

When an input face image x is presented to our system, we project the face image onto the eigenspace, the face image x is now represented by a M dimensional vector p. by a minimal distance classifier, we can get the vector p^* in the train database which is closest to p, then x is classified to the pose associated with p^*. In this way, we can simply divide face images into several classes, but we still can not get the accurate rotated degree of a face image.

3 Face Pose Estimation Algorithm

In order to get accurate pose of the input face image, a fuzzy C-means clustering algorithm is applied to the projections of the train set. The train set is divided into several classes and the class centers are decided. When a face image is presented to our system, we first project the face image to the eigenspace, the projection is then compared with the class centers to calculate its subordinate degree to each class. Finally, the accurate pose of the face image is calculated combining the subordinate degrees.

3.1 Fuzzy C-means Clustering Algorithm

The fuzzy clustering algorithm can not decide which class the test sample belongs to, instead, it gives the subordinate degree of the test sample to each class. The main target for a fuzzy clustering algorithm is to minimize the clustering loss function [10]. For a sample set $\{x_i, i = 1, 2, \cdots n\}$, suppose the sample set can be divided into c classes, $\{m_i, i = 1, 2, \cdots c\}$ are the centers of the classes, $\mu_j(x_i)$ is the subordinate degree of the ith sample to the jth class. The clustering loss function is defined as:

$$J_f = \sum_{j=1}^{c} \sum_{i=1}^{n} [\mu_j(x_i)]^b \|x_i - m_j\|^2 \tag{1}$$

Here $b > 1$ is a parameter which can control the fuzzy degree of the clustering result. For a fuzzy C-means clustering algorithm, the sum of subordinate degree of one sample to all the classes is constrained to 1.

$$\sum_{j=1}^{c} \mu_j(x_i) = 1, \quad i = 1, 2, \cdots n \tag{2}$$

By equation (3), (4), we can iteratively get the class centers of the train samples and their subordinate degrees to each class.

$$m_j = \frac{\sum_{i=1}^{n} [\mu_j(x_i)]^b x_i}{\sum_{i=1}^{n} [\mu_j(x_i)]^b}, j = 1, 2, \cdots c \tag{3}$$

$$\mu_j(x_i) = \frac{\left(1/\|x_i - m_j\|^2\right)^{1/(b-1)}}{\sum_{k=1}^{c}\left(1/\|x_i - m_j\|^2\right)^{1/(b-1)}}, i = 1, 2, \cdots n\, j = 1, 2, \cdots c \quad (4)$$

3.2 Pose Estimation by Subordinate Degrees

After applying the fuzzy C-means clustering algorithm, we can get the subordinate degrees of an input face image to each class using equation (4). The subordinate degree represents the similarity degree between the test face image and a face image class. Then the rotated angle of the input face image can be calculated using the following equation:

$$Angle(x) = \sum_{j, if \mu_j(x) > T} \mu_j(x) \times A_j \quad (5)$$

Here A_j is the angle of the jth class, T is a threshold, if $\mu_j(x)$ is smaller than T, we consider that the input is impossible to belong to the jth class. T can be determined through a train process.

4 Experiments and Conclusions

4.1 Face Image Collection Device

Currently, there are many face databases available for face recognition research, such as the CMU-PIE face database, the FERET face database, etc. Most of the face databases contain face images collected in different pose condition, different light condition, etc. But to the best of our knowledge, none of the face databases recorded accurate pose of face image. Considering this, we designed a face image collection device to collect face image with accurate pose. The device is shown in Fig. 1. The camera is fixed on a circle rail, when collecting face image, the volunteer sit at the center of the circle, the camera moves around the volunteer to collect face image, the rotated degree can be read from the marks on the circle rail. The precision of our face image collection device is within 1 degree.

4.2 Experiments

Our algorithm is tested on TH face database. TH face database includes face images of different pose from 305 volunteers. The accurate pose of each face image in the database is recorded, the angles of face images range from 45 degree left to 45 degree right. Before the experiments, we manually marked the two irises of the face images, and then the face images are normalized and cropped according to the location of the irises. One sample of cropped image is shown in Fig. 2.

The train set includes face images from 50 volunteers in TH face database and the poses of the train set are $\{0°, \pm10°, \pm20°, \pm30°, \pm40°\}$, the test set includes

Fig. 1. Accurate pose face image collection device

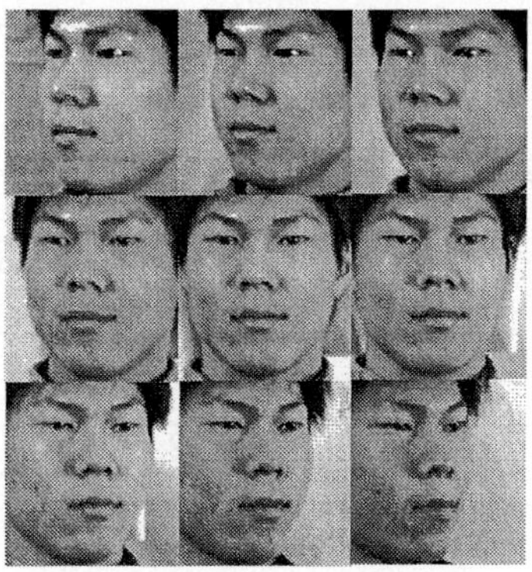

Fig. 2. Cropped image in TH face database

face images from the other volunteers in TH face database with pose ranging from 45 degree left to 45 degree right.

For comparison, we also implemented a pose estimation algorithm using direct PCA. The test results of our algorithm and direct PCA are shown in table 1. The average RMS of our pose estimation algorithm is 3.67° which is 5.34° when using direct PCA method. Blanz V. and Vetter T. [5] used 3D morphable model to estimate pose of the face image, their test results are shown in table 2. We can see that our method performs almost equivalent to their method, but the computation of their method is much more expensive, moreover, the RMS of estimation tends to increase with the true angle increasing in their results which can not be seen in the results of our method.

For the face images whose true angle is in the train set ($\{0°, \pm10°, \pm20°, \pm30°, \pm40°\}$), the RMS of estimation is a little smaller than the face images whose true angle is not in the train set ($\{\pm5°, \pm15°, \pm25°, \pm35°, \pm45°\}$).

Table 1. Pose estimation result of our method and direct PCA

True Angle (degree)	Our method		Direct PCA	
	Average estimate (degree)	RMS (degree)	Average estimate (degree)	RMS (degree)
0	0.17	3.14	0.10	4.10
5	3.70	4.53	4.70	6.23
10	9.90	3.05	9.93	4.32
15	14.23	4.32	14.52	6.29
20	19.57	2.56	19.49	4.36
25	25.56	4.06	25.32	6.62
30	29.36	2.98	29.51	4.68
35	34.11	4.40	34.31	6.40
40	38.86	3.98	39.20	5.08
Average RMS		3.67		5.34

Table 2. Pose estimation result reported in [5]

Angular Distance	Front-side (degree)	Front profile (degree)	Side-profile (degree)
Average Estimate	18.1	63.2	45.2
RMS	2.4	4.6	4.5
True Angle	16.5	62.1	45.6

4.3 Conclusion

In this paper, we present a new face pose estimation algorithm. By eigenspace analysis and fuzzy C-means clustering, we can calculate the accurate rotated degree of the face image. Our method performs much better than direct PCA method and almost the same with the 3D morphable model method. Our method is computationally inexpensive and hence very fast. Here, we only test face images rotated from left to right with no tilt. With more train samples, we believed that our algorithm can be extended to more generalized cases. The feature extraction process of our algorithm is based on eigenspace analysis, so inevitably our method is sensitive to the change of light conditions and background which is the main focus of our future work.

References

1. Gao, Y., Leung, M.K.H., Wang, W., Hui, S.C.: Fast face identification under varying pose from a single 2-D model view. IEE Proceedings of Vision, Image and Signal Processing, Vol. 148 , Issue: 4 , (2001) 248-253
2. Gee, A., Cipolla, R.: Estimating gaze from a single view of a face. International Conference on Pattern Recognition, Vol. 1. (1994) 758-760
3. Horprasert, Thanarat, Yacoob, Yaser; Davis, Larry S.: Computing 3D head orientation from a monocular image sequence. Proceedings of SPIE - The International Society for Optical Engineering, Vol. 2962, (1997) 244-252
4. Watta, P., Gandhi, N., Lakshmanan, S.: An eigenface approach for estimating driver pose. Proceedings of Intelligent Transportation Systems, (2000) 376-381
5. Volker Blanz, Thomas Vetter: Face recognition based on fitting a 3D morphable model. IEEE Transactions on pattern analysis and machine intelligence, Vol. 25, No. 9, (2003) 1063-1073
6. Elagin, E., Steffens, J., Neven, H.: Automatic pose estimation system for human faces based on bunch graph matching technology. Third IEEE International Conference on Automatic Face and Gesture Recognition, (1998) 136-141
7. Srinivasan, S., Boyer, K.L.: Head pose estimation using view based eigenspaces. 16th International Conference on Pattern Recognition, Vol. 4, (2002) 302-305
8. Saito, H., Watanabe, A., Ozawa, S.: Face pose estimating system based on eigenspace analysis. International Conference on Image Processing, Vol. 1, (1999) 638-642
9. Turk, M., Pentland, A.: Face recognition using eigenfaces. International Conference on Computer Vision Pattern Recognition, (1991) 586-59
10. Zhaoqi B., Xuegong Z.: Pattern recognition, Press of tsinghua university, P.R. China, (2000) 273-283

Chaotic Time Series Prediction Based on Local-Region Multi-steps Forecasting Model

Minglun Cai[1], Feng Cai[2], Aiguo Shi[2], Bo Zhou[2], and Yongsheng Zhang[2]

[1] Computer Science Dept., Ningbo Institute of Technology, Zhejiang University, China 315100
[2] Navigation Dept. of Dalian Naval Academy, China 116018
vipcaif@tom.com

Abstract. Large computational quantity and cumulative error are main shortcomings of add- weighted one-rank local-region single-step method for multi-steps prediction of chaotic time series. A local-region multi-steps forecasting model based on phase-space reconstruction is presented for chaotic time series prediction, including add-weighted one-rank local-region multi-steps forecasting model and RBF neural network multi-steps forecasting model. Simulation results from several typical chaotic time series demonstrate that both of these models are effective for multi-steps prediction of chaotic time series.

1 Introduction

Local-region forecasting method based on Takens embedding theory is a simple and effective method in predicting kinds of chaotic time series [1][2][3][4]. Common add-weighted one-rank local-region method (AOLM) is a single-step forecasting model. New predicted data obtained by AOLM should be added to original data to make multi-steps forecasting with AOLM. So large computational quantity and cumulative error are obvious shortcomings of AOLM for multi-steps forecasting. This paper presents two new local-region multi-steps forecasting models, including add-weighted one-rank local-region multi-steps method (AOLMM) and RBF neural network local-region multi-steps method (RBFNNLMM), both of which improve the multi-steps forecasting validity and avoid cumulative error.

2 AOLM and AOLMM

2.1 AOLM [5]

By phase-space reconstruction of time series x_1, x_2, $\cdots$, x_N based on Takens embedding theory, we can get the phase-space point $Y_i = \left(x_i, x_{i+\tau}, \cdots, x_{i+(m-1)\tau}\right)$, (Here, i = 1, 2, $\cdots$, M, M is the points number in reconstructed phase-space, and $M = N - (m-1)\tau$, m is the embedding dimension, τ is the time delay, and N is

F. Yin, J. Wang, and C. Guo (Eds.): ISNN 2004, LNCS 3174, pp. 418–423, 2004.

the length of time series). Assume neighbors of the center point Y_M in phase-space are Y_{Mi}, $i=1,2,\cdots,q$, the distance between Y_M and Y_{Mi} is d_i, and $d_{\min}$ is the minimum distance among d_i, the weight of Y_{Mi} is defined as:

$$P_i = \frac{\exp(-\alpha(d_i - d_{\min})}{\sum_{i=1}^{q} \exp(-\alpha(d_i - d_{\min})}, \tag{1}$$

α is a parameter and has the value of $\alpha = 1$ in common. Thus, the one-rank local-region linear fit is:

$$Y_{Mi+1} = ae + bY_{Mi}, i=1,2, \cdots, q, \tag{2}$$

Here, e is a m-dimensional vector, $e = (1,1,\cdots,1)^T$.

The add-weighted least square method for embedding dimension $m = 1$ requires

$$\sum_{i=1}^{q} P_i (x_{Mi+1} - a - bx_{Mi})^2 = \min$$

The model above is called AOLM.

2.2 AOLMM

The essence of single-step forecast based on AOLM is, to find the （m+1） most similar points (nearest neighbors of center point) in reconstructed phase-space first, make the single-step forecast according to the single-step evolving rule of these (m+1) neighbors. It's well known that the evolvement between a pair of nearest neighbor follows the index transition rule $e^{-\lambda t}$, here, λ is the largest Lyapunov exponent. In the same way as AOLM, we can make multi-steps forecast according to the multi-steps evolving rule of these (m+1) neighbors, which is the essence of local-region multi-steps method. The AOLMM for embedding dimension $m > 1$, and k (k>1)-steps forecast model is deduced in detail as follows:

Assume the reference vector set is $\{Y_{Mi}\}$, $i=1,2, \cdots, q$, which will be evolved as $\{Y_{Mi+k}\}$ after k-steps. According to the add-weighted least square method, the following equation exists:

$$\sum_{i=1}^{q} P_i \left[\sum_{j=1}^{m} \left(x^j_{Mi+k} - a_k - b_k x^j_{Mi}\right)^2 \right] = \min \tag{3}$$

The equation above is a function of dual variables a_k and b_k. By differentiating this equation, we can get:

$$\begin{cases} \sum_{i=1}^{q} P_i \sum_{j=1}^{m} \left(x_{Mi+k}^j - a_k - b_k x_{Mi}^j\right) = 0 \\ \sum_{i=1}^{q} P_i \sum_{j=1}^{m} \left(x_{Mi+k}^j - a_k - b_k x_{Mi}^j\right) x_{Mi}^j = 0 \end{cases},$$

which can be unwrapped as:

$$\begin{cases} a_k \sum_{i=1}^{q} P_i \sum_{j=1}^{m} x_{Mi}^j + b_k \sum_{i=1}^{q} P_i \sum_{j=1}^{m} \left(x_{Mi}^j\right)^2 = \sum_{i=1}^{q} P_i \sum_{j=1}^{m} x_{Mi+k}^j x_{Mi}^j \\ a_k m + b_k \sum_{i=1}^{q} P_i \sum_{j=1}^{m} x_{Mi}^j = \sum_{i=1}^{q} P_i \sum_{j=1}^{m} x_{Mi+k}^j \end{cases}$$

and the matrix form is:

$$\begin{pmatrix} coe1 & coe2 \\ m & coe1 \end{pmatrix} \begin{pmatrix} a_k \\ b_k \end{pmatrix} = \begin{pmatrix} e_k \\ f_k \end{pmatrix},$$

here, $coe1 = \sum_{i=1}^{q} P_i \sum_{j=1}^{m} x_{Mi}^j$, $coe2 = \sum_{i=1}^{q} P_i \sum_{j=1}^{m} \left(x_{Mi}^j\right)^2$, $e_k = \sum_{i=1}^{q} P_i \sum_{j=1}^{m} x_{Mi+k}^j x_{Mi}^j$, $f_k = \sum_{i=1}^{q} P_i \sum_{j=1}^{m} x_{Mi+k}^j$. Thus,

$$\begin{pmatrix} a_k \\ b_k \end{pmatrix} = \begin{pmatrix} coe1 & coe2 \\ m & coe1 \end{pmatrix}^{-1} \begin{pmatrix} e_k \\ f_k \end{pmatrix}.$$

Here, $\sum_{j=1}^{m} x_{Mi}^j$ in $coe1$ is the element sum of Y_{Mi}, $\sum_{j=1}^{m} \left(x_{Mi}^j\right)^2 = Y_{Mi} Y_{Mi}^T$, $\sum_{j=1}^{m} x_{Mi+k}^j x_{Mi}^j = Y_{Mi+k} Y_{Mi}^T$, and $\sum_{j=1}^{m} x_{Mi+k}^j$ in f_k is the element sum of Y_{Mi+k}. So the formula for calculating variables a_k and b_k is as:

$$\begin{pmatrix} a_k \\ b_k \end{pmatrix} = \begin{pmatrix} \sum_{i=1}^{q} P_i \sum_{j=1}^{m} x_{Mi}^j & \sum_{i=1}^{q} P_i Y_{Mi} Y_{Mi}^T \\ m & \sum_{i=1}^{q} P_i \sum_{j=1}^{m} x_{Mi}^j \end{pmatrix}^{-1} \begin{pmatrix} \sum_{i=1}^{q} P_i Y_{Mi+k} Y_{Mi}^T \\ \sum_{i=1}^{q} P_i \sum_{j=1}^{m} x_{Mi+k}^j \end{pmatrix} \quad (4)$$

With values a_k and b_k by formula (3) and the reference vector set $\{Y_{Mi+k}\}$, $i=1,2, \cdots, q$, we can get k th-step prediction is:

$$Y_{M+k} = a_k e + b_k Y_M \tag{5}$$

The detailed steps for AOLMM are as follows:

(1) Reconstruct the phase-space

First, calculate the correlation dimension d according to G-P method [6]. Second, select the embedding dimension $m \geq 2d+1$ according to Takens embedding theory. Third, get the average orbit period mtbp by FFT analysis, and calculate the time delay τ by $mtbp = (m-1)\tau$ [7]. Thus, the reconstructed phase-space is $Y_i = (x_i, x_{i+\tau}, \cdots, x_{i+(m-1)\tau})$, i=1,2, ···, M.

(2) Find the nearest neighbors of center point in phase-space

Calculate distances between each point and center point Y_M in phase-space, and find the q ($q = m+1$) number of nearest points as reference vector set Y_{Mi}, i=1,2, ···, q, then calculate the weights P_i of each nearest neighbor.

(3) Calculate coefficients a_k and b_k with formula (4)

(4) Forecast

Make multi-steps forecast with formula (5) and the value of a_k and b_k.

3 RBFNNLMM

AOLM(AOLMM) uses least square method to get phase points' one-step(multi-steps) evolving rule. We can also use the neural network technique to get such evolving rule. For RBF neural network has virtues of better ability of approach and classification and fast learning speed, we can select RBFNN instead of add-weighted one-rank method to build the local-region multi-steps forecasting model. The detailed model is as follows:

(1) Normalize the original chaotic time series;

(2) Reconstruct the phase-space, and find the nearest neighbors of center point (the same as AOLM);

(3) $k=1$, construct and train the RBF neural network;

Set the weighted sum of (m+1) nearest neighbors as sample input, and the corresponding weighted sum of these neighbors after one-step evolving as sample output, then call the NEWRBE function in MATLAB to design an accurate RBF neural network and train it.

(4) Forecast;

With the reference phase point Y_{Mi}, call the SIM (net, Y_{Mi}) function in MATLAB and the output is the single-step prediction value.

(5) $k>1$, repeat steps (2) and (3).

That is, to set the weighted sum of (m+1) nearest neighbors as sample input, and the corresponding weighted sum of these neighbors after k steps evolving as sample output, then call the NEWRBE function in MATLAB to design an accurate

RBF neural network and train it. With the reference phase point Y_{Mi}, call the SIM (net, Y_{Mi}) function and the output is k-step prediction value.

4 Simulation

We use Chen's chaotic system, Lorenz system and Logistic system as samples to test the validity of AOLMM and RBFNNLMM.

Chen's chaotic attractor is generated by the following differential system [8]:

$$\begin{cases} \dot{x} = a(y - x) \\ \dot{y} = (c - a)x - xz + cy \\ \dot{z} = y - bz \end{cases} \tag{6}$$

Here, $a = 35, b = 3, c = 28$.

Take the initial values as $x(0) = 0$, $y(0) = 1$, $z(0) = 0$, and the integral time interval is $h = 0.001$. Integrate equations (6) with four-rank Runge-Kutta method to get a chaotic time series of 15000 points, omit the first 1000 points, take 9000 points as learning sample, and take 5000 points as testing sample. The parameters for phase-space reconstruction are $m = 8$ and $\tau = 10$. Fig.1 and Fig.2 are forecast results with AOLMM and RBFNNLMM, the real line shows true values while the dashed line shows the predicted values. The latter result is normalized. Both of these two results show that the first 2100 predicted points have small errors.

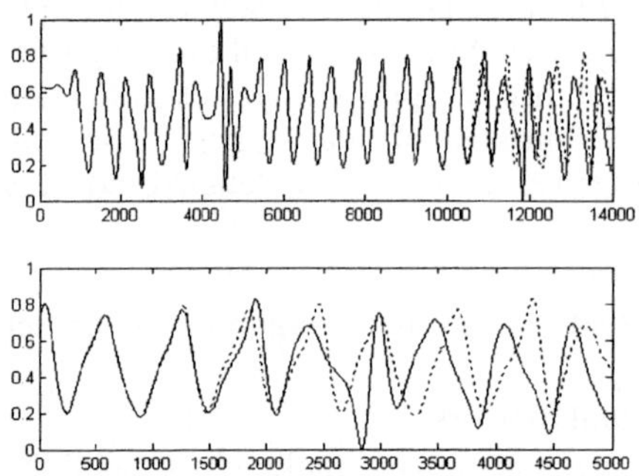

Fig. 1. Forecast Chen's result with RBFNNLMM

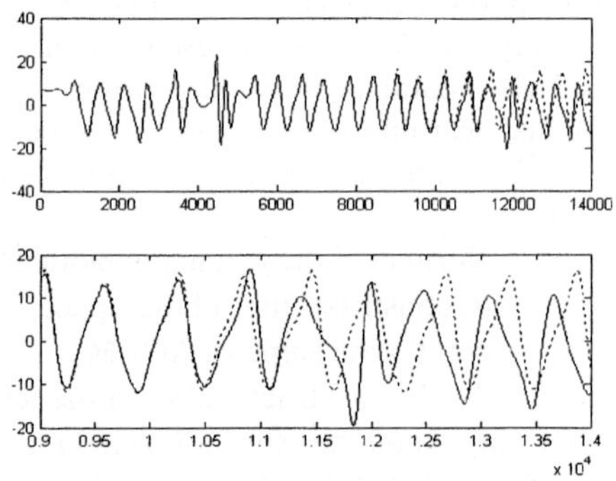

Fig. 2. Forecast Chen's result with AOLMM

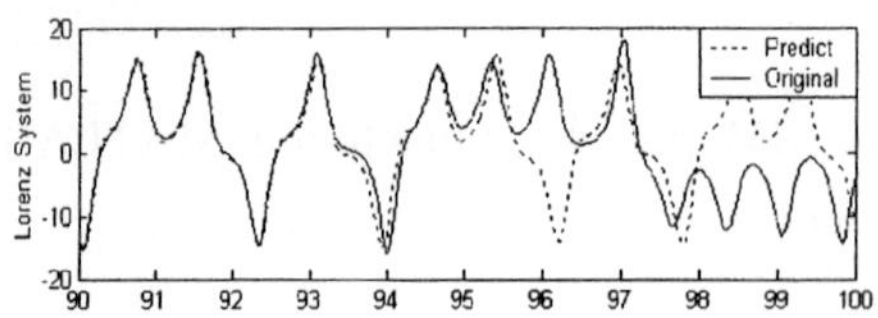

Fig. 3. Forecast result of Lorenz system

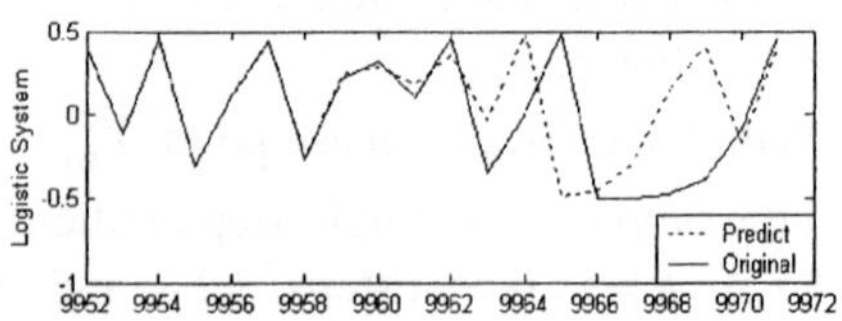

Fig. 4. Forecast result of Logistic System

Fig.3 is the forecast result of Lorenz chaotic system with AOLMM. Compared with results in reference [9]' the AOLMM is more effective for multi-steps forecast.

Fig.4 is the forecast result of Logistic system with AOLMM, which is better than the result in reference [10].

5 Conclusion

Simulation results from several typical chaotic time series, including Chen's system, Lorenz system and Logistic system, demonstrate that both of AOLMM and RBFNNLMM are effective for multi-steps prediction of chaotic time series.

References

1. Tianyun Li, Zifa Liu.: The chaotic property of power load and its forecasting. Proceedings of the CSEE (2000) 20(11): 36–40
2. Jinhu Lü, Suochun Zhang.: Application of adding-weight one-rank local region method in electric power system short-term load forecast. Control theory and applications (2002) 19(5): 767–770
3. Huaming Liu, Huan Qi, Zhiqiang Cai.: Nonlinear chaotic model of landslide forecasting. Chinese Journal of Rock Mechanics and Engineering (2003) 22(3): 434–437
4. Yuqiao Gu, Tianlun Chen, Wuqun Huang.: Forecasting of Heng Sheng index futures based on neural networks based on chaotic algorithm. Acta Scientiarum Naturalium Universitatis Nankaiensis (2000) 33(1): 119–123
5. Jinhu Lü, Jun'an Lu, Shihua Chen.: Chaotic time series analysis and its applications. Wuhan: Wuhan University Publication (2002)
6. Grassberger, P., Procaccia, I.: Characterization of strange attractors, Phys. Rev. Lett. 50 (1983) 346.
7. Shaoqing Yang, Chuanying Jia.: Two practical methods of phase space reconstruction. ACTA Physica Sinica (2002) 51(11): 2452–2458
8. Jun'an Lu, Jinhu Lü, Shihua Chen.: Chen's chaotic attractor and its characteristic quantity. Control theory and applications (2002) 19(2): 308–310
9. Mingyan Liu, Zongding Hu.: Chaos prediction of hydrodynamic behavior in gas-liquid two-phase bubble column with single orifice. Journal of Chemical Industry and Engineering (2000) 51(4): 475–479
10. Yadong Jiang, Jiangtao Shen, Bingru Yang.: A method of chaotic time series prediction based on wavelet neural network. Microcomputer development (2001) 37(3): 37–39

Nonlinear Prediction Model Identification and Robust Prediction of Chaotic Time Series

Yuexian Hou, Weidi Dai, and Pilian He

Department of Computer Science and Technology, Tianjin University,
Tianjin, 300072, China
{yxhou, plhe}@tju.edu.cn , davidy@126.com

Abstract. Although, in theory, the neural network is able to fit, model and predict any continuous determinant system, there is still an obstacle to prevent the neural network from wider and more effective applications due to the lack of complete theory of model identification. This paper addresses this issue by introducing a universal method to achieve nonlinear model identification. The proposed method is based on the theory of information entropy and its development, which is called as nonlinear irreducible autocorrelation. The latter is originally defined in the paper and could determine the optimal autoregressive order of nonlinear autoregression models by investigating the irreducible autodependency of the investigated time series. Following the above proposal, robust prediction of chaotic time series became realizable. Our idea is perfectly supported by computer simulations.

1 Introduction

The neural network has been of great research interest as a universal nonlinear fitting and prediction model for a long time. Compared to traditional linear models (ARMA [3]), the neural network predictor has some remarkable advantages. For example, MLPs with proper topological structure can approximate any continuous function with arbitrary precision; MLPs and certain types of recurrent networks (RNN[1~2], NARX[6], TLRN[4] etc.) have the ability of achieving NAR and NARMA respectively; MLPs and two types of recurrent networks with rational parameters have Turing-equivalent computational ability [5~6]. According to Church-Turing's thesis, the limitation of the computational ability of neural networks is the same as that of Turing algorithms. To sum up, in theory, neural networks possess universal modeling ability, which is superior to linear models, thus, it is possible for neural networks to model complex phenomena more precisely in reality.

The advantages in the modeling ability of neural networks are obtained at the cost of the analytical difficulty, which is an aftermath of the introduction of nonlinearity. Due to the nonlinearity, researchers could not develop a set of statistical methods to implement model identification that can determine the type and the order of a given model. But in practice, researchers often need to accomplish system identification and chaotic time series prediction at the absence of transcendent knowledge. For example, there might be a demand for predicting the future tendency of stock prices according to the historical time series of stock prices. In many similar cases, researchers only have a univariate or multivariate time series. Due to the lack of prior knowledge and mature model identification techniques, researchers have to compromise to adopt lin-

F. Yin, J. Wang, and C. Guo (Eds.): ISNN 2004, LNCS 3174, pp. 424–429, 2004.

ear models, or they have to rely on incomplete experience and arbitrary experiments to perform model identification. Obviously, these two ways are usually insufficient: the former can't be effectively used in the nonlinear time series; the latter can't be certain to choose the optimal model.

Based on information entropy theory, we put forward a method to define and measure the Nonlinear Irreducible Autocorrelation (NIA) of a time series, which is a quantitative index and is naturally corresponding to the nonlinear autoregressive order of NAR. Utilizing NIA to choose optimal scale of neural network predictors, we are able to achieve robust prediction of chaotic time series. The result of simulations supports our idea perfectly.

2 Nonlinear Irreducible Autocorrelation

Our method is a model-independent solution based on information entropy, in which only the statistical character of experimental time series and the underlaying thought of dynamical system (DS) are concerned, thus we might avoid dealing with the analytical difficulty brought in by nonlinearity.

Assume that we have a time series $X_t, t = 1,2,...$, in hands that leaks hints, e.g., the passing of test of the surrogate data [10], to be regarded as a chaotic time series. Then from the perspective of DS, we postulate that $X_t, t = 1,2,...$, is a discrete sampling of an observable dimension $x^{(i)}$ of certain differential equations with k free degrees:

$$\dot{\mathbf{x}} = f(\mathbf{x}), f \in C^1 \tag{1}$$

where $\mathbf{x} = [x^{(1)}, x^{(2)},..., x^{(k)}]^T$ is a $k \times 1$ column vector. Assuming that there are few very loose preconditions, we can indicate that there is a smooth function to determine $x_t^{(i)}$ by its finite history $x_{t-1}^{(i)}, x_{t-2}^{(i)},..., x_{t-n}^{(i)}$, where $x_t^{(i)}$ denote $x^{(i)}$ at the time of tT. Here, t should usually be an integer, T is the time delay of sampling, and n is a positive integer determined by the topological property of the attractor of DS (1).

Definition 1:If ϕ is a flow on an open subset U of R^k, which is determined by DS (1), $h: U \to R$ is a smooth function, and T is a positive number (called the delay), define the delay-coordinate map $F(h,\phi,T): U \to R^n$ by

$$F(h,\phi,T)(\mathbf{x}) = [h(\mathbf{x}), h(\phi_{-T}(\mathbf{x})),..., h(\phi_{-(n-1)T}(\mathbf{x}))]^T$$

Theorem 1 [8]: (Fractal Delay Embedding Prevalence Theorem). Let ϕ be a flow on an open subset U of R^k, and let A be a compact subset of U of box-counting dimension d. Let $n > 2d$ be an integer, and $T > 0$. Assuming that A contains at most a finite number of equilibria, no periodic orbits of ϕ of period T and $2T$, at most finite many periodic orbits of period $3T, 4T,..., nT$, and the linearization of those periodic orbits have distinct eigenvalues. Then for almost every smooth function h on U, the delay coordinate map $F(h,\phi,T): U \to R^n$ is a smooth diffeomophism from

A onto its image $F(A)$ (called the reconstructed set), that is, a smooth one-to-one map that has a smooth inverse.

Theorem 2 [7, 9]:Supposed that $F(h,\phi,T):U \to R^n$ is the delay coordinate map with the conditions of theorem 1, and $x_t^{(i)}, t=1,2,...$, is a T delay discrete sampling of an observable dimension $x^{(i)}$ of certain differential equations that determine the flow ϕ. Then, $x_t^{(i)}$ could be functionally determined by a C^1 function g

$$x_{t+1}^{(i)} = g(x_t^{(i)}, x_{t-1}^{(i)},..., x_{t-(n-1)}^{(i)}) \tag{2}$$

Following theorem 2, we get a physical explanation of NAR models. It utilizes finite historical information to fit g, moreover, NARMA could utilize moving average terms to improve the fitting precision additionally.

It is obvious that the goal of the identification of nonlinear autoregressive models is to choose the smallest n that lets g become a function. Traditionally, to estimate n, researchers have to estimate some dimensions of the underlaying attractor first, e.g., box-counting dimension [8] or correlation dimension [10]. According to the results of the first-step-estimating, researchers might follow the sufficient condition of embedding diffeomophism, which was proposed by Taken embedding theorem [12] originally, to get an interval of n. But the estimating of the dimension of the underlaying attractor is hard to be achieved precisely [14–15], and Taken theorem could only ascertain a rough interval of n even if the first step estimating have been done well. There are a few references that show it is easy to get a fine reconstruction under the condition of $n =$ *dimensions of the underlaying attractor* in spite of the sufficient condition of $n > 2\times$ *dimensions of the underlaying attractor* [11]. Thus, for high dimension attractors, conventional method couldn't work perfectly.

To get the optimal value of n, we utilize a quantitative index NIA to measure the irreducible dependency of time series, which is naturally corresponding to the smallest n that lets g be a function.

The NIA is derived from MI. The definition of MI of a time series $X_t, t=1,2,...$ is

$$I(X_{t+1} \mid X_t) \equiv H(X_{t+1}) - H(X_{t+1} \mid X_t) = H(X_{t+1}) + H(X_t) - H(X_{t+1}, X_t)$$

where $H(X_t) = -\sum_i P_{X_t}(x_{ti}) \log P_{X_t}(x_{ti})$, $H(X_{t+1} \mid X_t) \equiv -\sum_i P_{X_t}(x_{ti}) H(X_{t+1} \mid x_{ti})$, $H(X_{t+1} \mid x_{ti}) \equiv -\sum_j P_{X_{t+1}|X_t}[x_{(t+1)j} | x_{ti}] \log P_{X_{t+1}|X_t}[x_{(t+1)j} \mid x_{ti}]$, x_{ti} denotes a possible value of X_t.

According to the formal properties and physical sense of information entropy. $H(X_{t+1})$ is a measure of the uncertainty of X_{t+1}, i.e., its information capacity. $H(X_{t+1} \mid X_t)$ is a measure of the posterior uncertainty of X_{t+1} given the value of X_t in advance. Therefore, $I(X_{t+1} \mid X_t)$ reflects the dependency between X_{t+1} and X_t, and is suitable to be an index of nonlinear autocorrelation of time series. For the reasons discussed above, developing the index NIA to reflect nonlinear irreducible dependency of time series is required for the goal of nonlinear model identification.

For the sake of intuitiveness, we give its second order definition first and then expand it to general situations. Let

$$I(X_{t+1} \mid X_t, X_{t-1}) = H(X_{t+1}) - H(X_{t+1} \mid X_t, X_{t-1})$$
$$= H(X_{t+1}) + H(X_t, X_{t-1}) - H(X_{t+1}, X_t, X_{t-1})$$

Following the above definition, we get the second order definition of $\rho_{NIA}(2)$:

$$NIA(2) \equiv I(X_{t+1} \mid X_t, X_{t-1}) - I(X_{t+1} \mid X_t)$$

The physical sense of the definition is clear. It is a measure of independent contribution of X_{t-1} to determine X_{t+1}. Expanding the definition to general conditions, we have

$$I[X_{t+1} \mid X_t, X_{t-1}, \ldots, X_{t-(p-1)}]$$
$$= H(X_{t+1}) + H[X_t, X_{t-1}, \ldots, X_{t-(p-1)}] - H[X_{t+1}, X_t, X_{t-1}, \ldots, X_{t-(p-1)}] \quad p \geq 1$$

let $I(p) \equiv I[X_{t+1} \mid X_t, X_{t-1}, \ldots, X_{t-(p-1)}]$, $p \geq 1$. Thus, we get kth order definition: $\rho_{NIA}(1) = I(1)$, $\rho_{NIA}(k) = I(k) - I(k-1)$, $k > 1$

3 Robust Prediction of Chaotic Time Series

We hope the autoregressive order p of the NAR or NARMA model implemented by neural networks is the smallest one that can satisfy the function relation determined by formula (2). If p is too big, the computational cost and the number of the local optimal points in the parameter space of neural network increase, while the convergent possibility of neural network decreases. What's more important is over-fitting phenomena may appear, which will suppress the prediction precision of noisy data prediction [4].

We selected 4 sets of sampling time series. One is from a chaotic system of the real world: Nuclear Magnetic Resonance (NMR) laser time series [13]. The others are gotten by numeric simulation. Two of simulation time series are derived by discrete map, whose iterative formulas are given by Logistic Map $X_{t+1} = 4X_t(1 - X_t)$ and Triangle2 Map $X_{t+1} = 3(\sin(X_t) + \cos(X_{t-1}))$. Another simulation time series is created by Lorenz Equations [10], whose numeric solution adopts Euler method with step length of 0.01. We made sampling towards parameter X One sample data was taken every 4 times of iteration. During sampling, we omitted the points generated under transient states, which is corresponding to the beginning phase of the iteration. The learning and prediction sets of all time series include 200 points respectively, which were normalized in the interval [-1,1].

The Mutual Information (left) and Nonlinear Irreducible Autocorrelations (right) of the 4 sets of sampling time series are traced out in Figure 1. The prediction errors of different neural networks predictors are shown in Table 1. We used 3 types of neural networks predictors, i.e., MLPs, RBF with Gaussian kernel, and TLFN [4], with different topological structures to predict the above sampling time series. Each numeric value in the table grid is the smallest square error (SSE) of the prediction sets. MLP1 denotes MLPs that have 1 processing element in the input layers, and analogically do RBFs and TLFNs.

As for Logistic and Triangle2 sampling time series, we can conclude directly from their iterative formula that there exists 1st and 2nd nonlinear irreducible auto-dependency among their sampling points, and the conclusion tallies with the computational results shown in Figure 1 (right) perfectly. From Table 1, we can conclude that the Logistic time series can be predicted effectively by using NAR(1), which is implemented by MLP1, RBF1 or TLFN1 with only one processing element in its input layer. Increasing the order of the model cannot improve the quality of prediction markedly. From Figure 1 (right), the nonlinear irreducible autocorrelations with orders bigger than 1 are approximately 0. This explains theoretically the adequacy of models with order 1. The similar analysis is true to Triangle2 time series. As for Lorenz sampling time series, whose underlaying dynamical system has an attractor with box-counting dimension 2.06 [10]. Its nonlinear irreducible autocorrelation reflects the proper embedding dimension, which is necessary for a topologically equivalent reconstruction and a satisfactory prediction of underlaying attractor. As for NMR, the optimal NAR order is 2. It accords with the computational results of NIA. On the other hand, as a contrast, mutual information shown in Figure 1 (left) couldn't show any hint to reflect the above insight.

Table 1. Prediction errors (SSE)

	Logistic	Triangle2	Lorenz	NMR
MLP1	0.58	13.7	15.8	17.3
MLP2	0.58	2.86	1.41	1.85
MLP3	0.69	3.03	1.39	1.82
RBF1	0.41	12.8	14.7	16.8
RBF2	0.42	2.16	1.04	1.77
RBF3	0.42	2.25	1.13	1.75
TLFN1	0.47	13.9	15.5	17.0
TLFN2	0.49	2.75	1.29	1.91
TLFN3	0.50	2.82	1.33	1.89

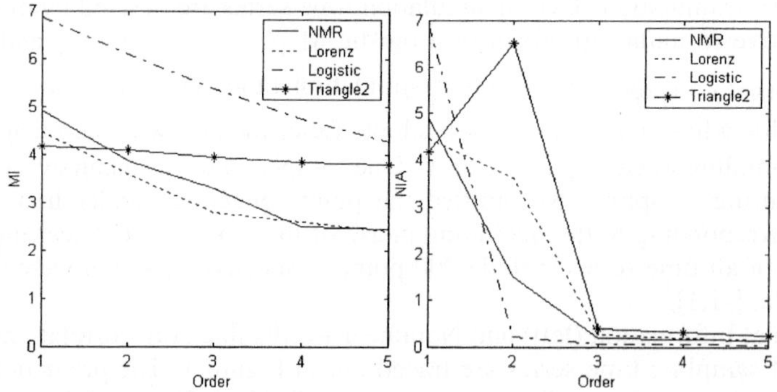

Fig. 1. Mutual information (left) and NIA (right) of 4 time serie

According to the above results, it is clear that the experimental results of optimal order of NAR are exactly corresponding to those of NIA. Therefore, we can conclude that nonlinear irreducible autocorrelation is reasonable as the measurement of irreducible auto-dependency of time series, which is suitable to determine the optimal order of universal NAR predictors.

4 Conclusion

We proposed a novel model identification technique based on nonlinear irreducible autocorrelation. Computer simulations indicated that the method could determine the optimal order of nonlinear auto-regression universally, thus improve the robustness of chaotic time series prediction effectively.

References

1. Connor, J.T. et al.: Recurrent Neural Networks and Robust Time Series Prediction. IEEE Trans. on Neural Networks, Vol. 5. No. 2. (1994) 240–254
2. Parlos, A.G. et al.: Application of the Recurrent Multiplayer Perception in Modeling Complex Process Dynamics. IEEE Trans. on Neural Networks, Vol. 5. No. 2. (1994) 255–266
3. Box, G. et al.: Time Series: Forecasting and Control Revised Edn. Holden-day, San Francisco (1976)
4. Principe, J.C. et al.: Neural Systems: Fundamentals through Simulations, NeuroDimension and Wiley & Son (2000)
5. Siegelmann, H.T. et al.: On the Computational Power Neural Networks. J. Comput. Syst. Sci., Vol. 50. No. 1. (1995) 132–150
6. Siegelmann, H.T. et al.: Computational Capabilities of Recurrent NARX Neural Networks. IEEE Trans. on Systems Man and Cybernetics, Vol. 27. No. 2. (1997) 208–215
7. Yuexian, Hou et al.: The Neural Network Predictor Identification Based on Prediction Complexity. Information and Control, Vol. 30. No. 1. (2001) (Chinese)
8. Sauer et al.: Embedology. J. Stat. Phys. Vol. 65. (1991) 579
9. Yuexian, Hou.: The Prediction and Control of Nonlinear System, (PhD degree dissertation). Tianjin University, Tianjin (2001)
10. Kaplan, D. et al.: Understanding Nonlinear Dynamics. Springer-Verlag, (1995)
11. Shinbrot, T. et al.: Nature. London (1993) 363–411
12. Taken, F.: Detecting Strange Attractors in Turbulence. Lecture Notes in Mathematics, No.898. Springer-Verlag, Berlin Heidelberg New York (1981)
13. Holger, Kantz. et al.: Nonlinear Time Series Analysis, Cambridge University Press (2000)

Wavelet Neural Networks for Nonlinear Time Series Analysis

Bo Zhou[1], Aiguo Shi[2], Feng Cai[2], and Yongsheng Zhang[2]

[1] Dep. of Basics, Dalian Naval Academy ,116018, China
judyever@sina.com
[2] Dep. of Navigation, Dalian Naval Academy, 116018, China

Abstract. A simple model based on the combination of neural network and wavelet techniques named wavelet neural network (WNN) is proposed. Thanks to the time-frequency analysis feature of wavelet, a selection method that takes into account the domain of input space where the wavelets are not zero is used to initialize the translation and dilation parameters. A proper choice of initialize parameters is found to be crucial in achieving adequate training. Training algorithms for feedback WNN is discussed too. Results obtained for a nonlinear processes is presented to test the effectiveness of the proposed method. The simulation result shows that the model is capable of producing a reasonable accuracy within several steps.

1 Introduction

The Wavelet Neural Network (WNN) belongs to a new class of neural networks with unique capabilities in system identification and classification. Wavelets are a class of basic functions with oscillations of effectively finite-duration that makes them look like “little waves”, the self-similar, multiple resolution nature of wavelets offers a natural framework for the analysis of physical signals. On the other hand, artificial neural networks constitute a powerful class of nonlinear function approximants for model-free estimation.

The concept of Wavelet Neural Network was inspired by both the technologies of wavelet decomposition and neural networks. In standard neural networks, the non-linearity is approximated by superposition of sigmoid functions. While in WNN, the nonlinearities are approximated by superposition of a series of wavelet functions.

We first present the concept of wavelets and the structures of WNN in section 2, in section 3, the initialization of parameters are presented. In section 4, the training systems and algorithms are described. For illustration purposes, the modeling of a nonlinear process by wavelet networks is presented in section 5.

F. Yin, J. Wang, and C. Guo (Eds.): ISNN 2004, LNCS 3174, pp. 430–435, 2004.

2 Basic Concepts of WNN

2.1 Wavelet Analysis

Wavelets are obtained from a single prototype wavelet $\psi(t)$ called mother wavelet by dilations and translations:

$$\Omega_c = \left\{ \psi_{a,b}(t) = \frac{1}{\sqrt{a}} \psi(\frac{t-b}{a}) \right\} \tag{1}$$

Where Ω_c is a family of wavelets, a is the dilation parameter and b is the translation parameter, they are real numbers in R and R^+ respectively.

The most important properties of wavelets are the admissibility and the regularity conditions and these are the properties, which gave wavelets their name. It is square integrable function $\psi(t)$ satisfying the admissibility condition:

$$C_\psi = \int_{-\infty}^{+\infty} \frac{|\Psi(w)|^2}{w} dw < \infty \tag{2}$$

$\Psi(w)$ stands for the Fourier transform of $\psi(t)$. Any function of $L^2(R)$ can be approximated to any prescribed accuracy with a finite sum of wavelets. Therefore, wavelet networks can be considered as an alternative to neural and radial basis function networks.

2.2 Structure of WNN

Employing the wavelets function $\psi_{a_j b_j}(t)$ as the activation functions of the hidden layer units, the output of the ith unit in the output layer of the proposed network is given as :

$$y_i(t) = \sum_{j=1}^{L} w_{ij} \psi_{a_j b_j} (\sum_{k=1}^{M} w_{jk} x_k)\ (i = 1,2,\ldots\ldots,N) \tag{3}$$

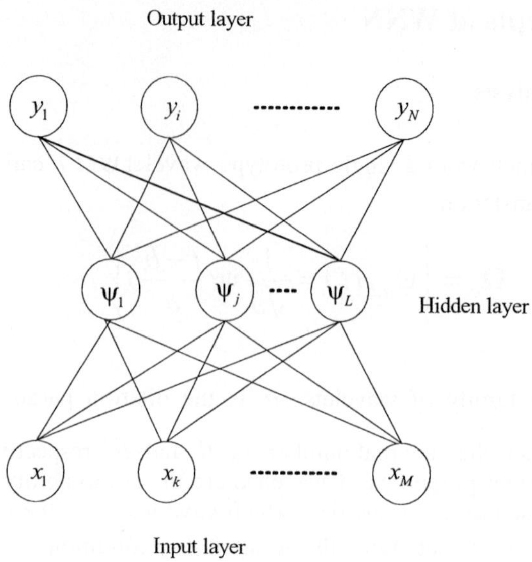

Fig. 1. The structure of wavelet neural network. x_k is the kth input vector, y_i is the ith output vector, w_{ij} is the synaptic weight connecting the output layer node i and hidden layer node j, w_{jk} is the synaptic weight connecting the input layer node k and hidden layer node j. a_j is the dilation parameter of the jth hidden node, b_j is its translation parameter ; $M(k=1,2,......M)$ is the number of neurons in the input layer, $n(j=1,2,......n)$ is the number of neurons in the hidden layer. $L(i=1,2,......,L)$ is the number of neurons in the output layer. y is the output of the network.

3 Initializing Wavelet Networks

Initializing the wavelet network parameters is an important issue. Similarly to Radial Basis Function networks (and in contrast to neural networks using sigmoid functions), we should not initialize the dilations and translations randomly, and we should take advantage of the input space domains where the wavelets are not zero[1][2].

Therefore, we propose an initialization for the mother wavelet the input domains defined by the examples of the training sequence.

We denote by $[x_{k\min}, x_{k\max}]$ the domain containing the input vector x_k. Let t_0 and σ_ψ be the center and the radius of the mother wavelet ψ in time domain, then the domain of $\psi_{a_j b_j}$ is given by $[b_j + a_j t_0 - a_j \sigma_\psi, b_j + a_j t_0 + a_j \sigma_{\hat{\psi}}]$.

In order that the wavelet $\psi_{a_j b_j}$ can cover the input space:

$$b_j + a_j t_0 - a_j \sigma_\psi = \sum_{k=1}^{M} w_{jk} x_{k\min} \tag{4}$$

$$b_j + a_j t_0 + a_j \sigma_\psi = \sum_{k=1}^{M} w_{jk} x_{k\max} \tag{5}$$

Dilations parameter is initialized to:

$$a_j = \frac{\sum_{k=1}^{M} w_{jk} x_{k\max} - \sum_{k=1}^{M} w_{jk} x_{k\min}}{2\sigma_\psi} \tag{6}$$

Translation parameter is initialized to:

$$b_j = \frac{\sum_{k=1}^{M} w_{jk} x_{k\max}(\sigma_\psi - t_0) + \sum_{k=1}^{M} w_{jk} x_{k\min}(\sigma_\psi + t_0)}{2\sigma_\psi} \tag{7}$$

These initializations guarantee that the wavelets extend initially over the whole input domain. The choice of the weights is less critical. They are initialized to small random values. This procedure is very simple and requires a small number of operations.

4 Training Wavelet Neural Networks

An error vector can be defined as the difference between the network output y and the desired output d. As usual, the training is based on the minimization of the following cost function:

$$E = \frac{1}{2}\sum_{i=1}^{N}(d_i - y_i)^2 \tag{8}$$

The minimization is performed by iterative gradient-based methods. The partial derivatives of the cost function with respect to the parameters θ are:

$$\frac{\partial E}{\partial \theta} = -\sum_{i=1}^{N}(d_i - y_i)\frac{\partial y}{\partial \theta} \tag{9}$$

Where θ denotes the parameters that should be trained, they are w_{ij}, w_{jk}, a_j and b_j. At each iteration , the parameters are modified using the gradient according to:

$$\theta(t+1) = \theta(t) - \eta \frac{\partial E}{\partial \theta} + u\Delta\theta \tag{10}$$

Where η is a constant known as the step-size or the learning rate, u is the momentum term.

As any gradient-descent based algorithm, the error back propagation algorithm suffers from slow convergence and high probability of converging to local minima. There are ways to improve the algorithm by using adaptive learning rates and momentum terms; the details are illustrated in reference [3]·

The algorithm is stopped when one of several conditions is satisfied: the variation of the gradient, or of the variation of the parameters, reaches a lower bound, or the number of iterations reaches a fixed maximum, whichever is satisfied first.

5 Simulation Results

The example is the approximation of a function of a variable function, given by:

$$f(x) = \begin{cases} 43.72x - 8.996 & for \quad x \in [0,0.4] \\ -84.92x + 42.46 & for \quad x \in [0.4,0.5] \\ 10\,exp(-x)\,sin(12x^2 + 2x - 4) & for \quad x \in [0.5,1] \end{cases} \tag{11}$$

A wavelet network approximation in the domain [0, 1] is to be found from a training sequence of 50 examples. We first predefined that there are 6 wavelets in the hidden layer, the maximum simulation steps is 200 and the maximum MSE is 0.001. Then we select the initiation value of a and b using the method mentioned above, at same time, randomly initiate the value of w_{ij} and w_{jk}. Finally, we trained the WNN using gradient-descent method.

Table 1. The changes of MSE relative to the change of steps. The MES is less than 0.001 only within 6 steps.

steps	Mean Square Error(MSE)
1	34.33712175
2	4.57306893
3	1.01194119
4	0.20106540
5	0.04185991
6	0.00763924

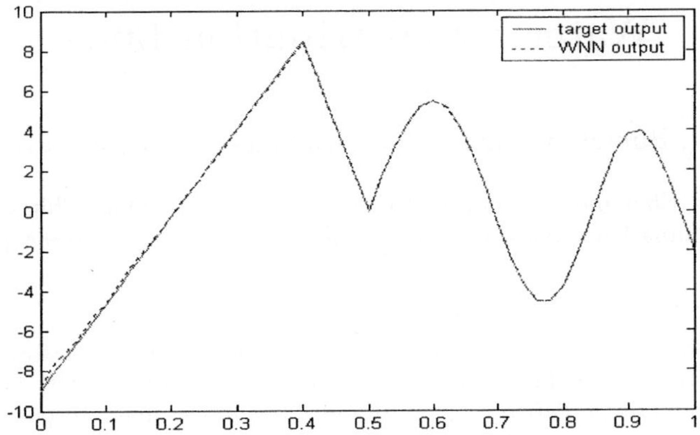

Fig. 2. The output of wavelet neural network (in dot line) and the target output (in real line)of the system characterized by formula (7).

6 Conclusion

In this paper, we introduce the use of wavelet networks for nonlinear function approximation. We show the training method through second order gradient descent implemented in a back propagation scheme, with appropriate initialization of the translation and dilation parameters. The training procedure is very simple and the result is reliable.

References

1. Yacine ,O., Gerard D.: Initialization by Selection for Wavelet Network Training. Neurocomputing, vol. 34. (2000) 131-143
2. Oussar, Y. , Rivals, I. , Personnaz, L.: Training Wavelet Networks for Nonlinear Dynamic Input-Output Modeling. Neurocomputing, vol.20. (1998) 173-188
3. Hornik, K. , Stinchcombe, M. , White, H.: Multilayer Feed Forward Networks Universal Approximators. Neural Networks, vol.2. (1989) 359-366

Neural Networks Determining Injury by Salt Water for Distribution Lines

Lixin Ma[1], Hiromi Miyajima[2], Noritaka Shigei[2], and Shuma Kawabata[2]

[1] Shanghai University of Electric Power, Shanghai, China
[2] Kagoshima University, 1-21-40 Korimomto, Kagoshima 890-0065, Japan

Abstract. Japan has so may solitary islands and wide coastal areas. In the air of these areas, there are so many grains of sea-salt. The salt grains adhere to distribution lines and damage the lines. In particular, the damage of covered electric wires is a serious problem. However, any method has not been proposed that judges if distribution lines are injured by salt water. This paper proposes a neural network method to determine injury of distribution lines by salt.

1 Introduction

Japan has so many solitary islands and wide coastal areas. In the air of these areas, there are so many grains of sea-salt blown by Typhoon and sea breeze. The salt grains adhere to distribution lines (service wires) and damage the lines. Hence,by adhering to distribution lines (service wires) of them, injury by salt for distribution lines occurs. In particular, the damage of covered electric wires (the connection between insulator and covered wire) is a serious problem. Much effort and money have been spent to prevent it. However, any effective method have not been found yet. Therefore, we consider to construct a diagnosis system that measure the rate of injury of distribution lines caused by salt. The diagnosis system alleviates the effects of salt on distribution lines and enables to replace seriously injured ones with new ones. As the first step to construct the diagnosis system, we have already measured the distribution of electric field around the connection between insulators and distribution lines by using an optical sensor[1].

In this paper, we construct a neural network method to determine injury of distribution lines by salt. The neural network method is based on higher order neural networks and back propagation. The effectiveness of the method is demonstrated on our actual measurement data.

2 Measurement of Injury by Salt for Distribution Lines

In the previous paper, we have measured the distribution of electric fields around insulators[1]. The Fig.1 is a conceptual figure to show the place of insulators. The distribution of electric field is measured at an impressed voltage of 1500[V]. Measuring points are assigned on lattice ones around insulators as shown in the Fig.2. In the Ref.[1], three solutions of salt with densities of 0.5, 1.0 and 2.0%

F. Yin, J. Wang, and C. Guo (Eds.): ISNN 2004, LNCS 3174, pp. 436–441, 2004.

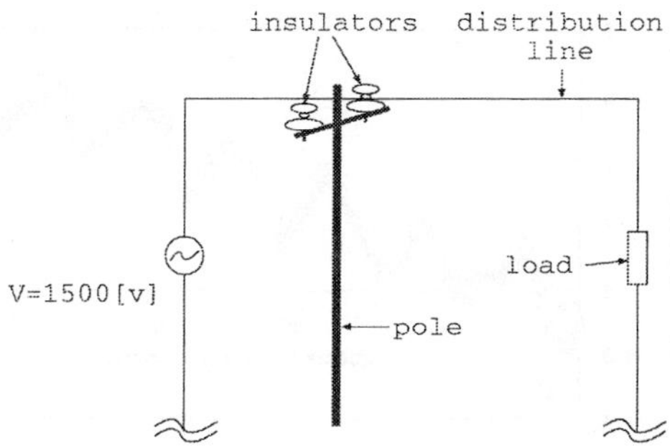

Fig. 1. A conceptual figure to show the place of insulators

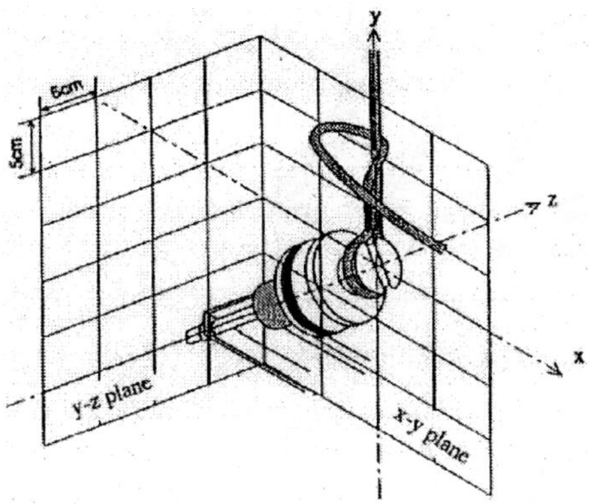

(a) The connection between an insulator and distribution lines

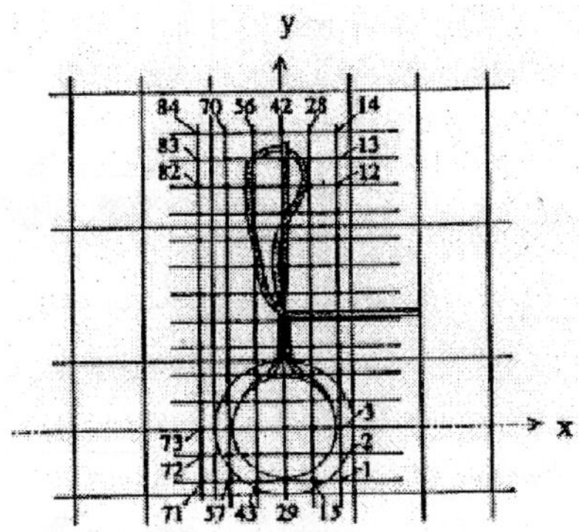

(b) Measuring points of electric field

Fig. 2. Measuring points of electric field.

were prepared and uniformly atomized to the measuring subject. Note that the density of sea water is about 3%. For each of three soil cases (solution densities of 0.5, 1.0 and 2.0 %) and the pure case, the measurement was performed repeatedly over several days. Fig.3 summarizes the measurement results. Fig.3(a), (b) and (c) show the average values of electric fields E_x, E_y and E_z in the direction of x-, y- and z-axes, respectively. The place of measurement is x-y plane with $z = 5$ cm. Note that the soil cases are not always separated from the pure case seemingly as shown in Fig.3, where $\diamond$ is for the pure case, and $\triangleleft$, $\circ$ and $\bullet$ are for soil cases with solution densities 0.5, 1.0, and 2.0%, respectively.

3 Higher Order Neural Networks and Back Propagation Method

First, we consider multi-layered higher order neural networks each of which consists of N_m ($m = 0, \cdots, M-1$) units, where the number of layers is M, the

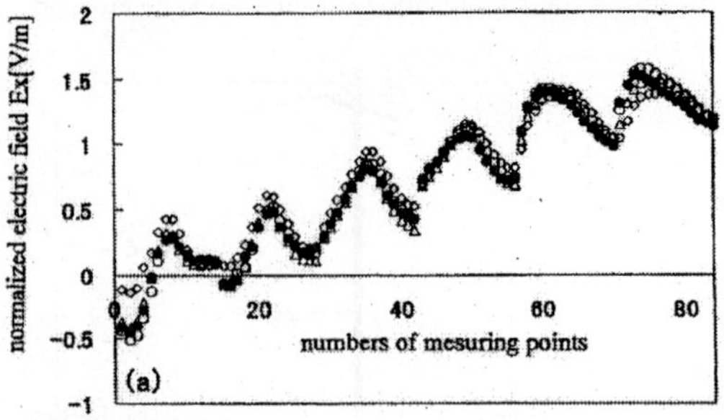

(a) Distribution of normalize electric field E_x

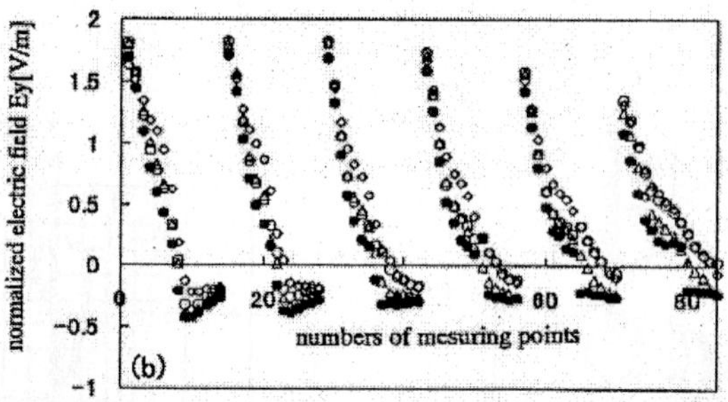

(b) Distribution of normalize electric field E_y

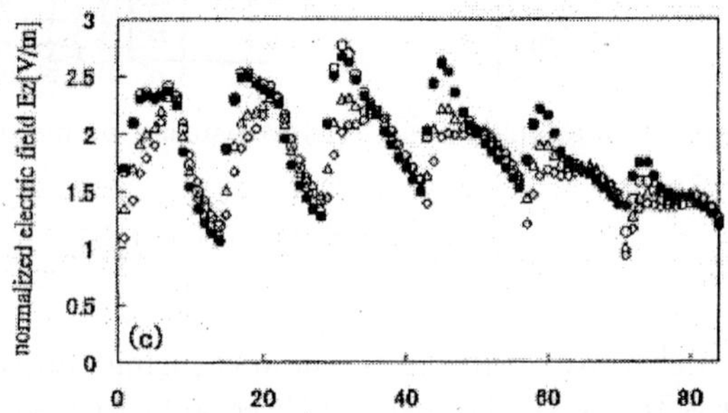

(c) Distribution of normalize electric field E_z

Fig. 3. Average values of the electric field around the connection between an insulator and distribution lines

0-th and the $(M-1)$-th layers are called input and output layers, respectively. The i-th (higher order neural) element of the m-th layer has N_{m-1} inputs and one output and each value of input takes any real number and output takes any real numbers from 0 to 1. The state of a unit (the output) is represented as a function of potential u. The potential u increases in proportion to the weighted $\sum_{[L_p]} W_{i[L_p]} y_{l_1} \cdots y_{l_p}$ of all combinations of products of p pieces of input variables y_i ,$1 \le i \le N_{m-1}$ and a threshold s_i, where

$[L_p] = l_1, \cdots, l_P$ and $\sum_{[L_p]}$ is defined as $\sum_{l_1=1}^{N_{m-1}-p+1} \sum_{l_2=l_1+1}^{N_{m-1}-p+2} \cdots \sum_{l_p=l_{p-1}+1}^{N_{m-1}}$ to exclude the overlapping of variables. $W_{i[L_p]}$ is called a weight of products of p pieces of input variables. Then, the output of the i-th unit of the m-th layer, $y_i^{(m)} (i = 1, \cdots, N_m, m = 1, \cdots, M-1)$, is determined as follows.

$$y_i^{(m)} = F[\sum_{p=1}^{k} w_{i[L_p]}^{(m)} y_{l1}^{(m-1)} \cdots y_{lp}^{(m-1)} + s_i^{(m)}], \qquad F[x] = \frac{1}{1+e^{-x}}, \tag{1}$$

where k means the maximum number of order for MHONNs. For data $\boldsymbol{y}^{(o)}$ to input layer, let the desirable value of output layer be represented by $\boldsymbol{d}$, where $\boldsymbol{y}^{(0)} = (y_1^{(0)}, \cdots, y_{N_0}^{(0)}), \boldsymbol{d} = (d_1, \cdots, d_{N_{M-1}})$. The output $y_i^{(m)}$ for $\boldsymbol{y}^{(0)}$ is computed from the Eq.(1). The square error E between the output $y_i^{(M-1)}$ and the desirable output d_i is defined as follows.

$$E = \frac{1}{2} \sum_{i=1}^{N_{M-1}} (y_i^{(M-1)} - d_i)^2 \tag{2}$$

In the following, we will deal with the case of $k = 2$, $M = 3$ and $N_2 = 1$ without loss of generality. The networks are called higher order neural networks (HONN). From the Eqs.(1) and (2), a back propagation method for HONN is shown as follows:

Let P be the number of input-output data and $r = 1, \cdots, P$.
[*Step* 1] Let $r = 1$.
[*Step* 2] The initial values of the weights are set to random real numbers. The input-output data $((y_1^{(0),r}, \cdots, y_{N_0}^{(0),r}, d^r)$ is given. For simplicity,the suffix r of $y_i^{(l),r}$ is omitted for $l = 0, 1, 2$.
[*Step* 3] For inputs $y_i^{(0)} (i = 1, \cdots, N_0)$, the output values in each layer are computed based on the Eq.(1) for $m = 1, 2$.
[*Step* 4] The objective function E over weights $w_{il_1}^{(m)}$ and $w_{il_1 l_2}^{(m)}$ is computed as follows.
(a) With output layer,

$$\begin{aligned} &\frac{\partial E}{\partial y_i^{(2)}} = y_i^{(2)} - d_i, \quad \frac{\partial E}{\partial x_i^{(2)}} = \frac{\partial E}{\partial y_i^{(2)}} y_i^{(2)} (1 - y_i^{(2)}) \\ &\frac{\partial E}{\partial w_{il_1}^{(2)}} = \frac{\partial E}{\partial x_i^{(2)}} y_{l_1}^{(1)}, \quad \frac{\partial E}{\partial w_{ill_2}^{(2)}} = \frac{\partial E}{\partial x_i^{(2)}} y_l^{(1)} y_{l_2}^{(1)} \\ &(i, l_1 = 1, \cdots, N_2; l = 1, \cdots, N_1 - 1, l_2 = l + 1, \cdots, N_1) \end{aligned} \tag{3}$$

(b) With hidden layer,

$$\frac{\partial E}{\partial y_l^{(1)}} = \sum_{i=1}^{N_2} \frac{\partial E}{\partial x_i^{(2)}} \frac{\partial x_i^{(2)}}{\partial y_l^{(1)}},$$
$$\frac{\partial x_i^{(2)}}{\partial y_l^{(1)}} = w_{il}^{(2)} + G(l) \sum_{l_1=1}^{l-1} w_{il_1 l}^{(2)} y_{l_1}^{(1)} + H(l) \sum_{l_2=l+1}^{N_1} w_{ill_2}^{(2)} y_{l_2}^{(1)} \tag{4}$$
$$\frac{\partial E}{\partial x_l^{(1)}} = \frac{\partial E}{y_l^{(1)}} y_l^{(1)}(1 - y_l^{(1)}), \quad \frac{\partial E}{\partial w_{lh}^{(1)}} = \frac{\partial E}{\partial x_l^{(1)}} y_h^{(0)}, \quad \frac{\partial E}{\partial w_{lpg}^{(1)}} = \frac{\partial E}{\partial x_l^{(1)}} y_p^{(0)} y_g^{(0)}$$
$$(i = 1, \cdots, N_2; p = 1, \cdots, N_1 - 1; l = 1, \cdots, N_1; \quad h = 1, \cdots, N_1; g = p + 1, \cdots, N_1),$$

where $G(l) = 0$ for $l = 1$ and 1 for $l \neq 1$,and $H(l) = 0$ for $l = N_2$ and 1 for $l \neq N_2$.

[*Step* 5] The weights $w_{il_1}^{(m)}$ and $w_{il_1 l_2}^{(m)}$are improved according to the following:

$$w_{il_1}^{(m)}[t+1] = w_{il_1}^{(m)}[t] - K_1 \frac{\partial E}{\partial w_{il_1}^{(m)}}, \quad w_{il_1 l_2}^{(m)}[t+1] = w_{il_1 l_2}^{(m)}[t] - K_1 \frac{\partial E}{\partial w_{il_1 l_2}^{(m)}} \tag{5}$$

[*Step* 6] If $r = P$, then the algorithm terminates, if $r \neq P$ then go to Step 2 with $r \leftarrow r + 1$.

Specifically, when $w_{il_1 l_2}^{(m)} = 0$ for $m = 1, 2$, $i = 1,\cdots,n$, $l_1 = 1,\cdots,N_2$ and $l_2 = l_1 + 1, \cdots, N_2$, the method is same as back propagation for the conventional neural networks (NN)[2].

4 Construction of Diagnosis Systems

By using back propagation method for HONN, learning and evaluation for diagnosis system are performed as follows:
(1) Measured data are separated to two groups of learning and testing one.
(2) Learning for each system of HONN is performed.
(3) By using testing data, generalization ability of each system is evaluated.

Let us explain each process easily. First, 84 learning data are used in order to construct a system. In this case, 42 data measured in the case (pure case) sprinkled nothing over distribution lines and 42 data measured in the case (soil case) sprinkled a solution of salt with density of 0.5, 1.0 or 2.0% over distribution lines are prepared. Three variables, two coordinates which are x- and y- ordinates and the absolute of electric field $\sqrt{E_x^2 + E_y^2 + E_z^2}$ as input variables, one variable as output ones and the appropriate number of the units for hidden layer are selected and learning is performed. Then, it is concluded that recognition of 100% which means that all the learning data are separated to pure and soil areas correctly is impossible in any case. It means that the number of input variables

Table 1. Result of diagnosis system by using data measured in the same day

Density		TNN	HONN
0.5%	Learning time [minute]	2	4
	The number of weights	12	7
	Rate of recognition [%]	86	90
1.0%	Learning time [minute]	5	10
	the number of weights	12	7
	Rate of recognition [%]	83	86
2.0%	Learning time [minute]	2	4
	The number of weights	12	6
	Rate of recognition [%]	90	95

is insufficiently. Therefore,we change three input variables to five variables, x-, y-coordinates, electric fields E_x. E_y and E_z. In this case, learning of 100% recognition is performed in all cases. Five variables as input are used in the following. Data used in computer simulation are normalized to the interval 0 to 1 . In determining of output data, the margin of 0.3 is allowed, where 0 and 1 of output value mean pure and soil cases, respectively. Table 1 shows a result of recognition for testing data, where the number of weights means ones of the hidden layer in the case where the best result has been obtained. The rate of recognition means one of data which are recognized correctly. In all cases in the Table 1, the value shows one of the best case. As a result, the recognition of 83-90% in NN and 86-95% in HONN are achieved.

Further, as compared with other soft computing method, we have constructed the systems by using Vector Quantization (VQ) and Fuzzy systems[3]. Numerical simulations are performed under the same condition as Table 1. As a result, the rates of recognition of 81–88% and 86–97% for VQ and FS are performed, respectively.

5 Conclusion

In this paper, we have constructed a diagnosis system to judge if distribution lines are soil (salty) or not. The system is based on higher order NNs. We have tested our method and a traditional one on actual measurement data. The result have demonstrated that our method is superior to the traditional one. In the future work, we will study to reduce the learning time and the number of sample points.

References

1. S. Kawabata et al., "Measurement of electric field around distribution lines by an optical sensor", National Convention Record of IEE.Japan, 1588, 1998 (in Japanese).
2. C.T. Lin and C.S.G. Lee, "Neural Fuzzy Systems", Prentice Hall PTR, 1996.
3. K. Kishida and H. Miyajima, "A learning method of fuzzy inference rules using vector quantization", Proc. ICANN, vol.2, pp.827–832, 1998.

EEG Source Localization Using Independent Residual Analysis

Gang Tan and Liqing Zhang*

Department of Computer Science and Engineering,
Shanghai Jiaotong University, Shanghai 200030, China
tangang@sjtu.edu.cn, zhang-lq@cs.sjtu.edu.cn

Abstract. Determining the location of cortical activity from electroencephalographic (EEG) data is important theoretically and clinically. Estimating the location of electric current source from EEG recordings is not well-posed mathematically because different internal source configurations can produce an identical external electromagnetic field. In this paper we propose a new method for EEG source localization using Independent Residual Analysis (IRA). First, we apply Independent Residual Analysis on EEG data to divide the raw signals into the independent components. Then for each component, we employ the least square method to locate the dipole. By localizing multiple dipoles independently, we greatly reduce our search complexity and improve the localization accuracy. Computer simulation is also presented to show the effectiveness of the proposed method.

Keywords: EEG, IRA, source localization, least square method

1 Introduction

Electroencephalography (EEG) is a technique for the non-invasive characterization of brain function. The localization of the neuronal activity responsible for measured EEG phenomena is one of the primary problems in EEG. The problem can be formulated in the framework of the *inverse problem* and the *forward problem* [1]. The inverse problem is an ill-posed problem. There is no unique solutions and solutions do not depend continuously on the data.

There exist several different approaches to solving the source localization problem. The dipolar approaches consist in searching one or few dipole locations and orientations the forward solution of which is best fit to measurements [4]. In the case of multiple dipoles overlapping, one must employ a spatio-temporal model. Dipoles are fit over the entire evoked potential epoch by using non-linear least-square error residual minimization [5]. The spatio-temporal model is developed into the spatio-temporal independent topographies model in MUSIC [6] and its extension, RAP-MUSIC [7]. Another approach is the class of

* The Project 60375015 supported by National Natural Science Foundation of China

F. Yin, J. Wang, and C. Guo (Eds.): ISNN 2004, LNCS 3174, pp. 442–447, 2004.

iterative focalization approaches. FOCUSS [2] for example, is based on a recursive minimum-norm approach. LORETA [8] is another example of an iterative re-weighting technique.

Although the above methods represent significant advances in source localization, they are usually very sensitive to noise. In this paper, we introduce a new method for spatio-temporal source localization of independent component. In our method, we first separate the raw EEG data into independent sources using our IRA algorithm. We then perform a separate localization procedure on each independent source. With precomputing the independent components of the EEG data and localizing on each component, we substantially reduce the complexity and increase the localization accuracy.

2 Method

2.1 Forward Problem

The forward equation, which gives scalp electric potentials as a function of the current sources, is:

$$\phi(t) = \mathbf{KJ}(t), \tag{1}$$

where $\phi(t) = [\phi_1(t), \phi_2(t), \ldots, \phi_N(t)]^T$ is comprised of measurements of scalp electric potentials at time instant t, N is the number of electrodes, and $\mathbf{J}(t) = [j_1(t), j_2(t), \ldots, j_M(t)]^T$ is comprised of M current dipoles within the brain volume at time instant t. The $N \times M$-matrix $\mathbf{K}$ is the so-called *lead field matrix* [3]. The problem of interest here is to compute the potentials at the electrodes from the known current sources $\mathbf{J}(t)$.

2.2 Inverse Solution

Solving the inverse problem consists in obtaining an estimate $\hat{\mathbf{J}}(t)$ from the EEG data $\phi(t)$:

$$\hat{\mathbf{J}}(t) = \mathbf{T}\phi(t), \tag{2}$$

where the $M \times N$-matrix $\mathbf{T}$ is some generalized inverse of the lead field matrix $\mathbf{K}$. The solution to this inverse problem can be formulated as the non-linear optimization problem of finding a least square fit of a set of current dipoles to the observed data over the entire time series, or minimization of the following cost function:

$$C(x, y, z, \theta, \psi, p) = \|\phi - \hat{\phi}\| = \sum_t \sum_{i=1}^{N} \left[\phi_i(t) - \hat{\phi}_i(t)\right]^2, \tag{3}$$

where $\phi_i(t)$ is the value of measured electric potential on the ith electrode at time instant t and $\hat{\phi}_i(t)$ is the result of the forward model computation for a particular source configuration; the sum extends over all channels and time frames. Each dipole in the model has six parameters: location coordinates (x, y, z), orientation (θ, ψ), and time-dependent dipole magnitude $p(t)$.

In equation (3), we need to deal with all of the time sequences, which is time-consuming. We will employ an effective method to reduce the search space in the next section.

2.3 Independent Residual Analysis

In order to extract statistically independent components from noisy EEG measurements, we can use Independent Component Analysis (ICA) [9], [1]. The ICA approach for EEG is implemented in the following way: ICA algorithm identifies temporally independent signal sources in multi-channel EEG data, and the extracted independent components are projected back to the scalp surface. These have been shown to be significantly more "dipole-like" than the raw EEG.

In ICA model, EEG measurements ϕ is considered as the linear mixture of the original sources $\mathbf{s} = [s_1, s_2, \ldots, s_N]^T$:

$$\phi = \mathbf{Hs}, \tag{4}$$

where $\mathbf{H}$ is an $N \cdot N$ unknown mixing matrix. In order to extract independent components of the EEG measurements ϕ, we establish a demixing model:

$$\mathbf{y} = \mathbf{W}\phi, \tag{5}$$

where $N \cdot N$ matrix $\mathbf{W}$ is so-called "unmixing" matrix, $\mathbf{y} = [y_1, y_2, \ldots, y_N]^T$ is the recovered version of the original sources $\mathbf{s}$.

There exist several different ways to estimate the $\mathbf{W}$ matrix [9], such as FastICA [10], the Bell-Sejnowski infomax algorithm [11] and its modification,the natural gradient algorithm [12]. When the statistics of data are sufficient, all these aforementioned algorithms generally perform almost equally well.

However, traditional ICA algorithms do not perform well on EEG data resulting from the ignorance of the temporal structure of the EEG signals. Therefore we formulate a new method for blind source separation named as Independent Residual Analysis (IRA) [13]. In this method, we describe the temporal structure of the sources by a stationary AR model:

$$s_i(t) = \sum_{p=1}^{L} a_p^i s_i(t-p) + \varepsilon_i(t), \tag{6}$$

where L is the degree of the AR model and $\varepsilon_i(t)$ is a zero mean independent and identically distributed time series called the residuals. By minimizing the mutual information of the residual signals, We can derive our gradient descent algorithm as follows [13]:

$$\Delta\mathbf{W} = \eta \left\{ \mathrm{I} - \sum_{p=0}^{L} \left[\mathbf{A}_p \varphi(\mathbf{r}) \mathbf{y}^T(t-p) \right] \right\} \mathbf{W}, \tag{7}$$

where $\mathbf{A}_0$ is the identity matrix, $\mathbf{A}_p = -diag(a_p^1, \ldots, a_p^N)$, $p \geq 1$ is a diago-

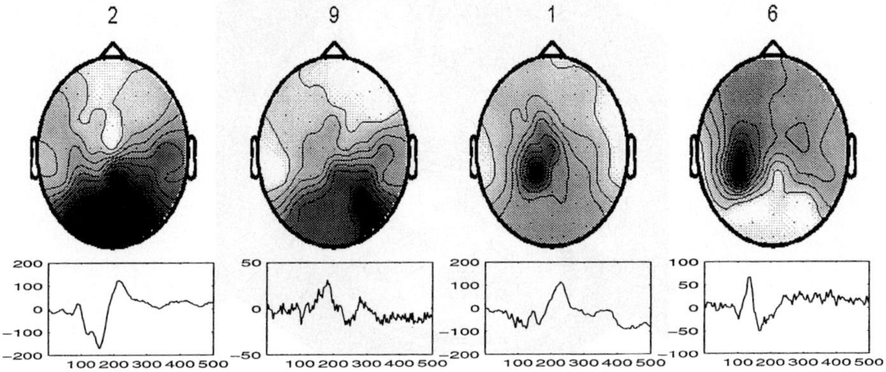

Fig. 1. The scalp maps of the separated components by IRA algorithm

nal matrix and $\mathbf{r}$ is the estimate residual signals. For non-linear function $\varphi(\mathbf{r})$, we can use the hyperbolic tangent function $\varphi(\mathbf{r}) = \tanh(\mathbf{r})$ for super-Gaussian sources and the cubic function $\varphi(\mathbf{r}) = \mathbf{r}^3$ for sub-Gaussian sources. The temporal structure might also affect the learning performance of the natural gradient algorithm. The gradient descent learning algorithm for the filter element a_p^i are:

$$\Delta a_p^i(k) = -\eta_k' \varphi_i \left[r_i(k)\right] y_i(k-p), \tag{8}$$

where η_k' is the learning rate.

After finding the weight matrix $\mathbf{W}$, we can use equation (5) to obtain the original sources. Projection of each independent component of original sources back onto the electrodes at time instant t can be done by:

$$\tilde{\phi}_i(t) = \mathbf{W}^{-1}\tilde{y}_i(t), \tag{9}$$

where $\tilde{\phi}_i(t)$ is the set of scalp potentials due to the ith source. For $\tilde{y}_i(t) = [0, \ldots, 0, y_i(t), 0, \ldots, 0]^T$ we zero out all rows of sources but the ith.

2.4 The Least Square Solution

In equation (9), $\tilde{y}_i(t)$ only scales the scalp potentials, so we need not compute the full time sequence to locate the dipole, but rather simply a single instant of activation. For this purpose, we set $\tilde{y}_i(t)$ to be unit. As a result, $\tilde{\phi}_i(t)$ elements are simply the corresponding columns of $\mathbf{W}^{-1}$ and the cost function can be simplified as:

$$C(x_i, y_i, z_i, \theta_i, \psi_i, p_i) = \|\tilde{\phi}_i - \hat{\phi}_i\| = \|\tilde{\phi}_i - \mathbf{k}_i \hat{\mathbf{j}}_i\|, \tag{10}$$

where $\mathbf{k}_i = (k_{1i}, k_{2i}, \ldots, k_{Ni})^T$ is the ith column of the lead field matrix $\mathbf{K}$. The orientation (θ_i, ψ_i) and strength p_i of the ith dipole can be thought of as

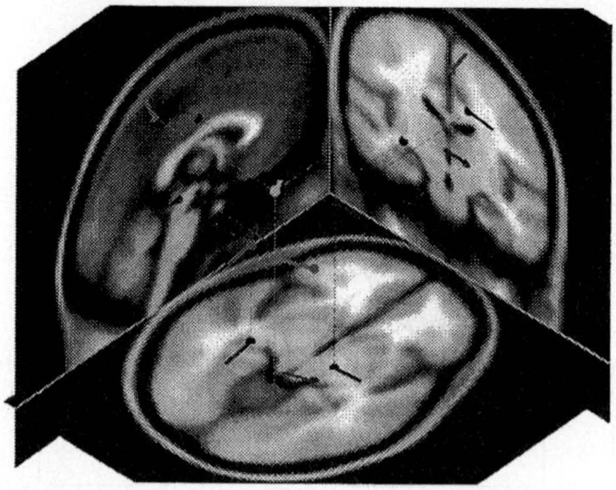

Fig. 2. Visualization of the dipole sources

components $(\hat{j}_{ix}, \hat{j}_{iy}, \hat{j}_{iz})$, the dipole strength in the x, y and z direction. Now we can obtain the dipole position (x_i, y_i, z_i) and dipole orientation (θ_i, ψ_i) by minimizing the cost function (10):

$$\begin{aligned}(x_i, y_i, z_i, \theta_i, \psi_i) &= \text{argmin}\|\tilde{\phi}_i - \mathbf{k}_{ix}\hat{j}_{ix} - \mathbf{k}_{iy}\hat{j}_{iy} - \mathbf{k}_{iz}\hat{j}_{iz}\| \\ &= \text{argmin}\sum_{j=1}^{N}(\tilde{\phi}_{ij} - k_{jix}\hat{j}_{ix} - k_{jiy}\hat{j}_{iy} - k_{jiz}\hat{j}_{iz})^2, \qquad (11)\end{aligned}$$

where $\tilde{\phi}_i = (\tilde{\phi}_{i1}, \tilde{\phi}_{i2}, \ldots, \tilde{\phi}_{iN})^T$ and $\mathbf{k}_{i\beta} = (k_{1i\beta}, k_{2i\beta}, \ldots, k_{Ni\beta})^T$ with $\beta = x, y, z$. The dipole position and dipole moment can be obtained by minimization of this cost function using an four-shell spherical model [14].

3 Results

The data used in the experiment is the time-dependent 20 EEG averaged measurements for 62 channels. We first perform IRA on the original EEG data to obtain the 62 independent components. We can visualize each component by ploting the scalp component maps. Fig. 1 shows the scalp maps of four components (they are dipole-like), which are related to the visual stimulus. The first component corresponds to the evoked potential in the primary visual cortex. The second component corresponds to the evoked potentials in the right side of the high visual cortex. And the third and fourth components are related to the face recognition. Last step is the source localization. We use the EEGLAB [15] plug-in DIPFIT to localize the dipole. Visualization of the obtained four dipoles is shown in Fig. 2.

4 Conclusions

In this paper, we present a new IRA method for EEG source localization that reduces the complexity. IRA considers the temporal structure of the sources and it can separate sources in a short time window. Projections of independent components obtained by IRA to the scalp surface have been shown to be more dipole-like. Localizing dipoles for each component separately dramatically increases the efficiency and accuracy of EEG source localization.

References

1. Zhukov, L., Weinstein, D., Jonson, C.: Independent component analysis for EEG source localization in realistic head models. IEEE Eng. Med. Biol. Mag. **19** (2000) 87-96
2. Gorodnistky, I., George, J., Rao, B.: Neuromagnetic imaging with FOCUSS: A recursive weighted minimum-norm algorithm. Electroencephalography and Clinical Neurophysiology. **95** (1995) 231-251
3. Burger, H., Van Millan, J.: Heart-vector and leads. Part I, Br. Heart J. **8** (1946) 157-161
4. Scherg, M., von Cramon, D.: Evoked dipole source potential of the human auditory cortex. Electroencephalography and Clinical Neurophysiology. **65** (1986) 344-360
5. Scherg, M., von Cramon, D.: Two bilateral sources of the late AEP as identified by a spatio-temporal dipole model. Electroencephalography and Clinical Neurophysiology. **62** (1985) 290-299
6. Mosher, J.C., Lewis, P.S., Leahy, R.M.: Multiple dipole modeling and localization from spatio-temporal MEG data. IEEE Trans Biomed Eng. **39** (1992) 541-557
7. Mosher, J.C., Leahy, R.M.: EEG and MEG source localization using recursively applied (RAP) MUSIC. Los Alamos National Laboratory Report. LA-UR-96-3889
8. Pascual-Marqui, R.D., Michel, C.M., Lehmann, D.: Low resolution electromagnetic tomography: a new method for localizing electrical activity in the brain. Int. J. Psychophysiol. **18** (1994) 49-65
9. Jung, T.P., Makeig, S., McKeown, M.J., Bell, A.J., Lee, T.W., Sejnowski, T.J.: Imaging brain dynamics using independent component analysis. Proceedings of the IEEE. **89** (2001) 1107-1122
10. Hyvarinen, A., Oja, E.: A fast fixed-point algorithm for independent component analysis. Neural Computation. **9** (1997) 1483-1492
11. Bell, A.J. Sejnowski, T.J.: An information-maximation approach to blind separation and blind deconvolution. Neural Computation. **7** (1995) 1129-1159
12. Amari, S., Cichocki, A., Yang, H.: A new learning algorithm for blind signal separation. Advances in Neural Information Processing System. **8** (1996) 757-763
13. Zhang, L., Cichocki, A.: Independent Residual Analysis for Temporal Correlated Signals. Computational Methods in Neural Modelling. Lecture Notes in Computer Sciences 2686, Eds: Mira, J., Alvarez, J. (2003) 158-165
14. Kavanagh, R., Darccey, T.M., Lehmann, D., Fender, D.H.: Evaluation of methods for three-dimensional localization of electric of electric sources in the human brain. IEEE Trans Biomed Eng. **25** (1978) 421-429
15. Delorme, A., Makeig, S.: EEGLAB Open Source Matlab Toolbox for Physiological Research. WWW Site: http://www.sccn.ucsd.edu/ scott/ica.html

Classifying G-protein Coupled Receptors with Support Vector Machine*

Ying Huang[1,2,3] and Yanda Li[1,2,3]

[1] Institute of Bioinformatics
[2] MOE Key Laboratory of Bioinformatics
[3] Department of Automation, Tsinghua University Beijing, 10084, China
hying99@mails.tsinghua.edu.cn

Abstract. G-protein coupled receptors (GPCRs) are a class of pharmacologically relevant transmembrane proteins with specific characteristics. They play a key role in different biological process and are very important for understanding human diseases. However, ligand specificity of many receptors remains unknown and only one crystal structure solved to date. It is highly desirable to predict receptor's type using only sequence information. In this paper, Support Vector Machine is introduced to predict receptor's type based on its amino acid composition. The prediction is performed to the amine-binding classes of the rhodopsin-like family. The overall predictive accuracy about 94% has been achieved in a ten-fold cross-validation.

1 Introduction

G-protein-coupled receptors (GPCRs) are a class of pharmacologically relevant proteins characterized by seven transmembrane (7TM) helices. Through their extracellular and transmembrane domains, GPCRs play a key role in a cellular signaling network that regulates many basic physiological processes: neurotransmission, cellular metabolism, secretion, cellular differentiation and growth, inflammatory and immune responses, smell, taste and vision [1]. According to their binding with different ligand types, GPCRs are further classified into different families. Many efforts in pharmaceutical research are current aimed at understanding their structure and function. Despite their importance, there is still only crystal structure solved to date [2]. And many known human GPCRs remain orphans (the activating ligand is unknown) [3]. In contrast, the sequences of thousands of GPCRs are known. So computational methods based on sequence information may be helpful to identify receptor's type [4]. Recently, Elrod and Chou [5] proposed a covariant discriminant algorithm to predict a GPCR's subfamily according to its amino acid composition. In the current study, we try

* This work was funded by the National Natural Science Grant in China (No.60171038 and No.60234020) and the National Basic Research Priorities Program of the Ministry of Science and Technology (No.2001CCA0). Y.H. also thanks Tsinghua University Ph.D. Grant for the support.

F. Yin, J. Wang, and C. Guo (Eds.): ISNN 2004, LNCS 3174, pp. 448–452, 2004.

to apply Support Vector Machine (SVM) method to approach this problem. To demonstrate our algorithm, we apply it to the amine-binding classes of the rhodopsin-like family of GPCRs. There are many medically and pharmacologically important proteins in this family. The results show that the prediction accuracy is significantly improved with this method.

2 Materials and Methods

2.1 Sequence Data

We used the same dataset as that of Erlod and Chou [5], which was taken from GPCRDB [6] (December 2000 release). There are 167 rhodopsin-like amine G-protein-coupled receptors classified into four groups, acetylcholine, adrenoceptor, dopamine and serotonin, as shown in Table 1. Other sub-families of rhodopsin-like amine GPCR are not included, for there are too few sequences in these sub-families to have any statistical significance. Redundancy was reduced so that pair-wise sequence identity is relative low. Accession numbers of these proteins in SWISSPROT can be obtained from Erlod and Chou [5].

Table 1. Summary of 167 receptors classified into four types

Group	Number of Sequences	Average Length
Acetylcholine	31	530.6
Adrenoceptor	44	448.4
Dopamine	38	441.4
Serotonin	54	433.7
Overall	167	457.3

2.2 Data Representation

Protein sequences are strings make up of 20 different amino acids (alphabets). To apply machine learning method such as SVM, we have to extract a fixed length feature vector from protein sequence with variable length. Following Chou [7], a protein is represented by its amino acid composition, corresponding to a vector in the 20-D (dimensional) space.

$$\mathbf{X_k} = \begin{bmatrix} x_{k,1} \\ x_{k,2} \\ \vdots \\ x_{k,20} \end{bmatrix} \qquad (k = 1, 2, \ldots, 167) \tag{1}$$

where $x_{k,1}, x_{k,2}, \ldots, x_{k,20}$ are the 20 components of amino acid composition for the kth protein X_k.

2.3 Support Vector Machine

SVM is a popular machine learning algorithm based on recent advances in statistical learning theory [8,9]. This algorithm first maps data into a high-dimensional feature space, and then constructs a hyperplane as the decision surface between positive and negative patterns. The actual mapping is achieved through a kernel function, making it easy to implement and fast to compute. Several popular kernel functions are:

$$\text{linear kernel: } K(x_i, x_j) = x_i^T x_j \tag{2}$$

$$\text{polynomial kernel: } K(x_i, x_j) = (\gamma x_i^T x_j + r)^d, \gamma > 0 \tag{3}$$

$$\text{RBF kernel: } K(x_i, x_j) = exp(-\gamma \|x_i - x_j\|^2), \gamma > 0 \tag{4}$$

In principle, SVM is a two-class classifier. With the recent improvements, the SVM can directly cope with multi-class classification problem now [10]. The software used to implement SVM was libSVM [11], which can be downloaded from http://www.csie.ntu.edu.tw/~cjlin/libsvm. LibSVM used "one-against-one" approach to deal with multi-class problem, in which $k(k-1)/2$ classifiers are constructed and each one trains data from two different classes.

2.4 Prediction Accuracy Assessment

The prediction performance was examined by the ten-fold cross-validation test, in which the dataset of 167 GPCRs for the four groups were divided into ten subsets of approximately equal size. We train our algorithm based on nine of these subsets, and then the remaining subset is used to estimate the predictive error of the trained classifier. This process is repeated ten times so that every subset is once used as the test data. The average of the ten estimates of the predictive error rate is reported as the cross-validation error rate. The prediction quality is then evaluated by the overall prediction accuracy and prediction accuracy for each group.

$$\text{overall accuracy} = \sum_{s=1}^{k} p(s)/N \tag{5}$$

$$\text{accuracy(s)} = p(s)/obs(s) \tag{6}$$

Where N is the total number of proteins in the data set (N=167), k is the number of groups (k=4), obs(s) is the number of sequences observed in group s, and p(s) is the number of correctly predicted sequences in group s.

3 Results

3.1 SVM Kernel and Parameters Selection

Experiments have been done on three popular kernel functions: the linear kernel, the polynomial kernel and the RBF kernel. We found that RBF kernel performed better than linear kernel and polynomial kernel. For RBF kernel, there are two parameters to be selected: kernel parameter γ and penalty parameter C. We choose $\gamma = 0.05$, which is the default value of libSVM software. As for C, various values ranging from 1 to 64 have been tested. The best result was achieved when $C = 4$.

3.2 Comparison with Existing Methods

The SVM prediction performance was compared covariant discriminant algorithm that also based on amino acid compositions. The results are summarized in Table 2. The overall success rate of SVM algorithm is 94.01%, which is about 10% higher than that of the covariant discriminant algorithm. We also observed that SVM improve the prediction performance for every group. Especially for acetylcholine type, SVM improve the accuracy significantly.

Table 2. Performance comparison between covariant discrimination and SVM algorithm.

Group	Cov[a] Accuracy(%)	SVM Accuracy(%)
Acetylcholine	67.74	100
Adrenoceptor	88.64	90.91
Dopamine	81.58	94.74
Serotonin	88.89	92.59
Overall	83.23	94.01

[a] covariant discrimination algorithm

4 Conclusion

In this paper, we introduced SVM method for recognizing the family of GPCRs. The rate of correct identification obtained in ten-cross validation is about 94% for four group classification, which is superior to existing algorithms. This result implies that we can predict the type of GPCRs to a considerably accurate extent using amino acid composition. It is anticipated that our method would be a useful tool for classification of orphan GPCRs and facilitate drug discovery for psychiatric and schizophrenic diseases.

References

1. Hebert, T., Bouvier, M.: Structural and functional aspects of g protein-coupled receptor oligomerization. Biochem Cell Biol **76** (1998) 1–11
2. Palczewski, K., Kumasaka, T., Hori, T., Behnke, C.A., Motoshima, H., Fox, B.A., Le Trong, I., Teller, D.C., Okada, T., Stenkamp, R.E., Yamamoto, M., Miyano, M.: Crystal structure of rhodopsin: A g protein-coupled receptor. Science **289** (2000) 739–45

3. Schoneberg, T., Schulz, A., Gudermann, T.: The structural basis of g-protein-coupled receptor function and dysfunction in human diseases. Rev Physiol Biochem Pharmacol **144** (2002) 143–227
4. Gaulton, A., Attwood, T.K.: Bioinformatics approaches for the classification of g-protein-coupled receptors. Curr Opin Pharmacol **3** (2003) 114–20
5. Elrod, D.W., Chou, K.C.: A study on the correlation of g-protein-coupled receptor types with amino acid composition. Protein Engineering **15** (2002) 713–715
6. Horn, F., Weare, J., Beukers, M.W., Horsch, S., Bairoch, A., Chen, W., Edvardsen, O., Campagne, F., Vriend, G.: Gpcrdb: an information system for g protein-coupled receptors. Nucleic Acids Res **26** (1998) 275–9
7. Chou, K.C.: A novel approach to predicting protein structural classes in a (20-1)-d amino acid composition space. Proteins Struct. Funct. Genet. **21** (1995) 319–344
8. Vapnik, V.N.: The Nature of Statistical Learning Theory. Springer, N.Y. (1995)
9. Vapnik, V.N.: Statistical Learning Theory. Wiley, New-York (1998)
10. Hsu, C.W., Lin, C.J.: A comparison of methods for multi-class support vector machines. IEEE Trans. Neural Networks **13** (2002) 415–25
11. Chang, C.C., Lin, C.J.: LIBSVM: a library for support vector machines. (2001) Software available at `http://www.csie.ntu.edu.tw/~cjlin/libsvm`.

A Novel Individual Blood Glucose Control Model Based on Mixture of Experts Neural Networks

Wei Wang [1,2], Zheng-Zhong Bian [1], Lan-Feng Yan [3], and Jing Su[2]

[1] Life Science and Technological School, Xi'an Jiaotong University, China, 710049
kinggreat@163.com

[2] Modern Technology Educational Center, Lanzhou Medical College, China, 730000

[3]Third Clinical Medical School, Lanzhou Medical College, China, 730000

Abstract. An individual blood glucose control model (IBGCM) based on the Mixture of Experts (MOE) neural networks algorithm was designed to improve the diabetic care. MOE was first time used to integrate multiple individual factors to give suitable decision advice for diabetic therapy. The principle of MOE, design and implementation of IBGCM were described in details. The blood glucose value (BGV) from IBGCM extremely approximated to training data (r=0.97±0.05, n=14) and blood glucose control aim (r=0.95±0.06, n=7).

1 Introduction

Different kinds of experts systems were developed to meet the needs of decision support in diabetes care [1]. It has been recognized that diabetic care needs a powerful decision support system to help patients to balance drugs, insulin, and diet plan [2], and a more advanced system aimed at support decisions and facilitating adjustment of insulin dosage [3]. Past studies based on decision support model of rule [4], case, knowledge and hierarchical decision model [5]. Some methods could be used to build diabetic decision support model (DDS) to adjust insulin accordingly [6]. These methods provided physicians with a reliable decision support tool to mainly adjust the dose of insulin, but could not deal with the multiple individual factors. The powerful DDS, which could advise suitable proportions of general drugs, insulin, diet, and exercises intensity, needs to be developed. The Mixture of Experts (MOE) neural networks were first time used in DDS. The MOE can divide a difficult task into appropriate subtasks, each of which can be solved by a very simple expert network [7]. This may offer robustness for an individual diabetic decision support model using multiple factors in the clinical practice and can extend to other decision fields.

2 Generic Model of Blood Glucose Control

Before setting up the individual blood glucose control model (IBGCM), the generic blood glucose control model must be built as the prototype of IBGCM. A simplified

F. Yin, J. Wang, and C. Guo (Eds.): ISNN 2004, LNCS 3174, pp. 453–458, 2004.

model is that the diabetic disease is treated as a partial or complete failure of the inherent pancreatic control loop (Fig. 1), which is replaced by the control loop of insulin or drugs therapy, diet plan and exercise intensity [1,4] to keep a physiological balance and recompense failure of the pancreatic function.

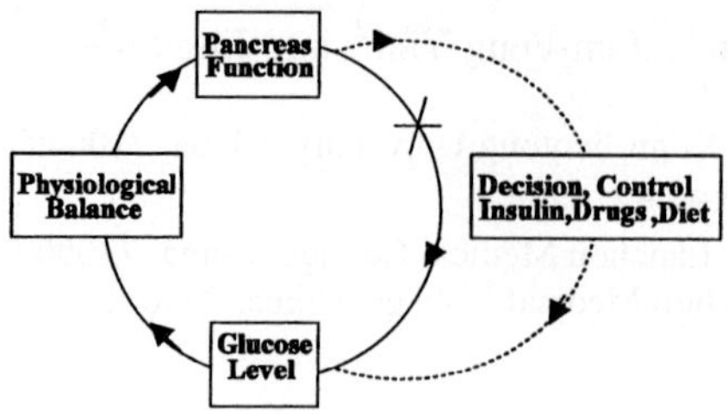

Fig. 1. Blood glucose control model treated as a recompensing control in pancreas failure

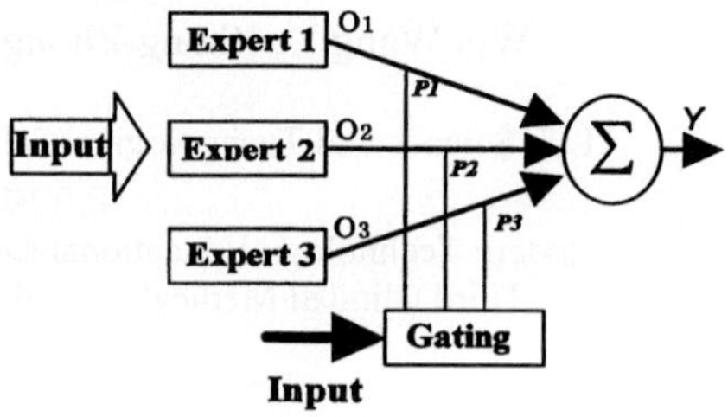

Fig. 2. The prototype of Mixture of Experts neural networks

3 Prototype of Mixture of Experts Neural Network

Before building the IBGCM, the principle of MOE must be discussed. MOE is a model that estimates the conditional probability distribution of a set of training patterns (Fig 2) [7]. It consists of multiple supervised neural networks, trained to specific regions of the input space. These networks use the idea of competitive learning where each Expert competes with the other Experts to generate the correct desired output. The gating network acts as a mediator to determine which Expert is best suited for that specific region of input space and assigns that portion to the Expert. The principle of this system is to divide a large complex set of data into smaller subsets via the gating networks that allows each Expert networks to better represent a specific subset. It, therefore, learns more quickly, since it is easier to learn multiple simple functions than it is to learn very large complex ones. It is also assumed that smaller networks can better generalize a smaller set of data than a complex network over a larger set of data. Expert networks consist of supervised linear or non-linear networks. The experts are assigned to a specific region of the input space, which allows them to model that space. The idea of Experts is that by implementing multiple networks in parallel, each Expert assigned to its own unique input space. In combining these networks, it combines all the Experts that generate the system output with a better performance. The equation for the output is:

$$Y = \sum_i p_i Y_i \tag{1}$$

Where p_i is the proportion that the i Expert contributed to the output, which determined by the gating network and Y_i is the output from the i Expert. Since the Expert networks can be either linear or non-linear networks, it is assumed that prior knowledge of Perceptrons and Multi-layered Perceptrons is known, the interaction between the last two layers in the network need to be discussed. For this layer, the values of all weights within the last layer were set to random values between -1 and 1.

In order to train the network, it is necessary to modify the weights between the last two layers of the networks based on the overall systems error. The basic sum-of-squares error function is:

$$E^c = \frac{1}{2}\sum_i p_i^c \| d^c - o_i^c \|^2 \tag{2}$$

Where d^c is the desired output and o^c is the actual output of the i Expert. The objective is to minimize the total error with respect to the output of the entire system. Using this equation, the system would assume that the prior probability, P_i generated by the Gating network was correct and would modify the weights based on the error from only the i Expert and not the error of the entire system. Therefore, the error function used is written as:

$$E^C = -\ln(\textstyle\sum_i p_i^C e^{-\| d^C - o_i^C \|^2}) \tag{3}$$

This relation is considered to be the negative log of the associative Gaussian mixture model. The derivative of this equation is:

$$\frac{\partial E^c}{\partial o_i^c} = -\left[\frac{p_i^c e^{-\frac{1}{2}\| d^c - o_i^c \|^2}}{\sum_j p_j^c e^{-\frac{1}{2}\| d^c - o_j^c \|^2}}\right](d^c - o_i^c) \tag{4}$$

The formula represents the amount of error that the i Expert contributed to the total system error. In training a network using gradient descent, the derivative of the error function with respect to the weights is used:

$$\frac{\partial E^c}{\partial w_{ij}} = \frac{\partial E^c}{\partial o_i^c} \times \frac{\partial o_i^c}{\partial w_{ij}} \quad \text{where} \quad \frac{\partial o_i^c}{\partial w_{ij}} = f'(\textstyle\sum_i w_{ij} b_i) b_i \tag{5}$$

The b_i equals to the input from the previous layer. The $f(x)$ is the activation function of the network. Therefore, the change in weight for the Expert networks, using gradient descent should be:

$$\Delta w_{ij} = n b_i \left[\left[\frac{p_i^c e^{-\frac{1}{2}\| d^c - o_i^c \|^2}}{\sum_j p_j^c e^{-\frac{1}{2}\| d^c - o_j^c \|^2}}\right](d^c - o_i^c)\right] f'(\sum_i w_{ij} b_i) \tag{6}$$

The gating network is the controller of the Expert networks. It determines what probability each Expert will generate the desired output. The gating network is considered to be a single layer Perceptron, although it is possible to implement into a Back-propagation system. The Gating network uses the same inputs as all the Expert systems, with the number of outputs equal to the number of Experts. The output from the network is the estimated probability, p_i, that the specific Expert will generate the desired output. For the gating network, the Softmax activation function shows as:

$$p_i = \frac{e^{x_j}}{\sum_j e^{x_j}} \qquad (\textstyle\sum p_i = 1) \tag{7}$$

The formula (7) is used to generate positive values with a sum equal to one. The Softmax function is a continuous differentiable function, which makes it perfect for gradient descent learning. Initially, each Expert should have the same probability of successfully generating the desired output; therefore the weights in the last layer of the gating network must be set to 0, which will cause the probabilities to be equally distributed. Similar to the Expert networks, the gating network is desirable to minimize the overall systems error in training. To substitute the Softmax function into the systems error function, we can obtain the error function written in the following form:

$$E^c = -\ln(\sum_i \left[\frac{e^x}{\sum_j e^{x_j}}\right] e^{-\|d^c - o_i^c\|^2}) = -\left[\ln(\sum_i e^x e^{-\|d^c - o_i^c\|^2}) - \ln(\sum_j e^{x_j})\right] \quad (8)$$

$$\frac{\partial E^c}{\partial u_i} = (h_i - p_i)x \quad (9)$$

The derivative of equation (8) yields the equation (9), where u_i represents the weights in the last layer of the gating network, h_i and x is the input. The p_i can be considered to be the prior probability of selected i Expert, while h_i is considered the posterior probability of selected i Expert. Therefore, when training by gradient descent learning, the change in weights of the gating network is defined by:

$$\Delta w_{ij} = n(h_i - p_i)x \quad (10)$$

4 Design of Individual Blood Glucose Control Model

It is difficult to completely control blood glucose value (BGV) with a single drug, diet and exercises. The past methods, such as time series analysis [2] are not an ideal for dealing with multiple parameters and nonlinear processing. MOE was applied to build the IBGCM with the data of BGV, drug and insulin dosage and exercises (Fig.3). In application, the output G of MOE considered as the control aim of BGV, and the inputs S_{ij} are the advices of IBGCM. The decision support process is a loop of learning. Regarding MOE, there are n individual Experts. Variable G is given by the summation of the values of G_i. Each G_i is weighted by weighting factor W_i, given in:

$$G = \sum_{i=1}^{n} G_i w_i \quad (11)$$

$$G_i = \sum_{j=1}^{m} a_{ij} S_{ij} + z_i \quad (12)$$

Expert variable G_i is a linear summation of m input parameters S_{ij} and constant Z_i, shown in equation (12). The weighting factor W_i is determined by another relationship of input parameters S_{ij}. The weighting factors W_i is optimized by Expectation Maximization (EM) [7,8], written in equation (14):

$$w_i = \frac{\exp(\phi_i)}{\sum_{i=1}^{n} \exp(\phi_i)} \quad (13)$$

$$\phi_i = \sum_{j=1}^{m} a_{ij} S_{ij} + \xi_i \tag{14}$$

In equation (13), Φ_i expressed as equation (14). ξ_i was a constant. There are n experts and m input parameters in a system, and coefficients (a_{ij}, z_i, s_{ij}, ξ_i) need to be determined [7,8]. The optimized process has two steps: In step one, the weighting factors associated with each output are fixed, and then the parameter values are optimized. In step two, the parameters values are fixed and then the weighting factors are optimized. For the whole process step one and step two are repeated until convergence is achieved. The coefficients (a_{ij}, z_i, ξ_i) are input coefficients. The variable G is the aim of blood glucose control; the data of diet, insulin, general drugs and intensity of exercise were directly inputted in variable S_{ij}. The weights W_i have a responding relationship with each output, which are optimized in the distribution of experts. In training, the coefficients S_{ij} could be initiated by the data of fat (*mg*), protein (*mg*), carbohydrate (*mg*), general drug dosage, insulin dosage (unit) and the intensity of exercises (min), and a_{ij}, z_i, ξ_i could be initiated with zero. In the DDS, G is initiated with BGV. The coefficients can be used to avoid the interference of other factors. The model uses normal glucose values as control aim, and insulin dosage, diet data and the intensity of exercise from input or historical data as the input to train the networks. According to the control aim, the IBGCM gets suitable proportion of fat, protein, carbohydrate, general drug dosage, insulin dosage and the intensity of exercises, finally giving out an optimal integrated advice to improve effect of diabetic therapy. It is important to note the calibration and learning: a). The normal blood glucose values must be the control aim. b). According to drug and insulin dosage, diet data, the intensity of exercise, the coefficients must be modified to respond the proportion of input data in the training. The value of coefficients must be limited to appropriate scope for learning speed.

5 Experiment Results

The blood glucose value following dinner two hours, diet, exercises, insulin dosages, general drugs dosages, and the intensity of exercises were collected from four diabetic patients as input S_{ij} for training IBGCM. Output G served as the control aim of BGV. Variable i was six responding to Experts, and j was fourteen according to the days of data collected. The data of fourteen days was collected to train IBGCM (Fig 3). After training, the ideal control aim of BGV was inputted to G, and then the advice of diet, exercises, general drugs dosage and insulin dosage were reversely given out. To validate IBGCM, the training data sets were re-inputted into the model, and the BGV from IBGCM extremely approximated to the training data sets (r=0.97±0.05, n=14). The BGV from the advices of other test were close to the control aim (r=0.95±0.06, n=7). In Fig. 4, the results demonstrated seven advice results from IBGCM to control BGV, and seven data sets of the doctor's experience compared with the results. It showed that the advice results from IBGCM were more balanced than doctor's experiential method. The control aim was 5.7 *mmol/L*. For BGV from the experience, standard deviation (SD) was 1.81. For BGV from IBGCM, SD was 0.21.

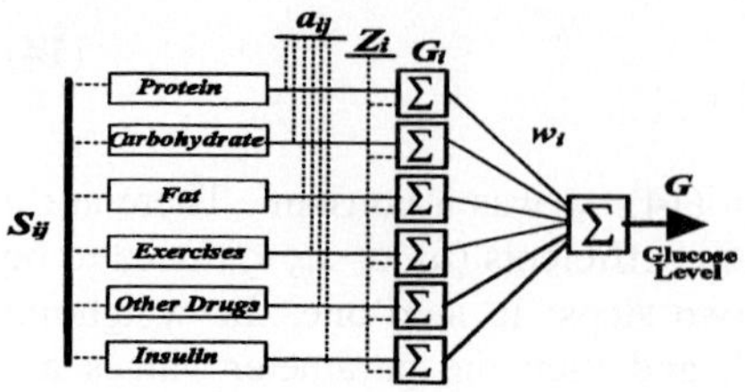

Fig. 3. Individual blood glucose control model based on MOE

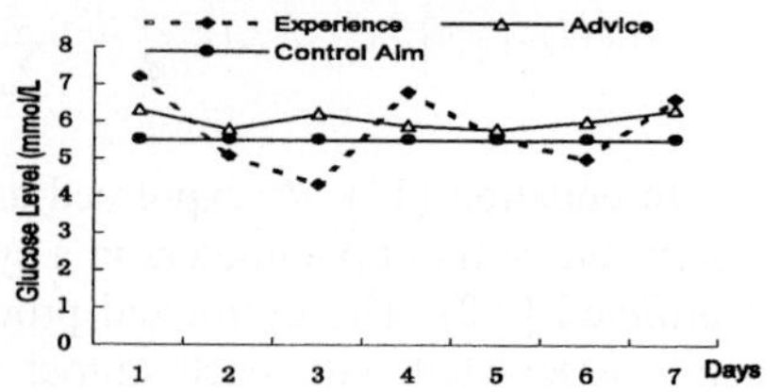

Fig. 4. BGV control results analysis from experience of doctor and IBGCM

6 Discussion

In the diabetic decision support, the suitable IBGCM must be built with robust methods, which can deal with multiple parameters from individual patient. The ideal model should include the advice of insulin and drug dosage, diet, and exercise according to individual and multi-factors. In this study MOE was first applied in IBGCM, clinical decision support, its training is simple. The IBGCM based on MOE has the advantage of serving any type of diabetes, and makes the advice more precise. The advice of DDS was improved with IBGCM. It was suggested that the advice results from IBGCM were more balanced than experience. The MOE can divides a large, difficult task into appropriate simple and each of subtasks can be solved by a very simple expert network [7], which is easier to implement in application than other methods. This method may offer robustness for an individual decision support system.

References

1. Carson, E.R., Fischer, U., Salzieder E.: Models and Computers in Diabetes Research and Diabetes Care. Comput. Methods Programs Biomed. 32 (1990) 171–356
2. Lehmann, E.D.: Application of Computers in Diabetes Care. Lancet. 344 (1994) 1010
3. Hovorka, R., Carson, E.R.: Computers in Diabetes. Comput. Methods Programs Biomed. 41 (1994) 151–303
4. Ambrosiadou, B.V., Goulis, D.G., Pazppas, C.: Clinical Evaluation of the Diabetes Expert System for Decision Support by Multiple Regimen Insulin Dose Adjustment. Computer Methods and Programs in Biomedicine. 49 (1996) 105-115
5. Montani, S., Magni, P., Bellazzi, R., Larizza, C., Roudsari, A.V., Carson E.R.: Integrating Model-based Decision Support in A Multi-modal Reasoning System for Managing Type 1 Diabetic Patients. Artificial Intelligence in Medicine. 29 (2003) 131–151
6. Montani, S., Bellazzi, R.: Supporting Decisions in Medical Applications: The Knowledge Management Perspective. International Journal of Medical Informatics. 68 (2002) 79-90
7. Jacobs, R.A., Jordan, M.: Adaptive Mixtures of Local-Experts. Neural Computation. 3 (1991) 79-87
8. Kurnik, R.T., Oliver, J.J., Waterhouse, S.R., Dunn, T., et al: Application of the Mixtures of Experts Algorithm for Signal Processing in A Noninvasive Glucose Monitoring System. Sensors and Actuators. b60 (1999) 19-26

Tracking the Amplitude Variation of Evoked Potential by ICA and WT

Haiyan Ding and Datian Ye

Department of Biomedical Engineering, Tsinghua University,
Beijing 100084, P. R. China
Dinghaiyan00@mails.tsinghua.edu.cn

Abstract. Evoked potential (EP) is non-stationary during the recording of electroencephalograph (EEG). This paper promotes a method to track the variation of EP's amplitude by the application of independent component analysis (ICA) and wavelet transform (WT). The utilization of the spatial information and multi-trial recording improves the signal-to-noise ratio (SNR) greatly. The variation trend of EP's amplitude across trials can be evaluated quantitatively. Our result on real auditory evoked potential shows a drop of about 40% on the amplitude of EP during 10 minutes recording. The present work is helpful to study the uncertainty and singularity of EP. Furthermore, it will put forward the reasonable experiment design of EP extraction.

1 Introduction

Evoked potential (EP) represents the gross electrical activity of specific regions of brain usually resulting from sensory stimulation. Like many neural signals, the measurement of EP is corrupted by noise as a result of the ongoing activity of other brain cells. The signal-to-noise ratio (SNR) of EP is typically quite low: the background potential distribution is of the order of 100 microvolt, whereas the evoked potential may be two orders of magnitude weaker. The traditional EP extraction method is ensemble averaging with the rationale that the single-trial electroencephalograph (EEG) data time locked to the stimulus consists of an average EP, whose time course and polarity is fixed across trials, plus other EEG processes whose time courses are completely unaffected by the experiment events. However, it has been proved that EP is non-stationary and has characteristics that vary across trials [1].

This paper aims to track the amplitude variation of EP during the experiment. With the application of independent component analysis (ICA), the spatial information and multi-trial matrix construction are employed to obtain the variation of EP across trials. Wavelet transform (WT) is used to extract the variation trend. Simulation and real auditory evoked potential data analysis shows that it is an effective method in tracking the amplitude variation of EP. The present work is helpful to study the uncertainty and singularity of EP during the extraction procedure. The characteristics of EP's variation may be used to optimize the weighted averaging method. Furthermore, it will put forward the reasonable EEG experiment design.

F. Yin, J. Wang, and C. Guo (Eds.): ISNN 2004, LNCS 3174, pp. 459–464, 2004.

2 Methodology

WT and ICA are highly effective at performing the de-noising, artifacts removing and feature extraction of EP in recent years [2, 3]. This work applies ICA to enhance SNR and extract EP's amplitude trial to trial. Being optimal resolution both in time and frequency domain, WT is employed to extract the variation trend of EP's amplitude. Further details about these algorithms can be obtained by the given references [4, 5].

As an unsupervised neural network learning algorithm, ICA finds a coordinate frame in which the data projections have minimal temporal overlap base on the assumptions of (1) the observation signals are the linear mixture of the statistically independent sources, (2) the propagation delays are negligible, and (3) the number of sources is the same as the number of sensors. Several different implementations of ICA are developed based on the different selection of statistically independent evaluation rules. In this paper, the fixed-point ICA [6] is applied to perform source separation.

We suppose that EP is fixed in waveform since it is responding to some certain neural information procedure triggered by stimulus and its amplitude varies across trials. Our task is to extract the transient amplitude features of EP. The key problem is to improve SNR to some extend that satisfies the requirement of tracking the amplitude variation. Two-step SNR enhancement procedures by ICA are applied as following:

(1) Through the utilization of spatial information, 'clearer' source activity of EP is picked out based on the comparison with the ensemble averaging EP waveform.

(2) Multi-trial signals of the 'clearer' source data are used to construct proper data matrix, run ICA again and obtain the indicator of EP's amplitude.

The first procedure is general in the application of ICA. And here we discuss the rationality of the multi-trial matrix construction.The active sources of ongoing EEG signal are statistic independent and both in localizations and temporal scale during the experiment. So if we construct an appropriate matrix out of a series of delay vectors (i.e. truncation by trial) from the 'cleared' source signal, the ICA may decompose the matrix into its underlying independent components (ICs) with an IC corresponding to the EP and others to the noise from ongoing EEG. According to the principle of ICA, the element of the mixing matrix represents the intensity of the corresponding IC. Therefore, we will obtain the relative amplitude variation of EP across trials.

The source activity separated from initial spatial observed EEG consists of the real EP and some other noise. We establish the signal model of single trial as following,

$$X_n(t) = a_n X^{EP}(t) + X_n^{EEG}(t) \tag{1}$$

Where n represents the number of recording trial, $X_n(t)$ is the source data, $X^{EP}(t)$ represents the EP with certain waveform, a_n is the constant amplitude coefficient of EP in the nth trial and $X_n^{EEG}(t)$ is the noise left.

We construct the N-trial signal matrix as below.

$$\begin{bmatrix} X_1(t) \\ X_2(t) \\ \vdots \\ X_N(t) \end{bmatrix} = \begin{bmatrix} a_1 X^{EP}(t) + X_1^{EEG}(t) \\ a_2 X^{EP}(t) + X_2^{EEG}(t) \\ \vdots \\ a_N X^{EP}(t) + X_N^{EEG}(t) \end{bmatrix} = \begin{bmatrix} A & B \end{bmatrix} \cdot \begin{bmatrix} X^{EP}(t) \\ S^{EEG}(t) \end{bmatrix} \tag{2}$$

Where $A_{N\times1} = \begin{bmatrix} a_1 \\ a_2 \\ \vdots \\ a_N \end{bmatrix}$ and $\begin{bmatrix} X_1^{EEG}(t) \\ X_2^{EEG}(t) \\ \vdots \\ X_N^{EEG}(t) \end{bmatrix} = B_{N\times N-1} \begin{bmatrix} S_1^{EEG}(t) \\ S_2^{EEG}(t) \\ \vdots \\ S_{N-1}^{EEG}(t) \end{bmatrix}$. $S^{EEG}(t)$ represents the spontaneous EEG sources corresponding to the 'noise' left in the EP source activity and matrix B represents the mixing matrix of $S^{EEG}(t)$. Generally we assume the sources are independent from each other. That makes this model satisfy the original assumption of ICA. As the effective preprocess of ICA, PCA may find out an outstanding principle component mainly responding to EP, and other small principle components responding to the ongoing EEG. PCA can cut down the dimension of matrix and speed the convergence of ICA. At the same time, it makes ICA focus 'attention' on the decomposition of EP that is very attractive in our work.

3 Simulation

It is no need for us to verify the efficiency of ICA in the spatial data decomposition. What is done in this section is to discuss the rationality of trial-matrix decomposition. Fig. 1 shows the simulate signal that consists of ten-period sine waves with additive random noise (SNR: 12.8dB) and lineally varied amplitude. The continuous signal is constructed into a 10 channels data matrix. Two main components are selected to do ICA. Fig. 2 shows the result of PCA in the left and that of ICA in the right.

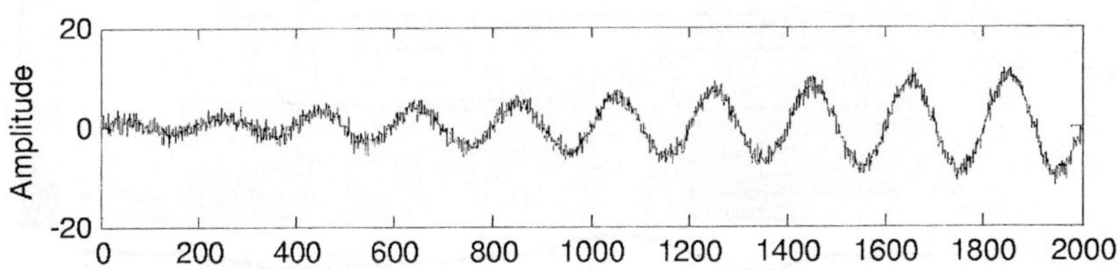

Fig. 1. The simulate signal

According to the discussion in section 2, the coefficients of mixing matrix reflect the intensity of ICs. Here the two ICs represent the signal and the noise respectively. Then we obtain the intensity variation of signal and noise in different period. As shown in Fig. 3, the amplitude variation calculated by ICA is linear to the real amplitude of signal, which reflects the variation of signal's amplitude faithfully.

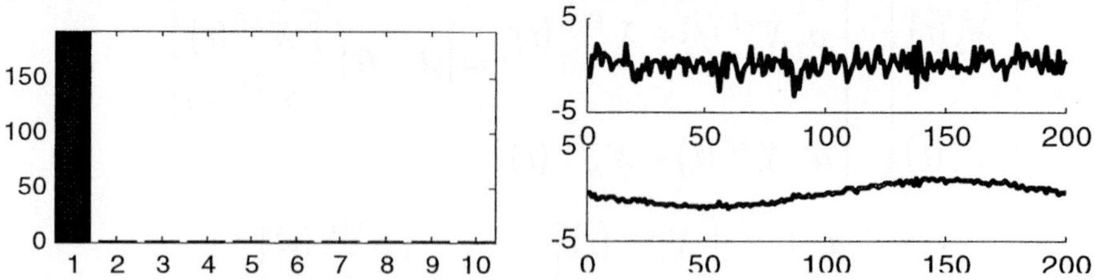

Fig. 2. The results of PCA and ICA on simulation signal

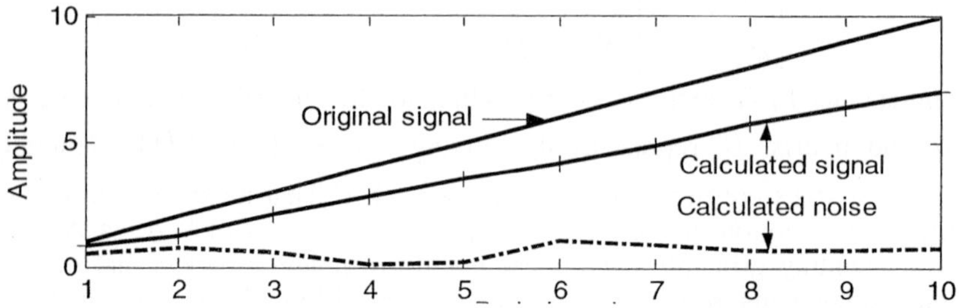

Fig. 3. The intensity variation of signal and noise in different period

4 Results

The variation of auditory evoked potential (AEP) is studied from actual recordings of a 29-year normal hearing subject on 32 electrodes affixed in the front, central and temporal areas in a quiet and electrical shielding cabin. The stimuli consist of three pure tones with different frequency in a pseudorandom sequence and present within about 10 minutes. The data corresponding to type I stimulus (800Hz, with 100ms duration and 700ms ISI) are analyzed.

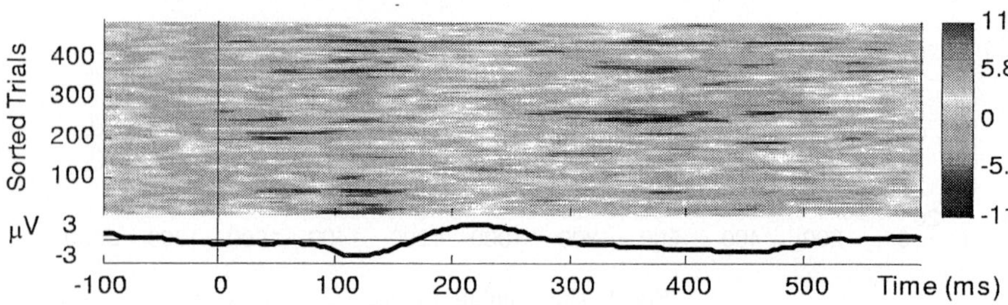

Fig. 4. Colored image of AEP and the averaging AEP in CFz

The original EEG data was re-referenced to the average potentials of ears and filtered by a 1~30Hz band pass filter. Trials with amplitude exceeding 50 μV from the 100 ms pre-stimulus baseline were rejected. We reserve 500 trials for further analysis with sorted number corresponding to the presentation time in experiment. Fig. 4 shows the colored image of AEP in CFz, in which each horizontal line represents a potentials time series during a single experimental trial and the changing color values indicate the potential at each time point in the trial. It shows that EP is non-stationary across trials. The moving average EP is illustrated at the bottom of Fig. 4, which works as the reference EP in the later analysis.

Fig. 5 shows the ICA result on the 32 channel recording. The left figure shows the eigenvalues of PCA, and the top 10 principle components (representing 93.7% of all) are applied for ICA. The result of ICA shows as the averaging unified waveform of 500 trials of each independent component (IC) in the right of Fig. 5.

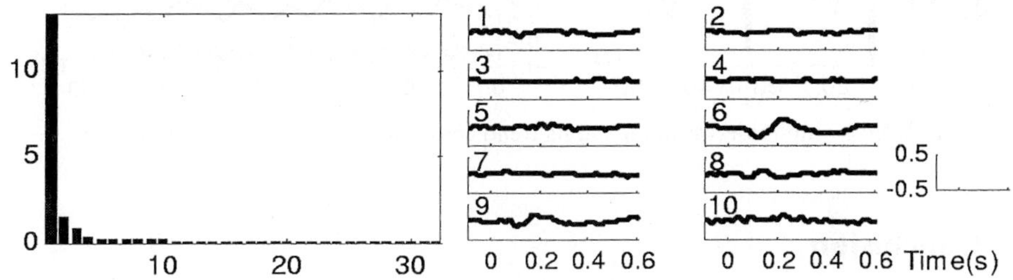

Fig. 5. The decomposition of 32 channels EEG and ICs' ensemble average

Here we compare the ensemble averaging ICs with the reference EP showed in Fig. 4. And the 'clearest' and 'strongest' IC 6 is picked out as the EP source activity. For the ambiguity of the amplitude of ICs, the relative amplitude of the EP is more important in this study. In order to improve the SNR, the 500 trials of IC 6 are averaged by every 10 trials before trial-matrix construction. 50 average trial data are used for PCA and the first 6 components are maintained for ICA with 72.6% power reserved. The eigenvalues of PCA and ensemble average results are showed in Fig. 6.

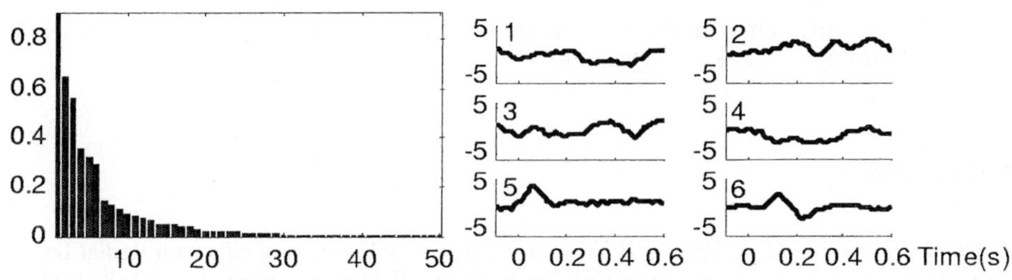

Fig. 6. The decomposition of IC 6 and the ICs' ensemble average result

Obviously, the last IC reserves the EP information most completely except for the reverse phase. The coefficients of the last IC reflect the variation of EP's amplitude. Fig. 7 plots the coefficients with phase reversed in the left. The EP's amplitude waves greatly and becomes weaker gradually.

In order to obtain the 'clear' variation trend, wavelet transform is applied to decompose the data into 4 scales by db5 wavelets. The reconstructions of 'approximation' and 'details' at different scales are shown in the right of Fig. 7

serially. The 'approximation' as illustrated in the first line shows the global trend of EP without any wave disturbance. The amplitude of EP is decreasing by time from the initial 0.1238 to 0.076 at the end point. It decreases by 38.61% after 10 minutes experiment. In the reconstructions of some 'details', as shown in the following four lines, the amplitude of EP waves across trials, that reflects the non-stationary of EP.

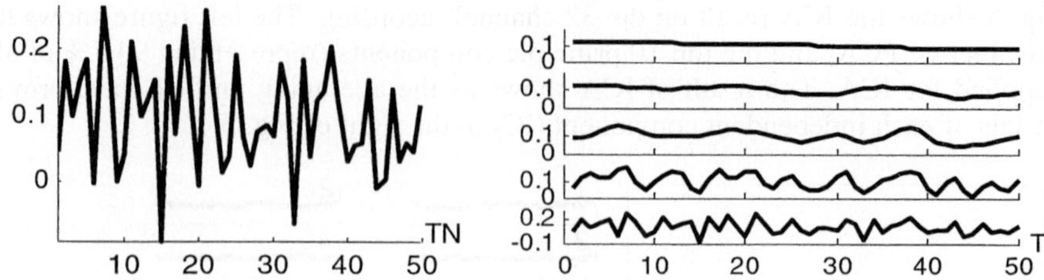

Fig. 7. EP's amplitude variation and its trend across trials (TN is the trial number)

5 Conclusion

In this paper, independent component analysis is used to improve the EP's SNR effectively by the utilization of the spatial and multi-trial information. The variation trend of EP during the EEG experiment is extracted by wavelet transform successfully. The result on real auditory EP shows a drop of about 40% on the amplitude of EP during 10 minutes. It proves that the brain fatigue cause by prolonging experiment will affect the EP significantly. The present work is helpful to the study on the uncertainty and singularity of EP during the extraction procedure. The description of EP's variation may used to optimize the weighted averaging method. Furthermore, it will put forward the reasonable EEG experiment design.

Acknowledgment. This study was supported by the National Natural Science Foundation of China contract 30270376

References

1. Wei Q., Kenneth S.M.F., Francis H.Y.C., et al.: Adaptive Filtering of EPs with Radial-basis-function Neural Network prefilter. IEEE Trans. Biomedical Engineering. 3 (2002) 224-232
2. Hyvarinen A.: Fast and Robust Fixed-pointed Algorithms for Independent Component Analysis. IEEE Trans Neural Networks. 10 (1999) 626-634
3. Bell A.J., Sejnowski T.J.: An Information Maximization Approach to Blind Separation and Blind Deconvolution. Neural Computation. 7 (1995) 1129-1159
4. Hyvarinen A., Karhunen J., Oja E.: Independent Component Analysis, John wiley & Sons Inc. (2001)
5. Mallat S.: A Wavelet Tour of Signal Processing, 2nd edn. Academic Press. (1999)
6. FastICA Matlab Package [On Line], available: http://www.cis.hut.fi/projects/ica/fastica

A Novel Method for Gene Selection and Cancer Classification

Huajun Yan and Zhang Yi

Computational Intelligence Laboratory, School of Computer Science and Engineering, University of Electronic Science and Technology of China, Chengdu 610054, People's Republic of China.
hearty@std.uestc.edu.cn; zhangyi@uestc.edu.cn
cilab.uestc.edu.cn

Abstract. Accurate diagnosis among a group of histologically similar cancers is a challenging issue in clinical medicine. Microarray technology brings new inspiration in solving this problem on genes level. In this paper, a novel gene selection method is proposed and a BP classifier is constructed for gene expression-based cancer classification. By testing on the open leukemia data set, it shows excellent classification performance. 100% classification accuracy is achieved when 46 informative genes are selected. Reducing genes number to 6, only a sample is misclassified. This study provides a reliable method for molecular cancer classification, and may offer insights into biological and clinical researches.

1 Introduction

Conventional cancer classification has been based primarily on macro- and microscopic histology and morphological appearance of the tumor. However, it has serious limitations: First, cancer with similar histopathological appearance can follow significantly different clinical courses and show different responses to therapy[1]; Second, many cancers lack distinctive morphological features for correct differential diagnosis; Third, human error or bias is inevitable during the subjective interpretation of the histopathology of a cancer specimen[2]. Accurate diagnosis among a group of histologically similar cancers is still a challenging issue in clinical medicine. Microarray technology simultaneously monitors the relative expression values of thousands of genes, and allows the genomic-scale study of biological processes. Since Gloub's[1] initial work, much research of cancer classification based on microarray data analysis has been done[2,3,4,5,6].

Gene selection and classifier design are two critical steps in gene expression-based cancer classification. In this paper, we propose an improved gene selection method, and construct a BP neural networks classifier. The open acute leukemia classification problem is taken as a test case. Computer simulation results show that our method meets higher classification accuracy.

F. Yin, J. Wang, and C. Guo (Eds.): ISNN 2004, LNCS 3174, pp. 465–469, 2004.

2 Material and Methods

2.1 Microarray Gene Expression Data

Acute leukemia is clinically classified into acute lymphoblastic leukemia(ALL) and acute myeloid leukemia(AML). The public leukemia data set is available at http://www.genome.wi.mit.edu/mpr. It consists of 72 tissue samples, each with 7129 gene expression values. It is divided into training set(27 ALL samples and 11 AML samples) and independent test set(20 ALLs and 14 AMLs). More detailed description about this data set can be acquired on the linkage.

2.2 Gene Selection

Gene expression data have unique characters: high dimension—usually containing thousands of genes; small sample size; most genes irrelevant to cancer distinction. Gene selection is to extract the most informative genes for cancer distinction and improve classification accuracy.

To leukemia data, each sample is expressed as: $s_j = (g_{1j}, g_{2j}, \cdots, g_{7129j})^T$, and each gene: $g_i = (g_{i1}, g_{i2}, \cdots, g_{iN})$, where g_{ij} denotes the expression value of g_i in sample s_j and N is the number of samples in the training set. Class distinction is represented as: $c = (c_1, c_2, \cdots, c_N)$, where $c_j = 1$ *or* 0 according to s_j belongs to ALL or AML class. For the training set, following values can be computed for every gene:

$$\mu 1_i = \frac{\sum_{j=1}^{N} g_{ij}, for\ c_j = 1}{num1}, \tag{1}$$

$$\mu 2_i = \frac{\sum_{j=1}^{N} g_{ij}, for\ c_j = 0}{num2}, \tag{2}$$

$$\sigma 1_i = \frac{\sqrt{\sum_{j=1}^{N} (g_{ij} - \mu 1_i)^2}, for\ c_j = 1}{num1 - 1}, \tag{3}$$

$$\sigma 2_i = \frac{\sqrt{\sum_{j=1}^{N} (g_{ij} - \mu 2_i)^2}, for\ c_j = 0}{num2 - 1}, \tag{4}$$

$$p_i = \frac{\mu 1_i - \mu 2_i}{\sigma 1_i - \sigma 2_i}, \tag{5}$$

where num1 and num2 are the numbers of ALL and AML samples in the training set respectively. $\mu 1_i$,$\sigma 1_i$ and $\mu 2_i$,$\sigma 2_i$ are the mean and standard squared variance of g_i for two classes. The bigger the absolute value of p_i, the more informative g_i is, and the sign of p_i being positive or negative denotes g_i being highly correlated with ALL or AML class. The value of p_i portrays between-class separation and within-class variance simultaneously, so it is a suitable correlation measure to some extent. Gloub etc.[1]'s gene selection method is based solely on it, however, adopting one and only correlation measure leads to some shortcomings.

We introduce two other correlation measures:

$$p1_i = |\mu 1_i - \mu_i| / \sigma 1_i, \tag{6}$$

$$p2_i = |\mu 2_i - \mu_i| / \sigma 2_i, \tag{7}$$

where μ_i is the average expression value of $g_i : \mu_i = \sum_{j=1}^{N} g_{ij}/N$. $p1_i$ and $p2_i$ reflect correlation degree of g_i with ALL or AML class.

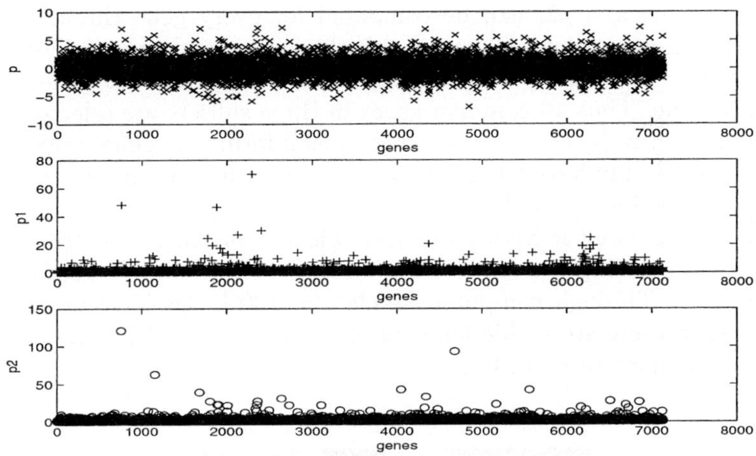

Fig. 1.

Fig.1 tells, while some genes have very large p1 or p2 value, their absolute value of p is not outstanding. No doubt, these genes are very informative too, but will be neglected according to [1]'s gene selection criterion. In contrast, our gene selection strategy considers three correlation measures comprehensively. The basic idea is to select informative genes using each measure independently, and then extract their common ingredients.

2.3 BP Classifier

To implement classification, many mathematical and statistical methods have been used: weighted voting of informative genes[1], Naive Bayes method[5], the nearest neighbor analysis[6], Fisher's linear discriminant function [2,3] and support vector machine[4].

Here, we adopt a full-connected single-hidden-layer BP neural networks to construct the classifier. Input nodes number is the number of informative genes selected, hidden neuron number is decided empirically and through computer simulation, and a single output neuron corresponds to either class. The transfer function between input and hidden layer is 'tansig', and between hidden layer and output is 'logsig'. The output is between 0 and 1, if it is greater than or equal to 0.5, the test sample will be judged as ALL ; else AML.

3 Results

Although much research has been done on this problem, 100% classification accuracy is rarely reached. In this section,first, we will use our method to make it.

Gene selection: we combine the original training set with independent test set. Then p_i, $p1_i$ and $p2_i$ can be computed for every gene through Eqn.1–7. Extract three gene subsets: 50 genes with the largest and the smallest p_i value, each 25 genes; 100 genes with the largest $p1_i$ value; and 100 genes with the largest $p2_i$ value. Thus 46 common genes in three subsets are selected out, i.e., they are informative genes. Compared with 50 informative genes selected out in [1], our 46 genes only have 23 common elements with it. It shows the difference of two gene selection methods.

Classifier: we adopt leave-one-out test, which is deemed as a stringent cross-validation test. 46 input nodes correspond to 46 informative genes, and one output neuron. Through computer simulation, 100% classification accuracy is obtained when there are 75 hidden neurons. The outputs of 72 times leave-one-out test are demonstrated in Fig.2.

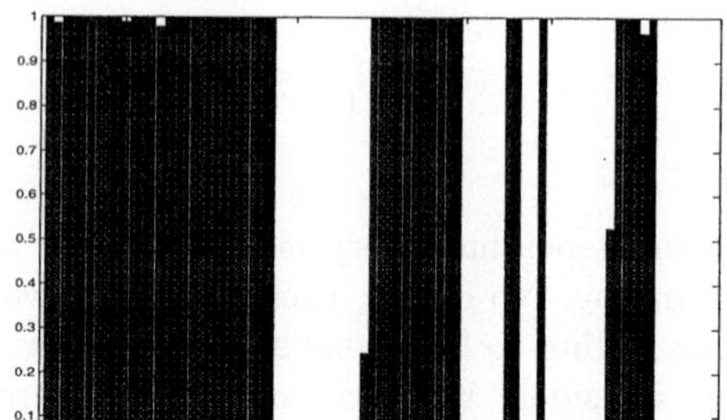

Fig. 2. The leave-one-out test outputs

Selecting less genes while keeping high classification accuracy is an important goal of concerned problem. We lessen informative genes to 6 by reducing the size of three subsets. When the number of hidden neurons is 15, through computer simulation, only a sample is misclassified. The GeneBank ID of six selected genes are M84526,M27891,D88270,M11722,K01911 and X82240. High classification accuracy (Table 1) and the coincidence of our results with others validate proposed method to certain extent.

Table 1. Classification Results Comparison

	number of misclassifications	number of genes	test method
Golub et al.[1]	4	50	independent test
Cho et al. [3]	2	6	cross validation
Fu et al. [4]	1	4	independent test
proposed	0 / 1	46 / 6	cross validation

4 Discussion

Gene selection is the most critical step in gene expression-based cancer classification. Golub et al.'s gene selection method has some limitations, we improve it and propose a novel one. BP classifier can efficiently recognize different acute leukemia patterns. High classification accuracy and the coincidence with other research justify the validity of proposed gene selection method and designed classifier.

Many researchers emphasize the coincidence of their research results(especially selected gene set) with others'. But at the same time, difference should not be neglected. Why many different compact gene sets can all achieve excellent classification performance? Prevalent answer is that selected gene set is the combination of some biological significance, and each biological significance can be fulfilled by a certain number of genes, so selected genes are those who can represent that significance well. The truth and rationale behind it will offer new insights into cancer research.

In addition, because of the small size of the samples, sometimes it is hard to evaluate a method objectively. At the same time, cancer-correlated genes digging is still a tough task on such a small data set.

References

1. Golub, T.R.,Slonim, D.K., Tamayo, P., Huard, C., Gaasenbeek, M., Mesirov, J.P., Coller, H., Loh, M.L., Downing, J.R., Caligiuri, M.A., Bloomfield, C.D., and Lander, E.S.: Molecular classification of cancer: Class discovery and class prediction by gene expression monitoring. (1999) Science **286** 531–537.
2. Xiong, M., Li, W., Zhao, J., Jin, L., and Boerwinkle, E.: Feature(Gene) selection in gene expression-based tumor classification. (2001) Molecular Genetics and Metabolism **73** 239–247.
3. Cho, J., Lee, D., Park, J.H. and Lee, I.: New gene selection method for classification of cancer subtypes considering within-class variation. (2003) FEBS. **551** 3–7.
4. Fu, L.M. and Fu-Liu, C.S.: Multi-class cancer subtype classification based on gene expression signatures with reliability analysis. (2004) FEBS. **561** 186–190.
5. Keller, A., Schummer, L., Hood, L., and Ruzzo, W.: Bayesian classification of DNA array expression data. (2000) Technical Report University of Washington. August.
6. Ben-Dor, A. etc.: Tissue classification with gene expression profiles. (2000) Proceedings of the Fourth Annual International Conference of Computational Molecular Biology,Tokyo,Japan.

Nonnegative Matrix Factorization for EEG Signal Classification

Weixiang Liu, Nanning Zheng, and Xi Li

Institute of Artificial Intelligence and Robotics
Xi'an Jiaotong University
Xi'an,Shaanxi Province 710049 P.R.China
{wxliu,xli}@aiar.xjtu.edu.cn, nnzheng@mail.xjtu.edu.cn

Abstract. Nonnegative matrix factorization (NMF) is a powerful feature extraction method for nonnegative data. This paper applies NMF to feature extraction for Electroencephalogram (EEG) signal classification. The basic idea is to decompose the magnitude spectra of EEG signals from six channels via NMF. Primary experiments on signals from one subject performing two tasks show high classification accuracy rate based on linear discriminant analysis. Our best results are close to 98% when training data and testing data from the same day, and 82% when training data and testing data from different days.

1 Introduction

Recognition and classification of Electroencephalogram (EEG) signals for Brain-Machine Interfaces (BCI) bring out many open problems from both theory and application; see [15] for a brief survey on the topic and [14] for more information. From the viewpoint of pattern recognition [1], one key problem is how to represent the recorded EEG signals for further analysis such as classification. In other words, it is, firstly, important to extract useful features from the EEG signals. There are many available representations from both time domain and frequency domain, such as AR model coefficients, Karhunen-Loéve transform [13] and maximum noise fraction (MNF) transform [15].

Recently nonnegative matrix factorization (NMF) [4], as a powerful feature extraction method for nonnegative data, has been successfully applied to music analysis [6,10,18,19], document clustering and information retrieval [7,8,20,21, 11,17], gene expression [9,22] and molecular analysis [23]. Inspired by the idea from above music analysis with magnitude spectra, we try to apply NMF to feature extraction for EEG signal classification. The basic idea is to decompose the magnitude spectra of EEG signals from six channels via NMF. We made our experiments on signals from one subject performing two tasks in two days and the results show high classification rate based on linear discriminant analysis. Our best results are approximately 98% when training data and testing data from the same day, and 82% when training data and testing data from different days.

F. Yin, J. Wang, and C. Guo (Eds.): ISNN 2004, LNCS 3174, pp. 470–475, 2004.

The rest of this paper is organized as follows. In section 2 we recall some basic theory of NMF. In section 3 we make our experiments and we give our conclusions and discuss some future work in section 4.

2 Nonnegative Matrix Factorization

NMF is a multivariate data analysis method. Given a nonnegative data set with m samples in R^d, denoted as $X_{d\times m}$, NMF finds two nonnegative matrix factors B and C (i.e. each element of B and C is nonnegative, denoted by $B \geq 0, C \geq 0$) such that [5]

$$X \approx BC \tag{1}$$

where B is a $d \times r$ basis matrix, each column of which is a basis vector, and C a $r \times m$ coefficient matrix, each column of which is a new feature vector. It leads to dimension reduction by choosing r smaller than d although it is an open problem to decide the optimal r. Two kinds of cost functions have been investigated [5] with multiplicative update rules which naturally preserve nonnegativity, i.e. the generalized Kullback-Leibler divergence

$$D_{KL}(B,C) = \sum_{i=1}^{d}\sum_{j=1}^{m}[X_{ij}\log\frac{X_{ij}}{(BC)_{ij}} - X_{ij} + (BC)_{ij}] \tag{2}$$

and the square Euclidean distance

$$D_2(B,C) = \sum_{i=1}^{d}\sum_{j=1}^{m}[X_{ij} - (BC)_{ij}]^2. \tag{3}$$

In this paper we adopt the multiplicative rules for minimizing eq. (2) as below [4]

$$C_{kj} \leftarrow C_{kj}\sum_{i} B_{ik}\frac{X_{ij}}{(BC)_{ij}} \tag{4}$$

$$B_{ik} \leftarrow B_{ik}\sum_{j} \frac{X_{ij}}{(BC)_{ij}}C_{kj} \tag{5}$$

$$B_{ik} \leftarrow \frac{B_{ik}}{\sum_{l} B_{lk}}. \tag{6}$$

For testing samples, it is convenient to get new features according to eq. (4) while fixing the learned basis matrix from training data.

With contrast to traditional principal component analysis (PCA) [3] and recent independent component analysis (ICA) [2], NMF has parts- based representation property because of nonnegativity [4].

NMF restricts its data with only nonnegativity. However, when we apply NMF for data analysis in magnitude spectra, it is not necessary to require that the source data is nonnegative. Some cases of NMF for music analysis [6,10,18, 19] provide promising results which inspire us to classify EEG signals via this method for representation.

As a new developed feature extraction method, NMF still has many open problems for investigation. See our technical report [12] for more information.

3 Experimental Results

3.1 EEG Data

The EEG data used here is available online[1], which has been discussed in detail [15,16]. We used the EEG signals from subject 1 performing two mental tasks, i.e. math multiplication and letter composing, in two days. Each tasks contains 5 trials on one day and another 5 trials on second day. Each trial is a 6×2500 matrix from six channels.

3.2 EEG Data Representation

Following [15], we first segmented each serial with 2500 times samples into 38 augmented data samples according to the procedure as below: each window has 128 time samples that overlap by 64 samples. Then we implemented discrete Fourier transform on each augmented sample and got the first 65 absolute versions of magnitude spectra. Finally each trial becomes a 38 data samples with 65×6 dimensions which can be reduced by NMF.

3.3 Classification Results

For classification, we adopted the discriminant analysis function *classify* in MATLAB[2] with linear method which fits a multivariate normal density to each group based on training data with a pooled estimate of covariance. We made three different initialization cases for dimension reduction by NMF and set $r = 6, 12, 18, \ldots, 66$ for comparison when testing the method.

Firstly we used the EEG signals from subject 1 on the first day. We used the first trial for training and the rest 4 trials for testing. The final results are shown in Table 1. We can see that from above table, different initializations and rs for NMF leads to different classification rate; the average accuracy of two tasks can get 98%. Our results are higher than those in [15] (90%) although we used 128 time samples per segmentation with overlap.

We also used the EEG signals from subject 1 from different days. We used the first trial on the first day for training and all 5 trials on the second day for testing. The final results are shown in Table 2. According to the average accuracy rate, the proposed method can get 82% or so as the best . Our results are higher than 75% reported in [15].

[1] http://www.cs.colostate.edu/eeg/index.html#Data

[2] http://www.mathworks.com/

Table 1. EEG signal classification results when training data and testing data from one day.

r	math	letter	average	r	math	letter	average	r	math	letter	average
6	0.770	0.875	0.822	6	0.829	0.868	0.849	6	0.770	0.855	0.813
12	0.822	0.914	0.868	12	0.921	0.928	0.924	12	0.895	0.928	0.911
18	0.914	0.954	0.934	18	0.947	0.928	0.938	18	0.908	0.895	0.901
24	0.947	0.895	0.921	24	0.895	0.934	0.914	24	0.908	0.895	0.901
30	0.954	0.947	0.951	30	0.974	0.947	0.961	30	0.987	0.934	0.961
36	0.928	0.941	0.934	36	0.928	0.908	0.918	36	0.947	0.954	0.951
42	0.961	0.961	0.961	42	0.993	0.941	0.967	42	0.934	0.928	0.931
48	0.947	0.914	0.931	48	**0.980**	**0.961**	**0.970**	48	0.934	0.941	0.938
54	0.921	0.967	0.944	54	0.947	0.961	0.954	54	0.921	0.928	0.924
60	0.987	0.954	0.970	60	0.974	0.947	0.961	60	0.947	0.914	0.931
66	**0.987**	**0.967**	**0.977**	66	0.908	0.868	0.888	66	**0.987**	**0.941**	**0.964**

Table 2. EEG signal classification results when training data and testing data from different days.

r	math	letter	average	r	math	letter	average	r	math	letter	average
6	0.647	0.642	0.645	6	0.674	0.489	0.582	6	0.695	0.516	0.605
12	0.905	0.542	0.724	12	0.884	0.695	0.789	12	0.863	0.616	0.739
18	0.784	0.737	0.761	18	0.911	0.647	0.779	18	0.932	0.595	0.763
24	0.858	0.637	0.747	24	0.789	0.784	0.787	24	0.889	0.674	0.782
30	**0.874**	**0.732**	**0.803**	30	**0.911**	**0.689**	**0.800**	30	0.900	0.579	0.739
36	0.895	0.653	0.774	36	0.842	0.716	0.779	36	0.847	0.574	0.711
42	0.868	0.711	0.789	42	0.889	0.647	0.768	42	0.895	0.674	0.784
48	0.921	0.563	0.742	48	0.853	0.679	0.766	48	**0.884**	**0.758**	**0.821**
54	0.879	0.632	0.755	54	0.916	0.537	0.726	54	0.884	0.653	0.768
60	0.895	0.511	0.703	60	0.821	0.753	0.787	60	0.853	0.716	0.784
66	0.916	0.621	0.768	66	0.847	0.526	0.687	66	0.900	0.668	0.784

Our results also indicate that it is difficult to select optimal r while reducing the dimension via NMF, which is an open problem for NMF [12].

4 Conclusions and Future Work

In this paper we first apply NMF to feature extraction for EEG signal classification. The basic idea is to decompose the magnitude spectra of EEG signals from six channels via NMF. We made our experiments on signals from one subject performing two tasks in the same day and the primary results show high classification rate based on linear discriminant analysis. Our best results are close to 98% when training data and testing data from the same day, and 82% when training data and testing data from different days.

Our results for EEG signal classification are promising. There are some directions for future work. Firstly we will further analyze the EEG signals based on NMF for several tasks. Secondly it is necessary to compare the proposed method with other methods such as MNF in [15], or with advanced classifiers such as support vector machines in [16]. Thirdly, it is possible to consider other time-frequency domain such as via Discrete Cosine Transform and Discrete Wavelet Transform. And finally, or the most important, it is potential to implement the method for real time analysis.

Acknowledgments. This work was supported by the NSF of China (No. 60205001). We would like to thank Lianyi Zhang for his introduction of the EEG data set and fruitful discussions.

References

1. Duda, R.O., Hart, P.E., Stork, D. G.: Pattern Classification. 2nd edn. John Wiley & Sons (2001)
2. Hyvärinen,A., Karhunen, J., Oja, E.: Independent Component Analysis. John Wiley & Sons (2001)
3. Jolliffe, I.T.: Principal component analysis. 2nd edn. Springer-Verlag, New York (2002)
4. Lee, D.D., Seung, H.S.: Learning the parts of objects with nonnegative matrix factorization. Nature **401** (1999) 788-791
5. Lee, D.D., Seung, H.S.: Algorithms for nonnegative matrix factorization.In: Leen, T., Dietterich, T., Tresp, V. (eds.): Advances in Neural Information Processing Systems 13. MIT Press, Cambridge, MA (2000)
6. Kawamoto, T., Hotta,K., Mishima,T., Fujiki, J., Tanaka, M., Kurita, T.: Estimation of Single Tones from Chord Sounds Using Non-Negative Matrix Factorization. Neural Network World **3** (2000) 429-436
7. Vinokourov, A.: Why Nonnegative Matrix Factorization Works Well For Text Information Retrieval. http://citeseer.nj.nec.com/458322.html.
8. Tsuge, S., Shishibori,M., Kuroiwa, S., Kita, K.: Dimensionality reduction using non-negative matrix factorization for information retrieval. In: IEEE International Conference on Systems, Man, and Cybernetics, Vol. 2. (2001) 960 -965
9. Seppänen,J. K. , Hollmén, J., Bingham, E., Mannila, H.: Nonnegative matrix factorization on gene expression data. Bioinformatics 2002, poster 49. (2002)
10. Smaragdis,P., Brown, J.C.: Non-negative matrix factorization for polyphonic music transcription. IEEE Workshop on Applications of Signal Processing to Audio and Acoustics. (2003) 177-180
11. Pauca,P., Shahnaz, F., Berry, M., Plemmons, R.: Text Mining using Nonnegative Matrix Factorizations. In: Proc. SIAM Inter. Conf. on Data Mining. (2003)
12. Liu, W.X., Zheng, N.N., Li, X.: Review on Nonnegative Matrix Factorization. Technical report, Institute of Artificial Intelligence and Robotics, Xi'an Jiaotong University. (2004)
13. Anderson, C., Devulapalli, S., Stolz, E.: EEG Signal Classification with Different Signal Representations. In: Girosi, F., Makhoul, J., Manolakos, E., Wilson, E. (des.): Neural Networks for Signal Processing V. IEEE Service Center, Piscataway, NJ. (1995) 475–483

14. Vaughan, T.M., Heetderks, W.J., Trejo, L.J., Rymer, W.Z., Weinrich, M., Moore, M.M., Kübler, A., Dobkin, B.H., Birbaumer, N., Donchin, E., Wolpaw, E.W. and Wolpaw, J.R.: Brain-computer interface technology: A review of the Second International Meeting. IEEE Transactions on Neural Systems & Rehabilitation Engineering **11** (2003) 94-109
15. Anderson, C.W., Kirby, M.: EEG Subspace Representations and Feature Selection for Brain-Computer Interfaces. In: Proceedings of the 1st IEEE Workshop on Computer Vision and Pattern Recognition for Human Computer Interaction (CVPRHCI). (2003)
16. Garrett, D., Peterson, D.A., Anderson, C.W., Thaut, M.H.: Comparison of Linear and Nonlinear Methods for EEG Signal Classification. IEEE Transactions on Neural Systems and Rehabilitative Engineering **11** (2003) 141-144
17. Xu,W., Liu,X., Gong, Y.H.: Document clustering based on non-negative matrix factorization. In: Proceedings of the 26th annual international ACM SIGIR conference on Research and development in informaion retrieval. (2003) 267 - 273
18. Plumbley,M. D. ,Abdallah,S. A., Bello,J. P., Davies, M. E., Monti,G., Sandler,M. B.: Automatic music transcription and audio source separation. Cybernetics and System **33** (2002) 603-627
19. Plumbley,M. D.: Algorithms for non-negative independent component analysis. IEEE Transactions on Neural Networks **14** (2003) 534- 543
20. Lu,J.J., Xu,B.W., Yang, H.J.: Matrix dimensionality reduction for mining web logs. In: Proceedings of IEEEWIC International Conference on Web Intelligence. (2003) 405 - 408
21. Xu,B.W., Lu,J.J.,Huang, G.S.: A constrained non-negative matrix factorization in information retrieval. In: IEEE International Conference on Information Reuse and Integration. (2003) 273 - 277
22. Kim,P.: Understanding Subsystems in Biology through Dimensionality Reduction, Graph Partitioning and Analytical Modeling. Phd thesis . (2003)
23. Brunet, J.P., Tamayo, P., Golub, T.R., Mesirov, J.P.: Metagenes and molecular pattern discovery using matrix factorization. Proc Natl Acad Sci U. S. A. **101** (2004) 4164-4169

A Novel Clustering Analysis Based on PCA and SOMs for Gene Expression Patterns

Hong-Qiang Wang[1,2], De-Shuang Huang[1], Xing-Ming Zhao[1,2], and Xin Huang[1]

[1] Hefei Institute of Intelligent Machines, Chinese Academy of Science,
P.O.Box 1130, Hefei, Anhui, 230031, China
{hqwang, dshuang, xmZhao, xhuang}@iim.ac.cn

[2] Department of Automation, University of Science and Technology of China,
Hefei, 230027, China

Abstract. This paper proposes a novel clustering analysis algorithm based on principal component analysis (PCA) and self-organizing maps (SOMs) for clustering the gene expression patterns. This algorithm uses the PCA technique to direct the determination of the clusters such that the SOMs clustering analysis is not blind any longer. The integration of the PCA and the SOMs makes it possible to powerfully mine the underlying gene expression patterns with the practical meanings. In particular, our proposed algorithm can provide the informative clustering results like a hierarchical tree. Finally, the application on the leukemia data indicates that our proposed algorithm is efficient and effective, and it can expose the gene groups associated with the class distinction between the acute lymphoblastic leukemia (ALL) samples and the acute myeloid leukemia (AML) samples.

1 Introduction

Microarray technology produces a large scale of gene expression measurements. But the analyses of these data collected over time and under different experimental conditions have become a formidable challenge. Gene clustering is an essential procedure in the gene analysis, because they can be used to understand the functional differences in cultured primary hepatocytes relative to the intact liver, on gene expression data for tumor and normal colon tissue probed by oligonucleotide arrays and to analyze temporal gene expression data during rat central nervous system development [1,2,3]. These applications were successful and worked well.

The clustering approaches often used include hierarchical clustering, Bayesian clustering, k-means clustering, and self-organizing maps (SOMs), etc. However, these clustering methods all have some shortcomings. In particlular, the common and big trouble is that there is not a formal and reasonable method for the choice of the number of the preset clusters with the actual meanings. The common approach to solve this dilemma is the manual trial-and-error scheme, but the large number of genes in the gene expression clustering analyses rends this approach infeasible.Though the automatic determination methods have been reported by some literatures [4,5,6], they have shortcomings individually.

This paper proposes the novel PCA-SOMs based clustering analysis algorithm for clustering the gene expression patterns. Because the principal components resulted from

F. Yin, J. Wang, and C. Guo (Eds.): ISNN 2004, LNCS 3174, pp. 476–481, 2004.

the PCA reflect the information about the distribution of gene expression levels, we can use the PCA method to deal with the dilemma of determining the number of the clusters, but not to transform the original data prior to the clustering analysis like the traditional applications of the PCA method.

This paper is organized as follows. Section II describes our proposed algorithm. The selection strategy of the number of the clusters for the gene expression patterns is given, and the full procedure of our proposed algorithm is presented. In Section III, the leukemia dataset is used to verify our proposed algorithm for discovering the underlying gene expression patterns. In addition, to assess the algorithm, we also use the clustering results of the different gene clusters to predict the unknown samples by two classifiers, i.e. the linear SVM classifier and the k-nearest neighbors classifier. Section IV provides the conclusion for this whole paper.

2 Main Results

2.1 The Determination of the Clusters *K* in the SOMs Clustering Analysis

In literatures [1,2,3] to discuss clustering analysis, the principal component analysis (PCA) is all used to capture the clustering structure of the original data to improve the performance of the clustering analyses. However, unlike the traditional application, the PCA method in our proposed algorithm is used to determine the number K of the clusters for the SOMs clustering analysis.

It is well known that the PCA method can reflect the distribution of all variables (genes) since there are maximal projection lengths for all variables along the directions of the principal components. In addition, the corresponding variance can indicate these information magnitudes along each principal component' direction. So, we can recognize the distribution of the gene expression patterns in the microarray data by the PCA processing, but not in detail. For examples, the traditional PCA can't specify the actual genes with the bigger projecting lengths in the principal component directions, however, they can be found by the SOMs based clustering analysis. In fact, these principal components by the PCA can be regarded as the centers of the gene expression patterns with the different confident values from the corresponding variances $Var_i, i = 1, 2, ..., N$. So, when the confident cutoff Θ is given, the number K of the gene expression patterns can be determined by counting the number of variances that are bigger than the cutoff Θ, i.e.

$$K = \sum_{i=1}^{N} \Gamma_i \tag{1}$$

where N is the number of principal components and

$$\Gamma_i = \begin{cases} 1 & Var_i \geq \Theta, \\ 0 & otherwise \end{cases} \tag{2}$$

2.2 The PCA-SOMs Based Clustering Analysis Algorithm

The goal of gene clustering analysis is to extract the fundamental patterns of gene expression levels inherent in microarray data. There have existed many clustering analysis approaches, such as hierarchical clustering, Bayesian clustering, k-means clustering, and self-organizing maps. Comparatively, SOMs have better computational properties [7]. In particular, they can obtain more features that make them well suitable to the clustering analysis of gene expression patterns [8].

In our algorithm, the SOMs based clustering analysis method (SOMs-CA) is adopted to cluster the gene expression patterns. However this method should be applied under the guide of the PCA, which is different from the previous methods where the number of the clusters is chosen empirically without actual meaning. In detail, prior to clustering the data, we perform the principal components analysis to extract the number of the patterns that take up most ratios in the pattern distribution as the number of the clusters in the unsupervised SOMs based pattern clustering analysis algorithm. So, our proposed clustering algorithm is endowed with the approximate pattern information, thus it is not blind any more in the practical analysis.

Commonly, if we can find the accurate variance cutoff Θ for determining the clusters, the following SOMs based clustering analysis can produce the ideal pattern clusters with the centralized distribution of the variances. But due to the difficult operation, the accurate cutoff $\Theta(0 < \Theta < 1)$ is very difficult to be chosen. So, to solve this problem, we perform the iteration of the PCA and SOMs-CA procedure recursively until all the satisfying patterns with the centralized distributions are found. That's to say, given a constant $\tau(\Theta < \tau < 1)$, which is also called as the stop condition parameter, these patterns need to meet the following condition:

$$\sum_{i=1}^{N} \ell_i = 1 \tag{3}$$

where N is the number of principal components and

$$\ell_i = \begin{cases} 1 & Var_i \geq \tau, \\ 0 & \text{otherwise.} \end{cases} \tag{4}$$

As a result, our proposed algorithm can unfold the gene expression patterns like a hierarchical tree.

Finally, our proposed novel PCA-SOMs based clustering analysis algorithm can be summarized as follows:

Begin

Step 1 Initialize the confident cutoff parameter Θ and the stop condition parameter τ.

Step 2 Preprocess the original microarray data to mean=0 across samples for each gene.

Step 3 Perform the principal components analysis on the preprocessed microarray data to attain the variance distribution and determine the clusters number K for the following SOMs based clustering analysis by $K = \sum_{i=1}^{N} \Gamma_i$.

Step 4 Perform the SOMs-CA algorithm to attain the gene expression patterns.

Step 5 Estimate each gene expression pattern by $\sum_{i=1}^{N} \ell_i$.
 if $\sum_{i=1}^{N} \ell_i > 1$,
 repeat from step.2.
Step 6 Draw the clustering results and the centroids of each gene expression pattern.

End

3 Experimental Results

In this section, we applied our proposed algorithm to the public available microarray data set, i.e., the leukemia data, which was obtained from cancer patients with two different types of leukemia first analyzed by Golub et al. This dataset contains 72 tissue samples divided into two categories, the ALL including 47 cases and the AML including 25 cases, each sample of which consists of expression values of 7129 gene variables that are absolute measurements from Affymetrix high-density oligonucleotide arrays. This dataset is available at http://www.genome.wi.mit.edu/MPR. In our experiment, we divided this dataset into two groups of the training set including 38 samples (27 ALLs and 11 AMLs) and the test set including 34 samples (20 ALLs and 14 AMLs) to assess our proposed algorithm by the performance of the clusters' classification of the samples.

3.1 The PCA-SOMs Clustering Analysis of the Leukemia Data

First, let us preset the two important parameters, the confident cutoff Θ and the stop condition τ, as 10% and 60%, respectively. Next, the iteration is started.

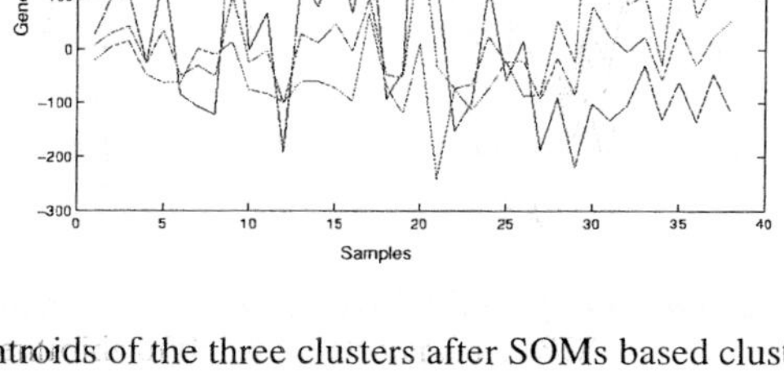

Fig. 1. The centroids of the three clusters after SOMs based clustering analysis.

As a result of the first iteration, the variances for the 38 principal components are worked out. From these variances, we can find that there are only the three leading principal components whose variances are bigger than the Θ=10%. This means that the original dataset perhaps contains three main gene expression patterns referring to the confident cutoff Θ. So, the number of the clusters for the SOMs based clustering analysis to follow should be defined as 3. After the SOMs based clustering analysis, the genes are assigned into three clusters. Fig.1 shows the centroids of the three clusters, where the horizontal axis denotes the total 38 training samples the first 27 of which belong to the ALL category but the remaining belong to the AML category. Because the leukemia dataset contains two categories: ALL and AML, the genes involved in the microarray chip should be in about three states, i.e., the up-regulated, the down-regulated and the non-significant between the two categories of samples. From Fig.1, we can note that the three centroids of the clusters are accordant with the three gene expression states. The pattern drawn by the blue curve is up-regulated in ALL samples but down-regulated in the AML samples, while the pattern by the red curve is down-regulated in the ALL samples but up-regulated in the AML samples. But the pattern by the green curve changes faintly across all 38 samples and is in the non-significant state between the two categories. Obviously, these three clusters are significantly distinct and responsible for the classification of the samples. So, this results are reasonable and practically meaningful.

Finally, after 15 iterations, our proposed algorithm gives the clustering result of the leukemia data, i.e., 19 clusters grouped. Fig.2 illustrates the centroids of all the clusters.

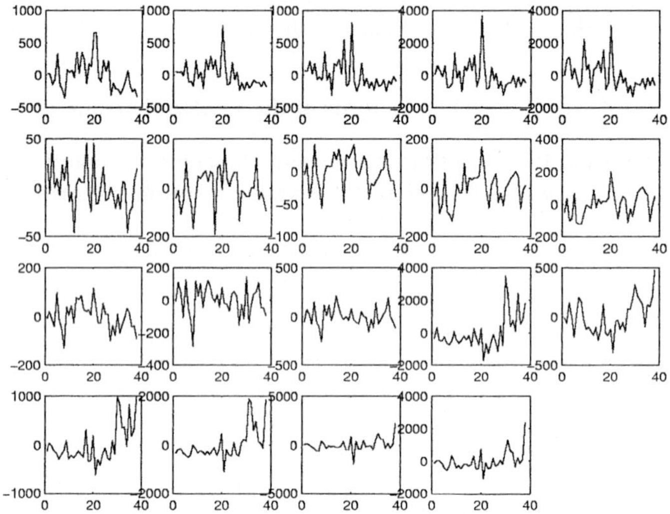

Fig. 2. The centroids of the final 19 clusters after the PCA-SOMs based clustering analysis.

3.2 The Assessment of Our Proposed Clustering Analysis Algorithm

The correct classification rates acquired by three gene expression patterns from the first iteration are listed in Table I. From Table I, we can find that both the first and the third clusters have similar classification rates that are much bigger than that of the second cluster. This result indicates that these three gene expression clusters completely correspond with the underlying class patterns in the dataset. Therefore, our proposed algorithm can reflect the actual gene expression patterns.

Table 1. The rates of the three clusters by the two classifiers: the linear SVM classifier and the k nearest neighbors classifier

Classifier	Cluster 1	Cluster 2	Cluster 3
The linear SVM classifier	0.79	0.70	0.79
The k nearest neighbors classifier	0.88	0.76	0.94

4 Conclusions

The PCA-SOM based clustering analysis algorithm proposed in this paper combines the merits of the PCA method and the SOM based clustering analysis method for clustering the gene expression patterns. It uses the result of the PCA to direct the determination of the clusters such that the SOMs based clustering analysis is not blind any longer. As a result, our proposed algorithm gives the informative clustering results in the hierarchical tree form. Finally, the application on the leukemia data indicates that our proposed algorithm is efficient and can expose the gene group associated with the class distinction between the ALL samples and the AML samples.

References

1. Baker, T. K., Carfagna, M. A., Gao, H., Dow, E. R., Li Q. Q., Searfoss, G. H., Ryan, T. P.: Temporal Gene Expression Analysis of Monolayer Cultured Rat Hepatocytes. Chem. Res. Toxicol, **14(9)**(2001) 1218-31
2. Alon, U., Barkai, N., Notterman, D. A., Gish, K., Ybarra, S., Mack, D., Levine, A. J.: Broad Patterns of Gene Expression Revealed by Clustering Analysis of Tumor and Normal Colon Tissues Probed by Oligonucleotide Arrays. Proc. Natl. Acad. Sci. USA, **96(12)** (1999) 6745-6750
3. Wen, X.l., Fuhrman, S., Michaels, G. S., Carr, D. B., Smith S., Barker, J. L., Somogyi, R.: Large-scale Temporal Gene Expression Mapping of Central Nervous System Development. Proc. Natl. Acad. Sci. USA, **95(1)** (1998) 334-339
4. Sharan, R., Shamir, R.: CLICK: A Clustering Algorithm with Applications to Gene Expression Analysis. Proccedings of the 2000 Conference on Intelligent Systems for Molecular Biology (ISMB00), La Jolla, CA. **8** (2000) 307-316
5. Tishby, N., Slonim, N.: Data Clustering by Markovian Relaxation and the Information Bottleneck Method. Neural Information Processing Systems, **13** (2001) 7
6. Kohonen, T.: Clustering, taxonomy, and topological maps of patterns. In Proc. 6ICPR, Int. Conf. on Pattern Recognition, (1982) 114-128
7. Kohonen, T.: The self-organizing map. Proceedings of the IEEE, **78(9)** (1990) 1464-1480
8. Kohonen, T.: Self-Organizing Maps. Springer, Berlin. (1997)

Feedback Selective Visual Attention Model Based on Feature Integration Theory

Lianwei Zhao and Siwei Luo

Dept. of Computer Science,
Beijing Jiaotong University,
Beijing 100044, P.R.China
lwzhao@sina.com

Abstract. In this paper the visual processing architecture is assumed to be hierarchical in structure with units within this network receiving both feed-forward and feedback connections. We propose a neural computational model of visual system, which is based on the hierarchical structure of feedback selectiveness of visual attention information and feature integration theory. The proposed model consists of three stages. Visual image input is first decomposed into a set of topographic feature maps in a massively parallel method at the saliency stage. The feature integration stage is based on the feature integration theory, which is a representative theory for explaining all phenomena occurring in visual system as a consistent process. At last stage through feedback selection, the saliency stimulus is localized in each feature map. We carried out computer simulation and conformed that the proposed model is feasible and effective.

1 Introduction

The most important function of selective visual attention is to orientate rapidly objects of interest in our visual environment. Attention allows us to break down the problem of understanding a visual scene into a rapid series of computationally less demanding, localized the visual analysis problems. So it is important to simulate the brain mechanism based on the physiological and psychological theories and experiments [1, 2].

In biological vision, visual features are computed in the retina, superior colliculus, lateral geniculate nucleus and early visual cortical areas. Neurons at the earliest stages are tuned to simple visual attributes such as intensity contrast, color, orientation, direction and velocity of motion, or stereo disparity at several spatial scales. Neurophysiologists have documented the existence of multiple cortical areas responsive to different visual features [3]. Through multiple neural networks early visual features are computed in a parallel manner across the entire visual field. The problem about how to bind the separated information is called binding problem. Among the models proposed to solve this problem, the feature integration theory proposed by Treisman [4] is has been considered as a representative theory. The next question is how to

F. Yin, J. Wang, and C. Guo (Eds.): ISNN 2004, LNCS 3174, pp. 482–487, 2004.

control a single attentional focus based on the multiple neural networks using multiple representations. To solve this problem, most models of bottom-up attention follow Itti and Koch [5] and hypothesize that the various feature maps feed into a unique saliency map. The saliency map is a scalar, two-dimensional map whose activity topographically represents visual saliency, irrespective of the feature dimension that makes the location salient.

But one important role of attention is to localize a stimulus in retinotopic, so that the interfering or corrupting signals are minimized, and the discriminability of a particular image subunit is increased. How does attention perform it? In this paper, we proposed a neural network model of visual attentive based on the feature integration theory. In this model we used a combination of a feed-forward bottom-up feature extraction and feedback selective mechanisms. The visual processing architecture is assumed to be hierarchical in structure with units within this network receiving both feed-forward and feedback connections. It is assumed that response strength of unit in the network is a measure of relative importance of contents of the corresponding receptive field in the scene.

Further more, this model is different from the top-down task-based attention model. The expression of the later is most probably controlled from higher areas. Such deployment of attention needs to move the eyes. Although certain features automatically attract attention, directing attention to other objects requires voluntary effort.

2 Binding Problem and Feature Integration Theory

The studies from neuroscience, computer science and psychology find that in our brain features such as form, color, orientation and motion are separated and processed in parallel. After this process, they are integration and recognized as one object. The process of reconstructing the image from the information that was separated is called binding. This problem is called binding problem, which is one of the most difficult problems in modeling the visual system. In recent years, the binding problem has gain widespread attention. The binding problem concerns the way in which we select and integrate the separated features of objects in the correct combinations.

The visual attention is one of the most important abilities for human beings to process much information. Some physiological experiments show that visual attention resembles the spotlight and this attention spotlight moves in visual space and select features. The feature integration theory proposed by Treisman [4] is based on the idea that spotlight of attention is focused on a small region in the master map of image, and the spotlight associates features in the different feature maps (form, color, orientation, motion, etc.) corresponding to the location of the spotlight. In this theory, different properties of the visual input are encoded in separate feature maps and are combined in perception by integrating separate feature maps though spatial attention, binding is the process of combining stimulus feature to form an objects representation. The mechanism of attention is considered as the integration function of features and the clue to solving the binding problem.

3 Feedback Selective Visual Attention Model

3.1 Architecture of Feedback Selective Visual Attention Model

The visual attention model proposed here (Fig.1) is based on Feature Integration Theory and other several models proposed to explain visual search strategies. This model includes three stages: the saliency stage, the feature integration stage and the feedback selective stage.

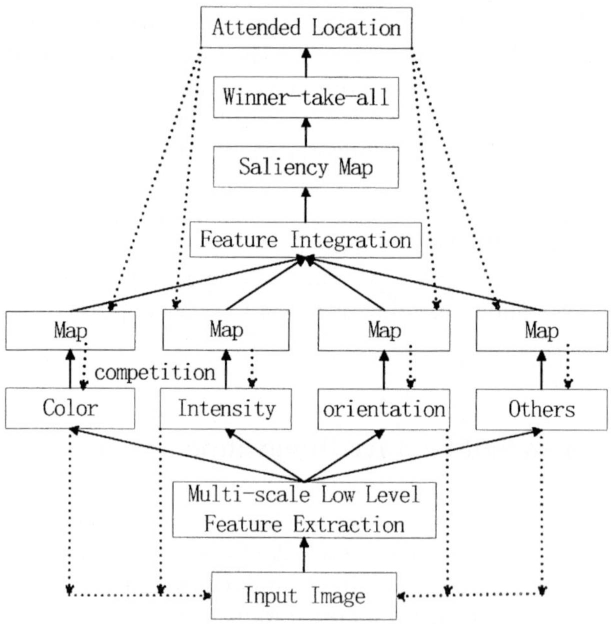

Fig. 1. The feedback selective visual attention model

In the first stage, visual image input is decomposed into a set of topographic feature maps in a massively parallel method, such as color, orientation, intensity and others. Different spatial locations then compete for saliency within each map, such that only one location that stands out from their surround can be the winner.

In the feature integration stage all feature maps feed into a master saliency map through feature integration in a bottom-up manner, which topographically codes for local conspicuity over the entire visual scene. In primates, such a map is believed to be located in the posterior parietal cortex as well as in the various visual maps. The model's saliency map is endowed with internal dynamics that generate attentional shifts.

The third stage is feedback selective stage. In this stage, the attended location for attention, which is selected on the basis of feed-forward activation at the second stage, is then propagated back through the activation of winner-take-all networks embedded within the bottom-up processing pyramid. This location will be the winning location of every saliency maps, and spatial competition for saliency is thus refined. The feed-forward paths that do not contribute to the winning location are pruned. Fig. 2 shows the feedback selective self-organization network.

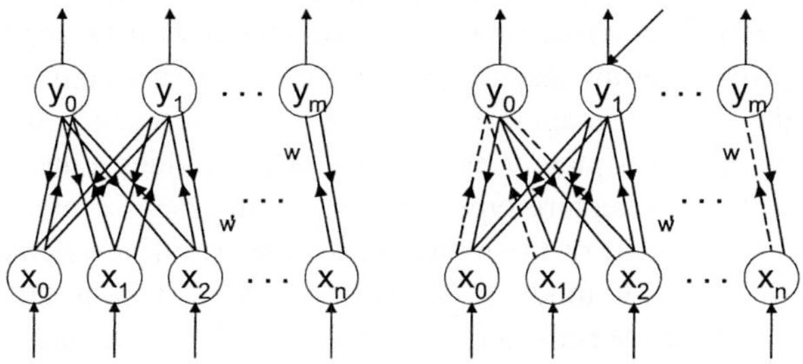

Fig. 2. Feedback selective Self-organization network

3.2 Feedback Selective Self-Organization Network

Like the general self-organization neural network, this feedback selective self- organization network also relies on a hierarchy of winner-take-all (WTA) process, and includes two layers: input and output layer. But between the two layers, there are connected by the weights of both bottom-up and top-down. That is, all the units in this network receive both feed forward and feedback connections.

When a stimulus is first applied to the input layer of the network in the first stage, it activates in a feed-forward manner all of the units within the network to which it is connected. The weights ***w*** of bottom-up represented the location of winner in the first stage, and then the weight ***w'*** of top-down represented the attended location (Fig. 2(left)). If these two locations are consistent with each other, then feedback selective need not implement. On the contrary, the feedback selectivity will be carried out, and all of the connections of the network that do not contribute to the winner are pruned (Fig. 2(right)). This strategy of finding the winners within successively smaller receptive fields and then pruning away irrelevant connections is applied recursively through the network. In this way the network learns attended location's information without destroying the old one.

4 Results

The bottom-up model of attention that we use has been shown to be effective with natural image [6]. Based on the biologically plausible architecture, this feedback selective computational model attempts to imitate the low-level, automatic mechanisms responsible for attracting our attention to the salient location in our environment, and closely follows the neuronal architecture of the earliest hierarchical levels and feature integration theory of visual processing.

To evaluation our proposed model, we ran an experiment based on the natural image (Fig. 3(a)), digitized at 256×256 resolution, and calculated the local contrasts for color, intensity and orientation features respectively. These feature types are computed in a center-surround fashion, providing a biologically plausible system that is sensitive to local spatial contrast. For the intensity and orientation features, a Gaussian filter is applied. The orientation feature is filtered with Gabor filter of angles 0, 45, 90, 135 degrees. Feature maps representing center-surround difference are then obtained from the filtered images. The feature maps for each feature are then combined respectively into conspicuity maps. Each map provides a measurement of scene area that pop out for that feature type. Combining the maps provide a unified measurement of pronounced part of the entire scene. This result a saliency map is the primary output of the attention model (Fig. 3(d)). The final stage in the bottom-up attention model is a winner-take-all network of simple integrate-and-fire neurons. The WTA network finds the attended location of the saliency map, which corresponds to the current most salient location in the scene.

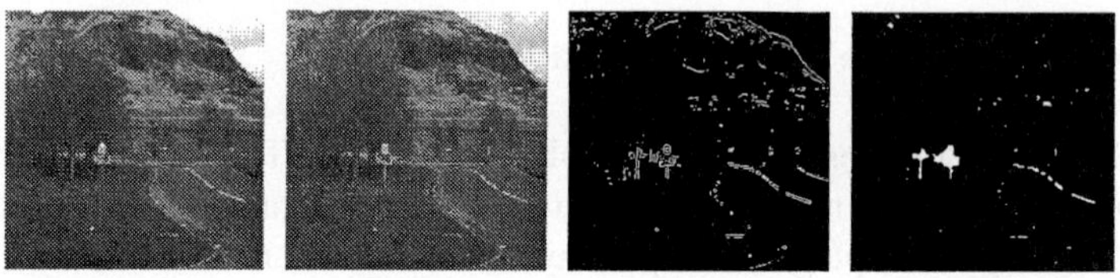

(a) (b) (c) (d)

Fig. 3. (a) Input image and attended location (b) Intensity image (c) Contour image (d) saliency map of intensity image

One of the feature maps in the first stage whose saliency area is different from the attended locations will be inhibited and implemented by reducing the chances of success of previous winners. This ensures that the focus of attention visits numerous parts of attended location through feedback selective self-organization network. Through this process, the spatial solution can also be enhanced. This result is consistent with the physiological study [7].

5 Discussions

We have discussed recent advances in the study of biologically plausible computational models of attention, and proposed a new model based on the bottom-up attention and the feature integration theory. The biological insight guiding its architecture proved efficient in the reproducing some of the performance of primate visual system. The efficiency of this approach for selective feedback has been tested, and by using feedback selective method, the multiple select of features in multi-scale can be performed in an integrated selection area, so the resolution of attention area can be increased.

However, controlling where attention should be deployed is not an autonomous feed-forward process. In order to provide a more scalable system, further work would be emphasized on the relation between the selective feedback attended locations and the task-based model. An important future direction for modeling work may include modeling of interaction between task-based and top-down cues, bottom-up cues.

Acknowledgements. The research is supported by national natural science foundations of china (60373029).

References

1. Reynolds, J.H., Desimone, R: The Role of Neural Mechanisms of Attention in Solving the Binding Problem. Neuron 24 (1999) 19–29
2. Hopfinger, J.B., Buonocore, M.H., Mangun, G.R: The Neural Mechanisms of Top-down Attentional Control. Nature Neurosci 3 (2000) 284–291
3. Corbetta, M., Kincade, J.: Voluntary Orienting Is Dissociated From Target Detection in Human Posterior Parietal Cortex. Nature Neurosci 3 (2000) 292–297
4. Treisman, A.M., Gelade, G.: A Feature-Integration Theory of Attention. Cogn. Psychol 12 (1980) 97–136
5. Itti, L., Koch, C.: A Saliency-Based Search Mechanism for Overt and Covert Shifts of Visual Attention. Vision Res. 40 (2000) 1489–1506
6. Itti, L., Koch, C., Niebur, E.: A Model of Saliency-Based Visual Attention for Rapid Scene Analysis. IEEE Trans. 20 (11) (1998) 1254–9
7. Yaffa, Y., Marisa, C.: Attention Improves or Impairs Visual Performance by Enhancing Spatial Resolution. Nature Neurosci 5 (1998) 72-75
8. Lee, D.K., Itti, L., Koch, C., Braun, J.: Attention Activates Winner-Take-All Competition Among Visual Filters. Nature Neurosci 2 (1999) 375–381
9. Kastner, S., de Weerd, P., Desimone, R.: Ungerleider Mechanisms of Directed Attention in the Human Extrastriate Cortex as Revealed by Functional MRI. Science 282 (1998) 108–111
10. Vanduffel, W., Tootell, R., Orban, G.: Attention-Dependent Suppression of Metabolic Activity in the Early Stages of the MacaqueVisual System. Cerebral Cortex (2000)
11. Brefczynski, J., DeYoe, E.: A Physiological Correlate of the 'Spotlight' of Visual Attention. Nat Neurosci. Apr. 2(4) (1999) 370-4

Realtime Monitoring of Vascular Conditions Using a Probabilistic Neural Network

Akira Sakane[1], Toshio Tsuji[1], Yoshiyuki Tanaka[1], Kenji Shiba[1], Noboru Saeki[2], and Masashi Kawamoto[2]

[1] Department of Artificial Complex Systems Engineering
Hiroshima University
Kagamiyama 1-4-1, Higashi-Hiroshima, Hiroshima, 739-8527 JAPAN
{sakane, tsuji, ytanaka, shiba}@bsys.hiroshima-u.ac.jp
http://www.bsys.hiroshima-u.ac.jp

[2] Department of Anesthesiology and Critical Care
Hiroshima University
Kasumi 1-2-3, Minami-ku, Hiroshima, 734-8551 JAPAN
{nsaeki,anekawa}@hiroshima-u.ac.jp

Abstract. This paper proposes a new method to discriminate the vascular conditions from biological signals by using a probabilistic neural network, and develops the diagnosis support system to judge the patient's conditions on-line. For extracting vascular features including biological signals, we model the dynamic characteristics of an arterial wall by using mechanical impedance and estimate the impedance parameters "beat-to-beat". As a result, this system can be utilized for the actual surgical operation, and the vascular conditions can be discriminated with high accuracy using the proposed method.

1 Introduction

A physician needs to judge patient's conditions accurately and take appropriate measures. Especially, blood pressure and electrocardiogram are used to judge the patient's conditions during surgical operations, and the waveforms of these biological signals change based on the vascular conditions. However, due to the inevitability of the human error factor in combination with inexperienced physicians, the judge of a patient's condition will most likely not be 100 percent accurate. Therefore, medical diagnosis computer assisted technology on vascular conditions is helpful for physicians in supporting complex diagnosis. As the development of a diagnosis support system based on vascular conditions, the estimation system of health condition and vascular aging have been reported [1],[2]. For example, Maniwa *et al.* estimated a acceleration plethysmogram and judged the conditions of health or illness quantitatively using a recurrence plot which is one of the chaos approach [1]. However, the proposed method does not correspond to long term-series behavior, and it is not applied to monitor the patient's conditions during operations.

F. Yin, J. Wang, and C. Guo (Eds.): ISNN 2004, LNCS 3174, pp. 488–493, 2004.

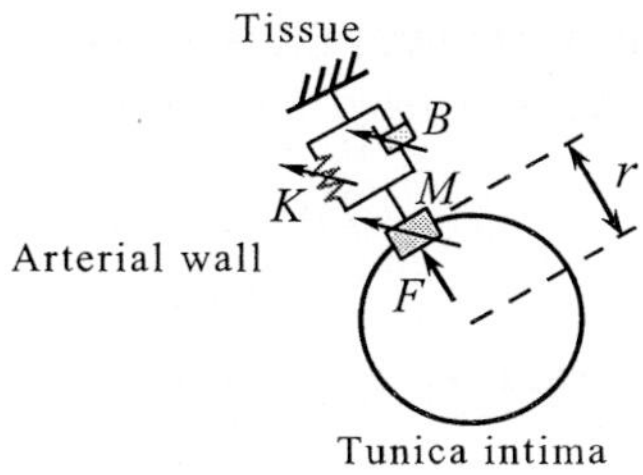

Fig. 1. An arterial wall impedance model

This paper aims to develop a diagnosis support system using a probabilistic neural network. In the many proposed techniques, we used Log-Linearized Gaussian Mixture Network (LLGMN) [3], based on the Gaussian mixture model (GMM) and the log-linear model of the probability density function. Specifically, about pattern discrimination of EMG signals, high discrimination capability is shown as compared with other neural networks [4]. We have discriminated vascular conditions from measured biological signals by using the neural network, thus the vascular conditions could be discriminated with high accuracy [5],[6]. However, the data used for discrimination were measured in advance, and the availability was not evaluated in actual surgical operation.

This paper proposes a method to discriminate vascular conditions using a probabilistic neural network on-line, and the availability of realtime monitoring of vascular conditions is verified.

2 Dynamic Characteristics of Arterial Wall

For extracting vascular features including biological signals, we estimated the arterial wall impedance [5]. Fig. 1 illustrates the proposed impedance model of the arterial wall. This model represents only the characteristics of the arterial wall in the arbitrary radius direction. The impedance characteristic can be described using an external force and a displacement of the arterial wall as follows:

$$dF(t) = Md\ddot{r}(t) + Bd\dot{r}(t) + Kdr(t) \tag{1}$$

where $F(t)$ is the force exerted on the arterial wall by blood flow; M, B, and K are the inertia, viscosity, and stiffness; $r(t), \dot{r}(t)$, and $\ddot{r}(t)$ are the position, velocity, and acceleration of the wall; the coefficient d means the variation from the time t_0; and t_0 denotes the start time just at moving arterial wall.

To estimate the impedance parameters given in (1), it is necessary to measure $F(t)$ and $r(t)$. Assuming that the force $F(t)$ is proportional to arterial pressure $P_b(t)$, the following equation can be obtained:

$$F(t) = k_f P_b(t) \tag{2}$$

where k_f is a proportional constant [5].

On the other hand, the vascular radius $r_v(t)$ is quite difficult to measure directly. So a plethysmogram is utilized instead of $r_v(t)$ as follows [5]:

$$r_v(t) = \frac{P_l(t) + A_D}{k_p} \tag{3}$$

where $P_l(t)$ is a plethysmogram, k_p is a proportional constant, and A_D is the absorbance.

The force exerted on the arterial wall is expressed by the arterial pressure $P_b(t)$ given by (2), and the vascular radius $r_v(t)$ is represented by the plethysmogram $P_l(t)$ in (3). Then, the arterial wall impedance is estimated by using $P_b(t)$ and $P_l(t)$ as follows:

$$dP_b(t) = \tilde{M} d\ddot{P}_l(t) + \tilde{B} d\dot{P}_l(t) + \tilde{K} dP_l(t) \tag{4}$$

where the parameter $\tilde{M}$ corresponds to the mass of the arterial wall existing in the measured part; $\tilde{B}$ and $\tilde{K}$ to the viscoelastic properties, respectively [5]. In the proposed method, the estimation period of the arterial wall impedance is the time between successive R-peaks (an RR interval), where the peak of the R wave can be detected from ECG signals. The impedance parameters $\tilde{M}$, $\tilde{B}$, and $\tilde{K}$ are estimated by substituting $dP_b(t)$ and $dP_l(t)$ into (4), where $dP_b(t)$ and $dP_l(t)$ are the variations of arterial pressure and plethysmogram from the detecting time of R wave t_0. Because the time needed for estimating arterial wall impedance is smaller than the RR interval, it is possible to estimate the arterial wall impedance "beat-to-beat" [5].

We proposed the impedance ratio to reduce the individual differential [5]. The impedance ratios $\tilde{M}_{\text{ratio}}$, $\tilde{B}_{\text{ratio}}$, and $\tilde{K}_{\text{ratio}}$ are calculated for $\tilde{M}$, $\tilde{B}$, and $\tilde{K}$ as follows:

$$\tilde{M}_{\text{ratio}} = \frac{\tilde{M}}{\tilde{M}_{\text{rest}}}, \tilde{B}_{\text{ratio}} = \frac{\tilde{B}}{\tilde{B}_{\text{rest}}}, \tilde{K}_{\text{ratio}} = \frac{\tilde{K}}{\tilde{K}_{\text{rest}}} \tag{5}$$

where $\tilde{M}_{\text{rest}}$, $\tilde{B}_{\text{rest}}$, and $\tilde{K}_{\text{rest}}$ are the nominal values of impedance ratios when patients are in a relatively rested condition. Similarly, the ratios of arterial pressure and plethysmogram are calculated as follows:

$$IBP_{\text{ratio}} = \frac{IBP_{\text{max}} - IBP_{\text{min}}}{IBP_{\text{rest}}}, PLS_{\text{ratio}} = \frac{PLS_{\text{max}} - PLS_{\text{min}}}{PLS_{\text{rest}}} \tag{6}$$

where IBP_{max}, IBP_{min}, PLS_{max}, and PLS_{min} are maximum and minimum values of arterial pressure and plethysmogram divided into beat-to-beat; IBP_{rest} and PLS_{rest} are maximum and minimum value differences of arterial pressure and plethysmogram in a relatively rested condition [5].

A neural network is used to discriminate the vascular conditions using the normalized ratios of impedance and biological signals given by

$$\tilde{M}'_{\text{ratio}} = \frac{\tilde{M}_{\text{ratio}}}{k_{\text{M}}}, \tilde{B}'_{\text{ratio}} = \frac{\tilde{B}_{\text{ratio}}}{k_{\text{B}}}, \tilde{K}'_{\text{ratio}} = \frac{\tilde{K}_{\text{ratio}}}{k_{\text{K}}},$$

$$IBP'_{\rm ratio} = \frac{IBP_{\rm ratio}}{k_{\rm I}}, PLS'_{\rm ratio} = \frac{PLS_{\rm ratio}}{k_{\rm P}} \tag{7}$$

where the gains for normalization, $k_{\rm M}$, $k_{\rm B}$, $k_{\rm K}$, $k_{\rm I}$, and $k_{\rm P}$ are determined by the maximum value of each estimated signal in advance [6].

The LLGMN receives the ratios of impedance and biological signals of (7).

3 Monitoring Experiment

3.1 Experimental Condition

The proposed method is applied to discriminate the vascular conditions. The subject was operated on using the endoscopic transthoracic sympathectomy for hyperhidrosis (Patient A : male 14 years). If a blood vessel contracts due to stimulation from sympathetic nerves, the palms and under arms will perspire. In this operation, the sympathetic nerves on the sides of the backbone are interrupted using a clip to stop the perspiration [7]. When the sympathetic nerve is interrupted, blood vessels become compliant on the spot. Therefore, if the vascular conditions can be identified on-line, it is possible to ascertain its success or failure during operation. Electrocardiogram ($ECG(t)$), arterial pressure ($P_b(t)$), and plethysmogram ($P_l(t)$) were measured at 125 [Hz] simultaneously for discriminating vascular conditions. The arterial pressure was measured through a catheter (24 gauge) placed in the left radial artery, and the plethysmogram was measured with the ipsilateral thumb (BSS-9800, NIHON KOHDEN Corp.).

We constructed the diagnosis support system equipped with a graphical display screen using LabVIEW (National Instruments Corp.) in order for a physician to ascertain the patient's conditions easily. This system was tested to determine whether realtime monitoring of vascular conditions was possible.

The learning data was created from that of the four patients (non-subjects) who operated the endoscopic transthoracic sympathectomy for hyperhidrosis patients. However, the learning data of the "Vasodilation" and "Shock" conditions were not available from the patients. Such unobservable facts were represented randomly by normal distribution with $N(\mu, 0.005)$. In this paper, means μ was set as follows: $\mu_{IBP'_{\rm ratio}} = 0.12, \mu_{PLS'_{\rm ratio}} = 0.75, \mu_{\tilde{M}'_{\rm ratio}} = 0.006, \mu_{\tilde{B}'_{\rm ratio}} = 0.006, \mu_{\tilde{K}'_{\rm ratio}} = 0.006$ for vasodilation; $\mu_{IBP'_{\rm ratio}} = 0.08, \mu_{PLS'_{\rm ratio}} = 0.1, \mu_{\tilde{M}'_{\rm ratio}} = 0.033, \mu_{\tilde{B}'_{\rm ratio}} = 0.033, \mu_{\tilde{K}'_{\rm ratio}} = 0.033$ for shock.

3.2 Experimental Result

Fig.2 shows the result of discrimination experiment. Time profiles of the ratio of arterial pressure, the ratio of plethysmogram, the ratio of inertia, the ratio of viscosity, the ratio of stiffness, the shock index (SI), the coefficient of determination, and the classification results are shown in order from the top. In this paper, we defined the four vascular conditions i.e., I) shock, II) normal, III) vasoconstriction, and IV) vasodilation [5]. The coefficient of determination is

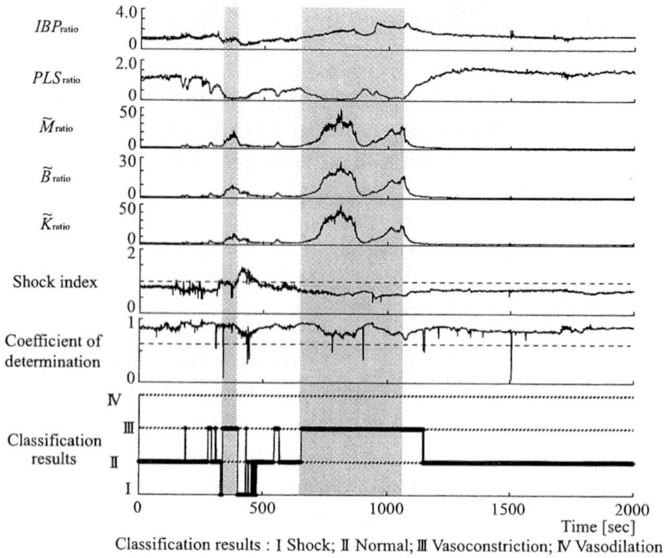

Fig. 2. Classification result of the vascular conditions during surgical operation

calculated from the measured arterial pressure and estimated arterial pressure using (4), and is used for judging whether discrimination should take place. Also, the shock index is used routinely for defining the "Shock" condition. This index is calculated from the ratio of heart rate to systolic arterial pressure; "Normal" condition at 0.6; while "Shock" condition over 1.0 [8].

The blood vessels become stiff in the shaded areas. The nominal values of impedance and biological signals are an average of consecutive 10 sample data after the patients are under anesthesia. The discrimination result is selected as vascular condition if the coefficient of determination $R^2 \geq 0.6$. On the other hand, the discrimination is suspended if $R^2 < 0.6$. Also, the normalization gains included in (7) are set as follows: $k_M = k_B = k_K = 30.0, k_I = 2.5, k_P = 2.0$.

In Fig.2, the estimated impedance shows that the blood vessels gradually became stiff because the doctor stimulated the patient's tissues to find the sympathetic nerves at 300-400 and 700-1100 [sec], and NN discriminated the vascular condition as "Vasoconstriction". After 400 [sec], the ratios of arterial pressure decreased rapidly, the vascular condition was discriminated as "Shock", and the corresponding shock index was indicated as equal or greater than 1.0 at the same time. Also, the sympathetic nerves are distributed on the two sides of body, the nerve that was concerned the left arm's blood vessels was interrupted at 1100 [sec]. Therefore, after 1100 [sec], the blood vessels became compliant, and vascular condition was discriminated as "Normal". After that, the operation that was interrupted the sympathetic nerve of the other side of the left arm measuring the biological signals was occurred at 1200-2000 [sec]. Then, the left arm's blood vessels were not constricted if the sympathetic nerve was interrupted properly. The discrimination result indicated "Normal" condition, and it meant that the sympathetic nerve was interrupted well. From this result, it was confirmed that the vascular conditions could be monitored in realtime during operation and judged more easily. However, the coefficient of determination showed low value because the phase difference between blood pressure and plethysmogram was large. It is considered to be easily solved by improving the model in contemplation of the pulse wave velocity.

4 Conclusion

This paper proposed a new method to discriminate the vascular conditions by using a probabilistic neural network, and developed the diagnosis support system to judge the patient's conditions on-line. This system could be utilized for the actual surgical operation, and the vascular conditions could be discriminated with high accuracy using the proposed method. Future research will test the method's validity using different surgery.

Acknowledgment. The authors would like to thank T.Ukawa and K.Tone of the NIHON KOHDEN Corp. for making available clinical data and for their precious suggestions. This work was supported by Grant-in-Aid for Scientific Research of Japan Society for the Promotion of Science (15008279).

References

1. Maniwa, Y., Iokibe, T., Koyama, M., Yamamoto, M., Ohta, S.: The Application of Pulse Wave Chaos in Clinical Medicine, in Proc. 17th FUZY System Symposium, Chiba (2001) 787-790
2. Takada, H., Mirbod, S.M., Iwata, H.: The relative vascular age derived from acceleration plethysmogram: A new attempt., Jpn. J. Appl. Physiol., Vol. 28, No. 2 (1998) 115-121
3. Tsuji, T., Fukuda, O., Ichinobe, H., Kaneko, M.: A Log-Linearized Gassian Mixture Network and Its Application to EEG Pattern Classification, IEEE Trans. Syst., Man, Cybern-Part C, Appl. and Rev., Vol. 29, No. 1 (1999) 60-72
4. Fukuda, O., Tsuji, T., Kaneko, M. Otsuka, A.: A Human-Assisting Manipulator Teleoperated by EMG Signals and Arm Motions, IEEE Trans. Robotics and Automation, Vol.19, No.2 (2003) 210-222
5. Sakane, A., Tsuji, T., Saeki, N., Kawamoto, M.: Discrimination of Vascular Conditions Using a Probabilistic Neural Network, Journal of Robotics and Mechatronics, Vol. 16, No. 2 (2004) 138-145
6. Sakane, A., Tsuji, T., Tanaka, Y., Saeki, N., Kawamoto, M.: DEVELOPMENT OF A DIAGNOSIS SUPPORT SYSTEM ON VASCULAR CONDITIONS USING A PROBABILISTIC NEURAL NETWORK, in Proc. 2nd International Symposium on Measurement, Analysis and Modeling of Human Functions (2004) (in press)
7. Drott, C., Gothberg, G., Claes, G.: Endoscopic transthoracic sympathectomy: An efficient and safe method for the treatment of hyperhidrosis, Journal of the American Academy of Dermatology, Vol. 33 (1995) 78-81
8. Rady, M.Y., Nightingale, P., Little, R.A., Edwards, J.D.: Shock index: a reevaluation in acute circulatory failure, Resuscitation, Vol. 23, No. 3 (1992) 227-234

Capturing Long-Term Dependencies for Protein Secondary Structure Prediction

Jinmiao Chen and Narendra S. Chaudhari

School of Computer Engineering, Nanyang Technological University, Singapore
pg05205549@ntu.edu.sg

Abstract. Bidirectional recurrent neural network (BRNN) is a non-causal system that captures both upstream and downstream information for protein secondary structure prediction. Due to the problem of vanishing gradients, the BRNN can not learn remote information efficiently. To limit this problem, we propose segmented memory recurrent neural network (SMRNN) and obtain a bidirectional segmented-memory recurrent neural network (BSMRNN) by replacing the standard RNNs in BRNN with SMRNNs. Our experiment with BSMRNN for protein secondary structure prediction on the RS126 set indicates improvement in the prediction accuracy.

1 Introduction

One of the most important open problems in computational biology concerns the computational prediction of the secondary structure of a protein given only the underlying amino acid sequence. During the last few decades, much effort has been made toward solving this problem, with various approaches including GOR (Garnier, Osguthorpe, and Robson) techniques [1], PHD(Profile network from HeiDelberg) method [2], nearest-neighbor methods [3] and support vector machines (SVMs) [4]. These methods are all based on a fixed-width window around a particular amino acid of interest. Local window approaches don't take into account the interactions between distant amino acids, which commonly occur in protein sequences, especially in the regions of beta-sheets.

The limitations of the fixed-size window approaches can be mitigated by using recurrent neural network (RNN), which is a powerful connectionist model for learning in sequential domains. Recently, RNNs have been applied to predict protein secondary structure. Gianluca Pollastri et. al proposed a *bidirectional recurrent neural network* (BRNN), which provides a noncausal generalization of RNNs [5]. The BRNN uses a pair of chained hidden state variables to store contextual information contained in the upstream and downstream portions of the input sequence respectively. The output is then obtained by combining the two hidden representations of context. However, learning long-term dependencies with gradient descent is difficult [6] and the generalized back propagation algorithm for training BRNN is gradient descent essentially. Therefor the error propagation in both the forward and backward chains is subject to exponential decay. In the practice of protein secondary structure prediction, the BRNN can

F. Yin, J. Wang, and C. Guo (Eds.): ISNN 2004, LNCS 3174, pp. 494–500, 2004.

utilize information located within about ±15 amino acids around the residue of interest. It fails to discover relevant information contained in even further distant portions [5].

In order to improve the prediction performance especially the recognition accuracy of β-sheet regions, we propose an alternative recurrent architecture called *Bidirectional Segmented-Memory Recurrent Neural Network*(BSMRNN), which is capable of capturing longer ranges of dependencies in protein sequences.

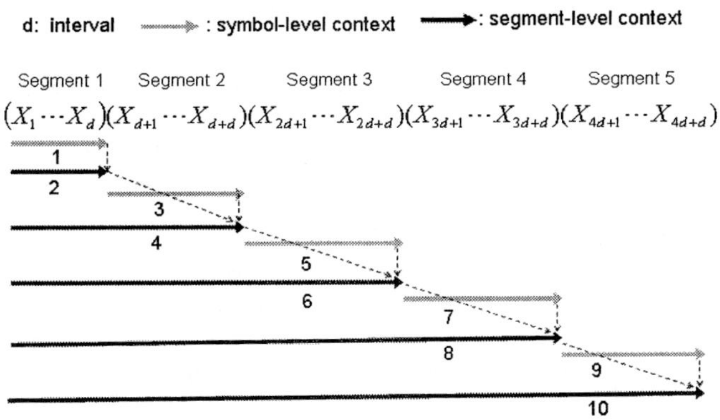

Fig. 1. Segmented memory with interval=d

2 Segmented-Memory Recurrent Neural Networks

As we observe, during the process of human memorization of a long sequence, people tend to break it into a few segments, whereby people memorize each segment first and then cascade them to form the final sequence. The process of memorizing a sequence in segments is illustrated in Figure 1. In Figure 1, gray arrows indicate the update of contextual information associated to symbols and black arrows indicate the update of contextual information associated to segments; numbers under the arrows indicate the sequence of memorization.

Based on the observation on human memorization, we believe that RNNs are more capable of capturing long-term dependencies if they have segmented-memory and imitate the way of human memorization. Following this intuitive idea, we propose *Segmented-Memory Recurrent Neural Network* (SMRNN) as illustrated in Figure 2.

The SMRNN has hidden layer H1 and hidden layer H2 representing symbol-level state and segment-level state respectively. Both H1 and H2 have recurrent connections among themselves. The states of H1 and H2 at the previous cycle are copied back and stored in context layer S1 and context layer S2 respectively. Most importantly, we introduce into the network a new attribute *interval*, which denotes the length of each segment.

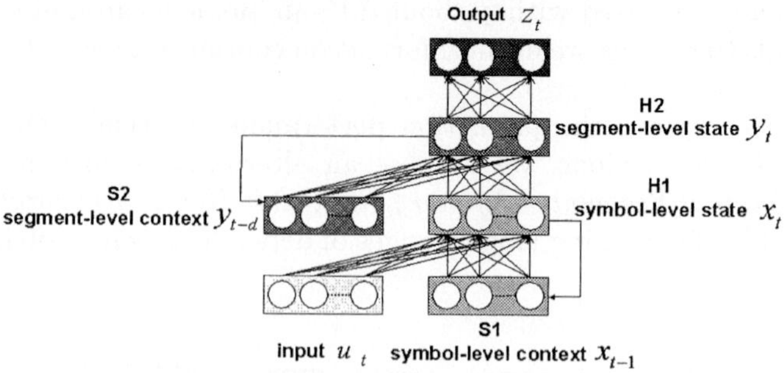

Fig. 2. Segmented-memory recurrent neural network with interval=d

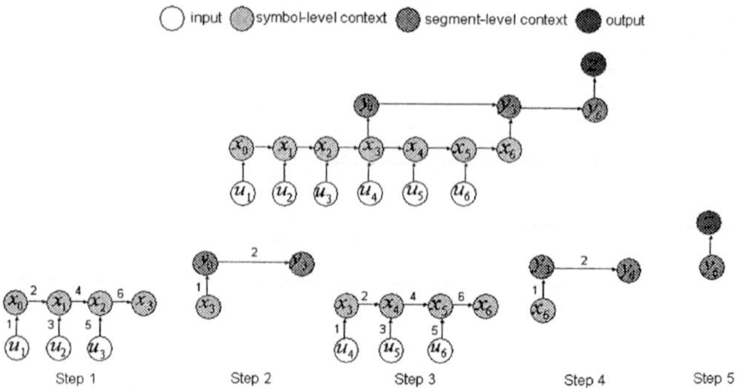

Fig. 3. Dynamics of segmented-memory recurrent neural network

In order to implement the segmented-memory illustrated in Figure 1, we formulate the dynamics of SMRNN with interval=d as below:

$$x_i^t = g(\sum_{j=1}^{n_X} W_{ij}^{xx} x_j^{t-1} + \sum_{j=1}^{n_U} W_{ij}^{xu} u_j^t) \tag{1}$$

$$y_i^t = g(\sum_{j=1}^{n_Y} W_{ij}^{yy} y_j^{t-d} + \sum_{j=1}^{n_X} W_{ij}^{yx} x_j^t) \tag{2}$$

$$z_i^t = g(\sum_{j=1}^{n_Y} W_{ij}^{zy} y_j^t) \tag{3}$$

We now explain the dynamics of SMRNN with an example(Figure 3). In this example, The input sequence is divided into segments with equal length. Then symbols in each segment are fed to hidden layer H1 to update the symbol-level context. Upon completion of each segment, the symbol-level context is forwarded to the next layer H2 to update the segment-level context. This process continues until it reaches the end of the input sequence, then the segment-level context is forwarded to the output layer to generate the final output. In other words, the network reads in one symbol per cycle; the state of H1 is updated at the coming of each single symbol, while the state of H2 is updated only after reading an entire segment and at the end of the sequence. The segment-level state layer behaves as if it cascades segments sequentially to obtain the final sequence as people often do: Every time when people finish one segment, they always go over the sequence from the beginning to the tail of the segment which is newly memorized, so as to make sure that they have remembered all the previous segments in correct order(see Figure 1).

The SMRNN is trained using an extension of the Real Time Recurrent Learning algorithm. Every parameter P is initialized with small random values then updated according to gradient descent:

$$\Delta P = -\alpha \frac{\partial E^t}{\partial P} = -\alpha \frac{\partial E^t}{\partial y^t} \frac{\partial y^t}{\partial P} \tag{4}$$

where α is the learning rate and E^t is the error function at time t.

Derivatives associated to recurrent connections are calculated in a recurrent way. Derivatives of segment-level state at time t depend on derivatives at time $t-d$ where d is the length of each segment.

$$\frac{\partial y_i^t}{\partial W_{kl}^{yy}} = y_i^t(1-y_i^t)(\delta_{ik} y_l^{t-d} + \sum_{j=1}^{n_Y} W_{ij}^{yy} \frac{\partial y_j^{t-d}}{\partial W_{kl}^{yy}}) \tag{5}$$

$$\frac{\partial y_i^t}{\partial W_{kl}^{yx}} = y_i^t(1-y_i^t)(\sum_{j=1}^{n_Y} W_{ij}^{yy} \frac{\partial y_j^{t-d}}{\partial W_{kl}^{yx}} + \delta_{ik} x_l^t) \tag{6}$$

where δ_{ik} denotes the Kronecker delta function(δ_{ik} is 1 if $i = k$ and 0 otherwise). Derivatives of symbol-level state at time t depend on derivatives at time t-1.

$$\frac{\partial x_i^t}{\partial W_{kl}^{xx}} = x_i^t(1-x_i^t)(\delta_{ik} x_l^{t-1} + \sum_{j=1}^{n_X} W_{ij}^{xx} \frac{\partial x_j^{t-1}}{\partial W_{kl}^{xx}}) \tag{7}$$

$$\frac{\partial x_i^t}{\partial W_{kl}^{xu}} = x_i^t(1-x_i^t)(\sum_{j=1}^{n_X} W_{ij}^{xx} \frac{\partial x_j^{t-1}}{\partial W_{kl}^{xu}} + \delta_{ik} u_l^t) \tag{8}$$

3 Bidirectional Segmented-Memory Recurrent Neural Network

In Gianluca Pollastri's BRNN, the forward subnetwork and the backward subnetwork are conventional RNNs. We replace the conventional recurrent subnetworks with SMRNNs and obtain a *Bidirectional Segmented-Memory Recurrent Neural Network* (BSMRNN) as illustrated in Figure 4. The BSMRNN is capable of capturing farther upstream context and farther downstream context.

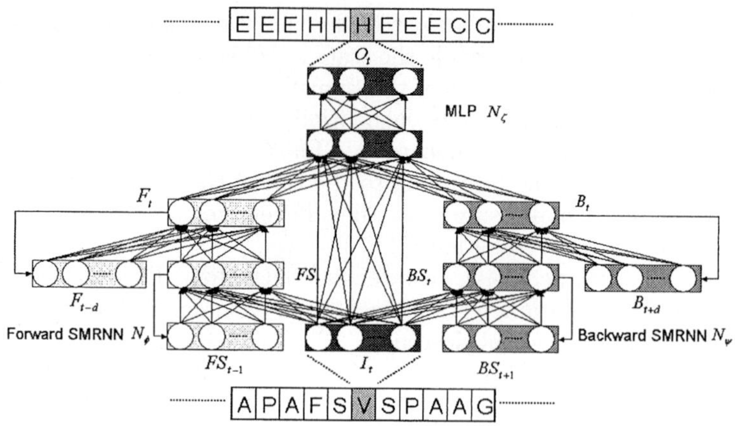

Fig. 4. Bidirectional segmented-memory recurrent neural network Architecture

The upstream context and downstream context are contained in vector F_t and B_t respectively. Let vector I_t encode the input at time t, vectors F_t and B_t are defined by the following recurrent bidirectional equations:

$$F_t = \phi(F_{t-d}, I_t) \tag{9}$$

$$B_t = \psi(B_{t+d}, I_t) \tag{10}$$

where $\phi()$ and $\psi()$ are nonlinear transition functions. They are implemented by the forward SMRNN N_ϕ and the backward SMRNN N_ψ respectively (left subnetwork and right subnetwork in Figure 4). The final output is obtained by combining the two hidden representations of context and the current input:

$$O_t = \zeta(F_t, B_t, I_t) \tag{11}$$

where $\zeta()$ is realized by MLP N_ζ(top subnetwork in Figure 4).

Usually the number of input neurons is equal to the size of input alphabet, i.e. $|\Sigma_i|$. Symbols in the sequences are presented to the input layer with one-hot coding. When a_l (the l-th symbol in the input alphabet) is read, the kth element of the input vector is $I_k = \delta(k,l)(\delta(k,l)$ is 1 if $k = l$ and 0 otherwise). Normally the synaptic input equals the sum of inputs multiplied by weights. Hence if an input is zero, the weights associated to that input unit won't be updated. Thus we apply a contractive mapping $f(x) = \epsilon + (1 - 2\epsilon)x$, with ϵ a small positive number, to input units. This mapping does not affect the essentials of the formalism presented here. The number of output units is equal to the size of output alphabet, i.e. $|\Sigma_o|$.

Given an amino acid sequence $X_1, \ldots, X_t, \ldots, X_T$, the BSMRNN can estimate the posterior probabilities of secondary structure classes for each sequence position t. Starting from $F_0 = 0$, the forward SMRNN reads in the preceding substring $X_1, \ldots, X_{t-1}$ from left to right and updates its states F_t, following eq. 9. Similarly, starting from $B_0 = 0$, the backward SMRNN scans the succeeding

Table 1. Comparison between BSMRNN and BRNN

	Q_3	Q_E	Q_H	Q_C
BSMRNN	66.7%	61.8%	52.1%	78.3%
BRNN	65.3%	57.1%	52.3%	77.3%

substring $X_{t+1}, \ldots, X_T$ from right to left and updates its states B_t, following eq. 10. After the forward and backward propagations have taken place, the output at position t is then calculated with eq. 11.

Learning in BSMRNN is also gradient-based. The weights of subnetwork N_ζ are adjusted in the same way as standard MLP. The derivatives of the error function with respect to states F_t and B_t are calculated and injected into N_ϕ and N_ψ respectively. Then the error signal is propagated over time in both directions and the weights of N_ϕ and N_ψ are adjusted using the same formulas as those of causal SMRNN (refer to section 2.4).

4 Experimental Evaluation

For the problem of PSS prediction, we use seven fold cross validation on the RS126 set(126 protein chains, 23,348 amino acids). Many PSS prediction methods were developed and tested on this set. With sevenfold cross validation, approximately six-sevenths of the dataset are used for training, and the remaining one-seventh is used for testing. In order to avoid the selection of extremely biased partitions that would give an inauthentic prediction accuracy, the RS126 set is divided into seven subsets with each subset having similar size and content of each type of secondary structure.

For each subnetwork of the BSMRNN, we use 10 hidden units. According to G.Pollastri's study, BRNN can reliably utilize information contained within about $\pm$ 15 amino acids around the predicted residue. Thus we set the interval of SMRNN to be 15. Recurrent neural network is a bit less stable than the feed-forward ones, so we shuffle the training set at each pass, so as to present it in different random order each time.

We obtained results illustrated in Table 1. Q_3 is the overall three-state prediction percentage defined as the ratio of correctly predicted residues to the total number of residues. Q_E, Q_H and Q_C are the percentage of correctly predicted residues observed in class E, H, C respectively. The results show that BSMRNN performs better on the problem of PSS prediction. Particularly, the higher prediction accuracy of beta-sheets indicates that BSMRNN captures longer ranges of dependencies in protein sequences than BRNN.

5 Concluding Remarks

The segmented memory recurrent neural networks are more capable of capturing long-term dependencies than conventional RNNs. From biology, we know protein secondary structure prediction is a long-term dependency problem. Therefore, BSMRNN can improve the prediction performance, especially the recognition accuracy of beta-sheets. However, there is a trade-off between the efficient training of gradient descent and long-range information latching. For BSMRNN, the training algorithm is also gradient descent essentially, hence BSMRNN does not circumvent the problem of long-term dependencies.

In practice, we found that the best prediction of testing data is obtained when the training error is not very small, and after that prediction accuracy begins to drop even though the error still keeps converging. One reason may be that the RS126 set is too small for BSMRNN to learn the complex mapping from amino acid sequence to secondary structure sequence. Hence a larger training set is required to achieve a satisfactory level of generalization.

A lot of research has shown that prediction quality can be improved by incorporating evolutionary information in the form of multiple sequence alignment [7, 8,9]. In the experiments, we performed prediction on single protein sequence only. The prediction accuracy can be further improved by using multiple alignments of homologous protein sequences.

References

1. Gibrat, J.F., Garnier, J., Robson, B.: Further Developments of Protein Secondary Structure Prediction Using Information Theory, J. Mol. Biol., vol. 198, (1987) 425-443
2. Rost, B., Sander, C.: Prediction of Protein Secondary Structure at Better than 70% Accuracy, J. Mol. Biol., vol. 232, (1993) 584-599
3. Yi, T.M., Lander, E.S.: Protein Secondary Structure Prediction Using Nearest-Neighbor Methods, J. Mol. Biol., vol. 232, (1993) 1117-1129
4. Hua, S., Sun, Z.: A Novel Method of Protein Secondary Structure Prediction with High Segment Overlap Measure: Support Vector Machine Approach, J. Mol. Biol., vol. 308, (2001) 397-407
5. Pollastri, G., Przybylski, D., Rost, B., Baldi, P.: Improving the Prediction of Protein Secondary Structure in Three and Eight Classes Using Recurrent Neural Networks and Profiles, vol. 47, (2002) 228-235
6. Bengio, Y., Simard, P., Frasconi, P.: Learning Long-Term Dependencies with Gradient Descent is Difficult, vol. 5, no. 2, (1994) 157-166
7. Di Francesco, V., Garnier, J., Munson, P.J.: Improving Protein Secondary Structure Prediction with Aligned Homologous Seqeunces, Prot.Sci., vol. 5, (1996) 106-113
8. Riis, S.K., Krogh, A.: Improving Prediction of Protein Secondary Structure Using Structural Neural Networks and Multiple Sequence Alignment, J. Comp. Biol., vol. 3, (1996) 163-183
9. Salamov, A.A., Solovyev, V.V.: Protein Secondary Structure Prediction Using Local Alignments, J. Mol. Biol., vol. 268, (1997) 31-36

A Method for Fast Estimation of Evoked Potentials Based on Independent Component Analysis

Ting Li, Tianshuang Qiu, Xuxiu Zhang, Anqing Zhang, and Wenhong Liu

School of Electronic and Information Engineering, Dalian University of Technology, 116024, Dalian, P. R China
tara2000@21cn.com, qiutsh@dlut.edu.cn

Abstract. Independent component analysis (ICA) is a new powerful tool for blind source separation. This paper proposes a new algorithm that combines two existent algorithms, the improved infomax algorithm and the fastICA algorithm. Utilizing the initial weights obtained by the improved infomax algorithm, we can not only reduce the length of data which fastICA algorithm needs, but also enhance the convergence stability of fastICA algorithm. The effectiveness of the algorithm is verified by computer simulations.

1 Introduction

The brain evoked potentials (EPs) are electrical responses of the central nervous system to sensory stimuli applied in a controlled manner. The EPs have a number of clinical applications [1]. The problem often encountered in the analysis of the EPs is that the signal-to-noise ratio (SNR) is often less than –10dB. Ensemble averaging method has been usually used to enhance the SNR. Such methods usually require a large number of sweeps to obtain a suitable estimate of the EP. The implicit assumption in the averaging is that the task-related cognitive process does not vary much in timing from trial to trial. However, it has been evident that in many cases this assumption is not valid. The observation of variation in the parameters of the EPs permits the dynamic assessment of changes in cognitive state. Thus, the goal in the analysis of EPs is currently the estimation from several potentials, or even from a single potential. This task is called fast estimation, or single-trial estimation. Recently, independent component analysis appeared as a promising technique in signal processing. ICA is based on the following principles. Assume that the original signals have been linearly mixed, and that these mixed signals are available. ICA finds in a blind manner a linear combination of the mixed signals which recovers the original signals, possibly rescaled and randomly arranged in the outputs.

The noiseless linear ICA model with instantaneous mixing may be described by

$$\mathbf{x} = \mathbf{A}\mathbf{s} \tag{1}$$

where $\mathbf{x} = [x_1, x_2, \cdots x_N]^T$ (T denoting the transpose) are observed signals, $\mathbf{s} = [s_1, s_2, \cdots, s_M]^T$ are the source signals, and $\mathbf{A}$ is an unknown mixing matrix.

F. Yin, J. Wang, and C. Guo (Eds.): ISNN 2004, LNCS 3174, pp. 501–506, 2004.

The goal is to estimate both unknowns from $\mathbf{x}$, with appropriate assumptions on the statistical properties of the source distributions. The solution is

$$\mathbf{y} = \mathbf{W}\mathbf{x} \tag{2}$$

where $\mathbf{W}$ is called the demixing matrix. The general ICA problem requires $\mathbf{A}$ to be an $N \times M$ matrix of full rank, with $N \geq M$. In this paper, we assume an equal number of sources and sensors to make calculation simple.

This paper is organized as follows. First, we introduce improved infomax algorithm in section 2. Second, we introduce fastICA algorithm in section 3. In section 4, we show the new algorithm proposed in this paper. In section 5, simulation results are shown, and in section 6, we present the conclusion.

2 Improved Infomax Algorithm

Bell [2] derives a self-organizing learning algorithm that maximizes the information transferred in a network of nonlinear units. We assume $\mathbf{W}$ is the demixing matrix, $\mathbf{x} = \mathbf{A}\mathbf{s}$, $\mathbf{u} = \mathbf{W}\mathbf{x}$, the output of neural network is $\mathbf{y} = f(\mathbf{u})$, $f(\cdot)$ is a nonlinear function. Maximizing the information transferred in a network is to maximize the mutual information of the input $\mathbf{x}$ and output $\mathbf{y}$ of the demixing system (a neural network). We choose a three-layer forward feedback neural network to construct demixing system, which includes an input layer, a linear hidden layer and a nonlinear output layer. The number of nodes of every layer is N. The block diagram of the neural network is shown in Fig. 1.

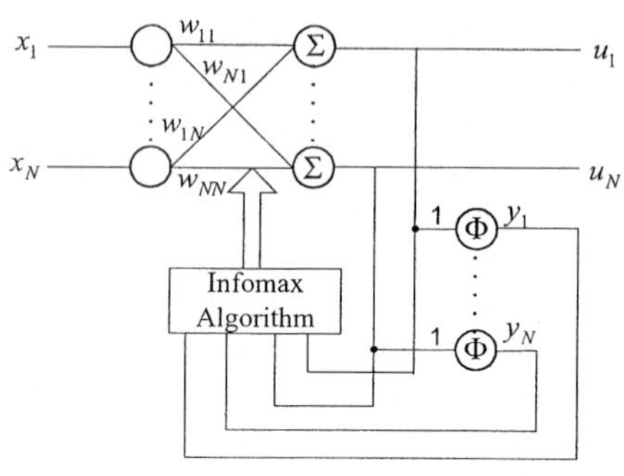

Fig. 1. The block diagram of the neural network

The weights between the input layer and the hidden layer construct the demixing matrix; the hidden layer is linear, the weights between it and the output layer are all 1s. We have noticed that the output of the hidden layer, $\mathbf{u}$, is what we want to separate. When we choose $\tanh(\cdot)$ function as a nonlinear unit, the output vector is

$$\mathbf{y} = \tanh(\mathbf{u}) \tag{3}$$

We use natural gradient instead of conventional Euclidean gradient to avoid calculating the inversion of matrix, and the iteration equation is

$$\mathbf{W} = \boldsymbol{\mu}(\mathbf{I} - \mathbf{y}\mathbf{u}^{\mathrm{T}})\mathbf{W} \tag{4}$$

where $\boldsymbol{\mu}$ is the learning rate. As we said before, it is important to choose the value of $\boldsymbol{\mu}$, because the convergence depends crucially on the correct choice of it. Barros et al. [3] suggested a method to calculate time-varying learning rate as

$$\boldsymbol{\mu} = \frac{\mathbf{2}}{\mathbf{y}^{\mathrm{T}}\mathbf{u} + \mathbf{1}} \tag{5}$$

where $\mathbf{1}$ is a matrix where elements are all 1s.

The process of improved infomax algorithm is:

1. Initiate demixing matrix W_1 with small random number;
2. Get the *i*th sample vector x_i, $i = 1, 2, \cdots, N$;

   (1) Calculate $u_i = W_i x_i$, $y_i = \tanh(u_i)$, and $\mu_i = \dfrac{2}{y_i^T u_i + 1}$;

   (2) Calculate $W_i = \mu_i(\mathbf{I} - y_i u_i^T)W_i$, and let $W_{i+1} = W_i + \Delta W_i$;
3. If $|W_{i+1} - W_i|$ is not less than ε (which is a small constant), let $i = i+1$, and go back to step 2. Otherwise, output W_{i+1}.

3 FastICA Algorithm [4,5,6]

FastICA algorithm, which is proposed by Hyvarinen et al., is one of the most popular ICA algorithms because of its fast convergence speed. A remarkable property of this algorithm is that a very small number of iterations, usually 5 to 10, seems to be enough to obtain the source signal. Before using fastICA algorithm, we must prewhiten or sphere the observed data $\mathbf{x}$. The observed data $\mathbf{x}$ is linearly transformed to a matrix $\tilde{\mathbf{x}} = \mathbf{M}\mathbf{x}$, which is to make the correlation matrix of $\mathbf{x}$ equal unity: $E[\mathbf{x}\mathbf{x}^T] = \mathbf{I}$. After the transformation, we have

$$\tilde{\mathbf{x}} = \mathbf{M}\mathbf{x} = \mathbf{M}\mathbf{A}\mathbf{s} = \mathbf{B}\mathbf{s} \tag{6}$$

where $\mathbf{M}$ is the whitening matrix, $\mathbf{B} = \mathbf{M}\mathbf{A}$ is an orthogonal matrix. The goal is to find the orthogonal demixing matrix $\hat{\mathbf{W}}$, which makes each component of $\mathbf{y}$ be independent. The output $\mathbf{y}$ is

$$\mathbf{y} = \hat{\mathbf{W}}^T\tilde{\mathbf{x}} = \hat{\mathbf{W}}^T\mathbf{M}\mathbf{x} = \hat{\mathbf{W}}^T\mathbf{M}\mathbf{A}\mathbf{s} \tag{7}$$

Through maximizing the kurtosis, we can extract source signals one by one. The process of fastICA algorithm is:

1. Take a random initial vector $\hat{w}(0)$ ($\hat{w}(k) = \hat{\mathbf{W}}_k$ is the *k*th row of it) of norm 1. Let $k = 1$;

2. Let $\hat{w}(k) = E[\tilde{x}(\hat{w}(k-1)^T \tilde{x})^3] - 3\hat{w}(k-1)$. The expectation can be estimated using a large sample of $\tilde{x}$ vectors. Divide $\hat{w}(k)$ by its norm;

3. If $\left|\hat{w}(k)^T \hat{w}(k-1)\right|$ is not close enough to 1, let $k = k+1$, and go back to step 2. Otherwise, output the vector $\hat{w}(k)$.

4 The New Algorithm

As is said before, both improved infomax and fastICA algorithm have advantages and disadvantages. We consider a new method combining these two algorithms, which can converge fast and does not need large size of data.

First, we deal with two periods of data with improved infomax algorithm, and get a demixing matrix. This is a preprocessing procedure; and the demixing matrix we get is used as an initial value of weights for fastICA algorithm. If fastICA algorithm receives an improper initial value of the demixing matrix, it converges very slowly, tens of times more than the usual. After the preprocessing procedure, we get a proper initial value to guarantee the fast convergence of fastICA algorithm. And the improved infomax algorithm converges comparatively fast at the beginning of the iteration, so two periods of data are enough to get an appropriate result.

Comparing eq. (2) with eq. (7), we have

$$\mathbf{W} = \hat{\mathbf{W}}^T \mathbf{M} \tag{8}$$

That is

$$\hat{\mathbf{W}} = (\mathbf{M}^{-1})^T \mathbf{W}^T \tag{9}$$

The whitening matrix $\mathbf{M}$ can be obtained by using principal component analysis (PCA). If we convert the demixing matrix $\mathbf{W}$ which is received from the improved infomax algorithm to $\hat{\mathbf{W}}$ according to eq. (9), we can get the initial value of weights of fastICA algorithm. Simulation results prove that EP signals can be extracted in only two periods using this method.

The process of the new algorithm:

1. Deal with two periods of signal with improved infomax algorithm, and get $\mathbf{W}$;
2. Calculate the initial weights of fastICA algorithm with $\hat{\mathbf{W}} = (\mathbf{M}^{-1})^T \mathbf{W}$;
3. Deal with the same two periods of observed signal using fastICA algorithm.

5 Simulation Results

Two independent sources are linearly mixed. One is the periodical EP signal, and the period is 128 points, the sampling frequency is 1000Hz. The other is a white Gaussian noise.

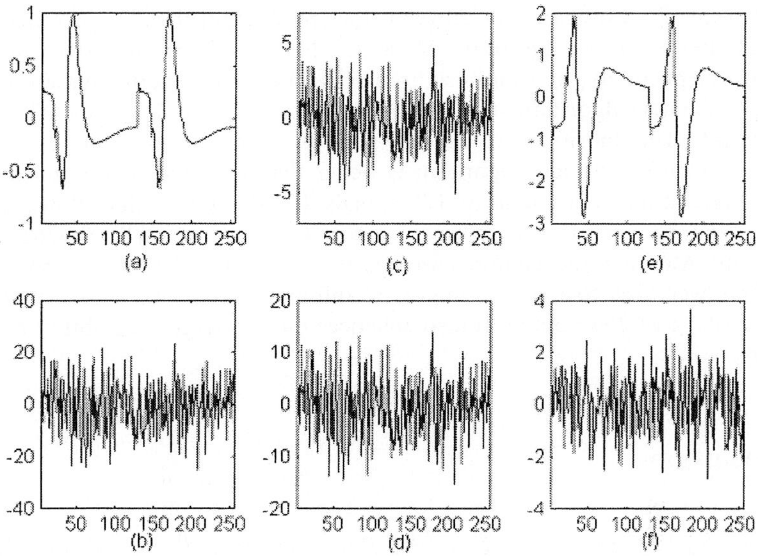

Fig. 2. Separating results (a)-(b) are the source signals. (c)-(d) are the mixed signals. (e)-(f) are the separating results.

Table 1. Comparison between three algorithms

The new algorithm	Iteration times	5	3	6	4	3	3
	Correlation coefficient of EP	0.9944	-1.0000	0.9847	-1.0000	-0.9864	-1.0000
	Correlation coefficient of noise	-0.9961	-0.9974	1.0000	0.9982	-0.9836	-1.0000
FastICA algorithm	Iteration times	14	4	158	5	3	44
	Correlation coefficient of EP	0.9943	1.0000	-0.9991	0.9334	0.9585	0.9943
	Correlation coefficient of noise	0.9949	-0.9993	0.9961	-0.9616	0.9523	-0.9999
Improved Infomax algorithm	Converge points	3150	1713	179	8895	871	517
	Correlation coefficient of EP	-1.0000	0.9957	0.9823	-0.9975	-1.0000	0.9997
	Correlation coefficient of noise	1.0000	0.9964	0.9914	1.0000	1.0000	0.9998

Experiment 1: We separate 2 sources with the new algorithm. Choose two periods of data, altogether 256 points. The simulation result is shown in Fig. 2. This algorithm separates independent sources effectively. The correlation coefficient between the separated and source EP signals is –0.9999, and the correlation coefficient between the separated and source white Gaussian noises is –0.9998. We only use two periods of EPs, and realize fast estimation of EPs.

Experiment 2: Comparison between the new algorithm, improved infomax algorithm and fastICA algorithm. Separate the mixed signals with these three algorithms respectively, and the results of 6 independent experiments are shown in Table 1. In the new algorithm and fastICA algorithm, we use 2 periods of data; in the improved infomax algorithm, we use 100 periods of data. From Table 1, we get that the performance of the new algorithm is better than the other two algorithms. The average correlation coefficient of EP signals for the three algorithms, the new, fastICA, and the improved infomax algorithms, are 0.9928, 0.9768, 0.9971, respectively. Although the average correlation coefficient of EP of the new algorithm is a little lower than that of the improved infomax algorithm, it uses much fewer points of data (just 256 points). It also enhances the convergence stability of fastICA algorithm.

6 Conclusions

In this paper, we combine two existent algorithms to construct a new algorithm, which has their advantages and without their disadvantages. Simulation results have verified that this new algorithm needs fewer data and has more stable convergence speed than the other two algorithms. And it can be used in blind separation for linearly mixed signals.

The improved infomax algorithm is relatively simple, and it is only suitable to separate sources with supergaussian distribution. In further research, we may choose the extended infomax algorithm, which can separate both sub- and super-gaussian sources, instead of it. Barros et al. suggested rICA algorithm (fastICA with reference signal), which can output the interested signal first. We can use this algorithm instead and output EP signal only to reduce calculation in later study.

References

1. Gharieb, R.R., Cichocki, A.: Noise Reduction in Brain Evoked Potentials Based on Third-Order Correlations. IEEE Trans. on Biomedical Engineering, vol. 48 (2001) 501-512
2. Bell, A. J., Sejnowski, T. J.: An Information-Maximization Approach to Blind Separation and Blind Deconvolution. Neural Computation, vol. 7 (1995) 1129-1159
3. Barros, A. K., Mansour, A., Ohnishi, N.: Removing Artifacts from Electrocardiographic Signals Using Independent Components Analysis. Neurocomputing, vol. 22 (1998) 173-186
4. Hyvarinen, A., Oja, E.: A Fast Fixed-Point Algorithm for Independent Component Analysis. Neural Computation, vol. 9 (1997) 1483-1492
5. Hyvarinen, A.: Fast and Robust Fixed-Point Algorithms for Independent Component Analysis. IEEE Trans. on Neural Networks, vol. 10 (1999) 626-634
6. Hyvarinen, A., Oja, E.: Independent Component Analysis: A Tutorial. http://www.cis.hut.fi/projects/ica/ (1999)

Estimation of Pulmonary Elastance Based on RBF Expression

Shunshoku Kanae, Zi-Jiang Yang, and Kiyoshi Wada

Kyushu University, 6-10-1 Hakozaki, Higashi-ku, Fukuoka, 812-8581 Japan
jin@ees.kyushu-u.ac.jp

Abstract. Building a precise respiration system model is very helpful for setting appropriate ventilation conditions to fit each patient when artificial respiration is performed on the patient. In this paper, a new respiration system model is proposed, which is a second order nonlinear differential equation including volume dependent elastic term described by RBF network. The model is able to describe the nonlinear dynamics of respiration. By using Sagara's numerical integration technique, a discrete-time identification model is derived. Then, off-line and on-line parameter estimation algorithms are presented. It is easy to obtain pulmonary elastance from identified model. The proposed model and the parameter estimation method are validated by clinical examples.

1 Introduction

If a patient cannot breathe autonomously, then artificial respiration will be performed on the patient. Artificial respiration is an emergency procedure used in almost all hospitals. When the artificial respiration is performed, it is necessary to appropriately set the parameters of respirator so that the ventilation conditions are suitable for each patient. For this purpose, pulmonary characteristics of each patient must be known. Especially, pulmonary elastance is generally considered as an important basis for deciding the pressure of ventilation.

Pulmonary elastance describes the relationship between pressure and volume of lung. There are various direct methods to measure the pulmonary elastance, for example, the super syringe technique[1], the interrupter technique[2,3], or the constant flow technique[4]. These techniques are sometimes not practical to perform or require the presence of a trained investigator. And the respirator settings may need to be changed for these techniques. On the other hand, several indirect methods have been proposed in the literature[5,6]. These methods adopt various respiration system models including a polynomial elastic term and a polynomial resistive term. The elastance can be determined from the parameters of estimated models. The most important advantage of these methods is that permit elastance to be obtained noninvasively without interfering with the ventilation pattern being employed. However, these model structures are still not efficient enough to model the clinical data, so sufficient accurate elastance estimate cannot be obtained based on these models.

In this paper, a second order nonlinear differential equation model including an elastic term expressed by RBF network is proposed for modeling respiration system, and estimation algorithms based on Sagara's numerical integration technique is derived.

F. Yin, J. Wang, and C. Guo (Eds.): ISNN 2004, LNCS 3174, pp. 507–512, 2004.

The pulmonary elastance can be easily obtained form estimated respiration model. The proposed model and the parameter estimation method are validated by clinical data.

2 Continuous-Time Model of Respiration System

Since 1950's, several simple models of respiration are proposed by Otis, Mead and Mount, respectively. These models can be generalized as a linear model[7]:

$$P(t) + a\dot{P}(t) = bV(t) + c\dot{V}(t) + f\ddot{V}(t), \tag{1}$$

where, $V(t)$ is the volume of lung, $\dot{V}$, $\ddot{V}$ the first and second order derivatives of the volume, $P(t)$ the pressure of the airway, $\dot{P}$ the first order derivative of the pressure. a, b, c, d are the coefficients of the model. In model (1), the pulmonary elastance and the airway resistance are expressed by a constant coefficient b and a constant coefficient c, respectively. But, in fact, the elastance and the resistance are not constant, values of the elastance and the resistance change nonlinearly with pulmonary volume. There exists essential problem that the model (1) is not able to describe nonlinearities.

In recent years, several models considering the nonlinearities of elastance and resistance have been proposed[6,8]. The expression of these models can be written as

$$P(t) = E(V)V(t) + R(V)\dot{V}(t), \tag{2}$$

where, the elastic coefficient and the resistive coefficient are described by polynomials of pulmonary volume, $E(V)$ and $R(V)$, respectively. So, a feature of model (2) is that the model has capability of describing the nonlinearities. However, the terms of $\dot{P}$ and $\ddot{V}$ are omitted in this model (in contrast to the model (1)), the model structure (2) is not sufficient to describe the dynamics of respiration. Consequently, the accurate respiration model cannot be acquired from the measurement data, and the sufficiently accurate estimates of $E(\cdot)$ and $R(\cdot)$ cannot be obtained.

In this paper, a new respiration system model is addressed:

$$P(t) + a\dot{P}(t) = f_E(V)V(t) + f_R(V)\dot{V}(t) + h\ddot{V}(t) + \epsilon(t), \tag{3}$$

$$f_E(V) = \sum_{i=1}^{n_E} b_i\psi_i(V(t)), \tag{4}$$

$$f_R(V) = c_0 + c_1|\dot{V}(t)|, \tag{5}$$

where, $\epsilon(t)$ contains model errors and measurement noises. In this model, the orders of derivatives are the same as model (1), and elastic coefficient is described by Radial Basis Function (RBF) network expressed by (4) whose input and output are respectively the volume of lung and the value of elastance. Here, n_E is the number of nodes, and b_i $(i = 1, \cdots, n_E)$ is the weight of i-th node. The RBF is as follows:

$$\psi_i(V) = \exp(-(V - V_{0i})^2/(2\pi\sigma^2)) \tag{6}$$

By the expressive power of RBF network, the model has the capability of describing the nonlinear dynamical characteristics of respiration.

Considering the relationship $Q(t) = \dot{V}(t)$ between the air flow $Q(t)$ and the volume of lung $V(t)$, the respiration model (3) can be written as:

$$P(t) + a\dot{P}(t) = f_E(V)V(t) + f_R(V)Q(t) + h\dot{Q}(t) + \epsilon(t). \tag{7}$$

Substituting (4) and (5) into (7), we have

$$P(t) = -a\dot{P}(t) + V(t)\sum_{i=1}^{n_E} b_i\psi_i(V) + Q(t)(c_0 + c_1|Q(t)|) + h\dot{Q}(t) + \epsilon(t). \tag{8}$$

Let data vector $\varphi(t)$ and parameter vector θ be

$$\varphi(t) = \begin{bmatrix} \dot{P}(t) \\ --- \\ V(t)\psi_1(V) \\ \vdots \\ V(t)\psi_{n_E}(V) \\ --- \\ Q(t) \\ Q(t)|Q(t)| \\ \dot{Q}(t) \end{bmatrix}, \quad \theta = \begin{bmatrix} -a \\ --- \\ b_1 \\ \vdots \\ b_{n_E} \\ --- \\ c_0 \\ c_1 \\ h \end{bmatrix}, \tag{9}$$

then simple expression of the model is obtained:

$$P(t) = \varphi^T(t)\theta + \epsilon(t) \tag{10}$$

3 Discrete-Time Model of Respiration System

In general, the measurement data that is obtained from the mechanical ventilation system is only the sampled data of pressure $\mathbf{P}$, air flow $\mathbf{Q}$, and volume $\mathbf{V}$.

$$\mathbf{P} = \begin{bmatrix} P(1) \\ P(2) \\ \vdots \\ P(N) \end{bmatrix}, \quad \mathbf{Q} = \begin{bmatrix} Q(1) \\ Q(2) \\ \vdots \\ Q(N) \end{bmatrix}, \quad \mathbf{V} = \begin{bmatrix} V(1) \\ V(2) \\ \vdots \\ V(N) \end{bmatrix} \tag{11}$$

In the continuous-time model (10) (or (8)), the differential terms are contained. Generally speaking, it is not desirable to calculate derivatives directly from the measurements, because it may make the noise effects worse. In this section, a discrete-time identification model for respiration system is derived based on Sagara's numerical integration technique[9].

Denote the sampling period of data collection as T. At time instant $t = kT$, integrate both sides of Equation (10) over the interval $[(k-\ell)T, kT]$. Let $y(k)$ be the left hand side of the resultant equation. Then $y(k)$ can be calculated as

$$y(k) = \int_{(k-\ell)T}^{kT} P(\tau)d\tau \doteq \sum_{j=0}^{\ell} g_j P(k-j), \tag{12}$$

where, ℓ is a natural number that decides the window size of numerical integration. The coefficients g_i $(i = 0, 1, \cdots, \ell)$ are determined by formulae of numerical integration. For example, when the trapezoidal rule is taken, they are given as follows:

$$\begin{cases} g_0 = g_\ell = T/2, \\ g_i = T, \quad i = 1, 2, \cdots, \ell - 1, \end{cases} \tag{13}$$

The integral of data vector $\varphi(t)$ can be calculated by

$$\phi(k) = \sum_{j=0}^{\ell} g_j \varphi(k-j) = \begin{bmatrix} P(k) - P(k-\ell) \\ --- \\ \sum_{j=0}^{\ell} g_j V(k-j)\psi_1(V(k-j)) \\ \vdots \\ \sum_{j=0}^{\ell} g_j V(k-j)\psi_{n_E}(V(k-j)) \\ --- \\ \sum_{j=0}^{\ell} g_j Q(k-j) \\ \sum_{j=0}^{\ell} g_j Q(k-j)|Q(k-j)| \\ Q(k) - Q(k-\ell) \end{bmatrix} \tag{14}$$

Here, analytical forms are taken for the terms where the integral can be calculated analytically. Get together the approximation error Δ_Σ caused by numerical integration and the integral of error term $\epsilon(t)$ of Equation (10) in $e(t)$. Namely, $e(k)$ is

$$e(k) = \Delta_\Sigma + \int_{(k-\ell)T}^{kT} \epsilon(\tau) d\tau. \tag{15}$$

Consequently, a discrete-time identification model of respiration system is derived:

$$y(k) = \phi^T(k)\theta + e(k). \tag{16}$$

4 Estimation of Pulmonary Elastance

From measurements of the airway pressure $P(k)$, air flow $Q(k)$ and pulmonary volume $V(k)$, it is easy to calculate $y(k)$ by equation (12) and $\phi(k)$ by equation (14) at each instant $k = \ell + 1, \cdots, N$, then $N - \ell$ regression equations can be derived as:

$$\mathbf{y} = \Phi\theta + \mathbf{e}, \tag{17}$$

where, $\mathbf{y} = [y(\ell+1) \ \cdots \ y(N)]^T$, $\Phi = [\phi(\ell+1) \ \cdots \ \phi(N)]^T$ and $\mathbf{e} = [e(\ell+1) \ \cdots \ e(N)]^T$, respectively.

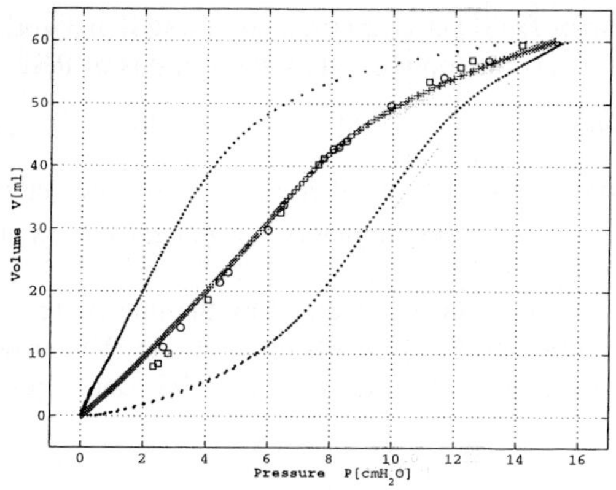

Fig. 1. Estimation of pulmonary elastance of a patient

The least squares estimate that minimizes the criterion function J defined as a sum of squared errors $J = ||\mathbf{y} - \Phi\theta||^2$ is given by

$$\hat{\theta} = (\mathbf{\Phi^T \Phi})^{-1}\mathbf{\Phi^T y} \tag{18}$$

provided the inverse exists. Then, the estimate of pulmonary elastance are obtained:

$$\hat{f}_E(V) = \sum_{i=1}^{n_E} \hat{b}_i \psi_i(V) \tag{19}$$

The above algorithm is an off-line algorithm in which the calculation is carried out after the data of length N are collected. But, in clinical cases, the data are recorded successively, and the state of lung may change, so on-line estimation algorithm is desired. The on-line algorithm for calculating the above LS estimate is as the follows[10]:

$$\begin{cases} \hat{\theta}(k) = \hat{\theta}(k-1) + L(k)(y(k) - \phi^T(k)\hat{\theta}(k-1)), \\ L(k) = \dfrac{S(k-1)\phi(k)}{\lambda + \phi^T(k)S(k-1)\phi(k)}, \\ S(k) = \dfrac{1}{\lambda}\left[S(k-1) - \dfrac{S(k-1)\phi(k)\phi^T(k)S(k-1)}{\lambda + \phi^T(k)S(k-1)\phi(k)}\right], \end{cases} \tag{20}$$

where, λ ($\lambda \leq 1$) is the forgetting factor, and the initial values of $\hat{\theta}$ and S are taken as $\hat{\theta}(0) = 0$, $S(0) = s^2 I$ (s is a sufficiently large real number).

5 Validation by Clinical Examples

In this section, to verify the applicability of our new model, the pulmonary elastance of a patient is estimated based on the proposed model and the on-line estimation algorithm. In this example, the sampling period is $T = 0.005$ second, the data length is $N = 470$. The measurement data of $P(k)$ and $V(k)$ are plotted by dotted line in Figure 1. In this example, the elastic term $f_E(V)$ is approximated by a RBF network which have $n_E = 5$ nodes. By the proposed estimation algorithm, the weights of RBF nodes are estimated:

$$[\,\hat{b}_1,\ \hat{b}_2,\ \hat{b}_3,\ \hat{b}_4,\ \hat{b}_5] = [1.444,\ -3.331,\ 4.496,\ -3.640,\ 1.662]$$

where, the integration window size is taken as $\ell = 20$. The static P/V characteristic curve calculated by the estimated parameters $\hat{b}_1, \hat{b}_2, \hat{b}_3, \hat{b}_4, \hat{b}_5$ is drawn in "+" symbols in Figure 1.

In figure 1, the static P/V curve obtained by the interrupter technique is drawn in "□" symbols (inspiration) and "o" symbols (expiration). We can see that the estimated elastance curve fits the practically measured static P/V curve quite well.

6 Conclusions

In this paper, a new respiration system model is proposed, which is a second order nonlinear differential equation with elastic term expressed by RBF network. The model is able to describe the nonlinear dynamics of respiration. By using Sagara's numerical integration technique, a discrete-time identification model is derived. Then, off-line and on-line parameter estimation algorithms are presented. It is easy to obtain pulmonary elastance from identified model. The proposed model and the parameter estimation method are validated by clinical examples.

References

1. Matamis, D., Lemaire, F., Harf, A., Brun-Buisson, C., Ansquer, J.C., Atlan, G.: Total Respiratory Pressure-Volume Curves in the Adult Respiratory Distress Syndrome. Chest, **86** (1984) 58–66
2. Gottfried, S.B., Rossi, A., Calverley, P.M.A., Zocchi, L., Milic-Emili, J.: Interrupter Technique for Measurement of Respiratory Mechanics in Anesthetized Cats. J. Appl. Physiol: Respir Environ Exercise Physiol **56** (1984) 681–690
3. Ranieri, V.M., Giuliani, R., Fiore, T., Damhrosio, M., Milic-Emili, J.: Volume-Pressure Curve of the Respiratory System Predicts Effects of PEEP in ARDS: "Occlusion" Versus "Constant Flow" Technique. Am. J. Respir. Crit. Care Med. **149** (1994) 19–27
4. Lu, Q., Vieira, S.R.R., Richecoeur, J., et. al.: A Simple Automated Method for Measuring Pressure-Volume Curves During Mechanical Ventilation. Am. J. Respir. Crit. Care Med. **159** (1999) 275–282
5. Uhl, R.R., Lewis, F.J.: Digital Computer Calculation of Human Pulmonary Mechanics Using a Least Squares Fit Technique. Comput. Biomed. Res. **7** (1974) 489–495
6. Wensley, D.F., Noonan, P., Seear, M.D., Werner, H., Pirie, G.E.: Pilot Study for the Development of a Monitoring Device for Ventilated Children. Pediatr. Pulmonol. **11** (1991) 272–279
7. Lorino, A.M., Lorino, H., Harf, A.: A Synthesis of the Otis, Mead, and Mount Mechanical Respiratory Models. Respiration Physiology, **97** (1994) 123–133
8. Muramatsu, K., Yukitake, K, Nakamura, M, Matsumoto, I., Motohiro, Y: Monitoring of Nonlinear Respiratory Elastance Using a Multiple Linear Regression Analysis. European Respiratory Journal **17** (2001) 1158–1166
9. Sagara, S., Zhao, Z.: Numerical Integration Approach to On-Line Identification of Continuous-Time Systems. Automatica **26** (1990) 63–74
10. Ljung, L.: System Identification Theory for the User. Prentice-Hall (1987)

Binary Input Encoding Strategy Based Neural Network for Globulin Protein Inter-residue Contacts Map Prediction

GuangZheng Zhang[1,2], DeShuang Huang[1], and Xin Huang[1]

[1] Intelligent Computing Lab, Hefei Institute of Intelligent Machines, Chinese Academy of Sciences, P.O.Box 1130, Hefei Anhui 230031, China
gzzhang@iim.ac.cn

[2] Department of Automation, University of Science and Technology of China Hefei, Anhui, 230026, China

Abstract. Inter-residue contacts map prediction is one of the most important intermediate steps to the protein folding problem. In this paper, we focus on protein inter-residue contacts map prediction based on radial basis function neural network (RBFNN), and propose a novel binary encoding scheme for the purpose of learning the inter-residue contact patterns, and get a better prediction results.

1 Introduction

Protein spatial structure predictions, including secondary, tertiary and quaternary structure prediction, from primary amino acids sequence are the fundamental and unresolved problems in molecular biology. Full molecular modelling to find the structures is at present intractable, and so intermediate steps, such as inter-residues contact prediction, [1]-[4], and residue spatial distance prediction, [5]-[7], have been developed rapidly recently. Our previous studies show that 2 amino acid residues which are far apart in sequence might be close together in spatial relation, and this will increases the contact probability of the residues. Therefore, the prediction of inter-residue contacts in proteins is an interesting problem whose solution may be useful in protein-folding recognition, secondary or tertiary structure prediction, de novo design and so on.

A contact map is a particularly useful two dimensional representation of a protein's three dimension structure. Within a protein sequence $S = \{S_1, S_2, \cdots, S_N\}$, two amino acid residues, S_i and S_j, are considered in contact if the 3D spatial distance is less than a contact threshold t. The distance involved in the different contact definitions can be that between the $C_\alpha - C_\alpha$ atoms of the two residues (Mirny and Domany, 1996), between the $C_\beta - C_\beta$ (Thomas et al., 1996; Lund et al., 1997; Olmea and Valencia, 1997) and the minimal distance between atoms belonging to the side chain or to the backbone of the two residues (Fariselli and CAsadio, 1999). But here, we use the distance between the two residues geometrical centers, determined by the coordinates of their atoms, to gain the contact map. Specifically, we use the spatial distance described in [7],

F. Yin, J. Wang, and C. Guo (Eds.): ISNN 2004, LNCS 3174, pp. 513–518, 2004.

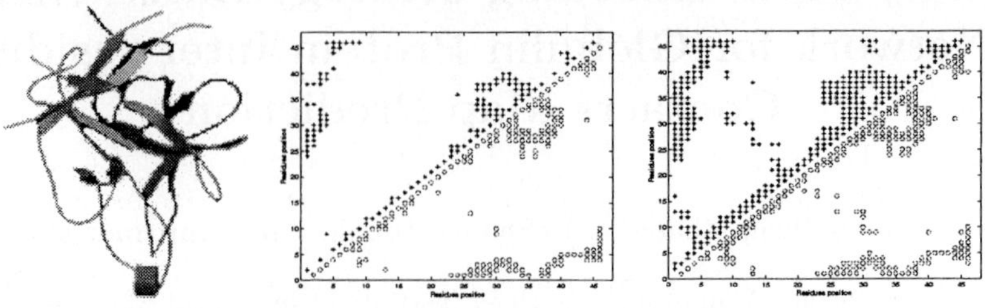

(a) Tertiary structure

(b) Desired maps of 5 and 6 Å

(c) Desired maps of 7 and 8 Å

Fig. 1. The tertiary structure and inter-residue contact maps of a soybean protein sequence (PDB code 1AVU). **(a)**: the sketch of the protein tertiary structure from PDB; **(b)**: blue asterisks in the upper left represent the contact map with the threshold of 5 Å, while red circles in the lower right denote the map of 6 Å; **(c)**: blue asterisks and red circles represent the maps of 8 and 7 Å, respectively.

$\sigma(S_i, S_j) = |r_i - r_j|$, where r_i and r_j are the geometrical coordinates of the 2 residues, to determine the residues are in contact or not.

The contact map of a protein with N amino acid residues ia an $N \times N$ matrix C, whose elements are defined as

$$C_{i,j} = \begin{cases} 1 \text{ if residues } S_i \text{ and } S_j \text{ are in contact} \\ 0 \text{ otherwise} \end{cases} \tag{1}$$

The contact between two residues can be defined in different ways. In particular, we will use the spatial distance described in [7] to determined the residues are in contact or not, so the definition 1 becomes

$$C_{i,j} = \begin{cases} 1 \text{ if } |r_i - r_j| < R_c \\ 0 \text{ otherwise} \end{cases} \tag{2}$$

where r_i and r_j are the geometrical coordinates of the 2 residues. As shown in Fig1, a soybean protein sequence (PDB code 1AVU) contact map and its tertiary structure are demonstrated. Here, we illustrate 4 inter-residue contact maps with different threshold values of 5, 6, 7 and 8 Å, respectively.

2 Radial Basis Function Neural Network

Generally, a radial basis function neural network consists three layers (see the RBFNN architecture which was shown in Fig.2): (1) the input layer, where all the feature vectors of the training samples are input; (2) the hidden layer, where at each node the input feature vector is put through a radial basis function (RBF) that is centered on a corresponding exemplar vector $\overrightarrow{V}$; (3) the output layer, where all the inputs from the hidden layer are combined to indicate the class of input sample. The weights between the hidden and the output layer are adjust during the training procedure for the purpose of minimizing the cost function.

Let we denote the training exemplar feature vectors set as $\{x^q : q = 1, \cdots, Q\}$, and x^q is an exemplar vector with N components, $x^q = (x_1^q, x_2^q, \cdots, x_N^q)$. Each actual output from the jth node, z_j^q is forced to match the target component, t_j^q by adjusting the weights. At the same time, a bias, b_j, is added onto the sum and also adjusted to help approximate the target value. Then the output from each of the J nodes for the input vector, x^q, has the following form:

$$\begin{aligned} z_j^q &= \frac{1}{M}\sum_{m=1}^{M} w_{mj} y_m^q + b_j \\ &= \frac{1}{M}\sum_{m=1}^{M} w_{mj} exp(-\frac{(x^q - x^m)^2}{2\sigma^2}) + b_j \end{aligned} \tag{3}$$

where b_j is a bias, w_{mj} is the weight between the hidden and the output layer, and the $exp(\cdot)$ the Gaussian radial basis function.

Table 1. The binary encoding scheme of the input sample space

Possible Pairwise	Binary (8 Bits)	Residue Classification	Binary (4 Bits)	Secondary Structure	Binary (3 Bits)
Ala-Ala	00000000	Np-Np	0000	α-α	000
Ala-Cys	00000001	Np-P	0001	α-β	001
Ala-Asp	00000010	Np-A	0010	α-Coil	010
Ala-Glu	00000011	Np-B	0011	β-β	011
Ala-Phe	00000100	P-P	0100	β-Coil	100
Ala-Gly	00000101	P-A	0101	Coil-Coil	101
⋮	⋮	P-B	0110		
Ala-Tpy	00010100	A-A	0111		
⋮	⋮	A-B	1000		
Tpy-Tpy	11010010	B-B	1001		

Np: nonpolar-hydrophobic; P: polar-hydrophilic; A: acidic and B: basic

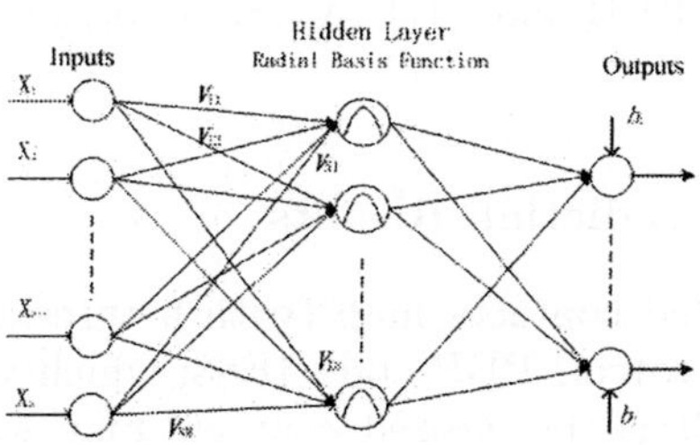

Fig. 2. The architecture of radial basis function neural network

3 Input Encoding Scheme

Here, we use binary encoding scheme to denote the input vectors for every possible residue pairwise. Specifically, each possible pairwise is encoded by a 15 bits binary, in which the first 8 bits denote the possible pair of 20 amino acids, the following 4 bits denote the residue classification information, while the last 3 stand for its secondary structure information.

Possible residues pairwise (8 bits for the 210 possible pairwises): As we known, the 20 different amino acid residues have 210 ($20 \times (20+1)/2 = 210$) possible couples (considering that each residue couple and its symmetric are coded in the same way)[3], such as the couple of Ala-Ala, Ala-Cys, Ala-Asp, and so on. So we can denote these 210 possible couples by a 8 bits binary which is shown in Table 1.

Residue classes (4 bits for the 10 possible class pairs): Residues may be classified into 4 classes, ie., nonpolar-hydrophobic, polar-hydrophilic, acidic or basic. So to a given pair of residues, there are totally 10 possible pair cases, such as both nonpolar-hydrophobic, nonpolar-hydrophobic to acidic, and so on. In this case, a given residue pair classification information can be denoted by a 4 bit binary as shown in Table 1.

Secondary Structure Information (3 bits for the 6 possible structure pairs): Studies show that the residue spatial structures, such as secondary structure and tertiary structure, are important factors of residue contact. A amino acid residue has three possible secondary structures:α-helix, β-sheet, and Coil. So, we can use a 3 bit binary to denote the 6 possible secondary structure pairs, ie., α-α, α-β, α-Coil, β-β, β-Coil and Coil-Coil. Table 1 gives the detailed encoding elements.

4 Simulation Results and Discussions

We use a large set of globulin proteins of known 3D structure to train the radial basis function neural network for the purpose of learning the inter-residue contact patterns. In Table 2, the 53 globulin protein sequences, which are got form PDB with the sequence identity lower than 90%, are grouped into 5 classes: $L_s < 100$, $100 \leq L_s < 200$, $200 \leq L_s < 300$, $300 \leq L_s < 400$ and $L_s \geq 400$, according to their residue length, and the five sequences(identified by boldface), named 1TTF, 1E88, 1NAR, 1BTJ_B and 1J7E_A, are used to test the RBFNN.

4.1 Contacts Map Prediction Results

Here, we give the detailed contacts map (with a threshold of 6 Å) prediction results of the globulin protein, PDB code 1E88, which was shown in Fig.3. In the figure, the upper-left is the desired contacts map with the threshold of 8 Å, and the lower-right area is the predicted contacts map based our proposed binary encoding scheme and an RBF network. Moreover, what should be stressed

Table 2. The PDB code of globulin proteins with different residue length used for training the radial basis probability neural network

$L_s < 100$		$100 \leq L_s < 200$		$200 \leq L_s < 300$		$300 \leq L_s < 400$		$L_s > 400$	
PDB Code	Length	PDB Code	Length	PDB Code	Length	PDB Code	Length	PDB Code	Length
1OWW	98	1DV9	162	1F5F	205	1LOT_B	376	1J78_A	458
1J8K	94	1KDK	177	**1NAR**	**290**	1OQH	337	1J78_B	458
1Q38	89	1KDM	177			1JQF	334	1H76	696
1QGB	93	1D2S	170			1D3K	329	1J7E_B	458
1FNA	91	1LHN	189			1BTJ_A	337	1LOT_A	458
1FBR	93	1LHO	189			1TFD	304	1KW2_A	458
1O9A_A	93	1LHU	189			1A8E	329	1JNF	676
1O9A_B	36	1LHV	189			1B3E	330	1OD5_A	492
2FN2	59	1LHW	189			1N7X	331	1OD5_B	492
1TTG	94	1MUP	166			1OQG	337	**1J7E_A**	**458**
1TTF	**94**	1QO6	101			1N84	331		
		1E8B	160			1BP5_A	337		
		1E88	160			1BP5_B	337		
		1E88	**160**			**1BTJ_B**	**337**		

Protein length L_S is the residue number of its amino acid, and the last 5 globulin protein (boldface) are used as test sequences.

Table 3. Numbers and accuracies of predicted contacts map

Globulin Sequence	5Å			6 Å			7 Å			8 Å		
PDB Code (Length)	N_p	N_d	A (%)	N_p	N_d	A (%)	N_p	N_d	A (%)	N_p	N_d	A (%)
1TTF (94)	145	482	30.08	245	783	31.29	346	1072	32.28	376	1421	26.46
1E88 (160)	424	1437	29.51	660	1955	33.76	772	2438	31.67	1006	3352	30.01
1NAR (290)	1394	4533	30.75	2025	6197	32.68	2583	7864	32.85	3346	10524	31.79
1BTJ_B (337)	1783	5944	30.00	2767	8630	32.06	3250	10745	30.25	3796	14283	26.58
1J7E (458)	2751	9732	28.18	4601	15688	29.33	5504	19645	28.02	6589	25026	26.33
SUM & Accuracy	6497	22128	29.36	10298	33253	30.96	12455	41764	29.82	15113	54606	27.67

is that all the residues pair contacts, whose sequence separation $|i-j| \leq 4$, are not shown in the figure.

4.2 Predicted and Desired Numbers

The five globulin protein sequences, named 1TTF, 1E88, 1NAR, 1BTJ_B and 1J7E_A, which are shown in Table 2, are used to test the proposed RBFNN. The correctly predicted contacts numbers and the corresponding accuracies are given in Table 3, from which we can see that the network overall performance with the contact threshold of 6 Å score higher than the other three, 5, 7 and 8 Å, and the average accuracy reaches about 30.96%. This demonstrates that the contact threshold with 6 Å is the best.

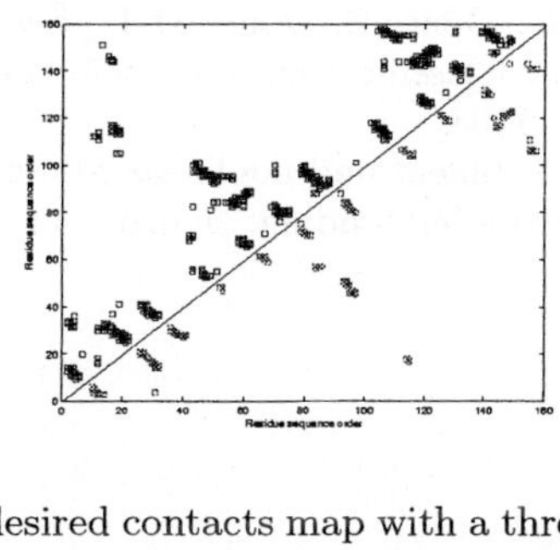

Fig. 3. The predicted and desired contacts map with a threshold of 6 Å for the globulin protein sequence 1E88, the red squares in the lower-right area denote the predicted contacts, while the blue squares in the upper-left stand for the desired contacts.

5 Conclusions

In this paper, we proposed a new binary input encoding scheme for the purpose of training radial basis function neural network to predict inter-residue contacts map within the globulin protein sequences, and got a better results. Future works will include protein inter-residue distance distribution analysis and tertiary structure prediction.

References

1. Kibeom Park, Michele Vendruscolo and Eytan Domany: Toward an Energy Function for the Contact Map Representation of Proteins. PROTEINS: Structure, Function, and Genetics. **40**(2000) 237-248.
2. Michlele Vendruscolo, Rafael Najmanovich and Eytan Domany: Can a Pairwise Contact Potential Stabilize Native Protein Folds Against Decoys Obtained by Treading?. PROTEINS: Structure, Function, and Genetics. **38** (2000) 134-148.
3. Piero Fariselli, Osvaldo Olmea, Alfonso Valencia and Rita Casadio: Prediction of contact maps with neural networks and correlated mutations. Protein Engineering. **Vol.14, No.11** (2001) 835-843.
4. Singer MS, Vriend G, and Bywater RP: Prediction of protein residue contacts with a PDB-derived likelihood matrix. Proetin Eng. **15 (9)** (2002) 721-725.
5. Steve Fairchild, Ruth Pachter and Ronald Perrin: Protein Structre Analysis and Prediction. The Mathematica Journal. **Volume 5, Issue 4** (1997) 64-69.
6. Albert J. Ketterman, Peerada Prommeenate, Chanikarn Boonchauy, Umnaj Chanama and etc.: Single amino acid changes outside the active site significantly affect activity of glutathione S-transferases insert. Biochemistry and Molecular Biology **31** (2001) 65-74.
7. Guang-Zheng Zhang and De-Shuang Huang: Radial Basis Function Neural Network Optimized by a Genetic Algorithm for Soybean Protein Sequence Residue Spatial Distance Prediction. 2004 IEEE Congress on Evolutionary Computation, Portland, Oregon, June 20-23, 2004(accepted).
8. Fariselli P, Olmea O, Valencia A, and Casadio R.: Progress in predicting inter-residue contacts of proteins with neural networks and correlated mutations. Proteins. **Suppl 5** (2001)157-162.
9. Fariselli P and Casadio R.: Neural network based prediction of residue contacts in protein. Protein Eng. **12** (1999)15-21.
10. Orengo C.A., Michie A.D., Jones S., Jones D.T., Swindells M.B., and Thornton J.M.: CATH- A Hierarchic Classification of Protein Domain Structures. Structure. **Vol 5. No 8** (1997)1093-1108.
11. D.S. Huang, and S.D. Ma: Linear and nonlinear feedforward neural network Classifiers: A comprehensive understanding. Journal of Intelligent Systems **Vol.9, No.1**(1999) 1-38.

Genetic Regulatory Systems Modeled by Recurrent Neural Network*

Xianhua Dai

Dept. of Electronic & Computer Engineering, School of Information Science and Technology
Sun Yat-sen University, Guangzhou 510275, China
issdxh@zsu.edu.cn

Abstract. In this paper, we focus on modeling and exploring the genetic regulatory systems (GRS) with an artificial recurrent neural network (ARNN). Based on the approximation capability of the ARNN, the proposed model can be used to express and analyze the genetic regulatory systems of genetic components, even the large-scale genetic systems. Unlike the conventional ARNN, the ARNN used in the paper is the echo state network (ESN), in which the connection weights of internal neurons are fixed and only the output weights are adjustable. Thus, there are no cyclic dependencies between the trained readout connections and, training the genetic regulatory system becomes a simple linear regression task. The experiment studies shows the new genetic regulatory system modeled by ESN and trained from the fluorescent density of reporter protein has a satisfactory performance in modeling the synthetic oscillatory network of transcriptional regulatory of *Escherichia coli* cells.

1 Introduction

Each cell of a (multi-cellular) organism holds the genome with the entire genetic material, represented by a large double-stranded DNA molecule with the famous double-helix structure. Cells are therefore the fundamental unit of living matter. They take up chemical substances from their environment and transform them. The functions of a cell are subject to regulation such that the cell acts and interacts in an optimal relationship with its environment. The interaction of genetic components or genetic regulatory is a complicated process, included 'Transcription' and 'translation'. Transcription is the process by which coding regions of DNA (called 'genes') synthesize RNA (m-RNA) molecules. This is followed by a process referred to as 'translation' synthesizing proteins using the genetic information in RNA as a template. Most proteins are enzymes and carry out the reactions responsible for the cell's metabolismthe reactions that allow it to process nutrients, to build new cellular material, to grow, and to divide. Thus, the genetic regulatory is a nonlinear dynamical process.

Modeling the genetic regulatory process as dynamical systems have been studied extensively for biochemical oscillations, the genetic toggle switch in *Escherichia coli*

* This work is supported by the NSFC 60272068.

F. Yin, J. Wang, and C. Guo (Eds.): ISNN 2004, LNCS 3174, pp. 519–524, 2004.

(*E.Coli.*) and λ phage, gene expression multi-stability in lactose utilization [1][5][8], and circadian clocks found in many organisms. All of these works stress the importance of feedback regulation of transcriptional factors. From the previous works, it is possible to design and construct an artificial neural network with new functional properties from genetic components to model the genetic regulatory process.

In this paper, we use ARNN, in particular echo state network, to model the genetic regulatory process, and then, the ESN is identified from the green fluorescent protein (GFP) of reporter protein, which represents the intensity of reporter proteins.

2 Genetic Regulatory Systems

Gene expression is a complex process regulated at several stages in the synthesis of protein, included 'Transcription' and 'translation'. There are several basic models for genetic networks [2][3][4][6][7]: 1) the Bayesian or logical (Boolean) model, 2) the ordinary (or partial) differential equation (ODE), called as the rate equations, also. By virtue of random graphs, the vertices or nodes represent the genes or other elements, and edges may represent the interaction relation. Under the Markov field assumption, the joint probability distribution of edges or multiple nodes, which may represents the interaction intensity of two or multiple genes, can be estimated from the expression data, such as the micro-arrays experiments. Figure 1(A) depicts an example of genetic regulatory network modeled by the graph models. Although the Bayesian networks and graph models are intuitive representation of genetic regulatory systems, they have the drawback of leaving dynamical aspects of gene regulatory implicit and coarse. In contrast, ODE or rate equations describes the concentrations of gene products such as mRNAs, protein and other small molecules, which are more accurate and can provide detailed understanding of the nonlinear dynamical behavior of biological systems, such as multi-stability, toggle switch and synthetic oscillatory. This paper will focus on the latter, i.e. the ODE based regulatory systems.

Figure 1 systemically shows genetic regulatory process, and the concentrations of gene products in Figure 1 can be modeled as

$$d\bar{\mathbf{m}}(t)/dt = \bar{\mathbf{f}}_m\left[\bar{\mathbf{m}}(t), \bar{\mathbf{p}}(t-\tau_p)\right] \tag{1}$$

$$d\bar{\mathbf{p}}(t)/dt = \bar{\mathbf{f}}_p\left[\bar{\mathbf{m}}(t-\tau_m), \bar{\mathbf{p}}(t)\right] \tag{2}$$

where $\bar{\mathbf{m}}(t) = [\bar{m}_1(t), \cdots \bar{m}_n(t)]^T \in R^N$ and $\bar{\mathbf{p}}(t) = [\bar{p}_1(t), \cdots \bar{p}_n(t)]^T \in R^N$ are the concentrations of mRNAs and proteins, respectively. $\tau_m = [\tau_{m1}, \cdots \tau_{mn}]^T \in R^N$ and $\tau_p = [\tau_{p1}, \cdots \tau_{pn}]^T \in R^N$ are the positive real vectors indicating the time-delays of mRNA and protein respectively.

$$\bar{\mathbf{f}}_m\left[\bar{\mathbf{m}}(t), \bar{\mathbf{p}}(t-\tau_p)\right] = \left[\bar{f}_{m1}\left[\bar{\mathbf{m}}(t), \bar{\mathbf{p}}(t-\tau_p)\right], \cdots \bar{f}_{mn}\left[\bar{\mathbf{m}}(t), \bar{\mathbf{p}}(t-\tau_p)\right]\right]^T \tag{3}$$

$$\bar{\mathbf{f}}_p\left[\bar{\mathbf{m}}(t-\tau_m), \bar{\mathbf{p}}(t)\right] = \left[\bar{f}_{p1}\left[\bar{\mathbf{m}}(t-\tau_m), \bar{\mathbf{p}}(t)\right], \cdots \bar{f}_{pn}\left[\bar{\mathbf{m}}(t-\tau_m), \bar{\mathbf{p}}(t)\right]\right]^T \tag{4}$$

are the continuous function vectors. By sampling procedure, the ODE (1)(2) can be rewritten in discrete form as

$$\mathbf{m}(k+1)=\mathbf{f}_m\left[\mathbf{m}(k),\mathbf{p}(k),\mathbf{p}(k-1),\cdots\mathbf{p}(k-Q_p)\right] \tag{5}$$

$$\mathbf{p}(k+1)=\mathbf{f}_p\left[\mathbf{m}(k),\mathbf{m}(k-1),\cdots\mathbf{m}(k-Q_m),\mathbf{p}(k)\right] \tag{6}$$

where $k=0,1,2,\cdots$ represents the sampling instant, $\mathbf{m}(k),\mathbf{p}(k)$ are the sampling values of $\mathbf{m}(t),\mathbf{p}(t)$. The vectors $\mathbf{f}_m[\bullet]$ and $\mathbf{f}_p[\bullet]$ are similar to their continuous forms. Q_p and Q_m are the positive integers.

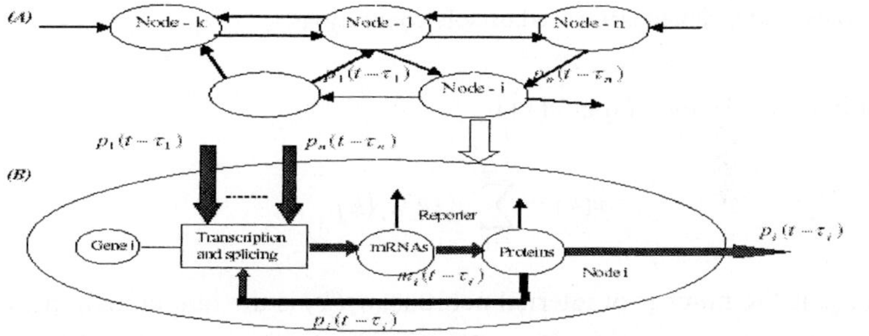

Fig. 1. Genetic regulatory network with feedback components for transcriptional and splicing processes. (A) genetic regulatory network, (B) structure of node-i or sub-network.

3 GRS Modeled by RNN and ESN

Certainly the ODE or rate equation can be modeled by an artificial recurrent neural network (ARNN). Figure 2 (A) depicts a standard ARNN, in which all weights denoted by dotted arrows, including input, output and internal (or hidden) neuron, are adjustable aim at minimizing the error $e(k)=d(k)-y(k)$.

Observing the ARNN, we can see that the output $y(k)$ and error $e(k)=d(k)-y(k)$ are apparently the nonlinear functions of the weight parameters apart from the output weights. As a result, the training is complicated and may converge to a local minima.

Instead of the standard ARNN, we use an ESN [9] to model the genetic regulatory networks. An ESN is also an ARNN, but its approach differs from the conventional ARNN in that a large RNN is used (on the order of 50 to 1000 neurons, the conventional ARNN typically use 5 to 30 neurons) and in that only the synaptic connections from the ARNN to the output readout (output layer) neurons are modified by learning. Figure 2 (B) illustrate an ESN, in which only the connection weights, dotted arrows, associated with the output neurons are adjustable by the learning. Because there are no cyclic dependencies between the trained readout connections and, training the genetic regulatory system becomes a simple linear regression task.

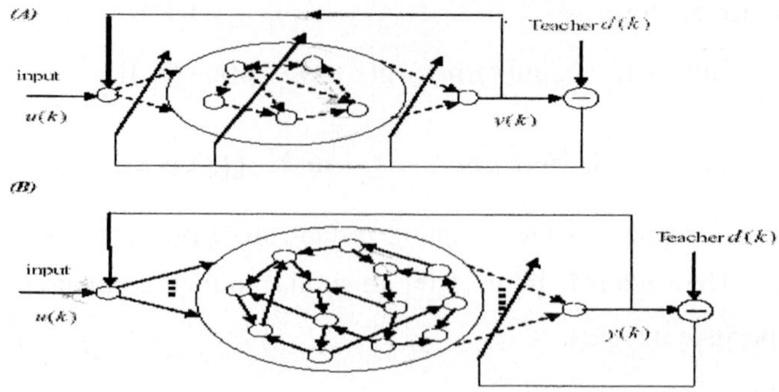

Fig. 2. (A) Conventional RNN model, (B) Echo State Networks. Sold bold arrows, fixed synaptic connections; dotted arrows, adjustable connections.

From Figure 2 (B), the output $y(k)$ is

$$y(k) = \sum_{i=1}^{N_{ESN}} w_i(k) x_i(k) \tag{7}$$

where N_{ESN} is the number of internal neurons, $w_i(k)$ is the output neuron weights, i.e. the adjustable parameters and $x_i(k)$ is the activation of the internal neuron at time k

$$x_i(k) = g\left[\sum_{l_1}\sum_{l_2} w_{i,l_1,l_2}^{(in)} u_{l_1}(k-l_2) + \sum_{l_1}\sum_{l_2} w_{i,l_1,l_2}^{(out)} y_{l_1}(k-l_2) + \sum_{j=1}^{N_{ESN}} w_{i,j}^{(inneuron)} x_j(k) + \beta_i\right] \tag{8}$$

with the fixed weights $w_{i,l_1,l_2}^{(in)}$, $w_{i,l_1,l_2}^{(out)}$ and $w_{i,j}^{(inneuron)}$. $g[\bullet]$ is the activation function and is chosen to be the Sigmoidal function, and β_i is a constant.

Because the connection weights of input and internal neurons are fixed, $x_i(k)$ is known in estimating $w_i(k)$. From (7), the adjustable weights can be easily estimated from the training set $\{d(k), y(k), x_i(k)\}$, $k = 1,2,\cdots$. More detail refers to [9].

4 GRS Identified by the Fluorescent Reporter Proteins

We consider a genetic regulatory system composed of three repressors in *E. coli.*, i.e. a plasmid illustrated in [1]. The first repressor protein, LacI from *E. coli.*, inhibits the transcription of the second repressor gene, *tet*R from the tetracycline-resistance transposon Tn10, whose protein product in turn inhibits the expression of a third gene, *cI* from λ phage. Finally, CI inhibits *lac*I expression, completing the cycle. The detail about the plasmid refers to [1]. By isolating single cells under the microscope, the transcription process is observed by monitoring their fluorescence intensity as they grew into small two-dimensional microcolonies consisting of hundreds of progeny cells. Figure 3 is copied from of the experiment measurements in [1].

We use ESN to model the synthetic oscillatory network, which is composed of three transcriptional repressors. Three repressor-protein concentrations, $p_i(k)$, and their corresponding mRNA concentrations, $m_i(k)$, are treated as the dynamical variables. In the initial phase, six variables are randomly perturbed to initiate the ESN, which simulates that the transcription process is initiated by the binding of several transcription factors to regulatory sites in the DNA, usually located in the promoter region of the gene. And then, ESN is trained by the GFP measurements.

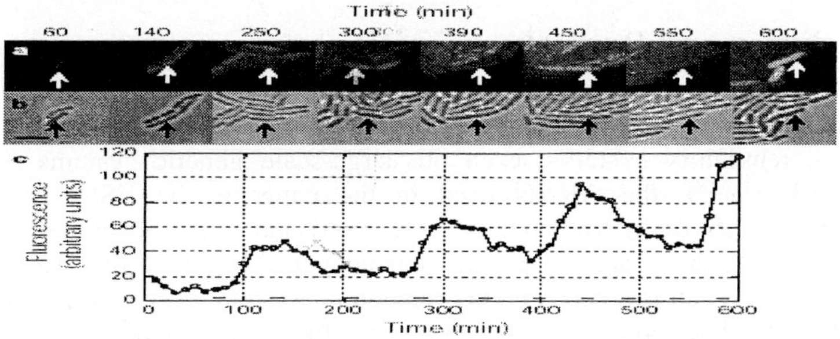

Fig. 3. Repressilation in living bacteria. a, b, The growth and time-course of green fluorescence protein (GFP) expression for a single cell of E. coli host strain MC4100 containing the repressilator plasmids. Snapshots of a growing micro-colony were taken periodically both in fluorescence (a) and bright-field (b). c. The pictures in a and b correspond to peaks and troughs in the time-course of GFP fluorescence density of the selected cell. Scale bar, 4 mm. Bars at the bottom of c indicate the timing of septation events, as estimated from bright-field images.

In ESN, we use 50 internal neurons, and the fixed weights, $w_{i,l_1,l_2}^{(in)}$ and $w_{i,l_1,l_2}^{(out)}$ take value as random variable uniformly distributed within $(0,1)$, and

$$w_{i,j}^{(inneuron)} = r_1(i,j)r_2(i,j) \tag{9}$$

where $r_1(i,j)$ is a random variable uniformly distributed within $(0,1)$, and $r_2(i,j)$ is

$$r_2(i,j) = \begin{cases} 1 & r < 0.1 \\ 0 & r \geq 0.1 \end{cases} \tag{10}$$

where r is a random variable uniformly distributed within $(0,1)$. Thus, the connectivity of internal neurons is only 10% controlled by (9)(10), a sparse interconnectivity.

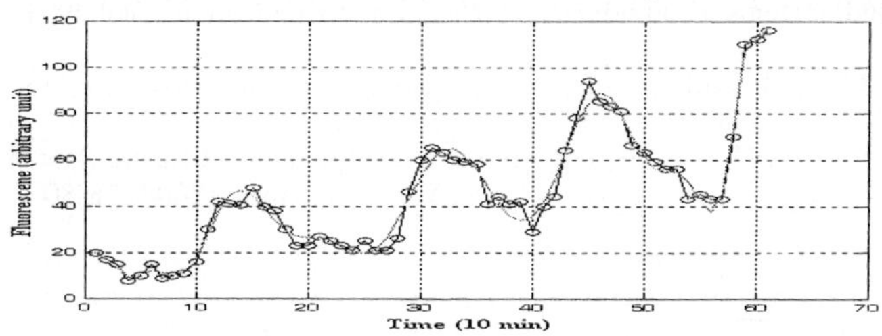

Fig. 4. The ESN outputs. (50 internal neurons, only 10% connectivity)

Figure 4 shows the original measurement and ESN outputs. Observing Figure 4, we can see that the ESN with sparse interconnectivity can better model the genetic regulatory system. Because only output weights need to estimated and adjusted, the learning process is simpler than the conventional ARNN. After ESN identification, it is easy to analyze the nonlinear dynamic property of the regulatory networks. Because of space limitation, the detail will not be discussed here.

5 Conclusion and Remarks

Based on the approximation capability, we have used an ARNN to model and analyze the genetic regulatory systems, even the large-scale genetic systems. Unlike the conventional ARNN, the ARNN used in the paper is the ESN, in which the connection weights of internal neurons are fixed and only the output weights are adjustable. Thus, there are no cyclic dependencies between the trained readout connections and, training the genetic regulatory system becomes a simple linear regression task. The experiment studies show the new genetic regulatory system modeled by ESN and trained from the GFP density of reporter protein has a satisfactory performance in modeling the synthetic oscillatory network of transcriptional regulatory of *E. coli* cells.

References

1. Elowitz, M. B. , Leibler, S.: A Synthetic oscillatory network of transcriptional regulators. Nature, Vol.403, 20, Jan. (2000) 335-338
2. Wolkenhauer, O. et al: Systems Biology. IEEE Control Systems Magazine, Aug. (2002) 38-48
3. Moreau, Y. et al: Functional Bioinformatics of Microarray Data: From Expression to Regulatory. Proc of the IEEE, Vol.90, 11 (2002) 1722-1743
4. De Jong, H.: Modeling and Simulation of Genetic Regulatory Systems: A Literature Review. J. of Computational Biology, Vol.29, 1 (2002) 67-103
5. Gardner, T. S. et al: Construction of a Genetic Toggle Switch in Escherichia Coli. Nature, Vol.403, 20, Jan. (2000) 339-342
6. Chen, L. et al: Stability of Genetic Regulatory Networks with Time Delay. IEEE trans. on Circuits and Systems –I. Fundamental Theory and Applications, Vol.49, 5 (2002) 602-608
7. Chen, L. et al: A Model of Periodic Oscillation for Genetic Regulatory Systems. IEEE trans. on Circuits and Systems –I. Fundamental Theory and Applications, Vol.49, 10 (2002) 1429-1436
8. Ozbudak, E. M. et al: Multistability in the lactose utilization network of Escherichia Coli. Nature, Vol.427, 19, Feb. (2004) 737-740
9. Jaeger, H. and Haas, H.: Harnessing Nonlinearity: Predicting Chaotic Systems and Saving Energy in Wireless Communication. Science, Vol.304, 2 April, (2004) 78-80

A New Computational Model of Biological Vision for Stereopsis*

Baoquan Song, Zongtan Zhou, Dewen Hu, and Zhengzhi Wang

Department of Automatic Control, National University of Defense Technology, Changsha, Hunan, 410073, P.R.C.
ranruomail@163.com

Abstract. This paper designs a new adaptive disparity filter which implements the binocular matching of disparity features in stereo images, and constructs a computational model as well as a stereopsis system based on the disparity filter along with mechanism of biological vision by integrating it with Grossberg's FACADE theory, to process real-world images of 3-D scenes. By using this stereopsis system, depth perception of surfaces in real-world stereo images is simulated and realized.

Keywords. Neural Networks, Depth Perception, Surface Perception, FACADE Theory.

1 Introduction

Although Marr's computational system of stereo-vision, which processes 3-D (three-dimensional) scenes through 2.5D representation, was the dominate theory in the research field of computer vision for several decades (see [1]) and succeeded in many applications, it is not a universal framework for processing images of complex and diverse 3-D natural scenes. As we all know, the fact that human's biological vision system can percept and understand natural scenes without any embarrassment, inspires lots of researchers to develop computer vision system by simulating mechanisms of biological vision (see [2], [3], etc.), and Grossberg's FACADE (Form-And-Color-And-DEpth) theory may be one of most successful and integrated framework among these efforts and work.

In 1980s, Grossberg proposed the concepts of BCS (Boundary Contour System) and FCS (Feature Contour System) as integrated pre-attentional processing models of biological vision system. BCS generates boundary segmentation by imposing neural mechanisms of short-range competition and long-range cooperation, while FCS obtains surface perception through filling and diffusing features such as brightness, color, within these segmentation boundaries from BCS. To explore more about the in-depth perception mechanism of binocular

* Supported by the Distinguished Young Scholars Fund of China (60225015), Natural Science Foundation of China (60171003), Ministry of Science and Technology of China (2001CCA04100) and Ministry of Education of China (TRAPOYT Project).

F. Yin, J. Wang, and C. Guo (Eds.): ISNN 2004, LNCS 3174, pp. 525–530, 2004.

vision inside biological system, Grossberg and his colleagues introduced the idea about FACADE model which combined BCS and FCS firstly in 1994 (see [3]), proposed the integrated FACADE model and then established FACADE theory, an integrated neural network theory of 3-D biological vision in 1997 (see [4]). Recently in 2003, Grossberg proposed the 3D LAMINART model (see [5]) to cope with data about perceptual development, learning, grouping and attention. The 3D LAMINART model, as Grossberg claimed, refined the FACADE model, introduced some new features, and can be employed to explain a much larger set of neurophysiological, anatomical and psychophysical data about stereopsis as well as 3-D surface perception than FACADE model had previously been possible.

FACADE theory and 3D LAMINART model have established primary frameworks of binocular vision system upon biological foundations. However, both models are primarily aiming at constructing mathematic depictions of biological vision system as closely as possible. They are chiefly applied to explain some visual phenomena such as illusions through simulating process using only simple images which contain ordinary object such as regular geometries, but it is quiet difficult for such models to analyze and process images of the real-world scenes in which there are lots of confusions, fuzzy edges and complex objects. Moreover, FACADE theory and 3D LAMINART model are both too complex to be practical as computational model for real-world image processing.

The aim of this paper is to implement a practical computational model and to construct an image processing computer system to be able to process real-world images of 3-D scenes based on biological mechanism. Inspired by the success of FACADE theory for their uniqueness and superiority with neural mechanisms such as competition and cooperation, we designed an adaptive disparity filter, and integrated it with FACADE theory to achieve a stereopsis system to process complex images with feasible computational load. By using this stereopsis system, depth perception of surfaces in real-world stereo images is effectively and efficiently realized.

2 Model of Biological Vision for Stereopsis

Full description of FACADE theory can be found in [3] and [4], this paper won't repeatedly discuss all the details, but introduce a similar model presented for this paper, whose diagram is shown in Fig.1, and explain the primary mechanisms. As what can be seen in the diagram, monocular preprocessing (MP) of left eye and right eye inputs generates parallel signals to BCS and FCS via pathways 1 and 2 respectively, which is carried out by the ON and OFF channels of lateral geniculate nucleus (LGN). Pathways 1 model monocular inputs to the inter-blobs in cortical area V1, and monocular inputs activate model simple cells which complete the function of oriented contrast filter (OC Filter). Pathways 2 model monocular inputs to blobs in cortical area V1 and thin stripe in cortical area V2. Pathways 3 support binocular combination of outputs from monocular simple cells at complex cells, which generate populations of disparity-sensitive cells (namely, multiple BCS copies are formed, each corresponding to certain depth plane) and realize disparity filtering. The outputs of complex cells reach the Cooperative-Competitive Loop (CC Loop) via pathways 4. CC Loop is constitutive of hyper-complex cells and bi-pole cells, in which each BCS copy implements boundary grouping and optimizing within a given depth range, its outputs are transported into cortical area V4 via pathways 5. While the FCS signals reach the cortical area V4 via pathways 6 that control the one-to-many topographic registration of the monocular FCS signals. In V4, firstly these FCS signals from left and right eye are binocularly matched, then the surviving FCS signals that can be selectively captured by the BCS boundary signals from pathways 5 are used to initiate diffusive filling-in of a surface representation. Finally, visible surface perception is gained by the diffusive filling-in in the corresponding depth plane. The following text will only present and discuss those significant processes of this stereopsis model, especially the disparity filter brought forward in this paper.

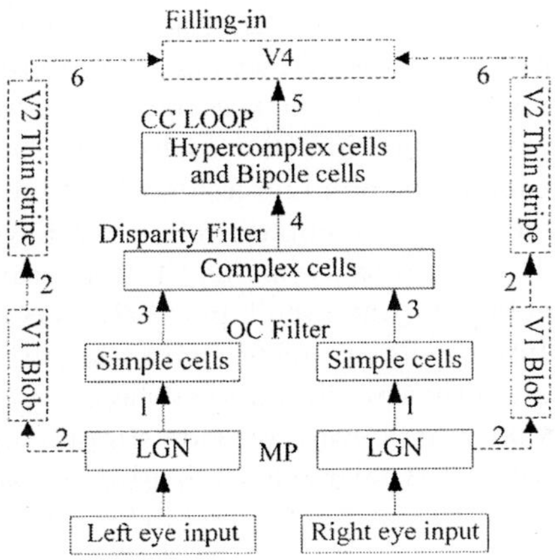

Fig. 1. Circuit diagram of the biological stereopsis model. Sold lines designate BCS, while dashed lines designate FCS

2.1 Disparity Filter

After MP has discounted the illuminant and OC filter has achieved boundary localization (see [4]), the complex cells in this stage combine the left and right monocular information for the first time, form multiple BCS copies to code different disparities and realize the function of disparity filter. Therefore, it is the key aspect of the stereopsis model. Considering the horizontal bias of eyes configuration in binocular vision system, we classify the binocular features received by complex cells into two categories: disparity features and non-disparity features, the latter only concern the horizontal contrast features and the former concern the others. To disparity features, the present model carries out the operation of disparity filtering to distribute them into the corresponding disparity-sensitive pools of cells (namely, the corresponding depth plan). On the contrary, to non-disparity features, this model directly transports them to the next stage.

In the FACADE model and the 3D LAMINART model , the membrane potential equations of complex cells that realize the function of disparity filter are the differential equations without analytical equilibrium solutions (see [4], [5]). So, a great number of iterative operations are needed to solve those differential equations. Ensuring inhibitory interactions are generated only when a cell is sufficiently active over its firing threshold, the two models introduce some half-wave rectifying operations with thresholds, which make their parameters less robust. However, an adaptive disparity filter designed by us overcomes the above shortcomings and achieves disparity tuning of complex cells rapidly and efficiently.

The adaptive disparity filter is composed of the binocular combination and the disparity competition. The binocular combination receives the outputs from the simple cells of left and right eye. When the complex cells receive the outputs with approximately the same magnitude of contrast from like-polarity simple cells of left and right eye, they register a high pattern match and are strongly activated, otherwise they register a less perfect match and aren't strongly activated. Then the following disparity competition suppresses false and weak binocular matches and implements the function that each disparity-sensitive pool of cells only codes the information corresponding its depth. Because, within the current implementation, the activities of complex cells that have been perfectly activated are approximately 2 times as strong as the activities of those that receive the common monocular inputs but aren't perfectly activated, in the binocular combination. Thus complex cells in each disparity-sensitive pool of cells can be inhibited by only those: (a) that are in different disparity-sensitive pools of cells; (b) that receive monocular inputs from common simple cells; (c) whose activities exceed half of their activities. Basing the above factors, we design the dynamics equation of disparity competition, the kernel of this adaptive disparity filter, as follows:

$$\frac{dJ_{ij\hat{k}d}}{dt} = -\alpha J_{ij\hat{k}d} + \left(U - J_{ij\hat{k}d}\right) F_{ij\hat{k}d} - \left(J_{ij\hat{k}d} + L\right) C \ . \tag{1}$$

where,

$$C = \sum_{e,p} g\left(F_{i+p,j,\hat{k},e}, \Gamma_{ij\hat{k}d}\right) G^{\mathrm{L}}_{pde} + \sum_{e,p} g\left(F_{i+p,j,\hat{k},e}, \Gamma_{ij\hat{k}d}\right) G^{\mathrm{R}}_{pde} \ . \tag{2}$$

$$\Gamma_{ij\hat{k}d} = 0.6 F_{ij\hat{k}d} \ . \tag{3}$$

$$g(x,y) = \begin{cases} x, & \text{for } x > y \\ 0, & \text{others} \end{cases} \ . \tag{4}$$

In (1), α is a constant decay rate, U/L bounds the upper or lower limit of cell activity, $\hat{k}$ designates the orientations corresponding to disparity features; $F_{ij\hat{k}d}$ is the total input to the complex cell centered on location (i, j), of orientation $\hat{k}$, and tuned to disparity d; $J_{ij\hat{k}d}$ is the output activity of complex cell. In (2),

G^{L}_{pde} and G^{R}_{pde} are the left and right inhibitory connections between complex cells that code different disparities. Equation (1) has an analytical equilibrium solution, (3) and (4) embody the adaptive threshold.

This adaptive disparity filter achieves the role of the disparity filter and generates the outputs in one step without iterative operations, which remarkably reduces computational load, and makes the present model's parameters more robust than other models'.

(a) Left image (b) Right image

Fig. 2. Real-world stereo image pair processed in the simulation of this paper

(a) Farther plane (b) Far plane (c) Near plane (d) Nearer plane

Fig. 3. The final outputs of the simulation of stereopsis model

2.2 Filling-in

The filling-in process in cortical area V4 is the last stage in this model. This stage receives not only the monocular FCS signals from left and right eye (via pathways 6) but also the binocular BCS signals from CC Loop (via pathways 5). Because only disparity features contain disparity information, non-disparity features don't, so in this model, only disparity features in binocular BCS are used to capture the outputs from the ON and OFF channels of LGN to select those FCS signals that are spatially coincident and orientationally aligned with binocular BCS boundaries in different depth planes. Then filling-in, which allows the brightness signals to diffuse spatially across the image except where gated by the presence of BCS signals (see [4]), is carried out in the corresponding depth plane using those captured FCS singles. In each depth plane, because there are two FCS signals (the outputs of ON channel and OFF channel) corresponding to one signal from BCS, there are two filling-in domains: ON filling-in domain and OFF filling-in domain. Ultimately, visible surface perception is generated by the competition between the outputs of ON filling-in domain and OFF filling-in domain.

3 Simulation and Realization

We have realized this stereopsis model in personal computer system and processed many stereo images of real-world scenes. Simulation and experimental results indicate that this stereopsis model is feasible and efficient. To demonstrate the capabilities of our realization in analyzing and processing 3-D real-world images, a sample everyday stereo image pair, as presented in Fig.2, is processed and exhibited here. To illustrate more expressly, only 4 separate disparity-sensitive pools of cells corresponding different depth planes are specified in this simulation.

Although there are fuzzy edges, shadows of the objects, and uneven illumination presented in the image pair shown in Fig.2, the processing result is satisfying. Figure 3 displays the final outputs of the stereopsis model, where the activities in the disparity-sensitive pools of cells reflect exactly the relative depth correlations of the objects in the stereo images, shadows of objects are suppressed, and figure-ground separation is also realized.

4 Discussion

The computational stereopsis model of this paper, as a realization of FACADE theory, circumvents the complex interactions between BCS and FCS to reduce model complexity and computational load. Moreover, the model is more efficient and parameter-robust by using the adaptive disparity filter proposed in this paper. Furthermore, it can process the stereo images of the real-world scenes rapidly and efficiently to realize 3-D surface perception, which is not achieved by other models that instantiate the FACADE theory.

However, original FACADE model simply divides the 3-D scene into a series of vertical planes, therefore the model can't realize 3-D perception of slanted or curved surfaces since these surfaces can't be filled out in any single depth plane. These situations will be further researched whether by designing a continuous disparity model or by constructing a filing-in mechanism across the disparity planes.

References

1. Marr, D.: Vision. 1st edn. W.H.Freeman, San Francisco (1982)
2. Harris, J.M., Parker, A.J.: Independent Neural Mechanisms for Bright and Dark Information in Binocular Stereopsis. Nature. **374** (1995) 808–811
3. Grossberg, S.: 3-D Vision and Figure-Ground Separation by Visual Cortex. Perception and Psychophysics. **55** (1994) 48–120
4. Grossberg, S., McLoughlin, N.P.: Cortical Dynamics of Three-Dimensional Surface Perception. Neural Networks. **10** (1997) 1583–1605
5. Grossberg, S., Howe, P.D.L.: A Laminar Cortical Model of Stereopsis and Three-Dimensional Surface Perception. Vision Research. **43** (2003) 801–829

A Reconstruction Approach to CT with Cauchy RBFs Network

Jianzhong Zhang and Haiyang Li

College of Computer and Information Engineering, Xiamen University
Xiamen, 361005, China
zjz98@sina.com

Abstract. A new reconstruction approach to computerized tomography(CT) with Cauchy Radial Basis Functions network is presented. The distribution of material parameters is represented by the weighting sum of Cauchy functions. The analytical formula of the line integral of Cauchy functions along any straight-line path is deduced, and the theoretical projection data along a bent ray is computed by optimization algorithm. The parameters in RBFs network are found by the learning rule based on the gradient decent method. The new reconstruction approach is suitable for the CT with a relatively small number of bent ray paths or projection dada, such as seismic tomography. Computer simulations show its good effects.

1 Introduction

Computerized Tomography (CT) is to visualize a spatial distribution of material parameters, such as attenuation rate or propagation velocity, from measured projection data. One of the key matters of CT is reconstruction algorithm. Filtered Back-Projection (FBP) of image reconstruction is in common use, but it is unfit for the case with a small number of projection dada. Seismic tomography is a method to determine the seismic wave propagating velocity in subsurface material from the seismic wave travel times measured in borehole or on ground, and has been used in geological and mineral prospecting applications. In seismic tomography, seismic travel times are given only; the goal is to infer the velocity, but the bent ray paths depend on the velocity distribution strongly. So the nonlinear inversion is required. One has proposed many iterative methods for traveltime inversion of seismic tomography, such as Algebraic Reconstruction Technique (ART), Least Squares Methods (LSM), and so on [1]. These methods divide the material into many small homogeneous pixels (or cells), and set up the discretized form of the equations between traveltimes at receivers and average velocities of all cells. Then the velocity value in each cell is reconstructed from traveltimes. In order to get a good reconstruction result, each pixel must be enough small, and be covered by a number of rays. When projection data are scarce it is uncertain whether each pixel covered by rays. If the pixels are large, the resolution of imaging must be lowered. Ichihashi et al (1993) proposed a neuro-fuzzy

F. Yin, J. Wang, and C. Guo (Eds.): ISNN 2004, LNCS 3174, pp. 531–536, 2004.

approach to CT for the reconstruction of cross section images from small number of projection data, assuming the straight ray-optic model of the propagation mechanism [2]. The Gaussian RBFs are adapted to the direct method of solution using iterative technique. The line integral formula of Gaussian RBFs along infinite straight-line can be obtained in a simple manner, but not for definite one. Consequently, the theoretical projection along a definite straight-line ray or a bent ray can not be computed accurately with this kind of RBFs, and the CT with bent ray, such as seismic traveltime tomography, can not be achieved. Gaussian and Cauchy functions have a same spatial distribution. The analytical formula of the line integral of Cauchy functions along any definite straight-line path can be deduced, so the Cauchy RBFs are used to express the propagating velocity distribution of material in this paper. A bent ray is approximated by a set of small straight segments. The theoretical projection along any bent ray path can be computed by summing the projection data along all segments. Moreover the proposed method with Cauchy RBFs in this paper can solve the CT problem with a relatively small number of bent ray paths.

2 Theory and Methodology

2.1 Cauchy RBFs Network

Cauchy distribution function in x-y plane can be expressed as

$$f(x,y)=\frac{1}{1+\alpha(x-a)^2+\beta(y-b)^2}, \tag{1}$$

where (a, b) is its central coordinate, and $-\infty<a$, $b<\infty$. α and β show the variation rate of the function along the radial direction, and $\alpha>0$ and $\beta>0$. The function is elliptical. If α equals β the function is circular.

Let the Cauchy function be kernel function of RBFs network. The final output of the network is defined as:

$$\mu(x,y)=\sum_{k=1}^{K}\omega_k f_k(x,y), \tag{2}$$

where K is number of neural cells in middle layer; ω_k is the kth weight, and $f_k(x,y)$ is kth Cauchy RBF.

2.2 Line Integral of Cauchy RBFs

Figure 1 shows the normalized region of imaging. 0 x 1, 0 y 1. The line EF represents straight path used by Ichihashi et al (1993), and SR represents bent path used in our method. The line from G (x_i, y_i) to H (x_{i+1}, y_{i+1}) can be written $y=y_i+\gamma_i(x-x_i)$, where $\gamma_i=(y_{i+1}-y_i)/(x_{i+1}-x_i)$. The line integral formula of the network output function $\mu(x,y)$ along straight line segment GH is obtained as:

$$I_i = \begin{cases} \sum_{k=1}^{K} \frac{\omega_k}{\sqrt{V_{i,k}}} \ln \frac{(2A_{i,k}x_{i+1} + B_{i,k} - \sqrt{V_{i,k}})(2A_{i,k}x_i + B_{i,k} + \sqrt{V_{i,k}})}{(2A_{i,k}x_{i+1} + B_{i,k} + \sqrt{V_{i,k}})(2A_{i,k}x_i + B_{i,k} - \sqrt{V_{i,k}})}, & if\ V_{i,k} > 0; \\ \sum_{k=1}^{K} \frac{4\omega_k A_{i,k}(x_i - x_{i+1})}{(2A_{i,k}x_i + B_{i,k})(2A_{i,k}x_{i+1} + B_{i,k})}, & if\ V_{i,k} = 0; \\ \sum_{k=1}^{K} \frac{2\omega_k}{\sqrt{-V_{i,k}}} \left(arctg \frac{2A_{i,k}x_{i+1} + B_{i,k}}{\sqrt{-V_{i,k}}} - arctg \frac{2A_{i,k}x_i + B_{i,k}}{\sqrt{-V_{i,k}}} \right), & if\ V_{i,k} < 0; \end{cases} \quad (3)$$

where

$$A_{i,k} = \alpha_k + \gamma_i^2 \beta_k , \quad (4)$$

$$B_{i,k} = 2[\gamma_i \beta_k (y_i - \gamma_i x_i - b_k) - \alpha_k a_k] , \quad (5)$$

$$C_{i,k} = \beta_k (y_i - \gamma_i x_i - b_k)^2 + \alpha_k a_k^2 + 1 , \quad (6)$$

$$V_{i,k} = B_{i,k}^2 - 4A_{i,k}C_{i,k} . \quad (7)$$

The projection data along broken line SR can be computed by summing the line integral value along each segment, that is

$$I = \sum_i I_i . \quad (8)$$

2.3 Computation of Bent Ray Paths and Projection Data

Seismic wave propagates in heterogeneous media along bent ray paths, which obey *Fermat's principle of least time*. We must trace the bent ray paths starting from the sources to the receivers for seismic tomography. Assuming the velocity distribution of media can be expressed as equation (2), and the parameters K, ω_k, α_k, β_k, a_k, and b_k are known, now we determine the bent ray path for each pair of sources and receivers. According to *Fermet's principle* we seek the least time path between the two points using trial and error. First we use many small straight segments to approximate the curve paths between S and R, as shown in Figure 1, and give the starting coordinates of the intersection points of the segments. The traveltime along each segment can be computed with equation (3), so the general traveltime from S to R can be obtained by simply summing. Then coordinates of the intersection points are corrected iteratively by optimization algorithm. Once the bent ray paths are determined the corresponding traveltimes (theoretical projection data) at receivers can easily obtained by equation (8).

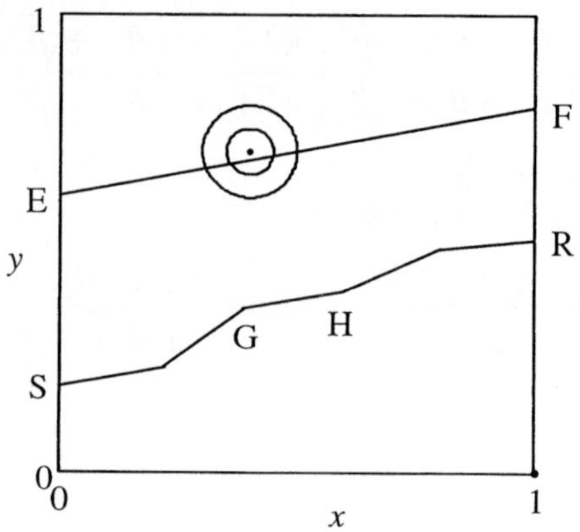

Fig. 1. A region of imaging and ray paths

2.4 Tomographic Inversion with Neural Network

If the information in very small number of projection data is not sufficient to reconstruct the velocity distribution the tomographic inversion is an ill-posed problem. It has an infinite number of solutions. The reguralization is a technique to transform an ill-posed problem into a well-posed one. Tabuchi et al (1995) consider the reconstruction method with regularization conditions from very small number of projection data on the assumption that the spatial distribution of media velocity is relatively smooth [3]. We let the cost function with regularizations be as

$$E=\frac{1}{2}\sum_{p=1}^{P}\left(I_p-I_{0p}\right)^2+\frac{1}{2}\lambda\sum_{(x,y)\in W}\left\{\left[\frac{\partial^2\mu(x,y)}{\partial x^2}\right]^2+\left[\frac{\partial^2\mu(x,y)}{\partial y^2}\right]^2\right\}, \tag{9}$$

where P is the number of rays; I_{0p} and I_p are measured projection data and computational projection data, respectively; W is the set of data points chosen in the region; λ is a positive weighting constant for regularization condition.

We apply the learning algorithm based on the steepest descent method to find the required Cauchy kernel function parameters α_k, β_k, a_k, and b_k, and weighting coefficient, ω_k in the RBFs network.

3 Computer Simulations

We demonstrate the theoretical modeling example of cross-borehole seismic tomography. The sources and receivers are located regularly on left and right edges of probing region, respectively, because of the limitation of probing direction. 36 propagation paths are chosen from 6 equally spaced transmitter and receiver locations. We have simulated three theoretic models, respectively shown in Fig.2a~Fig.3a. The velocity of background media is 1.2km/s, and that of local anomalous velocity objects is 4.8km/s. 16 Cauchy functions are distributed evenly in the probing region. The initial values of parameters α_k, β_k and ω_k all are 0.01. The corresponding reconstruction results are shown in Fig.2b~Fig.4b. We can see the proposed method have a good performance. If we discretize the region of imaging according to the size of a single anomalous velocity object in the theoretic model, there will be 100 pixels, while the number of projection data is only 36, so it's a serious ill-posed inverse problem. It's hard to reconstruct the velocity distribution with usual discrete algorithms, such as ART, LSM, and so on.

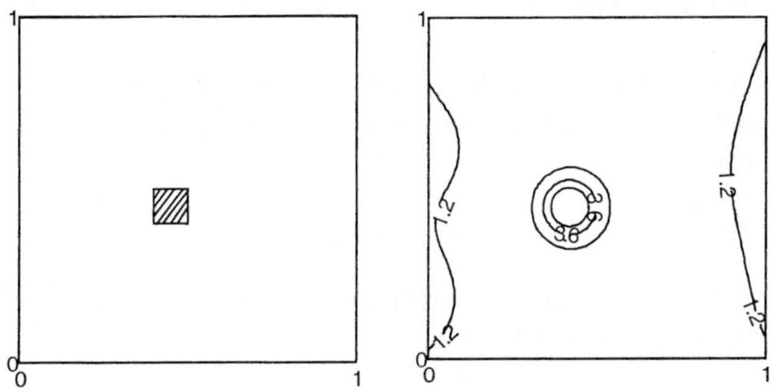

Fig. 2. Theoretic model I (left) and reconstructed result (right)

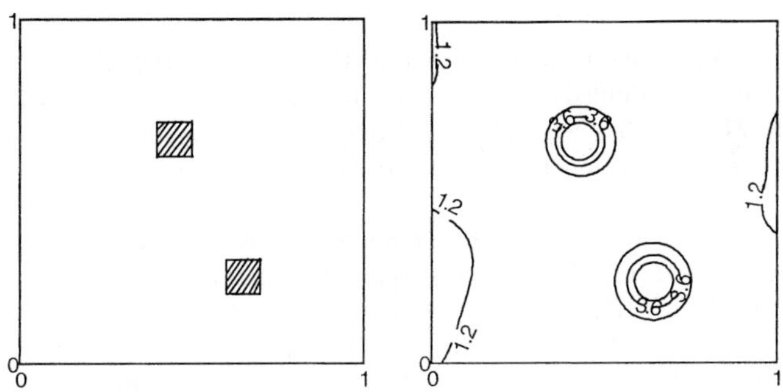

Fig. 3. Theoretic model I (left) and reconstructed result (right)

4 Conclusion

In the article we presented a reconstruction approach to CT with Cauchy RBFs Network. The formula to compute the projection data is deduced for the spatial distribu-

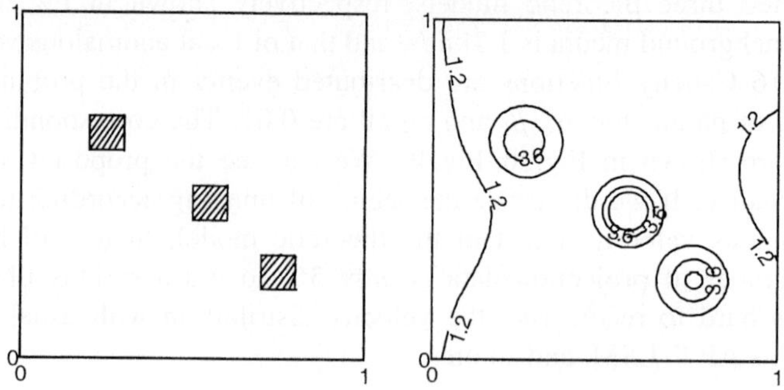

Fig. 4. Theoretic model I (left) and reconstructed result (right)

tion of material parameters as Cauchy RBFs. The bent ray paths are traced. The proposed method can obtain an effective reconstruction result with a relatively small number of bent ray paths.

Acknowledgement. This work is supported by Nature Science Foundation (No. D0310001) of Fujian province, China.

References

1. Herman, G. T., Langenberg, K. G., Sabatier, P. C.: Basic Methods of Tomography and Inverse Problems. IOP Publishing Limited and Individual Contributor (1987)
2. Ichihashi, H., Miyoshi, T., Nagasaka, K.: Computed Tomography by Neuro-Fuzzy Inversion. Proc.of 1933 International Joint Conference on Neural Networks, Vol. 1, (1993) 709-712
3. Tabuchi, H., Miyoshi, T., Ichihashi, H., Ohno, K.: Computerized Tomography with Radial Basis Functions Network: A Neuro-fuzzy Approach. IEEE Int. Conf. Neural Network Conf. Proc. Vol. 5 (1995) 2258-2263

Fault Diagnosis on Satellite Attitude Control with Dynamic Neural Network

HanYong Hao, ZengQi Sun, and Yu Zhang

State Key Lab of Intelligent Technology and Systems, Dept of Computer Science & Technology, Tsinghua University Beijing, 100084, P.R.China
{hhy01,szq-dcs,zhang-y-03}@mails.tsinghua.edu.cn

Abstract. With increasing demands on higher performance, more safety and reliability of dynamic systems, especially on safety-critical systems, fault diagnosis became a research interests in recent years. In this paper, a systematic approach to design fault diagnosis and accommodation with compound approach such as applying improved robust observer to fault diagnosis on linearized system aided by dynamic neural network, which is trained to bridge the gap between simulated system and real system on nonlinear attributes and modeling errors. Using an instance of fault diagnosis on attitude control system of satellite attitude, advantages of new scheme are tested to be effective.

1 Introduction

With modern control systems being more complicated, researches on how to make control systems cost effective and reliable are not only important to safety critical systems such as satellites, chemical plants, but also systems applied on vehicles and home electrical appliances which are close to our daily lives. Fault diagnosis and fault tolerance control approaches, which provide solutions for these demands, have been received more attention of many researchers [1,2].

Fault diagnosis can be classified as linear and nonlinear by characteristics of systems [2]. Research on fault diagnosis of linear system had made great improvements. Although there are some achievements on fault diagnosis, it is still hard to find a desirable general approach on nonlinear fault diagnosis [1]. [3] presents a fault diagnosis of satellite based on model-free approaches such as possibility theory and fuzzy set. Combination of neural network and dead-beat observer in [4] improves fault detection. Systems with weak nonlinear attributes can be linearized on several key points, a set of segmented linear systems can be used to approximate whole nonlinear system in wide operation region [5]. Systems with hard nonlinear attributes cannot be approximated by a set of linearized systems with desired precision due to modeling error. Both model-based and model-free approaches have their pros and cons, their merits can be applied and drawbacks be discarded. Robust observer approaches can be applied to tolerate modeling error and model-free approaches such as dynamic neural network can be applied to absorb hard-modeling factors. With inspiration of taking advantages of both model-based and model-free approaches, a project of fault diagnosis on attitude control of satellite is proposed.

F. Yin, J. Wang, and C. Guo (Eds.): ISNN 2004, LNCS 3174, pp. 537–542, 2004.

2 Scheme of Fault Diagnosis

A linearized model, which can model most dominant characteristics of nonlinear system, and a dynamic neural network, which is applied to compensate residual error between linearized model and real system, are applied together to simulate real system. Compound approach in Fig.1 is aimed to take advantages of linear fault diagnosis on nonlinear systems. Simulated model is in dotted zone. Linear fault diagnosis approach can be applied on dominant linear revised model. Common fault diagnosis schemes use different kinds of observers to generate diagnostic residual signals. After residual generation, different methods of residual analysis can be done.

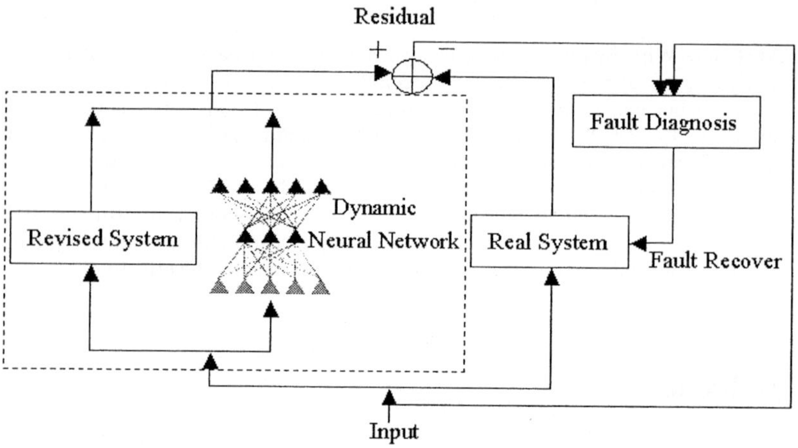

Fig. 1. Scheme of compound fault diagnosis approach

Linearized model can be described as (1) and real nonlinear model as (2). A dynamic neural network constructed by dynamic neurons partially interconnected to a function of their own output is trained off-line to bridge the gap between proximate linearized model and real system by data pairs of inputs and respective errors between dominant linearized model and real system. The general equation of a dynamic neural network can be described as (3). x is state variable. u is external input. θ is a vector of parameters. Function f and g are vector fields that define dynamics and output mapping of the designed dynamic neural network respectively [6]. Dynamic neural network, formed by many neurons, can be described as (4). $\beta_i, \omega_{i,j}, \gamma_{i,j}$ are adjustable weights. x_i is activation state of unit i. $\sigma(x_j)$ is activation function, $C_n = [I_{p\times n} 0_{p\times(n-p)}]^T$.

$$\dot{x}(t) = Ax(t) + Bu(t), \qquad y(t) = Cx(t) \tag{1}$$

$$\dot{x}(t) = f(x(t), u(t)), \qquad y(t) = g(x(t)) \tag{2}$$

$$\dot{x}(t) = f(x(t), u(t), \theta), \qquad y(t) = g(x(t), \theta) \tag{3}$$

$$\dot{x}_i = -\beta_i x_i + \sum_{j=1}^{n} \omega_{ij}\sigma(x_j) + \sum_{j=1}^{p} \gamma_{ij}u_j \text{,} \ y_n(t) = C_n x(t) \text{,} \ \sigma(x_j) = \tanh(x_j) \tag{4}$$

The form of dynamic neural network is very similar to affine system with same input and output states variables. Proved in [6], these similarities do ensure nice and stable approximation. After training dynamic neural network by error between real system and linear model with various random inputs and corresponding errors from different initial conditions, optimizing on structure of network, real system can be simulated by a well-trained dynamic neural network and dominant linear model. With *prior* knowledge on faults, most parts of faults can be modeled. This provides freedom as well as difficulty for design. Balance on complexity of linear model and performance of fault diagnosis is also an optimization issue. Define differences between real system and linear model as (5) with x_{real} and x_{lin}, $\dot{x}_{real}$ and $\dot{x}_{lin}$ being respective state and derivative of state of real and linear system. A performance index of approximation is defined as (6), N is the number of evaluation data.

$$e = x_{real} - x_{lin} \text{,} \ \dot{e} = \dot{x}_{real} - \dot{x}_{lin} \tag{5}$$

$$I = \lim_{N \to \infty} \frac{1}{N} \sqrt{\sum_{i=1}^{N} \left(\frac{x_{real} - x_{lin}}{x_{real}}\right)^2} = \frac{\left\|x_{real} - x_{lin}\right\|}{\left\|x_{real}\right\|} \tag{6}$$

It is not hard to choose a linear model, which can approximate system with $I \le 0.1$. The following theorem proves that compound approach with linear model and a dynamic neural network simulates the real system with higher precision.

Theorem 1: A linear model (1) and a dynamic neural network (3) simulate nonlinear system (2) with higher precision as (7). $\overline{(x_{real} - x_{lin})}$ *is simulated output of dynamic neural network.* Δ *is efficiency of approximation of dynamic neural network.*

$$\lim_{t \to \infty} \frac{\left\|x_{lin}\right\| + \left\|\overline{(x_{real} - x_{lin})}\right\|}{\left\|x_{real}\right\|} = (1 - I + I * \Delta) \tag{7}$$

Proof: Define modeling error between linear model and real system as (8), error between real system and compound system as (9). A dynamic neural network is trained to simulate the error between system and linear model. A theorem in [6] gives a boundary of simulation error between nonlinear model and dynamic neural network as (10). Q_0 is strictly positive definite matrix. ε^0 is bounded positive. Define efficiency of approximation as (12). $e(t)$ is approximate error of neural network. Compound system can simulate nonlinear system described as (2) with precision as (13). So, (7) is proven.

$$\ell = x_{real} - x_{lin} \text{,} \ \left\|\ell\right\| = \left\|x_{real} - x_{lin}\right\| = I * \left\|x_{real}\right\| \tag{8}$$

$$\partial = x_{real} - x_{lin} - x_{dnn} \tag{9}$$

$$\limsup_{t\to\infty}\|e\| \le \sqrt{\frac{\eta}{\lambda_{\min}(P)}} = \Omega \tag{10}$$

$$\eta = \max(V(e(0)), \frac{\varepsilon^0}{\lambda_{\min}(R_p)}),\ R_p = P^{-1/2}Q_0\ P^{-1/2},\ V(e(t)) = e(t)^T Pe(t) \tag{11}$$

$$\limsup_{t\to\infty}\left\|\frac{e}{x}\right\| \le \frac{\Omega}{\sup\|x\|} = \Delta \tag{12}$$

$$\lim_{t\to\infty}\frac{\|x_{lin}\| + \|(x_{real} - x_{lin})\|}{\|x_{real}\|} = \lim_{t\to\infty}(1 - I + \frac{\|(x_{real} - x_{lin})\|}{\|x_{real}\|}) = (1 - I + I * \Delta) \tag{13}$$

Compound approach can simulate nonlinear system within a higher precision as (7). Higher precision means better performance of fault diagnosis.

3 Project Description

Satellite is highly autonomous spacecraft. Once it is at fault on orbit, it must make self-diagnosis and reconfiguration to regain control and stability to avoid the loss. Both hardware and analytical redundancies can help satellite to prolong life with fault. This paper describes a project of satellite self-diagnosis and reconfiguration of attitude control. The approach of attitude stabilization is zero-momentum using 3 plus 1 momentum wheels to absorb disturbances by exchange momentum with satellite on X, Y, Z axes. The whole scheme can be shown as Fig.2. A controller is applied to change speeds of wheels to absorb disturbances to maintain precious attitude.

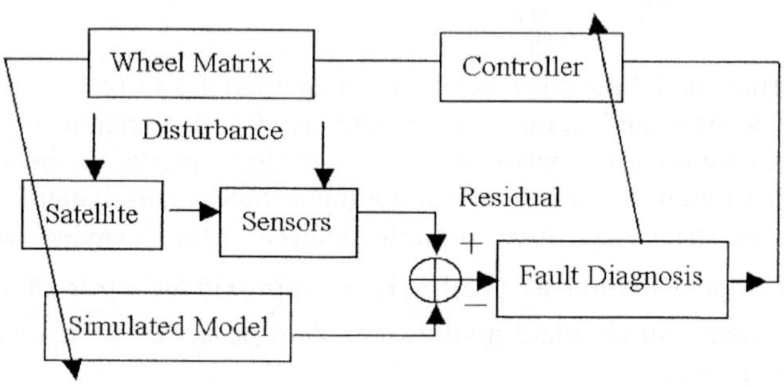

Fig. 2. Sketch of attitude control system

When one of three wheels fail and backup wheel starts up, the control law of attitude system must be reconfigured. The angular momentum equation of satellite can be described as (14). $\tilde{\omega}$ is skew symmetric matrix of ω, $\omega = [\omega_x \quad \omega_y \quad \omega_z]^T$. By leaving high order items and coupling effects between attitude and orbit out, dynamic equation of attitude can be simplified as (15). $x(s)$, $y(s)$, $z(s)$ are respective attitude angles of X,Y,Z axes. $\dot{h}_x(s)$, $\dot{h}_y(s)$, $\dot{h}_z(s)$ are control momentums on respective axis. $T_x(s)$, $T_y(s)$, $T_z(s)$ are disturbance moments on respective axis. Being simulated by a linear part and a dynamic neural network, linear fault diagnosis such as robust observer is applied in this paper. In design of robust observer, an improvement on precision is applied in [5]. After fault detection by observers, corresponding remediation can be applied to maintain health of system to the utmost extent, such as shutdowns fault components, starts backup, changes control law for reconfigured system. Linear Quadratic Regulator (LQR) is applied to realize optimal solution with constrain on energy of control signal and tracking error. This control law saves energy and avoids surges of control signal while minimize the tracking error. By applying LQR, state feedback law, $u = -Kx$, minimizes quadratic cost function (16) with weighted matrices Q,R,N.

$$\dot{H} = I\dot{\omega} + CI_s\dot{\Omega} + \tilde{\omega}H = 0 \tag{14}$$

$$s^2 \times I_s \times \begin{bmatrix} x(s) \\ y(s) \\ z(s) \end{bmatrix} = -\begin{bmatrix} \dot{h}_x(s) \\ \dot{h}_y(s) \\ \dot{h}_z(s) \end{bmatrix} + \begin{bmatrix} T_x(s) \\ T_y(s) \\ T_z(s) \end{bmatrix} \tag{15}$$

$$J(u) = \int_0^\infty (x^T Q x + u^T R u + 2x^T N u)dt \tag{16}$$

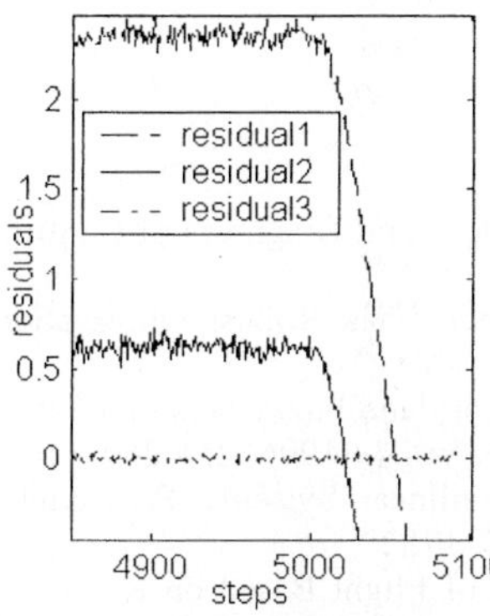

Fig. 3. Residuals of compound approach

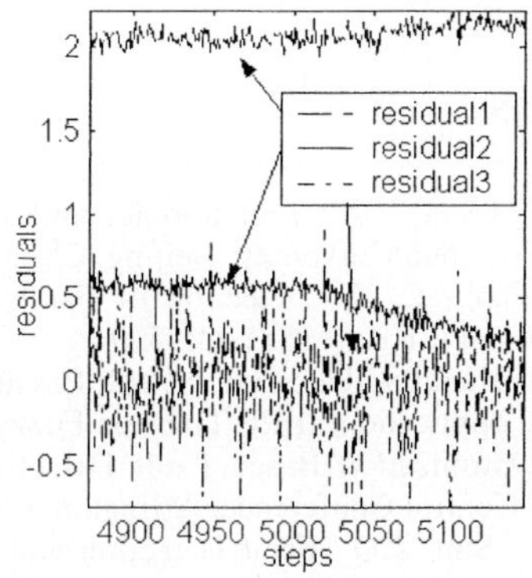

Fig. 4. Residuals of linear approach

4 Simulation of an Instance

In this instance, a certain type of satellite is adopted. Parameters in I_s are I_x=80kg, I_y=200kg, I_z=210kg, I_{xy}=8kg, I_{yz}=15kg, I_{xz}=16kg. Models of reaction wheels and sensors are defined with dead zones and saturations. In simulation, system is sampled every 0.05 second. A fault on motor of 3rd wheel, named fault3, defined as drifts 2% from normal point of reaction wheel with white noise of 1% occurred at 5000th step. Fig.3 and Fig.4 show results of improved approach and direct linearization approach respectively while fault3 occurred. In Fig.3, thresholds can be easily defined and response of fault detection is faster than case in Fig.4. After simulations at different working points, statistics conform that compound approach can keep fault alarm below 0.1% with desirable efficiency contrast to direct one about 1%, which uses about triple detection time. Aided by dynamic neural network, this compound strategy of fault diagnosis makes nice performance.

5 Conclusion

In this paper, a new scheme of fault detection and accommodation on attitude of satellite, with a compound approach based on improved robust observers and a well-trained dynamic neural network has been proposed, which can make better performance as well as make design easier, can be applied to nonlinear model for general cases. This new scheme also can be extended to relative application fields.

Acknowledgements. This work was jointly supported by the National Key Project for Basic Research of China (Grant No: G2003cb312205).

References

1. Zhou, D.H., Sun, Y.X.: Technologies on Fault Detection and Diagnosis of Control System. Press of Tsinghua University, Beijing, China (1994)
2. Ge, W., Fang, C. Z.: Detection of Faulty Components via Robust Observation. Int. J. Control, Vol.47. UK, (1988) 581-599
3. Didier, C.: Handling Uncertainty with Possibility Theory and Fuzzy Sets in a Satellite Fault Diagnosis Application. IEEE Trans on Fuzzy Systems, Vol.4, (1996) 251-269
4. Tor, F.: Optimization Based Fault Detection for Nonlinear Systems. Proceedings of the American Control Conference, Arlington, (2001) 1747-1752
5. Hao, H.Y., Sun, Z.Q.: Fault Detection and Isolation of Flight Based on Robust Observer. Proceedings of WCICA'04, Hangzhou, China, To be printed (2004)
6. Freddy, G., Victor, M.B.: Strategies for Feedback Linearisation: A Dynamic Neural Network Approach. Springer, London (2003)

A New Strategy for Fault Detection of Nonlinear Systems Based on Neural Networks

Linglai Li and Donghua Zhou

Department of Automation, Tsinghua University, Beijing 100084, China
lilinglai01@mails.tsinghua.edu.cn, zdh@mail.tsinghua.edu.cn

Abstract. Parity space is a well-known model-based scheme for fault detection and diagnosis. However, the construction of parity vector is strictly based on the formulation of linear systems, and can hardly be extended to nonlinear cases. From the view of analytical redundancy and the nature of parity space, we propose a new parity check scheme for nonlinear deterministic systems, which can be realized by neural networks and the condition that the parity relation exists are also given theoretically. Simulation studies on the model of the three tank system DTS200 demonstrate the effectiveness of the new strategy, which can detect faults fast and accurately.

1 Introduction

Fault detection and diagnosis (FDD) has received more and more attention due to the increasing demand for higher performance as well as for more safety and higher reliability of dynamic systems, in which an active research area is model-based FDD, and a classical method is parity space proposed by Chow and Willsky [1], which has received much attention.

However, the construction of parity vectors is strictly based on the formulation of linear systems, and can hardly be extended to nonlinear cases. Yu et al. [2] and Gertler et al. [3] extended the parity space to bilinear systems and a special type of nonlinear systems which can be transformed into an input-output form. Note that these two types of nonlinear systems are very special which are very similar to linear cases, so parity vectors can be constructed under the similar way. Staroswiecki et al. [4] investigated the parity check scheme for another special type of nonlinear system, namely polynomial systems. Krishnaswami et al. [5] proposed a nonlinear parity equation for FDD based on NARMAX models, but the design method is a bit complex and lacks theoretical analysis.

Inspired by the work of Moraal et al. [6] and Guo et al. [7], and based on analytical redundancy: the nature of parity space, a new strategy for FDD of nonlinear systems is proposed here which can be realized by neural networks due to its universal approximation capability [8]. On the basis of the observability of nonlinear systems, the condition for the existence of parity relation is also given theoretically. Simulation results on the three tank system DTS200 with different types of faults illustrate the effectiveness of our new FDD strategy.

F. Yin, J. Wang, and C. Guo (Eds.): ISNN 2004, LNCS 3174, pp. 543–548, 2004.

2 A New FDD Strategy Based on NN

2.1 Brief Review of Parity Space

Consider a linear discrete system as follows,

$$\begin{cases} x_{k+1} = Ax_k + Bu_k \\ y_k = Cx_k + Du_k \end{cases} \tag{1}$$

where $x \in \mathbb{R}^n$, $u \in \mathbb{R}^m$ and $y \in \mathbb{R}^p$. In the sampling period $[k, k+s]$, we have

$$\begin{bmatrix} y_k \\ y_{k+1} \\ \vdots \\ y_{k+s} \end{bmatrix} = \begin{bmatrix} C \\ CA \\ \vdots \\ CA^s \end{bmatrix} x_k + \begin{bmatrix} 0 & 0 & \cdots & 0 \\ CB & 0 & \cdots & 0 \\ \vdots & \ddots & \ddots & \vdots \\ CA^{s-1}B & \cdots & CB & 0 \end{bmatrix} \begin{bmatrix} u_k \\ u_{k+1} \\ \vdots \\ u_{k+s} \end{bmatrix} \tag{2}$$

Define

$Y_{k:k+s} = [y'_k \quad y'_{k+1} \quad \cdots \quad y'_{k+s}]'$, $U_{k:k+s} = [u'_k \quad u'_{k+1} \quad \cdots \quad u'_{k+s}]'$,

$$H_s^x = \begin{bmatrix} C \\ CA \\ \vdots \\ CA^s \end{bmatrix}, \text{ and } H_s^u = \begin{bmatrix} D & 0 & \cdots & 0 \\ CB & D & \cdots & 0 \\ \vdots & \ddots & \ddots & \vdots \\ CA^{s-1}B & \cdots & CB & D \end{bmatrix}$$

Under certain conditions [1], the left null space of H_s^x is not null, which is called parity space. Then we can find a matrix Ω such that $\Omega H_s^x = 0$ i.e. the rows of Ω which are called parity vectors span the parity space. Then the residual is constructed as follows [1],

$$r_k = \Omega(Y_{k:k+s} - H_s^u U_{k:k+s}) = \Omega H_s^x x_k \tag{3}$$

It is obvious that r_k is non-zero only if a fault occurs in the system. The existence of Ω denotes the parity relations in (2).

2.2 Extension to Nonlinear Case Based on Neural Networks

Consider a nonlinear deterministic discrete systems as follow,

$$\begin{cases} x_{k+1} = f(x_k, u_k) \\ y_k = h(x_k, u_k) \end{cases} \tag{4}$$

Let $f^u(x) := f(x,u)$ and $h^u(x) = h(x,u)$, then as is done in (2) we have,

$$\begin{bmatrix} y_k \\ y_{k+1} \\ \vdots \\ y_{k+s} \end{bmatrix} = \begin{bmatrix} h^{u_k}(x_k) \\ h^{u_{k+1}} \circ f^{u_k}(x_k) \\ \vdots \\ h^{u_{k+s}} \circ f^{u_{k+s-1}} \circ \cdots \circ f^{u_{k+1}}(x_k) \end{bmatrix} \text{ i.e. } Y_{k:k+s} = \mathcal{H}_s(x_k, U_{k:k+s}) \tag{5}$$

where "$\circ$" denotes the composition of functions. Different from the linear case, the unknown variables x_k can't be eliminated simply by orthogonality technique as is done in (3). So the parity space method proposed by [1] can't be extended to nonlinear systems intuitively.

However, it can be seen that the nature of parity space is analytical redundancy, the redundancy of the set of linear equations of (2). Without loss of generality it's assumed that the first n rows are linear independent. Then it is to say that other rows of H_x^s can be described as linear combinations of the first n rows and it's obvious that parity vectors in Ω could be obtained directly from the coefficients of linear combinations, which means that the residual in (3) is actually constructed from the redundancy of the set of linear equations of (2). Therefore it is intuitively to guess that we can also obtain the residual from the redundancy of nonlinear set of equations of (5). First a definition is given below.

Definition 1 (N-Observability). [6] The nonlinear system (4) is said to be N-observable at a point $\bar{x} \in \mathbb{R}^n$ ($N \geq 1$) if there exists $U_{k:k+s}$ such that $\bar{x}$ is the unique solution of (5) with $s = N-1$. The system is uniformly N-observable if the mapping $\mathcal{H}_{N-1}^* : \mathbb{R}^n \times (\mathbb{R}^m)^{N-1} \to (\mathbb{R}^p)^N \times (\mathbb{R}^m)^{N-1}$ by $(x_k, U_{k:k+N-1}) \to (\mathcal{H}_{N-1}(x_k, U_{k:k+N-1}), U_{k:k+N-1})$ is injective; it is locally uniformly N-observable with respect to $\mathcal{O} \in \mathbb{R}^n$ and $\mathcal{U} \in (\mathbb{R}^m)^{N-1}$ if $\mathcal{H}_s^*$ restricted to $\mathcal{O} \times \mathcal{U}$ is injective.

When the system (4) is (locally) uniformly N-observable, the equation (5) with $s = N-1$ can be uniquely solved for x_k i.e. $x_k = \Psi_{N-1}(Y_{k:k+N-1}, U_{k:k+N-1})$ where Ψ_{N-1} is the first n rows of the inversion of $\mathcal{H}_{N-1}^*$. Moraal et al. [6] solved x_k by Newton-Broyden's method, based on which an observer was also constructed. Guo et al. [7] proposed to use neural networks to approximate the inversion Ψ_{N-1} to get a rough estimation of state variables which are used as the initial states of EKF.

Then if we construct (5) for a N-observable system with $s \geq N$, there exist redundancy equations in (5), i.e. a definite mapping Φ_N^s, which denotes the parity relation in the set of equations in (5), exists such that

$$Y_{k+N:k+s} = \Phi_N^s(Y_{k:k+N-1}, U_{k:k+s}) \tag{6}$$

Note that (6) is always satisfied if there is no fault. i.e. the residual is simply constructed as $r_k = Y_{k+N+1:k+s} - \Phi_N^s(Y_{k:k+N}, U_{k:k+s-1})$.

2.3 Discussions

The parity relation and residual for fault detection has been constructed theoretically in the previous sub-section. However, for general cases, the mapping Φ_N^s can hardly be obtained explicitly. So it is proposed here to use neural networks as an approximation of Φ_N^s to get the parity vector. As indicated by Hornik et al. [8] the neural networks can approximate any continuous real function with arbitrary accuracy in a compact set.

The question is why we don't use neural networks as in [7] to get the estimation of states x directly? As is well-known, one of the main difficulties in model-based FDD is due to the presence of unknown and unmeasured states x. Two approaches can be used to deal with them: estimation and elimination. The estimation of x is usually known as observer or filter, which is also a difficult problem in nonlinear cases. The elimination of x, such as parity space, directly explores the analytical redundancy in the mathematical model. Note that our purpose is to detect if a fault occurs in the system, not to estimate the states! On the other hand there exists unavoidable error between the NN and the real function, though NN has universal approximation capability theoretically. Although there will also be errors in our new strategy, the threshold for fault detection can be decided referring to the maximum training error. But if the states x are not accurate, the property of residual which is transformed from x can hardly be analyzed.

Another problem is the training of NN. Theoretically NN should be trained all over the area $\mathcal{O}\times\mathcal{U}$ in which the nonlinear system (4) is N-observable, but it's usually a hard task, especially when N and s is large. Fortunately, in engineering we only need to train NN with the data under normal conditions. Bias from normal conditions can also be regarded as a fault, though there may not be any real faults in the plant, actuators or sensors.

3 Simulation Study

The simulation model is the DTS200 three-tank system, which is a benchmark problem in chemical engineering [9]. The dynamics are described as follows,

$$\begin{aligned} A\cdot dh_1/dt &= -Q_{13}+Q_1 \\ A\cdot dh_3/dt &= Q_{13}-Q_{32} \\ A\cdot dh_2/dt &= Q_{32}-Q_{20}+Q_2 \end{aligned} \quad \text{where} \quad \begin{aligned} Q_{13} &= az_1S_n sign(h_1-h_3)\sqrt{2g|h_1-h_3|} \\ Q_{32} &= az_3S_n sign(h_3-h_2)\sqrt{2g|h_3-h_2|} \\ Q_{20} &= az_2S_n\sqrt{2gh_2} \end{aligned} \tag{7}$$

The state vector is the level of three tanks: $x=[h_1,h_3,h_2]^T$; the input vector is the controlled pump flow $u=[Q_1,Q_2]^T$; and the output vector is $y=[h_1,h_2]^T$, i.e. the measurement equation is $y=Cx$, where $C=\begin{bmatrix}1&0&0\\0&0&1\end{bmatrix}$. The actual parameters are

$h_{max} = 62 \pm 1\text{cm}$, $Q_{1max} = Q_{2max} = 100\text{ml/s}$, $az_1^0 = 0.5$, $az_3^0 = 0.45$, $az_2^0 = 0.6$, $g = 9.81\text{m/s}^2$, $A = 0.0154\text{m}^2$, $S_n = 5\times10^{-5}\text{m}^2$. The levels of tank 1 and tank 2 are both controlled by PI controllers respectively and the parameters of the controllers are both $K_p = 0.001$ (gain constant) and $T_I = 5000\text{s}$ (integral time constant) in our simulations. By Euler discretization with sampling time $T = 0.1\text{s}$, the nonlinear discrete model (4) is obtained.

DTS200 is locally uniformly 2-observable, so we have $s = 2$. Without u in the measurement equation, the input of Φ_N^s in (6) is simplified to $(Y_{k:k+N-1}, U_{k:k+s-1})$, so there's eight inputs and two outputs in our example now. We choose three layer multilayer perceptron as our neural network in simulations with 16 perceptrons in the hidden layer. The normal setpoints of the levels of tanks 1 and 2 are $h_1^0 = 0.5\text{m}$ and $h_2^0 = 0.3\text{m}$, respectively. The training data are obtained from step response with change $\pm 0.05\text{m}$ in setpoints, and trained by the NN toolbox of MATLAB 6.5. In the following simulations with a fault at 100s, the setpoint of the level of tank 1 is $h_1^0 = 0.48\text{m}$ at the beginning and step to $h_1^0 = 0.46\text{m}$ at 50s and the setpoint of the level of tank 2 is $h_2^0 = 0.31\text{m}$ for all the simulation times, which are all different from the training data to test the generality of the NN. Consider three types of faults in the system: plant fault, actuator fault and sensor fault, respectively, and the simulation results are illustrated in Fig. 1.

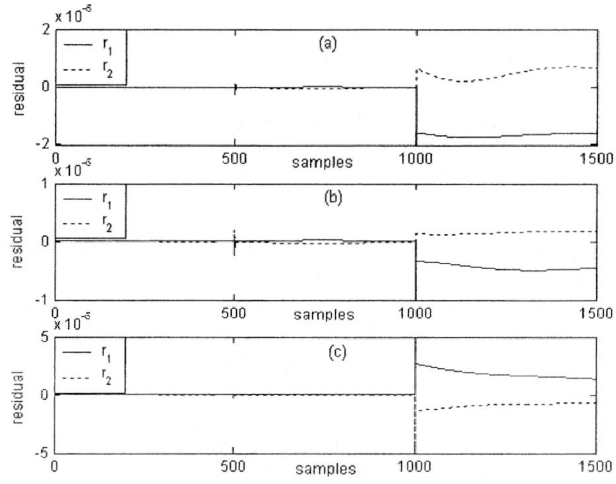

Fig. 1. The residual in simulations. (a) Plant fault: abrupt leakage in tank 1 with $Q_{leak}^1 = az_1 \cdot \pi r^2 \sqrt{2gh_1}$, where $r = 3\text{mm}$. (b) Actuator fault: the gain of pump1 step to 70% of normal value. (c) Sensor fault: the gain of sensor 2 step to 80% of normal value.

Fig. 1 shows that different types of faults in the system can be detected quickly by our new parity check strategy based on NN. Note that at 50s there is a small jump in the residual respect to the change of set-point of h_1, but compared to the jump of PV respect to the fault it's negligible. Then we can conclude that the training of NN is fairly good with some degree of generality, i.e. the residual is robust to the working condition and the change of set-point, but still sensitive to faults.

4 Conclusions

By exploring the nature of parity space, a new parity check strategy for nonlinear deterministic system is proposed in this paper. In literature, parity space method was proposed strictly based on the description of linear systems, which is very difficult to be extended to nonlinear cases. However, the existence of parity relation is due to the redundancy of the set of equations, from which we similarly obtain the parity check strategy for nonlinear systems. The proposed strategy is realized by neural networks. Simulations on a benchmark problem of chemical engineering show the effectiveness of the new strategy. The future work is to enhance the robustness of the method to uncertainties of models and keep sensitive to faults as well.

Acknowledgement. This work was supported by the NSFC (Grant No. 60025307, 60234010), and the national 973 program (Grant No. 2002CB312200) of China.

References

1. Chow, E.Y., Willsky, A.S.: Analytical Redundancy and the Design of Robust Failure Detection Systems. IEEE T. Automat. Contr. 29 (1984) 603-613
2. Yu, D., Shields, D.N.: Extension of the Parity-Space Method to Fault Diagnosis of Bilinear Systems. Int. J. Syst. Sci. 32 (2001) 953-962
3. Gertler, J., Staroswieeki, M.: Structured Fault Diagnosis in Mildly Nonlinear Systems: Parity Space and Input-Output Formulations. Preprints of IFAC 15th Triennial World Congress. Barcelona Spain (2002) 439-444.
4. Staroswiecki, M., Cometet-Varga, G.: Analytical Redundancy Relations for Fault Detection and Isolation in Algebraic Dynamic Systems. Automatica 37 (2001) 687-699
5. Krishnaswami, V., Luh, G.C., Rizzoni, G.: Fault Detection in IC Engines Using Nonlinear Parity Equations. Proc. 1994 American Contr. Conf. Baltimore Maryland (1994) 2001-2005
6. Moraal, P.E., Grizzle, J.W.: Observer Design for Nonlinear Systems with Discrete-Time Measurement. IEEE T. Automat. Contr. 40 (1995) 395-404
7. Guo, L.Z., Zhu, Q.M.: A Fast Convergent Extended Kalman Observer for Nonlinear Discrete-Time Systems. Int. J. Syst. Sci. 33 (2002) 1051-1058
8. Hornik, K., Stinchcombe, M., White, H.: Multilayer Feedforward Networks Are Universal Approximations. Neural Networks 2 (1989) 359-366
9. Xie, X.Q., Zhou, D.H., Jin, Y.H.: Strong Tracking Filter Based Adaptive Generic Model Control. J. Process Contr. 9 (1999) 337-350

Non-stationary Fault Diagnosis Based on Local-Wave Neural Network

Zhen Wang, Ji Li, Zijia Ding, and Yanhui Song

Institute of Vibration and Noise, Dalian University, Dalian, 116622, China
jswangzhen@163.com

Abstract. In view of non-stationary and non-linearity of vibration signal from machine surface, a new method, which is called Local-Wave Method (LWM), is presented to decompose it into number of Intrinsic Mode Weighs (IMW). Then improved RBF network model is constructed and trained using IMW as inputs. Taking the diesel fault diagnosis as an example, the method, which is checked through theory and practice, provides a power means for condition monitoring and fault diagnosis for the diesel engine.

1 Introduction

The vibration signal from machine surface as the multi-stimulation response, not only contains stimulation information in detail, but also contains the transferring characteristic and relative fault information. So it is an effective method to carry performance monitoring and fault diagnosis without disassembling by using the surface vibration signal. However, especially in the forepart of the diagnosis happening, the characteristic parameters selected and pick-up was very difficulty because the signal is non-stationary in machine fault diagnosis. For the traditional method, the characteristics, which are expressed by different diagnosis, can possibly be the same or similar fault, so only by these methods can't diagnose the type of the machine fault.

LWM may decompose non-stationary time-varying signal into some intrinsic-mode weights, and there is little influence among them. The intrinsic-mode weight has two kinds of information, one is fuzzy frequency band, the other is instantaneous frequency. As a method of nonlinear-mode identifying, RBF neural network has stronger capability of self-organizing and self-learning. On the basis of the two methods, a new method is represented. Namely network is construct through learning and training by intrinsic-mode weights.

2 Local-Wave Fundamental

Global wave analysis methods are suitable for stationary and linear signals. Local-wave Method[1] is a new method based on the development of the empirical-mode

F. Yin, J. Wang, and C. Guo (Eds.): ISNN 2004, LNCS 3174, pp. 549–554, 2004.

decomposition[2], and which can processes available analysis for time-varying or non-stationary signals.

A complex non-stationary signal may be include multi-oscillation modes, and the signal can have multi-instantaneous frequencies at any time. While the instantaneous frequency is the single-value function of the time, so there is only one frequency-value at any time. Thereby it can form the discrepancy in the signal analysis. Hence, it is necessary to process the complex signal in order to realize the separation of the multi-oscillation modes at some time. Based on the necessary condition of the instantaneous frequency on the physical meaning, the IMW must satisfy two conditions: (1) in the data set, the numbers of the extremum point and zero crossing point must either equal or differ at most by one; (2) at any time, the mean value of the envelope defined by the local maximum and envelop defined by the local minima is zero[2].

According to the discussed fundamental above, the Local-wave decompose of the original signal $X_i(t)$ may be shown as equation 1.

$$X(t) = \sum_{i=1}^{l} C_i + R_l(t) \ . \tag{1}$$

where $C_i(t)$ is the i intrinsic-mode weight, $R_l(t)$ is the trend weight, when its value is less than the desire value or is the monotone function, the decomposition of the signal is over. The reference [2, 3] proved the maturity and feasibility of the decomposition method. So the information included in the original signal may be adequately embodied through the intrinsic-mode weights.

Considering the simulating signal as equation 2, which is formed by three signals, one is sine signal which frequency is fixed, the other two is linearity frequency modulation signal which instantaneous frequency is different.

$$f(t) = a\sin(2\pi f_1 t) + b\sin(2\pi f_2 (t+1)t) + c\sin(2\pi f_3 (t+1)t) \ . \tag{2}$$

under this case, the frequency maybe mix in certain time. So the analysis error would be obviously, if the spectrum analysis way was adopt. However they can be separated by Local-wave decomposition because of instantaneous frequencies different.

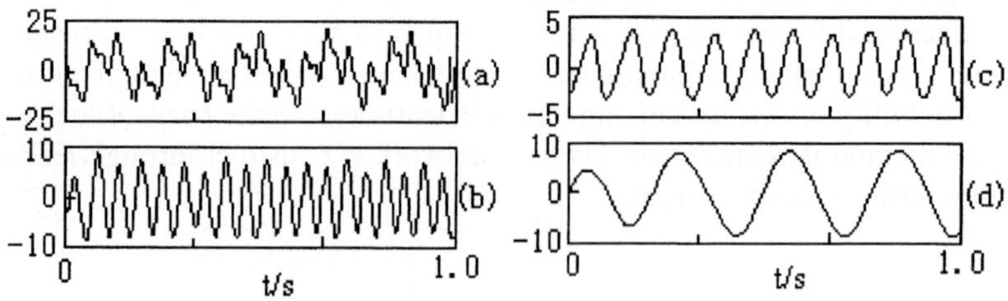

Fig. 1. Simulating signal and its weights, (a) is original data, figure 1(b) ~ (d) is three weights respectively

When *a* is 10, *b* is 5, *c* is 10, and f_1 is 5Hz, f_2 is 10Hz, f_3 is 25Hz, the time-domain waves of the simulating signal and the intrinsic-mode weight of the Local-wave decomposition are shown in Fig. 1. In order to show distinctly, here only gives partial data (sampling frequency is 1000Hz).

As fig. 1 shown, the complex multi-oscillation signal can be efficiently decomposed into every mode weight by Local-wave. In flowing content, the author choice intrinsic-mode weighs as characteristics, which are input in constructed network. By this way, local wave neural network could be obtained.

3 Construction of RBFN

3.1 Neural Networks Principle and Structure

RBF neural network shares the features of Back Propagation neural networks (BPNN) for pattern recognition. They are being extensively used for on-line and off-line nonlinear adaptive modeling and control applications. Now, the BP neural network is broadly and maturely used in pattern recognition domain. Most fuzzy neural networks use BP neural network to recognize. But there are some shortages in practical application. Such as learning convergence speed is slow and it is easy to appear local minimum. The initial data has great effect on the study performance, etc.

RBF neural network is one of the neuron-classified organs. Its neural center transformation function can form several segment-received domain. They can form complex structure decision field. RBF neural networks are presented with training set of input-output pairs and use a training algorithm to learn the nonlinear mapping form input to output. Thus they essentially carry out an approximation for nonlinear mapping from input to the output. RBF NNs have been used in signal processing, system identification and control applications. This paper uses the RBF neural network to train and recognize.

RBF NNs is one of three feedback neural networks; containing input node level, middle nonlinear dealing level and linear output level. The middle nonlinear dealing level adopts radial symmetry neuron function. The mostly common use is the Gaussian function. The expression is equation 3 [4].

$$R_i = \exp\left(-\frac{1}{\sigma_i^2}\|x - c_i\|^2\right) \quad i = 1,2,3,\cdots,N \ . \tag{3}$$

where: x is n-dimension input vector; c_i is the center of neural networks with the same dimension of x; σ_i is stands for the scalar quality value of width, m is number cell of middle level, $\| x\text{-}c_i \|$ is stands for Euclid distance between x and c_i.

Neural networks output is the linear combination of neuron function as equation 4.

$$y_k(x) = \sum_1^m w_{ik} R_i(x) \quad k = 1,2,3,\cdots, p \ . \tag{4}$$

w_{ik}-stands for authority value between the No. i node of latent level and the No. k node of output data p-stands for the numbers of output node.

3.2 Algorithm and Training Process

In fact, RBF algorithm is to chose classification core. The SAOSL (self-adaptive orthogonal least squares) learning algorithm is adopted in order to improve traditional approach. To eliminate the correlation among sample radically, the method translates samples into orthogonal samples type through matrix transform, then classification core is chosen from its using LS (Least Squares). SAOLS principle is introduced in detail as following.

The relation between network input and output can denote for equation 5 simply.

$$Y = XT \ . \tag{5}$$

where: Y is N-dimension output row vector, X is $N \times M$-dimension input matrix, T is M-dimension transfer row vector. By matrix transform method, X may be expressed orthogonal matrix as equation 6 shown.

$$X = CA \ . \tag{6}$$

Where: A is $M \times M$-dimension triangle matrix, C is satisfy with the equation of $C^T C = H$, and C^T is variation matrix of C, H is $M \times M$-dimension diagonal matrix. It shows that row vectors of C are orthogonal, so input samples are orthogonal. Based on equation (5) and (6), equation (7) is gained as following.

$$Y = CAT = Cg \ . \tag{7}$$

So the square of network output can express as equation 8.

$$N^{-1} Y^T T = N^{-1} \sum_{i=1}^{M} g_i C_i^T C_i \ . \tag{8}$$

g_i is i-th element of g, C_i is row vector of C matrix. During network training, original signal energy and its intrinsic-mode weights energy are characteristics of input samples. By this way, the ideal network will be constructed.

4 Application

4.1 Test Data and Parameters

Fig. 2 left is the time-domain waves under different stations of the 6BB1 diesel engine, which is measured when the diesel engine is at the same working condition (the rotate speed of the crankshaft is 1100rpm, moderate load, and guarantee the normal lubricating state.

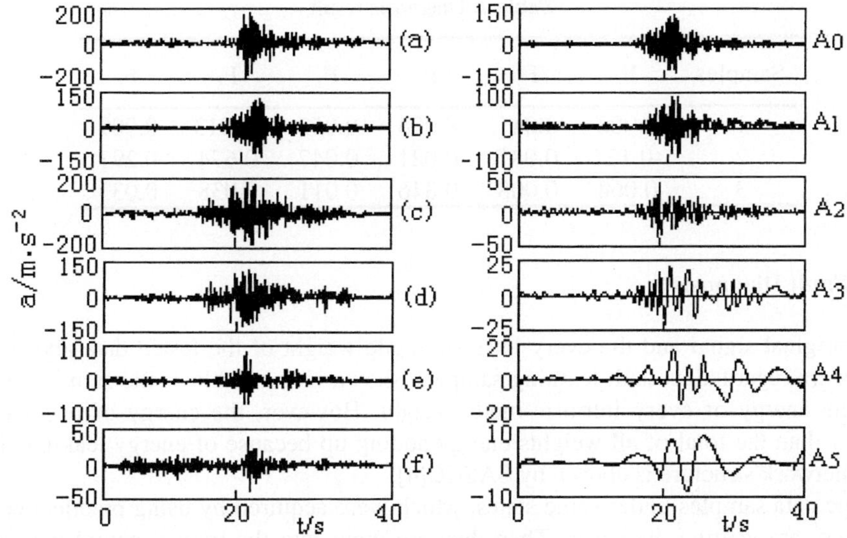

Fig. 2. Time domain wave of tested signals and its weights. Normal is (a). Injection starting pressure high (ISPH) is (b). Piston-liner wear (PLW) is (c). Exhausting gate leak (EGL) is (d). Nozzle leaking oil (NLO) is (e). Nozzle leaking oil badly (NLOB) is (f). The sampling frequency is 25.6kHz and 1024 points waiting for analyzing.

Fig. 2 right is the time-domain waves of the original data (A0) and intrinsic-mode weights (A1-A6). It can be seen that the number of the intrinsic-mode weight is decided by the complexity of the signal. According to the fundamental of the Local-wave, it can be known that every intrinsic-mode weight shows the information of different frequency region of the original signal and instantaneous frequency, so using the characteristic parameter of every weight as the input of the network, and carrying through training for it. By this way, the kind of fault and fault parts can determinate easily. The Local-wave decomposition of the other state signals is not given here because of the limit of the paper length.

Table 1. Six stations learning samples (unit: $m^2 \cdot s^{-4}$)

Faults		A0 E_0	A1 E_1	A2 E_2	A3 E_3	A4 E_4	A5 E_5
Normal	F_1	681.8	176.9	320.0	97.6	46.4	21.8
ISPH	F_2	620.7	150.3	311.6	84.2	40.7	18.2
PLW	F_3	770.3	208.2	356.4	110.7	60.2	27.6
EGL	F_4	645.3	219.8	248.0	97.4	48.7	20.2
NLO	F_5	641.5	142.2	301.3	89.6	69.4	19.8
NLOB	F_6	594.0	134.5	269.8	90.4	70.3	20.1

Table 2. Diagnosis result

Samples	F_1	F_2	F_3	F_4	F_5	F_6
1	0.908	0.041	0.003	0.052	-0.013	0.009
2	-0.122	0.087	0.021	0.047	0.674	0.293
3	0.064	0.008	0.846	0.011	0.038	0.033

4.2 Fault Diagnosis

The original signal and the every intrinsic-mode weight of the tested data above six states are regarded as fault learning samples, shown in the Table 1, the sign E stands for the energy of every intrinsic-mode weight. However, the energy of original is higher than the total of all weights energy adding up because of energy leaking. The best network structure is chosen by SASCC[4].

The data samples under some states, which were acquired by using practical engineering, are waiting diagnosis. Then they are input into the trained neural network, the result of the diagnosis shown as Table 2. Seeing from the result of the diagnosis, it is coincident with the factual stations of tested diesel engine.

5 Conclusion

Comparing with original signal, non-stationary level of each weight is lower than original signal, and each weight contains fuzzy frequency band and instantaneous frequency specially. These characteristics stand for different fault types and fault degree. Using them as training samples of neural network makes process converge quickly, meanwhile network model is stable. In fault diagnosis especially using non-stationary signals, the diagnosing result is very accurate proving in practice.

References

1. Ma, X.J., Yu B.: A New method for Time-Frequency---Local Wave Method. J. of Vibration Eng. 13 (2000) 219-224
2. Huang, N.E, Shen, Z., Long, S.R.: The Empirical Mode Decomposition and Hilbert Spectrum for Nonlinear and Non-stationary Time Series Analysis. Proc. Royal Soc. London. 454 (1998) 903-985
3. Narendra, K.G.: Application of Radial Basis Function (RBF) Neural Network for Fault Diagnosis in HVDC System. IEEE Tran. on Power Systems. 13(1998) 177-183
4. Wang, Z.P.: Research on Fault Diagnosis Method Based on Information Fusion and Application. J. of Dalian Univ. of Tech.. 4(2001) 128-132

Hybrid Neural Network Based Gray-Box Approach to Fault Detection of Hybrid Systems

Wenhui Wang, Dexi An, and Donghua Zhou

Department of Automation, Tsinghua University, Beijing 100084, China
wangwenhui99@mails.tsinghua.edu.cn, zdh@mail.tsinghua.edu.cn

Abstract. The fault diagnosis of hybrid systems is a challenging research topic at present. Model based fault diagnosis methods have been paid much attention in recent years, however, because of the complexity of hybrid systems, it is usually difficult to achieve a first principle model. To address this problem, this paper proposes a novel hybrid neural network, and based on it, a gray-box approach to fault detection of hybrid systems is presented, which combines some priori knowledge with neural networks instead of the common first principle model. Simulation results illustrate the effectiveness of the proposed approach.

1 Introduction

Hybrid systems have been used to describe complex dynamic systems that involve both continuous and discrete states. The continuous state evolves according to a differential/difference equation that depends on the discrete state (or mode), and the modes are described by a finite set. Mode transitions may occur when the certain conditions are satisfied. Due to the complexity of hybrid systems and the increasing demand for quality, safety and reliability, the fault diagnosis problem of hybrid systems becomes a crucial and challenging research topic. In terms of the type and amount of available knowledge, three levels of model can be defined. White-box models are derived from the first principle, black-box models are constructed from the observed data, and gray-box models incorporate some priori knowledge into observed information. Recently, white-box model based hybrid system fault diagnosis have been much studied, such as Petri nets [1], mixed logical dynamical [2], hybrid bond graph [3], particle filter [4]. The main disadvantage of white-box approaches is that the success is highly dependent on the quality of models. While the hybrid system is very complex, it is often difficult to achieve an accurate white-box model. In order to address this problem, this paper presents a novel hybrid neural network (HNN). Based on the HNN, a gray-box approach to fault detection of hybrid systems is proposed. This gray-box approach incorporates some priori knowledge about hybrid system modes and observed data into the HNN. The HNN consists of an adaptive resonance theory 2 (ART2) net [5], a probabilistic neural network (PNN) [6] and a bank of BP-L nets, where a BP-L net [7] consists of a backpropagation (BP) net and a linear dynamic system. The priori knowledge is embedded into the HNN construction in training stage. After training, the HNN can predict the hybrid system output

F. Yin, J. Wang, and C. Guo (Eds.): ISNN 2004, LNCS 3174, pp. 555–560, 2004.

accurately based on the system input. By comparing the observed output with the predicted output, the difference is used as a residual to detect faults. To the authors' knowledge, this is the first such approach to fault detection of hybrid systems.

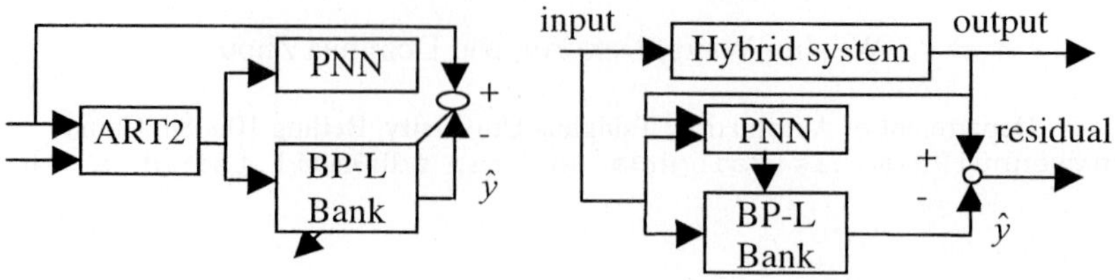

(a) HNN training stage (b) HNN recalling stage

Fig. 1. A hybrid neural network based gray-box approach

2 The Hybrid Neural Network Based Gray-Box Approach

The hybrid system studied in this paper has no well-developed model, the interior mechanism is unknown but some priori knowledge is available. Hence, it is necessary to develop a gray-box approach that combines the priori knowledge with the observed data for detecting faults in the hybrid system. A hybrid neural network based gray-box approach is proposed as shown in Fig.1.

In this gray-box approach, the priori knowledge about system modes is used at the selection of the adequate parameters of the ART2 net. ART2 [5] is known as a category learning net that self-organizes input patterns into various clusters. The parameters of ART2 net play a crucial role in determining how fine the categories will be. By adequate selecting the parameters, the ART2 net classifies the historical data into various categories while each category denotes a mode of the hybrid system through an unsupervised learning algorithm. The classification results determine the amount of the BP-L nets that construct the BP-L bank, and supervise the training of the PNN and the bank of BP-L nets.

According to the classification results, the PNN [6] is trained to identify the system modes. The PNN has four layers: an input layer, a pattern layer, a summation layer and an output layer. The input of PNN is the hybrid system's input-output, and the output is the current mode based on the Bayes optimal decision. The training and recalling stage of the PNN is described as follows (for detail, please refer to [6]).

In the training stage, the weight ω_{ij}^{1} between input node i and pattern layer node j is

$$\omega_{ij}^{1} = x_i^{(j)} \quad \text{for } i = 1,2,\ldots,n\,,\ j = 1,2,\ldots,p \tag{1}$$

where $x_i^{(j)}$ denotes the ith node input of the jth training pattern. The weight ω_{jk}^{2} between pattern layer node j and summation layer node k is

$$\omega_{jk}^{2} = \begin{cases} 1, & \text{pattern } j \in \text{mode } k \\ 0, & \text{pattern } j \notin \text{mode } k \end{cases} \quad \text{for } k = 1,2,\ldots,m\,,\ j = 1,2,\ldots,p \tag{2}$$

where p is the number of training patterns, n is the dimension of training pattern, m is the number of modes.

In recalling stage, the input vector is $\mathbf{x} = [x_1, ..., x_n]^T$, the output of pattern layer is

$$\mathrm{P_j} = \exp\left(-\sum_{i=1}^{n}\left(x_i - \omega_{ij}^1\right)^2 \Big/ 2\sigma^2\right) \quad \text{for } j = 1, 2, ..., p \tag{3}$$

where σ is the smoothing parameter. Then the output of summation layer nodes is

$$\mathrm{S_k} = \sum_{\mathrm{j=1}}^{\mathrm{p}} \omega_{\mathrm{jk}}^2 \cdot \mathrm{P_j} \Big/ \sum_{\mathrm{j=1}}^{\mathrm{p}} \omega_{\mathrm{jk}}^2 \quad \text{for } k = 1, 2, ..., m \tag{4}$$

The output layer makes a decision that which mode is active

$$\mathrm{O} = \arg\max_{\mathrm{k}} \mathrm{S_k} \quad \text{for } k = 1, 2, ..., m \tag{5}$$

For each mode, a BP-L net is trained to approximate the continuous nonlinear dynamic trajectory using the corresponding observed data. So a bank of BP-L nets is trained for the different modes of the hybrid system. All the BP-L nets have same structure but different parameters. A BP-L net [7] consists of a three-layer BP net, a linear dynamic system and a delay unit. The basic idea is to determine the parameters of the BP net using the corresponding historical data, so as to obtain the trained BP net that can drive the selected linear dynamic system to track the observed system output. The BP net has the sigmoid activation function for hidden layer and the linear activation function for output layer. The output of BP net is

$$v_i = \sum_j w_{ij}^{(2)} \sigma\left(\sum_k w_{jk}^{(1)} u_k - b_j^{(1)}\right) - b_i^{(2)} \tag{6}$$

where $\sigma(x) = 1/(1 + e^{-x})$, $w_{ij}^{(2)}$, $b_i^{(2)}$, $w_{jk}^{(1)}$ and $b_j^{(1)}$ are the parameters of the BP net, **u** and **v** are the input and output vector of BP net, respectively.

The linear dynamic system has a general form:

$$\begin{cases} \mathbf{x}(k+1) = \mathrm{A}\mathbf{x}(k) + \mathrm{B}\mathbf{v}(k) \\ \hat{\mathbf{y}}(k) = \mathrm{C}\mathbf{x}(k) \end{cases} \tag{7}$$

where the matrices A, B and C are chosen to make the linear dynamic system completely controllable.

In the training stage, the objective is to minimize the error

$$J = \left(\mathbf{y}(k) - \hat{\mathbf{y}}(k)\right)^T \left(\mathbf{y}(k) - \hat{\mathbf{y}}(k)\right)/2 \tag{8}$$

where $\mathbf{y}(k)$ and $\hat{\mathbf{y}}(k)$ are the observed and BP-L net's output, respectively. The gradient method for the optimization of parameters based on the partial derivative of the objective function $\boldsymbol{J}$ [8] is used to derive the off-line learning rule for the BP net. Let θ denotes the parameter to be optimized, and then the off-line learning rule is:

$$\theta(k+1) = \theta(k) - \lambda \times \partial J / \partial \theta \tag{9}$$

where λ is the learning velocity. $\partial J / \partial \theta$ can be calculated by (6), (7) and (8).

In fault detection, the PNN is used to discriminate the mode by classifying the observed data, track the mode changes and activate the corresponding BP-L net to track the continuous dynamic evolution. The output of the active BP-L net is used to predict the hybrid system output. The difference $\mathbf{r}(k) = \mathbf{y}(k) - \hat{\mathbf{y}}(k)$ is used as a residual to detect faults. So the fault detection law is

$$\Delta = \frac{1}{L} \sum_{i=k-L+1}^{k} \mathbf{r}^T(i)\mathbf{r}(i) \quad \begin{cases} \Delta \le \delta & \text{no fault in system} \\ \Delta > \delta & \text{fault occured} \end{cases} \tag{10}$$

where L is the size of the decision time window and δ is a predefined threshold.

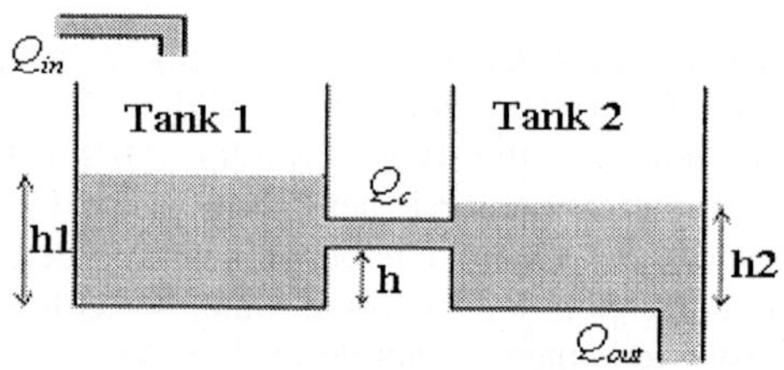

Fig. 2. Two-tank system

3 Simulation Study

The proposed approach is applied to a two-tank system that is a simple but typical hybrid system as shown in Fig.2. The system consists of two identical cylindrical tanks that are connected by a pipe at level h (20cm). Two tanks water levels h_1 and h_2 are measured via piezo-resistive pressure meters. The dynamic evolution is

$$\begin{aligned} A \cdot dh_1 / dt &= Q_{in} - Q_c \\ A \cdot dh_2 / dt &= Q_c - Q_{out} \end{aligned} \tag{11}$$

where A is the section of each tank (154cm^2). Q_{in} is the input flow provided by a pump. $Q_{out} = k_{out} \cdot \sqrt{2gh_2}$ is the output flow. $g = 980 cm/s^2$ and k_{out} is the linear gain (0.25). Q_c is the flow between the two tanks and has four modes that depend on the water levels h_1 and h_2 as follows:

$$Q_c = \begin{cases} 0 & \text{if } h_1 < h \text{ and } h_2 < h, \text{ Mode 1} \\ k_c\sqrt{2g(h_1 - h)} & \text{if } h_1 > h \text{ and } h_2 < h, \text{ Mode 2} \\ -k_c\sqrt{2g(h_2 - h)} & \text{if } h_1 < h \text{ and } h_2 > h, \text{ Mode 3} \\ \operatorname{sgn}(h_1 - h_2)k_c\sqrt{2g|h_1 - h_2|} & \text{if } h_1 > h \text{ and } h_2 > h, \text{ Mode 4} \end{cases} \tag{12}$$

where k_c is a linear gain (0.25). The continuous state is $x = [h_1, h_2]^T$ and the input is $u = Q_{in}$. Using the Euler discretization method with the sampling interval $T = 1\text{s}$, the discrete-time model is described by

$$\begin{cases} x(k+1) = f_q(x(k), u(k)) + \varpi(k) \\ y(k+1) = x(k+1) + \xi(k) \end{cases} \tag{13}$$

where $q \in \{1,2,3,4\}$ is the discrete mode as described in (12). f_q describes the continuous dynamic based on the mode q. Mode transition occurs when the water level exceeds h. $\varpi(k), \xi(k)$ are both white noise with mean zero and variance 0.01. The level of tank 2 is controlled by a discrete PI controller and the parameters are $K_p = 100$ (proportion gain) and $K_I = 0.1$ (integral gain) with the set-point $h_2^0 = 10\text{cm}$ In the simulation, the PNN has 3 input nodes and one output node. For each mode, a BP-L net consists of a 3-7-2 BP net and a linear dynamic system where the system matrices are all selected as identity matrices. Let $\delta = 0.1$ and $L = 10$.

Case 1: There is an abrupt leakage fault in tank 1 $Q_l = 0.075 \cdot \sqrt{2gh_1}$ after $t = 100\text{s}$. The result is shown in Fig.3 (a). We detect the fault at $t = 108\text{s}$. Case 2: There is an incipient clogging fault in the connection pipe $k_c' = (1 - 5\times10^{-4}(t - 100))k_c$ after $t = 100\text{s}$. The result is shown in Fig.3 (b). We detect the fault at $t = 130\text{s}$. It is shown clearly that the residual increases rapidly once the fault occurs. Either the abrupt fault or the incipient fault can be accurately detected.

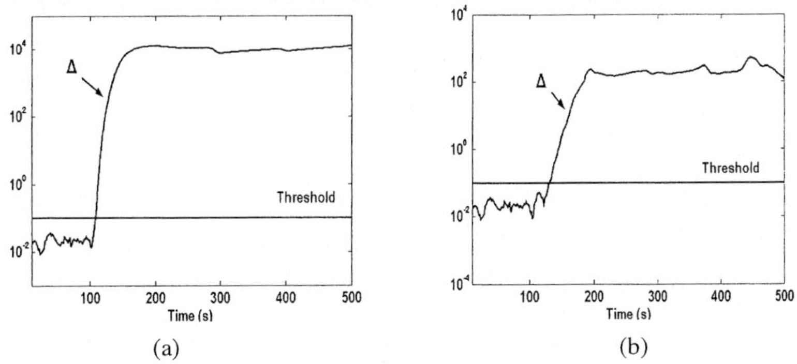

Fig. 3. The results of simulation: (a) an abrupt leakage in tank 1, and (b) an incipient clogging in connection pipe

4 Conclusions

This paper presents a novel gray-box approach based on the hybrid neural network for fault detection of hybrid systems The main advantages of this approach are: 1) it's feasible for fault detection of hybrid systems while first principle model is not available; 2) it's feasible for both abrupt and incipient faults; 3) it works well under closed-loop control. The simulation results show the effectiveness of the proposed approach. Further work will be focused on the fault isolation problem.

Acknowledgements. This work was supported by the NSFC (Grant No. 60025307, 60234010) and the national 973 program (Grant No. 2002CB312200) of China.

References

1. Tromp, L., Benveniste, A., Basseville, M.: Fault Detection and Isolation in Hybrid Systems, a Petri Net Approach. Proc. of 14th IFAC World Congress, Beijing, China (1999) 79-84
2. Bemporad, A., Mignone, D., Morari, M.: Moving Horizon Estimation for Hybrid Systems and Fault Detection. Proc. of 1999 American Control Conference, San Diego, USA (1999) 2471-2475
3. Narasimhan, S., Biswas, G.: An Approach to Model-Based Diagnosis of Hybrid Systems. In: Tomlin, C.J., Greenstreet, M.R. (eds.): Hybrid Systems: Computation and Control. Lecture Notes in Computer Science, Vol. 2289. Springer-Verlag, Berlin Heidelberg New York (2002) 308-322
4. Koutsoukos, X., Kurien, J., Zhao F.: Monitoring and Diagnosis of Hybrid Systems Using Particle Filtering Methods. Proc. of 15th International Symposium on Mathematical Theory of Networks and Systems, Notre Dame, USA (2002)
5. Carpenter, G.A., Grossberg, S.: ART2: Self-organization of Stable Category Recognition Codes for Analog Input Patterns. Applied Optics 26 (1987) 4919-4930
6. Specht, D.F.: Probabilistic Neural Networks. Neural networks 3 (1990) 109-118
7. Chen, G., Chen, Y., Ogmen, H.: Identifying Chaotic Systems via a Wiener-type Cascade Model. IEEE Control Systems 10 (1997) 29-36
8. Narendra, K.S., Parthasarathy, K.: Gradient Methods for the Optimization of Dynamical Systems Containing Neural Networks. IEEE Trans. on Neural Networks 2 (1991) 252-262

Application of Enhanced Independent Component Analysis to Leak Detection in Transport Pipelines*

Zhengwei Zhang, Hao Ye, and Rong Hu

Dept. of Automation, Tsinghua Univ
Beijing, 100084, P .R. China
zhangzw02@mails.tsinghua.edu.cn

Abstract. In this paper, a new Eigencurves method to detect leaks in oil pipelines is presented based on enhanced independent component analysis (EICA) and wavelet transform. EICA is used to derive Eigencurves from appropriately reduced principal component analysis (PCA) space of the training pressure images set. Wavelet transform de-noising (WTDN) is employed to preprocess measured pressure signals before getting the training and test images. In order to detect leaks, a classifier is designed to recognize negative pressure wave curve images by training set. The test results based on real data indicate that the method can detect many leak faults from a pressure curve, and reduce the ratio of false and missing alarm than conventional methods.

1 Introduction

Leak detection in oil transport pipelines is important for safe operation of pipelines, reducing oil loss and environmental pollution. This problem has drawn intensive attention. For all a lot of pipeline leak detection techniques, the negative pressure wave based approach has been widely adopted in practical applications.

The basic idea of the negative pressure wave based methods is that, when leak occurs, there is a negative pressure wave propagating from the leak point toward the upstream and downstream ends and so that we can detect leaks by detecting the pressure changes at the both ends [1].

Many approaches have been introduced for leak detection based on negative pressure wave, including Wavelet Transform [2], Pattern Recognition [3] etc.. However, these methods often have high possibility of false alarm when there are strong noises in the pressure measurement records and high possibility of miss detection when a leak is small or slow.

In this article, we propose a novel Eigencurves method to detect changes in a pressure curve, which uses wavelet transform to de-noise the pressure signals and

* Supported by the National Natural Science Fund of China (60274015) and the 863 Program of China

F. Yin, J. Wang, and C. Guo (Eds.): ISNN 2004, LNCS 3174, pp. 561–566, 2004.

uses Eigencurves extracted by Enhanced Independent Component Analysis (EICA) [4] to process leak detection. The improved performance has been tested by the experiment.

Compared with second-order method principal component analysis (PCA), ICA, an extension of the PCA, derives independent components by means of high order statistics. Different from normal ICA methods, the Enhanced ICA proposed by [4] is applied in this paper, which balances the representation and magnitude criterion and has better performance.

The paper is organized as follows. Section 2 gives out a three-step Eigencurves method which is composed of preprocessing pressure signals with wavelet transform de-noising (WTDN), extracting Eigencurves, and detection of negative pressure wave. The proposed method is demonstrated by history data of a real pipeline in section 3. Last, we draw a conclusion in section 4.

2 Eigencurves Method

This section details the three-step Eigencurves method: (i) De-noising; (ii) extracting eigencurves using EICA; (iii) projecting new pressure curves to be detected onto the eigencurves, then, using the nearest neighbor rule to realize on-line detection.

2.1 Preprocessing Pressure Signals with WTDN

In transport pipelines, noises are sometimes so large that the useful information is buried in them invisible. So the observed pressure signals often need to be de-noised before they are used to characterize the whole process. In this paper, we use discrete wavelet transform to decompose the pressure signals, and then de-noise them separately, and finally reconstruct the de-noised pressure signals [5].

The key of de-noising the signal is how to select the wavelet coefficient of signal decomposition. A threshold algorithm proposed by Donoho [5]

$$\delta_j = \lambda_0 \sigma_j \sqrt{2\log n} \Big/ \sqrt{n} \tag{1}$$

is adopted in this paper, where σ_j is the variance of the noise at the j th scale detail coefficients D_j, n is the size of the D_j, and λ_0 is a constant. In the de-noising procedure, any D_j smaller than σ_j is set to zero, while the others remain unchanged.

2.2 Extracting Eigencurves

Different from the conventional leak detection methods which were based on the idea of signal processing, in this paper we treat the pressure curve of some length as an image and use EICA, an image processing method, to realize leak detection in transport pipelines.

EICA is proposed in [4] to improve the generalization performance of ICA and reduce its computational complexity, which first chooses a proper low dimension space based on PCA by excluding small-valued trailing eigenvalues before extracting Eigencurves.

Let $\chi \in \Re^{N^2}$ be a vector representing a de-noised pressure image with $N \times N$ pixels. The vector is formed by concatenating the rows or the columns of the image. Let the training set of pressure images be $X = \{\chi_1, \chi_2, \ldots, \chi_M\}$, where M is the number of images in the training set. It is usually desirable to work with the standardized variables that are normalized to have unit norms. Standardization avoids the problems of having variables with large variance unduly.

Let $C \in \Re^{N^2 \times N^2}$ denote the covariance matrix of X. We can get the orthogonal eigenvector matrix $\Psi \in \Re^{N^2 \times N^2}$ and a diagonal eigenvalue matrix $\Lambda \in \Re^{N^2 \times N^2}$ with diagonal elements in decreasing order by using PCA. Now let $P \in \Re^{N^2 \times n}$ be a matrix whose column vectors are the first n ($n < N^2$) leading eigenvectors of C,

$$P = [\psi_1, \psi_2, \ldots, \psi_n] \tag{2}$$

where $P \in \Re^{N^2 \times n}$ is the loading matrix, n is determined by balancing the image representation and ICA generalization [4].

The projection of X on P is then adopted as a new matrix Y in the reduced PCA, which is defined as

$$Y = P^T X \tag{3}$$

After building the reduced PCA space, we can directly extract Eigencurves in it using the ICA method [4], i.e.

$$Y = FS \tag{4}$$

where $F \in \Re^{n \times m}$ is called the mixing matrix. The row vectors ($m \leq n$) of the new matrix $S \in \Re^{m \times M}$ are independent.

From Eq. (3) and (4), we can get,

$$S = (P\tilde{F}^T)^T X \tag{5}$$

where $\tilde{F} \in \Re^{n \times m}$ is the inverse (or pseudo inverse) matrix of the mixing matrix. (Conventionally, we often choose n=m). The column vectors of $P\tilde{F}^T$ are in analogy to the Eigenfaces [6], so we call these basis vectors the Eigencurves.

In this paper, we use the FastICA, introduced by Hyvärinen to extract Eigencurves [7]. FastICA is a method by which the independent components are extracted one after another by using Kurtosis. This method has high-speed convergence.

2.3 Online Detection

For an unknown pressure curve image f, we firstly normalize the pressure image to zero-mean and unit norm and get a normalized vector χ. Then we get the feature vector [see (5)] s of χ by projecting the normalized vector χ to the Eigencurves space of the training image set extracted with EICA.

$$s = (P\tilde{F}^T)^T \chi \tag{6}$$

Let $I_k, k = 1,2,\ldots,N$, be the mean of the training images for class ω_k in the EICA space (N is the number of classes), we can use Eigencurves extracted with EICA to realize classification based on the nearest neighbor (to the mean) rule. Let δ be some similarity measure

$$\delta(s, I_k) = \min_j \delta(s, I_j) \rightarrow s \in \omega_k \tag{7}$$

The feature vector s is classified to the images subset of the closest mean I_k. In this paper we use well-known cosine similarity measure $\delta_{\cos}$, which is defined as

$$\delta_{\cos}(\chi, y) = (-\chi^T y)/(\|\chi\|\|y\|) \tag{8}$$

where $\|.\|$ denotes the norm operator.

3 Experiment

This section assesses the performance of the Eigencurves method based on EICA for Recognition of the negative pressure waves. The effectiveness of the Eigencurves method is shown in terms of high performance.

3.1 Data Preparation and Sample Set

In this paper, the pressure curve is formed by ordinal connecting 2000 continually sampled pressure data. The sample time is 60 ms.

From the experience of the operator, we can detect occurrence of the negative pressure wave by search the shape of the pressure curve. Since the shape of the pressure curve is some local information, here we fragment the original pressure curve image into several 30×30 images and treat them as training and test images set.

Fig.1 shows some sample images we get by fragmenting one-hour pressure curve.

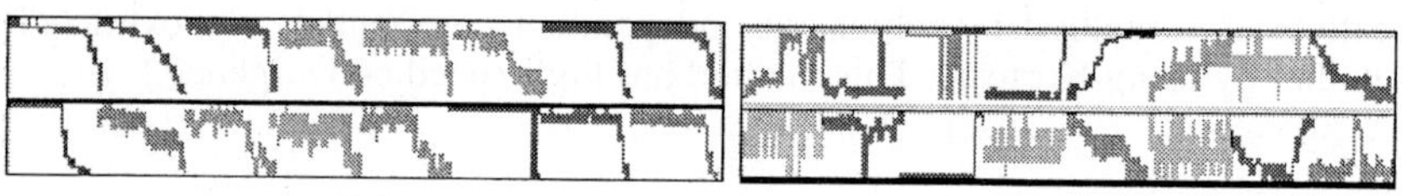

(a) (b)

Fig. 1. The sample images set reflecting the occurrence of the negative pressure wave. (b) The sample images set indicating normal condition

In this paper, we intercept 7,080 fractional images as sample set from the history data of a 32 Km real oil pipeline, in which 2,600 images contain negative pressure wave. We pick 3,200 samples images as training samples and the remnant 3,880 images are used as test images set.

3.2 Eigencurves

Since the condition of the transport pipeline is very complicated, the patterns of the sample images are diversiform. To achieve high accuracy of leak detection and reduce the computational complexity, we classify the training images set as 6 different patterns (negative pressure wave curves class, horizontal curve class, ascending curve class, fluctuant curves class, noisy curve case and sloping curves class respectively).

Fig.2 (a) shows the first 60 Eigencurves extracted with EICA from the training images. For comparison purpose, Fig.2 (b) shows the first 60 Eigencurves derived from the same training set with PCA. We can see the EICA-Eigencurves express more local feature.

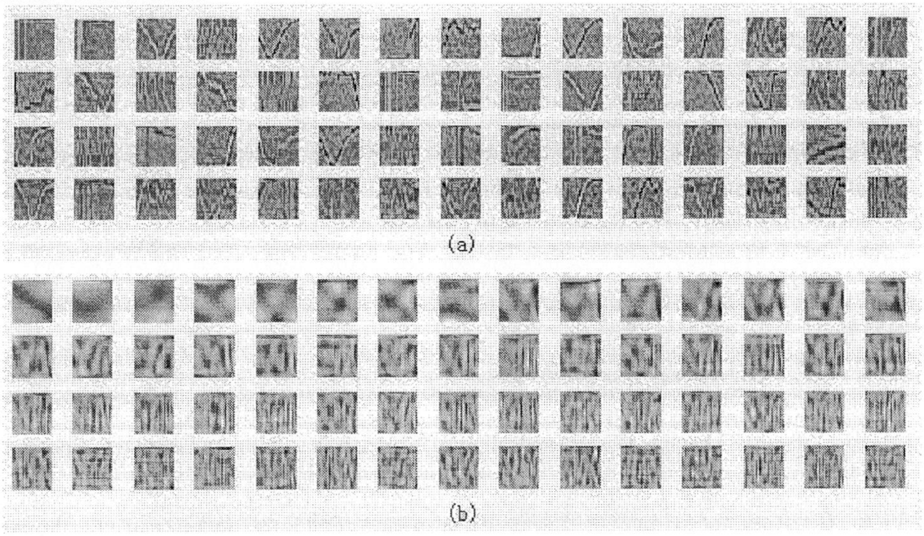

Fig. 2. (a) First 60 Eigencurves extracted with EICA from training pressure images set. (b) First 60 Eigencurves extracted with PCA from the same training pressure images set

3.3 Recognition of Negative Pressure Wave

For an unknown pressure curve image f, we firstly normalize the pressure image to zero-mean and unit norm and get a normalized vector. Then we get the feature vector [see (5)] s by projecting the normalized vector to the Eigencurves space. Finally, the On-line detection in section 2.3 is applied to realize the recognition of negative pressure wave.

$$\delta(s, I_1) = \min_{j=1,\ldots,6} \delta(s, I_j) \rightarrow \textit{the} \quad \textit{Negative} \quad \textit{Pressure} \quad \textit{Wave} \tag{9}$$

Table 1. Test results

Test set	Num. of samples	Num. of correct classif.	% of correct
Leak samples	1000	939	93.90
No-leak samples	2880	2838	98.54
Sum	3880	3777	97.35

3.4 Results

The results of detection by our approach are shown in Table 1. It can be seen that the Eigencurves method based on ICA can effectively detect the negative pressure wave with an acceptable miss detection rate 6.1% and false alarm rate 1.46% respectively.

4 Conclusion and Future Research

In this paper, a new Eigencurves method based on EICA to detect the negative pressure wave is proposed, in which the key step is to extract Eigencurves in a proper low dimension space achieved with PCA. The experimental results verify that the Eigencurves method can detect many leak defaults more effectively.

Some further work will be done. For example, the problem of operations on the upstream and downstream which might also lead to the pressure changes is not discussed in this paper. The localization of leak is another challenge.

References

1. Turner, N.C.: Hardware and Software Techniques for Pipeline Integrity and Leak Detection Monitoring. Proceedings of Offshore Europe 91, Aberdeen, Scotland (1991)
2. Ye, H., Wang, G.Z., Fang, C.Z.: Application of Wavelet Transform to Leak Detection and Location in Transport Pipelines. Engineering Simulation, Vol. 13 (1995) 1025-1032
3. Jin, S.J., Wang, L.N., Li, J.: Instantaneous Negative Wave Pattern Recognition Method in Leak Detection of Crude Petroleum Transported Pipeline. Journal of Electronic Measurement and Instrument, Vol.12 (1998) 59-64. (In Chinese)
4. Liu, C.J.: Enhanced Independent Component Analysis and Its Application to Content Based Face Image Retrieval. IEEE Trans. Systems, Man, and Cybernetics-part B: Cybernetics, vol. 34, no. 2 (2004) 1117-1127
5. Donoho, D.L.: De-noising via Soft-thresholding. IEEE Trans. Inform. Theory, vol. 41 (1995) 613-627
6. Turk, M., Pentland, A.: Eigenfaces for Recognition. J. Cognitive Neurosci., vol. 13, no. 1 (1991) 71-86
7. Hyvärinen, A.: Fast and Robust Fixed-point Algorithm for Independent Component Analysis. IEEE Trans. Neural Networks, vol. 10 (1999) 626-634

Internet-Based Remote Monitoring and Fault Diagnosis System

Xing Wu, Jin Chen, Ruqiang Li, and Fucai Li

The State Key Laboratory of Vibration, Shock & Noise, Shanghai Jiao Tong University, Shanghai 200030, P. R. China
{km_wx,fcli}@sjtu.edu.cn, jinchen@mail.sjtu.edu.cn, rqli163@163.com,

Abstract. This paper presents a novel monitoring-oriented and diagnosis-oriented software architecture model of Internet-based remote monitoring and fault diagnosis system based on Unified Modeling Language. This system combines rule-based reasoning and data mining techniques with case-based reasoning for fault diagnosis. A novel knowledge-based fuzzy neural network has been proposed and implemented to mine data stored in monitoring database. The applied technologies can be used for other research domains.

1 Introduction

Recently, the growth of equipment poses formidable challenges to cost-effective maintenance, such as distribution, high speed, and complexity. Personal computer, field bus, networking, and Internet technologies provide an easily implemented communications backbone for solving the above challenges. An Internet-based remote monitoring and fault diagnosis system (IRMFDS), as a feasible solution, has been designed and developed for key equipment in industries. The main advantages of an IRMFDS are *no-cost clients*, *low-cost maintenance*, *low-cost network investment*, *shared knowledge*, *shared experts*, *shared component packages*, and *cooperation diagnosis*. Data mining (DM) techniques provide an effective way to mine knowledge from machine fault data, which are combined with rule-based reasoning (RBR) and case-based reasoning (CBR) in IRMFDS. Section 2 introduces a novel monitoring-oriented and diagnosis-oriented software architecture (MODOSA) model based on Unified Modeling Language (UML); a novel DM approach for diagnosis is described in section 3; the prototype system implementation and conclusion are in sections 4 and 5, respectively.

2 MODOSA Model

Software technology now goes into an era of development based on component technology after structured design and object-oriented design [1] and software architecture (SA). Some domain-specific software architectures have been proposed, such as avionics control system, mobile robotics [2]. Developing MODOSA of

F. Yin, J. Wang, and C. Guo (Eds.): ISNN 2004, LNCS 3174, pp. 567–573, 2004.

IRMFDS is as essential as having a blueprint for large building. UML [3] was chosen to be the modeling language for description of MODOSA. The following sections will demonstrate our modeling works, which are typical parts of an iterative process consisting of the analysis, design, implementation and deployment of an IRMFDS.

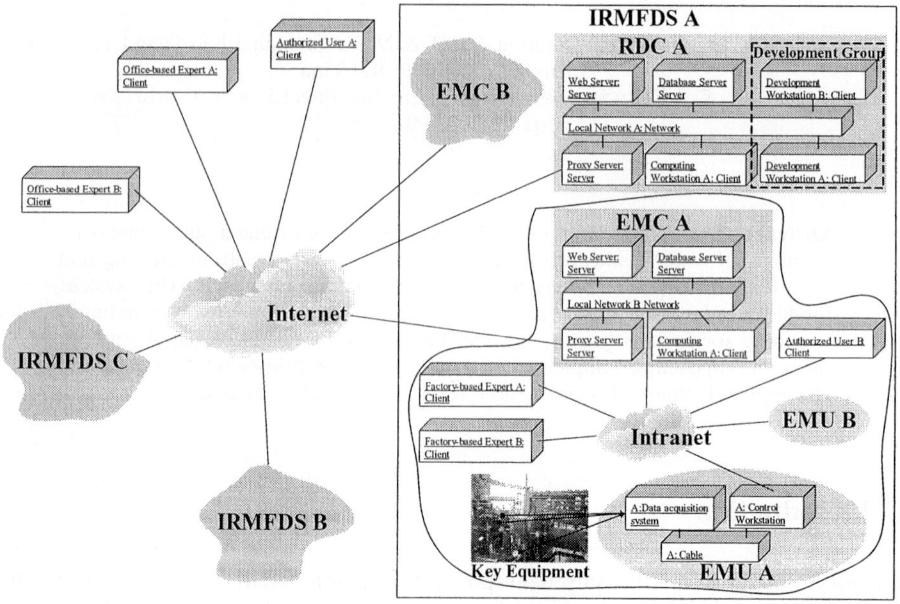

Fig. 1. Topological structure of IRMFDS

2.1 Topological Structure Model of IRMFDS

An IRMFDS is a combined hardware and software system, which provides enterprises with the capability to connect factory-based experts and office-based experts together over Internet. Heterogeneous architecture should be adopted for an IRMFDS, which consists of client-server style, layered style, object-oriented style and repository style. An IRMFDS consists of one Remote Diagnosis Center (RDC), several Enterprise Monitoring Centers (EMCs), many Enterprise Monitoring Units (EMUs), and a great deal of authorized experts and users (see Fig. 1).

- *EMU*. It collects data from sensors, preprocesses data and generates alarms based on defined operational limits from an EMC.
- *EMC*. It manages several EMUs in the same Intranet, displays real-time alarms and monitoring plots, extracts knowledge from fault data, and diagnoses machine faults.
- *RDC*. It mainly provides tech support about monitoring methods and diagnosis algorithms. Any authorized user can use remote analysis and remote diagnosis tools in a RDC.

From Fig. 1, a great cooperation diagnosis network (CDN) can be composed of the IRMFDS A, the IRMFDS B and the IRMFDS C. Also, authorized users can fuse diagnosis results from different RDCs using a result fusion tool in a RDC. The performance of the CDN will be continually improved and upgraded using new component packages from the development groups.

2.2 Static Class Model of EMC

A simplified static class model of an EMC is shown in Fig. 2. Some stereotype such as <<client page>>, <<server page>>, <<HTML Form>>, <<link>> and <<target link>> are described in [4]. A HTML formatted <<client page>> function1.html will be sent back to the client side when a user requests the <<server page>> function1. An instance of the <<boundary>> GUIObject1 class is embedded in the function1.html. The UIRMSController class manages the communications between the DBAccess class, <<boundary>> classes, and <<entity>> classes. The DBAccess class separates application logic from a database. <<Table>>, <<SP Container>> and <<derive>> are used to implement a database model.

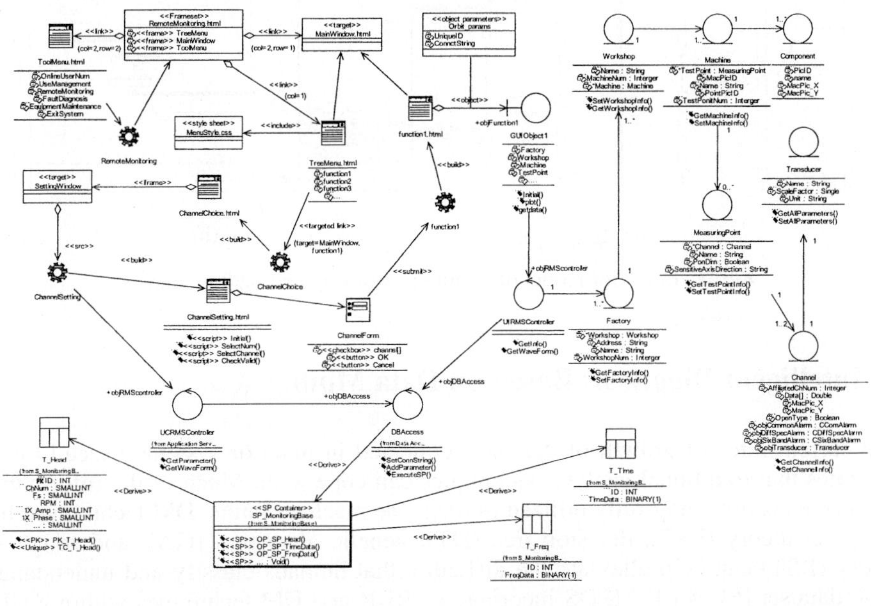

Fig. 2. Static class model of EMC

2.3 Dynamic Behavior Model of Monitoring and Diagnosis

Remote monitoring is one of key functions in an IRMFDS. Both the thin Web client pattern and the thick Web client pattern are based on HTTP. The inherently connectionless behavior of HTTP is not suitable for many real-time monitoring applications on the Internet. If there is effective control of client and network configurations, the Web delivery pattern is very suitable for quick and voluminous data exchanges between the client and the server [5]. A monitoring mechanism based on a mixed thick Web client pattern (see Fig. 3a) has been designed that combined the thick Web client pattern with the Web delivery pattern. An IRMFDS is a data-driven system. The knowledge learning process extracts expert knowledge from confirmed fault data using data mining techniques. These knowledge works within a CBR cycle (Fig. 3b) to support knowledge-based fault diagnosis of an EMC and a RDC.

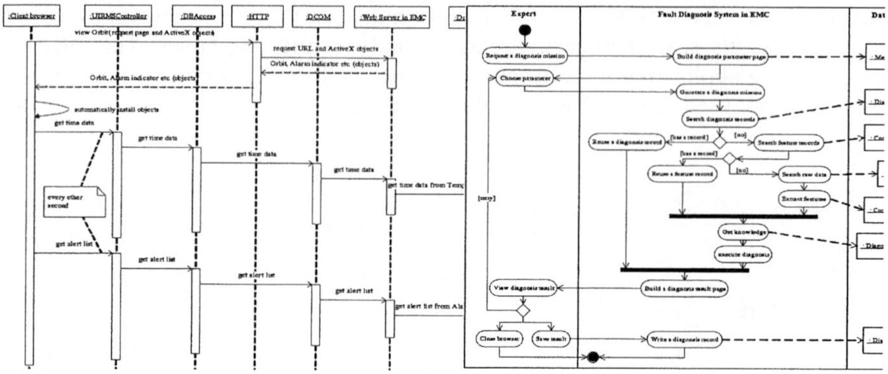

(a) (b)

Fig. 3. Behavior model of remote monitoring and fault diagnosis

3 Intelligent Diagnosis Based on Data Mining

The current rates of growth of data sets collected in monitor systems exceed by far any rates that traditional analysis techniques can cope with. Much of the information in this data is not being fully utilized to diagnose machine faults. DM techniques, like fuzzy set theory (FST), decision tree (DT), genetic algorithm (GA), and rough set theory (RST) can help alleviate the difficulty that humans classify and understand a large data set [6]. An IRMFDS incorporates RBR and DM techniques within a CBR cycle. A novel knowledge-based fuzzy neural network (KBFNN), as a DM approach for fault diagnosis, has been investigated that integrates RST, FST, NN and GA [7]. It takes advantages of the above theories and offset their shortages for fault diagnosis. Firstly, we obtain crude rules from training samples and calculate the dependency factors and antecedent coverage factors of the rules based on RST, and then a NN is constructed using the rules and configured using the factors mentioned above. The input and output parameters of the NN are also fuzzlized. At the same time, GA is utilized to optimize output parameters of those networks. Finally, a KBFNN are constructed based on the rules and factors. A five-layer ANN is used in the KBFNN which consists of the input layer, the fuzzification layer, the third and forth layers for describing the antecedent and consequential parts of the rules, and the unfuzzification layer. The implemented KBFNN component in an IRMFDS employed a digital signal processing (DSP) component, a RST component, and a GA component. All components are developed using Microsoft's Component Object Model (COM).

The classification performance of a KBFNN and a general back propagation fuzzy neural network (GFNN) is compared in terms of efficiency and accuracy (see Table 1). The structure of KBFNN is 11-3-7-5-5. The structure of GFNN is 11-33-10-5-5. Hardware used in the experiment is shown in Fig. 4. Some typical fault samples are obtained through simulating corresponding faults on Bently Rotor Kit 4 at certain constant speeds. All calculations were carried out on a 1.0GHz Pentium III server with 524MB RAM running under the Windows NT operating system. In summary, the KBFNN performs better than the GFNN in both the network training speed and accuracy.

Table 1. Performance comparison between KBFNN and GFNN

	KBFNN (11-3-7-5-5) (%) Train epoch times				GFNN (11-33-10-5-5) (%) Train epoch times			
	100	200	300	400	100	200	300	400
Normal	98.00	100.00	98.00	98.00	64.00	88.00	92.00	84.00
Unbalance	78.00	94.00	94.00	100.00	70.00	82.00	66.00	98.00
Radial Rubbing	95.83	97.50	92.50	96.67	77.50	86.67	84.17	80.00
Oil Whirl	100.00	100.00	100.00	100.00	95.00	100.00	98.75	100.00
Rotor Crack	99.00	99.00	99.00	99.00	93.00	94.00	90.00	97.00
Level	95.50	98.25	96.50	98.50	82.25	90.75	87.25	91.00
Train Time(s)	0.328	0.656	1.000	1.453	0.876	1.750	2.625	3.579

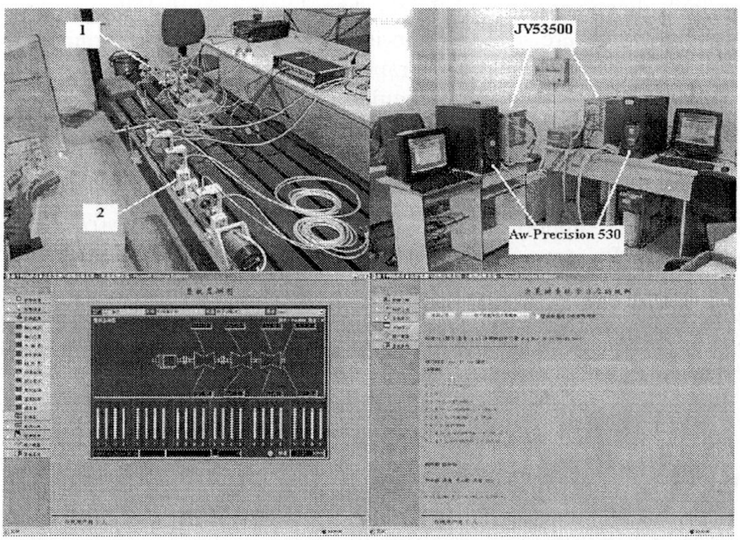

Fig. 4. IRMFDS hardware and software

4 System Implementation

A prototype system of an IRMFDS has been implemented using the MODOSA model and component-based programming. It consists of two EMUs, an EMC and a RDC. The basic functions of the EMU are data acquisition, data preprocessing, online alarm, and data backup. The basic functions of the EMC are user management (UM), equipment management, remote monitoring, and fault diagnosis. The basic functions of the RDC include remote analysis and remote diagnosis besides UM. A test environment of the IRMFDS has been built in our laboratory (see Fig. 4). There are two rotor test beds as monitored objects in top left corner of Fig. 4. The number 1 is Bently Rotor Kit 4. The EMU in top right corner consists of a 6 slot VXI-bus JV53500 and a DELL Aw-Precision 530. The picture in bottom left corner is a complete machine Web page to monitor the No.2 rotor test bed. The picture in bottom right corner displays the diagnosis knowledge mined from a confirmed fault base using a decision tree algorithm. This IRMFDS can steadily run and satisfy the requirements of online monitoring through continually monitoring Bently Rotor Kit 4.

5 Conclusion

This article presents the design and development of IRMFDS. A UML-based MODOSA model of an IRMFDS is proposed that has proven very useful after being used in a prototype system. As a direct result, this IRMFDS will easily able to suitably evolve in the fast changing Internet environment. This IRMFDS integrates RBR and DM techniques within a CBR cycle. A novel KBFNN approach has been designed and implemented to mine data stored in an EMC and an RDC for performing fault diagnosis. Three MODO shared component packages are published on the Internet, which are a basic algorithm package, a basic monitoring graphic package and an advanced analysis graphic package. This IRMFDS gives other researchers a paradigm to accomplish similar systems. It is of value to plant operators in order to better schedule maintenance and avoid damaging expensive plants.

Acknowledgements. Support for this project from the National "10.5" Science and Technology Key Project (Approved Grant: 2001BA204B05-KHKZ0009) and National Natural Science Foundation of China (No.50335030) is gratefully acknowledged. The thanks go to Haiming Qi, Yanxue Liang, Yiming Chen, Guangwei Guo, Liang Chen, Xun Chen and Yanxia Yuan for support and work during the project.

References

1. Shaw M., Garlan D.: Software Architecture: Perspectives on an Emerging Discipline Prentice Hall PTR, U.S.A (1996).

2. Katharine Whitehead.: Component-based Development. Addison-Wesley Longman Inc., U.S.A (2001).
3. Booch G., Rambaugh J., Jacobson I.: The Unified Modeling Language Reference Manual. Addison-Wesley Longman Inc., Boston, U.S.A (1999).
4. Conallen J.: Modeling Web Application Architectures with UML. Communications of the ACM, 42 (1999) 63-70.
5. Conallen J.: Building Web Applications with UML (Second Edition). Addison-Wesley Longman Inc., Boston, U.S.A (2002).
6. Usama Fayyad, Paul Stolorz. "Data mining and KDD: Promise and challenges". Future Generation Computer Systems, 13 (1997) 99-115.
7. Ruqiang Li, Jin Chen, Xing Wu: Fault Diagnosis of Rotor using Knowledge-based Fuzzy Neural Networks. Applied mathematics and mechanics (English Edition), in press.

Rough Sets and Partially-Linearized Neural Network for Structural Fault Diagnosis of Rotating Machinery

Peng Chen[1], Xinying Liang[1], and Takayoshi Yamamoto[2]

[1] Department of Environmental Science & Technology, Mie University
1515 Kamihama, Tsu, Mie Pref., 514-8507, Japan
chen@bio.mie-u.ac.jp

[2] Project of Machinery Diagnosis, Japan Science and Technology Agency

Abstract. Structural faults, such as unbalance, misalignment and looseness etc., are often occurring in a shaft of rotating machinery. These faults may cause serious machine accidents and bring great production losses. In order to detect faults and distinguish fault type at an early stage, this paper proposes a diagnosis method by using "Partially-linearized Neural Network (PNN)" by which the type of structural faults can be automatically distinguished on the basis of the probability distributions of symptom parameters. The symptom parameters are non-dimensional parameters which reflect the characteristics of time signal measured for diagnosis of rotating machinery. The knowledge for the PNN learning can be acquired by using the Rough Sets of the symptom parameters. The practical examples of diagnosis for rotating machinery are shown to verify the efficiency of the method.

1 Introduction

In the case of machinery diagnosis, the knowledge for distinguishing failures is ambiguous because definite relationships between symptoms and fault types cannot be easily identified. The main reasons can be explained as follows. (1) It is difficult to identify the symptom parameters for diagnosis by which all fault types can be distinguished perfectly. (2) In the early stages of a fault, effects of noise are so strong that the symptoms of a fault are not evident.

The conventional neural network (NN) cannot reflect the possibility of ambiguous diagnosis problem. Furthermore, the NN will never converge, when the symptom parameters put in the 1st layer have the same values in different states [1].

To overcome the problem and to detect faults of rotating machinery at the early stage, we propose the diagnosis method using "Partially-linearized Neural Network (PNN)" to identify the fault type of machinery.

The data (the diagnosis knowledge) for the PNN learning must be acquired in some way. Therefore, we also propose the knowledge acquisition method for the PNN by using the Rough Sets [2]. In this paper, practical examples of fault diagnosis of rotating machinery will verify that the method is effective.

F. Yin, J. Wang, and C. Guo (Eds.): ISNN 2004, LNCS 3174, pp. 574–580, 2004.

2 Partially-Linearized Neural Network

The neuron numbers of m-th layer of a NN is N_m. The set $X^{(1)} = \{X_i^{(1,j)}\}$ expresses the pattern inputted to the 1st layer and the set $X^{(M)} = \{X_i^{(M,k)}\}$ is the teacher data to the last layer (M-th layer). Here, $i = 1\ to\ P$, $j = 1\ to\ N_1$, $k = 1\ to\ N_M$, and,

$X_i^{(1,j)}$: The value inputted to the j-th neuron in the input (1st) layer;

$X_i^{(M,k)}$: The output value of k-th neuron in the output (M-th) layer; $k = 1\ to\ N_M$

Even if the NN converge by learning $X^{(1)}$ and $X^{(M)}$, it cannot well deal with the ambiguous relationship between new $X^{(1)*}$ and $X_i^{(M)*}$, which have not been learnt. In order to predict $X_i^{(M)*}$ according to the probability distribution of $X^{(1)*}$, partially linear interpolation of the NN is introduced as Fig. 1, we called it "Partially-linearized Neural Network (PNN)".

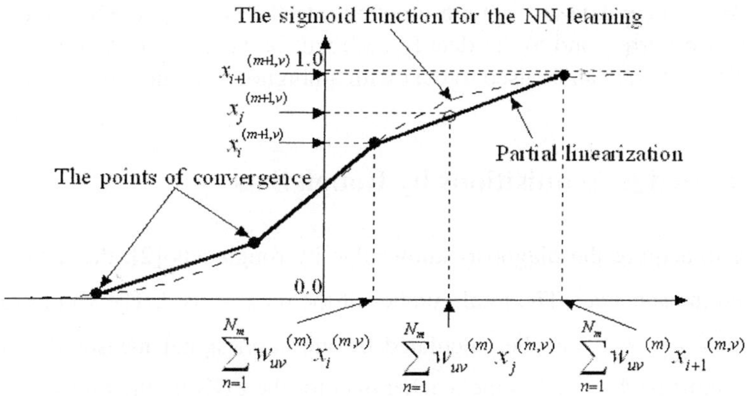

Fig. 1. The partial linearization of the sigmoid function

In the NN which has converged by the data $X^{(1)}$ and $X^{(M)}$, the symbols are used as follows.

$X_i^{(m,t)}$: The value of t-th neuron in the hidden (m-th) layer; $t = 1\ to\ N_m$

$W_{uv}^{(m)}$: The weight between the u-th neuron in the m-th layer and the v-th neuron in the $(m+1)$-th layer; $m = 1\ to\ M; u = 1\ to\ N_m; v = 1\ to\ N_{m+1}$

If these values are all remembered by computer, when new values $X_j^{(1,u)*}$ ($X_j^{(1,u)} < X_j^{(1,u)*} < X_{j+1}^{(1,u)}$) are inputted to the first layer, the predicted value of v-th neuron ($v=1$ to N_m) in the $(m+1)$-th layer ($m=1$ to $M-1$) will estimated by

$$X_j^{(m+1,v)} = X_{i+1}^{(m+1,v)} - \frac{\{\sum_{\mu=0}^{Nm} W_{uv}^{(m)}(X_{i+1}^{(m,u)} - X_j^{(m,u)})\}(X_{i+1}^{(m+1,v)} - X_j^{(m+1,v)})}{\sum_{\mu=0}^{Nm} W_{uv}^{(m)}(X_{i+1}^{(m,u)} - X_i^{(m,u)})} \tag{1}$$

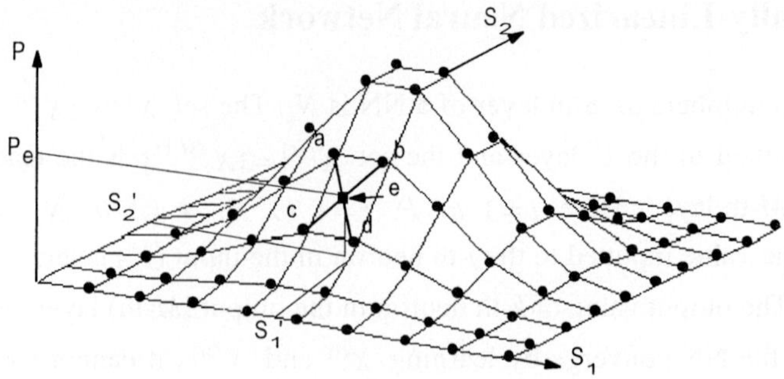

Fig. 2. Interpolation by the PNN

In the above way, the sigmoid function is partially linearized as shown in Fig. 1. If a function, such as Fig. 2, need to be learnt, the PNN will learn the points • shown in Fig. 2. When new data (s_1', s_2') are inputted into the converged PNN, the value symbolized by ■ correspond to the data (s_1', s_2') will be quickly identified as P_e shown in Fig. 2. So the PNN can be used to deal with ambiguous diagnosis problems.

3 Knowledge Acquisitions by Rough Sets

In order to acquire the diagnosis knowledge by rough sets [2], the data x_j must be measured in each state. The total number of the data x_j is J. The values of symptom parameters ${}^jp_{1s} \cdots {}^jp_{ms}$ can be calculated by vibration signal measured in each state. The ${}^jp_{is}$ must be digitized as the teacher data for the PNN by the following formula.

$$ {}^jp_{is} = 0\ to\ A_{pi} = \mathrm{int}[{}^jp_{is} / \{(\max\{{}^jp_{is}\} - \min\{{}^jp_{is}\}) / N_{pi}\} + 1] \tag{2} $$

Here, int[x] is the function which gives the integral values of x.

$$ P = \{p_1, p_2, \cdots, p_m\} \tag{3} $$

is the initial symptom parameter set. ${}^jP_S$ is the set of symptom parameter values measured in the state S.

$$ {}^jP_S = \{{}^jp_{1S}, {}^jp_{2S}, \cdots, {}^jp_{mS}\} \tag{4} $$

$[x_j]_{Ri}$ and A set are shown as follows :

$$ [x_j]_{Ri} = r_i = \{x_j \mid x_j \quad and \quad x_t \in [x_j]_{Ri} \rightarrow {}^jp_{i\bullet} = {}^tp_{i\bullet};\ j,t = 1\ \ to\ \ J\} \tag{5} $$

$$ A = \{r_i \mid i = 1\ to\ Q\} = \{[x_j]_{Ri}\} \tag{6} $$

The value sets (${}^{ri}P$ and ${}^{rj}P$) of symptom parameters of r_i and r_j are :

$$^{ri}P=\{^{ri}p_{1.},^{ri}p_{2.},\cdots,^{ri}p_{m.}\};^{rj}P=\{^{rj}p_{1.},^{rj}p_{2.},\cdots,^{rj}p_{j.}\} \tag{7}$$

The symptom parameters set P_{ij} selected from P shown in formula (3), which can discriminate between r_i and r_j, is :

$$P_{ij}=\{p_k \mid p_k\in P;\ p_k^* \text{ is the value of } P_k;\ p_k^*\in{}^{ri}P \text{ or } p_k^*\in{}^{ri}P\rightarrow P_k^*\in({}^{ri}P\cup{}^{rj}P)-({}^{ri}P\cup{}^{rj}P)\} \tag{8}$$

For distinguishing r_i(*i=1* to *Q)* from r_j (*j=1* to *Q, j i*), there may be redundant symptom parameters in the initial set *P* shown in formula (3). In order to remove the redundant symptom parameters, the following algorithm is proposed.

(1) Removing p_i from *P*;

(2) Calculating P_{ij} shown in formula (8).

(3) If $P_{ij}\neq\Phi$ (empty set), then p_i is the redundant symptom parameter. Removing p_i from *P*. Returning to (1) and repeating from (1) to (3) and from *i=1* to *i=Q*;

(4) After removing all of redundant symptom parameters, the new set of symptom parameters $P'=\{p_1,p_2,\cdots,p_l\}$ $(l\leq m)$ is obtained and the values set of P' of r_i is :

$$^{ri}P'=\{^{ri}p_1,^{ri}p_2,\cdots,^{ri}p_l\} \tag{9}$$

The possibility $^S\beta_{ri}$ of state *S* expressed by r_i can be calculated by

$$^s\beta_{ri}=\frac{card(^sx_j)}{card(x_i)}\% \tag{10}$$

Here, card(·) is the element number. $^Sx_j\in r_i$ is the x_j obtained from state *S*.

According to the above principle, the input data and teacher data (diagnosis knowledge) for PNN are as follows.

Input data : The value sets $^{ri}P'$ (formula (9)) of symptom parameters of r_i (*i=1* to *Q*), from which redundant symptom parameters have been removed.

Teacher data : The possibility $^S\beta_r$ (formula (10)) of state *S*.

4 Failure Diagnoses for Rotating Machinery

The above method has been applied to the diagnosis of rotating machinery. Here, we use the non-dimensional symptom parameters in time domain shown as follows for the diagnosis. The x_i is digital data of vibration signal for the diagnosis.

$$p_1=\frac{\sigma}{\overline{X}};\quad \text{Here, } \overline{X}=\frac{\sum_{i=1}^{N}|x_i|}{N} \tag{11}$$

$$p_2=\frac{\sum_{i=1}^{N}(|x_i|-\overline{X})^3}{\sigma^3};\quad \text{Here, } \sigma=\sqrt{\frac{\sum_{i=1}^{N}(|x_i|-\overline{X})^2}{N-1}} \tag{12}$$

$$p_3 = \frac{\sum_{i=1}^{N}(|x_i| - \overline{X})^4}{\sigma^4} \quad (13)$$

$$p_4 = \overline{X}_p / \overline{X} \quad (14)$$

Here, $\overline{X}_p$ is the mean value of peak values of $|x_i|$ ($i=1$ to N).

$$p_5 = \overline{|X_{max}|} / \overline{X}_p \quad (15)$$

Here, $\overline{|X_{max}|}$ is the mean value of 10 peak values (from maximum peak value to tenth value).

$$p_6 = \overline{X}_p / \sigma_p \quad (16)$$

Here, σ_p is the standard deviation of peak values of $|x_i|$ ($i=1$ to N).

$$p_7 = \overline{X}_L / \sigma_L \quad (17)$$

Here, x_L and σ_L are the mean value and the standard deviation of valley values of x_i respectively.

$$p_8 = \frac{\sum_{i=1}^{N}\sqrt{|x_i|}}{\sqrt{\sigma}} \quad (18)$$

$$p_9 = \frac{\sum_{i=1}^{N} x_i^2}{\sigma^2} \quad (19)$$

$$p_{10} = \frac{\sum_{i=1}^{N}\log(|x_i|+1)}{\log(\sigma)}; \; x_i \neq 0 \quad (20)$$

Fig. 3. Rotating machine

Fig. 3 shows the rotating machine for the diagnosis. Fig. 4 shows the time signals (1Hz to 320Hz low pass filtering) measured in each state. In this case, the values of the initial symptom parameters calculated by using the time signals of vibration are shown in Table 1. The redundant symptom parameters in the rough sets are removed so that we can distinguish each state by only using p_6 and p_{10}. The acquired knowledge of diagnosis for PNN learning is shown in Table 2.

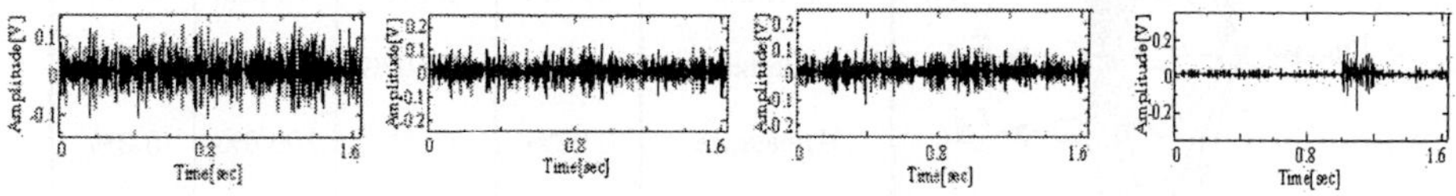

(a) Normal (b) Misalignment (c) Unbalance (d) Looseness

Fig. 4. Time signal of vibration measured in each state

Table 1. Values of symptom parameters in each state

	p_1	p_2	p_3	p_4	p_5	p_6	p_7	p_8	p_9	p_{10}
Normal:	1.39	0.34	1.01	2.28	2.61	1.72	-0.20	0.90	1.44	-0.01
	⋮	⋮	⋮	⋮	⋮	⋮	⋮	⋮	⋮	⋮
Misalignment :	1.25	0.55	1.88	1.59	4.41	1.39	-0.66	0.81	1.10	-0.01
	⋮	⋮	⋮	⋮	⋮	⋮	⋮	⋮	⋮	⋮
Unbalance:	1.24	0.58	2.14	1.50	4.34	1.36	-0.63	0.81	1.11	-0.01
	⋮	⋮	⋮	⋮	⋮	⋮	⋮	⋮	⋮	⋮
Looseness:	1.09	4.37	43.1	3.94	7.82	1.26	0.20	0.85	1.61	-0.00
	⋮	⋮	⋮	⋮	⋮	⋮	⋮	⋮	⋮	⋮

Table 2. Knowledge of diagnosis for PNN learning

	Input data (SP)		Teacher data			
	p_6	p_{10}	${}^N\beta_{ri}$	${}^M\beta_{ri}$	${}^U\beta_{ri}$	${}^L\beta_{ri}$
Normal :	8	5	1	0	0	0
	⋮	⋮	⋮	⋮	⋮	⋮
Misalignment :	1	3	0	1	0	0
	⋮	⋮	⋮	⋮	⋮	⋮
Unbalance :	5	5	0	0	1	0
	⋮	⋮	⋮	⋮	⋮	⋮
Looseness :	2	4	0	0	0	1
	⋮	⋮	⋮	⋮	⋮	⋮

Fig. 5 shows the PNN built on the basis of the method, and verification results. By learning the knowledge shown in Table 2, the PNN can quickly diagnose the faults with the possibility grade ${}^s\beta_{ri}$.

We sued the data measured in each state, which have not been learnt by the PNN, to verify the efficiency of the PNN. When the data of symptom parameters measured in the normal state (N), are inputted to the PNN, the probability of correct judgment is 87.5%. Similarly, the probabilities of correct judgment for M, U and L stets are all 100.0% respectively as shown in Fig. 5.

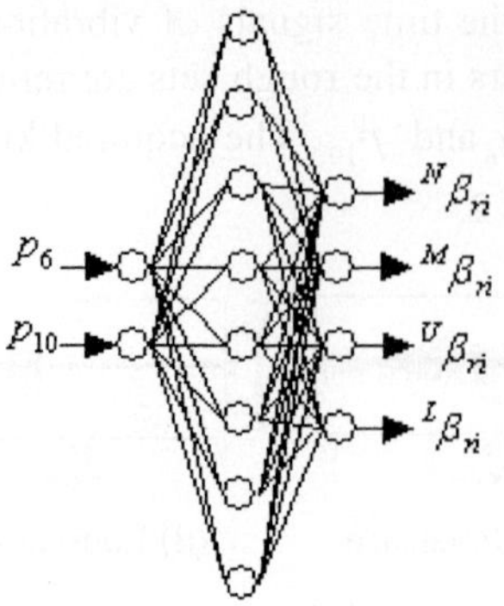

State by judgment ↓	Sate of signals for verifications			
	N	M	U	L
Normal	87.5%	0.0%	12.5%	0.0%
Misalignment	0.0%	100.0%	0.0%	0.0%
Unbalance	0.00%	0.0%	100.0%	0.0%
Looseness	0.00%	0.0%	0.0%	100.0%

Fig. 5. PNN for the diagnosis and the verifications

5 Conclusions

In order to diagnosing faults of rotating machinery and processing ambiguous information for diagnosis by neural network, the "Partially-linearized Neural Network (PNN)" and the knowledge acquisition method by rough sets have been proposed. The diagnosis results of structural faults of rotating machinery have been shown to verify the effectiveness of the methods discussed here.

The superiority of PNN can be explained as;
(1) It can deal with the ambiguous diagnosis for machinery fault;
(2) It is always convergent because the learning data is made consistent by rough sets;
(3) The convergence when learning is speedy.

References

1. Christopher Bishop M.: Neural Networks for Pattern Recognition. Oxford University Press (1995)
2. Pawlak Z.: Rough Sets. International Journal of Computer Information Science. Vol. 11 (1982) 344-356

Transient Stability Assessment Using Radial Basis Function Networks

Yutian Liu, Xiaodong Chu, Yuanyuan Sun, and Li Li

School of Electrical Engineering, Shandong University
250061 Jinan, P.R. China
liuyt@sdu.edu.cn

Abstract. A practical approach is proposed to power system transient stability assessment by monitoring and controlling active power flow on critical transmission lines. A radial basis function network is trained to estimate transient stability limit of active power flow on a critical line. Once the transient stability limit is violated, a preventive control decision is made in terms of generation rescheduling. Control amount is determined using first-order sensitivities of transient stability margin with respect to generator outputs, which can be derived directly from partial derivatives of the trained radial basis function network's output to inputs. Simulation results of a real-world power system demonstrate the effectiveness of the proposed approach.

1 Introduction

Electric power systems nowadays tend to be operated with smaller security margins due to inadequate transmission capacity caused by the lack of incentives for transmission expansion and operating uncertainties introduced by the deregulation of the electric power industry. To maintain power system security, on-line dynamic security assessment (DSA) plays an increasingly important role. The goal of on-line DSA is to analyze the dynamic state of the power system in an on-line manner and ensure power system security against a set of credible contingencies such as line and generator outages.

As a function of DSA, transient stability assessment (TSA) focuses on judging whether the power system loses synchronism under credible contingencies, evaluating the degree of transient stability or instability, and making control decisions to ensure stability [1]. A practical implementation of on-line TSA is to provide timely information on transient stability limits of critical operating parameters. Stability status of the power system can be derived from such information including whether the power system is stable or unstable and how large the stability margin is. For instance, it is a common practice to execute on-line calculation of transient stability limit of active power flow on a critical line. The line is at a transmission interface or a bottleneck constrained by transient stability for power transfer.

In this paper, an on-line TSA approach is proposed to monitor and control active power flow on critical lines. Periodically, active power flow on a critical line is

F. Yin, J. Wang, and C. Guo (Eds.): ISNN 2004, LNCS 3174, pp. 581–586, 2004.

monitored and its transient stability limit is computed. If current active power flow violates its transient stability limit, the power system is judged to be unstable but can be prevented from losing synchronism by generation rescheduling. First-order sensitivities of the stability margin with respect to changes of generator outputs are used to determine the generation schedule. Due to their learning ability, adaptability and fast execution, artificial neural networks have the potential to be applied to on-line estimation of transient stability limit of active power flow on a critical line. Also, first-order sensitivities can be derived directly from partial derivatives of the trained neural network's output to inputs [2]. Radial basis function (RBF) networks are applied because of their better convergence characteristics over commonly used multilayer perceptron networks [3].

2 RBF Network for Online TSA

In terms of active power flow on a critical line, transient stability margin is defined as the difference between its transient stability limit and current value. Instead of estimating transient stability limit directly, transient stability margin is approximated by a RBF network-based approach.

2.1 Estimation of Transient Stability Margin

Estimation of transient stability margin can be regarded as a function approximation problem mapping current operating condition to transient stability margin considering a set of credible contingencies

$$f : O \to M \tag{1}$$

where O denotes the current operating condition including network topology, load level, generation schedule, etc., M represents the minimum transient stability margin among all considered contingencies.

As general ANN-based approaches, the first step of our approach is to generate training examples off-line. To ensure the representativeness of training examples, a large number of time-domain simulations are executed covering various operating conditions under a set of credible contingencies. The operating conditions are combinations of different network topologies, load levels, generation schedules, etc. More importantly, they are all feasible cases, i.e., power flow equations can be solved under these conditions. For each case, generation shifts are made among generators with transient stability simulations executed under a contingency until the maximum value of active power flow on the line is found. The process is repeated for the whole set of contingencies. Each contingency corresponds to a constrained active power flow and transient stability limit is the minimum among them. Then transient stability margin is derived from the difference between transient stability limit and current value of active power flow on the line.

Three types of quantities are selected as input features including active power outputs of generators, load level and generation shift factors. These quantities represent operating conditions well and are measurable or can be derived easily. Particularly, it is flexible to take generation shift factors as representatives of network topologies. Generation shift factors are defined as [4]

$$a_{li} = \frac{\Delta f_l}{\Delta P_i} \tag{2}$$

where ΔP_i denotes change in generation at bus i and Δf_l represents change in active power flow on line l when a change in ΔP_i occurs. They are linear sensitivity factors and can be computed easily. Thus the input vector is composed of active power outputs of generators, load level and generation shift factors of the critical line with respect to generators

$$\boldsymbol{x} = [P_1, \cdots, P_{Ng}, a_{l1}, \cdots, a_{lNg}, L]^T \tag{3}$$

where Ng represents the number of generators in the power system, l is the critical line and L is load level.

Output is transient stability margin

$$y = M \; . \tag{4}$$

A RBF network is applied to such function approximation problem. RBF networks are three-layer networks with activation function of hidden nodes being radial basis function, e.g., Gaussian function. Input-output relationship of a RBF network with n inputs, m hidden nodes and one output is as

$$y = \sum_{k=1}^{m} w_k \exp\left(-\frac{\|x - c_k\|^2}{\sigma^2}\right) + w_0 \tag{5}$$

where w_k denotes the connecting weight between the k-th hidden node and the output node, w_0 is the bias, c_k and σ are center and width of the k-th hidden node, respectively, $\|\bullet\|$ represents the Euclidean norm.

The orthogonal least squares (OLS) algorithm is employed to train the RBF network [5].

2.2 Generation Rescheduling for Preventive Control

Periodically, the critical line is monitored with its transient stability margin estimated using the trained RBF network. Once the transient stability margin is estimated to be less than zero, i.e., active power flow on the line violates its transient stability limit; the power system would lose synchronism should occur a contingency. A preventive control decision must be made to maintain transient stability. Generation rescheduling, redispatching active power outputs among generators, is an effective control measure. First-order sensitivities of transient stability margin with respect to changes of generator outputs are used to determine the control amount.

Assume the sensitivity of transient stability margin with respect to changes of active power output of the i-th generator being

$$s_{li} = \frac{\partial M}{\partial P_i} . \tag{6}$$

By changing the active power output of the i-th generator with

$$\Delta P_i = -(s_{li})^{-1} M \tag{7}$$

the power system is pulled back to the stable status. First-order sensitivities can be derived directly from partial derivatives of the trained RBF network's output to inputs

$$s_{li} = \frac{\partial M}{\partial P_i} = -\frac{2}{\sigma^2} \sum_{k=1}^{m} w_k \exp\left(-\frac{\left\| \boldsymbol{x} - \boldsymbol{c}_k \right\|^2}{\sigma^2}\right)(P_i - \boldsymbol{c}_k(i)) . \tag{8}$$

In practice, generators are ordered according to their sensitivities with their active power outputs shifted one by one, i.e., generator with the maximum absolute sensitivity is redispatched first until the needed control amount is beyond its available control range.

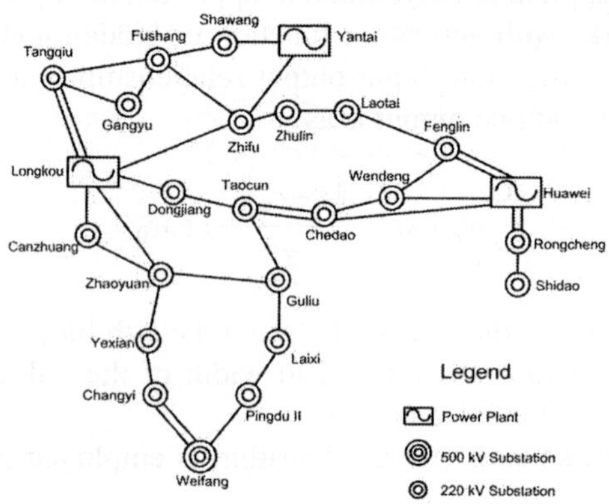

Fig. 1. Geographical diagram of Yantai-Weihai power system

3 Simulation Results

A regional power system in China, Yantai-Weihai power system is simulated. Geographical diagram of Yantai-Weihai power system is shown in Fig. 1. There are three power plants in the system including Huawei, Longkou and Yantai with about 2250 MW installed generating capacity. Constrained by transient stability, power output of Huawei power plant is suppressed below its rated value. Fenglin-Laotai line, responsible for power transfer from Huawei power plant to the load center, is a critical line vulnerable to transient stability limit. An on-line TSA scheme is designed with active power flow on Fenglin-Laotai line monitored and controlled.

3.1 Estimation of Transient Stability Margin

A RBF network is applied to the estimation of transient stability margin. To generate training examples for the RBF network, a large number of power flow cases are considered covering various operating conditions. Load level fluctuates in the range of 80% to 120% of the basic value (about 2234 MW). Power outputs of Huawei, Longkou and Yantai change randomly from 80% to 100% of the rated values. Two typical topologies are considered. Totally, there are 610 cases. For each case, a set of three phase-to-ground short-circuit faults on transmission lines are simulated and the most restricting one determines transient stability limit of active power flow on Fenglin-Laotai line.

In total, there are 610 input-output pairs generated. Since generators among the power plants usually behave coherently, the input vector is formed by quantities of power plants instead of individual generators. To be specific, the input vector is seven- dimensional composed of active power outputs and generation shift factors of Huawei, Longkou and Yantai power plants and load level. Output is transient stability margin, the difference between transient stability limit and current value of active power flow on Fenglin-Laotai line. Take randomly 400 input-output pairs as training examples and the other 210 ones as test examples. After training, average relative error is 0.05% for the training set and 1.77% for the test set, which verifies the RBF network is precise in estimating transient stability margin. Estimation results of part of the test set by the RBF network are shown in Fig. 2, compared with results by numerical computations.

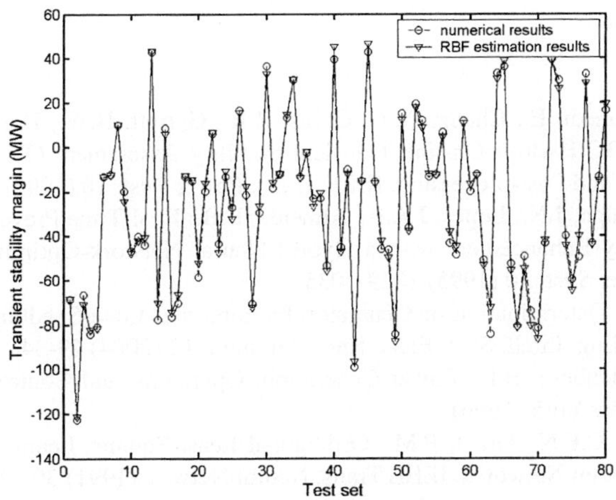

Fig. 2. Estimation results of RBF network compared with numerical results

3.2 Preventive Control

As listed in Table 1, two unstable scenarios are selected to test preventive control effects. Shift range of active power outputs for every power plant is ±5%. After generation rescheduling, the power system is stable.

Table 1. Preventive control results (Huawei, Longkou, Yantai)

Active power outputs/MW	First-order sensitivities	Generation rescheduling/MW
850, 1080, 320	-0.334, -0.031, 0	-39.7, 0, 0
850, 864, 256	-0.349, 0, 0	-37.7, 0, 0

4 Conclusions

A RBF networks is successfully applied to the estimation of transient stability limit of active power flow on a critical transmission line. Having first-order sensitivities of transient stability margin with respect to generator outputs, an appropriate preventive control decision can be made in terms of generation rescheduling whenever active power flow on the critical line violates its transient stability limit to maintain power system transient stability.

Acknowledgement. The authors would like to thank the financial support by State Power Corporation of China under the contract SP 11-2001-04-50-09.

References

1. Mansour, Y., Vaahedi, E., Chang, A.Y., Corns, B.R., Garrett, B.W., Demaree, K., Athay, T., Cheung, K.: BC Hydro's On-line Transient Stability Assessment (TSA) Model Development, Analysis and Post-processing. IEEE Trans. Power Syst. 10 (1995) 241-253
2. Miranda, V., Fidalgo, J.N., Lopes, J.A.P., Almeida, L.B.: Real Time Preventive Actions for Transient Stability Enhancement with a Hybrid Neural Network-Optimization Approach. IEEE Trans. Power Syst. 10 (1995) 1029-1035
3. Liu, Y., Chu, X.: Determination of Generator-Tripping and Load-Shedding by Compound Neural Network. Eng. Intell. Syst. Elect. Eng. Commun. 12 (2004) 29-34
4. Wood, A.J., Wollenberg, B.F.: Power Generation, Operation, and Control. 2nd edn. John Wiley & Sons, New York (1996)
5. Chen, S., Cowan, C.F.N., Grant, P.M.: Orthogonal Least Squares Learning Algorithm for Radial Basis Function Networks. IEEE Trans. Neural Netw. 2 (1991) 302-309

An Optimized Shunt Hybrid Power Quality Conditioner Based on an Adaptive Neural Network for Power Quality Improvement in Power Distribution Network

Ming Zhang, Hui Sun, Jiyan Zou, and Hang Su

Department of Electrical and Electronic Engineering, Dalian University of Technology, 116023 Dalian, China
v_zhang_cn@sohu.com
{dutshui, jyzou}@dlut.edu.cn

Abstract. This paper presents an optimized shunt hybrid power quality conditioner (SHPQC) for the compensation of harmonics and reactive power in power distribution network. A novel signal processing technique based on the adaptive neural network algorithm is applied to determine harmonics for generating the reference signals for the current controller of the SHPQC. The conventional hysteresis current controller for the three-phase insulated gate bipolar transistor (IGBT) voltage source inverter (VSI) is replaced by a current regulated space vector pulse width modulated controller. Simulations and experimental results are presented verifying the excellent power quality improvement performance of the proposed topology while keeping the apparent power of the SHPQC minimal.

1 Introduction

The wide application of nonlinear loads, such as static power converters and adjustable speed drives, etc., has resulted in a decreasing power quality due to the harmonic distortion. These harmonic current producing loads have caused various undesirable problems to both the power system and the equipment connected to these power lines.

Different harmonic power filters [1], [2] have been investigated in the past and have been installed in power systems to keep the harmonic distortion within acceptable limits. However, traditional passive filters have suffered from unstable filtering characteristics. A high cost for high-power application restricts the popularization of active power filters (APFs). In this paper, a full digital compensating system called a shunt hybrid power quality conditioner (SHPQC) is presented for the compensation of harmonics and reactive power in power distribution network, overcoming the shortcomings of the aforementioned filters.

Various harmonic analysis methods such as DFT, FFT, pq-transformation, etc., need additional updates and hence mostly complicated and difficult to implement and adjust [3]. Nowadays, neural networks [4] are widely applied in different fields of engineering and their application to estimation of harmonic components in power distribution network contributes to a solution to power quality improvement.

F. Yin, J. Wang, and C. Guo (Eds.): ISNN 2004, LNCS 3174, pp. 587–594, 2004.

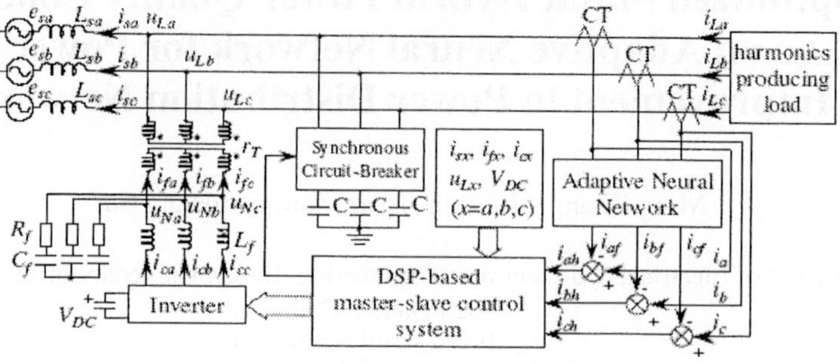

Fig. 1. Shunt hybrid power quality conditioner topology

In this paper, the estimation of the magnitude and phase angle of the current harmonics with an adaptive linear neural network is based on its training process [5]. The proposed algorithm is much simpler, highly adaptive, and capable of tracking the time varying harmonics accurately.

Simulations for a typical harmonic distorted power distribution network with reactive power compensation are presented as well as experimental results obtained with a laboratory prototype to verify the excellent performance of the SHPQC.

2 Shunt Hybrid Power Quality Conditioner

The proposed topology is illustrated in Fig. 1. A distribution network represented by the voltage source e_s and its inductance L_s, supplies a harmonics producing load.

The reactive power compensator consists of a chain of capacitor banks connected in a grounded Y connection and a medium-voltage synchronous circuit-breaker (SCB) with a magnetic drive [6]. Switching transients can be avoided or reduced by keeping the appropriate conditions of null initial current and null voltage difference between the capacitor banks and the power lines [7] with the help of the SCB which allows the closing and opening operations to be synchronized with the distribution network.

The shunt APF (SAPF) controlled as a current source compensates the harmonic currents of the nonlinear load, and the residual small amounts of load unbalance and power factor that the capacitor banks cannot eliminate due to their binary condition. In addition, the rating of the SAPF is notably reduced because most of reactive power compensation is done by the capacitor banks, and there is a sharp reduction of the troublesome problems caused by connecting capacitors to power systems.

An adaptive neural network is employed to determine the current harmonics of the nonlinear load for generating the reference signals for the current controller of the whole system. The total calculation and control of the SHPQC is implemented on a DSP-based master-slave system. It mainly comprises the synchronization of all signals, calculation of active and reactive power, and control of the SCB and the inverter.

2.1 Calculation of Harmonic Components

In the proposed adaptive neural network the nonlinear load current is sampled synchronously and the obtained values are used to extract the fundamental component of the distorted load current. The general form of a nonlinear load current with the fundamental angular frequency ω can be represented as

$$i_L(t) = I_0 e^{-\lambda t} + \sum_{n=1}^{N} I_n \sin(n\omega t + \phi_n) \ . \tag{1}$$

where the I_0 and λ are the initial value and decaying constant of the dc component, I_n and ϕ_n are the amplitude and phase of the nth (n=1, 2, ... , N) harmonic currents respectively, and N is the total number of the harmonics.

In general, a decaying dc component can be expressed using the first two terms of its Taylor series. If we transform (1) into the discrete time domain, we obtain

$$i_L(k) = I_0 - I_0 \lambda k T_s + \sum_{n=1}^{N} I_n \cos\phi_n \cdot \sin n\omega k T_s + \sum_{n=1}^{N} I_n \sin\phi_n \cdot \cos n\omega k T_s \ . \tag{2}$$

where T_s is the sample time and k is the time index or adaptation cycle number.

For the adaptive on-line estimation of the harmonics, a Fourier linear combiner called adaptive linear element or Adaline [4] is proposed. The implementation of the neural network circuit according to (2) is given by Fig. 2.

The input vector at time k can be written as

$$X(k) = [1, -kT_s, \sin\theta, \cos\theta, \ldots, \sin N\theta, \cos N\theta]^T \ . \tag{3}$$

The weight vector for the original Widrow-Hoff delta rule [4] is updated as

$$W(k+1) = W(k) + \frac{\alpha(k) e(k) X(k)}{\beta + X^T(k) X(k)} \ . \tag{4}$$

$$\alpha(k+1) = \rho \cdot \alpha(k) + \mu \cdot e^2(k) \ . \tag{5}$$

where $0 < \rho < 1$ and $\mu > 0$. $\alpha(k)$ is the adaptive learning parameter, the linear error $e(k)$ is the difference between the desired response input $\hat{i}(k)$ (training signal) and the linear output $i(k)$, and β is usually a small value avoiding a zero denominator.

The choice of the step size α plays an important role in the performance of the Adaline algorithm. A smaller value of the step size produces lower misadjustment, but it yields lower convergence speed of the algorithm and hence may not be able totrack nonstationary signals. While a bigger step size ensures faster convergence and better tracking capability at the cost of misadjustment. With the help of the fuzzy logic based algorithm using the error covariance, α is rewritten as

$$\alpha(k+1) = \rho \cdot \alpha(k) + \Delta\alpha(k) \ . \tag{6}$$

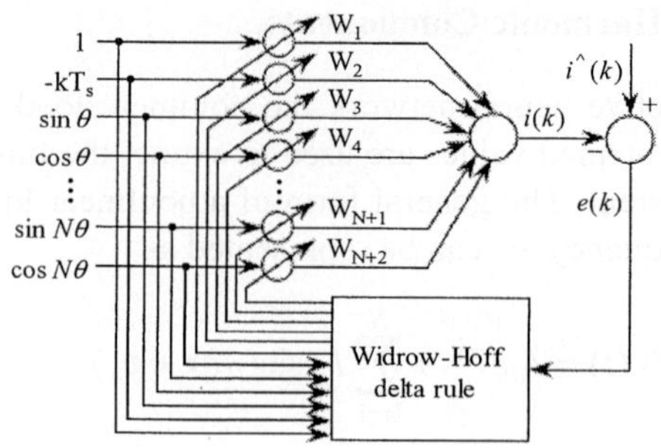

Fig. 2. Block diagram for estimation of harmonics using an Adaline ($\theta = kT_s$)

$$\Delta\alpha(k) = \sum_{i=1}^{6} \rho_i \Delta\alpha_i \Big/ \sum_{i=1}^{6} \rho_i \ . \tag{7}$$

where ρ_i and $\Delta\alpha_i$ are the firing level of the fuzzy rule and the center of gravity of the output fuzzy subset of the ith rule, respectively [8], [9].

Furthermore, the disadvantageous effect on the accuracy of the Adaline algorithm due to the fundamental frequency drift within the permitted range, can be effectively solved by the FFT algorithm with windows and interpolation [10].

The convergence of the learning rule for the neural estimator can be studied by Lyapunov Theorem [5]. During the training process, the present error is reduced by the factor of α when the weights are changed while keeping the input fixed until the next adaptation cycle starts. Then the next error is unceasingly reduced as such processes continue. The initial weight vector is usually set to be zero and the practical range of α is between 0.1 and 1 so that the tracking error $e(k)$ can converge to zero asymptotically. When a perfect training is obtained, the error will be brought to zero, and after the final convergence is reached the weight vector yields the Fourier coefficients in (2) as

$$W_0 = [I_0, I_0\lambda, I_1 \cos\phi_1, I_1 \sin\phi_1, \ldots, I_N \cos\phi_N, I_N \sin\phi_N]^T \ . \tag{8}$$

Finally, the amplitude and phase of the nth harmonic current are given by

$$I_n = \sqrt{W_0^2(2n+1) + W_0^2(2n+2)} \ . \tag{9}$$

$$\phi_n = \arctan[W_0(2n+2)/W_0(2n+1)] \ . \tag{10}$$

2.2 Discrete Current Controller

From Fig. 1, the total of all harmonics calculated by subtracting the fundamental component from the nonlinear load current, are used to generate the reference signals for the current controller. Since the filter voltage of a voltage source inverter (VSI) can be controlled directly [11], the previously obtained signals together with the supply voltage and inverter phase voltage are applied for the calculation of the command voltage space vector, and the gating signals of the VSI are determined by a space vector pulse width modulation (PWM) technique [12], [13], [14].

The optimization of connection configuration for power factor correction (PFC) capacitor banks can achieve a desirable performance of reactive power compensation. However, the residual small amounts of load unbalances and power factor cannot be eliminated due to the binary condition of the capacitor banks. When the reactive current component is considered simultaneously, we obtain the perfect reactive power compensation as well as harmonic compensation.

3 Simulations for a Power Distribution Network

In the simulation environment PSCAD and MATLAB, the proposed SHPQC has been examined in detail. The values used in the simulations are given in Table 1. The nonlinear load current containing the fundamental current, the most common harmonic currents, and a decaying dc component, is represented by

$$i_L = 0.12\sin(\omega t + \pi/3) + 0.02\sin(5\omega t) + 0.014\sin(7\omega + \pi/10)$$
$$+ 0.009\sin(11\omega t + \pi/7) + 0.03e^{-15t} \;. \tag{11}$$

The sampling frequency is set to 6.4 kHz and the learning parameter is chosen as 0.8 for this study. Fig. 3 shows the performance of tracking of the fundamental component. When full compensation is provided, the nonlinear load, compensating, and source currents are shown in Fig. 4. If a compensating current is applied as depicted in the second diagram below, the source current becomes nearly sinusoidal.

Table 1. Simulation values

Description	Symbol	Value
Source voltage	e_s	380 V
Fundamental frequency	f_1	50 Hz
Source inductance	L_s	2.68 mH
PFC capacitor	C	0.66 mF
Filter inductance	L_f	820 μ H
Filter resistance	R_f	4 Ω
Filter capacitor	C_f	2.5 μ F
Switching frequency	f_s	12.5 kHz
DC-link capacitor	C_{DC}	4.7 mF
DC-link voltage	V_{DC}	390 V
Load resistance	R_L	16 Ω
Load inductance	L_L	22.8 mH

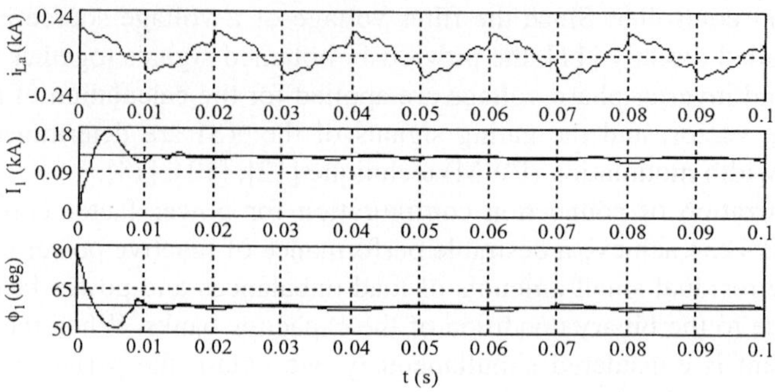

Fig. 3. Signal waveform (i_{La}), amplitude tracking of fundamental component (I_l), and phase tracking of fundamental component (ϕ_1)

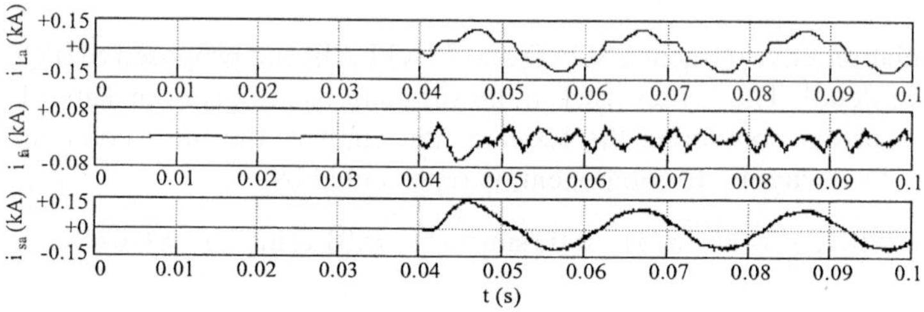

Fig. 4. Simulations of the load current (i_{La}), compensating current (i_{fa}), and source current (i_{sa}) with the optimized compensation

4 Experimental Results

A full-rated experimental prototype for harmonic and reactive power compensation was installed with a three-phase full-bridge rectifier supplying an inductance of 22.8 mH and a resistor of 16 Ω in series at the dc output of the rectifier. The source inductance has been measured at the output terminal and a value of 2.68 mH was obtained. The transformer with a magnetizing inductance of 0.12 mH referred to the secondary side and a ratio of ten, was used to reduce the voltage applied to the IGBT VSI so that the switching devices with a lower breakdown voltage of 1200 V can be used directly in the power distribution network. The dc-link capacitor of the VSI has a capacitance of 4.7 mF and the dc voltage set to 390 V can be controlled by injecting active power from or into the dc capacitor. The smoothing output filter of the VSI consists of a inductor of 0.82 mH, a resistor of 12 Ω, and a capacitor of 2.5 μF.

The experimental results are partly shown in Fig. 5. Particularly, the measuring values of the harmonic currents at 20 ms are given in Table 2. The reactive power compensator of 75 kVA with capacitors of 0.66 mF as well as the APF of 15 kVA can maintain the power factor around 0.95.

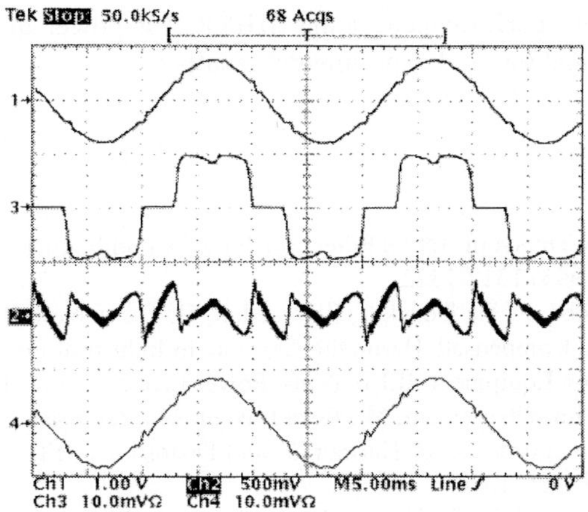

Fig. 5. Harmonic filtering performance of the SHPQC (from top to bottom): line voltage (400 V/div), load current (25 A/div), compensating current (15 A/div), and source current (25 A/div)

Table 2. Measuring values of harmonic currents at 20 ms

Harmonic order	Amplitude / Phase Theoretic value (p.u.)	Measuring value (p.u.)	Absolute error (p.u.)	Relative error (%)
1	1.3278 / 30	1.3251 / 30.11	-0.0027 / 0.11	-0.20 / 0.37
5	0.2213 / 150	0.2202 / 150.48	-0.0011 / 0.48	-0.50 / 0.32
7	0.1549 / 210	0.1574 / 211.03	0.0025 / 1.03	1.61 / 0.49
11	0.0996 / 330	0.1017 / 327.26	0.0021 / -2.74	2.11 / -0.83
13	0.0879 / -30	0.0863 / -29.02	-0.0016 / 0.98	-1.82 / -3.27
17	0.0695 / -90	0.0708 / -93.14	0.0013 / -3.14	1.87 / 3.49
19	0.0598 / -150	0.0612 / -144.05	0.0014 / 5.95	2.34 / -3.97

5 Conclusion

In today's competitive energy market, power quality issues are becoming an important performance indicator and receiving much more attention than in the past. In this paper, an optimized shunt hybrid power quality conditioner with a novel signal processor based on an adaptive neural network has been studied for the compensation of harmonics and reactive power, and thus improves the power quality in distribution network. The proposed algorithm of harmonic determination in the Adaline form is applied to provide the reference currents for the SHPQC by sensing the nonlinear load currents only and keeping adequate tracking of time varying harmonics. The conventional hysteresis current controller for the three-phase IGBT VSI is replaced by a space vector PWM controller to avoid subharmonics and minimize the inverter losses. An excellent performance of the SHPQC for power quality improvement is proved via all simulations and experimental results.

References

1. Akagi, H.: New Trends in Active Filters for Power Conditioning. IEEE Trans. Ind. Applicat., Vol. 32. (1996) 1312-1322
2. Detjen, D., Jacobs, J., De Doncker, R.W., Mall, H.-G.: A New Hybrid Filter to Dampen Resonances and Compensate Harmonic Currents in Industrial Power Systems with Power Factor Correction Equipment. IEEE Trans. Power Electron., Vol. 16. (2001) 821-827
3. Karimi, H., Karimi-Ghartemani, M., Reza Iravani, M., Bakhshai, A.R.: An Adaptive Filter for Synchronous Extraction of Harmonics and Distortions. IEEE Trans. Power Delivery, Vol. 18. (2003) 1350-1356
4. Widrow, B., Lehr, M.A.: 30 Years of Adaptive Neural Networks: Perceptron, Madaline, and Backpropagation. in Proc. IEEE, Vol. 78. (1990) 1415-1442
5. Dash, P.K., Swain, D.P., Liew, A.C., Rahman, S.: An Adaptive Linear Combiner for On-Line Tracking of Power System Harmonics. IEEE Trans. Power Systems, Vol. 11. (1996) 1730-1735
6. Cereda, C., Gemme, C., Reuber, C.: Synchronous MV Circuit-Breaker with Magnetic Drive and Electronic Control. ABB Review, (1999) 13-21
7. Dixon, J., del Valle, Y., Orchard, M., Ortúzar, M., Morán, L., Maffrand, C.: A Full Compensating System for General Loads, Based on a Combination of Thyristor Binary Compensator, and a PWM-IGBT Active Filter. IEEE Trans. Ind. Electron., Vol. 50. (2003) 982-989
8. Dash, P.K., Mishra, B., Panda, S.K., Salama, M.M.A.: Computation of Power Quality Symmetrical Components Using Fuzzy Logic Based Linear Combiners. in Proc. IEEE EMPD'98, Vol. 1. (1998) 17-22
9. Dixon, J.W., Contardo, J.M., Morán, L.A.: A Fuzzy-Controlled Active Front-End Rectifier with Current Harmonic Filtering Characteristics and Minimum Sensing Variables. IEEE Trans. Power Electron., Vol. 14. (1999) 724-729
10. Andria, G., Savino, M., Trotta, A.: Windows and Interpolation Algorithms to Improve Electrical Measurement Accuracy. IEEE Trans. Instrume. and Measurem., Vol. 38. (1989) 856-863
11. Malesani, L., Mattavelli, P., Buso, S.: Robust Dead-Beat Current Control for PWM Rectifiers and Active Filters. IEEE Trans. Ind. Applicat., Vol. 35. (1999) 613-620
12. Holtz, J.: Pulsewidth Modulation-A Survey. IEEE Trans. Ind. Electron., Vol. 39. (1992) 410-420
13. Bakhshai, A., Espinoza, J., Joos, G., Jin, H.: A Combined Artificial Neural Network and DSP Approach to the Implementation of Space Vector Modulation Techniques. in Conf. Rec. IEEE-IAS Annu. Meeting, Vol. 2. (1996) 934-940
14. Kazmierkowski, M.P., Malesani, L.: Current Control Techniques for Three-Phase Voltage-Source PWM Converters: A Survey. IEEE Trans. Ind. Electron., Vol. 45. (1998) 691-703

Cyclic Statistics Based Neural Network for Early Fault Diagnosis of Rolling Element Bearings

Fuchang Zhou, Jin Chen, Jun He, Guo Bi, Guicai Zhang, and Fucai Li

The State Key Laboratory of Vibration, Shock & Noise, Shanghai Jiao Tong University, Shanghai 200030, P. R. China
zhoufuchang@sjtu.edu.cn, jinchen@mail.sjtu.edu.cn

Abstract. This paper proposes a novel cyclic statistics based artificial neural network for early fault diagnosis of rolling element bearing, via which the real time domain signals obtained from a test rig are preprocessed by cyclic statistics to perform monitoring fault diagnosis. Three kinds of familiar faults are intentionally introduced in order to investigate typical rolling element bearing faults. The testing results are presented and discussed with examples of real time data collected from the test rig.

1 Introduction

Vibration monitoring has been widely reported as a useful technique for the analysis of the condition of rotating machines [1,2]. More efficient maintenance schedule can be planned if accurate information about the machine's condition is known and if on-line monitoring and diagnosing [3] is used, the machine can be left unattended for longer periods of time. Artificial neural networks have been used in a wide variety of pattern recognition applications including vibration monitoring [4] and consequently appear to be a possible solution to this problem. This paper presents the design of the cyclic statistics based artificial neural network diagnosis algorithm. The bearing vibration invariable frequency features extracted via cyclic statistics processing are applied to artificial neural networks to build an automatic bearing fault diagnosis machine. Experimental results indicate that a cyclic statistics based neural network bearing vibration diagnosis algorithm can give very accurate information about the condition of the rolling element bearing.

2 Cyclic Statistics and Artificial Neural Networks

2.1 Cyclic Statistics Theories

A random signal $x(t)$ is 1st-order cyclostationary, if its first order moment, i.e. the mean, is time periodical for all t

$$m_x(t) = E\{x(t)\} = m_x(t+T) \tag{1}$$

F. Yin, J. Wang, and C. Guo (Eds.): ISNN 2004, LNCS 3174, pp. 595–600, 2004.

where E is the mathematical expectation. The coefficients in the generalized Fourier series expansion of $m(t)$ is called the Cyclic Mean (CM)

$$M_x(\alpha) = \lim_{T\to\infty} \frac{1}{T} \int_{-T/2}^{T/2} m_x(t) \exp(-j2\pi\alpha t) dt = \langle m_x(t) \exp(-j2\pi\alpha t) \rangle_t \tag{2}$$

where α is the so called cyclic-frequency.

The signal $x(t)$ is 2nd-order cyclostationary, if its second order moment $R_x(t,\tau) = E\{x(t+\tau/2)x^*(t-\tau/2)\}$, i.e. autocorrelation function is time periodical for all t

$$R_x(t,\tau) = R_x(t+T,\tau) \tag{3}$$

where $*$ denotes complex conjugation. T is the cyclic period, and $\alpha = 1/T$ is the fundamental cyclic frequency. The coefficients in the generalized Fourier series expansion of $R_x(t,\tau)$ is called Cyclic Autocorrelation Function (CAF).

$$R_x^\alpha(\tau) \stackrel{\Delta}{=} \frac{1}{T} \int_{-T/2}^{T/2} R_x(t,\tau) \exp(-j2\pi\alpha t) dt = \langle R_x(t,\tau) \exp(-j2\pi\alpha t) \rangle_t \tag{4}$$

where $\alpha = nT, n = 1,2,3,\ldots$.

Fourier transforming $R_x^\alpha(\tau)$, then give the Cyclic Spectral Density (CSD) function. This is the cyclic version of the power spectral density known as Cyclic Wiener Relation and has the following equation

$$S_x^\alpha(f) = \int_{-\infty}^{+\infty} R_x^\alpha(\tau) \exp(-j2\pi f\tau) d\tau \tag{5}$$

2.2 Artificial Neural Networks

Artificial Neural Network has been found to be an effective tool for pattern recognition in many situations where data is fuzzy or incomplete. It is based upon models of human neurons and it has been used to perform tasks, which previously relied on human judgment to make a decision. The basic building block for an artificial neural network is neuron. Each neuron consists of many inputs and one output. The output is a non-linear transformation of the sum of the inputs scaled by weights and a bias.

$$y_m = \phi(\sum w_{mi} x_i + b) \tag{6}$$

where $\{x_i\}$ are the inputs, $\{w_i\}$ are the corresponding weights, b is the bias, y is the output of the neuron and ϕ represents the transformation. A pictorial representation of an artificial neuron is given in Fig.1. In this paper, the tangent sigmoid function

$$\phi(x)=\frac{\exp(x)-\exp(-x)}{\exp(x)+\exp(-x)} \tag{7}$$

is used for $\phi(x)$. We use the most common configuration, the feed-forward neural network, the neurons are arranged into 3 layers. The output of each layer is fed into the input of the next layer as shown in Fig.2.

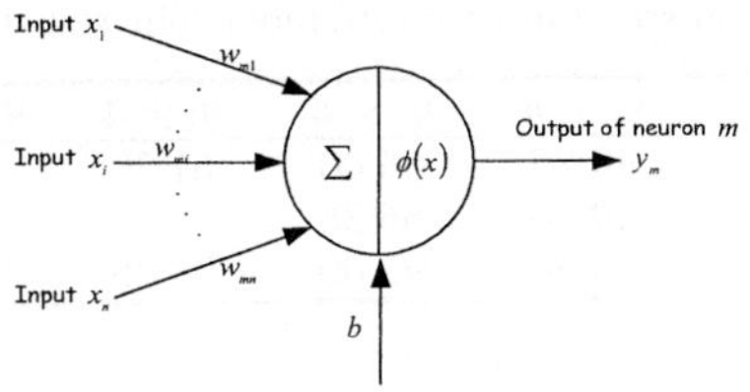

Fig. 1. Artificial neuron

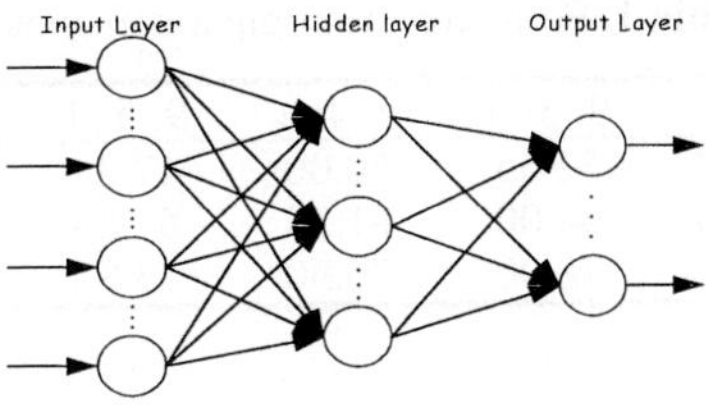

Fig. 2. Configuration of feed-forward ANN

2.3 Artificial Neural Network Based on Cyclic Statistics

The architecture of a feed-forward neural network is not suitable for direct classification from the time series data, as it has no memory of previous inputs. Therefore, it is necessary to extract time invariant features from the time series data to use as inputs to the network. The choice of features has a significant impact on the network's success.

If the vibrations of the bearings are stationary, the statistics such as the Mean, the Root Mean Square (RMS), the Peak Value, the Kurtosis and so on, will be invariant and classification based upon them can give an indication of the bearing's health. But unfortunately, in the fact, the vibration signals of rolling element bearing are produced by a combination of periodic and random processes due to the machine's rotation cycle and interaction with the real world. This phenomena appears prominently especially when faults occurred. The combination of such components can give rise to signals, which have periodically time-varying ensemble statistical and are best considered as cyclostationary, so it is much more appropriate to use cyclic statistics such as CM, CAF, CSD and so on to extract the almost time invariant features of the cyclostationary process and used as inputs. The cyclic statistics based artificial neural network diagnosis algorithm can be performed as shown in Fig.3.

The neural network used here consists of three layers. There are 9 neurons in the input layer, 3~10 neurons in the hidden layer and 4 in the output layer. The training data need 9 input nodes to correspond to the characteristic parameters extracted via cyclic statistics of the signal. The 4 output nodes are used for representing the probabilities rolling element bearing's conditions. We use 100 groups of training sets of each condition to train the ANN, and use 100 groups of testing sets of each condition to test the ANN based on cyclic statistics. The performance of ratio of calculation quantity of feature extracting and training of ANN about different structure for different cyclic statistics are compared, as shown in Table.1

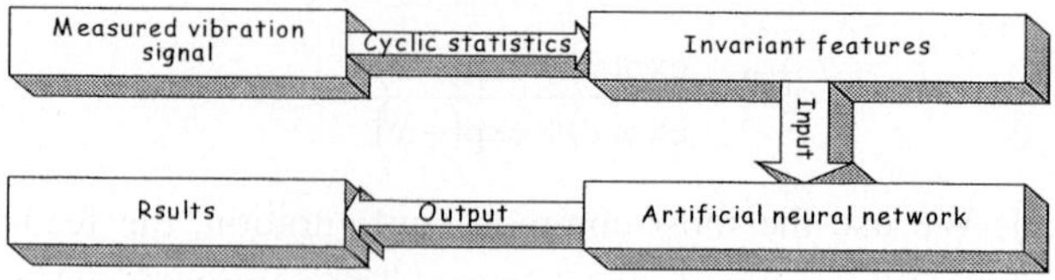

Fig. 3. Execute scheme of the cyclic statistics based artificial neural network diagnosis system

Table 1. Performance comparison about different structures for different cyclic statistics

	9: 3: 4	9: 4: 4	9: 5: 4	9: 6: 4	9: 7: 4	9: 8: 4	9: 9: 4	9: 10: 4
CM	52.65	50.00	47.25	45.00	49.85	56.40	61.20	68.75
CAF	64.00	61.35	57.28	54.30	59.50	66.36	75.68	86.25
CSD	78.00	76.50	73.80	70.00	83.50	92.00	98.75	100.00

3 Test Rig

The data used for this work are obtained from the following experimental test rig, as shown in Fig.4. Fig.4 (a) illustrates the plot of the rolling element bearing vibration-measuring system. Fig.4 (b) illustrates a view of implementation sketch of this experiment. The test rig can be divided into three subsystems: the machine set, the transducers and the computer hardware.

The machine set used is a rolling element bearing vibration-measuring machine ZST-1, consisting of a three-phase asynchronies AC motor, a belt transmission system, a shaft coupling, two back-up bearing systems, and a test bearing with load. The rotating frequency of the shaft f_r is about $11.67Hz$. A series GB203 rolling element bearing with local detects are used. The ball diameter is $6.747\ mm$, the pitch diameter is $28.5\ mm$, the contact angle $\theta = 0$, and the number of rolling elements is $n = 7$. The transducers used are a B&K4368 accelerometers positioned to measure

(a) Plots of a rolling element bearing vibration-measuring system

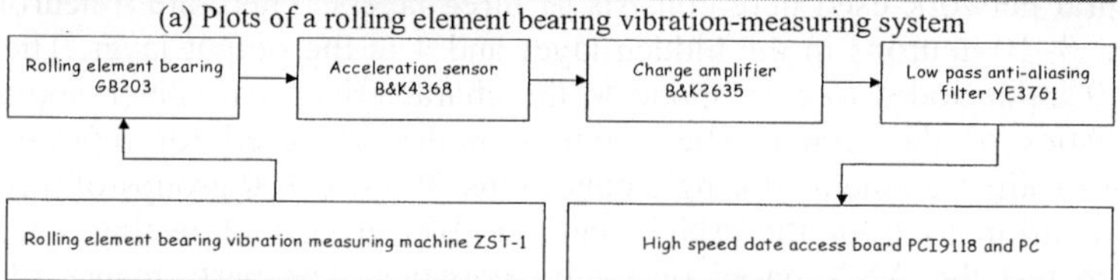

(b) Sketch of the measuring system for vibration signal of rolling element bearings

Fig. 4. Experimental test rig

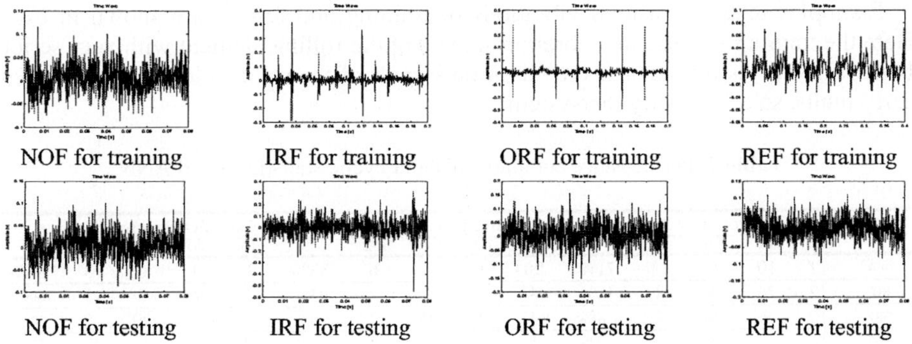

Fig. 5. Time waveform of training and testing signals

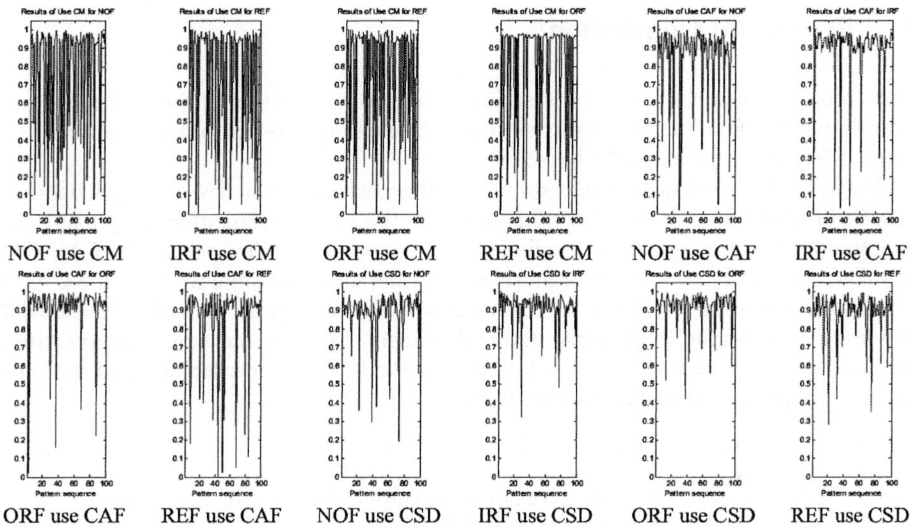

Fig. 6. Output of the cyclic statistics based ANN

the vibration signals of the outer-race, a charge amplifier B&K2635, and a low pass anti-aliasing filter YE3761. The computer hardware is a 2.4*G* Hz PⅣ PC with a high speed data access board PCI9118. The sampling frequency is $5120Hz$, and the number is 4096.

Using this equipment, it was possible to create four different conditions:

- *No fault applied (NOF)*: the rolling element bearing is in normal condition.
- *Only outer-race early fault applied (ORF)*: the rolling element bearing has a single early fault in outer-race.
- *Only inner-race early fault applied (IRF)*: the rolling element bearing has a single early fault in outer-race.
- *Only rolling element early fault applied (REF)*: the rolling element bearing has a single early fault on one rolling ball.

Examples of the measured vibrations of training and testing are shown in Fig.5. Here the training sets use the vibration signals of the rolling element with very serious faults, and the testing sets use the vibration signals of the rolling element bearing with early faults, so as to testify the system.

Table 2. Performance comparison about cyclic statistics based ANN

	Classification use CM				Ratio	Classification use CAF				Ratio	Classification use CSD				Ratio
	NOF	IRF	ORF	REF		NOF	IRF	ORF	REF		NOF	IRF	ORF	REF	
NOF	71	10	6	13	71%	88	4	2	6	88%	95	1	0	3	95%
IRF	12	76	4	8	76%	4	93	1	2	93%	1	98	0	1	98%
ORF	10	5	78	7	78%	3	1	95	2	95%	1	0	99	0	99%
REF	16	7	5	72	72%	5	3	2	90	90%	2	1	0	97	97%

4 Result and Conclusion

Fig.6 plots the results of output nodes for each testing condition of the rolling element bearing with different cyclic statistics. Table.2 illustrates the performance comparison about different cyclic statistics based ANN.

These results presented above show that the network trained with given data sets successfully performed early fault diagnosis for the rolling element bearing using pre-processed vibration signals by cyclic statistics processing. Comparing the CM, CAF and CSD, we find that it is much more appropriate and realizable to using the CAF to extract the invariable features of the vibration signal. Cyclic statistics based ANN condition monitoring and early fault diagnosis can replace some conventional methods currently in use. This work is also related to the representation of the input vector to multilayer feed-forward network.

Acknowledgements. The authors wish to acknowledge the financial support of the National Natural Science Foundation of China (Approved Grant: 50175068) and the Key Project supported by National Natural Science Foundation of China (No. 50335030).

References

1. Renwick, J.T.: Vibration Analysis-A Proven Technique as a Predictive Maintenance Tool. IEEE Transactions on Industry Applications, 21 (1985) 324-332
2. Cory, W.T.W.: Overview of Condition Monitoring with Emphasis on Industrial Fans. Processing of the Institution of Mechanical Engineers, Part A, 204 (1991) 225-240
3. Mayes, I.W.: Use of Neural Networks for On-Line Vibration Monitoring. Processing of the Institution of Mechanical Engineers, Part A, 208 (1994) 267-274
4. Lui, T.I., Mengel, J.M.: Intelligent Monitoring of Ball Bearing Conditions. Mechanical Systems and Signal Processing, Vol. 6, 5 (1992) 419-431

Application of BP Neural Network for the Abnormity Monitoring in Slab Continuous Casting

Xudong Wang, Man Yao, and Xingfu Chen

Department of Material Engineering, Dalian University and Technology, Dalian, 116023
Liaoning, People's Republic of China
hler@dlut.edu.cn

Abstract. The objective of this paper is to model the abnormal mould friction in continuous casting, using back propagation (BP) neural network. It is also discussed two capturing algorithms of abnormal characteristics, ample-change judgement and modified pass-zero-ratio judgement algorithms, which are designed to capture the abnormal modes, sharp pulse and big ramp. A set of software for mould friction abnormity has been developed, and the results of simulating prediction accord elementarily with the abnormal records in steel plant. Compared to the traditional technique, the BP neural network in combination with ramp and pulse judgements is much more convenient and direct, and can achieve a much better prediction effect.

1 Introduction

Approximately 80% of all steel produced today is continuously cast. In continuous casting, liquid steel is continuously cast into the mould which oscillates at a definite frequency and is strongly cooled by water, along with the mould length, newly forming shell will becomes thicker and the slab are drown out the mould. Mould behavior have a critical influence on the formation of an even shell and strand surface quality, while shell quality and breakout are closely related to the heat transfer and friction force between the newly forming shell and the oscillating mould. The interaction between shell surface and copper plates is expressed in terms of mould friction (MDF). Breakout and other MDF abnormities bring with much economic loss to continuous casting, so the method of abnormal prediction is a concerned question to steel plants and investigators, and much work has been done on these. Based on the analytical results of measuring MDF data by power method [1], it is found that the dramatic change in MDF can indicate a critical situation or an abnormity in the mould.

The object of this paper is to model and predict the abnormal mould friction in continuous casting by applying the method combining BP network and two capturing algorithms of abnormal characteristics, ample-change judgement and modified pass-zero-ratio judgement algorithms. Based on the method, a set of software for abnormity prediction is developed, and the results of simulating prediction for on-line measurement MDF data are discussed in the end of the paper.

F. Yin, J. Wang, and C. Guo (Eds.): ISNN 2004, LNCS 3174, pp. 601–606, 2004.

2 Model by Neural Network

Nowadays, the neural architectures have become more and more important as an effective learning technique in pattern recognition areas. The neural networks have strong abilities to learn and to self-organize information, and need only a few specific requirements and prior assumptions for modeling. These advantages have attracted much interest in the research on manufacturing process. Thus, in this study, the neural network analysis will be adopted to model the complicated MDF abnormities.

2.1 BP Neural Network

The most successful learning algorithm used to train multilayer neural networks is currently the BP scheme [2]. It is basically a gradient descent algorithm designed to minimize the error function in the weights space. And the BP network is also the most popular learning algorithm for multiplayer feedforward neural networks because of its simplicity, its power to extract useful information from examples, and its capacity of storing information implicitly in the connecting links in the form of weights [3]. In principle, it has been proved that any continuous function can be uniformly approximated by a neural network model with only one hidden layer. Therefore, a three-layer BP network is employed in our study.

2.2 Determination of Inputs and Output(s)

In order to construct a BP network, the number of input and output(s) should be determined. In learning and forecasting, *n* pairs of root mean square (RMS) of MDF and mould oscillation frequency ahead of *t-m* moment is assigned for the inputs, and the output is the forecast RMS of MDF at current moment *t* (shown in Figure.1). In view of the training time and the training precision, as is the case with m = 15, n = 15, these parameters might be chosen through many applications of try-and-error procedures. Therefore, there are thirty neurons in the input layer and one neuron in the output layer.

2.3 Decision of Number of Hidden Layer Neurons

It is not so easy to choose the appropriate number of neurons in the hidden layer because there is currently no definite rule to determine it except through

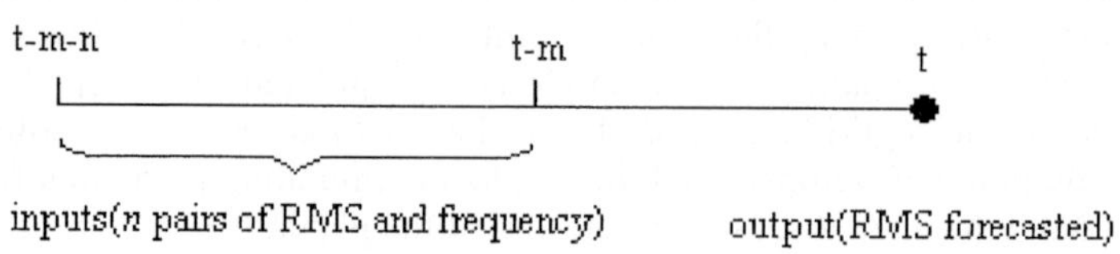

Fig. 1. Time schematic illustration for training and forecasting

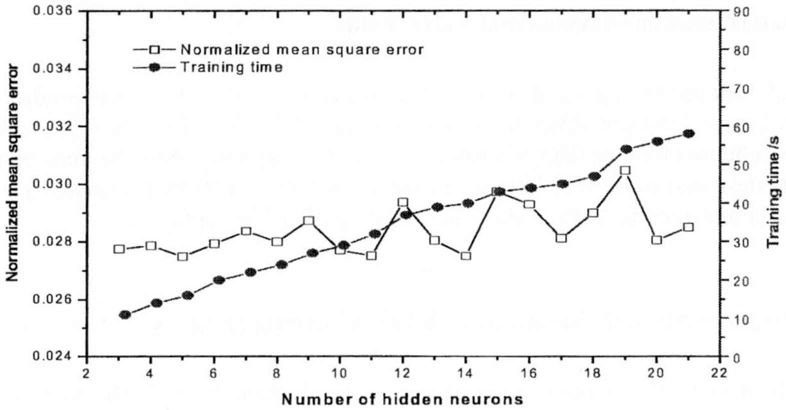

Fig. 2. NMSE and training time as a function of number of hidden neurons

experimentation. Too few neurons impairs the neural network and prevents the correct mapping of input to output., however, too many neurons impedes generation and increases training time. A common strategy, the one used in this study, is to replicate the training several times, starting with 3 neurons and then increasing the number while monitoring the normalized mean square error (NMSE) and training time. For all the BP models with different hidden neurons, they are trained with the same patterns using above BP algorithm, and the training stops after same iterations. Such experiment is carries out until there is no significant improvement in the NMSE with an increase in the hidden neurons. As shown in Fig.2 for this study, the optimized number of hidden neurons is 10.

2.4 The Data for Use in Training and Testing the BP Network

Data of typical normal and abnormal samples are gathered for use in training and testing the BP network, including the input patterns and the target output, one half of which are randomly chosen as training patterns and the other as testing patterns. These data basically contain the various distinct characteristics of the problem that the BP network is likely to encounter in the finished application. Preprocessing of the data is usually required before presenting the patterns to the BP network. This is necessary because the sigmoid activation function modulates the output of each neuron to values between 0 and 1. Various normalization strategies have been proposed. The following normalization procedure is commonly adopted and is used in this study.

$$Y = \frac{X - X_{\min}}{X_{\max} - X_{\min}} \times d_1 + d_2 \tag{1}$$

where X is the original value of RMS or frequency, Y is the normalized value, d_1 and d_2 are 0.8 and 0.1, respectively, and X_{min} and X_{max} *are* the minimum and maximum values in the data set.

2.5 Determination of Abnormal Valve Value

Through training by a great deal of on-line measuring data, in normal conditions, the maximal error between RMS forecasted and target RMS is less than 5%, while the error will be much more than 5% when abnormities happen. Thus, the abnormal valve value is designed to be RMS/20, when the error between RMS forecasted and target RMS is larger than the valve value, the abnormity is to be captured.

3 Judgements for Abnormal MDF Characteristics

It is sensitive to the abnormal conditions in production through the method which calculates the difference between measured value and forecasted value by BP network. At the same time, based on the analysis and summary of the abnormal time-region characteristics in the MDF curve before abnormities happened, MDF date often appear sharp pulse and big ramp fluctuation. In order to improve the prediction accuracy to the MDF abnormities, two capturing algorithms of abnormal characteristics, ample-change judgement and modified pass-zero-ratio judgement algorithms, are adopted to capture sharp pulse and big ramp abnormal modes [4].

3.1 Ample-Change Judgement

In order to capture the sharp pulse of MDF abnormities, the method based on range change judgement is applied in this study. The MDF of current moment is to be compared with the average value of continuous N entries of MDF current moment ahead, if the difference between them overruns the limitation set, the instance will be captured as a sharp pulse abnormity. This method is very sensitive to the state of ascending and descending of MDF, and the time lag is very small.

3.2 Modified Pass-Zero-Ratio Judgement

The modified pass-zero-ratio method is adopted to capture the abnormal ramp change of MDF. Firstly, each data of M entries of MDF current moment ahead will be subtracted by their average value, then the difference of symbol function is summed up, finally the pass-zero-ratio is calculated through dividing the sum by M. If the absolute value of pass-zero-ratio is closed to 1, it means that the MDF is in the ascending and descending trend.

4 Simulation Results

Based on the method which combines the BP network with two capturing algorithms of abnormal characteristics to predict MDF abnormity discussed above, a set of software for MDF abnormity prediction is developed using VC++ language. Based on

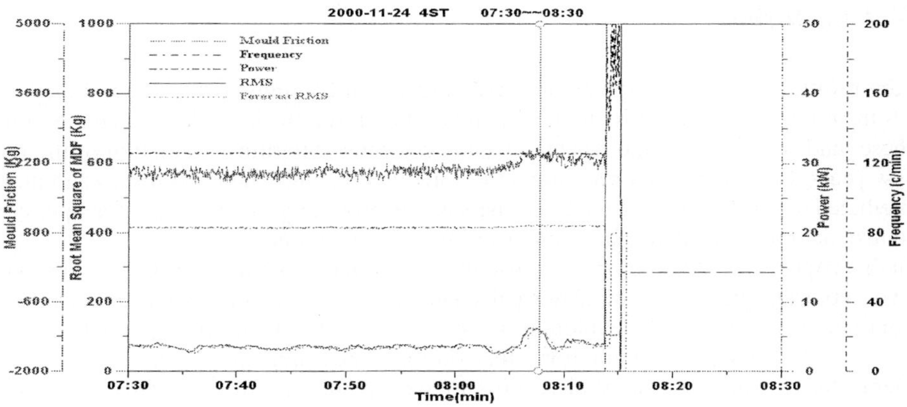

Fig. 3. The prediction result for breakout

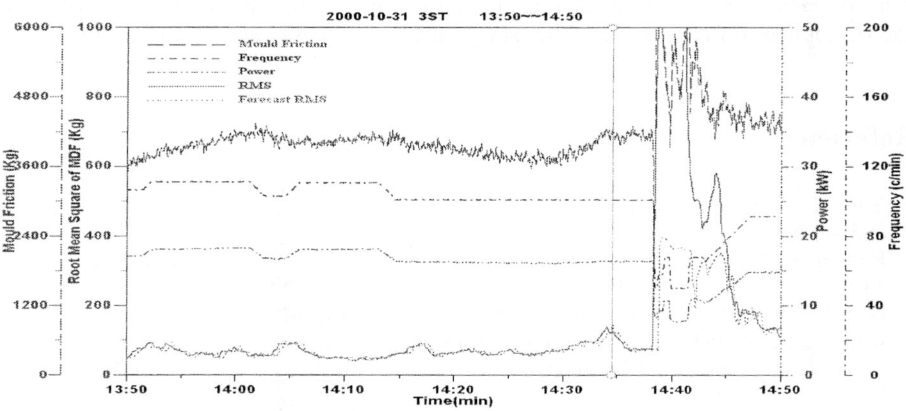

Fig. 4. The prediction result before submerged entry nozzle broken

the measuring data and abnormal records of continuous slab casting in steel plant, using the software to predict MDF abnormities, many simulations are carried out. Fig.3 and Fig.4 show two simulation results of prediction off-line. The vertical line in these figures describe that the abnormity is captured at that moment.

Fig.3 gives the fluctuation of measuring data before a breakout (8:14). It can be seen that RMS forecast is almost equal to the actual RMS in normal conditions, but the difference between them is increasing rapidly when abnormity happened. The software captures the abnormity of breakout at 8:08, which is 6 minutes ahead of breakout. The abnormal condition of submerged entry nozzle (through which molten steel flow into mould continuously) broken (14:48) is shown in Fig.4. The abnormity is captured in the middle of MDF ascending, and it is captured 4 minutes before abnormity happens.

5 Conclusion

Abnormities influence seriously on steady process in continuous casting, and they can result in huge economic loss to steel plants. Therefore, much work has been done on these, and some models have been developed for the abnormity prediction in the past few years. One of the important points for prediction is that the abnormities should be predicted several minutes ahead, for the sake of providing enough time for operators to take action to avoid the abnormities or alleviate economic loss.

In this paper, the authors propose to use the model that combines the BP network with two capturing algorithms of abnormal characteristics to predict MDF abnormity in continuous casting, and develop a set of software for abnormity prediction. The results of simulating prediction for on-line measurement MDF data accord elementarily with the abnormal records in steel plant, and the software can predict most abnormities such as breakout, submerged entry nozzle broken, mould level fluctuation and other abnormities. During simulation process, it is also found that the method can make prediction several minutes ahead of traditional temperature system warning. Thus, results of simulation predictions prove that the model is feasible and can be applied to future work on the abnormity prediction in continuous casting.

References

1. Yao, M.: On Line Measuring Method for Mould Friction in Continuous Casting. Ironmaking and Steelmaking, Vol. 23 (1996) 19-23
2. Rumelhart, D.E.: Learning Representation by BP Errors. Nature(London), Vol. 7 (1986) 149-154
3. Neural Application Corp.: Brimingham Steel. Intelligent Arc Furnace Operation at Birmingham Steel[J]. Steel Times International, Vol. 20 (1996) 20-21
4. Yang, J.H., Wang, G., Liu, W.Q.: Mould Friction Based Steel Leakiness Prediction of Oscillation Motor. Chinese Journal of Scientific Instrument., Vol. 23 (2002) 505-506

A TVAR Parametric Model Applying for Detecting Anti-electric-Corona Discharge

Zhe Chen, Hongyu Wang, and Tianshuang Qiu

School of Electronic and Information, Dalian university of technology,
Dalian, P.R. China 116023
eeyin@dlut.edu.cn
http://dsp.dlut.edu.cn

Abstract. It is well known that the time-varying parametric model based on wavelet neural network has excellent performance on modeling a signal. The spark discharge signal is a typical nonstationary one. The spark discharge has two types in a static dust catcher. one is normal discharge, another is anti-electric-corona discharge. A few simulations indicate that the performance of the TVAR model on modeling a discharge signal is fineness, especially, on distinguishing between normal discharge and anti-electric-corona discharge.

1 Introduction

The time-varying autoregressive (TVAR) parametric model is representative and used widely [1]. Assume $x(n)$ is a nonstationary random signal with zero-mean. Grenier proved that the TVAR parametric model does not always exist [2], In this paper, how to prove the model exists or not is not touched upon. it is supposed that the model exist is true, and its analytical expression is shown in (1).

$$x(n)+a(n,1)x(n-1)+\cdots+a(n,p)x(n-p)=e(n)\ . \tag{1}$$

Accordingly, its time-varying transform function which is denoted by $H(n, z)$ can be written as

$$H(n,z)=\frac{1}{1-\sum_{i=1}^{p}a(n,i)z^{-i}}\ . \tag{2}$$

Where, $a(n, i)$, $i = 0, 1, \ldots, p$, are the time-varying parameters of the model. $e(n)$ is the white noise with zero mean and σ^2 variance for exciting the model.

For some special random signals, Kalman filter algorithm, polynomial-algebraic algorithm etc. can be used to solve the parameters of model [1], [3]. But, decomposing $a(n, i)$ to linear combination of base function is a general method [1]. The base function can be chose freely. In general, They are at least linear independent, self-contained, may as well orthogonal to reduce the number of parameters (such as Walsh

F. Yin, J. Wang, and C. Guo (Eds.): ISNN 2004, LNCS 3174, pp. 607–612, 2004.

function). Assume that base function is $f_k(n)$, k = 0, 1, … and $s_i(k)$ are the weighted coefficients, then $a(n, i)$ can be written as

$$a(n,i) = \sum_{k=0}^{\infty} s_i(k) f_k(n) \ . \tag{3}$$

Put (3) in (2),

$$H(n,z) = \frac{1}{1 - \sum_{i=1}^{p} \sum_{k=0}^{\infty} s_i(k) f_k(n) z^{-i}} \ . \tag{4}$$

By this way, the problem with time-varying coefficients can be converted to the one with constant coefficients (all of $s_i(k)$ are constant). Certainly, other costs have to be suffered. Because of the number of unknown coefficients increasing, the chief cost is the scale of problem. When the time-varying problem is considered, the cost will be worth paying out. To get these coefficients(viz. $s_i(k)$), Recursive Least Square (RLS) algorithm or polynomial-algebraic method are available [1], [3]. For all practical purposes, limited precision $a(n, i)$ may be also accepted. Therefore, superior limit of k need not be up to ∞, Assume it is up to m-1. The "m" is named "base degree"[1].

The wavelet neural network (WNN) [4] for function approximation can be write as

$$f(t) = \sum_{k=0}^{m-1} w_k \varphi(\frac{t - g_k}{h_k}) + d \ . \tag{5}$$

Where, "m" is the number of wavelet function (viz. base degree); "g_k" is the kth shift factor; "h_k" is the kth scale factor; "w_k" is the kth weight coefficient; " $\psi(t)$" is a kind of mother wavelet. "d" is a constant for approximating to non-zero mean function. In expressions (1), substitute $a(n, i)$ by the WNN shown in expressions (5), and change wavelet to a discrete form. The TVAR based on WNN can be obtained as

$$x(n) + \sum_{i=1}^{p} [\sum_{k=0}^{m-1} w_{ik} \varphi(\frac{n - g_{ik}}{h_{ik}}) + d_i] x(n-i) = e(n) \ . \tag{6}$$

After the wavelet neural network is introduced into TVAR parametric model, How to estimate the parameters of model is converted to how to train the neural network. There are several training algorithms available to solve the model parameters in (6). Zhang used the random gradient algorithm in his WNN [4]. Szu used conjugation gradient algorithm to solve parameters [5]. Other algorithms, such as orthogonal Least Square algorithm are available [6]. To reduce computation cost, a new LMS algorithm based on input signal whitened is presented [7]. Because only part data is used in an iterative training, its computation cost will be very low.

2 The TVAR Parametric Model Based on WNN Applying for Detecting Anti-Electric-Corona Discharge

The electric sketch map of static dust catcher is shown in Fig.1(a). A high voltage charge source builts a static field between two pole board. When a dust particle passes the room between two pole board, ionization will happen. The ionization results in the dust particle is decomposed into a electron and a dust particle with positive charge. Any electriferous particle will be forced to move in electric field as shown in Fig.1(a). When a dust particle reach negative pole board, the electric neutralization makes it discharge its charge. When a dust particle passes through the median of two pole board, The minimum voltage to insure any dust will not escape from the dust catcher is

$$U_{\min} = \frac{md^2v^2}{qL^2} \, . \tag{7}$$

Where, d is the distance of two pole board. U is the voltage of two pole board. L is the length of pole board. m is the mass of a dust particle. q is the charge amount of a dust particle.

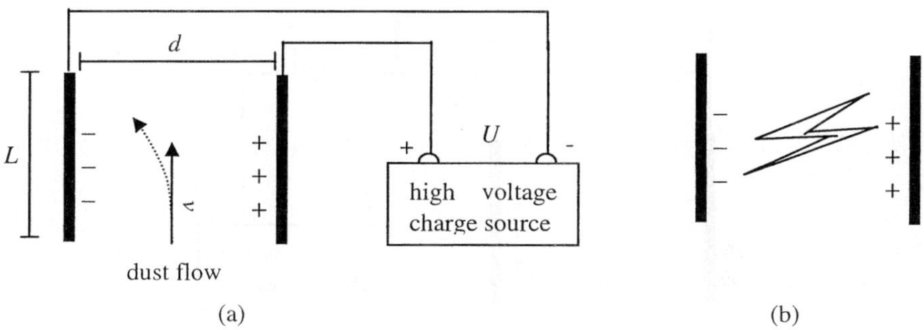

Fig. 1. Electric sketch map of static dust catcher

For a kind of dust, m and q are usually considered as a constant. When a dust catcher is made of, d and L are almost not changed. To process more dust particles, v is needed to increase. So U is also needed to increase. By limit of air discharge intensity, U does not increase unboundedly. Because the property of dust particles changes momently, discharge between two pole board may occur(shown in Fig.1(b)). The power of high voltage source is usually small. When a discharge occurs, the voltage between two board will be very low. A spark discharge will not transition an arc discharge. To avoid the discharge is very difficult, but, the times need to be controlled. If not, the efficiency of dust catcher will be very low.

After discharge, dust particles reveal neutral electricity character. But they still stay up in virtue of adhere force. Obviously, the discharge of dust particles on surface have to pass inner dust particles. While the resistance of dust layer is very high, it will results in discharging difficultly. In fact, dust layer on negative pole board comes into being a layer with positive charge, which is shown in Fig.2(a), and it is zoomed in

Fig.2(b). In Fig.2, the electric field largens the adhere force of dust particles, and dust is wiped off hardly by vibration. On the other hand, dust layer close with negative board, the discharge happens easily. (anyone can easily find the baby blue glimmer on negative board surface) This kind of discharge is called anti-electric-corona discharge. The anti-electric-corona discharge is usually expected, because it makes the electric field intensity descend, and results in dust can not be wiped off effectively.

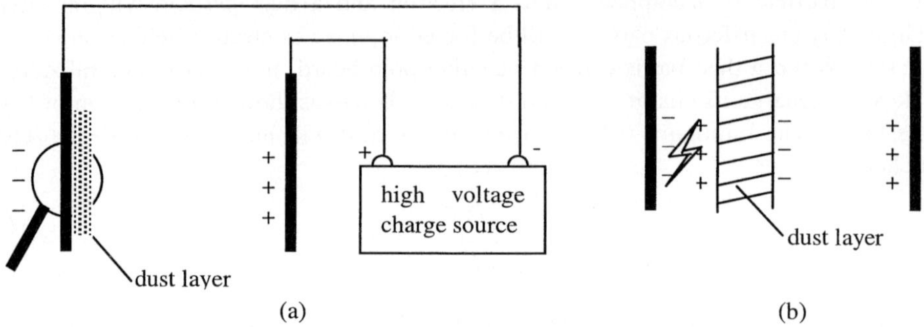

Fig. 2. Sketch map of anti-electric-corona discharge

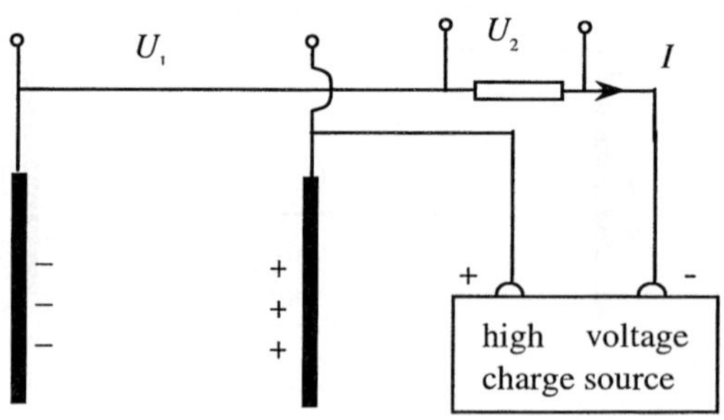

Fig. 3. Discharge signal collection sketch map

When a static dust catcher is working, its work state is needed monitor, especially, if anti-electric-corona discharge has occurred. If it is, all of steps must be adopted to wiped off dust on pole board. As we know, the mechanism of discharge is that there is a plasm channel between two electriferous object. Because plasm is very unsteady, the electric character of plasm is also nonstationary. How to monitor dust catcher also becomes very difficult. In this paper, The TVAR parametric model based on WNN is adopted for discharge signal modeling. By the TVAR parametric model, signal character can be extracted easily. In Fig.3, The signal collection sketch map is shown. In Fig.3, U_1 denotes the voltage between two board. U_2 denotes the current between two board by ohm law ($I = U_2/R$). They are collected by digital oscillograph. Usually, a broad way attenuation network is needed to avoid to damage the input device of oscillograph.

3 Simulation and Conclusion

For normal discharge and anti-electric-corona discharge, four sets data including voltage and current signal are collected. The total number of examples is twenty. The TVAR parametric model based on WNN is used to model the collection data, and the training algorithm in [7] is used to solve the parameters of model. In the TVAR model, p equals 2, base degree equal 3. The initial value of WNN parametrics are according to reference [5]. Because digital oscillograph is working in trigger-mode, The are a lot of noise data before-and-after valid data. To improve modeling effect, shorten the training time, All of original data are needed to preprocess properly. The preprocessing process includes seven step. They are (1)high pass filter with linear phase to wipe off DC component; (2) wipe off unuseful noise data by short time windowed energy; (3)obtain main component period (T_{main}) by classical power spectrum density; (4) extract valid data including about ten T_{main}; (5) high pass filter again with linear phase to wipe off DC component; (6)find out absolute maximum; (7) normalize all data by the absolute maximum.

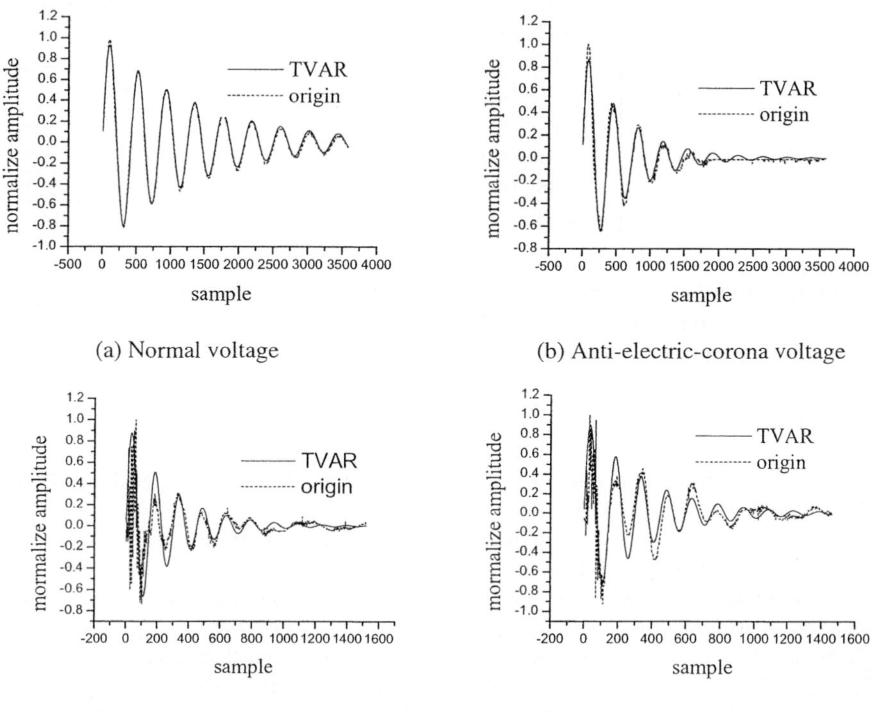

Fig. 4. A few modeling results (TVAR denotes the output of model)

A few results of modeling are shown in Fig.4. It can be found that the avail is acceptable. According to same procedure, every data example has its model. Based on these models, the equivalent attenuation coefficient of model can be calculated [8], and they are listed in table 1. The equivalent attenuation coefficients of normal discharge voltage fasten on 0.0007, and the ones of anti-electric-corona discharge voltage fasten on 0.0015. the equivalent attenuation coefficients of normal discharge current fasten on 0.0036, and the ones of anti-electric-corona discharge current fasten on 0.0033. They can be easily distinguish from each other, so, they can be parametric for monitoring dust catcher directly or input for other pattern sorter.

The wavelet transform has excellent character on time-frequency field, and it is also called microscope on math. The neural network has powerful learning ability, and it is fit for nonlinear problem. The WNN combine. When the WNN is introduced in the TVAR model, their virtue is exerted, and very fit for model of nonstationary signal.

Table 1. Equivalence attenuation coefficients by algorithm in this paper

discharge state(voltage)	coefficients	discharge state(current)	coefficients
normal	7.263e-004	Normal	3.649e-003
	7.120e-004		3.646e-003
	7.584e-004		3.827e-003
	7.193e-004		3.800e-003
	7.088e-004		3.538e-003
anti-electric-corona	1.623e-003	anti-electric-corona	2.946e-003
	1.587e-003		3.515e-003
	1.597e-003		3.346e-003
	1.548e-003		3.437e-003
	1.633e-003		3.257e-003

Acknowledgements. The research is supported by: National Natural Science Foundations of China (No. 60172072, No. 60372081).

References

1. Hongyu, W.: Nonstationary Random Signal Analysis and Processing. National Defence Industry Press of China (1999)
2. Grenier, Y.: Time-Dependent ARMA Modeling of Nonstationary Signals. IEEE Trans. on ASSP, Vol.31 (1983) 899–911
3. Mrad, R.B.(ed.): A Polynomial-Algebraic Method for Nonstationary TARMA Signals Analysis—Part I: the Method. Signal Processing, Vol.65 (1998) 1–19
4. Zhang, Q., Benveniste, A.: Wavelet Networks. IEEE Trans. on NN, Vol.3 (1992) 889–898
5. Szu, H.(ed.): Neural Network Adaptive Wavelets for Signal Representation and Classification. Optical Engineering, Vol.31 (1992) 1907–1916
6. Chen, S.(ed.): Orthogonal Least Squares Learning Algorithm for Radial Basis Function Networks. IEEE Trans. on NN, Vol.2 (1991) 302–309
7. Chen, Z.(ed.): A TVAR Parametric Model Based on WNN. Nanjing: ICNNSP (2003) 802–805
8. Taibin, C.: Circuit Analysis Tutorial. Beijing: Publish House of Electronic Industry of China (2003)

The Study on Crack Diagnosis Based on Fuzzy Neural Networks Using Different Indexes

Jingfen Zhang [1], Guang Meng [1], and Deyou Zhao[2]

[1] The State Key Laboratory of Vibration, Shock& Noise, Shanghai JiaoTong University, Shanghai 200240
jingfenzhang@sjtu.edu.cn, gmeng@sjtu.edu.cn
[2] Department of Naval Architecture, Dalian University of Technology, Dalian,116023

Abstract. Fuzzy neural networks fault diagnosis technology and diagnosis mode are used to diagnose cracks. It is trained with promoted BP arithmetic. The faults of cracked cantilever plate are diagnosed. Firstly the mode and frequency of numerical simulation intact plate and different cracked plates are calculated. Then five crack diagnosis indexes are calculated. Divide five indexes into three groups and create three fuzzy neural networks. The fuzzy neural networks are trained using these indexes, and diagnosis is taken to the crack in the end.

1 Introduction

Crack is a common form of structural damage. Plate and beam are the basic element of engineering structure [1]. Diagnosis to thin plate crack is taken as an example in this paper. Fault diagnosis based on fuzzy neural networks adapts to fault feature of fuzzy boundary, multi-levels and randomness. It can make good use of self-adaptive and self-learning ability of neural networks, and the strong encoding ability to scribe the expert language of fuzzy sort.

2 Fuzzy Neural Network Model [2]

A simplified fuzzy neural networks model is used. As to the structural crack, we use a half-liter concave convex function as the membership function:

$$\mu_{K_j}(x_i) = \begin{cases} 0, & 0 \le x_j \le a_j \\ a_j(x_j - a_j)K_j, & a_j < x_j \le a_j + 1/a_j^{1/k} \\ 1, & x_j > a_j + 1/a_j^{1/k} \end{cases} \qquad j = 1,2,\cdots 6 \tag{1}$$

F. Yin, J. Wang, and C. Guo (Eds.): ISNN 2004, LNCS 3174, pp. 613–618, 2004.

where K is the crack identifying index. a is the x value of crossing point between half lifter concave convex function and x axis. Each node of the second level represents a fuzzy rule to calculate the relevance grade of each rule:

$$\alpha_j = \mu_1^i \cdot \mu_2^{i2} \cdots \mu_n^{in} \tag{2}$$

The output of third level is fuzzilization value and normalizes it:

$$\overline{\alpha}_j = \frac{\alpha_j}{\sum_{i=1}^{m} \alpha_i} \qquad j = 1,2,\cdots m \tag{3}$$

Take separation angle cosine function as the output assorting membership grade:

$$\mu_{ij} = \frac{\sum_{k=1}^{n} x_{ik} . c_{jk}}{\sqrt{\sum_{k=1}^{n} x_{ik}^2} \cdot \sqrt{\sum_{k=1}^{n} c_{jk}^2}} \tag{4}$$

where c_{jk} is the k th eigenelement of the j th mode center c_j:

$$c_{jk} = \sum_{m=1}^{n_j} \frac{x_{mk}}{n_j} \qquad k = 1,2,\cdots,n \tag{5}$$

The following is learning ratio adjustment algorithms:

$$\begin{cases} \Delta R = R(n) - R(n-1) \\ \Delta = \mathrm{sgn}(\Delta R) \left| \ln[R(n) + 0.5] \right| \\ \lambda = +\varsigma\Delta \\ \eta(n+1) = \lambda\eta(n) \end{cases} \tag{6}$$

The weight adjustment equation is:

$$w_{ji}(k+1) = w_{ji}(k) + \eta\delta_j O_i + \alpha \cdot [w_{ji}(k) - w_{ji}(k-1)] \tag{7}$$

3 FEM Analysis Stimulation to Cantilever Thin Aluminum Plate Crack [3]

There is a cantilever thin aluminum plate with one side fixing, which is 0.4m long and 0.2 wide. Its Young's modulus $E = 68.5GPa$, Poisson's ratio $\mu = 0.34$, density $\rho = 2700Kg/m^3$ and thickness is 0.005m.The FEM model to calculate the cantilever plate is shown as figure 1, the element mesh size is 0.01m. We use the change of mode and natural frequency to diagnose the cracks.

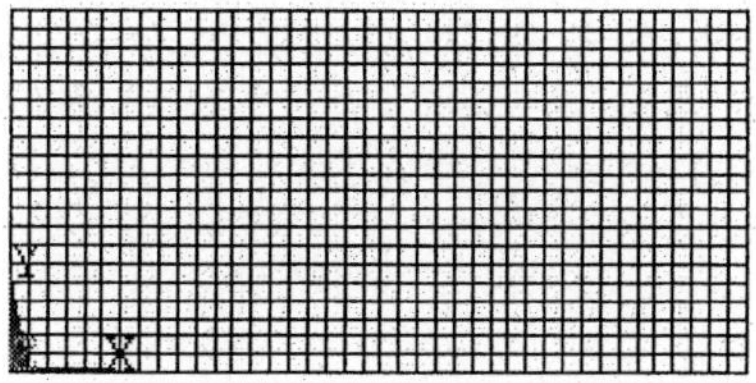

Fig. 1. FEM analysis mode of cantilever plate

4 Crack Diagnosis Index Choosing

4.1 Mode Tangential Angle Variance Ratio Index [4]

$$\{\Delta tg\alpha\}_{j,i} = \{tg\alpha\}_{j,i+1} - \{tg\alpha\}_{j,i} \tag{8}$$

$$\{tg\alpha\}_{j,i} = \frac{\{u_x\}_{j,i+1} - \{u_x\}_{j,i}}{\Delta l_x} \tag{9}$$

$$\{tg\alpha\}_{j,i} = \frac{\{u_y\}_{j,i+1} - \{u_y\}_{j,i}}{\Delta l_y} \tag{10}$$

where $\{u_x\}_{j,i+1}$ and $\{u_x\}_{j,i}$ is respective the ($i+1$)th and i th mode of j th range in y direction tangent line. Δl_x is the position distance between ($i+1$)th and i th.

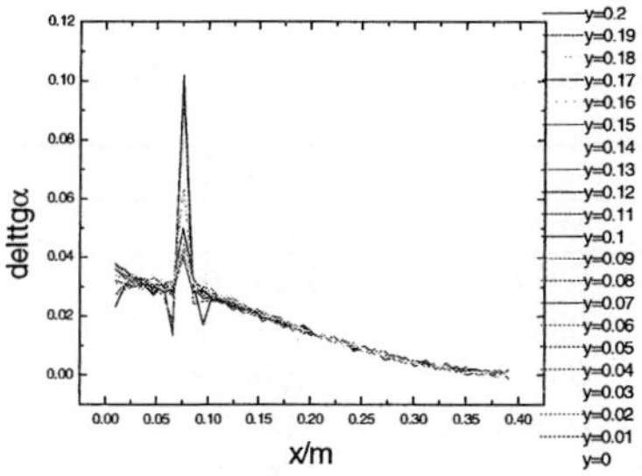

Fig. 2. $\Delta tg\alpha$ of a cracked plate

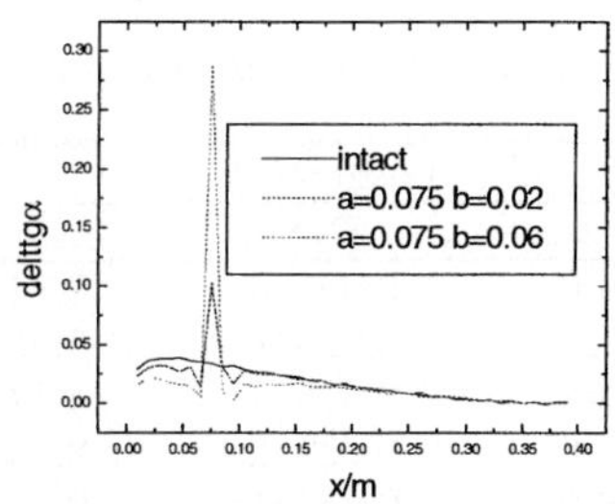

Fig. 3. $\Delta tg\alpha$ of different cracked plate

The above is the mode tangential angle variance ratio of a marginal straight crack of cantilever plate (see figure 2). Figure 3 is the mode tangential angle variance ratio of intact plate and cracked plate.

4.2 DSN Index [5]

$$\{DSN_j\}=\frac{DS_j}{\sum_{j=1}^{n}|DS_j|} \tag{11}$$

$$\{DS_i\}=\frac{\{\Delta\varphi_i\}}{\Delta f_i^2} \tag{12}$$

4.3 Rate Decline of The Natural Frequency

$$\Delta f=\frac{f_{i^o}^2-f_{i'}^2}{f_{i^o}^2}\times 100\% \tag{13}$$

where f_{i^o} is the i th natural frequency of intact plate. $f_{i'}$ is the i th natural frequency of cracked plate.

5 Crack Diagnosis Using Different Indexes

Take $\Delta tg\alpha$ and Δf, DSN and Δf, φ and f and Δf as different three groups to diagnose cracks based on FNN [6-7]. Choose 62 different crack as training samples. Another 10 different crack as the detection samples. The following tables show the diagnosis result. The analysis to the results is given in these tables at the same time.

Table 1. Diagnosis result to single crack based on $\Delta tg\alpha$ and Δf

Position/m			Length/m		
Actual	FNN	Error	Actual	FNN	Error
0.100	0.09266	7.34	0.005	0.00446	12.83
0.025	0.02642	5.66	0.068	0.07289	7.19
0.018	0.01879	4.37	0.092	0.09693	5.36
0.250	0.22955	8.18	0.078	0.08305	6.47
0.300	0.27231	9.23	0.052	0.04715	9.32
0.034	0.03182	6.41	0.041	0.03690	10.01
Depth/m			**Angle/degree**		
Actual	FNN	Error	Actual	FNN	Error
0.005	0.00479	4.14	90	83.916	6.76
0.005	0.00483	3.42	90	87.408	2.88
0.005	0.00485	3.07	45	47.763	6.14
0.005	0.00470	5.97	60	54.012	9.98
0.005	0.00468	6.35	90	80.172	10.92
0.003	0.00271	9.76	90	92.133	2.37

Table 2. Fault diagnosis result to single crack based on DSN and Δf

Position/m			Length/m		
Actual	FNN	Error	Actual	FNN	Error
0.100	0.08846	11.54	0.005	0.00431	13.72
0.025	0.02697	7.89	0.068	0.07435	9.34
0.018	0.01894	5.23	0.092	0.09846	7.02
0.250	0.22470	10.12	0.078	0.08464	8.51
0.300	0.26616	11.28	0.052	0.04584	11.84
0.034	0.03116	8.36	0.041	0.03580	12.69

Depth/m			Angle/degree		
Actual	FNN	Error	Actual	FNN	Error
0.005	0.00468	6.37	90	82.863	7.93
0.005	0.00476	4.81	90	85.851	4.61
0.005	0.00476	4.74	45	48.897	8.66
0.005	0.00465	7.09	60	52.812	11.98
0.005	0.00460	7.92	90	77.067	14.37
0.003	0.00258	14.05	90	95.769	6.41

Table 3. Diagnosis result to single crack based on φ, f and Δf

Position/m			Length/m		
Actual	FNN	Error	Actual	FNN	Error
0.100	0.08369	16.31	0.005	0.00413	17.49
0.025	0.02802	12.07	0.068	0.07744	13.88
0.018	0.01983	10.19	0.092	0.10319	12.16
0.250	0.20605	17.58	0.078	0.08897	14.07
0.300	0.24171	19.43	0.052	0.04319	16.95
0.034	0.02929	13.84	0.041	0.03357	18.11

Depth/m			Angle/degree		
Actual	FNN	Error	Actual	FNN	Error
0.005	0.00438	12.41	90	77.562	13.82
0.005	0.00460	7.99	90	81.927	8.97
0.005	0.00460	8.03	45	51.4485	14.33
0.005	0.00452	9.69	60	50.448	15.92
0.005	0.00450	9.96	90	72.675	19.25
0.003	0.00242	19.17	90	101.106	12.34

6 Conclusions

The diagnosis result indicates that crack diagnosis based on fuzzy neural networks is an effective intelligent technique. Different crack diagnosis indexes group bring different diagnosis precision. The diagnosis precision using $\Delta tg\alpha$ and Δf indexes group is much better than others. The diagnosis precision to longer crack is higher than that of small cracks. Full crack diagnosis precision is higher than that of partial cracks. The diagnosis precision to straight crack is higher than that of oblique cracks. The diagnosis precision to crack nearby fixing side is higher than that of far fixing side cracks.

Acknowledgements. The authors wish to acknowledge the financial support of the National 863 Plan (2002AA412410) and the Key Project supported by National Natural Science Foundation of China (No.50335030).

References

1. Zaihua Liu, Guoxing Dong, Wen'an Wang: Engineering Structure Breaking Resistance Design Basis. Wuhan, Huazhong University of Technology Press. (1990) 3-5
2. Liming Zhang: Artificial Neural Networks Model and Application. Xi'an, Xidian University Press. (1995) 25-28
3. Jingfen Zhang: Research on Intelligent Diagnosis to Equipment Fault and Structural Crack Based on Vibration. Dalian, Dalian University of Technology. (2003) 54-76
4. Grimes P J: Advancements in Structural Dynamic Technology Resulting from Saturn Programs. NASA CR-1539 and CR-1540. 1 (1970) 6
5. Fox C H J: The Location of Defects in Structures: A Comparison of Natural Frequency and Mode Shape Data. In Proceedings of the 10th International Modal Analysis Conference. USA. (1992) 522-528
6. Jingfen Zhang: Fault Diagnosis to Engineering Structure and Mechanical Equipment Based on Fuzzy Neural Network. Dalian, Dalian University of Technology. (2003) 20-35
7. Koenelija Zgonc, Jan D. Achenbach: A Neural Network for Crack Sizing Trained by Finite Element Calculations. NDT&E International. 29 (1996) 147-155

Blind Fault Diagnosis Algorithm for Integrated Circuit Based on the CPN Neural Networks

Daqi Zhu[1], Yongqing Yang[2], and Wuzhao Li[1]

[1] School of Communication and Control Engineering Southern Yangtze University
Wuxi, Jiangshu, Province, China, 214036
zdq367@yahoo.com.cn

[2] Department of Mathematics Southern Yangtze University
Wuxi, Jiangshu, Province, China, 214036
yangyongqing@sytu.edu.cn

Abstract. A blind diagnosis method of photovoltaic radar digital and analog integrated circuit based on the CPN neural networks is presented. By measuring the temperature and voltage of circuit component, the membership functional assignment of two sensors to circuit component is calculated, and the fusion membership functional assignment is gained by using the CPN neural networks, then according to the fusion data, the fault component is found. Comparing the diagnosis results based on separate original data with the ones based on CPN fused data, it is shown that the blind diagnosis method is more accurate.

1 Introduction

In analog circuit fault diagnosis, according to the different stages of testing process, we can divide the method of the analog circuit fault diagnosis into two types: one is before-testing simulation[1] and the other is after-testing simulation[2].In these methods, no matter before-testing simulation diagnosis or after-testing simulation method, generally, we must analyze the working principle and inner structure of the circuits, and know some information about the circuit, then undergo the diagnosis. However, in many cases, it's very difficult for us to get such information, and this limits the effectiveness of these diagnosis. For example, in an electronic component fault hunting system of a certain type airplane photovoltaic radar[3], although we know the fundamental functions of many electronic components, we don't understand their inner structure, and it's still difficult to find precisely fault components by using general diagnosis methods.

In this paper the multi-sensor information fusion technique[4-6] based on CPN(Counter propagation Network) is introduced for photovoltaic radar electronic components fault diagnosis. By using multi-dimension signal processing method of information fusion, it can diminish the uncertainty of analog electronic components fault diagnosis, and exactly orient the fault components. Moreover, it need not know the principle and structure of the circuit. It is a blind diagnosis.

F. Yin, J. Wang, and C. Guo (Eds.): ISNN 2004, LNCS 3174, pp. 619–626, 2004.

2 CPN Neural Networks Information Fusion Fault Diagnosis

The construction of multi-sensor information fusion system is shown in Figure1. The key voltage of each circuit component can be gotten by using probes. The temperature of each component is measured by using thermal image instrument (inframetrics 600). According to fuzzy logic theory, for each sensor, the fault possibility of the tested component can be described by a set of membership function values. Then two fuzzy sets of membership function values can be gotten, which may appear two cases: one is for the mutual affection of each component in an electric circuit, there has sometimes some misjudgement when using a signal sensor to distinguish a fault component; secondly if two sensors give different membership function values, it will be much harder to judge a fault component. An approach is to make a fuzzy neural network information fusion classify model, the model combines fuzzy logic with CPN neural network, and the input and output data are the meaningful membership function values. During fault diagnosis, consider the membership function values of voltage and temperature tested as input data of CPN network. And output data are the fusion fault membership function values. By using the fusion fault membership function values, the fault component can be determined on certain fault determination criterion.

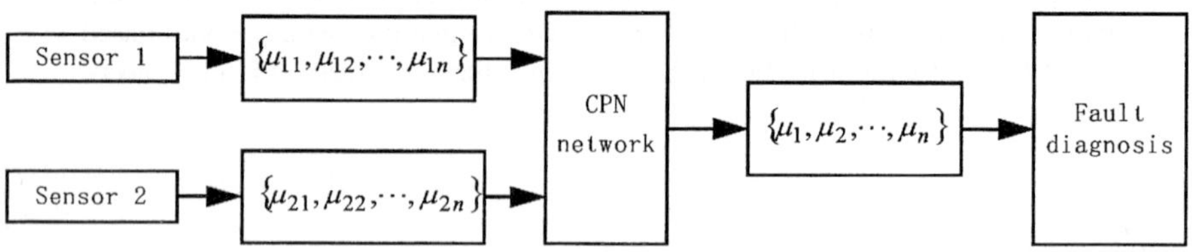

Fig. 1. Fault diagnosis based on multi-sensor information fusion

2.1 Forms of Membership Function

Membership function is designed by the working characteristics of the sensors and the characteristics of the parameters to be tested. For a certain component in the electric circuit system, when the system is working properly, the voltages of component key points should be stable and the temperature should be a fixed value. If there are some fault components in the system, generally the voltage values will deviate the normal range and the temperatures will have a change (increasing or decreasing).

The bigger the deviations are, the more fault possibility will be. For a convenience, we define the distribution of membership function $\mu_{ij}(x)$ as shown in Figure 2. The formula (1) is the distribution of membership function $\mu_{ij}(x)$. X_{oij} is the standard parameter value of the tested component when the electric circuit is working properly; e_{ij} is the normal changing range of the tested component parameters; t_{ij} is the margin deviation of the parameter of the component to be tested; $\mu_{ij}(x)$ is the membership function value of the component *j* tested by the sensor *i*; X_i is the practical measuring value of the sensor *i*.

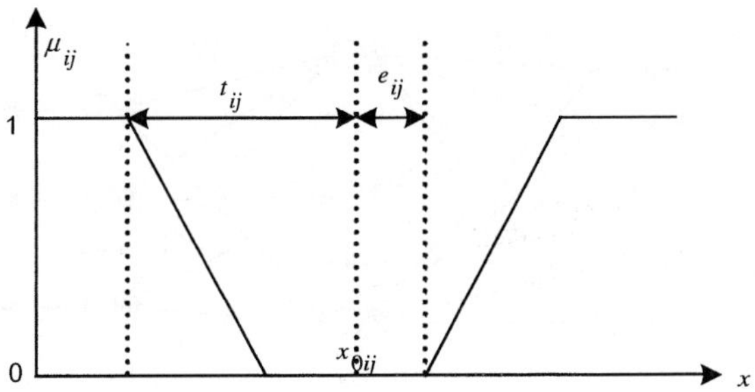

Fig. 2. Membership function distribution

$$\mu_{ij}=\begin{cases}1 & x_i\le 0\\ -\alpha(x_i-x_{0ij}+e_{ij})/(t_{ij}-e_{ij}) & 0<x_i\le x_{0ij}\\ 0 & x_i=x_{0ij}\\ \alpha(x_i-x_{0ij}-e_{ij})/(t_{ij}-e_{ij}) & x_{0ij}<x_i\le 2x_{0ij}\\ 1 & x_i>2x_{0ij}\end{cases} \tag{1}$$

2.2 CPN Neural Networks Model

CPN is a new mapping neural network[7-8] by combining Kohonen and Grossberg learning algorithm, and it's widely used in data classification, pattern recognition, associative memory and so on. In this paper, it was used in information fusion to recognize the fault components. The CPN has three layer neurons, the input layer: receives the input information; the middle layer is Kohonen competition layer and it is connected to the input layer through weight matrix V; Grossberg output layer is connected to the middle layer through weight matrix W. The structure of CPN is shown in Figure 3. When applying the CPN information fusion to recognize the fault pattern, the input and output of CPN are value that has certain physical meaning. In the figure, the input layer corresponds to the feature of fault(the membership function value of voltages and temperature) and the output layer corresponds to reasons(fault component). In practical data fusion fault diagnosis system of electric circuit, the applied network model has 2n input nodes to represent fault membership function values of the component tested by the two sensors. There are n nodes in the output layer to represent the membership function values of the fusion data(n is the number of component tested). And there is one hide layer.

2.3 Fault Component Judge Criterion

To each fused component, using methods based on rules to recognize the fault pattern, its specific contents are as follows:①Object patterns to be judged should have

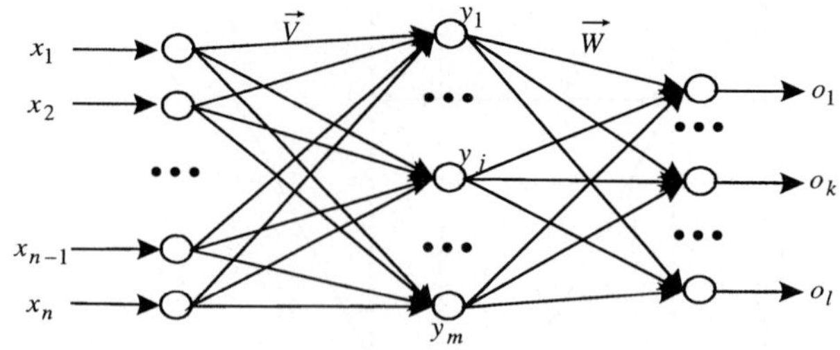

Fig. 3. CPN network stucture.

maximal membership function value; moreover, it should be bigger than a certain value. Generally, this value should be at least bigger than 1/n (n stands for the number of components to be tested), and the bigger the threshold value is and the more precise the judgment should be. However, if the threshold value is too big, membership function value from testing will not satisfy the requirement. Therefore, we should, according to the real situation, choose a moderate one, such as 0.7 in the diagnosis example given in this paper.②The difference of the function value between object patterns with other patterns should be bigger than a certain threshold value, such as 0.50 in the example of this paper.

3 Photovoltaic Radar Electronic Components Fault Diagnosis

In a certain type plane photovoltaic radar electronic components fault diagnosis system, the "fault tree" method is adopted to search fault components[3], Testing voltage of the components on specific circuit board and comparing it with the normal signals, we can make a proper judgment. Although this method is simple and convenient, its diagnosis accuracy is poor. The reason lies in two aspects: when a fault in an analog circuit occurs, not only the fault component's output signal but also the conjoint components corresponding signals will be distorted, because the conjoint components functions are affected. In other words, for the mutual influence of each component electric signal, it is hard to judge correctly whether a component has some fault by testing only the voltage or electric current of the component. On the other hand, for some imported components, we don't know their inner structure, the "fault tree" may not be well rationalized. So we introduce CPN to fault components hunting. We use two kinds of sensors to test electronic component fault information from different aspect, then conduct neural network information fusion and judge the fault components. This method need not know the knowledge of principle and structure of the electronic components, and it is a blind diagnosis method. Figure 4 is a picture of voltage-code changing circuit board for fault diagnosis experiment, and its main function is to change analog voltage into eleven bit digital signal. There are five electric component(A_1,A_2,A_3,A_4,A_5) are fault components doubted. U_1,U_2 ,U_3 ,U_4 and U_5 are the key voltages of the component 1,2,3,4 and 5 respectively. Firstly, the tem-

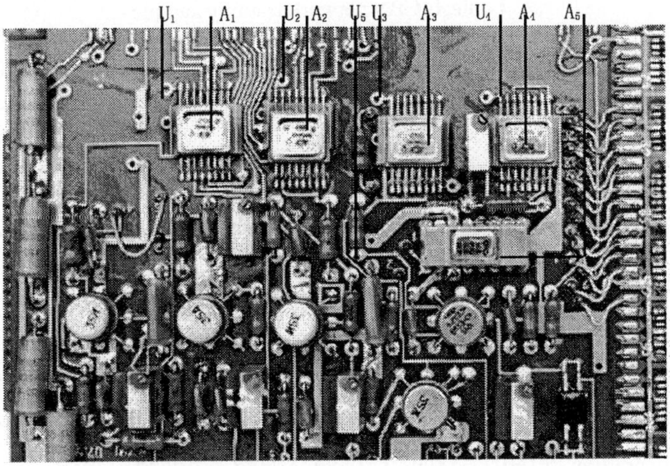

Fig. 4. The A/D circuit of photovoltaic radar diagnosed

perature of each component is measured by a thermal image instrument (inframetrics 600) when the circuit is working properly. Secondly, if there is some fault in the cir cuit, the fault components temperature will change (increase or decrease), so the new temperature of each component is measured, and the temperature fault membership function value can be calculated using the formula given before. In addition, the voltage of each key point can be gotten by using probes, and the voltage fault membership function values can be gotten also.

3.1 Structure of CPN and Training Sample

In this experiment, there are five object components and two sensors, so the neuron number of network input should be 10 and the neuron of output should be 5; the neuron number of the middle hidden layer is m=12; the initialized weight is a normalized random value. Membership function is defined mainly by the characteristics of sensors and object parameters, and for a certain component in electronic equipment, the distribution of its membership function $\mu_{ij}(x)$ can be illustrated by formula (1). To simplify the problem, and under the condition of not changing the fault characteristics, we choose $e_{ij} = 0$, $t_{ij} = x_{0ij}$, voltage sensor α=1/3, temperature sensor α=2. 8, and set different artificial fault, under different input conditions, calculate the membership function value of each component tested by the two sensors. Each group of sensor membership function value is normalized and then used as the input vector of network training sample; the output vector is defined by the fault components, that is, for the real fault components, their corresponding neuron output is 1, while for other components, the output is 0. Table 1 shows the membership function

Table 1. CPN network training sample

No.	Sensor	Input membership function value					Output value				
		A_1	A_2	A_3	A_4	A_5	A_1	A_2	A_3	A_4	A_5
1	Voltage	.2746	.0313	.3511	.0708	.2722	1	0	0	0	0
	Temperature	.5608	.2143	.0454	.0891	.0902					
2	Voltage	.0144	.5152	.2656	.0108	.1939	0	1	0	0	0
	Temperature	.1000	.6185	.1905	.0000	.0921					
3	Voltage	.0113	.0326	.6900	.2158	.0505	0	0	1	0	0
	Temperature	.0852	.0760	.6314	.1395	.0679					
4	Voltage	.0393	.3132	.2212	.3875	.0387	0	0	0	1	0
	Temperature	.1039	.0000	.0000	.7957	.1004					
5	Voltage	.0101	.2838	.2486	.0069	.4506	0	0	0	0	1
	Temperature	.1143	.0892	.0616	.1779	.5569					
6	Voltage	.3272	.0613	.2775	.0747	.2593	1	0	0	0	0
	Temperature	.6169	.1489	.0489	.0966	.0887					
7	Voltage	.0168	.5029	.2698	.0139	.1964	0	1	0	0	0
	Temperature	.0050	.7004	.1893	.0045	.1008					
8	Voltage	.0112	.0350	.7055	.1160	.1570	0	0	1	0	0
	Temperature	.0882	.0761	.6382	.1228	.0748					
9	Voltage	.0466	.3047	.2121	.3964	.0402	0	0	0	1	0
	Temperature	.1011	.0055	.0114	.7826	.0994					
10	Voltage	.0078	.2761	.2444	.0067	.4641	0	0	0	0	1
	Temperature	.0996	.0853	.0572	.1762	.5816					

value of each object component on the voltage-code changing circuit board, that is, input and output of CPN training sample.

3.2 Discussion of Fault Diagnosis Results

According to the above fusion algorithm and fault judging rules, the fault diagnosis result of the circuit is shown in Table 2. The first and the second groups of the table is respectively the membership function value of component tested by temperature sensor and voltage senor, the third group shows the membership function value of each component after fusion. The table shows distinctly that fault membership function values gotten by the two sensors respectively are sometimes very similar for the five components diagnosed in photovoltaic radar electric circuit system, the single sensor can not judge the fault component correctly using the judge criterion given before, but fusion membership function values can judge the fault component. For example, when component 1 is the real fault component, the membership function values of component 1 and ones of component 3 from voltage sensor are very similar, and we can not decide the fault component. But after fusion, the membership function

Table 2. Comparison of diagnosis results by single sensor and data fusion

Fault comp.	Sensor	Fault membership function value					Diagnosis results
		A_1	A_2	A_3	A_4	A_5	
A_1	Voltage	.3390	.0513	.3559	.0533	.2011	uncertain
	Temperature	.5677	.1378	.0579	.0971	.1394	uncertain
	Fusion	.7095	.0266	.1217	.0410	.1141	A_1 failure
A_2	Voltage	.0718	.5012	.2138	.0196	.1936	uncertain
	Temperature	.0091	.6822	.1804	.1032	.0251	uncertain
	Fusion	.0340	.8021	.0508	.0334	.0813	A_2 failure
A_3	Voltage	.0113	.0326	.6900	.2158	.0505	uncertain
	Temperature	.0852	.0760	.6314	.1395	.0679	uncertain
	Fusion	.0623	.0122	.7881	.0331	.1045	A_3 failre
A_4	Voltage	.0393	.3132	.2212	.3875	.0387	uncertain
	Temperature	.1039	.0000	.0000	.7957	.1004	A_4 failure
	Fusion	.0132	.0231	.0491	.8532	.0616	A_4 failure
A_5	Voltage	.0718	.2012	.2138	.0796	.4336	uncertain
	Temperature	.0901	.0782	.0804	.1932	.5581	uncertain
	Fusion	.0421	.0520	.1278	.0501	.7315	A_5 failure

value of component 1 increased greatly and is much higher than the values of other components, we can point out the fault component correctly. In other words, the information fusion algorithm increases the objective membership function value assignment and decreases the values of the other components. This makes the uncertainty of the diagnosis system decrease greatly and error diagnosis has been removed from the inadequacy of single sensor information. In the example, the correction of fault diagnosis can reach 100%. So multi-sensor information fusion based on CPN network and fuzzy logic improves the analyzable of the system and the judgment ability of fault models, and greatly increases the correction of fault component decision.

It can be seen that the data used in table 2 is not the sample data in table 1, which indicates that when applying the CPN to realize multi-sensors information fusion, it has certain generalization ability. So this method has a good adaptability. Additionally, in the whole fault diagnosis process, we needn't know too much about the inner principle and structure of the electronic components, and it's a blind diagnosis method for fault components hunting.

4 Conclusion

We can see from the experiment results that, as long as the components diagnosed are properly chosen, and the signals are precisely tested, then the fault components can be precisely recognized by using the multi-sensor CPN neural network information fusion.

References

1. Chakrabarti S.and Chatterjee A.: Compact fault dictionary construction for efficient isolation. Proc. 20th Anniversary Conf. Advanced Research in VLSI, (1999)327-341
2. Chen Y Q.: Experiment on fault location in large-scale analogue circuits. IEEE Trans. On IM. Vol.42. 1(1993)30-34
3. Daqi Zhu, Shenglin Yu, Wenbo Liu.: study of electronic components fault diagnosis based on fault tree and artificial apparatus. Journal of instrument and meter, Vol.16. 1(2002)16-19
4. Luo R.C. and Scherp R.S.: Dynamic multi-sensor data fusion system for intelligent robots. IEEE J. Robotics and Automation, Vol.4. 4(1998)386-396
5. Gao P., Tantum S. and Collins L.: Single sensor processing and sensor fusion of GPR and EMI data for landmine detection. Proc. SPIE Conf. Detection Technologies Mines Mine-like Targets, Vol.3710. Orlando(1999)1139-1148
6. Milisavljevic N. and Bloch I.: Sensor fusion in anti-personal mine detection using a two-level belief function model. IEEE Trans. On Systems, Man, and Cybernetics, Vol.33. 2(2003)269-283
7. Han Liqun.: the theory, design and application of neural networks, chemical industry press, Beijing(2002)76-80
8. Ho Y.C and Cassandras C.: A new approach to the analysis of discrete event dynamic systems. Automatica, Vol.19. 2(1983)312-323.

A Chaotic-Neural-Network-Based Encryption Algorithm for JPEG2000 Encoded Images

Shiguo Lian[1], Guanrong Chen[2], Albert Cheung[3], and Zhiquan Wang[1]

[1] Department of Automation, Nanjing University of Science & Technology
Nanjing 210094, P.R China, sg_lian@163.com

[2] Department of Electronic Engineering, City University of Hong Kong
Kowloon Tong, Hong Kong SAR, P.R. China

[3] Shenzhen Applied R&D Centres, Engineering, City University of Hong Kong
Kowloon Tong, Hong Kong SAR, P.R. China

Abstract. In this paper, a cipher based-on chaotic neural network is proposed, which is used to encrypt JPEG2000 encoded images. During the image encoding process, some sensitive bitstreams are selected from different subbands, bit-planes or encoding-passes, and then are completely encrypted. The algorithm has high security with low cost; it can keep the original file format and compression ratio unchanged, and can support direct operations such as image browsing and bit-rate control. These properties make the cipher very suitable for such real-time encryption applications as image transmission, web imaging, mobile and wireless multimedia communication.

1 Introduction

With the rapid development of multimedia technology, research on multimedia encryption, such as image, audio and video encryption, has become intensive recently. Due to its large-volume property with real-time transmission requirement, multimedia data cannot be well encrypted by traditional cryptographic techniques, at least not directly. Faster and better encryption algorithms are needed in order to meet these demands.

Neural networks can be used as a good random source based on its complicated and time-varying structures, if suitably designed [1]. This makes it suitable for data encryption. In the current literature, there exist some neural-network-based encryption algorithms reported [2-4]. The one proposed in [2] gives a good method to generate pseudo-random numbers; while the one proposed in [3] presents a method for secure communication by combining chaotic systems with neural networks. Moreover, the one proposed in [4] presents a block cipher based on the combination of a chaotic stream cipher and a neural network. All these algorithms have some properties suitable for the intended applications.

A novel still image compression standard, JPEG2000 [5, 6], has been widely used in today's markets. In this paper, we propose a novel encryption algorithm by combining chaotic neural networks with JPEG2000 codec. It encrypts images selectively for plaintexts of arbitrary sizes. Because the selected bitstream is often

F. Yin, J. Wang, and C. Guo (Eds.): ISNN 2004, LNCS 3174, pp. 627–632, 2004.

higherly sensitive to the images' understandability, the designed encryption algorithm has high security with low cost, and supports direct bit-rate control required by many imaging applications.

The rest of the paper is organized as follows. In Section 2, a chaotic-neural-network-based cipher is proposed. A JPEG2000 image encryption algorithm is presented in Section 3. In Section 4, the algorithmic performance is analyzed in detail. Finally in Section 5, some conclusions are drawn and future work is discussed.

2 A Cipher Based on Chaotic Neural Network (CNN)

For JPEG2000 compressed bitstream, a stream cipher [7] is more suitable in order to support direct bit-rate control. However, most random number generators are not truly random, which makes them theoretically predictable. In order to solve this problem, here we propose a Chaotic Neural Network (CNN)-based cipher, which is a modification of the one developed in [4]. The new cipher proposed here is described as follows.

1) Extension to Arbitrary Sizes

The encryption process is symmetric to the decryption process. Here, P and C are the plaintext and the ciphertext, respectively, and the weight ω and bias Θ are used as encryption/decryption keys. P is composed of N bits, and $P = [p_0, p_1, \ldots, p_{N-1}]^T$ ($p_i = 0$ or 1). Similarly, $C = [c_0, c_1, \ldots, c_{N-1}]^T$ ($c_i = 0$ or 1), $\bullet = [\bullet_0, \bullet_1, \ldots, \bullet_{N-1}]^T$ ($\bullet_i = \pm\frac{1}{2}$), and

$$\omega = \begin{bmatrix} \omega_{0,0} & 0 & \cdots & 0 \\ 0 & \omega_{1,1} & \ddots & \vdots \\ \vdots & \ddots & \ddots & 0 \\ 0 & \cdots & 0 & \omega_{N-1,N-1} \end{bmatrix} \quad (\omega_{i,i} = \pm 1, i = 0, \cdots, N-1).$$

Thus, the encryption process is

$$C = f(\omega P + \theta). \tag{1}$$

Here, $f(x)$ is a function taking value 1 if $x \geq 0$ and 0 otherwise. The parameters ω and Θ are determined by a chaotic sequence generator that will be described below. The decryption process is symmetric to the encryption one; that is, the plaintext P is replaced by the ciphertext C, and the ciphertext C by the plaintext P. In the proposed cipher, the size of the plaintext, N, is arbitrary.

2) The Chaotic Sequence Generator

A chaotic sequence generator based on the Logistic map [4] is not secure enough. This is because its control parameter is fixed at 4, which makes the relationship between adjacent states recoverable, and gives some help to the attacker. Observing this weakness, a more suitable chaotic map, the skew tent map, is used here, which is defined as

$$x_{j+1} = f(x_j, h) = \begin{cases} \dfrac{x_j}{h} & 0 < x_j \le h, \\ \dfrac{1-x_j}{1-h} & h < x_j \le 1. \end{cases} \tag{2}$$

Here, the parameter h ranges from 0 to 1, and together with x_0 they are used for the key. Thus, the chaotic binary sequence b(0)b(1)b(2)... generated by quantizing the chaotic sequence $\{x_0, x_1, \ldots x_n\}$ is used to determine the parameters ω and Θ. For example, if b(i) = 0, then $\omega_{i,I} = 1$; otherwise, $\omega_{i,I} = 0$.

3) The Chained Encryption Mode

A chained encryption mode is used here to enhance the cryptosystem's security. The encryption process is

$$C_i = E(P_{i-1} \oplus P_i, K_i). \tag{3}$$

Here, $\oplus$ means the bitwise operation, and $E(\cdot)$ is the encryption algorithm. The decryption process is

$$\begin{aligned} P_i &= D(C_i, K_i) \oplus P_{i-1} = D(C_i, K_i) \oplus D(C_{i-1}, K_{i-1}) \oplus P_{i-2} \\ &= D(C_i, K_i) \oplus D(C_{i-1}, K_{i-1}) \oplus \cdots \oplus D(C_1, K_1) \oplus D(C_0, K_0) \oplus P_{-1}. \end{aligned} \tag{4}$$

Seen from (4), the i-th ciphertext decryption depends on all the previous ones, but is not affected by the following ones. Thus, the encrypted data stream can be directly cut off at the end, which makes it suitable for applications with the bit-rate control requirement.

3 Encryption Algorithm for JPEG2000 Encoded Images

JPEG2000 codec is based on embedded block coding with optimized truncation (EBCOT) scheme [8]. The encoder consists of six steps: preprocessing, intercomponent transform, intracomponent transform, quantization, tier-1 encoder, and tier-2 encoder. Among them, Tier-1 encoder is the core, and it encodes each code block with three coding passes: significance pass, refinement pass, and cleanup pass. It is actually a context-based adaptive arithmetic coder named MQ [9].

Considering that image encryption emphasizes on content protection, encrypting only some sensitive data segments is recommended and studied here, which has both high security and high speed. A secure encryption scheme is proposed here, as shown in Fig. 1. The whole image is divided into three parts according to frequency ranges: low-frequency part (I), middle-frequency part (II), and high-frequency part (III). The entire encryption scheme is described as follows.

1) Low-frequency part encryption. For each code block, the six most significant bit-planes are encrypted. And, for each bit-plane, the three passes are all encrypted.

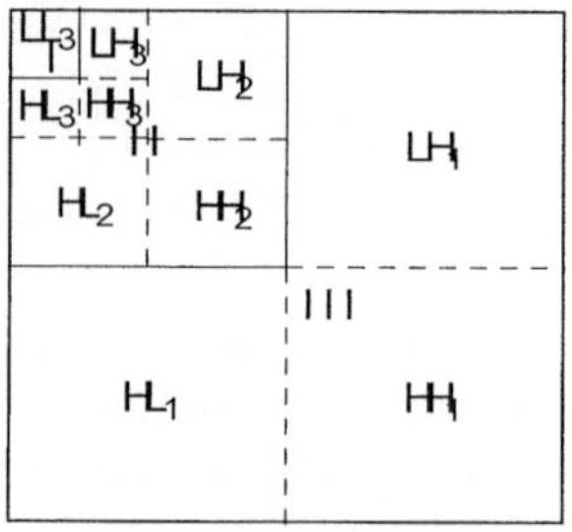

Fig. 1. Frequency band division. Taking 3-level wavelet transform for example, LL_3 belongs to low-frequency part, LH_3, HL_3, HH_3, LH_2, HL_2 and HH_2 belong to middle-frequency part, and LH_3, HL_3, HH_3 belong to high-frequency part.

2) *Middle-frequency part encryption.* For each code block, the 6 most significant bit-planes are encrypted. But, for each bit-plane, only the cleanup pass is encrypted.
3) *High-frequency part.* For each code block, the 3 most significant bit-planes are encrypted. And, for each bit-plane, only the cleanup pass is encrypted.
4) *The cipher to be used.* For all the encryption processes, the CNN-based cipher proposed in Section 2 is used.

Taking two images for example, the encryption results are shown in Fig. 2. As can be seen, the encrypted images are too chaotic to be understood by human eyes.

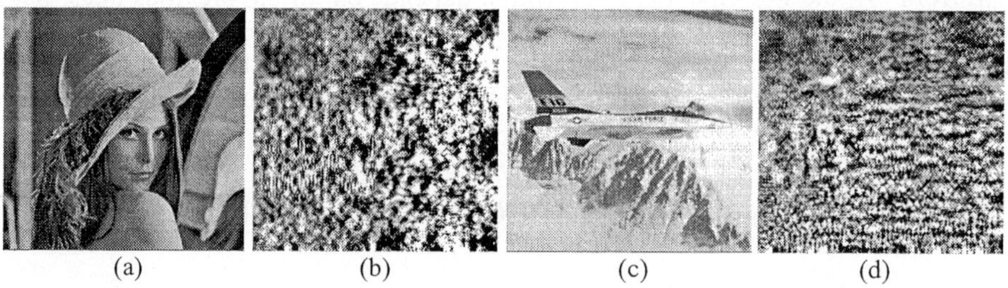

Fig. 2. Image encryption results. (a) and (c) are the original images; (b) and (d) are the encrypted images. The two images are Lena (256× 256, gray, 5-level wavelet transform) and Plane (512× 512, colorful, 6-level wavelet transform), respectively.

4 Performance Analysis

4.1 Security Analysis

Its security against brute-force attack is determined by the key in the CNN-based cipher. In this cipher, the initial condition x_0 and control parameter h are used as key, which are both composed of 64 bits. Thus, the key space is 2^{128}. If n different keys are used, then the key space is increased to 2^{128n}. Therefore, it can be secure enough against the brute-force attack.

For this cipher, the known-plaintext attack process is to solve Eq. 1. Taking the *i*-th plaintext-ciphertext couple for example, in order to obtain ω_i and Θ_i, 4 times of attempts should be carried out. Thus, the attacker should attempt for 2^{2N} times in order to obtain P. If N is bigger than 64, the difficulty is no smaller than the brute-force attack ($2^{2N} \geq 2^{128}$). This property keeps the cryptosystem secure enough against the known-plaintext attack.

Replacement attack [10] is often used to break multimedia encryption algorithms, which means to replace some of the encrypted data with others in order to reconstruct the plaintext. In wavelet transform, the unencrypted coefficients can give some help to replacement attackers, so most of the coefficients should be encrypted. For example, the reconstructed images, encrypted by different algorithms, are shown in Fig. 3. It can be seen that the algorithm proposed by in this paper is secure enough against the replacement attack.

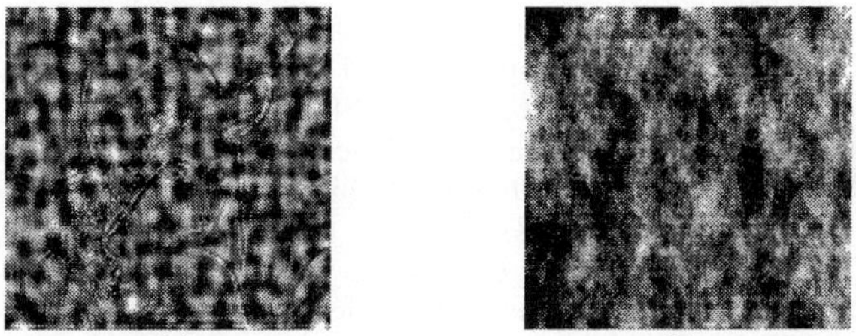

(a) (b)

Fig. 3. Replacement attack. (a) is the image reconstructed from the one whose low-frequency and middle-frequency parts are encrypted while the high-frequency part is left unencrypted; (b) is the image reconstructed from the one that is encrypted by the algorithm proposed in this paper.

4.2 Computational Complexity

Various images have been tested, on the encryption data ratio (Edr) and the encryption time ratio (Etr). The results are shown in Table 1. It shows that the Edr is often no more than 20% and the Etr is often smaller than 15%. This implies that the encryption process does not affect the encoding process significantly. Similar result is obtained on decryption process.

Table 1. Encryption speed test. Different gray and color images with different sizes are tested.

Image	Size	Color/Gray	Edr	Etr
Lena	128*128	Gray	13.4%	8.9%
Boat	256*256	Gray	10.8%	10.6%
Village	512*512	Gray	16.8%	11.1%
Lena	128*128	Color	14.7%	9.4%
Peppers	256*256	Color	12.1%	10.4%
Baboon	512*512	Color	15.3%	9.7%

5 Conclusions and Future Work

In this paper, a CNN-based encryption algorithm has been proposed. Both theoretical analysis and experimental results have shown that the algorithm has high security with low cost, and supports direct bit-rate control. It should be noted that the design methodology can be further extended to MPEG4 codec, which is left for future work.

Acknowledgements. This research was supported by the Shenzhen Applied R&D Centres, City University of Hong Kong under Grants 9410011 and 9620004.

References

1. Karras, D.A., Zorkadis, V.: On Neural Network Techniques in the Secure Management of Communication Systems through Improving and Quality Assessing Pseudorandom Stream Generators. Neural Networks, Vol.16, No.5-6 (2003) 899-905
2. Chan, C.-K., Cheng, L.M.: Pseudorandom Generator Based on Clipped Hopfield Neural Network. In: Proceedings of the 1998 IEEE International Symposium on Circuits and Systems, Vol.3. Monterey, CA (1998) 183–186
3. Mislovaty, R., Klein, E., Kanter, I., Kinzel, W.: Public Channel Cryptography by Synchronization of Neural Networks and Chaotic Maps. Physical Review Letters, Vol.91, No.11 (2003) 118701/1-118701/4
4. Yen, J.-C., Guo, J.-I.: A Chaotic Neural Network for Signal Encryption/Decryption and Its VLSI Architecture. In: Proc. 10th (Taiwan) VLSI Design/CAD Symposium. Taiwan (1999) 319-322
5. Rabbani, M., Joshi, R.: An Overview of the JPEG2000 Still Image Compression Standard. Signal Processing: Image Communication, Vol.17, No.1 (2002) 3-48
6. ISO/IEC 15444-1: Information Technology-JPEG 2000 Image Coding System-Part 1: Core Coding System (2000)
7. Buchmann, J.A.: Introduction to Cryptography. Springer-Verlag, New York (2001)
8. Taubman, D.: High Performance Scalable Image Compression with EBCOT. IEEE Trans. on Image Processing, Vol.9, No.7 (2000) 1158-1170
9. Witten, I.H., Neal, R.M., Cleary, J.G.: Arithmetic Coding for Data Compression. Communications of the ACM, Vol.30, No.6 (1987) 520-540
10. Podesser, M., Schmidt, H.P., Uhl, A.: Selective Bitplane Encryption for Secure Transmission of Image Data in Mobile Environments. In: CD-ROM Proceedings of the 5th IEEE Nordic Signal Processing Symposium (NORSIG 2002). Tromso-Trondheim, Norway (2002)

A Combined Hash and Encryption Scheme by Chaotic Neural Network

Di Xiao [1, 2] and Xiaofeng Liao [1]

[1] College of Computer Science and Engineering,
[2] College of Mechanical Engineering,
Chongqing University, 400044, Chongqing, People's Republic of China
xiaodi_cqu@hotmail.com
xfliao@cqu.edu.cn

Abstract. A combined hash and encryption scheme by chaotic neural network is proposed. With random chaotic sequences, the weights of neural network are distributed and the permutation matrix P is generated. The nonlinear and parallel computing properties of neural network are utilized to process hash and encryption in a combined mode. In-depth analyses of the performance indicate that the scheme is efficient, practicable and reliable, with high potential to be generalized.

1 Introduction

As the core of Cryptography, hash and encryption are the basic techniques for info-security. Currently, some novel directions have been widely exploited, for example, the chaos based cryptography [1]. In this paper, by investigating the nonlinear and parallel computing properties of neural network and integrating chaos with them, a combined hash and encryption scheme by chaotic neural network is proposed. The scheme can securely, efficiently complete hash and encryption in this combined mode. It is practicable and reliable, with high potential to be adopted for E-commerce.

2 Our Scheme

In this paper, a network is called a chaotic neural network if its weights and biases are determined by a chaotic sequence. Chaos is a kind of deterministic random-like process generated by nonlinear dynamic systems. Its properties include: sensitivity to initial conditions and parameters, random-like behavior, ergodicity and one-way, etc.

F. Yin, J. Wang, and C. Guo (Eds.): ISNN 2004, LNCS 3174, pp. 633–638, 2004.

2.1 Our Chaotic Neural Network Model and Its Theoretical Foundation for Hash and Encryption Application [2] [3]

We construct a 4-layer neural network model shown in Fig.1. The 1st layer is not actually a layer of neurons, but rather the input 640-bit plaintext themselves. The 2nd layer consists of 64 neurons. The 3rd layer consists of 128 neurons, and will generate the final hash result. The 4th layer consists of 640 neurons, and will output the ciphertext. The activation functions in each layer are the nonlinear Sigmoidal function: $f(x)=1/(1+exp(-\beta x)), \beta>0$, where β determines the slope of the function. When β become infinite, the Sigmoidal function is actually the Step function. In our scheme, $\beta=1$ in the 2nd layer, $\beta=1000$ in the 3rd layer, $\beta=\infty$ in the 4th layer.

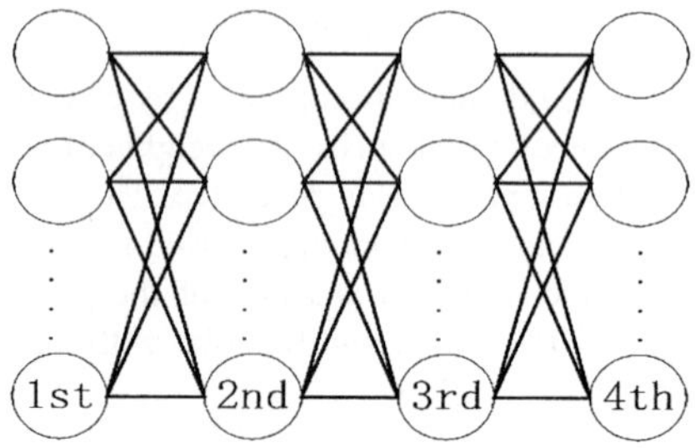

Fig. 1. Structure of the chaotic neural network

Nonlinearity is used in most existing algorithm. Neuron is a nonlinear component, while neural network consisting of Neurons has complicated nonlinear characteristics. Moreover, we utilize the chaotic system to distribute the weights and permute the corresponding relation between plaintext and ciphertext. All form the solid theoretical foundation for hash and encryption. The detailed analysis will be given later.

2.2 The Weight Distribution Algorithm

The chaotic map chosen in the scheme is logistic map: $X_{n+1}= bX_n(1-X_n)$, where b is the gain and $X_n\in[0,1]$. b=3.78 and X_0=0.232323 are chosen secretly as the key.

Calculates the chaotic sequence $X(1),X(2),\cdots, X(640)$, and transforms them to binary representation. Extracts the 16-bit values of $X(m)$ after the decimal point to the corresponding $b(16m-16),b(16m-15),b(16m-1)$; and gets a array $b(k)$, k=1,2,⋯,10240. To reduce the computation, on the precondition that the scheme security can be guaranteed, the value of the array can be reused so that a new array $b(k)$,k=1,2, ⋯,131072, can be generated for the following weight distribution.

The weights W_{ij}^{12} between 1st layer and 2nd layer are distributed as follows:

```
For i=1 To 640
  For j=1 To 64
    If i=j
```

$$W_{ij}^{12} = \begin{cases} 1, b(640*(i-1)+j) = 0 \\ -1, b(640*(i-1)+j) = 1 \end{cases}$$

```
    Else        //i≠j
```

$$W_{ij}^{12} = \begin{cases} b(640*(i-1)+j, i+j = odd \\ -b(640*(i-1)+j), i+j = even \end{cases}$$

```
    End
  End
End
```

The weights W_{ij}^{23} between 2nd and 3rd layer, W_{ij}^{34} between 3rd and 4th layer can be distributed similarly. We will not give detailed description due to page limit.

2.3 Generation of Permutation Matrix P

After section 3.1, using *X(640)* of iteration orbit as the new initial value to begin new iterations: *X(1),X(2),* ⋯, *X(i)*, ⋯. When *i*=1, the number *640*X(1)* is rounded to a integer $r \in [1, 640]$, and stored into permutation matrix P; when *i*=2, 3, ⋯, the new random integer gotten each time will be verified whether it has existed in P, if not, it will be stored into P. Similar operations will be done until there are 640 numbers in P. The ergodicity of chaotic orbit can ensure the smooth generation of P. The matrix P actually rearranges the integers from 1 to 640.

2.4 Realization of Hash Function

The input can be of variable bits long, similar to MD5, padding is done: A single '1' bit is appended followed by '0' bits until the length of the message is congruent to 448 bits. This is followed by a 64-bit representation of the original message length. The hash process proceeds by taking the inputting 512-bit data at a time, combines the input with the 128-bit output of the previous round, and obtains the 128-bit representation. The process stops when on more 512-bit data blocks are available. The last 128-bit output is the hash representation of the inputting plaintext. A known initialization vector is needed at the beginning of the process.

The hash function involves the 1st, 2nd, 3rd layers of our model. The jth neuron in i^{th} layer is represented as $S_i(j)$. The hash function is realized as follows: Equation (1), Equation (2) denotes the process in 1st –2nd and 2nd –3rd layers respectively, where *f(x)* is the corresponding Sigmoidal function in 2nd and 3rd layers. The output of each neuron is a real number between 0 and 1. We set a threshold of 0.5: if output>=0.5, it will be taken as 1; while output<0.5, it will be taken as 0. The output of the 3rd layer in the last round $S_3(1), S_3(2), \cdots, S_3(128)$ is the final hash value of plaintext.

$$\begin{cases} f(\sum_{j=1}^{640}(S_1(j) * W_{j1}^{12})) = S_2(1) \\ f(\sum_{j=1}^{640}(S_1(j) * W_{j2}^{12})) = S_2(2) \\ \cdots\ \cdots \\ f(\sum_{j=1}^{640}(S_1(j) * W_{j64}^{12})) = S_2(64) \end{cases} \tag{1}$$

$$\begin{cases} f(\sum_{j=1}^{64}(S_2(j) * W_{j1}^{23})) = S_3(1) \\ f(\sum_{j=1}^{64}(S_2(j) * W_{j2}^{23})) = S_3(2) \\ \cdots\ \cdots \\ f(\sum_{j=1}^{64}(S_2(j) * W_{j128}^{23})) = S_3(128) \end{cases} \tag{2}$$

2.5 Realization of Encryption and Decryption Function

The encryption and decryption functions involve the 1st, 3rd, 4th layers of our model. The 1st layer is the plaintext input layer, $S_1(j)$ represents the jth neuron in 1st layer, it is 0 or 1; the 4th layer is the ciphertext output layer, $S_4(j)$ represents the jth neuron in 4th layer; the permutation matrix P confuses the corresponding relations between neurons in 1st and 4th layer. The encryption and decryption is realized as follows:

For the encryption of the pi^{th} plaintext bit $S_1(pi)$, the corresponding index i to pi is looked up through the permutation matrix P, and the ith ciphertext bit $S_4(i)$ is:

$$S_4(i) = S_1(pi) \oplus f(\sum_{j=1}^{128}(S_3(j) * W_{ji}^{34}))\ ;$$

For the decryption of the kth ciphertext bit $S_4(k)$, the corresponding pk to k is looked up through the permutation matrix P, and the pk^{th} plaintext bit $S_1(pk)$ is:

$$S_1(pk) = S_4(k) \oplus f(\sum_{j=1}^{128}(S_3(j) * W_{jk}^{34}))\ .$$

Where $f(x)$ is the Sigmoidal function in 4th layer, actually a Step function.

3 Performance Analysis

3.1 Analysis of Hash Function

3.1.1 One-Way

One-way property means that it only allows a digest to be created from the original, yet the inverse is very hard [4]. Our final hash is gotten by equation (1) and (2). It seems possible to find the original from the final by performing Gauss-Jordan reduction. Suppose that a probability of 5% for the output to be 0.9999999999999999999, then β=-20ln((1/0.9999999999999999999)-1)=921≈1000.By setting β=1000 in the 3rd layer, the probability that 1-bit output is not equal to 1 or 0 will be 10%, and that for 128-bit output will be 0.1^{128}, which is so small that Gauss-Jordan reduction will no longer work.

3.1.2 Collision Resistance and Birthday Attacks Resistance

Collision resistance is the difficulty of finding two different inputs hashing to the same. Birthday attacks are similar [4]. Each neuron in a layer is connected to all in the next layer. This inherent structure expedites the avalanche effect; furthermore, we set β=1 in the 2nd layer, while the input can only consist of 0 or 1. These ensure that even a single bit change in input will definitely result in great changes in the output.

3.1.3 Flexibility

Simply modifying the number of neurons in every layer, the length of the final hash value will be changed. Compared with the traditional hash algorithm, it can satisfy the actual demand than that of the traditional hash algorithm better.

3.2 Analysis of Encryption Function

3.2.1 Exhaustive Attack Resistance

The key space of *X0* and *b* in logistic map is very large. Even if attacker knows all the weight values of the neural network, the matrix P with 640 number has still 640! possible arrangements, which is greatly larger than the key space 2^{56} of DES.

3.2.2 Auto-Correlation and Cross-Correlation Function with Least Different Keys

Repeated plaintext such as "*QQ*……" is encrypted and the corresponding auto-correlation function is analyzed. It is a δ function and has no repeated period. In Fig.2, for the sensitivity of chaos, least different keys will result in the great changes in the weights and the matrix P, which will totally change the ciphertext.

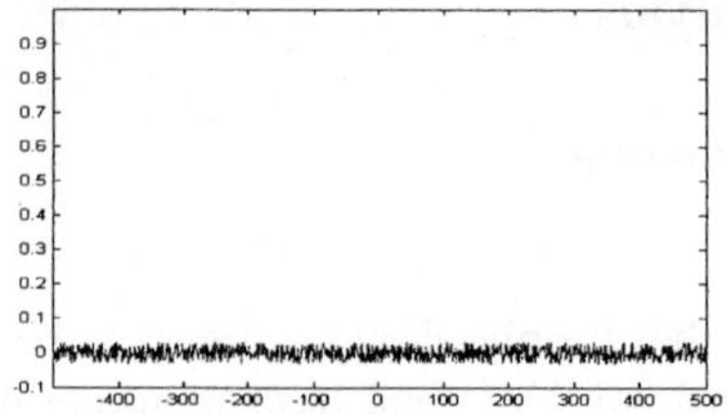

Fig. 2. Cross-correlation function

3.3 Analysis of the Process Speed and Its Applying Prospect

Neural network can process vast amount of data in parallel. Although the weight distribution and matrix P generation seem a little troublesome, yet they are only the pre-process before the true hash and encryption. The parallel data-process will inevi tably ensure the high efficiency of the complete scheme. Furthermore, we can only use chaos to determine part of the weights and fix the rest without bad effects to the scheme security. Finally, hardware can be utilized to realize the neural network. Our scheme can perform encryption, decryption and hash in a combined mode. Integrated with RSA, a simple system can be constructed to efficiently deal with lots of actual occasions where encryption and hash are required at the same time.

4 Conclusions

In this paper, we investigate the nonlinear and parallel computing properties of neural network and integrate chaos with them, propose a combined hash and encryption scheme by 4-layer chaotic neural network. The scheme can securely and efficiently complete hash and encryption. It is practicable, flexible and reliable.

References

1. Frank Dachselt, Wolfgang Schwarz: Chaos and Cryptography. IEEE Trans. Circuits and Systems-I, 48, (2001) 1498-1509
2. Liew Pol Yee , Liyanage C. De Silva: Application of Multi-layer Perception Network as One-way Hash Function. Proceedings of the IEEE-INNS-ENNS International Joint Conference on Neural Networks, Vol.2, (2002) 1459-1462
3. Toru Ohira: Encryption with Delayed Dynamics. Computer Physics Communications, 121, (1999) 54-56
4. B. Schneier: Applied Cryptography. 2nd edn. Wiley, New York (1996)

A Novel Symmetric Cryptography Based on Chaotic Signal Generator and a Clipped Neural Network

Tsing Zhou, Xiaofeng Liao, and Yong Chen

Department of Computer Science and Engineering, Chongqing University
400044, Chongqing, China
zq336@sina.com

Abstract. Recently, a novel encryption algorithm that integrates Haar wavelets transformation into chaotic signal generator, was proposed by R.Luo et al. In this paper, we first analyzed the merits and demerits of this algorithm. Then an improved scheme which uses a clipped neural network is proposed. Both theoretical analysis and computer simulations show our proposed scheme succeeds in overcoming the defects of Luo's algorithm while retaining all its merits. Moreover, the way that the clipped neural network evolves may present a new idea to the cryptography.

1 Introduction

Chaos-based encryption has received much attention in recent years. Recently, a hybrid symmetric cryptography[1] is proposed. In the scheme, a plaintext block p of 16-bit is first multiplied by a matrix H_{16} (i.e., $m=pH_{16}$), which is one permutation of the 16 subwavelet bases $\{h_0, h_1, \ldots, h_{16}\}$. Then chaotic signals (x, y, z) from the Chua's circuit[2] are generated. Finally, m is added to one chaotic signal, see, x to produce the ciphertext. For more security, the authors suggest that all the chaotic parameters be updated from time to time using a one-way function.

There are at least 4 virtues of this novel encryption scheme:

1) *New approach of encryption*
2) *Enhanced security for chaotic encryption*
3) *Fast implement by hardware*
4) *Feasible way to error-detection*

However, we also found some flaws of the scheme:

1) *Difficult for key management*
2) *Slow by software implement*
3) *Longer ciphertext than plaintext*
4) *Vulnerable to chosen-plaintext attack*

The fourth one is the foremost defect of the algorithm. The chosen-plaintext attack is based on the fact that, for some special plaintexts, the encoded data remain unchanged however the 16 wavelet bases are arranged. For example, in the extremely simple case, when we choose zero vector as the plaintext , the encoded data m must

F. Yin, J. Wang, and C. Guo (Eds.): ISNN 2004, LNCS 3174, pp. 639–644, 2004.

be zero vector in spite of the arrangement of the wavelet bases. Note that on such condition, the scheme is degraded to the pure chaotic synchronization encryption. Using the method proposed in [3], we could identify the private key with only 2000 blocks of ciphertexts averagely. Thus, the cryptography system is broken down with a simple trick. We noticed that the primary cause to its failure is the lack of confusion before combining with the chaotic signals. This makes us think of neural network naturally.

The content of this paper is organized as follows: Section 2 gives a detail description of our scheme. Section 3 shows some experiment results. In Section 4, we discuss the security issues, advantages and disadvantages of the scheme, and give some conclusion.

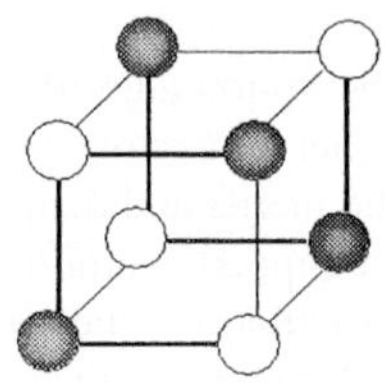

Fig. 1. Structure of the CNN: color of each vertex stands for neuron's state: black for 1 and white for -1, while the thickness of edge indicates the weight between neurons: the thicker one for 1 and thinner one for -1.

2 A Symmetric Cryptography Scheme

2.1 Structure of the Clipped Neural Network (CNN)

The structure of the neural network is depicted in Fig1. Obviously, neurons are organized as a cube. We clip the synaptic weight between two neurons to only two values {-1, 1}, and for this reason we name it as clipped neural network (CNN). Each neuron has a state, which could be also -1 or 1. Each time the neuron perceives states of its three neighbors, and then transforms its own states as a response. The transforming function is defined as:

$$f(s_i) = sign(\sum_{j=1}^{8} w_{ij} s_j) \tag{1}$$

where s_i is the state of ith neuron, w_{ij} is the synaptic weight between ith and jth neuron. $sign(x)$ returns the sign of x.

2.2 Evolutionary Mechanism of the CNN

The information of neuron network relies upon the synaptic weights and how they evolve. It is often helpful to emulate Nature to obtain complex and confused data. In our scheme, we regard CNN as a collect of interconnected neural cells, and chaotic signal as the environment that stimulate neurons. As a matter of fact, neuron changes its weight to adapt to the outside environment e_i. The weight changes as follow:

$$\text{IF } s_i = e_i \text{ THEN } w_{ij} \text{ and } w_{ji} \text{ do not change, ELSE } w_{ij} = -w_{ij},\ w_{ji} = -w_{ji} \tag{2}$$

2.3 Encryption/Decryption

The encryption system of our scheme consists of two main parts: the CNN and the chaotic tent map. The tent map is defined by (3):

$$T(x) = \begin{cases} rx & 0 < x \le 0.5 \\ r(1-x) & 0.5 < x < 1 \end{cases} \tag{3}$$

Where r is a real number very near to 2.

Now, we describe the detail procedure of encryption algorithm as follows:

Step 1) Both the transmitter and the receiver hold the same private key, which includes initial state of CNN S_0, initial value x_0 and parameter r of chaotic tent map.

Step 2) For higher security, both the transmitter and the receiver iterate the tent map and the CNN for 128 times before encryption.

Step 3) Plaintext is divided into blocks of 8 bits, i.e., one byte each block. For each block of plaintext:

i) CNN works to obtain its new state $s_0, s_1, \ldots, s_7$.

ii) The tent map iterates for 8 times, generating $x_0, x_1, \ldots, x_7$.

iii) Extract the 4th bit from each x_i, and substitute -1 for 0 to get e_i. As strictly proved by [4], these neuron's states are independent and identical distribution.

iv) The 8 states $s_0, s_1, \ldots, s_7$ is XORed with the plaintext byte, thus generates a new byte B. Transform the new byte to a decimal fraction between 0 and 1.

v) Combine B with the chaotic signal x_7 together to produce the final ciphertext byte C by $C = f(B + x_8)$ where $f(x)$ is an addition modulo 1 operation.

vi) Finally, change the CNN weights according to (2), to prepare for next plaintext block.

Step 4) The transmitter do step 3 with next plaintext block until the whole plaintext is exhausted.

The decryption process is similar to encryption, except iv of Step 3 where the states $s_0, s_1, \ldots, s_7$ is XORed with the ciphertext byte instead of plaintext, and vi of Step 3) where $C = f(B + x_8)$ is replaced by $B = f(C - x_8)$. Error-detection can also be done in decryption process.

3 Experiment Result

We implemented the proposed algorithm by Matlab software. In the experiment, we have S_0= [1 -1 1 -1 1 -1 1 -1], x_0=0.40 and r=1.99, and the gray image lena.gif

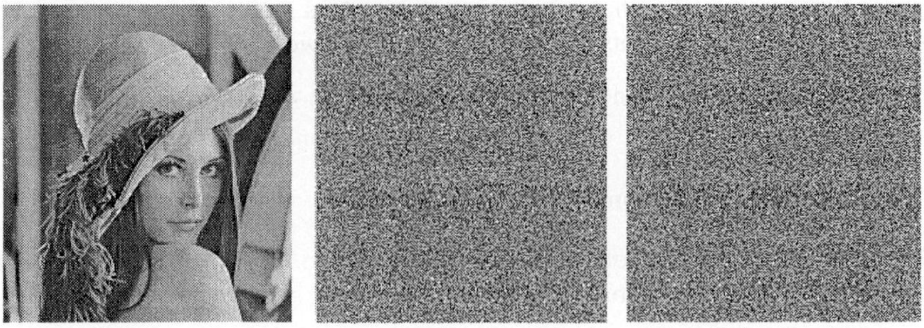

Fig. 2. Gray image Lena.gif (a) the source image (b)XORed with the CNN states (c) Encrypted image

Table 1. Statistic comparison of CNN and Uniform distribution

	Mean	STD
CNN State	127.500	74.045
Uniform	127.505	73.913

(Fig2(a)) with the size of 512*512 pixels and 8-bits/pixel as the plaintext. Both the chaotic signal and the ciphertext are stored with a precision of 52-bits. Fig2 depicts 3 states of the picture during the encryption process.

3.1 Statistic Characters of the CNN States

1) Statistic characters of the neuron's state

We generated first 2097152 states of CNN using the encryption scheme with parameters mentioned above, of which 50.027% are 1 and 49.973% are -1.

2) Statistic characters of the CNN's state

The CNN's state is the combination of the 8 neurons' states each time. To quantify it more evidently, we transform every 8 neurons' states to an integer between 0 and 255, then compare their statistic characters with those of ideal uniform distribution (Table 1).

From all the statistic data presented above, we can see that the CNN's states have the character of uniform distribution.

3.2 Sensitivity to the Outside Stimulation

A basic requirement of encryption algorithm is that the ciphertext must be sensitive to the key. In other words, a slightest variation of key should lead to ciphertext's obvious change. We have carried out 3 experiments to test the sensitivity of the CNN's state to the little change of the initial state S_0 of the CNN, the initial value x_0 and the parameter r of tent map respectively. The experiment results are depicted in Fig3 which show that the states of the neurons are sensitive to the key.

Fig. 3. The first 32 neurons' states (black for 1 and white for -1) driven by deferent keys with slightest change. On bottom line of each subplot, the corresponding key is S_0=[1 -1 1 -1 1 -1 1 -1], r=1.99 and x_0=0.40, and on the top line the corresponding key is:(a) S_0=[1 -1 1 -1 1 -1 1 1], r=1.99 and x_0=0.40 (b) S_0=[1 -1 1 -1 1 -1 1 -1], r=1. 98999999 and x_0=0.40 (c) S_0=[1 -1 1 -1 1 -1 1 -1], r=1.99 and x_0=0. 39999999

4 Discussion and Conclusion

Our scheme can resist the cipher-only attack, known-plaintext attack as proved by [1] in detail, due to the same aperiodic and unpredictable character of the chaotic signal and the CNN's state. Moreover, this new proposed scheme can resist the chosen-plaintext attack we proposed in Section I. Suppose the attacker could choose any plaintext p and acquire corresponding ciphertext C. To get the chaotic signal x, he must first guess the value of m which totally depends on CNN's state even the plaintext is a constant. Moreover, the CNN's state is sensitive to the chaotic signal x in turn. Therefore, if the attacker decides to use the attack in [3], he may have to try all possible states of the CNN, the number of which is 2^8. Suppose the attacker can figure the key out by a series of only 20 tent map chaotic signals (the average number presented by [3] is 400), there are total of 2^{160} possible states. For this reason, the attacker would prefer to guess the key, whose possible number is only about 2^{130}. Thus, it could be seen that our scheme is resistant to the chosen-plaintext attack in Section I.

It is easy to verify that our scheme retains 4 advantages of [1]. However, it can overcome almost all the disadvantages of [1]:

1) *Difficult for key management.* In our scheme, the secret key includes S_0, x_0 and r. The number of the parameters is much less than that of [1]. Therefore it is easier to generate and transmit the key.

2) *Slow by software implement.* Substitution of tent map for Chua circuit improves the speed of the software encryption greatly.

3) *Vulnerable to chosen-plaintext attack.* As discussed above, our scheme can resist such attack.

For scheme we proposed here, it still cannot avoid the 3th disadvantage of [1], i.e.,

4) *Longer ciphertext than plaintext*

Similar to [1], 8 bytes ciphertext is produced for each byte of plaintext. However, we found there may be a simple method to solve this problem, that is, replacing the cube structure of CNN by a 4-dimension hypercube. Hence, the CNN can generate 64-bits each time and the scheme produces 8 bytes ciphertext for 8 bytes plaintext. This new idea is still under experiment for security and efficiency issues.

From the discussion above, we can see that our scheme has overcome almost all the flaws we found of in [1], while retaining all its merits successfully. We noted that such good performance of our scheme originates mostly from the introducing of the CNN.

Acknowledgements. The work described in this paper was supported by National Nature Science Foundation of China (NO. 60271019) and the Doctorate Foundation of the Ministry of Education of China (No 2002061107)

References

1. Luo R., Chung L., and.Lien C,: A Novel Symmetric Cryptography Based on the Hybrid Haar Wavelets Encoder and Chaotic Masking Scheme. *IEEE trans. Industrial Electronics*, Vol.49, NO. 4, Aug. 2002
2. Chua L.O, Komuro M., Matsumoto T.: The Double Scroll Family. *IEEE Trans. Circuits Syst.*, Vol.33, NO.11, pp.1072-1118, Nov.1986.
3. Dedieu H. and Ogorzalek M.J.: Identifiability and Identification of Chaotic Systems Based on Adaptive Synchronization. *IEEE Trans. Circuits Syst.*, Vol.44, No10, pp.948-962, Oct.1997
4. Kohda T.: Information Sources Using Chaotic Dynamics. *Proceedings of the IEEE*, Vol.90, NO.5, pp.642-661, May. 2002

A Neural Network Based Blind Watermarking Scheme for Digital Images

Guoping Tang [1,2] and Xiaofeng Liao [1]

[1] Department of Computer Science and Engineering, Chongqing University, 400044 Chongqing, P. R. China
xfliao@cqu.edu.cn

[2] Logistical Engineering University of PLA, 400016 Chongqing, P. R. China
tgptgp@sina.com

Abstract. A new blind watermarking scheme by combining neural network with chaotic map is proposed. Using a chaotic sequence, the binary highpass watermark is generated. An encrypted watermark is to map to a selected pixel is modified according to the bit will be embedded. Embedding and extraction of watermark are based on the relationship between the pixel and its neighborhood in each block of image. A neural network and an adaptive embedding algorithm are adopted to enhance the characters of the watermarking system. Experimental results show that this watermarking scheme is very robust to common image processing and the extracted watermarks are readily recognizable.

1 Introduction

Over the past few decades, the amount of digital multimedia information distributed through communication networks has increased rapidly. Nevertheless, the illegal data access and unauthorized data reproduction have become easier and more prevalent. The design of robust schemes for copyright protection and content verification of multimedia data have become an urgent necessity. Such demand has been addressed lately by the emergence of a variety of watermarking method. The digital watermarking can be described as the process of embedding by means of a secret key, an imperceptible digital signal (the watermark) into multimedia content. The watermark may be visible or invisible in the embedded media.

A number of invisible watermarking techniques have been reported in recent years [1-4]. Among these schemes, many require that the original signal is available during the detection [1], whereas some schemes, i.e., the so-called blind methods [2–4], don't. In terms of the domain where the watermark signal is embedded, two techniques have been presented: (i) embedding watermark in the spatial domain [4] by modulating the intensity of pre-selected samples; (ii) embedding watermark in an appropriate transform domain, e.g. the DCT[1], DFT[2], or DWT[3] domain.

In this paper, a robust blind watermarking scheme over the spatial domain is proposed. The basic idea is as follows. First, the watermark is encrypted and its spectrum

F. Yin, J. Wang, and C. Guo (Eds.): ISNN 2004, LNCS 3174, pp. 645–650, 2004.

is spread using a chaotic sequence. Second, the host image is divided into 3×3 blocks, and a neural network is given to memorize the relationships among the pixels in each blocks. Finally, the masked binary watermark is embedded into the host image adaptive through changing the value of central pixel in a block of image. The extraction is the inverse processing.

2 The Chaotic Watermarks Generating

The chaotic sequence with such as pseudo-random and broad-spectrum properties makes it naturally suitable to secure communication and robust watermarking systems. The watermark generation procedure aims at generating a highpass chaotic watermark with a secret key *K*. As in [6] theoretically and experimentally evaluated, highpass chaotic watermarks have been proven to perform better than white ones, whereas lowpass watermarks have the worst performance.

Firstly, the skew tent map is used to generate a chaotic sequence based on the selected parameter α and initial value x_0 (used as the secret key *K*), which can be described as:

$$f(x)=\begin{cases} \frac{1}{\alpha}x & 0\le x\le\alpha \\ \frac{1}{\alpha-1}x+\frac{1}{1-\alpha} & \alpha<x\le 1 \end{cases} \tag{1}$$

where $x\in[0,1]$. When <0.5, the chaotic real-value sequence $\{f^n(x_0)\}_{n=0}^{\infty}$ is highpass[6]. Through a threshold function $\theta_t(x)$:

$$\theta_t(x)=\begin{cases} 1 & x\ge t \\ 0 & x<t \end{cases}, \quad (t>0) \tag{2}$$

we can obtain a binary sequence $\{\theta_t(f^n(x))\}_{n=0}^{\infty}$ from the sequence $\{f^n(x)\}_{n=0}^{\infty}$.

Secondly, by employing the classical chaotic masking technique [5], a highpass watermark $w^{'}\in\{0,1\}$ can be obtained:

$$w^{'}=w\oplus\theta_t(x) \tag{3}$$

3 The New Watermarking Scheme

3.1 Watermark Embedding

After the watermark generation we proceed to the watermark embedding by altering the pixels of the original (host) image *I*(**s**) according to the following formula:

$$I_w(\mathbf{s}) = \begin{cases} f_0(I(\mathbf{s}), N(\mathbf{s})) & \text{if } w'(\mathbf{s}) = 0 \\ f_1(I(\mathbf{s}), N(\mathbf{s})) & \text{if } w'(\mathbf{s}) = 1 \end{cases} \tag{4}$$

where f_0, f_1 are called embedding functions and $N(\mathbf{s})$ denotes a function that depends on the neighborhood of pixels **s**. The functions used in our scheme are based on superposition of real quantities in the pixels, which can be described as:

$$f_1(I(\mathbf{s}), N(\mathbf{s})) = \max\{I(\mathbf{s}), \sigma_1 + \delta\} \tag{5}$$

$$f_0(I(\mathbf{s}), N(\mathbf{s})) = \min\{I(\mathbf{s}), \sigma_0 - \delta\} \tag{6}$$

Where

$$\sigma_1 = \begin{cases} I(\mathbf{s}) & \text{if } B(\mathbf{s}) - I(\mathbf{s}) > \delta \\ B(\mathbf{s}) & \text{if } B(\mathbf{s}) - I(\mathbf{s}) \le \delta \end{cases} \tag{7}$$

$$\sigma_0 = \begin{cases} B(\mathbf{s}) & \text{if } B(\mathbf{s}) - I(\mathbf{s}) > \delta \\ I(\mathbf{s}) & \text{if } B(\mathbf{s}) - I(\mathbf{s}) \le \delta \end{cases} \tag{8}$$

here, $B(\mathbf{s})$ is the output of a neural network from the input $N(\mathbf{s})$ and δ is the embedding strength, its value determines the watermark power, which is defined in next sub-section.

3.2 Decision of Watermarking Strength

In a watermarking algorithm, it is noted that the bigger the watermarking strength, the better the robustness of watermark, but the less the invisibility. To achieve maximal watermarking while remaining invisible to the human eyes, we using an adaptive approach based on signal-noise-ratio (*SNR*) [7], which defined as:

$$SNR = 10\log_{10} \frac{\sum_x \sum_y p_{x,y}^2}{\sum_x \sum_y (p_{x,y} - p'_{x,y})^2} \tag{9}$$

where $p_{x,y}$, $p'_{x,y}$ is the pixel value of original and watermarked image, respectively. $x=1,2,\ldots,M_1$, $y=1,2,\ldots,M_2$ represent respectively the image width and height. For a given *SNR*, we can calculate the strength of embedding for each block:

$$\delta = \left| p_{x,y} - p'_{x,y} \right| = \sqrt{\sum_x \sum_y p_{x,y}^2 \times 10^{\frac{-SNR}{10}}} \tag{10}$$

Consequently, the embedding strength is adaptive. Thus, the properties of invisibility and robustness are both guaranteed.

The size of the region around **s** used for the calculation of $N(\mathbf{s})$ is important for the watermarking procedure. Moreover, the number of pixels used for calculation of $N(\mathbf{s})$ determines the upper bound of the number of watermarked pixels in an image. If a pixel to be selected is contained the neighboring region of another selected pixel, the related watermark detection may be affected by the neighboring pixel alterations, furthermore resulting in a false detection. To avoid such problem we should use small watermark embedding blocks (e.g., 3×3). Assume that a host image is $M_1 \times M_2$ dimensions and is divided into $(2r+1)\times(2r+1)$ blocks, Then the maximum number of pixels that can be selected in this image is given by [5] $n=(M_1 \times M_2)/(r+1)^2$.

3.3 The Neural Network Training

It is well-known that neural networks perform a highly adaptive nonlinear decision function from training examples. We establish the relationship among the pixels around **s** by using the back-propagation neural networks (BPNN) model. For a selected pixel $I_{i,j} = I(\mathbf{s})/255$, the network is trained with its 3×3 neighbors , i.e., let $I_{i-1,j-1}$, $I_{i-1,j}$, $I_{i-1,j+1}$, $I_{i,j-1}$, $I_{i,j+1}$, $I_{i+1,j-1}$, $I_{i+1,j}$, $I_{i-1,j+1}$ as input vector and the value of pixel $I(\mathbf{s})/255$ as output value. We construct three layer BPNN with 8, 10 and 1 neurons in the input, hidden and output layer respectively, and the tangent sigmoid, sigmoid transfer function are used for recognition. The mean absolute value of difference, for example, between simulation and true in Lena, Pepper, and Fishingboat is 2.101, 3.243 and 2.445, respectively. Generally, 2~4 differences in pixels value is invisible for human vision. This states that there is tight correlation among the pixels in a block of image, the BPNN can approach the relationship and memorize it between the original image and the watermarked image.

3.4 Watermark Extracting

According to the model of BPNN and the secret key K, the masked watermark can be retrieved as follows:

$$\tilde{w} = \begin{cases} 1 & \text{if } I_w(\mathbf{s}) > B_w(\mathbf{s}) \\ 0 & \text{otherwise} \end{cases} \tag{11}$$

where $I_w(\mathbf{s})$, $B_w(\mathbf{s})$ is the watermarked image and the output of BPNN, respectively. Performing as Eq.(3), the original watermark w can be obtained approximately.

4 Simulation Results

To demonstrate the effectiveness of the proposed algorithm, MATLAB simulation results are given in this section. In the simulation, host image of size 512×512 called "Lena" is used. The watermark is a image with size 64×64. The value of parameters in Eq.(1) is t=0.5, α=0.35, and the initial value of x is selected arbitrary x_0=0.670118. The original image and watermark are shown in Fig. 1 (a) and (b), respectively. The watermarked image with SNR=35 in Eq.(10) is shown in Fig. 1 (d) for comparison.

(a) (b) (c) (d)

Fig. 1. (a) the original image, (b) the watermark, (c) the masked watermark, (d) the watermarked image

(a) (b) (c) ρ=0.9568 (d) ρ=0.8681

Fig. 2. Degraded by noise with (a) *PSNR*=36.97, (b) *PSNR*=31.54, and the corresponding extracted watermark (c) and (d)

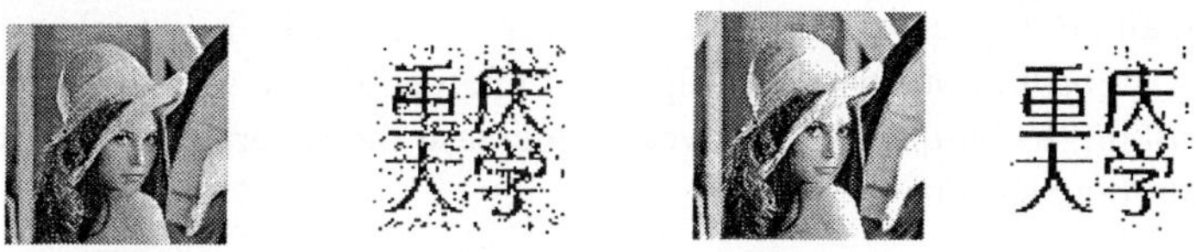

(a) (b) ρ=0.9401 (c) (d) ρ=0.9872

Fig. 3. (a) JPEG lossy compression, (b) the extracted watermark, and (c) Enhance contrast using histogram equalization, (d) the extracted watermark.

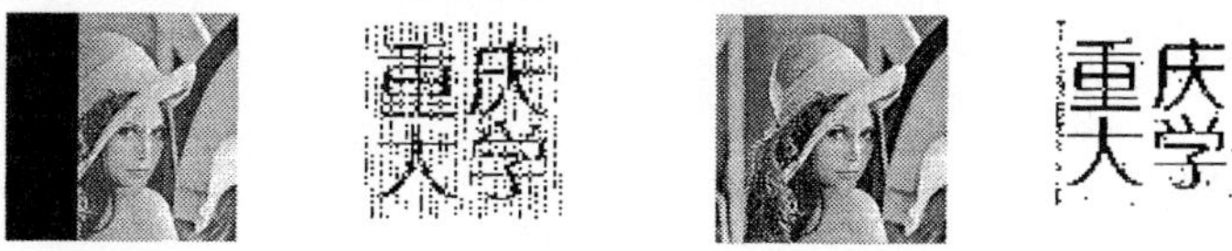

(a) (b) ρ=0.8990 (c) (d) ρ=0.9891

Fig. 4. (a) Cropped 31%, (b) the extracted watermark, and (c) Gaussian lowpass filtered, (d) the extracted watermark.

The similarity between the extracted watermark and the original watermark is defined as:

$$\rho(\widetilde{W},W)=\frac{\sum_{i=1}^{N}\widetilde{W}(i)W(i)}{\sqrt{\sum_{i=1}^{N}\widetilde{W}(i)^2}\sqrt{\sum_{i=1}^{N}W(i)^2}} \tag{12}$$

where W is the original watermark, $\widetilde{W}$ is the extracted watermark. The experimental results are shown as follows: Fig. 1 (a) and (d) show that the embedding watermark does not degrade the quality of the host image. In fact, we calculate *PSNR*=48.55. Fig. 2 (a)-(b) show that the watermarked "Lena" is degraded by Gaussian noise with *PSNR*=36.97, 31.54 respectively. Fig. 2 (c)-(d) present the corresponding extracted watermarks. Fig. 3 (a) and (b) show the watermarked "Lena"(768K) is compressed by JPEG to 58.7K and the corresponding extracted watermark. Fig. 3 (c) and (d) denote the watermarked "Lena", which is attacked by enhance contrast using histogram equalization with 256 discrete levels and the corresponding extracted watermark. Fig. 4 (a) and (b) show the watermarked image cropped 31% and the corresponding extracted watermark. Fig. 4 (c)-(d) denotes the watermarked "Lena" attacked by Gaussian lowpass filter with *n*=3, sigma=0.5 and the extracted watermark, respectively.

5 Conclusion

A blind watermarking scheme by combining neural network with chaotic map has proposed in this paper. It has the following characteristics: (i) Due to working in the spatial domain, the speed of this scheme is fast. (ii) The BPNN model and an adaptive embedding algorithm are used to improve imperceptibility and robustness. (iii) The chaotic masking technique is employed to increase the security and the robustness of the watermarking system.

Acknowledgements. The work described in this paper was supported by the National Nature Science Foundation of China (Grant No. 60271019) and the Doctorate foundation of the Ministry of Education of China (Grant No. 20020611007).

References

1. Cox, I. J. Kilian, J. Leighton, F. T. Shamoon, T.: Secure Spread Spectrum Watermarking for Multimedia, IEEE Trans. Image Processing, vol. 6, (1997) 1673–1687
2. Lin, C.-Y. Wu, M. Bloom, J. Cox, I. Miller, M. Lui, Y.: Rotation, Scale, and Translation Resilient Watermarking for Images, IEEE Trans. Image Processing, vol. 10, (2001) 767–782
3. Wang, Y. Doherty, J. Dyck, R. V.: A Wavelet-based Watermarking Algorithm for Ownership Verification of Digital Iimages, IEEE Trans. Image Processing, vol. 11, (2002) 77–88
4. Nikolaidis, A. Pitas, I.: Robust Watermarking of Facial Images Based on Salient Geometric Pattern Matching, IEEE Trans. Image Processing, vol. 2, (2000) 172–184
5. Schneier, B.: Applied Cryptography, 2nd Edition, Wiley, 1996
6. Tefas, A. Nikolaidis, A. Nikolaidis, N. Solachidis, V. Tsekeridou, S. Pitas, I. Performance Analysis of Correlation Based Watermarking Schemes Employing Markov Chaotic Sequences, IEEE Trans. signal processing, Vol. 51, No. 7, (2003) 1979-1994
7. Yu, Y. Wang, X.: An Algorithm of Adaptive Image Watermarking Based on SNR, Computer Engineering, In Chinese, Vol. 29, (2003) 70-73

Color Image Watermarking Based on Neural Networks

Wei Lu, Hongtao Lu, and Ruiming Shen

Department of Computer Science and Engineering,
Shanghai Jiao Tong University,
Shanghai 200030, P.R.China
luweisjtu@sjtu.edu.cn, {lu-ht,shen-rm}@cs.sjtu.edu.cn

Abstract. This paper presents a color image watermarking scheme based on neural networks. A binary image watermark is embedded into the spatial domain of the host image. A fixed binary sequence is added to the head of the payload image watermark as the samples to train the neural networks. Each bit of the watermark is embedded in multiple positions. Because of the good adaptive and learning abilities, the neural networks can nearly exactly extract the payload watermark. Experimental results show good performance of the proposed scheme resisting common signal processing and geometric attacks.

1 Introduction

With the rapid development of digital technology and multimedia networks, almost all kinds of works can be put on Internet, and be duplicated and spread quickly. A very crucial issue for copyright protection has emerged. As an effective method for digital works copyright protection and authentication, watermarking technique has been extensively studied and notable results have been obtained [1,2,3,4,5,6].

Some watermarking schemes based on neural networks have been developed in the literature [4,5,6]. In [3], Kutter proposed a method using amplitude modulation in spatial domain to embed watermark, Yu *et al.* improve the robustness of Kutter's method by using neural networks [4]. In this paper, we propose another method based on Kutter's scheme and use neural networks in a different way from [4] for watermark extraction to improve the robustness.

2 Watermark Embedding

Let I represent a RGB color image with size $M \times N$, where $I = \{V_r, V_g, V_b\}$, V_r, V_g and V_b are the red, green and blue channels information of image I, $V_r(i,j), V_g(i,j), V_b(i,j) \in \{0, 1, \ldots, 255\}$, $i \in \{0, 1, \ldots, M-1\}$, $j \in \{0, 1, \ldots, N-1\}$. We embed the watermark into the blue channel. The embedded watermark sequence W consists of two parts

$$W = P + S = p_0, p_1, \ldots, p_{k-1}, s_0, s_1, \ldots, s_{l-1} = w_0, w_1, w_2, \ldots, w_{m-1} \quad (1)$$

F. Yin, J. Wang, and C. Guo (Eds.): ISNN 2004, LNCS 3174, pp. 651–656, 2004.

where P is a fixed binary pattern sequence with length k, $p_0, p_1, \dots, p_{k-1} \in \{0,1\}$, S is a payload watermark sequence with length l, $s_0, s_1, \dots, s_{l-1} \in \{0,1\}$, which is obtained by a binary image watermark, and $m = k + l$, $w_0, w_1, \dots, w_{m-1} \in \{0,1\}$.

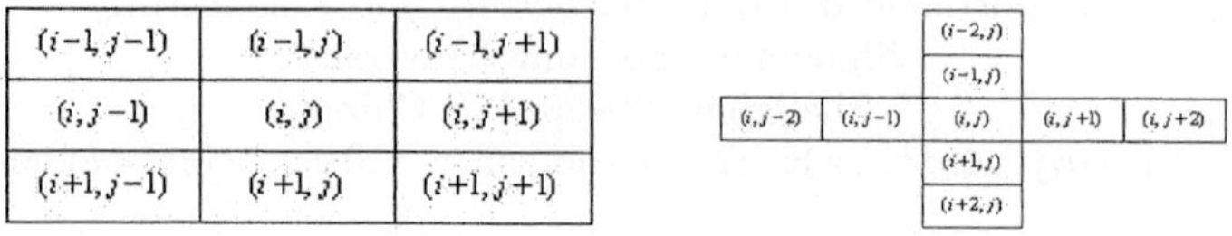

(a) (b)

Fig. 1. (a) The symmetric square-shaped window. (b) The symmetric cross-shaped window.

The proposed watermark embedding scheme can be described as follows

1) Generate a pseudo-random coordinate sequence $Z = \{(i_t, j_t) | t = 0, 1, \dots, m \times n\}$ according to a secret key, where $i_t \in \{0, 1, \dots, M-1\}$, $j_t \in \{0, 1, \dots, N-1\}$, n represents the repeated embedding times for a bit.

2) For each position $(i_t, j_t) \in Z$, firstly we compute the luminance by

$$L(i_t, j_t) = 0.299V_r(i_t, j_t) + 0.587V_g(i_t, j_t) + 0.114V_b(i_t, j_t) \tag{2}$$

then we compute the blue channel average value $\overline{V}_b(i_t, j_t)$ over the square-shape window centered at (i_t, j_t), as shown in Fig. 1(a), that is

$$\overline{V}_b(i_t, j_t) = \frac{1}{(2c+1)^2 - 1}\left(\sum_{x=-c}^{c}\sum_{y=-c}^{c} V_b(i_t + x, j_t + y) - V_b(i_t, j_t)\right) \tag{3}$$

where c is the width parameter of the window.

3) We embed each bit of W n times at n different locations selected sequentially from Z. Suppose a watermark bit w_s is embedded, $s \in \{0, 1, \dots, m-1\}$, the embedding formula is

$$V_b(i_t, j_t) = \overline{V}_b(i_t, j_t) + (2w_s - 1)L(i_t, j_t)\alpha, \quad t = sn, sn+1, \dots, (s+1)n - 1 \tag{4}$$

where α is the positive embedding strength.

4) Repeat step 3 m times until all watermark bits are embedded into the original image I. We then obtain the watermarked image with size of $M \times N$.

In our scheme, we use an adjustable k to control the pattern sequence length as the training sets of the neural network proposed in Section 3. We also consider the correlation between the embedded point and its square shaped neighborhood by using $\overline{V}_b(i_t, j_t)$ in step 3, which makes the scheme more robust than Kutter's, while in Kutter's scheme $V_b(i_t, j_t) = V_b(i_t, j_t) + (2w_s - 1)L(i_t, j_t)\alpha$.

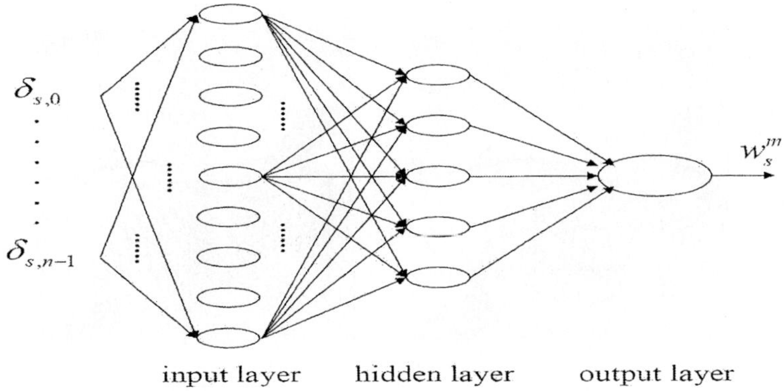

Fig. 2. Back Propagation Neural Network

3 Watermark Extraction

For watermark extraction from the input image $I^* = \{V_r^*, V_g^*, V_b^*\}$, the following steps are performed

1) Generate the same Z using the same secret key as the embedding process.

2) For each $(i_t, j_t) \in Z$, we compute the luminance $L^*(i_t, j_t)$ and the neighborhood average value $\overline{V}_b^*(i_t, j_t)$ as the step 2 in the embedding process.

3) In order to extract w_s, $s \in \{0, 1, \ldots, m-1\}$, we define

$$\delta_s(i_t, j_t) = V_b^*(i_t, j_t) - \overline{V}_b^*(i_t, j_t), \quad t = sn, sn+1, \ldots, (s+1)n-1 \tag{5}$$

then we get the input sets $\{\delta_s(i_t, j_t)\}$, $s = 0, 1, \ldots, m-1$, for the neural network.

4) In our scheme, we use the traditional Back Propagation Neural Network shown in Fig. 2, which is a 9-5-1 multiplayer perceptron. The training set is built from the first k input sets $\{\delta_s(i_t, j_t)\}$ and pattern sequence P as $\{\{\delta_s(i_t, j_t)\}, 2w_s - 1\}$, $s = 0, 1, \ldots, k-1$.

5) For extraction of the payload watermark sequence, the last l input sets are fed to the trained neural network, each bit s_t^* is determined by the output w_{k+t}^o of the neural network, the extracted s_t^* can be obtained by

$$s_t^* = \begin{cases} 1 & \text{if } w_{k+t}^o \geq 0 \\ 0 & \text{else} \end{cases} \qquad t = 0, 1, \ldots, l-1 \tag{6}$$

thus we obtain the extracted watermark sequence W^*.

In our scheme, because we use the correlation of square shaped neighborhood area in watermark embedding process, $\delta(i, j)$ that we compute in step 3 of section 2 reflect the composition of the luminance and embedding strength more accurately than the cross shaped neighborhood in Kutter's scheme [3] shown in Fig. 1(b). Furthermore, in step 4, we use the BP neural network to determine the adaptive watermark which can improve the robustness as shown in later experiments, while in [4], the train sets also do not consider the neighborhood correlation as in [3].

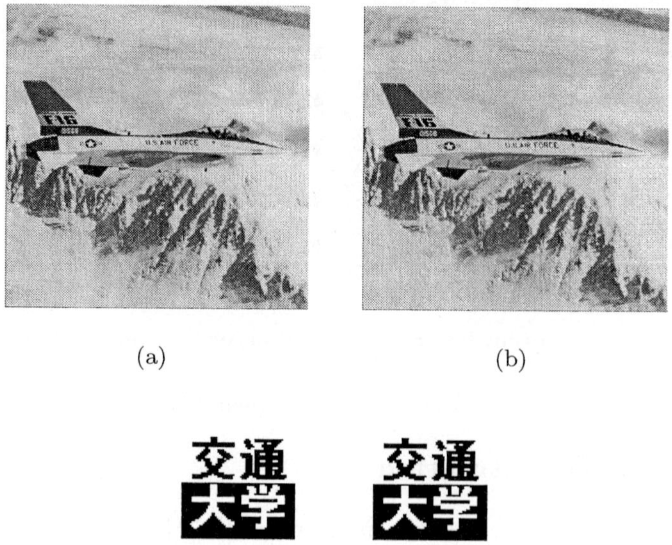

(a) (b)

(c) (d)

Fig. 3. (a) Original image. (b) Watermarked image. (c) Original watermark. (d) Extracted watermark from (b) (epochs=40).

4 Experimental Results

In our experiments, we use the color 512×512 F16 battleplane image with 24-bit color as the original image shown in Fig. 3(a). We use the $010101\ldots0101$ sequence as the pattern with length $k = 50$, The payload watermark is a 64×64 binary image shown in Fig. 3(c). The embedding strength $\alpha = 0.1$. The multiple embedding parameter $n = 8$, the square window width parameter $c = 1$, and the learning rate for BP network $lr = 0.2$. The watermarked image is shown in Fig. 3(b).

In order to describe the correlations between original watermark S and the extracted watermark S^* for making a binary decision on whether a given watermark exists or not, we calculate the normalized cross-correlation (NC) values [1] defined by

$$NC = \frac{\sum_i \sum_j S(i,j) \cdot S^*(i,j)}{\sum_i \sum_j [S(i,j)]^2} \tag{7}$$

Fig. 3(d) shows the extracted watermark from Fig. 3b, its NC is computed to be 1, which shows the exactly extraction.

We add Gaussian additive noise to the watermarked image is shown in Fig. 4(a), and the extracted watermark is shown in Fig. 4(c). Fig. 4(b) shows the blurred image with a 5×5 cross-shaped window filter, Fig. 4(d) shows the extracted watermark. Fig. 4(e-g) show the extracted watermarks from the attacked

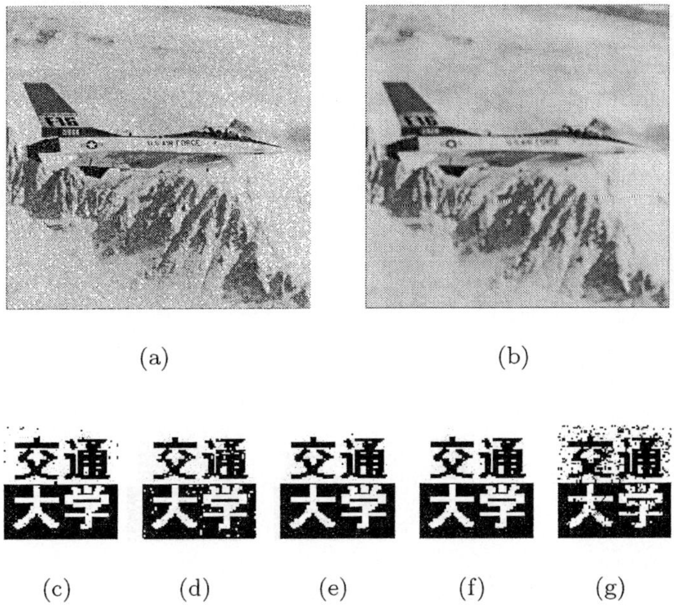

Fig. 4. (a) Image under additive Gaussian noise attack. (b) The watermarked image was blurred. (c) Extracted watermark from (a) with NC=0.9848(epochs=70). (d)Extracted watermark from (b) with NC=0.9663(epochs=70). (e) Extracted watermark from Gaussian low-pass filter with NC=0.9985(epochs=60). (f) Extracted watermark from high-pass filter with NC=0.9995(epochs=60). (g) Extracted watermark from JPEG quality 70 with NC=0.8677(epochs=80).

image under Gaussian low-pass filter, high-pass filter and JPEG compression with quality 70. All these show good robustness against common signal processing attacks.

To test the robustness of the proposed scheme against geometric attacks, the watermarked image has been rotated to the right by an angle 60° shown in Fig. 5(a). In order to decrease the search space, supposing it is a pure rotation attack, and Fig. 5(b) shows the rotation response to the watermark detector, we can see that the watermark can also be extracted, which is shown in Fig. 5(c). Fig. 5(d) shows the extracted watermark under attack of resizing with factor 1.5, while Fig. 5(e) with factor 0.5 which is less robust than Fig. 5(d), this is because zooming out image result in much more information loss than zooming in.

5 Conclusions

In this paper, we propose a color image watermarking scheme based on neural networks. The experimental results show that it is more robust to common image processing and geometric distortions. Since the watermark embedding is controlled by a secret key, our scheme is also secure.

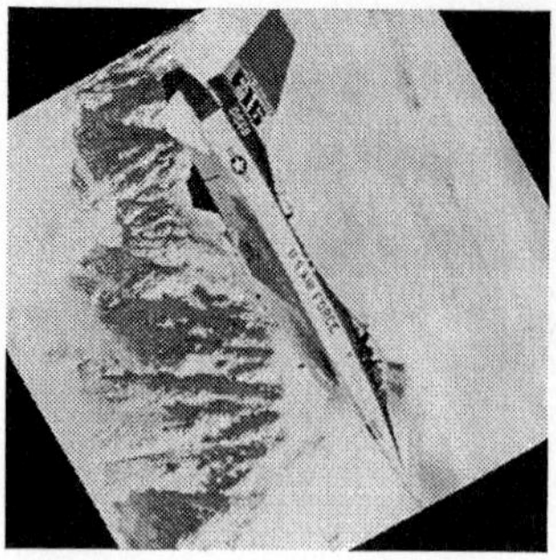

(a)

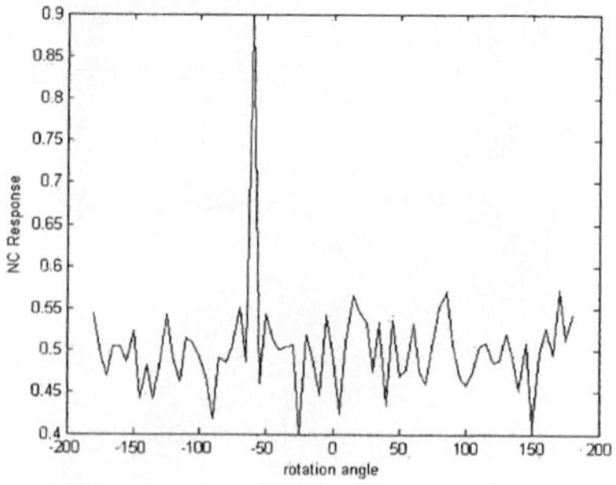

(b)

(c) (d) (e)

Fig. 5. Geometric attacks. (a) The watermarked image was rotate 60^0 to right. (b) NC detection response to the rotation space. (c) Extracted watermark from (a) with NC=0.9 (epochs=100). (d) Extracted watermark with resize factor 1.5 (epochs=100). (e) Extracted watermark with resize factor 0.5 (epochs=100).

Acknowledgments. This work is supported by Shuguang Program of Shanghai Education Development Foundation.

References

1. Petitcolas, F.A.P., Anderson, R.J., Huhn, M.G.: Information hiding - a survey. Proc. IEEE, **87(7)** (1999) 1062-1078
2. Cox, I.J., Killian, J., Leighton, T., Shamoon,T.: Secure Spread spectrum watermarking for multimedia. IEEE Trans. Image Processing, **6** (1997) 1673-1687
3. Kutter, M., Jordan, F., Bossen, F.: Digital Watermarking of Color Image Using Amplitude Modulation. J. Electronic Imaging, **7(2)** (April 1998) 326-332
4. Yu, P.T., Tsai, H.H., Lin, J.S.: Digital Watermarking based on Neural Networks for Color Images. Signal Processing, **81** (October 2001) 663-671
5. Fan, Y.C., Mao, W.L., Tsao, H.W.: An Artificial Neural Network-based Scheme for Fragile Watermarking. IEEE Int. Conf. Consumer Electronics, (2003) 210-211
6. Zhang, Z.M., Li, R.Y., Wang, L.: Adaptive Watermark Scheme with RBF Neural Networks. IEEE Int. Conf. Neural Networks & Signal Processing, (Dec 2003) 1517-1520

A Novel Intrusion Detection Method Based on Principle Component Analysis in Computer Security*

Wei Wang[1], Xiaohong Guan[1, 2], and Xiangliang Zhang[3]

[1] SKLMS (State Key Laboratory for Manufacturing Systems Engineering) and Research Center for Networked Systems and Information Security,
Xi'an Jiaotong University, 710049 Xi'an, China
{wwang, xhguan}@sei.xjtu.edu.cn

[2] Center for Intelligent and Networked Systems,
Tsinghua University, 100084 Beijing, China

[3] Department of electronic science and technology,
Xi'an Jiaotong University, 710049 Xi'an, China
zhangxl@mailei.xjtu.edu.cn

Abstract. Intrusion detection is an important technique in the defense-in-depth network security framework and a hot topic in computer security in recent years. In this paper, a new intrusion detection method based on Principle Component Analysis (PCA) with low overhead and high efficiency is presented. System call data and command sequences data are used as information sources to validate the proposed method. The frequencies of individual system calls in a trace and individual commands in a data block are computed and then data column vectors which represent the traces and blocks of the data are formed as data input. PCA is applied to reduce the high dimensional data vectors and distance between a vector and its projection onto the subspace reduced is used for anomaly detection. Experimental results show that the proposed method is promising in terms of detection accuracy, computational expense and implementation for real-time intrusion detection.

1 Introduction

Intrusion detection system (IDS) is an important component of the defense-in-depth or layered network security mechanisms [1]. In general, the techniques for intrusion detection fall into two major categories depending on the modeling methods used: misuse detection and anomaly detection. Since anomaly detection can be effective against new attacks, it has become a hot topic in research of computer security.

Many types of data can be used for anomaly detection, such as Unix commands, audit events, keystroke, system calls, and network packages, etc. Early studies [2, 3] on anomaly detection mainly focus on learning normal system or user behaviors from monitored system log or accounting log data. Examples of the information derived

* The research in this paper was supported in part by the National Outstanding Young Investigator Grant (6970025), National Natural Science Foundation (60243001) and 863 High Tech Development Plan (2001AA140213) of China.

F. Yin, J. Wang, and C. Guo (Eds.): ISNN 2004, LNCS 3174, pp. 657–662, 2004.

from these logs are: CPU usage, time of login, duration of user session, names of files accessed, etc. In recent years, many research in anomaly detection focus on learning normal program behavior. In 1996, Forrest et al. introduced a simple anomaly detection method based on monitoring the system calls issued by active, privileged processes [4]. This work was extended by various methods. Lee et al. used data mining approach to study a sample of system call data to characterize sequences occurring in normal data by a small set of rules [5]. Warrender et al. proposed Hidden Markov Model (HMM) method for modeling and evaluating invisible events based on system calls [6].

In practice, a protected computer system could produce massive data streams, for example, during the experiments of capturing the system calls on the *sendmail*, only 112 messages produced a combined trace length of over 1.5 million system calls [4]. Therefore, processing the high dimensional audit data in real time for online intrusion detection would be computationally expensive.

Principle Component Analysis (PCA, also called Karhunen-Loeve transform) is one of the most wildly used dimension reduction techniques for data analysis and compression in practice. In this paper, we discuss a novel intrusion detection method based on PCA, by which intrusion detection can be employed in a lower dimensional subspace and the computational complexity can be significantly reduced. Two types of data are used to verify the proposed method and the testing results show that the method is efficient and effective.

2 The Proposed Intrusion Detection Method Based on PCA

Suppose an observation dataset is divided into m blocks by a fixed length (e.g. divided consecutively by 100 in the command data) or by an appointed scheme (e.g. separated by processes in system call data), and there are totally n unique elements in the dataset, the observed data can be expressed by m vectors with each vector containing n observations. A $n \times m$ Matrix X, where each element X_{ij} stands for the frequency of i-th individual element occurs in the j-th block, is then constructed.

Given a training set of data vectors $\mathbf{x}_1, \mathbf{x}_2, \cdots, \mathbf{x}_m$, the average vector $\boldsymbol{\mu}$ and each mean-adjusted vector can be computed. m eigenvalue-eigenvector pairs $(\lambda_1, \mathbf{u}_1), (\lambda_2, \mathbf{u}_2), \cdots, (\lambda_m, \mathbf{u}_m)$ of the sample covariance matrix of vectors $\mathbf{x}_1, \mathbf{x}_2, \cdots, \mathbf{x}_m$ can be also computed [7, 8].

Several eigenvectors $\mathbf{u}_1, \mathbf{u}_2, \cdots, \mathbf{u}_k$ ($k << m$) forming the $n \times k$ matrix U, which can be used to represent the distribution of the original data, is decided by experiences. Any data vector of the training set can be represented by linear combination of the k eigenvectors such that the dimensions of the data are reduced.

Given a test data vector $\mathbf{t}$, it can be projected onto the k-dimensional subspace according to the rules [7]

$$\mathbf{y} = U^T(\mathbf{t} - \boldsymbol{\mu}) \tag{1}$$

The distance between the test data vector and its projection onto the subspace is simply the distance between the mean-adjusted input data vector $\mathbf{\Phi} = \mathbf{t} - \boldsymbol{\mu}$ and

$$\mathbf{\Phi}_f = U\mathbf{y} \tag{2}$$

If the test data vector is normal, the vector and its projection would be very similar and the distance between them would be very small and near to zero [9]. Based on this property, normal program and user behaviors are profiled for anomaly detection. In this paper, three measures, squared Euclidean distance, Cosine distance and Signal-to-Noise Ratio (SNR) measure, are applied to map the distance or similarity of this two vectors for comparison of the experimental results.

Squared Euclidean distance, Cosine distance and SNR measure, are defined respectively by the following rules for anomaly detection

$$\varepsilon_1 = \left\| \mathbf{\Phi} - \mathbf{\Phi}_f \right\|^2 \tag{3}$$

$$\varepsilon_2 = \frac{\mathbf{\Phi}^T \mathbf{\Phi}_f}{\left\| \mathbf{\Phi} \right\| \left\| \mathbf{\Phi}_f \right\|} \tag{4}$$

$$\varepsilon_3 = 10\log\left(\frac{\left\| \mathbf{\Phi} \right\|^2}{\left\| \mathbf{\Phi} - \mathbf{\Phi}_f \right\|^2} \right) \tag{5}$$

In the procedure of anomaly detection, ε_1, ε_2 and ε_3 are considered as *detection index*. If either ε_1, ε_2 is below or ε_3 is above a predetermined threshold, the test data $\mathbf{t}$ is then classified as normal, otherwise as anomalous.

3 Experiments

3.1 Experiments on System Call Data

The first data used in the experiments is from the data set collected by Warrender and Forrest [6]. Since a great number of traces are included in the *lpr* data, we use MIT *lpr* data to validate the proposed method in this paper. The data set is available for downloading at http://www.cs.unm.edu/~immsec/. The procedures of generating the data are also described in the website. Each trace of the *lpr* data is the list of system calls issued by a single process from the beginning of its execution to the end. The data set includes 2703 traces of the normal data and 1001 traces of the intrusion data. We use the former 600 traces of the normal data and the former 300 traces of the intrusion data in the experiments. The data descriptions in the experiments are shown as Table 1.

Table 1. Data descriptions in the experiments

Number of system calls	Number of unique system calls	Number of normal traces	Number of intrusion traces
842,279	41	600	300

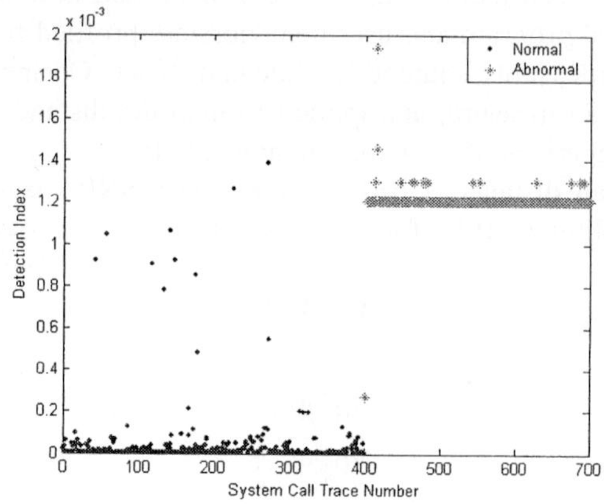

Fig. 1. Experimental results on the MIT *lpr* system call data. The y-axis represents the *detection index* and x-axis represents the system call trace number. The star (*) stands for abnormal data and dot (•) for normal data

Using PCA for intrusion detection, we can get good testing results. Figure 1 shows the experimental results using squared Euclidean distance measure with the former 200 traces of data for training and other 700 traces for testing. It is observed that abnormal data can be easily distinguished from normal data.

Using different number of the traces in the normal data for training and different distance or similarity measures for anomaly detection, we can get different detection rate and false alarm rate. We use principle component percentage of the total variation as 99.9% in the experiments and the results are summarized as table 2.

Table 2. False Alarm Rate (FAR) and Detection Rate (DR) with different conditions

Number of the normal training traces	Squared Euclidean distance measure		Cosine distance measure		SNR measure	
	FAR	DR	FAR	DR	FAR	DR
100	3.4%	100%	12.4%	100%	12.4%	100%
200	2.75%	100%	10.25%	100%	10.25%	100%
300	3%	100%	8%	100%	8%	100%
400	4%	100%	7%	100%	7%	100%

From table 2, we observe that the experimental results are the best for low alarm rate when the number of the training traces is 200 with squared Euclidean distance measure. Another observation is that the squared Euclidean distance is better than Cosine distance measure and SNR measure for intrusion detection.

3.2 Experiments on Unix Command Data

To further investigate the performance of intrusion detection using the proposed method, we use another data set which comes from a UNIX server at AT&T's Shannon Research Laboratory. User names and the associated command sequences make up the data. Fifty users are included with 15000 commands for each user, divided into 150 blocks of 100 commands. The first 50 blocks are uncontaminated while some masquerading command blocks are inserted into the command sequences of the 50 users starting at block 51 and onward. The goal is to correctly detect the masquerading blocks in the user community. The data are available at http://www.schonlau.net/intrusion.html, see [3] for more about the data descriptions. We revised the data and reconstructed them in the experiments. We selected two data sets of two users from the user community. The first 50 data blocks of the first user are used for training and other data, which contain 100 data blocks of the first user considered as normal and 150 blocks of the second user as abnormal, are used for testing. User 5 and user 32 are selected in the experiments.

We used principle component percentage of the total variation as 99.9% and squared Euclidean distance for anomaly detection in the experiments. Experimental results are shown as Fig.3.

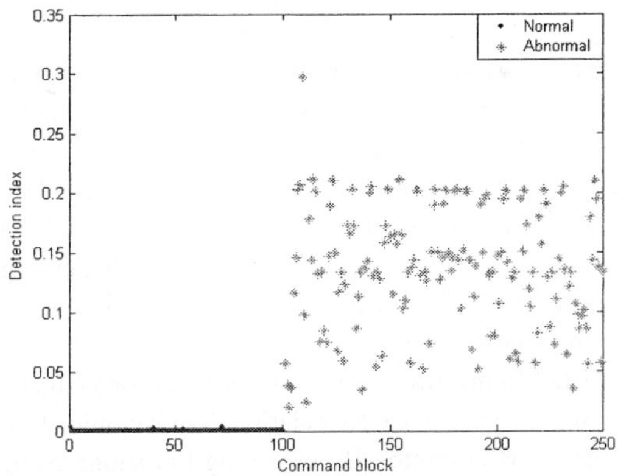

Fig. 2. The experimental results of the combining data of user 5 and user 32. All the data blocks of user 5 and 32 are uncontaminated, therefore the first 100 data blocks from user 5 are treated as normal (•) and blocks 101~250 from user 32 are considered as abnormal (*)

From Fig.2, it is easily observed that the abnormal data can be 100% distinguished from the normal data without false alarm by using PCA.

4 Conclusion

In this paper, a new intrusion detection method based on PCA is proposed. Instead of considering the transition information of the system calls or commands, the new method takes into account those of frequency property. Since there is no need to consider each system call in each trace or command in each block, the computational cost of the proposed method is low and suitable for real-time intrusion detection. Data found in intrusion detection problem are often high dimensional in nature. By using the proposed method, the high dimensional data can be greatly reduced by projecting them onto a lower dimensional subspace for intrusion detection so that the complexity of the detecting algorithm is significantly reduced.

The method is implemented and tested on the system call data from University of New Mexico and the Unix command data from AT&T Research lab. Experiment results show that the method is promising in terms of detection accuracy, computational expense and implementation for real-time intrusion detection.

Further research is in progress to mix the frequencies property with the transition information of system calls and commands so that lower false alarms and missing alarms can be achieved.

References

1. Lee, W., Xiang, D.: Information-Theoretic Measures for Anomaly Detection. Proceedings of the IEEE Symposium on Security and Privacy, IEEE Computer Society Press, Oakland, CA (2001) 130-143
2. Anderson, D., Frivold, T., Valdes, A.: Next-Generation intrusion Detection Expert System (NIDES): A Summary. Technical Report SRI-CSL-95-07, Computer Science Laboratory, SRI International, Menlo Park, California (1995)
3. Schonlau, M., Theus, M.: Detecting Masquerades in Intrusion Detection Based on Unpopular Commands. Information Processing Letters, Vol. 76 (2000) 33-38
4. Forrest, S., Hofmeyr, S.A., Somayaji, A., Longstaff, T.A.: A Sense of Self for Unix Processes. Proceedings of the IEEE Symposium on Security and Privacy, IEEE Computer Society Press, Oakland, CA (1996) 120-128
5. Lee, W., Stolfo, S.: Data Mining Approaches for Intrusion Detection. Proceedings of the 7th USENIX Security Symposium, Usenix Association, San Antonio, Texas (1998) 79-94
6. Warrender, C., Forrest, S., Pearlmutter, B.: Detecting Intrusions Using System Calls: Alternative Data Models. Proceedings of the IEEE Symposium on Security and Privacy, IEEE Computer Society Press, Oakland, CA (1999) 133-145
7. Duda, R.O., Hart, P.E., Stork, D.G.: Pattern Classification. 2nd Edn. China Machine Press, Beijing (2004) 568-570
8. Jolliffe, I.T.: Principal Component Analysis. 2nd Edn. Springer-Verlag, New York (2002)
9. Turk, M., Pentland, A.: Eigenfaces for Recognition. Journal of Cognitive Neuroscience. Vol. 3, No. 1 (1991) 71-86

A Novel Wavelet Image Watermarking Scheme Combined with Chaos Sequence and Neural Network

Jian Zhao[1,2], Mingquan Zhou[1], Hongmei Xie[3], Jinye Peng[2], and Xin Zhou[2]

[1] Computer Science Department, Northwest University,
Xian 710069, Shaanxi, PR China
[2] Electronic Science Department, Northwest University,
Xian 710069, Shaanxi, PR China
[3] Electronic Engineering Department, Northwestern Polytechnical University,
Xian 710072, Shaanxi, PR China

Abstract. Digital watermarks have been proposed in recent literature as a means for copyright protection of multimedia data. In the absence of standardization and specific requirements imposed on watermarking procedures, anyone can claim ownership of any watermarked images. In order to protect against these counterfeiting techniques, we examine the properties that are necessary for resolving ownership via invisible watermarking. A method for watermarking of chaos and neural network is proposed in this paper. The watermark is embedded in the wavelet descriptors. Watermarks generated by this nonlinear technique can be successfully detected even after rotation, translation, scaling. And watermarks of our scheme are good at protecting from many kind watermark attacks. The experimental results demonstrate that the watermark is useful and practical.

Keywords. Digital Watermarking, Chaotic Sequence, neural network, wavelet transform

1 Introduction

With the increasing importance and widespread distribution of digital media, the protection of the intellectual property rights of the owner for their media has become increasingly significant. One type of such media is digital imagery, which can be copied and widely distributed without any significant loss of quality. Protecting the property rights of these images is therefore more important. A straightforward way to protect this is to completely encrypt the data and thereby require the end user to have the encryption key for the decoding. Another means to protect this data is to apply a digital watermark.

Digital watermarking is the process of encoding an image with its owner's watermarks. It can be done in two general approaches. One is to transform the original image into its frequency domain representation and embed the watermark data therein. The second way is to embed the watermark into spatial domain of the original image directly. The general methods of watermark embedding and verifying are shown in Figure 1 and Figure 2.

F. Yin, J. Wang, and C. Guo (Eds.): ISNN 2004, LNCS 3174, pp. 663–668, 2004.

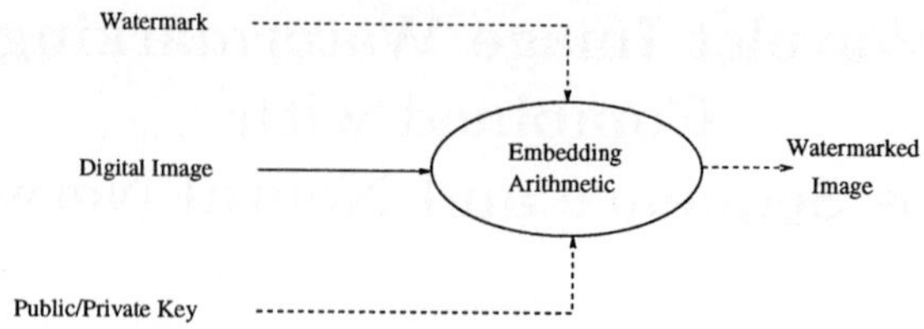

Fig. 1. General method of watermark embedding

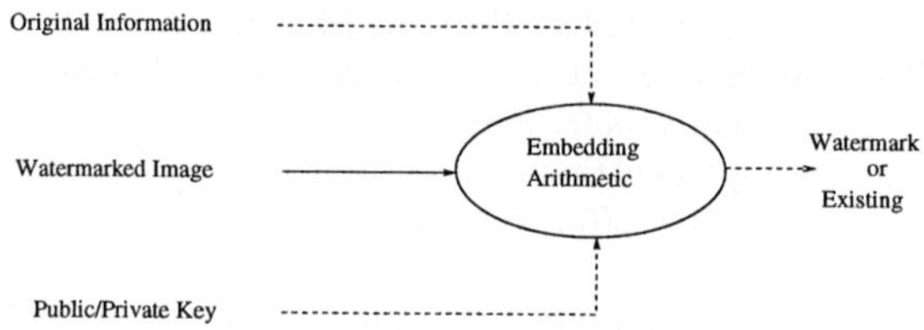

Fig. 2. General method of watermark verifying

2 Advantages of Wavelet Image Watermark

Embedding watermark at the frequency domain is better than doing it in the spatial domain. There are some virtues: people cannot distinguish the distinction between embedded images and original images; the energy of watermark is dispersed to all pixels; computing at the frequency domain matches the international standard (Such as JPEG2000). In addition, considering the character of a visual system, we can get the effective results in concealment and robustness. The wavelet basis function has better correlate to the broadband nature of images than the sinusoidal waves used in Discrete Cosine Transform (DCT). The Discrete Wavelet Transform (DWT) adapts HSV more than DCT, so it attracts more attention to searching in digital watermark. The new technologies used in JPEG2000 makes the watermark arithmetic, which is based on DWT, more practicable.

In the wavelet domain, the low frequency band owns much larger energy than the high frequency band. That means the most energy of image exists in the low frequency; and the energy of the image edge exists at high frequency. If we embed the watermark at the low frequency band, the information of image may be lost. While if we embed it in the high frequency band, some watermark information may disappear after image processing (such as code, compress). So, when we do the experiment of this paper, the image is disposed by the 3rd DWT function firstly, then it is divided as big as the watermark image size, and the sub-image have no any overlapped blocks. After that, we embed watermark to each sub-image. The area of the high frequency is bigger than that of the low frequency, so we get more sub-images at high frequency band. Thus the part of high frequency gets more watermarks while the part of low frequency gets fewer. Through this method, When image embeded watermark is transformed back to spatial domain, the watermark is widespread to whole image. And the watermark of image is more robust according to spread-frequency theory.

3 Application of Chaos and Neural Network

A deterministic system is a system which statistic result is completely determined by its initial conditions. On the contrary, a stochastic system's statistic result only partially determined by the initial condition, due to noise, or other external circumstances beyond our control. For a stochastic system, the present state reflects the past initial conditions and the particular realization of the noise generated. Chaos means that the system obeys deterministic laws of evolution, but the outcome is highly sensitive to small uncertainties in the specification of the initial state. If a deterministic system is locally unstable and globally mixing, it is said to be chaotic.

The chaos function can be superposed and extended, so it could not be forecasted. In the paper, we get the chaos sequence from the discrete dynamic Logistic mapping. We define:

$$x_{k+1} = \mu x_k(1 - x_k), \quad x_k \in (0, 1) \tag{1}$$

Where $0 \leq \mu \leq 4$. After transformation, the Logistic mapping is expressed at $[-1, 1]$ as follow:

$$x_{k+1} = (1 - \lambda x_k^2), \quad \lambda \in (0, 2) \tag{2}$$

With the value of λ increasing, the result of the function will be changed singularity and markedly. When $\lambda = 1.40115$, after several iterative steps, the system will be chaotic. The average value of the chaotic sequence is 0, and the sequence is the δ - like self-correlation sequence, and its cross correlation is 0. It has all the characters of white noise's statistical property. Here, we let $\lambda = 2$, initial value of x_0 is 0.3. After iterative computing of equation (2), we get the sequence X_K. We use chaos key as the user's key. Though the pirate picked-up the information of the watermark, they cannot resume the image of the watermark if they do not know the value of K. In the experiment, we define $k = 8$, and the precision is $10e(-4)$, the chaotic sequence is $X_K = \{0.3, 0.82, -0.3448, 0.7622, -0.1620, 0.9475, -0.7955, -0.2656\}$. We denote the watermark image as vector W_p, where $p = 1, 2, \cdots, M \times N$, and then encrypt the watermark image by X_K. We get the cryptograph watermark image W_v.

$$W_v = W_p \otimes X_K, \quad v = 1, 2, \cdots, M \times N \tag{3}$$

Here, "$\otimes$" denotes the XOR.

In order to make X_K more difficult to be interpreted, a nonlinear mapping by neural network of X_K is used. We use a simply 3 layers BP neural network, and X_K is nonlinear mapping to X'_K. User can design the X'_K sequence. Sometimes you can design X'_K more like some statistic sequence to puzzle the people who want to get watermark. So at last when we use function (3), we use more like some statistic sequence to puzzle the people who want to get watermark. So at last when we use function (3), we use X'_K instead of X_K. Fig 3 shows how 3 layers BP neural network make those nonlinear mapping.

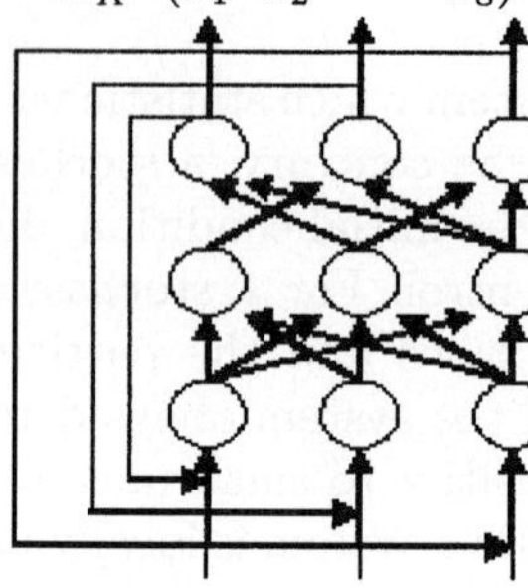

Fig. 3. Nonlinear Mapping (X_K to X'_K)

4 The Realization of the Scheme

Suppose X denote the original image whose size is $M \times M$, each pixel is s bits, then

$$X = \{w(m,n), \quad 0 \leq m,n \leq M\} \tag{4}$$

Here $w(m,n) \in \{0,1,\cdots,2^s-1\}$ is the value of the point (m,n) at the original image.

Suppose W is the watermark image, the size is $N \times N$, and each pixel is c bits.

$$W = \{w(m,n), \quad 0 \leq m,n \leq N\} \tag{5}$$

Here $w(m,n) \in \{0,1,\cdots,2^c-1\}$ is the value of the point (m,n) at the watermark image.

In general, suppose the size of watermark image is smaller than the size of the original image, and $M = 2^p \cdot N$ (where p is positive integer), then the algorithm process is:

1. Confusing the watermark image through Arnold transform, W_d denotes the confused image.
2. Transforming the original image by the 3rd DWT, we get some specifics sub-images X_j^k and an approach sub-image X_3^0 in different respective scale.

$$X_v = DWT(X) = \{X_j^k, j,k = 1,2,3; \quad \text{if } k=0, \text{ then } j=3\} \tag{6}$$

Where X is the wavelet-transformed image.

3. Separating all specifics sub-images to $2^{(p+1)-2(j-1)}$ blocks, which have the same size as the watermark image, and any blocks do not overlap each other.

$$X_b = \mathbf{Block}(X_v) = \{X_j^{k,i}, i = 1,2,\cdots,2^{(p+1)-2(j-1)}\} \tag{7}$$

4. Computing the data of the watermark image according to formula 3, we will get the encrypted watermark which processing by Logistic mapping chaotic sequence and neural network at cryptograph space.
5. High-frequency part should be embedded repeatedly, low frequency should be embedded only once. Embed the encrypted watermark to every sub-images with formula:

$$X_{bs}^{i} = X_{b}^{i} + \alpha(W_d) = \{X_{j}^{k,i}(m,n) + \alpha w_d(m,n), 0 \le m,n \le N\} \quad (8)$$

   Where α express the intensity of the watermark image. We should balance the relation between visibility and robustness of the watermarked image. If the value of α is large, then we'll get high robustness but low quality images, and vice versa.
6. Jointing all computed sub-images orderly. We will get the sub-image X_{bs}, which size is $M \times M$. Restore the sub-image by reverse DWT, then we'll get the embedded image:

$$X_s = IDWT(x_{bs}) = \{x_s(m,n), 0 \le m,n \le M\} \quad (9)$$

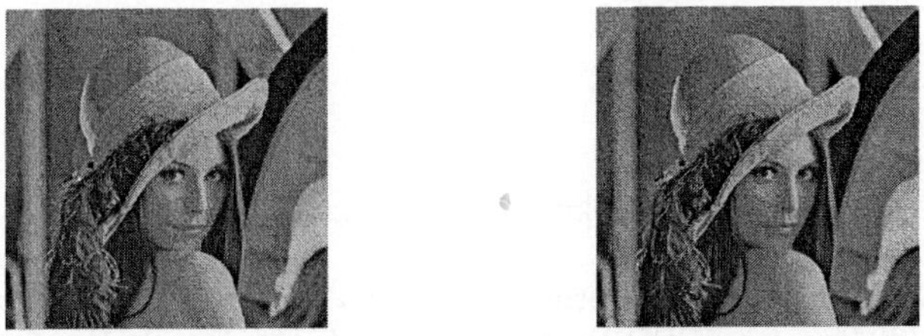

(a) original image (b) watermarked image

Fig. 4. Watermark performance

(a) original watermark image (b) chaotic watermark image

Fig. 5. Watermark image

5 Results and Conclusion

Figure 4 is the performances of the image, which is the watermark image processed by the chaotic sequence and neural network at wavelet domain. Figure 4(a) is the original image; figure 4(b) is the watermarked image. From figures we can tell that there is no much different on visual affection between watermarked image and original image. Figure 5(a) is original watermark; figure 5(b) is the watermark image processed by Logistic mapping chaotic sequence and neural network. From figure 5(b) we can see that original watermark has been obscured and can not be detected easily.

The afterward study shows the algorithm is good at the robustness. More simulations show that it can extract watermark from embedded image absolutely even the image suffers from pepper noise, Gauss noise, or is compressed by JPEG. As facts show above, we can draw conclusions of the algorithm:

1) Encryption key's Exclusivity. Different key gives different X_K and different W_v , so each image has its own key and watermark.
2) Irreversibility. We could not work out X_K according to W_v .
3) Watermark invisibility. The embedded digital image is almost the same as original image.
4) Robustness. The wavelet watermark image scheme has characters of spread spectrum system, which is robust to disturb and noise. Here we use neural network to improve robust of the algorithm. So our algorithm is more robust than before studied similar algorithm.

As the mentioned above, this algorithm is valuable and practical. The further study of choosing the chaotic sequence and choosing the key should improve the robust, exclusivity and credibility.

References

1. Tewfik A H: Digital watermarking. IEEE Signal Processing Magazine, 2000, 17(9):17–88.
2. Nikolaidis A.: Asymptotically optimal detection for additive watermarking in the DCT and DWT domains. IEEE Transac-tions on Image Processing, 2003, 12(5):563–571
3. N. Nikolaidis and I. Pitas: Robust image watermarking inthe spatial domain. Signal Processing, sp.issue on Copyright Protection and Access control, 1998, 66(3):385–403
4. I. J. Cox, J. Kilian, F. T. Leighton, and T. Shamoon: Secure spread spectrum watermarking for multimedia. IEEE Transactions on Image Processing, 1997, 6(12):1673–1687
5. Dawei Zhao, Guanrong Chen, and Wenbo Liu: A chaos-based robust wavelet-domain watermarking algorithm ,Chaos, Solitons and Fractals, 2004, 22(1): 47-54
6. Szu Harold, Noel Steven, and Yim Seong-Bin: Multimedia authenticity protection with ICA watermarking and digital bacteria vaccination, Neural Networks, 2003,16(5-6):907-914
7. Alexandre H. Paquet, Rabab K. Ward, and Ioannis Pitas: Wavelet packets-based digital watermarking for image verification and authentication, Signal Processing, 2003,83(10): 2117-2132

Quantum Neural Network for Image Watermarking

Shiyan Hu

Department of Computer and Information Science
Polytechnic University,
Brooklyn, NY 11201, USA
shu@cis.poly.edu

Abstract. In this paper, we investigate the application of *Quantum neural network* (QNN) to the document image watermarking problem. The method first divides the document into some weight-invariant partitions in the spatial domain. For better performance, we then reduce the watermarking problem to a classification problem, and use the Quantum neural network to solve it. QNN, characterized by the principles of quantum computing including concepts of qubits, superposition and entanglement of states, is a relatively new type of neural networks. Owning to the power of Quantum search, QNN is considered to have at least the same computational power as classical networks. We test the performance of QNN and the experimental results indicate the soundness of our method.

1 Introduction

Security issue about digital images has attracted a lot of research recently. Many methods such as [6,8] have been proposed for image watermarking and most of them focus on minimizing the distortion to the image while protecting the information. In particular, security concern about document images is important, since document images are distributed in large amount both electronically and physically. They are easily copied and the copyright information is difficult to identify. Most general image watermarking methods are based on "transform-domain" techniques and are less useful for document images because their modifications tend to be visible in the document and are easily removed by binarization [7]. Several methods specific for document watermarking have been proposed such as [1,2].

This paper presents a new method for document image watermarking using the Quantum neural network. The method first divides the document into some weight-invariant partitions in the spatial domain. For better watermarking results, we then reduce the watermarking problem to a classification problem, and use the *Quantum Neural Network* (QNN) to solve it. QNN, characterized by the principles of quantum computing including concepts of qubits, superposition and entanglement of states, is a relatively new type of neural networks. Owning to the power of Quantum search, QNN is considered to have at least the same

F. Yin, J. Wang, and C. Guo (Eds.): ISNN 2004, LNCS 3174, pp. 669–674, 2004.

computational power as classical networks. The experimental results indicate the soundness of our method.

The rest of the paper is organized as follows: Section 2 describes the method for embedding bits into a weight-invariant partition. Section 3 describes the Quantum neural network for improving the partition method. Section 4 shows the experimental results. In Section 5, we summarize the work.

2 Embedding Bits into Weight-Invariant Partitions

Our weight-invariant partition follows the one in [5]. For completeness, we briefly describe the method in [5] as follows. Suppose we want to embed the binary sequence τ into a document. To *partition* a document is to divide the whole document into pair-wise disjoint parts. In the following, we abuse the term "partition" a little: we also let partition mean the disjoint part in the above definition. Let $S(P)$ of cardinality n denote the set of all text lines in a partition P. The *weight* $w(l_i)$ of a text line l_i is the number of total pixels in l_i. The *sum weight* of P is the sum of the weight of all text lines in $S(P)$. The *average weight* $A(P)$ equals to the sum weight of P divided by n. We also call $A(P)$ the *weight of the partition* P. Let S_{in} of P denote the set of text lines with weight between $A(P) \pm \delta$, δ being a positive integer. Let S_{out} be $S - S_{in}$. The key observation of the method is that $A(P)$ of a partition P is not likely to be significantly changed due to noise.

Informally, a partition P is a *uniform partition* if half of lines in $S(P)$ have the weight very close to $A(P)$. If δ is appropriately chosen, $|S_{in}|$ will not be too small and we can embed enough bits into a partition as follows. We first modify the partition so that all the lines in S_{in} have weights of exactly $A + \delta$ or $A - \delta$. We call it *standardization process*. Suppose that we have added r pixels in a total (note that r can be negative) through the process. In order to maintain $A(P)$, we must accordingly remove r pixels from lines in S_{out}, which is called *flushing*. We use *late flushing* strategy, that is, we delay the flushing till the end of the embedding process. After standardization process, τ is sequentially embedded to each line in S_{in}. We will further modify the partition only when embedding 0: we add or remove $\delta/2$ pixels to l_i such that after modification, l_i has a weight of exactly $A + \delta/2$ or $A - \delta/2$.

Refer to [5] for further details about determining δ, flushing pixels and the principle for modifying a text line. The process of extracting watermarks from the embedded document is simple. Since the weight of a partition is not changed by the embedding process, we can simply search the embedded partition, compute $A(P)$, and then compute δ as follows: find the line l_i such that $|A - w(l_i)| = \min_k |A - w(l_k)|$, then $\delta = 2|A - w(l_i)|$. With δ, the extraction is straightforward. Since we will always face with small noise, we have to set an error-tolerant constant ξ for practical purpose, i.e., we treat $\delta \approx \delta \pm \xi$.

For robustness, we note that the average weight of a partition is unlikely to be changed due to noise, assuming that the partition itself is large enough. For higher robustness, we can incorporate the idea of fault tolerance, i.e., we can modify τ by using any *error correcting code* or simply the *repetition code* before the embedding process.

It remains to present a method to partition the whole document into smaller uniform partitions. The simplest way is that we first group a fixed number of consecutive pages as a partition and then check for the uniform condition. If the partition is not uniform, we divide it into smaller ones. This process is repeated until some partitions are uniform. The main assumption of the above strategy is that a reasonable number of physically consecutive lines can form a uniform partition, however, this is not always true. Even if it is the case, partitions formed in this way are usually small (while there are not many partitions), which greatly limits the capacity for embedding bits. Therefore, we propose an improved partition method in Section 3 without this assumption.

3 Partition Method Based on Quantum Neural Network

To compute the partition composed of logically close lines, which are not necessarily physically close, we first select some featured lines from the document as training examples, then use them to train a Quantum neural network. Finally, we use the network to classify all other lines. The lines then in the same class are logically close to each other.

Since the sum of the number of pixels in a line is important for the embedding process, it is natural to use this information to guide the classification. Precisely, suppose a text line contains m scan lines, then we will have m inputs, each of which is a sum of the number of pixels in its own scan line, to the neural network. After selecting some featured lines as training examples, each of which will be assigned to a unique class as the target output of the neural network for that training example, we are ready to describe the process of training the Quantum neural network.

The *Quantum Neural Networks* [4], characterized by the principles of quantum computing including concepts of qubits, superposition and entanglement of states, are relatively new members of neural networks. Several quantum neural networks have been proposed such as [9,10], some of which are shown to be more efficient than conventional neural networks. For our classification problem, we adopt the one proposed in [10], which excels especially in its efficiency of the Quantum training process. We begin with a fundamental introduction to Quantum computation.

As the smallest unit of information, a quantum bit or *qubit* is a quantum system whose states lie in a two dimensional Hilbert space. Note that a qubit can be in "1" state, "0" state or simultaneously in both (*superposition*). The state of a qubit can be represented as $|\Psi\rangle = \alpha|0\rangle + \beta|1\rangle$, where α and β specify the probability of the corresponding states, and $|\alpha|^2 + |\beta|^2 = 1$. The state of a qubit can be changed by unitary transformation (or *quantum operator*), which is of central importance in quantum mechanics for many reasons, e.g., the closed quantum mechanical systems transform only via unitary transformations and unitary transformation preserves quantum probabilities. The state $|\Psi\rangle$ of a general quantum system can be described by the linear superposition of the basis states $|\phi_i\rangle$ as $|\Psi\rangle = \sum_i c_i|\phi_i\rangle$, where c_i is a complex and $\sum_i |c_i|^2 = 1$. By quantum operators, an eigenvalue equation can be written as $A|\phi_i\rangle = a_i|\phi_i\rangle$, where A is an operator and a_i is the eigenvalue. The solutions $|\phi_i\rangle$ to such equations are called *eigenstates* and can be used to construct the above basis of the Hilbert space. In quantum computation, the target problem is first "translated" to the language of quantum states and then quantum operators are applied to drive the system to a final state where the solution can be identified with a high probability. For an in-depth introduction to Quantum computation, we refer the interested readers to [3]. We are now ready to describe the method proposed in [10].

Suppose the number of training examples is n. The Quantum neural network operates like a classical artificial neural network composed of several layers of perceptions. There are one input layer, one or more hidden layers and one output layer. Denote by $|\alpha\rangle_i, |\beta\rangle_k, |\Omega\rangle_j$ the quantum register for the input node i, hidden node k and target output node j, respectively. Note that the range of i, j, k denote the number of parameters, e.g., suppose there are five parameters in all n training examples, then these examples share five input Quantum registers. Each layer is fully connected to the previous layer. Each hidden layer computes a weighted sum $|\psi\rangle_t$ of the outputs of the previous layer. If the sum is above a threshold, the node goes high. Otherwise, it stays low. The output layer works similarly, in addition, it checks the accuracy of the output of QNN as follows. The QNN compares each computed output to its target output $|\Omega\rangle_j$, setting $|\varphi\rangle_j$ high if they are equivalent, where $|\varphi\rangle_j$ is a new Quantum register representing the ability of the network to classify the training examples. Then the performance of the network is denoted by $|\rho\rangle = \sum |\varphi\rangle_j$, which is the number of computed output equal to their corresponding target output.

A straightforward way for training the QNN is that we start off with $|\psi\rangle$ as a superposition of all possible weight vectors and all other registers except $|\alpha\rangle$ and $|\Omega\rangle$ are set to $|0\rangle$. The QNN then exhibits its power in the training process where we can by superposition classify the training examples with respect to every possible weight vector simultaneously. The training process terminates when the solution correctly classifies an acceptable percentage of the training examples.

The above algorithm is not efficient when carried out in our (classical) computers and can be improved as suggested in [10]. We start off as a conventional neural network. A node is randomly chosen and its weight vector $|\psi\rangle_i$ is put into superposition. Note that all other weight vectors start with the classical initial weights. Suppose this network correctly classifies a certain percentage of training examples. We then search for a new weight vector for this node that causes the network to correctly classify a higher percentage of training examples. Then the weight vector is fixed classically and the process continues for another randomly chosen node. Due to the nature of Quantum computing, the updating process for weight vectors incorporates with some amount of randomness to avoid being stuck in local optimum. For further improving the convergence, we add a small penalty to weight vectors which cause a hidden node to output the same value for all training examples. In this way, the selected weight vectors usually contains useful information for output nodes. Once again, the training process terminates when an acceptable percentage of the training examples is correctly classified.

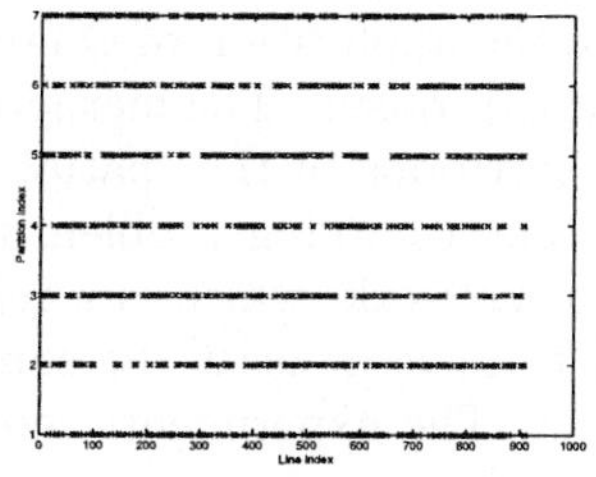

Fig. 1. Distribution of lines in partitions

Table 1. Average line weight and its standard deviation before/after the embedding phase

Original Document		Embedded Document	
Line Avg.	Standard Dev.	Line Avg.	Standard Dev.
54742	131836	54742	160422
56196	109865	56195	140998
53065	95433	53067	118032
56331	148903	56328	179138
53552	89932	53550	129054
55835	128043	55835	158965
55482	119050	55481	150783

4 Experimental Results

Testing the robustness of the watermarking scheme described in Section 2 lies outside the scope of this paper, we refer interested readers to [5] for the details. We apply the method based on QNN-partition to embed a 128-bit sequence τ into a twenty-three-page single-column paper where repetition code is used for high robustness. We first preprocess the document such that it contains only text by the standard block extraction method. We then use the QNN-based method to partition the document into seven parts. The average line weight and the standard deviation of the original document and the embedded document are summarized in Table 1. The average line weight remains, while the standard deviation differs. Figure 1 shows which partition the incrementally indexed lines of the document are classified into. From Figure 1 and Table 1, one sees that within each partition, the lines are not always physically close but the weights of them are close. Therefore, we can embed enough number of bits into the partitions. Finally, the extractor is applied and τ is successfully extracted from the scanned embedded document.

5 Conclusions

In this paper, we investigate the application of Quantum neural network to the document image watermarking problem. The method first divides the document into some weight-invariant partitions in the spatial domain. For better performance, we then reduce the watermarking problem to a classification problem, and use the Quantum neural network, which is a relatively new type of neural networks and is considered to have at least the same computational power as classical networks, to solve it. The experiments indicate the soundness of our new method.

References

1. Amano, T., Misaki, D.: A Feature Calibration Method for Watermarking of Document Images. Proc. 5th Int. Conf. Document Analysis and Recognition (1999) 91–94
2. Brassil, J., Low, S., Maxemchuk, N.: Copyright Protection for the Electronic Distribution of Text documents. Proc. IEEE 87 (1999) 1181–1196
3. Deutsch, D., Ekert, A.: Quantum Computation. Physics World (1998)
4. Ezhov, A., Ventura, D.: Quantum Neural Networks. In Future Directions for Intelligent Systems and Information Science (Ed. N. Kasabov), Physica-Verlag (2000)
5. Hu, S.: Document Image Watermarking Based on Weight-Invariant Partition Using Support Vector Machine. IAPR Int. Workshop on Document Analysis Systems (to appear)
6. Jiang, G., Yu, M., Shi, S., Liu, X., Kim, Y.: New Blind Image Watermarking in DCT Domain. Proc. 6th Int. Conf. Signal Processing (2002) 1580–1583
7. Low, S., Maxemchuk, N.: Capacity of Text Marking Channel. IEEE Signal Processing Letters 7 (2000) 345–347
8. Moulin, P., Mihcak, M., Lin, G.: An Information-Theoretic Model for Image Watermarking and Data Hiding. Proc. Int. Conf. Image Processing (2000) 667–670
9. Narayanan, A., Menneer, T.: Quantum Artificial Neural Network Architectures and Components. Information Sciences 128 (2000) 231–255
10. Ricks, B., Ventura, D.: Training A Quantum Neural Network. Advances in Neural Information Processing Systems 16 (2003)
11. Ventura, D., Martinez, T.: Quantum Associative Memory. Information Sciences 124 (2000) 273–296

An NN-Based Malicious Executables Detection Algorithm Based on Immune Principles*

Zhenhe Guo, Zhengkai Liu, and Ying Tan

Department of Electronic Engineering and Information Science,
University of Science and Technology of China, Hefei 230027, China
zhenhe@mail.ustc.edu.cn

Abstract. Detection of unknown malicious executables is one of most important tasks of Computer Immune System (CIS) studies. By using non-self detection, anomaly detection based on thickness, diversity of anti-body (Ab) and artificial neural networks, this paper proposes an NN-based malicious executables detection algorithm. This algorithm includes three parts, i.e., detector generation, anomaly information extraction and classification. At last, a number of experiments illustrate that this algorithm has high detection rate with very low the false positive rate.

1 Introduction

There doesn't exist a strict definition for the Malicious Executable so far. It is generally defined as a program that has some malicious functions that differ from benign executable, such as compromising a system's security, damaging a system or obtaining sensitive information without the permission of users. These programs include virus, trojan, worm and so on [3]. Some researchers has proposed some algorithms or methods to detect Malicious Executables [1,2,3,4,5].

Aiming at automation detection Malicious Executables, this paper proposes a novel Malicious Executable Detection Algorithm (MEDA) on the basis of the immune principle and artificial neural networks (ANN). Extensive experiments show that the algorithm has a better detection performance.

2 The Illumination of Immune System to Malicious Executable Detection Research

Non-self Detection Principles: To nature immune system, all cells of body are divided into two types of self and non-self. The immune process is to detect non-self from cell set.

---

* This work was supported by the Natural Science Foundation of China with grant no. 60273100, and partially supported by the Distinguished Youth Scientific and Technological Foundation of An Hui Province (2002-2003).

F. Yin, J. Wang, and C. Guo (Eds.): ISNN 2004, LNCS 3174, pp. 675–680, 2004.

Anomaly Detection Based on Thickness: In the process of non-self detection, the activation threshold of immune cells is determined by the thickness of immune cells matching antigens.
The Diversity of Detector Representation and Anomaly Detection Hole: The diversity of the representation of MHC cells determines the diversity of anti-body. This property is very useful to increase the power of detecting mutated antigens and decrease the anomaly detection hole.

3 Experimental Data Set

Dataset is composed of 4481 executables, which includes 915 benign executables and 3566 virus programs. All files are scanned then classified by a virus cleaner tool. We collected all *.exe files (benign executables) of a computer when a windows 2000 operation system and some other application programs were installed in it. The 3566 virus files are collected from Internet, and also scanned as virus files, i.e., Malicious Executables Set that mainly consists of DOS virus, Win32 virus, Trojan and Worm and so on, by a virus killer.

4 Malicious Executable Detection Algorithm

4.1 Data Structure Definition

$Seq(s,k,l)$: short sequence cutting operation. Suppose s be binary sequence, and $s=b(0)b(1)\ldots b(n\text{-}1)$, $b(i)\in B$, $B=\{0,1\}$, then $Seq(s,k,l)=b(k)b(k+1)\ldots b(k+l\text{-}1)$, k is the starting position of the short sequence in s.

E: executables set, $E= E(m)\cup E(b)$, m denotes malicious executable, b denotes benign executable.

$e(f_j,n)$: executable, $e(f_j,n)$ can be expressed as binary sequence that Its length is n, and f_j is executable identifier.

l_d: detector code length.

l_{step}: detector generation step length.

d_l: detector, $d_l = Seq(s,k,l)$.

D_l: detector set, $D_l =\{ d_l (0), d_l (1),\ldots, d_l (n_d\text{-}1)\}$, $|D_l|= n_d$.

4.2 Algorithm Structure

First of all, the algorithm constructs a gene (G) that is used to generate detectors. In this paper, G is constructed with $E_g(b)$, a subset of $E(b)$. Secondly, the algorithm extracts anomaly information according to the anomaly detection principle based on thickness. Thirdly, according to the diversity of detector representation, the algorithm detects malicious executables with superior classifier. This algorithm includes three parts, i.e., detector generation, anomaly information extraction and classification, whose chart is shown in Fig.1.

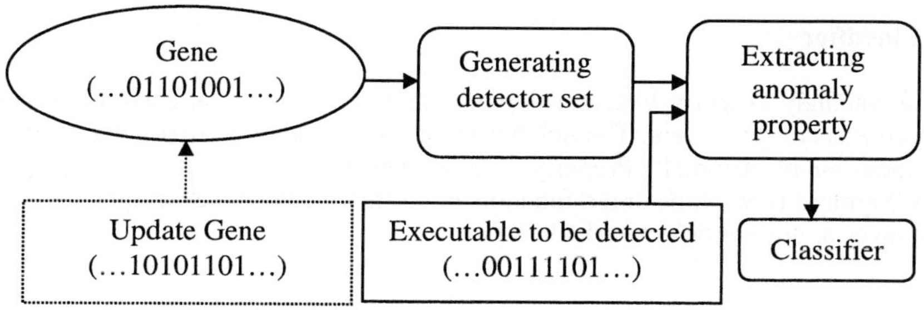

Fig. 1. Chart of the malicious executable detection algorithm (MEDA)

4.3 Generation of Detector Set

Detector is encoded as a binary sequence, $d_l = Seq(s,k,l)$. The diversity of detector is realized by changing the code length l. Through partitioning binary sequence of the program into equal length, we generate detectors. We must construct the Gene ($E_g(b)$, a subset of $E(b)$), and there are two parameters to be set before the detector set is generated. One is the generating step length l_{step}, the other is detector code length l_d.

Detector generating algorithm:
Step 1. Initialize l_{step}, l_d.
Step 2. Open one $e(f_k,n)$ from $E_g(b)$.
Step 3. If $e(f_k,n)$ isn't cut over, do $d:=Seq(e(f_k,n),i, l_d)$, add d to D_{ld}, and $i:=i+l_{step}$.
Step 4. Goto step5 if $E_g(b)$ is empty, else goto step 2.
Step 5. Output D_{ld}.

4.4 Extraction of Anomaly Characteristics

According to anomaly detection principle of immune system based on thickness, we define the percentage of non-self unit number to file binary sequence as Anomaly Property. Anomaly Property is named as Non-self Thickness (NTh) in this paper, and expressed as p_l. If detectors have m kinds, the file has an Anomaly Property Vector ($P=(p_{l1}, p_{l2}, \ldots, p_{lm})$). We set n_s is the number of self unit in one executable ($e(f_k,n)$), and n_n is the number of non-self unit.

NTh extraction algorithm:
Step 1. Initialize l_{step}, l_d.
Step 2. Open $e(f_k,n)$ to be detected.
Step 3. If $e(f_k,n)$ isn't cut over, do $s:=Seq(e(f_k,n),i, l_d)$. if $s \notin D_{ld}$, then $n_n := n_n+1$,else $n_s := n_s+1$.
Step 4. Compute p_{ld}, $p_{ld} := n_n/(n_n+n_s)$.

For different detector set, this algorithm computes a separate NTh of the file. Finally, all NThs are combined together to construct the Anomaly Property Vector.

4.5 Classifier

We use Anomaly Property Vector as input variable of a BP neural network classifier, which consists of two layers. The number of neural nodes in the first layer is equal to the dimension of Anomaly Property Vector. The transfer function of input layer selects Sigmoid-type while liner function is chosen as the transfer function of the output layer of our neural network classifier.

5 Experiment

5.1 Experimental Data Set Partition

Total data set is divided into benign and malicious executables subset ($E(b)$ and $E(m)$). $E(b)$ is composed with 915 program files, and is divided into generating detector data set ($E_g(b)$), which is the Gene used to generate detector data set, and testing data set ($E_{test}(b)$). $E_g(b)$ has 613 program files, and is constructed by randomly selection from $E(b)$. Other 302 program files of $E(b)$ is components of $E_{test}(b)$. $E(m)$ is composed with 3566 virus files.

5.2 Generating Detector Set

We select $E_g(b)$ as Gene of generating detector, $l_d \in \{16,24,32,64,96\}$, and l_{step}=8bits. With detector generating algorithm, we can get 5 detector sets, which are separately referred as D_{16}, D_{24}, D_{32}, D_{64}, and D_{96}.

5.3 Experimental Results

5.3.1 The Experimental Result Using Single Detector Set

At beginning, we separately detect malicious executables with single detector set of 5 kinds detector set. The experimental result is shown as Fig. 2.

When FPS = 0, we can get the best detection effect while l_d =96, and the DR will be worse as the length of l_d being shorter. DR is changing better as FPS being worse. When FPS 1%, the detection effect is best while l_d =24. At same time, DR will be worse as l_d being longer. On the other hand, when l_d =16, the DR is 0 because the detector number is 2^{16}. When l_d =96, to difference FPS, the DR changing is slow.

5.3.2 The Experimental Result Using Multi-data Set and BP Network

We don't use D_{16} data set because the DR is 0 when select D_{16}. At the same time, we set the upper limit of l_d equals to 96 because the DR is almost same when l_d =96. So, this experiment selects D_{24}, D_{32}, D_{64} and D_{96} four data sets as anomaly detection data set, and uses these four data sets to extract Anomaly Property Vector, and uses BP network to classify executables. In the process of classification, we randomly selects 30% files of $E_{test}(b)$ to train BP network, and use other data to test the anomaly detection effect. The experimental result is shown with ROC. It is shown as Fig.3.

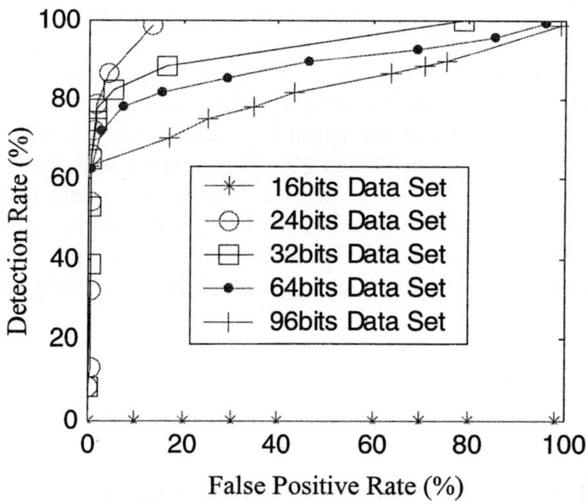

Fig. 2. Detection ROCs of five kinds of detector-sets

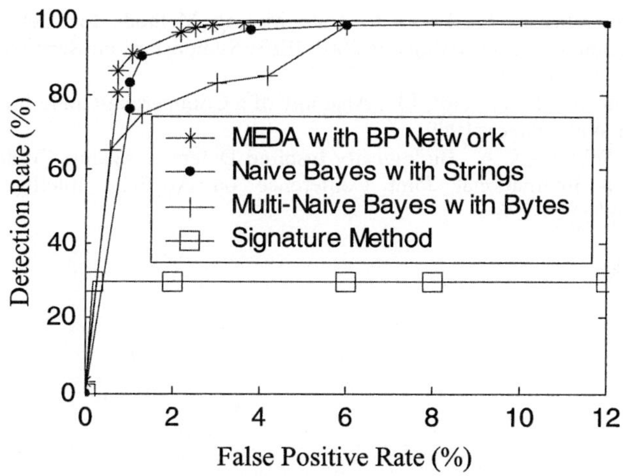

Fig. 3. The comparison of experiment results of several algorithms

5.4 Comparison with Matthew G. Schultz's Method

Known from Matthew G. Schultz's experimental result [3], using the Naive Bayes approach, the DR can reach 97.43% when FPR is 3.80%. To MEDA, While FPR is 2%, the DR can reach the value. While using Multi-Naive Bayes, the algorithm detection effect is the best, the DR can reach 97.76%. But to MEDA, the best DR can reach 99.50%. The detail comparison is shown as Fig.3.

6 Conclusions

We can conclude from experiments that the detector generated with executables binary short sequences can be used to detect malicious and benign executables. The best detection performance of our algorithm is obtained when a multi-detector set is used, and the DR is 97.46% when FPS is 2%. This verifies our clamed principle that diversity of detector representations can decrease the anomaly detection hole, and also tests the rightness and validity of non-self detection and detection based on thickness.

Because of the size of Detector Set, this algorithm is adaptive to statistically detect Malicious Executables. The real-time detection capability of our algorithm is our next research content.

References

1. MEF Group: Malicious Email Filter Group. Published as Online Publication. http://www.cs.columbia.edu/ids/mef
2. Jeffrey, O.K., William, C.A.: Automatic Extraction of Computer Virus Signatures. Proceedings of the 4th Virus Bulletin International Conference. Abingdon England (1994) 178-184
3. Matthew, G.S., Eleazar, E., Erez, Z.: Data Mining Methods for Detection of New Malicious Executables. Proceedings of 2001 IEEE Symposium on Security and Privacy. Oakland USA (2001) 38-49
4. Steve, R.W., Morton, S., Edward, J.P.: Anatomy of a Commercial-Grade Immune System. IBM Research White Paper (1999)
5. Jeffrey, O.K., William, C.A.: Biologically Inspired Defenses Against Computer Viruses. Proceedings of International Joint Conference on Artificial Intelligence. Morgan Kaufmann (1995) 985-996

The Algorithm for Detecting Hiding Information Based on SVM

JiFeng Huang[1,2], JiaJun Lin[1], XiaoFu He[2], and Meng Dai[2]

[1] Information Science College, East China University of Science and Technology, ShangHai, China
jfhuang@shnu.edu.cn , jjlin@ecust.edu.cn
[2] Department of Computer Science and Technology, Shanghai Normal University, ShangHai, China
{xfhe, damon}@pub.shnu.edu.cn

Abstract. Steganographic messages can be embedded into digital image while most methods in use today are invisible to an observer's senses. Mathematical analysis may reveal statistical anomalies in the stego images. Since the difference of adjacent DCT coefficients can better reflect the statistical difference between the cover image and the stego image, we select six differences of adjacent DCT coefficients as the feature vectors used by the classifier for training. Experimental results show that our approach based on SVM can determine the existence of hidden messages in JPEG images reliably and get excellent computational efficiency.

1 Introduction

The digital steganography is hiding information into a multimedia object (such as image, video, audio etc.). The main purpose of steganography is to convey messages secretly by concealing the very existence of messages. The embedded information called hidden information. We can't distinguish the difference between the the cover image and the stego image, which is the best property of steganography. So it can't attract the opponents'attention. Whether critical information embedded in a digital media or not is the steganalysis's main purpose. The relationship between the steganography and steganalysis is similar to the relationship between the encryption and decryption.

In this paper, we propose an novel algorithm of detecting the messages hidden in JPEG image using the SVM as the classifier, which has advantages of high speed and accurate detection.

The presence of embedded messages is often imperceptible to the human eye, so we can't figure out the difference between the cover image and the stego image. One of the efficient methods is to recognise the feature vector which can perfectly present the changes after hiding the information.

Farid[1] proposed a dection algorithm which can be used generally in natural images. First, after transformed by multi-scaling wavelet transformation, the image was decomposed into approximate, vertical, horizontal and diagonal subbands. Then the approximate subband is further decomposed using the same

F. Yin, J. Wang, and C. Guo (Eds.): ISNN 2004, LNCS 3174, pp. 681–687, 2004.

method. For each vertical, horizontal and diagonal subband coefficients, calculate four variables respectively, i.e mean, variance, skewness and kurtosis at each scale. For example, $3 \times 4 \times (n-1)$ feature vectors will be generated according to this calculation. With the prediction value of each property, there will be $24 \times (n-1)$ feature vectors. If $n = 4$, it will be 72.

Table 1. Frequency grouping

C_i ={i}, i∈{1,2,3...8,9,10}
C_{11} ={11,12}
C_{12} ={13,14}
C_{13} ={15,16}
C_{14} ={17,18}
C_{15} ={19,20}
C_{16} ={21,22,23}
C_{17} ={24,25,26}
C_{18} ={27...63}

Jeremiah, J.H.[2] used KFD as classifier to detect the presence of messages hidden in the JPEG images. Histogram is calculated according to the DCT coefficients, e.g $h_i[d]$, where k∈0...63, which value is got according to the Zig-Zag sequence. After quantization, the expected number of nonzero coefficients decreases as the frequency indices increase. To make use of these higher frequency coefficients, grouping the middle-high frequency coefficients is necessary, which refers to Table 1.

We calculate the histogram according to following equation

$$h_c[d] = \sum_{k \in c} h_k[d] \; , \tag{1}$$

For example $h_c[2]$, where c∈ $\{0 \ldots 63\}$ denote the total number of quantized coefficients with the value of 2 and $h_c[3]$, where $c = \{0\}$ is the total number of DC coefficients with the value 3.

Select the feature vectors according equation (2)

$$F = \{h_{c1}[-p] \cdots h_{c1}[0] \cdots h_{c1}[p] \cdots h_{c18}[-p] \cdots h_{c18}[0] \cdots h_{c18}[p]\} \; . \tag{2}$$

For instances, there are $9 \times 18 = 162$ feature vectors with p=4.

The number of the selected feature vectors of these two kinds of algorithms we mentioned before is too much and the calculation is too complex, which effects the detection effeciency.

In this paper, we propose a steganalytic technique based on the feature of histogram of DCT coefficients. We select the difference between two neighbourhood frequency coefficients as feature vector, which can determine the existence of hidden messages only with six feature vectors. We use the algorithm of JSteg and OutGuess to hide information in our experiment.

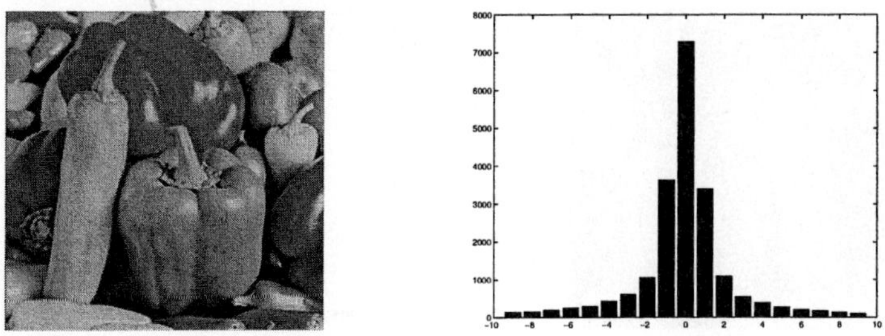

Fig. 1. Peppers(*JPEG80*) and Quantized DCT coefficients distribution(*left and right*)

2 Analysis of JPEG Coefficients

There are two processes in JPEG: lossy and lossless process. First, at the lossy process, the encoder divides an image into 8×8 blocks, and each block denoted $f(x, y)$ is transformed by DCT into a set of 64 values denoted $F(\mu, \nu)$.

$$F^{Q}(\mu, \upsilon) = IntegerRound\left(\frac{F(\mu, \upsilon)}{Q(\mu, \upsilon)}\right), \tag{3}$$

where, $Q(\mu, \nu)$is the 8×8 quantization table which is determined by the quantity factor. Figure.1.left shows peppers image after JPEG compression with quantity factor 80. Figure.1. right shows the histogram of JPEG coefficients.

JSteg[3] is the first publicity steganographic algorithm based on JPEG images. JPEG JSteg algorithm is a typical steganographic algorithm using JPEG file as cover-image proposed by Derek Upham. After quantization of DCT coefficients, JPEG-JSteg replaces the least significant bits(LSB) of the quantized DCT coefficients by the secret message bits. The embedding mechanism skips all coefficients with the values 0 or 1. The OutGuess0.1 algorithm [4]has some improvement to JSteg, which randomly select the DCT coefficients to embed the hidden message by pseudo-random generator.

The only difference between JSteg and OutGuess0.1 is that the former selects the hidden coefficients sequentially but the latter randomly. Figure.2.left shows the image after embedded 4000 bits data into the peppers(JPEG format). Figure.2.right shows quantized coefficients histogram of Figure.2.left.

For natural images, the distribution of DCT coefficients resemble Laplacian or generalized Gaussian distributions[5]. The model shows that DCT coefficients distribution satisfies symmetry.

The DCT coefficient frequency of cover image is denoted as $h_c(i)$ where i is the DCT coefficients value; $h_s(i)$ presents the DCT coefficient frequency after data embedded. Based on the feature of DCT coefficient distribution, we get:

$$h_c(i) > h_c(i+1) > h_c(i+2) \cdots \text{where } i > 0$$

and

$$h_c(i) > h_c(i-1) > h_c(i-2) \cdots \text{where } i < 0$$

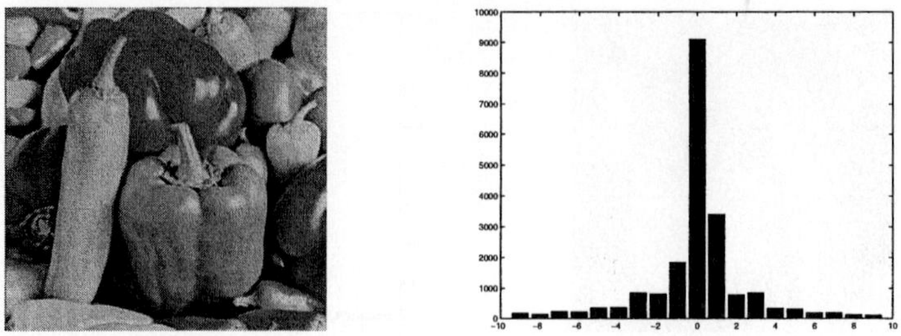

Fig. 2. Peppers after embedding messages and Quantized coefficient histogram of it (*left and right*)

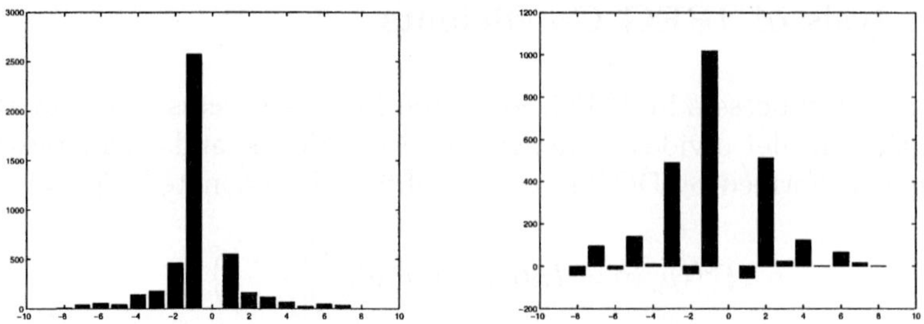

Fig. 3. Difference of adjacent coefficient of Figure.1.right and Difference of adjacent coefficient of Figure.2.right(*left and right*)

As the hidden message are even and random, after using JSteg and OutGuess algorithm it occurs that the numbers of $h_c(i)$ flipping to $h_s(i+1)$ are more than $h_c(i+1)$ flipping to $h_s(i)$ where i >0. Therefore, we can draw the conclusion:

$$|h_c(i) - h_c(i+1)| > |h_s(i) - h_s(i+1)| \, , \tag{4}$$

which is also satisfied where $i < 0$. According to the equation (4), we can know that the difference of adjacent DCT coefficients frequency in cover image is greater than which in hidden message image. Figure.3.left and Figure.3.right show the difference of adjacent DCT coefficient frequency before and after embedding message respectively.

According to the above analysis, we argue that the difference of adjacent coefficients can better reflect the statistical difference between the cover image and the stego image. We select six differences of adjacent coefficients as the feature vectors used by the classifier for training.

Before embedding information, the feature vectors are:

$$F_c = \{h_c(i) - h_c(i+t)\} \, , \tag{5}$$

where $i = \{-3, -2, -1, 1, 2, 3\}$, if $i > 0$ t=1, else $t = -1$;
After embedding information, the feature vectors become:

$$F_s = \{h_s(i) - h_s(i+t)\} \, . \tag{6}$$

Based on the equation (6) and (7), we can extract the features from cover images and stego images.

3 Support Vector Classifier

The main idea of SVM is to establish a hyperplane as the decision-making surface, which maximize the boundary between two classes.

First, we suppose that two classes is linear separable. The training sample is denoted as $\{(x_i, d_i)\}_{i=1}^{N}$ where x_i represents the input pattern and d_i represents the target value ($d_i \in \{+1, -1\}$). With the hyperplane

$$\omega x + b = 0 , \tag{7}$$

If the input pattern can be separated linearly, there would be exist (ω, b) which satisfy:

$$\omega x_i + b_0 > 0 \quad , \quad d_i = +1$$

$$\omega x_i + b_0 < 0 \quad , \quad d_i = -1$$

where, ω is the adjustable weight vector and b_0 is excursion.

Given a hyperplane, the margin of separation is the distance between the closest vector to the hyperplane. If the margin of separation is maximal, the separating hyperplane is optimal. Supposing that we now get the optimal hyperplane, namely, the root set of $\omega x + b = 0$. For this root set, the value of ω and b_0 are not exclusive. Consider a canonical hyperplane, where the minimum distance is $1/\|\omega\|$, i.e ω, b_0 are constrained by:

$$\min_{x_i} \left| \omega x_i + b_0 \right| = 1 , \tag{8}$$

The hyperplane that optimally separates the data is the one that minimizes following equation:

$$\Phi(\omega) = \frac{1}{2}\|\omega\|^2 , \tag{9}$$

Under the constraints of $y_i(\omega x_i + b_0) \geq 1$, $i = 1, \ldots, n$. The solution to the Equation (9) is given by the saddle point of Lagrange function, which is the classical solution to the problem of

$$J(\omega, b_0, \lambda) = \frac{1}{2}\|\omega\|^2 - \sum_{i=1}^{n} \lambda_i (y_i(\omega x_i + b_0) - 1) , \tag{10}$$

Calculating the differential coefficient of ω and b_0, we get the following optimal condition:

$$\omega = \sum_{i=1}^{n} \lambda_i y_i x_i , \tag{11}$$

$$\sum_{i=1}^{n} \lambda_i y_i = 0 , \tag{12}$$

To calculate the Lagrangian, using Equation (11) to expand the expression of $J(\omega, b_0, \lambda)$, we can get the expression

$$Q(\lambda) = \sum_{i=1}^{n} \lambda_i - \frac{1}{2} \sum_{i=1}^{n} \sum_{j=1}^{n} \lambda_i \lambda_j y_i y_j x_i x_j \ , \tag{13}$$

which is only depended on λ.

If λ_i^* is optimal, then

$$\omega^* = \sum_{i=1}^{n} \lambda_i^* y_i x_i \ , \tag{14}$$

From the Kunh-Tucker conditions, we know that at the saddle point the product of the constraint and the dual variant is zero, i.e:

$$\lambda_i(y_i(\omega x_i) + b) - 1 = 0 \quad , \quad i = 1, 2, \cdots, n$$

Hence, the discriminant can be expressed by

$$\begin{aligned} f(x) &= \mathrm{sgn}((\omega^* x) + b) \\ &= \mathrm{sgn}\left\{ \sum_{i=1}^{n} \lambda_i^* y_i (x_i x) + b \right\} , \end{aligned} \tag{15}$$

For the non-separable case, some samples can't satisfy the condition (9). So we introduce non-negative variables xi_i, where

$$\xi_i \geq 0 \ , \tag{16}$$

The constraints of Equation (9) are modified as

$$y_i \big[(\omega x_i) + b \big] - 1 - \xi_i \geq 0 \ , \tag{17}$$

Solving the non-separable problem is to minimize the function subject to the constraints of Equation (16) and (17)

$$\Phi(\omega, \xi) = \frac{1}{2} \|\omega\|^2 + C \sum_{i=1}^{n} \xi_i \ , \tag{18}$$

Where C is a given constant which controls the penalty level of errors. The solution to this optimization problem is identical to the optimal separating plane. After that, we can find the result is mostly similar to the expression for separable case except the former should satisfy more restrict constraints:

$$0 \leq \lambda_i \leq C \ , \tag{19}$$

For non-linear case, we map input vector to a high dimensional space by non-linear transform. Then get the optimal separating hyperplane through doing classification in this feature space. Substitute point-conduct with inner-conduct

$(K(x, y))$, which is equal to change the original space to a new feature space. The optimized function of equation (13) becomes:

$$Q(\lambda) = \sum_{i=1}^{n} \lambda_i - \frac{1}{2} \sum_{i=1}^{n} \sum_{j=1}^{n} \lambda_i \lambda_j y_i y_j K(x_i, x_j) , \tag{20}$$

also the equation (15) converts to

$$f(x) = \mathrm{sgn}\Big\{ \sum_{i=1}^{n} \lambda_i^* y_i K(x_i, x) + b \Big\} , \tag{21}$$

In this paper, we select the radial-base inner-conduct function as the core-function.

$$k(x, x_i) = \exp\left\{ -\frac{|x - x_i|^2}{256\sigma^2} \right\} . \tag{22}$$

4 Experimental Results and Conclusion

The training set consists of 1200 JPEG images which include 600 original images and 600 hidden message images. The test set consists of 800 JPEG images involving 400 original images and 400 hidden message images. All images are 256×256 pixels collected from Philip Greenspnn's photography [1].

According to the method of section 2, we extract six feature vectors from each image. After scale transform,the range of each vector value is between -1 and 1.

We embed messages into the images using JSteg and OutGuess0.1 algorithm and the length of those messages are 2000 bits, 4000 bits, 6000 bits. Our approach can produce excellent detection rates which averagely exceeds 97 percentage for all the kinds of embedding length.

References

1. Farid, H.: Detecting Hidden Messages Using Higher-Order Statistical Models. In International Conference on Image Processing, Rochester, New York (2002)
2. Jeremiah, J.H., William, A.P.: Kernel Fisher Discriminant for Steganalysis of JPEG Hiding Methods. Available: `http://www.cipr.rpi.edu/ harmsj/pubs`
3. Anderson, R.J., Petitcolas, F.A.P.: On the limits of steganography. IEEE Journal on Selected Areas in Communications, Vol.16, No.4 (1998) 474–481
4. Provos,N.: Defending Against Statistical Steganalysis. Proc.10th USENIX Security Symposium, Washington, DC(2001)323-335
5. Lam, E.J., Goodman, J.W.: A Mathematical Analysis of the DCT Coefficient Distributions for Images. IEEE Transactions on Image Processing, Vol.9, No.10 (2000) 1661–1666

[1] `http://www.photo.net/philg/digiphotos`

An E-mail Filtering Approach Using Neural Network

Yukun Cao, Xiaofeng Liao, and Yunfeng Li

Department of Computer Science,Chongqing UniversityChongqing,400044,P.R.China
marilyn_cao@sina.com

Abstract. The communication via electronic mail is one of the most popular services of Internet. The volume of emails that we get is constantly growing. In particular, unsolicited messages or *spam*, flood our email boxes, and result in causing frustration, and wasting bandwidth and time. The paper presents a novel schema to automatically filter spam emails by using the principal component analysis(PCA) and the Self Organized Feature Map (SOFM). In our schema, each email is represented by a series of textual and non-textual features. To reduce the number of textual features, PCA is used to select the most relevant features. Finally the output of the PCA and the non-textual features should be inputted into a well-trained SOFM to classify (*spam* or *normal*). In comparison with some traditional classification methods, the experimental result denotes that the scheme will increase the accuracy of filtering emails.

1 Introduction

Internet traffic has doubled almost every year since 1997, a growth rate set to continue for some time. The communication via electronic mail is one of the most popular services of the Internet. And the volume of e-mail that we get is constantly growing. In particular, unsolicited messages or *spam*, flood our email boxes, and result in causing frustration, and wasting bandwidth and time. They also cost money with connections and may expose minors to unsuitable content (e.g. when advertising pornographic sites). Future mailing system should require more capable filters to help users in the selection of what to read and avoid them to spend more time on processing incoming messages.

Most modern email software packages have provided some form of programmable filtering and classifying based rules, which involve in human beings observing emails and writing a set of logical rules. Moreover, as the characteristics of the spam email (e.g. topics, frequent terms) changes over time, these rule sets must be constantly tuned and refined by the user themselves, that is time-consuming and often tedious. A system that can automatically learn how to filter spam emails and classify non-spam emails into a set of folders is highly desirable.

It is well known that artificial neural networks have been applied successfully in a variety of fields. However, little studies have been presented using artificial neural networks (ANN) to filter and classify emails. After adequate training, the main ad-

F. Yin, J. Wang, and C. Guo (Eds.): ISNN 2004, LNCS 3174, pp. 688–694, 2004.

vantage of ANN is the accuracy they can achieve, that are sometimes more accurate than those of the symbolic classifiers.

In this paper, a system named Email Filtering Toolkit (EFT) are proposed. In Section 2, the structure of the system is expatiated in detail. Section 3 presents our intelligent and dynamic email filtering tool. Section 4 analyzes the experimental results. Finally, the conclusion and future work can be found in Section 5.

2 Architecture of Email Filter Toolkit

EFT, designed to filtering and classify emails efficiently, has several tasks to resolve as follows:

(1) Extracting non-textual features of emails and parsing the text content of emails.
(2) Representing the emails and user's interests.
(3) Evaluate the similarity between the emails and user's interests.
(4) Filtering and classifying process.
(5) Learning process.

To address those above mentioned problems, the architecture of EFT is composed of *four modules*: (1) a emails' non-textual features retrieving module;(2) a emails' content preprocessing module; (3) a emails' textual features retrieving module based on PCA, (4) a spam emails filtering module based on SOFM neural network, which is shown in Figure 1 .

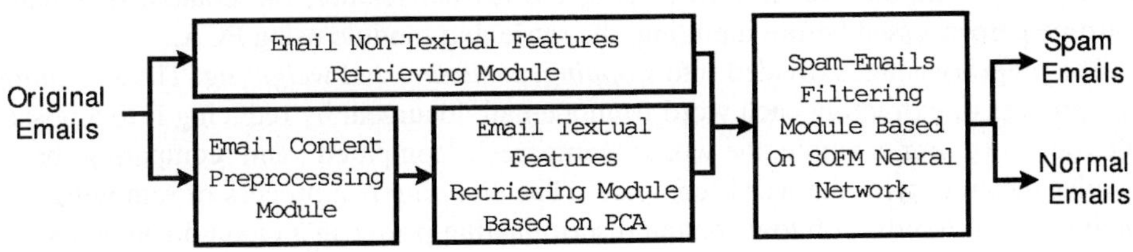

Fig. 1. Email Filter Toolkit Architecture

3 Intelligent and Dynamic Email Filter Toolkit

In EFT, original emails should be inputted into the emails' feature retrieving module first, whose results are as the input to the spam filtering module. Finally, the original emails will be divided into two categories: spam emails and normal emails.

3.1 Emails' Non-textual Features Retrieving Module

In the approach, a emails' non-textual feature retrieving module is designed to account the non-textual aspect as a knowledge model of emails to filter. In order to extract the non-textual feature, we collected a great deal of emails including *spam* and *non-spam*, and searched a set of factors that may discriminate emails and classify them best.

Email contains many non-textual characters, which provide a great deal of information allowing emails discrimination. In the approach, we consider particular non-textual features, such as the domain type of the sender. For example, *spam* email is virtually never sent from .edu domains. Another indicator (*recipient*) is found in examining if the message was sent by an individual user or via a mailing list. A number of other simple distinctions, such as whether a message has attached documents (most *spam* email doses not have attachment), or when a given message was received (most *spam* email is sent at night), are also powerful distinguishers between *spam* emails and *non-spam*. We also consider a number of other useful distinctions, such as the percentage of non-alphanumeric characters, abbreviations, or numerical characters. For example, *spam* email contains a high percentage of non-alphanumeric characters.

In the approach, the non-textual features of emails are: *domain type of the email sender (.com, .edu, .gov, etc.), header length, text content length, abbreviation, non-alphanumeric characters, attached documents, sentence length, date*, etc.

3.2 Emails' Content Preprocessing Module

In the approach, the textual features of a email are extracted from the content of the email. And parsing textual characters of a email is to represent the email as a term vectors in multidimensional space. Each word is assigned a weight, which represents its degree of importance. Before parsing the textual feature, the content of emails must be preprocessed before input into the retrieving module using PCA.

The preprocessing is divided into *stopping*, *stemming* and *weighting*. The *stemming* is a process of extracting each word from a email document by reducing it to a possible root word. For example, the words 'compares', 'compared', and 'comparing' have similar meaning with the word 'compare'. The *stopping* is a process of removing the high frequent words with low content discriminating power in a email document such as 'to', 'a', 'and', 'it', etc. Removing these words will save spaces for storing document contents and reduce time taken during the subsequent processes. The *weighting* is a scheme using TF-IDF algorithm to evaluate the importance of a term in a email.

TF-IDF is the one that has been well studied in the information retrieval literature. This scheme is based on the assumption that terms that occur in fewer documents are better discriminators. Therefore, if two terms occur with the same frequency in a email document, the term occurring less frequently in other documents will be assigned a higher value. Specifically, the importance of a term is proportional to the occurrence frequency of the term in each document, and inversely proportional to the total number of documents to which the term occurs in a given document collection. Let TF_i be *score* of term t_i in a document p where $i=1,...,m$, and let DF_i be the frequency of occurrence of document p in a collection. The weight of word i, denoted by $W_{i,p}$, is expressed as $W_{i,p} = \left(TF_i / |P| \right) \cdot \log\left(n / DF_i \right)$, where n is the number of documents in the collection and $|P| = \sum_j TF_j$ is used to normalize term frequency to [0,1] in order to avoid favoring long documents over short documents, where $j=1,..,n$.

After the page is automatically parsed ignoring common stop words, TF_i (a term's *score*) could be calculated according to three rules shown as follow:

(1) Each time a word occurs in the title its *score* is increased by ten.
(2) Each time a word occurs in a heading its *score* is increased by five.
(3) Each time a word occurs its *score* is increased by one.

3.3 Emails' Textual Features Retrieving Module Based On PCA

According to the length of emails' content, the high-dimensionality of the term-weight vectors representing emails has made the difficulty to filter and classify. To improve the accuracy of filtering and classifying, our approach uses the principal component analysis (PCA) technology, with a combination of statistical and natural language approaches, to reduce the original term-weight vectors with high dimensionality to a small number of relevant features.

The term-weight vector representing a email document is considered as inputs to a PCA system. PCA provides a means by achieving such a transformation where the feature space accounts for as much of the total variation as possible. Specifically, based on the PCA procedure the original feature space is transformed into another feature space that has exactly the same dimension as the original. However, the transformation is designed in such a way that the original feature set may be represented by a reduced number of 'effective' features and yet retains most of the intrinsic information content of the data. Therefore, using PCA we achieve not only the increasing of feature variation but also the decreasing of feature space dimensionality. Karhunen–Loeve transformation (KLT) is a well-known tool for PCA in multivariate analysis. A complete analysis of the PCA method used in this paper may refer to Refs. [2,3].

Briefly, in the approach the input vector of the PCA is the term-weight vector **x**, while the output is a reduced vector **y**, which is given by the relation: **y=Wx**, where the transformation matrix **W** contains the neural network coefficients. It is denoted that after training, **W** will approach the matrix whose rows are the first *d* eigenvectors of **C**, ordered by decreasing eigenvalues, where **C** is the covariance matrix of the training set. And the set principal components **y** is represented as $y_1 = e_1^T x, y_2 = e_2^T x, \ldots, y_d = e_d^T x$, where e_i^T is eigenvector *i* of **C**.

3.4 Spam Emails Filtering Module Based on SOFM Neural Network

The textual and non-textual features representing a email, obtained through the method mentioned previously, are as the input to the spam email filtering module. In the approach, the filtering module is represent by a SOFM neural network.

The Self Organized Feature Map (SOFM) is one of the most widely applied ANN first introduced by Kohonen [4]. It has successfully been used in a variety of applications including, data visualization, data clustering, pattern recognition and data mining. The SOFM's objective in document clustering is to group the documents, which appear similar, close to one another and place the very different ones distant from one another. The Kohonen model is composed of two layers neurons, input layer and output layer. And the output layer, called competition layer, is generally represented on a map (in two dimensions) or on a line(one dimension).

In our paper, the inputs of SOFM are the outputs of the PCA representing emails. The network combines its input layer with a competitive layer of neurons, and is trained by unsupervised learning. The two layers are fully interconnected since every input is connected to all of the neurons in the competitive layer. Kohonen self-organization has a result the representation of similar classes, by neighboring neurons on the competitive layer. Typically, the competitive layer is organized as a line.

Figure. 2 gives the Kohonen SOFM topology as the second part of the entire neural network. After training, one neuron in the competition layer represents the *spam* emails category to be filtered from the mailbox, the other is the non-spam emails class.

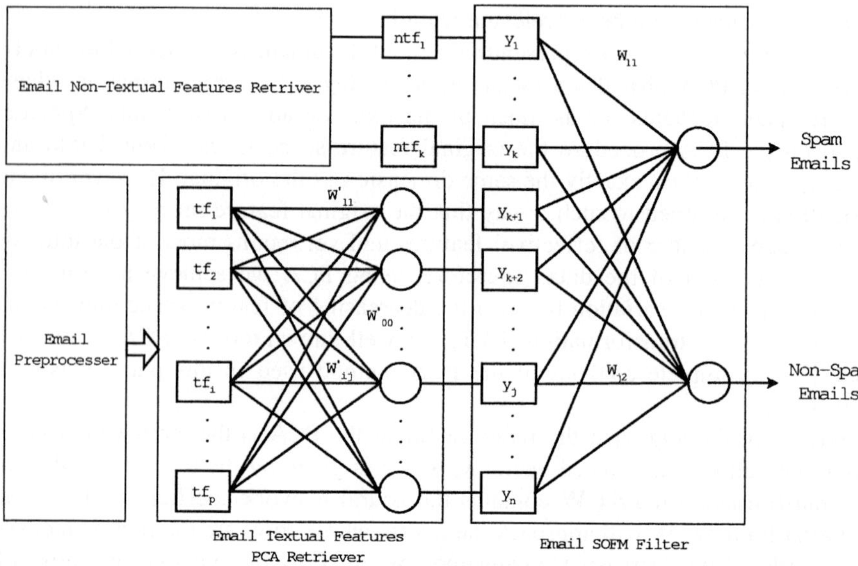

Fig. 2. The two Stage of a neural network

Suppose that the input $y = [y_1, y_2, ..., y_p]^T$, the weight vector of the neuron j in SOFM is $w_j = [w_{j1}, w_{j2}, ..., w_{jp}]^T$. There are two basic steps in SOFM, which are the search for the Best Matching Unit (BMU) i of weight vector w_i and y, and updating the BMU i with it's neighbours. The BMU i is found by computing the Euclidean distance between the input data vector y and the reference vector w_i as show as follows:

$$i(y) = \arg\min_{j}\{\ \| y - w_j \| \}\ ,\ j = 1,2,\ldots,\ n \tag{1}$$

where n is the number of neurons in the SOFM's feature map. Once we have found the BMU we update the BMU and it's neighbouring nodes using:

$$w_i(t+1) = w_i(n) + \Lambda_{i,j}(t)[x(t) - w_i(t)]\ \ ,\ t = 1,2,3,\ldots \tag{2}$$

Where:

$\Lambda_{i,j}(t)$ — the neighourhood function

t — a discrete time constant

The neighbourhood function $\Lambda_{i,j}$ used in equation (2), is a time decreasing function which determines to which extent the neighbours of the BMU will be updated. The extent of the neighbourhood is the radius and learning rate contribution, which should both decrease monotonically with time to allow convergence. The radius is simply the maximum distance at which the nodes from the BMU are affected. A typical smooth Gaussian neighbourhood kernel is given bellow in equation (3).

$$\Lambda_{i,j}(t) = \alpha(t) \cdot \exp\left(-\frac{\| r_i - r_j \|^2}{2\sigma(t)}\right) \tag{3}$$

where $\alpha(t)$ is the learning rate function, $\sigma(t)$ is the kernel width function, $\| r_i - r_j \|^2$ is the distance of BMU i unit to current unit j.

There are various functions used as the learning rate $\alpha(t)$ and the kernel width functions $\sigma(t)$. For further details about the SOFM please refer to [3] and [4].

4 Experiments

The proposed scheme was extensively tested on a great deal of emails. Due to space limitation, not to present additional examples in this paper. We have collected 800 emails from 10 mailboxes, including 300 spam emails and 500 normal emails. This collection is split into a training set of 400(200 spam and 200 normal emails) mails and a testing set of 400 mails (100 spam and 300 normal emails). The average precision of filtering email is 87.56%. And the average precision of Bayesian classification is 83.28%. In comparison with Bayesian classifying, the above-mentioned scheme, based on a combination of the PCA and SOFM, will increase the accuracy of filtering emails.

5 Conclusion and Future Work

Because filtering is closely related to information retrieval, most information filtering methods are based directly or indirectly on traditional techniques of information retrieval. Research in information filtering is still an open field. One of the most promising approaches is to use advanced natural language technology such as lexical semantic, terminology, shallow parsing, etc. An efficient filtering tool must use various strategies to retrieve, filter, or to infer information.

Our system is an intelligent and dynamic email filtering toolkit, which helps users in filtering emails and selecting what to read. The system contains two feature retrievers and a SOFM neural network to filter emails automatically. The feature selection in the system is based on the combination of the PCA and traditional statistical methods. The scheme qualifies it to be a real user assistant by reflecting its intentions during filtering operation and by having capability of auto-organization to act better.

References

1. Pandya, A.S., Macy, R.B.: Pattern Recognition with Neural Networks in C++. CRC Press and IEEE Press, (1995) 195–211
2. Sanger, T.D.: Optimal unsupervised learning in a single-layer linear feedforward neural network. Neural Networks 12 (1989) 459–473
3. Hykin, S.: Neural Networks a comprehensive Foundation. 2rd edn. Printice-Hall (1999)
4. Kohonen, T.: Self Organizing Maps Second Edition. Springer (1997)
5. Nouali, O., Blache, P.: A semantic vector space and features-based approach for automatic information filtering. Expert Systems with Applications, 26 (2004) 171-179
6. Clark, J.: A Neural Network Based Approach to Automated E-mail Classification. Proceedings of the IEEE/WIC International Conference on Web Intelligence, Halifax, Canada (2003) 702-705
7. Selamat, Ali: Web page feature selection and classification using neural networks. Information Sciences 158(2003) 69-88
8. Pui Y.Lee: Neural Networks for web content Filtering. IEEE Intelligent Systems, Vol.17, No.5(2002) 48-57
9. Chih-Ming Chen: Incremental Personalized Web Page Mining Utilizing Self-organizing HCMAC Neural Network. Proceedings of the IEEE/WIC international Conference on Web Interlligence, Halifax, Canada (2003)15-27
10. Soe-Tsyr Yuan: A personalized and integrative comparison-shopping engine and its applications. Decision Support Systems 34(2002)139-156
11. W. Cohen: Learning Rules that Classify E-Mail. AAAI symp. on Machine Learning in Inf. Access (1996)18-25
12. Sahami, M., Dumais, S. et al: A Bayesian Approach to Filtering Junk E-Mail. In Learning for Text Categorization: Papers from the 1998 workshop. AAAI Technical Report WS-98-05 (1998)
13. Jason D.M. Rennie: ifile: An Application of Machine Learning to E-Mail Filtering. Proceedings of the KDD-2000 Workshop on Text Mining(2000)

Topography-Enhanced BMU Search in Self-Organizing Maps

James S. Kirk[1] and Jacek M. Zurada[2,*]

[1] Union University, Jackson, Tennessee, U.S.A.
jkirk@uu.edu

[2] University of Louisville, Louisville, Kentucky, U.S.A.
jacek.zurada@louisville.edu

Abstract. Self-organizing maps (SOMs) have proven extremely useful because of their ability to provide a condensed representation of data. This is accomplished by creating a mapping from an often continuous data space to a discrete map grid, relationships on which ideally preserve much of the structure of the original data. When well-trained, Kohonen's original SOM successfully preserves continuity in the mapping from the SOM grid to the data space. When the dimension of a map approximately matches the dimension of the data, or when the folding of a map into higher-dimensional data space is controlled, mapping algorithms can preserve the organization of the data more comprehensively. In these cases, the structure of the map grid may reflect the data structure at several levels of granularity, allowing the search for the best-matching unit (BMU) of the trained map to speed up significantly.

1 Introduction

The basis of the usefulness of the self-organizing map (SOM) is its ability to condense the representation of a set of data. The vector quantization performed by the SOM contributes to this compression, as does the regression on the data that the trained SOM represents in the form of adjacencies on the output map lattice. This regression is intended to preserve certain data structure, which is necessary if an application is to rely upon a mapping for an accurate summary of the data it requires. Preservation of data structure is clearly important in many applications for which SOMs and similar mapping algorithms are commonly used. Such applications include the common uses of SOMs in data mining and compression, as well as the use of SOMs to represent high-dimensional data visually, sometimes as a user interface [2]. In addition to these benefits of structure-preserving mappings, it becomes apparent that, to the extent that a mapping can preserve data structure at multiple levels of granularity, the speed of finding the best-matching unit (BMU) for a given data point can be improved significantly.

---

* Work of this author was sponsored in part by the Systems Research Institute (IBS) of the Polish Academy of Sciences, 01-447 Warsaw, ul. Newelska 6, Poland.

F. Yin, J. Wang, and C. Guo (Eds.): ISNN 2004, LNCS 3174, pp. 695–700, 2004.

2 Topography-Enhanced BMU Search

Although a rigorous definition of *topology* preservation can be borrowed from mathematics, the term *topography*-preserving mapping has been either used informally, or defined flexibly as a mapping that preserves the data structures important for a particular application [2]. The current paper discusses one benefit of mappings that implicitly preserve a hierarchy of data structure. In this context then, a topography-preserving mapping is one that preserves data relationships at various levels of granularity.

Without the benefit of parallel search, performing the mapping $\mathbf{r}_i = \Omega(\mathbf{v}_k)$ from data point $\mathbf{v}_k \in V$ to SOM neuron $\mathbf{r}_i \in A$ generally requires computation on the order of dp, where d is the dimensionality of the data and p is the number of neurons in the map. A topography-preserving mapping allows faster search, closer to the order of $d \log(p)$. This improvement can be achieved by taking advantage of the implicit hierarchy that the output lattice imposes on prototype vectors $\mathbf{w}_i$ when the mapping is topography preserving. When this is the case, a single codebook vector can serve as the prototype vector not only of the data within its own receptive field or Voronoi cell, but also as a rough prototype for the data mapping to neurons nearby in the discrete output space A. In other words, $\mathbf{w}_k$ might serve as a representative for the cluster of neurons $\mathbf{r}_{k-2}$ $\mathbf{r}_{k-1}$, $\mathbf{r}_k$, $\mathbf{r}_{k+1}$, $\mathbf{r}_{k+2}$, where these neurons form a neighborhood in A. In this way, $\mathbf{w}_k$ can serve as prototypes simultaeously at more than one level of data generalization or abstraction. The property of topography preservation can therefore enable BMU search to approach the speed of a search of prototypes in a hierarchical map without the overhead of a hierarchical architecture.

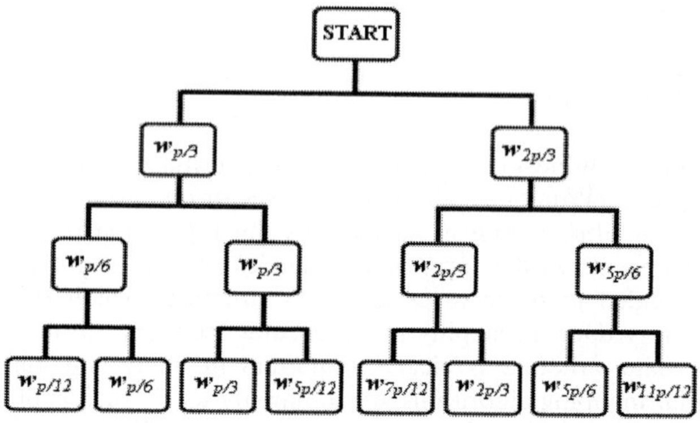

Fig. 1. BMU search tree for the first three comparisons in a one-dimensional topography-preserving map.

For example, for a one-dimensional map lattice with p neurons, a comparison of a new data point $\mathbf{v}$ with the pair of model vectors $\mathbf{w}_{p/3}$ and $\mathbf{w}_{2p/3}$ can narrow the BMU search to half the neurons of a topography-preserving mapping. When the better match for $\mathbf{v}$ is found in $\{\mathbf{w}_{p/3}, \mathbf{w}_{2p/3}\}$, the remaining search for the BMU can be restricted to half of the map represented at this highest level of abstraction by the better-matching prototype, assuming that the map preserves topography. With the next step of the search, $\mathbf{v}$ is compared with the pair of prototype vectors at one-third and two-thirds of the remaining half-map, either $\{w_{p/6}, w_{p/3}\}$ or $\{w_{2p/3}, w_{5p/6}\}$. The repeated elimination of half the remaining neurons continues until, at some point, an exhaustive search must be made of the remaining prototypes, since a lower-dimensional mapping cannot in general perfectly preserve the structure of a higher-dimensional data set. Ideally, this search of the prototype vectors of the one-dimensional map, especially at the highest, most abstract levels of the search tree, would require on the order of log(p) comparisons with each datapoint. In fact, as Fig. 1 shows for the first three comparisons, search at each level of the tree can reuse the measure of similarity calculated at the previous, "parent" comparison, allowing search with exactly log(p) comparisons. Of course, the situation is seldom ideal. The bifurcation of the data space will certainly not be so neat that datapoints that map to neurons near the boundary between submaps will always indicate a continued search in the correct submap. A possible solution is to continue search in both submaps whenever a datapoint excites both neurons beyond a certain threshold.

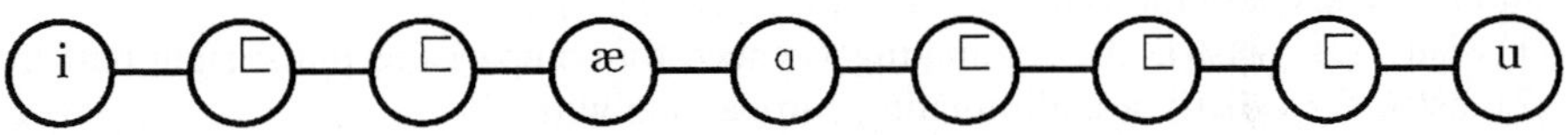

Fig. 2. One-dimensional topography-preserving mapping trained on six-dimensional vectors denoting articulatory features of English simple vowels

Table 1. Articulatory features describing simple vowels in a dialect of American English.

	i			æ					u
tongue high	1	1	0	0	0	0	0	1	1
tongue low	0	0	0	1	1	0	0	0	0
tongue back	0	0	0	0	1	1	1	1	1
lips round	0	0	0	0	0	0	1	1	1
lips spread	1	1	1	1	0	0	0	0	0
tense	1	0	0	0	0	0	0	0	1

To illustrate enhanced BMU search using a small and straighforward example, a nine-neuron one-dimensional map was trained using the Genetically-Trained Topographic Map (GTTM) [3][4]. Data were drawn from six-dimensional vectors representing articulatory features of the set of simple vowel phonemes for a dialect of American English. The features used to define each data point were adapted from a subset of those described in Chomsky and Halle's groundbreaking *The Sound Pattern of English* [1], where the discussion concerned universal distinctive features, i.e., features used across all natural human languages. The resulting trained mapping is represented in Fig. 2, and a list of the features describing each datapoint is given in Table 1. The distance relationships between the nine feature vectors are presented in Table 2, which depicts a dissimilarity matrix based on the Hamming distance between pairs of vectors. Vowel labels in the table are ordered by their position on the trained one-dimensional map, so small distances found near the main diagonal indicate near-continuity in the mapping, and a perfectly topography-preserving mapping would show increasing Hamming distances corresponding to greater distance from the main diagonal on each row and column. From the dissimilarity matrix, it is obvious that perfect topography preservation is impossible for a one-dimensional mapping of this data, since there is no way to order the vectors along one dimension that results in a total ordering of the Hamming distances between them. In other words, the data do not lie in a one-dimensional subspace of *V*.

If topography has been adequately preserved in the trained map, searching the nine-neuron map can be accomplished using the search tree displayed in Fig. 3. In this search tree, a data vector $\boldsymbol{v}$ is compared with pairs of prototype vectors that are siblings on the tree, eliminating half of the tree with the match at each level of abstraction. For example, searching for a vector representing the English phoneme / i / (corresponding to the vowel sound in the word *neat*) would use the following sequence of comparisons. Here, International Phonetic Alphabet symbols are used as abbreviations for feature vectors, and dist($\boldsymbol{v}_1$,$\boldsymbol{v}_2$) indicates the dissimilarity of vectors $\boldsymbol{v}_1$ and $\boldsymbol{v}_2$, in this case the Hamming distance between them.

1) dist(i, □) < dist(i, □), eliminating the need for comparisons in the right half-tree
2) dist(i, □) < dist(i, æ), eliminating comparison with □
3) dist(i, i) = 0

Table 2. Dissimilarity matrix showing Hamming distances between feature vectors representing English simple vowels

	i	□	□	æ	ɑ	□	□	□	u
i	0	1	2	3	5	4	5	4	3
□	1	0	1	2	4	3	4	3	4
□	2	1	0	1	3	2	3	4	5
æ	3	2	1	0	2	3	4	5	6
ɑ	5	4	3	2	0	1	2	3	4
□	4	3	2	3	1	0	1	2	3
□	5	4	3	4	2	1	0	1	2
□	4	3	4	5	3	2	1	0	1
u	3	4	5	6	4	3	2	1	0

It can be calculated that, when a new vector is presented that matches exactly one of the prototype vectors in this mapping, it will be found in 5 comparisons at most, with an average search of 3.56 comparisons. This compares favorably with the 9 comparisons for each new vector required by the typical exhaustive search. If a vector is presented to the trained map for which there is not an exactly matching prototype, the search must be pursued to the leaves of the search tree, requiring approximately 2 log(*p*) comparisons, each of which may involve *d* operations. This results in a complexity of *V* (*d* log(*p*)), as mentioned previously.

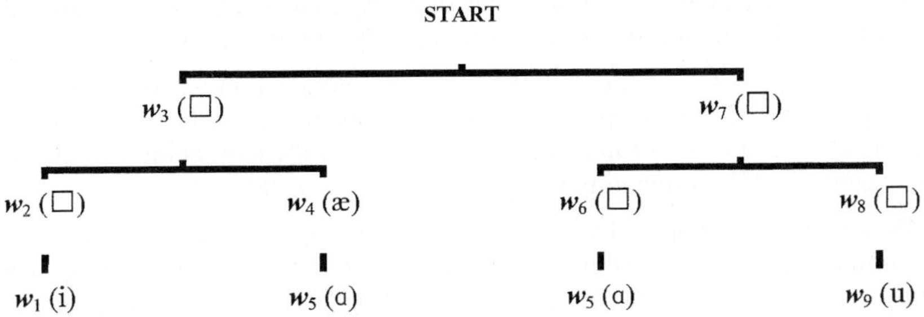

Fig. 3. Search tree for topography-enhanced BMU search of mapping of English vowel phonemes.

The example above of articulatory feature mapping to a one-dimensional phoneme map illustrates how topography-enhanced BMU search can operate. The procedure is easily adapted for a two-dimensional map lattice. If the output grid is almost square, a data vector $\boldsymbol{v}$ can be compared with four model vectors at each "level of abstraction." The U-Matrix in Fig. 4 shows an implicit and unnaturally regular hierarchy of data discovered by a 1024-neuron square map. In such an ideal example, one would expect that BMU search could be accomplished in very nearly $4 \times \log_2 (p^{0.5})$ comparisons for each new data vector $\boldsymbol{v}$. For this map, $\boldsymbol{v}$ will first be compared with four model vectors representing the four quadrants of the map that are separated by areas of greatest dissimilarity. After the first match, three-fourths of the map can be eliminated from the search. Comparison with four model vectors at the second level of abstraction eliminates three-fourths of the remaining one-fourth of the map, etc.

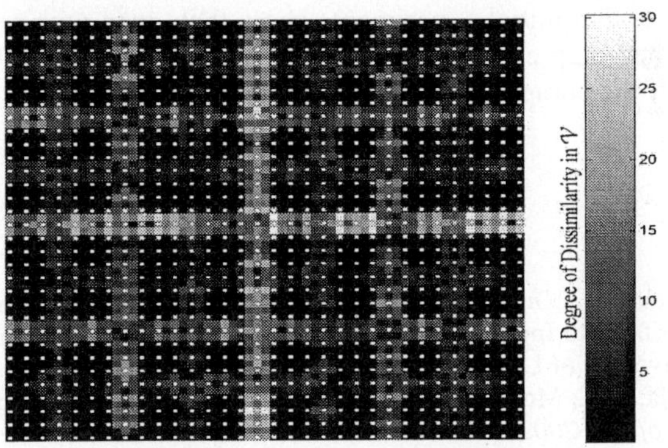

Fig. 4. U-Matrix for a 32x32 mapping revealing implicit hierarchy that can be used for topography-enhanced BMU search

For a final experiment, a 1024-neuron square map was trained on a three-dimensional artificial data set from which 2214 randomly-selected points were drawn to test the mapping. It was discovered that mapping a new data point to the best-matching neuron could be accomplished without error by making only 72 comparisons, rather than the 1024 comparisons required by exhaustive search. In this experiment, BMU search proceeded according to the following steps:

1) Data point ***v*** is compared with 4 model vectors representing the global structure of the 32 x 32 map. Matching ***v*** to a model vector at this most general level leads to reduction of the length of the sides of the map being examined by approximately 1/3, eliminating roughly 5/9 of the neurons.

2) Data point ***v*** is compared with 4 model vectors representing an abstraction of the remaining 22 x 22 submap. The length of the sides of the map under consideration is again reduced by 1/3, eliminating 5/9 of the remaining neurons.

3) Attention is now focused on the best-matching 15 x 15 submap, and ***v*** is again compared to 4 prototype vectors to reduce the scope of the search further.

4) The best-matching 11 x 11 submap is examined. After finding the best-matching region of the submap, an 8 x 8 region of the original map remains.

5) Exhaustive search finds the BMU among the 64 remaining neurons.

3 Conclusions

It is apparent that the success of topography-enhanced BMU search depends upon both the training of the map and the nature of the data. If the dimensionality of the map, δ_A, approximately matches the effective dimensionality of the data, δ_{eff}, and if the map is trained to preserve global as well as local relationships, then the enhanced BMU search can be applied effectively. However, when δ_{eff} is enlarged relative to δ_A, topography preservation becomes increasingly difficult. Despite this fact, it is encouraging to note that initial experiments show that, even with relatively high δ_{eff}, a globally well-trained map can benefit from topography-enhanced BMU search, often saving at least half the comparisons required by exhaustive search.

References

1. Chomsky, N., Halle, M.: *The Sound Pattern of English*. New York, Harper and Row (1968)
2. Kirk, J.S.: Algorithms to Improve Topography Preservation in Self-Organizing Maps. Ph.D. dissertation, University of Louisville (2003)
3. Kirk, J.S., Zurada, J.M.: Motivation for a Genetically-Trained Topography-Preserving Map. In *Proceedings of the 2002 International Joint Conference on Neural Networks*, held in Honolulu, HI, May 12-17 (2002) 394-9
4. Kirk, J.S., Zurada, J.M.: An Evolutionary Method of Training Topography-Preserving Maps. In *Proceedings of the 2001 INNS-IEEE International Joint Conference on Neural Networks*, held in Washington, D.C., July 14-19 (2001) 2230-4.
5. Kohonen, T.: *Self-Organizing Maps*, 3rd ed. Springer, Berlin Heidelberg New York (2001)
6. Zurada, J.M.: Introduction to Artificial Neural Systems. PWS, Boston (1992)

Nonlinear Optimal Estimation for Neural Networks Data Fusion

Ye Ma, Xiao Tong Wang, Bo Li, and JianGuo Fu

Dept. of automation, Dalian Naval Academy, 116018 Dalian, P. R. China
Anewday_021004@hotmail.com

Abstract. As the activation function of neural networks is nonlinear, an adaptive model and an algorithm are suggested for real-time calculation of the network weights using an Unscented Kalman Filter (UFK). Because the method can simplify the calculation of nonlinear systems, a global optimization of data fusion information can be achieved by real-time adjusting the weights of the networks according to the variations of the system inputs. An experiment was made on a vessel board equipped with DGPS/GPS/ROLANDC/COMPASS to testify the availability of the algorithm. Three algorithms are applied to the system: the 1st with fixed weights of the network, the 2nd adjusting the weights of the system with linear activation function by KF, and the 3rd adjusting the weights of the system with nonlinear activation function by UKF. Experimental results show that the proposed algorithm can improve the tracking accuracy of the system. This method can be applied to integrated navigation and other data fusion systems.

1 Introduction

The Kalman filtering theory has been widely used in the integrated navigation field. The famous federated filtering [1] is a two-stage data processing technique in which the outputs of local sensor-related filters are subsequently processed and combined by a larger master filter. The global optimal state estimation is produced based on various local state estimation and covariance. Literature [2] proposed a method of state estimation using neurons. The weights of NN in literature [2] are off-line trained and on-line applied. If a local state estimation has changed and the weights of NN are still fixed, its global estimation precision will be affected. Using UKF nonlinear optimal estimate theory[3][4], we developed a real-time learning model of multiplayer neural networks to fuse each local state estimates. Consequently, the neural networks weights can be adjusted in real-time via UKF. The model has the better fitting effect and self-adapting estimate capacity. The fusion model of multi-sensors with DGPS/GPS/ROLANDC/COMPASS in integrated navigation system[8][9] is implemented. The experimental results show that the proposed model and adaptive training algorithm are effective to enhance the precision and the operating speed of the overall system.

F. Yin, J. Wang, and C. Guo (Eds.): ISNN 2004, LNCS 3174, pp. 701–706, 2004.

2 NN-UKF Model Research

The data fusions with neural networks combines local filter outputs and will yield estimate that is globally optimal or near optimal. As suggested by the figure1, each local filter operates in parallel, processing unique data from their local sensor, and a destination data from a shared system. First, let the local filter #i solution be represented by the state vector $\hat{X}_i(k)$ and covariance matrix P_i, and the full state vector $\hat{X}_f(k)$. Globally optimal state relies on each local estimate $\hat{X}_i(k)$ and its covariance P_i, namely:

$$\hat{X}_f(k) = E\{X(k) \mid \hat{X}_i(k), P_i, i = 1,2,........ N\} \tag{1}$$

The maximum likelihood estimate $\hat{X}_i(k)$ and covariance P_i are obtained according to the reference [2]. $\hat{X}_f(k)$ can be simplified as:

$$\hat{X}_f(k) = W\mu \tag{2}$$

Obviously we may replace the formula (2) with the neural networks, where, W is the neural networks weight vector, μ is the input vector. This method avoids computing the condition covariance of each local filter, only needs to estimate the neural networks weight vector.

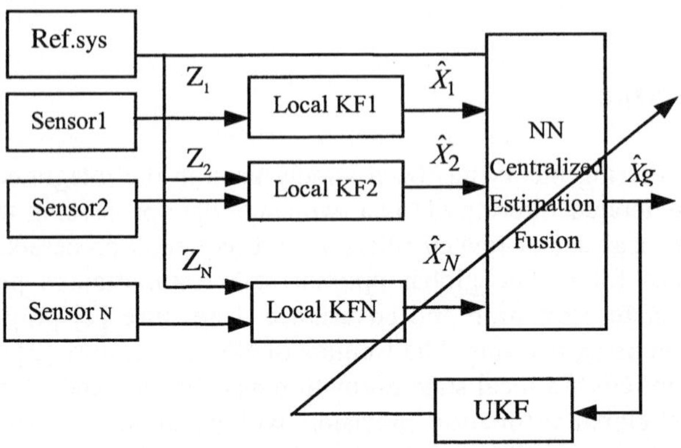

Fig. 1. NN-UKF Model

As each local filter output varies with time, and the network weights was trained offline[1][6]. Therefore an online adjustment is made to adapt each local state variation and to keep the state estimation optimal.

3 Algorithm Research

3.1 State Equation of Neural Networks

Supposition 1. The number of every layer neurons is N for n layers networks. Connection weight from the *j*th neuron of the *(n-1)* th layer to the ith neuron of the nth layer is w_{ij}^{n}, and its threshold value is θ_i^n;
Supposition 2. Nonlinear transformation $f(\cdot)$ of every neuron uses the sigmoid function. That is:

$$f(x_i^{\,n}(k)) = 1/(1-\exp(-x_i^{\,n}(k))) \tag{3}$$

Definition 1. Based on supposition 1 and supposition 2, transformation relation between the input and output is:

$$x_i^n(k) = f\Big(\sum_{j=1}^{N} w_{ij}^{\;n-1} x_j^{\;n-1}(k) + \theta_i^{n-1}\Big) = f\Big(\sum_{j=1}^{N+1} w_{ij}^{\;n-1} x_j^{\;n-1}(k)\Big) \tag{4}$$

If we use the network connection weight as the state vector and regard the object output vector as the observation value, then non-linear state equation can be written:

$$\begin{aligned} w(k+1) &= w(k) + \eta(k) \\ z(k) &= f(w(k), x(k)) + v(k) \end{aligned} \tag{5}$$

Where, $W_{(k)} \in R^n$ is the system state vector at sample k and $\eta_{(k)}$ is the process noise it is a zero-mean Gaussian white noise sequences and its covariance is $Q_{1.}$ $y_{(k)} \in R^m$ is measurement vector, $v_{(k),}$ is the measurement noise, it is also the zero-mean Gaussian white noise sequences, the covariance is R_1.

The formula (5) is the non-linear state equation, which is unable to use the standard Kalman filtering to carry on the state estimation. Here, we use UKF to carry on the filter.

3.2 Nonlinear Filter Algorithm

Assume the stable neural networks weight is the UKF initial value of state. Firstly carrying on the time update.

Supposes x_k is N dimension state vector, which mean is $x_{k/k}$ and covariance is $p_{k/k}$, the number of sigma-points needed to do this 2n + 1, where n is the dimension of X[5], that is:

$$\begin{aligned}
&\chi_0 = x_{k/k} && W_0 = k/(n+k) && \\
&\chi_i = x_{k/k} + (\sqrt{(n+k)P_{k/k}})_i && W_i = 1/2(n+k) && i = 1,.....2n \\
&\chi_{i+n} = x_{k/k} - (\sqrt{(n+k)P_{k/k}})_i && W_{i+n} = 1/2(n+k) && i = n,.....2n+1
\end{aligned} \tag{6}$$

Where $k \in R$, is a proportionality factor, $\sqrt{(n+k)P_{k/k}}$ the ith row or column of weighted matrix square root of the covariance $P_{k/k}$, W_i is the weight that is associated with the *ith* point. After giving out these sigma points, we may carry on the forecast to the state variable and covariance, the method is:

$$\hat{x}(k+1/k) = \sum_{i=0}^{2n} W_i \chi_i (k+1/k) \tag{7}$$

$$P(k+1/k) = \sum_{i=0}^{2n} W_i [\chi_i (k+1/k) - \hat{x}(k+1/k) \times [\chi_i (k+1/k - \hat{x}(k+1/k)]^T \tag{8}$$

According to the observation model, we can calculate the measurement value and the measurement estimate value:

$$Z_i(k+1/k) = f(\chi_i(k+1/k), k) \tag{9}$$

$$\hat{z}(k+1/k) = \sum_{i=0}^{2n} W_i Z_i (k+1/k) \tag{10}$$

measurement update:

$$P_{vv}(k+1/k) = \sum_{i=0}^{2n} W_i [\chi_i (k+1/k) - \hat{x}(k+1/k)] \times [\chi_i (k+1/k - \hat{x}(k+1/k)]^T \tag{11}$$

$$P_{xv}(k+1/k) = \sum_{i=0}^{2n} W_i [\chi_i (k+1/k) - \hat{x}(k+1/k)] \times [Z_i (k+1/k - \hat{z}(k+1/k)]^T \tag{12}$$

$$K_k = P_{xv} P_{vv}^{-1} \tag{13}$$

$$\hat{x}_k = \hat{x}(k+1/k) + K_k (z(k) - \hat{z}(k)) \tag{14}$$

$$P(k+1) = P(k+1/k) - K_k P_{vv} K_k^T \tag{15}$$

4 Experiments Results

To verify the usability of the proposed method, we integrate the navigation equipments of DGPS/GPS/LORANC/COMPASS to carry out the actual vessel experiment. The vessel sails at velocity ranged from 18 to 22 knots, heading in the direction of 20^0~40^0.The filter's state space consists of the longitude X_1 *(L)*, the latitude X_2 (λ), the north ocean current state X_3 (V_{CN}), east ocean current state X_4 (V_{CE}), the direction X_5 *(S)* and the speed X_6 *(H)*, namely, $X = (\varphi, \lambda, V_N, V_E, G, V)^T$. The state dynamics are:

$$\delta X(k+1) = \Phi(k+1/K)\delta X(K) + W \tag{16}$$

where:

$$\Phi = \begin{bmatrix} 1 & 0 & 0 & (1-e^{-\beta_c T})/\beta_c & \cos HT & -S\sin HT(\pi/180) \\ 0 & 1 & (1-e^{-\beta_c T})/\beta_c & 0 & \sin HT & S\cos HT(\pi/180) \\ 0 & 0 & e^{-\beta_c T} & 0 & O_{2\times3} & \\ O_{3\times4} & & & e^{-\beta_c T} & I_{3\times3} & \end{bmatrix}$$

The correlation model in the ocean current, the concealment supposition is the correlation time, that is: $1/\beta_c$.

To guarantee the KF running time is short enough, we suppose KF measurement vector always maintain 4x1 dimensional, that is: $Z(k)=(\varphi, \lambda, G, V)^T$. In actual system, the biases of filter state model and a measurement model are designed. The new measurement equation is:

$$\delta(Z(k)) = H(k)\delta X(k) + V(k) \tag{17}$$

Here, $\delta X(k) \in R^n$ is the system state vector at sample k and $W(k)$ is process noise, it is a zero-mean Gaussian white noise sequences, its covariance is R_2, $\delta Z(k) \in R^M$ is the measurement, and measurement noise $V(k)$ is the same as $W(k)$, its covariance is R_{2},. $H(k)$ is a constant matrix. We take DGPS localization information as the standard, and fuse various filters information with the neural networks. Because the system described by formula (5) is nonlinear, NN weights are on-line estimated with UKF. We compare this method with the linear networks. The simulation result showed that fitting precision is higher. The simulation curve and estimate value is showed in figure2.

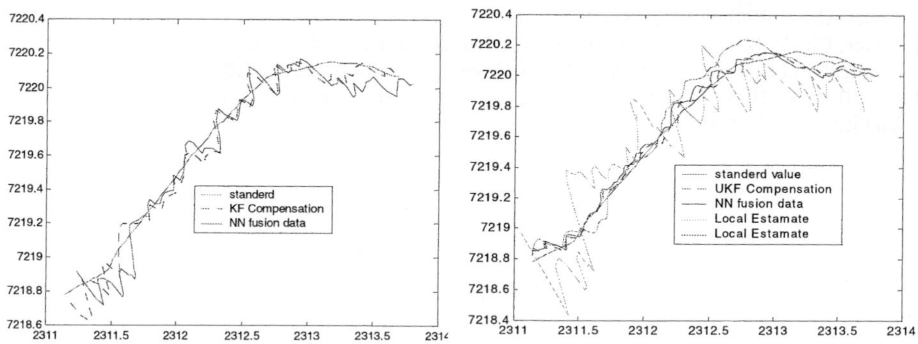

Fig. 2. Simulation curve (a) linear NN and KF (b) nonlinear NN and UKF

5 Conclusions

From the experimental results we can see that the nonlinear optimal estimation for neural networks data fusion [7] can effectively solve the problem of real-time information fusion. Using this method, as no calculation of various error covariance and inverse matrix is needed, so that the computational loads are reduced greatly and the computation precision is increased. Simultaneity, the global information can be feasibly allocated, according to the change of the estimation quality in local filters. The comparisons of its performance with those of the linear filtering in an integrated navigation system have been made. The theoretical analyses and experimental results show that this model and adaptive training algorithm are effective and practical.

References

1. Carlson N A: Federated Square Root Filter or Decentralized Parallel Processes. IEEE Trans Aerospace and Electronic Systems, AES-26. (1990) 517-525
2. Chowdhury F N: A Neural Approach to Data Fusion. Proc. American control Conf. (1995) 1693-1697
3. S.J.Julier, J.K.Uhlmann: A New Extension of the Kalman Filter to Nonlinear Systems. Proceedings of the SPIE, Orlando, FL: SPIE. (1997) 182-193
4. E.Awan, A.T.Nelson: Neural Dual Extended Kalman Filtering Applications in Speech Enhancement and Monaural Blind Signal Separations. Proc. Neural Networks for Signal Processing Workshop.IEEE (1997).
5. S.J.Julier: The Scaled Unscented Transformation. Proceedings of the 2002 American Control Conference, vol. 6, Anchorage (2002) 4555- 4559
6. J.K.Uhlmann: Algorithms for Multiple Target Tracking. American Scientist, vol.80. (1992) 128-141
7. Leonard C: Application of Neural Networks in Target Tracking Data Fusion. IEEE Trans.AES vol.30. (1994) 281-287
8. Meier, Ch.: A Multi-Sensor-System for Advanced Surface Movement Guidance and Control, Concept and first Results. IFAC, Braunschweig, Juni 2000, IFAC, (2000)
9. Farrell, J. A., Barth M.: The Global Positioning System and Inertial Navigation. New York: McGraw-Hill (1999)

Improvement of Data Visualization Based on SOM

Chao Shao and Houkuan Huang

School of Computer and Information Technology, Beijing Jiaotong University, Beijing, 100044, China
sc_flying@sina.com, hkhuang@center.njtu.edu.cn

Abstract. In order for the map to visualize the structure of input data more naturally, the positions of neurons of the SOM should be adjusted based on their similarities. We study these position-adjustable SOM algorithms using Himberg's contraction model, and then improve them into a uniform PASOM algorithm, which is easier to implement and control. In addition, by making use of the SOM's topological ordering, the side effect of randomly selected initial weight vectors and the excess contraction to one point can also be avoided mostly. Finally, the PASOM algorithm is verified by experimental results very well.

1 Introduction

The Self-Organizing Map (SOM), which was first introduced in 1982 by Kohonen[1], is a singular-layer neural network model based on competitive learning. As the combination of vector quantization and non-linear projection[2], the SOM can map high-dimensional input data onto a regular low-dimensional discrete lattice, i.e. a fixed grid of neurons, while preserving the topological relations of the input data as faithfully as possible. Therefore, the SOM can generate interesting low-dimensional representations of high-dimensional data, which makes it a popular clustering, visualization and abstraction tool.

Due to the topology preserving nature (although incomplete topology preservation[3][4]), the SOM is effective for the visualization of high-dimensional data. However, there are several drawbacks for the standard Kohonen's SOM to visualize high-dimensional data. First, the network structure has to be prespecified. To solve this problem, several dynamic SOMs such as GCS[5][6][7] and GSOM[3] have been presented recently. However, these approaches require more prespecified training parameters, which makes them much less robust than the standard Kohonen's SOM[8]. Second, due to the fixed grid of neurons, the projection is very rude[2], and the inter-neuron distances are not directly visible, which makes it necessary to use some color-coding schemes such as the U-matrix[2] to mark relative distances between neighboring neurons in the input space. Even so, the structure of input data may often appear in a distorted and unnatural form[9].

In order for the map to visualize the structure of input data more naturally, the distance quantity must be preserved on the map, along with the topology[9].

F. Yin, J. Wang, and C. Guo (Eds.): ISNN 2004, LNCS 3174, pp. 707–712, 2004.

One type of methods is to combine the distance-preserving mapping such as Multidimensional Scaling (MDS) with the SOM[4], or to minimize the MDS type quantitative measure of topographical agreement directly[10], which is computationally very intensive, and which cannot avoid the shortcoming of MDS too, e.g. no generalization capability. Another type of methods is to adjust the positions of neurons synchronously during the training process, we call position-adjustable SOMs, such as the grouping neuron (GN) algorithm[11], the adaptive coordinate (AC) algorithm[8] and the double self-organizing feature map (DSOM) algorithm[12][13]. Being add-ins to the standard Kohonen's SOM, the robustness of these methods is assured[8]. In addition, the adjustment rule is simple, so the additional computation is little. In this paper, we study these position-adjustable SOM algorithms using Himberg's contraction model[14], and then improve them into a uniform PASOM algorithm, which is easier to implement and control.

This paper is organized as follows: In Section 2, we recall the standard Kohonen's SOM. In Section 3, we improve those position-adjustable SOM algorithms into a uniform PASOM algorithm. Finally, experimental results and conclusions are given in Section 4 and Section 5 respectively.

2 The Standard Kohonen's SOM

The standard Kohonen's SOM consists of a singular-layer of neurons located on a regular low-dimensional lattice, usually 1-D or 2-D. Each neuron k is represented by an n-dimensional weight vector $m_k = \{m_{k1}, \cdots, m_{kn}\}$, i.e. the reference vector, where n is the dimension of input data. On each training step, a data sample x is selected randomly, and then the winner neuron (the best-matching unit, BMU) m_c is found according to the following rule:

$$\|m_c - x\| = \min_k \|m_k - x\| \tag{1}$$

The weight vectors of the BMU and its neighbors are updated towards x:

$$m_k(t+1) = m_k(t) + \alpha(t) \cdot h_{ck}(t) \cdot (x - m_k(t)) \tag{2}$$

Where $\alpha(t)$ is the learning rate and $h_{ck}(t)$ is the neighborhood kernel centered on the BMU c. To assure the convergence, both the learning rate $\alpha(t)$ and the neighborhood kernel radius (or its scalar parameter) $\sigma(t)$ should decrease monotonically with time.

The resulting weight vectors can be interpreted as conditional averages of the input data. Because not only the BMU but also its neighbors are updated, the weight vectors of neighboring neurons resemble each other[2]. Consequently, the BMUs of similar data samples are close to each other within the lattice, which is so-called SOM's topological ordering.

3 The PASOM Algorithm

To improve data visualization based on SOM, the distance quantity should be used to represent the similarities on the map directly, rather than use a color-coding scheme in a distorted lattice indirectly. However, the positions of neurons are fixed for the standard Kohonen's SOM, so several methods have been presented to adjust the positions of neurons synchronously during the training process, such as the GN algorithm[11], the AC algorithm[8] and the DSOM algorithm[12][13], we call position-adjustable SOMs, which don't affect the whole architecture of the standard Kohonen's SOM and then preserve its goodness.

In these position-adjustable SOM algorithms, each neuron k has an n-dimensional weight vector m_k and a 2-D position vector p_k attached, which are updated synchronously. p_k is initialized as the position of neuron k within the lattice. On each training step, m_k are updated according to formula (2), different method adjusts p_k differently (however, similar to formula (2)):

$$\textit{the GN algorithm}: \quad p_k(t+1) = p_k(t) + a(t) \cdot b_{ck}(t) \cdot (p_c(t) - p_k(t)) \tag{3}$$

$$\textit{the AC algorithm}: \quad p_k(t+1) = p_k(t) + \Delta Dist_k(t+1) \cdot (p_c(t) - p_k(t)) \tag{4}$$

$$\textit{the DSOM algorithm}: p_k(t+1) = p_k(t) + \eta_{ck}(t) \cdot h'_{ck}(t) \cdot (p_c(t) - p_k(t)) \tag{5}$$

Both $b_{ck}(t)$ in the GN algorithm and $h'_{ck}(t)$ in the DSOM algorithm decrease with the distance between the weight vectors of the BMU c and neuron k, so both methods can attract the neurons whose weight vectors are close together in the input space; however, both $\Delta Dist_k(t+1)$ (it can be proved easily.) in the AC algorithm and $\eta_{ck}(t)$ in the DSOM algorithm decrease with the distance between the positions of the BMU c and neuron k within the lattice, so both methods can attract the neurons that are adjacent to each other in the output map. Due to the SOM's topological ordering, these methods can reach similar results.

In fact, $b_{ck}(t)$, $\Delta Dist_k(t+1)$, $\eta_{ck}(t)$ and $h'_{ck}(t)$ can be used to represent the similarity between the BMU c and neuron k in some sense, therefore, these methods fall into the category of Himberg's contraction model[14] (similar to the hierarchical clustering), in which the neurons are contracted based on their similarities: the new position of a neuron is calculated as the weighted average of the positions of all the neurons, and the weights are the corresponding similarities.

Following the philosophy of Himberg's contraction model and its similarity to the SOM, these methods can also be unified (similar to formula (2)):

$$p_k(t+1) = p_k(t) + \alpha^p(t) \cdot h^p_{ck}(t) \cdot (p_c(t) - p_k(t)) \tag{6}$$

Where $\alpha^p(t)$ is the learning rate, and $h^p_{ck}(t)$ is the similarity kernel between the BMU c and neuron k, which can take any form of $h_{ck}(t)$, $b_{ck}(t)$, $\Delta Dist_k(t+1)$, $\eta_{ck}(t)$, $h'_{ck}(t)$ or their combinations, etc. Like the standard Kohonen's SOM, the parameters which are used to adjust the positions in these methods can be classified into the learning rate $\alpha^p(t)$ and the similarity kernel scalar parameter $\sigma^p(t)$ respectively, both of which are monotonically decreasing functions of time.

For the convenience of implementation, and to converge synchronously with the weight vectors, we can specify simply that the scalar parameter $\sigma^p(t)$ of the similarity kernel $h^p_{ck}(t)$ in (6) is identical to the scalar parameter $\sigma(t)$ of the neighborhood kernel $h_{ck}(t)$ in (2), and that the learning rate $\alpha^p(t)$ in (6) is constant times of the learning rate $\alpha(t)$ in (2):

$$\sigma^p(t) = \sigma(t) \tag{7}$$

$$\alpha^p(t) = cf \cdot \alpha(t) \tag{8}$$

Where cf is the contraction factor to control the extent of one contraction step.

These methods don't take the side effect of randomly selected initial weight vectors into consideration. During the forepart of iterations, i.e. the ordering phase, the neurons are still unordered. To solve this problem, we should adjust the positions not from the beginning but after the proper ordering takes place, i.e. during the fine-adjustment phase. Thus, we can also diminish the time cost by making use of the SOM's topological ordering, What's more, the excess contraction to one point is avoided mostly, because the number of iterations is reasonably large in order to achieve the better statistical accuracy of the final mapping. So, the PASOM algorithm can be described as follows:

```
Specify the network structure;
Initialize the weight matrix randomly;
Initialize the position matrix whose elements are identical
to the corresponding positions within the lattice;
t=0; Specify alpha(0) and sigma(0);
Specify one form of the similarity kernel;
Specify cf and threshold;
WHILE the stop condition is not met
    Initialize all the input data samples still unselected;
    Do
        Select an input data sample x unselected randomly;
        Find the BMU c according to formula (1);
        Update the weight vectors according to formula (2);
        IF t>threshold THEN
            Adjust the position vectors according to formula (6);
    Until all the input data samples are selected;
    Decrease alpha(t) and sigma(t); t=t+1;
END
```

4 Experimental Results

We apply the PASOM algorithm described above on the butterfly dataset (Fig. 1, 2-D for comparison between the results of PASOM and the original dataset; we can also apply PASOM on other high-dimensional datasets.) using different contraction factors and similarity kernels. In the experiments, we specify that the neighborhood kernel is a Gauss function, and the parameters are defined as follows: the number of neurons m=10×10, $\alpha(0)$=0.9,

$\sigma(0)$=66.666, $\alpha(t+1)$=0.99×$\alpha(t)$, $\sigma(t+1)$=0.98×$\sigma(t)$, the number of iterations is 1000, *threshold*=200. The results are given in Fig. 3-8. From these results, we can see that the structure of input data can be visualized more naturally and clearly by using PASOM (Fig. 3-8) than by using the standard Kohonen's SOM (Fig. 2), especially by selecting the more proper contraction factor (e.g. cf=0.1), which is easier to control.

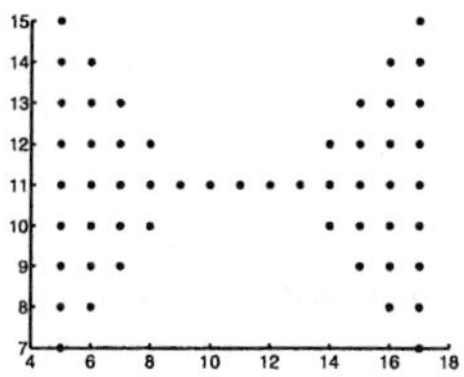

Fig. 1. The butterfly dataset

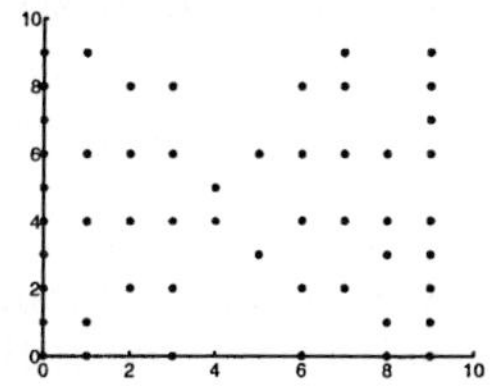

Fig. 2. Standard SOM's result

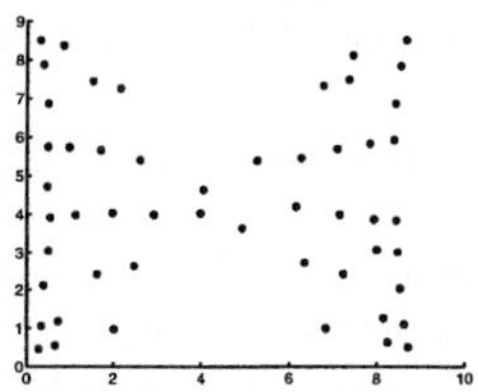

Fig. 3. cf=0.05,$h^p_{ck}(t)$=$h'_{ck}(t)$

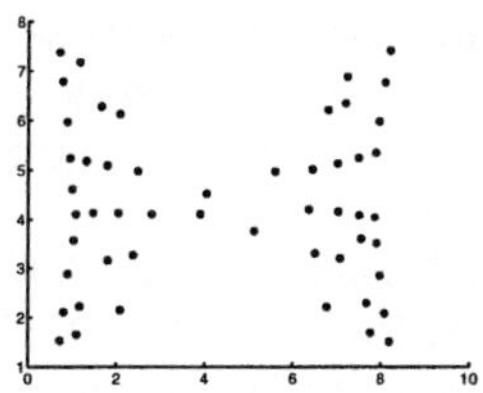

Fig. 4. cf=0.05,$h^p_{ck}(t)$=$b_{ck}(t)$

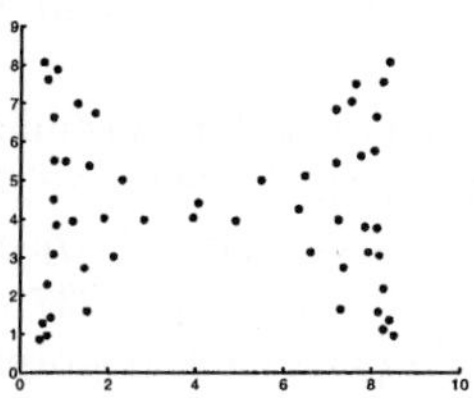

Fig. 5. cf=0.10, $h^p_{ck}(t)$=$h'_{ck}(t)$

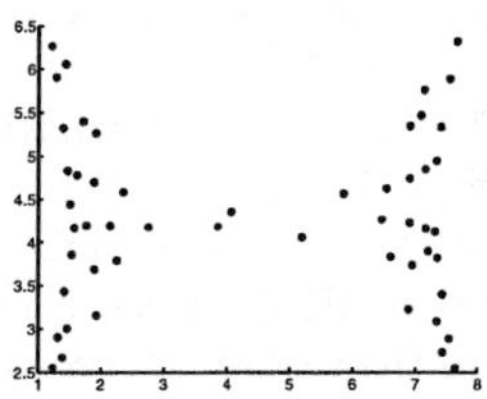

Fig. 6. cf=0.10, $h^p_{ck}(t)$=$b_{ck}(t)$

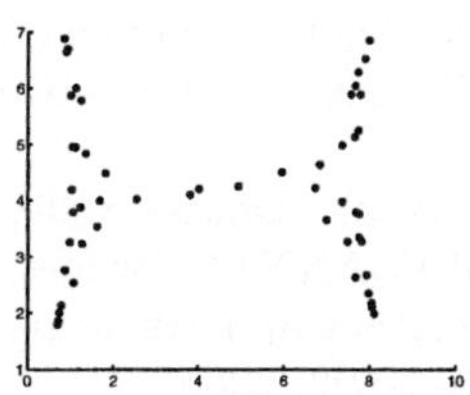

Fig. 7. cf=0.25, $h^p_{ck}(t)$=$h'_{ck}(t)$

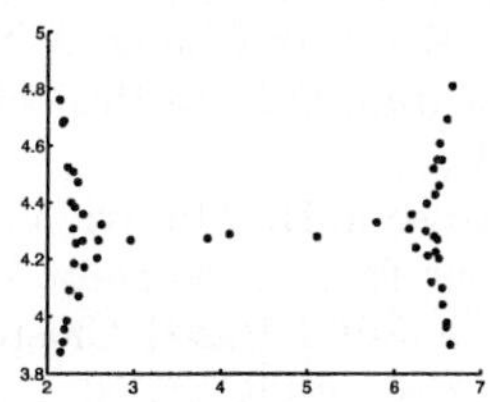

Fig. 8. cf=0.25, $h^p_{ck}(t)$=$b_{ck}(t)$

Following the philosophy of Himberg's contraction model (similar to the hierarchical clustering), Our PASOM algorithm can improve the visualization of input data greatly.

5 Conclusions

In this paper, we adjust positions of all the neurons of SOM adaptively according to their similarities, by which we improve those position-adjustable SOM algorithms into a uniform PASOM algorithm, which is easier to implement and control. What's more, by making use of the SOM's topological ordering, the side effect of randomly selected initial weight vectors and the excess contraction to one point can also be avoided mostly.

References

1. Kohonen, T.: Self-Organized Formation of Topologically Correct Feature Maps. Biol. Cybern., vol. 43, pp. 59-69, 1982
2. Vesanto, J.: SOM-Based Data Visualization Methods. Intelligent Data Analysis, Vol. 3, pp. 111-126, 1999
3. Alahakoon, D., Halgamuge, S. K., Srinivasan, B.: Dynamic Self-Organizing Maps with Controlled Growth for Knowledge Discovery. IEEE trans. Neural networks, vol. 11, no. 3, pp. 601-614, 2000
4. Duch, W., Naud, A.: On Global Self-Organizing Maps. In Proc. of ESANN'96, Bruges, pp. 91-96, 1996
5. Fritzke, B.: Kohonen Feature Maps and Growing Cell Structures - a Performance Comparison. In Proc. of NIPS'92, Denver, pp. 123-130, 1992
6. Fritzke, B.: Growing Cell Structures - a Self-Organizing Network for Unsupervised and Supervised Learning. Neural Networks, vol. 7, no. 9, pp. 1441-1460, 1994
7. Wu, Z., Yen, G. G.: A Som Projection Technique with the Growing Structure for Visualizing High-dimension Data. In Proc. of IJCNN'03, Portland, pp. 1763 - 1768, 2003
8. Merkl, D., Rauber, A.: Alternative Ways for Cluster Visualization in Self-Organizing Maps. In Proc. of WSOM'97, Espoo, pp. 106-111, 1997
9. Hujun Yin: ViSOM - a Novel Method for Multivariate Data Projection and Structure Visualization. IEEE trans. Neural networks, vol. 13, pp. 237 - 243, 2002
10. Duch, W.: Quantitative Measures for the Self-Organizing Topographic Maps. Open Systems & Information Dynamics, vol. 2, pp. 295-302, 1995
11. Rubio, M., Gimnez, V.: New Methods for Self-Organising Map Visual Analysis. Neural Comput & Applic, vol. 12, pp. 142-152, 2003
12. Mu-Chun Su, Hsiao-Te Chang: A New Model of Self-Organizing Neural Networks and Its Application in Data Projection. IEEE trans. Neural networks, vol. 12, pp. 153-158, 2001
13. Dali Wang, Ressom, H., Musavi, M., Domnisoru, C.: Double Self-Organizing Maps to Cluster Gene Expression Data. In Proc. of ESANN'02, Bruges, pp. 45-50, 2002
14. Himberg, J.: A SOM Based Cluster Visualization and Its Application for False Coloring. In Proc. of IJCNN'00, Como, pp. 587-592, 2000

Research on Neural Network Method in GPS Data Transformation

Min Han[1], Xue Tian[1], and Shiguo Xu[2]

[1] Sch. of Electronic and Information Eng., Dalian U. of Technology, Dalian, 116023, China
minhan@dlut.edu.cn
[2] Sch. of Civil and Hydraulic Eng., Dalian U. of Technology, Dalian, 116023, China

Abstract. This paper discusses an important problem emerged in applying GPS to wetland study--transformation method between coordinates system of local electrical maps and global coordinates system of GPS data. Traditional method is complex and not precise enough for large area. In this paper, a feed-forward neural network is build and the back propagation algorithm is used. The Levenberg-Marquardt algorithm and a modified error function are adopted to improve convergence rate and generalization ability. Neural network model, learning principle, simulation curves and results comparation are discussed in detail. The study is undertaken on Songhua River Basin. Practice results prove that compared with traditional method, neural network method succeeds in solveing complex problem and is effective in dynamic data analysis of wetland.

1 Introduction

It has been recognized nowadays that wetland is a kind of important multi-functional resource. Applying sufficient methods and reliable information technology to wetland protection study has improved the work very much [1, 2].

Any point's three-dimensional position on the earth can be obtained by receiving GPS (Global Positioning System) signals. To link local electrical maps with GPS data, coordinates transformation is necessary [3]. Concerning transformation, traditional method is complex and not precise enough for large area because of over many calculations and similarities. New tools or better methods should be developed. NN (Neural Network) has been widely applied in nonlinear function approximation. Application of NN in GPS data transformation is a new trial attempt. The results of practice prove that NN method is easier and more precise than traditional method. It is expected to be an efficient and feasible method in transformation of GPS data.

2 Traditional Method of Transformation

Coordinates system of data derived form GPS is WGS84 which is a global uniform geographic coordinates with the best approximation in the world as a whole [4]. Local coordinates system such as BJZ54 is usually used as local electrical map's system and projected to *x-y* coordinates system. Therefore during the process of identifying posi

F. Yin, J. Wang, and C. Guo (Eds.): ISNN 2004, LNCS 3174, pp. 713–718, 2004.

tions on the electrical maps as illustrated in Fig. 1, *x-y* should be used and latitude-longitude data should be transformed into *x-y*. Then it is possible to show moving positions on electrical maps by getting the real latitude and longitude from GPS message and transform them to the scale on the projected maps.

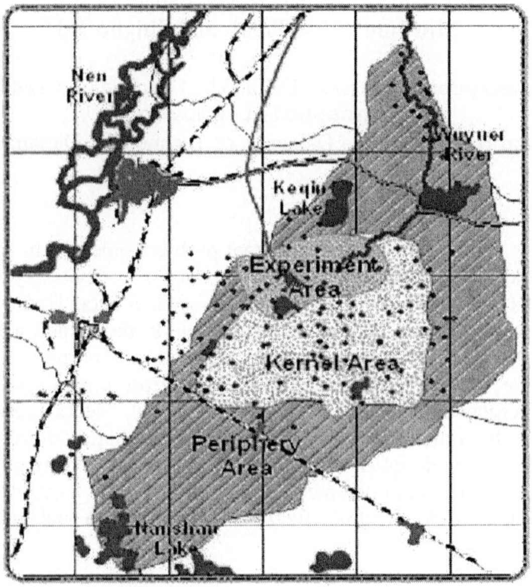

Fig. 1. Local electrical maps of Zhalong Wetland

The transformation from geodetic coordinates (*B*, *L*, *H*) to three-dimensional Cartesian coordinates (*x*, *y*, *z*) is given by the following equations:

$$\begin{cases} x = (N+H)\cos B\cos L \\ y = (N+H)\cos B\sin L \\ z = [N(1-e^2)+H]\sin B \end{cases} . \tag{1}$$

Where $N = a/\sqrt{1-e^2\sin^2 B}$, the principal radius of curvature along the prime vertical, $a = 6378137\,m$, $e^2 = 2f - f^2$ and $1/f = 298.257$, the parameters describing size (semimajor axis) and shape (flattening) of the adopted reference ellipsoid.

Prior to GPS and WGS-84, there were already many reference ellipsoids and datums chosen to fit particular areas of the world. Each of these is typically referred to as "local datum". Coordinates can be transformed from one geocentric conventional terrestrial datum to another local reference datum, and seven parameters similarity transformation formulae is one of the common transformation models [5].

Local mapping projection is another necessary work. One of the conversion formulae, Gauss-Kruger projection formulae, can be found in reference [6].

3 Neural Network Method of Transformation

The neural network method of transformation includes taking latitude and longitude of geodetic coordinates derived from GPS as input, training the neural network according to certain rules, finally getting *x*-*y* values in local coordinates system as output. A three layers feed-forward neural network is adopted to build the transformation model. Hidden layer neurons' transfer function is tan-sigmoid. Output layer neurons' transfer function is linear function.

In almost all kinds of neural network algorithms, back propagation algorithm has gained nearly full development. However, it still has several disadvantages for example slow convergence rate and over fitting problem. The algorithm performance is very sensitive to the proper setting of learning rate, so the optimal learning rate should be adjusted on line during the learning process to quicken the convergence rate. Another problem during neural network learning is called over fitting. The error on learning set is driven to a very small value, but when new data is presented to the network the error is large. The network has memorized the learning set, but it has not learned to generalize to new situations.

In this paper, a modified performance function is adopted in neural network learning process to solve the over fitting problem and Levenberg-Marquardt algorithm which is proved suitable to the transformation work is adopted to make the learning process become faster. The principle of the algorithm is introduced as following. The standard performance function used for training feed-forward neural networks is mean sum of squares of the network errors. The standard performance function is modified by adding a term that consists of sum of network weights squares as $F=\beta\sum_{i=1}^{n}e_i^2+\alpha\sum_{i=1}^{n}\omega_i^2$, where F is the performance function, α and β are the objective function parameters, e is the vector of network errors, ω is the network weights, and n is the number of learning patterns. The values of α, β are optimized according to Bayesian's rule as following equations [7]:

$$\alpha=\frac{\gamma}{2\sum_{i=1}^{n}\omega_i^2},\ \beta=\frac{n-\gamma}{2\sum_{i=1}^{n}e_i^2}. \tag{2}$$

Where $\gamma=N-\alpha\, tr(H)^{-1}$ is the effective number of parameters, N is the total number of parameters in the network, H is the Hessian matrix determined by the equation $H=\beta J^TJ+\alpha I_N$ and J is the Jacobian matrix of performance function with respect to the weights. This performance function forces network response to be smooth and less likely to be over fit. L-M algorithm is used to approach second-order learning rate without having to compute the Hessian matrix. The Hessian matrix can be approximated as $H=\beta J^TJ+\alpha I_N$ and the gradient can be computed as $g=\beta J^Te+\alpha\omega$. L-M algorithm uses the approximation to Hessian matrix in the following weights update equation:

$$\omega_{k+1}=\omega_k-[H+\mu I_N]^{-1}g=\omega_k-[\beta J^TJ+\alpha I_N+\mu I_N]^{-1}(\beta J^Te+\alpha\omega_k). \tag{3}$$

Where μ is the scalar. When μ is zero, this is Newton's method. When μ is large, this becomes gradient descent with a small step size. The aim is to shift towards Newton's method as quickly as possible.

4 Simulation Examples

The study area is Songhua River Basin in northeast China, and Zhalong Wetland is taken as the central research area. GPS application on wetland guarantees the precision of positional information and spatial data. To display and track points' positions in local electrical maps, coordinates transformation between GPS data and local electrical maps is necessary. This paper collects sample data from Songhua River Basin, and conducts simulation experiments to validate the efficiency of neural network method.

Table 1. Parts of the learning points

Points	Latitude	Longitude	y (units: m)	x (units: m)
1	49.333	121.50	13510177.25	5485613.66
2	49.167	122.75	13649170.81	5467081.23
3	48.833	124.25	13815963.29	5430016.15
⋮	⋮	⋮	⋮	⋮

200 points is selected to train the network and 100 points to test. All these points are collected from the study area which takes Zhalong Wetland as center and 45°30 to 48°20 degrees in north latitude, 122°00 to 126°15 degrees in east longitude. Parts of the learning points are shown in Table 1. Before learning process, it is important to scale the inputs and targets so that they always fall within the range [-1, 1].

The number of input neurons is two according to latitude and longitude in geodetic coordinates system and the number of output neurons is two according to x-y in local coordinates system. The number of hidden layers is one and the number of hidden neurons is ten obtained by simulation experiments. The network structure is described as following: two input neurons, single hidden layer, ten hidden layer neurons and two output neurons. Weights of network are initialized as random numbers. The objective function parameters, α and β, are initialized as 0 and 1 respectively. The scalar μ is initialized as 0.005.

Neural network method of transformation is mainly realized by two steps. First, training the network according to samples, and second transforming test data using neural network with learning results. The learning and testing error curves of neural network with L-M and modified performance function are shown in Fig. 2 (a), where error is computed by mean sum of squares of network errors. To compare, gradient descent and standard performance function with the same learning data, testing data and network structure is adopted. The learning rate is set to 0.01. The learning and testing error curves of neural network with gradient descent and standard performance function are shown in Fig. 2 (b), where error is computed by mean sum of squares of network errors. Furthermore, L-M and standard performance function with the same learning data, testing data and network parameters is adopted. The learning and testing error curves of neural network with L-M and standard performance function are shown in Fig. 2 (c), where error is computed by mean sum of squares of network errors. Performance index of different methods is shown in Table 2.

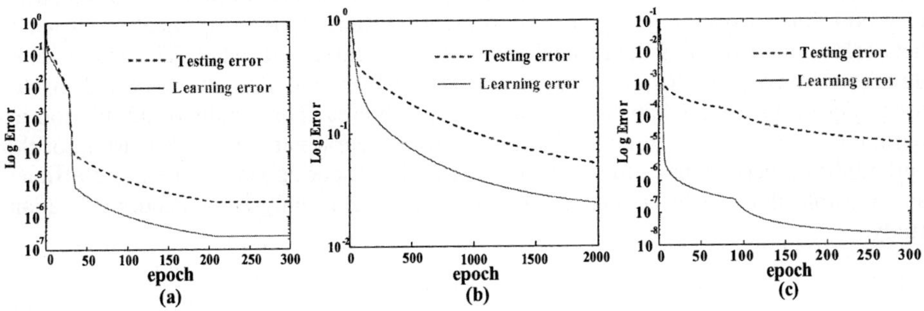

Fig. 2. Learning and testing error curves with (a) L-M and modified performance function; (b) gradient descent and standard performance function; (c) L-M and standard performance function

Table 2. Performance index of different methods

Methods	Testing error	Learning error	Computing time (units: s)	Epoch
(a)	1.91e-6	1.82e-7	0.54	300
(b)	0.35e-1	0.18e-1	3.23	2000
(c)	1.41e-5	1.58e-8	0.35	300

The error curves in Fig. 2 and performance index in Table 2 show that L-M algorithm is suitable to the GPS data transformation with fast convergence and the modified performance function can solve the over fitting problem.

Table 3. Transformation results of different methods (units: m)

Points	BJZ-54 local mapping grid system	L-M and modified performance function	Gradient descent and standard performance function	L-M and standard performance function	Traditional method
1	*x*:13621372.12 *y*:5485613.66	*x*:13621903.66 *y*:5483433.74	*x*:13651120.35 *y*:5407389.53	*x*:13645396.49 *y*:5448476.39	*x*:13652375.99 *y*:5539287.31
2	*x*:13621372.12 *y*:5467081.23	*x*:13621794.02 *y*:5465498.47	*x*:13654591.80 *y*:5387579.37	*x*:13637427.64 *y*:5441768.38	*x*:13835562.33 *y*:5499440.19
3	*x*:13649170.81 *y*:5430016.15	*x*:13649312.21 *y*:5429192.14	*x*:13645484.64 *y*:5365089.40	*x*:13651524.25 *y*:5425060.43	*x*:13530419.38 *y*:5389504.22
⋮	⋮	⋮	⋮	⋮	⋮

Table 3 is several points' transformation results with different methods including L-M and modified performance function, gradient descent and standard performance function, L-M and standard performance function and traditional method. Table 3 summarizes L-M and modified performance function can get better transformation results.

5 Conclusions

This paper focuses on the research of transformation method between local electrical map coordinates system and global coordinates system of GPS. Application of neural network method in GPS data transformation is discussed in detail. The study is undertaken on Songhua River Basin in northeast China and Zhalong Wetland in Heilongjiang Province is taken as the central research area. The results of simulation experiments show that the three layers feed-forward neural network with L-M algorithm and modified performance function adopted in this paper appears to be more efficient and feasible than traditional method in dynamic data analysis of Zhalong Wetland study.

Acknowledgements. This research is supported by the project (50139020) of the National Natural Science Foundation of China. It is also supported by the project (60374064) of the National Natural Science Foundation of China. All of these supports are appreciated.

References

1. Chapman, H.P., Noort, R.V.: High-Resolution Wetland Prospection, Using GPS and GIS: Landscape Studies at Sutton Common (South Yorkshire) and Meare Village East (Somerset). Journal of Archaeological Science, **28** (2001) 365-375
2. Nemenyi, M., Mesterhazi, P.A., Pecze, Zs., Stepan, Zs.: The Role of GIS and GPS in Precision Farming. Computers and Electronics in Agriculture, **40** (2003) 45-55
3. Muro-Medrano, P.R., Infante, D., Guillo, J., Zarazaga, J., Banares, J.A.: A CORBA Infrastructure to Provide Distributed GPS Data in Real Time to GIS Applications. Computers, Environment and Urban Systems, **23** (1999) 271-285
4. Soler, T.: A Compendium of Transformation Formulas Useful in GPS Work. Journal of Geodesy, **72** (1998) 482-490
5. Zhan, S.B., Zhang, Q.S.: The Algorithm and Implementation of Coordinate Transformation in GPS and DRM. Journal of Beijing University of Aeronautics and Astronautics, **22** (1996) 530-534
6. Xiang, H.K., Liu, X.M.: Data Communication, Transformation and Display in GPS, GIS and GSM. Traffic and computer, **19** (2001) 20-23
7. Foresee, F.D., Hagan, M.T.: Gauss-Newton Approximation to Bayesian Learning. Proceedings of the 1997 International Joint Conference on Neural Networks, (1997)

Dynamic File Allocation in Storage Area Networks with Neural Network Prediction

Weitao Sun, Jiwu Shu, and Weimin Zheng

Department of Computer Science & Technology, Tsinghua University
Beijing 100084, China
sunwt@tsinghua.edu.cn

Abstract. Disk arrays are widely used in Storage Area Networks (SANs) to achieve mass storage capacity and high level I/O parallelism. Data partitioning and distribution among the disks is a promising approach to minimize the file access time and balance the I/O workload. But disk I/O parallelism by itself does not guarantee the optimal performance of an application. The disk access rates fluctuate with time because of access pattern variations, which leads to a workload imbalance. The user access pattern prediction is of great importance to dynamic data reorganization between hot and cool disks. Data migration occurs according to current and future disk allocation states and access frequencies. The objective of this paper is to develop a neural network based disk allocation trend prediction method and optimize the disks' file capacity to their balanced level. A Levenberg-Marquardt neural network was adopted to predict the disk access frequencies with the I/O track. History. Data reorganization on disk arrays was optimized to provide a good workload balance. The simulation results proved that the proposed method performs well.

1 Introduction

Over the last decade, there has been an explosive growth of the Internet and data, which has led to a rapidly increasing demand for storage. Much effort has focused on improving distributed storage to provide better performance and scalability. The Storage Area Network (SAN) represents a new scheme to transfer data directly between the clients and the network storage system [1]. Disk arrays are used in the SAN to provide mass storage and high I/O performance. Data striping spreads data to multiple disks for parallel I/O processing and load balancing.

However, partitioning huge data objects into small chunks and distributing them onto storage devices do not guarantee the optimal performance of an application. Dynamic file data allocation in SAN disk arrays is being considered as a promising approach. Striping data across multiple disks has originally been proposed in [2] and [3]. The work in [4] and [5] presented file systems for dynamic data creation and reorganization in disk arrays. But none of the mentioned work has considered the fact that the disk allocation state and access frequencies tend to vary. Future disk load imbalance status prediction is of great importance for I/O optimization. For example,

F. Yin, J. Wang, and C. Guo (Eds.): ISNN 2004, LNCS 3174, pp. 719–724, 2004.

if there is an overloaded disk A with decreasing access frequency and an underloaded disk B with increasing access frequency, dynamic data migration from disk A to disk B may cause a greater load imbalance.

This work aims at predicting future disk status and access frequencies by neural network so as to guide the data reorganization. The look-ahead heuristic algorithm provides a good way to maximize the I/O performance and minimize the response time difference. In this study, a load generator and a simulated I/O system were implemented to provide more insights into the performance of this new algorithm.

2 Dynamic File Allocation Problem in the SAN

Static file allocation with balanced response time on a distributed multi-server system has been studied in [6]. The dynamic file allocation algorithm in [4] is based on the current allocation state and disk access frequencies. This paper proposes an extended file allocation algorithm with neural network prediction for dynamic data migrations in SAN storage systems. Fig. 1 shows the basic elements of the SAN framework and the data organization on the disks. The smallest data unit *block* is the minimal transfer unit and is required to be the same for all disks. A physically contiguous collection of logically consecutive blocks is a *run*. A physically contiguous collection of one or more runs on one disk is called an *extent*. An extent is described by its disk number, start address, and its size in blocks. Detailed information of the above definitions can be found in [4]. Our object was to allocate files dynamically to multiple disks so as to minimize the variation in disk response time.

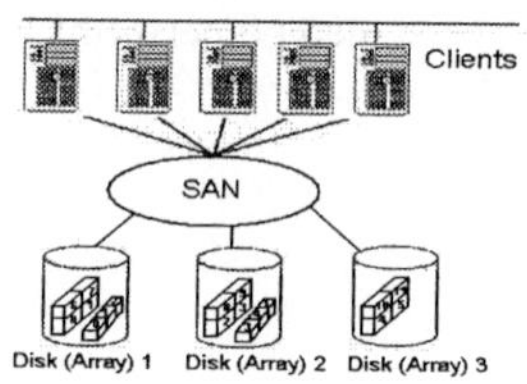

Fig. 1. The Storage Area Network (SAN) framework

Assume that files have been distributed and there is no replication of extents on the disks. The static file allocation response time analysis in [6] is extended to the dynamic data migration. Data requests are assumed to arrive at the disk according to a Poisson distribution. Each individual disk is modeled by an M/M/1 queue. When the response times are balanced among the disks, the mean response time R for any given disk i and disk j are equal. The optimal total file access rate that disk D_k can accommodate without exceeding a balanced response time is:

$$\lambda_{k\,opt} = s_k - \frac{s_{total} - f_{total}}{n_d}\,. \tag{1}$$

Here s_k is service rate of disk k. s_{total} is sum of the service rates of all disks. f_j is access rate of extent j. f_{total} is sum of the access rates of all extents.

The *file capacity* that disk D_k can further accommodate is defined as:

$$c_k = \lambda_{k\,opt} - \lambda_k \,. \tag{2}$$

Thus, $c_k > 0$ means that the disk can still accommodate more extents before reaching the optimal balanced response time; and $c_k = 0$ means that the server has reached the allocation limit; $c_k < 0$ means that the server is overloaded. If the disk response time is balanced, $c_k = 0$ for all *k*. λ_k is access rate of all extents assigned to disk *k*.

From the above analysis, the dynamic file allocation in a SAN storage system can be defined as a bounded optimization problem as: find a dynamic allocation, $X = \{x^i_{jk} \mid i = 1,...,n_r; j = 1...,n^i_e; k = 1,...,n_d\}$, where

$$x^i_{jk} = \begin{cases} 1 & \text{if extent } E_j \text{ is allocated to disk } D_k \\ 0 & \text{if extent } E_j \text{ is not allocated to disk } D_k \end{cases}, \tag{3}$$

so as to minimize

$$\max_{k=1,...,n_d} |c_k|, \tag{4}$$

subject to:

$$\sum_{k=1}^{n_r \cdot n^i_e} \lambda_k = f_{total}, \tag{5}$$

where $\lambda_k = \sum_{i=1}^{n_r} \sum_{j=1}^{n^i_e} x^i_{jk} f_j$. n_d is total disk number. n_r is total region number. n^i_e is the number of extents belong to region i.

Here λ_k and capacity factor c_k are dependent on both x^i_{jk} and f_j. At the same time, f_j and x^i_{jk} are influenced by the dynamic data migration caused by λ_k. Whether the dynamic data migration should be implemented is not only based on the current file capacity c_k but is also based on the predicted c_k' for the next time step. When the c_k<0 and $c_k' < c_k$, the disk *k* is overloaded and will be more overloaded in the future. In this case the data should be moved from this overloaded disk to other under-loaded disks. The file capacity c_k' and disk access rate λ_k' for the next time step is predicted by a neural network method. With the aid of NN prediction, file capacities of the disks are more close to zero, which provides a more balanced storage system.

3 Disk Status Prediction by Neural Network (NN)

Neural networks have been studied since 1943 [7]. Hopfield and Tank first used a neural network to solve an optimization problem [8]. The original description of the Levenberg-Marquardt algorithm was given in [9]. The application of Levenberg-Marquardt to neural network training was described in [10]-[12]. This algorithm appeared to be the fastest method for training moderate-sized feedforward neural networks (up to several hundred weights). In the present study, a Levenberg-Marquardt Neural Network (NN)-based predictor is implemented to provide the look-ahead parameters, such as the disk allocation state and disk access frequencies in the future. The disk status variation trend is one of the most important factors for dynamic file allocation. The outstanding advantage of NN prediction in this problem is that the unnecessary data migrations are prevented according to the file allocation trend and all the disks are utilized at their optimal performance level.

The NN based dynamic file allocation consists of three stages. First, the NN is trained through simulated problem sets of many different random access patterns and generates the look-ahead parameters. Secondly, the NN model is used to simulate the real time variation of the future disk access trend. Then the real disk access frequency becomes known as time passed, and the differences between the predicted and the real disk status values are used to retrain the NN. As a result, the NN model is dynamically corrected so as to match different user access patterns all the time.

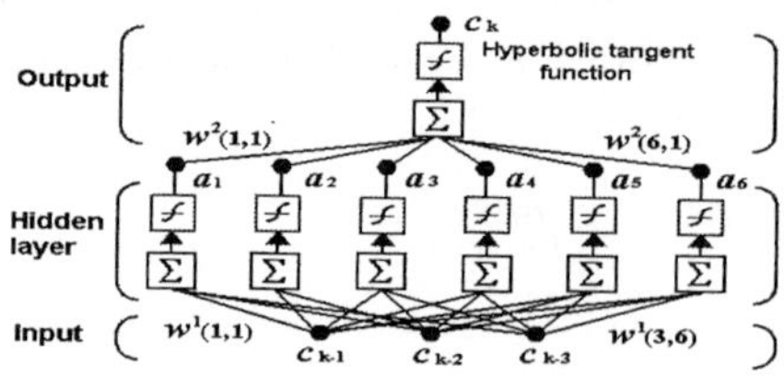

Fig. 2. The Levenberg-Marquardt neural network with one hidden layer. The disk file capacities at the three previous time steps *k*-3, *k*-2, *k*-1 are used as input values. The disk file capacity at the next time step *k* is predicted to guide the dynamic file allocation.

In order to predict the file capacity c_k, application of the Levenberg-Marquardt neural network was adopted (Fig. 2). The file capacity samples are passed forward from the input layer to the output layer through a hidden layer with six nodes. The number of the input nodes is three for the last three c_k in the history. A hyperbolic tangent function was used as transfer function at each node of the network. An explanation of the training and simulating process is provided in the following section.

4 Test of Dynamic File Allocation with Neural Network

A SAN storage system with five disks is simulated. The disk number and service rates were fixed to test the effect of data distribution and system load on the response time. The service rates of the five disks are listed in Table 1.

Table 1. Service rates of the five disks

Disk ID	D_1	D_2	D_3	D_4	D_5
Service rate (transaction/sec)	200	150	100	50	125

The total number of extents was fixed at 30 and the file access rates were simulated by a distribution pattern (Fig. 3 (a)) with a constant time shift for each extent. The extent access rate at each time point represents the mean rate within a time window with a certain length. First, the file allocation was fixed after the creation. The variation trend of the file capacity with time was simulated by neural network. With 300 times training, the simulation provided a very satisfactory file capacity prediction. The predicted file capacities of disk D_1 through current f_j and x_{jk} are shown in Fig. 3 (b).

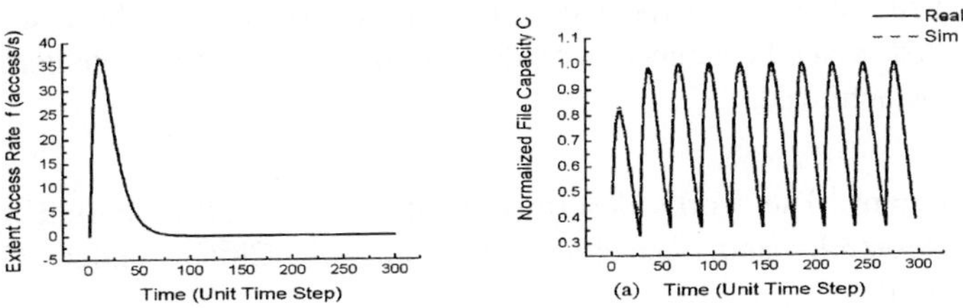

Fig. 3. (a) The extent access rate distribution pattern; **(b)** Comparison of real D_1 file capacity curve with the NN simulated D_1 file capacity curve for fixed file allocation method

Secondly, the dynamic file allocation was used to balance the load during the simulation. Extents were reorganized according to the disk file capacity variation at each time step. Fig. 4 (a) shows a good match of the fluctuation trend between the real file capacity and the NN predicted results.

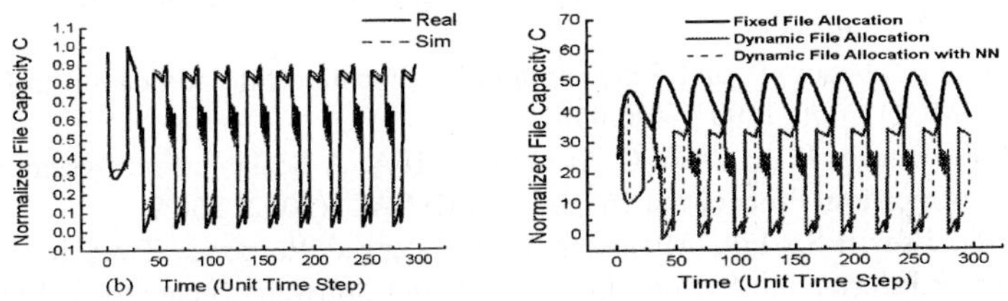

Fig. 4. (a) Comparison of real D_1 file capacity curve with the NN simulated D_1 file capacity for dynamic file allocation; **(b)** Comparison of simulated D_1 file capacities by fixed, dynamic and NN based dynamic file allocation algorithms.

Finally, the dynamic file allocation with NN prediction was used to minimize the response time difference between disks. The time that data immigration occurred was not only dependent on whether the disk was overloaded, but was also dependent on the criteria of c_k introduced in section 2. The NN based dynamic file allocation shows a better performance than the one without future disk status information. In Fig. 4 (b) the fixed, dynamic and NN based dynamic file allocation methods are compared. When the disk file capacity curve of fixed file allocation algorithm increases, the dynamic file allocation algorithm moves data from other overloaded disks to this disk and causes a decrease of the file capacity value. The neural network prediction algorithm provides a file capacity curve with the lowest values, which means that the disk load is further balanced.

5 Conclusions

A neural network based dynamic file allocation algorithm was proposed to minimize the difference of response time between disks in SAN storage systems. The disk allocation and file access information in the next time step was simulated by neural network. Data migration usually occurs only when a disk is overloaded, but it should also happen when a disk is expected to become overloaded in the near future. By predicting a disk's status, unnecessary data migration can be prevented, and more balanced can be achieved.

Acknowledgement. The works described in this paper are supported by National High-Tech Research and Development Plan of China under Grant (No. 2001AA111110).

References

1. Wilson, S.: Managing a Fibre Channel Storage Area Network. Storage Networking Industry Association White Paper (1998)
2. Kim, M. Y.: Synchronized Disk Interleaving. IEEE Trans. on Computers, Vol. C-35, No. 11 (1986) 978-988
3. Salem, K., and Garcia-Molina, H.: Disk Striping. Proc. of the 2nd Int. Conf. on Data Engineering, (1986) 336-342
4. Weikum, G., Zabback, P. and Scheuermann P.: Dynamic File Allocation in Disk Arrays. ACM SIGMOD Conf., (1991) 406-415
5. Tsai, H. P., Huang, J. L., Chao, C. M., et al.: SIFA: A Scalable File System with Intelligent File Allocation. Proc. of the 26th Int. COMPSAC, (2002) 26-29
6. Cai, W.T., Lee, B.S.: File Allocation with Balanced Response Time in a Distributed Multi-server Information System. Information and Software Technology, Vol. 40 (1998) 27-35
7. McCulloch, W.S., Pitts, W.H.: A Logical Calculus of Ideas Imminent in Nervous Activity. Bullet. Math. Biophys., Vol. 5, No.1 (1943) 15-33
8. Hopfield, J., Tank, D.W.: Neural Computation of Decision in Optimization Problems. Biol. Cybernet. Vol. 52, No.1 (1985) 41-52
9. Marquardt, D.: An Algorithm for Least-Squares Estimation of Nonlinear Parameters, SIAM Journal Applied Math. Vol. 11 (1963) 431-441
10. Hagan, M. T., Menhaj, M.: Training Feedforward Networks with the Marquardt Algorithm, IEEE Trans. on Neural Networks, Vol. 5, No. 6 (1994) 989-993
11. Ampazis, N., Perantonis, S. J.: Two Highly Efficient Second-Order Algorithms for Training Feedforward Networks. IEEE Trans. on Neural Networks, Vol. 13, No.5 (2002) 1064-1074
12. Peng, H., Ozaki, T., Haggan-Ozaki, V., et al.: A Parameter Optimization Method for Radial Basis Function Type Models. IEEE Trans. on Neural Networks,Vol.14, No.2 (2003) 432-438

Competitive Algorithms for Online Leasing Problem in Probabilistic Environments

Yinfeng Xu[1,2] and Weijun Xu[2]

[1] School of Management, Xi'an Jiaotong University,
Xi'an, Shaan'xi, 710049, P.R. China
xuweijun75@163.com

[2] The State Key Lab for Manufacturing Systems Engineering,
Xi'an, Shaan'xi, 710049, P.R. China
Yfxu@mail.xjtu.edu.cn

Abstract. We integrate probability distribution into pure competitive analysis to improve the performance measure of competitive analysis, since input sequences of the leasing problem have simple structure and favorably statistical property. Let input structures be the characteristic of geometric distribution, and we obtain optimal on-line algorithms and their competitive ratios. Moreover, the introducing of interest rate would diminish the uncertainty involved in the process of decision making and put off the optimal purchasing date.

1 Introduction

In recent years, competitive analysis has been gaining recognition for being a complementary approach in the analysis of algorithmic decision-making under uncertainty. With the emergence of many on-line financial papers, the competitive approach is shown to be productive for a variety of financial problems. For on-line leasing problem, the prototype was the well-known "Ski-Rental" example put forward by Karp in the field of theoretical computer science in 1992 [9]. Subsequently, a series of research has been carried out on the basic model. In 1994, Karlin et al. made on-line analysis for what they called "the ski-rental-family of problem" [8]. In 1998, S. Irani et al. studied the situation which the purchasing price varies but the rental cost stays at a fixed price [7]. In 1999, R. El-Yaniv et al. investigated the leasing problem with interest rate [4]. In the same year, S. al-Binali developed a famous Risk–Reward Framework to analyze the rental problem and unidirectional trading problem [2]. In 2001, S. Albers et al. introduced and explored natural delayed information and action models to investigate several well-known on-line problems inclusive of the rental problem [1]. However, the previous research always avoids probabilistic assumption intentionally. In 2002, H. Fujiwara et al. firstly integrated probabilistic distribution into pure competitive analysis to study on-line leasing problem, while they suppose that input sequences are drawn form the exponential distribution [6].

F. Yin, J. Wang, and C. Guo (Eds.): ISNN 2004, LNCS 3174, pp. 725–730, 2004.

2 Our Contributions

The purpose of this study is to improve the performance measure of competitive analysis by integrating distribution information into pure competitive analysis for the leasing problem. Similar to the results in [6], we also obtain several interesting results as follows. If the average cost of always leasing is less than the purchasing cost, the optimal strategy for an investor is to lease the equipment forever. Otherwise, the optimal strategy is to purchase the equipment after leasing several periods, which the optimal purchasing data would be determined by using the dichotomous search algorithm in the polynomial time. Furthermore, we introduce the nominal interest rate on the market into the model. It could be found that the introducing of interest rate would diminish uncertainty involved in the process of decision making and put off the purchasing date. Although this is only one step toward a more realistic solution of the problem, the introduction of this parameter considerably complicates the analysis and also arises some new issues that do not exist in the Fujiwara's model with no interest rate.

Moveover, the reasons based on such consideration are as follows. Pure competitive analysis always assumes that an investor has no information for input sequences. Indeed, whenever a decision-maker does have some side information or partial (statistical) knowledge on the evolution of input sequences it would be a terrible waste to ignore it, which is precisely what the competitive ratio does. In this case the use of competitive algorithms may lead to inferior performance relative to Bayesian algorithms. Moreover, it is hardly true that competitive analysis of the worst case intentionally emphasizes on difficulty to estimate the input distribution. Other than many combinatorial problems with more complicated input structures, there do exist a number of interesting problems with relatively simple and tractable input structures. We can characterize accurately their input structures by using statistical theory. Hence, stochastic competitive analysis in this paper, as well as in [6], might help to overcome these difficulties.

The reason that we consider the on-line rental problem with geometric distribution comes from the literature [10] which analyzes a class of optimal stopping of geometric distribution and from the "tossing coin" idea that the leasing doesn't cease until the purchasing appears. However, the literature [6] highlights the continuous model with the exponential distribution, while we study a discrete model that assumes the evolution of input sequences subject to the characteristic of geometric distribution. Maybe, we could also amend several aspects of the literature [6] as follows. (i) The continuous model could be unnecessarily equivalent to the discrete case because the leasing problem is essentially discrete; (ii) Geometric distribution may be more reasonable than exponential distribution to depict input structures of the leasing problem, because the leasing activity every period is similar to doing the Bernoulli trial what on earth to rent continuously or to purchase immediately; (iii) In our discrete model, the immediately purchasing at the beginning, i.e. the strategy $A(0)$, could be also an optimal strategy in practice, and the competitive ratio $C(0)$ is a finite value, while it could be not the case that the literature [6] pointed out the competitive ratio $c(k)$ diverging to $+\infty$ as k approaching to zero.

3 Optimal Analysis of Online Algorithms

For on-line problem, we consider the following deterministic on-line strategies $A(k)$ $(k = 0, 1, 2, \cdots)$: rent up to k times and then buy. Thus let $Cost_{ON}$ be the optimal cost of on-line algorithms, and let $Cost_{OPT}$ be the optimal cost of off-line algorithms. In this paper, we consider the discrete model. Therefore the conception of the stochastic competitive ratio is defined as follows.

Definition 1. Let the number of the leasing be a stochastic variable X which is subject to some type of probability distribution, and the probability function is $P(X = t)$, where t is the number of actual leasing. Then the discrete stochastic competitive ratio is defined as

$$C(k) = E_X \frac{Cost_{ON}(X,k)}{Cost_{OPT}(X)} = \sum_{t=0}^{\infty} \frac{Cost_{ON}(t,k)}{Cost_{OPT}(t)} P(X = t), \tag{1}$$

where $P(X = t)$ is a probability function that investors approximately estimate for input structures. We consider that the inputs are drawn from the geometric distribution, and let the hazard rate of continuous renting in every period activity be θ, and then the probability function is $P(X = t) = (1 - \theta)\theta^{t-1}$ $(t = 0, 1, 2, 3, \cdots)$.

Note that there is an essential difference of definitions between the stochastic competitive ratio in this paper and the randomized competitive ratio in the literature [9], [4]. The former indicates that input structure may be subject to some probability distribution, i.e. investor has certain information distribution, while the latter means that on-line players or adversary players choose some strategy randomly in the strategy space set. Thereby we use different terms to imply different meanings.

3.1 Leasing in a Market Without Interest Rate

Let the costs of renting and purchasing equipment be 1 and positive integer s, respectively. Obviously, optimal off-line decision-making cost is

$$Cost_{OPT}(t) = \min\{s, t\}. \tag{2}$$

Based on the strategy set $A(k)$, on-line decision-making cost is

$$Cost_{ON}(t,k) = \begin{cases} t & t \le k, \\ k+s & t > k. \end{cases} \tag{3}$$

Obviously, the optimal strategy is immediately purchasing if s were equal to 1, so s is at least 2. We also make the assumption that the player needs the equipment throughout n contiguous time periods.

According to (1), (2), and (3), we could obtain that, for k=0, 1, 2, 3, $\cdots$, s,

$$C(k) = (1 - \theta^k) + (k+s)(1-\theta) \sum_{t=k+1}^{s} \frac{1}{t}\theta^{t-1} + \frac{k+s}{s}\theta^s, \tag{4}$$

and for $k = s+1, s+2, s+3, \cdots$,

$$C(k) = (1-\theta^s) + \frac{1-\theta}{s}\sum_{t=s+1}^{k} t\theta^{t-1} + \frac{k+s}{s}\theta^k. \quad (5)$$

Then we obtain the following result by theoretical analysis.

Theorem 1. 1). If $\frac{1}{1-\theta} < s$, then the average cost of always leasing is less than the purchasing cost s. The optimal strategy for an investor is to lease the equipment forever, and the competitive ratio is $1 + \frac{\theta^s}{s(1-\theta)}$;

2). If $\frac{1}{1-\theta} = s$, then the average cost of always leasing is equal to the purchasing cost s. The optimal strategy for an investor is to purchase the equipment after leasing $s-1$ periods, and the competitive ratio is $1 + (1-\frac{1}{s})^s$;

3). If $\frac{1}{1-\theta} > s$, then the average cost of always leasing is greater than the purchasing cost s. The optimal strategy for an investor is to purchase the equipment after leasing k_0 periods, and the competitive ratio is $1 - [1 - \frac{k_0 s(1-\theta)}{k_0+1} - \frac{s^2(1-\theta)}{k_0+1}]\theta^{k_0}$, where k_0 satisfies $(1-\theta)s^2 - 0.09s - 1 < k_0 < (1-\theta)s^2 - 1$, which the decision-making data k_0 must be determined by using the dichotomous search algorithm in the polynomial time $O(\log s)$;

4). If $\frac{1}{1-\theta} \to \infty$, then the average cost of always leasing limits to ∞, and the optimal competitive ratio of any strategy $A(k)$ is $1 + \frac{k}{s}$. The optimal strategy for an investor is to purchase the equipment at the very beginning, and the competitive ratio approaches to 1.

Whichever case to consider, the competitive ratio that the investor takes the strategy $A(s-1)$, even when there is a large deviation for the hazard ratio θ to be estimated, also is better than the deterministic competitive ratio $2 - \frac{1}{s}$ in [9], and the randomized competitive ratio in [4]. For example, if $s = 10$ and $\theta = 0.95$, then the competitive ratio 1.56722 in this paper is better than the competitive ratio 1.9 in [9], and better than the randomized competitive ratio 1.582 in [4].

3.2 Leasing in a Market with Interest Rate

When considering alternative financial decisions, an agent must consider their net present value, that is, accounting for the market interest rate is an essential feature of any reasonable financial model. Hence, let i be the nominal interest rate in the financial market. Without loss of generality we assume that $\frac{1}{s} > \frac{i}{1+i}$. This is a reasonable assumption for any practical use because the purchase price of the equipment must be less than the present discount value of the alternative of always leasing ($s < \sum_{j=0}^{\infty} \frac{1}{(1+i)^j}$). Otherwise, the online player can attain a competitive ratio of 1 by simply never purchasing the equipment. Set $\beta = \frac{1}{1+i}$, and then $\frac{1}{s} + \beta - 1 > 0$. From economic point of view, $\frac{1}{s} + \beta - 1$ is relative opportunity cost to purchase equipment. In addition, let $\triangle = (\frac{1}{s} + \beta - 1)^{-1}$.

Clearly, the adversary player will never purchase the equipment after leasing it for some time (as in [4]). Therefore, for any n optimal offline decision-making cost is

$$Cost_{OPT}(t) = \begin{cases} \frac{1-\beta^t}{1-\beta} & t \le n^*, \\ s & t > n^*, \end{cases} \tag{6}$$

where n^* is the number of rentals whose total present value is s. In other words, n^* is the root of $\frac{1-\beta^n}{1-\beta} = s$. That is $n^* = \frac{\ln(1-s(1-\beta))}{\ln \beta}$. Based on the strategy set $A(k)$, on-line decision-making cost is

$$Cost_{ON}(t,k) = \begin{cases} \frac{1-\beta^t}{1-\beta} & t \le k, \\ s\beta^k + \frac{1-\beta^k}{1-\beta} & t > k. \end{cases} \tag{7}$$

According to (1), (6), and (7), we could obtain that, for k=0, 1, 2, 3, $\cdots$, n^*,

$$C(k) = (1-\theta^k)+(1-\theta)(1-\beta^{n^*+k}) \sum_{t=k+1}^{n^*} \frac{1}{1-\beta^t}\theta^{t-1} + (\beta^k + \frac{1-\beta^k}{s(1-\beta)})\theta^{n^*}, \tag{8}$$

and for $k = n^*+1, n^*+2, n^*+3, \cdots$,

$$C(k) = (1-\theta^{n^*}) + \frac{1-\theta}{s(1-\beta)} \sum_{t=n^*+1}^{k} (1-\beta^t)\theta^{t-1} + (\beta^k + \frac{1-\beta^k}{s(1-\beta)})\theta^k. \tag{9}$$

Then we obtain the following result by theoretical analysis.

Theorem 2. 1). If $\frac{1}{1-\theta} < \frac{\triangle}{1+i}$, then the average cost of always leasing without interest rate is less than the present discount value of the reciprocal of relative opportunity cost. The optimal strategy for an investor is to lease the equipment forever, and the competitive ratio is $1 + \frac{1+\beta(1-2\theta)}{\triangle(1-\beta)(1-\beta\theta)}\theta^{n^*}$;

2). If $\frac{1}{1-\theta} = \frac{\triangle}{1+i}$, then the average cost of always leasing without interest rate is equal to the present discount value of the reciprocal of relative opportunity cost. The optimal strategy for an investor is to buy the equipment after n^*-1 periods, and the competitive ratio is $1 + (\frac{\theta}{1+i})^{n^*}$;

3). If $\frac{1}{1-\theta} > \frac{\triangle}{1+i}$, then the average cost of always leasing without interest rate is greater than the present discount value of the reciprocal of relative opportunity cost. The optimal strategy for an investor is to buy the equipment after k_0 periods, and the competitive ratio is $1 + \frac{s\beta(1-\theta)(1-\beta^{n^*+k_0})-\beta^{n^*}(1-\beta^{k_0+1})}{\beta^{n^*}(1-\beta^{k_0+1})}\theta^{k_0}$, where the decision-making data k_0 is established by using the dichotomous search algorithm in the polynomial time $O(\log n^*)$;

4). If $\frac{1}{1-\theta} \to \infty$, i.e. the average cost of always leasing without interest rate limits to $+\infty$, then the optimal competitive ratio of any strategy $A(k)$ is $\frac{1}{s(1-\beta)} + (1 - \frac{1}{s(1-\beta)})\beta^k$. The optimal strategy for an investor is to purchase the equipment at the very beginning, and the competitive ratio approaches to 1.

Note that Theorem 2 is the further extension of Theorem 1. If $i \to 0$, then $n^* \to s$, and $\frac{\triangle}{1+i} \to s$. Similar to Theorem 1, if $s = 10$, and $\theta = 0.95$, and $i = 0.01$, then the competitive ratio of non-optimal strategy $A(9)$ that is 1.4983 is better than the competitive ratio 1.9 in [9], and better than the randomized competitive ratio 1.582 in [4], and better than the stochastic competitive ratio

1.56722 without interest rate. Moreover, it could be found that the entrance of interest rate diminishes uncertainty involved in financial decision making and puts off optimal purchasing date. For example, if $s = 19$ and $\theta = 0.98$, the introducing of interest rate $i = 0.02$ results in that the competitive ratio reduces, and that the optimal strategy $A(6)$ with no interest rate is postponed to become the optimal strategy $A(11)$ with interest rate.

4 Concluding Remarks

Although Ran El-Yaniv proposed the axiom set of the competitive ratio, the conception has still inherent and insurmountable limitations [3]. It is still the subject of worthy attention how to improve the performance measure of the competitive ratio by combining other methods. Furthermore, it also is the subject of research how to depict input information under uncertainty, while we think that the theory of Rough set and possibility distribution may be an useful tool to be integrated into pure competitive analysis to improve the performance measure of on-line algorithms in the future.

Acknowledgements. We would like to express our appreciation to Prof. Lili Wei and Dr. Jianjun Wang for the helpful discussion during the research. Moreover, we would also like to acknowledge the support No. 10371094, 70121001 from the NSF of China.

References

1. Albers, S., Charikar, M., Mitzenmacher, M.: On Delayed Information and Action in On-line Algorithms. Information and Computation. **170** (2001) 135–152
2. al-Binali, S.: A Risk–Reward Framework for the Competitive Analysis of Financial Games. Algorithmica. **25** (1999) 99–115
3. Borodin, A., El-Yaniv, R. (eds.): On-line Computation and Competitive Analysis. Cambridge University Press (1998)
4. El-Yaniv, R., Kaniel, R., Linial, N.: Competitive Optimal On-line Leasing. Algorithmica. **25** (1999) 116–140
5. Fiat, A., Woeginger, G.J. (eds.): On-line Algorithms: the State of Art. Springer (1998)
6. Fujiwara, H., Iwama, K.: Average-case Competitive Analyses for Ski-rental Problems. ISAAC'02 (2002) 476–488
7. Irani, S., Ramanathan, D.: The Problem of Renting versus Buying. Personal communication (1998)
8. Karlin, A.R., Manaees, M.S., McGeogh, L., Owichi, S.: Competitive Randomized Algorithms for Nonuniform Problems. Algorithmica. **11** (1) (1994) 542–571
9. Karp, R.: On-line Algorithms versus Off-line Algorithms: How Much Is It Worth to Know the Future? Proc. IFIP 12th world computer congress. **1** (1992) 416–429
10. Wei, L.L., Zhang, W.X.: A Class of Bayesian Stopping and Decision Rule of Geometric Distribution. Acta Mathematicae Applicatae Sinica. **26** (1) (2003) 181–185

Wavelet Network for Nonlinear Regression Using Probabilistic Framework

Shu-Fai Wong and Kwan-Yee Kenneth Wong

Department of Computer Science and Information Systems
The University of Hong Kong, Hong Kong
{sfwong, kykwong}@csis.hku.hk

Abstract. Regression analysis is an essential tools in most research fields such as signal processing, economic forecasting etc. In this paper, an regression algorithm using probabilistic wavelet network is proposed. As in most neural network (NN) regression methods, the proposed method can model nonlinear functions. Unlike other NN approaches, the proposed method is much robust to noisy data and thus over-fitting may not occur easily. This is because the use of wavelet representation in the hidden nodes and the probabilistic inference on the value of weights such that the assumption of smooth curve can be encoded implicitly. Experimental results show that the proposed network have higher modeling and prediction power than other common NN regression methods.

1 Introduction

Regression have long been studied in statistics [1]. It receives attention from researchers because of its wide range of applications. These applications include signal processing, time series analysis and mathematical modeling etc.

Neural network is a useful tools in regression analysis [2]. Given any input signal $\{x_i, t_i\}_{i=1}^{N}$, neural network was found to be capable to estimate the nonlinear regression function $f(\cdot)$ such that the equation, $x_i = f(t_i) + \epsilon$, hold, where ϵ is the model noise. It outperforms many linear statistical regression approaches because it makes very few assumptions, such as linearity and normality assumptions, as other statistical approaches do [3]. It was also found that nonlinear neural network exhibits universal approximation property [4].

Researchers have proposed many neural networks for solving regression problem during past decades. For instance, multilayer network (MLP) [5], and Radial basis function networks (RBFN) [6] have been explored in previous research. Although the universal approximation property seems to be appealing, it may cause problem in regression especially when data contains heavy noise. Noisy data is quite common in applications such as financial analysis, signal processing etc.

Wavelet denoising and regression have been proposed to handle the noisy data [7]. The idea of using wavelet in noisy data regression is that the signal is firstly broken down into constituent wavelets, and those important wavelets are then chosen to reconstruct back the denoised signal. The signal without irrelevant wavelets or pulses is then used to approximate the nonlinear function

F. Yin, J. Wang, and C. Guo (Eds.): ISNN 2004, LNCS 3174, pp. 731–736, 2004.

$f(\cdot)$. Such approach has been integrated with neural network to form wavelet network by some researchers such as [8]. However, the problem of over-fitting still remains because the number of hidden nodes is unknown before regression. If the number of hidden nodes is more than enough, over-fitting still occurs.

Inspired by the probabilistic framework of neural network [9] and Bayesian model for wavelets [10] presented recently, probability framework for wavelet network (wavenet) is proposed in this paper. Under this framework, the weights in the wavenet will be updated according to the prior assumption of the model simplicity and smooth curve, instead of minimizing the square error as in other regression methods. Due to the above assumption, the final curve will be denoised in certain sense such that it is smooth and is also best fitted to all data points although the number of hidden nodes can be infinite initially.

2 Probabilistic Wavenet

As described in previous section, the proposed probabilistic wavenet aims at modeling the nonlinear function or the series of input data without the problem of over-fitting. In order to model the nonlinear function, simple wavelet analysis is performed to find out the constituent wavelets of the input series or signal. To be noise tolerant, wavelet denoising will be performed to remove those unimportant wavelets. By using the proposed probabilistic wavenet, these two steps can be performed automatically at once with high accuracy and in short time.

2.1 Wavelet Network

Wavelet network has been used for regression since its introduction [8]. The data set is in the form of time series data $D = \{x_i, t_i\}_{i=1}^{N}$, the wavenet will estimate the regression function $f(\cdot)$ for $X = \{x_i\}$ and $T = \{t_i\}$ by the regression model: $X = f(T) + \epsilon$.

In wavenet, the function is indeed represented by wavelet composition: $f(t_i;\omega) = \sum_{j=1}^{K} \omega_j \psi_j(t_i)$, where ω_j and ψ_j are the wavelet coefficients and the wavelet functions respectively.

To limit the number of wavelets to be used, dyadic wavelet is used. In other words, the wavelet functions are constructed by translating and scaling the mother wavelet as: $\psi_{m,n}(t) = 2^{-m/2}\psi(2^{-m}t - n)$, where m and n are the scaling and translating factor respectively. Given the signal size N, m is ranged from 1 to $log_2(N)$ and n is ranged from 0 to $2^{-m}N$.

In the proposed system, the mother wavelet function is set as the "Mexican Hat" function: $\psi(t_i) = (1 - (\frac{||t_i-\mu_\psi||}{\sigma_\psi})^2)exp(\frac{-||t_i-\mu_\psi||^2}{2\sigma_\psi{}^2})$, where μ_ψ and σ_ψ are the transition and scale factor of the mother wavelet.

Given a finite number of wavelet functions $\psi_{m,n}$, regression is done by finding optimal solution set of wavelet coefficients (ω_j). In most neural network applications, the value of such wavelet coefficients or weights are estimated by minimizing the total error as shown:

$$\omega^* = min_\omega\{\sum_{i=1}^{N}(x_i - \sum_{j=1}^{K}(\omega_j\psi_j(t_i)))^2\} \tag{1}$$

Though it is usually possible to estimate the value of ω using common optimization methods, over-fitting may occur. To overcome such problem, probabilistic inference is proposed to estimate the value of the wavelet coefficients.

2.2 Probabilistic Framework

In order to handle the problem of over-fitting because of noisy data, probabilistic framework is adopted to estimate the value of the wavelet coefficients (or weights in wavenet) and thus estimate the original signal (or smoothened signal).

As described in previous subsection, wavenet perform regression analysis on time series data. Using the same set of notation as above, the regression model and the wavelet composition model can be combined in matrix form as $X = \Psi\omega + \epsilon$, where Ψ is the $N \times K$ design matrix formed from the wavelet functions and ω is the weight vector.

With reference to the regression model, at any time t_i, the probability of having signal value x_i or the likelihood of the model is given by:

$$p(x_i|t_i) \sim N(f(t_i), \sigma^2) \tag{2}$$

where $\epsilon \sim N(0, \sigma^2)$ in the regression model.

To be more specific, the matrix form of the regression model can be used in expressing the probability of having data X given certain wavenet:

$$p(X|\omega, \sigma^2) = (2\pi\sigma^2)^{\frac{-N}{2}} exp\{-\frac{1}{2\sigma^2}||X - \Psi\omega||^2\} \tag{3}$$

In common neural network, the weights (ω) are found by error minimization and thus may cause over-fitting. In contrast, hyperparameter (α) is introduced in probabilistic wavenet to limit the number of wavelets with large weight value. The distribution of the value of weights is proposed as the following:

$$p(\omega|\alpha) = \Pi_{i=1}^{K} N(\omega_i|0, \alpha_i^{-1}) \tag{4}$$

From above, it is clear that the mean value of all weights is set to be zero. Thus, there is a preference to have fewer constituent wavelets. Those constituent wavelet should have large variance (α_i^{-1}). This conditional probability is served as the prior probability in estimating optimal wavelet coefficient such that the preference of fewer number of constituent wavelets is included in the estimation.

Given the weight (ω), the hyperparameter(α), the system noise (σ) and the time series data $D = (x_i, t_i)$ (or $X = \{x_i\}$), the prediction can also be done. Prediction is represented as the following through marginalization:

$$p(x^*|X) = \int\int\int p(x^*|\omega, \alpha, \sigma^2)p(\omega, \alpha, \sigma^2|X)d\omega d\alpha d\sigma^2 \tag{5}$$

Estimation of the value of weights, hyperparatmeters and system noise is indeed done by maximizing the prediction power of the regression model.

The second term can be expressed as:

$$p(\omega, \alpha, \sigma^2|X) = p(\omega|X, \alpha, \sigma^2)p(\alpha, \sigma^2|X) \tag{6}$$

From above, the posterior probability of having certain weight can be now represented as:

$$p(\omega|X, \alpha, \sigma^2) = \frac{p(X|\omega, \sigma^2)p(\omega|\alpha)}{p(X|\alpha, \sigma^2)} \tag{7}$$

Suppose the normalizing term above follows Gaussian distribution, the posterior probability can be simplified using Equation (3) and Equation (4):

$$p(\omega|X, \alpha, \sigma^2) = (2\pi)^{-\frac{N+1}{2}}|\Sigma|^{-\frac{1}{2}}exp\{-\frac{1}{2}(\omega-\mu)^T\Sigma^{-1}(\omega-\mu)\} \tag{8}$$

where $\Sigma = (\sigma^{-2}\Psi^T\Psi + A)^{-1}$, $\mu = \sigma^{-2}\Sigma\Psi^T X$ and $A = diag(\alpha_1, ..., \alpha_N)$. This gives expected value of weights, μ.

On the other hand, the second term in Equation (6) can be expressed as:

$$p(\alpha, \sigma^2|X) \propto p(X|\alpha, \sigma)p(\alpha)p(\sigma^2) \tag{9}$$

It is possible to consider the likelihood function alone to obtain the optimal values of α and σ. Here is the likelihood expression:

$$\begin{aligned} p(X|\alpha, \sigma^2) &= \int p(X|\omega, \sigma^2)p(\omega|\alpha)d\omega \qquad (10) \\ &= (2\pi)^{-\frac{N}{2}}|\sigma^2 I + \Psi A^{-1}\Psi^T|^{-\frac{1}{2}}exp\{-\frac{1}{2}X^T(\sigma^2 I + \Psi A^{-1}\Psi^T)^{-1}X\} \end{aligned}$$

According to [9], the optimal value (α_{MP} and σ_{MP}) can be obtained by iteration:

$$\begin{aligned} \alpha_i^{new} &= \frac{\gamma_i}{\mu_i^2} \\ (\sigma^2)^{new} &= \frac{||X - \Psi\mu||^2}{N - \sum_i \gamma_i} \end{aligned} \tag{11}$$

where $\gamma_i = 1 - \alpha_i \Sigma_{ii}$.

After estimated the value of ω, α and σ at each stage, the prediction of the trend of the signal can be made as the following:

$$\begin{aligned} p(x^*|X, \alpha_{MP}, \sigma^2_{MP}) &= \int p(x^*|\omega, \sigma^2_{MP})p(\omega|X, \alpha_{MP}, \sigma^2_{MP})d\omega \\ &\sim N(x^*|\mu_*, \sigma^2_*) \qquad (12) \end{aligned}$$

where $\mu_* = \mu^T\psi(t_{N+1})$ and $\sigma^2_* = \sigma^2_{MP} + \psi(t_{N+1})^T\Sigma\psi(t_{N+1})$. The predicted value ($\mu_*$) and its variance ($\sigma^2_*$) is thus obtained.

In regression, the whole inference process repeats with the new values of wavelet coefficients, the new values of both hyperparameters and system noise are evaluated using Equation (8) and Equation (11) respectively until the equilibrium state is achieved. In making prediction, the predicted value and its variance can be obtained using Equation (12).

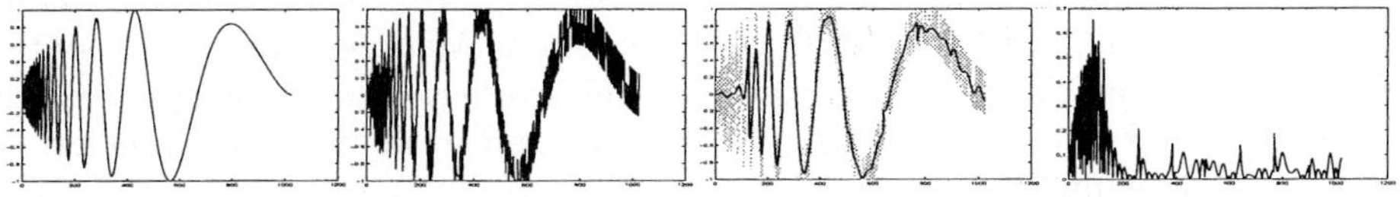

(a) (b) (c) (d)

Fig. 1. In (a), it shows the original Doppler function. In (b), it shows the Doppler function contains 10dB noise. In (c), it shows the denoised signal in black solid line compare with the noisy signal in cyan dotted line. In (d), it shows the relative square error with peak value 0.67 and average value 0.153.

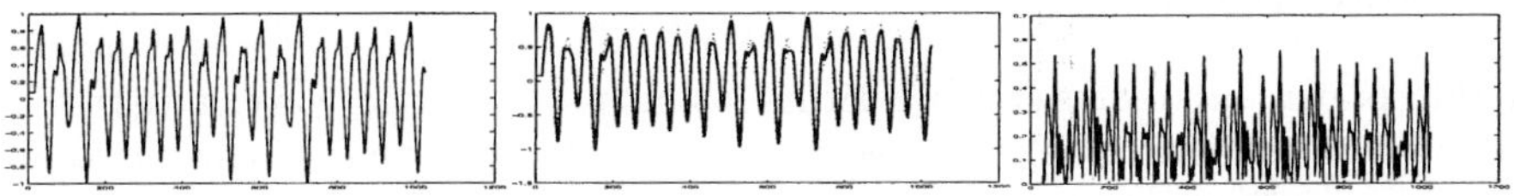

(a) (b) (c)

Fig. 2. In (a), it shows the Mackay-Glass series. In (b), it shows the prediction curve in black solid line compare with the noisy signal in cyan dotted line. In (c), it shows the relative square error with peak with 0.56 and average value 0.317.

3 Experimental Result

The proposed system was implemented using Visual C++ under Microsoft Windows. Two experiments was performed to test the system. The experiments were done on a P4 2.26 GHz computer with 512M ram running Microsoft Windows.

In the first experiment, the denoising and modeling power of the proposed network was tested. A Doppler function ($f(t) = \sqrt{t(1-t)}sin(2\pi(1+a)/t+a))$ with $a = 0.05$) which contains additive noise (10dB) was used in this experiment. This function has been commonly used in research in wavelet denoising such as [10]. Noisy signal generated from such function which with signal size 1024 was analysed by the network. The result of Doppler function modeling is shown in Figure 1. The average relative square error $((|x_{original} - x_{denoised}|^2)/(x^2_{original}))$ is 0.153. The processing time is around 4 seconds. Except the highly oscillated region in the begining of the signal (up to signal point 180), the relative square in later part is usually lower than 2. It shows that the modeling power of the proposed network is good without susceptible to noise.

In the second experiment, the prediction power of the proposed network was tested. Mackay-Glass chaotic time series were used in this experiment. The Mackay-Glass series ($x_{t+1} = (0.2x_{t-\Delta})/(1+(x_{t-\Delta})^{10}) + 0.9x_t$ with $\Delta = 17$, $x_t = 0.9$ for $0 \le t \le 17$) has been used in prediction test for a long time and a comparative study in regression methods using such series can be found in [11]. In the experiment, a range of data of size 64 was analysed by the network each time and the prediction is made at the time slot 65. Prediction result is obtained by performing predictions using the appropriate range of data shifting along the input signal (with size 1024) generated from Mackey-Glass series stated above. The result is shown in Figure 2. The average relative square error is 0.317. The processing time is usually less 1 second in making each prediction. The normalized prediction error ($\varepsilon = (\sum_t (x_t - x_t^{predict})^2)/(\sum_t (x_t - x_{mean})^2)$) is 0.4777%. This result is relatively lower than the result given by other neural network approaches or linear approach as indicated in [11] (normalized error of MLP is 1.0%, those of RBF is 1.1%, those of polynomial fitting is 1.1%, those of local linear fitting is 3.3%). It shows that the prediction power of the proposed system is better than those of the other common neural network approaches.

4 Conclusions

In this paper, an regression algorithm using probabilistic wavelet network is proposed to perform nonlinear regression reliably such that it can be applied to real life applications such as economic forecasting. Experimental results show that the proposed network have relatively high modeling power and prediction power compare with common neural network regression methods. The proposed method can model nonlinear functions reliably without susceptible to data noise.

References

1. Fox, J.: Multiple and Generalized Nonparametric Regression. Sage Publications, Thousand Oaks CA (2000)
2. Stern, H.S.: Neural networks in applied statistics. Technometrics **38** (1996) 205–214
3. Bansal, A., Kauffmann, R., Weitz, R.: Comparing the modeling performance of regression and neural networks as data quality varies: A business value approach. Journal of Management Information Systems **10** (1993) 11–32
4. Cybenko, G.: Approximation by superposition of a sigmoidal function. Mathematics of Control, Signals and Systems **2** (1989) 303–314
5. Hornik, K., Stinchcombe, M., White, H.: Multilayer feedforward networks are universal approximators. Neural Networks **2** (1989) 359–366
6. Park, J., Sandberg, I.W.: Universal approximation using radial-basis function networks. Neural Computation **3** (1991) 246–257
7. Donoho, D., Johnstone, I.: Ideal spatial adaption by wavelet shrinkage. Biometrika **81** (1994) 425–455
8. Zhang, Q.: Using wavelet network in nonparametric estimation. IEEE Trans. Neural Networks **8** (1997) 227–236
9. MacKay, D.J.C.: Bayesian methods for backpropagation networks. In Domany, E., van Hemmen, J.L., Schulten, K., eds.: Models of Neural Networks III. Springer-Verlag, New York (1994)
10. Ray, S., Chan, A., Mallick, B.: Bayesian wavelet shrinkage in transformation based normal models. In: ICIP02. (2002) I: 876–879
11. Lillekjendlie, B., Kugiumtzis, D., Christophersen, N.: Chaotic time series - part ii: System identification and prediction. Modeling, Identification and Control **15** (1994) 225–243

A Neural Network Approach for Indirect Shape from Shading

Ping Hao, Dongming Guo, and Renke Kang

MOE Key Laboratory for Precision & Non-traditional Machining, Dalian University of Technology, Dalian, P. R. China
haoping@student.dlut.edu.cn
{guodm, kangrk}@dlut.edu.cn

Abstract. For the reason that the conventional illumination models are empirical and non-linear, the traditional shape from shading (SFS) methods with conventional illumination models are always divergent in the process of iteration and are difficult to initialize the parameters of the illumination model. To overcome these disadvantages, a new approach based on the neural network for indirect SFS is proposed in this paper. The new proposed approach applies a series of standard sphere pictures, in which the gradients of the sphere can be calculated, to train a neural network model. Then, the gradients of the reconstructed object pictures, which are taken in the similar circumstances as that of the standard sphere pictures, can be obtained from the network model. Finally, the height of the surface points can be calculated. The results show that the new proposed method is effective, accurate and convergent.

1 Introduction

Many surface reconstruction methods, including the stereovision, the shape from shading (SFS), the structured light ranging and the photometric stereo, etc., are developed rapidly recently. Especially, the method of SFS, which takes advantage of the illumination model information to reconstruct the surfaces only by a single image, has been paid great attention in surface reconstruction field. However, there are three disadvantages of the method of SFS with illumination models. Firstly, for the reason that the illumination models, generally are Lambert model, Torrance-Sparrow model [1] and Cook-Torrance model [2], are empirical models, with a low accuracy of reconstructed surface obtained [3]. Secondly, because the illumination models applied in SFS are non-linear equations, a divergent process of iteration for surface reconstruction with the method of SFS always takes place [4]. Finally, parameters of the conventional illumination models are difficult to be estimated and have to be initialized by experience. Recently, with the development of the neural network technology, some scholars focus on the research of SFS based on the neural network model. Wei [5] developed a multiplayer feedforward network for the parameterization of the object surface. Cho and Chow [6] apply the neural network as a general illumination model to take the place of the conventional illumination models in SFS. However, similarly, they convert the SFS to an ordinary extremum problem in which the cost function is to be minimized with respect to the network weights. Problems of

F. Yin, J. Wang, and C. Guo (Eds.): ISNN 2004, LNCS 3174, pp. 737–742, 2004.

the initialization of model parameters and the convergence problem are not resolved effectively. Cheung [7] applies a self-organization neural network for direct shape from shading. A linear feedforward network is applied by Jiang [8] to calculate the height directly. Nevertheless, their methods lack the consideration of the integrability and the smoothness for the reconstructed surface as a whole.

To overcome the disadvantages of the traditional SFS methods, a new neural network approach for indirect SFS is proposed in this paper, which does not have to initialize the parameters of the illumination model as traditional SFS methods. Meanwhile, two experiments with this approach are provided. The new approach for indirect SFS shows a high accuracy in surface reconstruction and a high efficiency. Moreover, without using the non-linear illumination model to reconstruct surface, the problem of divergence in iteration is avoided.

2 Neural Network Approach for Gradient Calculation

2.1 Traditional SFS Method

Generally speaking, SFS techniques are to reconstruct a surface from only a single gray image with some necessary constraints. Take the most classical approach, the minimization approach, for example, some constraints, such as the brightness constraint, the smoothness constraint and the integrability constraint, have been considered to ensure the reverse problem resolvable by minimizing an energy function. The general energy function can be represented as:

$$F = \iint_{\Omega} \left\{ \left(I(x,y)-R(p,q)\right)^2 + \lambda\left(p_x{}^2+p_y{}^2+q_x{}^2+q_y{}^2\right) + \mu\left[\left(z_x-p\right)^2+\left(z_y-q\right)^2\right]\right\} d\Omega \ , \tag{1}$$

where I and R are the real gray level and the calculated gray level, respectively, p and q are the gradients of a point on reconstructed surface, z is the height of a point, λ and μ are the coefficients of the smooth constraint and the integrability constraint, respectively, Ω is the effective area of an image. Especially, the important parameter R, calculated by the illumination model, is empirical and non-linear, represented as: $R_p = k_d I_l \cos\theta_i$, where I_l is the effective brightness for diffuse reflection when the light irradiates the surface perpendicularly, k_d is the coefficient of diffuse reflection and θ_i is the incident angle. In addition, a discretization formula about gradient and height is represented as:

$$\begin{cases} (k\lambda'+\mu)p_{ij} = \left(k\lambda'\bar{p}_{ij}+\mu z_x\right) + \left(I(x,y)-R(p,q)\right)R_p \\ (k\lambda'+\mu)q_{ij} = \left(k\lambda'\bar{q}_{ij}+\mu z_y\right) + \left(I(x,y)-R(p,q)\right)R_q \ , \\ \dfrac{k}{\varepsilon^2} z_{ij} = \dfrac{k}{\varepsilon^2}\bar{z}_{ij} - \left(p_x+q_y\right) \end{cases} \tag{2}$$

where $\lambda' = \lambda/\varepsilon^2$, $\bar{p}_{ij}$ is the average value of p corresponding to the pixel of (i, j), ε is the distance between neighboring pixels, k is the discretization coefficient of $\bar{p}_{ij}$.

To prevent divergence in the process of iteration, the non-linear component of the irradiance equation has to be dealt with by linearization. Based on Taylor formula, a linear iteration formula of Eq. (2) is obtained[4]:

$$\begin{cases} \left(\lambda''+{R_p^{(k)}}^2\right)\delta p_{ij}+R_p^{(k)}R_q^{(k)}\delta q_{ij}=\left(k\lambda'\delta\bar{p}_{ij}+\mu\delta z_x\right)+\left(I(x,y)-R^{(k)}\right)R_p^{(k)} \\ \left(\lambda''+{R_p^{(k)}}^2\right)\delta q_{ij}+R_p^{(k)}R_q^{(k)}\delta p_{ij}=\left(k\lambda'\delta\bar{q}_{ij}+\mu\delta z_y\right)+\left(I(x,y)-R^{(k)}\right)R_q^{(k)} \end{cases}, \tag{3}$$

where $R^{(k)}{}_p$ and $R^{(k)}{}_q$ are $\partial R^{(k)}/\partial p^{(k)}$ and $\partial R^{(k)}/\partial q^{(k)}$, respectively, $\lambda''=k\lambda'+\mu$, δ is variance relative to the reference point. Then the value of (p,q), corresponding to the direction of a surface point, can be derived as:

$$\begin{cases} p_{ij}^{(k)}=p_{ij}^{(k-1)}+\delta p_{ij} \\ q_{ij}^{(k)}=q_{ij}^{(k-1)}+\delta q_{ij} \end{cases}. \tag{4}$$

Simultaneously, the formula for surface height z_{ij} is also similar to Eq. (2).

2.2 Training the Network

As well known, the method of SFS with illumination model is to resolve a reverse problem of the partial differential equations, which are derived by the non-linear illumination models. To overcome the disadvantages brought by the non-linear illumination models, an indirect approach based on the neural network to reconstruct surface is proposed. The new approach take the place of the conventional illumination model with a network model, which is trained with a series of standard sphere pictures taken in the similar circumstances as the reconstructed surface, to calculate the gradient of each point in the surface. Then, according to the gradients obtained, the surface height of each point can be calculated.

To reduce the errors brought by the network, the circumstances of the standard sphere pictures and the reconstructed surface picture have to be similar, including the locations of the visual point and the light source, the reflection characters of the surface and the camera parameters, such as the shutter speed, the f-number and the white balance, etc. Meanwhile, the standard sphere pictures provided to train the network should be complete.

The network model we use here is a backpropagation based neural network with five layers: an input layer, an output layer and three hidden layers as shown in Fig. 1 (a). The nodal points of hidden layers are 8, 10 and 12, respectively. The output layer has two parameters: the gradient (p,q). Assume that the gray level, the reflected light direction and the incident light direction of a point $P(i,j)$ on the reconstructed surface is $I(i,j)$, (V_x,V_y) and (L_x,L_y), respectively, the input layer has nine parameters: $I(i,j)$, $I(i+1,j)$, $I(i-1,j)$, $I(i,j+1)$, $I(i,j-1)$, V_x, V_y, L_x and L_y. The parameters $I(i+1,j)$, $I(i-1,j)$, $I(i,j+1)$ and $I(i,j-1)$ are to ensure the smoothness and the integrability of the reconstructed surface. The parameters $I(i,j)$, V_x, V_y, L_x and L_y are the critical factors to calculate the gradients of the points on reconstructed surface. Additionally, for the reason that the pictures of the standard sphere and the reconstructed surface are taken in similar circumstances, some other factors, which also contribute to the light reflection, are approximately equal and are ignored. Moreover, the process of the training goes point by point except for the boundary points of the picture.

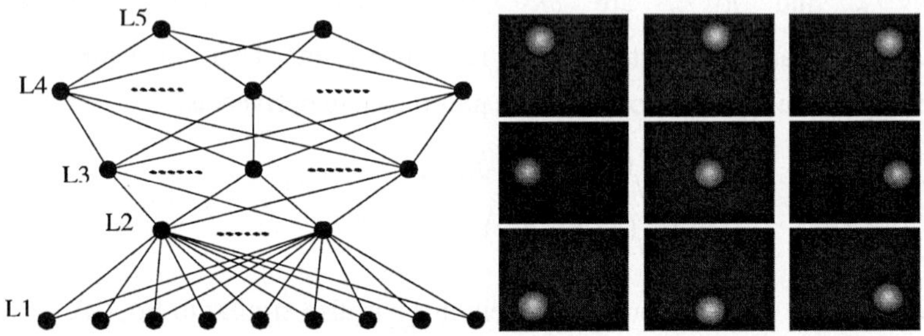

(a) (b)

Fig. 1. The image (a) is the network structure used in this paper with size $9{\times}8{\times}10{\times}12{\times}2$. The images (b) (each with size $30{\times}30$ pixels) are the standard sphere images to train the network. The origin of the Cartesian coordinate in image (b) is in the center of each image; the x axis is toward and vertical to the bottom margin and the y axis is parallel to the bottom margin and toward the right margin; locations of the light source and the visual point are L (-70, 0, 720) and V (0, 0, 720), respectively

For considering a complete circumstance of the input parameters and the gradient of a point in the surface, nine standard sphere pictures with different locations are provided to train the network model, as shown in Fig. 1 (b). To assign appropriate nodal points in the hidden layers, the process of training is convergent rapidly.

3 Calculation for Height

Generally speaking, for the discontinuous boundary conditions, the gradients obtained from the network might be discontinuous in the boundary between the background and object surface. Therefore, for reconstructing more accurate surface, the gradients have to be filtered. By plenty of experiments, the averaging filter shows more efficient and the gradients obtained from the trained network are filtered before reconstruction.

According to the representation of the surface gradient: $p=\partial z/\partial x$ and $q=\partial z/\partial y$ [3], the height of a surface point $p(i,j)$ can be obtained by a discretization method:

$$z(i,j)=\frac{(\varepsilon p(i,j)+z(i-1,j))+(\varepsilon q(i,j)+z(i,j-1))}{2}, \quad (5)$$

where ε is the distance between neighboring pixels. Additionally, the heights of the boundary points are initialized with zero. By calculating the height of the reconstructed surface point by point, a point cloud of the final surface can be obtained and the reconstructed surface can be formed with B-spline surface.

4 Using the Trained Network to Reconstruct Complex Surface

Firstly, we reconstruct one of the nine standard sphere images (size 30×30 pixels) by the trained network. For the reason that the neural network is trained by the standard sphere images, the reconstructed surface for one of the standard sphere image is perfect except for some boundary points, as shown in Fig. 2 (a).

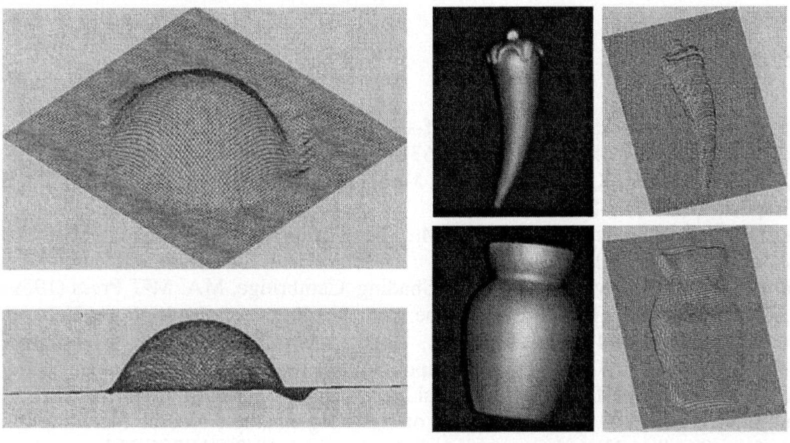

(a) (b)

Fig. 2. The images (a) are the result for hemisphere reconstruction with different view points. The images (b) are the original images and reconstruction results for more complex surfaces, a capsicum and a vase

Additionally, two more complex surface pictures, a capsicum picture and a vase picture, which are taken in the similar circumstances as the nine standard sphere images, are reconstructed by the trained network model, as shown in Fig. 2 (b). Because there is little gray information in highlight area, the reconstructed surface of the highlight area shows a little error. However, as a whole, the reconstructed results of the complex surfaces are perfect. This new SFS approach, based on the neural network model, shows effective in complex surface reconstruction and overcomes the disadvantages of the iterative SFS method with conventional illumination model, such as the low accuracy of the illumination model, the divergence in iterations and the difficulty in initializing the parameters of the illumination model.

5 Conclusions

In this paper, a new indirect SFS method based on the neural network model is proposed. This method overcomes the disadvantages of the traditional SFS method with conventional illumination models, such as the low accuracy of the empirical illumination model, the divergence in iterations and the difficulty in initializing the parameters of the conventional illumination model. The purpose of the neural network model adopted in this paper is to learn the gray information from the standard pictures and derive the gradients of the points on the reconstructed surface. Then, the heights of the surface points can be calculated from the discretization representation of the height and the gradient. The reconstructed results show that the proposed method is effective and robust. In addition, this approach can be extended to the fields of object recognition, orientation tracking and non-contact measurement with low accurate requirement.

References

1. Torrance, K., Sparrow, E.: Theory for Off-specular Reflection from Rough Surfaces. Journal of the Optical Society of America 57 (1967) 1105–1114
2. Cook, R.L., Torrance, K.E.: A Reflection Model for Computer Graphics. ACM Transactions on Graphics 1 (1982) 7-24
3. Zhang, R., Tsai, P.S., Cryer, J.E., Shah, M.: Shape from Shading: A Survey. IEEE Trans. On PAMI 21 (1999) 690-705
4. Horn, B.K.P., Brooks, M.J.: Shape from Shading. Cambridge, MA: MIT Press (1989)
5. Wei, G.Q., Hirzinger, G.: Learning Shape from Shading by A Multiplayer Network. IEEE Trans Neural Network 4 (1996) 985-995
6. Cho, S.Y., Chow, T.W.S.: Neural Computation Approach for Developing A 3-D Shape Reconstruction Model. IEEE Trans Neural Network 5 (2001) 1204-1214
7. Cheung, W.P., Lee, C.K.: Application of Self-Organization Neural Network for Direct Shape from Shading. Neural Computing & Applications 10(2001) 206-213
8. Jiang, T., Liu, B., Lu, Y.L., Evans, D.J.: A Neural Network Approach to Shape from Shading. Intern. J. Computer Math. 4 (2003) 433-439

A Hybrid Radial Basis Function Neural Network for Dimensional Error Prediction in End Milling

Xiaoli Li, Xinping Guan, and Yan Li

School of Electric Engineering, Yanshan University, Qinhuangdao, 066004, P. R. China
xlli@ysu.edu.cn

Abstract. This paper presents an approach to predict dimensional errors in end milling by using a hybrid radial basis function (RBF) neural network. First, the results of end milling experiments are discussed and the effects of the cutting parameters on dimensional errors of machined surfaces are analyzed. The results showed the dimensional errors are affected by the spindle speed, the feed rate, the radial and axial depths of cut. Then, a hybrid RBF neural network is applied. This neural network combines regression tree and an RBF neural network to rapidly determine the center values and its number, and the radial values of the radial basis function. Finally, the prediction models of dimensional errors are established by using the RBF neural network, the ANFIS (adaptive-network-based fuzzy inference system), and the hybrid RBF neural network for end milling. Compared with the predicted results of the above three models, the performance of the hybrid RBF neural network-based method is shown to be the best.

1 Introduction

End milling is widely used for rough or finish cutting processes. An end mill induced by cutting forces, however, often results in a deflection, which is defined as tool deflection, similar to a cantilevered beam at a free end. The deflection will result in the dimensional errors on the machined surfaces. The dimensional errors on the machined surfaces are directly affected by cutting parameters (i.e., spindle speed, feed rate, the axis depth of cut AD, the radial depth of cut), the size and characteristics of the end mill, and the material properties of the workpiece and cutter. Therefore, the determination of dimensional errors in end milling is not an easy task, which includes two main problems: the cutting force estimation or measurement and the tool deflection calculation.

Dimensional errors estimated by using cutting force models have already been widely investigated, such as [1] developed a model for predicting dimensional errors by summing up the cutting forces generated on chip load elements and mimicking an end mill as a slender cantilever beam. [2] presented a dimensional-accuracy model, which took a dimensional error as a function of cutting conditions by using a cutting force model derived from solid modeling and tool-workpiece interaction. In [3] a theoretical model for improving the accuracy in end milling by combining the static

F. Yin, J. Wang, and C. Guo (Eds.): ISNN 2004, LNCS 3174, pp. 743–748, 2004.

stiffness of an end mill and instantaneous cutting forces was developed. However, all the methods above have a serious shortcoming: they are based on chip thickness, cutting force coefficients, tool stiffness and equivalent tool diameter, etc. It is known that the determination of the above parameters would be a very difficult task in practical end milling. To overcome the difficulty derived from the estimation of cutting forces, dynamometer sensors have been developed and directly applied to measure cutting forces in the end milling. Commercial dynamometers provide accurate measurements of cutting forces, and different types of dynamometers are available for different cutting types. However, some limitations of using dynamometers have been reported in the some literatures, which include reducing the stiffness of machine tools, leading to chatters and dimensional errors, lack of overload protection and high cost [4]. Therefore, the dynamometers are suited for fundamental studies on the cutting processes, but not for practical processes in the industry.

Neural networks are parallel and distributed models for modeling complex systems [5]. In recent years, neural networks have been applied in manufacturing [6-7] based on their learning capability. The objection of this paper is to apply a neural network to build a prediction model of the dimensional errors based on the cutting parameters. Then, a hybrid RBF neural network that combines the regression tree and an RBF neural network is addressed and applied to model the relationship between the cutting parameters and the dimensional errors in the end milling. Finally, the trained neural network is tested by additional cutting tests; the results show that the method above is useful to predict the dimensional errors in the end milling operations.

2 Experiments and Analysis

Before the neural network is used, we should firstly analyze the effect of cutting parameters on the dimensional errors. From these results, we will find that it is very necessary to use neural network to map the nonlinear relationship between the cutting parameters and the dimensional errors; and how to apply neural network, such as the selection of inputs of neural network. Forty eight cutting tests were performed, which are associated with various spindle speeds, feed rates, the radial depths and axial depths of cut, are performed on the vertical machine center Mazada with a 6 mm diameter, 30° helix angle, and 4 fluted high speed steel end mill, dry and down milling conditions. The end mill with 30 mm overhang length is clamped in the chuck tool holder. Workpiece is aluminum alloys. 1-18 cutting tests are taken as the testing samples, the rest of cutting tests are taken as the training samples. Before cutting, the surface of workpiece is prepared by light milling. After each cutting test, the dimensional error is determined with the prepared surface by using a dial gage. The cutter is normal condition, without heavy wear. In this paper, we will not discuss the effect of cutter wear on the dimensional errors.

Cutting forces are directly related to the tool deflection that results the dimensional errors on the machined surfaces, so the factors that influence cutting forces in the end milling should be taken into account. First, let us analyze the effect of spindle speed on the cutting force and the dimensional error in the end milling. It is found that the dimensional errors decrease with the increase of spindle speed when the spindle speed is less 1500 rpm; when the spindle speed goes on increasing, the dimensional error will rise again because of the increase of cutting temperature. Feed rate is one of the important influential cutting parameters during cutting process. The feed rate is generally taken as a controlled parameter in the adaptive cutting system. For reducing dimensional error, the feed rate is also selected as the key adjusted parameter in end milling. The feed rate has an effect of the dimensional errors in the end milling. Another two influence factors are the radial depth of cut and the axial depth of cut on the dimensional errors in the end milling. The two values result in the change of the chip thickness and the cutting forces. The dimensional errors linearly increase with the increase of the radial depth of cut or axial depth of cut.

Based on above analysis, it is clear that the dimensional errors are affected by the spindle speed, the feed rate and the radial and axial depths of cut in the end milling. The dimensional errors can be taken as a function of these cutting parameters. However, it is not easy to represent this function by using a simple mathematical method because of the nonlinear relation between the cutting parameters and the dimensional errors. We should note that the dimensional errors on the machined surface more or less affect the cutting parameters. In this paper, these cutting parameters are mainly considered when using neural network to predict the dimensional error in the end milling. In the following section, the feed-forward neural network will be used to map the relation of the cutting parameters and the dimensional error.

3 Hybrid RBF Neural Network

Radial basis function (RBF) neural networks offer some advantages over multiplayer perceptions (MLP) in some applications, because RBF neural networks are easier to train than MLP neural networks [8]. Meanwhile, RBF neural networks can transform the n-dimensional inputs nonlinearly to an m-dimensional space like the MLP neural network, and then estimate a model by using a linear regression. The non-linear transformation is controlled by a set of m basis functions each characterized by a position or center $\boldsymbol{c}_j$ in the input space and a width or radius vector $\boldsymbol{r}_j$, $j \in \{1,2,\cdots,m\}$ [9]. The basis functions are typically Gaussian-like functions, such as

$$h_j(\mathrm{x}) = \exp(-\sum_{k=1}^{n} \frac{(x_k - c_{jk})^2}{r_{jk}^2}) \quad . \tag{1}$$

Where $\boldsymbol{c}_{jk}$ -center in the input space; $\boldsymbol{r}_{jk}$ is a radius vector. Clearly, the output of basis functions respond most strongly to the inputs nearest to the center $\boldsymbol{c}_j$, in the metric determined by the radius $\boldsymbol{r}_j$. The output of the RBF neural network, $f(\mathbf{x})$, which is a model based on linear regression, is expressed as follows:

$$f(\mathrm{x}) = \sum_{j=1}^{m} w_j h_j(\mathrm{x}) \quad . \tag{2}$$

In brief, to build RBF neural network often meet four problems, including (1) the number of nodes for RBF neural network; (2) the identification of Gaussian center $\boldsymbol{c}_j$;

(3) the proper setting of the radius $\boldsymbol{r}_j$; (4) the output-layer weights. For the fourth problem, it is relatively simple to determine the output-layer weights w_j by using the delta rule or some traditional statistical approach such as the pseudoinverse matrix. For the first problem, it is involved in the second problem because the identification of Gaussion center $\boldsymbol{c}_j$ includes its size and its number. Therefore, the more difficult problems are the identification of Gaussian center $\boldsymbol{c}_j$ and the proper setting of the radius $\boldsymbol{r}_j$. To overcome the above problems, various algorithms have been presented [10-13]. However, most of these techniques are computationally expensive. To summarize, the design of a successful RBF neural network involves several nontrivial problems, and so far there does not seem to exist any simple and general algorithms, or heuristic. [14] first suggested combining the decision tree and RBF neural network for overcoming the drawbacks of the other algorithms. Each terminal node of the decision tree contributes one unit to RBF neural network, the center and radius of which are determined by the position and size of the corresponding hyper rectangle. The decision tree can sets the number, positions and sizes of all RBF centers in the network. Because the decision tree nodes were converted into RBF unites, the only level of model complexity control was the tree pruning provided by C4.5 software and the choice of RBF scale, α. Unfortunately, Kulbat did not present a method to optimize either the degree of pruning or the value of the parameter α. Mark Orr (2000) presented a new method by combining regression tree and RBF networks, and added an automatic method for the control of model complexity through the selection of RBFs, the detail can be found in [15].

4 Dimensional Error Prediction

In this paper, the feed-forward neural networks, which are RBF neural network, ANFIS, and hybrid RBF neural network, are used to map the relationship between the cutting parameters and the dimensional errors. The input layer is made up of four nodes corresponding to the spindle speed, the feed rate, and the radial and axial depth of cut. The one output layer node corresponds to the dimensional error.

Firstly, the RBF neural network has been used to map the relationship between the dimensional errors and the cutting parameters for the end milling. In the training phase, the number of nodes in the hidden layer is automatically determined by the trained samples during learning, the result is that the hidden layer has 8 nodes in the trained RBF neural network; the training time is 2.7 minute, the minimal error is 0.27mm. The trained RBF neural network is used to estimate the dimensional errors, the test samples are shown in Table 1, which is from cutting tests 1 to 18. The results are shown in Fig.1 (a).It is found that the predictions of the model differed from the measured data to within 15%.

Secondly, an adaptive neuro-fuzzy inference system (ANFIS) has been used to map the relationship between the cutting parameters and the dimensional errors. ANFIS fuses the fuzzy system and neural network into a learning system, which would combine the benefits of the fuzzy system and neural network [16]. The resulting neuro-fuzzy system, a hybrid, has fuzzy system architecture, but uses neural network learning techniques so that it can be trained automatically. For a given input/output data set, ANFIS can construct a fuzzy inference system whose membership functions are tuned by using either a back propagation algorithm alone, or in combination with the least square method. Here, the trained ANFIS had 112 nodes, 46 fuzzy rules and 345 linear and 36 nonlinear parameters (345 in total), applied on 30 training data pairs (from 19 to 48). Three bell-type membership functions were used while aiming for the training error of 0.1 (goal), with the initial step size of 0.5 (0.9 step size decrease, and 1.2 increase rates). The training completed at epoch 1000, with the minimal training RMSE of 0.45 mm. The training time is 3.12 minute. The trained ANFIS is used to estimate the dimensional errors; the results are shown in Fig. 1 (b). It is found that the predictions of the model differed from the measured data to within 25%.

Finally, the hybrid RBF neural network is used to map the relationship between the cutting parameters and the dimensional errors. In the training phase, owing to hybrid RBF neural network combine regress tree and RBF neural network, the center values and its number, and radius values of radial basis function can be rapidly determined by the trained samples (from cutting tests 19 to 48) during learning. The hidden layer of hybrid RBF neural network has 11 nodes, the training time is 1.81 minute, and minimal training error is 0.2662mm. The trained hybrid RBF neural network is used to estimate the dimensional error, the test samples are from cutting tests 1 to 18. The calculated results are shown in Fig. 1. It showed that the predictions of the model differed from the measured data to within 5%. From these results, it is verified that the model with hybrid RBF neural network, whose results are the best by comparison with RBF neural network and ANFIS, is useful in the prediction of dimensional error for end milling.

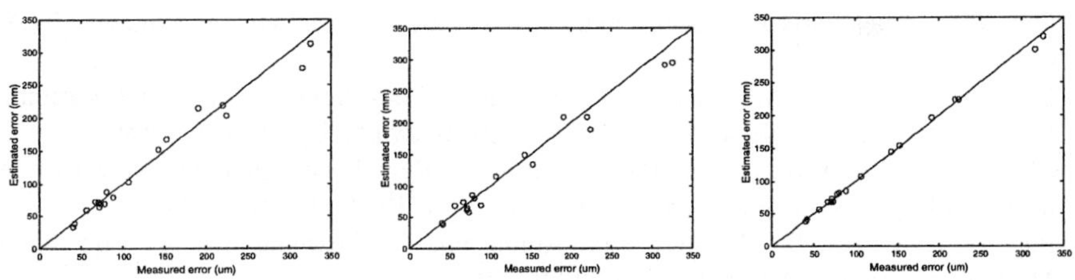

(a) (b) (c)

Fig. 1. Comparison of the dimensional errors measured and predicted by the neural network; (a) RBF NN (b) ANFIS; (c) hybrid RBF NN.

5 Conclusions

For the given end mill and workpiece, the dimensional error is generally determined by the selection of the cutting parameters corresponding to the spindle speed, the feed rate, and the radial and axial depth of cut in the end milling. In this paper, a hybrid

RBF neural network is discussed and applied to calculate the dimensional errors associated with the spindle speed, the feed rate, and the radial and axial depth of cut for the end milling. The dimensional errors can be clearly modeled by utilizing the hybrid RBF neural network and the experiments under the given cutting environments. By comparison with the RBF neural network and ANFIS, the hybrid RBF neural network is capable of predicting the dimensional errors accurately according to cutting parameters before machining.

References

1. Kline, W.A., DeVor, R.E., Lindberg, J.R.: The Prediction of Cutting Forces in End Milling with Application to Cornering Cuts. Int. J. Mach Tool Des. Res. 22 (1982) 7-22
2. Budak, E., Altintas, Y.: Peripheral Milling Conditions for Improved Dimensional Accuracy. Int. J. Mach. Tools Manufact. 34 (1994) 907-918
3. Matsubara, T., Yamamoto, H., Mizumoto, H.: Study on Accuracy in End Mill Operations. Bull. Japan Soc. Prec. Eng. 21 (1987) 95-100
4. Byrne, G., Dornfled, D., Inasaki, I., Ketteler, G., Konig, W., Teti, R.: Tool Condition Monitoring (TCM) - The Statue of Research and Industrial Application. Annals of the CIRP, Vol. 44 (1995) 541-567
5. Hanna, M.M., Buck, A., Smith, R.: Fuzzy Petri Nets with Neural Networks to Model Products Quality from a CNC-milling Machining Centre. IEEE T. Syst. Man. Cy. A, 26 (1996) 638-645
6. Cho, S., Cho, Y., Yoon, S.: Reliable Roll Force Prediction in Cold Mill Using Multiple Neural Networks. IEEE T. Neural Networks 8 (1997) 874-882
7. Li, X., Djordjevich, A., Patri, K.V.: Current Sensor-based Feed Cutting Force Intelligent Estimation and Tool Wear Condition Monitoring. IEEE T. Ind. Electron. 47 (2000) 697-702
8. Orr, M.J.L.: Regularisation in the Selection of RBF Centres. Neural Comput. 7 (1995) 606-623
9. Chen, S., Cowan, C.F.N., Grant, P.M.: Orthogonal Least Squares Learning Algorithm for Radial Basis Function Networks. IEEE T. Neural Networks 2 (1991) 302-309
10. Cheng, Y.H., Lin, C.S.: A Learning Algorithm for Radial Basis Function Networks: With the Capability of Adding and Pruning Neurons. Proc. IEEE (1994) 797-801
11. Poggio, T., Girosi, F.: Regularization Algorithms for Learning that are Equivalent to Multilayer Networks. Science 247 (1990) 987-982
12. Haykin, S.: Neural Networks: A Comprehensive Foundation. Maxmillan, New York (1994)
13. Lowe, D.: Adaptive Radial Basis Function Nonlinearities and the Problem of Generalization. 1th Int. Conf. Artificial Neural Networks, London, U.K. (1989) 171-175
14. Kubat, M.: Decision Trees Can Initialize Radial-Basis Function Networks. IEEE T. Neural Networks 9 (1998) 813-824
15. Orr, M.J.L.: Recent Advances in Radial Basis Function Networks. Technical Report, Institute for Adaptive and Neural computation, Edinburgh University (1999)
16. Jang, J.S.R.: ANFIS: Adaptive-Network-Based Fuzzy Inference System. IEEE Transactions on Systems, Man and Cybernetics 23 (1993) 665-685

Bayesian Neural Networks for Life Modeling and Prediction of Dynamically Tuned Gyroscopes

Chunling Fan[1,2], Feng Gao[3], and Zhihua Jin[2]

[1] College of information and Control Engineering, Qingdao University of Science and Technology, Qingdao 266042, China
[2] Instrument Engineering Department of Shanghai Jiaotong University, Shanghai 200030, China
[3] Mitsubishi Heavy Industries Haier (Qingdao) Air conditioners Co.Ltd, Qingdao 266100, China
fanchunling@sjtu.edu.cn

Abstract. In this paper we apply Bayesian neural networks to life modeling and prediction with real data from a dynamically tuned gyroscopes (DTG). The Bayesian approach provides consistent way to inference by integrating the evidence from data with prior knowledge from the problem. Bayesian neural networks can overcome the main difficulty in controlling the model's complexity of modeling building of standard neural network. And the Bayesian approach offers efficient tools to avoid overfitting even with very complex models, and facilitates estimation of the confidence intervals of the results. In this paper, we review the Bayesian methods for neural networks and present results of case study in life modeling and prediction of DTG.

1 Introduction

Due to high cost and small batch of some mechanical-electric components, such as flywheels and gyroscopes, how to estimate their reliability and life has become a difficulty. In addition, these products have fewer samples and are lack of practical criterion of accelerated life test comparing with electric parts. Moreover, there is no evidence to prove that the life of gyroscopes and flywheels has obvious relations with environmental parameters, such as temperature, pressure etc. Therefore, under the request of increasing life for these products, real-time life test of one to one is high cost, which induces waste on the material and financial resources and becomes a key factor of prolonging schedule of developing spacecraft. Thus doing research and applying new method to model and predicting the life of mechanical-electric components are very significant.

Usually, two strategies are applied to estimate the life of such mechanical-electric components. One is to employ accelerated-life test to shorten the time of life test [1]; the other is to use parameter extrapolation to estimate the life of such mechanical-electric components. In this paper, we adopt parameter extrapolation to estimate the life of gyroscopes. And because of the advantages of Bayesian neural networks, such as automatic complexity control, possibility to use prior information and hierarchical models for the hyperparameters and predictive distributions for outputs [2], hence

F. Yin, J. Wang, and C. Guo (Eds.): ISNN 2004, LNCS 3174, pp. 749–754, 2004.

Bayesian neural networks [3] is applied to life modeling and prediction for DTG in this paper.

2 Bayesian Principle

The key principle of Bayesian approach is to construct the posterior probability distributions for all the unknown entities in a model, given the data sample.

Consider a regression problem involving the prediction of a noisy vector $\mathbf{y}$ of target variables given the value of a vector $\mathbf{x}$ of input variables. The process of Bayesian learning is started by defining a model and prior distribution $p(\theta)$ for the model parameters θ. Prior distribution expresses our initial beliefs about parameter values, before any data has been observed. After observing new data $D=\{(\mathbf{x}^{(1)},\mathbf{y}^{(1)}),\cdots,(\mathbf{x}^{(n)},\mathbf{y}^{(n)})\}$, prior distribution is updated to the posterior distribution using Bayes' rule

$$p(\theta \mid D)=\frac{p(D\mid\theta)p(\theta)}{p(D)}\propto L(\theta)p(\theta)\,. \tag{1}$$

where the likelihood function $L(\theta)$ gives the probability of the observed data as function of the unknown model parameters.

To predict the new output $\mathbf{y}^{(n+1)}$ for the new input $\mathbf{x}^{(n+1)}$, predictive distribution is obtained by integrating the predictions of the model with respect to the posterior distribution of the model parameters

$$p(\mathbf{y}^{(n+1)}\mid \mathbf{x}^{(n+1)},D)=\int p(\mathbf{y}^{(n+1)}\mid \mathbf{x}^{(n+1)},\theta)p(\theta\mid D)d\theta\,. \tag{2}$$

where θ denotes all the model parameters and hyperparameters of the prior structures. It is the same as taking the average prediction of all the models weighted by their goodness.

3 Bayesian Learning for Neural Networks

In the following, we give a short review of the Bayesian learning for neural networks. We concentrate on model, prior and Markov chain Monte Carlo numeric simulation method for computing the integrations.

3.1 Model

The posterior predictive distribution of output $\mathbf{y}^{(n+1)}$ for the new input $\mathbf{x}^{(n+1)}$ given the training data $D=\{(\mathbf{x}^{(1)},\mathbf{y}^{(1)}),\cdots,(\mathbf{x}^{(n)},\mathbf{y}^{(n)})\}$, is obtained by integrating the predictions of the model with respect to the posterior distribution of the model using Eq. (2).

The probability model for the measurements, $p(\mathbf{y} \mid \mathbf{x}, \theta)$, contains the chosen approximation functions and noise models. It defines also the likelihood part in the posterior probability term, $p(\theta \mid D) \propto p(D) p(\theta)$. The probability model in a regression problem with additive error is

$$y = f(\mathbf{x}; \theta_{\mathbf{w}}) + e \ , \tag{3}$$

where $f()$ is the MLP function

$$f(\mathbf{x}; \theta_{\mathbf{w}}) = \mathbf{b}^2 + \mathbf{w}^2 \tanh(\mathbf{b}^1 + \mathbf{w}^1 \mathbf{x}) \ . \tag{4}$$

where θ_w denotes all the parameters $\mathbf{w}^1, \mathbf{b}^1, \mathbf{w}^2, \mathbf{b}^2$, which are the hidden layer weights and biases, and the output layer weights and biases, respectively. The random variable e is the model residual.

3.2 Prior

We have to define the prior information about our model parameters, before any data has been observed. Usual prior is that the model has some unknown complexity but the model is neither constant nor extremely flexible. To express this prior belief we can set hierarchical model specification.

Parameters w define the model $f(\mathbf{x}, \theta_{\mathbf{w}})$. Because the complexity of the MLP can be controlled by mastering the size of the weights w. This can be achieved by using, e.g., Gaussian prior distribution for weights w given hyperparameter α

$$p(w \mid \alpha) = (2\pi)^{-m/2} \alpha^{m/2} \exp(-\alpha \sum_{i=1}^{m} w_i^2 / 2) \tag{5}$$

3.3 Prediction

After defining the model and prior information, we can integrate the evidence from the data to get the posterior distribution for the parameters

$$p(w, \alpha \mid D) \propto L(w, \alpha \mid D) p(w, \alpha) \tag{6}$$

Predictive distribution for new data is then obtained by integrating over this posterior distribution

$$p(\mathbf{y}^{(n+1)} \mid \mathbf{x}^{(n+1)}, D) = \int p(\mathbf{y}^{(n+1)} \mid \mathbf{x}^{(n+1)}, w, \alpha) p(w, \alpha \mid D) dw\alpha \ , \tag{7}$$

We can also evaluate expectations of various functions with respect to the posterior distribution for parameters. For example in regression we may evaluate the expectation for a component of $\mathbf{y}^{(n+1)}$

$$\hat{\mathbf{y}}_k^{(n+1)} = \int f_k(\mathbf{x}^{(n+1)}, w) p(w, \alpha \mid D) dw\alpha \,. \tag{8}$$

which corresponds to the best guess with squared error loss.

The posterior distribution for the parameters $p(w,\alpha \mid D)$ is typically very complex, with many modes. Evaluating the integral of Eq. (8) is therefore a difficult task. The integral can be approximated with Markov chain Monte Carlo numerical approximation.

3.4 Markov Chain Monte Carlo Numerical Approximation Method

Neal [4] has introduced implementation of Bayesian learning for MLPs in which the difficult integration of Eq. (8) is performed using Markov chain Monte Carlo (MCMC) methods.

The integral of Eq. (8) is the expectation of function $f_k(\mathbf{x}^{(n+1)}, w)$ with respect to the posterior distribution of the parameters. This and other expectations can be approximated by Monte Carlo method, using a sample values $w^{(t)}$ drawn from the posterior distribution of parameters

$$\hat{\mathbf{y}}_k^{(n+1)} \approx \frac{1}{N} \sum_{t=1}^{N} f_k(\mathbf{x}^{(n+1)}, w^{(t)}) \,. \tag{9}$$

Note that samples from the posterior distribution are drawn during the 'learning phase' and predictions for new data can be calculated quickly using the same samples and Eq.(9).

Neal has used the hybrid Monte Carlo (HMC) algorithm for parameters and Gibbs sampling for hyperparameters. HMC is an elaborate Monte Carlo method, which makes efficient use of gradient information to reduce random walk behavior [4]. The gradient indicates the direction in which one should go to find states with high probability. Application of Gibbs sampling for hyperparameters helps to minimize the amount of tuning that is needed to obtain good performance in HMC.

4 Application

This section mainly describes application of Bayesian neural network in life modeling and prediction for DTG. Firstly, we should observe some parameters, which are related with DTG's life. These parameters are vibration, x-axis drift, environmental temperature and power. After analyzing these parameters, and according to characteristics of life prediction for DTG, we choose drift and vibration as modeling parameters and collect 106-day data for modeling.

These original data should be normalized before putting into the neural networks. The structure of MLP is chosen as 4-30-2. Number of training data points is 20, the prior of weight is 0.001 and error coefficient of data is 100. The weights of network are drawn from a zero mean, unit variance isotropic Gaussian, with varied scaled by the fan-in of the hidden or output units as appropriate. And number of samples omitted at start of chain is 200.

The modeling output of Bayesian neural networks for such two parameters is shown in the Fig.1, where the days are normalized between 0 and 1. Then we should to implement inverse-normalization to process modeling output. After converting, the modeling output and real detected data of such two parameters are also shown in the Fig.1.

Then we should evaluate the precision of this Bayesian neural network. The modeling error is listed in table 1.

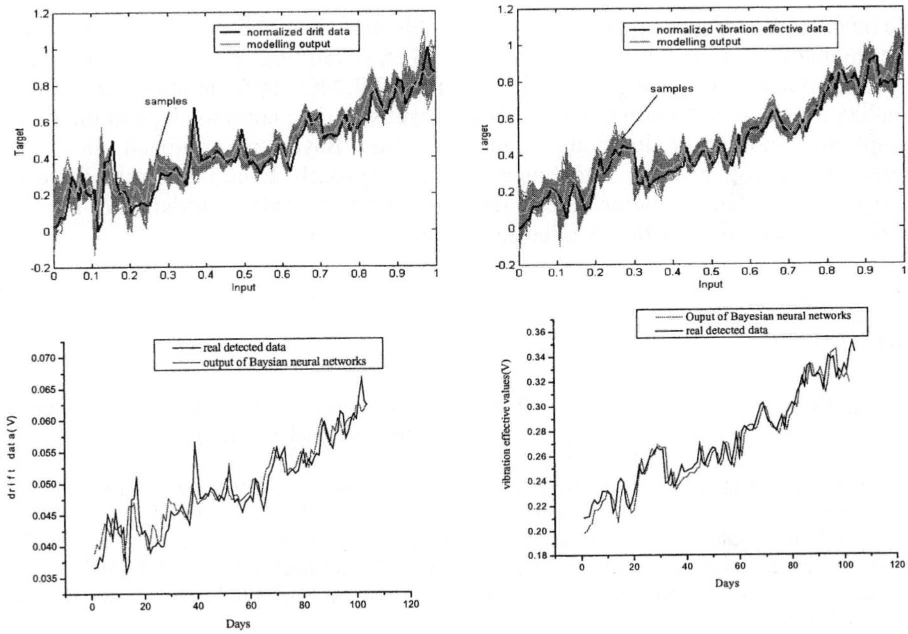

Fig. 1. Modeling output of two parameters

Table 1. Modeling error of Bayesian neural network

Output	Mean square error (MSE) (V)	Average percent absolute error (APAE) (%)
Drift	2.6332×10^{-4}	3.15 %
Vibration effective value	9.7517×10^{-4}	2.48 %

Table 2. Comparison of predicted and detected values for 2 parameters

Output	Predictive values (V)	Real detected values (V)
Drift	0.083	0.08
Vibration effective value	0.410	0.4

From the table 1, we can see the modeling MSE can reach 10^{-4}, and it's the largest of APAE is 3.15 %, so the modeling precision is 96.85%. Finally, we should estimate the predictive performance. We use a real detected value as reference to evaluate the predictive values. Comparison of predicted and detected values for 4 parameters is listed in table 2.

From the table 2, we can calculate their APAE error value and get the largest error is 3.75%. Thus the predictive precision can reach 96.25%.

5 Conclusions

We have applied Bayesian neural network in the modeling and prediction for a DTG. The modeling and predictive precisions are 98.38% and 96.25%, respectively. Bayesian neural networks gave good results by using MCMC approximation for the marginalization. It must be emphasized that the results of data analysis depend on the assumptions and approximations made-thus the Bayesian approach does not automatically give better results than a classical approach. Even though the Bayesian models do not need validation data to set the model complexity, validation of the final models is essential, as with any other modeling approach.

References

1. E.A., Elsayed, Algon, C.K. Chen: Recent Research and Current Issues in Accelerated Testing, Systems, Man, and Cybernetics, IEEE International Conference on, Vol.5. IEEE Electronic Library (1998) 4704-4709
2. Aki, Vehtari, Jouko, Lampinen: Bayesian MLP Neural Networks for Image Analysis, Pattern Recognition Letters, Vol. 21. Elsevier, Science Direct Digital Library (2000) 1183-1191
3. Jouko, Lampinen, Aki, Vehtari: Bayesian Approach for Neural Networks-Review and Case Studies, Neural networks, Vol. 14. Elsevier, Science Direct Digital Library (2001) 257-274
4. Radford, M., Neal: Probabilistic Inference Using Markov Chain Monte Carlo Methods, Technical Report CRG-TR-93-1, (1993) 47-86

Furnace Temperature Modeling for Continuous Annealing Process Based on Generalized Growing and Pruning RBF Neural Network

Qing Chen[1], Shaoyuan Li[1], Yugeng Xi[1], and Guangbin Huang[2]

[1] Institute of Automation, Shanghai Jiao Tong University, Shanghai, 200030, P.R. China
{king, syli, ygxi}@sjtu.edu.cn

[2] School of Electrical and Electronic Engineering, Nanyang Technological University, Nanyang Avenue, 639798, Singapore

Abstract. Dynamic modeling for the quality of large-scale process is studied in this paper combined with continuous annealing process. This kind of process is constituted with several sub-processes. There is complex nonlinear mapping between the sub-process set points and the final quality. The quality model should be constructed and updated based on the new data from the real process. To meet this demand, a novel generalized growing and pruning RBF (GGAP-RBF) network is used to establish the quality model. GGAP-RBF is a sequential learning algorithm so that we can establish the model dynamically. Last, we do some on-line application study on the continuous annealing furnace in a steel factory. The quality model between the furnace temperature of each zone in the furnace and the exit strip temperature is constructed.

1 Introduction

In some industrial processes, the final product quality is influenced by the overall process, which is made up of several sub-processes. There are complex nonlinear mapping functions between the set points of each sub-process and the final quality. The optimal set points of each sub-process should be given dynamically according to the quality model and performance index in real-time optimization layer. So it is very important to construct and update the quality model using the input-output data obtained from the real plant.

A continuous annealing process is a highly efficient heat treatment process for cold-rolled strips in steel works. It consists of heating, soaking, and cooling furnaces. The strip temperature is measured and controlled at the outlet of each furnace. Among these furnaces, the heating furnace (HF), which is constituted with several zones, has large effects on production rate, strip quality, stability of operation, etc. The temperatures of each zone in HF affect the strip temperature at the exit side. Some research work has been proposed on the modeling, control and optimization of strip temperatures for the HF in continuous annealing [1].

Neural networks, especially Radial Basis Function (RBF) networks have gained much popularity in recent times due to their ability to approximate complex nonlinear

F. Yin, J. Wang, and C. Guo (Eds.): ISNN 2004, LNCS 3174, pp. 755–760, 2004.

mappings directly from the input-output data with a simple topological structure. Several learning algorithms have been proposed in the literature for training RBF networks [2]-[4]. In practical online applications, sequential learning algorithms are generally preferred over batch learning algorithms, as they do not require retraining whenever a new data is received and the model can be updated.

This paper applied the GGAP-RBF algorithm in [5] on modeling the strip temperature in the heating furnace of continuous annealing process. The algorithm introduces the concept of "significance" for the hidden neurons and directly links the required learning accuracy to the significance of neurons in the learning algorithm so as to realize a compact RBF network. The algorithm is sequential and can be used for on-line learning in real time applications

2 The HF Plant and Quality Model

In continuous annealing process, the material for annealing is a cold-rolled strip coil. The strip runs through the process with a certain line speed. The heating furnace is a very important part of the continuous annealing process. It is made up of several zones and each zone has its temperature set point. When the line speed is constant and the dimension of the strip is certain, the furnace temperatures of each zone determine the strip temperature at the delivery side. The nonlinear mapping from the furnace temperatures of several zones to the strip temperature is the quality model we want to get in this paper.

In the real industry, the quality model should be constructed and updated based on the new data measured from the plant with the time going. In this paper, a novel neural network, GGAP-RBF is proposed to construct the model. This sequential learning algorithm is better to use the new input-output data for model updating.

3 GGAP-RBF Modeling Method

3.1 Definition of Significance of Neurons

This section first introduces the notion of *significance* for the hidden neurons based on their statistical average contribution over all inputs seen so far, although those inputs are discarded and not stored in the system after being learned. In the GGAP-RBF algorithm, the significance is used in growing and pruning strategies. The output of a RBF network with K neurons for an input vector $\mathbf{x} = (x_1, \cdots x_l)^{\mathrm{T}} \in X \subseteq \mathbf{R}^l$, where l is the dimension of input observation space, is given by:

$$f(\mathbf{x}) = \sum_{k=1}^{K} \alpha_k \phi_k(\mathbf{x}) \tag{1}$$

where α_k is the weight connecting the k-th hidden neuron to the output neuron and $\phi_k(\mathbf{x})$ is the response of the k-th hidden neuron for an input vector $\mathbf{x}$:

$$\phi_k(\mathbf{x}) = \exp(-\frac{\|\mathbf{x}-\mu_k\|^2}{\sigma_k^2}) \tag{2}$$

where $\mu_k = (\mu_{k,1},\cdots,\mu_{k,l})^{\mathrm{T}}$ and σ_k are the center and width of the k-th hidden neuron, respectively, $k = 1,\cdots,K$. Let a series of training samples $(\mathbf{x}_i, \mathbf{y}_i)$, $i = 1,2,\cdots$,be drawn sequentially and randomly from a rang X with a sample density function of $p(\mathbf{x})$, where X is a subset of an l-dimensional Euclidian space. After sequentially learning n observations, assume that a RBF network with K neurons has been obtained. The network output for input $\mathbf{x}_i$ is given by f_1. If the neuron k is removed, the output of the RBF network with the remaining neurons for the input $\mathbf{x}_i$ is f_2. Thus, for an observation $\mathbf{x}_i$ the error resulted from removing neuron k is given by:

$$E(k,i) = \|f_1 - f_2\|_q = \|\alpha_k\|_q \,\phi_k(\mathbf{x}_i), i = 1,\cdots,n \tag{3}$$

where $\|\bullet\|_q$ is the q-norm of vectors, indicating the L_q-distance between two points in Euclidian space. The q-norm of the error E_q for all n sequentially learned observations caused by removing the neuron k is:

$$E_q(k) = \|(E(k,1),\cdots,E(k,n))^{\mathrm{T}}\|_q = \|\alpha_k\|_q \left(\sum_{i=1}^{n} \phi_k^q(x_i) \Big/ n \right)^{\mathrm{T}} \tag{4}$$

Let the sample range X be divided into N small spaces Δ_j, when the number of input observations n is large and Δ_j is small, we can obtain the definition of *significance* of a hidden neuron proposed in [5]:

$$E_{sig}(k) = \|\alpha_k\|_q \left(\int_X \exp(-\frac{q\|x-\mu_k\|^2}{\sigma_k^2}) p(x)dx \right)^{1/q} \tag{5}$$

This is the statistical contribution of neuron k to the overall output of the RBF network. If the distributions of the l attributes $(x_1,\cdots,x_i,\cdots,x_l)^{\mathrm{T}}$ of observations $\mathbf{x}$'s are independent from each other, "significance" (5) can be re-written as:

$$E_{sig}(k) = \|\alpha_k\|_q \prod_i^l \left(\int_{a_i}^{b_i} \exp(-\frac{q(x-\mu_{k,i})^2}{\sigma_k^2}) p_i(x)dx \right)^{1/q} \tag{6}$$

where (a_i, b_i) is the interval of the i-th attribute x_i of observations, $i = 1,\cdots,l$. The input samples which are measured from real plant are often uniformly drawn a range X, the sampling density function $p(\mathbf{x})$ is given by $p(\mathbf{x}) = 1/S(\mathbf{x})$, where $S(\mathbf{x})$ is the size of the range X. In fact, as we do in some of our applications, one may just normalize the inputs to the range $[0\ \ 1]^l$ and get $S(X)=1$.

3.2 GGAP-RBF Learning Algorithm

The learning process of GGAP-RBF involves the allocation of new hidden neurons as well as adaptation of network parameters. A new neuron is only added if the input data is sufficiently far from the existing neurons. At the same time, it should be ensured that the significance of the newly added neuron is greater than the required approximation accuracy. One needs only to adjust parameters of the neuron nearest to the most recently received input if no new neuron is added and only needs to check the nearest (most recently adjusted) neuron for pruning. As for the parameter adjustment, the adjustment will be done for the parameters of the nearest neuron using the EKF algorithm in [3]. The GGAP-RBF algorithm can be generalized as:

Step 1: For each observation $(\mathbf{x}_n, \mathbf{y}_n)$ presented to the network, compute the overall network output:

$$f(\mathbf{x}_n)=\sum_{k=1}^{K}\alpha_k \exp(-\frac{1}{\alpha_k^2}\left\|\mathbf{x}_n-\mu_k\right\|^2) \tag{7}$$

where K is the number of hidden neurons.

Step 2: Select the threshold ε_n in growing criterion appropriately and calculate the error:

$$e_n = y_n - f(x_n) \tag{8}$$

Step 3: Apply the criterion for adding neurons, if

$$\begin{cases} \left\|\mathbf{x}_n-\mu_{nr}\right\| > \varepsilon_n \\ \left\|e_n\right\|_q \left(\int_X \exp(-\frac{q\left\|\mathbf{x}-x_n\right\|^2}{\kappa\left\|\mathbf{x}_n-\mu_{nr}\right\|^2} p(\mathbf{x})d\mathbf{x} \right)^{1/q} > e_{\min} \end{cases} \tag{9}$$

allocate a new hidden neuron $K+1$, where κ is an overlap factor that determines the overlap of the responses of the hidden neurons in the input space, μ_{nr} is the center of the hidden neuron nearest to the new coming data, $e_{\min}$ is the expected approximation accuracy. The parameters for the new add hidden neuron is:

$$\begin{cases} \alpha_{K+1} = e_n \\ \mu_{K+1} = \mathbf{x}_n \\ \sigma_{K+1} = \kappa\left\|\mathbf{x}_n-\mu_{nr}\right\| \end{cases} \tag{10}$$

Else, if (9) does not hold, adjust the network parameters α_{nr}, μ_{nr} for the nearest neuron only. (Refer to [3] for the adjustment algorithm)

Step 4: Check the criterion for pruning the adjusted hidden neuron, if

$$E_{sig}(nr) = \|\alpha_{nr}\|_q \left(\int_X \exp(-\frac{q\|\mathbf{x}-\mu_{nr}\|^2}{\sigma_{nr}^2}) p(\mathbf{x}) d\mathbf{x} \right)^{1/q} < e_{\min} \tag{11}$$

Remove the nr-th hidden neuron, and $E_{sig}(nr)$ is the significance of the nearest neuron.

Table 1. Experiment results in three work conditions

Dimension $Thickness \times Width$ $mm \times mm$	Line Speed mm/s	Training Error		Testing Error	
		MAE	RMSE	MAE	RMSE
800×1173	250	0.0478	0.0599	0.0469	0.0587
600×823	330	0.0545	0.0680	0.0541	0.0682
706×1147	200	0.0552	0.0769	0.0550	0.0760

4 Application Study

On-line application study has been done on the continuous annealing furnace in a steel factory. The heating furnace is made up of eight zones and the furnace temperatures of each zone can be measured. We construct the nonlinear mappings from the temperatures of eight zones to the exit strip temperature when the dimension and line speed unchanged. The sampling time is $10s$. Two learning accuracy measurements are used in all experiments. Mean Arithmetic Error (MAE) is defined as:

$$MAE = \frac{\sum_{i=1}^{n} \|f(\mathbf{x}_i) - \mathbf{y}_i\|}{n} \tag{12}$$

The Root Mean Square Error (RMSE) is defined as:

$$RMSE = \sqrt{\frac{\sum_{i=1}^{n} \|f(\mathbf{x}_i) - \mathbf{y}_i\|_2^2}{n}} \tag{13}$$

where, $(\mathbf{x}_i, \mathbf{y}_i)$ is sample data and $f(\mathbf{x}_i)$ is the output of the network. We do experiments in three work conditions and the experiment results are showed in Table 1.

In the experiments, we found that most error between the output of the testing sample and the network is less than $3^{\circ}C$, which can meet the demand of application in real industry process. In the first work condition, we use 660 training data and

220 testing data. In the second work condition, 750 training data and 250 testing data and in the third work condition, 960 training data and 320 testing data, respectively.

5 Conclusion

Dynamic modeling for the quality of large-scale process is studied in this paper combined with continuous annealing process. First, introduce the continuous annealing furnace and the modeling problem for the heating furnace temperature. Then, a novel generalized growing and pruning RBF neural network is used to establish the quality model which is the mapping from the furnace temperatures of each zone to the strip temperature at delivery side. GGAP-RBF is a sequential learning algorithm so that we can establish the model and update it using the new input-output data measured from the real plant. Last, the experiment results in a steel factory show that the model accuracy can meet the demand of applications in industry process.

Acknowledgement. The authors would like to acknowledge the financial support of the High Technology Research and Development Program of China (Grant No.2002AA412130) and the Specialized Research Fund for the Doctoral Program of Higher Education of China (Grant No.20020248028).

References

1. Naoharu, Y., Akihiko, H.: Model-based Control of Strip Temperature for The Heating Furnace in Continuous Annealing. IEEE Transactions on Control Systems Technology, 6, (1998) 146-156
2. Derong, L., Hohil, M., Stanley, S.: N-bit Parity Neural Networks: New Solutions Based on Linear Programming. Neurcomputing, 48, (2002) 477-488
3. Yingwei, L., Sundararajan, N., Saratchandran, P.: A Sequential Learning Scheme for Function Approximation Using Minimal Radial Basis Function (RBF) Neural Networks. Neural Computation, 9, (1997) 461-478
4. Rojas, I., Pomares, H., Bernier, J.L., Ortega, J., Pino, B., Pelayo, F.J., Prieto, A.: Time Series Analysis Using Normalized PG-RBF Network With Regression Weights. Neurocomputing, 42, (2002) 267-285
5. Guangbin, H., Saratchandran, P., Sundararajan, N.: A Generalized Growing and Pruning RBF (GGAP-RBF) Neural Network for Function Approximation, Accepted by IEEE Transactions on Neural Network. (2004)

A Learning-Based Contact Model in Belt-Grinding Processes

Xiang Zhang, Bernd Kuhlenkötter, and Klaus Kneupner

University Dortmund, Department of Assembly and Handling System
Leonhard-Euler-Strasse 2,44227 Dortmund, Germany
{xiang.zhang,bernd.kuhlenkoetter,klaus.kneupner}@uni-dortmund.de

Abstract. It is crucial to calculate the Signorini contact problem at high speed in real-time simulation of the belt grinding process. Finite Element Method (FEM) is the traditional way to solve such a contact problem. However, FEM is too time-consuming to fulfill the real-time simulation requirement. This paper demonstrates a new approach to model the Signorini contact problem based on learning. This new model averts doing the expensive optimization problem for each contact by FEM; hence dramatically reduces the calculating time. Multi-Layers Neurons (MLN) and Support Vector Regression (SVR) are adopted as a learning machine. The testing errors of these two learning machines are also compared. SVR showed superior performance than MLN in this application.

1 Bottleneck in Belt Grinding Process Simulation

Recently, industrial robots are introduced in the free-formed surface grinding process to promote the surface quality and raise producing efficiency as well. In order to automatize the process, a simulation of the real time grinding status becomes necessary. Through the simulation results it turns possible to plan the robot path automatically and adjust robot reaction dynamically.

The simulation of the grinding process is more difficult than those operating processes by precise cutting edges like milling, because the removal of surface can not be obtained directly by CAD data and tool shape. Quite several parameters can simultaneously work on the final removal from the workpiece surface, for example material of the grinding belt, the elasticity of the grinding wheel, temperature and so on. Thus, the removal from workpiece surface can be written as a function of some factors. To model this, Hammann [1,2] presented a linear experiential formula

$$r = K_A . k_t . \frac{V_b}{V_w, l_w} . F_A \tag{1}$$

where r is the removal; K_A, a combination constant of some static parameters; k_t, grinding belt wear factor; V_b, grinding velocity; V_w, workpiece moving velocity; l_w, length of grinding area, F_A, the acting force. Among those factors, the most important one is the resulting force F_A between the workpiece surface

F. Yin, J. Wang, and C. Guo (Eds.): ISNN 2004, LNCS 3174, pp. 761–766, 2004.

and the elastic grinding wheel. Roughly speaking the global removal is linearly related to the global force given that other factors are unvaried according to formula (1). This linear global relation can be applied to grind the surface with a sample shape. But for free-formed surface grinding, one should get the thorough removals in entire grinding area which are not homogeneously distributed on the workpiece surface. The different removals on the workpiece surface result from the different local contact forces between the workpiece surface and the elastic grinding wheel. Therefore, force distribution should be determined before the local removals are considered.

The traditional way to get force distribution is to consider it as a Signorini contact problem and then calculating the forces according to an energy minimization principle. Blum [3,4] and Suttmeier [5] worked out a Finite Element model that deals with Signorini contact problem with an optimized mesh discretization. Nevertheless, it still requires about 15 minutes for calculating one contact situation. This is far away from the demand for real time simulation of belt grinding process, not to mention adjusting the robot's reaction in time. Calculating the force distribution becomes a bottleneck of the whole simulation flow.

To accelerate the calculation, two branches are under research nowadays. The first branch is to optimize the mesh division; another one is to improve the convergent rate and stability of the optimization algorithm. Both can't avert doing the iteration steps each time when a new contact situation is presented. To overcome this, a learning machine is introduced to approximate the well-established FEM model. Although an optimization process is also necessary in the training phase, it can finish calculation of one contact situation in a very short time because the time-consuming transaction is put in the training phrase, no longer in the run time. Multi-Layer Neurons (MLN) and Support Vector Regression (SVR) are tested as the learning machine.

2 Contact Digitization and Feature Selection

Fig (1) illustrates a contact status between the workpiece and the grinding wheel. Imagine that the elastic grinding wheel deforms equally according to the workpiece surface [1]. In the same words, the grinding wheel and workpiece cling together without gaps between them. Then, the distances of grinding wheel surface to a base surface totally define the workpiece surface geometry within contact area, see right part of fig (1). The base surface is vertical to the grinding wheel and is fixed with a constant relative position to the grinding wheel. Using an imaginary grinding wheel surface instead of the workpiece surface aims to achieve a unified measure standard. This contact representation method is called Height Model, which is specially put forward to belt grinding contact problem for our destination. The distances between the imaginary grinding wheel surface and the base surface (the length of pillars) are called Heights. For simplification, the

[1] Not really so, it is only a method to represent the surface geometry of workpiece surface

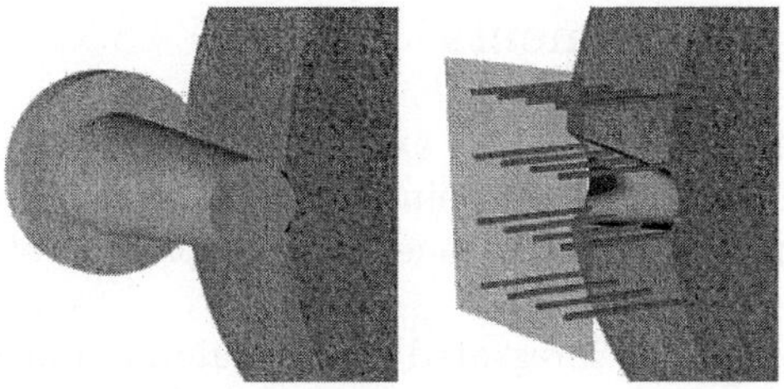

Fig. 1. Height model

contact area is limited to a $50mm \times 50mm$ square area and is discretized into 50×50 mesh evenly spaced with $1mm$ interval. Therefore, the initial contact boundary condition can be digitized into a 50×50 matrix.

It is not a good idea to impose the whole contact (Heights matrix) directly as a input on learning machine, just like what is done by FEM. Normally, one can't expect good results or generalization with such a high input dimension due to the so-called curse of dimensionality. According to one assumption, the contact problem can be localized to reduce the input dimension. The assumption is that **the force on one mesh point is affected only by contact situation (Heights) of its surrounding points inside a finite size area**. That is to say, the force at one contact point is a function of Heights of its surrounding points inside an area named function area. Obviously, if the function area is large enough or is the whole contact area, the assumption is undoubtedly correct. The small function leads to a low input dimension, but weakens the correctness of the assumption. Thus, the function area size must not be too small to guarantee the assumption's correctness. Through training experiments, the best function area size is $11mm \times 11mm$. So, all the experiments results listed in later part are adopting this function area size.

The input dimension is 121, which is still high, when the function area size is 11mm. Two methods can be used to further reduce the input dimension. One is Part Point Selection (PPS); another one is Principle Component Analysis (PCA) [6]. **PPS**, as its name implies, takes only a couple of points in function area instead of all points with preferences. PPS can easily lower the input dimension without losing much information because the workpiece surface is assumed to be continuous in every direction and varied smoothly, not sharply. Only 41 points, which locate on four lines(vertical, horizontal and two diagonals), are selected out from all 121 points in the $11 \times 11mm^2$ function area. Compared to the PPS that is enlightened by the practical case, **PCA** is a pure mathematic method that is intended to reduce correlation of components in input data. It first orthogonalizes the components of the input vector so that they are uncorrelated with each other; Then eliminates those components in the input vector that contribute to variation less than specified ratio. PCA can largely elevate the performance of MLN, but demands additional computation.

3 Numerical Experiments

180 characteristic contact situations are defined for training and 64 for testing and there are over 100,000 contact points in training situations and about 50,000 in testing situations. There exists one training pair according to one contact point.

For MLN, PCA must be integrated in the training process, otherwise results will be very poor generalization. As mentioned above, because PCA need additional computer resource, only a part of the training sets can be involved in the training of MLN. One hidden layer is used because more than one hidden layer didn't show any advantages in experiments. In comparison to MLN, SVR doesn't rely much on the PCA. So it is possible to get more training sets into training. We have tried linear kernel, RBF kernel, polynomial kernel of degree two and three. Except that the linear SVM is apparently incapable of the task, other three kernels performed slightly different from each other depending on the range of tube width ε. The difference can be neglected when the value of ε is in normal work range. However, the RBF executed much faster than polynomial kernels. The sigmoid kernel is not included in evaluation for two reasons. The one is that sigmoid kernel does not always fulfill the Mercer Condition [7, 8]; another is that the sigmoid would perform similarly to RBF kernel when the phase item θ is a very small value [9]. Therefore, in the experiments we only consider RBF kernel in the model.

Table 1. One model for all training sets

MLN			SVR		
Training Pairs	Neurons	Mean Relative Simulation Error	Training Pairs	nSV	Mean Relative Simulation Error
942	8	11.2%	5160	540	8.6%
1570	10	10.5%	5160(PPS)	606	8.8%
2355	11	10.4%	6880	887	7.7%
3140	11	10.2%	6880(PPS)	675	6.5%
4710	15	10.8%	10264	933	8.2%
			10264(PPS)	880	8.2%

Table (1) shows the best results of different batches of training pairs. All models are managed to simulate the 64 contact situations after they are trained. Mean Relative Simulation Error indicates the average simulation error of 64 testing contact situations. For MLN, the simulation error is not likely to keep on declining when the training sets are over 3000. The error can't fall under 10% even when getting more training sets in training and more neurons serving in hidden layer. For SVR, only PPS is adopted for feature selection. It is very remarkable that the results with PPS are not worse (even better) than that without PPS. The best result of SVR is 6.5%, which is still required to be cut down though excels 10.2% of MLN.

To do this, a training set classification strategy is applied by close analysis of the simulation result. Fig (2) is an example of simulation result of the SVR model created above. It is apparent that errors are relatively bigger in the boundary and corner of the contact area. The mapping from local Heights to forces is dependent on the subarea, in which the calculation point locates. Therefore, training sets (contact points) are divided into some categories according to local contact situation (16 possibilities all together) and then train a model for each category to further reduce the simulation error.

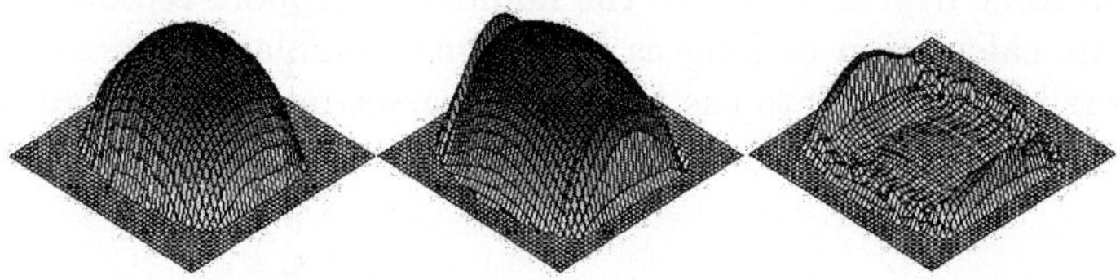

Fig. 2. Simulation result of one model for all. The left is the output of the FEM model; middle is the output from SVR and right is the error between them.

Table 2. One model for each category

	MLN		SVR	
Type	Neurons	Relative Testing Error	nSV	Relative Testing Error
1	3	6.2%	421	3.0%
2	6	5.2%	657	4.0%
3	8	5.4%	1023	3.0%
4	10	2.7%	1081	2.0%
5	8	5.0%	659	4.0%
6	2	4.0%	657	1.9%
7	10	2.8%	1112	2.0%
8	8	2.6%	1233	2.8%
9	4	5.1%	1027	3.0%
10	10	3.1%	1092	1.9%
11	5	6.9%	680	2.4%
12	8	2.6%	1303	1.8%
13	8	2.7%	1119	1.9%
14	8	3.1%	1277	2.8%
15	8	2.6%	1273	1.8%
16	4	1.5%	1159	2.3%
MRSE	4.9%		4.1%	

This time PPS and PCA are combined in MLN training and SVR uses only PPS for feature selection. Table (2) are training results of MLN and SVR.

SVR excels MLN once again in performance. Except type 16, SVR generalizes a smaller testing error than MLN. The MRSE in Table (2) is the mean relative simulation error of 64 testing contact situations. The error of SVR is 4.1% compared to MLN 4.9%. Additionally, the force distribution given by SVR looks smoother than that by MLN. However, MLN conducts calculation much faster than that of SVR. Operating complexity for MLN is approximate 2×10^6 times multiplications in contrast to that of SVR, 7×10^7 multiplications plus 1.5×10^6 exponential operation. SVR requires about 1 second for one contact situation calculation on a PC (CPU AMD Athlon 2600, 512MB memory). The $\varepsilon-tube$ can be relaxed in some degree to reduce the number of support vectors and hence to accelerate the calculation as long as the average simulation error is below 5%. It is also possible to calculate the forces on coarser mesh (less than 50 knots in two directions) and then apply interpolation to reduce the calculation time.

4 Conclusion

This paper demonstrates a new way to model the Signorini problem using SVR and MLN to learn the map relation rather than solve the optimization problem each time when a new contact situation is presented. The experiments show that both SVR and MLN can approximate the traditional FEM model with error below 5%. SVR has a relatively better approximating precision than that of MLN, but longer calculating time than MLN. Calculating time is reduced to 1 second compared to 15 minutes of original FEM model. This makes a real-time simulation of belt grinding process possible and imposes the control on robots as well.

References

1. Hammann, G.: Modellierung des Abtragsverhaltens elastischer robotergefuehrter Schleifwerkzeuge. PhD thesis, University Stuttgart (1998)
2. Schueppstuhl, T.: Beitrag zum Bandschleifen komplexer Freiformgeometrien mit dem Industrieroboter. PhD thesis, University Dortmund (2003)
3. Blum, H., Suttmeier, F.T.: An adaptive finite element discretisation for a simplified signorini problem. Calcolo **37** (2000) 65–77
4. Blum, H., Schroeder, A., Suttmeier, F.T.: A posteriori error bounds for finite element schemes for a model friction problem, Witten-Bommerholz (2003)
5. Suttmeier, F.T.: Error Analysis for Finite Element Solutions of Variational Inequalities. PhD thesis, Dortmund University (2001)
6. Jolliffe, I.: Principal Component Analysis. Springer, New York (1986)
7. Burges, C.J.C.: A tutorial on support vector machines for pattern recognition. Data Mining and Knowledge Discovery **2** (1998) 121–167
8. Smola, A., Schoelkopf, B.: A tutorial on support vector regression (1998)
9. Lin, H.T., Lin, C.J.: A study on sigmoid kernels for svm and the training of non-psd kernels by smo-type methods (2003)

Application of General Regression Neural Network to Vibration Trend Prediction of Rotating Machinery

Zhipeng Feng[1], Fulei Chu[1], and Xigeng Song[2]

[1]Department of Precision Instruments and Mechanology, Tsinghua University, Beijing 100084, China
{fengzp, chufl}@mail.tsinghua.edu.cn
[2]Department of Power Engineering, Dalian University of Technology, Dalian 116024, China
xuedx@dlut.edu.cn

Abstract. The General Regression Neural Network (GRNN) is briefly introduced. The BIC method for determining the order of Auto Regression (AR) model is employed to select the number of input neurons, and the Genetic Algorithm is applied to calculate the optimal smoothing parameter. The GRNN is used to predict the vibration time series of a large turbo-compressor, and its performance is compared with that of Radial Basis Function Neural Network (RBFNN), Back Propagation Neural Network (BPNN), and AR. It is indicated that the GRNN is more appropriate for the prediction of time series than the others, and is qualified even with sparse sample data.

1 Introduction

It is difficult to accurately model complicated machinery for vibration trend prediction due to the dynamic complexity and various influences. Therefore, the methods for time series prediction such as Auto Regression, Auto Regression and Moving Average, Grey Model, and Prony Exponential Model are usually employed to solve this problem [1]. Besides, neural networks are applicable for their ability to the approximation of any map. In this paper, GRNN is applied to vibration condition prediction of a large turbo-compressor, and is compared with RBFNN, BPNN, and AR.

2 General Regression Neural Network

The general regression neural network was presented by D. F. Specht in 1991 [2]. Based on nonlinear kernel regression in statistics, it can approximate the map inherent in any sample data, and the estimate can converge to the optimal regression surface even with sparse sample. At present, it has been applied in a variety of fields such as system identification, adaptive control, pattern recognition, and time series prediction [2-6].

F. Yin, J. Wang, and C. Guo (Eds.): ISNN 2004, LNCS 3174, pp. 767–772, 2004.

2.1 General Regression

From the Parzen's nonparametric probability density estimate, the fundamental formulation of the GRNN is deduced as follows.

$$\hat{Y}(X) = \frac{\sum_{i=1}^{n} Y_i \exp[-\frac{(X-X_i)^T(X-X_i)}{2\sigma^2}]}{\sum_{i=1}^{n} \exp[-\frac{(X-X_i)^T(X-X_i)}{2\sigma^2}]} \tag{1}$$

where X, X_i, and Y_i are the sample observations, $\hat{Y}(X)$ is the estimate associated with the sample X, σ is the smoothing parameter, i.e. the kernel width of the Gaussian function, and n is the number of samples. It can be noted that the estimate $\hat{Y}(X)$ is the weighted average of all the sample observations Y_i, where the weight for each observation is the exponential of the squared Euclidean distance between the sample X and X_i.

From formula (1), the general regression neural network is constructed as shown in Fig. 1.

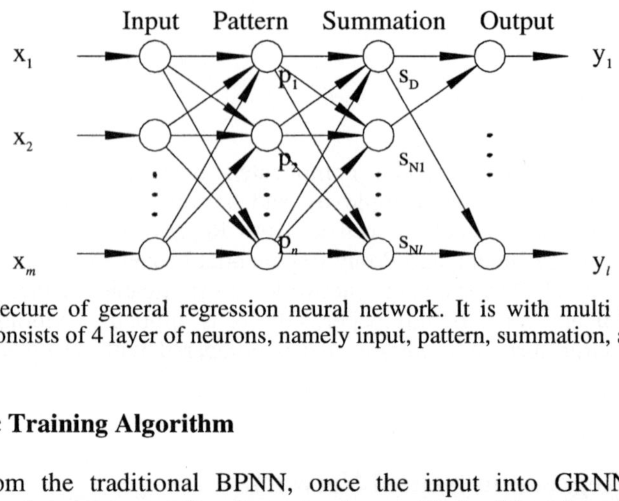

Fig. 1. Architecture of general regression neural network. It is with multi inputs and multi outputs, and consists of 4 layer of neurons, namely input, pattern, summation, and output layer.

2.2 Genetic Training Algorithm

Different from the traditional BPNN, once the input into GRNN is given, its architecture and weights are also determined accordingly. Therefore, training GRNN is essentially to optimize the smoothing parameter, and thereby modify the transfer functions of the neurons in pattern layer to obtain the optimal regression estimate. Presently, various optimization methods are employed to improve the training algorithm, such as the simple climbing up and the conjugate gradient [3,4].

Compared with those methods, genetic algorithm is more suitable to solve complex optimization problem. It is originated from biological evolutions and genetics. Because of the computation complexity, the difficulty in computing the derivatives, and the uncertain existence of the convex derivative, the genetic algorithm is adopted to find the optimal smoothing parameter for its global optimization in this paper. The genetic algorithm attempts to maximize the fitness defined as

$$f(\sigma) = \frac{1}{1+e} \tag{2}$$

subject to $\sigma \in [\sigma_{\min}, \sigma_{\max}]$, where e is the mean squared error obtained with the holdout method, i.e. remove a sample from the training data, construct a GRNN with the rest to estimate this sample observation, and calculate the resulting estimate error. Repeat the same process for each sample, and finally obtain the mean squared value of the error series

$$e = \frac{1}{n}\sum_{i=1}^{n}[\hat{Y}_i(X_i) - Y_i]^2 \tag{3}$$

The lower and upper limits of the smoothing parameter, $\sigma_{\min}$ and $\sigma_{\max}$, are determined respectively as follows [3].

$$\sigma_{\min} = (-\frac{D_{\min}^2}{2\ln\varepsilon})^{\frac{1}{2}} \tag{4}$$

$$\sigma_{\max} = \frac{D_{\max}}{2} \tag{5}$$

where $D_{\min}$ and $D_{\max}$ are the minimum and maximum Euclidean distance of any two samples respectively, and $\varepsilon > 0$ is the minimum value dependent on the PC hardware.

It is recommended that each sample be assigned with different smoothing parameter respectively, resulting in a better estimate [3]. While for time series prediction, the samples are with the same attribute, therefore the smoothing parameter is taken as a unique value in this paper.

It validates the neural network as well as optimizes the smoothing parameter, thereby it is unnecessary to constitute the training data with additional validation data.

3 Vibration Trend Prediction of Rotating Machinery

In mathematic sense, time series prediction is a map from the field of historic record to the field of future tendency. The mathematic model to predict the future trend in l step ahead with m historic data can be described as

$$(x(i), \cdots, x(i+l)) = p(x(i-1), \cdots, x(i-m)) \tag{6}$$

where $i = m+1, m+2, \cdots, m+(N-m-l)/s$, $(N-m-l)/s$ is the number of subsamples, N is the length of overall sample, and s is the number of sliding points (usually given with 1).

GRNN is capable of approximating any map inherent in the sample data, therefore it is applied to the vibration condition time series prediction of a large turbo-compressor in this paper.

3.1 Selection of Input Neuron Number

It is obvious that the number of the input neurons corresponds to the length of subsample for training GRNN, namely the dimension of the input pattern. With the sample data given, if the input neurons are too many, the subsample will be few. Otherwise, if the input neurons are few, the subsample will be too many. The output of GRNN is the weighted average of all the observed samples, therefore the number of the input neurons affects its prediction performance.

Referring to the determination of the order of AR model [1], a method to select the number of the input neurons is proposed. Firstly, initialize the number of input neurons according to the empirical formula for selecting the order of AR model

$$m = \begin{cases} N/2 & N = 20 \sim 40 \\ N/3 \sim N/2 & N = 40 \sim 100 \\ N/\ln N & N = 100 \sim 200 \end{cases} \tag{7}$$

Then, search the optimal number of the input neurons in a range around the initial m. After optimizing the smoothing parameter, compute the prediction error e according to (3) for each m. Finally, determine the optimal number of the input neurons according to the minimum BIC value

$$\mathrm{BIC}(m) = N \ln e + m \ln N \tag{8}$$

In practice, the prediction error e can also be used to evaluate the GRNN. When it becomes small enough, the search can be terminated.

3.2 Vibration Trend Prediction

The GRNN is applied to predicting the vibration trend of a large turbo-compressor in petrochemical industry. The rated speed and power of the machine are 12300 rpm and 2500 kw respectively. The radial displacement vibration signal of the rotor is acquired in-field with the sampling frequency being 8000Hz. The important index for the evaluation of vibration condition, namely peak-peak value is employed to predict the vibration trend of the machine. The obtained peak-peak time series is shown as Fig. 2, which is composed of 46 observations with the interval being 1 hour. The first 40 data are used as the training data, and the rest as the test data.

Firstly, according to the minimum BIC (BIC vs. number of the input neurons is shown as Fig. 3), the number of input neurons is determined as 12, and 17 subsamples are resulted consequently. Then, by virtue of genetic optimization, the optimal smoothing parameter is found to be 0.671. Finally, given these parameters, the GRNN is determined, and the time series is predicted, with the results denoted by Fig. 4 and Table. 1.

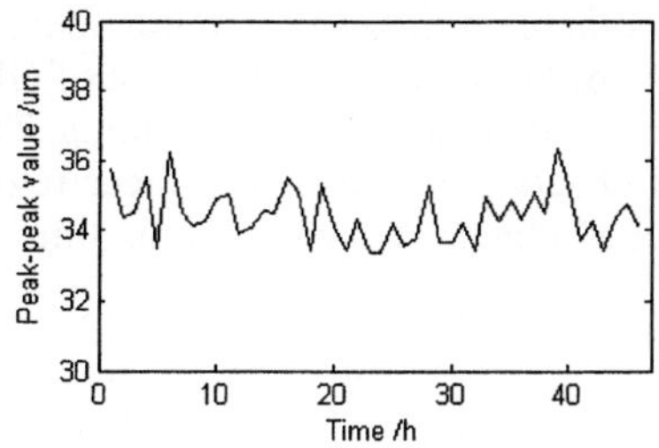

Fig. 2. Vibration peak-peak time series. It is composed of 46 observations. The first 40 data are used as the training data, and the rest as the test data

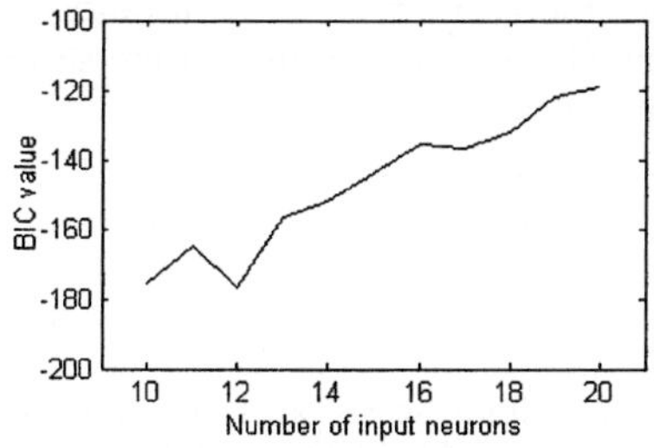

Fig. 3. BIC vs. number of input neurons. According to the minimum BIC, 12 input neurons are set for the GRNN

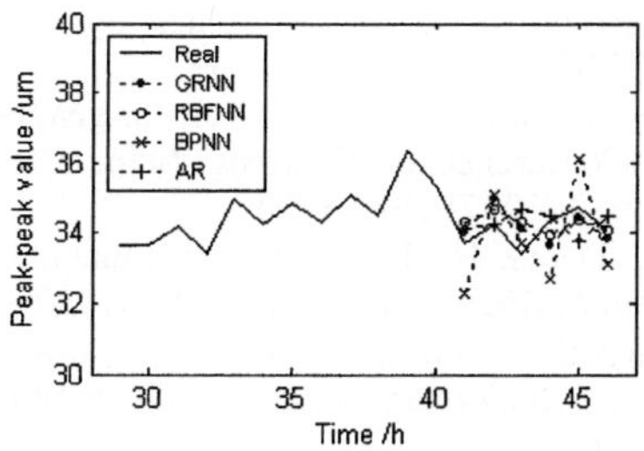

Fig. 4. Prediction of test data. The last 6 observations are predicted with GRNN, RBFNN, BPNN, and AR respectively

Table 1. Results of vibration trend prediction

Method	Architecture	Smoothing parameter	MSE of sample	MSE of test
GRNN	12-17-2-1	0.671	0.299	0.249
RBFNN	12-17-1	0.853	0.399	0.293
BPNN	12-10-1	-	0.885	1.409
AR	-	-	0.367	0.480

For performance comparison of various methods, the predictions based on the RBFNN, BPNN, and AR are also studied, and the results are shown as Fig. 4 and listed in Table. 1.

It is obvious that the prediction by GRNN reflects the trend and fluctuation of the machine vibration condition even with sparse sample data as few as 40, and its accuracy is higher than the others for either the training data or the test data. Although the prediction by RBFNN is of the same level with that of GRNN, its training with the error back propagation algorithm is more time consuming. The prediction by BPNN is far different from the real observations. AR is a linear regression model, so that its prediction cannot reflect the nonlinear dynamics of the time series accurately.

4 Conclusions

GRNN can approximate any map inherent in the observation samples with one-pass learning. By virtue of the proposed method for selection of the input neurons, and the genetic optimization of the smoothing parameter, it outperforms RBFNN, BPNN, and AR for the prediction of machine vibration trend time series, and is qualified for this task even with sparse sample data.

References

1. Yang S. Z., Wu Y., et al, Times series analysis in engineering application, Wuhan: Huazhong University of Science and Technology, (1991)
2. Specht D. F., A general regression neural network, IEEE Transactions on Neural Networks, 2 (6) (1991) 568-576
3. Tomandl D., Schober A., A modified general regression neural network (MGRNN) with new, efficient training algorithms as a robust 'black box'-tool for data analysis, Neural Networks, 14(4) (2001) 1023-1034
4. Masters T., Land W., A new training algorithm for the general regression neural network, 1997 IEEE International Conference on Systems, Man, and Cybernetics: Computational Cybernetics and Simulation, 3 (1997) 1990-1994
5. Leung M. T., Chen A. S., Daouk H., Forecasting exchange rates using general regression neural networks, Computers & Operation Research, 27(4) (2000) 1093-1110
6. Lee W. M., Lim C. P., Yuen K. K., Lo S. M., A hybrid neural network model for noisy data regression, IEEE Transactions on System, Man, and Cybernetics-Part B: Cybernetics, 34(2) (2004) 951-960

A Novel Nonlinear Projection to Latent Structures Algorithm

Shi Jian Zhao[1], Yong Mao Xu[1], and Jie Zhang[2]

[1] Department of Automation, Tsinghua University, Beijing 100084, China
zhaoshj00@mails.tsinghua.edu.cn, xym-dau@mail.tsinghua.edu.cn

[2] School of Chemical Engineering and Advanced Materials, University of Newcastle, Newcastle upon Tyne NE1 7RU, UK
jie.zhang@newcastle.ac.uk

Abstract. Starting from an equivalent presentation of projection to latent structures (PLS), a novel nonlinear PLS approach is presented where both nonlinear latent structures and nonlinear reconstruction are obtained straightforwardly through two consecutive steps. First, an radial basis functions (RBF) network is utilized to extract the latent structures through linear algebra methods without the need of nonlinear optimization. This is followed by two feed-forward networks (FFN) to reconstruct both the original predictor variables and response variables. The proposed algorithm exhibits fast convergence speed and its efficiency is assessed through both mathematical example and modelling of a pH neutralization process.

1 Introduction

PLS is a popular linear regression method based on projecting the information in a high dimensional space down onto a low dimensional space defined by some latent variables. It is successfully applied in diverse fields to deal with noisy and highly correlated data whereas only a limited number of observations available [1]. However, data from practical situations such as industrial plants generally exhibits strong nonlinear behavior. In order to address the inherent nonlinearity, a number of approaches have been proposed to extend PLS to the nonlinear case (c.f. [2,3,4] and refer to [5] for a survey).

However, these approaches suffer from being capable of presenting linear latent structures only or demanding optimization of complex networks. In this paper, based on an equivalent presentation of PLS which illustrates that linear PLS can be performed by a projection step and a reconstruction one consecutively, a nonlinear PLS algorithm is presented along the same line by employing one RBF network to extract the latent structures and two FFNs to reconstruct the original variables. The potential benefits of the proposed approach is assessed through both a mathematical example and a pH neutralization process.

F. Yin, J. Wang, and C. Guo (Eds.): ISNN 2004, LNCS 3174, pp. 773–778, 2004.

2 An Equivalent Presentation of Linear PLS

PLS is originally characterized as an iterative bootstrap process and its related mathematical properties can be found in [1]. Assuming that both $\mathbf{X}$ and $\mathbf{Y}$ assemble the variables in their columns and have zero-mean and unit-variance, statisticians have presented a perspective of PLS in favor of more interpretable version such as [6]. In fact, it can be proven that finding the i-th linear latent variable $\mathbf{t}_i = \mathbf{X}\mathbf{w}_i$ of data matrices $\mathbf{X}$ and $\mathbf{Y}$ can be reformulated as an optimization problem [7]:

$$\mathbf{w}_i = \arg\max \sum_{j=1}^{q} \langle \mathbf{t}_i, \mathbf{y}_j \rangle^2 = \sum_{j=1}^{q} \langle \mathbf{X}\mathbf{w}, \mathbf{y}_j \rangle^2 \tag{1}$$

subject to

$$\mathbf{w}^T\mathbf{w} = 1 \text{ and } \mathbf{w}^T \Sigma_{\mathbf{X}} \mathbf{W}_{i-1} = \mathbf{0}^T \tag{2}$$

where $\Sigma_{\mathbf{X}} = \mathbf{X}^T\mathbf{X}$ and $\mathbf{W}_{i-1} = [\mathbf{w}_1, \dots, \mathbf{w}_{i-1}]$. Furthermore, regressing between the obtained $\mathbf{t}_i$ and the original $\mathbf{X}$ and $\mathbf{Y}$ will result in the similar decomposition:

$$\mathbf{E}_i = \mathbf{E}_{i-1} - \mathbf{t}_i\mathbf{p}_i^T \text{ with } \mathbf{p}_i = \frac{\mathbf{X}^T\mathbf{t}_i}{\mathbf{t}_i^T\mathbf{t}_i} \tag{3}$$

$$\mathbf{F}_i = \mathbf{F}_{i-1} - \mathbf{t}_i\mathbf{r}_i^T \text{ with } \mathbf{r}_i = \frac{\mathbf{Y}^T\mathbf{t}_i}{\mathbf{t}_i^T\mathbf{t}_i} \tag{4}$$

Note that in both (3) and (4) the regression coefficients $\mathbf{p}_i$ and $\mathbf{r}_i$ associate with the original $\mathbf{X}$ and $\mathbf{Y}$.

According to the presented equivalent presentation, PLS can be regarded as, roughly speaking, two related but relatively independent steps:

1. *Projection* step: to calculate the latent variables which have maximum covariance with the response variables under certain regularization constraints.
2. *Reconstruction* step: providing the latent variables $\mathbf{T}$ are obtained, it can then be utilized to reconstruct the original data $\mathbf{X}$ and $\mathbf{Y}$ or, equivalently, to deflate them.

3 Nonlinear Projection to Latent Structures

Although different presentations of linear PLS are mathematically equivalent, the extensions of linear PLS may have different possibilities providing distinct performance rules are employed for nonlinear methodologies. Recognizing that the predictor variables can correlate nonlinearly with each other and with the response variables, in this paper, both the projection and reconstruction steps are identified through nonlinear neural networks.

3.1 Nonlinear Optimization Objective Functions

As aforementioned, the main idea of PLS is to perform the *projection* step to attain the latent variables $\mathbf{T}$ and then the *reconstruction* step to reproduce the original data. Intuitively, if $\mathbf{T}$ is related to the original process variables $\mathbf{X}$ through certain nonlinear relationship, the *projection* part of PLS is then extended to the nonlinear case. Analogously, if the original $\mathbf{X}$ and $\mathbf{Y}$ are reproduced by $\mathbf{T}$ nonlinearly, the *reconstruction* part will then also be nonlinear.

Nonlinear projection. For the i-th latent structure $\mathbf{t}_i$, analogous to (1), assuming that $\mathbf{t}_i = \mathbf{g}_i(\mathbf{X})$ where function $\mathbf{g}_i(\cdot)$ is certain nonlinear mapping satisfying specific regularization constraints, the search of $\mathbf{t}_i$ can then be formulated as:

$$\underset{\mathbf{t}_i}{\text{maximize}} \ J(\mathbf{t}_i) = \sum_{j=1}^{q} \langle \mathbf{t}_i, \mathbf{y}_j \rangle^2 \tag{5}$$

where $\mathbf{Y} = [\mathbf{y}_1, \ldots, \mathbf{y}_q]$. After $\mathbf{t}_i$ is obtained, it can then be employed to reconstruct the original data $\mathbf{X}$ and $\mathbf{Y}$.

If it is required that A latent variables $\mathbf{T} = [\mathbf{t}_1, \ldots, \mathbf{t}_A] = \mathbf{g}(\mathbf{X})$ should be extracted simultaneously, (5) can be modified correspondingly:

$$\underset{\mathbf{t}_i}{\text{maximize}} \ J(\mathbf{t}_i) = \sum_{k=1}^{q} \langle \mathbf{t}_i, \mathbf{y}_k \rangle^2, \ \text{for } i = 1, \ldots, A \tag{6}$$

subject to the orthogonality constraint $\mathbf{t}_i^T \mathbf{t}_j = 0$, for $j < i$.

Suppose that an RBF network illustrated in Fig.1 is employed for the extraction of latent structures. Denote $\mathbf{C}$ and σ the centers and the corresponding widths of the RBF function, respectively. For input patterns $\mathbf{X} \in \Re^{n \times p}$, let $\mathbf{f}(\mathbf{X}) \in \Re^{n \times h}$ be the output of neurons in the RBF hidden layer where h is the number of neurons in the hidden layer. The k-th latent variable $\mathbf{t}_k = \mathbf{f}(\mathbf{X})\mathbf{w}_k$ can then be formulated as

$$\underset{\mathbf{w}_k}{\text{maximize}} \ J(\mathbf{t}_k) = \sum_{i=1}^{q} \langle \mathbf{f}(\mathbf{X})\mathbf{w}_k, \mathbf{y}_i \rangle^2 \tag{7}$$

subject to $\mathbf{w}_k^T \mathbf{w}_k = 1$ and $\mathbf{t}_i^T \mathbf{t}_j = 0$, for $j < i$. Let

$$\mathbf{B} = \sum_{i=1}^{q} \mathbf{f}^T(\mathbf{X})\mathbf{y}_i \mathbf{y}_i^T \mathbf{f}(\mathbf{X}) = \mathbf{f}^T(\mathbf{X})\mathbf{Y}\mathbf{Y}^T \mathbf{f}(\mathbf{X})$$

Eqn.(7) can thus be represented as

$$\underset{\mathbf{w}_k}{\text{maximize}} \ J(\mathbf{t}_k) = \sum_{i=1}^{q} \langle \mathbf{f}(\mathbf{X})\mathbf{w}_k, \mathbf{y}_i \rangle^2 = \mathbf{w}_k^T \mathbf{B} \mathbf{w}_k \tag{8}$$

It is apparent that this is a linear PLS problem or equivalently, to find the first A linear latent structures between data $\mathbf{f}(\mathbf{X})$ and $\mathbf{Y}$. It can be solved using linear algebra methods directly.

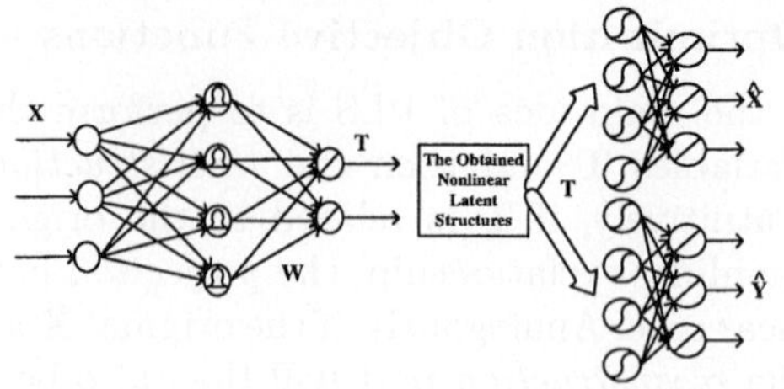

Fig. 1. Architecture of networks to perform nonlinear PLS. Nodes with curves represent nonlinear activation functions and straight lines the linear ones

Nonlinear reconstruction. The step aims at minimizing the reconstruction errors $\tilde{\mathbf{X}} = \mathbf{X} - \hat{\mathbf{X}}$ and $\tilde{\mathbf{Y}} = \mathbf{Y} - \hat{\mathbf{Y}}$. Because a neural network with one hidden layer of sigmoidal units can model any nonlinear function with bounded number of hidden units [8], two separate networks are employed to reconstruct **X** and **Y** from the obtained **T**, respectively.

Therefore, the proposed nonlinear PLS algorithm can be regarded as the combination of two steps:

1. Extract the latent structures by the RBF network through linear algebra without nonlinear optimization introduced. The centers **C** and widths σ of the Gauss RBF network are determined independently prior to the calculation of weights **W** through clustering methods such as k-means and the corresponding width determination algorithm [9].
2. Reproduce or reconstruct the original data **X** and **Y** using two independent networks as shown in Fig.1, whose structures are determined by cross-validation, through extensively investigated methods such as error back-propagation and Levenberg-Maquardt algorithm [10].

4 Results

4.1 Example 1: A Simulation Problem

A mathematical example is devised to assess its efficiency. It is defined as

$$\begin{aligned} x_1 &= t^2 - t + 1 + \epsilon_1 \\ x_2 &= \sin t + \epsilon_2 \\ x_3 &= t^3 + t + \epsilon_3 \\ y &= x_1^2 + x_1 x_2 + 3\cos x_3 + \epsilon_4 \end{aligned}$$

where t is uniformly distributed over [-1, 1] and $\epsilon_i, i = 1, 2, 3, 4$, are noise components uniformly distributed over [-0.1, 0.1].

The generated data with 300 samples is segmented into training, validation and testing data sets consecutively with equal length. One latent structure is retrieved. The number of neurons in the hidden layers for the RBF network with Gaussian functions and two feed-forward networks with sigmoidal functions to reconstruct $\mathbf{X} = [x_1, x_2, x_3]$ and $\mathbf{Y} = y$ are finally chosen as 9, 8 and 15 respectively through cross-validation. Mean-squared-errors (MSE) of response variables $\mathbf{Y}$ is used to determine the network architecture. The results are listed in Table 1. Results using linear PLS are also included for comparison. It is plain to see that the proposed approach provides a better presentation of the underlying problem than the linear PLS method.

Table 1. The MSE for training, validation and testing data sets

Data	1 LV NLPLS		1 LV PLS		2 LVs PLS		3 LVs PLS	
Set	MSE_X	MSE_Y	MSE_X	MSE_Y	MSE_X	MSE_Y	MSE_X	MSE_Y
Train	0.0190	0.0543	0.0337	0.2984	0.0051	0.0899	0.0000	0.0818
Validation	0.0211	0.0821	0.0304	0.3225	0.0053	0.1227	0.0000	0.1088
Test	0.0186	0.0850	0.0257	0.2664	0.0056	0.1258	0.0000	0.1070

4.2 Example 2: Modelling of a Neutralization Process

In this example consider a well-known highly nonlinear pH neutralization process taking place in a CSTR (Continuous Stirred-Tank Reactor) with two input streams of acetic acid and sodium hydroxide. The data sets are generated by adding multi-level random perturbations to the flow rate of acetic acid $u(t)$ while keeping the other inputs to the reactor constant. The output pH measurements $y(t)$ are corrupted with random noise with the distribution $N(0, 0.2)$. The dynamic models are of the form $y(t) = f[y(t-1), y(t-2), u(t-1), u(t-2)]$ and refer to [11] for details.

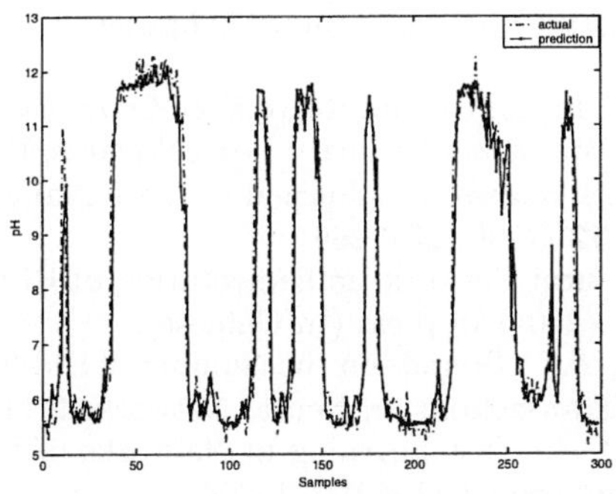

Fig. 2. Predictions of the model on testing data set

The training data set is therefore 4-dimension and contains totally 298 samples. Finally three latent structures are extracted and the number of neurons in the hidden layers for the RBF network with Gaussian functions and the reconstruction networks with sigmoidal functions for **X** and **Y** are chosen as 4, 10 and 20 respectively through cross-validation. The corresponding prediction result for unseen testing data shown in Fig.2 is very satisfactory.

5 Conclusions

Based on an equivalent presentation of PLS, a novel nonlinear PLS approach is presented where both nonlinear latent structures and nonlinear reconstruction are obtained straightforwardly through two consecutive steps. Its efficiency is demonstrated through both mathematical example and a pH neutralization process.

Acknowledgements. The authors greatly appreciate the financial support of both the UK EPSRC through the grant GR/R10875 and the National High Technology Research and Development Program of China (863 Program) (No. 2001AA413320).

References

1. Höskuldsson, A.: PLS Regression Methods. Journal of Chemometrics **2** (1988) 211–228
2. Qin, S.J., McAvoy, T.J.: Nonlinear PLS Modeling Using Neural Network. Computers & Chemical Engineering **16** (1992) 379–391
3. Wilson, D.J.H., Irwin, G.W., Lightbody, G.: Nonlinear PLS Using Radial Basis Functions. Transactions of the Institute of Measurement and Control **19** (1997) 211–220
4. Malthouse, E.C., Tamhane, A.C., Mah, R.S.H.: Nonlinear Partial Least Squares. Computers & Chemical Engineering **21** (1997) 875–890
5. Baffi, G., Martin, E.B., Morris, A.J.: Non-linear Projection to Latent Structures Revisited: the Quadratic PLS Algorithm. Computers & Chemical Engineering **23** (1999) 395–411
6. Stone, M., Brooks, R.J.: Continuum Regression: Cross-validated Sequentially Constructed Prediction Embracing Ordinary Least Squares, Partial Least Squares and Principal Components Regression. Journal of the Royal Statistical Society. Series B (Methodological) **52** (1990) 237–269
7. Zhao, S., Xu, Y.: Neural Network Implementation of PLS Algorithm. Journal of Tsinghua University (2004) In press (In Chinese).
8. Huang, S.C., Huang, Y.F.: Bounds on the Number of Hidden Neurons in Multilayer Perceptrons. IEEE Transactions on Neural Networks **2** (1991) 47–55
9. Moody, J., Darken, C.J.: Fast Learning in Networks of Locally-tuned Processing Units. Neural Computation **1** (1989) 281–294
10. Hagan, M.T., Menhaj, M.B.: Training Feedforward Networks with the Marquardt Algorithm. IEEE Transactions on Neural Networks **5** (1994) 989–993
11. Zhang, J.: Developing Robust Neural Network Models by Using Both Dynamic and Static Process Operating Data. Industrial & Engineering Chemistry Research **40** (2001) 234–241

Inversing Reinforced Concrete Beams Flexural Load Rating Using ANN and GA Hybrid Algorithm

Zeying Yang[1], Chengkui Huang[1], and Jianbo Qu[2]

[1] State Key Laboratory of Coastal and Offshore Engineering, Dalian University of Technology, Dalian 116024, China
yangbridge2003@yahoo.com.cn
huangck@dlut.edu.cn

[2] Shandong shipping (group) CO., LTD. Jinan 250014, China
jb@qu.com.cn

Abstract. This paper presents a new method of using ANN (Artificial Neural Networks) to solve the inverse problem: predicting reinforced concrete (RC) beams flexural load rating (ratio of flexural load to ultimate bearing capacity) given apparent damage parameter (crack width, crack height, deflection). A hybrid algorithm (G-Prop) that combines a genetic algorithm (GA) and BP (back-propagation) is used to train ANN with a single hidden layer. The GA selects the initial weights and changes the number of neurons in the hidden layer through the application of specific genetic operators. The case study shows that ANN based on GA and BP is a powerful instrument for predicting flexural load rating and further for evaluating RC beams safety.

1 Introduction

Up to now, a proof load test [1] is the more appropriate method to assess the actual load carrying capacity of a bridge. A successful proof load test demonstrates immediately that the resistance of the bridge is greater than the proof load. However extensive load testing programs are very costly. It should be recognized also that there is a risk that the structure will be damaged or not survive a proof load test (referred to herein as test risk) and so proof load testing may not always be cost-effective. So a new method to predict load rating (ratio of bearing load to ultimate bearing capacity) and further to evaluate safety of RC beams of bridges without carrying out load testing is cried for in practical engineering. It is a fact that the relative size between load and resistance appears through structural element's apparent damage parameter (e.g. crack width, crack height, deflection etc.) [2,3]. China standard of "Technical Specification of Maintenance for Highway" [4] specified that maximum of vertical crack width near the main steel is *0.25mm* for reinforced concrete beams, but witch always is considered from durability and can't be used as standard to evaluating the safety of RC beams.

F. Yin, J. Wang, and C. Guo (Eds.): ISNN 2004, LNCS 3174, pp. 779–785, 2004.

There are seldom study concerning predicting flexural load rating of RC beams through their apparent damage parameter seen in last literatures. Literature [5] proposed two simple formula by carrying out dead-load test and regression analysis of testing results of RC beams. But their calculating errors are too big to be used in practice.

The relationship between flexural load rating and apparent damage parameter of RC beams is too complex to be represented by explicit mathematical function. This paper presents a new method of using ANN (Artificial Neural Networks) to solve the inverse problem: predicting load rating given apparent damage parameter. Witch aims to provide engineering assessor with a useful approach.

2 BP Neural Networks

An artificial neural network [6] is a massively parallel distributed processor made up of simple processing units (neurons), witch has a natural propensity for storing experiential knowledge and making it available for use. It resembles the brain in two respects: ① Knowledge is acquired by the network from its environment through a learning process. ② Inter neuron connection strength ,known as synaptic weights, are used to store the acquired knowledge. Any kinds of nonlinear mapping can be implemented by a three layer feedforward neural network. In an implicit sense, the learning process must decide witch features of the input pattern should be represented by the hidden neurons. The learning process is therefore made more difficult because the search has to be conducted in a much large space of possible functions, and a choice has to be made between alternative representations of the input pattern. The development of the back-propagation algorithm represents a landmark in neural networks in that it provides a computationally efficient method for the training of multiplayer ANN. For detail, refer to [6].

3 Genetic Algorithms

Genetic algorithms (GAs) [7,8] are new global optimization techniques based on the principles of natural evolution. GA operates on the population of potential solutions (also called chromosomes) to a problem. A notion of fitness is used in GA to measure the "goodness" of a candidate solution (chromosome). Genetic operators of selection, crossover, and mutation are repeatedly applied to the population to increase the fitness of chromosomes.

The success of employing GA to solve a given optimization problem greatly depends on correctly choosing the fitness function. Fitness function must provide positive values and must be maximized. Representation of candidate solution is another important issue in GA-based optimization. Traditionally, GA use binary encoding to build chromosomes. Therefore, an adequate mapping of problem variables to binary strings and vice versa is required.

Selection operator randomly chooses chromosomes from the population to produce a new population with the probabilities proportional to the values of fitness of chromosomes. Two other genetic operators: crossover and mutation are applied to the new population.

The crossover operator randomly picks two chromosomes and mates them producing two child chromosomes. To this end, first the split point in chromosomes is randomly chosen, then child chromosomes are generated, each of them inheriting the genes up to the split point from one parent, and from that point on from the other parent. Mutation randomly picks one of the bits in a chromosome and flips it. Fig.1 expresses how to generate offspring (an elite preservation strategy is adopted).

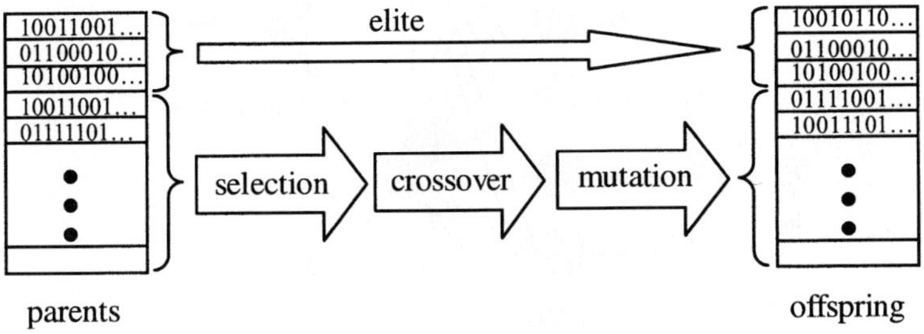

Fig. 1. Generate a new population. This shows how to generate offspring (an elite preservation strategy is adopted).

4 ANN Based on GA and BP

4.1 Back-Propagation ANN

In this paper, a three-layer feedforward neural network [9,10] is established to map the nonlinear relationship between apparent damage parameter (including crack width, crack height, deflection) and load rating of RC beams. Recently, literature [5] has carried out a dead-load test of reinforced concrete beams and has done regression analysis of the test data. Based on witch, he then pointed out that crack intervals couldn't be used to deduce load rating because of dependency relation between them is not distinct enough. So according to literature [2,3,5], the input layer to the network consists of nine neurons: crack width, relative crack height, deflection, concrete compression strength, reinforcement ratio, steel bar yield strength, steel bar diameter, bond stress, cover thickness, they are vector $X : x_1,\ x_2,\ x_3,\ldots,x_9$ respectively. The number of neurons in the output layer is one, which corresponds to the number of output: load rating, it is vector y. The reader is reminded that in this paper relative crack height refers to the ratio a/h and load rating refers to m/m_u (where a is the crack height, h is the beam depth, m is the moment load, m_u is the flexural ultimate load-bearing capacity, respectively). Fig.2 shows the architecture of the neural network.

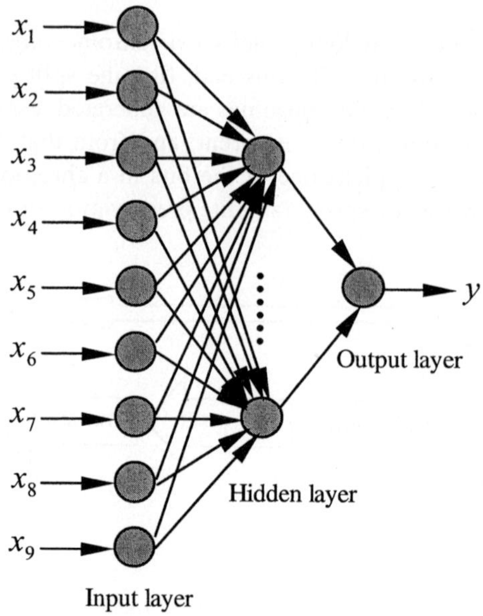

Fig. 2. Architecture of back-propagation neural network. This shows the architecture of the neural network.

4.2 Hybrid Algorithm Based on GA and BP

Even as BP methods are successfully used in many fields, due to its leaning ability, it does encounter certain difficulties in practice [11]: ① convergence tends to be extremely slow and very dependent on the initial weights; ② convergence to the global optimum is not guaranteed; ③ learning constants and hidden layer size must be guessed heuristically or through systematic tests. Evolutionary neural networks are a more efficient way of searching, but still, they search over a subset of all possible parameters. The hybrid approach of GA and BP to train ANN makes use of the capacity of both algorithms: the ability of the GA to find a solution close to the global optimum, and the ability of the BP to tune a solution and reach the nearest local minimum by means of local search from the solution found by the GA.

GA is used to modify the initial weights only, while BP is used to train from those weights. This makes a clean division between global and local search. The hidden layer size is searched throughout the application of some new GA operators: substitution, addition and elimination. These operators are developed to perform incremental (adding hidden neurons) or decremental (pruning hidden neurons) learning. For detail, refer to [11]. This paper used MATLAB [12,13] to develop the calculating program.

5 Application, Results, and Evaluation

The objective of this study is to investigate the feasibility of using neural networks to estimate flexural load rating given apparent damage parameter. Currently there is no simple solution available to solve this inverse problem. In order to use neural networks, a reliable comprehensive dataset is needed for training and testing the networks. Dataset used in this paper is come from literature [5]. In literature [5], all beams were designed to fail in bending and tested in flexure under third point loading. The load was applied following successive increments. For each increment load, measurements of strains, deflection and crack openings were measured. The patterns in the dataset were divided into two sets. The first set is used for training the network. The second set consisted of the remainder patterns and was used for testing the trained network. Overall procedure of this study is shown in fig.3.

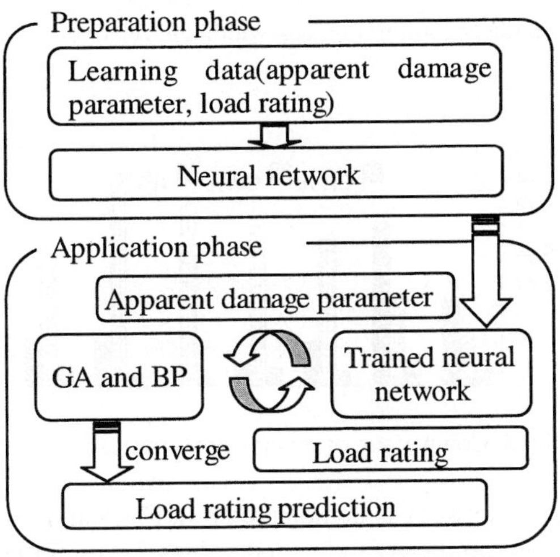

Fig. 3. Load rating prediction procedure. There is preparation phase and application phase.

In the preparation phase, firstly, neural network was trained using training data. The regression error computed according to crack-width and crack-height in literature [5] are 20.4923%~72.3193, 0.5200~39.5293% respectively. However, the error of prediction of network is only 0.0351~9.9.82%. Error comparison is demonstrated in Fig. 4. Fig. 4 shows that, for training data, the error between prediction and actual values is less than the error between regression results according to literature [5] and actual values dramatically.

In the application phase, after the neural network was trained using the training data, it was applied to test data. The regression error computed according to crack-width and crack-height in literature [5] are 20.7%~72.7, 0.5~61.5% respectively. However, the error of prediction of network is only 4.35~9.36%. Error comparison is demonstrated in Fig. 5. Fig.5 shows that, for testing data, the error between prediction and actual values is less than the error between regression results according to literature [5] and actual values dramatically.

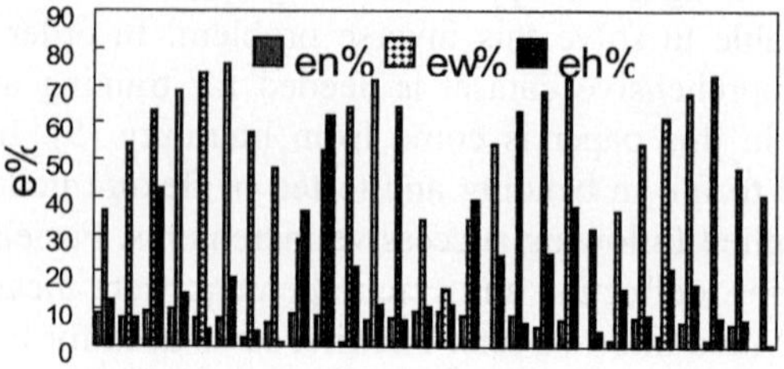

Fig. 4. Comparison of relative error for training data. It is noted that the relative error $e_n = |r_n - r|/r$, $e_w = |r_w - r|/r$, $e_h = |r_h - r|/r$. r is the actual value of flexural load-bearing condition of RC beams; r_n is the ANN prediction value; r_w and r_h are the calculated values by regression formula according to crack width and relative crack height respectively in literature [5].

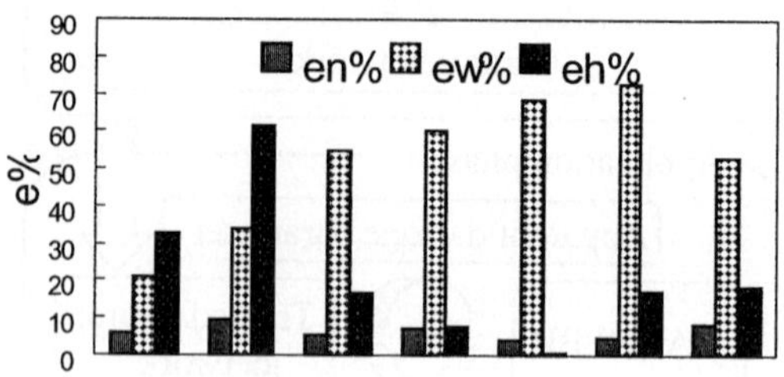

Fig. 5. Comparison of relative error for testing data.

The real power of the network is demonstrated in Fig.4 and Fig.5, witch show very good agreement between predicted and actual values for both the training and testing data. The ANN trained by the hybrid algorithm of GA and BP obtains a much high degree of generalization and can be used in practice.

6 Conclusion

Based on fact that the relative size between load and resistance of reinforced concrete beams appears through apparent damage parameter, a neural network was designed and trained for solving the inverse problem of flexural load rating: given crack width, relative crack height, deflection, concrete compressive strength, steel ratio, steel yield strength, steel bar diameter, steel bar bond coefficient, cover thickness. G-Prop, an algorithm to train ANN based on GA and BP, is used to obtain a much higher degree of generalization and minimize the number of hidden units. At the same time, GA is a good strategy to avoid local minima.

A load test results dataset has been used to test the proposed methodology. The results show that ANN can predict load rating quite accurately. Hence, the feasibility and effectiveness of using this technology to solve real world problems is demonstrated.

References

1. China ministry of communications standards.: Identification of Load carrying capacity for existing highway bridge (trial implementation) (in chinese). People's Communications Publishing House, Beijing (1989)
2. Dajun, Ding.: Crack resistance, crack and stiffness of reinforced concrete element (in chinese). Nanjing Technical college press, Nanjing (1986)
3. Park, R., Paulak, T.: Reinforced Concrete Structures. John Wiley & Sons, Canada (1975)
4. China standard (JTJ 073-96).: Technical Specifications of Maintenance for Highway (in chinese). People's Communications Publishing House, Beijing (1996)
5. Guanhua, Zhang.: Flexural crack parameter, inspection and evaluation of existing RC bridges (in chinese). Dalian university of technology, Dalian (2003)
6. Simon, H.: Neural Networks: a comprehensive foundation. Tsinghua University Press, Beijing (2001)
7. Goldberg, D.E.: Genetic Algorithms in search, Operatimization, and Machine Learning. Addison-Wesley (1989)
8. Yong, Liu., Lishan, Kang., Yuping, Chen.: Genetic algorithms. Non-numerical parallel algorithms, Vol. 2. Science Press, Beijing (2000)
9. Mohamed, A., Mahmoud, M.A., Kiefa, A.: Neural network solution of the inverse vibration problem. NDT&E. Int. 32 (1999) 91-99
10. Suh, M.-W., Shim, M.-B.,: Crack identification of using hybrid neuro-genetic technique. J. Sound and Vibration. Vol. 238 (4) (2000) 617-635
11. Castillo, P.A., Merelo, J., Prieto, A., Rivas, V., Romero, G.: G-Prop: Global optimization of multiplayer perceptrons. Neurocomputing. 35 (2000) 149-163
12. Shuntian, Lou., Yang, Shi.: Systematic analysis and design based on MATLAB: neural networks (in chinese). Sian electronic and technical university press, Sian (1999)
13. Shi Yang.: MATLAB essence and dynamic emulation tool SIMULINK (in chinese). North-West industry press, Sian (1997)

Determining of the Delay Time for a Heating Ventilating and Air-Conditioning Plant Using Two Weighted Neural Network Approach

Mengdi Hu[1], Wenming Cao[2], and Shoujue Wang[3]

[1]Department of Civil Engineering, Zhejiang University Of Science and Technology, Hangzhou, China
humengdi1958@163.com

[2]Information College, Zhejiang University of Technology, Hangzhou 310014, China
[3]Institute of Semiconductors, Chinese Academy of Science, Beijing 100083, China

Abstract. This paper presents an two weighted neural network approach to determine the delay time for a heating, ventilating and air-conditioning (HVAC) plan to respond to control actions. The two weighted neural network is a fully connected four-layer network. An acceleration technique was used to improve the General Delta Rule for the learning process. Experimental data for heating and cooling modes were used with both the two weighted neural network and a traditional mathematical method to determine the delay time. The results show that two weighted neural networks can be used effectively determining the delay time for AVAC systems.

1 Introduction

Since Wang shoujue proposed two weighted neural network[1], it have widely applied pattern recognition[2]、control theory[3] and approximation theory[4] . In my paper, we will apply two weighted neural network approach to determine the delay time for a heating, ventilating and air-conditioning (HVAC) plan to respond to control actions.

Figure 1 denotes the control system of a general heating, ventilating and air-conditioning. $u(t)$, $y(t)$, $r(t)$, $e(t)$ denote the controlling parameter, controlled parameter, defining parameter and error. Given a signal (as figure 2), the system will responds as figure 2. In this figure, the interval BC at which y(t) ascending from 5% to 95% is defined as ascending time (Ts). And the interval AB at which system does not respond is defined as delay time (Td). The interval AC is defined as responding time (Tr). Ascending time is relevant with system and controller, but delay time is only relevant with system.

F. Yin, J. Wang, and C. Guo (Eds.): ISNN 2004, LNCS 3174, pp. 786–791, 2004.

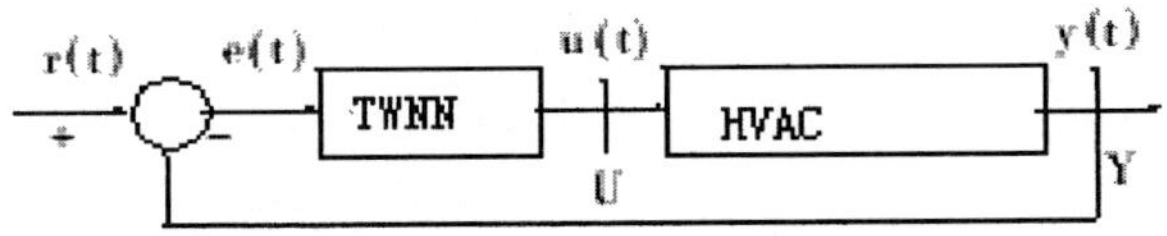

Fig. 1. Two weighted neural networks feedback control system.

Delay time can also be defined by Dirac pulse. In theory, a Dirac pulse would respond at Y when a Dirac pulse happened at U (as figure 3b). But, in fact, the system responds as figure 3c. The interval of the pulse is defined as delay time. In general, the output of a controller can be parsed as a serial of Dirac pulse (as figure 4a). The system will produce a responding curve $y(t)$, which looks like $u(t)$ (as figure 4b). Every response has its own delay time. The comprehensive response of linear system to continuous control signal has the same delay time. Then, we can determine the delay time by comparing $u(t)$ and $y(t)$.

In this paper, we use two weighted neural network to determine the delay time of system. At first, get some control signal: $u(0), u(Tc), u(2Tc), \cdots, u(nTc), \cdots$, and the corresponding responses:

$$y(0), y(Tc), y(2Tc), \cdots, y(nTc), y([n+1]Tc), y([n+2]Tc), \cdots, y([n+k]Tc), \cdots,$$

(Tc is the sample interval). If the response of control signal $u(nTc)$ is $y([n+k]Tc)$, then the delay time T_d is the product of sample times and sample interval Tc:

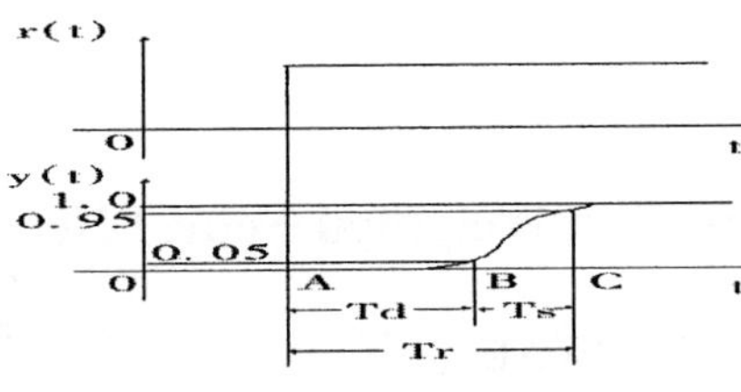

Fig. 2. The response of system.

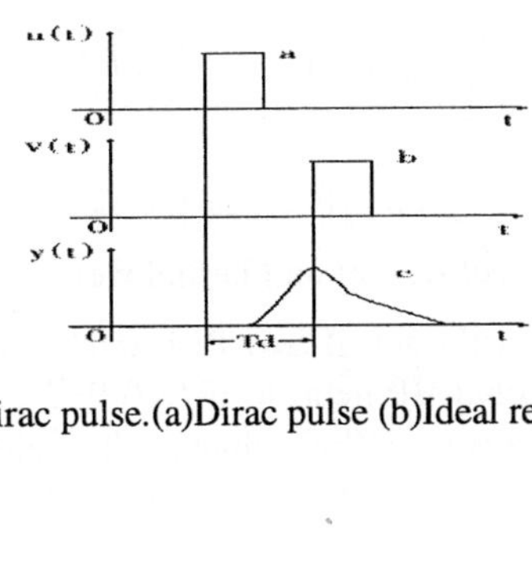

Fig. 3. The response of Dirac pulse.(a)Dirac pulse (b)Ideal response (c)Practical response

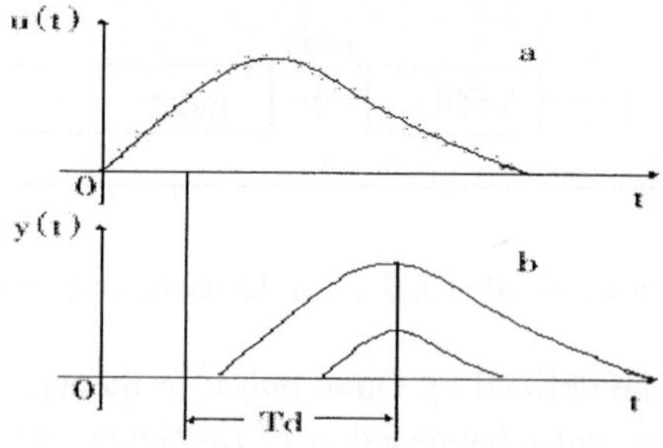

Fig. 4 (a). The practical response of control. (**b**) Resemble coefficient.

$$T_d = K \bullet T_C \ . \tag{1}$$

The array $u(Tc), u(2Tc), \cdots, u(nTc), \cdots$ is much is the array $y([k+1]Tc), y([k+2]Tc), \cdots, y([k+n]Tc), \cdots$. We can determine the delay time by check whether $u(iTc)$ resemble $y([k+i]Tc)$ $(i = 1,2,\cdots,n)$. Shown as Figure 5, when $u(iTc) = y([k+i]Tc)$, the output of z is 1, else the output of z is less than 1.0. The more $u(iTc)$ resembles $y([k+i]Tc)$, the more the value of z approaches 1.0. Here, we define z as resemble coefficient. In the paper, two weighted neural network has two inputs, the one is $u(iTc)$, and the other is $y([k+i]Tc)$. The time corresponding to k which make the largest output of two weighted neural network is delay time (as formula (1)).

2 The Design of the Two Weighted Neural Network

2.1 Two Weights Neural Network

The model is $Y = f\left[\sum_{j=1}^{n}\left(\frac{W_j}{|W_j|}\right)^{s} | \ W_j(X_j - W_j') \ |^{P} - \theta\right]$

In which:Y is the output of neuron ; f(…) is the neuron activation function; X_j is neuron jth input; W_j W_j' are direct weight and kernel weigh; θ is the threshold; n is the dimension of input.We can easily get: if S=1,P=1 and all W are zero, then the above equation becomes the traditional BP neuron; if S=0,P=2 and all the W are one, then the neuron becomes RBF neuron; if S=0, change the value of W and P, and if the basis of function f (…) is zero, the track of input X becomes the following closing hypersurfaces Fig 5.

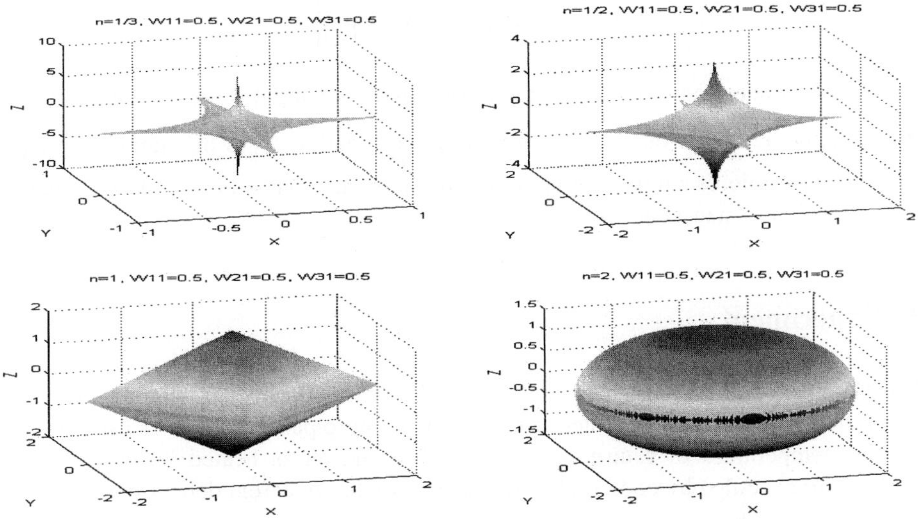

Fig. 5. Closing hypersurfaces of Two weighted neural

2.2 Topology Structure of Two Weighted Neural Network

The two weighted neural network's topology is shown in figure 6. The first layer is input layer which has two input node. The forth layer is the output layer which only has one neuron. The second and third layer both has five neurons. The output of controller and the response (temperature) which is normalized is the input of the two weighted neural network. Every neuron is a Sigmoid function. The output of two weighted neural network is between 0 and 1. The weights of the two weighted neural network were adjusted by the two weighted neural networks algorithm[3].

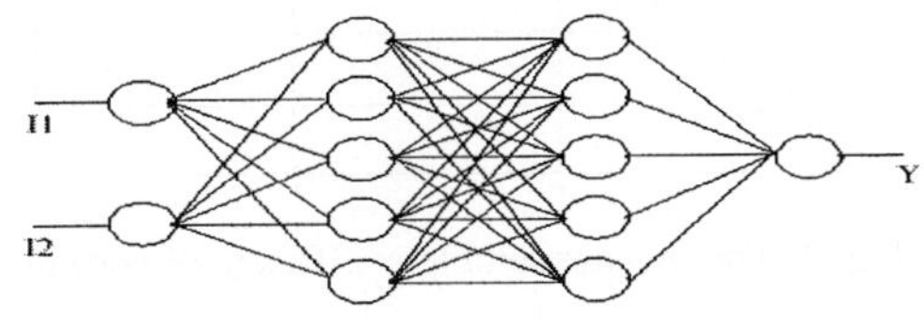

Fig. 6. The topology of the weighted neural network.

We adapt two two weighted neural network in our experiment, one of them has three layers, and the other has four layers. Every hide layer of the two weighted neural network both has five neurons. The simulation was shown in figure 7 and figure 8. The two criterions, mean square error (Ea) and the largest error (Em), are used to assess our two weighted neural network. The criterions were defined as follows:

$$E_a = \frac{1}{N_b} \sum_{b=1}^{N_b} (D_b - Y_b)^2 \tag{2}$$

$$E_m = \max\{abs(D_b - Y_b)\} \tag{3}$$

D_b is between 0 and 1 which is the expect output of input state b in one circle of study; Y_b is the actual output, which is between 0 and 1. N_b is the number of input state in one circle of study. The learning algorithm also has a powerful influence on the two weighted neural network[4].

3 Experiment

The HVAC in our experiment was composed with 2000-cfm fan-winding-pipes, refrigeration, cold water, hot water and steam-winding-pipes, a controller used to control the temperature of return wind. When heating, air is heated by steam; when cooling, air is cooled by cooler. Different operation has different delay time.

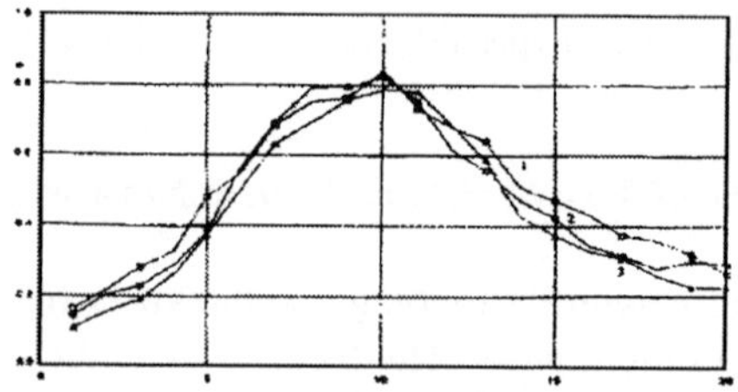

Fig. 7. The mean resemble coefficient of heating.

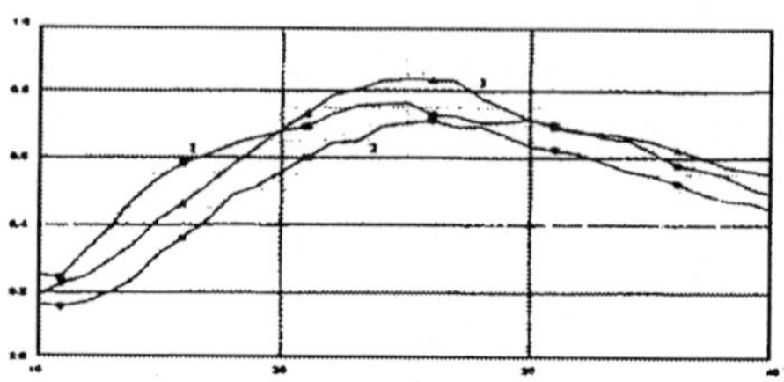

Fig. 8. The mean resemble coefficient of cooling

We get signals (control signal and the temperature of return wind) from controller with data sample system. And then the signals were normalized between 0 and 1. The two weighted neural network can get resemble coefficients with interval K. In figure 7, curve 1 depicts the result of heating. In figure 7 curve 1 depicts the result of cooling. In figure 7, the largest resemble coefficient is 0.5 at $k = 18$, the interval of sample is 2.12 minute, so, the delay time of heating is 86. 1 minute. In figure 8, the largest resemble coefficient is 0.84 at $k = 15$, so the delay time of cooling is 50 minute.

4 Conclusion

In this paper, a two weighted neural network with four layers was used to determine the delay time of HVAC system. The simulation and experiment denote that two weighted neural network is efficient in determining the delay time of HVAC system. Compared with conventional algorithm, two weighted neural network can eliminate the interrupt of input noise by adopting the bias and coefficient of the neuron. Then, two weighted neural network is more precise than conventional algorithm. It can be used on controller to predict system delay time. When learning, the acceleration algorithm can improve the conventional learning algorithm, it lower the output error and accelerate the convergence.

References

1. Wang ShouJue etc.: Discussion on the Basic Mathematical Models of Neurons in General Purpose Neuro computer. ACTA ELECTRONICA SINICA, Vol.29 No.5, 2001
2. Wang Shoujue etc.: The Sequential Learning Ahead Masking (SLAM) Model of Neural Networks for Pattern Classification. Proceedings of JCIS98. Vol.IV. pp199-202. Oct. 1998. RTP. NC. USA
3. Caoyu,Wang Shoujue: Two Weighted Neural Networks and Its Approximation, Thesis, Institute of Semiconductors. Chinese Academy of Science, Beijing, China,2000

Intelligent Forecast Procedures for Slope Stability with Evolutionary Artificial Neural Network

Shouju Li and Yingxi Liu

State Key Laboratory of Structural Analysis for Industrial Equipment, Dalian University of Technology, Dalian, 116024, China
{Lishouju,Yxliu}@dlut.edu.cn

Abstract. Evolutionary artificial neural network is applied for the prediction of the slope stability from the geotechnical material properties and the slope geometries. Coupling the genetic algorithm with artificial neural network, an effective forecast procedure is presented to analyze slope stability. In order to deal with the local minimal problem of artificial neural network with Back-Propagation rule, the connection weights of the artificial neural network are changed by using the genetic algorithm during the iteration process. The practical application demonstrates that the forecast of slope stability using artificial neural network is feasible and a well trained artificial neural network reveals an extremely fast convergence, a better generalization and a high degree of accuracy in the intelligent forecast for the slope stability.

1 Introduction

The majority of slope stability analyses performed in practice still use traditional limit equilibrium approaches. A difficulty with all the equilibrium methods is that they are based on the assumption that the failing soil mass can be divided into slices[1]. This in turn necessitates further assumption relating to side force directions between slices, with consequent implications for equilibrium. The economic and social values of slope stability analysis are very high. As a result, the problem has attracted many researchers in the area of computational intelligence recently. A powerful technique, artificial neural networks, has begun to be used in industry. Neural networks are nonlinear dynamic systems that have important features, such as self-learning, adaptive recognition, nonlinear dynamic processing and associative memory. They have the ability to learn knowledge from historical data and bring forth new knowledge and generalization. The most frequently used Back-Propagation neural networks have a set of problems: dependence on initial parameters, long training time, lack of problem-independent way to choose appropriate network topology and incomprehensive (Black box) nature of ANN. This paper uses the genetic algorithm to address some of these problems. In order to improve global convergent characteristics of ANN, the genetic algorithm is employed to evolve the connecting weights of neural network.

F. Yin, J. Wang, and C. Guo (Eds.): ISNN 2004, LNCS 3174, pp. 792–798, 2004.

2 Artificial Neural Networks

An artificial neural network is a biologically inspired computational model formed from hundreds of single units, artificial neurons, connected with coefficients (weights) which constitute the neural structure[2]. The Back-propagation networks, shown as Fig. 1, consist of an input layer, one or more hidden layers and an output layer.

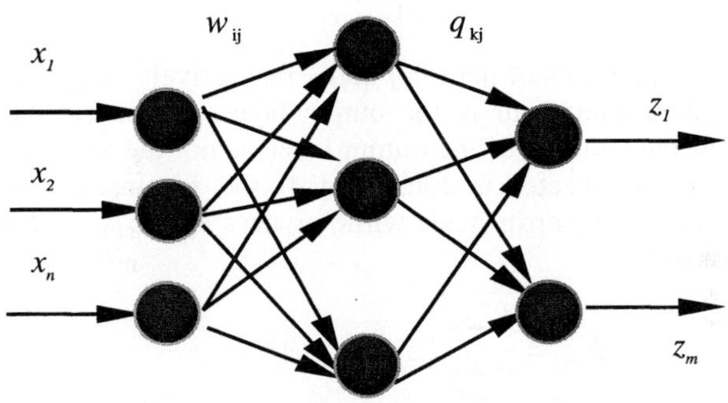

Fig. 1. Schematic diagram of multilayer neural network

For the hidden layer or output layer, each neuron forms a weighted sum, which is the input of every neuron. It can be described as follows:

$$u_{jp} = \sum_{i=1}^{M} w_{ij} x_{ip} . \tag{1}$$

$$v_{kp} = \sum_{j=1}^{J} q_{kj} y_{jp} . \tag{2}$$

Where u_{jp} and v_{kp} are the net input of neuron j and k of the pth training pair in the hidden layer and output layer. x_{ip} and y_{jp} are the net output of neuron i and j of the pth training pair in the input layer and hidden layer. w_{ij} and q_{kj} are the weights linking input layer and hidden layer and the weights linking hidden layer and output layer. The artificial neuron is the building component of the ANN designed to simulate the function of the biological neuron. The arriving signals, called inputs, multiplied by the connection weights(adjusted) are first summed (combined) and then passed through a transfer function to produce the output for that neuron. The activation function is the weighted sum of the neuron's inputs and the most commonly used transfer function is the sigmoid function. The output of the neuron j or k of the pth training pair in the hidden layer or in the output layer is calculated using the logarithmic sigmoid transfer function

$$y_{jp} = \frac{1}{1+\exp\{-(\sum_{i=1}^{M} w_{ij}x_{ip} + \theta_j)\}} . \tag{3}$$

$$z_{kp} = \frac{1}{1+\exp\{-(\sum_{j=1}^{J} q_{kj}y_{jp} + \theta_k)\}} . \tag{4}$$

Where θ_j and θ_k are the biases of neuron j and k respectively. Z_{kp} is the output of the neuron k of the pth training pair in the output layer. In general the output vector, containing all z_{kp} of the neurons of the output layer, is not the same as the true output vector. This true output vector is composed of the summation of o_{kp}. The error between these vectors is the error made while processing the input-output vector pair is calculated as follows

$$E_p = \frac{1}{2}\sum_{k=1}^{L}(z_{kp} - o_{kp})^2 . \tag{5}$$

Where z_{kp} and o_{kp} are the actual and desired outputs of neuron k respectively.

3 Training Neural Networks with Genetic Algorithm

Since error surfaces of objective function for neural networks can be quite complex with many local optima, the genetic algorithm seems to be better suited for this type search. The genetic algorithm searches from one population of points to another, focusing on the area of the best solution so far, while continuously sampling the total parameter space. When a network is trained with a database containing a substantial amount of input and output vector pairs the total error E can be calculated. Evolutionary neural network was widely applied for breast cancer diagnosis，pattern recognition, expert system and electric load forecasting [3,4,5]. The algorithm used to train network makes use of the genetic algorithm. The algorithm is more powerful than the common used gradient descent methods because the genetic algorithm makes training more accurate and faster near global minima on the error surface. For the model, the architecture of the network was determined to be a three-layer fully connected feed-forward network and the weights between the layers were allowed to adjust according to the constraints given and the function to be optimized[6].

The weights between the layers were constrained to vary between -4 and 4. Notably, the values for nodes in the input layer are usually normalized to the range between 1 and-1 in training an ANN. To reduce the training error, the connection weights are changed during a completion of a forward and backward pass by adjustments($\Delta w,\Delta q$) of all the connections weights. Equations (6) and (7) calculate those adjustments by using genetic algorithm. This process will continue until the training error reaches a predefined target threshold error. By using the genetic algorithm to minimize the mean squared difference between the desired and actual network outputs, the connecting weights are adjusted by the following

$$w_{ij}^{t+1} = w_{ij}^{t} + \Delta w_{ij}^{t} \ . \tag{6}$$

$$q_{ij}^{t+1} = q_{ij}^{t} + \Delta q_{ij}^{t} \ . \tag{7}$$

Where Δw_{ij}^{t} and Δq_{ij}^{t} are the weight adjustments based on the computation of genetic algorithm. The main operations of genetic algorithm include the recombination, mutation and selection. Recombination is a process by which information contained in two candidate solutions is combined. In the recombination, each individual is first paired with an individual at random. Let a pair of present individuals be given by[p_{α}^{t}, p_{β}^{t}]. A new pair [p_{α}^{t+1} , p_{β}^{t+1}] is then created in terms of a phenomenological recombination formula:

$$p_{\alpha}^{t+1} = (1-\mu)p_{\alpha}^{t} + p_{\beta}^{t} \ . \tag{8}$$

$$p_{\beta}^{t+1} = (1-\mu)p_{\beta}^{t} + \mu p_{\alpha}^{t} \ . \tag{9}$$

Where μ is defined by a normal distribution with mean 0 and standard deviation σ.

$$\mu = \mathrm{N}(0,\sigma^2) \ . \tag{10}$$

A new pair similar to the present pair can be created, provide that μ is closed to zero, and they are different as far as μ is not equal to zero.

Mutation is a process by which vectors resulting from selection and recombination are perturbed. The intuition behind the mutation operation is the introduction of some extra variability or diversity into the population. The main role of mutation is to provide genes not present in the initial population so as to prevent stagnation at local optima. This may be important if the initial population size is a small subset of all possible solutions. The mutation is conducted with only a small probability by definition. An individual, after this mutation, p_i^{t+1} , is described as

$$p_i^{t+1} = rand\{p_{idown}^{t}, p_{iup}^{t}\} \ . \tag{11}$$

Where p_{up} and p_{down} are the upper and the lower bounds of the parameter, respectively. Selection is simply the copying of quality solution in proportion to their effectiveness. The selection operator is used to determine which population members are chosen as parents that will create offspring for the next generation. The essential idea of all these strategies is that the above-average chromosomes are picked from the current population in a probabilistic manner. The evaluation of the fitness can be conducted with a linear scaling, where the fitness of each individual is calculated as the worst individual of the population subtracted from its objective function value.

$$f_j = 1/E_j \,. \tag{12}$$

Where f_j is the fitness of j-th individual. In the selection process, the reproduction probabilities of individuals are given by their relative fitness:

$$pro_j = \frac{f_j}{\sum f_j} = \frac{f_j}{f_{sum}} \,. \tag{13}$$

The main steps for the weight adjustments of neural networks with genetic algorithm are shown as follows: Step 1: determine the domains of weights and biases of neural networks. Step 2: randomly generate an initial population of candidate solutions. Step 3: compute the fitness of each individual. Step 4: execute recombination operation. Step 5: execute mutation operation. Step 6: execute select operation. Step 7: execute stopping criterion. If stopping criterion can not be reached, then, go to Step 3; otherwise, the weight computation stops. When the ANN produces the desired output the weighted links between the units are saved. These weights are then used as an analytical tool to predict results for a new set of input data. This is a recall or prediction phase when network works only by forward propagation of data and there is no backward propagation of error. The output of a forward propagation is the predicted model for the validation data.

4 Application of Neural Networks to Slope Stability Analysis

As shown as Figure 2, the factors influencing slope stability include the unit weight of the soil γ, cohesion c, internal fraction angle φ, slope angle α, slope height H and pore pressure ration γ_u.

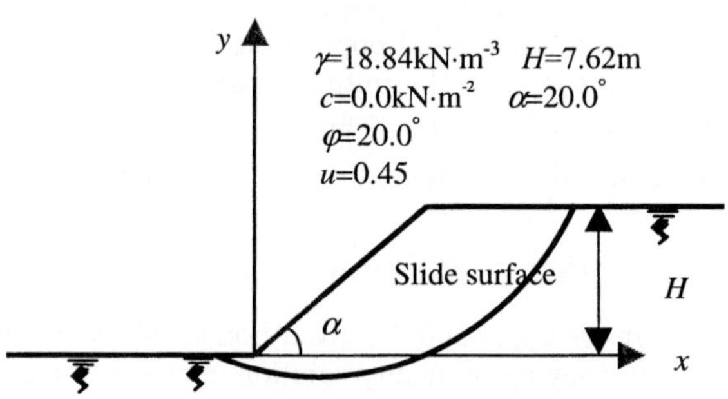

Fig. 2. Definition of geometric parameters of a typical slice circle

As a result, there were six neutrons in the input layer of ANN, 32 neutrons in the hidden layer and 1 neutron in the output layer. Designing network architecture requires more than selecting a certain number of neurons, followed by training only. Especially phenomena such as over-fitting and under-fitting should be recognized and avoided in order to create a reliable network. And 46 input-output samples were gotten from reference[7]. After training ANN with the practical engineering samples, the prediction of slope stability for the other 10 cases of the slope failure are done with well trained ANN. Basic character parameters of geotechnical slopes and comparison between the practical states and the forecasting states are shown in Table 1. The practical application of proposed intelligent forecast procedure for slope stability shows that a well trained artificial neural network reveals an extremely fast convergence, a better generalization and a high degree of accuracy.

Table 1. Basic character parameters of geotechnical slopes and the comparison between the practical states and forecasting states

No.	γ/kN·m^{-3}	c/kN·m^{-2}	φ/°	α/°	H/m	u_c	Slope states#
1	18.84	0.00	20.00	20.0	7.62	0.45	Failed/ Failed
2	28.44	29.42	35.00	35.0	100.0	-	Stable/ Stable
3	28.44	39.23	38.00	35.0	100.0	-	Stable/ Stable
4	21.43	0.00	20.00	20.0	61.0	0.50	Failed/ Failed
5	19.06	11.71	28.00	35.0	21.0	0.11	Failed/ Failed
6	18.50	25.00	0.00	30.0	6.0	-	Failed/ Failed
7	21.51	6.94	30.00	31.0	76.8	0.38	Failed/ Failed
8	31.30	68.00	37.00	47.0	213.0	-	Failed/ Failed
9	25.00	46.00	35.00	50.0	284.0	-	Stable/ Failed
10	27.00	35.00	35.00	42.0	359.0	0.25	Stable/ Stable

#Note: Practical states/Forecasting states

5 Conclusion

Using artificial neural networks to predict the slope stability does have some benefits compared to the statistical predictions and the limit equilibrium methods. First, the predictions of slope stability with ANN are more accurate than the statistical predictions. Furthermore, and more important, the ANN is able to generalize much better than statistical models do. This makes the applicability of ANN for the predictions of the slope stability more reliable than statistical models. The proposed approach has better generalization and lower computational cost than traditional back propagation approaches. A major advantage to the evolutionary approach over traditional learning algorithms such as BP is the ability to escape a local optimum. More advantages include robustness and ability to adopt in a changing environment.

References

1. Al-Karni A.: Study Of The Effect Of Soil Anisotropy On Slope Stability Using Method Of Slices. Computer and Geotechnics. 26(2000)83-103
2. Kustrin S.: Basic Concepts Of Artificial Neural Network (ANN) Modeling And Its Application In Pharmaceutical Research. J. of Pharmaceutical and Biomedical analysis. 22(2000)717-727
3. Abbass H A.: An Evolutionary Artificial Neural Networks Approach For Breast Cancer Diagnosis. Artificial Intelligence in Medicine. 25(2002)265-281

4. Maniczzo V.: Genetic Evolution Of The Topology And Weight Distribution Of Neural Networks. IEEE Trans. Neural Networks. 5(1994)39-53
5. Yao X.: A New Evolutionary System For Evolving Artificial Networks. IEEE Trans. Neural Networks. 8(1997)694-713
6. Srinivasan D.: Evolving Artificial Neural Networks For Short Term Load Forecasting. Neurocomputing. 23(1998)265-276
7. Sah N K.: Maximum Likelihood Estimation Of Slope Stability. Int. J. Rock Mech. Min. Sci. & Geomech. Abstr. 31(1994)47-53

Structural Reliability Analysis via Global Response Surface Method of BP Neural Network

Jinsong Gui[1,2], Hequan Sun[1], and Haigui Kang[1]

[1] State Key Lab. of Coastal and Offshore Engineering, Dalian University of Technology, Dalian 116024, P. R. China
{hqsun, hgkang}@dlut.edu.cn

[2] College of Civil Engineering, Dalian Fisheries University, Dalian 116023, P. R. China
guijs@dlfu.edu.cn

Abstract. When the performance function cannot be expressed exactly, response surface method is often adopted for its clear thought and simple programming. The traditional method fits response surface with quadratic polynomials, and the accuracy can not be kept well, which only the area near checking point coincides well with the real limit state surface. In this paper, a new method based on global response surface of BP neural network is presented. In the present method, all the sample points for training network come from global area, and the real limit state surface can be fitted well in global area. Moreover, the examples and comparison are provided to show that the present method is much better than the traditional one, the amount of calculation of finite element analysis is reduced quite a lot, and the accuracy is increased.

1 Introduction

Structural reliability covers structural security, structural applicability and structural persistence, and the scale of structural reliability can be measured by the implementation of scheduled performance probability under prescribed time and condition. When the performance function is known, the first-order second-moment method is adopted. But the real structure is very complicated, the numerical analysis including finite element method (FEM) has to be adopted, even in some determinant analysis. Meanwhile, Monte Carlo method with FEM, random FEM and the response surface method would be nice for the structural reliability analysis when the performance function is unknown. Monte Carlo method is the simplest one with high accuracy, but it needs the large computation, thus it is usually applied to checking the accuracy of other approximation methods. The random FEM is modified from the determinant finite element program, but the analysis program is difficult to construct for describing all the possible random factors owing to complicacy of theory and programming. The response surface method can be implemented by using a series of the finite element calculations to fit a response surface instead of the unknown real limit state surface, and the existent FEM program can be applied to the reliability analysis directly. This method is used widely in practice for its clear thought and easy program-

F. Yin, J. Wang, and C. Guo (Eds.): ISNN 2004, LNCS 3174, pp. 799–804, 2004.

ming [1], [2]. Among all the response surface methods, the iterative sequence method in quadratic polynomial is used generally based on finding checking point. But this method does not work well in the real limit state surface beyond the places around the checking point($\pm\ \sigma$). The method specified in this paper is based on global response surface method of BP neural network. The FEM computation can be reduced greatly, the accuracy and efficiency can be increased, and the real limit state surface can be fitted very well in the global area by using the present method.

2 Mathematics

2.1 BP Neural Network

BP neural network is adopted widely with three layers of input, hidden and output, which can be trained to learn the relationships between inputs and outputs. Each layer has a number of nodes, linked by weighted connections. The input layer can be used for feeding patterns into the network, and the output layer is for producing the results. BP network can be trained by providing an input pattern with a corresponding target pattern. The error caused by the difference is propagated back to hidden layer, and all weights should be adjusted. A set of training patterns are learned for many times until the mean squared error is relatively small, then the network can be used in practice.

With the development of neural networks, some applications have been implemented in structural analysis [3], [4]. The theory of structural approximation analysis has been developed based on Kolmogorov theory. The neural network is very flexible for interpolation with many sigmoid functions, and can be used to fit well the real limit state function rather than quadratic polynomial function. Response surface method with three layers of neural network is presented in this paper. The structure of neural network is illustrated in Fig. 1.

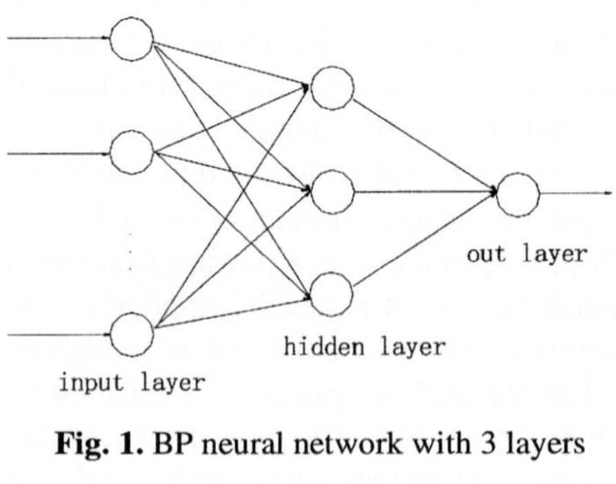

Fig. 1. BP neural network with 3 layers

2.2 Mathematic Model of Reliability Index by Optimization Method

When the first-order second-moment method is employed to structural reliability analysis, central point method, JC method, mapping transformation method, practical analysis method, and optimization method can be adopted. In this paper, the optimization is adopted for the structural reliability, and the performance function is taken in the form of BP neural network.

The limit state formula $Z=g(X_1, X_2, \ldots, X_n)=0$ is expressed by n independent random variables $X_1, X_2, \ldots, X_n$ with arbitrary distribution in structural reliability analysis. The non-normal variables can be normalized by R-F method, and the mean value ($\mu_{X_i'}$), the standard deviation ($\sigma_{X_i'}$) and the reliability index (β) of equivalent normal distribution can be obtained by the following equations:

$$\sigma_{X_i'} = \Phi\{\Phi^{-1}[F_{X_i}(x_i^*)]\}/ f_{X_i}(x_i^*) \tag{1}$$

$$\mu_{X_i'} = x_i^* - \Phi^{-1}[F_{X_i}(x_i^*)]\sigma_{X_i'} \tag{2}$$

$$\beta = \{\sum[(x_i^* - \mu_{X_i'})/\sigma_{X_i'}]^2\}^{1/2} \tag{3}$$

When the checking point is unknown at the beginning, β can be taken as the function of the point $P(X_1, X_2, \ldots, X_n)$ on the limit state surface. The minimum β value is found by optimization method, as well as reliability index β and checking point $P^*(x_1^*, x_2^*, \ldots, x_n^*)$. Finding the reliability index can be summarized as the following constrained optimization model.

$$\begin{cases} \text{Min} \quad \beta^2 = \sum_{i=1}^{n} [(x_i - \mu_{Xi}')/\sigma_{Xi}']^2 \\ \text{s.t.} \quad g(x_1, x_2, \ldots, x_n) = 0 \end{cases} \tag{4}$$

The performance function $g(x_1, x_2, \ldots, x_n)$ can be expressed by either quadratic polynomial or BP neural network. Each can be substituted into the optimization model as the constraining condition.

3 The Response Surface Method of Quadratic Polynomials

When the performance function can not be expressed exactly, the response surface method is often adopted to solve the reliability problem, which replaces an inexplicit function with a suitable and explicitly expressed one. For the structural reliability analysis, the structural FEM is often employed to fit a response surface to replace the real unknown limit state surface, and the first-order second-moment method can be used for this purpose. The iterative serial response method via finding checking points is used in common among the methods. The response surface function of this method is described by quadratic polynomials without intercross terms as Eq. (5). The number to-be-determined in this function is 2*n*+1. The calculation procedure has been provided in [5] in detail.

$$Z = g(X) = a + \sum_{i=1}^{n} b_i X_i + \sum_{i=1}^{n} c_i X_i^2 \tag{5}$$

4 The Methods Used in This Article

It is time-consumable to make structural analysis by ANSYS or SAP, and each sample point is very valuable. Thus all the points should be used sufficiently. For the quadratic polynomial response surface method, only 2*n*+1 sample points are required for each step. Because all the points come from the area close to the checking point ($\pm\sigma$), the conformance with real limit state is fairly satisfied only near the checking points. The present method uses 2*n*+*k* sample points in every step for network training. In fact, the checking point obtained after the first step of iteration is close to the ultimate one, and it is enough to increase one point which is the checking point obtained in the current step. It is not necessary to increase 2*n*+1 sample points in every step as usual. In this way, 2*n*+1 finite element calculations are required in the first step, only one calculation is required in the following steps, and the calculation of finite element is reduced greatly. That is, only 2*n*+*k* calculations are required when converging is reached at the *k*th step. Without getting new expanding points $x_M^{(k)}$ by linear interpolation, the checking point $x^{*(k)}$ obtained in the previous step is used for the next calculation in the present method. It is clear that calculation of structural finite element is reduced with no negative influence on converging speed. The calculation procedure is operated as follows:

(1) Select the mean values $\boldsymbol{x}^{(1)}=(x_1^{(1)},x_2^{(1)}, \cdots x_i^{(1)}, \cdots ,x_n^{(1)})$ as the initial values.

(2) Calculate the result $g(x_1^{(1)}, x_2^{(1)}, \cdots, x_i^{(1)}, \cdots, x_n^{(1)})$ and $g(x_1^{(1)},,x_2^{(1)}, \cdots ,x_i^{(1)}\pm f\sigma_i, \cdots ,x_n^{(1)})$ by structural FEM, where *f* equals 3.0 for the first step and 1.0 for the following steps.

(3) Train the BP neural network by using the 2*n*+*k* sample points obtained through the previous k steps.

(4) Find checking point $x^{*(k)}$ and reliability index $\beta^{(k)}$.

(5) Output β if $|\beta^{(k)} - \beta^{(k-1)}|$ < the given accuracy (0.001 in this paper), or return to step (2) to continue the next iterative step until converging.

If the number of random variables is *n* and converging is reached at the *k*th step, there will be 2*n*+*k* iterative steps in the process.

5 Examples and Results

To save the computation time, some examples with known performance function is chosen.

Example 1: To find the checking point and the structural reliability index when the performance function is Z=18.46-7.48X_1/X_2^3, where X_1 and X_2 are given by X_1~N(10,2) and X_2~N(2.5,0.375).

Example 2: To find the checking point and the structural reliability index when the performance function is Z=1+X_1X_2-X_2, where X_1 and X_2 are given by X_1~LN(2,0.4), X_2~LN(4,0.8).

Example3: To find the checking point and the structural reliability index when the performance function is Z=0.567X_1X_2-0.0005X_3^2, where X_1, X_2 and X_3 are given by X_1~N(0.6,0.0786), X_2~N(2.18,0.0654), X_3~LN(32.8,0.984).

Example4: To find the checking point and the structural reliability index when the performance function is Z=X_1X_2-X_3, where X_1, X_2 and X_3 are given by X_1~N(0.5472,0.0274), X_2~N(3.8,0.304), X_3~N(1.3,0.91).

Table 1. Comparison of calculation results

	Computing item	The first-order second-moment method	Quadratic polynomial response surface method	Global neural network response surface method
1	Reliability index	2.330	2.331	2.330
	Checking point	(11.186,1.655)	(11.012,1.647)	(11.183,1.655)
	Times of iterative		5	6
	Times of FEM		29	10
2	Reliability index	4.69	4.690	4.691
	Checking point	(0.797,4.933)	(0.798, 4.949)	(0.800,5.00)
	Times of iterative		6	6
	Times of FEM		35	10
3	Reliability index	1.965	1.965	1.965
	Checking point	(0.456,2.159,33.418)	(0.456,2.159,33.419)	(0.456,2.159,33.428)
	Times of iterative		3	6
	Times of FEM		23	12
4	Reliability index	3.795	3.795	3.798
	Checking point	(0.505,2.893,1.461)	(0.505,2.894,1.461)	(0.505,2.900,1.463)
	Times of iterative		3	9
	Times of FEM		23	15

As shown in Table 1, the present method improves efficiency greatly by achieving the same computing accuracy. The reason is that our method needs only one more sample point for each step after first step in the iterative process, thus computation efficiency can be increased remarkably.

6 Conclusions and Prospects

BP neural network is used for structural reliability analysis in the paper. A new method based on the global response surface is developed for the inexplicit performance function, and a global response surface of BP neural network is constructed. The results by the present method fit well with the real limit state surface in the global area. Comparing with the traditional response surface method of quadratic polynomials, it is shown that the present method is suitable, and the quantity of calculation of finite element analysis is reduced quite a lot with high calculation accuracy.

Acknowledgements. The research is sponsored by the National Science Fund for Distinguished Young Scholars under Grant No.50125924 and by the National Natural Science Foundation of China under Grant Nos. 50379001 and 10332050.

References

1. Bucher, C.-G., Bougund, U.: A Fast and Efficient Response Surface Approach for Structural Reliability Problem. Structural Safety, Vol. 5, No. 1. (1990) 57-66
2. Faravelli L.: A Response Surface Approach for Reliability Analysis. Journal of Engineering Mechanics, Vol. 115, No. 12. (1989) 2763-2781
3. Parvin A., Serpen G.: Recurrent Neural Networks for Structural Optimization. Computer-Aided Civil and Infrastructure Engineering, Vol. 14, No. 6. (1999) 445-451
4. Tashakori A., Adeli H.: Optimum Design of Cold-formed Steel Space Structures Using Neural Dynamics Mode. Journal of Structural Engineering. (2001) 127
5. Zhao G. F.: The Reliability Principles and Applications for The Engineering Structure. Dalian University of Technology Press (1996) 145-158 (in Chinese)

Modeling Temperature Drift of FOG by Improved BP Algorithm and by Gauss-Newton Algorithm

Xiyuan Chen

Department of Instrument Science and Engineering , Southeast University, Nanjing city, 210096, China
Chxiyuan@263.net, chxiyuan@seu.edu.cn

Abstract. The large temperature drift caused by variation of environmental temperature is the main factor affecting the performance of fiber optical gyroscope (FOG). Based the advantages of artificial neural network and the fact that the temperature drift of FOG is a group of multi-variable non-line time series related with temperature, this paper presents modeling temperature drift of fiber optical gyro rate by improved back propagation (BP) training algorithm and by Gauss-Newton training algorithm, comparison between the modeling results of by improved BP algorithm and by gauss-newton algorithm is presented. Modeling results from measured temperature drift data of FOG shows that Gauss-Newton algorithm has higher training precision and shorter convergence time than improved BP algorithm on the same training conditions for application of modeling temperature drift of FOG.

1 Introduction

The concept of Fiber optical gyros (FOG) was first proposed by Pircher and Hepner in 1967 and was given the first experimental demo in American Utah University By Vali and Shorthill in 1976. So far, FOGs have been widely used because of unique advantages such as high rotation-rate resolution, and a high zero-point stability, maintainance free, reliability, short warming-up time and so on[1]-[3].With the extension of research, noises and biasing drifts of a FOG represents the major error source of inertial navigation systems that are required to operate over long time intervals. Up to the present, causes of many noises and drifts have been discovered and resolved [4]. However the problems of FOG's sensitivity to environment still need to be further studied. Among the environmental factors, the variation of environmental temperature is one of important elements to influence the performance of FOG. A mathematical model or more practical and effective signal processing techniques for temperature drift of FOG must be taken account in performing an error analysis or optimization study of a FOG inertial navigation system.

Efforts have been made to obtain a math model for the temperature drift of FOG. Researchers studied the complex mechanism of the temperature effect but found the influence to temperature drift from each component is a complex equation without

F. Yin, J. Wang, and C. Guo (Eds.): ISNN 2004, LNCS 3174, pp. 805–812, 2004.

any prior assumption [5]. Conventionally, nonlinear time-sequence auto-regression model was usually used to evaluate and compensate the temperature drift like modeling random drift of dynamic tuning gyro [6][7]. Modeling temperature drift of FOG by artificial neural network (ANN) is studied in the recent years[8][9]. For raising the FOG's measurement precision, the intelligent signal-processing strategy which includes a series single-layer network (SSLN), an advanced learning scheme, and an RBF network (RBFN) was proposed[8]. [9] introduces a projection pursuit learning network (PPLN) for modeling temperature drift of FOG theoretically. However the complex of FOG's performance, modeling FOG's drift still to be further studied and the signal processing strategy still need to be verified .

This paper presents modeling temperature drift of fiber optical gyro rate by improved back propagation (BP) training algorithm and by Gauss-Newton training algorithm based the advantages of artificial neural network and the fact that the temperature drift of FOG is a group of multi-variable non-line time series related with temperature ,and presents comparison between the modeling results of by improved BP algorithm and by gauss-newton algorithm.

2 Improved BP Algorithm and Gauss-Newton Algorithm

2.1 Structure of Neuron and Model of an ANN[10]

In this paper, a three-layer error feed-forward neural network is used as a network model for modeling , its structure of a neuron is shown in Fig.1. It consists of an input, a single hidden layer and an output layer. Each node represents a neuron. The number of nodes on each layer is determined according to practical situation. It is assumed that the input vector is $X(x_1, x_2, \cdots, x_N)$, and the output vector is $Y(y_1, y_2, \cdots, y_l)$. The numbers of hidden nodes are M. ω_{ij}, v_{ki} ($\omega_{ij} \in R^{N\times M}, v_{ki} \in R^{M\times L}$), are respectively, the connection weight matrix between the input layer and hidden layer, and that between the hidden and output layer. $\theta_i, \beta_i (\theta_i \in R^M, \beta_i \in R^L)$ is the threshold vector of the hidden layer and the output layer separately. Then the above interrelationship forms a non-linear map from the input space R^N to the output space $R^L : X \rightarrow Y$.

Here, the output vector of the hidden layer is

$$O_i = f(\sum_{j=1}^{N} \omega_{ij} x_j + \theta_i) \quad (i = 1, 2, 3, \ldots, M) \tag{1}$$

in which $f(\bullet)$ refers to the activation function. Being a non-linear function, the most widely used activation function is the sigmoid function. And the output vector of the output layer is

$$y_k = g(\sum_{i=1}^{M} v_{ki} O_i + \beta_i) \ (i = 1, 2, 3, \ldots, L) \tag{2}$$

in which $g(\bullet)$ refers to the activation function. Here the activation function is assumed as proportional and linear function for the output vector of the output layer.

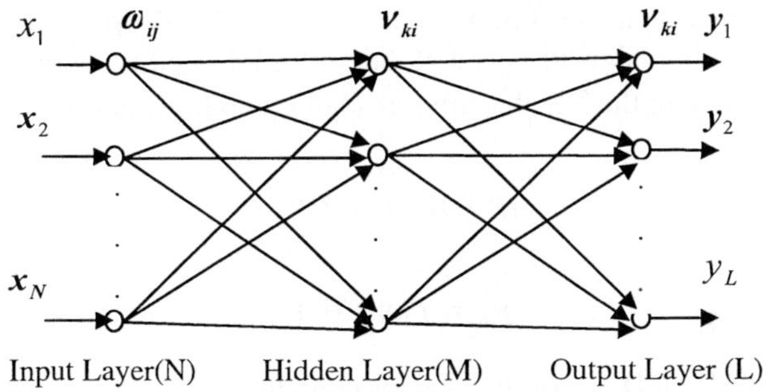

Fig. 1. Three-layer feed-forward ANN topology

2.2 Improved BP Learning Algorithm and Gauss-Newton Learning Algorithm

According conventional BP learning algorithm, the network is trained with a gradient-descent technique.

At the successful completion of the training, the root mean square (RMS) error is minimum for all of the training samples ,and can be defined as follows:

$$E = \sqrt{\frac{1}{2}\sum_{p=1}^{P} e_p} = \sqrt{\frac{1}{2}\sum_{p=1}^{P}\sum_{k=1}^{L}(t_{kp} - y_{kp})^2} = \min(E) \tag{3}$$

Here, P is the number of training samples, and t_{tp} is the desired output, y_{kp} is the output vector of the output layer .

By modifying the learning parameters of $\boldsymbol{\omega}_{ij}$, θ_i and $\boldsymbol{v}_{ki}$, we can meet the above objective of this investigation, ie equation (3) .

$$\Delta W = -\eta \frac{\partial e_p}{\partial W} = \eta \sum_{k=1}^{L} (d_{kp} \frac{\partial y_{kp}}{\partial W}) \tag{4}$$

In which,, $\Delta W = W(t) - W(t-1)$ $\eta \in (0,1)$ is the learning rate. $d_{kp} = t_{kp} - y_{kp}$. Here,

$$\Delta W = (\Delta \omega_{ij}, \Delta \theta_i, \Delta V_{ki}), \tag{5}$$

Above conventional BP algorithm as some advantages but convergence speed is very slow. For improving training speed, equation (4) is changed as follows:

$$\Delta W^{(n)} = -\eta \frac{\partial e_p}{\partial W} + \alpha \Delta W^{(n-1)} = \eta \sum_{k=1}^{L} (d_{kp} \frac{\partial y'_{kp}}{\partial W}) + \alpha \Delta W^{(n-1)} \tag{6}$$

Here, $\alpha \in (0,1)$ is damping parameter. Considering training speed and to avoid diverging , usually α =0.90~0.98, η=0.15~0.5.

BP algorithm is non-line optimal method with local optimum and low training speed. Gauss-Newton algorithm is a global optimum method with fast training speed. Major difference with the BP algorithm is about weight vector W , for Gauss-Newton algorithm,

$$W = W(0) + \Delta W \tag{7}$$

Then, for

$$y'_p = f(x'_p, W) \tag{8}$$

Expansion at the $W(0)$ according Taylor equation,

$$f(x'_{ip}, W) = f_0(x'_{ip}, W) + \frac{\partial f_0(x'_{ip}, W)}{\partial W} \Delta W \tag{9}$$

Here, $f_0(x'_{ip}, W) = f[x'_{ip}, W(0)]$ $\quad \frac{\partial f_0(x'_{ip}, W)}{\partial W} = \left.\frac{\partial f(x'_{ip}, W)}{\partial W}\right|_{W=W(0)}$

For meeting the same objective of this investigation as BP algorithm, ie equation (3).then

$$\frac{\partial E}{\partial W} = 0 \tag{10}$$

Vector ΔW is solved by equation (12).the other steps are the same as the BP algorithm .

3 FOG's Temperature Drift Model and Analysis Based Feed Forward ANN

3.1 ANN Topology of FOG's Temperature Drift[11]

ANN topology of FOG's non-line temperature drift is shown as Fig.2, the architecture the ANN in this paper is adopted 1-3-1(Fig.3), for kth sample, identified error function $d(k)$ is defined as:

$$d(k) = y(k) - \hat{y}(k) \tag{11}$$

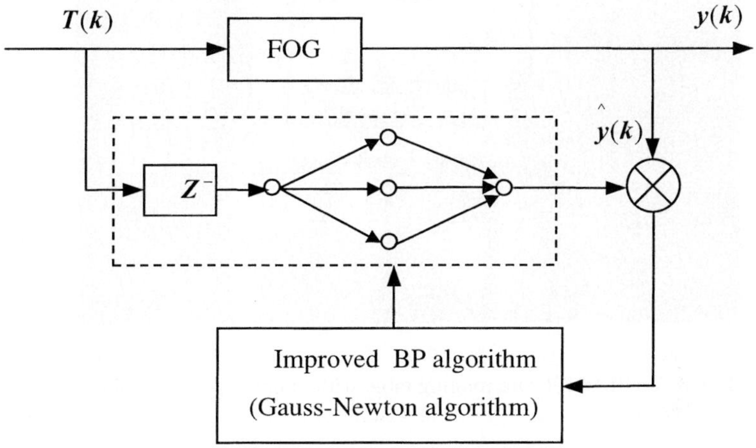

Fig. 2. FOG's temperature drift ANN topology

for kth sample, net error function $e(k)$ is defined as

$$e(k) = \frac{1}{2}[y(k) - \hat{y}(k)]^2 \tag{12}$$

Assume number of the network input training samples and output training samples is P, network target function E is defined as follows:

$$E = \sum_{k=1}^{P} e(k) = \frac{1}{2}\sum_{k=1}^{P}[y(k) - \hat{y}(k)]^2 \tag{13}$$

3.2 Collecting the Experimental Data

The FOG's temperature drift model is identified by off-line method. On the different temperature condition, we collect FOG's drift. The experimental device is rate rotating table with temperature controlled produced by Co. WUILFERT in France (Fig.3). Main technical parameters is as follows: minimum rate is ±0.0001°/s, maximum rate is ±800°/s, revolution is 0.0001°/s , temperature range is -50°C~60°C.

With the Measurement temperature increasing from 18°C to 43°C at the temperature interval of 0.5°C, the observations of the temperature drift of FOG is considered as output samples of identified system as table 1.

Fig. 3. WUILFERT rate rotating table with temperature controlled

Table 1. Output samples of identified system (t_p unit: °/s)

$t_1 \sim t_7$	$t_8 \sim t_{14}$	$t_{15} \sim t_{21}$	$t_{22} \sim t_{28}$	$t_{29} \sim t_{35}$	$t_{36} \sim t_{42}$	$t_{43} \sim t_{45}$
0.0012223	0.0013086	0.0012518	0.0012598	0.0012450	0.0012166	0.0012420
0.0011198	0.0013633	0.0013955	0.0012821	0.0013426	0.0012059	0.0012141
0.0011636	0.0011769	0.0011465	0.0013337	0.0012915	0.0012614	0.0012127
0.0012429	0.0012356	0.0012413	0.0012779	0.0011709	0.0012824	
0.0013746	0.0012374	0.0012759	0.0011644	0.0013128	0.0011227	
0.0012725	0.0012818	0.0011763	0.0012645	0.0012163	0.0011775	
0.0011972	0.0011444	0.0012267	0.0012305	0.0012601	0.0012464	

3.3 Applying the Improved BP Learning Algorithm and Gauss-Newton Algorithm for Evaluation and Analysis of Temperature Drift Model

Applying the improved BP learning algorithm to evaluate the drift, we choose initial vector $\boldsymbol{W}$ =[-0.6410908, -0.9218503,0.3532885,-0.7065259,0.6347983,-0.9050612] M=3 (M is numbers of hidden nodes). η=0.3, α=0.9, β=1. The procedure of the target function $\boldsymbol{E}$ decreasing is shown in table 2.

Using the Gauss-Newton learning algorithm to evaluate the presented drift, the other parameters such as initial vector, threshold values of neuron are same as BP algorithm parameter except β=2. The procedure of the target function $\boldsymbol{E}$ decreasing is shown in table 3.

Table 2. The procedure of the target function E decreasing applying the improved BP algorithm

Epoch TT	0	1	3	5	7	9
E	7.2973	0.4337	0.0086	0.0080	0.0073	0.0067
Epoch TT	50	100	200	300	400	500
E	0.0013	1.902e-4	3.433e-6	4.516e-7	2.238e-7	1.076e-7

Table 3. The procedure of the target function E decreasing applying the Gauss-Newton algorithm

Epoch TT	0	1	3	5	7	9
E	9.0733	8.42e-4	1.118e-4	1.864e-6	1.163e-7	8.878e-8
Epoch TT	50	100	200	300	400	500
E	8.834e-8	8.834e-8	8.834e-8	8.834e-8	8.834e-8	8.834e-8

Comparing with the results by improved BP learning algorithm and Gauss-Newton learning algorithm, training converging procedures of ANN by both algorithms are steady. But converging speed of ANN by improved BP algorithm is slower than by Gauss-Newton learning algorithm. Modeling precision of ANN by latter is higher than former algorithm for the same epochs.

4 Conclusions

Modeling results from measured temperature drift data of FOG shows that Gauss-Newton algorithm has higher training precision and shorter convergence time than improved BP algorithm on the same training conditions for application of modeling temperature drift of FOG.

Acknowledgments. The work was supported by the Southeast University excellent young teacher foundation (4022001002). The author would like to thank Dr. Daqing Zhang and Prof. Dejun Wan of the Department of instrument science and engineering at Southeast University for helpful suggestions.

References

1. Hotate, K.: Fiber Optic Gyros. A Chapter in Optical Fiber Sensors, Vol. IV. Norwood, MA. Artech House (1997) 167-206
2. Lefevre, H.C.: The Fiber Optic Gyroscope. Norwood, MA. Artech House (1993)
3. Hotate, K.: Future Evolution of Fiber Optic Gyros. In Proc. SPIE Fiber Optic Gyros.20th Anniversary Conf., Vol. 2837, Denver, CO. (1996) 33-44

4. Saida, T., Hotate, K.: General Formula Describing Drift of Interferometer Fober-Optic Gyro- due to Faraday Effect Reduction of the Drift in Twin-Depo-I-FOG. Journal of Lightwave Technology, Vol. 17 (1999) 222-228
5. Fengping, Y., Huijuan, L., Shuisheng, J.: Investigation of the Temperature Compensated Method for Fiber Optic Gyros. Acta Optica Sinica, Vol. 19 (1999) 968-974
6. Xiyuan, C.: Modeling Random Gyro Drift by Time Series Neural Networks and by Traditional Method. Proceedings of IEEE Int. Conf. Neural Networks & Signal Processing, Vol. 1 (2003) 810-813
7. Sudhakar, M.P., Weibang, Z.: Modeling Random Gyro Drift Rate by Data Dependent Systems. IEEE Transactions on Aerospace and Electronic Systems, Vol. 22 (1986) 455-459
8. Rong, Z., Yanhua, Z., Qilian, B.: A Novel Intelligent Strategy for Improving Measurement Precision of FOG. IEEE Transactions on Instrumentation and Measurement, Vol. 49 (2000) 1183-1188
9. Hongwei, B., Zhihua, J., Weifeng, T.: A Projection Pursuit Learning Network for Modeling Temperature Drift of FOG. Proceedings of IEEE Int. Conf. Neural Networks & Signal Processing, Vol. 1 (2003) 87-90
10. Daqing, Z.: Study on Application of All-Digital IFOG in New Style Strapdown Attitude and Heading Reference System [D]. Southeast University (2002)
11. Colin, O.B., Shengchai, C., Tarek, G., Catherine, A.R.: Comparing BP and ART II Neural Network Classifiers for Facility Location. Computers and Engine, Vol. 28 (1995) 43-50

A Novel Chaotic Neural Network for Automatic Material Ratio System

Lidan Wang[1] and Shukai Duan[1,2]

[1] School of Electronic Information Engineering, Southwest China Normal University 400715, Chongqing, China
duansk@swnu.edu.cn, duanshukai@sina.com

[2] Department of Computer Science and Engineering of Chongqing University 400044, Chongqing, China

Abstract. Aiming at improving automatic level of traditional material ratio control system, we propose a novel chaotic neural network (NCNN) for automatic material ratio control system. The NCNN controller has following properties: (1) separation of superimposed ratio patterns, (2) learning unknown ratio patterns successively. As for the first feature, we utilize the feature of chaotic neural network proposed by K.Aihara [1]. As for the first property, when a stored pattern is given to the network, the network searches around the input pattern by chaos, and it recalls the ratio patterns from superimposed patterns. As for the second one, when an unknown input pattern is given, a different response will be received. So it can distinguish unknown patterns from the known patterns and learn the unknown patterns successively. A series of computer simulations demonstrate the effectiveness and stability of the proposed method.

1 Introduction

In the traditional material ratio control system, it can not realize automatically separating superimposed patterns and successive learning. It is reported that chaotic behavior existing in olfactory neural systems in physiological experiments plays an important role in memory and recalling in human brain [1]. In order to mimic the real neurons, Aihara et al. proposed a chaotic neural networks (CNN) model in 1990 [2]. In the model, chaos is considered having three properties of biological neurons: (1) graded response,(2) relative refractoriness, (3) spatio-temporal summation.

On the other hand, many neural network models on chaotic associative memory have been made [2-7]. Most of them cannot realize many-to-many associations and successive learning which are very important to both mimic real neuron and apply in the engineering field.

In this paper, we propose a NCNN for automatic material ratio control system. It can deal with separation of superimposed ratio patterns and successive learning. In the proposed model, we replace whole spatio-temporal summation with a part of it.

F. Yin, J. Wang, and C. Guo (Eds.): ISNN 2004, LNCS 3174, pp. 813–819, 2004.

After a training set which including a common term is stored, the network can recall them by chaos when the common term being given as an external input. In addition, The NCNN can distinguish an unknown pattern from the known patterns and learn the unknown pattern successively. This behavior is similar to the physiological facts in the olfactory bulb of a rabbit [2].

2 Chaotic Neural Network

In the history, Mcculloch-Pitt neuron model was proposed firstly [1]. In 1971, Nagumo and Sato modified Mcculloch-Pitt neuron model and proposed the Nagumo-Sato neuron model [1]. In this section, we briefly review chaotic neural networks (CNN) proposed by K.Aihara et al in 1990. The dynamics of the ith chaotic neural in a neural network composed of M neurons can be modeled as

$$x_i(t+1) = f_i\left(\sum_{j=1}^{M} W_{ij} \sum_{r=0}^{t} k^r h_j(x_j(t-r)) + \sum_{j=1}^{N} V_{ij} \sum_{r=0}^{t} k^r I_j(t-r) - \alpha \sum_{r=0}^{t} k^r g_i(x_i(t-r)) - \theta_i \right. \quad (1)$$

where $x_i(t+1)$ is the output of the ith chaotic neuron at the discrete time t+1; M is the number of the chaotic neurons ; N is the number of externally applied inputs. W_{ij} shows the connection weight between the ith chaotic neuron and the jth chaotic neuron; V_{ij} is the connection weight from the jth externally applied input to the ith chaotic neuron; $I_j(t\text{-}r)$ is the strength of the jth externally applied input at the time $t\text{-}r$; h_j is the transfer function; g_i is refractory function of the ith chaotic neuron; k^r is the damping factor of the refractoriness; α is a the scaling factor of the refractoriness and θ_i is the threshold of the ith chaotic neuron. The f_i is the continuous output function of the ith chaotic neuron, the $f_i(.)$ is represented by the following equation

$$f_i(y) = \frac{1}{1+\exp(-y/\varepsilon)} \quad (2)$$

where ε is the steepness parameter.

3 NCNN for Automatic Material Ratio System

In this section, we propose a novel chaotic neural network (NCNN) for automatic ratio system based on the novel successive learning chaotic neural network proposed in our previous work [8-9]. Here, we explain the structure of the NCNN. Furthermore, we will discuss how the proposed model realizes its functions such as separation of superimposed ratio pattern and learning unknown ratio patterns successively.

3.1 Structure of NCNN for Automatic Ratio System

The NCNN for Automatic Ratio System is based on NCNN model which can separate superimposed pattern and realize successive learning. The NCNN model replaces whole spatio-temporal summation with a part of it. The dynamics of the ith chaotic neuron in the αth layer can be modeled as,

(1)when $t>t_0$,

$$x_i^{(\alpha)}(t+1)=\psi[v\sum_{d=t-t_0}^{t}k_s^d I_i^{(\alpha)}(t-d)+\sum_{\substack{\beta=1\\ \beta\neq\alpha}}^{L}\sum_{j=1}^{N^{(\beta)}}\omega_{ij}^{(\alpha\beta)}(t)\sum_{d=t-t_0}^{t}k_m^d x_j^{(\beta)}(t-d)-\gamma\sum_{d=t-t_0}^{t}k_r^d x_i^{(\alpha)}(t-d)+\theta_i] \tag{3}$$

(2) when $t<=t_0$, replace $d=t-t_0$ with $d=0$ in equation (3).

Where $x_i^{(\alpha)}(t+1)$ is the output of the network. $N^{(\alpha)}$ and are the number of chaotic neurons in the αth layer and the βth layer; L is the number of layers; $I_i^{(\alpha)}(t)$ is the externally applied input of the ith chaotic neuron in the αth layer at the time t; v is the connection weight from the input to the neurons; $\omega_{ij}^{(\alpha\beta)}(t)$ shows the weight between the ith chaotic neuron in the αth layer and the jth chaotic neuron in the βth layer; k_s, k_m and k_r are the damping factor of the refractoriness; r is a the scaling factor of the refractoriness and θ_i is the threshold. The ψ (.) is defined as,

$$\psi(y)=\frac{2}{1+\exp(-y/\varepsilon)} \tag{4}$$

In the proposed model, we use continuously external input instead of initial input and replace whole spatio-temporal summation with a part of it.

3.2 Recalling Stored Patterns for Automatic Ratio System

Here, we explain the reason why the many-to-many associations can be realized. Let us consider the following training set (A1, B1, C1), (A1, B2, C2), (A3, B3, C3), which including a common term A1.

When A_1 is given to the first layer of the NCNN continuously, the second layer receives the superimposed pattern (B_1+B_2) and searches around B_1 and B_2 by chaotic itinerancy, and in the third layer, the network searches around C_1 and C_2. So, we can receive (A_1, B_1, C_1) and (A_1, B_2, C_2) at different time and realize many-to-many associations. Obviously, if the training set does not include common set, the proposed model can recall them respectively.

While a known pattern is given to the NCNN, the following two forces will work to same direction: (1) a force approaching a minimum of the energy function; (2) a force approaching the input pattern. The stored patterns are recalled by two forces which working to a same direction immediately. However, while an unknown pattern is given to the NCNN as an externally input, the two forces will work to different directions and a chaotic itinerancy among the stored patterns appears. The NCNN can make use of the different response and learn the unknown pattern.

In the proposed model, when an input is considered as an unknown pattern, the unknown pattern is learned by Hebbian learning. The connection weights are updated by the following equation,

$$\omega_{ij}^{(\alpha\beta)}(t_0+1)=\omega_{ij}^{(\alpha\beta)}(t_0)+\lambda x_i^{(\alpha)}(t_0)x_j^{(\beta)}(t_0) \tag{5}$$

Where $x_i^{(\alpha)}(t_0)$ is output in the α th layer at t_0 and λ is the learning rate; t_0 is the time chaotic itinerancy appears.

4 Automatic Material Ratio System Including NCNN Controller

In this part, we build up an automatic material ratio system using a NCNN controller discussed in section 3.

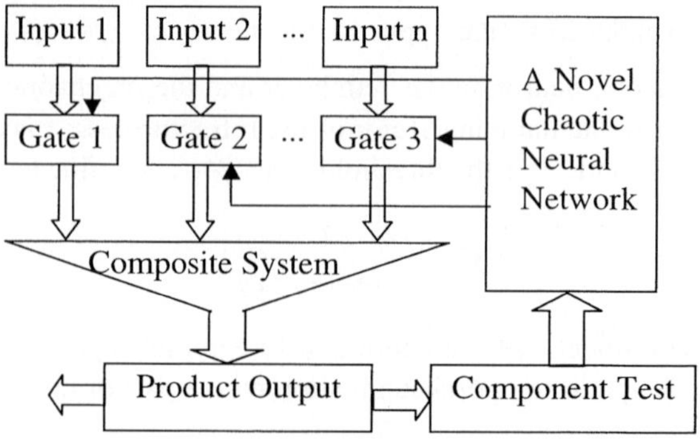

Fig. 1. Structure of the Automatic Material Ratio Control System

Fig. 1 shows the structure of the system. It consists of five parts called material input and product output, gate of control, composite system, component test and NCNN controller. In this section we will discuss the working process of the system in details.

4.1 Separation Stored Material Ratio Patterns

As it is shown in fig. 1, when the system begin to work, the NCNN controller controls different material flowing into ratio system by different ratios according to gates. After been mixed up in the composite system, the product will be sent out. And a part of product will be feedback by component test controller, then the analyzing result of component test controller should feedback to NCNN controller and it can make the system more stability.

While only one inputting material of a ratio pattern being sent into input equipment, the NCNN controller will separate the whole ratio pattern from the stored patterns immediately by the inputting material being given continuously which produced by component test controller. Hence, the system will accomplish the work accurately.

4.2 Learning New Ratio Patterns Successively

In traditional material ratio system, we must store different ratio patterns in the controller. While in the automatic ratio control system we proposed, the NCNN controller can distinguish new patterns and learn they successively by inputting material of new patterns into ratio system continuously.

While a new pattern being given continuously, the component test controller analyzes the components and sends the result into the NCNN controller successively. Then, the two forces in the NCNN controller will work to different directions and a chaotic itinerancy among the stored patterns appears. The NCNN controller makes use of the different response and learns the new material ratio pattern. The system may learn various material ratio patterns successively by repeating above process.

So, there are two ways the system learning a new pattern as follows:(1) storing a new pattern in the NCNN controller directly, (2) giving a new pattern as an external input continuously. The two way shows advantage of the system we proposed.

5 Computer Simulations

We design a NCNN controller and prove its effectiveness by series of computer simulations. The parameters are: k_s= 0.90, k_m=0.1, k_d =0.90, r=9.8, v=200, λ =1, θ=0, ε = 0.02, t_0=3. Bipolar data (-1 or 1) are used.

5.1 Recalling Patterns for the NCNN Controller

Table 1 shows the patterns of automatic ratio control system which been stored in the NCNN controller.

Table 1. The patterns of automatic ratio control system

Ratio patterns	Input material		
	Input 1	Input 2	Input 3
pattern 1	A	a	1
pattern 2	B	b	2
pattern 3	C	c	3

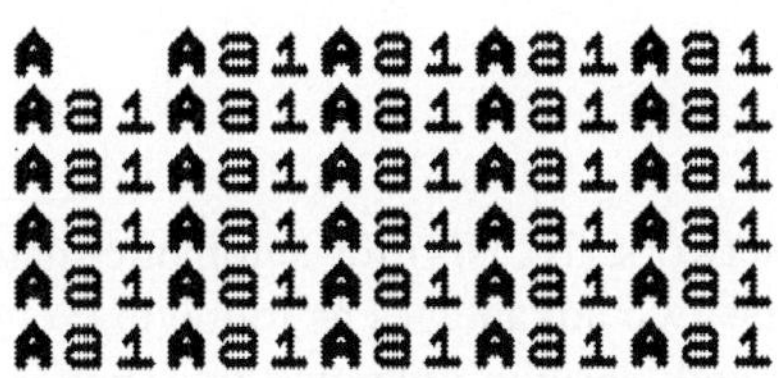

Fig. 2. No. A material is given to NCNN

Fig. 3. No. b material is given to NCNN

As it is shown in fig. 2, while No. A material of pattern 1 being given as an externally input, the NCNN will soon recalled the whole pattern 1 exactly. In the same way, while No. b material of pattern 2 being given, the pattern 2 appears immediately (fig. 3).

Obviously, the computer simulations shows perfect effectiveness of the NCNN controller we proposed .

5.2 Successive Learning for NCNN Controller

We also design successive learning simulations for the NCNN. In the simulation, (*A*, *a*, 1) is stored in the NCNN initially.

Fig. 4. (*A*,*a*,1) is given to NCNN

Fig. 5. (*C*,*b*,2) is given to NCNN

Fig. 6. (*A*,-,-) is given to NCNN

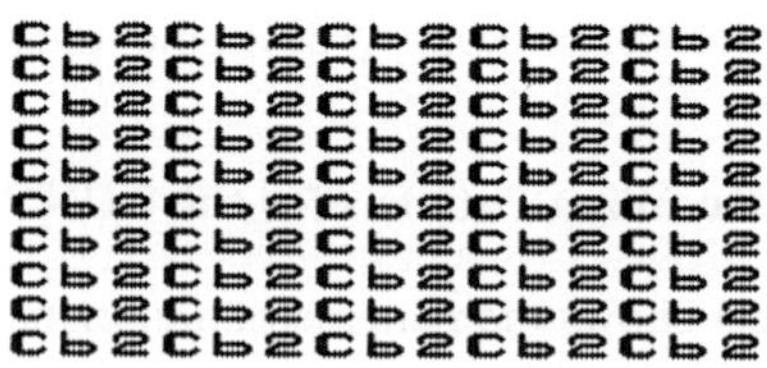

Fig. 7. (*C*,-,-) is given to NCNN

Fig. 4 shows (*A*,*a*,1) is given as an external input, it is recalled at once. While (*C*,*b*,2) is given as shown in figure 5, (*C*,*b*,2) is considered as an unknown pattern. The network appears a chaotic itinerancy and the pattern is stored. Fig. 6,7 show the learning result of the NCNN controller. When *A* and *C* being given, (*A*,*a*,1) and (*C*,*b*,2) are recalled, respectively.

Clearly, the automatic material ratio system we proposed shows the effectiveness, it can realize separation the stored patterns and learn new patterns successively.

6 Conclusions

In this paper, we propose a new chaotic neural network (NCNN) for automatic material ratio control system. The NCNN controller has two advantages as following: (1)it can realize recalling superimposed ratio patterns and learning unknown ratio patterns successively. (2) it can adjust itself by the feedback information of the output. Computer simulations prove its effectiveness and stability of the proposed method.

Acknowledgement. The work is supported by Fund Project of Chongqing Natural Science, China.

References

1. Aihara, K., Takabe, T., Toyoda, M.: Chaotic Neural Networks. Physics Letters A, Vol. 144. (1990) 333 –340
2. Yao, Y., Freeman, W.J.: Model of Biological Pattern Recognition with Spatially Chaotic Dynamics Neural Networks. Neural Networks, Vol. 3. USA (1990) 153-170
3. Osana, Y., Hagiwara, M.: Separation of Superimposed Pattern and Many-to-Many Associations by Chaotic Neural Networks. IEEE International Joint Conference on Neural Networks Proceedings, Vol. 1. Anchorage (1998) 514-519
4. Kaneko, K.: Clustering, Coding, Switch, Hierararchial Ordering and Control in a Network of Chaotic Elements. Physics Letters D, Vol. 41. (1990) 137-172
5. Ishii, S., Fukumizu, K., Watanabe, S.: A Network of Chaotic Elements for Information Processing. Neural Networks, Vol. 1 (1996) 25-40
6. Osana, Y., Hattori, M., Hagiwara, M.: Chaotic Bidirectional Associative Memory. International Conference on Neural Networks, Vol. 2. Houston (1997) 816 –821
7. Osana, Y., Hagiwara, M.: Successive Learning in Chaotic Neural Network. IEEE International Joint Conference on Neural Networks Proceedings, Vol. 2. Anchorage (1998) 1510-1515
8. Liu, G.Y., Duan, S.K.: A Chaotic Neural Network and its Applications in Separation Superimposed Pattern and Many-to-Many Associative Memory. Computer Science, Vol. 30. Chongqing, China (2003) 83-85
9. Duan, S.K., Liu, G.Y., Wang, L.D., Qiu, Y.H.: A Novel Chaotic Neural Network for Many-to-Many Associations and Successive Learning. IEEE International Conference on Neural Networks and Signal Processing, Vol. 1. Nanjing, China (2003) 135-139

Finite Element Analysis of Structures Based on Linear Saturated System Model

Hai-Bin Li[1,2], Hong-Zhong Huang[1], and Ming-Yang Zhao[2]

[1] School of Mechanical Engineering, Dalian University of Technology
Dalian, 116023, China
hzhhuang@dlut.edu.cn

[2] Shengyang Institute of Automation, Chinese Academy of Sciences
Shengyang, 110015, China
{hbli, myzhao}@sia.ac.cn

Abstract. According to the property that global stiff matrix is positive definite after being adjusted and specific formation of elastomer potential energy function, linear saturated system model (LSSM) is introduced into finite element neurocomputing. Based on the neural network, a circuit implementation of an example is given and the time, error characteristic and simulation of the circuit are analysed.

1 Introduction

Finite element method was developed quickly after being brought forward in 1953 and it has become a most efficient structural analysis method. With structural analysis becoming more and more complicated, it is impossible to implement real-time calculation of complicated structure with finite element method because of the limitation of memory capacity and running speed of computer. So some structural analyst and mathematicians began to study parallel process of finite element structure after parallel computer came into being in the 1970's.

Neural network is a complicated non-linear dynamic system with great parallel computation ability. Neural network has been widely used in many fields such as optimization, pattern recognition, automatic control and economic prediction and so on since it revived in 1980's. Optimization problem can be mapped into a dynamic circuit with proper neural network, so solution to the problem can be obtained in circuit reaction time. The essence of neurocomputing is to construct proper network structure and learning method which can minimum the energy during running time and make the objective function value minimum when the system attains equilibrium. At present, widely used neural networks of optimization are Hopfield model and its modified models.

Neural network of optimization can be used to resolve structural analysis problems of elastic mechanics because elastic mechanics problems can be summed up to quadratic optimization under equality constraint. Some scholars have done a lot of works on the subject [1, 2]. At present, all the optimization neural network algorithms are all based on gradient descent algorithm of network energy, so it is inevitable that

F. Yin, J. Wang, and C. Guo (Eds.): ISNN 2004, LNCS 3174, pp. 820–825, 2004.

the optimization get into local minimum value. A new method that combines simulated annealing with gradient descent algorithm to searching global optimum solution was developed in [3]. Simulated annealing is a general random searching method that corresponds to the principle of mental annealing in physics. The searching direction can be chosen not only the better one but also the worse one in every step, so it can realize global searching. But simulated annealing is a global searching method based on probability, it can not converge to the global minimum value in complete probability which will bring great economic loss to important structural component. In the other hand, there is not a uniform theorem to choose the parameters in simulated annealing, such as initial temperature, internal cycle time criterion and stop criterion, which will affect the algorithm's performance. Moreover those parameters are chosen according to experience, which can not realized with fixed circuit. For the reasons mentioned above Linear Saturated System Model (LSSM) is introduced into finite element neural network according to the property that global stiff matrix are positive definite after being adjusted and specific expression of elastomer potential energy function. The finite element neural network can obtain solution without error in theorem. Circuit implementation to an example is given and the computation time and computation error are discussed.

2 Theorem Analysis

Finite element calculation of elastic mechanics can be summed up to quadratic optimization as following [1]:

$$\min_{x\in\Omega} \Pi = \frac{1}{2}\delta^{\mathrm{T}}\mathrm{K}\delta - \mathrm{q}^{\mathrm{T}}\delta \qquad (1)$$

$$\text{s.t} \quad \mathrm{A}\delta = \overline{\delta}(\text{in } S_u)$$

where

$$\mathrm{K} = \sum_{e=1}^{n} \int_{\Omega} (\mathrm{B}^{\mathrm{T}} D\mathrm{B}) \mathrm{d}\Omega$$

$$\mathrm{q} = \sum_{e=1}^{n} \int_{\Omega} (N^{\mathrm{T}} \mathrm{d}b) \mathrm{d}\Omega + \sum_{e=1}^{n_s} \int_{S} (N^{\mathrm{T}} \mathrm{d}\overline{\mathrm{q}}) \mathrm{d}S$$

N is displacement function, B is strain matrix, K is global stiffness matrix, q is node loading array, A is constraint matrix.

Above optimization can be expressed with linear equations:

$$\mathrm{K}\delta = \mathrm{q} \qquad (2)$$

where $\delta_{x_i} = \overline{\delta_i}$, $i = 1, 2, \cdots, s$.

Proper row transformation is applied on Eq. (2), i.e. known constraint variables are moved to the bottom of the equations:

$$\begin{bmatrix} K'_{(n-s)\times(n-s)} & K'_{(n-s)\times s} \\ K'_{s\times(n-s)} & K'_{s\times s} \end{bmatrix} \begin{bmatrix} \delta_{(n-s)\times 1} \\ \delta_{s\times 1} \end{bmatrix} = \begin{bmatrix} \hat{q}_{(n-s)\times 1} \\ \hat{q}_{s\times 1} \end{bmatrix}$$

where $\delta_{s\times 1}$ is constraint vector, $\delta_{s\times 1} = \overline{\delta}_{s\times 1}$.

According to matrix operation theorem:

$$K'_{(n-s)(n-s)}\delta_{(n-s)\times 1} = \hat{q}_{(n-s)\times 1} - K'_{(n-s)\times s}\overline{\delta}_{s\times 1}$$

let $\hat{q}_{(n-s)\times 1} - K'_{(n-s)\times s}\overline{\delta}_{s\times 1} = q'$, $K'_{(n-s)(n-s)}$ is denoted as K', then Eq. (1) can be transformed into unconstrained optimization problem:

$$\min_{x\in\Omega} \Pi = \frac{1}{2}\delta^{T} K' \delta - q'^{T}\delta \tag{3}$$

Building a neural network whose status function is:

$$\frac{d\delta_i}{dt} = -\frac{1}{\tau}\frac{dE}{d\delta_i} = -\frac{1}{\tau}\left(\sum_{j=1}^{m} K'_{ij}\delta_j - q'_i\right) \tag{4}$$

where τ is integral constant of integrator.

For the integrated circuit is restricted by maximum output voltage k i.e. $\delta_i \le k$ during the circuit implement, the neural network system is called Linear Saturated System Model, in brief LSSM.

According to matrix theorem, the analytical solution of Eq. (4) of neural network is:

$$\delta(t) = e^{-\frac{1}{\tau}K'(t-t_0)}\delta(t_0) + \int_{t_0}^{t} e^{-\frac{1}{\tau}K'(t-s)} q' / \tau ds \tag{5}$$

Suppose the input of neural network is $\mathbf{x}_0$ at the beginning, then solution to the differential equation is:

$$\delta(t) = e^{-\frac{1}{\tau}K't} x_0 + e^{-\frac{1}{\tau}K't} q' K^{-1}(e^{\frac{1}{\tau}kt} - 1) = e^{-\frac{1}{\tau}K't} x_0 + q' K'^{-1}(1 - e^{-\frac{1}{\tau}K't}) \tag{6}$$

When t approximate to $+\infty$, $\delta(t)$ will approximate to a solution vector which has no business with initial value x_0, i.e. $\lim_{t\to\infty}\delta(t) = q'K^{-1}$. The equilibrium point of the neural network is the analytical solution to the finite element equations.

Mentioned as above that the integrated circuit is restricted by maximum input voltage k in LSSM, $\delta(t)$ must be bounded in the running process of neural network. The proving of boundness is given as following. According to Eq. (6) and matrix analysis theorem:

$$\|\delta(t)\| \le \left\| e^{-\frac{1}{\tau}K't} \right\| \|x_0\| + \left\| q'K'^{-1} \right\| \left\| 1 - e^{-\frac{1}{\tau}K't} \right\|$$

$$\le \left\|x_0\right\| + \left\|q'K'^{-1}\right\|$$

$$\le \left\|x_0\right\| + \left\|q'\right\|\left\|K'^{-1}\right\|$$

The result shows that $\delta(t)$ is bounded in the running process of neural network. In practical application, if $\left\|x_0\right\| + \left\|q'\right\|\left\|K'^{-1}\right\| \ge k$, let $q' = q'/n$ (where $q' = q'/n$) at first, then run the neural network. After obtaining calculation result δ, let $\delta = \delta n$, then the solution to the problem can be obtained.

3 Circuit Implement of Neural Network

Dynamic system of neural network with property of Eq. (4) is realized through circuit as following. Neurocomputation circuit system is composed of integrator, adder, feedback loop and reverse controller which realizes negative-resistance. Integrator is composed of simulated operation amplifier (LM324), capacitance and resistance. The I/O relationship is:

$$u_o = \frac{1}{RC}\int u_i \, dt$$

Adder is a reverse adder that is composed of simulated operation amplifier (LM324) and resistance. The I/O relationship is:

$$u_{oi} = -k_{i1}u_{i1} - k_{i2}u_{i2} - \cdots - k_{in}u_{in}$$

where k_{ij} can be obtained through adjusting ratio of node resistance in the adder. Reverse controller is a special case of adder when the relationship of I/O is $u_{oi} = -u_i$. Neural network system to resolve finite element equations can be obtained through connecting adder, integrator, and reverse controller with lead.

4 Example

An example is given to show the application of neural network in finite element. Suppose a sheet whose thickness t = 0.1mm, Yangs' modulus E, poisson ratio μ = 0, load F = 10kN/m. Neglect gravity, try to find each node displacement with finite element and neural network method.

The 6 orthogonal points are input into neural network and the network is run. Calculation results are shown in Table 1.

The results in Table 1 show that whatever the initial points are chosen the neurocomputing dynamic system will convergence to the same point. But it is easy to find that there is little difference between the calculation result and analytical result (relative error is about 2%) because of the impact of maladjustment of input voltage and input current in integrated circuit. The phenomenon that the input values of neurocomputing are zero but the outputs are not zero are called zero-drift, shown as in Table 2. For improving the calculation accuracy, output should be adjusted. The adjusted value equal real output value subtract zero-drift value because that the network system is a linear one and the I/O property of linear system obeys superposition principle. Adjusted values are listed in Table 3. So it is very close between the adjusted values and analytical values, the maximum relative error is within 0.5%.

Table 1. Circuit simulation results of neurocomputing

Input variable	V_1	V_2	U_3	V_3	U_5	U_6
$\mathbf{X}_1$	-3.255	-1.253	-0.08865	-0.3742	0.1748	0.1726
$\mathbf{X}_2$	-3.255	-1.253	-0.08865	-0.3742	0.1748	0.1726
$\mathbf{X}_3$	-3.255	-1.253	-0.08865	-0.3742	0.1748	0.1726
$\mathbf{X}_4$	-3.255	-1.253	-0.08865	-0.3742	0.1748	0.1726
$\mathbf{X}_5$	-3.255	-1.253	-0.08865	-0.3742	0.1748	0.1726
$\mathbf{X}_6$	-3.255	-1.253	-0.08865	-0.3742	0.1748	0.1726
Analytical solution	-3.2527	-1.2527	-0.0879	-0.3736	0.1758	0.1758

Table 2. Zero-drift values of neural network

Output variable	V_1	V_2	U_3	V_3	U_5	U_6
Zero-drift value	-3.104×10^{-3}	-9.762×10^{-4}	-7.608×10^{-4}	-7.608×10^{-4}	-9.762×10^{-4}	-3.104×10^{-3}

Table 3. Adjusted results of neurocomputing

Output value	V_1	V_2	U_3	V_3	U_5	U_6
Adjusted value	-3.252	-1.253	-0.08857	-0.3734	0.1758	0.1757
Analytical solution	-3.2527	-1.2527	-0.0879	-0.3736	0.1758	0.1758

For comparing the calculation results with the results of Hopfield neural network and the results of M-TH neural network in [4], E=200MPa is adopted and the calculation results are listed in the same table, shown as in Table 4.

Table 4. Results of different calculation methods

Method	V_1	V_2	U_3	V_3	U_5	U_6
M-Hopfield	-0.016263	-0.006263	-0.000440	-0.001868	0.000879	0.000879
M-TH	-0.016257	-0.006261	-0.000439	-0.001862	0.000878	0.000878
LSSM(Protel 99)	-0.01626	-0.006265	-0.00044285	-0.001867	0.000879	0.0008785
Analytical solution	-0.016264	-0.006264	-0.000440	-0.001868	0.000879	0.000879

It can be seen from the Table 4 that the errors between the simulation results of LSSM on Protel 99/sim workstation and that of Hopfield neural network and M-TH neural network is within 0.5%.

According to the definition of circuit constant time, the constant time of analytical expression (5) of neural network status equation (4) can be defined as $\frac{1}{\tau}\|K\|$. To specific problems, K is known, only $\tau = RC$ is adjustable. It is easy to obtain different computation speed of neurocomputing system only through adjusting R and C. Stabilizing time is listed in Table 5. It can be seen through Table 5 that network's convergence speed increases proportionally with the decrease of product of R and C.

Table 5. Stabilizing time of neural network

Circuit parameter	R=10, C=1uF	R=10, C=10uF	R=100, C=1uF
Stabilizing time	250us	2.5ms	2.5ms

5 Conclusions

Circuit simulation shows that questions can be resolved (error within 0.5%) in circuit constant time with finite element neural network based on LSSM system. The method developed in the paper and finite element neural network method based on Hopfield network both can obtain right result. Due to the need of circuit design, additional energy items, don't exist in object optimization energy function, come out in the energy function of optimization neural network during the optimization process of Hopfield neural network and its modified form (M-TH neural network). To obtain correct result, high gain ideal operational amplifier is usually chosen in circuit implementation of neural network. The amplifier can make the redundant energy approximate to zero. However, it is difficult to realize the ideal operational amplifier and the solution of the circuit will appear periodical oscillation when the gain of amplifier is big enough, which will result in calculation error and solution instability [5]. So finite element neurocomputing based on LSSM system is more practical. Moreover the number of parameters in finite element equations has no effect on computation speed of the neural network because neural network is parallel system.

References

1. Sun, D., Hu, Q., Xu, H.: Real Time Neurocomputing Theory and Numerical Simulation on Elastic Mechanics. Acta Mechanica Sinica 30 (1998) 348-352
2. Chen, X., Huang, H.-Z..: On the Neurocomputing Time and Error of Structural FEM. Chinese Journal of Mechanical Engineering 36 (2000) 22-26
3. Zhao, Z., Huang, H.-Z..: Method of Stochastic Neural Network Method for Structural Optimization Problem and Its Computer Simulation. Chinese Journal of Computational Mechanics 18 (2001) 242-245
4. Xie, M.: Development of Large Scale Integration Technology. World Science-Technologic Research and Development 23 (2001) 40-44
5. Jiao, L.: Neural Network Computation. Xidian University Press Sian (1996)

Neural Network Based Fatigue Cracks Evolution

Chunsheng Liu[1], Weidong Wu[1], and Daoheng Sun[2]

[1] Department of Mechanical Engineering, Heilongjiang Institute of Science and Technology
Harbin 150027, China
{Liu_chunsheng, Wu-weidong}@163.com
[2] Department of Electromechanical Engineering, Xiamen University
Xiamen, 361005, China
Sundh1339@hotmail.com

Abstract. The crack density and crack growth rate are important parameters, which are used to describe the fatigue damage and predict the fatigue life of a material. There are many researches on the quantitative description of the fatigue cracks density and the crack growth rate, and several models are proposed, but these models cannot be widely used. In this paper, the BP network is used to describe the evolution of the fatigue crack density and the crack growth rate. It can be seen that the proposed method is feasible. The proposed method does not need to determine the interface between the long and short crack, and overcome the shortcoming of traditional models in which physical background of the parameters are uncertain, so it is difficult to determine in engineering.

1 Introduction

The crack density and crack growth rate are two important parameters that are used to describe the fatigue damage and predict the fatigue life of a material. The research on the quantitative description of the smooth sample surface fatigue crack density, evolution and the fatigue crack growth law has always been stressed. The evolution model of the fatigue crack density has ever been proposed successively from different perspectives by Suh [1], She [2], etc. Paris formula successfully describes the fatigue crack growth behavior of the crack object, which satisfies LEFM constrained conditions. Miller [3-6], Hobson [7], Polak [8] and Dowling [9] did a lot of work on the description of the fatigue short crack growth law and they posed their own models respectively. Due to the lack of the comparable experimental results and the unified standard, it is difficult to evaluate the advantages and disadvantages of these models. On the one hand, there are some parameters in these models and the physical significances of these parameters are unclear and hard to define practically; on the other hand, every model bases on a certain material structural characteristics, specific experiment conditions and it is proposed on a certain phase of the crack growth. As a result, their practical scope is constrained. Up to now, there is not extensively accepted the quantitative model of the crack evolution.

The neural network is a complicated nonlinear dynamic system with the ability of real time and pairing collective operation. In recent years, the neural network has been successfully applied in the field of automatic control, pattern recognition, and expert

F. Yin, J. Wang, and C. Guo (Eds.): ISNN 2004, LNCS 3174, pp. 826–831, 2004.

system, etc. But the research on applying the neutral network to describe the fatigue crack evolution law is rare in the literature. There are four main types of the neutral network, which are multilayer forward, feedback, self-organizing and random network. Among them, the multilayer forward network is fit to process a large amount of discrete, noise-bearing and incomplete data collection, and the law can be extracted from the process without analytical expression. It is a good tool to process the fatigue crack evolution experimental data and describe the fatigue crack evolution behavior.

2 Factors Affecting the Fatigue Crack Evolution

Many factors can affect the fatigue crack evolution of the material. These factors include the material features, load features and environment, etc. The following equation can describe the evolution of the smooth sample surface fatigue crack density:

$$n(N) = f_n(K_c, d, \Delta\sigma, R, C, \cdots) \tag{1}$$

where n is the crack density; N is the load cyclic frequency; K_c is the fracture toughness of the material; d is the microstructure scale of the material; $\Delta\sigma$ is the applied stress amplitude; R is the stress proportion; C is the environmental temperature.

Only through experiments can we know how these factors affect the fatigue crack density. At present, the influence of a few factors can be studied through experiments. In addition, even if the experiment studies the function of every factor, it is difficult to embody the results on the quantitative model. For example, the model of AlSi304 stainless steel fatigue crack density under the temperature of 538°C set by Suh, Lee, Kang, Ahn, and Woo [1] only embodies the function of the stress amplitude.

Concerning the problem of the fatigue crack growth, it is widely thought that the growth behavior of the long crack is different from that of the short crack and models set up to describe their laws. It is reasonable to apply Paris formula to describe the long fatigue crack. In the improved Paris formula, the influence of the stress proportion R is considered, which makes the result of the prediction more conform to the reality. The behavior of the short fatigue crack is very complicated. Besides the affecting factors of the long crack, the microstructure of the material, such as, the crystallite dimension, the shape and orientation of the defect, the local directional properties and so on will affect the growth of the short crack. Miller [3-6], Hobson [7], Polak [8] and Dowling [9] did a lot of work to describe the evolution law of the fatigue short crack and all of them set up their own models.

It is difficult to actually define the boundary between the long crack and the short crack, so the above-mentioned model can hardly be applied practically. With the aid of the neural network, the long crack and the short crack can be united and described, the general expression is:

$$\frac{\mathrm{d}a}{\mathrm{d}N} = f_a(K_c, d, \Delta K \text{ or } \Delta\sigma, a, R, C, \cdots) \tag{2}$$

where ΔK is the stress intensity factor variation amplitude; $\Delta\sigma$ is the stress variation amplitude; R is the stress proportion; a is the length of the crack.

The factors which affect the evolution law of the fatigue crack are: the crack ductility of the material, the size of the microstructure, the stress variation amplitude, the length of the crack, the stress proportion and environmental temperature, etc. Among them, the performance of the material and the applied stress amplitude are more important.

3 The Neural Network Model of the Fatigue Crack Evolution

To extract the evolution law of the fatigue crack from the fatigue experimental data through the application of the BP network, at first, the network structure should be defined according to the practical problem, that is, the number of the neuron in the input, output and hidden layer. And then, train the network repetitively by way of the samples made up of the experimental data and the BP algorithm until it converges, so far, the establishment of the model accomplishes and the working phase begins.

3.1 Smooth Sample Surface Fatigue Crack Density

Eq. (1) shows that the crack ductility of the material K_c, the size of the microstructure of the material d, the applied stress amplitude $\Delta\sigma$, the stress proportion R and the environmental temperature C will affect the number of the crack at a certain recycle times in varying degree. For this reason, both the above-mentioned parameters and loaded cyclic numbers are regarded as the input of the network, the number of the neuron in the input layer corresponds to it. The unit of the output layer corresponds to the crack density n. To the experiments on the same material with the same stress proportion and environmental conditions, the unit of the input layer of the network corresponds to the stress amplitude and loaded recycle times. 1Cr18Ni9Ti smooth sample surface fatigue experiment is carried out on the MTS material fatigue experimental machine. The stress control is adopted, the monopodium tensile and compressive stress proportion is $R=-1$, the loaded frequency is 2.0Hz, the experimental environment is indoor temperature 17° C and the medium is air. The replication method is adopted to observe the evolution of the sample surface crack. At first, duplicate the change of the surface state in the sample fatigue process discontinuously, and then, put the replication thin film under the optical microscope to amplify to many times and observe it, the observing order is inversed. When replicating, it is guaranteed that the sample is proceeding under the bearing of 10kN tensile load, moreover, two roughly symmetrical plates are duplicated every time in order to ensure that the interested area is not left out. Table 1 shows the experimental data of 1Cr18Ni9Ti smooth sample surface fatigue crack density.

Double hidden layers are taken; all the units of the hidden layers are five. When training, the learning factor η=0.9, the inertia term coefficient α=0.7 are taken. Through 30 000 learning, the network converges; the population error is 3.4×10^{-5}. The selection of the number of the hidden layer, the learning factor and the inertia term coefficient is empirical. But they just affect the speed of convergence, not the result of convergence.

Table 1. Experimental data of 1Cr18Ni9Ti smooth sample surface fatigue crack density

σ_a (MPa)	N/N_f n	N/N_f n	N/N_f n	N/N_f n	N/N_f n	N/N_f n	N/N_f n	N/N_f n
340	0.981	0.921	0.809	0.698	0.587			
	203	41	17	4	1			
350	0.965	0.89	0.816	0.742	0.668	0.594	0.519	0.445
	302	276	194	118	57	2	1	1
370	0.913	0.783	0.652	0.522	0.391	0.261		
	219	122	68	45	4	1		
320	0.966	0.875	0.783	0.692	0.646	0.601	0.556	
	82	47	27	15	2	2	1	
310	0.936	0.869	0.803	0.736	0.669	0.602	0.535	
	73	40	25	12	2	2	2	

3.2 Smooth Sample Surface Fatigue Crack Growth Rate

Comparing Eq. (1) with (2), the results show that all the parameters, which affect the fatigue crack density, will also affect the fatigue crack growth rate, in addition, the growth rate of the crack varies with its length. Consequently, to the experiments to the same material with the same stress proportion and environmental conditions, the input units of the network corresponds to the stress amplitude and the crack length; the output unit corresponds to the crack growth rate. Table 2 shows the experimental data (training sample) of 1Cr18Ni9Ti smooth sample surface crack growth rate.

Table 2. Experimental data (training sample) of 1Cr18Ni9Ti smooth sample surface crack growth rate

σ_a (MPa)	a da/dN	a da/dN	a da/dN	a da/dN	a da/dN	a da/dN	a da/dN	a da/dN	a da/dN	a da/dN
340	25	64	180	520	660	1640	4200			
	0.034	0.052	0.0928	0.272	0.112	0.784	3.776			
350	4	16	24	88	200	280	580	1080	2080	
	0.0053	0.016	0.0107	0.0853	0.0747	0.1067	0.4	0.6667	1.3333	
370	45	88	192	224	560	2200				
	0.075	0.0717	0.1733	0.0533	0.56	2.7333				
320	10	12	12	15	32	68	92	140	400	1240
	0.004	0.0008	0.0001	0.0012	0.0068	0.0144	0.0096	0.0096	0.052	0.168
310	5	12.5	17.5	20	60	60	84	112	160	220
	0.001	0.0015	0.001	0.0005	0.008	0.0001	0.0048	0.0056	0.0096	0.012
400	24	68	88	112	140	180	200	200	460	580
	0.0274	0.176	0.16	0.192	0.224	0.32	0.16	0.0001	1.04	0.48

The hidden layer of the network is double-layer; the number of the neurons in the first and second hidden layer is 5 and 7 respectively. In the training process, the learning factor $\eta = 0.9$, the inertia coefficient $\alpha = 0.7$ are taken. Through 30 000 learning, the total error is 1.89×10^{-4}.

Before training the network, the normalization of the samples must be carried out, that is, adjusting all the sample values to the interval [0,1].

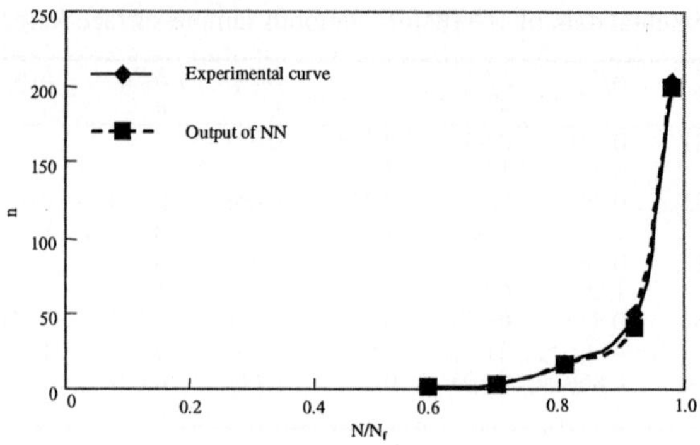

Fig. 1. Curve of 1Cr18Ni9Ti material smooth sample surface crack quantity

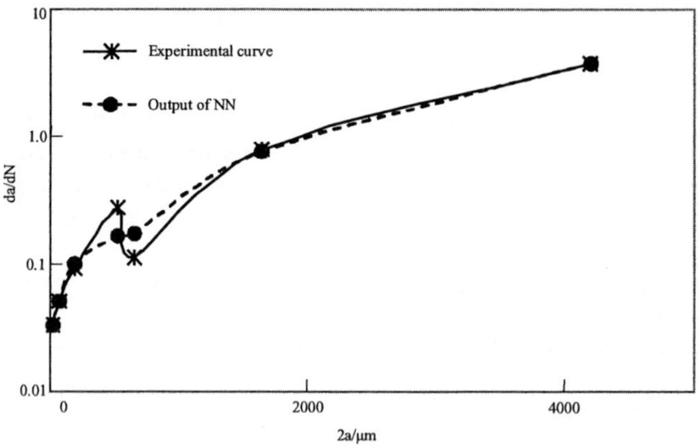

Fig. 2. Curve of 1Cr18Ni9Ti material smooth sample surface crack growth rate

Figure 1 and 2 show the experimental curve of 1Cr18Ni9Ti material smooth sample surface crack quantity and the crack growth rate and the neural network model calculating curve under the acting of the stress amplitude $\alpha = 340$MPa respectively (there is a conspicuous depression on the experimental curve in Figure 3, which is caused by the deficiency of the data and that the measured data is random to certain extent). The figures reflect the experimental curve and the network output curve of the crack quantity and the crack growth rate are identical.

4 Conclusions

The neural network is a complicated nonlinear dynamic system with the ability of real time and pairing collective operation. It combines calculation and storage into one. Recently the neural network has been applied in many fields and the attempt is successful. The multilayer forward network is suitable to process a large number of discrete, noise-bearing and incomplete data collection, from which the law can be extracted without analytical expression. This is a good tool to process the fatigue crack evolution experimental data and describe the fatigue crack evolution behavior.

When applying the neural network to describe the fatigue crack growth rate, it is unnecessary to distinguish the long and short crack. Therefore, it overcomes the shortcomings of traditional models in which it is difficult to determine the interface between the long and short crack, and the physical background of the parameter in analytical models is uncertain and hard to define practically. The reason of the method is very simple and it is fit for the engineering application.

A network can be trained with the fatigue crack evolution experimental data to different materials and under different stress levels, thus, a data base of the material fatigue crack evolution can be set up. The similar method can be adopted to process the experimental data of the material performance.

This paper adopts the neural network to describe the evolution laws of the fatigue crack density and crack growth rate. The analysis to the examples has proved that the method is feasible.

References

1. Suh, C.M., Lee, J.J., Kang, Y.G., et al.: A Simulation of the Fatigue Crack Process in Type 304 Stainless Steel at $538^\circ C$. Fatigue Fract. Engin Mater. Struct. 15 (1992) 671-684
2. She, S., Landes, J.D.: Statistical Analysis of Fracture in Graphite. Int. J. of Fract. 63 (1993) 189-200
3. Miller, K.J., Mohammed, H.J., de los Rios, E.R. In: Miller, K.J., de los Rios, E.R. (eds.): The Behaviour of Short Fatigue Cracks. EGP Publication 1, Mechanical Engineering Publications, London (1986) 491-511
4. Miller, K.J. In: Bilby, B.A, Miller, K.J., Willis, J.R. (eds.): Fundamentals of Deformation and Fracture. Cambridge University Press, Cambridge (1985) 477-500
5. Miller, K.J. In: Van, Elst, H.C., Bakker, A. (eds.): Proceedings Fracture Control of Engineering Structures ECF6. Amsterdam, 3 (1986) 2149-2167
6. Miller, K.J., Mohammed, H.J., Brown, M.W., de los Rios, E.R. In: Ritchie, R.O., Lankford, J. (eds.): Small Fatigue Cracks. Pennsylvania, A Publication of the Metallurgical Society, Inc., (1986) 639-656
7. Hobson, P.D., Brown, M.W., de los Rios, E.R. In: Miller, K.J., de los Rios, E.R. (eds.): The Behaviour of Short Fatigue Cracks. EGP Publication 1, Mechanical Engineering Publications, London (1986) 441-459
8. Obrtlik, K., Polak, J., Hajek, M., Vasek, A.: Short Fatigue Crack Behaviour in 316L Stainless Steel. Int. J. Fatigue 19 (1997) 471-475
9. Dowling, N.E., In: Cyclic Stress-Strain and Plastic Deformation Aspects of Fatigue Crack Growth, ASTM STP637, Philadelphia (1977) 97-121

Density Prediction of Selective Laser Sintering Parts Based on Artificial Neural Network

Xianfeng Shen[1], Jin Yao[1], Yang Wang[2], and Jialin Yang[2]

[1] School of Manufacture Science and Engineering, Sichuan University, Postalcode 61 00 65, 86028 Chengdu, China
shxfmail@21cn.com, jinyao163@163.com

[2] Institute of Machinery Manufacturing Technology, China Academy of Engineering Physics, Postalcode 62 19 00, 860816 Mianyang, China
yang_wang_mail@yahoo.com.cn, myyangjialin@sohu.com

Abstract. Selective laser sintering (SLS), one of rapid prototyping technologies, employs laser beam to selectively fuse fully powder into a solid object layer by layer. However, density prediction of SLS parts using finite elements analysis (FEA) having been reported, heavily depends on the precision of the FEA model. An Artificial neural network (ANN) approach presented in this paper has been developed for density prediction of SLS parts. Two-layer supervised neural networks are used, and the inputs to the neural network are known SLS process parameters such as laser power, scan speed, scan spacing and layer thickness. Orthogonal experimental method is employed for collection of experimental training and test sets. The construction of network is also investigated. Comparison of predicted and experimental data has confirmed the accuracy of the ANN approach.

1 Introduction

Global competition has dramatically driven manufacturing companies to respond to customers' requirements at low cost in the shortest time. The need to reduce time and cost in the product development process has brought up a wide range of different technologies for rapid prototyping (RP), rapid tooling (RT) and rapid manufacturing (RM). Among them, selective laser sintering (SLS) can be used to produce three-dimensional parts directly from a CAD model by the selective sintering of successive layers of polymer, metallic or ceramic powder. In addition, SLS need not any frock mould, directly changes the CAD model into prototyping parts, and is not confined by parts complexity. Because of the unique advantage in the diversification of the SLS material, material saving, prototyping without support, high intensity parts and rapid metal parts manufacturing, SLS can be widely used and it attracts attention from all works of industry.

The aim of SLS is to attain high density parts, which need suitable SLS process parameters. Many parameters affect the quality of sintered parts, including sintering

F. Yin, J. Wang, and C. Guo (Eds.): ISNN 2004, LNCS 3174, pp. 832–840, 2004.

material properties, powder properties, the specifications of SLS machine and process parameters, such as laser power, scan speed, scan spacing and layer thickness. But once the SLS machine is certain and the sintering powder has been selected, the main factors affecting part density are the process parameters. It must be pointed out that sintered part from powder is porous and its density presents its quality. Though part quality should be evaluated by combination of tensile strength, elongation, and hardness, sintered part density is the most important factor. Comparison of half-density part to full-density part, the former is inferior to the latter in tensile strength, hardness and surface quality. Therefore, density prediction of SLS parts provides suitable guidelines in the selection of appropriate process parameters.

But, nowadays most of the research methods on density prediction of SLS parts are based on finite elements analysis. Several studies dealing with density prediction of sintered polymers produced by SLS processes have been reported. For selective laser sintering of amorphous polymers, Nelson et al. predicted density of a sintered polycarbonate part, using 1-D finite element and 1-D finite difference methods [1]. As to ABS, Weissman and Hsu used a 1-D finite element method [2]. Considering polycarbonate, a 2-D model of sintering has been developed in order to predict density and to accommodate edge effects [3], [4]. Similarly, a 2-D model applied to ABS material has also been reported [5]. Tontowi et al. discussed density prediction of crystalline polymer sintered parts at various powder bed temperatures using method of a 2-D finite element simulation [6].

It must be pointed out that FEA strongly depends on the precision of SLS model and operators must be familiar with the FEA modeling and have good mathematical skills. Furthermore, the process of SLS not only has physical effects but also chemical changes. The relationship between process parameters and part density is quite complicated and nonlinear. Thus, heretofore researchers have not found a systematical theory and satisfying method for SLS part density prediction. Therefore, we can not theoretically establish a function to represent part density by means of process parameters. To change the present situation a new method must be found to make sure that the high precision and high intensity parts can be produced directly in SLS process.

An artificial neural network (ANN) method, which has characters of human intelligence and need not any precise mathematical model, is suitable for the part density prediction of SLS. ANN after successfully training can rapidly reason, and associate. It can filter the inputted noise, map from 'n-dimensional input space' to 'm-dimensional output space' and approach any nonlinear function precisely. An ANN can be trained iteratively from examples to learn and represent the complex relationships implied within the data. In fact the general applicability of the technique has been demonstrated by the proof that feedforward networks with a single hidden layer, having sufficient number of neurons, using threshold or sigmoid activation function, are universal approximators [7]. The part density predict in SLS can be thought of as a continuous non-linear mapping from '3-dimensional process parameters space' to '1-dimensional density space'. In this paper, a 2-layer supervised neural network is employed to predict part density of SLS. Inputs to the neural network are known SLS process parameters such as scan speed, scan spacing, laser power, and layer thickness.

Mathematical modeling, the structure design, training data selection, training, and results analysis of ANN are also discussed in this paper. Comparison of predicted and experimental data has demonstrated the validity of the ANN approach.

2 Problem Representation and Modeling

As mentioned above, sintered part quality can be evaluated by its density, and SLS process parameters include laser power, scan speed, scan spacing and layer thickness. So, the problem can be tread as non-linear mapping from '3-dimensional process parameters space' to '1-dimensional density space'.

The following discusses are based on the same material powder and the same SLS machine. If other material powder or SLS machine are needed, the neural network should be trained respectively using the particular training set based on experiments. The artificial neural network structure need be altered only if the generalization error is beyond desired target.

The SLS process parameters have a non-linearity profound effect upon the final part density. When employing an ANN approach, it is relatively easy to incorporate a large number of system inputs to accommodate these effects. Since the modeling is directly incorporated within the weights of the ANN connections, any non-linearity or inter-dependence within the relationships is necessarily incorporated within the output predictions. Hence Artificial Neural Networks (ANN) is identified as being particularly suitable for modeling the final part density under various SLS processing conditions.

The mapping from 'n-dimensional input space' to 'm-dimensional output space' can be represented as following.

Input vector:

$X=[x1,x2,...,xn]^T$

Target vector:

$Y=[y1,y2,...,yn]^T$

One group training set:

$x(k)=[x1(k),x2(k),...,xn(k)]^T$

$y(k)=[y1(k),y2(k),...,yn(k)]^T$

The aim of transformation is to find out the underlying function of data set between X (n-dimension) and Y (m-dimension).

Sintered part density prediction is nonlinear. The inputs to the neural network comprise SLS process parameters such as laser power (P), scan speed (v), scan spacing (d), and layer thickness (h). The output is the sintered density (ρ). Then the model can be represented as:

$$\rho=F(P, h, v, d). \tag{1}$$

Backpropagation (BP) is employed in learning the function F. It has been proved that a network of two layers, where the first layer is sigmoid and the second layer is linear, can be trained to approximate any function with a finite number of discontinuities arbitrarily well. So the ANN described here employs BP network with just one hidden layer and hyperbolic tangent sigmoid transfer function.

3 An Algorithm and Its Implementation

The generalized steps of ANN modeling can be divided several steps. First, ascertain problem domain and ensure the mapping from input to target space. Second, collect training data through experiments. Third, design network structure and determine layers count, neurons number and their connection mode. Finally, train the network, then the network can learn the underlying rule from input to output. Hence the modeling is finished.

Figure 1 shows the steps involving in the algorithm, such as enter of initial parameters, initialization, training, test of the artificial neural network.

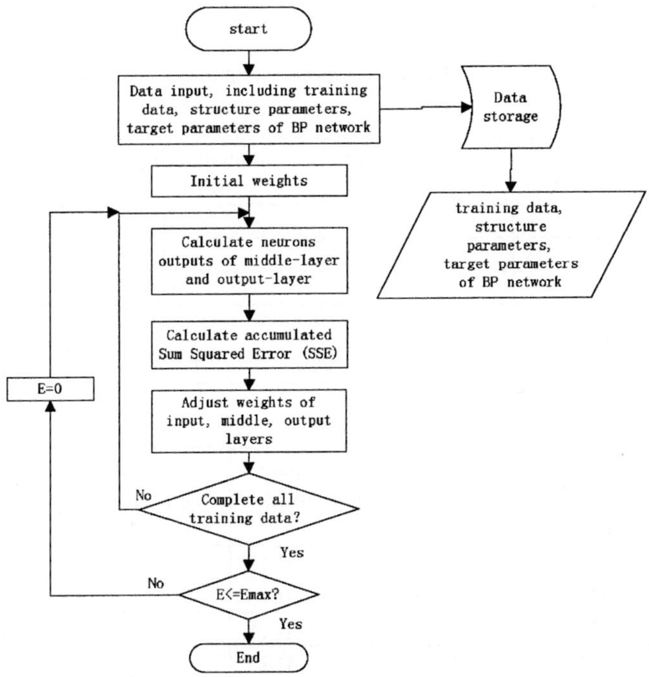

Fig. 1. Flow chart of the BP neural network algorithm

Batch training was adopted in this BP network. Batch training of a network proceeds by making weights and biases changes based on an entire set (batch or epoch) of input vectors. Incremental training changes the weights and biases of a network as needed after presentation of each individual input vector.

In this flow chart, sum-squared error (SSE) is the sum of squares due to error.

$$E = (Tp - Yp)^2/2. \quad (2)$$

where E is SSE, Tp is the target output, and Yp is the network output for *p*th pattern [8].

This statistic measures the deviation of the responses from the fitted values of the responses. A value closer to 0 indicates a better fit.

3.1 Collection of Training Data

It is well known that insufficient or less representative training data lead that BP network can not fully learn the law underlying training data, whereas excessive training data result in over fitting BP network. Over fitting BP network leads to too complicated ANN structure and larger generalized error because it memorizes specific features of particular data, hence precise model can not be attained. In addition, collection of a lot of training data is a time and energy consuming job, especially through experiments.

Table 1. Orthogonal table

No.	Factors			
	Laser power (W)	Scan speed (m/s)	Scan spacing (mm)	Layer thickness (mm)
1	15	0.5	0.12	0.2
2	20	0.75	0.15	0.3
3	25	1.0	0.18	0.4

Table 2. Orthogonal experimental scheme and experimental data

No.	Factors				Experimental Results
	Laser power (W)	Scan speed (m/s)	Scan spac-ing (mm)	Layer thick-ness (mm)	part den-sity(kg/m3)
1	15	0.5	0.12	0.2	696
2	20	0.5	0.15	0.3	599
3	25	0.5	0.18	0.4	636
4	15	0.75	0.15	0.4	793
5	20	0.75	0.18	0.2	764
6	25	0.75	0.12	0.3	663
7	15	1.0	0.18	0.3	954
8	20	1.0	0.12	0.4	823
9	25	1.0	0.15	0.2	766

Based on statistic principle, orthogonal experimental method can chooses adequate and representative points from lots of available trial points, and arrange multi-factor trails by employing standard orthogonal table. Equilibrium decentralization of orthogonal experimental method ensures representative samples. Furthermore, the outstanding advantage of orthogonal experimental method is that it can reduce trial times while hardly harms to trial results. Thus, trial efficiency is dramatically improved.

Training data are measured through experiments [6]. In the experiments, the material used is a nylon-12 crystalline polymer powder, known as Duraform polyamide. Specimens with the simple shape, whose 3-D are 20mm, 20mm and 10mm, have been made in a SLS machine.

Laser power, scan speed, scan spacing and layer thickness compose testing factor and 4-factor and 3-row orthogonal table are established which is listed in Table 1.

3.2 Data Normalization

Before training the network, the training data was normalized in the range of 0 and 1:

$$x' = (x - Xmin) / (Xmax - Xmin). \quad (3)$$

where x' is the normalized value for the variable, and Xmin and Xmax are the minimum and maximum of each variable 'x'.

BP network outputs should be in the range 0 and 1, so the neurons of target matrix must be normalized by applying (3). Once the network training has been finished, the real values of ANN outputs can be obtained from:

$$x = Xmin + x' (Xmax - Xmin). \quad (4)$$

Normalized network weights abate network training difficulties because they fall in a short range. In addition, if sigmoid transfer function is adopted, normalization can prevent neuron outputs from saturation, which improves network accuracy.

3.3 Network Training and Test

Training error is referred as the sum of squared differences between the network targets and actual outputs for a given input vector or set of vectors, which shows network precision against training set, whereas generalization error is termed as a symbol of network precision against test set. It is obvious that the smaller generalization error, the better generalization performance the network has.

The neurons number of the hidden layer largely affects network performance. The training error shrinks whereas the generalization error increases with the increase of the neurons number. Therefore, enough neurons number of the hidden layer is very essential to network. The dimension of input vector is labeled as P, and then the neurons number of the hidden layer can be initialized as 2P+1. The neurons number of the hidden layer here is 9.

Decision of ANN training times can be obtained by setting of maximum training steps m_{ne} and sum-squared error s_{se}. ANN training times before stopping learning increases with the larger m_{ne} and the smaller s_{se}. Commonly, training error and generalization error decreases along with increased training times. But too large training times will lead to over fitting, which can be described as a case in which the error on the training set is driven to a very small value, but when new data is presented to the network, the error is large. To avoid over fitting, the proper training times should suit the requirement of ANN. Here we took m_{ne}=2000 and s_{se}=0.001.

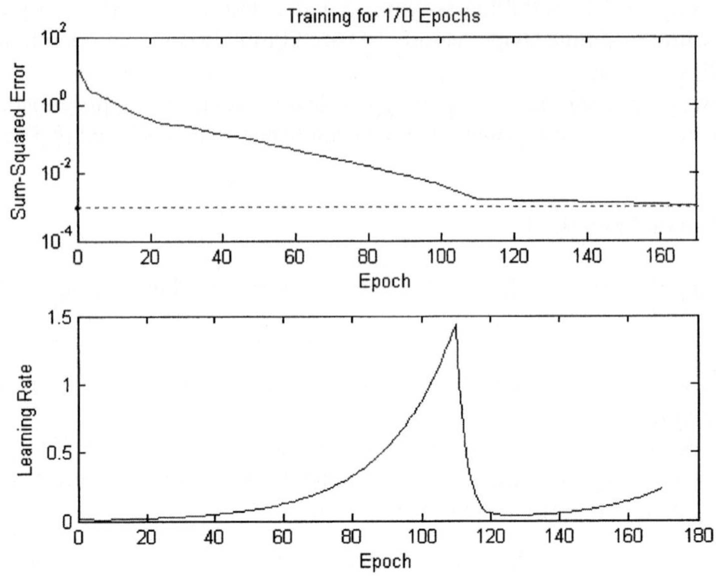

Fig. 2. Curve of network sum-squared error and learning rate vs. training times (Epoch)

Batch training, in which weights and biases are only updated after all of the inputs and targets are presented, is taken as training method for better network performance.

All training is done using backpropagation networks (BP) with both adaptive learning rate and momentum method. An adaptive learning rate requires some changes in the training procedure. During training, the learning rate increases, but only to the extent that the network can learn without large error increases. Thus, a near-optimal learning rate is obtained for the local terrain. When a larger learning rate could result in stable learning, the learning rate is increased. When the learning rate is too high to guarantee a decrease in error, it gets decreased until stable learning resumes. As with momentum, if the new error exceeds the old error by more than a predefined ratio, the new weights and biases are discarded. In addition, the learning rate is decreased. Otherwise, the new weights are kept. If the new error is less than the old error, the learning rate is increased. The trend can be seen from Figure 2, in which the learning rate increases to about 1.48, then decreases sharply. It can be seen clearly from Figure 2 that training converges quickly after 110 epochs. This mainly causes by applying adaptive learning rate and momentum methods.

4 Results

The comparison of predicted and experimental data in Figure 3 show the accuracy 93 percent of the ANN approach. Post-processing has been employed for converting normalized outputs back into the same units that were used for the original targets using (4).

Table 3. Experimental data versus corresponding ANN predictions for sintered density

No.	Factors				Experimental Results	Predicted Results	Per-cent error
	Laser power (W)	Scan speed (m/s)	Scan spacing (mm)	Layer thick-ness (mm)	part density (kg/m^3)	part density (kg/m^3)	Percent error (%)
1	17.5	0.73	0.12	0.25	933	878.6	-6.2
2	17.5	0.58	0.15	0.2	911	937.2	2.8
3	22.5	0.64	0.16	0.35	821	811.3	-1.2
4	22.5	0.89	0.14	0.30	649	660.7	1.7
5	17.5	0.97	0.15	0.4	621	592.4	-4.8
6	22.5	0.89	0.15	0.25	650	682.5	4.7

5 Conclusion

From the experiments and predicted results, it can be concluded that density prediction of SLS parts by applying ANN approach is rather accuracy. This method has an outstanding advantage that it need not know the precise model. Employment of orthogonal experimental method makes trail more reasonable with less trail times and almost no sacrificed performance. Training can be accelerated by applying a combination of batch training, adaptive learning rate and momentum method.

Acknowledgements. This work was financially supported by Scientific Foundation of China Academy of Engineering Physics under Grant No. 2002Z0303.

References

1. Nelson, J.C., Xue, S., Barlow, J.W., Beaman, J.J., Marcus, H.L. and Bourell, D.L.: Model of the Selective Laser Sintering of Bisphenol-A Polycarbonate. Ind. Eng. Chem. Res., Vol. 32 (1993) 2305-17

2. Weissman, E.M. and Hsu, M.B.: A Finite Element Model of Multi-layered Laser Sintered Part. Solid Freeform Fabrication Proceedings, University of Texas at Austin, TX. (1991) 86-93
3. Berzins, M., Childs, T.H.C. and Ryder, G.W.: Selective Laser Sintering of Polycarbonate. Annals of CIRP., Vol. 45. No. 1 (1996) 187-90
4. Nelson, J.C., Xue, S., Barlow, J.W., Beaman, J.J., Marcus, H.L. and Bourell, D.L.: Model of the Selective Laser Sintering of Bisphenol-A Polycarbonate. Solid Freeform Fabrication Proceedings, University of Texas at Austin, TX. (1995) 196-203
5. Bugeda, G., Cervera, M. and Lombera, G.: Numerical Prediction of Temperature and Density Distribution in Selective Laser Sintering Process. Rapid Prototyping Journal, Vol. 5. No. 1 (1999) 21-6
6. Tontowi, A.E. and Childs, T.H.C.: Density Prediction of Crystalline Polymer Sintered Parts at Various Powder Bed Temperatures. Rapid Prototyping Journal, Vol. 7. No. 3 (2001) 180-184
7. Hornik K.M., Stinchcombe M. and White, H.: Multi-layer Feedforward Networks Are Universal Approximators. Neural Networks, Vol. 2. No. 5 (1999) 359-66
8. Cherian R.P., Smith L.N. and Midha P.S.: A Neural Network Approach for Selection of Powder Metallurgy Material and Process Parameters. Artificial Intelligence in Engineering, 14 (2000) 39-44

Structure Optimization of Pneumatic Tire Using an Artificial Neural Network

XuChun Ren and ZhenHan Yao

Department of Engineering Mechanics, Tsinghua University, Beijing 100084, People's Republic of China
rxc00@mails.tsinghua.edu.cn, demyzh@tsinghua.edu.cn

Abstract. An application of neural networks to tire optimization designs is presented to alleviate the stress concentration of toe opening. As well known, it is either uncertain or time-consuming to obtain the global optimum solution by using classical local search methods when objective function of optimization is both nonconvex and implicit. In addition, it is infeasible to use local search method based on iteration to optimize tire mechanical property because analysis of tire mechanical responses is involved with material nonlinearity, geometry nonlinearity and boundary nonlinearity. In this paper, a GRNN is constructed to optimize the stress of toe opening by looking at an optimum Young's modulus and cord direction of tire body rubber-cord composite material layer.

1 Introduction

Pneumatic tire is one of the important components of an automobile driver system. Recently, with the development of automobile, the mechanical response of tire has been extensively researched in various respects such as rolling resistance, maneuverability, durability, noise and riding comfort. To improve these performances, a lot of effort should be made on profile contour, construction, material and tread pattern.

According to the investigation on tire failure, toe opening damage, which results from the stress concentration of toe opening, is very popular. In order to reduce the stress of toe opening, the designing of the Young's modulus and the pave direction of cords in tire body composite layers should receive much more attention.

Traditionally, optimality criterion approach, mathematical programming approach and sensitivity analysis are extensively applied to mechanical structure optimization [1]. Involving with complex nonlinearity, traditional methods are time-consuming for tire optimization. So the screening method is used alternatively [2]. On the other hand, artificial neural networks have recently been used to optimize mechanical response of pneumatic tire. Then the method will be used in this paper to reduce the stress of toe opening.

In the following sections, the optimization model and its FEA model are firstly built for reducing the stress concentration. A radial basis function (RBF) neural network is then constructed for this constrained optimization problem. Training samples are designed by finite element method (FEM). At last, mechanical responses of the new design are compared with the original ones.

F. Yin, J. Wang, and C. Guo (Eds.): ISNN 2004, LNCS 3174, pp. 841–847, 2004.

2 Optimization Model and Tire Mechanical Responses

2.1 Optimization Model

$$
\begin{aligned}
&\text{Minimize}: \ w_{\max}(E,\theta) \\
st. \quad & \left\|\mathbf{u}_j(E,\theta)\right\| \le \overline{u} \qquad j=1,\ldots,m \\
& \left\|\boldsymbol{\sigma}_k(E,\theta)\right\| \le \overline{\sigma} \qquad k=1,\ldots,l \\
& E^l \le E \le X^u \\
& \theta^l \le \theta \le \theta^u
\end{aligned} \tag{1}
$$

where E^l and E^u are the lower and the upper bound of Young's modulus respectively, θ^l and θ^u the lower and the upper bound of cord direction. The inequality on $\mathbf{u}_j$ is displacement constraint of the *j*th point, and the inequality on $\boldsymbol{\sigma}_k$ displacement constraint of the *k*th point. Design variables include Young's modulus E and direction θ of cords in profile layers. Object function is to minimize the maximal strain energy density $w_{\max}(E,\theta)$ around toe opening. $w_{\max}(E,\theta)=\int_0^{\mathbf{E}} \mathbf{S}:d\mathbf{E}$ is determined by equation:

$$
\nabla_0 \cdot \left(\mathbf{S}\cdot\mathbf{F}^T\right) + \rho_0 \mathbf{f} = \rho_0 \ddot{\mathbf{u}} \tag{2}
$$

Where $\mathbf{S}$ is the second Piola-Kirchhoff (PK2) stress tensor and $\mathbf{E}$ is Green strain tensor. Because of the nonlinear constitutive equations, the geometry equations of large deformation and the nonlinear boundary condition, the object function is implicit and nonconvex.

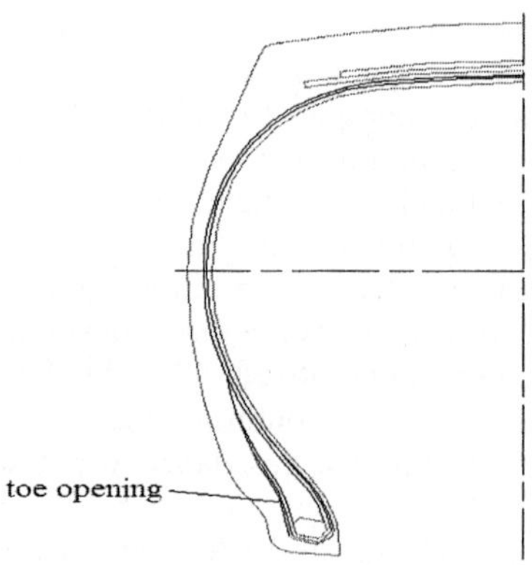

Fig. 1. Material distribute (cross-section)

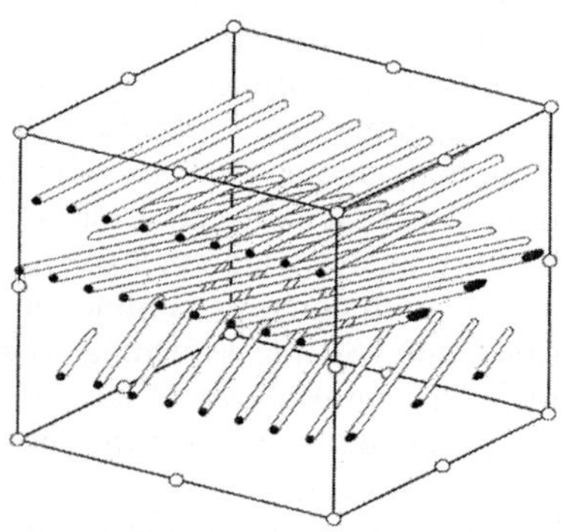

Fig. 2. Rubber-cord composite element

2.2 FEM Simulation of Mechanical Responses

There are more than ten kinds of different components in 10020R tire, which include eleven kinds of rubbers, three belt rubber-steel cord layers, steel and two kinds of polyester cord-rubber layers, as shown in Fig. 1. Carbon black-filled rubber, a hyper-elastic material, is described by moony-rivlin constitutive equation. Contacts happen between steel rim and tire toe as well as between tread and ground. Large deformation occurs with full loads applied.

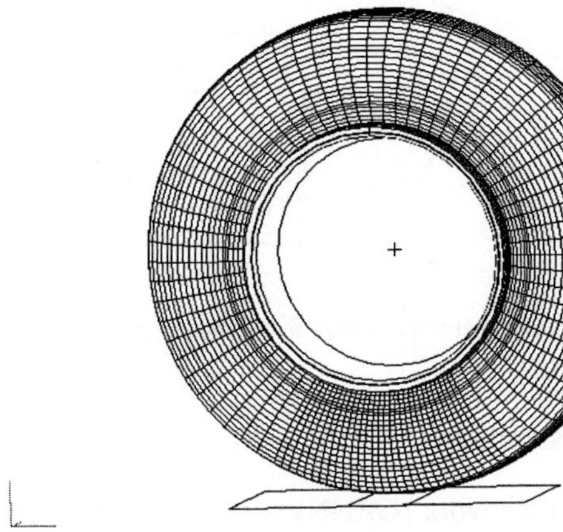

Fig. 3. FEA model of tire: eleven kinds of rubber materials and five kinds of rubber-cord composites. The contact between rim and tire and the contact between tire and ground are considered. In order to improve the numerical stability and accuracy, fine meshes are created in the parts where the contact may occur. There are more than 16,000 elements in the model, and a full simulation procedure consumes about 7 hours on a PC with an Intel 1.6Ghz CPU and 1Gb memory

To solve fomula (1), the local search algorithm must iterate from multi initial values to avoid local optimal pseudo-solution. However, a single FE analysis procedure consumes large CPU time, and hundreds of iterations required will make the local search algorithm infeasible.

3 RBF Neural Network for Optimization of Tire Toe-Opening

Many kinds of neural networks have been established over the past 20 years. They include multilayer feedforward neural network (MLFN), recurrent neural network (RNN), self-organized neural network and Hopfield network. Radial basis function (RBF) network belong to the group of kernel function network that utilize simple kernel functions, distributed in different neighborhoods of the input space, whose responses are essentially local in nature. The architecture of RBF network consists of one hidden and one output layer. This shallow architecture has great advantage in terms of computing speed compared to multiple hidden layer nets. The good approximation ability of RBF multilayer network was proved in reference [3-5].

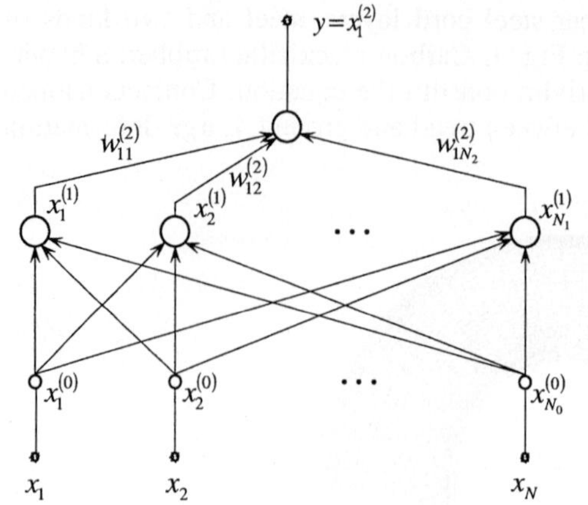

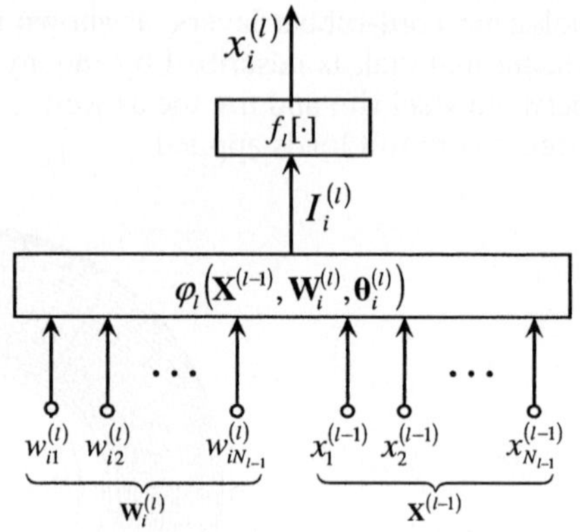

Fig. 4. Two layers and one output RBF network **Fig. 5.** A RBF neuron in GRNN hidden layer

3.1 GRNN with Multi Inputs and One Output

Generalized Regression Neural Network (GRNN) is special RBF network with a linear second layer. A GRNN, in which the hidden layer employs RBF and output layer employs linear function, is constructed by using E and θ as the input vector components and $w_{\max}$ as output. We employ Gaussian RBF in the hidden layer (Fig 4, 5):

$$f_l\left[I_i^{(l)}\right] = \left(\sqrt{2\pi}\sigma_i^{(l)}\right)^{-1} \exp\left(-I_i^{(l)}\right), \quad \sigma_i^{(l)} > 0 \tag{3}$$

$$I_i^{(l)} = \varphi_l\left(\mathbf{X}^{(l-1)}, \mathbf{W}_i^{(l)}\right) = \frac{\left\|\mathbf{X}^{(l-1)} - \mathbf{W}_i^{(l)}\right\|^2}{2\left(\sigma_i^{(l)}\right)^2} \tag{4}$$

And the mapping function from input $\mathbf{X}$ to output y is

$$y = \sum_{i=1}^{N1} w_i^{(2)} \left(\sqrt{2\pi}\sigma_i^{(1)}\right)^{-1} \exp\left(\frac{-\left\|\mathbf{X} - \mathbf{W}_i^{(1)}\right\|^2}{2\left(\sigma_i^{(1)}\right)^2}\right) \tag{5}$$

where $\mathbf{W}_i^{(1)}$ is the weight vector from the input layer to the i th neuron of the hidden layer, and $w_i^{(2)}$ is the weight vector from the i th neuron of the hidden layer to the output layer.

3.2 Treating Implicit Constraint Conditions of Optimization Problem Using GRNN, Training Samples, and Neural Network Parameters

In optimization problem (1) displacement constraint conditions and stress constraint conditions are implicit. If only the objective function is approximated with GRNN, the feasible domain of the optimization problem will be hardly determinated. Therefore, the dimension of output vector is expanded to $m+l+1$ to contain stresses and displacements. Consequently, the optimization problem (1) is transformed into a solvable optimization problem with the objective function and constraint variables both approximated by GRNN. Hence, the required floating-point operations to get the objective function and constraint variables are $O(100)$ order and can be accomplished in several milliseconds. Finally the whole design variable space is discreted and an enumeration algorithm is employed to get an approximate solution of optimization problem (1).

Training samples are obtained using FEA software MSC.MARC to analyze 50 different models in which Young's modulus E and cord direction are taken from ticked data in Table 1. E^l and E^u in fornula (1) are taken as 1.6×10^{11}MPa and 2.5×10^{11}MPa respectively. θ^l and θ^u in formula (1) are taken as 0 degree and 20 degree respectively. The spread $\sigma_i^{(l)}$ (formula (3)) of GRNN is 0.08. And the input vector and the output vector are normalized so that all E values fall in interval $[0,0.9]$ and all θ values fall in interval $[0,2]$.

Table 1. Traing samples. The first column is discrete values of Young's modulus (unit: Pa). The first row is discrete values of cord direction (unit: degree). The ticks indicate the corresponding value pairs taken to bulid a model analysed with FEA software MSC.MARC

	0	2.5	5	7.5	10	12.5	15	17.5	20
1.6e11		√		√		√		√	
1.7e11	√		√		√		√		√
1.8e11		√		√		√		√	
1.9e11	√		√		√		√		√
2.0e11		√		√		√		√	
2.1e11	√		√		√		√		√
2.2e11		√		√		√		√	
2.3e11	√		√		√		√		√
2.4e11		√		√		√		√	
2.5e11	√		√		√		√		√

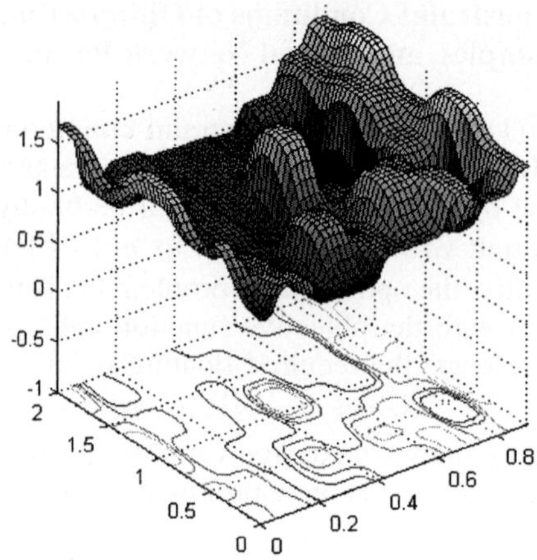

Fig. 6. Approximation function of max strain energy density (objective function): there are two independent variables i.e. E and θ. The bottom curves show the contours of the max strain energy density. Obviously, the objective function is nonconvex

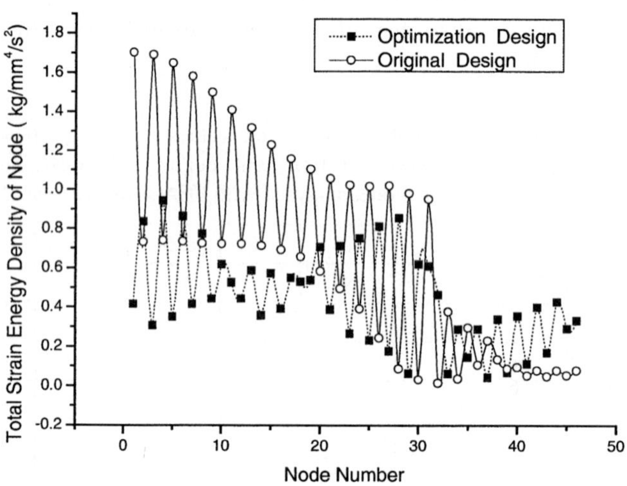

Fig. 7. Comparation between the optimization design and the original design. The total strain energy density value are obtained using MSC.MARC. Nodes path including the max total strain energy density node is set along the circumference of toe-opening

4 Numerical Results and Conclusions

The objective function approximated by GRNN is as shown in Fig 6. After discreting the whole design variable space and using enumeration algorithm, an approximate solution of optimization problem (1) is achieved as $E = 1.798\times10^{11}$ and $\theta = 17.6$.

For the convenience of engineering use, we round-off the result as $E = 1.8 \times 10^{11}$ and $\theta = 18$. This optimization design is modeled and analyzed by MSC.MARC. Comparing with the original design ($E = 2.1 \times 10^{11}$, $\theta = 0$), there is 10~70 percent reduction of strain energy density for almost all the nodes, particularly the max total strain energy density decreases over 80 percent (Fig 7). The values near the lower bound of the total strain energy density increase by more than 100 percent however they are still in the low level region (Fig 7). These differences show the optimization design carry loads in a more effective way.

4.1 Conclusions

As a relatively efficient model to optimize tire designs, the GRNN model is found to save hundreds of CPU hours and satisfy required accuracy. For the optimization problem with implicit noncovex objective function and implicit constraint conditions, the new method is also very superior as the constraint variables contained in the GRNN output vector. And corresponding optimization problem could be solved with a considerable small CPU cost, even using an enumeration algorithm.

References

1. Sui, Y. K.: Modelling Transformation and Optimization New Developments of Structural Synthesis Method. Dalian University of Technology Press (1996)
2. Weiss, M., Tsujimoto,S., Yoshinaga, H.: Belt Construction Optimization for Tire Weight Reduction Using the Finite Element Method. Tire Science and Technology, TSTCA, Vol. 21, No. 2, April-June (1993) 120-134
3. Hartman, E. J., et al: Layered Neural Networks with Gaussian Hidden Units as Universal Approximation, Neural Computation, Vol. 2, No. 2 (1990) 210-215
4. Park, I., et al: Universal Approximation Using Radial Basis Function Networks. Neural Computation, Vol. 3 (1991) 247-257
5. Park, I., et al: Approximation and Radial Basis Function Networks. Neural Computation, Vol. 5 (1993) 305-316

A Multiple RBF NN Modeling Approach to BOF Endpoint Estimation in Steelmaking Process*

Xin Wang [1], Shaoyuan Li [1], Zhongjie Wang[2], Jun Tao [3], and Jinxin Liu [4]

[1] Institute of Automation, Shanghai Jiao Tong University, 200030
wangxin26@sjtu.edu.cn
[2] Depart. of Control Science&Engineering, Tongji University, 200092
[3] Baosight Software Corporation, Shanghai, 201900
[4] Research center of Automation, Northeastern University, 110004

Abstract. In order to estimate the bath endpoint phosphorus [P] content and manganese [Mn] content for Basic Oxygen Furnace (BOF) steelmaking process, three BOF endpoint estimation models are given according to the fast speed case, the slow speed case and the middle speed case of analyzing the sublance in-blowing sample. The first one is modeled from metallurgic mechanism, the second is with RBF NN and the last one is modeled using least square method. With theses models, steelmaker can get the estimation of BOF endpoint [P] and [Mn] based on the process information and the sublance measurement. The industrial experiment shows that these models are helpful and powerful.

1 Introduction

Basic Oxygen Furnace (BOF) steelmaking is one of the key processes in the iron & steel industry. The process of BOF steelmaking is a complex physical-chemical process, which takes hot metal (HM), pig iron and scrap as materials [1]. In order to decrease impurity compositions level (such as carbon, [P] *etc.*) and raise bath temperature to catch the tapping aim, it ejects oxygen from oxygen lance into the bath [2]. Because of the poor condition of BOF process, bath compositions can't be measured on line continuously [3]. So a sublance is designed to measure the bath. However, it can only provide the bath carbon consistence, the temperature, samples *etc.* at BOF in-blowing and endpoint stages. As to bath [P], [Mn] level, the sublance can't do any measurement directly. In practice, steelmakers have to wait for the analysis result from the endpoint sublance sample to determine whether bath [P], [Mn] has caught the aim requirement, which lowers the productivity. On the other hand, most BOF models are used to estimate the carbon content [C], the temperature [4] and can't do any estimation about bath [P], [Mn] level [5].

In this paper, multiple models method is introduced into the field of BOF endpoint [P]&[Mn] estimation. Three intelligent models are presented according to the analyzing speed of in-blow sublance sample by using metallurgical mechanism, RBF

* This work is supported by 863 High Technology Project (2002AA412130, 2003AA412310).

F. Yin, J. Wang, and C. Guo (Eds.): ISNN 2004, LNCS 3174, pp. 848–853, 2004.

NN and least square weighting method respectively. With the aid of these models, steelmaker can get the estimation of BOF endpoint impurity composition based on the process and sublance information, and needn't wait for the sample analysis.

2 Oxidation Theory of Bath [P]&[Mn]

In the BOF process, bath [P]&[Mn] enter the slag phase as different compositions after the oxide production. The more oxide production enters the slag, the more capability of decreasing impurity can be got with the slag. The capacity (C_i) can be expressed with the equilibrium status to the oxidation reaction. The distribution ratio (L_i) is defined as the result of the concentration in steel phase divided by the concentration into slag phase to each element, which can be denoted as

Table 1. Denotations of C_i & L_i to Element [P]&[Mn]

	Capacity	Distribution Ratio
P	$C_P = \dfrac{(\%P)}{a_P \cdot a_O^{2.5}}$	$L_P = C_P \cdot a_O^{2.5} \cdot f_P = \dfrac{K_1^{\theta} \cdot a_{O^{2-}}^{1.5}}{f'_{PO_4^{3-}}} \cdot a_O^{2.5} \cdot f_P$
Mn	$C_{Mn} = \dfrac{(\%Mn)}{a_{Mn} \cdot a_O}$	$L_{Mn} = C_{Mn} \cdot a_O \cdot f_{Mn} = \dfrac{K_2^{\theta}}{f'_{Mn^{2+}} \cdot a_{O^{2-}}} \cdot a_O \cdot f_{Mn}$

where (%P), (%Mn) are percentage concentrations for [P], [Mn] in slag phase, K_1^{θ}, K_2^{θ} are equilibrium coefficients for reaction [P], [Mn], a_i, f_i are activities and activity coefficients for compositions i, respectively. The BOF endpoint [P]&[Mn] estimation model in this paper is just based on the essential theory above.

3 BOF Endpoint [P]&[Mn] Estimation Models

Because BOF dynamic process is much smoother and last less time, the estimation precision of the model will be enhanced greatly if the chemical analysis result of the in-blowing sample is available in estimating BOF endpoint [P]&[Mn] content. But in practical BOF process, BOF dynamic process lasts for 1 to 2 minutes and the analyzing process of sublance sample always takes about 3 minutes. To solve this problem, three BOF endpoint estimation models are presented according to the period taken in analyzing the sublance in-blowing sample.

3.1 Model I

Model I is used in the case of low sample analyzing speed. In this case, the BOF endpoint estimation model is required to describe the whole steelmaking process for each heat. But the relation between endpoint [P]&[Mn] and the process conditions is complex and most BOF model can not be used to estimate the composition of slag. So the equation of capacity (C_i) and distribution ratio (L_i) to each element above couldn't be used directly, which must be simplified to be utilized.

Under the condition of chemistry reaction equilibrium is reached, the distribution ratios (L_i) for element [P]&[Mn] between slag phase and steel phase could be simplified as [6]

$$L_{\mathrm{P}} = F_P\left(\sum_{i=1}^{m}\left(a_{P,i} \cdot x_{P,i}\right)\right), \tag{1}$$

$$L_{\mathrm{Mn}} = F_{Mn}\left(\sum_{i=1}^{n}\left(a_{Mn,i} \cdot x_{Mn,i}\right)\right), \tag{2}$$

where F_P , F_{Mn} are functions for the distribution ratio for element [P], [Mn] between slag and steel, $a_{P,i}$, $a_{Mn,i}$ are coefficients of factor x_i on element [P], [Mn], $x_{P,i}$, $x_{Mn,i}$ are factors that affect the distribution ratio for element [P], [Mn] between slag and steel, such as endpoint [C] content, endpoint temperature *etc.*, respectively.

It is proved that the data identification result of the coefficient of factor x_i on element [P]&[Mn] is poor with the least squares parameter estimates method for the influence from sample noise. In Model I, robust regress method is used to identify the model coefficients, which causes the result of coefficient identification is much more robust because it is obtained by weighing the sample data based on the sample noise.

Generally, Model I for estimation BOF endpoint [P]&[Mn] content is characterized as follows:

- The model originates from metallurgic mechanism and each coefficient presents obvious physical or chemical meanings. So operators can modify it according to the change of steelmaking conditions.
- Model coefficients can be identified and adapted online, which improve the ability of model tracking and adaptation to the change of steelmaking condition.
- Combined with BOF dynamic simulation model and replaced the endpoint [C] and temperature with the dynamic model outputs [C] and temperature as input variables, the model can simulate the change of bath [P]&[Mn] content real-timely.

3.2 Model II

Model II is used in the case of fast sample analyzing speed. It utilizes material data, process data, endpoint sublance data and in-blowing sample analysis data, and finally obtains the BOF endpoint [P]&[Mn] content.

Because there is some phosphorus returned into steel bath from slag and the mechanism is complex, it is difficult to build a mechanism model like Model I. As a result, the BOF endpoint [P]&[Mn] content is estimated using NN technology.

In this paper Radial Basis Function Neural Network (RBF NN) is applied. As a three-layer forward-feed neural network, RBF NN excel over Back Propagation (BP) neural network in approximation ability, classification ability and learning speed, *etc.* Its basic ideas is as follows: RBF, as the "basis" of the mid-layer hidden cell, makes up hidden layer space and maps the input vector onto the hidden space. The output of the mapping from the hidden space to the output space is linear; however, the mapping from input to output is nonlinear.

The approximation function of RBF NN is as follows:

$$F(x)=\sum_{i=1}^{c} w_i G(\|x-v_i\|), \tag{3}$$

where v_i is the center; $\|\cdot\|$ is Euclidean norm; $G(x)$ is Green Function (Gauss Function is selected here), respectively.

The process for the solution of function F is to search for a suitable set of NN weight $\{w_i | i=1,2,...c\}$ to make the functional $\xi(F^*)$ reach the minimum. The critical functional in this paper is given in (4), where the first item is the standard error item, and the second item is standardized item, which demonstrates that the index function conceals the requirement for the smoothness and continuity.

$$\xi(F^*)=\sum_{k=1}^{n}[d_k-F(x_k)]^2+\lambda\|PF^*\|^2=\sum_{k=1}^{n}[d_k-\sum_{i=1}^{c} w_i G(\|x_k-v_i\|)]^2+\lambda\|PF^*\|^2 . \tag{4}$$

By minimizing the item $\xi(F^*)$ with the vector W, the iterative learning procedures of the weight W is obtained

$$W_{i+1}=W_i-\eta\frac{\partial\xi(F)}{\partial W_i}=W_i-\eta\cdot[(G^{\tau}G+\lambda G_0)W_i-G^{\tau}d], \tag{5}$$

where $d=[d_1,d_2,...,d_n]^{\tau}$, $W=[w_1,w_2, \ldots, w_c]^{\tau}$ and

$$G=\begin{bmatrix} G(x_1;v_1) & G(x_1;v_2) & \ldots & G(x_1;v_c) \\ G(x_2;v_1) & G(x_2;v_2) & \ldots & G(x_2;v_c) \\ \vdots & \vdots & & \vdots \\ G(x_n;v_1) & G(x_n;v_2) & \ldots & G(x_n;v_c) \end{bmatrix}.$$

Relatively, Model II for estimation BOF endpoint [P]&[Mn] content is characterized as:

- The output is modified with the RBF cluster centers and the connection weights of the output layer, which avoids the complex modeling process.
- The analysis result of sublance in-blowing sample is utilized as input variable, which improves the model estimation precision.

3.3 Model III

Model III is used in the case of middle sample analyzing speed. In this case, the sample analyzing period is between that of the Model I and II, which causes the mechanism model like the Model I and identification model like the Model II can not be used individually. To improve the BOF endpoint [P]&[Mn] estimation precision, part information but not the whole information of Model I and II should be used. So Model III is set up by weighting and combining the models above. The weighting parameters can be got using a least square method to minimize the estimation error.

4 Industrial Experiment

Material data, process data, endpoint [C] and endpoint temperature are considered as input variables in Model I. There are two steps in building this model, one is determining the input variable and model structure, and the other is identifying the model parameter.

Table.2 presents the input and output variables to the RBF NN in Model II, and table.3 give the selection of some parameters to the RBF NN.

Table 2. Input and Output Variables for Model II

Model	Input	Output
Endpoint [Mn] Estimation Model	In-blowing analysis [Mn] content The amount of oxygen used Endpoint [C] content of steel bath Endpoint temperature of steel bath	Endpoint [Mn] content
Endpoint [P] Estimation Model	[Si] content in hot metal In-blowing analysis [P] content The amount of lime used The amount of fluorite used The amount of dolomite used The amount of oxygen used Endpoint [C] content of steel bath Endpoint temperature of steel bath	Endpoint [P] content

Table 3. Some Parameters for Model II

Item	[Mn]	[P]
Number of RBF Cluster Centers	20	40
Learn Speed of Connection Weights	0.001	0.001

In this section the industrial experiment results with the data from BOF steelmaking process are given to verify the effectiveness of the estimation models. Two sets of data (sample set and test set) are included in the simulation to each furnace. Sample set of data is used to identify the structure and parameter to Model I, to train the RBF cluster centers and connection weights of Model II, to identify the weighing parameters in Model III. While the test set of data is used to verify the effectiveness of these models.

It shows the results of the three estimation models in Table 4. The result information includes the hit ratios and the correlation coefficients of the three models for sample set and test set. The hit aim slots are selected as $\pm 2.5\times10^{-3}\%$ for [P] estimation and $\pm 2.5\times10^{-2}\%$ for [Mn] estimation respectively. Statistically, both for [P] estimation and for [Mn] estimation, the hit ratios of Model I and III exceed 70% and those of Model II exceed 80%. Thus it is proved that the models presented in this paper is effective in estimating the BOF endpoint [P] and [Mn] content.

Table 4. Model Test Result

Model I	Sample Set (226 Heats)		Test Set (250 Heats)	
	P	Mn	P	Mn
Hit Ratio (%)	79.20	79.65	76.40	81.60
Correlation Coefficient	0.6382	0.7025	0.5577	0.6568
Model II	Sample Set (184 Heats)		Test Set (160 Heats)	
	P	Mn	P	Mn
Hit Ratio (%)	85.33	90.22	84.38	86.88
Correlation Coefficient	0.8212	0.8464	0.7487	0.7687
Model III	Sample Set (231 Heats)		Test Set (254 Heats)	
	P	Mn	P	Mn
Hit Ratio (%)	79.63	74.13	71.23	75. 51
Correlation Coefficient	0.7156	0.6467	0.725	0.6377

5 Conclusion

Three intelligent models are presented according to the analyzing speed of in-blow sublance sample are presented in this paper. Model I is modeled from metallurgic mechanism, Model II is with RBF NN and Model III is modeled using least square method, respectively. The industrial experiment verified the effectiveness of these models. With the aid of these models, steelmaker can get the estimation of BOF endpoint [P] and [Mn] based on the process information and the sublance measurement.

References

1. Sumi, I., Kawabata, R., *etc.*: Technique of Controlling Dust Generation During Oxygen Top Blowing in BOF, Steel Research, 74 (2003) 14–18
2. Swift, T.: BOF Bath Level Measurement for Lance Height Control at The Sparrows Point Plant, Iron and Steelmaker, 29 (2002) 37–40
3. Nirschel, W.F., Stone R.P., Carr, C.J.: Overview of Steelmaking Process Control Sensors for The BOF, Ladle and Continuous Casting Tundish, Iron and Steelmaker, 28 (2001) 61–65
4. Tao, J., Wang X., *etc.*: Intelligent Control Method and Application for BOF Steelmaking Process, Proceedings of the IFAC World Congress, (2002)
5. Yang L.H., Liu L., He P.: [P] Prediction and Control Model for Oxygen-Converter Process at The End Point Based on Adaptive Neutro-Fuzzy System, Proceedings of the World Congress on Intelligent Control and Automation, (2002) 1901–1905
6. Tao J., Qian W.D.: Intelligent Method for BOF Endpoint [P]&[Mn] Estimation, Proceedings of the IFAC workshop on new technologies for automation of metallurgical industry, (2003)

Application of Neural Network on Wave Impact Force Prediction

Hongyu Zhang, Yongxue Wang, and Bing Ren

State Key Lab. of Coastal and Offshore Engineering,
Dalian University of Technology, Dalian, 116024, P. R. China
missmean@etang.com
{wangyx, bren}@dlut.edu.cn

Abstract. This paper investigates the regular wave impact force on open-piled wharf deck by Artificial Neural Network. A three-layered neural network is employed and the units of input layer are wave period, T, incident wave height, H, and relative clearance, s/H. The unit of output layer is the maximum wave impact force, F_{max}. It is shown that the neural network with parameters determined through self-learning can predict wave impact force reasonably.

1 Introduction

In coast and offshore engineering, the safety of the structures whose superstructures located in the splash zone such as piling wharf, shore trestles, oil-drilling platforms, etc, have great relation to wave slamming. Under confused sea condition, the structures may subject to very strong wave impulsive load due to slamming by the wave with significant crest when waves propagate underneath the structure and surge up to its surface. Previous studies indicate that impact pressure are characterized by initial peak pressure of considerable magnitude but of short duration, and which typically is first positive then decreases to zero and becomes negative. In hostile sea state, the peak pressures may cause the damage of the horizontal members of the structures or make the whole superstructure collapsed. However, due to the complicated phenomena of wave slamming problem, the most research work on the impact wave force is conducted by model tests. And the limited mathematical models meet considerable difficulty to compute the wave impact force under complicated states [1], [2].

Artificial Neural Network is a kind of calculative technique and belongs to artificial intelligence. It can deal with the nonlinear dynamic system effectively and realize the complex nonlinear mappings. For some complicated engineering problems with no available mathematical model, ANN is usually the most effective approach. Furthermore, with the development of ANN, it has been used successfully to dispose the variable problem with many parameters. In the area of ocean engineering, Maus et al. [3] employ neural network to predict occurrence of impact wave force, etc. The aim of this study is to discuss the application of neural network to calculate wave impact force and to provide a new approach to study the complex nonlinear relation between

F. Yin, J. Wang, and C. Guo (Eds.): ISNN 2004, LNCS 3174, pp. 854–859, 2004.

wave impact forces and influence factors, such as wave period, incident wave height and relative clearance etc.

2 Back Propagation Neural Network

Back Propagation Neural Network [4] is organized in layers, in generally, input, hidden and output layer. Through training and learning with sample, we can determine the correlation parameter of net. Actually the training of BP neural network includes two alternant iteration procedures that are forward propagation of the signal and backward propagation of the error.

By defining $W_{ji}^{(k-1)}$ as the weight between unit of i in layer k-1 and unit of j in layer k, θ_j^k as the threshold of unit of j in layer k, $f(x)$ as the transfer function, n_k as the number of units in layer k, M as the total number of net layers, the forward propagation is described by

$$y_j^k = f_j^k\left(\sum_{i=1}^{n_{k-1}} W_{ji}^{(k-1)} \cdot y_i^{(k-1)} - \theta_j^k\right), j = 1, 2, \cdots, n_k; k = 1, 2, \cdots, M. \tag{1}$$

The backward propagation is error propagation from output layer to input layer as well as the emendation of correlation parameter. The amount of change of the weights and thresholds are given by

$$\Delta \mathrm{x}_k = -[\mathrm{J}^T(\mathrm{x}_k)\mathrm{J}(\mathrm{x}_k) + \mu_k \mathrm{I}]^{-1} \mathrm{J}(\mathrm{x}_k)\mathrm{V}(\mathrm{x}_k). \tag{2}$$

$$\mathrm{J(x)} = \begin{bmatrix} \frac{\partial v_1(\mathrm{x})}{\partial x_1} & \frac{\partial v_1(\mathrm{x})}{\partial x_2} & \cdots & \frac{\partial v_1(\mathrm{x})}{\partial x_n} \\ \frac{\partial v_2(\mathrm{x})}{\partial x_1} & \frac{\partial v_2(\mathrm{x})}{\partial x_2} & \cdots & \frac{\partial v_2(\mathrm{x})}{\partial x_n} \\ \vdots & \vdots & & \vdots \\ \frac{\partial v_N(\mathrm{x})}{\partial x_1} & \frac{\partial v_N(\mathrm{x})}{\partial x_2} & \cdots & \frac{\partial v_N(\mathrm{x})}{\partial x_n} \end{bmatrix}. \tag{3}$$

In which, $\mathrm{V}(\mathrm{x}_k)$ is the error vectors, J(x) is jacobian matrix.

The aim of learning is to make learning error less than the value that is defined by user previously. For this study, we choose the sum of squares error to express learning error.

The sum of squares error, E, between the value from the output unit of k, a_k, and the teach signal, t_k, is expressed by

$$E = \frac{1}{2}\sum_k (t_k - a_k)^2. \tag{4}$$

3 Prediction Model for Computing Wave Impact Force

3.1 Sample

The sample data used in this paper are obtained by experiments in the large wave-current tank in the state key Laboratory of coastal and offshore Engineering, Dalian University of Technology. The tank is 69m in length, 2m in width, 1.8m in depth. The wharf deck is made of organic glass, whose length along the direction of wave propagation is 1m and the width is 0.65m,thickness is 0.02m.In experiment, wave period, T, is 1.0s, 1.2s, 1.5s, 2.1s; incident wave height, H, is 0.1m, 0.15m, 0.2m, 0.25m; the ratio of the surface level of wharf deck to the incident wave height, s/H, that is relative clearance is -0.1, 0, 0.1, 0.2, 0.3, 0.4. 60 groups of experimental data are chosen. In which, 54 groups are used training data, 6 groups are used test data.

The hidden layer and output layer all employ log-sigmoid transfer function. In order to speed up the rate of convergence and avoid the saturation state, firstly, this paper make the data locate the range from 0 to 1 through the following formula.

$$X' = \frac{X - X_{min}}{X_{max} - X_{min}}. \tag{5}$$

In which, X' show disposed data, X show original data, X_{min}, X_{max} show the minimum and maximum of original data, respectively.

3.2 The Number of Layers

The imitational quality has positive relation to the number of neural network layers. Although more neural network layers can improve the imitational quality, the net complexity and the training time are increased. Hecht-Nielsem [5] proved a three-layered feedforward Neural Network can approach any function with many variable. So, for this study, we adopt a three-layered net to predict wave impact force. Both hidden and output layer employ log-sigmoid transfer function. The expression is

$$f(u) = \frac{1}{1+e^{-u}}. \tag{6}$$

3.3 The Number of Neurons

The neuron number of input and output layer is related to the practical problem. It is decided by the number of input and output variable. For regular wave impact force on open-piled wharf deck, the main influence factors are T, H, s/H. Thus we choose the number of input and output units as 3 and 1, respectively.

Because the method deciding the number of hidden units has not theory basis at present, we choose hidden layer neurons as 10 and then add the number in turn, when the aim error is 0.08. We determine hidden units from the many training results in terms of the principle of better training time, training epoch, the average value of relative error of the training specimen (ARET for short) and the average value of relative error of the verifying specimen (AREV for short). For each kind of condition, we employ 10 different initial weight matrixes to train the net. Table 1 lists the average value of 10 training. From the table, when the number of hidden layer neurons is 17, comparing with other case, the training result is better.

Table 1. Training result of the different number of hidden units

Marquardt algorithm						
The number of hidden units	10	15	16	17	18	20
Training time	2.4461	1.5992	1.6153	1.8076	1.9750	2.2573
Training epoch	19	11.3	10.5	11.2	13	13.1
ARET	0.1035	0.0954	0.0904	0.0854	0.0847	0.0808
AREV	0.2861	0.2221	0.2264	0.2001	0.2023	0.2914

3.4 Aim Error

The aim error is determined in terms of experience in the practical problem. We also can choose partial sample to train the net with different aim error and employ the rest to test it, then we can determine the better aim error in terms of the prediction quality. When the number of hidden units is 17, for different aim error, we employ 10 different initial weight matrixes to train the net. Table 2 lists the average value of 10 training. From the table, we can see, when both ARET and AREP are better, the aim error can be chosen as 0.08.

Table 2. Training result of the different number of hidden units

Marquardt algorithm, 17 hidden units							
Aim error	0.005	0.05	0.07	0.08	0.09	0.4	0.8
ARET	0.0256	0.0775	0.0811	0.0854	0.0885	0.1827	0.2168
AREV	0.3225	0.2839	0.2115	0.2001	0.2019	0.1930	0.2239

3.5 Learning Algorithm

In order to choose an appropriate BP algorithm for the wave impact force problem, the conventional and evolutionary back-propagation algorithms are tested to train the network. For the net adopting different algorithm, we choose the appropriate parameter and employ 10 different initial weight matrixes to train. Table 3 lists the average value of 10 training. It is seen that from the table, the marquardt algorithm is the best, according to the training time, training epoch, ARET and AREP. Because marquardt algorithm is not sensitive to parameters, μ_k=0.001, θ=10 are set.

Table 3. Training result of different learning algorithm

Learning algorithm	BP	BPX	CGBP	LMBP
Brief introduce	Conventional Back Propagation Caculation,3 0 hidden units	Momentum calculation combine with Variable Learning Rate,35 hidden units	Conjugate Gradient,19 hidden units	Marquardt algorithm,17 hidden units
Training time	348.6965	83.0052	42.0565	1.8076
Training epoch	7959	1458	256	11
ARET	0.1076	0.1000	0.1089	0.0854
AREV	0.2018	0.2022	0.2211	0.2001

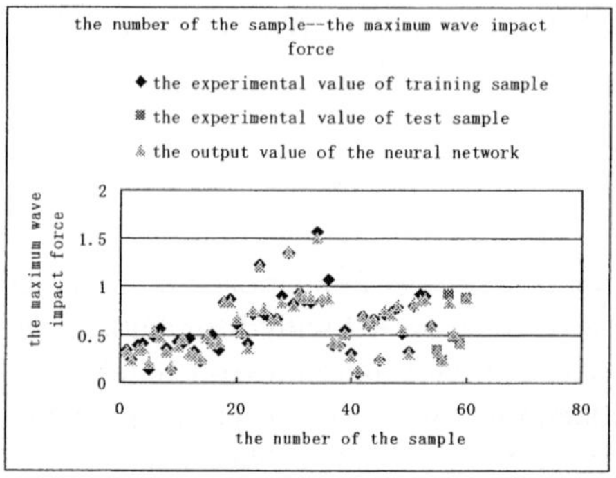

Fig. 1. Comparison of the experimental value and the output value of neural network

3.6 Result

In this paper, the training procedure of BP neural network is programmed and the prediction model has three layers: 3 input units, 17 hidden units, 1 output unit. The marquardt algorithm and the aim error of 0.08 are adopted. 10 different initial weight matrixes are employed to train. The net whose prediction quality is the best in the 10 training is chosen as the prediction model. The ARET of the net is 8.61%. Table 4 shows the prediction result. In this table, the AVAE and AVRE stand for the absolute value of absolute error and the absolute value of relative error, respectively. Figure 1 shows the Comparison of the experimental value and the output value of neural network.

4 Conclusions

In this study, the neural network is applied to predict wave impact force and the application of neural network in the complex hydrodynamics problems is discussed.

Table 4. Prediction result

Specimen		T(s)	H(m)	s/H	F_{max} (experimental value)	Marquardt algorithm			
						F_{max} (the output value of the net)	AVAE	AVRE (%)	The average of relative error (%)
Test specimen	1	1.0	0.10	0.1	0.339	0.3197	0.0193	5.68	4.56
	2	1.2	0.10	0.3	0.243	0.2332	0.0098	4.02	
	3	1.5	0.20	0.2	0.910	0.8240	0.0860	9.45	
	4	2.0	0.10	0.1	0.478	0.5113	0.0333	6.96	
	5	2.0	0.15	0.3	0.401	0.4031	0.0021	0.51	
	6	2.0	0.25	0.0	0.886	0.8796	0.0064	0.72	

It is shown that the model of neural network is better from the prediction result. The average value of relative error of the training specimen is 8.61%; the average value of relative error of the probative specimen is 4.56%. The wave impact forces can be predicted by present model although the choice of initial weight matrix and the determination of network architecture need a further study.

References

1. Wang, Y., Ren, B.: Study on Wave Slamming by Turbulent Model. Journal of Hydrodynamics, Ser. A, Vol. 14, No. 4 (1999) 409–417
2. Wang, Y., Ren, B.: Nonlinear Wave Impact on Structures. Journal of Dalian University of Technology, Vol. 39, No. 4 (1999) 562–566
3. Mase, H., Kitano, T.: Prediction Model for Occurrence of Impact Wave Force. Ocean Engineering, Vol. 26 (1999) 949–961
4. Hagan, M., Demuth, H., Beale, M.: Neural Network Design. China Machine Press (2003) ISBN: 7-111-10841-8
5. Hecht-Nielsen: R Komogorovs Mapping Neural Network Existence Theorem. Proceedings of the International Conference on Neural Networks, New York, IEEE (1987) 11–13

RBF Neural Networks-Based Software Sensor for Aluminum Powder Granularity Distribution Measurement*

Yonghui Zhang, Cheng Shao, and Qinghui Wu

Institute of Advanced Control Technology, Dalian University of Technology,
Dalian, China, 116024
zhyhemail@sohu.com

Abstract. For aluminum powder nitrogen atomizing process, it is important but difficult to establish the aluminum powder granularity distribution using real-time measurements of exiting online sensors. In this paper, a novel software sensor model based on RBF Neural Networks is presented to estimate the granularity distribution of aluminum powder by means of measurements of melted aluminum level and temperature, atomizing nitrogen temperature and pressure, and recycle nitrogen temperature and pressure combined with granularity statistics distribution of the aluminum powder. The software sensor model can be obtained by training the RBFNN offline or online iteratively. An error analysis is carried out to illustrate effectiveness of the proposed software sensor model of aluminum powder granularity distribution.

Keywords. Aluminum powder nitrogen atomizing, Aluminum powder granularity distribution, RBF Neural Networks, Software sensor

1 Introduction

Nitrogen atomizing is a widely applied technique to produce the super-tiny spherical aluminum powder. During the atomizing process, the aluminum ingots are heated, melted and atomized to become aluminum powder. For nitrogen atomizing process, the granularity distribution of aluminum powder is a very important parameter required for automatic control, but can't be directly measured real-time using exiting online sensors. Two ways dealing with this type of non-direct measurable problems made in the literature are to develop new hardware sensor for monitoring the desired parameters, and to employ software sensors, one indirect method, by means of relating measurements available on-line.

There are two types of modeling approaches in developing software sensors. One is a deterministic model from process mechanism [1, 2] and the other is of the black-box approach using only the observed values [3]. Nitrogen atomizing process is with

* Supported by State Science & Technology Pursuing Project (2001BA204B01) of China and Foundation for University Key Teacher by the Ministry of Education of China.

F. Yin, J. Wang, and C. Guo (Eds.): ISNN 2004, LNCS 3174, pp. 860–865, 2004.

nonlinearities, large time delay, strong coupling and severe uncertainty, and thus it is difficult to obtain the deterministic model by mechanistic method.

Recently, Neural Networks (NN) has received special attention from many researchers because of its approximating ability for complex nonlinear systems [4]. Compared by BP Neural Networks and any others, Radial Basis Function Neural Networks (RBFNN) has many advantages, such as global approaching, sample structure, training quickly, etc., in modeling nonlinear systems. In this paper, a novel software sensor model based on RBFNN is presented to estimate the granularity distribution of aluminum powder. The software sensor model is trained by combining measurements of melted aluminum level and temperature, atomizing nitrogen temperature and pressure, and recycled nitrogen temperature and pressure with granularity distribution character of the aluminum powder. The simulation results illustrate the effectiveness of the proposed software sensor model of aluminum powder granularity distribution.

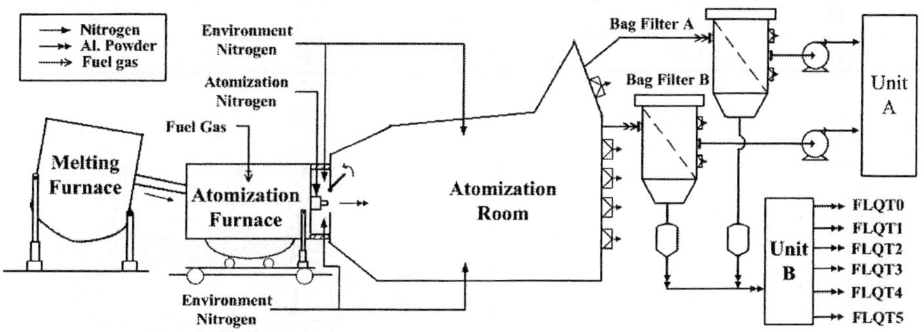

Fig. 1. Aluminum powder nitrogen atomizing production process flow chart. Unit A: Nitrogen Recycle Unit; Unit B: Aluminum Classification Unit

2 Aluminum Powder Nitrogen Atomizing Production Process

The flowchart of aluminum powder nitrogen atomizing production process is shown in Fig. 1. Aluminum ingots are heated in ***Melting Furnace*** up to melting at a certain temperature, and then the melted aluminum is delivered to ***Atomization Furnace*** through a guide channel and heated again in ***Atomization Furnace*** continuously to keep a higher temperature(about 900•). Through an atomizing nozzle towards ***Atomization Room*** in the front of the ***Atomization Furnace***, the melted aluminum is spurted to ***Atomization Room*** and the aluminum powders are produced by action of high-pressured atomizing nitrogen. The atomizing speed of melted aluminum and the size of aluminum powder can be affected by melted aluminum level and atomizing nitrogen from circular Venruri distributor. With the protection and cooling of recycle nitrogen, aluminum powder solidified rapidly. The mixtures of aluminum powder and nitrogen are pulled into the two ***Bag Filters***, where the aluminum powders are separated with nitrogen. The purified aluminum powders are delivered to the inferior store tanks, and then classified to six different degrees of powders in classification unit (***Unit B***). The whole atomizing process takes place in the ***Atomization Room*** filled fully by recycle nitrogen. With dust removing and cooling, nitrogen separated from ***Bag Filters*** is used repetitively. Some of the nitrogen is dried, compressed, heated, and sent to circular Venruri distributor as atomizing nitrogen.

Aluminum powder nitrogen atomizing production is followed by a national military criterion"*Super-tiny spherical aluminum powder criterion*" (GJB1738-93). According to the size of middle diameter (D_{50}), there are six degrees of aluminum powder named from FLQT0 to FLQT5 (see Table 1). The percentage of each degree of aluminum powder in gross weight is different as shown in Fig.1 at one sampling of producing in atomization device. In this paper, the aluminum powder granularity distribution infers to the percentage distribution of each degree of aluminum powder.

Table 1. The degrees of aluminum powder

Degrees	FLQT 0	FLQT 1	FLQT 2	FLQT 3	FLQT 4	FLQT 5
D_{50}(μm)	40±5	29±3	24±3	13±2	6±1.5	2±1

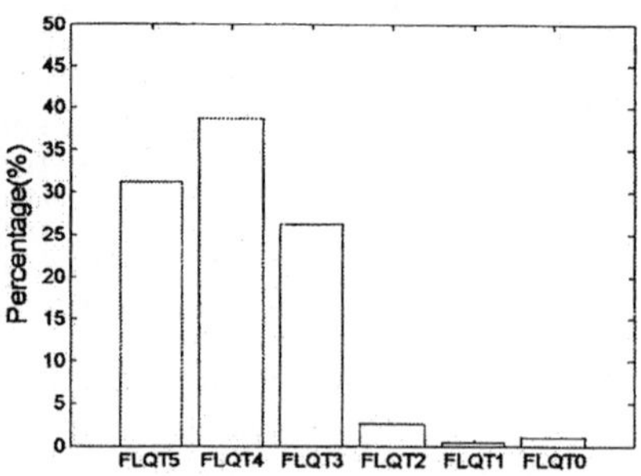

Fig. 2. The percentages of six degrees of aluminum powder

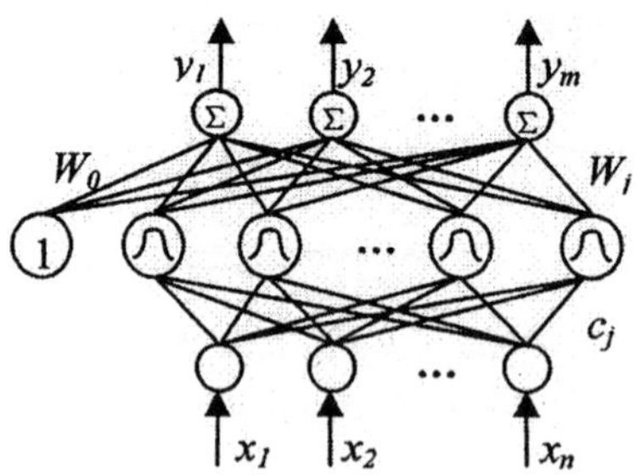

Fig. 3. The structure of MIMO RBFNN

3 The Software Sensor Model Based on RBFNN

Among all types of the Neural Networks, BP and RBF Networks should be the two often applied in software sensor modeling [4]. Compared by BP Networks, RBFNN has the feature of global approximating, well classifying, and simpler structure and faster training.

3.1 MIMO Radial Basis Function Neural Networks

RBFNN is a kind of 3-layer forward Networks (see Fig. 3). The first layer is a signal input unit; the second is hidden layer, the units of hidden layer are nonlinear; the third is output layer, the units of this layer are linear. The transformation function in hidden layer is varied from different occasions. Theoretically, RBFNN has the ability of approaching arbitrary nonlinear functions. The MIMO RBF Networks shown in Fig. 3 has n input units, a hidden layer including L units, and m linear output units. The mapping of the RBF Networks f_r•R^n •R^m, is defined by

$$\mathbf{y} = f_r(\mathbf{x}) = \mathbf{W}_0 + \sum_{j=1}^{L} \mathbf{W}_j \phi(\|\mathbf{x} - \mathbf{c}_j\|) \tag{1}$$

Where $\mathbf{x} \in R^n, \mathbf{y} \in R^m$, $\phi(\bullet)$ is the mapping from R^n to R, $\|\bullet\|$ is the Euclid norm, $\mathbf{W}_j$ is the weight vector $\mathbf{W}_j \in R^m$, $(j = 1, 2, \cdots, L)$, $\mathbf{W}_0$ is bias vector $\mathbf{W}_0 \in R^m$, $\mathbf{c}_j$ is the center vector, and $\mathbf{c}_j \in R^n$, $(j = 1, 2, \cdots, L)$, respectively. Here $\phi(\bullet)$ is chosen as Gaussian function of the form

$$\phi(v) = \exp(-v^2 / \sigma^2) \tag{2}$$

The width of Gaussian function is σ. Let $\mathbf{c}_0 = \mathbf{x}$, then

$$\phi(\|\mathbf{x} - \mathbf{c}_0\|) = \phi(0) = 1 \tag{3}$$

Let $\mathbf{W}_j = [W_{1j}, W_{2j}, \cdots, W_{mj}]$, $(j = 0, 1, 2, \cdots, L)$, eqn. (1) can be rewritten as

$$\mathbf{y} = f_r(\mathbf{x}) = \sum_{j=0}^{L} \mathbf{W}_j \phi(\|\mathbf{x} - \mathbf{c}_j\|) \tag{4}$$

3.2 Modeling of the MIMO Software Sensor Based on RBFNN

By the analysis of aluminum powder atomizing process, it is indicated that the percentage of each degree of aluminum powder is affected by many parameters. In this paper, six parameters directly affecting the aluminum powder granularity distribution, the level and temperature of melted aluminum, the temperature and pressure of atomizing nitrogen, are chosen as the inputs of software sensor, and the temperature and pressure of recycle nitrogen, and the percentage contents prediction of six degrees of aluminum powder as the outputs. Then the software sensor model can be described as

$$[y_0(k), y_1(k), \cdots, y_5(k)]^T = f(L(k), T_L(k), T_A(k), P_A(k), T_E(k), P_E(k)) \tag{5}$$

Where $y_0(k), y_1(k), \cdots, y_5(k)$ are corresponding to the percentages of the FLQT0, FLQT1,…, FLQT5 aluminum powder, $L(k)$ is the level of melted aluminum, $T_L(k)$ is the temperature of melted aluminum, $T_A(k)$ is the temperature of atomizing nitrogen, $P_A(k)$ is the pressure of atomizing nitrogen, $T_E(k)$ is the temperature of recycle nitrogen, and $P_E(k)$ is the pressure of recycle nitrogen, respectively. Let $\mathbf{x}(k) = [L(k), T_L(k), T_A(k), P_A(k), T_E(k), P_E(k)]$, $\mathbf{y}(k) = [y_0(k), y_1(k), \cdots, y_5(k)]^T$ Then from eqns. (4) and (5) one obtains the RBFNN model of the software sensor

$$\mathbf{y}(k) = f_r(\mathbf{x}(k)) = \sum_{j=0}^{L} \mathbf{W}_j \phi(\|\mathbf{x}(k) - \mathbf{c}_j\|) \tag{6}$$

3.3 Training and Verifying for RBFNN Model

One group of historical sample data of aluminum powder nitrogen atomizing production is chosen for RBFNN training; another group of sample data of recent two months is chosen for extensive verifying. Before training, the sample data have been processed by digital filtering, eliminating delay and deleting invalid data etc. With the RBF Neural Networks tool box of MATLAB, the software sensor model for aluminum powder granularity distribution is established, the offline training is done, and the well-trained Networks model is verified. The online batch training is considered. The sampling data of last month are used to train the software sensor model online every month.

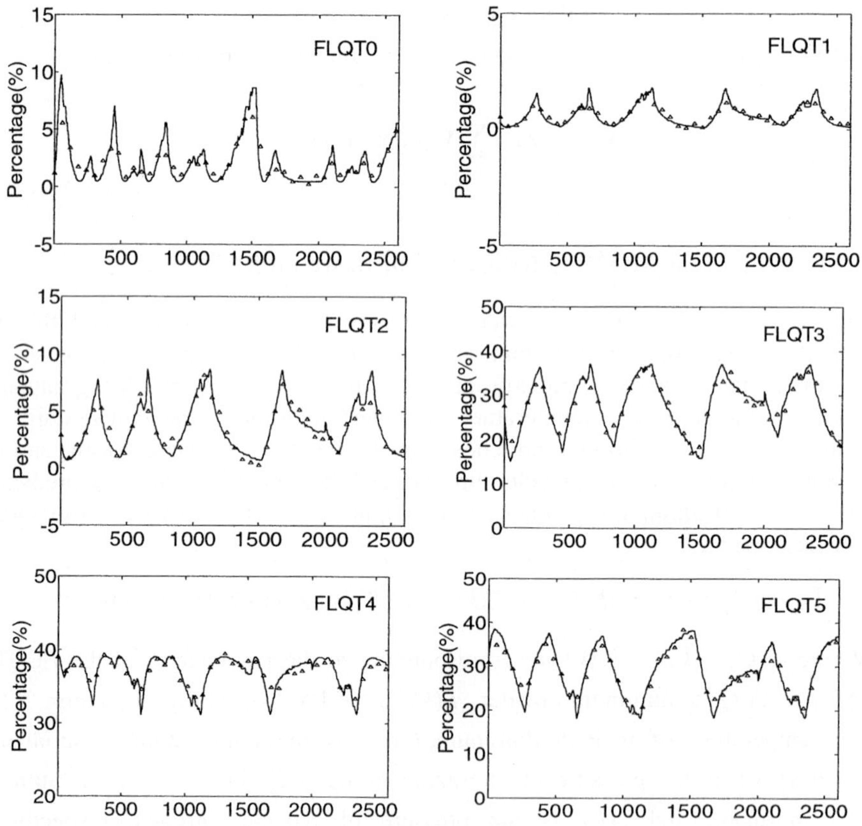

Fig. 4. The comparison between the prediction of RBFNN model (—) and verifying data(△)

Table 2. The errors between the detections and the predictions of software sensor of aluminum powder granularity distribution

Degrees	FLQT0	FLQT1	FLQT2	FLQT3	FLQT4	FLQT5
e_{max}	4.1569	1.9828	3.7933	1.5465	1.3179	3.4829
e_{min}	0.0029	0.0014	0.0021	0.0027	0.0006	0.0037
e_{mse}	0.9124	0.5335	0.7020	0.4799	0.1587	0.4348

The maximum absolute error e_{max}, the minimum absolute error e_{min} and square root e_{mse} are listed in Table 2. As shown, in all the degrees, the maximum e_{max} is 4.1569; the minimum e_{min} is 0.0006; the maximum e_{mse} is 0.9124. It is followed from Fig. 4 and above computing results that the prediction of software sensor model match verifying data very well and can track accurately the varying of aluminum powder granularity distribution real time, therefore the software sensor model of aluminum powder granularity distribution based on RBFNN is effective, and its measurement accuracy can meet the needs of applying advanced control in production process well.

4 Conclusions

The granularity distribution of aluminum powder is an important parameter for aluminum powder nitrogen atomizing process, but it is difficult to measure by exiting online sensors. In this paper, a software sensor based on RBFNN is presented to estimate the granularity distribution of aluminum powder. The error analysis shows that the software sensor is well-matched to the verifying data. By applying of the software sensor, the granularity distribution of aluminum powder can be monitored real time, which gives technical support to realize closed-loop control in the aluminum powder nitrogen atomizing process.

References

1. Acha V., Meurens M., Naveau H., Dochain D., Bastin G.and Agathos S. N.: Model-based Estimation of an Anaerobic Reductive Dechlorination Process via an Attenuated Total Reflection-Fourier Transform Infrared Sensor. Water Sci. Technol. Vol. 40, No. 8, (1999) 33–40.
2. Chéruy A.: Software Sensors in Bioprocess Engineering. Journal of Biotechnology. Vol. 52, (1997)193–199.
3. Farza M., Busawon K. and Hammouri H.: Simple Nonlinear Observers for On-line Estimation of Kinetic Rates in Bioreactors. Automatica. Vol. 34, No.3, (1998) 301–318.
4. Lee D. S. and Park J. M.: Neural Network Modeling for On-line Estimation of Nutrient Dynamics in a Sequentially-operated Batch Reactor. J. Biotechnol. Vol. 75, (1999) 229–239.

A Technological Parameter Optimization Approach in Crude Oil Distillation Process Based on Neural Network

Hao Tang[1], Quanyi Fan[2], Bowen Xu[2], and Jianming Wen[3]

[1] Department of Automation, Tsinghua University, Beijing, 100084, China
th01@mails.tsinghua.edu.cn

[2] Department of Automation, Tsinghua University, Beijing, 100084, China
{fqy-dau, yjb-dau}@mail.tsinghua.edu.cn

[3] Fujian refinery Petrochemical Co. Ltd., Quanzhou, Fujian, 362117, China
wenjianming@sina.com

Abstract. Applying data analysis-based intellectual optimization approach to the industrial process optimization may improve the weakness of the mechanical model-based optimization method which is mostly based on the incomplete knowledge and understanding of a system. Therefore, it shows great application prospects. In this paper, an operating parameter optimization model of crude oil distillation process is developed using BP neural network. Then genetic algorithm is applied to search for the optimal technological parameters. Simulation results indicate that this approach is effective and practical.

1 Introduction

Process optimization is a well-concerned problem for enterprises. In the fierce modern market competition, furthest optimizing the process and ensuring the production to be effective and stable are very important for enterprises to enhance their profits. Technological parameter optimization in process industry can not only increase the production, but also take great effect in saving energy sources, protecting environments and enhancing gross economic benefits for enterprises. In China, the optimization strategies of atmospheric and vacuum distillation devices in refinery are mostly based on the processes mechanical models. In this method, algebraic and differential equations are used to describe the running status of production devices. Then according to the optimization objectives, a specific optimization algorithm is used to search for the optimal operating parameters. But due to the complexity of production devices, processes and techniques, there are usually many difficulties in optimization object modeling. Then the optimized technological parameters obtained from solving the model are usually not the best ones for the system. Besides, due to the fact that different devices may have different mechanisms and production techniques, the objects mathematical models may not totally be the same as others. Thus the applica

F. Yin, J. Wang, and C. Guo (Eds.): ISNN 2004, LNCS 3174, pp. 866–873, 2004.

bility of the models is poor, the optimization method and optimized parameters for one device are difficult to be used in others.

With the management information systems (MIS) being established and complemented in petroleum refinery, importing computer technology into production management brings enterprises plenty of data about production processes, while alleviates the work intension as well. The data reflect the production circumstances directly and factually. By collecting, systematizing and analyzing the data, people can get evidences of how to adjust the production processes and make management decisions. Here, based on the operating data collected from field site, artificial neural network (ANN) and genetic algorithm (GA) are synthetically used to optimize the light oils yield ratio of an atmospheric and vacuum distillation device in a refinery, and satisfying results are obtained.

2 System Model Based on Neural Network

2.1 Optimization Objective

Crude oil distillation is in the first place of petroleum refining. Improving the light oils yield ratio can not only increase the production, but also decrease operating expenses of the following treatment processes. So the light oils yield ratio is the most important economic target for petroleum refineries to maximize their profits. In general, crude oil distillation process includes atmospheric distillation and vacuum distillation. Here the light oils yield ratio optimization problem under the circumstances of atmospheric pressure distillation is considered.

Light oils yield ratio of the atmospheric distillation column is that, while ensuring the products quality, the ratio of the total amount of products from the top to the third side-stream of the distillation column to the feeding crude oil. The mathematical description is,

$$f_G = \frac{\sum_k P_k}{\sum_j F_j} \times 100\% \ . \tag{1}$$

where P_k is the output of product k and F_j is the jth feedstock. Optimization of light oils yield ratio is just to maximize f_G. In the process of crude oil distillation, while the feeding flow and components of crude oil are stable, the influence of feedstock can be omitted, then maximizing f_G is equivalent to optimize the light oils production yields. That is,

$$max \ f_L = \sum_k P_k \ . \tag{2}$$

It is meant to maximize the total amount of products from the top to the third side-stream of the column. This is the objective function of light oils yield ratio optimization of atmospheric pressure distillation column, where P_k (k=0,1,2,3) are the optimization decision variables.

2.2 Establishment of System Model

Since atmospheric distillation is a complex process, the back-propagation (BP) neural network [1] is used here to establish the system model.

Having got the optimization objective function, on the basis of analysis of the technological mechanism [2], the following 23 operating parameters that reflect the actual conditions of the production process are collected. They are: returned top reflux, top circulating reflux, first intermediate circulating reflux, second intermediate circulating reflux, stripping steam in the base, top temperature, return temperature of cooled top reflux, return temperature of top circulating reflux, withdrawing temperature of top circulating reflux, return temperature of first intermediate circulating reflux, withdrawing temperature of first intermediate circulating reflux, return temperature of second intermediate circulating reflux, withdrawing temperature of second intermediate circulating reflux, temperature of stripping section, temperature of stripping steam in the base, temperature of first side-stream draws, temperature of second side-stream draws, temperature of third side-stream draws, temperature of fourth side-stream draws, base temperature, feedstock temperature, top pressure and pressure of stripping section. The outputs of the atmospheric distillation column are also collected; they are top draws and the first, second and third side-stream draws. Since the problem considered is restricted to the fixation of the input material, the test analysis data of components and properties of the feeding crude oil are not collected; that is to say, the influences of these two factors are not considered while doing optimization calculation.

The nonlinear reflection between the main operating parameters of atmospheric distillation process and its productions is established using BP neural network. It is just the technological parameter optimization model based on the field operating data, and the optimization calculation will be conducted on it.

The number of nodes of BP network's input layer is decided by the dimensions of source data. Here, the input layer's nodes are set to 23, each of which represents one of the 23 main operating parameters listed above, respectively. The output layer's nodes are set to 4, which represent the 4 light oils output of atmospheric distillation column.

It's always a hard problem that how to determine the number of hidden layers and each hidden layer's nodes in a BP neural network. A simplified approach is used here. For small neural networks, since one with two hidden layers is not always better in performance than the one with one hidden layer, the BP neural network used to establish optimization model is designed with only one hidden layer. Due to the 23 inputs and 4 outputs, the effect of the neural network can be treated as a feature compression process. Let the number of hidden layer's nodes to meet the relationship that $23 : N = N : 4$, then $N = 9.6$. Repeated experiments indicate that while the hidden layer's nodes are set to 12, it is faster for the neural network to converge while being trained, and the performance is much stable as well.

When the neural network model is established, it is trained using the sampled operating data collected from field site.

3 Optimization Calculation

Genetic algorithm (GA) [3] is used here to search for the optimal technological parameters on the neural network model.

GA is a self-adaptive and iterative random search algorithm. It comes from natural selection and evolution mechanism of biology. It begins with the problem's potential solution population, which consists of a certain number of gene-encoded individuals. Individuals are initialized at random while calculation begins, and each individual's fitness is calculated. Thus the first generation or initial population is generated. If this generation does not meet the optimization criteria, calculation is conducted to generate the next generation. Generations evolve to create better approximate solutions in terms of the principle of the survival of the fittest. In each generation, individuals are selected according to their fitness values, and then crossover and mutation are conducted on them in terms of certain probability. Thus the child population is generated. The individuals' fitness values of this child population are recalculated. The procedure leads the new generation to fit environments better, just like biology evolution. This process is repeated until the optimization criteria are met. The best individual of the last population can be treated as the approximate best solution of the problem by gene decoding.

The basic operators of GA include selection and reproduction based on fitness values, crossover and mutation. To the optimization problems of some complex nonlinear, multi-goal functions, conventional linear programming and nonlinear programming will meet difficulties in the course of calculation; while GA can conveniently get better results for its extensive adaptability.

According to the artificial neural network model established just now, the 23 input variables are encoded with real number. Set the population size to 30, crossover rate to 0.8, and mutation rate to 0.1. Aiming at improving the light oils yield ratio, GA is conducted to search for the optimal technological parameters on the neural network model.

4 Simulation Results and Discussions

Taking the atmospheric distillation column operating data of Oct. 22, 2003 from a petrochemical refinery as an example, the technological operating parameters optimization is conducted using NN and GA approach mentioned above.

First of all, for the reason of influence of some uncertain factors to the production process, the data collected from field site usually contain more or less noises. These may cause data distortion, incompletion and inconsistency. So it is necessary to filter and clean up the data before calculation. This is so-called preprocessing. Put the sample data, which are the main technological parameters collected when the system is in the steady state condition, into the form that the BP neural network can accept, and kick out the obviously wrong data and the data that don't meet the quality requirements. Besides, all the input and output data are preprocessed by normalization to avoid the influence of dimension differences among data to the performance of the neural network.

On the basis of data preprocessing, BP neural network is used to establish the optimization model.

Then GA-based calculation is conducted on the well-trained neural network to search for the optimal technological operating parameters. Each element variable of the input vector is restricted to the sample data's range so that the products quality can be guaranteed to the utmost extent while optimizing the light oils yield ratio. The process of GA is shown in Figure 1, and the average results of several simulations are shown in Table 1. The simulation results of one optimization calculation are shown in Table 2.

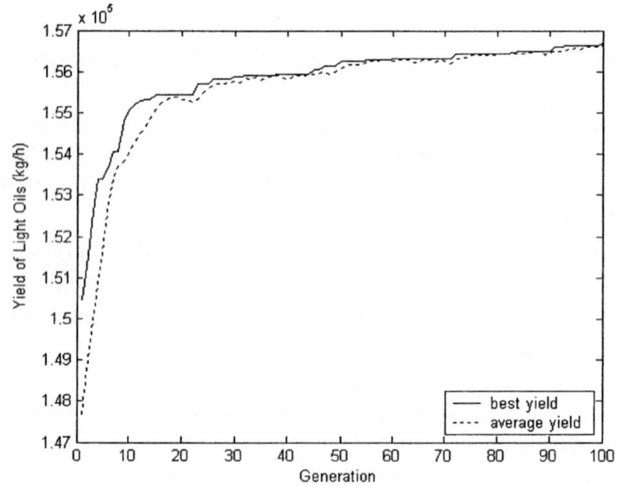

Fig. 1. The Process of GA

Table 1. Optimization results of light oils yield

Number of sample data	Maximum light oils yield in sample data (t/h)	Average optimized light oils yield (t/h)	Yield increasing (%)
721	152.8	156.6	2.49%

The above figures indicate that the light oils yield will increase of 3.8 tons per hour if the atmospheric distillation column runs according to the optimized technological operating parameters. In the case that the crude oil is processed 10,000 tons a day, for an atmospheric distillation device with 4 million tons annual process capability, the light oils yield ratio will increase of 0.91%. It will bring tremendous profits to the enterprise.

Table 2. Simulation results of one optimization calculation

Technological operating parameters	Units	Lower limits of sample data	Upper limits of sample data	Optimized values
Returned top reflux	kg/h	4886	10359	6979
Top circulating reflux	kg/h	69955	75045	73349
First intermediate circulating reflux	kg/h	87801	88195	88054
Second intermediate circulating reflux	kg/h	97862	98096	98018
Stripping steam in the base	kg/h	3803	3865	3817
Top temperature	°C	112.0	114.4	113.0
Return temperature of cooled top reflux	°C	34.4	46.6	39.5
Return temperature of top circulating reflux	°C	83.8	86.7	84.4
Withdrawing temperature of top circulating reflux	°C	136.9	140.6	139.7
Return temperature of first intermediate circulating reflux	°C	107.6	111.3	108.5
Withdrawing temperature of first intermediate circulating reflux	°C	192.0	202.5	200.1
Return temperature of second intermediate circulating reflux	°C	197.8	215.6	211.5
Withdrawing temperature of second intermediate circulating reflux	°C	280.9	289.2	285.4
Temperature of stripping section	°C	349.9	352.7	350.3
Temperature of stripping steam in the base	°C	374.3	385.0	378.0
Temperature of first side-stream draws	°C	166.2	172.8	171.3
Temperature of second side-stream draws	°C	241.0	250.5	246.6
Temperature of third side-stream draws	°C	310.7	314.8	311.6
Temperature of fourth side-stream draws	°C	347.9	350.1	348.7
Base temperature	°C	346.2	346.9	346.4
Feedstock temperature	°C	359.9	360.1	359.9
Top pressure	MPa	0.0285	0.0307	0.0293
Pressure of stripping section	MPa	0.0579	0.0607	0.0599

Side-stream draws	Top draws	1st side-stream draws	2nd side-stream draws	3rd side-stream draws	Total
Average light oils yield in sample data (kg/h)	22406	38587	41484	45453	147930
Optimized light oils yield (kg/h)	24622	41265	41409	49407	156703
Yield increasing (%)	9.89%	6.94%	-0.18%	8.70%	5.93%

The advantage of this approach is that, the BP neural network used to establish the optimization model does not need to be trained online. It can be trained only once as long as the production process is stable in a span of time. Thus the time expense can be greatly saved while the optimization calculation runs. Yet there are two major weaknesses of this approach.

(1) To guarantee the products quality. Although all the operating variables are restricted to the sample data's range during the optimization process, and the sample data that don't meet the products quality are kicked out, it is still hard to ensure fully that the products quality is satisfied and stable when the distillation process runs with the optimized technological operating parameters. A feasible solution [4] is that, on the basis of collecting the device technological operating parameters and the products quality analysis data, a quality estimation model is established first between the products quality and technological operating parameters using neural network approach. While the optimized parameters are obtained, they can be imported into the quality estimation model to estimate the products quality. Those parameters that will lead rejected products will be discarded while doing optimization searching. Thus the products quality can be guaranteed. Besides, this approach can avoid the complex mechanism analysis of how these operating parameters influence the products quality; and the optimization calculation, which is based on the input-output data, is much simplified in comparison with the mechanical model-base optimization method.

(2) Training of the neural network may influence the optimization results. Simulation shows that, when the BP neural network is initialized at random, there may be great differences among the optimization results obtained by GA, and the light oils yield may fluctuate conspicuously. Since the weight and threshold values of the BP neural network needn't to be frequently updated, an experimental method can be adopted while training, and a BP neural network model with good performances among the several network models obtained from experiments is to be reserved.

5 Conclusions

Implementation of CIMS in process industry is the main competitive means for large-scale continuous production enterprises to increase their profits and market competition capabilities. On the basis of application of databases, the process optimization approach based on operating data breaks the original objectives and levels of application of management information systems in the production process management. It shows great significance for enterprises to make their production management decisions, and has brilliant application prospects. Furthermore, the implementation of online data analysis and intelligent optimization systems will make real-time observation of the production processes more efficient, so that the online operating instructions and optimum production may come true.

Acknowledgements. Financial support for this work was provided by 863 High Tech Program (Contract No. 2002AA414610).

References

1. Nascimento, C.A.O., Giudici, R., Guardani, R.: Neural Network Based Approach for Optimization of Industrial Chemical Processes. Computers & Chemical Engineering, Vol. 24, 9-10 (2000) 2303-2314
2. Speight, J.G., Özüm, B.: Petroleum Refining Processes. New York: Marcel Dekker (2002)
3. Goldberg, D.E.: Genetic Algorithms in Search, Optimization and Machine Learning. Mass: Addison-Wesley (1989)
4. Ganguly, S., Sadhukhan, J., Saraf, D.N.: Artificial Neural Network Based Estimation of Petroleum and Product Properties. Ind. Chem. Eng. Section A, January March 44 (2002) 294–299

A Neural Network Modeling Method for Batch Process

Yi Liu[1], XianHui Yang[1], and Jie Zhang[2]

[1] Institution of Process Control Engineering, Department of Automation, Tsinghua University, Beijing, 100084, China
liuyi00@mails.tsinghua.edu.cn,
yangxh@mail.tsinghua.edu.cn

[2] Centre for Process Analytics and Control Technology, School of Chemical Engineering and Advanced Materials, University of Newcastle, Newcastle upon Tyne, NE1 7RU, U. K.
jie.zhang@ncl.ac.uk

Abstract. For the production of high quality chemicals batch manufacture plays an important role. Since some parameters of a batch chemical reaction are difficult to measure with conventional instruments, Artificial Neural Network (ANN) model has become a useful tool for the estimation of chemical reaction indexes. Data restructuring is an effective method for improving properties and accuracy of the ANN model. This paper constructed an ANN model for estimating the number-average molecular weight and polydispersity in batch polymerization process and presented the method of training data restructuring. The results were satisfactory.

1 Introduction

Batch processes are suitable for the manufacturing of high value added products such as pharmaceuticals, specialty chemicals, and bio-chemicals. With process manufacturing increasingly being driven by market forces and customer needs, there is a growing necessity for flexible and agile manufacturing [1]. Customer oriented highly responsive production has resulted in a tendency for the chemicals industries to move towards batch or mixed batch-continuous production where chemical effect are a major product quality issue.

Unlike continuous processes, batch processes are inherently transient and typically also nonlinear. To maintain consistently high quality production, close monitoring of batch processes is required. In this paper, a batch polymerization process is used as a cased study. Since batch polymerization processes are usually highly non-linear and the associated quality indexes are generally difficult to be measured on-line, ANN technique is used here to provide inferential estimations of polymer quality indexes.

Because of the difference in recipe and environment within different batches, there are many problems in the ANN modeling. For example, the obvious dispersity of data will affect fitting accuracy. This paper studies a useful data restructuring method. The simulation indicates that it is an efficient way of improving ANN model properties.

F. Yin, J. Wang, and C. Guo (Eds.): ISNN 2004, LNCS 3174, pp. 874–879, 2004.

2 A Batch Polymerization Process

In the production of high quality chemicals, batch processes play an important role. In a batch polymerization process, polymer molecular grows during certain time period. The batch polymerization reactor studied here is a one-gallon well-stirred vessel with a jacket. A cascade control system is used to keep the polymerization temperature constant. The reactor temperature is the primary controlled variable, while the jacket outlet temperature is taken as the secondary controlled variable. The manipulated variables are the cold and hot water flow rates. The detail of the polymerization system is discussed by Zhang [2]. A rigorous simulation program is developed to generate polymerization data under different batch operating conditions. These data are used to train and test neural network models.

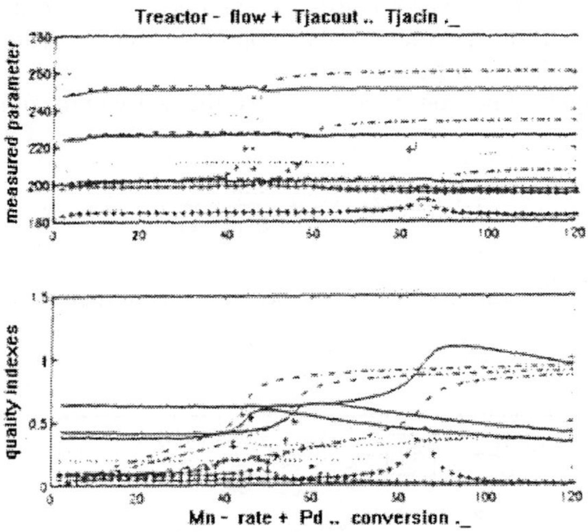

Fig. 1. Quality index and measurable variable curves in three different batches

Number-average molecular weight (Mn) and polydispersity (Pd) are polymer product quality indexes. Reaction rate and monomer conversion, as reaction indexes, represent reaction progress.

They are usually difficult to measure. Some researchers have studied the inferential estimation of polymer quality based on state estimations [3,4]. This approach, however, requires an accurate model based on the knowledge of polymerization kinetics. The development of such an accurate model could be very time consuming and effort demanding.

In this paper, neural networks based inferential estimation is studied. In a soft sensor model, the task of ANN is to estimate these indexes from certain parameters measured on-line. Fig. 1 shows how these parameters vary in three batches under different conditions. It can be seen that polymer particle growth is not proportional. It barely changes in some regions, but grows quite fast in other regions. In different batches, there are great differences in the growth time and growth values. The measurable variables, such as reactor temperature, jacket inlet and outlet temperature and flow rate through the jacket, related only partially to variations of Mn, Pd, reaction rate, and conversion. The curve shapes of Mn and Pd are very different from the curve shapes of the measurable parameters. Consider these indexes to be functions of the measured variables as follow:

$$Index = f(\text{var}\,1, \text{var}\,2, \cdots) \tag{1}$$

3 Quality Estimation Model 1 and Its Problems

This is an elementary function approximation trial where a recurrent neural network is used. In a recurrent network, the outputs depend on the current inputs, the past inputs, and the past outputs. The network consists of two layers. The hidden layer contains sigmoidal neurons, while the output layer contains a linear neuron. This recurrent network structure is shown in fig. 2.

Hyperbolic tangent function was taken as the activation function for hidden neurons:

$$y = \tan sig(x) = \frac{1-\exp(-x)}{1+\exp(-x)}, \tag{2}$$

Where $x = w^T * p + b$, and p is the input vector, w is a vector of network weights, and b is a bias.

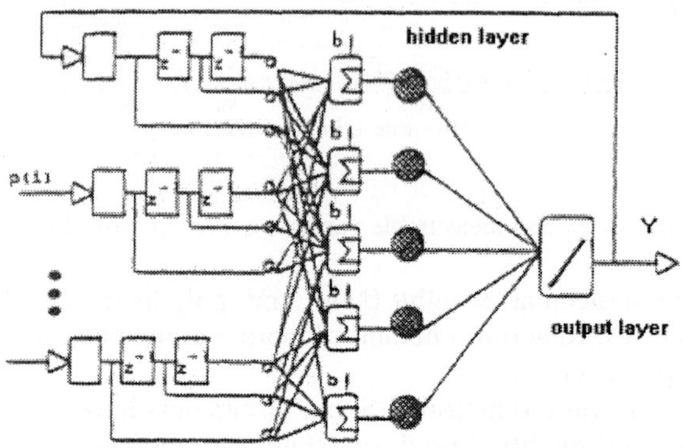

Fig. 2. Recurrent Architecture of Model 1

The training rule is back propagation with momentum and adaptive learning rate. They can be used to adjust the weights and biases of networks in order to minimize the sum of squared errors of the network. This is done by continuously changing the values of the network weights and biases in the direction of steepest descent of the sum of squared errors with respect to network weights. Starting from the network output layer, derivatives of error are calculated, and then back propagated through the network.

Fig. 3 shows the test results of the trained ANN model 1. It can be seen that the performance of the ANN model 1 is not satisfactory. Since quality index Mn has more complex relationship with measurable parameters, their estimated curves are worse than the estimated rate curves. As a soft sensor, it should have much higher accuracy.

From theoretical point of view, the network with a hidden layer of sigmoidal neurons and a linear output layer should be capable of approximating any function and obtain a satisfactory accuracy. Unfortunately training results were not satisfactory, although during training we've managed to use various ways to improve training results, such as changing input dimensions, changing neuron number in the hidden layer, selecting training data, canceling output recurrence, pre-processing input data etc.

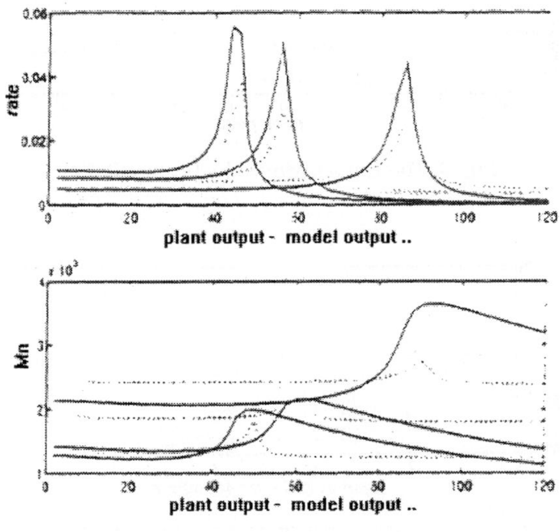

Fig. 3. Test curves of model 1

4 Data Restructuring via Moving Time Referential Point

By carefully observing Fig. 1, it can be discovered that the distribution of quality index data along the time axis is dispersed. At the same time in different batches, the dispersity of data distribution usually is very obvious. This dispersed data distribution will also affect fitting accuracy.

If the curves of Mn, Pd and reaction rate are moved by some feature point, e.g. the peak point of Mn or reaction rate, i.e. if the peak point is taken as referential point of time axis, the data distribution will decrease its dispersity. Fig. 4 shows the moved curves and their data distribution via new time axis. It can be seen that the data distribution dispersity at the same time in different batches becomes much smaller.

The fitting relationships hence get simplification. This is another way for improving network accuracy.

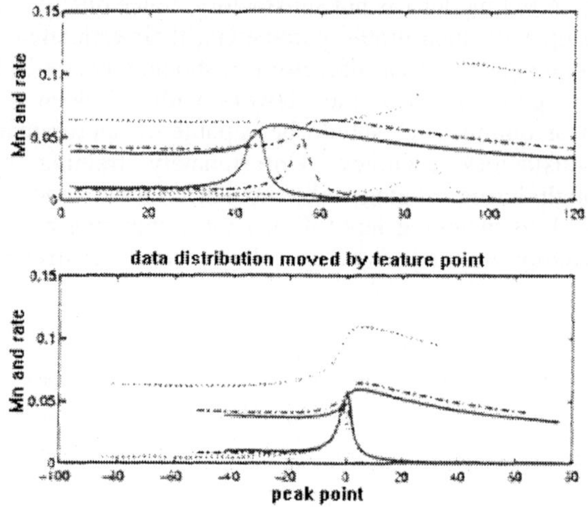

Fig. 4. The Curves Moved by Feature Point

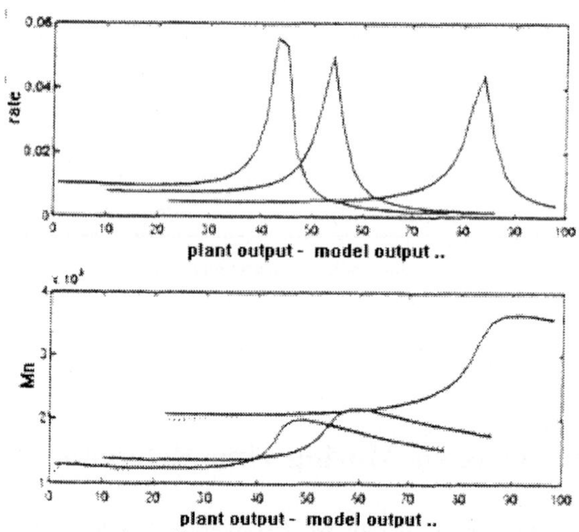

Fig. 5. Test Curve of synthetic model 2

To complete the simplification, a network is first trained to find the feature points under different batches. When reaction initial conditions and relative parameters are put into the network as inputs, the trained network model can accurately find the feature points. The estimated error is within ±1 sampling step.

The sample data are reorganized by the new time referential point. The new input matrix is formed and is used to train quality index model. The combination of the two networks synthesizes the quality estimation model. Test results show that the estimation model has satisfactory accuracy.

Fig. 5 shows test curves of the synthetic model 2. The network for estimating feature point consists of radial basis neurons, while the other network consists of sigmoidal neurons. For fitting problems of many sets of data, the radial basis networks usually show obvious advantages.

5 Conclusions

Since batch polymerization processes are usually highly non-linear and the associated quality indexes are generally difficult to be measured on-line, ANN technique is used here to provide inferential estimations of polymer quality indexes. Data restructuring via moving time referential point is an effective method for improving properties and accuracy of a neural network model. Data restructuring has no definite rules and should be based on practical needs and possibilities. To avoid pitfalls in using neural networks, it's very necessary to analyze the problem to be solved and to find some features of the plant.

Acknowledgements. The work was supported by the UK EPSRC through the Grant GR/R10875 - Intelligent Computer Integrated Batch Manufacturing.

References

1. Benson, R.: Process Control - the Future. Computing & Control Engineering Journal, Vol. 8 (1997) 161-166
2. Zhang, J.: Inferential Estimation of Polymer Quality Using Bootstrap Aggregated Neural Networks. Neural Networks, Vol.12 (1999) 927-938
3. Dimitrators, J., Georgaskis, C., El-Asser, M. S., Klein, A.: Dynamic Modeling and State Estimation for an Emulsion Copolymerization Reactor. Computers and Chemical Engineering, Vol. 13 (1989)21-33
4. Kozub, D. J., MacGregor, J. F.: State Estimation for Semi-Batch Polymerization Reactors. Chemical Engineering Science, Vol. 47 (1992) 1047-1062
5. Zhang, J., Morris, A. J.: A Sequential Learning Approach for Single Hidden Layer Neural Networks. Neural Networks, Vol.11 (1998) 65-80

Modelling the Supercritical Fluid Extraction of Lycopene from Tomato Paste Waste Using Neuro-Fuzzy Approaches

Simon X. Yang[1,2], Weiren Shi[1], and Jin Zeng[2]

[1] College of Automation, Chongqing University, Chongqing, 400044, China
[2] School of Engineering, University of Guelph, Guelph, ON N1G 2W1, Canada
syang@uoguelph.ca

Abstract. Industrial production of lycopene and β-carotene from tomatoes appears to be in highly demand in recent years by food industry and pharmaceutical companies for the development of functional foods. In this paper, a novel neuro-fuzzy model is first proposed to model the supercritical CO_2 extraction of lycopene and β-carotene from tomato paste waste by considering several process parameters. It can help to achieve the optimal control, and to develop a simulation and auto-control system for quality assurance in the industrial production. The effectiveness of the proposed model is demonstrated by simulation study.

1 Introduction

Lycopene and β-carotene are important members of the carotinoid families, they are the strongest carotinoid for neutralising free-radicals, help to resist-oldness, resist-tired, prevent cancer and anticancer. Lycopene and β-carotene can be used in medicines, cosmetics, health protection profession and colour matching, etc. They are principal pigments found in tomatoes and are responsible for the characteristic deep-red colour of ripe tomato fruits and tomato products.

In recent years, the food industry has dramatically increased their interest and awareness of the health benefits of lycopene and β-carotene from tomatoes. Industrial production of lycopene and β-carotene from tomatoes appears to be in high demand by food and pharmaceutical companies for the development of functional foods [1]. The most commonly used method is supercritical fluid extraction technique that is used to extract many natural products from a wide variety of substrates [2,3,4,5,6]. Currently, extract lycopene and β-carotene by supercritical CO_2 extraction is still at the laboratory scale. Hence, a proper model to predictive control the extraction on an industrial scale with minimal loss of bioactivity is highly desirable for food and pharmaceutical applications.

In this paper, a novel neuro-fuzzy model is first proposed to model and optimise the lycopene and β-carotene extraction by supercritical CO_2 extraction procedure in laboratory. It will helps to optimise the extraction process, and pave the way to develop a simulation and auto-control system for quality control in the industrial production. This model can be also used to optimise the extraction processes of other nutrient and functional foods from plant material.

F. Yin, J. Wang, and C. Guo (Eds.): ISNN 2004, LNCS 3174, pp. 880–885, 2004.

2 The Proposed Neuro-Fuzzy Model

In order to model an industry process, the commonly used conventional approach is the mathematics method. First determine the relative variables and parameters, after analyse the experiment data then establish a series differential equations by physical laws to describe the non-linear relationships of the whole process. Finally, use the statistics to evaluate the results. The mathematics approach is the commonly used approach, but it only suits for some simple processes. When the non-linear relationships of the whole process become more and more complex, it is extremely hard for the conventional method to find the accurate parameters and equations to describe the complex non-linear process.

Artificial neural networks are well known for their capacity to learn from real patterns and to generalise the input-output relationship, especially for those high non-linear process [7]. They have been widely used in task domains ranging from concept learning to function approximation. But the neural network approach still has some shortcomings, it is an blind simulation process, can not provide the explanations for the simulation results. Hence, the neuro-fuzzy systems are introduced to overcome this problem, where in the neuro-fuzzy system, the neural networks are used to adjust the membership functions to make the fuzzy system complies to the input-output requirements. Hence, in this paper, the neuro-fuzzy models are proposed to model the extraction of lycopene and β-carotene from tomato paste waste.

The proposed neuro-fuzzy model for modelling of the supercritical CO_2 extraction of lycopene and β-carotene from tomato paste waste is illustrated in Fig. 1. This proposed model use ANFIS architecture [8] respect to each output, five parameters (temperature, pressure, co-solvent rate, extraction time and CO_2 flow rate) are selected as the inputs, the recovery rate of lycopene and β-carotene are the outputs, respectively.

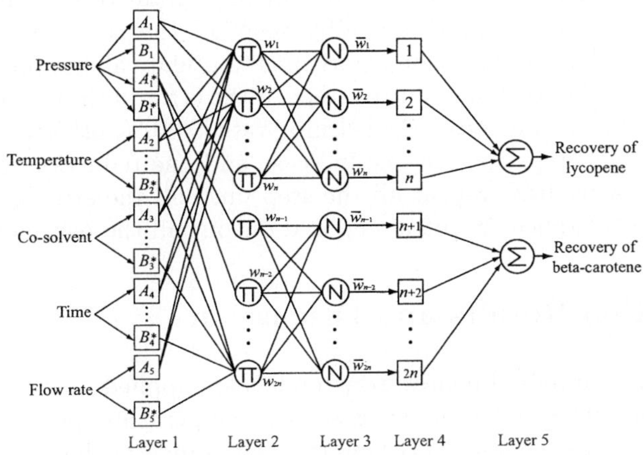

Fig. 1. Proposed neuro-fuzzy model for supercritical CO_2 extraction of lycopene and β-carotene from tomato paste waste.

For the proposed neuro-fuzzy model, it contains 64 rules, 4 membership functions assigned to each input variables, and in total, there are 60 premise parameters and 384 consequent parameters. As show in Fig. 1, in Layer 1, each unit contain a bell-shaped membership function to represents a linguistic term, given as

$$\mu_j^{(i)}(x_i) = \frac{1}{1 + \left[(\frac{x_i - c_i}{a_i})^2 \right]^{b_i}}, \tag{1}$$

where a_i, b_i and c_i are premise parameters to determine the shape of bell functions, x_i is the input to unit i. Each unit in Layer 2 multiplies all the signals come from Layer 1 then send the firing strength ω_i as the output to the next layer, given as

$$\omega_i = \prod_j \mu_j^{(i)}(x_i). \tag{2}$$

In Layer 3, the normalised firing strengths of all rules are calculated by

$$\bar{\omega}_i = \frac{\omega_i}{\sum_i \omega_i}. \tag{3}$$

Then the fuzzy *if-then* rules are calculated in Layer 4, given as

$$O_{R_i} = \bar{\omega}_i(\alpha_0^{(i)} + \alpha_1^{(i)} x_1 + \alpha_2^{(i)} x_2 + \ldots + \alpha_n^{(i)} x_n), \tag{4}$$

where $\alpha_0^{(i)}, \alpha_1^{(i)}, \ldots \alpha_n^{(i)}$ are the consequent parameters. Finally, the output is computed in Layer 5 by

$$output = \sum_i \bar{\omega}_i(\alpha_0^{(i)} + \alpha_1^{(i)} x_1 + \alpha_2^{(i)} x_2 + \ldots + \alpha_n^{(i)} x_n). \tag{5}$$

To training the proposed neuro-fuzzy models, firstly, keep the premise parameters fixed, then determine the consequent parameters by iterative least square estimations techniques; secondly, keep all the consequent parameters fixed, the use the backpropagation algorithms to update all the premise parameters; thirdly, consider four consecutive steps, if the error in these four consecutive steps are reduced, then increase the learning rate by 10%, otherwise, if the error in two consecutive steps contain one increase and one reduction, then decrease the learning rate by 10%; repeat all the step until all the error is smaller than the pre-defined tolerance, stop training and output all the parameters.

3 Simulation Results and Discussion

In this study, the proposed neuro-fuzzy models are applied to the available data sets in literature [9]. Five factors are assessed: temperature, pressure, co-solvent ethanol, extraction time and CO_2 flow rate, the range of those parameters are 20^oC to 100^oC, $100bar$ to $300bar$, 1% to 20%, $0.5h$ to $3.5h$ and $1kg/h$ to $8kg/h$, respectively. The performance for the neuro-fuzzy models are compared to the

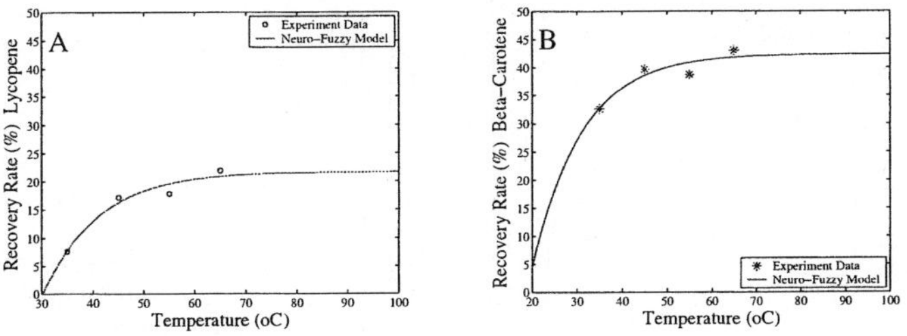

Fig. 2. Effects of temperature on SFE for lycopene (A) and β-carotene (B), respectively. *(pressure: 300bar; flow rate: 4kg/h; extraction time: 2h; co-solvent: 0%)*

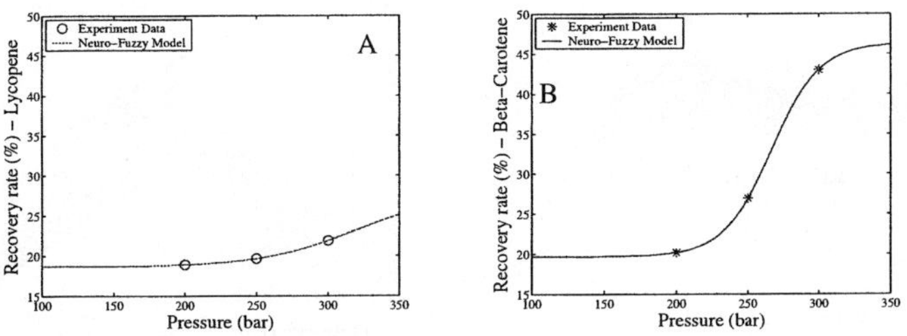

Fig. 3. Effects of pressure on SFE for lycopene (A) and β-carotene (B), respectively. *(temperature: 65°C; flow rate: 4kg/h; extraction time: 2h; co-solvent: 0%)*

experiment data by least mean square error (E) and correlation coefficient (R): the smaller E is, the better; the closer R is to 1, the better.

The relationship between the extractor temperature and recovery rate of lycopene and β-carotene by the neuro-fuzzy model are given by Figs. 2A and 2B, respectively. The circle and star dot denote the experiment data, the solid lines represent the neural-fuzzy model results for recovery rate of lycopene and β-carotene. Analyse Figs. 2, the proposed neuro-fuzzy model can fit and correlate the experimental data very well, it shows when the temperature over $55°C$, the recovery rate of lycopene and β-carotene reach to the saturated status, so the optimal temperature for the extraction process given by the proposed model is $55°C$ to $60°C$. With respect to temperature, the R and the E for the whole system are 0.9773 and 0.4332, respectively.

The relationships between extractor pressure and yield rate of lycopene and β-carotene, extraction time and yield rate of lycopene and β-carotene are shown in Figs. 3A and 3B, Figs. 4A and 4B, respectively. Similarly, experiment data are denoted by circle and star dot respectively, the neuro-fuzzy model results are denoted by solid line. Respect to extractor pressure, the R and the E for the neuro-fuzzy model are 0.9993 and 0.0917; while respect to extraction time, the R and the E are 0.9984 and 0.2114. The results show that the proposed neural-fuzzy model can give the accurate prediction to the extraction process, the optimal pressure ranges are 300 to $350bar$ and the optimal extraction time ranges are 2 to $2.5h$, respectively.

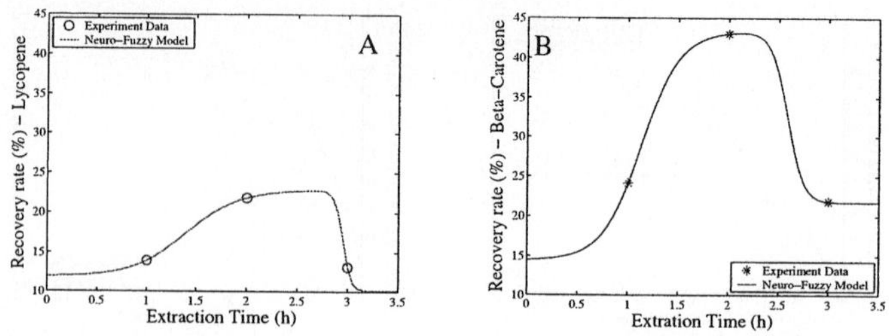

Fig. 4. Effects of extraction time on SFE for lycopene (A) and β-carotene (B), respectively. *(pressure: 300bar; temperature: 65°C; flow rate: 4kg/h; co-solvent: 0%)*

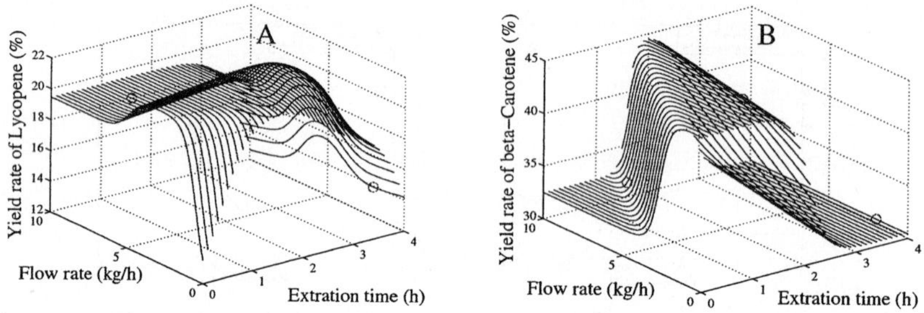

Fig. 5. Effects of extraction time and CO_2 flow rate on SFE for lycopene (A) and β-carotene (B), respectively. *(pressure: 300bar; temperature: 65°C; co-solvent: 0%)*

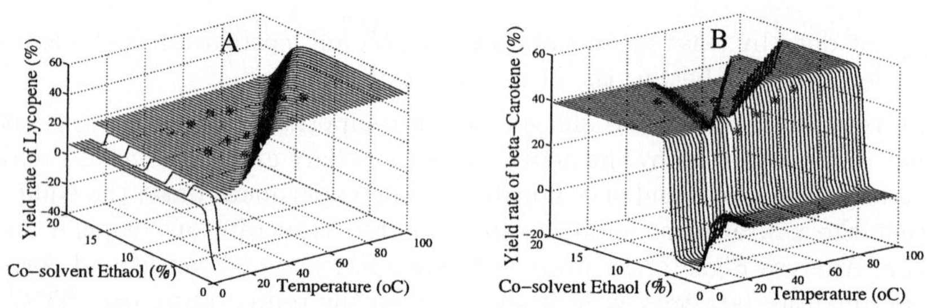

Fig. 6. Effects of temperature and co-solvent on SFE for lycopene (A) and β-carotene (B), respectively. *(pressure: 300bar; extraction time: 2h; co-solvent: 0%)*

Figs. 5A and 5B give out the results of relationship between yield rate and two variables, extraction time and CO_2 flow rate, for lycopene and β-carotene respectively, experiment data are denoted by circle dot, the results by neural-fuzzy model are represented by line surface. Analyse Fig. 5, the neuro-fuzzy model can fit the experiment data very well, the line surface give the accurate predictions to the extraction process, it can helps to predict and find the optimal point for the whole process and hence to improve the productivity. The R and the E for the whole system are 0.9998 and 0.3454, respectively.

When consider the relationship between yield rate and other two variables, extraction temperature and co-solvent ethanol, for lycopene and β-carotene, the results shown in Figs. 6A and 6B, respectively. In Fig. 6, experiment data are denoted by star dot, the results by neural-fuzzy model are represented by line surface. Analyse those figures, obviously, the proposed neuro-fuzzy model can always fit the experiment data, the line surface give the accurate predictions to the extraction process. The R and the E for the whole system are 0.9989 and 0.5476, respectively.

From above results, obviously, the proposed neuro-fuzzy model can successful model the extraction process of lycopene and β-carotene from tomato paste waste and helps to optimise the extraction process and highly improve the productivity

4 Conclusion

In this paper, a novel neuro-fuzzy model is proposed to model the supercritical CO_2 extraction of β-carotene and lycopene from tomato paste waste. Generally the proposed neuro-fuzzy model can provide the accurate prediction to the whole extraction process. The effectiveness of the proposed neuro-fuzzy model is demonstrated by simulation studies. It is believed that the proposed novel neuro-fuzzy approach will be able to be applied to practical areas and highly improve the productivity.

References

1. Shi, J., Maguer, M.L.: Lycopene in tomatoes: Chemical and physical properties affected by food processing. Critical Reviews in Food Science and Nutrition **40** (2000) 1–42
2. Taniguchi, M., Tsuji, T., Shibata, M., Kobayashi, T.: Extraction of oils from wheat germ with supercritical carbon dioxide. Agric. Biol. Chem. **49** (1985) 2367–2372
3. Kaihara, A., Yoshii, K., Tsumura, Y., Ishimitsu, S., Tonogai, Y.: Multiresidue analysis of 18 pesticides in fresh fruits, vegetables and rice by supercritical fluid extraction and liquid chromoatography-electrospray ionization mass spectrometry. Journal of Health Science **48** (2002) 173–178
4. Lucas, A.D., Rincon, J., Gracia, I.: Influence of operating variables on yield and quality parameters of olive husk oil extracted with supercritical carbon dioxide. Journal of the American Oil Chemists' Society **79** (2002) 237–243
5. Oliveira, R., Rodrigues, M.F., Bernardo, M.G.: Characterization and supercritical carbon dioxide extraction of walnut oil. Journal of the American Oil Chemists' Society **79** (2002) 225–230
6. Rozzi, N.L., Singh, R.K., Vierling, R.A., Watkins, B.A.: Supercritical fluid extraction of lycopene from tomato processing byproducts. Journal of Agricultural and Food Chemistry **50** (2002) 2638–2643
7. Karayiannis, N.B.: Artificial Neural Networks Learning Algorithms, Performance Evaluations and Application. Kluwer Academic, Boston (1993)
8. Jang, J.S.R.: ANFIS: Adaptive-Network-Based Fuzzy Inference System. IEEE Transactions on Systems, Man, and Cybernetics **23** (1993) 665–684
9. Baysal, T., Ersus, S., Starmans, D.: Supercritical co_2 extraction of β-carotene and lycopene from tomato paste waste. J. Agric. Food Chem. **48** (2000) 5507–5511

Detection of Weak Targets with Wavelet and Neural Network

Changwen Qu, You He, Feng Su, and Yong Huang

Department of Electronic Engineering, Naval Aeronautical Engineering Institute
Yantai, Shandong, 264001, P. R. China
qcwwby@sohu.com

Abstract. A fundamental task of radar is the detection of targets in noise and clutter. It is very difficult to detect weak targets in heavy sea clutter. In this paper we apply the discrete wavelet transform (WT), along with multilayer feedforward neural networks, to the detection of radar weak targets. The wavelet transform employs the Haar scaling function, which is well matched to the signals from targets. The hard threshold filter is adopted to remove sea clutter. The neural networks are trained with the back-propagation rule, which are used to detect weak targets. The simulated results show that the proposed method is very effective for radar detection of weak targets.

1 Introduction

Radar is a critical element of many surveillance systems that exist in modern military and civilian environments where its roles include target detection and identification, navigation and mapping, target tracking and weapons guidance. Radar detects targets by emitting electromagnetic energy, then processing the signal from the reflected radiation. There are many kinds of clusters in radar signals. For marine radar, one of the major sources of noise is random reflection from the ocean's surface. It is very difficult to detect weak targets in heavy sea clutter. It is the important task of marine radar or coast-based radar to suppress sea clutter. In our work, we are particularly interested in the detection of weak targets within heavy sea clutter.

The WT is a tool for carving up radar signals into different frequency components, allowing one to study each component separately. We use the WT with the Haar wavelets to separate the radar targets from sea clutter. The hard threshold filter is applied to remove some high frequency coefficients, which result from sea clutter. Then we reconstruct the radar signal, in which sea clutter is suppressed. Using a multilayer feedforward network, we detect weak targets from the reconstructed radar signal finally.

This paper is organized as follow. The next section shows characteristics of echoes from targets and sea clutter. In section 3, the Harr wavelet and the multiresolution analysis are reviewed for radar signals. The method to suppress sea clutter is given in section 4. In section 5, the detection of radar targets based on the multilayer feedforward network is studied. Finally, conclusions are given in section 6.

F. Yin, J. Wang, and C. Guo (Eds.): ISNN 2004, LNCS 3174, pp. 886–891, 2004.

2 Features of Echoes from Targets and Sea Clutter

After the pulse is transmitted by pulse radar, the system tries to receive any echoes from the pulse. As shapes and states of targets are steady, returned signals from targets are pulse-type and their envelopes vary very slowly. The irregularity of the ocean surface arises through the interaction of wind and water. The system is open because of the external forcing of the sun's heat, and is dissipative because of the heat loss due to friction. Also, the interaction of wind and water is nonlinear. Signals resulting from open dissipative systems with nonlinear dynamics are known to be fractal. Thus sea clutter is fractal [1]. In addition, sea clutter is faster fluctuation than echoes from targets. For low-resolution radar, sea clutter has many characteristics like those of thermal noise. It is randomly fluctuation in both amplitude and phase, and in many case the probability density function is of the Rayleigh type, like that of thermal noise. But the spectrum of sea clutter is often narrower than that of white noise.

Echoes from targets last long intervals and are steady, while sea clutter is fractal and fast fluctuation. Echoes from targets are low frequency components, and sea clutter is high frequency component. According characteristics of echoes from targets and sea clutter, we will choose the Haar wavelet to decompose echoes into low and high components.

3 Discrete Wavelet Transform of Radar Signals

The Haar scaling function $\phi(t)$ is defined by

$$\phi(t) = \begin{cases} 1, & 0 \le t < 1 \\ 0, & otherwise \end{cases} \tag{1}$$

The scaling function is a low pass filter. A low pass filter suppresses the high frequency components of a radar signal and allows the low frequency components through. The Haar scaling function calculates the average, which results in a smoother, low pass signal. One prescription used to generate a function family $\phi_{j,k}(t)$ from the scaling function $\phi(t)$ is

$$\phi_{j,k}(t) = 2^{-j/2}\phi(2^{-j}t - k) \tag{2}$$

where $j,k \in Z$. The index *j* refers to *dilation* and *k* refers to *translation*. Define the vector space V_j as

$$V_j = span\{\phi_{j,k}(t)\} \tag{3}$$

where span denotes the linear span. It is shown that $j,k \in Z$ forms an orthonormal basis.

The Haar wavelet function $\psi(t)$ is defined by

$$\psi(t)=\begin{cases}1, & 0\le t<1/2\\ -1, & 1/2\le t<1\\ 0, & otherwise\end{cases} \tag{4}$$

The wavelet function is a high pass filter. A high pass filter allows the high frequency components of a radar signal through while suppressing the low frequency components. As above, we define the vector space W_j as

$$W_j = span\left\{\psi_{j,k}(t)=2^{-j/2}\psi(2^{-j}t-k)\right\} \tag{5}$$

It is also shown that the wavelet family $\left\{\psi_{j,k}(t)\right\}$ forms an orthonormal basis. The relationship between the Haar wavelet function and scaling function is given

$$V_j = V_{j+1}\oplus W_{j+1} \tag{6}$$

Furthermore, one may get

$$V_0 = \cdots V_j \oplus W_j \oplus W_{j-1}\cdots\oplus W_1 \tag{7}$$

Then $\left\{V_j\right\}$ constructs a multiresolution analysis. From Eq. (10), one can see that every signal $s(t)$ may be represented as a series of the form

$$s(t)=\sum_{k=-\infty}^{\infty}a_m(k)\varphi_{m,k}(t)+\sum_{j=1}^{m}\sum_{k=-\infty}^{\infty}d_j(k)\psi_{j,k}(t) \tag{8}$$

The first term is low frequency component, which provides the "general picture". The second term is high frequency component, which provides the "details". The above equation represents one way of decomposing the radar signal $s(t)$ into low and high components.

4 Method to Suppress Sea Clutter

The low frequency components of the WT of a radar signal result from echoes of targets and the high frequency components correspond to sea clutter. Wavelet-domain filtering produces wavelet shrinkage estimates. That is, certain wavelet coefficients are reduced to zero. There are two predominant thresholding schemes. The soft threshold filter shrinks coefficients above and below the threshold. Soft thresholding reduces coefficients toward zero. However, performs better for detection of weak targets in sea clutter. In the hard threshold filter, we compare wavelet coefficients of high frequency components with a threshold value q_j. If wavelet coefficients of high frequency components are less than the threshold, they are reduced to zero in order to remove the sea clutter. That is

$$d_j(k)=\begin{cases} d_j(k), & \left|d_j(k)\right|\geq q_j \\ 0, & \left|d_j(k)\right|<q_j \end{cases} \tag{9}$$

Wavelet coefficients over the threshold are used to reconstruct the signal that mainly contains echoes from targets. The chose threshold value is important. In our method, the threshold value is computed as the standard deviation of wavelet coefficients of high frequency components.

The processing steps to suppress sea clutter are performed as follows.

Step 1: Calculate the discrete WT of the radar signal.

Step 2: Calculate the threshold and remove the wavelet coefficients of the high frequency components based on the hard threshold filter.

Step 3: Reconstruct the radar signal.

We simulate the signal from a target in sea clutter to demonstrate the above method. In the simulation, sea clutter is the Rayleigh type and the signal-to-clutter ratio is –5 dB. The Haar wavelet is used to decompose the radar signal with the seventh level. Fig. 1 shows the echo from a target without sea clutter. Fig. 2 shows the echo from a target within sea clutter. Fig. 3 shows the reconstructed signal using the proposed method. It is shown that the method is powerful to suppress sea clutter from simulated results.

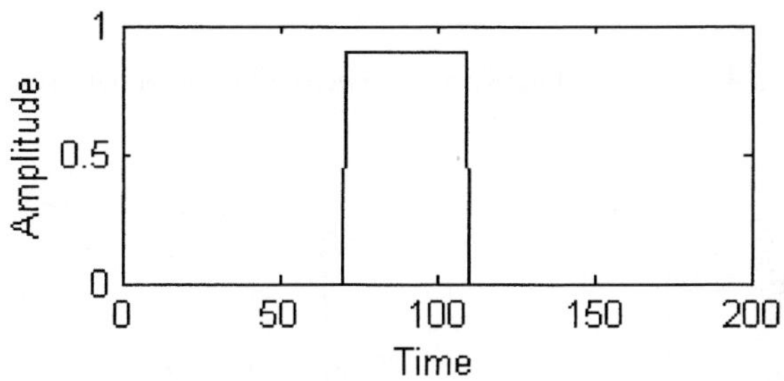

Fig. 1. The echo from a target without sea clutter

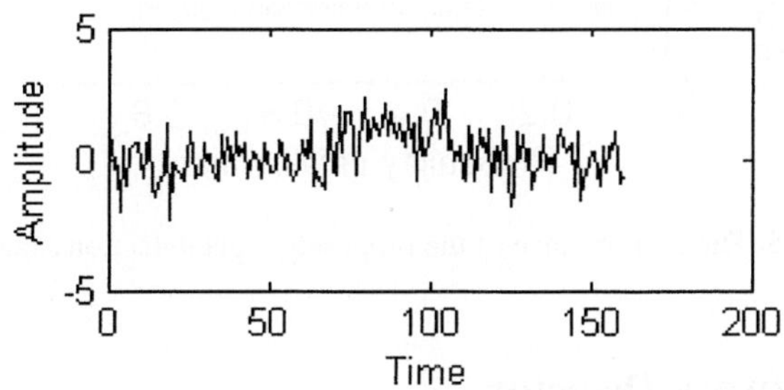

Fig. 2. The echo from a target within sea clutter

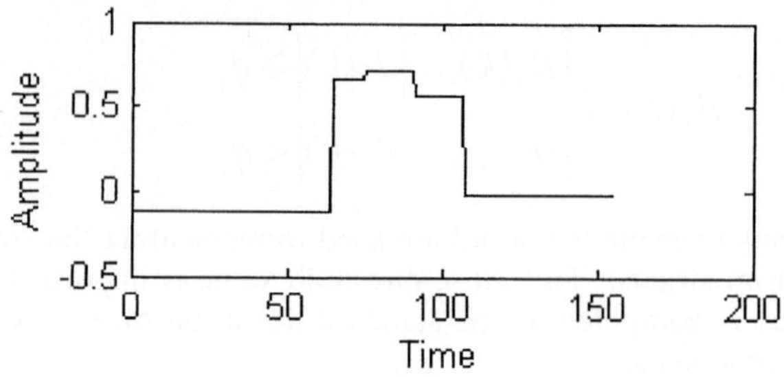

Fig. 3. The reconstructed signal

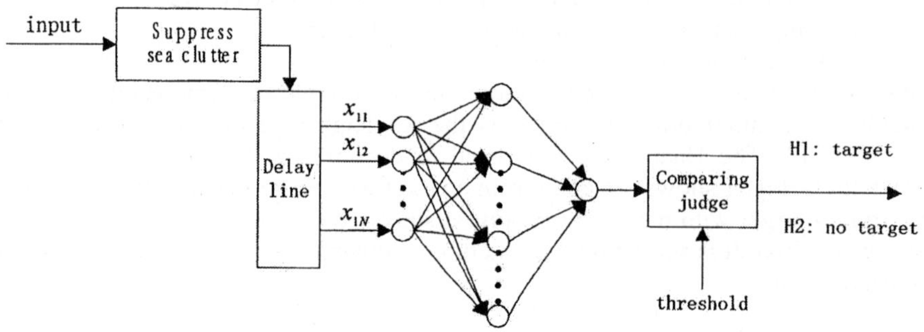

Fig. 4. The Neural network architecture for target detection

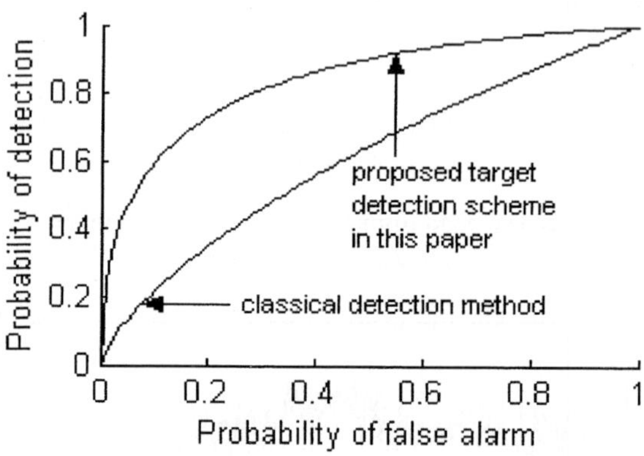

Fig. 5. The performance of the proposed target detection scheme

5 Neural Network Detector

After sea clutter is suppressed by WT along with the hard threshold filter, we need to design a detector to detect targets. Although many classical detection algorithms are available for this purpose, neural networks are very promising due to their advantages over conventional methods. The advantages of neural networks as pattern classifiers have been well investigated, and a detailed description of the architecture and training of neural networks can be found in [2]. The main reason for the feasibility of neural networks is their ability to generalize after learning from training data. That is, they can identify a test pattern even when that particular pattern is not used for training.

Fig. 4 shows the neural network architecture we employ for target detection. The network is comprised of 3 layers of artificial neurons: an input layer, a hidden layer, and an output layer. Signals flow forward through the network, that is from input layer to hidden layer to output layer. Such an architecture is known as a multilayer feedforward network or multilayer perceptron. The input neuron layer performs no processing. It merely provides means for coupling the input vectors to the hidden layer. The neurons in the middle layer sum the weighted network inputs, along with an internal bias for each neuron, then apply the nonlinear sigmoidal activation function. The single output neuron computes the weighted sum of the outputs of the hidden neurons, along with its internal bias. This architecture can represent an arbitrary function arbitrarily well, given a sufficient number of neurons in the hidden layer.

The particular function mapping that the network performs is determined by the values of the weights between neuron layers and the internal neuron biases. Various learning algorithms exist for computing the network weights and biases for a given problem. The back-propagation learning algorithm is applied to train the neural network to achieve the best detection accuracy.

To demonstrate the performance of the proposed target detection scheme, 100 target signals in the Rayleigh type sea clutter are simulate. The signal-to-clutter ratios are 0dB. The performance of the proposed target detection scheme is shown in Fig. 5. Clearly, the performance of the proposed target detection scheme is better than the classical detection method.

6 Conclusions

We choose the Haar wavelet to decompose radar signals into low and high components. Using a hard threshold filter to suppress sea clutter. Based on a multilayer feedforward network, we detect weak targets from the reconstructed radar signal. Simulated results illustrate that the method can perform the detection of weak targets in sea clutter beautifully.

References

1. Noel, S., Szu, H.: Wavelets and Neural Networks for Radar. Progress in Unsupervised Learning of Artificial Neural Networks and Real-World Applications, Russian Academy of Nonlinear Sciences (1998)
2. Haykin, S.: Neural Networks. 2nd ed. NJ: Prentice-Hall, Englewood Cliffs (1999)

An Artificial Olfactory System Based on Gas Sensor Array and Back-Propagation Neural Network

Huiling Tai, Guangzhong Xie, and Yadong Jiang

School of Opto-electronic Information
University of Electronic Science & Technology of China, Chengdu, 610054, P.R. China
taihuiling@163.com

Abstract. An artificial olfactory system is constructed to determine the individual gas concentrations of gas mixture (CO and H_2) with high accuracy. And a Back-Propagation (BP) neural network algorism has been designed using MATLAB neural network toolbox. And an effective study to enhance the parameters of the neural network, including pre-processing techniques and early stopping method is presented in this paper. It is showed that the method of BP artificial neural improves the selectivity and sensitivity of semiconductor gas sensor, and is valuable to engineering application.

1 Introduction

It is important to analyze gas mixture for science research and environmental detection. However, a single gas sensor has a poor selective and sensitivity, so it is very difficult to process the quantitative analysis of the mixed gases using a single sensor. But gas sensor arrays that are made up of a number of sensors with different selectivity can make use of the over-lapping sensitivity, and each sensor responds to a range or class of gases rather than to a specific one. We can extract this information from the combined response of all sensors in sensor array, and in order to process such a partially overlapping response, pattern recognition (PARC) techniques are generally employed [1]. Pattern recognition using artificial neural networks (ANN) has received considerable interest during recent years. ANN is especially valuable because of their adaptability to almost any mathematical transform within a given region. The models obtained can be both linear and nonlinear.

An artificial olfactory system [2], which is composed of a gas sensor array with the pattern recognition technique of a feed forward neutral network, is designed to determine the concentrations of the gas mixture (H_2 and CO). And a BP artificial neural network has been designed by using MATLAB neural network toolbox [3] to process the signals of the gas sensor arrays. In order to develop the training efficiency and the generalization of the network, early stopping method supplied by MATLAB neural network toolbox is chosen to process the nonlinear problem. And we will process the concentrations of the mixed gases and the output signals of the gas sensor array in advance to improve the performance of the ANN.

F. Yin, J. Wang, and C. Guo (Eds.): ISNN 2004, LNCS 3174, pp. 892–897, 2004.

2 Experiment

2.1 The Experimental System

In order to have rapid, accurate and reliable measurements of the response of the gas sensor array to the test vapors, a measurement and characterization system has been set up, which can produce certain concentrations of the tested gases, control the working conditions of the gas sensor array and the environment conditions (such as temperature and humidity) through a personal computer and collect the input and output data of the gas sensor array by using a keithley 2700 meter. The gas chamber is provided with the necessary electrical connections, gas inlet and outlet valves for sensor characterization in presence of test gases. Gas is injected into the chamber through three mass-flow controllers (MFCS), which control the flow of two different test gases and nitrogen gas. The whole system is shown in Fig.1.

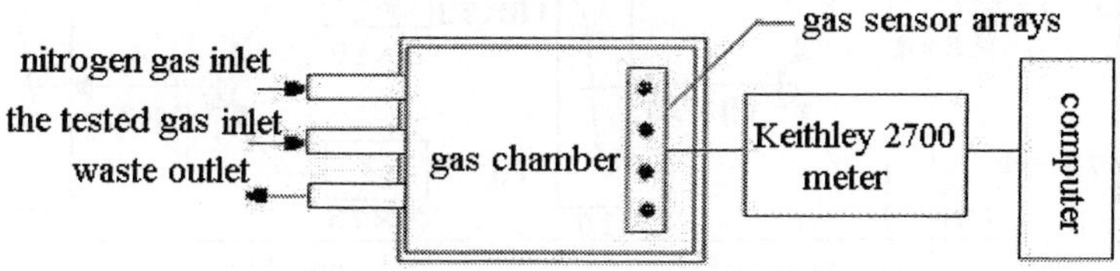

Fig. 1. Structure of the measurement and characterization system

In the gas sensor array, we employed 4 commercial gas sensors sensitive to CO and H_2, whose names and parameters are presented in Table 1. The sensors' conductivity changes with the gas concentrations. Every sensor has a different heater voltage, which can be offered by the programmable power.

According to the requirement of the analysis, the concentrations of CO and H_2 are obtained every 50ppm in the finite range from 100ppm to 500ppm (1ppm=10^{-6}), so 81 different gas mixtures have been obtained, in which eight samples are selected as the test set, and the rest of them are used as the training set in the experiment.

Table 1. Gas sensors

Commercial name	Measuring range (ppm)	Heater voltage
TGS2104	10~100(H_2) 50~1000(CO)	7.0±0.35DC
TGS821	30~1000(H_2) 1000~10000(CO)	5.0±0.2DC
TGS2442	30~1000(CO) 500~1000(H_2)	4.8±0.2DC
TGS203	50~1000(CO)	5.0±0.01DC

2.2 Designed BP Network Model

In the experiment, a three-layer ANN shown in Fig.2 is designed. The structure of the used network is 4: N: 2, because we use an array of 4 sensors, and we want to obtain two gas concentrations in the mixture (H_2 and CO), N is the number of neurons in the hidden layer. It is a three-layer structure, so we can adapt our model to any function. A training algorithm called "Levenberg-Marquardt back-propagation" is used to minimize the neural network error, in which trainlm function is chosen as the training function, and tansig function is chosen as the transfer function. The network's training goal is set up as 0.00001. In order to develop the training efficiency and the generalization of the network, early stopping method supplied by MATLAB neural network toolbox is chosen to process the nonlinear problem.

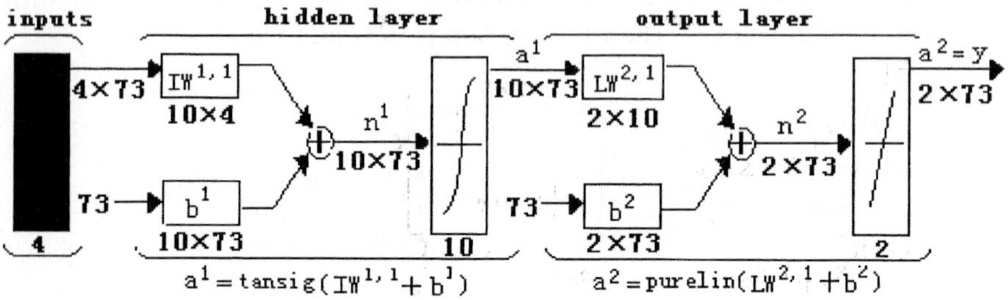

Fig. 2. Structure of Artificial Neural Network

The neural network model, before application, requires a previous calibration or training in order to minimize the error of the model. A way to define this error is SSE: sum of the squares of the differences between the output of the network and the corresponding target value, summed over all test cases, which is also the performance function of the network. The influence of the number of neurons in hidden layer on the ANN error is shown In Fig.3. From Fig.3, we can see that the network error is the smallest when the number of the neurons is equal to 10. So, we choose 10 neurons in hidden layer as a proper value of N for our experiment.

2.3 Pre-process Method

In order to optimize the neural network and its training and achieve a convenient performance of ANN, it is necessary to pre-process all data before using it in the neural network. In this paper, the pre-processing procedure is composed of the processing of the output data of gas sensor array and the concentrations of the gas mixture.

The pre-processing algorithm of the concentrations of the gas mixture is as follows:

$$C_i' = C_i / C_{\max} \qquad i = 1,2 . \tag{1}$$

where C_i' is the new concentration, C_i is the old one, and C_{max} is the largest one, which is equal to 500ppm. Thus, the new concentrations will lie between 0 and 1.

The raw input data of ANN is the sensor resistance or conductance measurements in the gas mixtures. The pre-processing step used over it consists of:

$$X_i' = X_i \Big/ \left(\sum_{i=1}^{4} X_i^2 \right)^{1/2} . \tag{2}$$

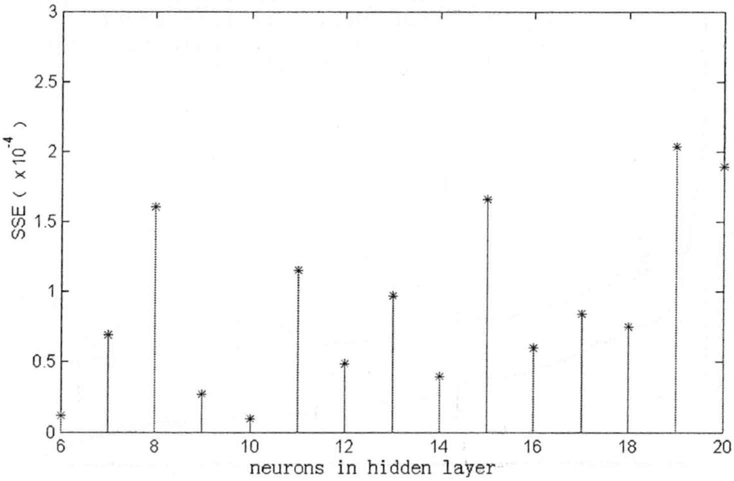

Fig. 3. Effect of the number of neurons in hidden layer on the performance of the network

3 Results

3.1 The Training and Regress Analysis of ANN

The whole input data of ANN is divided into three parts: training samples, validation samples and test samples because of the early stopping method employed. The change curves of error are shown in Fig.4. We can observe that the training of network stops early only with 315 epochs, and the performance of the trained network is 1.03076e-005.

In order to prove the performance of the trained network future, the regress analysis can be made using postreg function. In the regress analysis, R is the correlative coefficient of the output vector of the network and the target vector. The results are satisfactory because both R are equal to 1, which show that BP network has had good performance.

3.2 Measuring Results of CO and H_2 Concentrations

The trained BP neuron network is tested with 8 test samples. The tested results of CO and H_2 are shown in Table 2, Table 3, Table 4 and Table 5 respectively.

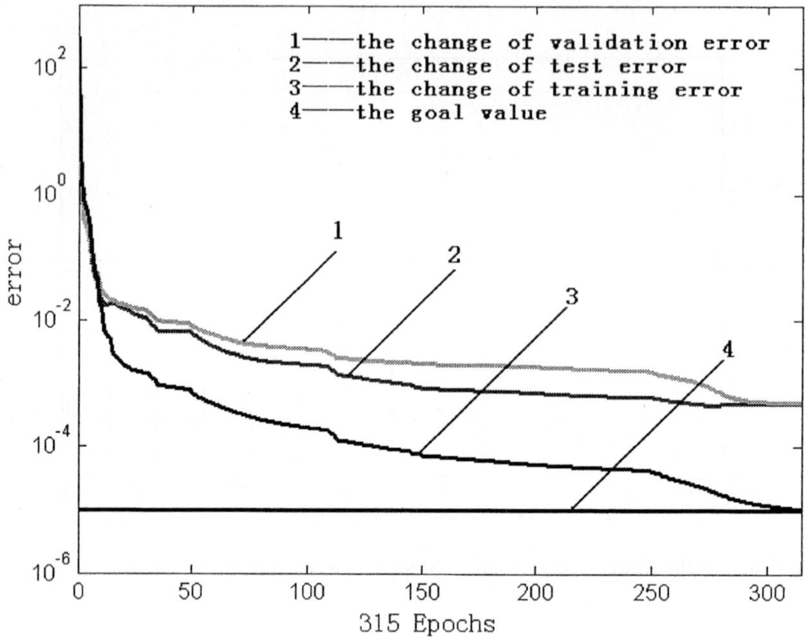

Fig. 4. Change curves of error

Table 2. Predicted results of CO (1~4)

Test sample	1	2	3	4
True value(ppm)	100	150	200	250
Predicted value(ppm)	99.80	150.18	200.25	249.83
Error (ppm)	0.2041	-0.1833	-0.2459	0.1686
Relative error (%)	0.20	0.12	0.12	0.07

Table 3. Predicted results of CO (5~8)

Test sample	5	6	7	8
True value(ppm)	300	350	400	450
Predicted value(ppm)	299.84	350.32	399.95	450.40
Error (ppm)	0.1605	-0.3243	0.0476	-0.4015
Relative error (%)	0.05	0.09	0.01	0.09

Table 4. Predicted results of H_2 (1~4)

Test sample	1	2	3	4
True value(ppm)	150	200	250	300
Predicted value(ppm)	150.13	201.62	250.06	299.87
Error (ppm)	-0.1277	-1.6191	-0.0611	0.1310
Relative error (%)	0.09	0.81	0.02	0.04

Table 5. Predicted results of H_2 (5~8)

Test sample	5	6	7	8
True value(ppm)	350	400	450	500
Predicted value(ppm)	349.83	399.88	449.92	500.23
Error (ppm)	0.1696	0.1225	0.0800	-0.2313
Relative error (%)	0.05	0.03	0.02	0.05

4 Conclusions

An artificial olfactory system including a gas sensor array and an Artificial Neural Network (ANN) is constructed, and a Back-Propagation (BP) neural network algorism has been designed using MATLAB neural network toolbox, in which the number of neurons in the hidden layer is chosen as 10 by experiment. And the concentrations of CO and H_2 can be obtained by the designed ANN system with better precision and fast response, which shows that an artificial olfactory system composed of the gas sensor arrays and BP artificial neural network can process quantitative analysis of the mixed gases. And an effective study to enhance the parameters of the neural network, including pre-processing techniques and early stopping method is presented in this paper. It is showed that the method of BP artificial neural improves the selectivity and sensitivity of semiconductor gas sensor, and is valuable to engineering application.

References

1. Gardner, J.W.: Detection of Vapors and Odors from a Multisensor Array Using Patter Recognition. Part 1.Principal Component and Cluster Analysis. Sens.Actuators B 4 (1991) 109-115
2. Gardner, J.W., Hines, E.L., Wilkinson, M.: Application of Artificial Neural Networks to an Electronic Olfactory System. Meas. Sci. Technol.1 (1990) 446-451
3. Demuth, Howard, Beale, M.K.: Neural Network Toolbox for Use with Matlab. The Math Works Inc. (1993) Chapter 5

Consumer Oriented Design of Product Forms

Yang-Cheng Lin[1], Hsin-Hsi Lai[1], and Chung-Hsing Yeh[2]

[1] Department of Industrial Design, National Cheng Kung University,
Tainan, 701, Taiwan
p3890104@ccmail.ncku.edu.tw
hsinhsi@mail.ncku.edu.tw

[2] School of Business Systems, Monash University, Clayton,
Victoria 3800, Australia
ChungHsing.Yeh@infotech.monash.edu.au

Abstract. This paper presents a new approach for designing form elements of a product based on consumers' perception of images of the product. Neural network (NN) models are used to suggest the best combination of product forms for matching a given set of product images. An experimental study on the form design of mobile phones is conducted to evaluate the performance of four NN models, which are developed by linking 9 form elements with 3 product images of mobile phones. The evaluation result shows that NN models can be used to help product designers determine the best combination of form elements for matching a set of desirable product images.

1 Introduction

Product image plays an important role in consumers' preference and choice of a product [1]. Whether consumers choose a product depends largely on their perception of images of the product, while designers design a product by considering only physical elements or characteristics of the product. To best meet consumers' need of a product from a design perspective, the physical elements of the product require being linked to the consumers' perception of the product. Design support models [1, 3] and consumer oriented technologies [5] have been developed as "translating technology of a consumer's feeling and image of a product into design elements". They have been applied successfully in the product design field [4, 5] to explore the relationship between the feeling (perception of product image) of the consumers and the design elements of the product.

The process of product form design or the perception of the consumers is a often black box, involving uncertain information. As such, we present a new approach using neural network (NN) techniques [7] to formulate the process of consumer oriented design of product forms. NN models are developed to transform form elements of a product into a set of product images perceived by consumers. To illustrate how the new approach works, we conduct an experimental study on mobile phones for their popularity as a consumer product and their wide variety of product form elements.

F. Yin, J. Wang, and C. Guo (Eds.): ISNN 2004, LNCS 3174, pp. 898–903, 2004.

2 Experimental Samples and Form Elements of Mobile Phones

In the experimental study of collecting numerical data for building NN models, we focused on investigating and categorizing of various mobile phones in the market. 33 mobile phones were used as representative experimental samples for training and test the NN models.

In this study, the product form of mobile phones is represented by both the outline shapes and the product elements. We use the morphological analysis to extract the form elements of the 33 mobile phones. The morphological analysis involves two steps, which are described in our prior study [2]. As a result of morphological analysis, Table 1 shows the 9 form elements extracted from the 33 mobile phones, together with their associated types. Each form element has different types of its own, ranging from 2 to 4, as indicated by the type number 1, 2, 3 or 4 in Table 1.

Table 1. Extracted form elements of mobile phones

	Items	Type 1	Type 2	Type 3	Type 4
1.	Top shape	Line	Curve	Arc	Irregular
2.	Body shape	Parallel line	Raised curve	Concave curve	
3.	Bottom shape	Line	Curve	Arc	
4.	Length and width ratio of body	Wide ratio 2:1	Middle ratio 2.5:1	Slender ratio 3:1	
5.	Function buttons style	With large button	Symmetry style	Other style	
6.	Number buttons arrangement	Regular	Irregular		
7.	Screen size	TV ratio 4:3	Movie ratio 16:9	Other ratio	
8.	Screen mask and function buttons	Independence	Function buttons dependence on screen mask	Interdependence	
9.	Outline division style	Normal division	Rim division	Special division	

This morphological analysis result can then be used to build and test the NN models. A group of 15 product design experts are involved in the experiment to evaluate the degree to which the 33 mobile phone samples match a given product image. To describe a specific image of mobile phones in terms of consumers' ergonomic and psychological estimation, a pair of image words is often used. Based on our prior study [2], we use three image word pairs for describing the images of mobile phones, which are Simple-Complex (S-C), Modern-Classic (M-C), and Young-Mature (Y-M). And the design elements used are the form elements of mobile phones as identified in Table 1. To obtain the evaluation value for the three image word pairs, a 7-point scale (1-7) of the semantic difference (SD) method [8] is used. As such, the 15 subjects are asked to assess the form elements of mobile phones on a simplicity-complexity scale of 1 to 7, where 1 is most simple and 7 is most complex. The last three columns of Table 2 show the evaluation results of the 33 experimental samples, including 28 samples in the training set and 5 samples in the test set (asterisked). For each selected mobile phone in Table 2, the first column shows the mobile phone number and Columns 2-10 show the corresponding type number for each of its 9 form elements, as given in Table 1. Table 2 provides a numerical data source for applying the neural network techniques.

Table 2. Consumer perception for 33 selected mobile phones

Phone No	X1	X2	X3	X4	X5	X6	X7	X8	X9	S-C	M-C	Y-M
1	2	2	1	3	2	2	1	2	1	4.467	5.030	4.874
2	2	3	1	2	3	2	1	2	3	3.133	3.370	3.319
3	1	3	2	1	2	1	2	3	1	2.867	5.859	5.237
4	3	3	1	2	2	1	3	1	1	5.533	5.756	4.615
5	4	1	1	2	3	2	3	1	2	5.533	4.874	3.941
6*	1	2	1	2	2	1	1	3	1	6.400	3.630	5.081
7	2	2	1	2	2	1	2	2	3	6.867	2.904	3.578
8	4	1	2	3	2	2	1	2	3	6.800	3.267	4.822
9	2	2	1	1	1	1	2	1	2	4.600	4.667	4.407
10	2	3	1	2	2	1	1	1	1	6.200	6.741	6.637
11*	3	2	3	3	2	2	1	1	2	6.600	2.541	2.074
12	2	1	1	2	2	1	2	3	1	4.533	6.689	6.741
13	1	2	1	2	1	1	3	3	1	3.933	5.652	5.859
14	3	2	2	3	1	1	2	3	3	6.933	5.652	5.807
15	2	2	1	3	1	1	1	3	1	5.600	5.237	4.822
16*	1	1	1	3	1	1	1	1	1	2.733	3.681	2.281
17	1	1	1	2	3	1	2	2	3	4.333	5.081	5.030
18	2	2	1	2	1	1	2	3	3	2.533	4.200	3.059
19*	2	3	2	2	1	1	2	1	2	4.533	3.163	1.815
20	2	2	2	2	1	1	1	2	1	3.200	1.867	2.178
21	2	2	2	2	2	1	1	2	1	5.533	2.956	3.837
22	2	2	2	2	3	1	1	1	1	6.867	3.630	3.422
23	2	2	1	2	2	2	1	2	3	4.467	2.956	3.993
24	2	2	1	2	2	1	2	2	1	3.667	4.304	3.578
25	3	3	1	2	1	1	2	2	3	3.067	2.956	1.763
26	3	3	3	3	2	1	2	3	1	4.533	4.200	2.074
27	3	2	1	3	2	2	1	2	1	3.600	3.889	3.578
28*	1	2	2	2	1	2	1	2	3	4.600	1.089	1.711
29	2	2	2	3	1	1	1	1	1	4.933	2.178	1.763
30	1	3	1	3	1	1	1	1	1	5.333	1.711	2.437
31	2	3	3	1	1	1	1	3	3	3.733	3.526	1.556
32	2	3	3	2	1	2	2	3	2	3.533	4.511	3.215
33	2	1	3	2	2	1	2	2	2	3.733	3.422	2.437

3 Experimental Analysis of Neural Network Models

Neural networks (NNs) are non-linear models and are widely used to examine the complex relationship between input variables and output variables [7]. In this paper, we use the multilayered feedforward NNs trained with the backpropagation learning algorithm, as it is an effective and popular supervised learning algorithm [6, 7]. In our experimental study, we develop four NN models for the training and test process.

3.1 Building and Training the Four NN Models

A combined NN model is developed by combining all three images as three output neurons of the model, using the average values of three image word pairs. The other three models are developed using each of three image word pairs (i.e. the S-C value, the M-C value, and the Y-M value) as the output neuron respectively. Each NN model has 27 input neurons, which are the whole 27 types of the 9 form elements given in Table 1. For each model, if a mobile phone has a particular form element type, the value of the corresponding input neuron is 1; otherwise the value is 0.

In this study, we apply the following four most widely used rules [7] to determine the number of hidden neurons in the single hidden layer for each model. Each model is associated with the rule used, such as -HN1, -HN2, -HN3, or -HN4.

(The numbers of input neurons + the numbers of output neurons) / 2 (1)

(The numbers of input neurons * the numbers of output neurons) ^ 0.5 (2)

(The numbers of input neurons + the numbers of output neurons) (3)

(The numbers of input neurons + the numbers of output neurons) * 2 (4)

The 28 samples in the training set (given in Table 2) were used to train the four NN models. Each model trained 5,000 epochs at each run with the root of mean square (RMS) errors being decreased significantly when these models were trained for about 4,000-5,000 epochs. To find the best training result, we continued the training process up to 50,000 epochs for all models. Table 3 shows the lowest RMS error of each model using a given hidden neuron number.

As shown in Table 3, the RMS error of the combined NN model using the HN1 rule in (1) is the lowest (0.0254), as compared to the other three rules. However, the lowest RMS error (0.0237) of the Y-M NN model is the one using the HN4 rule in (4). This result is in line with the notion that there is no best rule for determining the number of neurons of the hidden layer and it largely depends on the nature of the problem. In Table 3, the Y-M NN model has the lowest average value of RMS errors (0.0374). The result indicates that the four NN models have the highest training consistency on the "Young-Mature" image.

3.2 Performance Evaluation of the Four NN Models

To evaluate the performance of the four NN models in terms of their prediction ability in determining the best design combination of mobile phone form elements for matching given images, the 5 samples in the test set are used. Another group of 15 product design experts are involved in the evaluation process, using the SD method [8] with a 7-point scale (1-7). Rows 2-4 of Table 4 show the average values of the three image word pairs on the 5 test samples evaluated by the 15 subjects, which are used as a comparison base for the performance evaluation. With the 5 test samples as the input, Table 4 shows the corresponding three image values predicted by using the four NN models respectively. The last two columns of Table 4 show the (lowest) RMS errors of the four models in comparison with the evaluated image values.

Table 3. RMS errors of the four models for the training set

	The combined NN model				The S-C NN model			
	-HN1	-HN2	-HN3	-HN4	-HN1	-HN2	-HN3	-HN4
Epochs	45000	50000	40000	40000	45000	30000	25000	40000
RMS errors	0.0254	0.0258	0.0267	0.0262	0.0430	0.0457	0.0622	0.0648
	The M-C NN model				The Y-M NN model			
	-HN1	-HN2	-HN3	-HN4	-HN1	-HN2	-HN3	-HN4
Epochs	35000	45000	45000	50000	50000	30000	50000	50000
RMS errors	0.0505	0.0461	0.0461	0.0471	0.0472	0.0301	0.0487	0.0237

Table 4. Predicted image values and RMS errors of the four models for the test set

Phone no.		6	11	16	19	28	RMS errors	
Evaluated Image value	S-C value	6.400	6.600	2.733	4.533	4.600		
	M-C value	3.630	2.541	3.681	3.163	1.089		
	Y-M value	5.081	2.074	2.281	1.815	1.711		
The combined NN model	-HN1 S-C	6.793	7.286	3.583	5.537	2.492		
	M-C	4.404	2.345	2.452	3.336	1.153	0.1325	
	Y-M	6.525	2.491	2.637	4.008	2.810		
	-HN2 S-C	6.026	7.329	3.254	5.358	2.140		
	M-C	5.397	3.337	3.505	5.076	1.121	0.1911	
	Y-M	6.996	3.844	3.353	5.638	2.898		
	-HN3 S-C	6.805	7.339	4.131	5.877	2.897		*0.1325*
	M-C	5.553	2.688	3.366	4.807	1.524	0.1871	
	Y-M	7.136	3.090	3.639	5.380	3.586		
	-HN4 S-C	6.930	7.315	3.748	5.592	3.324		
	M-C	5.622	2.199	2.829	3.810	1.905	0.1585	
	Y-M	7.132	2.410	2.950	4.360	4.075		
The S-C NN model	-HN1	6.529	7.158	4.227	5.364	2.184	0.1839	
	-HN2	5.960	6.969	4.107	5.275	1.967	0.1899	*0.1647*
	-HN3	6.437	7.299	4.310	5.262	2.655	0.1647	
	-HN4	6.562	7.313	4.392	5.303	2.592	0.1715	
The M-C NN model	-HN1	5.156	2.424	4.104	4.011	0.909	0.0859	
	-HN2	4.764	2.233	3.922	3.645	1.100	0.0614	*0.0614*
	-HN3	5.194	2.630	3.969	4.042	0.910	0.0868	
	-HN4	5.388	2.628	3.778	4.185	1.270	0.0971	
The Y-M NN model	-HN1	6.684	2.274	3.952	4.696	2.307	0.1940	
	-HN2	6.032	2.624	4.624	4.007	2.081	0.1765	*0.1765*
	-HN3	6.694	2.865	4.032	4.351	3.070	0.1975	
	-HN4	7.237	2.512	3.323	5.183	3.999	0.2455	

As shown in Table 4, the M-C NN model has the lowest RMS error (0.0614). This suggests that the M-C NN model has the highest prediction ability in determining the best design combination of form elements. The results obtained from the NN models can be used to help product designers to work out the best combination of form elements for a particular design concept represented by a set of product images. The product designer can focus on the determination of the desirable product images, and the NN models can help determine what combination of form elements can best match the desired product images. These design-supporting results provide useful insights in designing form elements of a product for reflecting the images of the product. To illustrate, Table 5 shows the best form combinations for a set of product images represented by the S-C, M-C, and Y-M values of 4. The combination of form elements given in Table 5 has its S-C value being 4.086, M-C value being 4.066, and Y-M value being 3.943, which is the closest among all form combinations.

Table 5. The best combination of product form elements

X_1	X_2	X_3	X_4	X_5	X_6	X_7	X_8	X_9

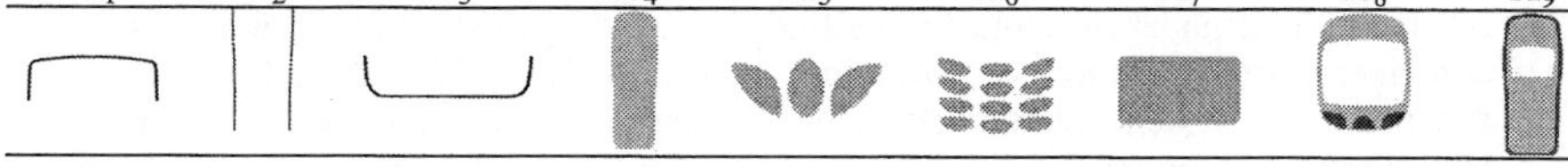

4 Conclusion

In this paper, we have presented a consumer oriented approach using NN models for transforming product forms into a set of product images perceived by consumers. To illustrate how such NN models can be built to help determine the best design combination of form element for matching given product images, we have conducted an experimental study on mobile phones. The NN models built can help product designers better understand consumers' perception of form elements, and work out the best combination of form elements with respect to desirable product images. Although mobile phones are chosen as an illustration of the approach, the approach can be applied to other products with various design elements.

References

1. Chuang, M.C., Chang, C.C., Hsu, S.H.: Perceptual Elements Underlying User Preferences Toward Product Form of Mobile Phones. International Journal of Industrial Ergonomics **27** (2001) 247-258
2. Guan, S.S., Lin, Y.C.: A Study on the Color and Style Collocation of Mobile Phones Using Neural Network Method. Journal of the Chinese Institute of Industrial Engineers **18** (2001) 84-94
3. Hsiao, S.W., Liu, M.C.: A Morphing Method for Shape Generation and Image Prediction in Product Design. Design Studies **23** (2002) 497-513
4. Kashiwagi, K., Matsubara, Y., Nagamachi, M.: A Feature Detection Mechanism of Design in Kansei Engineering. Human Interface **9** (1994) 9-16
5. Nagamachi, M.: Kansei Engineering: A New Ergonomics Consumer-Oriented Technology for Product Development. International Journal of Industrial Ergonomics **15** (1995) 3-10
6. Negnevitsky M.: Artificial intelligence. Addison-Wesley, New York (2002)
7. Nelson, M.: Illingworth WT. A Practical Guide to Neural Nets. Addison-Wesley, New York (1991)
8. Osgood, C.E., Suci, C.J.: The Measurement of Meaning. Urbana: University of Illinois Press (1957)

An Intelligent Simulation Method Based on Artificial Neural Network for Container Yard Operation

Chun Jin, Xinlu Liu, and Peng Gao

School of Management, Dalian University of Technology,
116023 Dalian, P.R.China
jinchun@dlut.edu.cn

Abstract. This paper presents an intelligent simulation method for regulation of container yard operation on container terminal. This method includes the functions of system status evaluation, operation rule and stack height regulation, and operation scheduling. In order to realize optimal operation regulation, a control architecture based on fuzzy artificial neural network is established. The regulation process includes two phases: prediction phase forecasts coming container quantity; inference phase makes decision on operation rule and stack height. The operation scheduling is a fuzzy multi-objective programming problem with operation criteria such as minimum ship waiting time and operation time. The algorithm combining genetic algorithm with simulation is developed. A case study is presented to verify the validity and usefulness of the method in simulation environment.

1 Introduction

Container yard operation is one of the important factors to keep container flow smoothly through the container terminal. In the past, researches analyzed the storage capacity of container yard and verified the capacity of operation machines [1]. But, considering the uncertainty and stochastic container flow, the un-manned, automation operation system is necessary for container yard operation to meet increasing container logistics [2].

Recently, intelligent control method has become effective to control dynamic operation system, and has combined with fuzzy artificial neural network (FANN), simulation and heuristic algorithm to enhance the operation efficiency and quality adaptively.

For this reason, it is significant to establish an intelligent method to analyze and control the handling operation of container yard, which makes both ship owners and terminals attain the maximum benefit by the minimum cost.

This paper aims to present an intelligent simulation method based on FANN to regulate container yard operation and determine operation scheduling with combination of genetic algorithm (GA) and discrete event system simulation. In addition, a case study is carried out to verify the validity and usefulness of our method.

F. Yin, J. Wang, and C. Guo (Eds.): ISNN 2004, LNCS 3174, pp. 904–911, 2004.

2 Architecture and Method of Intelligent Operation Control

2.1 Container Yard Operation

Yard Operation. The typical container yard operation of container terminal is shown in Fig. 1. The container operation is divided into container ship operation, container yard operation, and container gate operation.

Container ship operation means that quayside cranes load containers into (or unload containers from) a container ship staying on berth. The operation cycle time obeys *m*-Erlang distribution [1]. Here, *m* is the number of quayside cranes in operation.

Container yard operation includes container handling operation by transfer cranes and transit operation by trailers. The operation cycle time is the sum of container handling time from loading to unloading and its transit time.

Container gate operation means container arrival to the gate by truck. At the range of random operation days, the relationship between arrival time interval and the arrival quantity obeys exponential distribution [1].

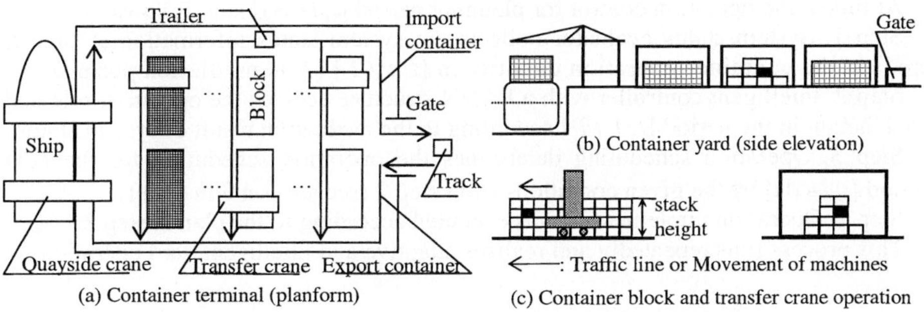

Fig. 1. Container yard operation on container terminal

Operation Regulation. For a ship staying in port, its turn-round time mainly includes waiting time and service time. For the benefit objective of the terminal, the waiting time should be reduced as much as possible. But to the service time of ship, or operation time for terminal, because it is concerned with operation capacity of port, i.e., the cost factor, it should be balanced in an acceptable level where both the operation cost can be reduced and the service for ship can be kept in good level. For this purpose, the regulation of operation is necessary.

To realize the operation objectives above in uncertainty and dynamic environment, the methods of operation regulation focus mainly on: (a) regulating operation rule including working shifts, scheduling rules of material handling machines in operation (e.g. FIFO), etc.; (b) regulating easiness for operation efficiency, such as regulation of stack height.

2.2 Architecture and Functions of Intelligent Operation Control

Fig. 2 shows the architecture of the intelligent operation control, which consists of four parts: system status evaluator, intelligent controller, operation scheduling module and operation implementation module.

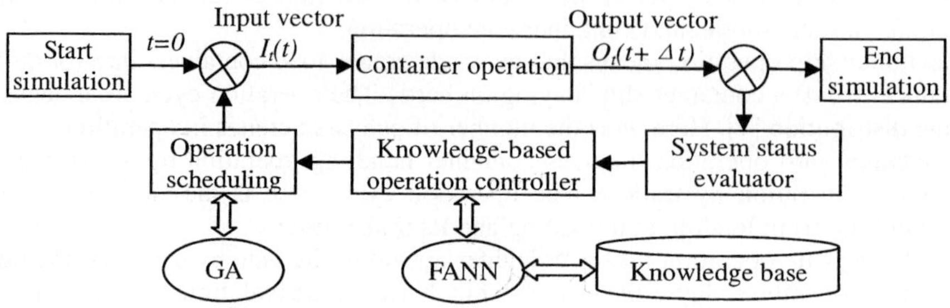

Fig. 2. Diagram of intelligent operation control

At time t, the operation control for planning period $[t, t+\Delta t]$ is as follows.

Step 1. System status evaluator collects the system status information at time t, predicts the container operation quantity in $[t, t+T_y]$ (T_y is prediction period).

Step 2. Intelligent controller with a FANN structure decides the operation rule and stack height in the period $[t, t+T_y]$ according to the evaluation result of system status.

Step 3. Operation scheduling determines the operation scheduling for the next period $[t, t+\Delta t]$ by the given conditions from step 2 (see section 3 in detail).

Step 4. Operation implementation is executed according to the plan at step 3.

This process runs repeatedly and realizes adaptive and intelligent yard operation.

2.3 Knowledge-Based Intelligent Operation Controller with FANN

The overall structure of the FANN is shown in Fig. 3. The control process is divided into two phases. Predicting phase: at time t, predicts container stock quantity at time $t+T_y$; Inference phase: determines the stack height and operation rule in the period $[t, t+T_y]$. There are two parallel ANNs in the second phase: stack height control ANN and operation rule control FANN [3]. The learning rule of these ANNs and FANNs is Back Propagation method.

Prediction ANN is used to forecast container operation quantity at time $t+T_y$, which composed by one input layer, two hidden layers and one output layer. The input layer includes eight units ($u_1, u_2 \ldots u_8$) whose values are computed by the system status evaluator. The output layer has one unit which produces the prediction result Q_t. Q_t is also the one of input units in the two other ANNs stated below.

Stack height control ANN is used to decide container stack height on yard until time $t+Ty$, which composed of one input layer, two hidden layers and one output layer. Input layer includes three units (Q_t, v_1, v_2): Q_t is the output of the prediction ANN; v_1 and v_2 are computed from the system status evaluator. The output layer has one unit which produces the decision result of stack height h_t.

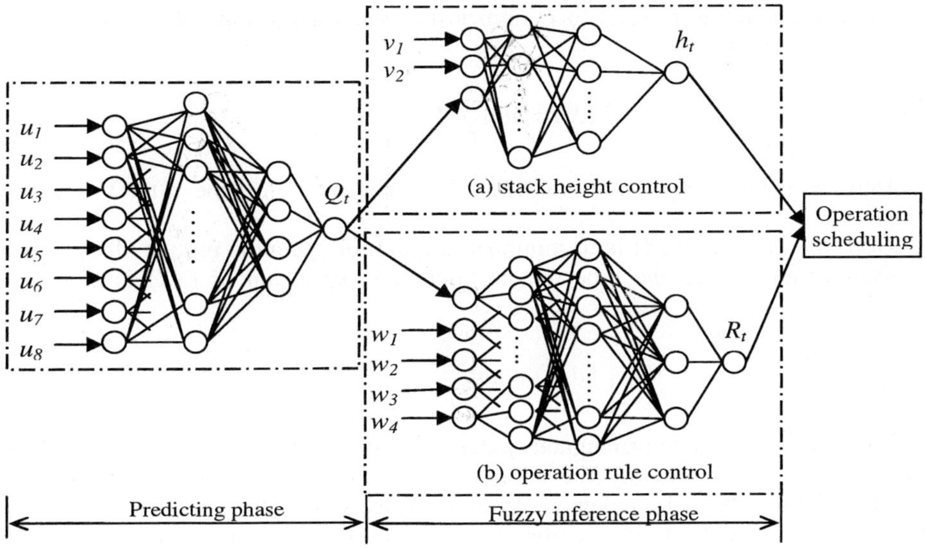

Fig. 3. Architecture of the whole FANN

Operation rule control FANN is used to decide operation rule in period $[t, t+T_y]$, which contains 5 layers: input layer, fuzzified layer, fuzzy rule layer, defuzzified layer and output layer. The input layer includes five units $(Q_t, w_1, \ldots, w_4)$. Q_t is from the output of the prediction ANN; others are from the system status evaluator. The output layer has one unit producing the decision result of R_t. R_t is used to determine parameters for operation scheduling in the next planning period.

3 Operation Scheduling with GA and Simulation

According to the operation rule and stack height decided in 2.3, the yard operation plan for next planning period can be scheduled. This operation scheduling problem is to determine operation series of yard machines, concerning with different combination of machines at different time. In this case, the following factors are considered: (a) uncertainty: there are many random factors in the operation; (b) computation complexity: this problem is a combinational optimization problem with NP-hard computation complexity; (c) ambiguousness: it concerns with the principles selecting feasible optimal results from a solution set. Here, simulation model is used to deal with (a); GA is used to deal with (b); and fuzzy set theory is adopted to select feasible solutions.

3.1 Formulation of the Optimization Problem

Decision variables are x_i, y_i, i=1,…,24, x_i represents the number of quayside cranes at time i by hour; y_i represents the number of transfer cranes at the same time. The decision vector $X = (x_1, \ldots, x_{24}, y_1, \ldots, y_{24})$ is used to show the scheduling result, i.e., the number series of operating quayside cranes and transfer cranes during the period time.

Objective function $f_1(X)$ is to minimize the total waiting time of arrival ships as shown in equation (1).

$$f_1(X) = \min \sum_{k=1}^{Ns} T_{ws}(k) \tag{1}$$

Here, Ns is the number of arrival ships, $T_{ws}(k)$ is the waiting time of the k-th arrival ship .

Objective function $f_2(X)$ is to minimize the sum of operation time of all operating transfer cranes. When the time unit is hour, the expression of $f_2(X)$ is shown in equation (2).

$$f_2(X) = \min \sum_{t=1}^{24} y_i \tag{2}$$

For $f_1(X)$ and $f_2(X)$, the ambiguous judgment is considered for the uncertain factors in decision-making process. We adopt a fuzzy satisfaction degree from selected Pareto solution set as the judgment on optimal decisions. Here, objective function $f_1(X)$ is transformed into its membership function (MF) $\mu_{f1}(X)$ shown in equation (3), and $f_2(X)$ is also transformed like so.

$$\mu_{f1}(X) = 1 - \frac{f_1(X) - f_{11}}{d_{f1}} \tag{3}$$

Here, f_{11} is the value of $f_1(X)$ when the $\mu_{f1}(X)$ is equal to 1, d_{f1} is the interval of $f_1(X)$ between the value of $\mu_{f1}(X)$ is 1 and 0.

Constraint function $g_1(X)$: the operating number of quayside cranes is not more than that of transfer cranes.

Constraint function $g_2(X)$: the truck waiting time for loading a container is less than the expected value.

Constraint function $g_3(X)$: the stocking quantity of containers is lower than yard storage capacity.

Constraint function $g_4(X)$: the handled quantity of containers should be equal to the planed quantity.

The four constraint functions $g_j(X)$ (j=1, 2, 3, 4) are transformed into their membership functions $\mu_{gj}(X)$ by means of the formulation shown in equation (3).

This fuzzy multi-objective programming (FMOP) problem with two fuzzy objective functions and four constraint functions is formulated into the following equations.

$$\lambda = \mu_D(\mu_f(X)) = \min\{\mu_{f1}(X), \mu_{f2}(X)\} \tag{4}$$

$$\left.\begin{array}{ll} \textit{Maximize} & \lambda \\ \textit{subject to} & \mu_{gj}(X) \geq \mu_{fi}(X) \end{array}\right\} (0 < \lambda \leq 1) \tag{5}$$

Here, i=1,2, j=1,2,3,4, λ is the value of integrated MF $\mu_D(X)$, and $\mu_D(X)$ is the minimum value of $\mu_{f1}(X)$ and $\mu_{f2}(X)$.The priority for objective functions is set to be $f_1(X)$, $f_2(X)$.

3.2 FMOP with GA and Simulation

For the FMOP problem, GA is used to compute the global integer solutions of scheduling. The GA construction is shown as follows [4].

For coding design, the decision variables $x_1,\ldots, x_{24}, y_1 ,\ldots, y_{24}$ are coded into 48 chromosomes and combined into one individual. The bit string structure of an individual is shown in Fig. 4.

0011	0101	……	1001	0111	1011	……	1111
x_1	x_2	……	x_{24}	y_1	y_2	……	y_{24}

Fig. 4. Example for bit string of an individual in GA

For GA operators, population with 100 individuals is set up. In fitness evaluation, fuzzy satisfaction degrees are computed from simulation results. Tournament selection, uniform crossover and random mutation strategies are used.

Programming algorithm based on GA and simulation is developed. The yard operation model is a discrete event system simulation model without the feature of differential functions, which is embedded into fitness evaluation stage in GA. According to the results from simulation for one individual X, the values of objective functions $f_1(X)$, $f_2(X)$ are computed. Then these values are transformed into their fuzzy satisfaction degrees $\mu_{f1}(X)$ and $\mu_{f2}(X)$.

4 Case Study

The case study of a container yard operation is to verify validity and usefulness of the intelligent operation regulation model (model A) in simulation environment. The main data of container yard and analysis conditions are shown in Table 1.

Table 1. Main parameters for simulation analysis of container yard

Parameter	Value	Parameter	Value
Yard storage capacity (1000TEU)	13.5	Number of berths	2
Past quantity: min-max (1000TEU)	7.8-12.1	Number of blocks	64
Quayside crane: num./cycle time(s)	6/102	Max. stocking time (*day*)	7
Transfer crane: num./cycle time(s)	8/90	Mean stocking time (*day*)	4
Trailer: num./speed (*m/min*)	36/175	Planning time unit (*hr*)	1
Slot number per block (row×col.)	4×15	Planning period Δt (*hr*)	24
Stack height scope (layer)	2-5	Prediction period T_v (*day*)	3

In Fig.5, regulation result of stack height and variation of container stocking quantity in analysis period of 30 days are shown. In order to analyze the regulation result of model A, Fig.6 shows variation of handling quantity of containers on ship side and gate side on the 17-th day. Fig.7 shows variation of the optimal arrangement number of transfer cranes by hour on the same day.

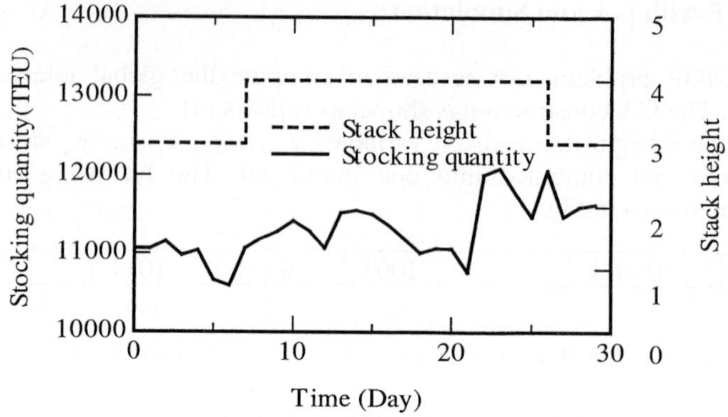

Fig. 5. Container stocking quantity and regulation result of stack height

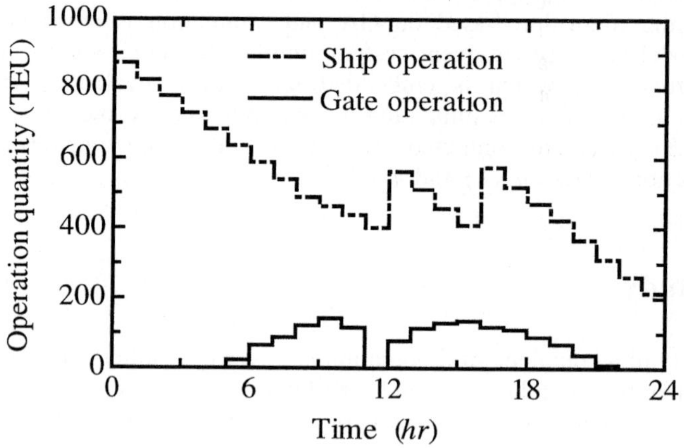

Fig. 6. Variation of ship operation and gate operation during one day

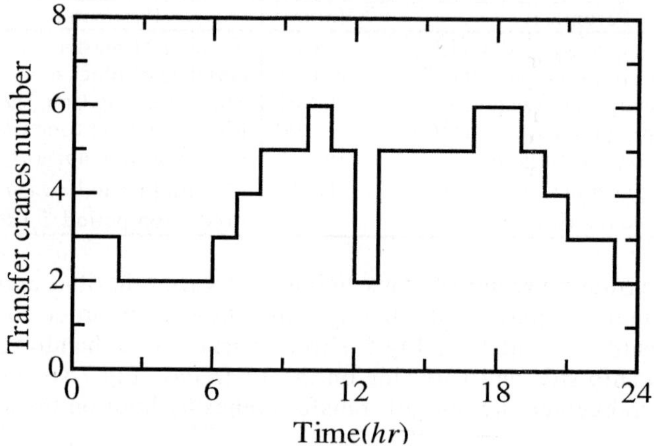

Fig. 7. Optimal number of transfer cranes for yard operation in model A

The planned number means the necessary number of machines meeting operation demand. It is seen that at 11 and 18 o'clock, with the increase of containers being handled on ship side and gate side, the operating number of transfer cranes is regulated to maximum value.

The comparison between model A and actual operation in 30 days has shown that at the same handled quantity, model A can shorten the total ship waiting time from 64 hours to 46 hours. It is indicated that there is still room to improve on the operation efficiency for present container yards.

5 Conclusions

In this research, the following conclusions can be drawn.

In our intelligent simulation model, in order to realize the objectives reducing ship waiting time and total operation time of transfer cranes, FANN is effective to forecast the operation quantity of containers, determine operation rule and control the container stack height adaptively.

It is clear that the algorithm combining GA with simulation is effective for optimal container yard operation scheduling problems with the features of uncertainty, computation complexity and ambiguousness.

By case study, it is testified and verified that the handling operation model with the function of intelligent control and optimal planning is effective to improve the container yard operation.

This intelligent control method can be applied to planning material handling system of container yard, and to regulating an actual yard operation for optimal schedule in real time environment.

References

1. Nisisaki, J., Kuwabara, A.: Container Yard Operation Programming Based on Simulated-annealing Algorithm. Proceedings of 7th Conference of Transportation and Logistics. Tokyo, Japan(1998)165-169
2. Takehara, T., Urasaki, G.: Automation Planning Method for Container Terminals (1st Report, Concept of Automation Container Terminal). Mitsui Shipbuilding Technique Report .164(1998)12-19
3. Schikoraa, P. F., Godfrey, M. R.: Efficacy of End-user Neural Network and Data Mining Software for Predicting Complex System Performance. Int. J. Production Economics. 84 (2003) 231–253
4. Yu, H., Liang, W.: Neural Network and Genetic Algorithm-based Hybrid Approach to Expanded Job-shop Scheduling. Computer & Industrial Engineering. 39 (2001) 337–356

A Freeway Traffic Incident Detection Algorithm Based on Neural Networks

Xuhua Yang, Zonghai Sun, and Youxian Sun

National Laboratory of Industrial Control Technology, Institute of Modern Control Engineering, Zhejiang University, Hangzhou, Zhejiang, P.R. China. 310027
xhyang@iipc.zju.edu.cn

Abstract. This paper proposes a novel freeway traffic incident detection algorithm. Two stages are involved. First, get the freeway traffic flow model based on BP neural networks and use the model to obtain the output prediction. The residual signals will be gotten from the comparison between the actual and prediction states. Second, a SOM neural networks is trained to classify characteristics contained in the residuals. Hence, based on the classification given by the SOM neural networks, traffic incidents can be detected. Both theory analysis and simulation research show that this algorithm is effective.

1 Introduction

This paper proposes a novel freeway traffic incident detection algorithm based on BP neural networks and SOM neural networks. The BP neural networks can approximate any nonlinear mapping relationship and the SOM neural networks have the ability to extract a system feature. Combining these two neural networks to detect the traffic incident can simply the modeling greatly and classify the traffic incident information quickly.

2 The Principle of Traffic Incident Detection

The principle of traffic incident detection is shown as figure 1.The input is the traffic parameters which can affect freeway's vehicle flow speed and vehicle flow density. (In this paper, vehicle flow speed is the vehicle flow space average speed). The output is the vehicle flow speed and the vehicle flow density. y is the actual output and $\hat{y}$ is the traffic flow model output prediction. This paper use the BP neural networks to get the freeway traffic flow model and use the model to obtain the output prediction.

Suppose both y and $\hat{y}$ are affected by noises, then

$$y = y_{real} + \xi \tag{1}$$

$$\hat{y} = \hat{y}_{real} + \hat{\xi} \tag{2}$$

F. Yin, J. Wang, and C. Guo (Eds.): ISNN 2004, LNCS 3174, pp. 912–919, 2004.

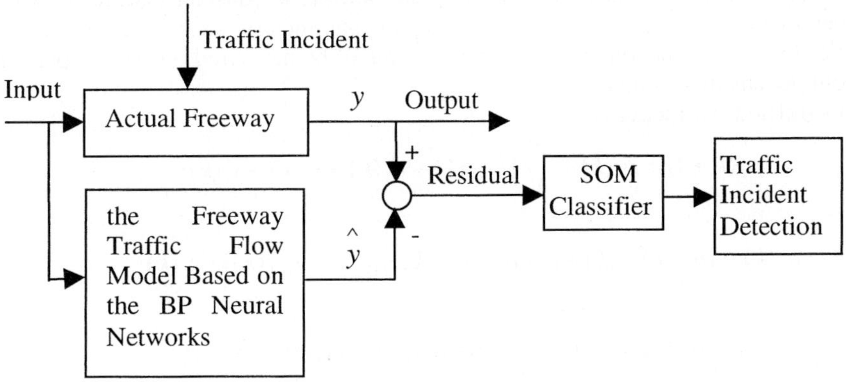

Fig. 1. The Principle of Traffic Incident Detection

In the formula (1) and (2), y_{real} is the real output, ξ is the measure noise, $\hat{y}_{real}$ is the traffic flow model's desired output prediction, $\hat{\xi}$ is the model noise. Suppose ξ and $\hat{\xi}$ are all noises with 0 mean value.

Define residual as formula (3)

$$\varepsilon = y - \hat{y} = (y_{real} - \hat{y}_{real}) + (\xi - \hat{\xi}) \tag{3}$$

The principle of traffic incident detection: When the freeway is in the normal traffic state, the traffic flow model can simulate the actual freeway primarily and $y_{real} \approx \hat{y}_{real}$,then the residuals should meet the distribution with 0 mean value. When there is traffic incident on the freeway, the traffic capacity will fall and the traffic model can not simulate the actual freeway and $y_{real} \neq \hat{y}_{real}$,then the residuals will mutate and should not meet the distribution with 0 mean value. The traffic incident often causes the accidental congestion and different level congestion corresponds to different level residuals. Therefore, using an SOM neural network to classify characteristics contained in the residuals can not only detect the traffic incident but also detect the level of the accidental congestion caused by the traffic incident.

3 The Freeway Traffic Flow Model Based on BP Neural Networks

The neural networks model is a black box model. For the purpose of using the system's information fully, we firstly analyze the traffic flow analytic model[1] which was proposed by Markos Papageorgiou:

Divide a freeway into N sections and each section's length Δ_i is several hundreds meters. In each section, the traffic state can be considered uniform approximately. Namely, the traffic volume, the vehicle flow speed and the vehicle flow density in a section are constant. In each section, the lane number is constant and the maximum number of entrance or exit is one. The traffic parameters, such as traffic volume, vehicle flow speed and vehicle flow density, can all be measured. Detection period T is about dozens of seconds.

This traffic flow model is

$$\rho_i(k+1) = \rho_i(k) + \frac{T}{\Delta_i}[q_{i-1}(k) - q_i(k) + r_i(k) - s_i(k)] \tag{4}$$

$$q_i(k) = \alpha\rho_i(k)v_i(k) + (1-\alpha)[\rho_{i+1}(k)v_{i+1}(k) - r_{i+1}(k)] - s_i(k) \tag{5}$$

$$v_i(k+1) = v_i(k) + \frac{T}{\tau}\{v[\rho_i(k)] - v_i(k)\} + \frac{T\xi}{\Delta_i}v_i(k)[v_{i-1}(k) - v_i(k)] - \frac{\gamma T}{\tau\Delta_i}\frac{\rho_{i+1}(k) - \rho_i(k)}{\rho_i(k+1) + \lambda} \tag{6}$$

where

$$v(\rho) = v_f \exp[-(1/b)(\rho/\rho_{cr})^b] \tag{7}$$

The formula (4)—(7) involves the following variables:

T sampling period;

$v_i(k)$ the vehicle flow speed of i th section at kT time;

$\rho_i(k)$ the vehicle flow density of i th section at kT time;

$q_i(k)$ the traffic volume from i th section to $(i+1)$ th section at kT time;

Δ_i the length of i th section;

$r_i(k)$ the on-ramp traffic volume for i th section at kT time;

$s_i(k)$ the off-ramp traffic volume for i th section at kT time;

Model parameters are $v_f, \rho_{cr}, b, \tau, \gamma, \xi, \lambda, \alpha$.

After analyzing the above analytic model, we can see that there exist nonlinear mapping relationship about vehicle flow speed and vehicle flow density as following:

$$v_i(k+1) = f_1[q_{i-1}(k), \rho_i(k), \rho_{i+1}(k), v_{i-1}(k), v_i(k), v_{i+1}(k), r_i(k), r_{i+1}(k), s_i(k)] \tag{8}$$

$$\rho_i(k+1) = f_2[q_{i-1}(k), \rho_i(k), \rho_{i+1}(k), v_{i-1}(k), v_i(k), v_{i+1}(k), r_i(k), r_{i+1}(k), s_i(k)] \tag{9}$$

f_1 and f_2 are relevant nonlinear mapping relationships.

Construct a BP neural networks with 9 inputs and 2 outputs. In this BP networks, Input variables are

$$q_{i-1}(k), \rho_i(k), \rho_{i+1}(k), v_{i-1}(k), v_i(k), v_{i+1}(k), r_i(k), r_{i+1}(k), s_i(k),$$

and output variables are $v_i(k+1)$, $\rho_i(k+1)$. Using the freeway's historical traffic data under the normal traffic state to train the BP networks can get the traffic flow model based on the BP neural networks. (This paper adopts the BP neural networks' learning algorithm in reference [2].)

4 SOM Neural Networks Classify the Residuals

In this paper, the residual vector is the input of SOM and the output layer adopt 2 dimensions neuron lattice. (This paper adopts the SOM neural networks' learning algorithm in reference [3].)SOM classify the residual vector's space into several subspace and each subspace correspond to one winning neuron in the output neuron lattice. So, the classification to the residuals can be achieved and the traffic incident can be detected.

The neighboring residual vector's subspace correspond to the neighboring neuron in the lattice because of the property of "topological ordering". Therefore, the SOM can help us to achieve "visual traffic state". Namely, the following two aspects:

(1) Observing the winning neuron in the lattice can detect not only the traffic incident but also the level of accidental congestion caused by the traffic incident.

(2) Observing the running trajectory of winning neurons can real-time trace the traffic state of the freeway. Namely, we can real-time observe traffic state transformation which is from the normal state to the incident state or from the incident state to the normal state.

4.1 Generating the Residuals

Construct a two dimensions vector which is composed of vehicle flow speed and vehicle flow speed density. Suppose the actual measure vector is $[v \quad \rho]^T$ and the traffic flow model prediction vector is $[\hat{v} \quad \hat{\rho}]^T$,then the residual vector is

$$[\varepsilon_v \quad \varepsilon_\rho]^T = [v \quad \rho]^T - [\hat{v} \quad \hat{\rho}]^T \tag{10}$$

4.2 SOM Neural Networks Classify the Residuals

Let the residual vector as the input of the SOM networks and the output layer adopt 10×10 two dimensions lattice. The classification involves two stages.

First, train the SOM and get the SOM model.

If we want to detect traffic incident of a freeway section, we should firstly train the SOM networks to get the SOM model. The training raw data is from this freeway section's traffic historical data including normal states and incident states. We can get the residual vector sequence by using the method in 4.1 and train the SOM .So, we can get the SOM model.

Second, apply the SOM model to the classification.

The traffic normal state and incident state correspond to different domain in the output neuron lattice. In practical application, substitution of the residual into the SOM model yields a winning neuron. From the neuron's position in the lattice, we can detect not only the traffic incident but also the level of accidental congestion caused by the traffic incident. Furthermore, observing the running trajectory of winning neurons can real-time trace the traffic state of the freeway. Namely, we can real-time observe traffic state transformation which is from the normal state to the incident state or from the incident state to the normal state.

5 Simulation Results

Consider a 1800 meters freeway which is divided into 3 sections and each section's length is 600 meters. There are on-ramp and off-ramp on the first and the second section. Suppose this freeway' traffic flow model is the formula (4)-(7) and each section has the same model parameters which are as following:

$\Delta_i = 0.5km$, $\alpha = 0.95$, $T = 12s$, $\xi = 1$, $\tau = 19.4s$, $\lambda = 1veh/km$

$\gamma = 34.7km^2/h$, $v_f = 98km/h$, $b = 3$, $\rho_{cr} = 32veh/km$

This paper researches the traffic detection on the second section.

5.1 Getting the Traffic Flow Model Based on BP Neural Networks

Generate sample data with the traffic flow analytic model (formula (4)-(7)) and use them to train the BP neural networks. So, we can get the traffic flow model based on the BP neural networks.

5.2 Getting Residual Vector Sequence

We need the residual vector sequence including normal traffic state's and incident traffic state's to train the SOM model.

(1) Getting the residuals of the normal traffic state

On the normal traffic state, he residual vector sequence should meet the distribution with 0 mean value. Let the vehicle flow speed residuals meet normal distribution with 0 mean and 1 variance and the vehicle flow density residuals meet normal distribution with 0 mean and 0.01 variance.

(2) Getting the residuals of the traffic incident state

Suppose the freeway's normal traffic state is in dynamic balance before the traffic incident and the balance traffic parameters are

$\rho_1(0) = 32.7veh/km$, $v_1(0) = 66.9km/h$

$\rho_2(0) = 33.7veh/km$, $\rho_3(0) = 32veh/km$

$v_3(0) = 68.3km/h$, $r_1 = 100veh/h$, $s_1 = 50veh/h$,

$r_2 = 610veh/h$, $s_2 = 500veh/h$

When there is traffic incident on the freeway, the traffic capacity will fall and the parameters of formula (7) will be changed. At the same time, traffic capacity falling means the effective traffic length of this freeway will fall. Namely, the Δ_i of formula (4) and (6) will be decreased.

Suppose the traffic incident can cause 3 different level of accidental congestion and the corresponding parameters' transformation is as following:

1) slight congestion

$$v_f = 70km/h\,,\ \rho_{cr} = 25veh/km\,,\ \Delta_i = 0.55km$$

2) middle congestion

$$v_f = 30km/h\,,\ \rho_{cr} = 15veh/km\,,\ \Delta_i = 0.50km$$

3) serious congestion

$$v_f = 15km/h\,,\ \rho_{cr} = 10veh/km\,,\ \Delta_i = 0.45km$$

After Updating corresponding parameters, we can use the traffic flow analytic model (formula (4)-(7)) to get the vehicle flow speed and the vehicle flow density of the incident state. So, we can obtain the residuals according to the principle of the figure 1.

5.3 Get the SOM Model

In this paper, the SOM's input is the 2-dimensional residual vector and the output layer adopt 10×10 2 dimensions lattice. Using the residual vector sequence to train the SOM can get the SOM networks model.

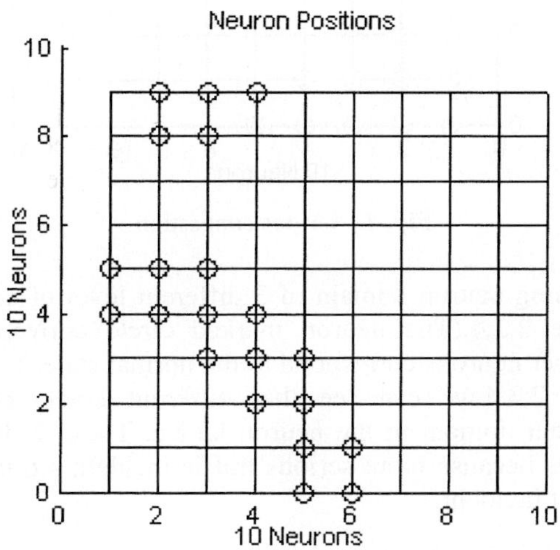

Fig. 2. Light congestion

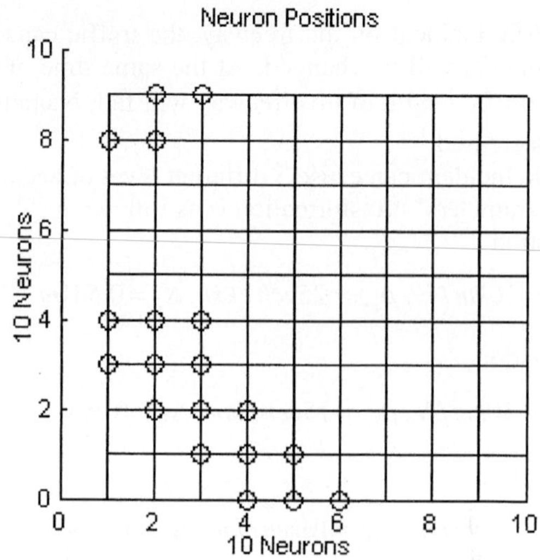

Fig. 3. Middle congestion

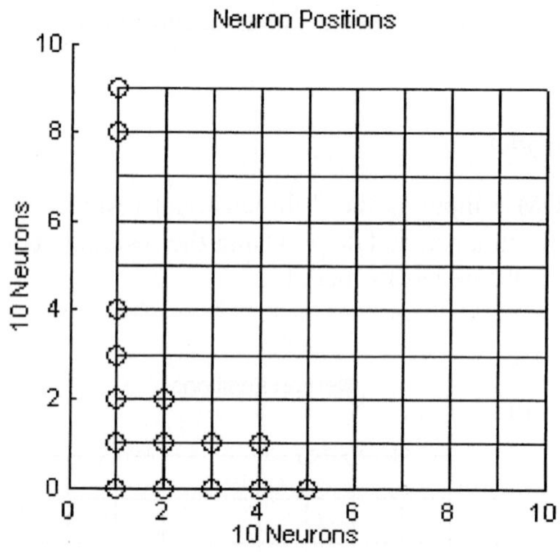

Fig. 4. Serious congestion

The possible winning neuron domain of 3 different level of accidental congestion are shown in figure 2,3,4.(The neuron marked circle corresponds to the traffic incident state and other neurons correspond to the normal state.)

Observing figure 2,3,4,we can see that different level accidental congestion correspond to different domain in the neuron lattice. These 3 domains have a little overlap in their edges because more serious traffic incident's transition state perhaps behave as more slight incident.

5.4 Practical Application

Substitution of the residual into the SOM model yields a winning neuron. From the neuron's position in the lattice, we can detect not only the traffic incident but also the level of accidental congestion caused by the traffic incident. Observing the running trajectory of winning neurons can real-time trace the traffic state of the freeway.

6 Conclusion

This paper proposes a novel freeway traffic incident detection algorithm based on the BP networks and SOM networks. Using this algorithm, we can detect not only the traffic incident but also the level of accidental congestion caused by the traffic incident. Furthermore, we can real-time trace the traffic state of the freeway by observing the running trajectory of winning neurons in the SOM neuron lattice. Both theory analysis and simulation research show that this algorithm is effective.

References

1. Shi Zhongke, Huang Huixian, Qu Shiru, Chen Xiaofeng: Traffic Control System Introduction, Science Publishing House, Beijing (2003)
2. Simon Haykin: Neural Networks: A Comprehensive Foundation. Tsinghua University Press, Pages:161 - 175
3. Kohonen T.: Self-Organized Formation of Topologically Correct Feature Maps, Biol Cyber (1982) 43:59-69

Radial Basis Function Network for Chaos Series Prediction

Wei Chi[1], Bo Zhou[2], Aiguo Shi[1], Feng Cai[1], and Yongsheng Zhang[1]

[1] Dep. of Navigation, Dalian Naval Academy, 116018, China
dl_tiger@163.com
[2] Dep. of Basics, Dalian Naval Academy, 116018, China
judyever@sina.com

Abstract. A method is described for using Radial Basis Function (RBF) network to predict the chaos time series. The structure of RBF network is introduced first, then the two-step training procedure for the network is proposed, that is the unsupervised learning in first layer using K-means clustering algorithm and the supervised learning in second layer using gradient-based methods. Results obtained for a chaos logistic map time series analysis is presented to test the effectiveness of the proposed method. This approach is shown to improve the overall reliability of chaos time prediction.

1 Introduction

Radial Basis functions emerged as a variant for artificial neural network in late 1980's. RBF's are embedded in a two layer neural network, where each hidden unit implements a radial activated function, the output units implement a weighted sum of hidden unit outputs. The input into an RBF network is nonlinear while the output is linear, due to these nonlinear approximation properties, RBF networks are able to model complex mappings, while perceptron neural networks can only model by means of multiple intermediary layers.

The paper is structured as follows. In Section 2, we explain the network topology. In section 3, we describe the training algorithms for RBF networks. In section 4, a experimental results for chaos time series prediction are provided, the conclusions of this article are presented in section 5.

2 Radial Basis Function Network Overview

When approximating time-series, the network input represent data samples at certain past time-laps, while the network has only one output representing a signal value as shown in Fig.1.

F. Yin, J. Wang, and C. Guo (Eds.): ISNN 2004, LNCS 3174, pp. 920–924, 2004.

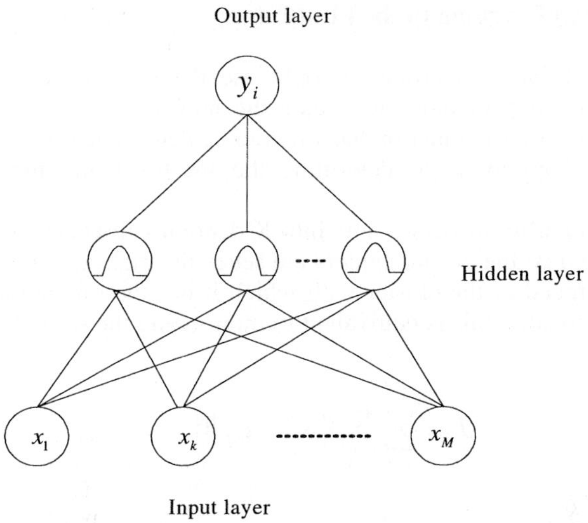

Fig. 1. A simplified architecture of a Radial Basis Function Network in time series modeling

For Radial Basis Function, the basis function is radial symmetry with respect to input, and the value is determined by the distance between data point and the RBF center. For instance, the Gaussian RBF:

$$\phi_j(\mathbf{x}) = \exp(-\|\mathbf{x}-\mathbf{c}_j\|/2\sigma^2) \tag{1}$$

Where $\mathbf{c}_j$ represents the center, σ the width, $\|\mathbf{x}-\mathbf{c}_j\|$ is the distance measure. For Euclidean distance:

$$\|\mathbf{x}-\mathbf{c}_j\| = \sum_{n=1}^{M}(x_n - (\mathbf{c}_j)_n)^2 \tag{2}$$

For RBF neural network, we expect that the function to be learned can be expressed as a linear superposition of a number of RBF.

3 Two-Step Training Strategies

In a RBF network, different layers perform different tasks. Therefore, it is useful to separate the optimization of the hidden unit and output layers of the network by using different techniques [1].

The parameters of the RBF networks are the centre and the influence field of the radial function and the output weight (between the intermediate layer's neurons and those of the output layer). So we can take a two-step training strategies to train them respectively.

3.1 Unsupervised Learning in the First Layer

This is to fix the basis functions by only use the knowledge of input data. For Gaussian RBF, it often includes to decide the number, locations and the width of RBF. The number of basis function has often to be determined by trials, the key issue in unsupervised learning is to determine the locations and the widths of basis functions.

The algorithm partitions data points into K disjoint subsets (K is predefined). The clustering criterion is: the cluster centers are set in the high-density regions of data; a data point is assigned to the cluster with which it has the minimum distance to the center ; Mathematically, this is equivalent to minimizing the sum-of-square clustering function,

$$J = \sum_{j=1}^{K} \sum_{n \in S_j} \| \mathbf{x}^n - \mathbf{c}_j \|^2 \tag{3}$$

Where S_j is the jth cluster containing N_j data points, $\mathbf{c}_j = \frac{1}{N_j} \sum_{n \in S_j} \mathbf{x}^n$ is the mean of the data points in cluster S_j.

3.2 Supervised Learning in the Second Layer

In order to determine the network weights in the second layer, supervised learning is adopted here. If we choose the sum-of-square error, it becomes a quadratic function optimization, which is easy to solve.
The network output after clustering:

$$\phi_j(\mathbf{x}) = \exp(-\| \mathbf{x} - \mathbf{c}_j \|^2 / 2\sigma^2) \tag{4}$$

For any $j \neq 0$, $\phi_j(\mathbf{x})$ is the Gaussian RBF, $\mathbf{c}_j$ is the centers obtained by clustering, $\phi_0(\mathbf{x}) = 1$ is the bias term.
The output of the RBF network is:

$$y^n = \sum_{j=0}^{M} w_j \phi_j^n \tag{5}$$

The sum-of-square error cost function:

$$E(\mathbf{w}) = \frac{1}{2} \sum_{n=1}^{N} (y^n - t^n)^2 \tag{6}$$

Where t^n is the target output at instant n. The minimization of the error is performed by iterative gradient-based methods. The partial derivative of the cost function with respect to the parameters is:

$$\frac{\partial E}{\partial w_j} = -(t - y)\phi_j(x) \tag{7}$$

In each step, the parameters are modified using the gradient, according to:

$$w_j(t+1) = w_j(t) - \eta \frac{\partial E}{\partial w_j} \tag{8}$$

Where η is a constant known as the step-size or the learning rate, as any gradient-descent based algorithm, the error back propagation algorithm suffers from slow convergence and high probability of converging to local minima. By using a momentum term u, the performance can be improved.

$$w_j(t+1) = w_j(t) - \eta \frac{\partial E}{\partial w_j} + u\Delta w_j \tag{9}$$

The two-step training avoids supervised learning simultaneously in two layers, and hence greatly simplifies the learning process [2].

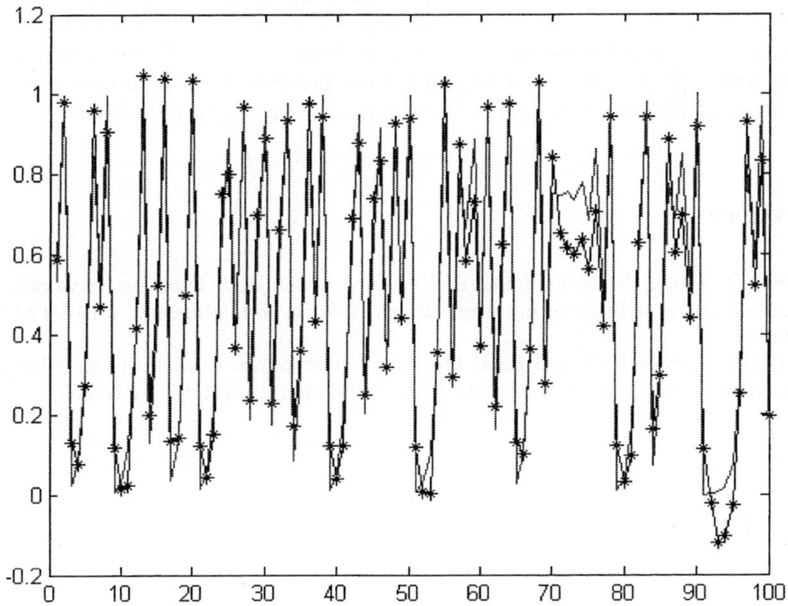

Fig. 2. The output of wavelet RBF network (in star line) and the target output (in real line) of the logistic system characterized by formula (10).

4 An Example of Chaos Time Series Prediction

We will show an example of using RBF for time series prediction. Time series prediction is to predict the system behavior based on its history.

Suppose the time course of a system is denoted as $\{x_1, x_2, \ldots\ldots, x_n\}$, where x_n is the system state at time step n. The task is to predict the system behavior at $n+1$ based on the knowledge of its history. This is possible for many problems in which system states are correlated over time.

Consider a simple example, the logistic map, in which the system state *x* is updated iteratively according to:

$$x_{n+1} = rx_n(1 - x_n) \tag{10}$$

Our task is to predict the value of *x* at any step based on its values in the previous two steps, i.e., to estimate x_n based on x_{n-1} and x_{n-2}. Choose *r=4* and the initial value of *x* to be *0.3.*

5 Conclusions

We have presented in this article an application of the RBF network on a Logistic Map time series prediction problems. Because there is only one hidden layer and the combination between the output layer and the hidden layer is linear, the structure of RBF is very simple. The two-step training process avoids supervised learning simultaneously in two layers, and hence the training process time is relatively short.

References

1. Ryad, Z., Daniel, R., Noureddine, Z.: Recurrent Radial Basis Function Network for Time-Series Prediction. Engineering Applications of Artificial Intelligence, vol.16. (2003) 453–463
2. Krzy,B. , Linder, T.:Radial Basis Function Networks and Complexity Regularization in Function Learning. IEEE Trans. on Neural Networks, vol. 9. (1998) 247-256

Feature Extraction and Identification of Underground Nuclear Explosion and Natural Earthquake Based on FM^{m}let Transform and BP Neural Network

Xihai Li, Ke Zhao, Daizhi Liu, and Bin Zhang

Section 602, Xi'an Research Institute of High Technology, Hongqing Town, Baqiao, Xi'an 710025, People's Republic of China
xihai_li@163.com

Abstract. Identification of underground nuclear explosion events and natural earthquake events is always worth to study because of the great military importance. The earthquake signal is non-stationary and FM^{m}let atom can delineate this signal more subtly, so we apply FM^{m}let Transform to extract the features of the nuclear explosion and natural earthquake signals, and then these features are recognized by the BP neural network. Experimental results indicate that the feature extraction and classification methods are effective and get good recognition results.

1 Introduction

In July and August 1958, a Committee of Experts met in Geneva to consider the design of an inspection system for nuclear tests[1]. In September 1996, a treaty banning underground nuclear test was subscribed in Geneva, seismic methods were the first chosen detection methods to monitor compliance with this treaty, and this led immediately to the problem of distinguishing between the signals from underground nuclear explosions and those of natural earthquakes. In order to distinguish the two kinds of signals, the first work to do is to extract features from these signals. There are different feature extraction methods to this identification problem, such as physical features, structural features and mathematical features. C.H.Chen[2] points out that mathematical features, especially mathematical features of frequency domain should be used to explain different characteristics of the earthquake. However, because the earthquake signals are non-stationary and non-linear, the traditional analysis methods (such as Fourier transform) cannot describe the time-frequency characteristics completely, the joint time-frequency method is first chosen to extract the features of earthquake signals[5]. FM^{m}let transform is a new time-frequency representation presented by Hongxing Zou in 2001[3], and it can delineate the earthquake signals more subtly than other time-frequency analysis methods, therefore, we first apply FM^{m}let transform to analyze earthquake signal to get the mathematical features of frequency domain, then we use BP neural network to identify the underground nuclear explosions and natural earthquakes.

F. Yin, J. Wang, and C. Guo (Eds.): ISNN 2004, LNCS 3174, pp. 925–930, 2004.

2 FM^{m}let Transform and Feature Extraction

2.1 FM^{m}let Transform

FM^{m}let transform is a parametric time-frequency representation, and some of the existing transforms like Fourier transform, STFT(including Gabor transform), wavelet transform, and chirplet transform may each be found to be a special case of the FM^{m}let transform with specific parameters [4]. Since many interesting physical and artificial process encountered in the real world yield linear and nonlinear time-varying phenomena, the FM^{m}let transform is therefore more flexible for delineating their time-varying structures. The FM^{m}let atom in a five-dimension (5-D) space is defined by[4]

$$q_{t_c, f_c, log(\sigma), r, m}(t) = \frac{1}{\sqrt{\sigma}} g(\frac{t-t_c}{\sigma}) \exp(j2\pi(1 + r(\frac{t-t_c}{\sigma})^m f_c(\frac{t-t_c}{\sigma}))) \quad (1)$$

where $g(\cdot) \in L^2(R)$denotes the window function. Clearly, an FM^{m}let atom, besides the window function, is governed by 5 parameters: time center t_c, frequency centerf_c, duration σ, chirplet r, and frequency modulation exponent m.

If window $g(\cdot)$ is a normalized Gaussian function, we will have the Gaussian FM^{m}let

$$g_{t_c, f_c, log(\sigma), r, m}(t) =$$

$$(\pi\sigma_t)^{\frac{-1}{4}} \exp(-\frac{1}{2}(\frac{t-t_c}{\sigma_t})^2) \exp(j2\pi(1 + r(\frac{t-t_c}{\sigma})^m f_c(\frac{t-t_c}{\sigma})) \quad (2)$$

Once the FM^{m}lets are used as atoms, the FM^{m}let transform of any square-intergrable signal $s(t) \in L^2(R)$ may be readily defined as

$$FM^m(t_c, f_c, log\sigma, r, m) = \langle s(t), q_{t_c, f_c, log(\sigma), r, m}(t) \rangle$$

$$= \frac{1}{\sqrt{\sigma}} \int_{-\infty}^{+\infty} s(v) \cdot g^*(\frac{v-t_c}{\sigma}) \cdot \exp(-j2\pi(1 + r(\frac{v-t_c}{\sigma})^m f_c(\frac{v-t_c}{\sigma}))) dv \quad (3)$$

where$\langle , \rangle$ denotes the inner product.

Due to the high dimension of parameter space of FM^{m}let, direct implementation of FM^{m}let transform, although straightforward conceptually, is not trivial in general. In order to circumvent this difficulty, an *ad hoc* decomposition algorithm for FM^{m}let transform based on so-called "matching-pursuits" is designed[4]. After each iteration, an FM^{m}let atom that best matched the dominating component of residual signal is selected. The decomposition process will stop when certain criteria such as the ratio between the energy of remainder and that of original signal(or the predetermined number of total iterations) is satisfied.

2.2 Feature Extraction

Because an FM^{m}let atom is governed only by five parameters, therefore FM^{m}let transform has great ability of feature extraction, and the features extracted by FM^{m}let transform have obvious physical characteristics comparing with the features obtained by AR model, MA model and ARMA model.

To each short-period sample of underground nuclear explosion and natural earthquake sets, 8 decompositions are carried out by FM^{m}let transform, and 8 components are obtained (the ratio between the energy of remainder and that of original signal is trivial after 8 decomposition), each component has five parameters:amplitude of signal a, variance of Gaussian window σ_0, frequency center f_0, chirplete r, frequency modulation exponent m.

3 Recognition Based on BP Neural Network

Artificial neural network (ANN) is a kind of physical simulation of human brain. Of all the ANN models, BP network is one of the widely used models. In 1989, Robert Hecht-Nielson proved that a 3-layer BP neural network can fulfill the map from n dimension to m dimension. To reduce the structure redundancy of BP network, we choose a 3-layer BP network to recognize underground nuclear explosions and natural earthquakes. The neuron number of hidden layer of BP network is always worth to study, in general case, it has essential relation with the problem to be solved, the number of training samples and the neuron number of input and output layer. In this paper, we choose the neuron number of hidden layer according to the formula of paper [6]. The formula is $J = \sqrt{I+K} + b$, where J is the neuron number of hidden layer, I the input neuron number,K output neuron number, b a constant between [1,10].

For nuclear explosion sample set and natural earthquake sample set, after FM^{m}let transform, each sample has 8 components and each component has 5 parameters, thereafter, there are 40 parameters totally. We divide these parameters into 3 cases to discuss.

(1) Use all the parameters as features to train the BP network. At this time, the input layer neuron number is 40, output is 1, then the neuron number of hidden layer $J = \sqrt{40+1} + b = 6.4 + b$.

(2) Use 5 parameters of each decomposition as features to train the BP network. At this time, the input layer neuron number is 5, output is 1, then the neuron number of hidden layer $J = \sqrt{5+1} + b = 2.45 + b$.

(3) Use 5 variants of 8 decomposition respectively as features to train BP network, that is using amplitude of signal a, variance of Gaussian window σ_0,frequency center f_0, chirplete r, frequency modulation exponent m as features respectively. At this time, the input layer neuron number is 8, output is 1, then the neuron number of hidden layer $J = \sqrt{8+1} + b = 3 + b$.

To compare the influence of different b, we suppose b equals to 1, 3, 5, 7, 9 respectively, and then calculate the neuron number of hidden layer J of 3 cases. Choose 80 samples from underground nuclear explosion set and natural earthquake set respectively to train BP neural network, and the rest 20 samples (40 samples totally) are tested as testing samples. The recognition results are listed in Table 1, 2, 3 (where N denotes nuclear explosion, E denotes natural earthquake, T denotes total recognition rate).

Table 1. Testing identification rate of the first case

J	4	6	8	10	12
$N\%$	75	80	85	80	85
$E\%$	75	80	85	80	85
$T\%$	75	75	75	75	75

Table 2. Testing Identification of the second case

D	J	4	6	8	10	12
1	$N\%$	70	70	75	70	75
1	$E\%$	85	85	85	90	85
1	$T\%$	77.5	77.5	80	80	80
2	$N\%$	75	80	75	80	75
2	$E\%$	70	70	65	65	70
2	$T\%$	72.5	75	70	72.5	72.5
3	$N\%$	70		75		75
3	$E\%$	85		60		75
3	$T\%$	77.5		67.5		75
4	$N\%$	30		25		45
4	$E\%$	80		85		65
4	$T\%$	55		55		55

From Table 1, we find that when all parameters are chosen as features, the testing recognition rate is not high, and at the same time the training recognition rate is almost 100%. This means that overfitting phenomenon occurs, the reason is that training samples are few, but input feature number is too large. So, all parameters are chosen as features is not a favorite method for recognition.

From Table 2, we find that only the first five parameters of the first decomposition are chosen as features, the testing recognition rate is relatively high, and the b has little influence to the recognition. when the parameters of the second decomposition, the third, the forth decomposition are chosen as features, the testing recognition rate is not good, the reason is that the major energy and shape of the earthquake signal is governed by the five parameters of the first decomposition, and these reflect primary characteristics of these two kinds of earthquake signals. So, the five parameters of the first decomposition are favorite features for the recognition.

Table 3. Testing Identification rate of the third case

para	J	4	6	8	10	12
a	$N\%$	85	95	95	90	90
a	$E\%$	80	65	75	70	75
a	$T\%$	82.5	80	85	80	82.5
σ_0	$N\%$	95	80	75	85	95
σ_0	$E\%$	55	65	75	65	65
σ_0	$T\%$	72.5	72.5	75	75	80
f_0	$N\%$	90	55	65	70	60
f_0	$E\%$	60	80	75	75	75
f_0	$T\%$	75	67.5	70	72.5	67.5
r	$N\%$	55	65	55	70	65
r	$E\%$	90	80	85	85	75
r	$T\%$	72.5	72.5	70	77.5	70

From Table 3, we find that only four parameters can calculate the recognition results, the parameter m can not calculate the recognition result validly because this parameter basically equals zero.We also find that the influence of b is little to the recognition. To these four parameters, only when the amplitude of signal a is chosen as features, the recognition rate is relatively high, the reason is that the amplitude of nuclear explosion signal and natural earthquake signal has great difference and the difference of other parameters is not obvious as the amplitude of the signals.

4 Concluding Remarks

In this paper, we first obtain some new parameters by FM^{m}let transform, then we analyze the performance of these parameters as the features of underground nuclear explosions and natural earthquakes. Experimental results and the corresponding analysis provided suggest that amplitude of the earthquake signals and the five parameters of the first decomposition are good remedies for seismological pattern recognition pathology.

Acknowledgments. The authors would like to express their gratitude to Dr. Hongxing Zou for his valuable advice. Thanks also to the National Natural Science Foundation for its support(No. 40274044).

References

1. Douglas, A.: Seismic Source Identification: A Review of Past and Present Research Efforts. Lecture Notes of Dr. David Booth ,Zhang Jiajie (2001)
2. C.H. Chen: Application of Pattern Recognition in Earthquake interpretation. In: K.S. Fu. (ed.): Application of Pattern Recognition, CRC Press, Inc., Boca Raton, Florida (1982) 157-175

3. Hongxing Zou, Qinghai Dai, Renming Wang, and Yanda Li: Parametric TFR via Windowed Exponential Frequency Modulated Atoms, IEEE Signal Processing Letters, Vol.8 (2001) 140-142
4. Hongxing Zou: Automatic Recognition of Underground Nuclear Explosion and Natural Earthquake(in Chinese), The Second Artillery Institute of Engineering, Xi'an (1995)
5. Ke Zhao, Daizhi Liu, et.al.: Feature Extraction on Time-frequency Plane for Seismic Pattern Recognition of Nuclear Explosion, Nuclear Electronics & Detection Technology(in Chinese), Vol.20 (2000) 272-278
6. Liming Zhang: Model of Artifical Neural Network and its Application(in Chinese), Fudan University (1992)

A Boosting-Based Framework for Self-Similar and Non-linear Internet Traffic Prediction

Hanghang Tong[1], Chongrong Li[2], and Jingrui He[1]

[1] Department of Automation, Tsinghua University, Beijing 100084, China
{walkstar98, hejingrui98}@mails.tsinghua.edu.cn

[2] Network Research Center of Tsinghua University, Beijing 100084, China
licr@cernet.edu.cn

Abstract. Internet traffic prediction plays a fundamental role in network design, management, control, and optimization. The self-similar and non-linear nature of network traffic makes highly accurate prediction difficult. In this paper, a boosting-based framework is proposed for self-similar and non-linear traffic prediction by considering it as a classical regression problem. The framework is based on Ada-Boost on the whole. It adopts Principle Component Analysis as an optional step to take advantage of self-similar nature of traffic while avoiding the disadvantage of self-similarity. Feed-forward neural network is used as the basic regressor to capture the non-linear relationship within the traffic. Experimental results on real network traffic validate the effectiveness of the proposed framework.

1 Introduction

Internet traffic prediction plays a fundamental role in network design, management, control, and optimization [11, 12, 14]. Prediction on large time scale is the base of long term planning of Internet topology. Dynamic traffic prediction is a must for predictive congestion control and buffer management. Highly accurate traffic prediction helps to make the maximum use of bandwidth while guaranteeing the quality of service (QoS) for Variable-Bite-Rate video which has a stringent requirement on delay and cell rate loss (CRL). Traffic prediction on different links can also be used as a reference to route. Layered-encoded video applications can determine whether to transmit the enhanced layer according to prediction results.

Essentially, the statistics of network traffic itself determines the predictability of network traffic [2, 9, 11, 12, 14]. Two of the most important discoveries of the statistics of Internet traffic over the last ten years are that Internet traffic exhibits self-similarity (in many situations, also referred as long-range dependence) and non-linearity. Since Will E. Leland's initiative work in 1993, many researchers have dedicated themselves to proving that Internet traffic is self-similar [8, 12]. On the other hand, Hansegawa et al in [5] demonstrated that Internet traffic is non-linear by using

F. Yin, J. Wang, and C. Guo (Eds.): ISNN 2004, LNCS 3174, pp. 931–936, 2004.

surrogate method [16]. The discovery of self-similarity and non-linearity of network traffic has brought challenges to traffic prediction [9, 11, 12, 14].

In the past several decades, many methods have been proposed for network traffic prediction. To deal with the self-similar nature of network traffic, the authors in [15] proposed using FARIMA since FARIMA is a behavior model for self-similar time series [3]; the authors in [17] proposed predicting in wavelet domain since wavelet is a natural way to describe the multi-scale characteristic of self-similarity. While these methods do improve the performance of prediction for self-similar time series, however, they are both time-consuming. To deal with the non-linear nature of network traffic, Artificial Neural Network (ANN) is probably the most popular method. While ANN can capture any kind of relationship between the output and the input theoretically [5, 7, 10], however, it might suffer from over-fitting [4]. Another kind of prediction method for non-linear time series is support vector regression (SVR) [10] which is based on structural risk minimization. However, the selection of suitable kernel functions and optimal parameters is very difficult [5].

In this paper, we propose a boosting-based framework for predicting traffic which exhibits both self-similarity and non-linearity (BBF-PT), by considering traffic prediction as a classical regression problem. On the whole, BBF-PT is based on AdaBoost [13] and follows R. Bone's work [1]. To take advantage of self-similarity while avoiding its disadvantage, Principle Component Analysis (PCA) is adopted as an optional step. BBF-PT takes feed-forward neural network (FFNN) as the basic learner to capture the non-linear characteristic of network traffic and makes use of the boosting scheme to avoid over-fitting. BBF-PT itself is straightforward and can be easily incorporated with other kind of basic regressors. Experimental results on real network traffic which exhibits both self-similarity and non-linearity demonstrate the effectiveness of the proposed framework.

The rest of this paper is organized as follows: in section 2, we present our framework (BBF-PT) in detail; experimental results are given in section 3; finally, we conclude the paper in section 4.

2 Boosting-Based Framework for Predicting Traffic (BBF-PT)

2.1 Problem Definition

By denoting network traffic as a time series: $X = (x_i : i = 0, 1, 2, \cdots)$. The prediction problem can be defined as follows [2]: given the current and past observed values $X_i = (x_{i-p+1}, \cdots, x_{i-1}, x_i)$, predict the future value x_{i+q}, where p is the length of history data used for prediction and q is the prediction step.

In practice, the traffic prediction problem can be considered as a classical regression problem under the assumption that $x_i \in \zeta_2$ $(i = 1, 2, \cdots)$ [2]. That is, the prediction of x_{i+q} can be written as:

$$\hat{x}_{t+q} = \underset{y \in \xi_2}{\operatorname{argmin}} E\{(y - x_{t+q})^2\} . \tag{2}$$

2.2 Flowchart of BBF-PT

As a general method for improving the accuracy of some weak learning algorithms, boosting has been shown to be very effective for classification [1, 6, 13]. When applied to a regression problem, there are mainly two classes of methods: 1) modifying some specific steps of the existing boosting algorithms for classification so that they are suitable for a regression problem [1, 6]; 2) converting a regression problem to a classification problem by introducing an additional variable [13]. In this paper, we adopt the first method and leave the latter for future work.

How to exploit the correlation structurc is the key problem in traffic prediction [11]. For self-similar traffic, it means: 1) containing more past observed values will be helpful to improve the prediction accuracy, however, it also requires more processing time; 2) traffic exhibits sustained burstiness, generally resulting in bigger peak prediction error [17]. To address these problems, we propose using Principle Component Analysis (PCA) as an optional step so that: 1) the same length of history data can contain more "information" about the past observed values and thus help to improve the prediction accuracy while not increasing the processing time drastically (here, the additional processing time is for PCA); 2) De-correlation on the input data might help to reduce the peak prediction error. Based on Ada-Boost [13] and following R. Bone's work [1], the flowchart of BBF-PT can be given as follows:

The flowchart of BBF-PT

1. Prepare a training set for regression. For every example in the training set, the input is $X_i = (x_{i-p+1}, \cdots, x_{i-1}, x_i)$ ($X_i \in R^p$), and the output is $Y_i = x_{i+q}$ ($Y_i \in R$), where $i = 1, 2, \cdots, N$ and N is the total number of examples.
2. Initialize weight distribution for all examples in the training set: $D_1(i) = 1/N$.
3. Iterate until the stopping criterion is satisfied. In every iteration $t\ (t = 1, 2, \cdots, T)$:
   - Train a weak regressor h_t using distribution D_t on the training set;
   - Get a weak regressor $h_t : R^{N_p} \to R$;
   - Evaluate the error $l_t(i)$ for every example i in the training set using the weak regressor h_t;
   - Compute the training error ε_t of h_t;
   - Choose $\alpha_t \in R$ to measure the importance of h_t;
   - Update the weight distribution of the examples in the training set: $D_t \to D_{t+1}$;
   - Judge whether the stopping criterion is satisfied.
4. Combine the weak regressors: $\{h_t, t=1, 2, \cdots, T\} \to H$.

Optional: Perform PCA on the input data in the training set; store the corresponding principle axis $u_i (i = 1, 2, \cdots, N_p)$; and take the projection of X_i on $u_i (i = 1, 2, \cdots, N_p)$ as the actual input for both training and testing.

2.3 The Details of BBF-PT

The necessary modifications to make Ada-Boost for classification suitable for a regression problem include: 1) the form of the basic learner (here the basic learner is the basic regressor); 2) the way to evaluate the error information for every example; 3) the scheme of combining weak learners.

The basic regressor: essentially, BBF-PT can use any kind of basic regressors. In this paper, we adopt feed-forward neural network (FFNN) [5, 7, 17] to capture the non-linearity within the network traffic.

Evaluating the error information: There are three basic forms to evaluate the error information for examples in the training set, namely linear, squared and saturated [1, 6]. We adopt the linear form in this paper.

Computing ε_t : the training error ε_t of h_t can be computed as $\varepsilon_t = \sum_{i=1}^{N} D_t(i) \cdot L_t(i)$.

Choosing α_t and updating weight distribution: BBF-PT takes the same scheme as Ada-Boost [13] to measure the importance h_t and update weight distribution.

Stopping criterion: In Ada-Boost, the iteration will stop when $\varepsilon_t \geq 0.5$. In BBF-PT, we set a maximum iteration number $T_{\max}$ to avoid slow convergence. That is, if $\varepsilon_t \geq 0.5$ or $t > T_{\max}$, the iteration process will stop.

Combining weak regressors: There are mainly two methods to combine the weak regressors: weighted mean and weighted median. Since the weighted median method is less sensitive than the weighted mean method [1, 6], it is adopted in BBF-PT.

3 Experimental Results

The network traffic that we use is the JPEG and MPEG version of the "Star Wars" video which is widely used to examine the performance of the network traffic prediction algorithms [7]. In our experiment, we divide the original traffic into some traces with equal length 1000. Then we make use of variance-time [8] and surrogate method [16] to test self-similarity and non-linearity of a given trace, respectively. Those exhibiting both self-similarity and non-linearity are selected. Each trace is normalized to [0,1] for comparison simplicity.

There are a set of parameters and operations that need to be set in BBF-PT:

- If PCA is not performed, the length of history data p used for predicting is set to be 10; else p is set to be 20 and N_p is 10;
- At the current stage, we are concerned with one-step prediction so q is 1;
- We used the first 50% data in each trace to compose the training set;
- The FFNN contains three layers and composes of 10 hidden neurons and 1 extra neuron for the output layer;
- The transfer functions for the hidden layer and the output layer in FFNN are sigmoid and linear respectively;

- ♦ The weights of FFNN are initialized randomly;
- ♦ FFNN is trained by standard back-propagation algorithm;
- ♦ The maximum iteration number T_{max} is set to be 30.

The mean-square-error (MSE) of the prediction results obtained by BBF-PT with and without PCA is listed in figure 1. It is compared with the results obtained by Feed-Forward Neural Network (mean results over 30 runs) without boosting. From figure 1, it can be seen that: 1) boosting does improve the prediction accuracy; and 2) PCA further improves the prediction performance.

We use the maximum absolute prediction error on the testing set of a given trace to measure the peak prediction error. The results obtained by BBF-PT with and without PCA are listed in figure 2. It can be seen that in the most cases, the PCA step can help to reduce the peak prediction error.

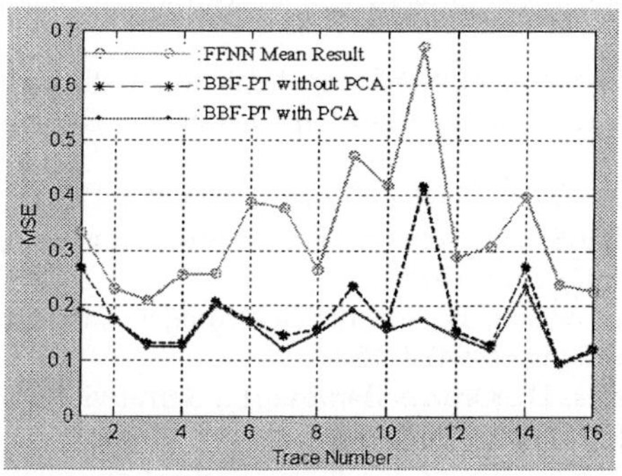

Fig. 1. MSE of prediction results

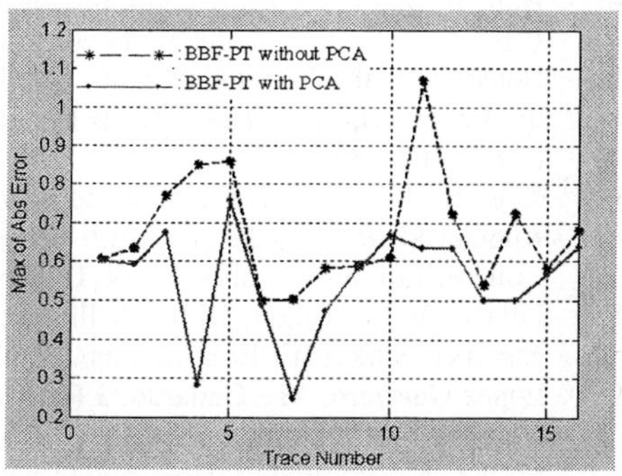

Fig. 2. Max. error of prediction results

4 Conclusion

In this paper, we have proposed a boosting-based framework for self-similar and non-linear network traffic prediction by considering it as a classical regression problem. The framework is based on Ada-Boost on the whole and makes use of FFNN to capture the non-linearity within traffic at the current stage. We also propose using PCA as an optional step to take advantage of self-similarity while trying to reduce the peak prediction error. Experimental results demonstrate the effectiveness of the proposed scheme. Future work includes: 1) incorporating other kinds of basic regressors into our framework; 2) extending the current method to multi-step prediction; 3) considering network traffic prediction from a classification point of view.

Acknowledgements. This paper is supported by National Fundamental Research Develop (973) under the contract 2003CB314805.

References

1. Bone, R., Assaad, M., Crucianu, M.: Boosting Recurrent Neural Networks for Time Series Prediction. Proceedings of the International Conference on Artificial Neural Networks and Genetic Algorithms (2003)
2. Brockwell, P., Davis, R.: Time Series: Theory and Methods. Springer-Verlag, New York, 2nd edition (1991)
3. Gripenberg, G., Norros, I.: On the Prediction of Fractional Brownian Motion. Journal of Applied Probability (1996) 400-410
4. Hall, J., Mars, P.: The Limitations of Artificial Neural Networks for Traffic Prediction. IEE Proceedings on Communications (2000) 114-118
5. Hansegawa, M., Wu, G., Mizuno, M.: Applications of Nonlinear Prediction Methods to the Internet Traffic. The 2001 IEEE International Symposium on Circuits and Systems (2001) 169-172
6. Kegl, B.: Robust Regression by Boosting the Median. COLT/Kernel (2003) 258-272
7. Khotanzad, P., Sadek, N.: Multi-Scale High-Speed Network Traffic Prediction Using Combination of Neural Network. IJCNN (2003) 1071-1075
8. Leland, W.E., Taqqu, M.S., Willinger, W., Wilson, D.: On the Self-Similar Nature of Ethernet Traffic. IEEE/ACM Tran. on Networking (1994) 1-15
9. Lopez-Guerrero, M., Gallardo, J.R., Makrakis, D., Orozco-Barbosa, L.: Sensitivity of a Network Traffic Prediction Algorithm. PACRIM (2001) 615-618
10. Muller, K.: Predicting Time Series with Support Vector Machines. Proceedings of the International Conference on Artificial Neural Network (1997) 999-1004
11. Ostring, S., Sirisena, H.: The Influence of Long-rang Dependence on Traffic Prediction. IEEE ICC (2001) 1000-1005
12. Sang, B., Li, S.: A Predictability Analysis of Network Traffic. Proceeding of INFOCOM (2000) 343-351
13. Schapire, R.E.: The Boosting Approach to Machine Learning: an Overview. MSRI Workshop on Nonliear Estimation and Classification (2002)
14. Shah, K., Bohacek, S., Jonckheere, E.: On the Predictability of Data Network Traffic. Proceedings of the 2003 American Control Conference (2003) 1619-1624
15. Shu, Y., Jin, Z., Zhang, L., Wang, L.: Traffic Prediction Using FARIMA Models. IEEE ICC (1999) 891-895
16. Small, M., Yu, D., Harrison, R.G.: Surrogate Test for Pseudoperiodic Time Series Data. Physical Review Letter (2001)
17. Wang, X., Shan, X.: A Wavelet-based Method to Predict Internet traffic. IEEE International Conference on Communications, Circuits and Systems and West Sino Expositions (2002) 690-694

Traffic Flow Forecasting Based on Parallel Neural Network*

Guozhen Tan and Wenjiang Yuan

Department of Computer Science and Engineering
Dalian University of Technology, 116024, China
gztan@dlut.edu.cn

Abstract. In Intelligent Transportation Systems (ITS), traffic flow forecasting is important to Traffic Flow Guidance System (TFGS). However most of the traffic flow forecasting models cannot meet the requirement of TFGS. This paper presents a traffic flow forecasting model based on BP neural network according to the correlation theory. This model greatly reduces the size of input patterns. Meanwhile, a new parallel training algorithm based on training set decomposition is presented. This algorithm greatly reduces the communication cost. Experiment results show that the new algorithm converges faster than traditional one, and can meet practical requirement.

1 Introduction

As an important aspect of Intelligent Transportation Systems (ITS), traffic flow guidance is considered as an optimum way to improve traffic efficiency and mobility. The basis of traffic flow guiding is real-time traffic flow forecasting. Scholars have advanced many methods for short-term traffic flow forecasting [1], [2]. At present, the algorithms for traffic flow forecasting can be divided into two categories: accurate mathematical model based algorithm and non-model algorithm [1]. Because of the urban traffic flow's time-varying volatility and nonlinearity, it's hard to be described by an accurate mathematical expression, all of these results in most algorithms of the upper two categories such shortcomings: slow convergent speed, low precision and bad adaptability, consequently cannot meet practical requirements.

But as one of the non-model algorithm, artificial neural network has many virtues such as good adaptability, high precision and needlessness of accurate mathematical expression, and has been widely used in many fields [3], [4], [5]. Error back-propagation algorithm is an effective algorithm for training neural network. The neural network trained by this algorithm is named as BP neural network. BP neural network is widely used in information forecasting. But in

* This work was supported in part by Grand 60373094 of National Natural Science Foundation of China and Grand 2002CB312003 of High Tech. Research and Development (973) Programme, China.

F. Yin, J. Wang, and C. Guo (Eds.): ISNN 2004, LNCS 3174, pp. 937–942, 2004.

practical traffic flow forecasting, because of the mutual influence between different road sections, the size of neural network's input pattern is greatly expanded [6], consequently the training time of the whole network is greatly increased, and can't meet the requirement of practical traffic flow forecasting.

Parallel processing has been proved to be a good method to reduce training time [4], [7]. Neural network's parallel training algorithms can be divided into two categories: node (neuron) parallelization and data parallelization [8]. Exploring from neural network' structure, there are five levels of parallelism: network level, training set level, layer level, neuron level and synapse level [9]. The paper [4] presented a parallel training algorithm based on search space partition, and effectively conquered the shortcomings of immersion into local vibration. The paper [5] used parallel neural network to forecast short-term load, the result is better than the one gained by other methods.

This paper presents a traffic flow forecasting model based on BP neural network, and chooses neural network's inputs according to the correlation theory. This method greatly reduces the size of input patterns. Meanwhile, a new parallel training algorithm based on training set decomposition is presented to accelerate the convergence.

2 The Traffic Flow Forecasting Model Based on BP Neural Network

In practical traffic flow forecasting, traffic flow has certain relations with the past ones of the current road section. Also, as a part of the road net, traffic flow is certainly influenced by the ones of upstream and downstream road sections. So future road section's traffic flow can be forecasted by the past ones of the following road sections: the current road section, the upstream road section and the downstream road section. Owing to the difference of different road section's traffic conditions, choosing the traffic flow of different road sections in different time segments, which has strong correlation with the being forecasted ones as the neural network's inputs is good for improving forecasting precision [10]. The different time segment's traffic flow is stochastic, so it can be treated as stochastic variable. According to the correlation coefficient between two stochastic variables, the correlation between different road sections' traffic flow can be calculated. Assume $u_i(t)$ is the traffic flow of road section i in future time segment t, $u_j(t-n)$ is the traffic flow of road section j in time segment $t-n$. $u_i(t)$ is the neural network's output, and chooses $u_j(t-n)$ which has strong correlation with $u_i(t)$ as the neural network's inputs. The correlation coefficient ρ between $u_i(t)$ and $u_j(t-n)$ is:

$$\rho = \frac{E((u_i(t) - \overline{u_i(t)})(u_j(t-n) - \overline{u_j(t-n)}))}{\sqrt{E(u_i(t) - \overline{u_i(t)})^2 E(u_j(t-n) - \overline{u_j(t-n)})^2}} \tag{1}$$

In equation (1), $E(X)$ is the variable X's mathematical expectation, $\overline{u_i(t)}$ and $\overline{u_j(t-n)}$ are the mathematical expectations of $u_i(t)$ and $u_j(t-n)$. The range of ρ is $[-1, 1]$. The bigger the value of $|\rho|$ is, the stronger correlation between $u_i(t)$ and $u_j(t-n)$ is. $|\rho| = 1$ denotes that $u_i(t)$ and $u_j(t-n)$ are linear correlative, $|\rho| = 0$ denotes that $u_i(t)$ and $u_j(t-n)$ are not correlative. By this way, this paper chooses $u_j(t-n)$ which has strong correlation with $u_i(t)$ as the BP neural network's inputs, and founds a traffic flow forecasting model based on BP neural network.

3 New BP Neural Network's Parallel Training Algorithm Based on Training Set Decomposition

3.1 Traditional Parallel Training Algorithm

The Master/Slave model is often used in traditional parallel training algorithm based on training set decomposition. The whole process is shown in Fig. 1:

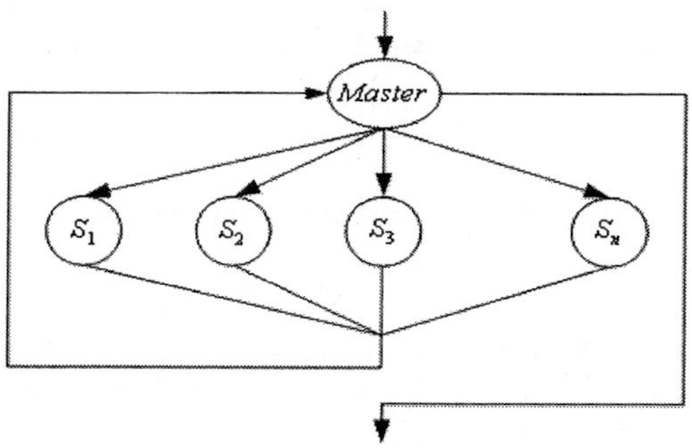

Fig. 1. Traditional parallel training algorithm based on training set decomposition

In Fig. 1, S_1, S_2, S_3, $\cdots\cdots$, S_n are slave tasks. This algorithm has the following shortcomings (Assume this parallel algorithm has n slaves):

(1) High communication cost. In each iteration, the master must receive weight updates from all slaves, then sends new weight to all slaves. So each iteration needs $2n$ times information transition. Assume the whole learning process needs N iterations, this parallel training algorithm needs $2Nn$ times information transition. Along with the increasing of N, the communication cost increases rapidly. Meanwhile, assume the time of once information transition is t, so each iteration needs $2nt$, and the whole parallel algorithm needs $2Nnt$.

(2) Mutual wait between master and slaves. In training process, the master must wait to receive weight updates. Then the salves must wait for new weight set. This mutual wait greatly increases the training time.

In order to conquer upper shortcomings, this paper presents a new parallel training algorithm based on training set decomposition.

3.2 New Parallel Training Algorithm Based on Training Set Decomposition

The new parallel training algorithm also uses Master/Slave model. But instead of the traditional communication profile, this new parallel training algorithm uses a new one. At the beginning, the master divides the whole training set into several subsets, each subset is used to train a copy of the network. Each slave consists of one subset and one copy of the network. Weight initialization and weight updating are done by slaves. After received the training subset, the slaves do error back-propagation learning, and calculate weight updates. Instead of sending weight updates to the master, the weight updates are sent between slaves. After receiving the weight updates, the slave updates the weights, then a new iteration starts. The whole training process can't stop until one slave estimates that the neural network has converged. The new communication profile used in this algorithm is shown in Fig. 2:

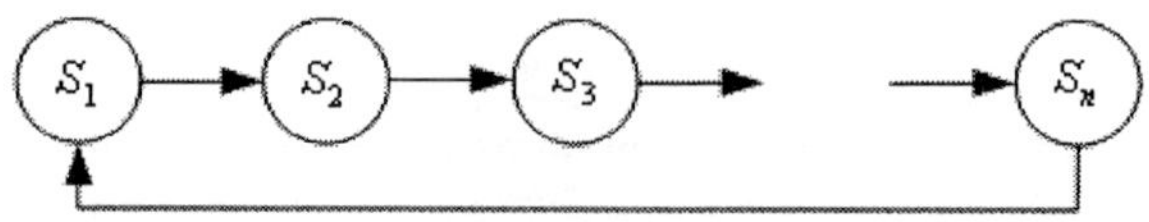

Fig. 2. New communication profile

In this communication profile, the slave $S_i (i < n)$ sends the weight updates to slave S_{i+1}, slave S_n sends to slave S_1. All of the slaves form a ring communication structure. Compare with traditional algorithm, the new algorithm has the following virtues (Assume this parallel algorithm has n slaves):

(1) Low communication cost. In this communication profile, each slave sends weight updates to its right neighbor, so each iteration only needs n times information transition. This is half of the traditional ones. Assume the time of once information transition is t. In order to avoid data losing, when one slave is in sending status, its destination slave is in receiving status. After completed this information transition, these two slaves change their status, sending changes into receiving, and receiving changes into sending. So at one time, there are $\frac{n}{2}$ slaves in sending status and $\frac{n}{2}$ slaves in receiving status. In parallel computer, these information transitions are not disturbing each other, and can be done simultaneously, so each iteration only needs $2t$ time, and greatly reduced the time spent on information transition.

(2) Mutual wait between different slaves is greatly reduced. In the new algorithm, the slaves need not to communicate with master, so the mutual wait between master and slaves is conquered. Meanwhile, in the new communication profile, the information transitions between slaves are not disturbing each other, and can be done simultaneously, so this new algorithm effectively conquers the mutual wait of the traditional parallel training algorithm.

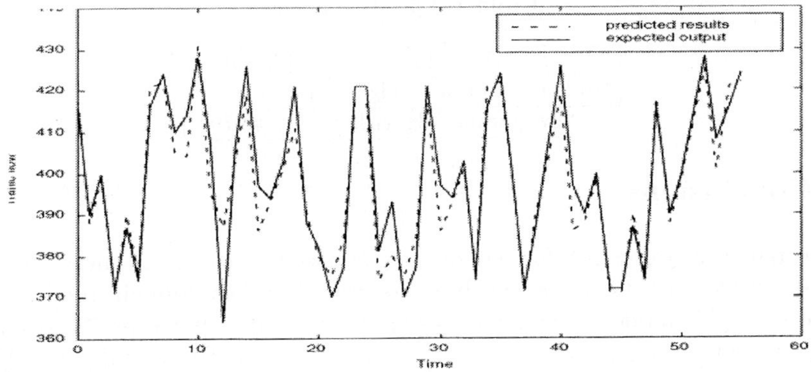

Fig. 3. Results forecasted by the trained neural network

4 Experiment Analysis

By the SCOOT system, this paper gets real traffic data of DaLian, and chooses three neighbor road sections' traffic data of Gaoerji Road from 7 : 30 to 8 : 30 in one month (30 days per month). The traffic data is partitioned based on time segment of 5 minutes, so 360 records of traffic data can be got. After correlation analysis, the following conclusion can be drawn: the traffic flow to be forecasted has strong correlation with the ones in the past first and second time segments of the current road section, has strong correlation with the ones in the past second and third time segments of the upstream road section, and has strong correlation with the ones in the past fourth and fifth time segments of the downstream road section. Assume to forecast certain road section's traffic flow in time segment $t \sim t+5$, the neural network's inputs are the current road section's traffic flow in time segments $t-5 \sim t$ and $t-10 \sim t-5$, the upstream road section's traffic flow in time segments $t-10 \sim t-5$ and $t-15 \sim t-10$, and the downstream road section's traffic flow in time segments $t-20 \sim t-15$ and $t-25 \sim t-20$. Using the traffic flow data got before, 60 input patterns can be formed. This paper using 30 input patterns to train the neural network, and uses the remaining 30 input patterns to verify the neural network.

The parallel algorithm's speed-up is obtained using the expression (2):

$$Speed-up = \frac{Sequential\_execution\_time}{Parallel\_execution\_time} \quad (2)$$

Table 1 shows the speed-up gained by the SHU-GUANG parallel computer. It can be clearly noticed that the new parallel training algorithm gains faster convergent speed than traditional one. And because of the bad communication profile, the traditional parallel training algorithm's convergent speed is much slower than the sequential one.

The results forecasted by the trained neural network are shown in Fig. 3.

Table 1. Speed-up gained

Algorithm	Speed-up
Traditional parallel algorithm	0.497
New parallel algorithm	4.979

5 Conclusions

In this paper, a traffic flow forecasting model based on BP neural network is presented, by the way of calculating the correlation between different road section's traffic flow, this model greatly reduces the size of input patterns. Meanwhile, a new parallel training algorithm is presented to train the BP neural network. The convergent speed gained by the new algorithm is much faster than traditional ones, and can meet the requirements of practical problems.

References

1. He, G.G., Li, Y., Ma, S.F.: Discussion on Short-Term Traffic Flow Forecasting Methods Based on Mathematical Models. System Engineering Theory & Practice, Vol.12 (2000) 51-56
2. Han, C., Song, S.: A Review of Some Main Models for Traffic Flow Forecasting. 2003 IEEE Intelligent Transportation Systems Proceedings, Vol.1 (2003) 216-219
3. Xia,Y.S., Wang, J.: A Discrete-Time Recurrent Neural Network for Shortest-Path Routing. IEEE Transactions on Automatic Control, Vol.45(11) (2000) 2129-2134
4. Li, J., Li, Y.X., Xu, J.W., Zhang, J.B.: Parallel Training Algorithm of BP Neural Networks. Proceedings of the 3rd World Congress on Intelligent Control and Automation, Vol.2 (2000) 872-876
5. Kalaitzakis, K., Stavrakakis, G.S., Anagnostakis, E.M.: Short-Term Load Forecasting Based on Artificial Neural Networks Parallel Implementation. Electric Power Systems Research, Vol.63 (2002) 185-196
6. Yang, Z.S., Gu, Y.L.: A Study on the Model for Real Time Dynamic Traffic Flow Forecasting. Journal of Highway and Transportation Research and Development, Vol.15 (1998) 4-7
7. Phua, P.K.H., Ming, D.: Parallel Nonlinear Optimization Techniques for Training Neural Networks. IEEE Transactions on Neural Networks, Vol.14(6) (2003) 1460-1468
8. Yasunaga, M., Yoshida, E.: Optimization of Parallel BP Implementation: Training Speed of 1056 MCUPS on the Massively Parallel Computer CP-PACS. IEEE International Joint Conference on Neural Networks Proceedings, Vol.1 (1998) 563-568
9. Nordstrom, T., Svensson, B.: Using and Designing Massively Parallel Computers for Artificial Neural Network. Journal of Parallel and Distributed Computing, Vol.14(3) (1992) 260-285
10. Tan, G.Z., Gao, W.: Shortest Path Algorithm in Time-Dependent Networks. Chinese J. Computers, Vol.25(2) (2002) 165-172

A High Precision Prediction Method by Using Combination of ELMAN and SOM Neural Networks*

Jie Wang and Dongwei Yan

School of Electrical Engineering of Zhengzhou University, 450002 Zhengzhou, PR China
wj@zzu.edu.cn
yandwemail@263.net

Abstract. This paper presents a combination of ELMAN and SOM neural networks in order to enhance the prediction precision. A new method of training and predicting of samples is developed. In this method, the training and predicting are divided into two steps, clustering at first and then training and predicting the samples at the clustered areas. As examples, this method is applied to weather broadcasting and disaster prediction. Simulation results show that this method can enhance the ability of local generalization of the network. The prediction precision of combined network presented in this paper is higher than that with normal BP network or just one of ELMAN or SOM network.

1 Introduction

It has been proved that some Artificial Neural Network (ANN) models, such as BP, RBF, ELMAN, etc., have the capacity of infinitely approaching any function. ANN was originally used to the prediction fields by Lapedes [1] and Weigend etc. [2]. At present, prediction using ANN method was widely used in stock, weather, disease and ecology, etc. The sample data obtained for training are practically less than needed for high precision prediction. To solve this problem, a characteristics of ANN can be used: the more similar the sample data are, the more higher the prediction precision in the neighboring field is. This paper presents a new prediction method with high precision. In this method, the predicting processes are divided into two steps, clustering sample data at first and then predicting the samples at the clustered areas. There is a pillar in the nerve cell model, which can pre-classify the complex information, thus make the sequential information processing and mapping more accurate and rapid. In this paper, SOM(Self-organizing feature map) network was chosen to pre-classify the samples for the object to be predicted since SOM network can successfully realize the clustering function to some extent [3]. Simulation results show that this method enhances the ability of local generalization of the network and the prediction precision with the method presented in this paper is higher than that with normal BP network or just one of ELMAN and SOM network.

* This research is supported by the Natural Science Foundation of Henan Province under grants of 0411010200

F. Yin, J. Wang, and C. Guo (Eds.): ISNN 2004, LNCS 3174, pp. 943–949, 2004.

2 Introduction of ELMAN and SOM Neural Network

2.1 ELMAN Neural Network

The structure of ELMAN neural network is shown in Fig.1. The output of the hidden layers of ELMAN neural network is interconnected with the input of the hidden layers by delay and storage. The interconnected mode makes it sensitive to the data of history state. The ability of processing the dynamic information is added to the inner feedback network and it is of advantage to the modeling for the dynamic process. ELMAN neural network can store information for further use and thus it can both learn space mode and time mode.

As for ELMAN neural network, its nonlinear space can be expressed as

$$y(k) = g(w^3 x(k)) \tag{1}$$

$$x(k) = f(w^1 x_c(k) + w^2 u(k-1)) \tag{2}$$

$$x_c(k) = x(k-1) \tag{3}$$

Wherein *u,x, u,* and x_c denote output, state, input and feedback state vectors with m, n, r and n dimensions respectively. $f(\bullet)$ and $g(\bullet)$ denote activation functions of the output and hidden nodes. w^3, w^2, w^1 denote the weights that link the output, the hidden, the input and the successive layers. According to equation (2) and (3),

$$x_c(k) = x(k-1) = f(w_{k-1}^1 x_c(k-1) + w_{k-1}^2 u(k-2)) = \cdots = \Phi(u^k, w^{1k}, w^{2k}) \tag{4}$$

where $u^k = \{u_0, u_1, \cdots, u_{k-2}\}$, $w^{1k} = \{w_0^1, w_1^1, \cdots, w_{k-1}^1\}$, $w^{2k} = \{w_0^2, w_1^2, \cdots, w_{k-1}^2\}$, u^k, w^{1k} and w^{2k} are information about previous input data and weights.

The Euclidean norm of the error is given by

$$E(w) = \sum_{p-1}^{m} [y_p(w) - \tilde{y}_p(w)]^2 / 2 \tag{5}$$

The weight is always be adjusted by BP algorithm:

$$w(k+1) = w(k) + \eta(k)(-\frac{\partial E(w)}{\partial w}) \tag{6}$$

$$\Delta w_{ij}^3 = \eta \delta_i^0 x_j(k) \qquad i = 1,2,\cdots,m, j = 1,2,\cdots n \tag{7}$$

$$\Delta w_{jq}^2 = \eta \delta_j^h u_q(k-1) \qquad j = 1,2,\cdots n, q = 1,2,\cdots,r \tag{8}$$

$$\Delta w_{jl}^1 = \eta \sum_{i-1}^{m} (\delta_i^0 w_{ij}^3) \frac{\partial x_i(k)}{\partial w_{jl}} \qquad j = 1,2,\cdots,n, l = 1,2,\cdots,n \tag{9}$$

Wherein η is a variable that controls the process of the adjustment.

$$\delta_i^0 = [y_i(k) - \tilde{y}_i(k)]g'(\bullet), \delta_j^h = \sum_i^m (\delta_i^0 w_{ij}^3) f'(\bullet) \tag{10}$$

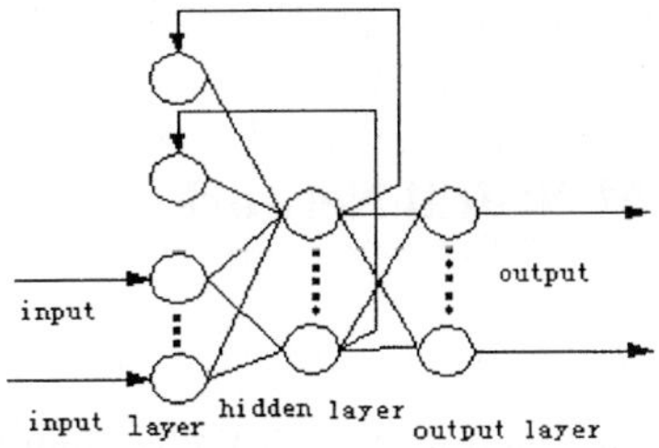

Fig. 1. ELMAN Neural Network

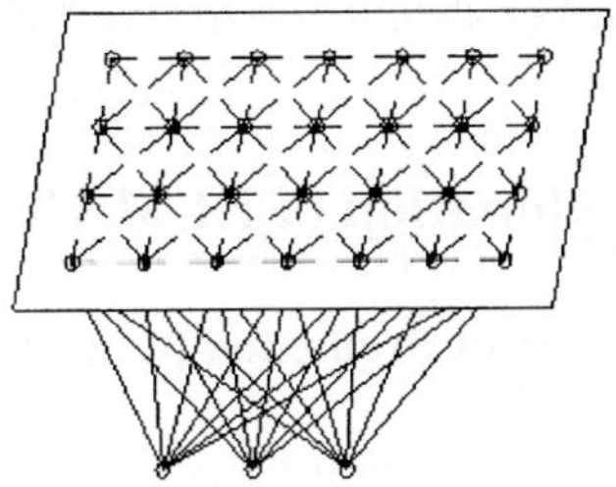

Fig. 2. SOM Neural Network

2.2 SOM Neural Network

The 1-dimensional SOM neural network (Fig.2.) utilized in this paper typically consists of two layers of nodes: input and Kohonen layer. The SOM network performs unsupervised learning. Its primary use is for tasks such as clustering, pattern recognition and various optimization problems. It is designed to capture topologies and hierarchical structures of higher dimensional input spaces. The learning algorithm is summarized as follows.

(1)Initialization. Randomly assign values to weight w_{ij}; set learning rate $\eta(0)$, topological neighborhood parameters $N_g(t)$ and maximum number of clusters T.

(2)Input randomly one of the q modes P_k into the input layer and normalize it.

(3)Normalize the weight vector $W_j = (w_{1j}, w_{2j}, \cdots, w_{jN})$ and compute Euclidean distance d_j between $\overline{W}_j$ and $\overline{P}_k$.

(4)Find the minimum distance d_g and locate the winning unit.

(5)Update weights for all units within a specified neighborhood $N_g(t)$:

$$\overline{w_{ji}(t+1)} = \overline{w_{ji}(t)} + \eta(t)[\overline{p}_l^k - \overline{w_{ji}(t)}] \tag{11}$$

$$j \in N_g(t), \qquad j = 1,2,\cdots,M \qquad (0 < \eta(t) < 1)$$

(6)Choose one of the learning patterns for the input layer and repeat step (3) until all inputs have been presented to the SOM once.

(7)Update learning rate and its neighborhood parameter $N_g(t)$: $\eta(t) = \eta(0)(1 - \frac{t}{T})$.

Suppose a neuron g in competitive layer, its coordinate value in a 2-dimentional array is (x_g, y_g), the neighborhood is a square constructed by left and right triangle parts restricted by ($x_g + N_g(t), y_g + N_g(t)$) and ($x_g - N_g(t), y_g - N_g(t)$) respectively. The neighborhood $N_g(t)$ could be modified by $N_g(t) = INT[N_g(0)(1-\frac{t}{T})]$, wherein the $INT[x]$ means taking integer of x and $N_g(0)$ is the initial value of $N_g(t)$.

(8)Let t=t+1 and go to step (2), stop when t=T.

3 Combination of ELMAN and SOM Neural Networks

3.1 Theory of Prediction by ANN

In a sense, prediction by ANN is to establish a function $Y = f(X)$ by inputting all influencing factors, $X = \{x_1, x_2, \cdots, x_n, \cdots\}$, and obtaining expected output Y. The function $Y = f(X)$ can give the output accurately if all the influencing factors could be given. But due to the complexity of matter itself, only main influencing factors can be taken as the input factors. That is, establish the function $Y = f'(X')$. X' is subspace of X by subtracting the factors that are not so important. As a result, the predicting function of system is now only decided by X' and thus cannot reflect the regulations entirely and systematically.

The common method of prediction by ANN is taking the main influencing factors as input, train the network with sample data until the error index is satisfied. To avoid the 'dimension disaster', the input of ANN should be as few as possible. The input should take only the main influencing factors. These may result in that the obtained rules will be not accordant with realities, especially when the subordinate factors are excessive and their influence cannot be neglected any more. But if all sample data are used to train the network with equal weight, the prediction error will be also increase because of the decentralization of the sample data.

3.2 Theory of Combined ANN

Practically, the rules that the system obeys are more similar when some of the influencing factors are similar. Take the agricultural harvest prediction as an example, the regulation for the south rainy region is much different from that for the north droughty region. But the regulation for different north regions is similar due to the similar weather characteristic. Take the national economy prediction as another example, the regulation for developing countries is much different from that for developed countries, but the regulation for different developing countries is always similar. According to the above fact, it is proposed that if the dispersive samples are preprocessed and pre-classified, the predicting functions for various similar situations will be obtained. Thus, the precision of prediction will be enhanced when it is put into practice.

Now, the problem lies in how to classify the training samples into different clusters that obey different regulations. In this paper, a new predicting method is proposed by considering properties of ANN that the generalization at a low-sample-density region is weaker than that at a high-sample-density region and the strong classifying ability of SOM neural networks and the powerful function-approaching ability of ELMAN neural networks.

4 Prediction by Combined ANN

At first, classify the sample data into different clusters which obey different regulations by SOM neural network, then train one ELMAN neural network for each cluster. As for samples to be predicted, the first step is to judge which cluster they belong to by using SOM neural networks, then predict by using the trained relative ELMAN neural network. The modeling of prediction by combined ANN can divided into several steps as follows:
(1) Investigation. Analyze the properties and the influencing factors and classify the main factors and the subsidiary factors.
(2) Take the number of dimensions of sample sequence as number of input nodes. The number of input layer (competing layer) nodes is no more than the number of input samples. Thus a SOM neural network is constituted.
(3) Classify the main factors by using SOM neural network and train the network. Get the classification number by the number of the clustering center of the wining samples, subtract the nodes of the completing layer which is distant from the clustering center and the output that approximates to zero to decrease the nodes of the completing layer. Retrain until convergence
(4) Establish an ELMAN neural network and initialize its weight values. The nodes in the input layer for the ELMAN neural network are composed of the nodes in the output layer for the SOM neural network, which has been trained successfully; and the nodes in the output layer for the ELMAN neural network depend on the number of variables that the prediction needs to output.
(5) Connect the output layer for the SOM neural network with the input layer for the ELMAN neural network, and train this combined ANN with the same input samples used in SOM neural network and the expected output samples (the expected prediction results). In this course, the weight values of SOM network keeps unchanged while the weight values of ELMAN network are trained using BP algorithm.
(6) After the training for the network converged, a new ANN, which is the combination of SOM neural network and ELMAN neural network and has the pre-categorized function, can be set up.

5 Illustrative Examples

5.1 Weather Forecasting

The experiment data for weather forecasting are taken from reference[3]. In this article, Yuan took the data in Apr.1~15 and May14~29(31 days) as the training examples for the neural network and took the data June 1 ~28 (28 days) as the testing samples to verify the prediction ability of the meteorological data testing system.

Firstly, the training samples are pretreated with SOM neural network whose initial topology architecture is composed of 12 inputs (the number of meteorological data per day) and 12 outputs. Thus, two clustering centers can be formed when the training for the SOM neural network is over and an improved SOM neural network with 12 inputs and 6 outputs, which is arranged in the form of one-dimensional vector, can be established. Secondly, ELMAN network with 6-8-3 topology architecture is set up, in which the 3 outputs represent the results of the weather forecasting for the present day, the second day and the third day respectively. Finally, the above-mentioned method is employed to train and test the combined neural network to verify the prediction ability. The testing results are shown in table 1. As comparison, the prediction results with common BP method are also listed.

Table 1. Correctness of different prediction methods for the weather forecasting

Model	The 1st day	The 2nd day	The 3rd day
Combined ANN	85.7%	71.4%	64.3%
Common BP	82%	67.9%	60.7%

It can be seen from table 1 that the correctness of the weather forecasting for the present day, the second day and the third day with combined ANN is higher than that with common BP.

It can be seen from table 1 that the correctness of the weather forecasting for the present day, the second day and the third day with combined ANN is higher than that with common BP.

Table 2. Results of different prediction methods for the disease incidence and the prediction error

	Actual Incidence	Incidence predicted by combined ANN	Prediction error of combined ANN	Incidence predicted by normal BP	Prediction error of normal BP
Combined ANN	5.3%	5.6%	5.7%	5.9%	11.3%
Normal BP	6.3%	6.8%	7.9%	6.9%	9.5%

5.2 Disease Prediction

The data for the disease prediction are obtained from reference [4]. The commonly used methods for disease prediction at home and abroad at present are gray model method, plural linear regression method, curve fitting method and Markov chain method etc. In this paper, prediction with the combined ANN was carried on in the army about the incidence of the acute infectious diarrhea disease.

The training samples include 12 samples in Table 1 and the No.1 and No.4 samples in Table 2, totally 14 samples, in reference [4]. The 2 testing samples are the No.2 and

No.4 samples in Table 2 in reference [4]. The testing results with the same method used in illustrative example 4.1 are shown in Table 2. As comparison, the prediction results with common BP method are also listed.

It can be seen from table 2 that it is effective for ANN to predict the disease and the prediction precision with combined ANN is higher than that with common BP.

6 Conclusions

The scatter of samples increases the error of prediction model based on common neural network. Classify the samples by clustering and then train the combination of ELMAN and SOM Neural Networks, and the precision of prediction obtained will be higher than that of normal BP neural network. By comparing two examples, the weather predicting and the disease-infecting rate, it is proved that the combining ANN improves the ability of local generalization of the networks and the prediction precision is higher than normal BP networks or just one of ELMAN and SOM networks.

References

1. Lapedes, A., Farber, R.: How Neural Nets Work. Proceedings of IEEE Conference on Neural Information Processing Systems—Nature and Synthetic. Denver (1987) 442-456
2. Andreas, W.S., Huberman, B.A., Rumelhart, D.E.: Predicting the Future: A Connectionist Approach. Int. J. of Neural Systems. 1(1990) 193-209
3. Zengren, Y.: Artificial Neural Networks and Applications. Publishing House of Tsinghua University, Beijing (1999) 319-326
4. Shouyi, Y., Qing, C., et al.: A Scroll Model for Predicting the Incidence of a Disease of Acute and Epidemic Diarrhea in Army, Journal of Prophylactic Medicine of PLA, 5 (1998) 349-351

Short-Term Traffic Flow Forecasting Using Expanded Bayesian Network for Incomplete Data

Changshui Zhang, Shiliang Sun, and Guoqiang Yu

State Key Laboratory of Intelligent Technology and Systems
Department of Automation, Tsinghua University, Beijing, China, 100084
zcs@mail.tsinghua.edu.cn
{suns102, ygq01}@mails.tsinghua.edu.cn

Abstract. In this paper expanded Bayesian network method for short-term traffic flow forecasting in case of incomplete data is proposed. Expanded Bayesian network model is constructed to describe the causal relationship among traffic flows, and then the joint probability distribution between the cause and effect nodes with dimension reduced by Principal Component Analysis (PCA) is approximated through Gaussian Mixture Model (GMM). The parameters of the GMM are learned through Competitive EM algorithm. Experiments show that the expanded Bayesian network method is appropriate and effective for short-term traffic flow forecasting with incomplete data.

1 Introduction

Short-term traffic flow forecasting, which is to determine the traffic volume in the next time interval usually in the range of five minutes to half an hour, is an important problem in the research area of Intelligent Transportation Systems (ITS). In the past years, many theories and methods on this theme were proposed including those based on time series models (including ARIMA, seasonal ARIMA), Kalman filter theory, neural network approaches, non-parametric methods, simulation models, local regression models and layered models [1]~[6]. Although these theories and methods have alleviated difficulties in traffic flow modelling and forecasting to some extent, they can do little when the data used for forecasting is partially missing or unavailable, while this case of incomplete data often occurs in practice. When this circumstance happens, the historical average (fill up the incomplete data with their historical average) method is often applied to cope with this issue; nevertheless, the forecasting performance is quite limited.

The Bayesian network approach, as studied comprehensively in the community of information science, gives us some inspiration on traffic flow forecasting. Considering the nature of short-term traffic flows, we can draw the conclusion that the traffic flow at a given road link is closely related to those of its neighbors, and thus in order to forecast as accurately as possible, we should also take into

F. Yin, J. Wang, and C. Guo (Eds.): ISNN 2004, LNCS 3174, pp. 950–955, 2004.

account the information provided by neighbor links. Bayesian network is such a model that can fully take into account the causal relationship between random variables statistically, and it has a natural capacity to deal with incomplete data. Our main contribution of the paper is that we introduce the concept and approach of Bayesian network in information science to the area of ITS for the first time and effectively solve the traffic flow forecasting problem with incomplete data. For a given road net, we focus on constructing rational expanded Bayesian networks, learn the statistical relations between the cause and effect nodes, and then based on the statistical relationship carry out forecasting. Experiments for real-world short-term vehicular flow forecasting in case of incomplete data are carried out to validate our method. Comparison with AR model and the historical average method shows that our expanded Bayesian network method is appropriate and effective for this kind of traffic flow forecasting problems.

2 Expanded Bayesian Networks

A Bayesian network, also known as casual model, is simply a directed graphical model for representing conditional independencies between a set of random variables. In a Bayesian network, an arc from node A to B can be informally interpreted as indicating that A "causes" B [7]. Suppose we have several random variables denoted by $x_1, x_2, ..., x_m, y_1, y_2, ..., y_n$ and z respectively. $x_1, x_2, ..., x_m$ are used to forecast y_1 and $y_1, y_2, ..., y_n$ are used to forecast z in turn. Then considering the causal relations in variable forecasting, we can construct two Bayesian networks as shown in Fig. 1.a where arrows start from the cause nodes and point to the effect nodes.

Now suppose data for random variable y_1 is missing while data for $x_1, x_2, ..., x_m, y_2, ..., y_n$ is complete(intact), then how can we forecast z? We can construct another Bayesian network to model the new causal relationship, which is given in Fig. 1.b. In the graph, $x_1, x_2, ..., x_m, y_2, ..., y_n$ serve as the cause nodes of z . Since the newly constructed Bayesian network often has more nodes than either of the original graphs, we call it Expanded Bayesian Network (EBN).

3 Related Methods

3.1 Dimension Reduction

When using Bayesian networks, the joint probability distribution of related variables should be estimated. Usually the dimension of the joint probability distribution using Bayesian network is high and the data is not enough comparatively. So there might be a large error coming from parameter learning. However, if we carry out parameter learning on a lower dimension with the same data, the estimation will be more accurate and efficient. Principal Component Analysis (PCA) is such an effective tool for linear dimension reduction [8]. By using PCA, we select a few principal components corresponding to the largest eigenvalues to represent data space.

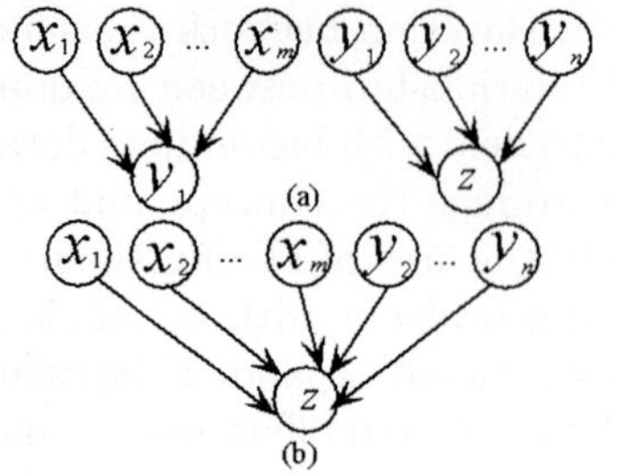

Fig. 1. (a) Two Bayesian networks. (b) Expanded Bayesian network (EBN)

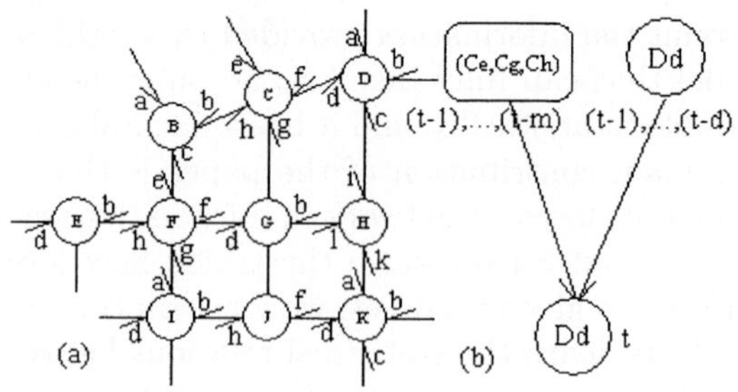

Fig. 2. (a) A patch taken from the whole map where UTC/SCOOT systems are placed. (b) The Bayesian network between the object road link and its neighbors

3.2 Parameter Learning of Joint Probability Distribution and Forecasting Formulation in Bayesian Networks

In this article, the joint probability density function (PDF) between cause nodes and effect node in a Bayesian network is approximated through Gaussian Mixture Model (GMM). Let x denote one random variable or multidimensional random vector, and then the GMM form of its probability distribution with M mixture components can be represented as:

$$p(x|\Theta) = \sum_{l=1}^{M} a_l G(x|\theta_l) \tag{1}$$

where the parameters are $\Theta = (\alpha_1, ..., \alpha_M, \theta_1, ..., \theta_M)$ and M, s.t. $\sum_{l=1}^{M} a_l = 1$. Each $G(.)$ is a Gaussian PDF parameterized by $\theta_l = (\mu_l, \Sigma_l), l = 1, ..., M$.

Usually we use Maximum Likelihood Estimation (MLE) to carry out parameter learning with given data. The Competitive Expectation Maximization (CEM) algorithm proposed by Zhang et al is an effective algorithm to carry out MLE, which is a useful variant of EM algorithm [9][10]. In this article, the parameters of a GMM which describe the joint PDF of the cause nodes and effect node in a Bayesian network are learned through CEM algorithm.

Traffic flow forecasting here can be considered as an inference problem in a Bayesian network. The main goal of inference in a Bayesian network is to estimate the values of hidden nodes, given the values of observed nodes. We use this mechanism to implement forecasting of traffic flows. Suppose (E, F) be a partitioning of the node indices of a Bayesian network into disjoint subsets, and (x_E, x_F) be a partitioning of the corresponding random variables. Under the rule of Minimum Mean Square Error (M.M.S.E.), the optimal estimation of x_F from x_E can be given as [11]:

$$\hat{x}_F = E(x_F|x_E) \ . \tag{2}$$

To deduce the representation of the optimal forecasting $\hat{x}_F$ under the GMM, we employ the following lemma.

Lemma [12] Let $G(x;\mu,\Sigma)$ denote a multidimensional normal density function with mean $\mu^T = (\mu_1^T, \mu_2^T)$ and covariance matrix $\Sigma = \begin{pmatrix} \Sigma_{11} & \Sigma_{12} \\ \Sigma_{21} & \Sigma_{22} \end{pmatrix}$. $x^T = (x_1^T, x_2^T)$ is a random vector. Then we have:

$$p(x) = G(x_1;\mu_1,\Sigma_{11})G(x_2;\mu_{x_2|x_1},\Sigma_{x_2|x_1}) \ ,$$

where $\mu_{x_2|x_1} = \mu_2 - \Sigma_{21}\Sigma_{11}^{-1}(\mu_1 - x_1), \Sigma_{x_2|x_1} = \Sigma_{22} - \Sigma_{21}\Sigma_{11}^{-1}\Sigma_{12}$.

If we rewrite $x^T = (x_F^T, x_E^T), \mu_l^T = (\mu_{lF}^T, \mu_{lE}^T)$, $\Sigma_l = \begin{pmatrix} \Sigma_{lFF} & \Sigma_{lFE} \\ \Sigma_{lEF} & \Sigma_{lEE} \end{pmatrix}$, we have:

$$p(x_F, x_E) = \sum_{l=1}^{M} a_l G(x;\mu_l,\Sigma_l) = \sum_{l=1}^{M} a_l G(x_E;\mu_{lE},\Sigma_{lEE})G(x_F;\mu_{lF|E},\Sigma_{lF|E}) \ .$$

Finally, we can get the optimal forecasting $\hat{x}_F$ under the criterion of M.M.S.E. as follows:

$$\hat{x}_F = E(x_F|x_E) = \sum_{l=1}^{M} \beta_l \mu_{lF|E} \ , \tag{3}$$

where $\beta_l = \frac{a_l G(x_E;\mu_{lE},\Sigma_{lEE})}{\sum_{j=1}^{M} a_j G(x_E;\mu_{jE},\Sigma_{jEE})}, \mu_{lF|E} = \mu_{lF} - \Sigma_{lFE}\Sigma_{lEE}^{-1}(\mu_{lE} - x_E)$, $\Sigma_{lF|E} = \Sigma_{lFF} - \Sigma_{lFE}\Sigma_{lEE}^{-1}\Sigma_{lEF}$.

4 Experiments

The experimental data for analysis is the vehicle flow rates of discrete time series recorded every 15 minutes on many road links by the UTC/SCOOT system in Traffic Management Bureau of Beijing, whose unit is vehicles per hour (vph). The data is from Mar. 1 to Mar. 25, 2002 and 2400 sample points totally. To evaluate our approach objectively, 2112 points (training set) of them are employed to learn parameters of GMM and the rest (test set) are employed to test the forecasting performance. Fig. 2.a shows the analyzed patch. Each circle node denotes a road link. Each arrow shows the direction of the traffic flow on the corresponding road.

We take road link D_d as an example to show our approach. From the view point of Bayesian Network, vehicle flows of C_e, C_g and C_h should have causal relations with vehicle flow of D_d. Similarly, vehicle flows of B_a and B_c should have causal relations with vehicle flow of C_h. Furthermore, considering the time factor, to predict the vehicle flow of D_d at time t (denoted by $D_d(t)$)we should use values $D_d(t-1), D_d(t-2), ..., D_d(t-d)$ as well. That is, some historical values of C_e, C_g, C_h and D_d could be regarded as the cause nodes of $D_d(t)$ in a Bayesian network. The causal relation is shown in Fig. 2.b.

Suppose traffic flow data $C_h(t-m)$ is missing, we can use EBN method to forecast $D_d(t)$. The EBN for forecasting $D_d(t)$ with missing data $C_h(t-m)$ is shown in Fig. 3. The flow chart of our forecasting procedure can be described as follows: 1). Construct an EBN between input (cause nodes) and output (effect node) for a given road link using the methods explained in section 2; 2). Approximate the joint PDF of all nodes in the EBN by PCA and GMM using methods explained in section 3; 3). Carry out the optimal estimation of flow rates of the object road link in the form of equation (3).

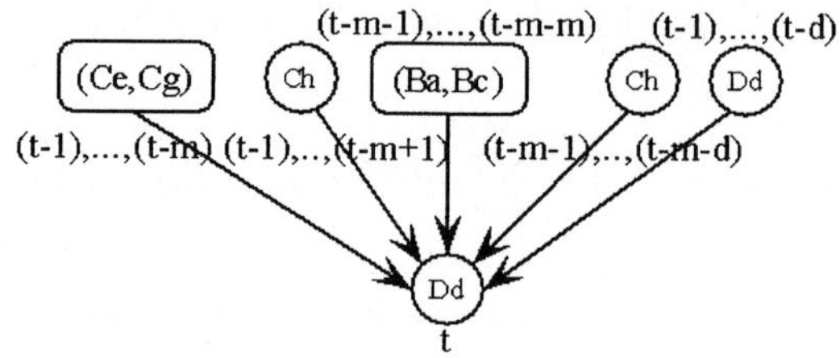

Fig. 3. The expanded Bayesian network for object road link D_d

In the experiment, the forecasting orders from the object road link and from the neighbor links are respectively taken as 4 and 5 empirically (parameters $d = 4, m = 5$). Then for Fig. 3 the joint PDF is: $p(C_e(t-j), C_g(t-j), B_a(t-j-5), B_c(t-j-5), C_h(.), D_d(t-j+1), j = 1, ..., 5)$, where $C_h(.) = (C_h(t-l), C_h(t-l-5), l = 1, ..., 4)$. We can see the dimension of the joint PDF is very high (dimension=33). So we use PCA to carry out dimension reduction before the parameter learning of GMM. For other road links, we also employ PCA for dimension reduction before the parameter learning of GMM. The final forecasting performances using the EBN method are listed in Table 1 evaluated by root Mean Square Error (RMSE). Since the parameters d and m is just selected empirically, utilizing AR model with the same parameter d for comparison is reasonable (AR model is comparable which only uses historical flow rates of the object road link to forecast). The forecasting results through AR model and the historical average method (using the the average value of the historical flow rates at the corresponding time to forecast) are also given in Table 1. For a given road link of Table 1, the smaller RMSE corresponds to the better forecasting performance (accuracy). From the experimental results we can see the outstanding improvements of forecasting capability brought by using EBN. For each of the four road links analyzed, the performances of the EBN outperforms the other two methods significantly.

5 Conclusion

In this paper, we first successfully introduce the concept and approach of expanded Bayesian network to the community of ITS for the application problem of incomplete data forecasting. Experiments with real-world data also show that

Table 1. Performance comparison of three methods for short-term traffic flow forecasting of four different road links

Methods	D_d	J_f	G_d	C_f
Historical Average	84.20	140.69	213.39	112.50
AR	66.14	123.65	155.20	90.76
Expanded Bayesian network(EBN)	*57.44*	*110.88*	*138.39*	*86.31*

EBN is appropriate and effective for incomplete short-term traffic flow forecasting. However, there are still some problems to be discussed in the future, e.g. how to effectively consider the periodical information of the traffic flows for our EBN method, etc.

Acknowledgements. This work was supported by National High Technology Research and Development Program of China (863 Program) under contract No.2001AA114190.

References

1. William, B.M.: Modeling and Forecasting Vehicular Traffic Flow as a Seasonal Stochastic Time Series Process. Doctoral dissertation. University of Virginia, Charlottesville (1999)
2. Okutani, I., Stephanedes, Y.J.: Dynamic Prediction of Traffic Volume through Kalman Filter Theory. Transportation Research, Part B, Vol.18B (1984) 1-11
3. Edwards, T., Transley, D.S.W., Frank, R.J., Davey, N.: Traffic Trends Analysis using Neural Networks. http://homepages.feis.herts.ac.uk/~nngroup/pubs/papers/ed-wards-iwannt97.pdf (1997)
4. Davis, G.A., Nihan, N.L.: Non-Parametric Regression and Short-Term Freeway Traffic Forecasting. Journal of Transportation Engineering (1991) 178-188
5. Roland, C., Joachim, W., Michael, S.: Traffic Forecast Using Simulations of Large Scale Networks. IEEE Intelligent Transportation Systems Conference Proceedings (2001)
6. Davis, G.A.: Adaptive Forecasting of Freeway Traffic Congestion. Transportation Research Record, No.1287. TRB, National Research Council, Washing, D.C. (1990)
7. Kevin, P.M.: An Introduction to Graphical Models. http://www.ai.mit.edu/~murphyk/Papers/intro_gm.pdf (2001)
8. Jolliffe, I.T.: Principal Component Analysis. Springer-Verlag, New York (1986)
9. Jeff, A.B.: A Gentle Tutorial of the EM Algorithm and its Application to Parameter Estimation for Gaussian Mixture and Hidden Markov Models. International Computer Science Institute, Berkeley CA, TR-07-021 (1998)
10. Zhang, B.B., Zhang, C.S., Yi, X.: Competitive EM Algorithm for Finite Mixture Models. Pattern Recognition, Volume: 37, Issue: 1, January (2004) 131-144
11. Andrew, H.J.: Stochastic Processes and Filtering Theory . Academic Press, Inc., New York and London (1970)
12. Rao, C.R.: Linear Statistical Inference and Its Applications (Second Edition). John Wiley&Sons, Inc. (1973)

Highway Traffic Flow Model Using FCM-RBF Neural Network*

Jian-Mei Xiao and Xi-Huai Wang

Department of Electrical and Automation, Shanghai Maritime University, Shanghai 200135, China
jmxiao@cen.shmtu.edu.cn, wxh@shmtu.edu.cn

Abstract. A highway traffic flow model using a distributed radial basis function (RBF) neural network based on fuzzy c-means (FCM) clustering is presented. FCM clustering is used to classify training data into a couple of clusters, each cluster is trained by a sub-RBF neural network, and membership values are used for combining several RBF outputs to obtain the final results. A highway traffic flow model with four segments is studied. The training data for traffic flow modeling were generated using a well-known macroscopic traffic flow model at different densities and average velocities. The simulation result proves the effectiveness of this method.

1 Introduction

Highway macroscopic traffic flow model is the basis of control, analysis, design, and decision-making in intelligent transportation system, and is often described by a set of non-linear, dynamic equations [1-2]. However, some of the existing macroscopic models have been found to exhibit instabilities in their behavior and often do not track real traffic data correctly. On the other hand, some microscopic traffic flow models can yield more detailed and accurate representations of traffic flow but are computationally intensive, and typically not suitable for real time implementation. Such implementations are likely to be necessary for development and application of advanced traffic control in intelligent transportation systems.

The neural network is paid great attention to the system dynamic modeling, because it has the capabilities of self-learning and approaching any nonlinear functions. In this paper, we present a distributed radial basis function (RBF) neural network with fuzzy c-means (FCM) clustering to set up the highway macroscopic traffic flow model. An FCM-RBF neural network model has been developed to emulate an improved version of a well-known discrete traffic flow dynamic model. Simulation results show that the FCM-RBF model can capture the traffic dynamics quite closely.

---

* This work was supported by Science Foundation of Shanghai Education (03IK09) and Key Subject Foundation (No. 4) of Shanghai Education Commission

F. Yin, J. Wang, and C. Guo (Eds.): ISNN 2004, LNCS 3174, pp. 956–961, 2004.

2 FCM-RBF Neural Network

The proposed algorithm consists of three steps. The first step is classifying training objects into several clusters using FCM clustering. The second step is training each cluster with a sub-RBF neural network. The third step is combining several RBF outputs to obtain the final result using membership values.

A RBF neural network has a three-layer architecture with no feedback [3-4]. The hidden layer consists of H hidden neurons (radial basis units), with radial activation functions. A typical choice for this function is the Gaussian function. So, the output of the h-th hidden neuron, z_h, is a radial basis function that defines a spherical receptive field in R^H given by the following equation:

$$z_h = \Phi(\| x - c_h \|) = \exp(-\frac{(x-c_h)^T(x-c_h)}{2\sigma_h^2}), \quad \forall h \tag{1}$$

In other words, each neuron in the hidden layer has a substantial finite spherical activation region, determined by the Euclidean distance between input vector, x, and the center, c_h, of the function z_h normalized with respect to the scaling factor σ_h.

From (1) we can know that each hidden neuron is associated with $H+1$ internal parameters; the H components of vector c_h that represents the H dimensional position of the radial function, and σ_h that determines the receptive field of the neuron. The receptive field is the region of the input space over which the neuron has an appreciable response. The set of hidden neurons is designed so that they cover all the significant regions of the input vector space.

The network output y is given by the following equation:

$$y = \sum_{h=1}^{H} w_h z_h \tag{2}$$

Fuzzy c-means clustering is an unsupervised classification algorithm [5], which uses a certain objective function, described in (3), for iteratively determining the local minima.

$$J_m(U,V) = \sum_{i=1}^{N}\sum_{j=1}^{C} u_{ij}^m d_{ij}^2(X_i, V_j) \tag{3}$$

where C is the number of clusters, N is the number of training objects $X=\{X_i|i=1, 2, \ldots N\}$. The objective function is a weighted within-groups sum of distances d_{ij}. An important parameter in this approach is the exponent m. This exponent is always greater than unity.

Cluster centers are given by the following equation:

$$V_j = \sum_{i=1}^{N} u_{ij}^m X_i \Big/ \sum_{i=1}^{N} u_{ij}^m, \quad \forall j \tag{4}$$

Membership values for objects are calculated using (5). An important feature in FCM is that the sum of an object's values over all the clusters equals unity as (6).

$$u_{ij} = 1 \Big/ \sum_{l=1}^{C} \left(\frac{d_{ij}}{d_{il}}\right)^{\frac{2}{m-1}}, \quad \forall i \tag{5}$$

$$\sum_{j=1}^{C} u_{ij} = 1, \qquad u_{ij} \in [0,1], \qquad \forall i, j \tag{6}$$

Different types of distance measures can be used to measure the distance between an object i and a prototype j. A Euclidean distance is given by (7).

$$d_{ij}^2(X_i, V_j) = (X_i - V_j)(X_i, V_j)^T, \quad \forall i, j \tag{7}$$

Equations (3)-(7) mainly describe the formulas which are used during the training stage. In online stage, new values of memberships are calculated using (8) along with the chosen distance measure. The membership values for a new vector is:

$$u_{N+1,j} = 1 \Big/ \sum_{l=1}^{C} \left(\frac{d_{N+1,j}}{d_{N+1,l}}\right)^{\frac{2}{m-1}}, \quad \forall i \tag{8}$$

The output y of the distributed FCM-RBF is as (9) shows:

$$y = \sum_{j=1}^{C} u_{N+1,j} f_j(X) \tag{9}$$

where f_j is the output of the *j-th* RBF sub-network, C is the number of sun-networks in the distributed network, i.e. the number of clusters of FCM.

3 Simulation

We use this FCM-RBF to set up the highway traffic flow model. Here we discuss a highway with four segments, which are same length. There is an on-ramp at the first segment, and an off-ramp at the third segment. The road structure of highway is show as Fig.1.

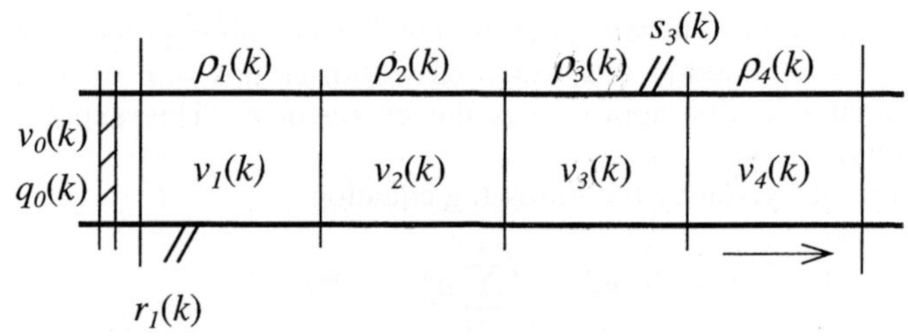

Fig. 1. Structure of highway traffic flow system

After analyzing the traffic flow, The FCM-RBF structure consisted of 12 inputs and 8 outputs, which correspond to the number of input and output variables, respectively. The inputs of FCM-RBF are traffic flow $q_0(k)$ and average velocity $v_0(k)$ at first segment beginning, on-ramp flow $r_1(k)$ at first segment, and off-ramp flow $s_3(k)$ at the third segment. To set up a dynamic model, we delay the density $\rho_i(k)$, average velocity $v_i(k)$ as $\rho_i(k-1)$ and $v_i(k-1)$ which are also used as the inputs of FCM-RBF. We use the density $\rho_i(k)$, average velocity $v_i(k)$ as the outputs of FCM-RBF. The number of neurons in the hidden layer depended on the number of training objects.

To model the traffic flow, input and output data must be collected first. There are different forms of microscopic traffic flow dynamic model, but main difference is how to describe the relations among flow, speed and density. Based on the dynamic speed-density equations provided by Payne, Papagergiou improved on momentum equation and got the model as follows [1-2]:

$$q_i(k) = \alpha\rho_i(k)v_i(k) + (1-\alpha)\rho_{i+1}(k)v_{i+1}(k) \tag{9}$$

$$q_i(k) = \alpha\rho_i(k)v_i(k) + (1-\alpha)\rho_{i+1}(k)v_{i+1}(k) \tag{10}$$

$$v_i(k+1) = v_i(k) + \frac{T}{\tau}(v(\rho_i(k)) - v_i(k)) + \frac{T\xi}{\Delta_i}v_i(k)(v_{i-1}(k) - v_i(k)) - \frac{\gamma T}{\tau\Delta_i}\frac{\rho_{i+1}(k)-\rho_i(k)}{\rho_i(k)+\lambda} \tag{11}$$

$$v(\rho) = v_f \exp(-1/b(\rho/\rho_{cr})^b) \tag{12}$$

where T is the sample time, Δ_i is the length of i-th segment, $\rho_i(k)$ is the traffic flow density of i-th segment sampled at k points, $v_i(k)$ is the average velocity of i-th segment at k points, $q_i(k)$ is the flow from i-th segment to $(i+1)$-th segment at the time of kT, $r_i(k)$ is the traffic flow of on-ramp at the time of kT, $s_i(k)$ is the flow of off-ramp at the time of kT. In above equations, v_f is free velocity of the vehicles, ρ_{cr} is the critical density, τ is the time constant, γ is the constant of ramps related to ramp geography shapes, ξ, b, λ and α are adjustable coefficients.

(12) is a stable model of traffic flow, it describes the relationship between velocity and density at the stable traffic status.

We simulate the traffic flow data with (9)-(12) for 120 min. The sample time is 30second, totally 240 sample data are used as training data for FCM-RBF. Supposed the flow at first segment beginning is 1600 veh/h, average velocity at the beginning is 80 km/h with white noise. The flow at the on-ramp changes from 0 to 1500veh/h at random, The flow at the off-ramp changes from 0 to 1200 veh/h at random. The parameters in the model are v_f=98(km/h), ρ_{cr}=32(veh/km), b=3, τ=19.4/3600, γ=34.7, ξ=1, λ=1, α=0.95, and Δ=0.8 km [2].

After training the network, we use traffic flow model (9) to (12) to get another group of data in 120 minute. T is still 30 seconds. Supposed the flow at first segment beginning is 1600 veh/h, average velocity at the beginning is 80 km/h. Flow at the on-ramp increases gradually from 0 to 1500veh/h in 60 minutes, then decreases from 1500 veh/h to 0 in 60 minutes. The flow at the off-ramp changes from 0 to 1200 veh/h in 60 minutes gradually, then decreases form 1200veh/h to 0 in 60 minutes. Fig.2 to Fig.5 are the density and velocity from segment 1 to segment 4 respectively, real line is the output of the traffic flow dynamic model, and dashed line is the outputs of FCM-RBF. The results show that the FCM-RBF neural network can catch up highway traffic flow model effectively.

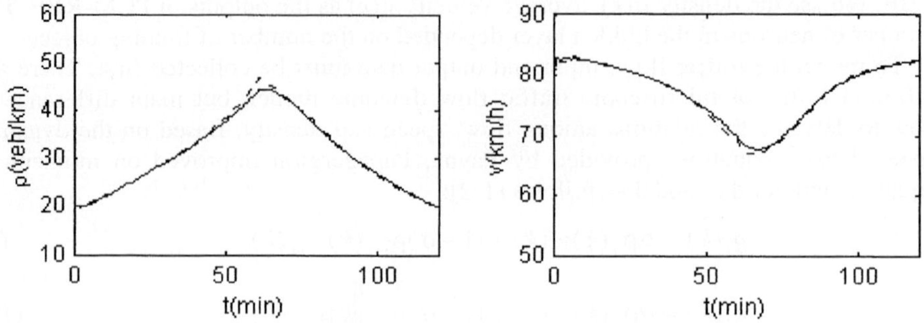

Fig. 2. Density and velocity of segment 1

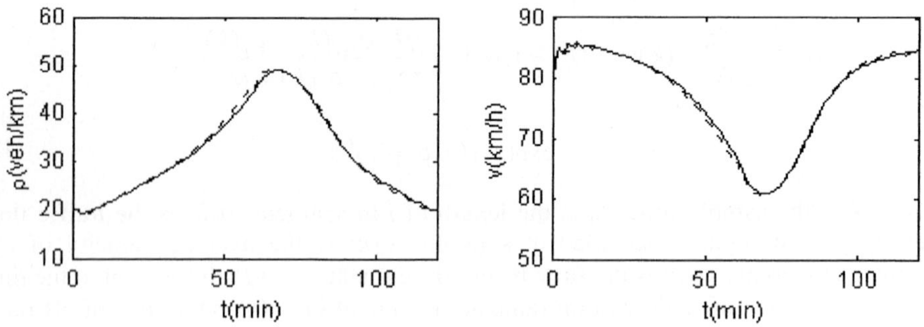

Fig. 3. Density and velocity of segment 2

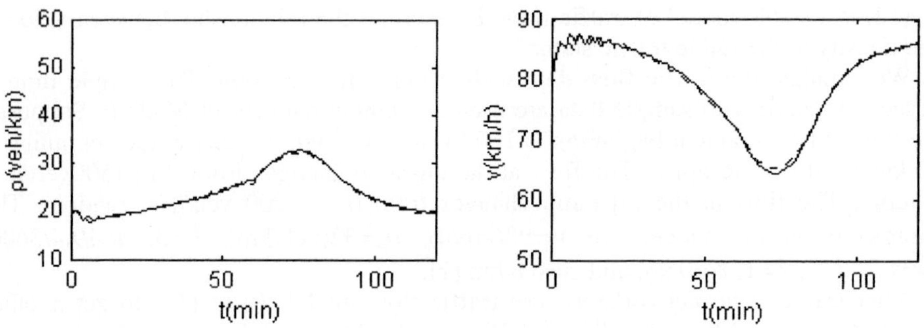

Fig. 4. Density and velocity of segment 3

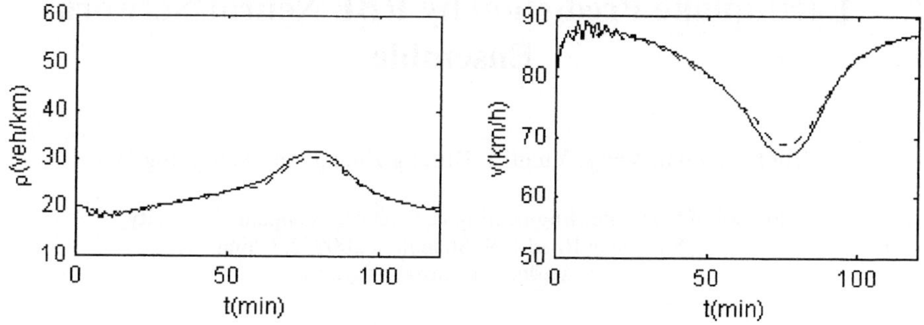

Fig. 5. Density and velocity of segment 4

4 Conclusion

The highly non-linear, dynamic characteristics of the macroscopic traffic flow problem require a modeling approach that is capable of dealing with the complex non-linear relationships between the speeds, flow and density. In this paper, we present the method of the highway traffic flow model by means of FCM-RBF neural network. The simulation shows FCM-RBF has a lot of superior peculiarities in highway traffic flow modeling. This algorithm has characteristic of training fast and is strong in network practicality, also can approach more accurately actual system. Future research will attempt to attain similar levels of performance using real traffic data.

References

1. Papageorgiou, M.: Multi-Layer Control System Design Applied to Freeway Traffic. IEEE Trans. Automation Control. vol. 29. (1984) 36–45
2. Luo, Z.W., Wu, Z.J., Han, Z.J.: Application of Extended Kalman Filter to the Freeway Traffic Flow Model. Acta Automatica Sinica. vol. 28. (2002) 90–96
3. Leonard, J.A., Kramer, M.A., Ugar, L.H.: Using Radial Basis Functions to Approximate Function and its Error Bounds. IEEE Trans. Neural Networks. vol. 3. (1991) 624–627
4. Catelani, M., Fort, A.: Fault Diagnosis of Electronic Analog Circuits Using A Radial Basis Function Network Classifier. Measurement. vol. 28. (2000) 147–158.
5. Bezdek, J.C, Pal, N.R.: An Integrated Approach to Fuzzy Learning Vector Quantization and Fuzzy C-Means Clustering. IEEE Trans. Fuzzy Systems. vol. 5. (1997) 622–628

Earthquake Prediction by RBF Neural Network Ensemble

YueLiu, YuanWang, YuanLi, Bofeng Zhang, and Gengfeng Wu

School of Computer Engineering & Science, Shanghai University,
Yanchang Road 149, Shanghai, 200072, China
yliu@mail.shu.edu.cn

Abstract. Earthquake Prediction is one of the most difficult subjects in the world. It is difficult to simulate the non-linear relationship between the magnitude of earthquake and many complicated attributes arising the earthquake. In this paper, RBF neural network ensemble was employed to predict the magnitude of earthquake. Firstly, the earthquake examples were divided to several training sets based on Bagging algorithm. Then a component RBF neural network, which was optimized by Adaptive Genetic Algorithm, was trained from each of those training sets. The result was obtained by majority voting method, which combined the predictions of component neural networks. Experiments demonstrated that the prediction accuracy was increased through using RBF neural network ensemble.

1 Introduction

Earthquake Prediction is one of the most difficult subjects in the world. It is difficult to give an overall and objective prediction because the earthquake is aroused by many complicated factors, and their relationship is non-linear. With the proposal of expert system, many researchers applied it to earthquake engineering. From 1980s, Chinese researchers have developed three generations of 'Expert System for Earthquake Prediction'. But there existed some limitations: 1) Bottleneck in knowledge acquirement. The knowledge can be only drawn by seismologist but cannot be acquired automatically. 2) High complexity and low efficiency because the operations in the old expert system were all serial. Especially, it was difficult to predict correctly when the input data were noisy, dirty and incomplete.

In order to overcome these limitations, neural network was applied because of its applicability in virtual situations in which a relationship between the predictor variables and predicted variables existed, even when that relationship was very complex and not easy to articulate in the usual terms of "correlations" or "differences between groups." However, the generalization ability of single neural network was not powerful. Neural network ensemble means that the generalization ability of a neural network system can be significantly improved through ensemble of a number of neural networks, i.e. training many neural networks and then combining their predictions. After originally proposed by Hansen and Salamon [1], the neural network system has been successfully applied across an extraordinary range of problem domains, in areas as diverse as face recognition [2,3], optical character recognition [4,

F. Yin, J. Wang, and C. Guo (Eds.): ISNN 2004, LNCS 3174, pp. 962–969, 2004.

5, 6], seismic signals classification [7], medical diagnosis [8,9], text and speech categorization [10], etc.

In this paper, a new approach based Bagging algorithm was proposed and realized in Chinese new generation of Expert System for Earthquake Prediction. RBF neural network was used to simulate the relationship between exception attributes and the magnitude of earthquake. We used Bagging on the training set to generate a number of RBF neural networks, in which adaptive genetic algorithm was employed to evolve the weights so that they can characterize the fitness of the neural networks in constituting an ensemble. Then we combined the predictions of component RBF neural networks obtained by the majority voting method, which was one of the most prevailing methods.

The rest of this paper was organized as follows. In Section 2, the whole expert system was shortly described in which neural network ensemble was realized. In section 3, the RBF neural network ensemble approach and procedure were proposed. In section 4, some experimental comparisons were presented. Finally, conclusions were drawn and several issues for future works were indicated.

2 Expert System for Earthquake Prediction

Expert system for earthquake prediction was composed of Knowledge Base, Inference Engine, Interpreter, and Knowledge Acquirement Machine as shown in Fig.1.

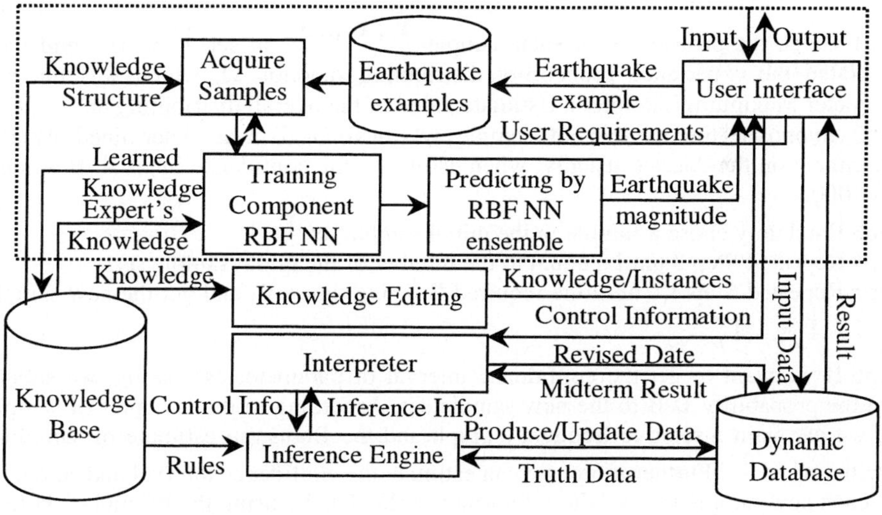

Fig. 1. Block diagram of the expert system for earthquake prediction

Here, we mainly discussed the top of the block diagram in details, which indicated the process of earthquake magnitude prediction by using RBF neural network ensemble.

3 Earthquake Prediction Using RBF Neural Network Ensemble

3.1 Preparation of the Samples

An earthquake example is composed of a number of exception attributes, such as earthquake belt, earthquake gap, etc, and the actual earthquake magnitude. In the traditional earthquake prediction approaches by using artificial neural network, the input data are values of some fixed exception attributes selected by seismologists according to their experiences. In this paper, we proposed a new exceptions-driven approach to prepare the training set. We only selected those values of the high-frequency exception attributes in the detected exceptions as the samples from the earthquake examples. We say now we get the initial samples, Γ. There will be another paper to depict it in details.

3.2 Division of the Samples

As for training component neural networks, the most prevailing approaches are Bagging and Boosting. Boosting was proposed by Schapire [11] and improved by Freund et al. [12,13]. Bagging was proposed by Breiman [14] based on bootstrap sampling [15]. It generates several training sets from the original training set and then trains a component neural network from each of those training sets. Here, we used the bootstrap to sampling.

Step1 Given the probability of each sample $x_1, x_2, \ldots x_n$ in set Γ is 1/n, and then calculated the experience probability distributing function Ω, which was the no-parameter Maximum Likelihood Estimate value of the overall distributing ω.
Step2 Supposed the number of sample sets was B. B was determined by the experiment on the dataset, usually, when estimated the confidence interval. B at least was 1000.
Step3 Randomly chose a sample in the initial sample set Γ.
Step4 Repeated Step 3 until n equal to the number of the initial sample set.
Step5 Repeated Step 3,4 until the required Bagging times, B, and get the new sample set, $\Gamma_1^*, \Gamma_2^*, \ldots, \Gamma_B^*$.
Step6 If we want to get the confidence interval of parameter estimating, we should give the probability 1/ B to the new sample set. According to the B copies of sample set, we can gain a relative frequency graph and the Bootstrap estimate of sampling distributing of Γ. Further more, we can estimate the confidence interval and so on.
For each new sample set, we should normalize the data by using the Formula 1 before training the neural network.

$$v' = \frac{v - \min_A}{\max_A - \min_A}(new\_\max_A - new\_\min_A) + new\_\min_A \tag{1}$$

3.3 Training the Component RBF Neural Network

Theory has been verified, for a given nonlinear function, using the radial based function (RBF) neural network can approximate it with any accuracy. And more important thing was that RBF neural network could avoid tedious redundancy computing of reverse direction propagation between the input layer and the hidden layer. And the speed of learning was 10^3-10^4 times faster than BP neural network [23]. So we used RBF as the component neural network of the ensemble.

In our RBF network, the radical function was Gauss function, which was most often used, for any input vector $X \in R^N$ (R^N was the input sample set).

$$R_i(x) = \exp[-\|X - C_i\|^2 / (2\sigma_i^2)] \quad \mathrm{i}=1,2,...,\mathrm{m} \tag{2}$$

Where $R_i(x)$ was the output of unit i in the hidden layer; X was an input vector with n dimensions, $X = \{X_p | X_p \in R^N, p = 1,2,...,K\}$; C_i was the Gauss function center of unit i in the hidden layer nodes; σ_i was the normalized parameter of the layer node i; m was the number of the hidden layers; P was the number of the samples.

The output layer can linear map $R_i(x)$ to y, which can be defined as:

$$y_j = \sum_i \varpi_{ij} R_i(x), j = 1,2,...q \tag{3}$$

Learning algorithm of RBF here was same to Moody and Darken algorithm [20]. But the centers were decided by using the nearest neighbor-clustering algorithm instead of K-Means Algorithm. The detailed description has been reported in [21].

In order to optimize the RBF, we employed the Adaptive Genetic Algorithm. The process was shown in Fig.2.

Coding. In Genetic Algorithm, each solution of an optimization problem is usually encoded as a bit string. But for the numerical optimization problem, encoding as real number is better than encoding as bit string. In this paper, the weights between the RBF nodes were all real. So we chose the real number encoding approach.

Fitness. It is important to choose a fitness function. The goal of Training RBF neural network was to make the overall error function of the network minimum, so we chose the network's error function as the fitness function of GA. The value was given by

$$E = \sum_{k=1}^{m} E_k = \sum_{k=1}^{m} \sum_{t=1}^{q} (y_t^k - C_t)^2 / 2 \tag{4}$$

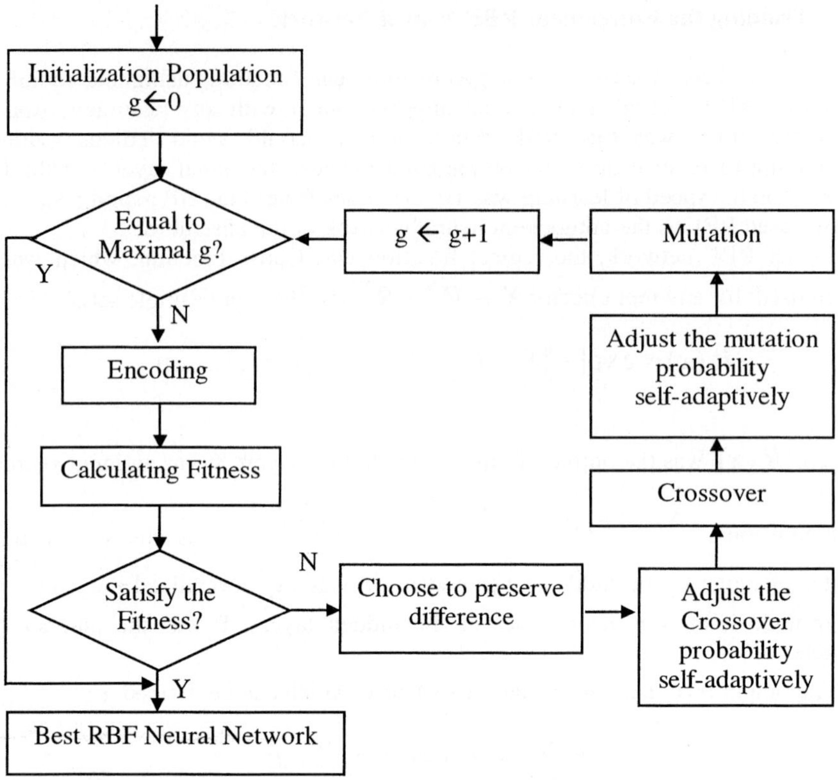

Fig. 2. The flow chart of training the component RBF neural network

3.4 Combination of the Results

Simple averaging and weighted averaging are prevailing methods for combining individual predictions of regression estimators. Majority voting and plurality voting are prevailing methods for combining individual predictions of classifiers. Here, we used majority voting method, which judged a prediction to be the final prediction earthquake magnitude if more than half of the individual networks voted to the prediction.

4 Experiment

We selected 17 attributes, mov_gra\dur1, l_level\dur1, ani_water\dur1, stre_meter\dur1, water\dur1, radon\dur1, app_res\dur1, s_level\dur1, Vp/Vs\dur, b_time\dur, frequency\fre_M, frequency\dur, train release\dur, gap\major axis, gap\dur, belt\dur, and the number of the exceptions, as the input attributes of RBF neural network. The earthquake examples from 1966 to 1998 [24, 25, 26] comprised

160 samples among which 44 samples made up the training set and 30 samples made up the test set after data preprocessing.

We used Bagging on the training set to generate 20 RBF neural networks. Each component RBF neural network had been trained for 5000 times. To compare the performance of earthquake prediction by ensemble of RBF neural networks, single RBF neural network based on the same training set had been trained for 15000 times. The results were shown in Table 1, where $\|\Delta M_1\|$ was the absolute value of the difference between the earthquake magnitude predicted by single RBF neural network and the actual earthquake magnitude, $\|\Delta M_2\|$ was the absolute value of the difference between the earthquake magnitude predicted by RBF neural network ensemble and the actual earthquake magnitude.

Table 1. Experimental comparison between single RBF NN and RBF NN ensemble

No.	Earthquake Name (Time)	Actual	Single RBF	RBF Ensemble	$\|\Delta M_1\|$	$\|\Delta M_2\|$
1	HeLinGeEr	6.4	6.18	6.28	0.22	0.12
2	Haicheng	7.4	7.01	7.07	0.39	0.33
3	LiYang(1974)	5.5	5.49	5.46	0.01	0.04
4	LongLin	7.4	7.29	6.99	0.11	0.41
5	BaYinBuRen	6.2	5.75	5.57	0.45	0.63
6	LiYang (1979)	7.1	5.67	6.79	1.43	0.31
7	FengZhen	6.2	5.54	5.68	0.66	0.52
8	JingTai	6.2	5.73	5.75	0.47	0.45
9	HeZe	6.1	5.89	5.69	0.21	0.41
10	South Yellow Sea(1984)	6.2	6.16	6.44	0.04	0.24
11	LingWu	5.3	5.61	6.21	0.31	0.91
12	LuQuan	6.3	6.01	6.05	0.29	0.25
13	DieBu	5.9	5.16	5.52	0.74	0.38
14	SeYang	5.1	5.22	5.39	0.12	0.29
15	ShuNan	5.7	5.33	5.68	0.37	0.02
16	DaTong	6.0	7.08	6.62	1.08	0.62
17	ChangShu	5.1	5.5	5.52	0.40	0.42
18	HaiYuan	7.6	5.6	6.63	2.00	0.97
19	ALaShan	5.3	5.56	5.45	0.26	0.15
20	JiaYuGuan	5.4	5.57	5.71	0.17	0.31
21	QiLian	5.0	5.26	5.43	0.26	0.43
22	Puer	6.3	5.32	5.36	0.98	0.94
23	QingHai	5.2	5.27	5.36	0.07	0.16
24	TaiWan Strait	6.3	6.94	6.03	0.64	0.27
25	BeiBuWang	6.1	6.22	5.84	0.12	0.26
26	FuShuNan	5.5	5.37	5.69	0.13	0.19
27	YongDeng	5.8	6.22	5.72	0.42	0.08
28	ChangShan	5.2	6.9	5.56	1.70	0.36
29	JiJiang	7.0	6.92	6.71	0.08	0.29
30	South Yellow Sea(1996)	6.1	6.36	6.49	0.26	0.39

Table 1 indicated the average error of single RBF NN was 0.09 where that of RBF NN ensemble was 0.07. If $\|\Delta M\| \leq 0.5$, then shown the prediction was correct. Based on this hypothesis, the prediction accuracy of single RBF NN was 73.3% where that of RBF NN ensemble was 80.0%. The prediction accuracy was improved.

5 Conclusion

It is a difficult thing to build a successful artificial neural network based application. The artificial neural network ensemble is a recently developed technology, which has the ability to significantly improve the performance of a system where a single artificial neural network is used. Since very easy to be used, it may potentially benefit the experts in artificial neural network research as well as the engineers in development of real world applications.

In this paper, we proposed a RBF neural network ensemble approach based on Bagging algorithm, which utilized artificial neural network ensemble to predict the earthquake magnitude in expert system for earthquake prediction. Firstly, an exceptions-driven sample acquirement method was devised to acquire samples automatically from the earthquake examples. Then the samples were divided to a number of training sets by a proposed approach based on Bootstrap algorithm. The component neural network was RBF neural network, which was optimized by Self-Adaptive Genetic Algorithm to increase the difference among the component RBF neural networks. Finally, the majority voting method was employed to combine the prediction results from each trained RBF neural network. Through adopting these techniques, the RBF neural network ensemble approach proposed in this paper achieved a high rate of prediction accuracy. In course of developing the Expert System for Earthquake Prediction, we found that employing strong pattern recognition techniques such as artificial neural network ensemble was only one key to improve the performance of the whole system. Interactive interpreter was another key. Moreover, we are glad to apply our ensemble approach to more real-world domains.

Acknowledgements. This research was supported by the National Natural Science Foundation of P. R. China (Grant No. 60203011) and the Science and Technology Development Foundation of Shanghai Municipal Education Commission (Grant NO. 02AK13).

References

1. Hansen, L.K., Salamon, P.: Neural Network Ensembles. IEEE Trans Pattern Analysis and Machine Intelligence 12 (1990) 993-1001
2. Gutta, S., Wechsler, H.: Face Recognition Using Hybrid Classifier Systems. In: Proc. ICNN-96. Washington, DC, IEEE Computer Society Press, Los Alamitos, CA (1996) 1017-1022
3. Huang, F., Zhou, Z., Zhang, H., Chen, T.: Pose Invariant Face Recognition. In: Proc. 4th IEEE International Conference on Automatic Face and Gesture Recognition, Grenoble, France, IEEE Computer Society Press, Los Alamitos, CA (2000) 245-250

4. Drucker, H., Schapire, R.: Improving Performance in Neural Networks Using a Boosting Algorithm. In: Hanson, S.J., Cowan, J.D., Giles, C.L. (eds.), Advances in Neural Information Processing Systems 5, Denver, CO, Morgan Kaufmann, San Mateo, CA (1993) 42-49
5. Hansen, L.K., Liisberg, L., Salamon, P.: Ensemble Methods for Handwritten Digit Recognition. In: Proc. IEEE Workshop on Neural Networks for Signal Processing, Helsingoer, Denmark, IEEE Press, Piscataway, NJ (1992) 333-342
6. Mao, J.: A Case Study on Bagging, Boosting and Basic Ensembles of Neural Networks for OCR. In: Proc. IJCNN-98, Vol. 3. Anchorage, AK, IEEE Computer Society Press, Los Alamitos, CA (1998) 1828-1833
7. Shimshoni, Y., Intrator, N.: Classification of Seismic Signals by Integrating Ensembles of Neural Networks. IEEE Trans Signal Processing 46 (1998) 1194-1201
8. Cunningham, P., Carney, J., Jacob, S.: Stability Problems with Artificial Neural Networks and the Ensemble Solution. Artificial Intelligence in Medicine 20 (2000) 217-225
9. Zhou, Z., Jiang, Y., Yang, Y., Chen, S.: Lung Cancer Cell Identification Based on Artificial Neural Network Ensembles. Artificial Intelligence in Medicine 24 (2002) 25-36
10. Schapire, R.E., Singer, Y.: BoosTexter: A Boosting-based System for Text Categorization. Machine Learning 39 (2000) 135–168
11. Schapire, R.E.: The Strength of Weak Learnability. Machine Learning 5 (1990) 197-227
12. Freund, Y.: Boosting a Weak Algorithm by Majority. Information and Computation 121 (1995) 256-285
13. Freund, Y., Schapire, R.E.: A Decision-theoretic Generalization of On-line Learning and an Application to Boosting. In: Proc. EuroCOLT-94, Barcelona, Spain, Springer-Verlag, Berlin (1995) 23-37
14. Breiman, L.: Bagging Predictors. Machine Learning 24 (1996) 123-140
15. Efron, B., Tibshirani, R.: An Introduction to the Bootstrap. Chapman & Hall, New York (1993)
16. Zhou, Z., Wu, J., Jiang, Y., Chen, S.: Genetic Algorithm Based Selective Neural Network Ensemble. In: Proc the 17th International Joint Conference on Artificial Intelligence, Seattle, WA 2 (2001) 797-802
17. Bauer, E., Kohavi, R.: An Empirical Comparison of Voting Classification Algorithms: Bagging, Boosting, and Variants. Machine Learning 36 (1999) 105-142
18. Perrone, M.P., Cooper, L.N.: When Networks Disagree: Ensemble Method for Neural Networks. In: Mammone, R.J. (ed.): Artificial Neural Networks for Speech and Vision, New York. Chapman & Hall (1993) 126-142
19. Wolpert, D.H., Macready, W.G.: An Efficient Method To Estimate Bagging's Generalization Error. Machine Learning 35 (1999) 41-55
20. Moody, J., Darken, C.: Learning with Localized Receptive Fields. In: Touretzky, D., Hinton, G., Sejnowski, T. (eds.): Proceedings of the 1988 Connectionist Models Summer School, San Mateo, CA (1989)
21. Han. J., Kamber, M.: Data Mining: Concepts and Techniques. Morgan Kaufmann Publishers Inc. (2000)
22. Wang, X., Cao, L.: Genetic Algorithm-Theory, Application and Implementation. Xian Jiaotong University Press (2002) 79-85
23. Robert, J., James, J.: Approximation of Nonlinear Systems with Radial Basis Function Neural Networks. IEEE Transactions on Neural Networks 12 (2001)
24. Zhang, Z. (ed.): China Earthquake Examples (1966-1985). Earthquake Press, Beijing (1990)
25. Zhang, Z. (ed.): China Earthquake Examples (1986-1991). Earthquake Press, Beijing (2000)
26. Chen, Q. (ed.): China Earthquake Examples (1992-1999). Earthquake Press, Beijing (2002)

Rainfall-Runoff Correlation with Particle Swarm Optimization Algorithm

Kwokwing Chau

Department of Civil and Structural Engineering, Hong Kong Polytechnic University, Hunghom, Kowloon, Hong Kong, People's Republic of China
cekwchau@polyu.edu.hk

Abstract. A reliable correlation between rainfall-runoff enables the local authority to gain more amble time for formulation of appropriate decision making, issuance of an advanced flood forewarning, and execution of earlier evacuation measures. Since a variety of existing methods such as rainfall-runoff modeling or statistical techniques involve exogenous input and different assumptions, artificial neural networks have the potential to be a cost-effective solution, provided that their drawbacks can be overcome. Usual problems in the training with gradient algorithms are the slow convergence and easy entrapment in a local minimum. This paper presents a particle swarm optimization model for training perceptrons. It is applied to forecasting real-time runoffs in Siu Lek Yuen of Hong Kong with different lead times on the basis of the upstream gauging stations or at the specific station. It is demonstrated that the results are both more accurate and faster to attain, when compared with the benchmark backward propagation algorithm.

1 Introduction

Precise prediction of rainfall-runoff is an important research topic in hydrologic engineering since it enables the local authority to gain more amble time for formulation of appropriate decision making, issuance of an advanced flood forewarning, and execution of earlier evacuation measures. However, the relationship between rainfall and runoff is not definite due to many pertinent factors such as ambient conditions, soil infiltration capacity, evapo-transpiration, etc. Existing rainfall-runoff modeling or statistical techniques require exogenous input and embrace different assumptions. In numerical modeling, the physical problem is represented by a highly coupled, non-linear, partial differential equation set. The involving processes are highly complex and uncertain which may demand huge computing cost and time. The representation by a deterministic or statistical model is not completely satisfactory.

Recently, owing to various advantages (built-in dynamism, data-error tolerance and no need to have exogenous input), artificial neural networks (ANN), and in particular, the feed forward back-propagation (BP) perceptrons, have been widely applied in water resources engineering [1]. However, the commonly used BP algorithm has the drawbacks of slow training convergence speed and easy entrapment in a local minimum.

F. Yin, J. Wang, and C. Guo (Eds.): ISNN 2004, LNCS 3174, pp. 970–975, 2004.

In this paper, the particle swarm optimization (PSO) algorithm is employed to train multi-layer perceptrons for rainfall-runoff prediction in Shatin catchment of Hong Kong with different lead times and input precipitation data at adjacent or that stations.

2 PSO Algorithm

PSO algorithm is initially developed as a tool for modeling social behavior and is able to optimize hard numerical functions [2-3]. It is currently adapted as a computational intelligence technique intimately related to evolutionary algorithms [4]. It is an optimization paradigm that mimics the ability of human societies to process knowledge. It has roots in two main component methodologies: artificial life on bird swarming; and, evolutionary computation.

Its principle is founded on the assumption that potential solutions will be flown through hyperspace with acceleration towards more optimum solutions. PSO is a populated search method for optimization of continuous nonlinear functions resembling the movement of organisms in a bird flock or fish school. Each particle adjusts its flying according to the flying experiences of both itself and its companions. In doing so, it keeps track of its coordinates in hyperspace which are associated with its previous best fitness solution, and also of its counterpart corresponding to the overall best value acquired thus far by any other particle in the population. Vector, as a convenient form for optimization problems, is used as the variable presentation to represent particles.

Its major advantages are relatively simple coding and hence computationally inexpensive. A similarity between PSO and a genetic algorithm is the initialization of the system with a population of random solutions and the employment of the fitness concept. However, the evolution of generations of a population of these individuals in such a system is by cooperation and competition among the individuals themselves. The population is responding to the quality factors of the previous best individual values and the previous best group values. The allocation of responses between the individual and group values ensures a diversity of response. The principle of stability is adhered to since the population changes its state if and only if the best group value changes. It is adaptive corresponding to the change of the best group value. The capability of stochastic PSO algorithm to determine the global optimum with high probability and fast convergence rate has been shown in other cases. In the following, it is adopted to train the multi-layer perceptrons.

3 Paradigm for Training of Network

If a three-layered preceptron is considered, $W^{[1]}$ and $W^{[2]}$ represent the connection weight matrix between the input layer and the hidden layer, and that between the hidden layer and the output layer, respectively. During training of the multi-layer preceptrons, the i-th particle is denoted by $W_i = \{W^{[1]}, W^{[2]}\}$ whilst the velocity of particle i is denoted by V_i. The position representing the previous best fitness value of any particle is denoted by P_i whilst the best matrix among all the particles in the population is recorded as P_b. Let m and n represent the index of matrix row and column, respectively, the following equation represents the computation of the new velocity of the particle based on its previous velocity and the distances of its current position from the best experiences both in its own and as a group.

$$V_i^{[j]}(m,n) = V_i^{[j]}(m,n) + r\alpha[P_i^{[j]}(m,n) - W_i^{[j]}(m,n)] + s\beta[P_b^{[j]}(m,n) - W_i^{[j]}(m,n)] \tag{1}$$

where j = 1, 2; m = 1, …, M_j; n= 1, …, N_j; M_j and N_j are the row and column sizes of the matrices W, P, and V; r and s are positive constants; α and β are random numbers in the range from 0 to 1. In the context of social behavior, the cognition part $r\alpha[P_i^{[j]}(m,n) - W_i^{[j]}(m,n)]$ represents the private thinking of the particle itself whilst the social part $s\beta[P_b^{[j]}(m,n) - W_i^{[j]}(m,n)]$ denotes the collaboration among the particles as a group. The new position is then determined based on the new velocity as follows.

$$W_i^{[j]} = W_i^{[j]} + V_i^{[j]} \tag{2}$$

The following equation is used to determine the fitness of the i-th particle in term of an output mean squared error of the neural networks

$$f(W_i) = \frac{1}{S}\sum_{k=1}^{S}\left[\sum_{l=1}^{O}\{t_{kl} - p_{kl}(W_i)\}^2\right] \tag{3}$$

where f is the fitness value, t_{kl} is the target output; p_{kl} is the predicted output based on W_i; S is the number of training set samples; and, O is the number of output neurons.

4 Application Case

The usefulness and applicability of any modeling system can only be affirmed by verifying its capability to mimic a particular case study with accurate depiction of real phenomena. This system has been verified and validated by applying to study the rainfall-runoff correlation in the Shatin catchment of Hong Kong [5-12]. Discharge at Siu Lek Yuen is forecasted with a lead time of 1 and 2 days based on the measured daily precipitations there and at Tate's Cairn. The data comprises continuous precipitations from 1998 to 2002 with 1460 pairs of daily records, of which two-third and one-third were used for training and validation, respectively. Data preprocessing is performed so that high and low discharge periods of the year and also rapid changes in runoffs are contained in both data sets.

Figure 1 shows the perceptron which has an input layer with one neuron, a hidden layer with three neurons, and output layer with two neurons. The input neuron represents the rainfall at the current day whilst the output nodes include the runoffs after 1 day and 2 days, respectively. All source data are normalized into the range between 0 and 1, by using the maximum and minimum values of the variable over the whole data sets. The number of population is set to be 30 whilst the maximum and minimum velocity values are 0.3 and -0.3 respectively.

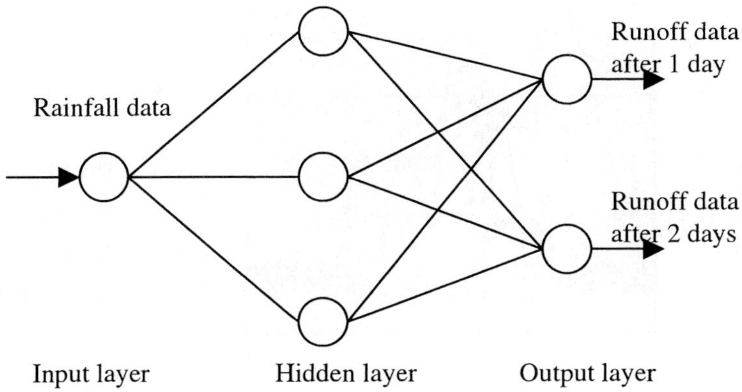

Fig. 1. Forecasting schema of PSO-based perceptrons network

Table 1. Normalized mean square errors at various fitness evaluation times during training for PSO-based and BP-based perceptrons

Fitness valuation time	Algorithm	Normalized MSE
5000	BP-based	0.21
	PSO-based	0.12
10000	BP-based	0.14
	PSO-based	0.09
20000	BP-based	0.11
	PSO-based	0.09

5 Analysis of Results

The performance of the PSO-based multi-layer ANN is evaluated in comparison with the benchmarking standard BP-based network. In order to provide a fair and common initial ground for comparison purpose, the training process of the BP-based perceptron commences from the best initial population of the corresponding PSO-based perceptron. Table 1 shows the normalized mean square errors (MSE) at various fitness evaluation times during training for PSO-based and BP-based perceptrons. The fitness evaluation time here for the PSO-based perceptron is equal to the product of the population with the number of generations. It can be observed that the PSO-based perceptron exhibits much better and faster convergence performance in the training process as well as better prediction ability in the validation process than those by the BP-based perceptron.

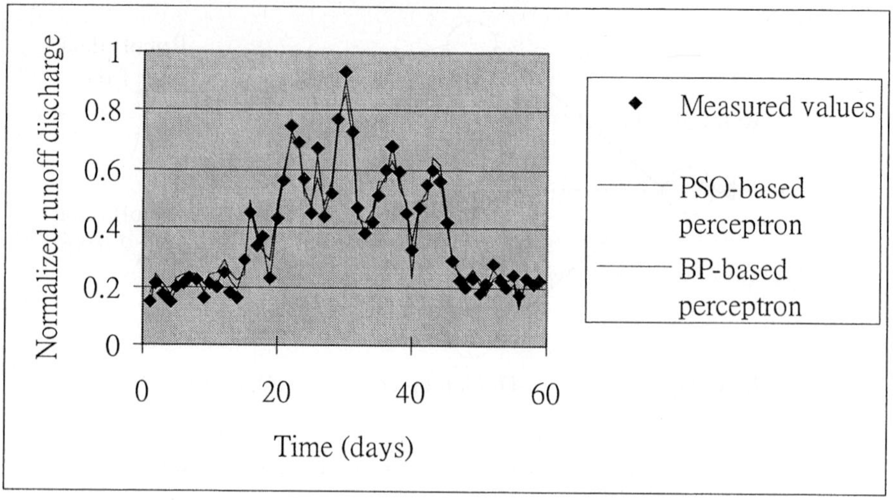

Fig. 2. 1 day lead time water discharge prediction by both perceptrons in the validation process

Figure 2 shows the 1 day lead time normalized water discharge prediction by both perceptrons in the validation process. Table 2 shows comparisons of the results for runoff forecasting at Siu Lek Yuen with both 1 day and 2 day lead times based on precipitation data at the same station (Siu Lek Yuen) and adjacent station (Tate's Cairn). It should be noticed that runoff forecasting at Siu Lek Yuen made by using the data collected at Tate's Cairn is generally better compared to the data collected at Siu Lek Yuen. From these analyses, as a final remark, it can also be observed that the performance of PSO-based perceptron for both training and verification simulations is better than its counterparts of BP-based perceptron.

Table 2. Results for runoff forecasting at Siu Lek Yuen based on precipitation data at the same and adjacent stations

Input data	Algorithm	Coefficient of correlation			
		Training		Validation	
		1 day ahead	2 day ahead	1 day ahead	2 day ahead
Siu Lek Yuen	BP-based	0.956	0.911	0.937	0.893
	PSO-based	0.975	0.964	0.953	0.941
Tate's Cairn	BP-based	0.973	0.945	0.957	0.907
	PSO-based	0.991	0.981	0.985	0.977

6 Conclusions

In this paper, a perceptron approach based on particle swarm optimization (PSO) paradigm is employed for real-time prediction of runoff discharge at Siu Lek Yuen in Shatin catchment with different lead times based on precipitation gauging stations at Tate's Cairn or at Siu Lek Yuen. The algorithm is shown to be capable to furnish model-free estimates in deducing the runoff output from the precipitation input, and hence is demonstrated to be a robust forewarning and decision-support aid. It is noticed from the training and verification simulation that, when compared with the benchmarking BP-based perceptron, the rainfall-runoff prediction results are apparently more accurate and at the same time consume less computational cost.

Acknowledgement. This research was supported by the Internal Competitive Research Grant of Hong Kong Polytechnic University (A-PE26).

References

1. Chau, K.W., Cheng, C.T.: Real-time Prediction of Water Stage with Artificial Neural Network Approach. Lecture Notes in Artificial Intelligence, 2557 (2002) 715-715
2. Kennedy, J., Eberhart, R.: Particle Swarm Optimization. Proceedings of the 1995 IEEE International Conference on Neural Networks. Perth (1995) 1942-1948
3. Kennedy, J.: The Particle Swarm: Social Adaptation of Knowledge. Proceedings of the 1997 International Conference on Evolutionary Computation. Indianapolis (1997) 303-308
4. Clerc, M., Kennedy, J.: The Particle Swarm—Explosion, Stability, and Convergence in a Multidimensional Complex Space. IEEE Transactions on Evolutionary Computation 6(1) (2002) 58-73
5. Chau, K.W.: Manipulation of Numerical Coastal Flow and Water Quality Models. Environmental Modelling & Software 18(2) (2003) 99-108
6. Chau, K.W.: Intelligent Manipulation of Calibration Parameters in Numerical Modeling. Advances in Environmental Research 8(3-4) (2004) 467-476
7. Chau, K.W.: A Prototype Knowledge-Based System on Unsteady Open Channel Flow in Water Resources Management. Water International 29(1) (2004) 54-60
8. Chau, K.W., Chen, W.: A Fifth Generation Numerical Modelling System in Coastal Zone. Applied Mathematical Modelling 25(10) (2001) 887-900
9. Chau, K.W., Chen, W.: An Example of Expert System on Numerical Modelling System in Coastal Processes. Advances in Engineering Software 32(9) (2001) 695-703
10. Chau, K.W., Cheng, C., Li, C.W.: Knowledge Management System on Flow and Water Quality Modeling. Expert Systems with Applications 22(4) (2002) 321-330
11. Chau, K.W., Lee, J.H.W.: Mathematical Modelling of Shing Mun River Network. Advances in Water Resources 14(3) (1991) 106-112
12. Chau, K.W., Yang, W.W.: Development of an Integrated Expert System for Fluvial Hydrodynamics. Advances in Engineering Software 17(3) (1993) 165-172

A Study of Portfolio Investment Decision Method Based on Neural Network*

Yongqing Yang[1,2], Jinde Cao[1**], and Daqi Zhu[2]

[1] Department of Mathematics, Southeast University, Nanjing 210096, China
[2] School of Science, Southern Yangtze University, Wuxi 214063, China
{yongqingyang,jdcao}@seu.edu.cn

Abstract. In the paper, a multi-objective programming of portfolio is proposed according to the assumption that total risk loss can be measured by the maximum of risk loss in all securities. After analyzing the risk preference of the investor and taking transaction cost function's linear approximation, the multi-objective programming model is transformed into simple-objective linear programming model. Based on neural network, a differential dynamical system for solving linear programming is constructed, and optimal portfolio decision is obtained.

1 Introduction

To decentralize the investment risk and obtain well-investment profit, an investor usually adopts the portfolio investment means, which is to invest some selected negotiable securities. The security investment theory proposed by H. M. Markowitz established the foundation for the modern portfolio investment theory. In this theory, H. M. Markowitz proposed two indexes about venture security appraisal: the expected profit rate and profit rate variance, and mean-variance model on portfolio [1]. In the paper, A new risk definition is given in another different view. By the new risk definition, we propose the multi-objective programming model of portfolio and compute the optimal solution with some methods by neural network.

2 Multi-objective Programming Model for Portfolio

In order to obtain the optimal portfolio decision, the key step is to construct an appropriate model. Assume an investor owns a great number of fund to be used as investment in a period, and the securities $S_i (i = 1, 2, \cdots, n)$ have been

* This work was jointly supported by the National Natural Science Foundation of China under Grant 60373067, the Natural Science Foundation of Jiangsu Province, China under Grants BK2003053 and BK2003001, Qing-Lan Engineering Project of Jiangsu Province and the Foundation of Southeast University, Nanjing, China under Grant XJ030714.

** Corresponding author

F. Yin, J. Wang, and C. Guo (Eds.): ISNN 2004, LNCS 3174, pp. 976–981, 2004.

selected and evaluated. The investor estimated that the average profit rate of security S_i is r_i and the risk loss rate is q_i by careful analysis. Obviously, if the investor buys security, he shall have to pay transaction cost. Assume transaction cost rate of security S_i is p_i, and the threshold of transaction cost in security S_i is u_i. Of course, not buy, not pay any. In addition, the bank deposit interests rate is r_0, no any risk and no any transaction cost. Due to the uncertainty of expected return, the investment risk will be taken place. To obtain net profit as much as possible and bring about risk loss as little as possible, a resolution is portfolio.

Let x_0,x_i,y_i denotes the proportion of fund deposit in bank and investment and transaction cost respectively, then the relation between y_i and $x_i$is as follows:

$$y_i(0) = 0, \quad y_i(x_i) = \frac{u_i}{M} p_i \ (0 < x_i \leq \frac{u_i}{M}), \quad y_i(x_i) = x_i p_i \ (x_i \geq \frac{u_i}{M}),$$

where, $i = 1, 2, \cdots, n$, M is the total sum of fund to be invested.

Generally, the investor always wish to obtain the maximum of expected return under the condition of the restricted risk level, or pursue the minimum of risk loss under the condition of expected profit target. Thus we can construct two simple-objective programming model (1) and (2).

$$\max R = r_0 x_0 + \sum_{i=1}^{n} (r_i x_i - y_i)$$

$$s.t. \begin{cases} x_0 + \sum\limits_{i=1}^{n} (x_i + y_i) = 1, \\ \sum\limits_{i=1}^{n} q_i x_i \leq Q_0 \\ x_i \geq 0, y_i \geq 0, \end{cases} \tag{1}$$

and

$$\min Q = \sum_{i=1}^{n} q_i x_i$$

$$s.t. \begin{cases} x_0 + \sum\limits_{i=1}^{n} (x_i + y_i) = 1, \\ r_0 x_0 + \sum\limits_{i=1}^{n} (r_i x_i - y_i) \geq R_0, \\ x_i \geq 0, y_i \geq 0, \end{cases} \tag{2}$$

In the two models, due to the risk level or profit target given in advance, it is difficult to decide the optimal investment. Moreover, risk relies on individual factors.In some cases, when the investor invests securities, he always hopes the loss in every security is not too much. So we give a new definition about total risk loss.

Definition: Suppose risk loss of security S_i is $q_i x_i$,then $Q = \max_{1 \le i \le n} \{q_i x_i\}$ is said to be total risk loss of portfolio. Namely, total risk loss can be measured by the maximum of risk loss of all securities.

According to new definition of total risk, we propose a multi-objective programming model (3).

$$\min\{-R, Q\} = \min\{-[r_0 x_0 + \sum_{i=1}^{n}(r_i x_i - y_i)], \max_{1 \le i \le n}\{q_i x_i\}\}$$

$$s.t. \begin{cases} x_0 + \sum\limits_{i=1}^{n}(x_i + y_i) = 1, \\ x_i \ge 0, y_i \ge 0, \end{cases} \tag{3}$$

As we are known, a core problem in the risk investment is to balance the profit and the risk, which involves individual risk preference. Optimizing security investment is to seek for the optimal combination between return and risk. Some investors are cautious and conservative. But other investors are willing to take on risk and purse the biggest profit. In portfolio, the investor always compares the profit with the loss, and selects a well-content proportion among all securities. Thus, we have to take risk preference of the investor by weight coefficient.

Let $\lambda(0 \le \lambda \le 1)$denotes the risk preference degree.

If the investor risks more, weight coefficient λ will be bigger. In particular, $\lambda = 0$ denotes the investor only purse profit disregard of risk. $\lambda = 1$ denotes the investor fears risk such that he will deposit all of his money in the bank.

According to the risk weight, we can transform model (3) into a simple-objective programming model (4).

$$\min Z = \lambda Q + (1 - \lambda)(-R),$$

$$s.t. \begin{cases} x_0 + \sum\limits_{i=1}^{n}(x_i + y_i) = 1, \\ x_i \ge 0, y_i \ge 0, \end{cases} \tag{4}$$

Considering variable y_i is piecewise function, total risk loss is measured by the maximum of risk loss of all securities. Obviously, model (4) is a nonlinear programming and is very difficult to be solved. Therefore we will again transform model (4).

1) Transaction Costs Function's Linear Approximation:

Because the threshold of investment is much small, it is almost impossible that the investment sum of a security is less than its threshold. Thus, we can assume $y_i = x_i p_i$. Although it is possible to bring out error among those securities whose investment sum are less than those thresholds. But the error cannot make the obvious influence in optimum result when the total investment sum is much bigger.

2) Risk Function's Transformation:

Let $x_{n+1} = Q$, we have $q_i x_i \le x_{n+1} (i = 1, 2, \cdots, n)$. Because the object is to optimize objective function $Z = \lambda x_{n+1} + (\lambda - 1)[r_0 x_0 + \sum_{i=1}^{n}(r_i - p_i)x_i]$, the optimal solution can let $\max\limits_{1 \le i \le n} \{q_i x_i\}$ achieve x_{n+1}. Therefore the model (4) can be transformed into a simple-objective linear programming model (5).

$$\min Z = \lambda x_{n+1} + (\lambda - 1)[r_0 x_0 + \sum_{i=1}^{n}(r_i - p_i)x_i],$$

$$s.t. \begin{cases} x_0 + \sum\limits_{i=1}^{n}(x_i + y_i) = 1, \\ q_i x_i - x_{n+1} \le 0, \quad (i = 1, 2, \cdots, n), \\ x_i \ge 0, y_i \ge 0. \end{cases} \tag{5}$$

3 The Neural Network Model for Portfolio

When solving linear programming problems, the traditional methods are the simplex method and the Karmarker method. But these methods involve an complex iterative process and longer computational time, and cannot satisfy the practical demand. Thus, it is urgent to develop a high-performance algorithm.

The neural network is a self-feedback complex system made of simple processing elements locally connected and has the characteristics of massive parallel computing and less computing time. In high-performance computation, neural network method has more superiorities than the traditional simplex method and the Karmarker method. In recent years, developing linear programming algorithm by neural network has widely attracted interests and made large progress. Many computing model were proposed by some authors. These models have their own advantages and disadvantages in reliability, stability, complexity and construction [2,3,4].

In the paper, using the dual theory, we propose the dual and mutual-feedback neural network system based on Hopfield neural network model. The stability of neural network model is analyzed by applying a energy function. And we also prove the theorem that the equilibrium point of neural network model is the optimal solution of linear programming problem and the minimum of the energy function.

Considering the linear programming problem as follows:

$$\min Z = C^T X,$$

$$s.t. \begin{cases} AX \ge b, \\ X \ge 0, \end{cases} \tag{6}$$

By the dual theory, [5] the dual problem of (6) is as follows:

$$\max W = b^T Y$$

$$s.t. \begin{cases} A^T Y \leq C, \\ Y \geq 0, \end{cases} \tag{7}$$

where $C \in R^m, X \in R^m, A \in A^{n\times m}, b \in R^n, Y \in R^n$,

Considering a new neural network model for solving problem (6) and (7), we design a differential dynamical system as follows:

$$\begin{cases} \frac{dX}{dt} = -[A^T(AX-b)^- + C(C^T X - b^T Y)], \\ \\ \frac{dY}{dt} = -[A(A^T Y - C)^+ - b(C^T X - b^T Y)], \end{cases} \tag{8}$$

where $X^- = (x_1^-, x_2^-, \cdots, x_m^-)^T \in R^m, x_i^- = \min(0, x_i), (i = 1, 2, \cdots, m)$ $Y^+ = (y_1^+, y_2^+, \cdots, y_n^+)^T \in R^n, y_i^- = \max(0, y_i), (i = 1, 2, \cdots, n)$

The energy function is in the following,

$E(X,Y) = \frac{1}{2}\|(AX-b)^-\|_2^2 + \frac{1}{2}\|(A^T Y - C)^+\|_2^2 + \frac{1}{2}(C^T X - b^T Y)^2$

Theorem X^*,Y^* are the optimal solution of the linear programming problem (6) and (7) respectively if and only if (X^*,Y^*) is the globally stable stability equilibrium point of system (8).

Proof (⇒)If X^*,Y^* are the optimal solution of the linear programming problem (6) and (7) respectively,then $C^T X^* = b^T Y^*, AX^* \geq b, A^T Y^* \leq C$.

So, we have

$$[A^T(AX^* - b)^- + C(C^T X^* - b^T Y^*)] = 0,$$
$$[A(A^T Y^* - C)^+ - b(C^T X^* - b^T Y^*)] = 0.$$

Namely, (X^*,Y^*) is equilibrium of model (8). In the following, we prove that (X^*,Y^*) is he globally stable stability equilibrium point of model (8).

Considering the positive define function

$$V = \frac{1}{2}\left\|\begin{bmatrix} X - X^* \\ Y - Y^* \end{bmatrix}\right\|_2^2.$$

Let $AX^* = b + U_1, A^T Y^* = C - U_2, U_1 \geq 0, U_2 \geq 0,$ then

$$\begin{aligned} \frac{dV}{dt} &= \frac{\partial V}{\partial X}\frac{dX}{dt} + \frac{\partial V}{\partial Y}\frac{dY}{dt} \\ &= -(X - X^*)^T[A^T(AX-b)^- + C(C^T X - b^T Y)] \\ &\quad -(Y - Y^*)^T[A(A^T Y - C)^+ - b(C^T X - b^T Y)] \\ &= [(AX-b)^T(AX-b)^- + (A^T Y - C)^T(A^T Y - C)^+ + C^T X - b^T Y)^2 \\ &\quad -U_1(AX-b)^- + U_2(A^T Y - C)^+] \\ &\leq -[\|(AX-b)^-\|_2^2 + \|(A^T Y - C)^+\|_2^2 + (C^T X - b^T Y)^2] \leq 0, \end{aligned}$$

so the equilibrium point of model (8) is global stable.

(⇐)If (X^*,Y^*) is the equilibrium point of model (8), then $\nabla E(X^*, Y^*) = 0$, so function $E(X,Y)$ can obtain the minimum in X^*,Y^*. Then $E(X^*, Y^*) = 0$, and $C^T X^* = b^T Y^*, AX^* \geq b, A^T Y^* \leq C$. By the dual theory, X^*, Y^* are the optimal solution of the linear programming problem (6) and (7), respectively.

4 An Example

Assume an investor owns one million dollars and has selected four securities to be invested in a period. By careful analysis, the relation data about these securities have been obtained. Namely, the average profit rate of securities are $r = (0.28, 0.21, 0.23, 0.25)$, and the risk loss rate are $q = (0.025, 0.015, 0.055, 0.026)$, and transaction cost rate are $p = (0.01, 0.02, 0.45, 0.65)$. Assume the threshold of transaction cost in security S_i is $u = 2500$ and the bank deposit interests rate is $r_0 = 0.05$ and risk preference coefficient $\lambda = 0.8$. We devise the optimal portfolio plan in the following.

Firstly, to devise the optimal portfolio plan, we construct multi-objective programming model (9) as follows:

$$\max R = 0.05x_0 + 0.28x_1 + 0.21x_2 + 0.23x_3 + 0.24x_4 - y_1 - y_2 - y_3 - y_4$$

$$\min Q = \max\{0.025x_1, 0.015x_2, 0.055x_3, 0.026x_4\}$$

$$s.t. \begin{cases} x_0 + \sum_{i=1}^{4}(x_i + y_i) = 1, \\ x_i \geq 0, y_i \geq 0, (i = 1, 2, 3, 4), \end{cases} \tag{9}$$

$$y_i(0) = 0, \quad y_i(x_i) = \frac{1}{400}p_i \ (0 < x_i \leq \frac{1}{400}), \quad y_i(x_i) = x_i p_i \ (x_i \geq \frac{1}{400}),$$

According to Section (2) and Section (3), model (9) can be transformed into simple-objective linear programming problem and a neural network system can be constructed. Its solution is (0.369, 0.615, 0, 0, 0).

5 Conclusions

In the paper, we construct the multi-objective model of portfolio according new risk loss definition and transform the multi-objective model into simple-objective linear programming model by risk preference of the investor. Using the neural network method, we are able to devise the optimal portfolio plan.

References

1. Markowitz, H.M.: Portfolio Selection. Journal of Finance, **7** (1952) 77–91
2. Wu, X.Y., Xia, Y.S., Li, J., Chen, W.K.: A High-Performance Neural Network for Solving Linear and Quadratic Programming Problems. IEEE Trans. Neural Networks, **7** (3) (1996) 643–651
3. Zhang, S., Constantinides, A.G.: Lagrange Programming Neural Networks. IEEE Transactions on Circuits and Systems: Analog and Digital Signal Processing, **39** (7) (1992) 441–452
4. Xia, Y., Wang, J.: Global Exponential Stability of Recurrent Neural Networks for Solving Optimization and Related Problems. IEEE Trans. Neural Networks, **11** (4) (2000) 1017–1022
5. Gass, S.I.: Linear Programming Methods and Application. Fifth Edition. Mc Graw Hill Book Company (1984)

Portfolio Optimization for Multi-stage Capital Investment with Neural Networks

Yanglan Zhang[1] and Yu Hua[2]

[1]Business School, Wuhan University, P.R. China
yhuastarmpls@hotmail.com
[2]Computer School, Wuhan University, P.R. China
yhuastar@hotmail.com

Abstract. Portfolio optimization has been researched for several years in financial engineering. The problem can be described as maximization of the expected growth rate of a portfolio with relatively small variance of the portfolio growth rate. In this paper, novel method of multi-stage portfolio optimization with partial information is introduced in detail. Investors can redistribute wealth in multiple periods in order to maximize the returns and minimize the risks in financial market. The experiments with random data can prove the method efficient and proper.

1 Introduction

In the financial markets, the models with deterministic coefficients are only good for a relative short period of time and cannot respond to changing conditions [1]. As for the investors, risk control over bankruptcy is crucial to a successful investment via dynamic portfolio selection, but there are not feasible and efficient ways solve the problem of risk control over bankruptcy and dynamic portfolio selection. In this paper, an analytical model is proposed to realize an optimal return with a mean-variance tradeoff and a proper risk control over bankruptcy. The Markowitz's mean-variance portfolio selection model in a single period aims to maximize the final wealth with the mean time to minimize the risk. And the criterion of the risk was the variance in order to enable investors to seek highest return with the acceptable risk level.

In [2], the authors introduced a class of dynamic Markowitz's mean-variance portfolio selection problem. Zhou and Li [3] presented a stochastic linear-quadratic control framework to study the continuous-time version of the Markowitz's problem. However, there are certain limitations when the stochastic differential equations and geometric Brownian motion models were applied into the continuous-time formulation of the mean-variance problem. Because the parameters, such as the interest rate and the stock volatility rates, are assumed to be insensitive to real markets changes. In particular, some intermediate states added could provide more accurate description for drastic changes in markets than two states of extremes. The Markowitz's mean-variance model [4] has been recently extended in [5] to a multi-period setting. The analytical expression of the efficient frontier for the multi-period portfolio selection is derived. Recently, the artificial intelligence techniques, such as genetic algorithm, ge-

F. Yin, J. Wang, and C. Guo (Eds.): ISNN 2004, LNCS 3174, pp. 982–987, 2004.

netic programming and reinforcement learning, have been applied into the fields of financial engineering. Moody *et al.* examine a recurrent reinforcement-learning algorithm that seeks to optimize an online estimate of the Sharpe ratio. Pictet *et al.* employ a GA to optimize a class of exponentially weighted moving average rules, but run into serious over fitting and poor out-of-sample performance. The portfolio optimization can be considered in multiple stages in order to redistribute the current wealth in an optimal way to realize the final wealth maximized.

2 Analytical Model

2.1 Problem Description

A capital market has $(n+1)$ risky securities with random rates of returns when investors enter the market at time 0 with an initial wealth x_0. Let x_t be the wealth of the investor at the beginning of the t th period, u_t^i, $i = 1,2,\cdots,n$, be the amount invested in the i th risky asset at the beginning of the t th time period. The amount invested in the 0th risky asset at the beginning of the t th time period is equal to $x_t - \sum_{i=1}^{n} u_t^i$. The wealth dynamics can be thus given as

$$x_{t+1} = \sum_{i=1}^{n} e_t^i u_t^i + (x_t - \sum_{i=1}^{n} u_t^i) e_t^0 \tag{1}$$

where $u_t = [u_t^1, u_t^2, \cdots, u_t^n]'$ and $P_t = [P_t^1, P_t^2, \cdots, P_t^n]'$. The mean-variance formulation for multi-period portfolio selection can be posed as follows when security 0 is taken as a reference: $\max\limits_{u} E(x_T) - w_T Var(x_T)$ where $w_T \in (0,\infty)$ represents the tradeoff between the expected terminal wealth and the associated risk represented by the variance of the terminal wealth. Furthermore, dynamic portfolio policy derived from maximizing a terminal objective may result in some very aggressive policies in the early stages, which are not penalized in the mathematical formulation. In real implementation, however, this negligence of risk control will be penalized by a high possibility of a miscarriage of an investment plan.

2.2 Method with Markov Decision Process

The Markov model can be described as: $X = \{1,2,\cdots,N_X\}$ is the set of finite states, $Y = \{1,2,\cdots,N_Y\}$ is the set of outputs, and $U = \{1,2,\cdots,N_U\}$ is the set of decisions and control. In addition, $P(u) = p_{ij}(u)$ is the $N_X \times N_X$ state transition matrix, and $Q(u) = q_{xy}(u)$ is the probability of state y when the state is x and action u has been taken.

When time is at t, the system is in the state $x_t = i$, and the state can be converted into state $x_{t+1} = j$ with probability $p_{ij}(u)$, when the decision $u_t = u$ has been taken. Then, the probability of y_{t+1} can be delineated as $q_{xy}(u) = \{y_{t+1} = y \mid x_{t+1} = i, u_t = u\}$ and new decision may be produced based on $W_t = (y_0, u_0 y_1, \cdots, u_t, y_{t+1})$.

Thus, in general, the purpose of risk-sensitive control problem is to look for the sequence of control policy $Z = \{z_0, z_1, \cdots, z_{T-1}\}$ with the finite horizon T. The process is $\xi^{\gamma}(p_0, z) = \lim_{T\to\infty} \sup \frac{\gamma}{T} \log M_{p_0}^{z} \exp(C_T / \gamma)$, where γ is the risk factor, M is the expectation, and C_T is the total cost for the corresponding horizon. We may present the average cost:

$$j_{x_0}(u) = \lim_{N\to\infty} \sup \frac{1}{N} M_u \left[\sum_{n=1}^{N} c(X_n, u_n) \mid X_0 = x_0 \right] \tag{2}$$

The aim is to find the optimal policy $u^* \in U$ that may has the minimum cost for all initial state $x_0 \in S$. According to [2], the optimal policy u^* is always of the form of a regular Markovian policy to some extent and can be realized by selecting the action to be taken at each state. Especially, at each stage with state i and control action u, the long-term expected value of the average-cost corresponding to policy τ is

$$\theta^{\tau} = \lim_{N\to\infty} \frac{1}{N} E\left\{ \sum_{n=0}^{N-1} f[X_n, \iota(X_n)] \right\} \tag{3}$$

The potential vector g is defined by Poisson equation $g = f/(I - P + e\pi)$, where I is the identity matrix, π is the steady state probability vector, $e = (1, \cdots, 1)^T$. At the same time, $L_i(j) = \min\{n : n \geq 0, X_n = j\}$ with starting state $X_0 = i$ in Markov chain $X = \{X_n, n \geq 0\}$. Thus, the difference between the system performances of Markov chain starting from state i and state j can be measured as follows:

$$E\left\{ \sum_{n=0}^{L_i(j)-1} [f(X_n - \eta) \mid X_0 = i] \right\} \tag{4}$$

Then, simple and reliable algorithms can realize the iterative optimization to decide the optimal control policy.

The long-run average cost of the Markov chain under policy τ is given by

$$\eta^{\tau} = \lim_{N\to\infty}\frac{1}{N+1}\sum_{n=0}^{N} E^{X_0,\tau}\{f(X_n,\tau(X_n))\} = \pi^{\tau} f^{\tau} \tag{5}$$

where $E^{X_0,\tau}$ is the expectation with the initial state X_0 under policy τ. And π^{τ} is the row of Cesaro-limiting probability matrix with identical rows. Furthermore, the MDP problem can be solved in polynomial time with respect to the sizes of the state-space and the action-space. At the same time, the preferred algorithms to solve MDP problem are value iterations and policy iterations based on the dynamic programming [1].

The term on the k th row and i th column of the matrix N is denoted by n_{ki}. There are several actions that the investor may choose during the wealth assignment. When a new period arrives, the current wealth may be assigned depending on the chosen action. Then, the current state may move to the corresponding next state. The main aim is to achieve the actions that may bring optimal results.

So, we describe the action state space as $N \in S$ by a M-vector $u = [u_1, u_2, \ldots, u_M]$ where $u_k \in \{0,1,2,\ldots, C\}$. And, the value of u_k denotes the number of units of wealth assigned. Then, the action space may be denoted as

$$K_N = \left\{ u : u_k \in \{0,1,\ldots,C\} and \sum_{k=1}^{M}\sum_{i=1}^{C} n_{ki} + \max\{u_1, u_2, \ldots, u_M\} \le C \right\} \tag{6}$$

The action space is used to express the set of all possible actions. When action u is the chosen action, u_k units of wealth will be assigned. Thus, the state will move to corresponding next state. Of course, if the tasks of certain wealth are completed, the units of wealth will be released naturally.

In addition, we describe the transition rate of the process of the assignment as a MDP. The transition rate from state N to state Q when action $u \in K_N$ is chosen. We describe the process with the parameter $a^{u}_{N,Q}$. And ∂_{ki} is defined in order to express the $M \times C$ matrix with zeros in all but the (k,i) th entry where it has a one.

Thus, we describe the whole process of the assignment comprehensively. When a new period arrives at the state N, the state N chooses a certain action and moves to a new state at the rate of λ_k. Hence, the current state has one more unit than the previous state. Likewise, the process of the wealth assignment will move to a new state that has one less wavelength at the rate of $n_{ki} u_{ki}$. The process of the assignment can be described as the MDP showed in Fig.1.

Thus, the algorithm with MDP in multiple stages, i.e. MSPO (Multiple Stages Portfolio Optimization) can be introduced:

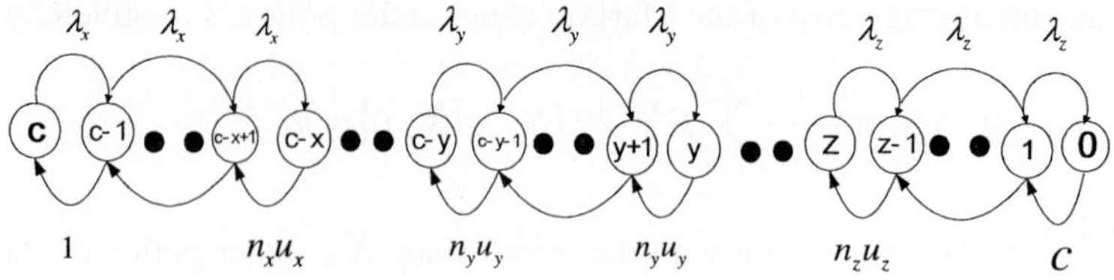

Fig. 1. The assignment process of the units of wealth

- Step1. Choose N and state X_0
- Step 2. Decide the initial policy τ_0 and set $k = 1$
- Step 3. Compute the length of each stage and estimate the potential average cost according to the transition matrix in N transitions.
- Step 4. If the risk is higher than the threshold, goto step 5
- Else accumulate the units of wealth $k := k + 1$, goto step 3.
- Step 5. Done and show the results.

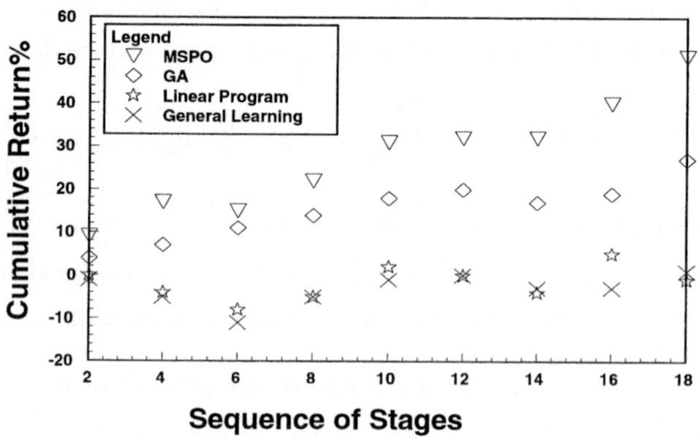

Fig. 2. Cumulative monthly returns

3 Performance Study

In order to prove the proposed model efficient and valid, the experiment was made with monthly returns of four stocks from January 1964 until July 1999. To obtain a performance estimate for each model, we used the sequential validation procedure, by first training on 60 months and thereafter retraining every 12 months, each time testing on the 12 months following the last training point. Assume that an investor invests his wealth with an initial wealth of 1 unit. The planning horizon is five months, and one time period is one month.

In Fig.2, cumulative monthly returns with different slippages were shown and MSPO can achieve better cumulative returns than GA, Linear Programming and General Learning. And the fluctuations of MSPO were also rather smaller and then the investors could achieve satisfactory returns. The average of the cumulative returns in multiple stages were more than 40.25%.

4 Conclusion

Bankruptcy control is pivotal to be addressed in dynamic portfolio optimization for the volatility of financial markets. In this paper, Portfolio optimization with stochastic market data was realized through the MSPO algorithm based on MDP method. Then, an integration of bankruptcy control and dynamic portfolio selection has been considered in this paper. Furthermore, the multi-stage description with mean-variance formulation was also introduced and an optimal investment policy can be generated to help investors not only achieve an optimal return in the sense of a mean-variance tradeoff, but also have a good risk control over bankruptcy. At last, the results of corresponding experiments could prove the ideas and method efficient and proper.

References

1. Postolos, P.A., Referes, N., Burgess, A.N., Bentz, Y.: Neural Networks in Financial Engineering A Study in Methodology. IEEE Transaction on Neural Networks, Vol. 8. No. 6. 11 (1997) 1222-1267
2. Yin, G., Zhou, X.Y.: Markowitz's Mean-Variance Portfolio Selection with Regime Switching From Discrete-Time Models to Continuous-Time Limits. IEEE Transactions on Automatic Control, Vol. 49. No. 3. 3 (2004) 349-360
3. Zhou, X.Y.: Continuous-Time Mean-variance Portfolio Selection A Stochastic LQ Framework.. Appl. Math. Optim, Vol. 42 (2000) 19-33
4. Markowitz, H.M.: Portfolio Selection Efficient Diversification of Investment. New York Wiley (1959)
5. Li, D., Ng, W.L.: Optimal Dynamic Portfolio Selection Multi-Period Mean Variance Formulation. Math. Finance, Vol.10 (2000) 387-406

Efficient Option Pricing via a Globally Regularized Neural Network

Hyung-Jun Choi, Hyo-Seok Lee, Gyu-Sik Han, and Jaewook Lee

Department of Industrial Engineering,
Pohang University of Science and Technology,
Pohang, Kyungbuk 790-784, Korea.
{chj,ronalee,swallow,jaewookl}@postech.ac.kr

Abstract. Nonparametric approaches of option pricing have recently emerged as alternative approaches that complement traditional parametric approaches. In this paper, we propose a novel neural network learning algorithm for option-pricing, which is a nonparametric approach. The proposed method is devised to improve generalization and computing time. Experimental results are conducted for the KOSPI200 index daily call options and demonstrate a significant performance improvement to reduce test error compared to other existing techniques.

1 Introduction

Since a major breakthrough in the pricing of stock options made by Black, Scholes, and Merton in the early 1970s, Black-Scholes model has had a huge influence on the way that traders price and hedge options. The celebrated original Black-Scholes pricing formula has a closed form based on a dynamic-hedging argument and a no arbitrage condition, but has shown systematic and substantial bias because of its unrealistic assumptions. To improve pricing performance, Black-Scholes formula has been generalized to various parametric option pricing models while closed form expressions are not available. However, as shown in [7], these parametric option pricing models didn't succeed and the resulting models are too complex, have poor out-of-sample performance. Even the most complex modern parametric models are imperfect and are outperformed by simple, less general models. They often produce parameters inconsistent with underlying time series and inferior hedging and retain systematic price bias they were intended to eliminate [2].

Prompted by shortcomings of modern parametric option-pricing, nonparametric approaches, which do not rely on pre-assumed models but instead try to uncover/induce the model, have recently emerged as alternative approaches. Especially, it is demonstrated in ([6]) that neural networks can be used successfully to estimate an option pricing formula with good out-of-sample pricing and delta-hedging performance. Since overfitting problems influence option-pricing models very much, many researches on neural networks' learning algorithms for option pricing have been focused on reducing overfitting, most of them relying

F. Yin, J. Wang, and C. Guo (Eds.): ISNN 2004, LNCS 3174, pp. 988–993, 2004.

on homogeneity assumption ([3], [6]), which can be inconsistent with complex asset price dynamics.

By relaxing this homogeneity assumption, in this paper, a new learning algorithm for option pricing is proposed to reduce overfitting. The performance of the proposed algorithm is verified concerned with improving the option-pricing accuracy and efficiency by applying it to KOSPI200 call options.

The rest of this paper is organized as follows. In section 2, an option and Black Scholes model are introduced and the problem for option pricing is formulated. Section 3 presents a description of the proposed method for option-pricing with empirical results shown in Section 4. Section 5 contains a conclusion.

2 Problem Formulation

2.1 Black-Scholes Model

To verify the merits of nonparametric and computational methods of option pricing, the performance of these methods needs to be measured against a widely used alternative model. In this paper, Black-Scholes model is used for this purpose. Black and Scholes model provides a closed-form formula for pricing call options using the assumption that the underlying follows a random diffusion process, which is described by

$$\begin{aligned} p^{(BS)} &= f^{(BS)}(\mathbf{x}) = SN(d_1) - Xe^{-r(T-t)}N(d_2), \\ d_1 &= [\ln(S/X) + (r + \sigma^2/2)(T-t)]/\sigma\sqrt{T-t}, \\ d_2 &= d_1 - \sigma\sqrt{T-t}, \end{aligned} \tag{1}$$

where S is the underlying asset price, X is the strike price of option, $T-t$ is the time to maturity, σ is the volatility (standard deviation) of the underlying asset, r is the continuously compounded risk-free interest rate, and $N(\cdot)$ is the cumulative normal distribution function. Unless otherwise specified, in this paper, we use these set of inputs normally given by $\mathbf{x} = (S, X, T-t, \sigma, r)$ as the training input data for a fair comparison with the BS.

2.2 Neural Network Model

The goal of a nonparametric method using a neural network is to model a mapping of some input variables onto the observed option prices. Our model is a typical three-layer perceptron. This one consists of an input layer, a hidden layer, and an output layer, interconnected by modifiable weights, represented by links between layers:

$$f^{(NN)}(\mathbf{w};\mathbf{x}) = \phi\left(\sum_{j=1}^{n_H} w_j \phi\left(\sum_{i=1}^{d} w_{ji}x_i + w_{j0}\right) + w_0\right) \tag{2}$$

where $\phi : \Re \rightarrow \Re$ is a nonlinear activation function, n_H denotes the number of hidden units and $\mathbf{w}$ denotes a synaptic weight vector. For a given training sample data $\{\langle \mathbf{x}_k, p_k \rangle : k = 1, ..., n\}$, these outputs $f^{(NN)}(\mathbf{w}; \mathbf{x}_k)$ are compared to the target values p_k, the past closing option market prices; any difference corresponds to an error (called a training error). This error or criterion function is some scalar function of the weights and is minimized when the network outputs natch the desired outputs. Thus the weights are adjusted to reduce this measure of error.

One drawback of this neural network approach is the virtual absence of inferential procedures to determine the best model specification; the neural network may overfit in which cases the training errors can be made very small if we increase the model complexity enough, but the test errors (out-of-sample errors) may be large, resulting in poor generalization of the network. To produce a network that generalizes well, we will propose a method based on a form of "regularization" [5], which will be presented in the next section.

3 The Proposed Method

Our proposed method is based on viewing the supervised learning of an MLP as an unconstrained minimization problem as follows;

$$\min_{\mathbf{w}} E_\lambda(\mathbf{w}) = \frac{1}{2} \sum_{i=1}^{N} \left(p_k - f^{(NN)}(\mathbf{w}; \mathbf{x}_k) \right)^2 + \lambda \|\mathbf{w}\|^2 \qquad (3)$$

where λ is a regularization paramter, which represents the relative importance of the complexity-penalty term with respect to the performance-measure term and $f^{(NN)}(\cdot)$ is the neural network model given in Eq. (2). To minimize Eq.(3), the proposed method consists of two phases: a Levenverg-Marquardt based local search phase and a weight reduction tunneling phase. In the first phase, a Levenverg-Marquardt algorithm is employed for fast convergence to a local minimum. After converging to a local minimum, the second phase, a weight reduction tunneling technique is invoked. If the second phase finds a new weight vector $\mathbf{w}$ with lower cost function value than those of previously obtained weight vectors, then the first phase is initiated again. These processes are repeated iteratively until some terminating condition is met.

Phase I: a Levenverg-Marquardt Based Local Search: The basic idea of a Levenverg-Marquardt based local search method [4] is to minimize the second-order Taylor approximation of E_λ in Eq.(3) around the current point $\mathbf{w}$. By solving this problem, the weight update rule at the current point $\mathbf{w}$ can be obtained as follows

$$\Delta\mathbf{w} = -\left[\mathbf{J}^T(\mathbf{w})\mathbf{J}(\mathbf{w}) + \alpha\mathbf{I}\right]^{-1} \mathbf{J}^T(\mathbf{w})\mathbf{e}(\mathbf{w}) \qquad (4)$$

where $\mathbf{J}(\mathbf{w}) \in \Re^{(n+m)\times m}$ is the Jacobian matrix at $\mathbf{w} = (w_1, ..., w_m)$ of

$$\mathbf{e}(\mathbf{w}) = (\mathrm{p}_1 - f^{(NN)}(\mathbf{w}; \mathbf{x}_1), ..., \mathrm{p}_n - f^{(NN)}(\mathbf{w}; \mathbf{x}_n), \sqrt{\lambda}w_1, ..., \sqrt{\lambda}w_m)^T \qquad (5)$$

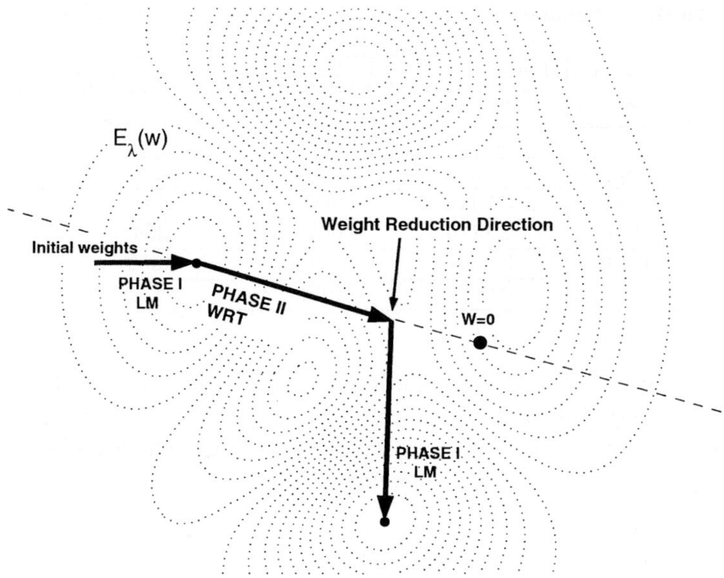

Fig. 1. Basic Idea of Tunneling Scheme

If $\alpha \to 0$, then it approaches the classical Newton method whereas if $\alpha \to \infty$, then it approaches the gradient decent method. After some iterations with a proper value of α, it will converge to a local minimum point. One nice property of the Levenberg-Marquardt based local search is that it retains the rapid rate of convergence of Newton's method, but is also generally applicable and globally convergent with proper choices of α [8].

Phase II: Weight Reduction Tunneling: Despite of its rapid convergence property, the Levenberg-Marquardt based local search would get trapped at a local minimum. To handle this problem, we adopt a tunneling strategy ([1], [9]), which attempts to compute a weight vector of next descent. Many implemented tunneling methods, however, often failed to find an appropriate tunneling direction which helps us to escape from a local minimum in high dimensional cases such as neural networks.

To overcome this difficulty, we propose a so-called *weight reduction tunneling* technique to enhance tunneling performance and efficiency. The proposed technique tunnels into the zero point of the weight space until it encounters the first local minimum point along the sequence of points generated by

$$\mathbf{w}(k+1) = \gamma \mathbf{w}(k), \qquad 0 < \gamma < 1 \tag{6}$$

Here γ represents the step length of tunneling. Then by invoking Phase I from this point, it obtains another local minimum.

The proposed tunneling technique has several advantages. First, since the obtained weight vector from Eq. (6) is located normally outside the convergent

Table 1. Simulation Result using six neural network techniques

Algorithm	Training Error	Test Error	Learning time(sec.)
BS model	1.7681	2.2684	-
EBP	0.5267	1.0668	765
R-EBP	0.4504	0.9568	4614
LM	0.2440	0.9664	438
BR	0.1776	1.0613	264
DTR	0.4467	0.9514	4659
Proposed	0.0096	0.0119	58.35

region of the initiated poor local minimum, successful tunneling happens more frequently than existing tunneling techniques. Second, during the tunneling process, the size of the weight vector becomes very small, which makes it easier to find a new weight vector of next descent with a lower E_λ.

4 Experimental Results

In this section, we verify how well the proposed method works compared with Black-Scholes (BS) model and other neural network algorithms which include

- Error Backpropagation(EBP)[6]
- Regularized EBP(R-EBP)[5]
- Levenberg-Marquadt Algorithm(LM)[4]
- Bayesian Regularization Algorithm(BR)[3]
- Dynamic Tunnelling Regularized Algorithm(DTR)[9]

The data used in the empirical analysis are daily closing quotes of the KOSPI200, during the time periods January 1, 2000 - December 31, 2000. The KOSPI200 index is a value-weighted combination of the 200 most traded securities in Korea Stock Exchange market. To avoid nonsynchronicity caused by trading effect, we use the closing prices of KOSPI call options as price data. We train a network using 2972 samples(80%) that are randomly selected. And the remaining 743 samples(20%) are used to test each neural network. Based on the universal approximation property of feedforward neural networks that no more than two hidden layers are needed to approximate an arbitrary function, we've used three-layer perceptrons as our neural network architecture and the network structures are chosen to the ones that show high performance for each algorithm. The neural network model for the proposed method is '4-10-1' with $S, X, T-t, \sigma$ as the input vector. Here we used σ available in the market, which was given according to the general implied volatility formula. We excluded r from the input vector because for the KOSPI200 data, it changes little during the time periods.

Simulation results are shown in Table 1. The network pricing perfomance measure is the test error and learning time which is becoming an important factor to choose the proper method for option-pricing. Not surprisingly, EBP and R-EBP provide the poorest performance because they take too much time and error is relatively severe. LM and BR algorithm have comparatively less train error and learning time, but shows more test error compared to other algorithms. DTR algorithm has comparatively less test error but shows intensive learning time. The proposed method shows the superior performance compared to the others in terms of both efficiency and accuracy. Especially, it reduces train error and test error by 99.4%, 99.5% respectively, compared with BS model.

5 Conclusion

In this paper, we've proposed a novel neural network training algorithm to mitigate overfitting and improve generalization for option pricing. The proposed method is the two-phase learning algorithm of an MLP using a Levenberg-Marquardt based local search and a weight reduction tunneling technique. To show the effectiveness of the proposed algorithm, we've conducted an experiment with daily KOSPI200 Index European options obtained from the Korea Stock Exchange for the period January 2000 to December 2000. The experimental results demonstrate that the new algorithm not only successfully improves generalization but also is substantially faster than other existing learning algorithms. An application of the method to other options remains to be further investigated.

Acknowledgement. This work was supported by Com^{2}Mac-KOSEF.

References

1. Barhen, J., Protopopescu, V., Reister, D.: TRUST: A Determionistic Algorithm for Global Optimization. Science, Vol. 276 (1997) 1094-1097
2. Bakshi, G., Cao, C., Chen, Z.: Empirical Performance of Alternative Option-Pricing Models. The Journal of Finance, Vol. 52, No. 5 (1997) 2003-2049
3. Gencay, R., Qi, M.: Pricing and Hedging Derivative Securities with Neural Networks: Bayesian Regularization, Early Stopping, and Bagging. IEEE Transactions on Neural Networks, Vol. 12, No. 4 (2001) 726-734
4. Hagan, M.T., Menhaj, M.: Training Feedforward Networks with the Marquart Algorithm. IEEE Transactions on Neural Networks, Vol. 5, No. 6 (1994) 989-993
5. Haykin, S.: Neural Networks: A Comprehensive Foundation. Prentice-Hall, New York (1999)
6. Hutchinson, J.M., Lo, A.W., Poggio, T.: A Nonparametric Approach to Pricing and Hedging Derivative Securities via Learning Networks. Journal of Finance, 49, (1994) 851-889
7. Lajbcygier, P.: Literature Review: The Problem with Modern Parametric Option Pricing. Journal of Computational Intelligence in Finance, (1999)
8. Nocedal, J., Wright, S.J.: Numerical Optimization. Springer, New York (1999)
9. RoyChowdhury, P., Singh, Y., Chansarkar, A.: Dynamic Tunneling Technique for Efficient Training of Multilayer Perceptrons. IEEE Transactions on Neural Networks, Vol. 10, No. 1 (1999) 48-55

Effectiveness of Neural Networks for Prediction of Corporate Financial Distress in China

Ji-Gang Xie, Jing Wang, and Zheng-Ding Qiu

Institute of information science, Beijing Jiaotong University, Beijing 100044, China
xiejigang823@sohu.com, zdqiu@center.njtu.edu.cn

Abstract. The study examines the effectiveness of two types of neural networks in predicting corporate financial distress in China. Back-propagation and LVQ neural networks are considered. The neural networks are compared against Logistic Regression, which is the most popular one among the traditional methods. The results show that the level of Type I and Type II errors varies greatly across techniques. The neural networks have low level of Type I error and high level of Type II error, while Logistic Regression has the reverse relationship. Since the cost of Type I is more expensive than that of Type II error in this field. We demonstrate that the performance of the neural networks tested is superior to Logistic Regression.

1 Introduction

Financial distress does not only affect the corporate itself, but it also affects the overall economy. Since the economic cost of corporate failure is large, there is a great need for models giving timely and robust prediction for this particular event. Therefore, the prediction of corporate financial distress is one of the key issues in modern finance.

Since the seminal paper of Altman [1], the prediction of corporate financial distress has been studied for about 40 years in the developed countries. Multiple Discriminant Analysis (MDA) and Logistic Regression (LR) had been widely used in practice and in many academic studies during 1960s-1980s [2]. The two models are essentially linear models that are used under some assumptions, such as the ratios being independent and the two classes having Gaussian distributions with equal covariance matrices. However, many research results show that the assumptions do not accord with the financial data [3][4]. In the 1990s, the limitation of "traditional statistical tools" was challenged, as Neural Networks (NNs) started to appear in the financial distress prediction [5][6]. Almost in all of these publications the conclusion has been that the prediction accuracy can be increased by using NNs instead of traditional MDA or LR.

The research in this area just begins in China. And mostly studies limit to using traditional statistic models, such as MDA and LR [7][8][9]. This study focused on the effectiveness of NNs in the prediction of financial distress in Chinese listed companies.

F. Yin, J. Wang, and C. Guo (Eds.): ISNN 2004, LNCS 3174, pp. 994–999, 2004.

2 Methods

2.1 Logistic Regression

LR has been used to investigate the relationship between binary response probability and explanatory variables. The method fits linear logistic regression model for binary response data by the method of maximum likelihood. LR weights the independent variables and assigns a z score in a form of failure probability to each company in a sample, the conditional probabilities or z scores lying between 0 and 1(on a sigmoidal curve) are determined with the next formula :

$$P(y=1|X)=P_1(X)=\frac{1}{1+e^{(-Z)}}=\frac{1}{1+e^{-(w_0+w_1x_1+\cdots+w_nx_n)}} \tag{1}$$

The exponent in formula (1) expresses the so-called 'Z scores'. The coefficients are estimated with the maximum likelihood method. Therefore, the likelihood function in formula (2) is maximized:

$$L(b)=\prod_{i=1}^{n} P_1(X_i)^{y_i}[1-P_1(X_i)]^{1-y_i} \tag{2}$$

with

$P_1(X_i)$ =probability of failure of ith company,

w =vector with n estimable parameters $w_1,\ldots,w_n$,

X_i =vector with characteristics of ith company,

y_i =1 if ith company fails, 0 if it doesn't fail.

2.2 Neural Networks

Neural networks (NNs) are attractive for classification problems because they are capable to learn from noisy data and to generalize, the effectiveness of neural network classification has been tested empirically. NNs have been successfully applied to variety of real world classification tasks in industry, business and science [10]. In this study we apply two popular types of NNs: feed-forward networks with back-propagation (BP) and learning vector quantization (LVQ) as the learning algorithms.

The feed-forward network consists of three layers, the input layer, the hidden layer with a number of hidden neurons, and the output layer with a single neuron. The hidden layer uses the hyperbolic tangent sigmoid activation function, while the output layer uses the log-sigmoid activation function. BP algorithm attempts to minimize the least-squares error function defined as the squared of the differences between the actual network output and the targeted output.

LVQ is a pattern classification method. Each category to be learned is assigned a number of reference vectors, each of which has the same number of dimensions as the input vectors of the categories. In the recognition stage, an unknown input vector is classified as the category of the reference vector that is closest to that input vector. Kohonen suggested three different ways of adjusting the position of the reference vectors resulting into algorithms LVQ1, LVQ2 and LVQ3. The details of LVQ training algorithms can be found in [11].

3 Data Set

Considering the background of Chinese stock market, we define being "specially treated" (ST) received by listed companies due to their abnormal financial performance as the indicator of financial distress. The reasons are as follows:

1. Though it is a common practice for Western researchers to use bankruptcy as an indicator of financial distress, up to now there has been no listed company that has gone bankrupt on the Chinese stock market. Since qualifications for public trading are still regarded as some sort of precious shell resources at the moment, listed firms will be taken over by other firms even in they are faced with bankruptcy. Therefore, it is almost impossible for a Chinese listed firm to file for bankruptcy. So at least for the time being, it is inappropriate to use bankruptcy to define financial distress in Chinese listed firms.
2. ST, as an objective event, has a high measurability.
3. From the experience of those firms that have got out of the shadow of ST, most firms have to resort to extensive assets reorganization before they can get rid of the ST label. This indicates that ST reflects corporate financial distress to a certain degree.

In order to access the predictive performance, we use the financial statements in fiscal year t-2 to predict the corporate status in year t. e.g. a company was specially treated in 1998, the study use its financial data stated in 1996. Our data set consists of 549 Chinese listed companies, including 115 ST and 434 non-ST companies during 1998-2002. We select 80 companies of ST and non-ST respectively as training set. The rest is used for testing.

In our study we do not apply any pre-classification system for the data, i.e., we do not use any technique to ensure that all non-ST companies in our sample are sound. Thus, the non-ST companies are selected randomly from the basic data. This means that in our sample of sound companies there are companies suffering from financial distress, maybe even companies that are going to fail within a year. This is done on purpose because we wanted to have all types of non-ST companies in our data. The absolute prediction accuracy may suffer from this starting point, but the comparison between LR and NNs is more realistic when the data is not manipulated ex ante.

4 Selected Ratios

As there is a lack of economic theoretic guidance to decide which ratios should be used for financial distress prediction, in this study, we have selected 16 ratios that previous domestic research has found useful in financial distress predictions as a starting point [7][8][9]. These ratios represent each of the five major ratio categories: profitability, solvency, liquidity, efficiency and growing ability (see Table 1 for details).

Table 1. Ratios

	Ratios	Ratio categories
R1	Cash/Net Sales	Liquidity
R2	Current Assets/Current Liabilities	Liquidity
R3	Current Assets/Total Assets	Liquidity
R4	Growth Rate of Total Assets	Growing Ability
R5	Growth Rate of Prime Activity Revenue	Growing Ability
R6	Long Term Debt/Equity	Efficiency
R7	Net Income/Total Assets	Profitability
R8	Profit of Prime Activity/Total Assets	Profitability
R9	Profit of non-Prime Activity/Total Assets	Profitability
R10	Quick Assets/Total Assets	Liquidity
R11	(Accounts) Receivable Turnover	Efficiency
R12	Retained Earnings/Total Assets	Profitability
R13	Stock Turnover	Efficiency
R14	Total Debt/Equity	Solvency
R15	Total Debt/Total Assets	Solvency
R16	Working Capital/Total Assets	Liquidity

We select the ratios by conducting single variable tests on the differences of five categories of financial ratios between ST and non-ST companies. If it is significant at the specified level, in our study 0.05, the ratio is entered into the models. Finally, six ratios are selected from the original list of 16 ratios: R16, R14, R11, R8, R9, R12.

5 Prediction Results

For BP networks, the number of input nodes is six. The number of output node is one, and the target output of the network was set 0.9 or 0.1, which represent ST and non-ST respectively. The threshold value is specified as 0.5 when network is used for classification testing; if output is larger than 0.5, then the result is regarded as 0.9, otherwise 0.1. As for how to determine the number of hidden nodes, we employ so-called trial and error by starting from a relatively larger network, and then pruning the number gradually till get an appropriate one. The selected network topology(input nodes-hidden nodes-output) included 6:20:1, 6:10:1, 6:8:1, 6:6:1, 6:4:1, 6:3:1, 6:2:1, 6:1:1. In addition, the initial learning rate is 0.3, and initial weights are random numbers between –0.3 and 0.3.

Regarding LVQ, the number of input nodes is six, and the number of output nodes is two that stand for ST and non-ST respectively. Since the number of hidden layer nodes has deep influence on its ability to classify. Here we construct various networks by selecting the number of hidden nodes as 20, 60 and 120, in order to find a LVQ with the best performance. Then three algorithms were used to test the effects with the initial learning rate a=0.1.

Empirical results (see Table 2 for details) show that BP(6-3-1) and LVQ(6-120-2) are superior to LR both in prediction accuracy and in Type I error. LR has the highest Type I error rates and lowest Type II error rates, while most NNs having low Type I error rates. So NN is an encouraging method since evidences show that the Type I error rates could be 35 times more costly than Type II error rates[12].

Table 2. Empirical results

Models	Network topology	Prediction accuracies	Type I error	Type II error
BPNN	6-1-1	69.41%	51.43%	28.53%
	6-2-1	81.23%	45.71%	16.10%
	6-3-1	83.55%	25.71%	15.54%
	6-4-1	82.52%	31.43%	16.10%
	6-6-1	81.49%	31.43%	17.23%
	6-8-1	80.46%	37.14%	17.80%
	6-10-1	81.49%	31.43%	17.23%
	6-20-1	79.18%	37.14%	19.21%
LVQ1	6-20-2	80.72%	22.86%	18.93%
	6-60-2	82.78%	28.57%	16.10%
	6-120-2	83.80%	22.86%	15.54%
LVQ2	6-20-2	68.12%	34.29%	31.64%
	6-60-2	82.26%	22.86%	17.23%
	6-120-2	83.29%	20.00%	16.38%
LVQ3	6-20-2	82.01%	22.86%	17.51%
	6-60-2	82.52%	25.71%	16.67%
	6-120-2	84.83%	22.86%	14.41%
LR		82.42%	81.8%	4.4%

6 Conclusions

In this study the group of original ratios is formed by selecting those ratios which in previous central studies have been found good predictors of financial distress. With the ratios subset selected by conducting single variable tests, the empirical results show that Type I error rates of NNs are much lower than that of LR. And the prediction accuracies of BP(6-3-1) and LVQ(6-120-2) are slightly better than that of LR. All this results show the effectiveness of NNs in the prediction of financial distress in Chinese listed companies. Next, we will attempt to improve the classification performance of NNs by selecting the optimal ratios set from the original ratios group.

References

1. Altman, E.I.: Financial Ratios, Discriminant Analysis and the Prediction of Corporate Bankruptcy. *J. Finance*, Vol. 39, No. 4 (1968) 589-609
2. Jones, F.L.: Current Techniques in Bankruptcy Prediction. *Journal of Accounting Literature*, Vol. 6 (1987)
3. Eisenbeis, R.A.: Pitfalls in the Application of Discriminant Analysis in Business, Finance, and Economics. *Journal of Finance*, Vol. 32 (1977) 875-900
4. Tam, K.Y., Kiang, M.: Managerial Applications of Neural Networks: the Case of Bank Failure Predictions. *Management Science*, Vol.381 (1992) 927-947
5. Charalambous, C., Charitou, A., Kaourou, F.: Comparative Analysis of Artificial Neural Network Models: Application in Bankruptcy Prediction. *American Accounting Association Conference* (1999)
6. Atiya, A.F.: Bankruptcy Prediction for Credit Risk Using Neural Network: a Survey and New Results. *IEEE Trans. NN*, Vol.12, No. 4 (2001)
7. Chen Jing: Study on Forecast of Financial Distress in China's Listed Companies. *Accounting Research*, (1999) 31-38
8. Wu Shi-Nong, Lu Xian-Yi: A Study of Models for Predicting Financial Distress in China's Listed Companies. *Economic Research Journal* (2001) 46-55
9. Jiang Xiu-Hua, Ren Qing, Sun Zheng: A Forecasting Model of Financial Distress for Listed Companies. *FORECASTING*, Vol.21; No.3 (2002) 56-61
10. Widrow, B., Rumelhard, D.E., Lehr, M.A.: Neural networks: Applications in industry, business and science. Commun. ACM, Vol.37 (1994) 93-105
11. Kohonen, T.: The Self-Organizing Map. *Proc. IEEE*, Vol. 78, No. 9 (1990)
12. Altman, E.I., Halderman, R., Narayaman, P.: Zeta Analysis. *Journal of Banking and Finance* (1977)

A Neural Network Model on Solving Multiobjective Conditional Value-at-Risk*

Min Jiang[1], Zhiqing Meng[2], and Qiying Hu[3]

[1] School of Economics and Management,
Xidian University, Xi'an, 710071,China
j_yx@263.net

[2] College of Business and Administration,
Zhejiang University of Technology, Hangzhou, 310032,China
mengzhiqing@zjut.edu.cn

[3] College of International Business & Management,
Shanghai University, Shanghai, 201800,China
huqiying@sina.com

Abstract. Conditional Value-at-Risk (CVaR) is a new approach for credit risk optimization in the field of finance engineering. This paper introduces the concept of $\boldsymbol{\alpha}$-CVaR for the case of multiple losses under the confidence level vector $\boldsymbol{\alpha}$. The problem of solving the minimal $\boldsymbol{\alpha}$-CVaR results in a multiobjective problem (MCVaR). In order to get Pareto efficient solutions of the (MCVaR), we introduce a single objective problem (SCVaR) and show that the optimal solutions of the (SCVaR) are Pareto efficient solutions of (MCVaR). We construct a nonlinear neural networks model with an approximate problem (SCVaR)′ of (SCVaR). We may get an approximate solution (SCVaR) by solving this nonlinear neural networks model.

Keywords: Credit risk, Loss functions, $\boldsymbol{\alpha}$-CVaR, Pareto efficient solutions

1 Introduction

Value-at-Risk (VaR) is a measure for the potential loss in the financial market. With respect to a specified probability level α, the α-VaR of a portfolio is the lowest amount y such that, with probability α, the loss will not exceed y. VaR has achieved its great success in practice. However, research shows that VaR has undesirable properties both in theory and in practice [1].

The concept of Conditional Value at Risk (CVaR) ([1]) was presented mainly to solve the problem of sub-additivity. With a specified probability α, the α-CVaR is the conditional expectation of losses above the VaR. CVaR overcomes several limitations of VaR and has some good properties, especially its good computability. Recently, there are several studies on CVaR area([2-5]). However, the

* The project was supported by the National Natural Science Foundation of China with grant 72072021.

F. Yin, J. Wang, and C. Guo (Eds.): ISNN 2004, LNCS 3174, pp. 1000–1006, 2004.

losses that should be considered in practical risk management are often multiple, such as those due to interest risk, exchange risk, shares risk, commercial risk. So, the risk management problems are of multiobjective. Krokhmal, Palmquist and Uryasev [5] solve efficient frontier problems with three losses under the framework of CVaR. However, they were not concern in theoretical studies.

The Hopfield neural networks is a very important tool to solve optimization problems [7-9]. This paper will construct a Hopfield neural networks model for the multiple conditional VaR. We present the optimization problem of minimizing the multiple $\boldsymbol{\alpha}$-CVaR as a multiple programming. We prove that under some conditions the multiple $\boldsymbol{\alpha}$-CVaR problem can be transformed into a single objective nonlinear programming problem. Based on this, we construct a nonlinear neural networks model as its an approximate problem.

2 CVaR with Multiple Losses

Let $f_i(\boldsymbol{x}, \boldsymbol{\xi}) \in R^n \times R^m \to R^1 (i = 1, 2, \cdots, I)$ be loss functions which depend upon the decision vector $\boldsymbol{x} \in X \subset R^n$ and the random vector $\boldsymbol{\xi} \in R^m$. X is the set of possible all portfolios. For simplicity, we assume that $\xi \in R^1$ is a random variable with the probability density function $p(z)$. However, this assumption is not critical for our discussions in the following. Denote by $\Psi_i(\boldsymbol{x}, \cdot)$ the distribution function of the loss $f_i(\boldsymbol{x}, \xi)$, i.e.,

$$\Psi_i(\boldsymbol{x}, y) = P\{f_i(\boldsymbol{x}, \xi) \leq y\} = \int_{f_i(\boldsymbol{x}, z) \leq y} p(z)dz, \qquad i = 1, 2, \cdots, I. \tag{1}$$

Definition 2.1 Given $\alpha_i \in (0, 1)$, $i = 1, 2, \cdots, I$, and $\boldsymbol{x} \in X$, the $\boldsymbol{\alpha}$-VaR (of the loss) associated with $\boldsymbol{x}$ under the confidence level vector $\boldsymbol{\alpha} = (\alpha_1, \alpha_2, \cdots, \alpha_I)$ is defined by

$$y^*(\boldsymbol{x}) = \min\{y \mid \Psi_i(\boldsymbol{x}, y) \geq \alpha_i, \quad i = 1, 2, \cdots, I\}. \tag{2}$$

When $I = 1$, the $\boldsymbol{\alpha}$-VaR is exactly the α-VaR for one loss defined by [1]. For $i = 1, 2, \cdots, I$,

$$y_i^*(\boldsymbol{x}) = \min\{y \mid \Psi_i(\boldsymbol{x}, y) \geq \alpha_i\}$$

is the α_i-VaR of x for the ith loss. We have $y^*(\boldsymbol{x}) = \max\{y_1^*(\boldsymbol{x}), y_2^*(\boldsymbol{x}), \cdots, y_I^*(\boldsymbol{x})\}$.

For each $i = 1, 2, \cdots, I$, we define as in [1] that

$$\phi_{\alpha_i}(\boldsymbol{x}, y) = (1 - \alpha_i)^{-1} \int_{f_i(\boldsymbol{x}, z) \geq y} f_i(\boldsymbol{x}, z)p(z)dz. \tag{3}$$

Definition 2.2 For given $\alpha_i \in (0, 1)$, $i = 1, 2, \cdots, I$, and $\boldsymbol{x} \in X$, we call the vector $(\phi_{\alpha_1}(\boldsymbol{x}, y^*(\boldsymbol{x})), \phi_{\alpha_2}(\mathbf{x}, y^*(\boldsymbol{x})), \cdots, \phi_{\alpha_I}(\boldsymbol{x}, y^*(\boldsymbol{x})))$ the $\boldsymbol{\alpha}$-CVaR associated with $\boldsymbol{x}$ under the confidence level vector $\boldsymbol{\alpha} = (\alpha_1, \cdots, \alpha_I)$.

It is not easy to compute $\phi_{\alpha_i}(\boldsymbol{x}, y)$. Thus we introduce another loss function as in [1]. For $i = 1, 2, \cdots, I$, we define

$$F_{\alpha_i}(\boldsymbol{x}, y) = y + (1 - \alpha_i)^{-1} \int_{z \in R} [f_i(\boldsymbol{x}, z) - y]^+ p(z) dz, \tag{4}$$

where $[f_i(\boldsymbol{x}, z) - y]^+ = \max(0, f_i(\boldsymbol{x}, z) - y)$. By [1], we have the following result under the condition that all the loss functions satisfy $P\{f_i(x, \xi) = y\} = 0$ for each $i = 1, 2, \cdots, I$ and each $Y \in R^1$.

Lemma 2.1 For each $i = 1, 2, \cdots, I$, $F_{\alpha_i}(\boldsymbol{x}, y)$ is a continuous differential and convex function and

$$\min_{y \in R} F_{\alpha_i}(\boldsymbol{x}, y) = \phi_{\alpha_i}(\boldsymbol{x}, y_i^*(\boldsymbol{x})), \tag{5}$$

$$\frac{\partial F_{\alpha_i}(\boldsymbol{x}, y)}{\partial y} = (1 - \alpha_i)^{-1} [\Psi_{\alpha_i}(\mathbf{x}, y) - \alpha_i]. \tag{6}$$

3 Main Results

We need to consider the minimal $\boldsymbol{\alpha}$-CVaR of $\boldsymbol{x} \in X$. That is, we face the following multiobjective problem:

$$\text{(MCVaR)} \min \left(\phi_{\alpha_1}(\boldsymbol{x}, y^*(\boldsymbol{x})), \phi_{\alpha_2}(\boldsymbol{x}, y^*(\boldsymbol{x})), \cdots, \phi_{\alpha_I}(\boldsymbol{x}, y^*(\boldsymbol{x}))\right)$$
$$\text{s. t. } \boldsymbol{x} \in X.$$

For any given $\boldsymbol{\alpha} = (\alpha_1, \alpha_2, \cdots, \alpha_I)$, if $\boldsymbol{x}^*$ is a Pareto efficient solution to (MCVaR)([6]), then $(\phi_{\alpha_1}(\boldsymbol{x}^*, y^*(\boldsymbol{x}^*)), \phi_{\alpha_2}(\boldsymbol{x}^*, y^*(\boldsymbol{x}^*)), \cdots, \phi_{\alpha_I}(\boldsymbol{x}^*, y^*(\boldsymbol{x}^*)))$ is called a Pareto-$\boldsymbol{\alpha}$-CVaR and $\boldsymbol{x}^*$ is called a Pareto-$\boldsymbol{\alpha}$-CVaR efficient solution. The set of all Pareto-$\boldsymbol{\alpha}$-CVaR efficient solutions is denoted by $E(\boldsymbol{\alpha})$. In the following, we want to find out such an efficient solution.

Let $\Lambda = \{\lambda = (\lambda_1, \lambda_2, \cdots, \lambda_I) \mid \lambda_i \in [0, 1], i = 1, 2, \cdots, I, \sum_{i=1}^{I} \lambda_i = 1\}$.

Theorem 3.1 Let $\boldsymbol{x} \in X$. Suppose that $\overline{y}$ is an optimal solution to the problem $\min_{y \in R} \sum_{i=1}^{I} \lambda_i F_{\alpha_i}(\boldsymbol{x}, y)$ and satisfies the condition

$$P\{f_i(\boldsymbol{x}, \xi) = \overline{y}\} = 0, \qquad i = 1, 2, \cdots, I. \tag{7}$$

Then

$$\sum_{i=1}^{I} \lambda_i \phi_{\alpha_i}(\boldsymbol{x}, \overline{y}) = \min_{y \in R} \sum_{i=1}^{I} \lambda_i F_{\alpha_i}(\boldsymbol{x}, y). \tag{8}$$

Proof. Because $\overline{y}$ is an optimal solution to the problem $\min_{y \in R} F(\boldsymbol{x}, y)$, we have from Lemma 2.1 that

$$\sum_{i=1}^{I} \lambda_i (1 - \alpha_i)^{-1} (\Psi_i(\boldsymbol{x}, \overline{y}) - \alpha_i) = 0. \tag{9}$$

With (3), we have

$$\sum_{i=1}^{I} \lambda_i F_{\alpha_i}(\boldsymbol{x}, \overline{y}) = \sum_{i=1}^{I} \lambda_i \phi_{\alpha_i}(\boldsymbol{x}, \overline{y}) + \overline{y} \sum_{i=1}^{I} \lambda_i (1-\alpha_i)^{-1} (\Psi_i(\mathbf{x}, \overline{y}) - \alpha_i),$$

which together with (9) implies (8). □

Now, suppose that $\Psi_i(\boldsymbol{x}, y)$ is strictly increasing in y. We consider the following single objective problem:

$$\begin{aligned} \text{(SCVaR)} \min & \sum_{i=1}^{I} \lambda_i F_{\alpha_i}(\boldsymbol{x}, y), \\ \text{s. t. } & y \in R, \boldsymbol{x} \in X, \\ & 0 \le \lambda_i \le 1, \sum_{i=1}^{I} \lambda_i = 1. \end{aligned}$$

Here, y and λ_i are decision variables.

Theorem 3.2 Suppose that $\Psi_i(\boldsymbol{x}, y)$ is strictly increasing in y. If $(\overline{\boldsymbol{x}}^*, \overline{y}^*, \overline{\lambda}^*)$ is an optimal solution to (SCVaR) and satisfies condition (7), then $\overline{\boldsymbol{x}}^*$ is a Pareto-$\boldsymbol{\alpha}$-CVaR efficient solution to (MCVaR).

Proof. For $\lambda \in \Lambda$, let

$$y^*(\boldsymbol{x}, \lambda) = \min\{y \mid \sum_{i=1}^{I} \lambda_i (1-\alpha_i)^{-1} \Psi_i(\boldsymbol{x}, y) \ge \sum_{i=1}^{I} \lambda_i \alpha_i (1-\alpha_i)^{-1}\}.$$

It is easy to see that

$$y^*(\boldsymbol{x}) = \max_{\lambda \in \Lambda} y^*(\boldsymbol{x}, \lambda). \tag{10}$$

Therefore, we have

$$\min_{\boldsymbol{x} \in X} \sum_{i=1}^{I} \lambda_i^* \phi_{\alpha_i}(\boldsymbol{x}, y^*(\boldsymbol{x})) = \min_{\boldsymbol{x} \in X} \min_{\lambda \in \Lambda} \min_{y \in R} \sum_{i=1}^{I} \lambda_i F_{\alpha_i}(\boldsymbol{x}, y). \tag{11}$$

From (11), if $(\overline{\boldsymbol{x}}^*, \overline{y}^*, \overline{\lambda}^*)$ is an optimal solution to (SCVaR), we know that $\overline{\boldsymbol{x}}^*$ is a Pareto-α-CVaR efficient solution to (MCVaR). □

Therefore, in order to solve the multiobjective problem (MCVAR), we need only to solve the single objective problem (SCVAR). When (4) can be approximately computed, we can use a nonlinear programming to solve (SCVaR). By [4], if we can obtain some points $z_j, j = 1, 2, \cdots, J$, sampled from $p(z)$, then

$$(1-\alpha_i)^{-1} \int_{z \in R} (f_i(\boldsymbol{x}, z) - y)^+ p(z) dz \approx (1-\alpha_i)^{-1} J^{-1} \sum_{j=1}^{J} (f_i(\boldsymbol{x}, z_j) - y)^+,$$

and the (SCVaR) can be approximated by

$$\text{(SCVaR)}' \min \sum_{i=1}^{I} \lambda_i \left(y + (1-\alpha_i)^{-1} J^{-1} \sum_{j=1}^{J} z_{ij} \right)$$
$$\text{s. t. } y \in R, \boldsymbol{x} \in X,$$
$$0 \le \lambda_i \le 1, \sum_{i=1}^{I} \lambda_i = 1.$$
$$z_{ij} \ge f_i(\boldsymbol{x}, z_j) - y, z_{ij} \ge 0, i = 1,2,\cdots,I, \quad j = 1,2,\cdots,J.$$

Now, we construct a nonlinear neural networks of like- Hopfield type for the (SCVaR)$'$, when $X = R^n$. Define an energy function by

$$\begin{aligned} F(y,\lambda,\mathbf{x},\mathbf{z},Z) = &\sum_{i=1}^{I} \lambda_i \left(y + (1-\alpha_i)^{-1} J^{-1} \sum_{j=1}^{J} z_{ij} \right) \\ &+\rho \sum_{i=1}^{I} (\max\{-\lambda_i, 0\}^2 + \max\{\lambda_i - 1, 0\}^2) + \rho(\sum_{i=1}^{I} \lambda_i - 1)^2 \\ &+\rho \sum_{i=1}^{I} \sum_{j=1}^{J} (\max\{f_i(\boldsymbol{x}, z_j) - y - z_{ij}, 0\}^2 + \max\{-z_{ij}, 0\}^2) \end{aligned} \tag{12}$$

and a dynamics system as follows:

$$\begin{aligned} \frac{dy}{dt} &= -\sum_{i=1}^{I} \lambda_i + 2\rho \sum_{i=1}^{I} \sum_{j=1}^{J} (\max\{f_i(\boldsymbol{x}, z_j) - y - z_{ij}, 0\}, \\ \frac{d\lambda_i}{dt} &= -y - (1-\alpha_i)^{-1} J^{-1} \sum_{j=1}^{J} z_{ij} + 2\rho(\max\{-\lambda_i, 0\} \\ &\quad -2\rho(\max\{\lambda_i - 1, 0\} - 2\rho(\sum_{i=1}^{I} \lambda_i - 1), \quad i = 1,2,\cdots,I, \\ \frac{dx_k}{dt} &= -2\rho \sum_{i=1}^{I} \sum_{j=1}^{J} (\max\{f_i(\boldsymbol{x}, z_j) - y - z_{ij}, 0\} \frac{\partial f_i}{\partial x_k}, \quad k = 1,2,\cdots,n, \\ \frac{dz_j}{dt} &= -2\rho \sum_{i=1}^{I} (\max\{f_i(\boldsymbol{x}, z_j) - y - z_{ij}, 0\} \frac{\partial f_i}{\partial z_j}, \quad j = 1,2,\cdots,J, \\ \frac{dz_{ij}}{dt} &= -\lambda_i (1-\alpha_i)^{-1} J^{-1} + 2\rho \max\{f_i(\boldsymbol{x}, z_j) - y - z_{ij}, 0\} \\ &\quad +2\rho \max\{-z_{ij}, 0\}, \quad i = 1,2,\cdots,I, j = 1,2,\cdots,J, \end{aligned}$$

where $\mathbf{z} = (z_1, z_2, \cdots, z_J), Z = (z_{i,j}), i = 1,2,\cdots,I, j = 1,2,\cdots,J$. Let $\mathbf{u} = (y, \lambda, \mathbf{x}, \mathbf{z}, Z)$. By the definition of the energy function, the above dynamics system is equalent to

$$\frac{d\mathbf{u}}{dt} = -\nabla F(y, \lambda, \mathbf{x}, \mathbf{z}, Z). \tag{13}$$

Theorem 3.3 If $\mathbf{u}^*$ is an equilibrium point of the dynamics system (13) under the parameter ρ, if $F(\mathbf{u}) \neq 0$ for $\mathbf{u} \neq 0$, then $\mathbf{u}^*$ is a stable point of the dynamics system (13).

Proof. According to hypothesis, if $\mathbf{u} \neq 0$, we have $F(\mathbf{u}) \neq 0$, because

$$\frac{\mathrm{d}F(\mathbf{u})}{\mathrm{d}t} = -\nabla F(\mathbf{u})^T \nabla F(\mathbf{u}) \leq 0.$$

According to the Lyapunov stable theorem, $\mathbf{u}^*$ is a stable point of the dynamics system (13). □

The following theorem is clear.

Theorem 3.4 If $\mathbf{u}^*$ is an optimal solution to the problem $\min_{\mathbf{u}} F(\mathbf{u})$, then $\mathbf{u}^*$ is an equilibrium point of the dynamics system (13). □

We know that an optimal solution to the problem $\min_{\mathbf{u}} F(\mathbf{u})$ is an approximate solution to (SCVaR)$'$. When the parameter ρ is large enough. From Theorem 3.3 and Theorem 3.4, we may obtain approximate solutions to (SCVaR)$'$ by finding out an equilibrium point of the dynamics system (13).

4 Conclusions

This paper discusses the CVaR problem with multiple losses. We introduce the concepts of $\boldsymbol{\alpha}$-VaR and $\boldsymbol{\alpha}$-CVaR with the confidence level vector $\boldsymbol{\alpha} = (\alpha_1, \alpha_2, \cdots, \alpha_I)$. It is shown that obtaining Pareto efficient solutions of the minimal $\boldsymbol{\alpha}$-CVaR can be transformed into solving a single objective problem under some mild conditions. Hence, we construct a nonlinear neural networks for multiobjective conditional value-at-risk problem. This paper resolves the problem of multiobjective CVaR in theory, and provides base to numerical calculation. This may help us to study multiobjective CVaR problems in theory and calculation under others definitions of $\boldsymbol{\alpha}$-VaR.

References

1. Rockafellar, R. T. and Uryasev, S.: Optimization of Conditional Value-at-Risk, Journal of Risk 2(2000)21-41
2. Chernozhukov, V. and Umantsev, L.: Conditional Value-at-Risk: aspects of modeling and estimation, Empirical Economics, 26(2001) 271-292
3. Andersson, F., Mausser, H., Rosen, D. and Uryasev, S.: Credit risk optimization with Conditional Value-at-Risk criterion, Math. Program. 89(2001)273-291
4. Rockafellar R. T. and Uryasev, S.: Conditional Value-at-Risk for general loss distributions, Journal of Banking & Finance 26(2002) 1443-1471
5. Krokhmal, P., Palmquist, J. and Uryasev, S.: Portfolio optimization with Conditional Value-at-Risk objectives and constraints, Journal of Risk, 2(2002)124-129

6. Sawragi, Y. , Nakayama, H. and Tanino, T.: Theory of multiobjective optimization, Academic Press, New York (1985)
7. J.J. Hopfield, DW.Tank. Neural computation of decision in optimization problems. Biological Cybernetics 58(1985)67-70
8. Youshe Xia and Jun Wang. A General Methodology for Designing Globally Convergent Optimization Neural Networks, IEEE Trans. On Neural Networks, 9(6)(1998)1331-1444
9. G.Joya, MA.Atencia, F.Sandoval. Hopfield neural networks for optimizatiom: study of the different dynamics. Neurocomputing 43(2002)219-237

DSP Structure Optimizations – A Multirate Signal Flow Graph Approach

Ronggang Qi[1], Zifeng Li[2], and Qing Ma[3]

[1] EMS Satellite Networks, 2341 Boul. Alfred-Nobel, Saint Laurent, Quebec H4S 2A9, Canada,
qi.r@emssatnet.com

[2] Dept. of Electrical & Computer Engineering, Concordia University, 1455 de Maisonneuve Blvd. W., Montreal, Quebec H3G 1M8, Canada

[3] Shenzhen Jiu Li Trading Ltd., Shenzhen, China

Abstract. A systematic approach for Digital Signal Processing (DSP) structure optimizations is introduced. The method is based on the Multirate Signal Flow Graph (MSFG) representation and transformations. It reduces the optimization of dsp algorithms/structures to a series of MSFG transformations that makes automation of dsp structure optimization possible. As an example, an efficient Digital Down Converter structure is derived through MSFG transformations. It is applicable to linear time-invariant or periodically time-varying systems.

1 Introduction

Multirate digital signal processing techniques have been widely used in areas where extensive digital processing is needed. In digital communications, for example, they have been used to design low-complexity and power-efficient communication equipment. Typical applications of multirate signal processing include sampling rate conversion, sub-band coding, and transmultiplexing.

1.1 Multirate System Optimization Approaches

It is well understood that multirate digital processing techniques provides ways of simplifying digital signal processing structures [1], [2]. Except for some simple and regular dsp algorithm and structures, simplifying and optimizing multirate dsp systems is generally not an easy task. Frequently used methods for developing efficient multirate digital signal processing systems include

a) Direct application of some existing efficient structures. Systems optimized in this way may not be optimal because the optimization may not be global.

b) Use mathematical representation and manipulations. This method is difficult to use because the structural information may not be clearly presented to the designer during the reduction process.

F. Yin, J. Wang, and C. Guo (Eds.): ISNN 2004, LNCS 3174, pp. 1007–1012, 2004.

c) Use conventional signal flow graph method. This method is valid only for linear time invariant systems. It has difficulties to handle multirate operations, which are periodical time varying in nature.

These methods, though effective in some situations, are very often ad hoc and difficult to use. In many cases, they can not be systematically and directly used to derive efficient multirate systems.

2 Multirate System Optimization Through MSFG Transformations

To overcome the difficulties of the conventional methods, a dsp structure optimization method based on the Multirate Signal Flow Graph (MSFG) representation was proposed in [3]. With this method, the multirate system optimization is achieved by applying a series of MSFG transformations that progressively minimize the overall system computational complexity. The structural information is maintained and explicitly shown throughout the optimization process.

2.1 Multirate Signal Flow Graph

MSFG operators. The most commonly used basic operators are adder, multiplier (multiplication by a signal), scalar (multiplication by a constant), delay element, down-sampler, and up-sampler. With basic operators, some composite operators that are frequently encountered in digital signal processing are also defined in MSFG. They include linear filter, Serial-to-Parallel and Parallel-to-Serial commutators, Permutator, Sampling & Hold, Integral & Dump, Sampler, Discrete Fourier Transform, and Switch. For details about these MSFG operators, see [3].

MSFG transformations. A multirate signal processing system can be transformed into various alternative structures without changing the functionality. These alternative structures can be obtained via MSFG transformations. Some fundamental and very useful relationships for multirate operators (referred to as the noble identities) are given in [1] and [2]. Many more useful MSFG transformations have been developed and summarized in [3].

2.2 Optimization of Signal Processing Structures

DSP algorithms can be represented as interconnected networks consisting of basis processing elements (e.g., adders, multipliers, filters etc.) and/or sub-systems depending on the level of abstraction. An efficient dsp structure requires less hardware resources or less processing time in implementation. The efficiency of a dsp algorithm can be measured by how much computation is involved to complete the intended task(s). In practice, we often use the implementation complexity, processing latency, and the required memory to evaluate the efficiency of a dsp algorithm. If the dsp algorithm or structure is linear, time invariant, or periodically time-varying, it can always be represented by a MSFG. Optimizing dsp structure is thus equivalent to transforming its MSFG under some given optimization criteria.

A convenient measure for complexity is the sum of arithmetic operations (typically, multiplication and addition) weighted by the sampling rate at which the operation is performed. The processing latency is the delay between input and the output signals of the algorithm. The number of memories required to implement a dsp algorithm is another important factor that may affect the choice of one dsp structure over the other. In summary, the criteria for dsp structure optimization can be the minimization of a) complexity, latency, and memory at the same time (if possible); b) any one of the three measures without concerning the other two; and c) one of the three measures under given constraints of the other two.

Another objective for dsp algorithm/structure optimization problem is to maximize the throughput of the dsp structure under given physical constraints (e.g., complexity and clock rate) of the system. This objective can be achieved by increasing the concurrency of the signal processing, i.e., through parallelization of the algorithm.

3 Example: An Efficient Digital Down Converter

Digital Down Converter

Digital down-converters (DDCs) have been widely used in digital receivers in wireless communications. Fig. 1 shows the functionality of a flexible DDC [4]. This DDC sees a wide sampled bandwidth within which the desired communication signal is located. Before down conversion, a digital quadrature mixer is used to tune the DDC to the desired signal. The spectral decimation is performed by seven programmable decimation filter stages. The first stage has a programmable decimation factor of 12, 6, 4, 3, or 2. The 5 middle stages are coefficient programmable and can be bypassed for increased flexibility. The DDC has therefore a decimation factors programmable from 4 to 768. The DDC shown in Fig. 1 is not efficient in terms of both complexity and power consumption because the filtering in each stage is performed at the high sample rate side and the digital mixer is running at the highest sampling rate.

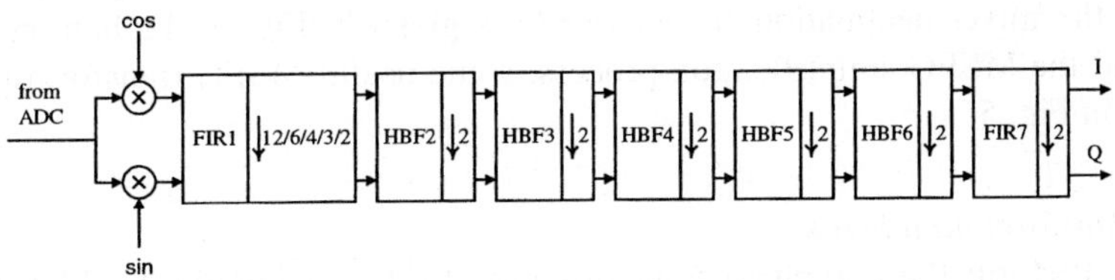

Fig. 1. The Digital Down Converter

Optimization of the DDC

The above DDC can also be represented by the MSFG shown in Fig. 2 where a rate change ratio is always expressed in terms of an interpolation ratio. The mixer takes a complex sinusoidal sequence of the center frequency of the desired signal, which is defined by complex tone $W_N^{kn} = \exp\left(j\frac{2\pi k}{N}\cdot n\right)$ where N and k are integers. The criterion to be used in this dsp structure optimization problem is to minimize the computational complexity without sacrificing the flexibility of the multistage architecture. We therefore optimize each of the decimation filter stages independently.

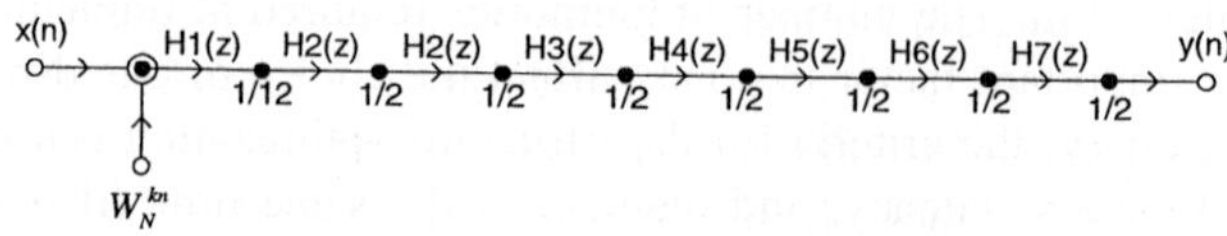

Fig. 2. DDC MSFG: conceptual model

Optimal structure for decimation-by-2 FIR filters. Structures for stages 2 through 7 are identical because they have the same decimation factor though their coefficients may be different. It is well known that an integer decimation FIR filter has an optimal dsp structure known as the polyphase decimation filter [1]. This relationship can be represented by a MSFG transform pair and the resulting optimal decimation-by-2 FIR filter is shown in Fig. 3.

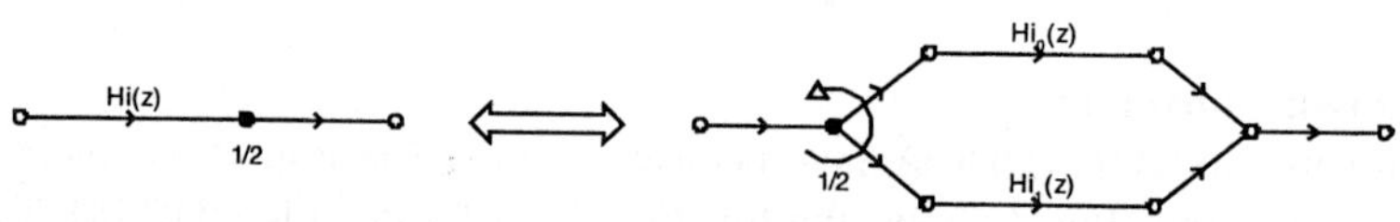

Fig. 3. The optimal decimation-by-2 FIR filter

Optimization of mixer-decimation filter cascade. Although we can use the same polyphase filter transformation for the first stage decimation filter, the resultant DDC structure is still not very efficient in computational complexity because the mixer has to operate at the highest sample rate. This can be a problem if the input sample rate is too high. High multiplication rate often leads to high power consumption in reality. It is thus desirable to move the mixing operations to a lower sample rate side (after the first decimation filter stage, for instance). To this end, we need to look into the optimization of the mixer-decimation-by-12 FIR filter cascade. A step-by-step simplification of the mixer-decimation filter cascade is given in Fig. 4. To help readers to understand the MSFG simplification process, some useful MSFG transform pairs [3] are listed in Fig. 5.

MSFG Transformation Steps

Step 1. Perform the polyphase transformation to the decimation-by-12 FIR filter. The result is shown in Fig. 4 (b).

Step 2. Move the mixer into the parallel branches of the Serial-to-Parallel commutator. This involves a simple MSFG transform pair given in Fig. 5 (a), where $W(N,-km) \equiv W_N^{-km}$ is a constant complex scalar. Another important transform that has been used here is given in Fig. 5 (b). The result of this step is shown in Fig. 4 (c).

Step 3. Move mixers after the FIR sub-filters. This step is the direct application of the transformation of Fig. 5 (c). As a result, the k-th real polyphase sub-filter of H1(z), H1k(z) becomes a complex filter H1'k(z). The complex phasers are also moved after the polyphase sub-filters and the result of this step is given in Fig. 4 (d).

Step 4. Replace the bank of identical mixers in Fig. 4 (d) with a shared one after the summation of polyphase branches. The resulting MSFG shown in Fig. 4 (e) is the final optimal dsp structure for the mixer-decimation filter cascade.

The structure given in Fig. 4 (e) not only reduces the computational complexity of the decimation FIR filter to its theoretical minimum, it also has the minimum multiplication rate by moving the mixing operation to the lower sample rate side.

The optimized DDC. Through separate optimizations of decimation filter stages and the mixer-decimation filter cascade, the optimized structure of the DDC is derived and is shown in Fig. 6. The mixer could be moved further to the output of the next defimation-by-2 FIR filter stage. But we may not gain much in reducing computational complexity because the polyphase decimation-by-2 filter becomes complex, which is equivalent to four real polyphase decimation filters when the input is also complex (this is different from the first stage whose input is real).

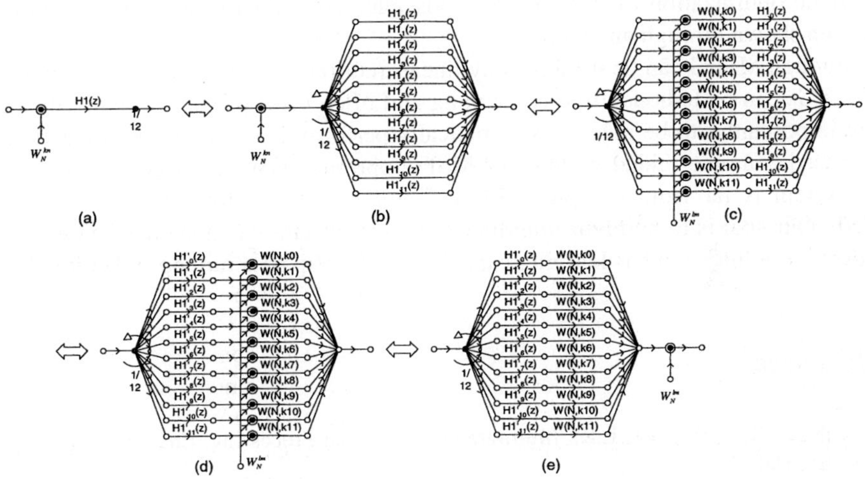

Fig. 4. Step-by-step MSFG transformation of mixer-decimation filter cascade

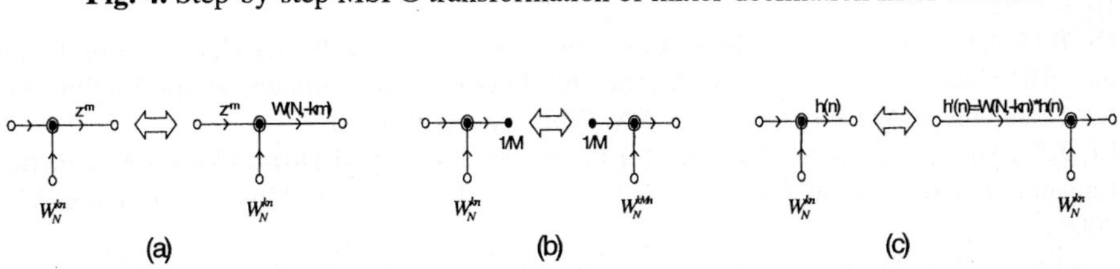

Fig. 5. Some useful mixer-related MSFG transform pairs

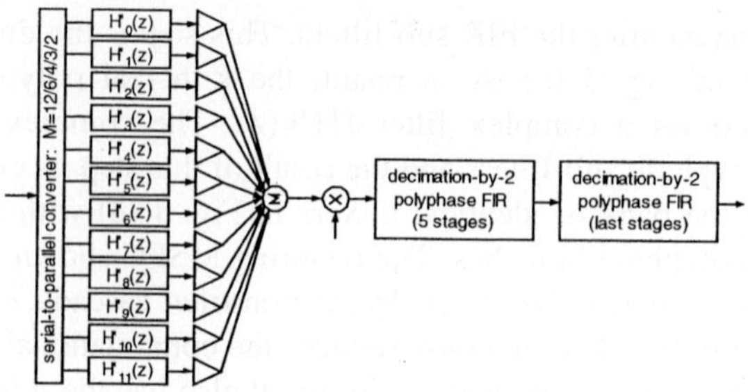

Fig. 6. Optimized DDC

4 Conclusions and Future Work

The MSFG approach proves to be a viable dsp structure design and optimization method. Compared to conventional methods, this approach provides better and more structural information and it can be easily represented in computers hence making the automatic optimization of dsp structure possible. The effectiveness of the approach has been shown through an example.

Although MSFG can be used for any linear multirate system, manual manipulation of MSFGs can be tedious, time-consuming and often error-prone. To avoid this and, more importantly, to be able to systematically perform dsp structure optimizations, a rule-based computer-aided system for MSFG optimizations has been developed [5]. The system is far from complete. More features and transformation rules will be added. Our goal is to build an 'intelligent' expert system which should be capable of adaptive searching and has learning capability in the dsp structure optimization.

References

1. Crochiere, R.E., Rabiner, L.R.: Multirate Digital Signal Processing, Englewood Cliffs, NJ: Printice Hall, 1983
2. Vaidyanathan, P.P.: Multirate Systems and Filter Banks, Englewood Cliffs, NJ: Printice Hall, 1993
3. Qi, R.G., Coakley, F., Evans, B.G.: Low Complexity and Low Power Consumption Design for OBP Multicarrier DEMUX VLSI, the 16th International Communications Satellite Systems Conference, Washington, DC, USA, February 25-29, 1996
4. Li, Z.F., Ma, Q., Qi, R.G.: Design of a Programmable Digital Down-Converter Structure, Canadian Conference on Electrical and Computer Engineering, Montreal, Canada, May, 2003
5. Ma, Q., Li, Z.F., Qi, R.G.: A Computer Aided Design System for Transformation and Optimization of Multirate Signal Processing Systems, GSPx & International Signal Processing Conference (ISPC), Dallas, Texas, USA, March 31-April 3, 2003

Author Index

Adachi, Masaharu I-217, I-395
Aguirre, Carlos I-55
Aknin, Patrice I-524
An, Dexi II-555
An, Jubai I-977
Asari, Vijayan K. I-870

Bai, Ling I-1002
Bao, Zheng I-531
Beadle, Patch I-726, II-350, II-356
Bentoumi, Mohamed I-524
Bi, Guo II-595
Bi, Weixing I-338, I-425
Bi, Yingwei II-405
Bian, Zheng-Zhong II-453
Bloch, Gérard I-524
Bo, Liefeng I-264
Brown, Carl I-977
Bu, Nan I-834

Cai, Feng II-418, II-430, II-920
Cai, Minglun II-418
Campos, Doris I-55
Cao, Guang-yi II-150
Cao, Jinde I-78, II-976
Cao, Li II-218
Cao, Wen I-965
Cao, Wenming II-786
Cao, Xiu II-188
Cao, Yijia II-188
Cao, Yukun II-688
Cao, Zongjie II-399
Cardon, Pierre I-864
Chai, Lin II-381
Chan, Francis I-732
Chan, Philip C.H. I-696
Chau, Kwokwing II-970
Chaudhari, Narendra S. I-362, II-494
Chen, Chunrong I-983
Chen, Gang I-971
Chen, Guanrong II-627
Chen, Huaidong II-399
Chen, Jin I-906, II-567, II-595
Chen, Jinmiao I-362, II-494
Chen, Junying I-317
Chen, Nian-Yi I-389
Chen, Peng II-574
Chen, Qing II-755
Chen, Quanshi I-512
Chen, ShaoBai I-565
Chen, Shuang I-199
Chen, Songcan I-180, I-768, I-858
Chen, Tianping I-38, I-144
Chen, Xiaoming I-401, I-425, I-430
Chen, Xiaoping I-311
Chen, Xingfu II-601
Chen, Xiyuan II-805
Chen, Yaqin I-786
Chen, Yen-Wei I-186
Chen, Yong I-102, I-120, I-512, II-639
Chen, Yuehui I-211
Chen, Yumin I-906
Chen, Zhe II-607
Cheng, Guojun II-110
Cheng, Hong I-792
Cheng, Xiefeng I-702
Cheung, Albert II-627
Chi, Wei II-920
Chicharo, Joe F. I-672
Choi, Hyung-Jun I-239, II-988
Chou, PenChen II-83
Chu, Fulei II-767
Chu, Xiaodong II-581
Cui, Pingyuan II-129

Dai, Meng II-681
Dai, Weidi I-840, II-424
Dai, Xianhua II-519
Dang, Chuangyin I-406
Deng, Beixing I-876
Deng, Bing II-324
Deng, Feiqi I-571
Ding, Haiyan II-459
Ding, Liu I-598
Ding, Zijia II-549
Doki, Shinji I-714
Dominguez, David I-14, I-20
Dong, Jiwen I-211
Du, Cheng II-411
Du, Hai II-163

Du, Haifeng I-299
Du, Minghui II-269
Du, Ruxu II-169
Duan, Shukai II-813

Elhabian, Tarek II-248, II-281
Ertunc, Metin II-254

Fan, Chunling II-749
Fan, Jian II-35
Fan, Quanyi II-866
Fan, Zhi-Gang I-804
Fang, Qun II-381
Fang, Yonghui I-276
Feng, Chun-Bo I-32
Feng, Guori I-235
Feng, Xiaoning I-846
Feng, Xin I-454
Feng, Zhipeng II-767
Fernández-Redondo, Mercedes I-223, I-229, I-852, I-912
Fu, JianGuo II-701
Fu, Xiao-wei II-150
Fu, Yonggang I-552

Gao, Feng II-749
Gao, Hangshan I-126
Gao, Jianfeng I-311
Gao, Jianming I-786
Gao, Peng II-904
Gielen, Georges I-305, I-876
Gong, Jianya I-906
Gong, Shengguang II-110
Gosztolya, Gábor I-996
Gou, Shuiping I-918
Gu, Wenjin II-117
Gu, Xiaodong I-26, I-413
Guan, Huanxin I-96
Guan, Xiaohong II-657
Guan, Xinping II-71, II-169, II-743
Gui, Jinsong II-799
Guo, Chengan I-370
Guo, Chuangxin II-188
Guo, Dongming II-737
Guo, Jun I-474
Guo, Lin I-648
Guo, Ping I-616, I-1002, II-369, II-393
Guo, Shide II-293
Guo, Shu-Mei II-375
Guo, Ying II-405
Guo, Zhenhe II-675
Guo, Zhi-Gang I-971

Hacker, Roger I-383
Han, Gyu-Sik II-988
Han, Jiuqiang I-622
Han, Liu I-598
Han, Min II-200, II-713
Hao, HanYong II-537
Hao, Pengwei I-168
Hao, Ping II-737
Hao, Zhifeng I-558
Hao, Zhihua I-744
He, Daihai I-138
He, Fan I-822
He, Hanlin I-44
He, Jingrui II-931
He, Jun II-595
He, Junfeng II-138
He, Mi I-138
He, Pilian I-840, II-424
He, Wuhong I-299
He, XiaoFu II-681
He, You II-886
He, Zhenya I-660
Hernández-Espinosa, Carlos I-223, I-229, I-852, I-912
Horzyk, Adrian I-150
Hotta, Hirohisa I-714
Hou, Yuexian II-424
Hsu, Chin-Yuan II-375
Hu, Bao-Gang I-7
Hu, Chang II-218
Hu, Dewen I-199, I-708, II-25, II-525
Hu, Gensheng I-571
Hu, Huibo I-738
Hu, Lingyun I-286
Hu, Mengdi II-786
Hu, Qinglei II-42
Hu, Qiying II-1000
Hu, Rong I-558, II-561
Hu, Shiyan II-669
Hu, Xuelei I-828
Hu, Yunan II-129
Hua, Yu II-982
Huang, Chengkui II-779
Huang, Degen I-983
Huang, De-Shuang I-114, I-648, I-774, I-1008, II-476, II-513
Huang, Gaoming I-660

Huang, Guangbin II-755
Huang, Hong-Zhong II-820
Huang, Houkuan II-707
Huang, Jianhuan I-894
Huang, Jie II-59
Huang, JiFeng II-681
Huang, Liangwei II-381
Huang, Xin II-476, II-513
Huang, Xuanjing I-1025
Huang, Xuming I-720
Huang, Yan-Xin II-337
Huang, Ying II-448
Huang, Yong II-886
Hwang, TsenJar II-83

Jang, Min-Soo I-642
Jang, Sung-Whan II-65
Ji, Ce I-96
Ji, Guangrong I-1015
Ji, Ye I-332
Jia, Ying I-761
Jiang, Chun-di II-19
Jiang, Hairong II-163
Jiang, Min I-406, II-1000
Jiang, Minghu I-305, I-876
Jiang, Qiuhao I-108
Jiang, Quanyuan II-188
Jiang, Yadong II-892
Jiang, Yuan I-356
Jiang, YuGang II-369
Jiao, Licheng I-258, I-264, I-299, I-605, I-629, I-918, II-236, II-275
Jin, Chun II-904
Jin, Lianwen I-988
Jin, Mingxi I-941
Jin, Wuyin I-138
Jin, Xiaoyi I-299
Jin, Yuqiang II-129
Jin, Zhihua II-749

Kanae, Shunshoku II-507
Kang, Chunyu I-930
Kang, Haigui II-799
Kang, Renke II-737
Kang, Zhiyu II-381
Kavak, Adnan II-254
Kawabata, Shuma II-436
Kawamoto, Masashi II-488
Kim, Hyun-Ki I-156
Kim, Sang-Jun I-642
Kim, Yong-Guk I-642
Kim, Yong-Soo I-174
Kim, Yongkab I-162
Kirk, James S. II-695
Kneupner, Klaus II-761
Kong, Fan-Sheng I-323
Koroutchev, Kostadin I-14, I-20
Kotani, Makoto I-395
Kuh, Anthony I-370
Kuhlenkötter, Bernd II-761
Kwan, Chiman I-543

Lai, Hsin-Hsi II-898
Lan, Xiang I-144
Lauer, Fabien I-524
Lee, Chang-Shing II-375
Lee, Dae-Won I-239, I-350, II-287
Lee, Dong-Yoon II-65
Lee, Hyo-Seok II-988
Lee, Hyuk-Soon II-287
Lee, Hyun-Gu I-642
Lee, Jaewook I-239, I-350, II-287, II-988
Lee, Shu Tak II-387
Lee, Soek-Joo I-642
Lei, Xusheng II-1
Leung, Hofung I-454
Li, Baoqing I-468
Li, Bi-Cheng I-965, I-971
Li, Bo I-332, II-701
Li, Chongrong II-931
Li, Chuandong I-61, I-102
Li, Chung Lam II-387
Li, Chunhung I-518
Li, DongSheng I-882
Li, Fucai II-567, II-595
Li, Gang I-786, II-362
Li, Guo-Zheng I-389
Li, Hai-Bin II-820
Li, Haiyang II-531
Li, Hongzhai II-393
Li, Huaqing I-487
Li, Ji II-549
Li, Jiandong I-654, I-690
Li, Jianmin I-462
Li, Kaili II-224
Li, Li II-581
Li, Linglai II-543
Li, Lishuang I-983
Li, Muguo II-163
Li, Ping II-181

Li, Qi I-598
Li, Qing I-792
Li, Ruqiang II-567
Li, Sen I-678
Li, Shaoyuan II-175, II-755, II-848
Li, Shouju II-792
Li, Shuchen II-181
Li, Songsong I-430
Li, Ting II-501
Li, Wuzhao II-619
Li, Xi II-150, II-470
Li, Xiaobing II-405
Li, Xiaoli II-169, II-743
Li, Xiaoou II-77, II-212
Li, Xiaoxia II-362
Li, Xihai I-959, II-925
Li, Xueming I-132
Li, Xunming I-32
Li, Yan II-743
Li, Yanda I-537, II-448
Li, Yong I-401, I-419, I-425, I-442
Li, Yuan II-962
Li, Yunfeng I-611, II-688
Li, Zhancheng II-350, II-356
Li, Zhengxue I-235
Li, Zhonghua I-565
Li, Zifeng II-1007
Lian, Han I-924
Lian, Shiguo II-627
Liang, Hua-Lou I-755
Liang, Xinying II-574
Liang, Yanchun I-558, II-104, II-337
Liang, Yanming I-598
Liao, Xiaofeng I-61, I-102, I-120, I-132, II-633, II-639, II-645, II-688
Liao, Xiaoxin I-44
Lin, Bin I-977
Lin, Chih-Chung I-377
Lin, Fuzong I-462
Lin, Hongwen I-930
Lin, JiaJun II-681
Lin, Ling I-786, II-362
Lin, Qiu-Hua I-755
Lin, Wei I-822
Lin, Xiaokang I-780
Lin, Yang-Cheng II-898
Lin, Yiping I-67
Lin, Zhonglin I-900
Linnell, Bruce I-543
Liu, Bing I-270
Liu, Chengming I-720
Liu, Chunsheng II-826
Liu, Dafan I-876
Liu, Daizhi I-959, II-925
Liu, Daxin I-846
Liu, Fang I-293, II-236, II-275
Liu, Fei I-798, II-123
Liu, Guangyuan I-276
Liu, Haiyuan I-578
Liu, Hongbo I-332
Liu, Hongwei I-531
Liu, Jiafeng I-894
Liu, Jian II-35
Liu, Jinxin II-848
Liu, Ju I-702, I-738, II-311, II-317
Liu, Jun I-768
Liu, Lijun I-282
Liu, Ming II-91
Liu, Peng II-91
Liu, Qun I-636
Liu, Weixiang II-470
Liu, Wenhong II-501
Liu, Wenling I-840
Liu, Wenqi II-144
Liu, Xinlu II-904
Liu, Yaqiu II-42
Liu, Yi II-156, II-874
Liu, Yingxi II-792
Liu, Yu I-317
Liu, Yue II-962
Liu, Yuliang II-362
Liu, Yutian II-581
Liu, Zengrong I-67
Liu, Zhendong II-163
Liu, Zhengkai II-675
Liu, Zhigang I-959
Long, Weijiang I-244
Lu, Bao-Liang I-480, I-804, I-888
Lu, Hongtao I-552, II-651
Lu, Jun I-389
Lu, Naijiang I-900
Lu, Wei II-144, II-651
Lu, Wen-Cong I-389
Lu, Wenkai I-500, I-537
Lu, Wenlian I-38
Lu, Yinghua II-104
Luo, Siwei II-482
Luo, Zhiwei II-7
Lv, Jiancheng I-810

Ma, Fei II-35
Ma, Guangfu II-42
Ma, Lin I-792
Ma, Lixin II-436
Ma, Qing II-1007
Ma, Runnian I-126
Ma, Xiaobin I-988
Ma, Xiaojiang I-744
Ma, Xiuli I-258
Ma, Ye II-701
Mao, Weihua I-90
Mao, Zongyuan I-565
Matsugu, Masakazu I-864
Mei, Fei I-654, I-690
Mei, Tiemin I-666, I-672
Meng, Guang II-613
Meng, Zhiqing I-406, II-1000
Millerioux, Gilles I-524
Miyajima, Hiromi II-436
Murahashi, Yoshimitsu I-714
Murray-Smith, Roderick II-52

Nakao, Zensho I-186
Nan, Dong I-235
Nath, Baikunth I-636
Nishi, Tetsuo I-474
Nishino, Fumihito I-1025

Oh, Sung-Kwun I-156, I-162, I-174, II-65
Okamoto, Masaru I-834
Okuma, Shigeru I-714
Ortiz-Gómez, Mamen I-229
Otten, Lambert I-383
Ou, Zongying I-611
Ouyang, Gaoxiang II-169

Pan, Daru II-269
Pan, Leilei I-383
Pan, Qielu II-7
Pang, Guibing II-262
Park, Byoung-Jun I-174
Park, Daehee I-162
Park, Gwi-Tae I-642
Park, Ho-Sung I-156, I-162
Pascual, Pedro I-55
Pedrycz, Witold I-156, I-162, I-174
Peng, Jinye II-663
Peng, Tian-Qiang I-965

Qi, Feihu I-487
Qi, Ronggang II-1007
Qian, Tao I-543
Qiao, Jianping I-702
Qin, Bo I-1015
Qiu, Tianshuang I-678, II-299, II-405, II-501, II-607
Qiu, Yuhui I-276
Qiu, ZhengDing I-882, II-994
Qu, Changwen II-886
Qu, Fuzheng I-750
Qu, Jianbo II-779

Ren, Bing II-854
Ren, Guanghui I-936
Ren, XuChun II-841
Rodríguez, Francisco B. I-14, I-20
Roh, Seok-Beom II-65

Saeki, Noboru II-488
Sakane, Akira II-488
Sato, Norihisa I-395
Sbarbaro, Daniel II-52
Seow, Ming-Jung I-870
Serrano, Eduardo I-14, I-20, I-55
Shang, Ruiqiang II-242
Shao, Chao II-707
Shao, Cheng II-860
Shao, Dingrong II-248, II-281
Shao, Zhiqiong I-235
Shen, Dongsheng I-988
Shen, Lixin I-678
Shen, Minfen I-726, I-732, II-350, II-356
Shen, Ruimin I-552
Shen, Ruiming II-651
Shen, Xianfeng II-832
Shen, Zhang-Quan I-323
Shi, Aiguo II-418, II-430, II-920
Shi, Haixiang II-230
Shi, Weiren II-880
Shi, Xiangquan II-344
Shi, Zhenwei I-684
Shi, Zhewen I-317
Shi, Zhiwei II-200
Shiba, Kenji II-488
Shigei, Noritaka II-436
Shu, Jiwu II-719
Song, Baoquan II-525
Song, Xigeng II-767
Song, Xin I-344
Song, Yanhui II-549

Song, Yibin II-224
Su, Feng II-886
Su, Guangda II-411
Su, Hang II-587
Su, Jianbo I-293, II-1, II-7
Su, Jing II-453
Sulistiyo I-186
Sun, Bing-Yu I-648
Sun, Changyin I-32
Sun, Daoheng II-826
Sun, Fuchun II-13
Sun, Hequan II-799
Sun, Hui II-587
Sun, Jiancheng I-578
Sun, Jiande I-738
Sun, Lisha I-732
Sun, Mingliang I-941
Sun, Shiliang II-950
Sun, Weitao II-719
Sun, Yongmei I-678
Sun, Youxian II-194, II-912
Sun, Yuanyuan II-581
Sun, Zengqi I-286, II-537
Sun, Zhan-Li I-774
Sun, Zonghai II-194, II-912

Tadeusiewicz, Ryszard I-150
Tai, Huiling II-892
Takahashi, Norikazu I-474
Tan, Gang II-442
Tan, Guozhen II-937
Tan, Xiaoyang I-858
Tan, Ying II-675
Tanaka, Yoshiyuki II-488
Tang, Fang I-448
Tang, Guoping II-645
Tang, Hao II-866
Tang, Haoyang I-61
Tang, Hong I-678
Tang, Huanwen I-684
Tang, Lixin I-454
Tang, Xianglong I-894
Tang, Yiyuan I-332, I-684
Tang, Zhenan I-696
Tang, Zheng I-338, I-401, I-419, I-425, I-430, I-442
Tao, Haihong II-236, II-275
Tao, Jun II-848
Tao, Ran II-324
Tao, Yewei I-702
Tian, Fengchun I-383
Tian, Shengfeng I-592
Tian, Xue II-713
Tong, Hanghang II-931
Torres-Sospedra, Joaquín I-223, I-229, I-852, I-912
Tóth, László I-996
Tsaih, Ray I-377
Tsuji, Toshio I-834, II-488

Valdes, Arturo II-52

Wada, Kiyoshi II-507
Wan, Anhua I-90
Wan, Chunru I-270
Wan, Yong II-311
Wan, Youchuan I-906
Wang, Bin II-331
Wang, Dan II-59
Wang, Gang I-49, I-708
Wang, Guangxing II-242
Wang, Haijun I-72
Wang, Han I-344
Wang, Hong-bin II-19
Wang, Hong-Qiang II-476
Wang, Hongrui II-19, II-206
Wang, Hongyu II-607
Wang, Jiahai I-401, I-419, I-425, I-430, I-442
Wang, Jianjun I-1, I-924
Wang, Jie II-943
Wang, Jing II-163, II-994
Wang, Jinkuan I-344
Wang, Jun I-1025
Wang, Junping I-512
Wang, Lidan II-813
Wang, Lin I-84
Wang, Ling I-264, I-448, II-236, II-275
Wang, Lipo I-270, II-230
Wang, Peijin II-224
Wang, Renming I-959
Wang, Ronglong I-419, I-442
Wang, Shaoyu I-487
Wang, Shoujue II-786
Wang, Wei II-200, II-453, II-657
Wang, Wenhui II-555
Wang, Xi-Huai II-956
Wang, Xia II-206
Wang, Xiao Tong II-701
Wang, Xicheng I-436

Wang, Xin I-750, II-175, II-848
Wang, Xiong II-97
Wang, Xiukun I-332
Wang, Xudong II-601
Wang, Xuening II-25
Wang, Xuexia I-936
Wang, Xugang I-338
Wang, Yan II-337, II-362
Wang, Yang II-832
Wang, Yong I-622
Wang, Yonggang I-605
Wang, Yongxue II-854
Wang, Yuan II-962
Wang, Yukun II-269
Wang, Yuwen II-399
Wang, Zengfu I-114, I-1008
Wang, Zhen II-549
Wang, Zheng I-924
Wang, Zhengzhi II-525
Wang, Zhiquan II-627
Wang, Zhongjie II-175, II-848
Wang, Zhongsheng I-44
Wang, Zhuo I-846
Wei, Guangfen I-696
Wei, Lixin II-206
Wei, Xiaopeng I-72
Wen, Jianming II-866
Wen, Xian-bin I-822
Wen, Yi-Min I-480
Wen, Yingyou II-242
Weng, Shifeng I-205, I-900
Wong, Fai I-205
Wong, Kwan-Yee Kenneth II-731
Wong, Shu-Fai II-731
Wu, Gengfeng II-35, II-962
Wu, Hao II-13
Wu, Jianping I-565
Wu, Lide I-1025
Wu, Qinghui II-156, II-860
Wu, Tao I-494
Wu, Wei I-235, I-282
Wu, Weidong II-826
Wu, Xiang I-500
Wu, Xing II-567
Wu, Yan I-941
Wu, Ying I-138
Wu, Zhilu I-936

Xi, Jiangtao I-672
Xi, Youmin I-126
Xi, Yugeng I-586, II-755
Xia, Guang'an I-419
Xia, Guangpu I-401, I-419, I-430, I-442
Xia, Shunren I-953
Xia, Zhimin I-900
Xiang, Tao I-132
Xiao, Di II-633
Xiao, Huanqin I-537
Xiao, Jian-Mei II-956
Xie, Guangzhong II-892
Xie, Hongmei II-663
Xie, Hua I-953
Xie, Ji-Gang II-994
Xing, Fei I-616
Xing, Hong-Jie I-7
Xiong, Wenjun I-108
Xiong, Zhangliang II-344
Xiong, Zhihua II-97
Xu, Bowen II-866
Xu, Hongji II-311, II-317
Xu, Hua I-1015
Xu, Jianhua I-252
Xu, Jianxue I-138
Xu, Lei I-828
Xu, Roger I-543
Xu, Shiguo II-713
Xu, Weidong I-953
Xu, Weijun I-1, II-725
Xu, Weiling I-726, I-732
Xu, Wenji II-262
Xu, Xin II-25
Xu, Xinhe II-181
Xu, Xinshun I-401, I-425, I-430
Xu, Xu II-104
Xu, Yinfeng II-725
Xu, Yongmao II-97, II-773
Xu, Yuesheng I-235
Xu, ZongBen I-1
Xue, Jin II-399

Yamamoto, Takayoshi II-574
Yan, Dongwei II-943
Yan, Huajun II-465
Yan, Lan-Feng II-453
Yang, Bo I-211, II-71
Yang, Chan-Yun I-506
Yang, Jialin II-832
Yang, Jianhua II-144
Yang, Jie I-235, I-389
Yang, Jinyao I-726, II-350

Yang, Luxi I-660
Yang, Shangming I-193
Yang, Simon X. I-383, II-880
Yang, XianHui II-874
Yang, Xiaowei I-558
Yang, Xuhua II-194, II-912
Yang, Yongqing II-619, II-976
Yang, Yu-Jiu I-7
Yang, Yuansheng I-983
Yang, Zeying II-779
Yang, Zi-Jiang II-507
Yao, Jin II-832
Yao, Kaifeng I-537
Yao, Man II-601
Yao, ZhenHan II-841
Ye, Datian II-459
Ye, Hao II-561
Ye, Mao I-810, I-815
Ye, Yisong I-941
Yeh, Chung-Hsing II-898
Yi, Hongjie I-1015
Yi, Zhang I-810, I-815, II-465
Yigit, Halil II-254
Yin, Chuanhuan I-592
Yin, Fuliang I-666, I-672, I-755, II-299
You, Lan I-1025
Young, Rebecca I-543
Yu, Daoheng I-26, I-413, II-293
Yu, Ensheng I-436
Yu, Guoqiang II-950
Yu, Hao I-1025
Yu, Jian I-168, I-592
Yu, Jin I-598
Yu, Jinyong II-117
Yu, Jun I-696
Yu, Qilian I-786
Yu, Shu I-558
Yu, Wen II-77, II-212
Yuan, Jianping II-381
Yuan, Kun I-78
Yuan, Wenjiang II-937
Yuan, Yanbo II-269
Yuan, Yuman II-117
Yue, Heng II-175

Zeng, Fanming II-110
Zeng, FanZi I-882
Zeng, Jin II-880
Zeng, Zhigang I-114, I-1008
Zha, Daifeng I-678
Zhai, Xiaobing II-262
Zhang, Anqing I-930, II-501
Zhang, Bin II-925
Zhang, Bo I-462, II-248, II-281
Zhang, Bofeng II-962
Zhang, Changshui I-205, I-900, II-950
Zhang, Daoqiang I-180
Zhang, Dianjun II-156
Zhang, Fuyan I-858
Zhang, GuangZheng II-513
Zhang, Guicai II-595
Zhang, Hande I-977
Zhang, Hao II-13
Zhang, Hongyu II-854
Zhang, Huaguang I-49, I-96
Zhang, Jianwei I-462
Zhang, Jianzhong II-531
Zhang, Jie II-97, II-773, II-874
Zhang, Jing I-636
Zhang, Jingfen II-613
Zhang, Li I-586
Zhang, Liming I-26, I-413, I-720, I-924, II-331
Zhang, Liqing I-235, II-442
Zhang, Ming II-587
Zhang, Naimin I-235
Zhang, Qiang I-72
Zhang, Qun II-163
Zhang, Shanwen I-537
Zhang, Taiyi I-578
Zhang, Tianqi I-780
Zhang, Weidong II-110
Zhang, Wenxiu I-244
Zhang, Xian-Da I-761
Zhang, Xiang II-761
Zhang, Xiangliang II-657
Zhang, Xinhua I-930
Zhang, Xiuling II-169
Zhang, Xuegong I-252
Zhang, Xuxiu II-501
Zhang, Yanglan II-982
Zhang, Yi I-193
Zhang, Yonghui II-156, II-860
Zhang, Yongsheng II-418, II-430, II-920
Zhang, Youyun I-518
Zhang, Yu II-537
Zhang, Yunfeng II-362
Zhang, Yuzheng II-350
Zhang, Zengke II-138
Zhang, Zhao I-1021

Zhang, Zhengwei II-561
Zhang, Zhiqiang II-169
Zhao, Chun I-90
Zhao, Deyou II-613
Zhao, Dongfang I-605
Zhao, Feng I-558
Zhao, Hai I-888
Zhao, Han-Qing I-494
Zhao, Jian II-663
Zhao, Jianli II-242
Zhao, Jianye II-293
Zhao, Ke II-925
Zhao, Li I-947
Zhao, Lianwei II-482
Zhao, Ming-Yang II-820
Zhao, Shi Jian II-773
Zhao, Xing-Ming II-476
Zhao, Yaqin I-936
Zhao, Zhijin I-654, I-690
Zhao, Zhong-Qiu I-648
Zhao, Zhonggai I-798
Zheng, Chunhong I-629
Zheng, Guiwen I-629
Zheng, Nannig I-792
Zheng, Nanning II-470
Zheng, Qin I-317
Zheng, Rirong I-565
Zheng, Tian I-822
Zheng, Weimin II-719
Zheng, Wenming I-947
Zheng, Xiaoshen I-840
Zheng, Xufei I-276
Zheng, Yong-Rui I-755
Zhou, Bo II-418, II-430, II-920
Zhou, Chun-Guang II-337
Zhou, Donghua II-91, II-543, II-555
Zhou, Dongsheng I-72
Zhou, Fuchang II-595
Zhou, Gengui I-406
Zhou, Hao II-331
Zhou, Hong II-305
Zhou, Jin I-144
Zhou, Jinjin II-262
Zhou, Mingquan II-663
Zhou, Tsing II-639
Zhou, Xiaoyan I-947
Zhou, Xin II-663
Zhou, Yunshui I-654
Zhou, Zhengzhong I-780
Zhou, Zhi-Hua I-180, I-356, I-768, I-858
Zhou, Zongtan I-199, II-525
Zhu, Daqi II-619, II-976
Zhu, Hui I-894
Zhu, Ruijun II-299
Zhu, Xiao-Long I-761
Zhu, Xin-jian II-150
Zhu, Yihua I-406
Zhu, Yongsheng I-518
Zong, Ziliang I-338
Zou, Cairong I-947
Zou, Dali I-750
Zou, Jiyan II-587
Zou, Shu-Xue II-337
Zurada, Jacek M. II-695

Lecture Notes in Computer Science

For information about Vols. 1–3058

please contact your bookseller or Springer

Vol. 3177: Z.R. Yang, H. Yin, R. Everson (Eds.), Intelligent Data Engineering and Automated Learning – IDEAL 2004. VXIII, 852 pages. 2004.

Vol. 3174: F. Yin, J. Wang, C. Guo (Eds.), Advances in Neural Networks - ISNN 2004, Part II. XXXV, 1021 pages. 2004.

Vol. 3173: F. Yin, J. Wang, C. Guo (Eds.), Advances in Neural Networks - ISNN 2004, Part I. XXXV, 1041 pages. 2004.

Vol. 3172: M. Dorigo, M. Birattari, C. Blum, L. M.Gambardella, F. Mondada, T. Stützle (Eds.), Ant Colony, Optimization and Swarm Intelligence. XII, 434 pages. 2004.

Vol. 3158: I. Nikolaidis, M. Barbeau, E. Kranakis (Eds.), Ad-Hoc, Mobile, and Wireless Networks. IX, 344 pages. 2004.

Vol. 3157: C. Zhang, H. W. Guesgen, W.K. Yeap (Eds.), PRICAI 2004: Trends in Artificial Intelligence. XX, 1023 pages. 2004. (Subseries LNAI).

Vol. 3156: M. Joye, J.-J. Quisquater (Eds.), Cryptographic Hardware and Embedded Systems - CHES 2004. XIII, 455 pages. 2004.

Vol. 3153: J. Fiala, V. Koubek, J. Kratochvíl (Eds.), Mathematical Foundations of Computer Science 2004. XIV, 902 pages. 2004.

Vol. 3152: M. Franklin (Ed.), Advances in Cryptology – CRYPTO 2004. XI, 579 pages. 2004.

Vol. 3150: G.-Z. Yang, T. Jiang (Eds.), Medical Imaging and Virtual Reality. XII, 378 pages. 2004.

Vol. 3148: R. Giacobazzi (Ed.), Static Analysis. XI, 393 pages. 2004.

Vol. 3146: P. Érdi, A. Esposito, M. Marinaro, S. Scarpetta (Eds.), Computational Neuroscience: Cortical Dynamics. XI, 161 pages. 2004.

Vol. 3144: M. Papatriantafilou, P. Hunel (Eds.), Principles of Distributed Systems. XI, 246 pages. 2004.

Vol. 3143: W. Liu, Y. Shi, Q. Li (Eds.), Advances in Web-Based Learning – ICWL 2004. XIV, 459 pages. 2004.

Vol. 3142: J. Diaz, J. Karhumäki, A. Lepistö, D. Sannella (Eds.), Automata, Languages and Programming. XIX, 1253 pages. 2004.

Vol. 3140: N. Koch, P. Fraternali, M. Wirsing (Eds.), Web Engineering. XXI, 623 pages. 2004.

Vol. 3139: F. Iida, R. Pfeifer, L. Steels, Y. Kuniyoshi (Eds.), Embodied Artificial Intelligence. IX, 331 pages. 2004. (Subseries LNAI).

Vol. 3138: A. Fred, T. Caelli, R.P.W. Duin, A. Campilho, D.d. Ridder (Eds.), Structural, Syntactic, and Statistical Pattern Recognition. XXII, 1168 pages. 2004.

Vol. 3136: F. Meziane, E. Métais (Eds.), Natural Language Processing and Information Systems. XII, 436 pages. 2004.

Vol. 3134: C. Zannier, H. Erdogmus, L. Lindstrom (Eds.), Extreme Programming and Agile Methods - XP/Agile Universe 2004. XIV, 233 pages. 2004.

Vol. 3133: A.D. Pimentel, S. Vassiliadis (Eds.), Computer Systems: Architectures, Modeling, and Simulation. XIII, 562 pages. 2004.

Vol. 3131: V. Torra, Y. Narukawa (Eds.), Modeling Decisions for Artificial Intelligence. XI, 327 pages. 2004. (Subseries LNAI).

Vol. 3130: A. Syropoulos, K. Berry, Y. Haralambous, B. Hughes, S. Peter, J. Plaice (Eds.), TEX, XML, and Digital Typography. VIII, 265 pages. 2004.

Vol. 3129: Q. Li, G. Wang, L. Feng (Eds.), Advances in Web-Age Information Management. XVII, 753 pages. 2004.

Vol. 3128: D. Asonov (Ed.), Querying Databases Privately. IX, 115 pages. 2004.

Vol. 3127: K.E. Wolff, H.D. Pfeiffer, H.S. Delugach (Eds.), Conceptual Structures at Work. XI, 403 pages. 2004. (Subseries LNAI).

Vol. 3126: P. Dini, P. Lorenz, J.N.d. Souza (Eds.), Service Assurance with Partial and Intermittent Resources. XI, 312 pages. 2004.

Vol. 3125: D. Kozen (Ed.), Mathematics of Program Construction. X, 401 pages. 2004.

Vol. 3124: J.N. de Souza, P. Dini, P. Lorenz (Eds.), Telecommunications and Networking - ICT 2004. XXVI, 1390 pages. 2004.

Vol. 3123: A. Belz, R. Evans, P. Piwek (Eds.), Natural Language Generation. X, 219 pages. 2004. (Subseries LNAI).

Vol. 3121: S. Nikoletseas, J.D.P. Rolim (Eds.), Algorithmic Aspects of Wireless Sensor Networks. X, 201 pages. 2004.

Vol. 3120: J. Shawe-Taylor, Y. Singer (Eds.), Learning Theory. X, 648 pages. 2004. (Subseries LNAI).

Vol. 3118: K. Miesenberger, J. Klaus, W. Zagler, D. Burger (Eds.), Computer Helping People with Special Needs. XXIII, 1191 pages. 2004.

Vol. 3116: C. Rattray, S. Maharaj, C. Shankland (Eds.), Algebraic Methodology and Software Technology. XI, 569 pages. 2004.

Vol. 3114: R. Alur, D.A. Peled (Eds.), Computer Aided Verification. XII, 536 pages. 2004.

Vol. 3113: J. Karhumäki, H. Maurer, G. Paun, G. Rozenberg (Eds.), Theory Is Forever. X, 283 pages. 2004.

Vol. 3112: H. Williams, L. MacKinnon (Eds.), Key Technologies for Data Management. XII, 265 pages. 2004.

Vol. 3111: T. Hagerup, J. Katajainen (Eds.), Algorithm Theory - SWAT 2004. XI, 506 pages. 2004.

Vol. 3110: A. Juels (Ed.), Financial Cryptography. XI, 281 pages. 2004.

Vol. 3109: S.C. Sahinalp, S. Muthukrishnan, U. Dogrusoz (Eds.), Combinatorial Pattern Matching. XII, 486 pages. 2004.

Vol. 3108: H. Wang, J. Pieprzyk, V. Varadharajan (Eds.), Information Security and Privacy. XII, 494 pages. 2004.

Vol. 3107: J. Bosch, C. Krueger (Eds.), Software Reuse: Methods, Techniques and Tools. XI, 339 pages. 2004.

Vol. 3106: K.-Y. Chwa, J.I. Munro (Eds.), Computing and Combinatorics. XIII, 474 pages. 2004.

Vol. 3105: S. Göbel, U. Spierling, A. Hoffmann, I. Iurgel, O. Schneider, J. Dechau, A. Feix (Eds.), Technologies for Interactive Digital Storytelling and Entertainment. XVI, 304 pages. 2004.

Vol. 3104: R. Kralovic, O. Sykora (Eds.), Structural Information and Communication Complexity. X, 303 pages. 2004.

Vol. 3103: K. Deb, e. al. (Eds.), Genetic and Evolutionary Computation – GECCO 2004. XLIX, 1439 pages. 2004.

Vol. 3102: K. Deb, e. al. (Eds.), Genetic and Evolutionary Computation – GECCO 2004. L, 1445 pages. 2004.

Vol. 3101: M. Masoodian, S. Jones, B. Rogers (Eds.), Computer Human Interaction. XIV, 694 pages. 2004.

Vol. 3100: J.F. Peters, A. Skowron, J.W. Grzymała-Busse, B. Kostek, R.W. Świniarski, M.S. Szczuka (Eds.), Transactions on Rough Sets I. X, 405 pages. 2004.

Vol. 3099: J. Cortadella, W. Reisig (Eds.), Applications and Theory of Petri Nets 2004. XI, 505 pages. 2004.

Vol. 3098: J. Desel, W. Reisig, G. Rozenberg (Eds.), Lectures on Concurrency and Petri Nets. VIII, 849 pages. 2004.

Vol. 3097: D. Basin, M. Rusinowitch (Eds.), Automated Reasoning. XII, 493 pages. 2004. (Subseries LNAI).

Vol. 3096: G. Melnik, H. Holz (Eds.), Advances in Learning Software Organizations. X, 173 pages. 2004.

Vol. 3095: C. Bussler, D. Fensel, M.E. Orlowska, J. Yang (Eds.), Web Services, E-Business, and the Semantic Web. X, 147 pages. 2004.

Vol. 3094: A. Nürnberger, M. Detyniecki (Eds.), Adaptive Multimedia Retrieval. VIII, 229 pages. 2004.

Vol. 3093: S.K. Katsikas, S. Gritzalis, J. Lopez (Eds.), Public Key Infrastructure. XIII, 380 pages. 2004.

Vol. 3092: J. Eckstein, H. Baumeister (Eds.), Extreme Programming and Agile Processes in Software Engineering. XVI, 358 pages. 2004.

Vol. 3091: V. van Oostrom (Ed.), Rewriting Techniques and Applications. X, 313 pages. 2004.

Vol. 3089: M. Jakobsson, M. Yung, J. Zhou (Eds.), Applied Cryptography and Network Security. XIV, 510 pages. 2004.

Vol. 3087: D. Maltoni, A.K. Jain (Eds.), Biometric Authentication. XIII, 343 pages. 2004.

Vol. 3086: M. Odersky (Ed.), ECOOP 2004 – Object-Oriented Programming. XIII, 611 pages. 2004.

Vol. 3085: S. Berardi, M. Coppo, F. Damiani (Eds.), Types for Proofs and Programs. X, 409 pages. 2004.

Vol. 3084: A. Persson, J. Stirna (Eds.), Advanced Information Systems Engineering. XIV, 596 pages. 2004.

Vol. 3083: W. Emmerich, A.L. Wolf (Eds.), Component Deployment. X, 249 pages. 2004.

Vol. 3080: J. Desel, B. Pernici, M. Weske (Eds.), Business Process Management. X, 307 pages. 2004.

Vol. 3079: Z. Mammeri, P. Lorenz (Eds.), High Speed Networks and Multimedia Communications. XVIII, 1103 pages. 2004.

Vol. 3078: S. Cotin, D.N. Metaxas (Eds.), Medical Simulation. XVI, 296 pages. 2004.

Vol. 3077: F. Roli, J. Kittler, T. Windeatt (Eds.), Multiple Classifier Systems. XII, 386 pages. 2004.

Vol. 3076: D. Buell (Ed.), Algorithmic Number Theory. XI, 451 pages. 2004.

Vol. 3075: W. Lenski, Logic versus Approximation. VIII, 205 pages. 2004.

Vol. 3074: B. Kuijpers, P. Revesz (Eds.), Constraint Databases and Applications. XII, 181 pages. 2004.

Vol. 3073: H. Chen, R. Moore, D.D. Zeng, J. Leavitt (Eds.), Intelligence and Security Informatics. XV, 536 pages. 2004.

Vol. 3072: D. Zhang, A.K. Jain (Eds.), Biometric Authentication. XVII, 800 pages. 2004.

Vol. 3071: A. Omicini, P. Petta, J. Pitt (Eds.), Engineering Societies in the Agents World. XIII, 409 pages. 2004. (Subseries LNAI).

Vol. 3070: L. Rutkowski, J. Siekmann, R. Tadeusiewicz, L.A. Zadeh (Eds.), Artificial Intelligence and Soft Computing - ICAISC 2004. XXV, 1208 pages. 2004. (Subseries LNAI).

Vol. 3068: E. André, L. Dybkjær, W. Minker, P. Heisterkamp (Eds.), Affective Dialogue Systems. XII, 324 pages. 2004. (Subseries LNAI).

Vol. 3067: M. Dastani, J. Dix, A. El Fallah-Seghrouchni (Eds.), Programming Multi-Agent Systems. X, 221 pages. 2004. (Subseries LNAI).

Vol. 3066: S. Tsumoto, R. Słowiński, J. Komorowski, J.W. Grzymała-Busse (Eds.), Rough Sets and Current Trends in Computing. XX, 853 pages. 2004. (Subseries LNAI).

Vol. 3065: A. Lomuscio, D. Nute (Eds.), Deontic Logic in Computer Science. X, 275 pages. 2004. (Subseries LNAI).

Vol. 3064: D. Bienstock, G. Nemhauser (Eds.), Integer Programming and Combinatorial Optimization. XI, 445 pages. 2004.

Vol. 3063: A. Llamosí, A. Strohmeier (Eds.), Reliable Software Technologies - Ada-Europe 2004. XIII, 333 pages. 2004.

Vol. 3062: J.L. Pfaltz, M. Nagl, B. Böhlen (Eds.), Applications of Graph Transformations with Industrial Relevance. XV, 500 pages. 2004.

Vol. 3061: F.F. Ramos, H. Unger, V. Larios (Eds.), Advanced Distributed Systems. VIII, 285 pages. 2004.

Vol. 3060: A.Y. Tawfik, S.D. Goodwin (Eds.), Advances in Artificial Intelligence. XIII, 582 pages. 2004. (Subseries LNAI).

Vol. 3059: C.C. Ribeiro, S.L. Martins (Eds.), Experimental and Efficient Algorithms. X, 586 pages. 2004.